T0231101

Digital Signal Processing for Multimedia Systems

Signal Processing

K. J. Ray Liu
Series Editor
University of Maryland
College Park, Maryland

1. Digital Signal Processing for Multimedia Systems, *edited by Keshab K. Parhi and Takao Nishitani*

Additional volumes in preparation

Multimedia Systems, Standards and Networks, *edited by Dr. Atul Puri and Dr. Tsuhan Chen*

Compressed Video Over Networks, *edited by Dr. Ming-Ting Sun and Dr. Amy Riebman*

Blind Equalization and Identification, *Dr. Zhi Ding and Dr. Ye (Geoffrey) Li*

Interprocessor Communication Strategies for Application Specific Multiprocessors, *Dr. Sundararajan Sriram and Dr. Shuvra S. Bhattacharyya*

Digital Signal Processing for Multimedia Systems

edited by

Keshab K. Parhi
University of Minnesota
Minneapolis, Minnesota

Takao Nishitani
NEC Corporation
Sagamihara, Japan

CRC Press
Taylor & Francis Group
Boca Raton London New York

CRC Press is an imprint of the
Taylor & Francis Group, an **informa** business

CRC Press
Taylor & Francis Group
6000 Broken Sound Parkway NW, Suite 300
Boca Raton, FL 33487-2742

Visit the Taylor & Francis Web site at
http://www.taylorandfrancis.com

and the CRC Press Web site at
http://www.crcpress.com

Series Introduction

Over the past 50 years, digital signal processing has evolved as a major engineering discipline. The fields of signal processing have grown from the origin of fast Fourier transform and digital filter design to statistical spectral analysis and array processing, and image, audio, and multimedia processing, and shaped developments in high-performance VLSI signal processor design. Indeed, there are few fields that enjoy so many applications—signal processing is everywhere in our lives.

When one uses a cellular phone, the voice is compressed, coded, and modulated using signal processing techniques. As a cruise missile winds along hillsides searching for the target, the signal processor is busy processing the images taken along the way. When we are watching a movie in HDTV, millions of audio and video data are being sent to our homes and received with unbelievable fidelity. When scientists compare DNA samples, fast pattern recognition techniques are being used. On and on, one can see the impact of signal processing in almost every engineering and scientific discipline.

Because of the immense importance of signal processing and the fast-growing demands of business and industry, this series on signal processing serves to report up-to-date developments and advances in the field. The topic of interests include but are not limited to the following:

- Signal theory and analysis
- Statistical signal processing
- Speech and audio processing
- Image and video processing
- Multimedia signal processing and technology
- Signal processing for communications
- Signal processing architectures and VLSI design

I hope this series will provide the interested audience with high-quality, state-of-the-art signal processing literature through research monographs, edited books, and rigorously written textbooks by experts in their fields.

K. J. Ray Liu

Preface

Within a few years, multimedia will become part of everyone's life. There has been tremendous growth in the field of communications and computing systems in this decade. Much of this growth has been fueled by the promise of *multimedia*- an exciting array of new products and services, ranging from video games, car navigators and web surfing to interactive television, cellular and video phones, personal digital assistants, video on demand, desktop video conferencing, entertainment and education. In addition to conventional user interfaces and I/O operations, these multimedia applications demand real-time digital signal processing (DSP) to support compression and decompression of audio, speech, and video signals, speech recognition and synthesis, and character recognition. These operations require intensive multiply and multiply-accumulate operations unique to DSP such as filtering, motion detection and compensation, discrete cosine transforms (DCTs), and fast Fourier transforms (FFTs). Although many of these algorithms have been known for decades, it is only in the last few years that VLSI technology has advanced to the point where the enormous computing power required for multimedia applications can be met at very low costs. It is predicted that in the next decade the CMOS technology will scale down to 0.1 micron devices with corresponding improvements in density and speed. The number of transistors on single chips is expected to increase from current $6M$ to $200M$ by the year 2010.

Digital TV transmission will begin in 1999. Consumers can then use the same device either as a digital TV or as a PC (in non-interlaced or progressive mode). The same device, along with a set-top box, can be used to watch television, listen to music CDs, access videos stored on CD-ROMs and DVD-ROMs, and for VCRs and video games. This setup has been termed *PC theater.* The PC theater can allow the user to read several email messages or access the Internet in the middle of watching a TV program (for example during a commercial). This convergence of the TV and the PC will be the key in accessing multimedia information from homes.

This book addresses applications of multimedia; programmable and custom architectures for implementation of multimedia systems; and arithmetic architectures and design methodologies.

The first part (Chapters 1 to 8) begins with applications of multimedia. Video and audio compression are main enabling technologies for multimedia. Chapters 2 and 3 present overview of video and audio compression needed for multimedia applications. Chapter 4 addresses system synchronization aspects. Chapter 5 presents digital versatile disks (DVDs) which will be the medium of storage for multimedia information. To receive multimedia information at home using the telephone connections, subscriber loops will play an important role. In particular, the information will be received by using asymmetric digital subscriber loops (ADSL) and very-high-data-rate digital subscriber loops (VDSL). Chapter 6 presents architectures for equalizers and modems for ADSL and VDSL applications. Chapter 7 presents cable modems while Chapter 8 presents an overview of wireless systems.

One of the important design considerations when building new silicon architectures is the choice between programmable and dedicated hardware. While dedicated design methodology was popular for high-throughput applications (such as video), the power of current microprocessors and (programmable DSP based) media processors encourages use of these systems to meet the flexibility needed in various forms of computing. It may, however, be noted that the dedicated (or vertically-integrated) design style is still important for low-power applications such as personal digital assistants (PDAs). The second part of the book (Chapters 9 to 19) addresses programmable and custom architecture design of various multimedia components. Chapter 9 begins with programmable digital signal processors (PDSPs) which were first developed in the early 1980s to support iterative and computation-intensive real-time digital signal processing. PDSPs provide an interesting tradeoff between performance and flexibility. These DSP processors deviate from general-purpose microprocessors by introducing specialized data paths to speed up the computations most common in DSP applications, i.e., fast multiply-accumulate operations, and by using multiple-access memories, etc. The first generation DSPs developed for high-throughput audio signal processing exhibited a 50:1 (5M vs. 100K multiply-add/s) performance advantage over the microprocessor of the day (M68000), for about the same cost and power.

Multimedia computation deals with various forms of data ranging from video, audio, speech, graphics, and text. The type of computing and precision needed also differs for various forms of data. For example, 8-bit precision is sufficient for video data while 16-bit precision may be needed for audio data. Computer graphics operations need even higher precision. The varying precision requirements also add to the complexity of the design of the media processors. Multimedia processing will continue to be the driving force for the evolution of both microprocessors and DSPs. Recently the introduction of efficient audio and video compression techniques and their standardization has created a new category of devices referred to as *media processors* which are discussed in Chapter 10. These processors is a special group of DSP processors which are equipped with audio, video and graphics accelerators suitable for multimedia signal processing applications. The media processors can achieve high throughput using moderate clock speed by exploiting parallelism in the form of SIMD or sub-word parallelism (also referred to as split-word parallelism). Most systems also contain coprocessors for computation-intensive tasks such as motion estimation, I/DCT and variable length coders (VLCs). Another approach to implementation of media processors are based on a generalized MMX instruction set for the microprocessors. These are also discussed in Chapter 10. Chapter 11 describes low-power PDSPs for wireless applications.

Chapters 12 to 19 present various custom architectures for multimedia algorithms. Chapter 12 presents design of motion estimation systems. Chapter 13 describes architectures for wavelets. Chapter 14 addresses implementation of DCTs. Chapters 15 and 16, respectively, present lossless coders and Viterbi decoders used in many decoding applications. Chapter 17 presents an overview of watermarking applications of multimedia which are needed for security in media systems. Chapters 18 and 19 present advanced recursive least square (RLS) adaptive filtering approaches which are essential for implementation of equalizers.

The third part of the book (Chapters 20 to 27) addresses arithmetic architectures which form the building blocks and design methodologies for implementations of media systems. Both high-speed and low-power implementations are considered. Chapter 20 addresses division and square-root architectures, Chapter 21 addresses finite field arithmetic architectures which are used for implementation of error control coders and cryptography functions. Chapter 22 presents CORDIC rotation architectures which are needed for implementation of space-time adaptive processing systems and orthogonal filtering applications. Chapter 23 presents advanced systolic architectures. Reduction of power consumption is important for media sytems implemented using scaled technologies. Low power consumption increases battery life in portable computers and communications systems such as personal digital assistants. Power consumption reduction also leads to reduction of cooling and packaging costs. Chapter 24 addresses low power design methodologies while Chapter 25 presents approaches to power estimation. Chapter 26 addresses power reduction methodologies through memory management. Chapter 27 addresses hardware description based synthesis based on custom as well as FPGA implementations which will form the main medium of system implementations in future decades.

This book is expected to be of interest to application, circuit and system designers of multimedia systems. No book brings together such a rich variety of topics on multimedia system design as does this one.

The editors are most grateful to all coauthors for their contributing excellent chapters. This book could not have been possible without their efforts. They are grateful to Ru-Guang Chen for his help in compiling this book. Thanks are also due to the National Science Foundation (NSF) and the NEC Corporation. A Japan Fellowship to KKP by the NSF was instrumental in bringing the editors together. The editors thank Dr. Ed Murdy and Dr. John Cozzens of NSF, Dr. Mos Kaveh of the University of Minnesota for their support and encouragement. The editors thank Graham Garratt, Rita Lazazzaro and Brian Black of Marcel Dekker, Inc. It was truly a pleasure to work with them.

Keshab K. Parhi
Takao Nishitani

Contents

Part II Programmable and Custom Architectures and Algorithms

Contributors

Florin Balasa, Ph.D.*
Engineer
IMEC
Kapeldreef 75
B-3001 Leuven
Belgium

David H. Bartley, M.A. (Comp. Sc.)
Distinguished Member, Technical Staff (DMTS)
Texas Instruments Incorporated
10235 Echo Ridge Court
Dallas, TX 75243
bartley@ti.com

Francky Catthoor, Ph.D.
Head, System Exploration for Memory and Power Group
VLSI System Design Methodology Division
IMEC
Kapeldreef 75
B-3001 Leuven
Belgium
catthoor@imec.be

Ingemar J. Cox, Ph.D.
Senior Research Scientist
NEC Research Institute
4 Independence Way
Princeton, NJ 08540
ingemar@research.nj.nec.com

Herbert Dawid, Dr.-Ing.
Member of Technical Staff
Research
Synopsys, Inc.
Digital Communication Solutions
Professional Services Group
Kaiserstrasse 100
D-52134 Herzogenrath, Germany
dawid@synopsys.com

Eddy De Greef, Ph.D.
Research Engineer
IMEC
Kapeldreef 75
B-3001 Leuven
Belgium
degreef@imec.be

Tracy C. Denk, Ph.D.
Staff Scientist
Broadcom Corporation
16251 Laguna Canyon Road
Irvine, CA 92618
tdenk@broadcom.com

Sachin G. Deshpande, M.S.
Ph.D. Candidate
Department of Electrical Engineering, Box 352500
University of Washington
Seattle, Washington 98195
sachind@ee.washington.edu

Wanda K. Gass, M.S.
Manager, DSP Architecture Definition
Texas Instruments Incorporated
P.O. Box 660199, MS 8723

* *Current affiliation:* Senior Design Automation Engineer, Rockwell Semiconductor Systems, Newport Beach, CA 92660.

Dallas, TX 75266
gass@ti.com

Alan Gatherer, Ph.D.
Manager/Senior Member of Technical Staff
Wireless Communications Branch
DSPS R & D Center
Texas Instruments
P.O. Box 655303, MS 8368
Dallas, TX 75265-5303
gatherer@ti.com

Hidenobu Harasaki, M.E.
Research Manager
C&C Media Research Labs., NEC Corporation
1-1, Miyazaki 4-chome, Miyamae-ku, Kawasaki, 216-8555, Japan
harasaki@ccm.cl.nec.co.jp

Chiung-Yu Hung, Ph.D.
Member of Technical Staff
8330 LBJ Freeway MS 8374
Dallas, Texas 75243
cy-hung@ti.com

Jenq-Neng Hwang, Ph.D.
Associate Professor
Department of Electrical Engineering
Box # 352500
University of Washington
Seattle, WA 98195
hwang@ee.washington.edu

Masahiro Iwadare, M.S.
Principal Researcher
Digital Signal Processing Technology Group
C&C Media Research Laboratories
NEC Corporation
1-1, Miyazaki 4-chome
Miyamae-ku, Kawasaki 216-8555, Japan
iwadare@dsp.cl.nec.co.jp

Olaf J. Joeressen, Dr.-Ing.
R&D Project Leader
Nokia Mobile Phones, R&D Center Germany
Meesmannstr. 103
D-44807 Bochum
Germany
Olaf.Joeressen@nmp.nokia.com

Ton Kalker, Ph.D.
Research Scientist
Philips Research Laboratories
Bldng WY 8.41, Pbox WY82
Prof. Holstlaan 4, 5656 AA
Eindhoven, The Netherlands
kalker@natlab.research.philips.com

Masayuki Kozuka, M.S.
Manager
Multimedia Development Center
Matsushita Electric Industrial Co., Ltd.
Ishizu Minami-machi 19-1-1207
Neyagawa, Osaka 572, Japan
mk@isl.mei.co.jp

Ichiro Kuroda, B.E.
Research Manager
NEC Corporation
1-1, Miyazaki, 4-chome, Miyamae-ku
Kawasaki, Kanagawa 216-8555, Japan
kuroda@dsp.cl.nec.co.jp

Tadahiro Kuroda, B.S.E.E
Senior Specialist
Toshiba Corp., System ULSI
Engineering Lab., 580-1, Horikawa-cho, Saiwai-ku, 210-8520, Japan
tadahiro.kuroda@toshiba.co.jp

Dominique Lavenier, Ph.D.
CNRS Researcher
IRISA
Campus de Beaulieu
35042 Rennes cedex
France
lavenier@irisa.fr

Junsoo Lee, M.S.
Ph.D. Candidate
Dept. Electrical Engineering
University of Minnesota
200 Union Street S.E.
Minneapolis, MN 55455
jlee@ece.umn.edu

Jean-Paul M. G. Linnartz
Natuurkundig Laboritorium WY8
Philips Research
5656 AA Eindhoven,
The Netherlands
linnartz@natlab.research.philips.com

K. J. Ray Liu, Ph.D.
Associate Professor
Systems Research Center
University of Maryland
A.V. Williams Building (115)
College Park, MD 20742
kjrliu@isr.umd.edu

Lori E. Lucke, Ph.D.
Senior Design Engineer
Minnetronix, Inc.
2610 University Ave., Suite 400
St. Paul, MN 55114
lelucke@minnetronix.com

Heinrich Meyr, Dr.-Ing.
Professor
Institute of Integrated Systems for Signal Processing (ISS)
Aachen University of Technology (RWTH Aachen)
Templergraben 55
D-52056 Aachen
Aachen, Germany
meyr@ert.rwth-aachen.de

Matthew L. Miller, B.A.
Senior Scientist
Signafy Inc.
4 Independence Way
Princeton, NJ 08540
mlm@signafy.com

Takao Nishitani, Ph.D.
Deputy General Manager
NEC Corporation
1-20-11-206, Minami-hashimoto,
Sagamihara, 223-1133, Japan
takao@mel.cl.nec.co.jp

Yasushi Ooi, M.S.
Principal Researcher
C&C Media Research Laboratories
NEC Corporation
4-1-1, Miyazaki, Miyamae,
Kawasaki, 216-8555, Japan
oioi@dsp.cl.nec.co.jp

Keshab K. Parhi, Ph.D.
Edgar F. Johnson Professor
Department of Electrical & Computer Engineering
University of Minnesota
200 Union St. S.E.
Minneapolis, MN 55455
parhi@ece.umn.edu

Patrice Quinton, Ph.D.
Professor University of Rennes 1
IRISA, Campus de Beaulieu,
35042 Rennes cedex, France
Patrice.Quinton@irisa.fr

K. J. Raghunath, Ph.D.
Member of Technical Staff
Lucent Technologies
Bell Laboratories
184 Liberty Corner Road, Room 1SC125
Warren, NJ 07059
raghunath@lucent.com

Sanjay Rajopadhye, Ph.D.
Senior Researcher, CNRS
IRISA
Campus Universitaire de Beaulieu
35042 Rennes cedex, France
rajopadhye@irisa.fr

Takayasu Sakurai, Ph.D.
Center for Collaborative Research, and Institute of Industrial Science,

University of Tokyo
7-22-1 Roppongi, Minato-ku, Tokyo, 106-8558 Japan
tsakurai@iis.u-tokyo.ac.jp

Janardhan H. Satyanarayana, Ph.D.
Member of Technical Staff
Bell Laboratories, Lucent Technologies
101, Crawfords Corner Road, Room 4D509
Holmdel, NJ 07733
jana@lucent.com

Naresh R. Shanbhag, Ph.D.
Assistant Professor, ECE Department
Coordinated Science Laboratory, Rm. 413
University of Illinois at Urbana-Champaign
1308 West Main Street, Urbana, IL 61801
shanbhag@uivlsi.csl.uiuc.edu

Peter Slock, Dip. Eng.*
IMEC
Kapeldreef 75
B-3001 Leuven
Belgium

Leilei Song, M.S.
Ph.D. Candidate
Department of Electrical & Computer Engineering
University of Minnesota
200 Union St. S.E.
Minneapolis, MN 55455
llsong@ece.umn.edu

Elvino S. Sousa, Ph.D.
Professor
Dept. of Electrical and Computer Engineering
University of Toronto
Toronto, Ontario, Canada
M5S 3G4
sousa@comm.utoronto.ca

H. R. Srinivas, Ph.D.
Team Leader
Lucent Technologies
Bell Laboratories
Room 55E-334
1247 S. Cedar Crest Boulevard
Allentown, PA 18103
hsrinivas@lucent.com

Akihiko Sugiyama, Dr. Eng.
Principal Reseacher
C&C Media Research Laboratories
NEC Corporation
1-1, Miyazaki 4-chome
Miyamae-ku, Kawasaki 216-8555
Japan
sugiyama@dsp.cl.nec.co.jp

Ming-Ting Sun, Ph.D.
Associate Professor
Department of Electrical Engineering, Box 352500
University of Washington
Seattle, Washington 98195
sun@ee.washington.edu

Shin-ichi Tanaka, B.S.
General Manager
Device Development Group
Optical Disk Systems Development Center
Matsushita Electric Industrial Co., Ltd.
1-42-14 Yamatehigashi, Kyotanabe
Kyoto, Japan 710-0357
stanaka@drl.mei.co.jp

Mihran Touriguian, M.Sc.
Manager, System Design
Atmel Corp.
2150 Shattuck Blvd., 3rd floor
Berkeley, CA 94704
tourigui@berkeley.atmel.com

* *Current affiliation:* M.S. candidate, K.U. Leuven, Gent, Belgium.

Kazuhiro Tsuga, M.S.
Manager
Visual Information Group
Multimedia Development Center
Matsushita Electric Industrial Co., Ltd.
9-33 Hanayashiki-tsutsujigaoka, Takarazuka
Hyogo, Japan, 665-0803
tsuga@hdc.mei.co.jp

Ingrid Verbauwhede, Ph.D.
Associate Professor
Electrical Engineering Department
University of California, Los Angeles
7440B Boelter Hall
Los Angeles, California 90095-1594
ingrid@JANET.UCLA.EDU

An-Yeu (Andy) Wu, Ph.D.
Associate Professor
Electrical Engineering Dept., Rm. 411
National Central University
Chung-Li, 32054
Taiwan
andywu@ee.ncu.edu.tw

Sven Wuytack, Ph.D.
Research Engineer
IMEC - VSDM Division
Kapeldreef 75
B-3001 Leuven
Belgium
wuytack@imec.be

Chapter 1

Multimedia Signal Processing Systems

Takao Nishitani
NEC Corporation
Sagamihara, Kanagawa, Japan
takao@mel.cl.nec.co.jp

1.1 INTRODUCTION

Multimedia is now opening new services that support a more convenient and easy to use environment, such as virtual reality for complex systems and for education systems, multiple-view interactive television services and three dimensional home theater. It is not too much to say that the introduction of audio and video into communications worlds, computer worlds and broadcasting worlds formed the beginning of the multimedia. Adding audio and visual environment to the conventional text-base services makes them vivid and attractive for many users. Therefore, realizing a seamless connection and/or fusion among computer, communication and broadcasting worlds, as shown in Fig. 1.1, leads to the possibility of dramatic changes in our lives. The key function, here, is efficient digitization of video and audio, because digital video and audio can be easily fed into the digital computers and digital communication networks.

However, these three worlds impose different requirements on the digitization of audio and video, due to their long history in each world. Also, information concerning the audio and video in direct digitization results in a much larger file than the conventional text-based files. Therefore, high capacity storage, high speed network and compression technologies for audio and video play an important role in the multimedia world. In addition, in order to encourage the multimedia world establishment, low cost implementation of such compression and transmission/storage hardware is inevitable. In this sense, VLSI design methodology for low power and low cost implementation is an important issue. In the following, the background of this area, multimedia signal processing and its hardware implementation, will be briefly reviewed, by focusing on the above mentioned issues.

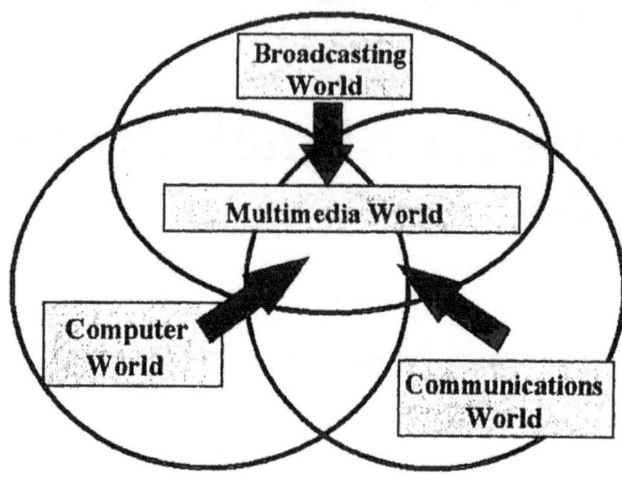

Figure 1.1 Generation of multimedia world.

1.1.1 Computer World

Data in the computers was originally composed of processing data and its transactions. Currently, document data including text, tables and figures are part of computer data. Percentage of document data in computer systems increases day after day, especially after personal computers have become very popular. For example, it is very hard now to publish documents and books without word processing software for PCs or workstations. Recently, audio and video signals as well as photographs were introduced into the computer data. This meant that audio and video can be treated just like text in a word processing system. Editing audio and video by "cut and paste" became possible, and one could easily make an attractive multimedia presentation of materials on a personal computer (PC). This was the beginning of multimedia computation. Examples of this can be seen in a "multimedia home-page" on the world wide web.

Another important fact is that recent high-end processors have reached the level of real-time software decompression (decoding) of compressed Video. Although real-time software Video compression is still far from microprocessor processing capability; ever-improving VLSI technology with advanced architecture based on low power and high speed circuit design surely enables down-sizing of super-mini computers to PC levels and this fact accelerates multimedia applications. Therefore, processor architectures for multimedia is one of the hot topics in this book.

1.1.2 Communications World

Although the above innovation has occurred in the computer world, the word "multimedia" was first introduced in communication area, where PCM (pulse code modulation) speech and computer data are transmitted through the same digital communications lines which are well established in many countries. Digital speech coding in PCM has a long history, because digitization gives high quality speech even in a long distance call [1][2]. This fact is hardly realized in an analog transmission due to signal attenuation and contamination by thermal noise and cross-talk. When digital speech transmission becomes popular, it is natural to send computer data through digital speech communication channels, instead of new computer network establishment. Thus, multimedia multiplexing systems support flexible and low cost data transmission.

This fact indicates two important points for multimedia networks. The first point is the inclusion of computer data on a digital speech transmission line; that is co-existence of time dependent speech data and time independent data. This is the beginning of seamless connection and/or fusion among computer, communications and broadcasting. The boundary of these separate worlds have disappeared, but the difference between time dependent data and time independent data causes some difficulty in real-time operations. Examples can be seen in video transmission through internet, where required bandwidth for real-time video transmission is hard to reserve. Constant Quality-of-Service (QOS) transmission is a barrier to the next generation multimedia networks.

The other point is low cost network implementation for communications, for computer and for broadcasting through a single communication line, especially bandwidth expansion on a subscriber line. When we make analog telephone subscriber line a digital multimedia line, this low cost implementation leads to personal applications on multimedia. Indeed, the bit rate on a conventional single subscriber line is increasing rapidly, due to the employment of advanced signal processing technology to a voiceband modem. Recent PCs have a built-in modem at the bit rate of 28.8Kb/s, 33.6Kb/s or 57.6Kb/s for telephone line transmission. When we ask the telephone carrier company for ISDNi (integrated services digital network) service, we can enjoy transmission at the rate of 144Kb/s on a conventional subscriber line. However, in order to get entertainment quality video, at least the bit rate of 1Mb/s is required. This is also an important issue and later we will explore the developments in this area. More precise technology information can be found in this book in Chapter 2. Due to these personal applications, multimedia has led to revolutionary changes in all of our lives.

1.1.3 Broadcasting World

Audio and video are the main contents in broadcasting. However, digitization of these signals has started recently in this world except for editing and storing these data in broadcast stations. The reason is that very high quality signal broadcasting is required for high quality commercial advertisement films, in order to attract support from many sponsors. Slight degradation incurred by digitization may cause to loss of sponsors. However, CATV and satellite television belong to different category, where they can collect subscription fee from many subscribers. Digitization has started in these areas when world standard MPEG-2 algorithms were completed. Motivation of digitization is to increase the number of channels on CATV or Satellite with reasonable quality. This is because MPEG-2 can achieve rather high compression ratio and the price of high speed digital modem for coaxial cables has reached the level of consumer products. Also, analog CATV quality is quite different at the subscriber's location, due to tree structure video networks, but digital transmission enables insuring the quality all over the network. Due to these digital broadcasting merits, terrestrial broadcasting is also going to be digitized. In ATV (advanced TV) project, HDTV (high definition TV) transmission in digital form is also scheduled from this year of 1999.

Cable modem implementation and new services called VOD (video on demand) together with its terminal STB (set-top box) are further discussed later. Cable modem is also addressed in Chapter 7 of this book.

1.2 DIGITIZATION OF AUDIO AND VIDEO

1.2.1 Information Amount

The essential problem of the digital audio and video processing lies in the huge amount of information which they require. Let us consider about the information in every media. One alphabet letter is represented in one byte of an ASCII code. Then, one page, consisting about 60 letters $\times$ 50 lines, requires 3 Kbytes. Therefore, one book of 330 pages requires storage of about 1 M byte. This volume is almost equal to that of a standard floppy disk of 1.44 Mbytes. On the contrary, Hi-Fi audio is composed of two channel signals (left and right) for stereo playback. Each channel signal is sampled at the sampling rate of 44 KHz in CD (compact disk) applications or at 48 KHz in DAT (digital audio tape) applications. These sample rates ensure up-to 20 KHz band audio signal reconstruction. Every sample is then converted into digital forms of 16 bits: 2 bytes for a sample. Therefore, one second stereo playback requires about 200 Kbytes. This means that in every 5 seconds, hi-fi audio signals generate information, comparable to a 330 page book. In the same way, consider video signals. In every one second, NTSC television processes 30 pictures (frames). One picture in the NTSC format is composed of 720 $\times$ 480 pixels. Every pixel is then converted into 24 bit R/G/B signals (an 8 bit signal for each component) or 16 bits of luminance/chrominance (an 8 bit luminance of full samples and two 8 bit chrominance signals by alternative sampling). As a result, NTSC information in one second requires at least 20 Mbytes. It is comparable to 20 contents of books in a second. Furthermore, HDTV signals in ATV have a picture of 1920 x 1080 pixels with 60 frames per second. In this case, total information amount reaches 240 Mbytes per second. Fig. 1.2 summarizes the above information amounts. It clearly shows that audio and video demand more than a few magnitudes larger memory capacity, compared with other text data. Now, we can say that in order to handle audio and video signals just like text data, compression technologies for these signals are essential.

Note that storing or play-back of digital audio signals has been available in the consumer market in the form of compact disk since early 1980s, but digital video storage availability had been limited to only professional use for a long time. Video disc and digital versatile disc (DVD), now available in consumer market, employ MPEG compression technology which will be described later (see also Chapter 5). In general, digital video signals without compression do not economically overcome their analog counter parts, although digital video and audio have the advantage of robustness to external noise. Advances in compression technology and large capacity storage as well as high speed communication networks enable realization of the multimedia world. All of these technologies are based on Digital Signal

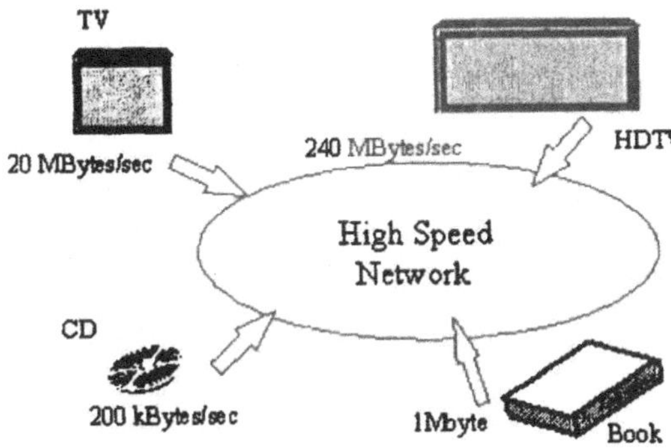

Figure 1.2 Required information amount of every media.

Processing (DSP), and therefore, multimedia signal processing systems and their VLSI implementation are of great interest.

1.2.2 Compression Technology

Fig. 1.3 shows video and audio requirements of seamless connection and/or fusion among computer, communications and broadcasting worlds. As the multimedia is supported by these three different worlds, and as these worlds have been independently developed until today, there are a lot of conflicts among them. These conflicts mainly come from digital video formats employed and required functions for video. These problems are examined below by considering the encoding algorithm in chronic order.

Compression technology itself started in the communications fields to send digital speech in PCM (pulse code modulation) form in 1960s, where nonlinear companding (compression and expanding) of sampled data was studied. Still picture was also compressed in the same time period for sending landscape of the moon surface from the NASA space rocket of lunar-orbiter to the earth through the space.

After these activities, Video compression appeared to realize television program delivery through 45 Mb/s high speed PCM backbone network from a station to another station in real time. Therefore, the most important requirement from the broadcasting world is to achieve NTSC quality as much as possible. This means that every video signal should have 30 frames in a second and every frame picture should have 720 $\times$ 480 pixels. On the contrary, teleconferencing systems and telephony systems started solely in the communications area. Therefore, their primary concern is the communication cost, rather than the picture quality. A single 64 Kbit/sec PCM channel or a primary multiplexed 1.544 Mb/s PCM line is acceptable for television telephony systems or teleconferences in terms of running cost. Therefore, the compression algorithms for these purposes employ lower resolution pictures and lower number of frames (pictures) in a second. The world wide standard on video compression algorithm of the recommendation H.261 from ITU-T (International Telecommunication Union, Telecommunication standardiza-

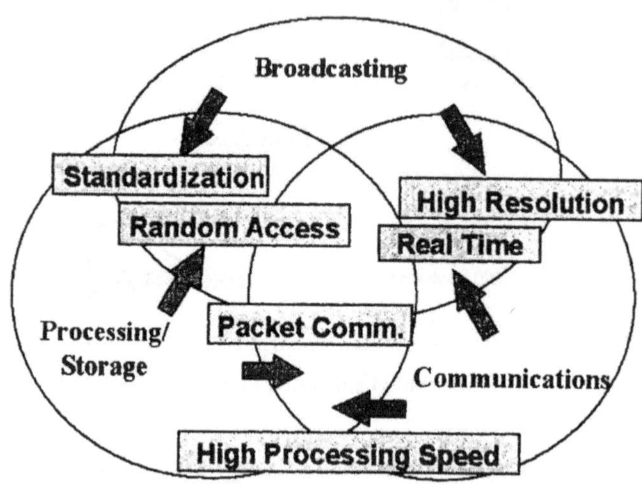

Figure 1.3 Requirements from every domain.

tion sector) employs CIF (common intermediate format) and QCIF (quarter CIF) which require a quarter and 1/16 resolution of NTSC, respectively. It also employs fewer frame rate than conventional 30 frames/sec. For example, a motion picture having 7.5 frame/sec with QCIF can be transmitted at 64 Kbits/s. Then, video signals in this format become only 1/54 of original NTSC information. Another important factor in the communication systems is the requirement on short coding delay. During the H.261 standardization period, the difference of the specification between communications and broadcasting became clear. The ITU-R (International Telecommunication Union, Radio communication sector) decided to make a broadcasting standard and this activity resulted in Recommendation 723, although the basic approach in this compression algorithm is almost the same: hybrid coding between DPCM (differential PCM) with MC (motion compensation) and DCT (discrete cosine transform) coding with variable bit-length coding [3][4].

Consider the area between broadcasting and computer worlds. Widely accepted standard compression algorithms are essentially required for wide distribution of Video programs from the viewpoint of broadcasting area. From computer side, the mandatory requirement is random access capability of video and audio files. This is because computer users want to access a certain period of a video sequence with audio, instead of the period starting from the beginning. Unfortunately, as this kind of functionality is not considered in the ITU-T recommendation H.261 and in ITU-R recommendation 723, the ISO (International Standards Organization) and IEC (International Electro-technical Commission) have decided to collaborate to make the world standard which covers the requirement from broadcasting, communications and computer worlds. The MPEG-1/2 algorithms have been standardized based on the forerunner algorithms of H.261 and G.723 with expanding functionality. They call MPEG a generic coding. As MPEG is designed to be generic, several parameters are specified for different applications. For example, the picture resolution is selected from several "levels" and minor modification in the algorithm is set by several "profiles".

The importance of MPEG activities can be seen in the following facts. The MPEG-2 standard which is originally based on the computer world is employed in the communications standard of ITU-T as H.262 which is a common text of MPEG specification. In 1996, the MPEG activities received the Emmy award from broadcasting world. These facts indicates that MPEG has become the glue of these three worlds and ties them together. This is the reason why this book highlights audio and video compression algorithms and implementation approaches for compression functions such as DCT, motion compensation and variable bit-length encoder (lossless coding).

In addition to compression algorithms for audio and video, error correcting encoding/decoding is mandatory to put them into multimedia storage systems or multimedia communications networks. This is because storage systems and communications networks are not always perfect and such systems give some errors in the compressed data sequence, although the error rate is very low. As the compressed data is a set of essential components, only single bit error may cause significant damages in the decoding process. Fig. 1.4 shows relationship between compression and error correction clearly. In Fig. 1.4, compression part is denoted as source coding and error correction part is as channel coding, because the compression function removes redundant parts from source data and error correction function adds some information to protect the compressed audio and video from the errors due to channels. Error correction should be effective for both random errors caused by external noise and burst errors caused by some continuous disturbance. Source coding is also referred to as low bit-rate coding.

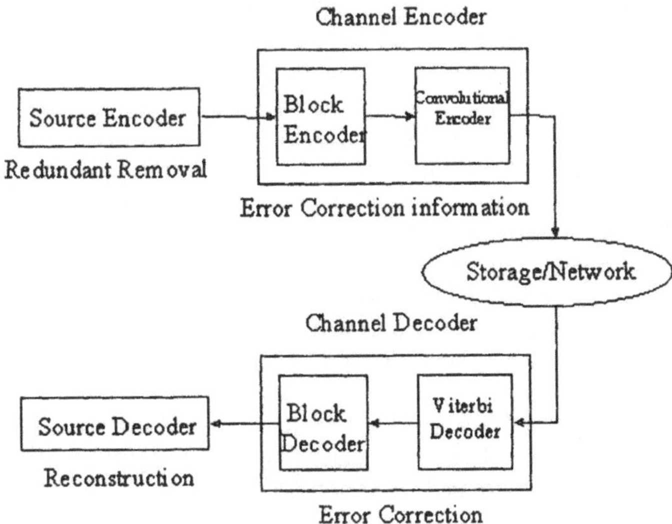

Figure 1.4 Error correction encoder/decoder location in multimedia systems.

In many cases, two different error correcting encoders are employed in a tandem connection form. The first one is a block code where error correction range is limited in a certain block. For example, compressed data should be put in a format for easy handling. ATM (asynchronous transfer mode) cell format or a packet format in high speed networks are examples. Block coding is effective only in the format areas. Reed-Solomon code is used for this purpose which is based on finite field arithmetic (see the Chapter 21).

After compressed data are formatted and coded by block coding, they are stored or transmitted serially. In this case, storage system hardware and/or transmission system hardware do not care for their contents. Contents are nothing but a single bit stream. In such cases, convolutional coding protects errors from channels. As error protection information is convolved into serial data, decoding process is a deconvolution process, and therefore, it becomes quite complex. However, in this field, Viterbi decoder efficiently decodes a convolved bit stream. Chapter 16 of this book covers theoretical background and implementation approaches of Viterbi decoders.

Note that in source coding DCT (discrete cosine transform) of Viterbi decoders is often used in standard encoding algorithm. This is because the basis functions in the cosine transform are very similar to that of the optimal KL transform for pictures. However, a new transformation, called the wavelet transform, has a similar function of human eye system: multi-resolution decomposition. In MPEG-2 standardization periods, some institutes and companies proposed this new transform instead of DCT. Although this proposal was rejected in MPEG-2 for the reason that this transform inclusion disturbs smooth transition from MPEG-1, MPEG-4 is going to accept this transform in the fusion area of computer graphics. Therefore, wavelet transform is also important and is described in Chapter 13.

1.2.3 Storage For Multimedia Applications

CD-ROM or its video storage application of video CD, and newly introduced DVD (digital versatile disk) also bridge the computer and the broadcasting worlds by storing movies and video in a large capacity. CD-ROM storage capacity has increased to 780 Mbytes in a disc of 12 cm diameter by employing optical data pick-up. Bit-rate from CD-ROM is normally set to 1.128 Mb/s excluding overhead information of error correction, which is described in the former section. Since the early 1980s, CD-ROM access speed has been improved and 16 or 32 times faster CD-ROM drive is available now, but the capacity itself has remained unchanged. MPEG-1 requirements on audio and video, specified in the beginning of MPEG-1 standardization, have been determined so that the normal CD-ROM can save one hour playback of television programs with a quarter resolution of NTSC called SIF (standard image format, almost equal to CIF), where 1 Mb/s for video and 128 Kbit/sec for audio are allocated.

DVD (the original abbreviated form of digital video disk and recently modified to digital versatile disk) specification is now available as a standard and its storage capacity has increased to about 4.7 giga byte which is same as the size of CD-ROM. This large capacity storage is a result of laser diode with shorter wave length and accurate mechanism control based on digital signal processing. DVD system employs MPEG-2 video compression algorithm which promises full NTSC compatible resolution. The employed video compression bit-rate is variable:

4 Mbits/s in average and upto 9 Mbits/s are allowable. The reason why video CD and DVD employ MPEG standard on audio and video is that pseudo-random access capability and fast forward play-back capability are embedded in the basic coding process. Chapter 5 of this book covers DVD and systems.

1.2.4 Multimedia Communications

Let us go into the overlapped area between communications and computers. Computer systems employ packet communications, when computers are connected to each other with a local area network such as *Ethernet*^{TM} [5][6]. In wide band communication networks, ATM is introduced which uses a set of cells, similar to packets. In the packet and cell communications networks, real time communications are sometimes frozen for a moment, when traffic becomes heavy. For example, packets are automatically held in the output buffer of a system, when the traffic congestion occurs in the network. In the ATM, cells are automatically dropped off, when the buffer in the switching system becomes full. Current hot topics in packet and cell based communications networks address video transmission over the internet. Three important issues, there, are bandwidth reservation of video transmission through the networks, high volume continuous video data, and correct 30 frame/sec synchronization in video reconstruction.

However, MPEG algorithms are robust to cell/packet loss problem. Quasi-random access capability of the video frame structure in MPEG algorithms terminates packet/cell loss error propagation. Also, the MPEG transport layer supports precise timing recovery through ATM network by incorporating digital phase lock mechanism. The systems aspect on the MPEG transport layer is addressed in Chapter 2.

A much more convenient traffic dependent approach is now discussed in MPEG-4, where encoding process is carried out by objects in a video sequence. Every picture in video is first structured by objects, and then, objects are encoded. When network traffic becomes heavy, only the most important objects in a video are transmitted.

In wide band communications networks, ATM is used in backbone networks which are composed of optical fiber transmission lines. However, direct connection to such optical networks from small offices or from home is still far from actual use with reasonable cost. Digitalization of existing subscriber lines is a good way to go and this has already led to the ISDN (Integrated Services Digital Network) standard. However, ISDN supports only 128 Kb/s data plus 16 Kb/s packet channels. Low bit rate multimedia terminals using such as H.263 Video codec and G.723 speech codec from ITU-T or MPEG-4 are available in this bit-rate range, but this bit-rate is too slow for sending and receiving MPEG-1/2 quality video, which are included in www contents.

The possible approach to increase available bit-rate over existing subscriber lines is called xDSL. This technology employs advanced modem technology of multi-carrier orthogonal frequency division multiplexing with water-filling bit allocations. This technology overcomes transmission line impairments such as cross-talks and non-flat frequency characteristics of subscriber lines to achieve high speed transmission of a few Mb/s bit-rate. MPEG-1 real-time down-loading during conversation, as an example, can be realizable through this approach.

Another approach to increase bit-rate with reasonable investment for users is to employ coaxial cables used for CATV. When video and audio become digitized, it is natural to employ digital modulation in CATV. As digital video and audio are normally compressed, digital CATV can carry more channel signals than analog CATV. Some of additional digital channels, generated by cable digitization, can be used for communication purpose or www applications. As co-axial cable transmission characteristics are much more natural, the employed technique in cable modem is QAM (Quadrature Amplitude Modulation) approach which was originally used in digital microwave transmission or satellite communications.

The area of digital wireless communications, including satellite and microwave communications, is one of the latest topics all over the world. Among them, digital cellular systems is of great interest. Digital cellular systems cover their service area with small cells, where weak electromagnetic carrier waves are employed. Due to their weakness, the same frequency carriers are repeatedly used in cells which are not adjacent to each other. As the cell coverage is small, a subscriber terminal need not send high power electromagnetic waves from its antenna. Therefore, recent terminals for digital cellular systems become very small and fit in a pocket. In order to reduce call congestion in a cell, CDMA (Code Division Multiple Access) approach is superior, where excessive calls causes only S/N degradation of multiple channels. S/N degradation of receiving signals slightly affects increase of bit error rate, but is free from congestion.

In a small office, internal local area network should be simple enough for lowering implementation cost. Twisted pair line transmission can also carry high capacity digital information, if the coverage area is within a few hundred meters.

All the modulation schemes described above are highly related to digital signal processing and, therefore, these topics are covered in other chapters.

1.3 MULTIMEDIA SERVICES

The multimedia world requires service or applications, which effectively employ multimedia environment. Most of the explanation until now contains some ideas on such service. Let us summarize these services, which are described separately in different sections. Fig. 1.5 shows locations of new multimedia systems and services on the domains shown in Fig. 1.1. Internet or world wide web is now very popular in all over the world which supports fusion among computer, communication and broadcasting. The internet was started to support the message transmission through computer networks with the standardized internet protocol. Most applications of interest had been e-mail and file transfer. Since the introduction of the www (world wide web), the internet has become the leader of the multimedia world. www provides simplified and unified command system by introducing URL (unified resource locator) and also hyperlink capability embedded in a document written in HTML (hyper text makeup language). In addition to text, graphics, and photographs, video and audio can be also included in HTML document. As everybody wants to enjoy www services on the internet more comfortably, the www results in accelerating ISDN (integrated services digital network) and xDSL modems on telephone lines which have already been described. Similarly, www expands browser market from PCs and workstations to consumer areas, due to the sky rocketing needs of internet browsers. Wireless communications also link to the internet, although the available bit rate is rather low: 9.8 Kbits/s to 64 Kbits/s. These wireless channels are going to be combined with palm-top computers and PDA (personal data assistants). This has led to the beginning of mobile computing.

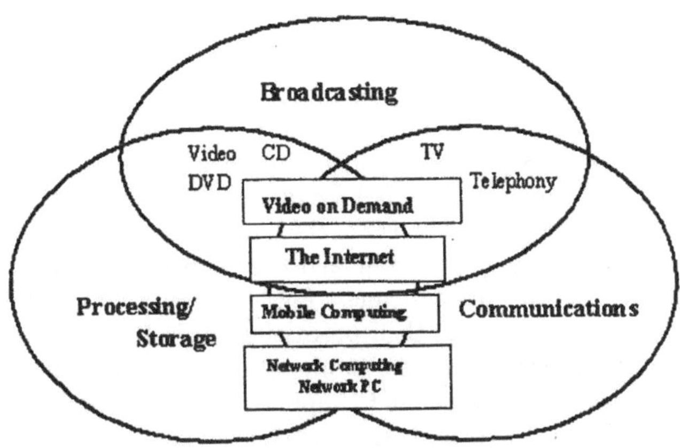

Figure 1.5 Multimedia systems and services.

On broadcasting side, the introduction of digital video and audio has created new business opportunities. A single channel of analog NTSC TV with 6 MHz bandwidth on terrestrial broadcasting can carry around 20 Mb/s by using digital modems, while MPEG-2 compression requires 4 to 9 Mb/s for a single video channel. Additional 3-4 channels on average become available on a single television channel bandwidth of conventional analog broadcasting. Many new video channels become available by using digital compression. In the same way, satellite transmitter can send around 30 Mb/s. Therefore, satellite CATV makes sense when digital transmission is employed. Furthermore, as the transmission has been carried out in digital form, HDTV program broadcasting becomes easier by using several digital television channels.

VOD (video on demand) addresses the new services in the overlapped areas of the three worlds in Fig. 1.5. In the system, a server machine is a computer system itself and manages video libraries stored in a set of large storage systems. Digital wide band transmission channels connect between the server (sometimes referred to head-end) and clients. The Video server (VOD head-end) sends a selected video by a client when it is requested. As the channel connected between a client terminal and the server is for sole use, client terminals can ask the server for the operations widely used in Video Cassette Recorders such as pause, rewind, fast forward, search and so on.

One big problem on multimedia services is the protection of author's copyright. Digital video and audio are going to be delivered through multimedia network and DVD storage. PCs in near future will become powerful enough to edit video and audio. Then, it is natural to use a part of existing materials to create new multimedia materials. When someone asks for originality in their multimedia material, they want to put some marks in it. Employing watermark in video and audio become important issue for protecting author's copyright. Recent signal processing technology enables watermarking without noticeable degradation.

The signature of the authors is added into the video and audio by using spread spectrum technique, for example. Although watermark technology is important, this technology is still in the infant phase. This book also covers up-to-date research activities on the watermark (see Chapter 17).

1.4 HARDWARE IMPLEMENTATION

Low cost implementation of multimedia terminals is the key issue, as described in Section 1.2 [7][8]. Thanks to the VLSI technology progress, the hardware cost is quickly decreasing and the establishment of the multimedia world is going to become a reality. However, even today, only high end microprocessors have a capability to decode MPEG-2 bit-stream in real time. Fig. 1.6 shows classification of recent programmable chips together with their processing capability. The upper direction shows general purpose RISC chips for workstations and the upper-left direction shows general purpose CISC chips for PC applications. The lower-left direction is embedded RISC chips for PDAs and Game machines. The lower direction indicates programmable DSP chips. The lower-right direction is for PCs which assist engine chips, called media processors.

After the introduction of the pentium chip, the difference between RISC chips and CISC chips has become small. This is because pentium employs pipeline arithmetic units and out-of-order super-scalar approach, both of which were first introduced to RISC processors for improving processing capability. The penalty of such approaches is the complex and huge control units on a chip. More than 50 % of die area is used for these units. As a result, power dissipation is around 20-30 watts.

The real-time MPEG-2 decoding requires around 1 giga operations per second and therefore it is impossible to decode it by using a less than 1 GIPS microprocessor with conventional architectures. Some of these chips employ the split ALU approach for real time MPEG-2 decoding, where a 64 bit ALU is divided into 4 different 16 bit ALUs under SIMD control. One of the recent embedded RISC chips also employ this split ALU approach for realizing real-time MPEG-1 encoding or MPEG-2 decoding. As complex control units are not employed in the embedded chip, the power dissipation can be reduced to about 1.5 watts. DSP chips are mainly employed in wireless communication terminals for realizing low bit-rate speech coding. This is because these DSP chips have extremely low power dissipation of less than 100 mW. Unfortunately, however, their processing capability does not reach the level of real time MPEG-2 decoding. For PDA use, video communications using MPEG-4 is reasonable in terms of compact and low cost realization. Wireless communications discussed in IMT-2000 (future mobile communications system) are considered to be around 64 Kb/sec. Therefore, QCIF (1-16th resolution of NTSC) and 7.5 to 15 frames/sec video can be compressed into this bit rate. Then, low power DSP chips can provide the MPEG-4 decoder function, due to the small amount of information. Media processors in the lower-right direction is an expansion of programmable DSP chips for PC support, including real-time MPEG-2 decoding. They have employed multiple number of processing units which

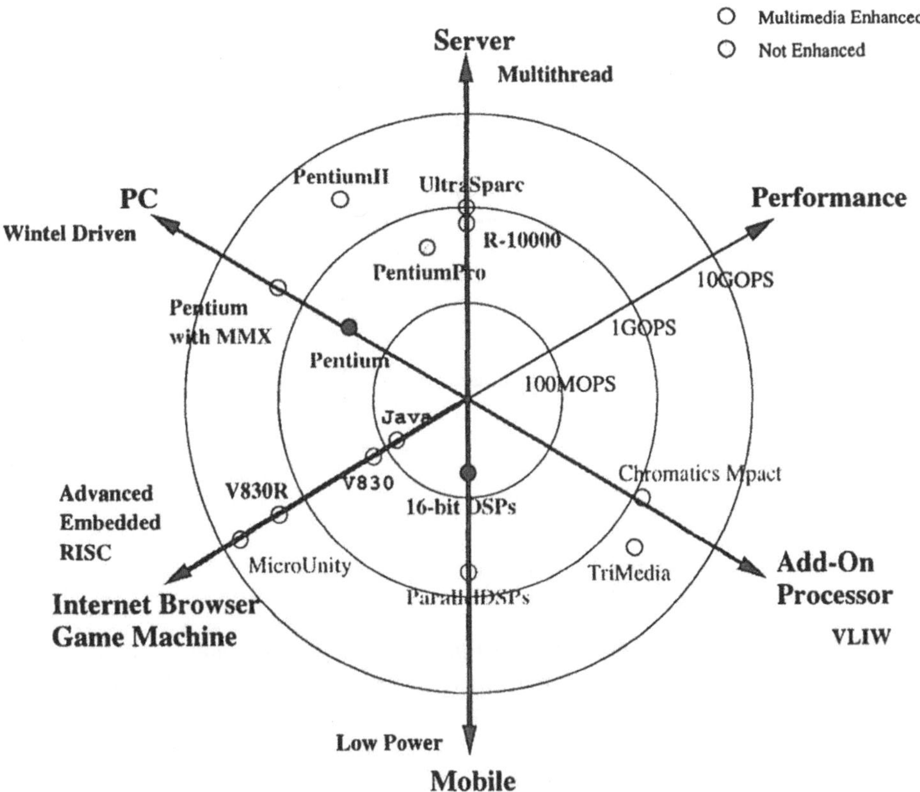

Figure 1.6 Multimedia processor classification.

are controlled by a VLIW (very long instruction word) approach. As their clock frequency is relatively slow, their power dissipation is around 4 Watts.

Although programmable chips have around 1GOPS processing capability which is adequate for real-time MPEG-2 decoding, real-time MPEG-2 encoding is far beyond their capability. Let us evaluate the complexity of motion estimation in a video encoding. The compression coding is carried out by extracting a past picture information from the current picture. Only residual components between two pictures are encoded. The motion estimation is used for improving extraction of the past picture information by compensating pre-estimate motion in the past picture, if there is some movement.

The motion estimation function is a set of pattern matching processes between a 16 × 16 pixel segment in the current frame picture and the whole reference frame picture (a past picture in many cases) to find out the most similar segment of 16 × 16 pixels. The motion information is then obtained from the location distance between the current segment and the detected segment. From the 720 × 480 pixel current frame, there exists 1350 different 16 × 16 pixel segments: 45 segments in the horizontal direction and 30 in the vertical direction. In Fig. 1.7, the search area in the past picture is limited to the square region covering the motion of -16 to +16 pixel positions in both horizontal and vertical directions for

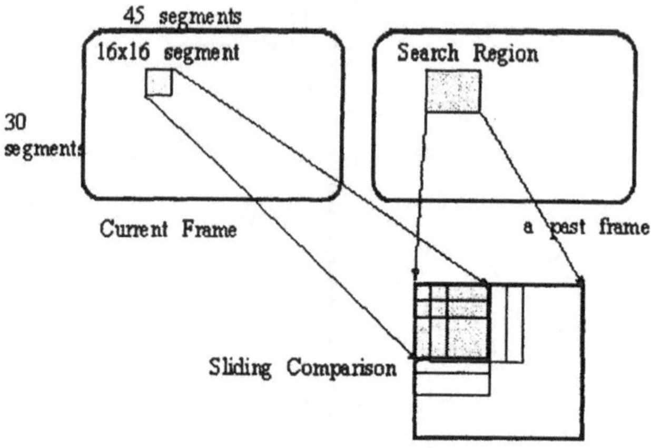

Figure 1.7 Motion estimation approach.

each 16×16 segment. This motion area limitation maybe reasonable, if the video is concerned with face-to-face communication. As MPEG-2 allows motion of half pixel positions, 16 × 16 current segment should compare with 64 different positions in both horizontal and vertical directions. The L1 distance measure, where absolute difference between corresponding pixels over the segments is accumulated, is used for best match criteria. Therefore, 16x16 absolute operations are required for one possible motion evaluation. As a result, the total number of operations per second can be calculated as

$$1350 \times (64 \times 64) \times (16 \times 16) \times 30 = 40\ GOPS. \quad (1)$$

This means that the above limited region motion estimation still requires more than 40 GOPS, as operations of DCT/IDCT and variable bit-length encoder should be included. According to the MPEG-2 specification, search regions are expandable to the whole frame pictures. Then, the required processing exceeds more than 10 Tera operations per second. These facts demand use of application-specific systems.

When an application-specific LSI is designed, for example, for an MPEG-2 encoder, the required processing capability is highly dependent on the employed algorithm. There are a lot of simplified algorithms for motion compensation, but the simplification of the algorithm generates some degradation in the reconstructed image. The trade-off between the hardware complexity and quality is an important issue with application-specific design methodology.

In the architecture design, pipelining hardware increases processing capability with small penalty of register insertion, where one processing is divided into two parts, for example, by inserting registers.. Then, first processing can be activated just after the second processing starts. Register insertion enables doubling the processing speed. In case of motion estimation described above or a set of Matrix and Vector products, a parallel pipeline structure, called Systolic Array, can be used. Many motion compensation chips and/or codec chips including motion compensation have employed systolic array, due to their regularity and simplicity in high speed processing. The advanced systolic array approaches are summarized in Chapter 23 in this book.

For channel coding part in Fig. 1.4, some processing is carried out in finite field and operations required there are quite different from the conventional ALUs. Therefore, this book shares hardware implementation of such arithmetic units in Chapter 21. Viterbi decoders are also included in Chapter 16.

Low power design is another important issue for hardware implementation, because low power realization of multimedia functions enable long life battery support for portable systems (see Chapter 24). Probably the most important market for the multimedia world in next generation is this portable terminal area which had not existed before. Fig. 1.8 shows our experimental multimedia equipment which enables watching video news in a commuter, every morning. In this system, video news is stored in a PCMCIA card in the form of MPEG-1. Audio quality is comparable to that of compact disk and MPEG-1 quality is quite reasonable for small screens. Although this system does not include wireless communication capability at the moment, future versions will surely provide functions for browsing the internet by WWW. When such a compact terminal has to send and receive video with audio of Mb/s digital information, adaptive antenna systems, for example, should catch the desired signal among noisy environment. This is the reason why high-speed VLSI implementations of adaptive algorithm are included in this book (see Chapter 18 and 19). More basic arithmetic operations such as division and square root, and primary functions such as sin, cos, log and exp functions might be included in such a system and these functions are also important for computer graphics. Therefore, design of division and square root arithmetic units and design of CORDIC algorithms for primary functions are described in Chapters 20 and 22, respectively.

Figure 1.8 Silicon view: a compact video terminal with MPEG-1 compressed video in a PCMCIA card.

As low power implementation approaches become very important, power estimation during design phases in architecture, logic, and circuit levels in an efficient way is important. In the book, some estimation approaches are included in Chapter 25 for this purpose.

Although many chapters are concerned with the technology of today's technologies, all of the chapters are related to basic technology to develop multimedia worlds. Recognition of Video/Image/Speech as an user friendly interface and biometric user identification will surely join this new world, but the basic technologies described here can support these functions. Moreover, processor architectures and low power VLSI implementations should be advanced based on the basic technologies. However, technologies in different areas are also under fusion in the multimedia world. For example, MPEG-4 uses a set of structured pictures based on objects. However, at the moment, this structure is given by manual instructions, because MPEG is an activity on decoder specifications. Object extraction from arbitrary pictures is now a hot topic in these days, but this area has also some overlap with image recognition.

REFERENCES

[1] N. S. Jayant and Peter Noll, *Digital Coding of Waveforms*, Prentice-Hall Signal Processing Series, Prentice-Hall, Inc. 1984.

[2] J. G. Proakis, *Digital Communications*, (Third Edition), McGraw-Hill, Inc.

[3] V. Bhaskaran and K. Konstantinides, *Image and Video Compression Standards*, Kluwer Academic Publishers 1995.

[4] Peter Noll, "Wideband Speech and Audio Coding," *IEEE Communications magazine*, vol. 31, no. 11, Nov. 1993.

[5] S. Okubo *et al.*, "ITU-T standardization of audiovisual communication systems in ATM and LAN environments," *IEEE Journal on Selected Areas in Communications*, vol. 15, no. 6, August 1997

[6] Takao Nishitani, "Trend and perspective on domain specific programmable chips," *Proc. of IEEE SiPS Workshop*, 1997.

[7] R.V. Cox *et al.*, "On the applications of multimedia processing to communications," *Proc. IEEE*, vol. 86, no. 5, May 1998.

[8] K.K. Parhi, *VLSI Digital Signal Processing Systems: Design and Implementation,* John Wiley and Sons, 1999.

Chapter 2

Video Compression

Keshab K. Parhi
Department of Electrical and Computer Engineering
University of Minnesota, Minneapolis, Minnesota
parhi@ece.umn.edu

2.1 INTRODUCTION

Digital video has many advantages over analog video. However, when the video signal is represented in digital format, the bandwidth is expanded substantially. For example, a single frame in high-definition TV (HDTV) format (with a frame size of 1920×1250 pixels and a frame rate of 50 frame/sec) requires storage size of 57.6 mega bytes and a video source data rate of 2.88 Giga bits per sec (Gb/sec). A two-hour HDTV movie requires about 414 giga bytes. Even with a capable storage device, there is no existing technology which can transmit and process motion video at such a high speed. In order to alleviate the bandwidth problem while taking advantage of digital video, various video compression techniques have been developed. This chapter summarizes some of the key concepts and provides hardware designers with the basic knowledge involved in these commonly used video coding techniques.

This chapter is organized as follows. Section 2.2 reviews the basic concepts of *lossless* coding schemes such as Huffman coding, arithmetic coding and run length coding. The compression ratios achievable by lossless schemes is limited. In contrast *lossy* compression schemes, discussed in schemes 2.4 and 2.5, give up exact reconstruction but achieve a significant amount of compression. Transform-based coding techniques are addressed in Section 2.3; these include discrete cosine transform, wavelet transform, vector quantization and reordering of quantized transform coefficients. The key scheme used in video compression, motion estimation and compensation, is addressed in Section 2.4. Section 2.5 provides and overview of some key features of the MPEG-2 video compression standard. Finally, the design challenges rendered by these sophisticated video coding schemes are discussed in Section 2.6.

2.2 ENTROPY CODING TECHNIQUES

The *first-order entropy* H of a memoryless discrete source containing L symbols is defined as follows:

$$H = \sum_{i=1}^{L} -p_i \log_2 p_i, \tag{1}$$

where p_i is the probability of the ith symbol. The entropy of a source has the unit *bits per symbol*, or bits/symbol, and it is lower bounded by the average codeword length required to represent the source symbols. This lower bound can be achieved if the codeword length for the ith symbol is chosen to be $-\log_2 p_i$ bits, i.e., assigning shorter codewords for more probable symbols and longer codewords for less probable ones. Although $-\log_2 p_i$ bits/symbol may not be practical since $-\log_2 p_i$ may not be an integer number, the idea of *variable length coding* which represents more frequently occurred symbols by shorter codewords and less frequently occurred symbols by longer codewords can be applied to achieve data compression. The data compression schemes which use source data statistics to achieve close-to-entropy bits/symbol rate are referred to as *entropy coding*. Entropy coding is *lossless*, since the original data can be exactly reconstructed from the compressed data.

This section briefly reviews the two most-frequently-used entropy coding schemes, *Huffman coding* [1] and *arithmetic coding* [2]. This section also includes another type of lossless source coding scheme, the *run-length coding*. It converts a string of same symbols into an intermediate length indication symbols called *run-legnth codes* and is often used together with entropy coding schemes to improve data compression rate.

2.2.1 Huffman Coding

When the probability distribution of a discrete source is known, the Huffman coding algorithm provides a systematic design procedure to obtain the optimal variable length code. Two steps are involved in design of Huffman codes: symbol merging and code assignment; these are described as follows:

1. **Symbol merging**: formulate the Huffman coding tree.
   (a) Arrange the symbol probabilities p_i in a decreasing order and consider them as leaves of a tree.
   (b) Repeat the following merging procedure until all branches merge into a root node:
      i. Merge the two nodes with smallest probabilities to form a new node with a probability equal to the sum probability of the two merged nodes;
      ii. Assign '1' and '0' to the pair of branches merging into one node.
2. **Code assignment**: The codeword for each symbol is the binary sequence from the root of the tree to the leaf where the probability of that symbol is located.

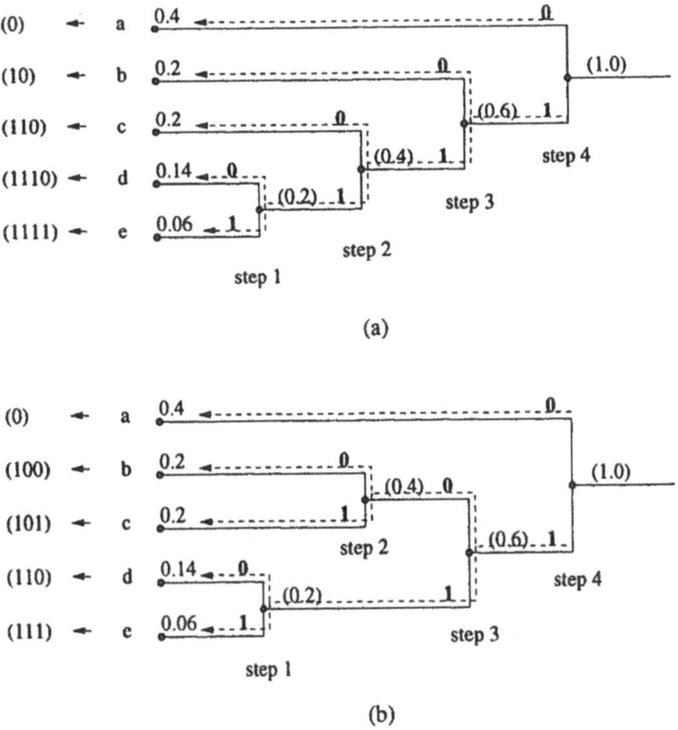

Figure 2.1 Huffman coding examples.

Example 2.2.1 *Consider a discrete source containing 5 symbols* $\{a, b, c, d, e\}$ *with probability distribution* $\{0.4, 0.14, 0.2, 0.2, 0.06\}$. *The Huffman coding procedure and the resulting Huffman codes are illustrated in Fig. 2.1. Note that there may be a tie of probability during the merging process. For example, in step 2 in Fig. 2.1(a), the merged probability of symbols d and e equals the probabilities of symbols b and c. In case of a tie, the choice of merging can be arbitrary, and the resulting codes may be different but have the same average bit rate and hence compression rate, as can be verified using the two code examples in Fig. 2.1(a) and (b).*

Huffman code is uniquely decodable. Once the codebook is generated, the encoding procedure can be carried out by mapping each input symbol to its corresponding codeword which can be stored in a look-up table (the codebook). The decoding procedure includes parsing codewords from a concatenated codeword stream and mapping each codeword back to the corresponding symbol using the Huffman codebook. One important property of Huffman codes is that no codes and any code combinations are prefix of any other codes. This *prefix condition* enables parsing of codewords from a concatenated codeword stream and eliminates the overhead of transmitting parsing positions. Conceptually, the codeword parsing can be carried out bit-by-bit by traversing the Huffman coding tree. The parsing starts from the root of the tree; at each intermediate node, a decision is made according to the

Table 2.1 Huffman Codebook in Example 2.2.1

symbols	*a*	*b*	*c*	*d*	*e*
codeword	*0*	*10*	*110*	*1110*	*1111*

received bit until the terminal node (the leaf) is reached; then a codeword is found and the corresponding bits are parsed from the bit stream.

Example 2.2.2 *This example illustrates the encoding and decoding procedures using the Huffman codes generated in Example 2.2.1 (Fig. 2.1(a)) whose codebook is shown in Table 2.1. Consider the source data sequence dbaaec. Using the codebook table, the corresponding codeword stream is computed as* 111010001111110. *At the decoder side, this bit stream can be parsed as* 1110, 10, 0, 0, 1111, 110 *and mapped back to the symbol sequence dbaaec.*

2.2.2 Arithmetic Coding

In arithmetic coding, the symbol probabilities p_i should also be known *a priori* or estimated on the fly. With known source data probability distribution, the arithmetic coding partitions the interval between 0 and 1 into sub-intervals according to symbol probabilities, and represents symbols by the mid-points of the sub-intervals.

Consider single-symbol based arithmetic coding of the ordered symbol set $\{a_i,\ 1 \leq i \leq L\}$ with probability distribution $\{p_i\}$. Let P_i denote the accumulative probability from the 1st symbol to the ith symbol, i.e., $P_i = \sum_{k=1}^{k=i} p_k$. In arithmetic coding, the interval $[0,1]$ is partitioned into L sub-intervals, $\{[0, P_1], [P_1, P_2], \cdots, [P_{L-1}, P_L = 1]\}$, and the ith interval $I(a_i) = [P_{i-1}, P_i]$ is assigned to the ith symbol a_i (for $1 \leq i \leq L$), as illustrated in Fig. 2.2(a). The binary representation of the mid-point of the ith interval is then computed, and the first $W(a_i)$-bits (after the point) is the arithmetic codeword for the symbol a_i (for $1 \leq i \leq L$), where $W(a_i) = \lceil \log_2(1/p_i) \rceil + 1$.

Example 2.2.3 *For the symbol set $\{a, b\}$ and the probability distribution of $p_0 = 1/4$, $p_1 = 3/4$, the interval $[0,1]$ is partitioned into two sub-intervals, $I(a) = [0, 1/4]$ and $I(b) = [1/4, 1]$. With $W(a) = [\lceil \log_2 4 \rceil + 1 = 3$ and $W(b) = \lceil \log_2 4/3 \rceil + 1 = 2$, the arithmetic codes for the symbols a and b are computed as* 001 *and* 10, *which are the first 3 bit of the binary representation of* 1/8 *(the mid-point of the interval $I(a)$) and the first 2 bits of the binary representation of* 5/8 *(the mid-point the interval $I(b)$), respectively. This is illustrated in Fig. 2.2(b).*

Arithmetic coding processes a string of symbols at a time to achieve better compression rate. It can usually outperform the Huffman code. The arithmetic coding of a symbol string of length l, $S = \{s_1, s_2, \cdots, s_l\}$, is carried out through

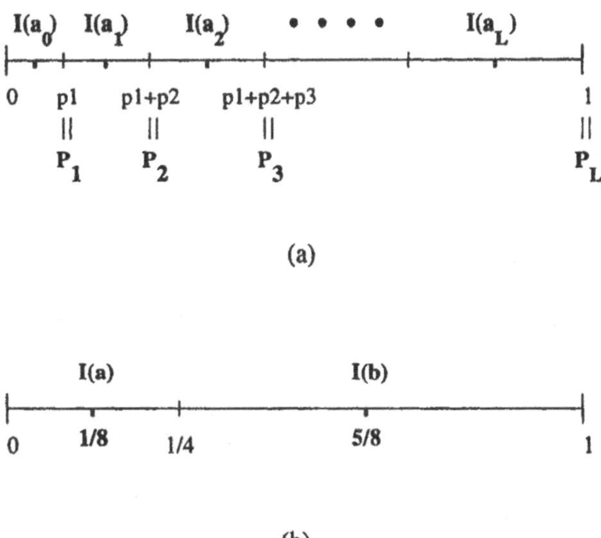

Figure 2.2 Sub-interval partitions in single-symbol based arithmetic coding for (a) a general example, (b) the Example 2.2.3.

l iterative sub-interval partitions based on the statistics of the symbol set, i.e., the probability distribution and conditional probabilities. The length of each sub-interval equals the probability of its corresponding symbol string. The arithmetic codeword of a symbol string S is the first W bits in the binary representation of the mid-point value of its corresponding sub-interval $I(S)$, where $W = \lceil \log_2 1/|I(S)| \rceil + 1$ and $|I(S)|$ denotes the length of the interval $I(S)$.

Example 2.2.4 *This example illustrates the arithmetic coding process for a symbol string from the symbol set in Example 2.2.3. Assume that the symbols in the source sequence are independent and identically distributed (iid). Consider the 4-symbol string $S = bbab$. Its arithmetic coding includes five steps, as shown in Fig. 2.3. In step 1, the interval $[0, 1]$ is partitioned into two sub-intervals based on the probabilities of a and b, and $I(a) = [0, 1/4]$ and $I(b) = [1/4, 1]$. Since the first symbol in string S is b, the second sub-interval is retained and passed to the next iteration. In step 2, the sub-interval $I(b)$ is partitioned into two sub-intervals, $I(ba) = [1/4, 7/16]$ and $I(bb) = [7/16, 1]$, based on the conditional probabilities $p(a|b)$ and $p(b|b)$, which, respectively, equal to $p(a)$ and $p(b)$ for iid source. According to the value of the second symbol, the sub-interval $I(bb)$ is retained and passed to the next iteration. Similarly, in step 3 the sub-interval $I(bba) = [7/16, 121/256]$ is retained and passed to the 4th iteration; the sub-interval $I(bbab) = [121/256, 37/64]$ obtained in step 4 is the final sub-interval for the symbol string $S = bbab$. Finally, in step 5, the binary representation of the mid-point of the sub-interval $I(bbab) = [121/256, 37/64]$, which equals to 269/512, is computed and the first bits, 10000, constitute the arithmetic codeword for symbol string $S = bbab$.*

$$\lceil \log_2(1/|I(bbab)|\rceil + 1 = \lceil \log_2 256/27 \rceil + 1 = 5 \quad (2)$$

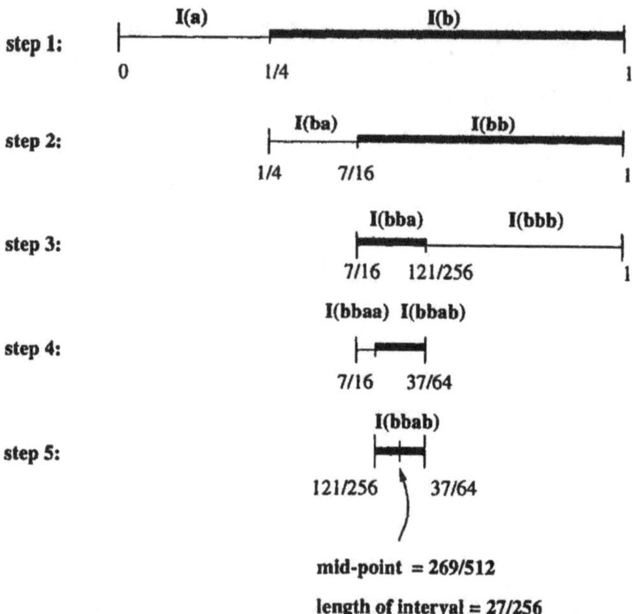

Figure 2.3 Arithmetic coding process for symbol string *bbab* in Example 2.3.

2.2.3 Run Length Coding

In run length code, a string of same symbols are represented using one length indication symbol and one value indication symbol. For example, the run length code for the source symbol sequence $\{0,0,0,0,0,3,0,0,0,5,6\}$ is $\{(\#5,0), (\#1,3), (\#3,0), (\#1,5), (\#1,6)\}$, where the value after # is the length indicator. These length and value indication symbols in run length codes can be coded using entropy coding schemes. For binary sequence, the consecutive strings have alternate values of 0 and 1 and these values need not be explicitly shown. Hence, only the length indication symbol and the first value of the whole sequence are required in the rub length code of a binary sequence. For example, the binary sequence $\{0,0,0,0,1,1,1,0,0,1,1,1\}$ can be run-length coded as $\{0, \#4, \#3, \#2, \#3\}$.

For data sequences corresponding to digital images, there are some high-probability symbols that always occur consecutively, such as zeros. In this case, only these symbol strings are run-length coded into intermediate symbols, and these intermediate symbols and the rest of the original source symbols are then coded using entropy coding schemes. For example, the sequence $\{0,0,0,0,0,3,0,0,0,5,6\}$ can first be run-length coded as $\{(\#5,3), (\#3,5), (\#0,6)\}$, where the second value in the little bracket is the value of successive non zero symbols and the first value in the little bracket indicates the number of its preceding consecutive zeros.

2.3 TRANSFORM CODING TECHNIQUES

Transform coding techniques have the tendency of packing a large fraction of average energy of the image into a relatively small component of the transform coefficients, which after quantization contain long runs of zeros. A transform-based coding system contains the following steps: transform (or decomposition) of the image blocks (or the image), quantization of the resulting coefficients, reordering of the quantized coefficients and formulating the output bit streams; these techniques are addressed respectively in this section. Two transforms are considered, including the discrete cosine transform and the wavelet transform.

2.3.1 Discrete Cosine Transform

The discrete cosine transform (DCT) was first introduced for pattern recognition in image processing and Wiener filtering [3]. DCT is an orthogonal transform which decorrelates the signals in one image block and compacts the energy of the whole image block into a few low frequency DCT coefficients. It has been incorporated into both still image and video compression standards due to its energy-compaction, and each of implementation. This section introduces the derivation of even symmetrical 1-D DCT.

Consider a N-point sequence $x(n)$, i.e., $x(n) = 0$ for $n < 0$ and $n > N - 1$. The N-point DCT and IDCT (inverse DCT) pair for this sequence is defined as:

$$X(k) = e(k) \sum_{n=0}^{N-1} x(n) \cos[\frac{(2n+1)k\pi}{2N}], \; k = 0, 1, \cdots, N-1 \tag{3}$$

$$x(n) = \frac{2}{N} \sum_{k=0}^{N-1} e(k) X(k) \cos[\frac{(2n+1)k\pi}{2N}], \; n = 0, 1, \cdots, N-1, \tag{4}$$

where

$$e(k) = \begin{cases} \frac{1}{\sqrt{2}}, & if \; k = 0, \\ 1, & otherwise. \end{cases} \tag{5}$$

The N-point DCT and IDCT pair can be derived using a $2N$-point discrete Fourier transform (DFT) pair. Construct a $2N$-point sequence $y(n)$ using $x(n)$ and its mirror image as follows:

$$y(n) = x(n) + x(2N - n - 1) = \begin{cases} x(n), & 0 \leq n \leq N-1 \\ x(2N-n-1), & N \leq n \leq 2N-1. \end{cases} \tag{6}$$

Hence $y(n)$ is symmetric with respect to the mid-point at $n = N - 1/2$. Fig. 2.4 shows an example for $N = 5$.

The $2N$-point DFT of $y(n)$ is given by

$$\begin{aligned} Y_D(k) &= \sum_{n=0}^{2N-1} y(n) e^{-j\frac{2\pi}{2N}kn} \\ &= \sum_{n=0}^{N-1} x(n) e^{-j\frac{2\pi}{2N}kn} + \sum_{n=N}^{2N-1} x(2N-n-1) e^{-j\frac{2\pi}{2N}kn}, \end{aligned} \tag{7}$$

for $0 \leq k \leq 2N - 1$. Substituting $n = 2N - n' - 1$ into the second summation in (7), we obtain

$$\sum_{n=N}^{2N-1} x(2N-n-1) e^{-j\frac{2\pi}{2N}kn} = \sum_{n'=N-1}^{0} x(n') e^{-j\frac{2\pi}{2N}k(2N-n'-1)}$$

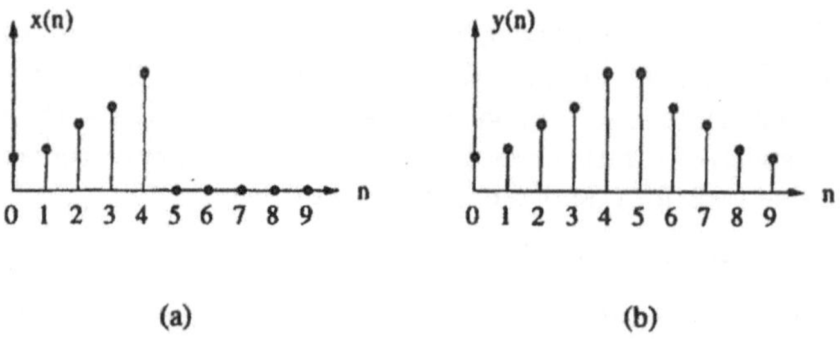

Figure 2.4 Relation between (a) N-point sequence $x(n)$ and (b) $2N$-point sequence $y(n) = x(n) + x(2N - n - 1)$.

$$= \sum_{n'=0}^{N-1} x(n') e^{j\frac{2\pi}{2N}kn'} e^{j\frac{2\pi}{2N}k}. \tag{8}$$

With (8), (7) can be rewritten as

$$\begin{aligned} Y_D(k) &= \sum_{n=0}^{N-1} x(n) e^{-j\frac{2\pi}{2N}kn} + \sum_{n=0}^{N-1} x(n) e^{j\frac{2\pi}{2N}kn} e^{j\frac{2\pi}{2N}k} \\ &= e^{j\frac{k\pi}{2N}} \left(\sum_{n=0}^{N-1} x(n) e^{-j\frac{(2n+1)k\pi}{2N}} + \sum_{n=0}^{N-1} x(n) e^{j\frac{(2n+1)k\pi}{2N}} \right) \\ &= e^{j\frac{k\pi}{2N}} \sum_{n=0}^{N-1} 2x(n) \cos\left(\frac{(2n+1)k\pi}{2N}\right). \end{aligned} \tag{9}$$

Define

$$\hat{X}(k) = \begin{cases} Y_D(k) e^{-j(\frac{k\pi}{2N})}, & 0 \le k \le N-1 \\ 0, & \mathit{otherwise}. \end{cases} \tag{10}$$

Then the N-point DCT can be expressed as $X(k) = e(k)\hat{X}(k)/2$.

The inverse DCT is derived by relating $Y_D(k)$ to $X(k)$, computing $y(n)$ from $Y_D(k)$ using the inverse DFT, and reconstructing $x(n)$ from $y(n)$. Although $Y_D(k)$ is a $2N$-point sequence and $X(k)$ is a N-point sequence, the redundancy (symmetry) in $y(n)$ enables $Y_D(k)$ to be expressed using $X(k)$. For $0 \le k \le N-1$, $Y_D(k) = e^{j\frac{k\pi}{2N}} \hat{X}(k)$; $Y_D(N) = 0$. For $N+1 \le k \le 2N-1$, $1 \le 2N-k \le N-1$. Therefore,

$$Y_D(2N-k) = e^{j\frac{(2N-k)\pi}{2N}} \hat{X}(2N-k) = -e^{-j\frac{k\pi}{2N}} \hat{X}(2N-k). \tag{11}$$

On the other hand, from (9),

$$\begin{aligned} Y_D(2N-k) &= e^{j\frac{(2N-k)\pi}{2N}} \sum_{n=0}^{N-1} 2x(n) \cos\left(\frac{(2n+1)(2N-k)\pi}{2N}\right) \\ &= -e^{j\frac{2N\pi}{2N}} e^{-j\frac{k\pi}{2N}} \sum_{n=0}^{N-1} 2x(n) \cos\left(\frac{(2n+1)k\pi}{2N}\right) \end{aligned}$$

$$
\begin{aligned}
&= e^{-j\frac{2k\pi}{2N}} e^{j\frac{k\pi}{2N}} \sum_{n=0}^{N-1} 2x(n)\cos(\frac{(2n+1)k\pi}{2N}) \\
&= e^{-j\frac{2k\pi}{2N}} Y_D(k). \qquad (12)
\end{aligned}
$$

Hence,

$$
\begin{aligned}
Y_D(k) &= e^{j\frac{2k\pi}{2N}} Y_D(2N-k) \\
&= -e^{j\frac{2k\pi}{2N}} e^{-j\frac{k\pi}{2N}} \hat{X}(2N-k) \\
&= -e^{j\frac{k\pi}{2N}} \hat{X}(2N-k), \qquad (13)
\end{aligned}
$$

for $N+1 \le k \le 2N-1$. Therefore, we have

$$
Y_D(k) = \begin{cases} e^{j\frac{k\pi}{2N}} \hat{X}(k), & 0 \le k \le N-1 \\ 0, & k = N \\ -e^{j\frac{k\pi}{2N}} \hat{X}(2N-k), & N+1 \le k \le 2N-1. \end{cases} \qquad (14)
$$

Taking the inverse DFT of $Y_D(k)$, we have

$$
\begin{aligned}
y(n) &= \frac{1}{2N} \sum_{k=0}^{2N-1} Y_D(k) e^{j\frac{2\pi}{2N}kn} \qquad (15) \\
&= \frac{1}{2N}(\sum_{k=0}^{N-1} \hat{X}(k) e^{j\frac{(2n+1)k\pi}{2N}} + \sum_{k=N+1}^{2N-1} (-e^{j\frac{k\pi}{2N}} \hat{X}(2N-k)) e^{j\frac{2\pi}{2N}kn}). \qquad (16)
\end{aligned}
$$

After change of variable in the second term and some algebraic manipulation, and using $1/e(0) = 2e(0)$ and $1/e(k) = e(k)$ for $k \neq 0$, (16) can be rewritten as

$$
\begin{aligned}
y(n) &= \frac{1}{2N}(\sum_{k=0}^{N-1} \hat{X}(k) e^{j\frac{(2n+1)k\pi}{2N}} + \sum_{k=1}^{N-1} \hat{X}(k) e^{-j\frac{(2n+1)k\pi}{2N}}) \\
&= \frac{1}{2N}(\hat{X}(0) + 2\sum_{k=1}^{N-1} \hat{X}(k)\cos(\frac{(2n+1)k\pi}{2N})) \\
&= \frac{2}{N}(\ e(0)X(0) + \sum_{k=1}^{N-1} X(k)e(k)\cos(\frac{(2n+1)k\pi}{2N})\ ), \qquad (17)
\end{aligned}
$$

for $0 \le n \le 2N-1$. The inverse DCT, obtained by retaining the first N points of $y(n)$, is given by

$$
x(n) = y(n) = \frac{2}{N} \sum_{k=0}^{N-1} e(k)X(k)\cos(\frac{(2n+1)k\pi}{2N}), \qquad (18)
$$

for $0 \le n \le N-1$.

Express the N-point sequences $x(n)$ and $X(k)$ in vector form as

$$
\mathbf{x} = \begin{bmatrix} x(0) \\ x(1) \\ \cdots \\ x(N-1) \end{bmatrix}, \quad \mathbf{X} = \begin{bmatrix} X(0) \\ X(1) \\ \cdots \\ X(N-1) \end{bmatrix}, \qquad (19)
$$

and the DCT transform in (3) in matrix form as

$$\Lambda = \begin{bmatrix} 1/\sqrt{2} & 1/\sqrt{2} & \cdots & 1/\sqrt{2} \\ \cos(\frac{\pi}{2N}) & \cos(\frac{3\pi}{2N}) & \cdots & \cos(\frac{(2N-1)\pi}{2N}) \\ & \cdots & \cdots & \\ \cos(\frac{(N-1)\pi}{2N}) & \cos(\frac{3(N-1)\pi}{2N}) & \cdots & \cos(\frac{(2N-1)(N-1)\pi}{2N}) \end{bmatrix}. \quad (20)$$

The DCT and IDCT coefficients can be computed using

$$\mathbf{X} = \mathbf{\Lambda x}, \quad \mathbf{x} = \frac{2}{N}\mathbf{\Lambda}^T\mathbf{X}. \quad (21)$$

This leads to $\mathbf{\Lambda\Lambda}^T = \frac{N}{2}\mathbf{I}_{N\times N}$, where $\mathbf{I}_{N\times N}$ is the identity matrix of dimension $N \times N$. Therefore, DCT is an orthogonal transform.

In image processing, one image frame is divided into $N \times N$ blocks and a separable two-dimensional DCT (2D-DCT) is applied to each $N \times N$ image block. An N-point one-dimensional DCT in (3) requires N^2 multiplications and additions. Direct computation of 2D DCT of length-N requires N^4 multiplications and additions. On the other hand, by utilizing the separability of 2D-DCT, it can be computed by performing N 1D DCT's on the rows of the image block followed by N 1D-DCT's on the resulting columns [4]. With this simplification, $N \times N$ 2D-DCT requires $2N^3$ multiply-add operations, or $4N^3$ arithmetic operations.

2.3.2 Wavelet-Based Image Compression

Wavelet transform is a multiresolution orthonormal transform [5]-[7]. It decomposes the signal to be represented into a band of energy which is sampled at different rates. These rates are determined to maximally preserve the information of the signal while minimizing the sampling rate or resolution of each subband.

In wavelet analysis, signals are represented using a set of basis functions (wavelets) derived by shifting and scaling a single prototype function, referred to as "mother wavelet", in time. One dimensional discrete-wavelet-transform (DWT) of $x(n)$ is defined as

$$\begin{aligned} y_i(n) &= \sum_{k=-\infty}^{\infty} x(k)h_i(2^{i+1}n - k), \ for \ 0 \le i \le m-2, \\ y_{m-1}(n) &= \sum_{k=-\infty}^{\infty} x(k)h_{m-1}(2^{m-1}n - k), \ for \ i = m-1, \end{aligned} \quad (22)$$

where the shifted and scaled versions of the mother wavelet $h(n)$, $\{h_i(2^{i+1}n - k),\ for\ 0 \le i \le m-1, -\infty < k < \infty\}$ are the basis functions, and $y_i(n)$ are the ***wavelet coefficients***. The inverse transform is computed as follows:

$$\begin{aligned} x(n) &= \sum_{i=0}^{m-2} \sum_{k=-\infty}^{\infty} y_i(k)f_i(n - 2^{i+1}k) \\ &+ \sum_{k=-\infty}^{\infty} y_{m-1}(k)f_{m-1}(n - 2^{m-1}k), \end{aligned} \quad (23)$$

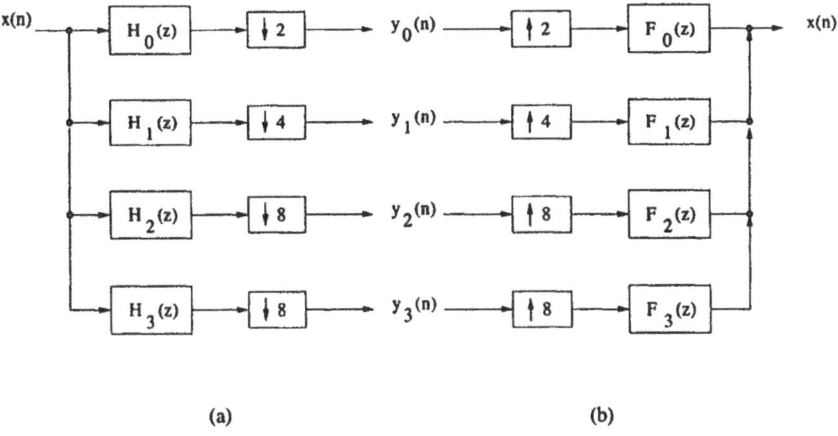

Figure 2.5 Analysis and synthesis filter banks for DWT and IDWT.

where $\{f_i(n - 2^{i+1}k)\}$ is designed such that (23) perfectly reconstructs the original signal $x(n)$. Note that the computations in DWT and IDWT are similar to convolution operations. In fact, the DWT and IDWT can be calculated recursively as a series of convolutions and decimations, and can be implemented using filter banks.

A digital *filter bank* is a collection of filters with a common input (referred to as the analysis filter bank) or a common output (referred to as the synthesis filter bank). Filter banks are generally used for subband coding, where a single signal $x(n)$ is split into m subband signals in the analysis filter bank; in the synthesis filter bank, m input subband signals are combined to reconstruct the signal $y(n)$.

Consider the computation of the discrete wavelet transform for $m = 4$ using filter banks. The wavelet coefficients

$$\begin{aligned} y_0(n) &= \sum_{k=-\infty}^{\infty} x(k)h_0(2n-k), \\ y_1(n) &= \sum_{k=-\infty}^{\infty} x(k)h_1(4n-k), \\ y_2(n) &= \sum_{k=-\infty}^{\infty} x(k)h_2(8n-k), \\ y_3(n) &= \sum_{k=-\infty}^{\infty} x(k)h_3(8n-k) \end{aligned} \tag{24}$$

can be computed using the analysis filter bank with decimators in Fig. 2.5(a). The signal $x(n)$ can then be reconstructed through inverse wavelet transform using interpolators and synthesis filter bank, as shown in Fig. 2.5(b). In practice, the discrete wavelet transform periodically processes M input samples every time and generates M output samples at various frequency bands, where $M = 2^m$ and m is the number of bands or levels of the wavelet. It is often implemented using a *tree-structured* filter bank, where the M wavelet coefficients are computed through $\log_2 M$ octave levels and each octave performs one lowpass and one highpass filtering

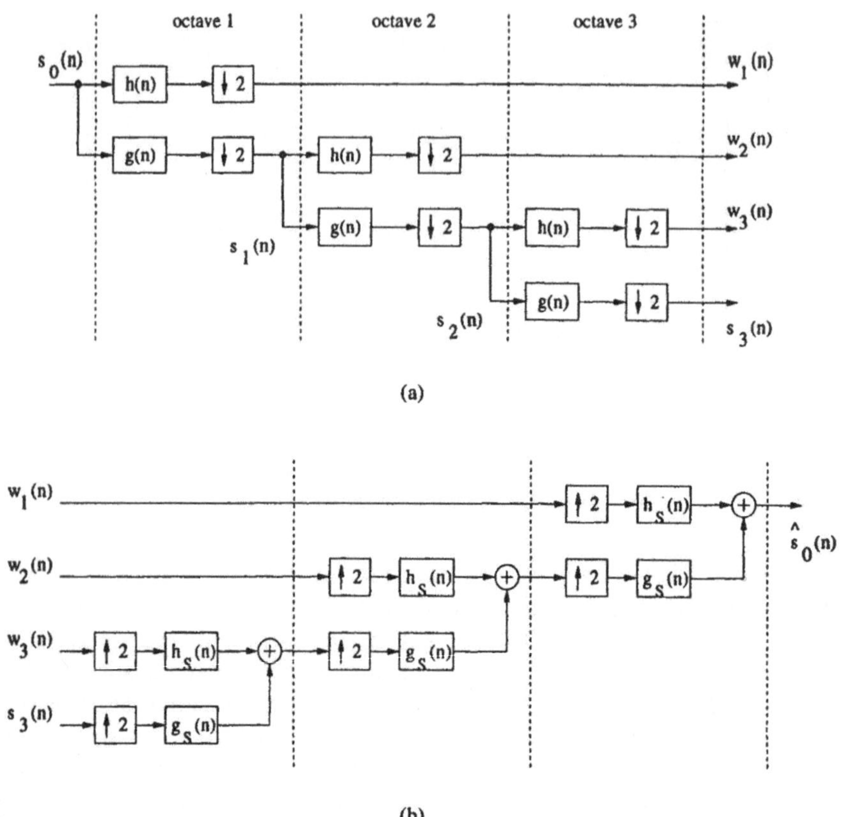

Figure 2.6 Block diagrams of tree-structured analysis and synthesis filter banks for DWT and IDWT.

operations. At each octave level j, an input sequence $s_{j-1}(n)$ is fed into lowpass and highpass filters $g(n)$ and $h(n)$, respectively. The output from the highpass filter $h(n)$ represents the detail information in the original signal at the given level j, which is denoted by $w_j(n)$, and the output from the lowpass filter $g(n)$ represents the remaining (coarse) information in the original signal, which is denoted as $s_j(n)$. The computation in octave j can be expressed as follows:

$$\begin{aligned} s_j(n) &= \sum_k s_{j-1}(k)g(2n-k) = \sum_k g(k)s_{j-1}(2n-k) \\ w_j(n) &= \sum_k s_{j-1}(k)h(2n-k) = \sum_k h(k)s_{j-1}(2n-k), \end{aligned} \quad (25)$$

where n is the sample index and j is the octave index. Initially, $s_0(n) = x(n)$. Fig. 2.6 shows the block diagram of a 3-octave tree-structured DWT.

Two-dimensional discrete wavelet transform can be used to decompose an image into a set of successively smaller orthonormal images, as shown in Fig. 2.7. The total size of all the smaller images is the same as the original image; however, the energy of the original image is compacted into low frequency small images at the upper left corner in Fig. 2.7.

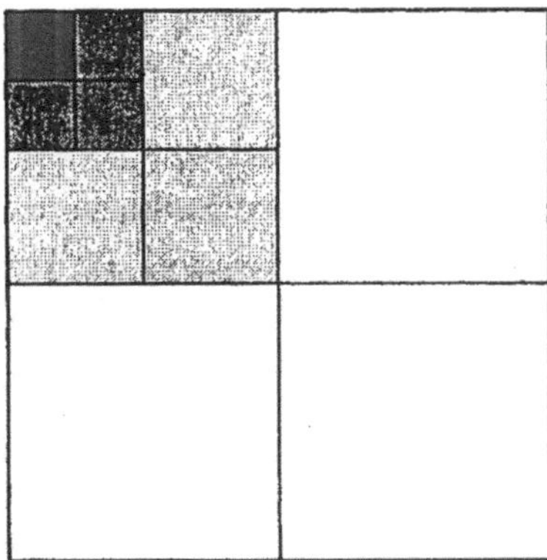

Figure 2.7 One image is decomposed into a set of successively smaller orthonormal images using two-dimensional discrete wavelet transform.

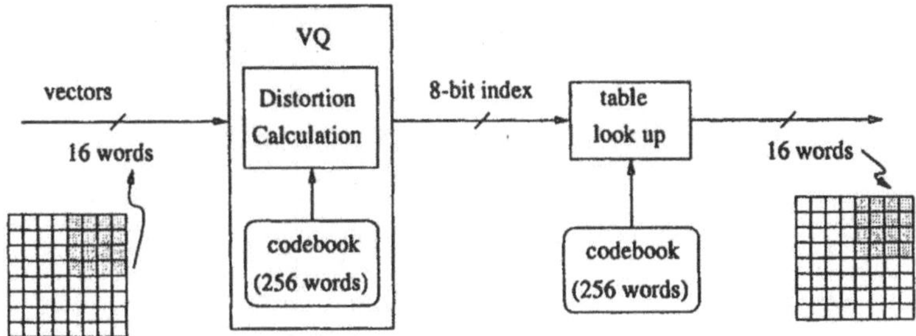

Figure 2.8 VQ-based vector compression and decompression.

2.3.3 Vector Quantization

The quantization process projects the continuous values of resulting transformed coefficients into a finite set of symbols, each of which best approximates the corresponding coefficient's original value. This single coefficient based quantization process is referred to as scalar quantization. In contrast, vector quantization maps sets of values (in the form of vectors), into a predefined set of patterns. Vector quantizer outperforms scalar quantizer in terms of performance; however, it is much more complicated to implement. The fundamental algorithms and implementation requirements of vector quantization are addressed in this section.

In a VQ system, a common definition-of-pattern codebook needs to be predefined and stored at both the transmitter (containing vector quantizer or encoder) and the receiver side (containing vector dequantizer or decoder). The vector quantizer transmits the index of the codeword rather than the codeword itself. Fig. 2.8 illustrates the VQ encoding and decoding process. On the encoder side, the vector quantizer takes a group of input samples (transformed coefficients), compares this input vector to the codewords in the codebook and selects the codeword with minimum *distortion*. Assume that vectors are k-dimensional and the codebook size is N. If the word-length of the vector elements is W and $N = 2^m$, then the m-bit address of the codebook is transmitted as opposed to the kW bits. This leads to a compression factor of m/kW. The decoder simply receives the m-bit index as the address of the codebook and retrieves the best codeword to reconstruct the input vector. In Fig. 2.8, each vector contains $k = 16$ pixels of word-length $W = 8$. The codebook contains $N = 256$ codewords, hence $m = 8$. Therefore, the vector quantizer in Fig. 2.8 achieves a compression factor of 1/16.

The encoding algorithm in the vector quantizer can be viewed as an exhaustive search algorithm, where the computation of distortion is performed sequentially on every codeword vector in the codebook, keeping track of the minimum distortion so far, and continuing until every codeword vector has been tested. Usually, the Euclidean distance between two vectors (also called square error)

$$d(\mathbf{x}, \mathbf{y}) = \|\mathbf{x} - \mathbf{y}\|^2 = \sum_{i=0}^{k-1} (x_i - y_i)^2 \tag{26}$$

is used as a distortion measurement. In practical implementations, the distortion between the input vector $\mathbf{x}$ and the j-th codeword vector $\mathbf{c_j}$ $(0 \leq j \leq N-1)$ is computed based on their inner product, instead of direct squaring operations [8]. By expanding (26), we get

$$d(\mathbf{x}, \mathbf{c_j}) = \|\mathbf{x}\|^2 - 2(\mathbf{x} \cdot \mathbf{c_j} + e_j), \tag{27}$$

where

$$e_j = -\frac{1}{2}\|\mathbf{c_j}\|^2 = -\frac{1}{2}\sum_{i=0}^{k-1} c_{ji}^2, \tag{28}$$

and the inner product is given by

$$\mathbf{x} \cdot \mathbf{c_j} = \sum_{i=0}^{k-1} x_i c_{ji}. \tag{29}$$

Since e_j depends only on the codeword vector $\mathbf{c_j}$ and is a constant, it can be precomputed and treated as an additional component of the vector $\mathbf{c_j}$. Therefore, for a fixed input vector $\mathbf{x}$, minimizing the distortion in (27) among the N codeword vectors is equivalent to maximizing the quantity $\mathbf{x} \cdot \mathbf{c_j} + e_j$, where $0 \leq j \leq N-1$. Therefore, the search process in VQ can be described as follows:

$$ind_n = (\min_{0 \leq j \leq N-1} d_j)^{-1} = (\max_{0 \leq j \leq N-1} \sum_{i=0}^{k-1} (x_i^n c_{ji} + e_j))^{-1}, \tag{30}$$

where the inverse means "output the index ind_n which achieves the minimum or maximum" and n represents the time instance. The search process can also be described equivalently in a matrix-vector multiplication formulation followed by comparisons as follows [9]:

$$\begin{aligned} \mathbf{D} &= [d_0 \; d_1 \; \cdots \; d_{N-1}]^T = \mathbf{Cx} + \mathbf{e} \\ ind_n &= (\mathbf{Max}\{d_i\})^{-1}, \end{aligned} \tag{31}$$

where $\mathbf{C} = \{c_{ji}\}$ is a $N \times k$ matrix with the j-th codeword vector $\mathbf{c_j}^T$ as its j-th row, $\mathbf{x}$ is the input vector of dimension k, and $\mathbf{e} = [e_0 \; e_1 \; \cdots \; e_{N-1}]^T$.

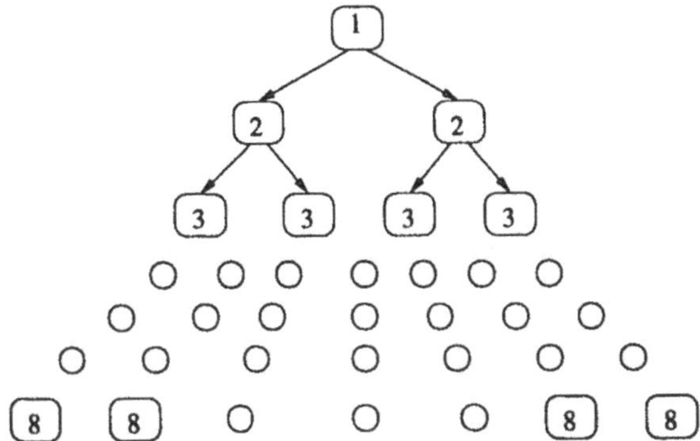

Figure 2.9 Tree-structured vector quantization.

The above searching algorithm is a brute-force approach where the distortion between the input vector and every entry in the codebook is computed, and is called ***full-search vector quantization.*** Every full-search operation requires N distortion computations and each distortion computation involves k multiply-add operations. Hence, computing the index for one k-dimensional input vector requires Nk multiply-add operations and $N-1$ comparisons, without including memory access operations. This algorithm may be a bottleneck for high performance for large N. For these cases, a ***tree-structured vector quantization*** scheme can be used whose complexity is proportional to $\log_2 N$. The basic idea is to perform a sequence of binary search instead of an exhaustive search, as shown in Fig. 2.9. At each level of the tree, the input vector is compared with the two codeword vectors and two distortion computations are carried out. This process is repeated until the leaf of the tree is reached. For the example in Fig. 2.8, the tree-search requires 16 distortion calculations, as compared to 256 in full search. The tree-search VQ is a sub-optimal quantizer which typically results in performance degradation. However, for carefully design codebook, the degradation can be minimized.

2.3.4 Reordering of the Quantized Transform Coefficients

Since most of the non zero values are at low frequency positions, the quantized transform coefficients can be reordered such that the resulting sequence contains long runs of zeros, which can be compressed efficiently using run length and entropy coding schemes. The most popular coefficient reordering for DCT coefficients is incremental zig-zag scanning starting from zero-frequency (DC) value to its highest frequency component.

2.4 MOTION ESTIMATION/COMPENSATION

For highly correlated input samples, a relatively accurate estimation of the current sample can be made based on past samples; or alternately, the past samples can be used to predict the current sample. This property enables predictive coding scheme. Fig. 2.10 shows the block diagram of a basic generic predictive coding system, where the predictive difference is coded and transmitted. The most popular form of predictive coding for image is the differential pulse code modulation

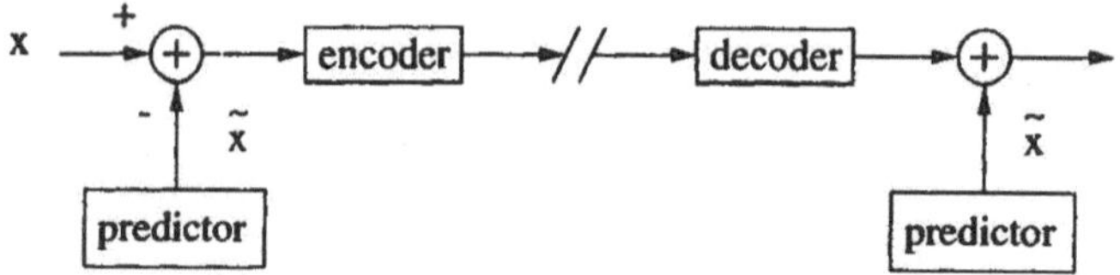

Figure 2.10 A basic predictive coding system.

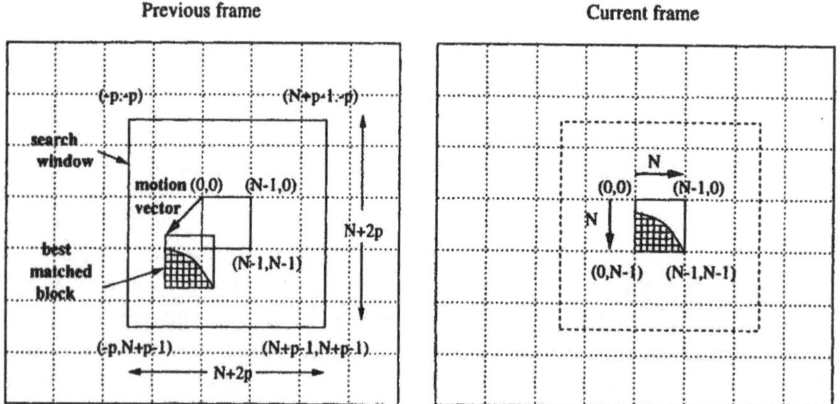

Figure 2.11 Block matching algorithm for motion estimation.

(DPCM). For still image compression, it uses neighboring pixels as a predictor and exploits the spatial correlation to achieve compression. In video sequences, consecutive frames contain very high similarities. These temporal redundancies can be removed through DPCM based interframe coding, which uses previous frames as the predictor, and is based on motion-compensated prediction between successive frames.

Motion-compensated prediction includes motion estimation and compensation, out of which motion estimation is the most computation-intensive part. In motion estimation, successive frames of a video sequence are analyzed to estimate the motion (or displacement) vectors of pixels or blocks of pixels. The motion vectors and the difference between the motion compensated frame and the original frame are further coded and transmitted.

Block-matching algorithms (BMA) are the most preferred schemes for motion estimation due to their relative simplicity. In BMA, each frame is partitioned into N-by-N macro reference blocks and it is assumed that all the pixels in one block have the same motion. Each reference block in the current frame is compared with displaced candidate blocks in the previous frame and the offset between the best fitting candidate block and the reference block is defined as its motion vector. The search range in the previous frame is called *search window* and is constrained to $+/-$ p pixels in both horizontal and vertical directions relative to the position of the reference block. Thus, the search window contains $(N + 2p)^2$ pixels. The block-matching algorithm is illustrated in Fig. 2.11.

Several search criteria can be used to define the best match, including cross-correlation function (CCF), mean-square error (MSE) and mean-absolute-difference (MAD). The MAD function is most widely used in practical implementations due to its simplicity and satisfactory performance. The displaced block difference $s(m, n)$ with displacement (m, n) using MAD is defined as

$$\begin{aligned} s(m,n) &= \sum_{i=0}^{N-1} \sum_{j=0}^{N-1} |x(i,j) - y(i+m, j+n)|, \\ & \quad for \ -p \leq m, \ n \leq p, \end{aligned} \tag{32}$$

where $x(i, j)$ and $y(i + m, j + n)$ correspond to the pixel values in the reference block in the current frame and the candidate block in the search window in previous frame, respectively. Note that (32) requires $3N^2$ computations (one subtraction, one absolute value and one accumulation are needed for each absolute difference computation). Several strategies can be used to determine the best matched block, out of which *full search* is the most straight-forward method. It searches all the $(2p + 1)^2$ positions in the search window and computes the motion vector $\mathbf{v}$ as

$$\begin{aligned} u &= min_{(m,n)}\{s(m,n)\}, \ for \ -p \leq m, \ n \leq p, \\ \mathbf{v} &= (m,n)|_u. \end{aligned}$$

Hence, for a $N_h \times N_v$ frame (N_h pixels per line, N_v lines per frame), the full-search BMA involves

$$\frac{N_h \times N_v}{N^2} \cdot (2p+1)^2 \cdot 3N^2 = 3(2p+1)^2 N_h N_v$$

computations per frame. Assume a frame rate of F frames/sec, the computation load of full-search BMA is $3(2p+1)^2 N_h N_v F$ operations/sec.

2.5 MPEG-2 DIGITAL VIDEO CODING STANDARD

Generally speaking, video sequences contain significant amount of spatial and temporal redundancy within a single frame and between consecutive frames. MPEG, a video communication standard developed by the Moving Picture Experts Group, reduces the bit-rate by exploiting both the spatial and temporal redundancies through intra- and inter-frame coding techniques. The ultimate goal of MPEG standard is to optimize image or video quality for a specified bit rate subject to "objective" or "subjective" optimization criteria [10]. Fig. 2.12 shows the block diagram of the MPEG-2 encoding process, where a motion-compensated prediction is followed by transform coding of the remaining spatial information; the transformed coefficients are then quantized and entropy coded [11].

This section overviews some of the key concepts in MPEG-2 standard; these include subsampling of color difference signals, intraframe and interframe coding of I, P and B frames, interlaced and progressive scanning techniques. Finally, the generic structure of MPEG-2 standard is briefly addressed and its profiles and levels are listed at the end of this section.

2.5.1 Subsampling

A color digital image consists of picture elements (pixels), which are represented using three primary color elements, red (R), green (G) and blue (B). The

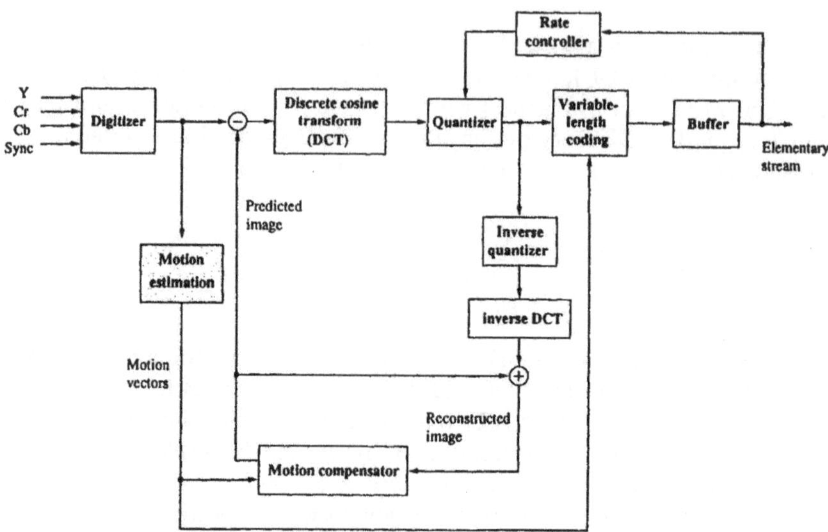

Figure 2.12 Block diagram of MPEG-2 encoder.

RGB representation is converted to YUV representation based on human visual systems, where Y stands for the luminance information and U and V are the color differences between Y and blue, Y and red, respectively, and are called chrominances. A full-sampling of YUV is represented as $4:4:4$ sampling and the resulting pixel is represented using 24 bits, 8 bits for each variable. With $4:4:4$ sampling, a CIF (Common Intermediate Format) frame with a frame size of 288×352 pixels and a frame rate of 30 frame/sec requires storage of size 2.433 Mega bits (Mb) and the video source data rate is 72.99 Mb/sec for a single frame. For high-definition TV (HDTV) video with a frame size of 1920×1250 pixels and a frame rate of 50 frame/sec, one frame requires storage size of 57.6 Mb and the video source data rate of 2.88 Giga bits per sec (Gb/sec). With a sequence of video containing hundreds and thousands of frames, storage and transmission in real-time is impossible with today's technology.

In reality, video frames are downsampled and quantized prior to coding by making use of specific physiological characteristics of human eyes and removing subjective redundancy contained in the video sequence. This can be considered as one of the elementary compression technique. Human eyes have fewer receptors with less spatial resolution for color than those for luminance. Hence the chrominance can be downsampled to reduce the source data rate and frame storage size. Typically, a $4:2:2$ or $4:2:0$ sampling is used. In $4:2:2$, the luminance Y is sampled for every pixel, while the chrominances U and V are sampled every other pixel horizontally resulting in 33% savings. In $4:2:0$, U and V are downsampled by a factor of 2 both horizontally and vertically, resulting in 50% savings.

2.5.2 Intraframe and Interframe Coding : the I, P and B Frames

MPEG-2 compression defines three types of image frames for encoding, including the I frame, P frame and B frame. The coding schemes for these three frame types are illustrated in Fig. 2.13 [12]. The I frames are intraframe coded as

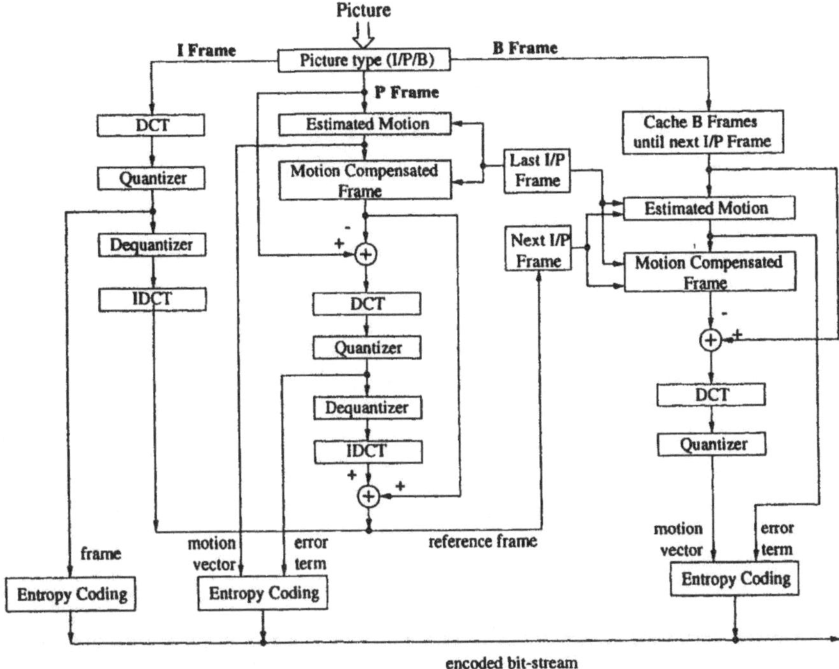

Figure 2.13 Coding procedures for I, P and B frames in MPEG.

an independent images. It is broken into macroblocks, each of which is compressed using DCT transformation followed by quantization and entropy coding. The P frame is coded based on forward motion-prediction. The difference between the current input picture and the last I/P frame is compressed using DCT; the quantized DCT coefficients and the motion vectors are entropy coded and transmitted. The B frame is coded based on bidirectional (forward and backward) motion-prediction, where the last I/P frame and the next I/P frame are used as references for motion estimation and compensation. It can be seen from Fig. 2.13 that both the coded I and P frames are converted back (de-quantized and inverse DCT transformed) to be used as reference frames for prediction; the B frames are never used as references for prediction.

Use of P and B frames increases the level of compression dramatically; however, they also create some inconvenience for display and random access of video sequence. Since the B frames are encoded with reference to both past and future frames, the video frames are encoded and transmitted in an order different from the original, and need to be re-ordered for display, as illustrated in Fig. 2.14. Moreover, decoding of P frames requires decoding of at least two frames including the reference I frame and itself, and decoding of B frames requires decoding of at least three frames including two reference frames and itself. Applications which require access to any part of the video sequence at random result in much more computation complexity, hence an increase in latency. In order to decode the coded bit stream at any arbitrary point, it is necessary to use a certain number of intraframe coded I frames in the video sequence.

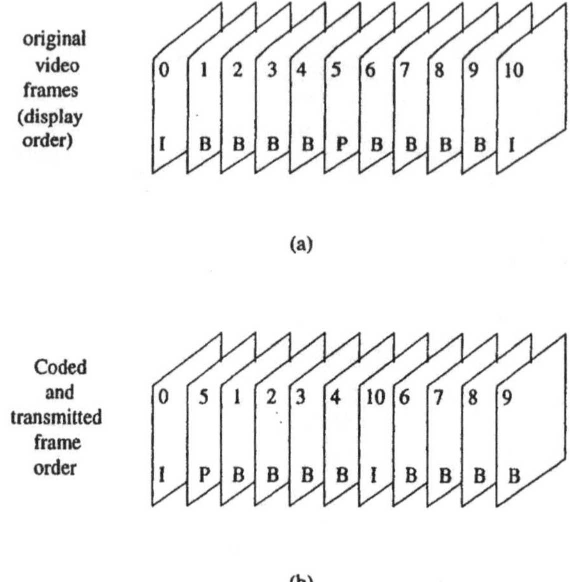

Figure 2.14 (a) Original and display order of video frames. (b) Coding and transmitting order of I, P and B video frames.

2.5.3 Bit Stream Generation of Quantized DCT Coefficients

The actual compression of each picture is carried out based on macroblocks, each of which contains n 8×8 blocks of data. For $4:2:0$ sampling, these n 8×8 blocks include $2m$ luminance blocks and m chrominance blocks where $n = 3m$. These 8×8 blocks are compressed individually using DCT, quantization and entropy coding. As discussed in Section 2.3.1, most of the energy in one 8×8 image block tends to be stored in the low frequency DCT coefficients, which are located at the upper left corner in the 8×8 block. After quantization, the non zero coefficients are concentrated in the upper left corner, as shown in Fig. 2.15. The quantized 8×8 DCT coefficients are "zig-zag" scanned such that the resulting sequence contains long runs of zeros, which can be compressed efficiently using run length coding and entropy coding schemes.

2.5.4 Interlaced/Non-Interlaced Scanning

An image display/recording system scans the image progressively and uniformly from left to right and top to bottom. Two scanning formats are generally used, including *interlaced scanning* and *non interlaced (progressive) scanning*. Interlaced scanning technique is used in camera or television display, where each frame is scanned in two successive vertical passes, first the *odd field*, then the *even field*, as shown in Fig. 2.16. On the other hand, computer video images are scanned in progressive format, where one frame contains all the lines scanned in their proper order, as shown in Fig. 2.17.

For processing motion images and design of image displays, temporal aspects of human visual perception is very important. It is observed that human eyes can

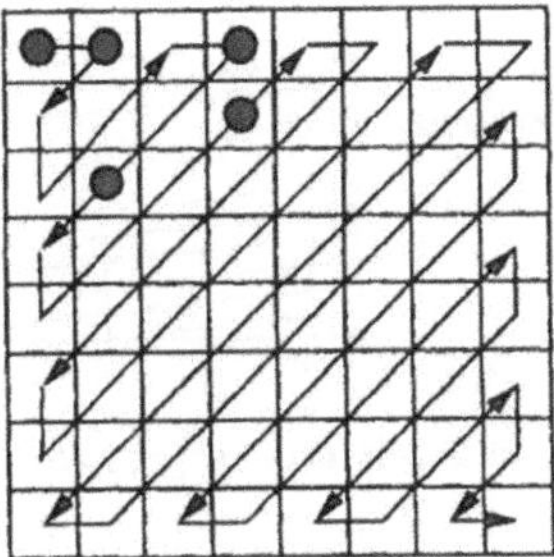

Figure 2.15 Zig-zag scanning of quantized DCT coefficients in an 8×8 block. Since most of the energy is concentrated around the lower DCT coefficients, zeros have high-probability and appear in consecutive strings in the output sequence, which can be compressed efficiently using run length coding and entropy coding schemes.

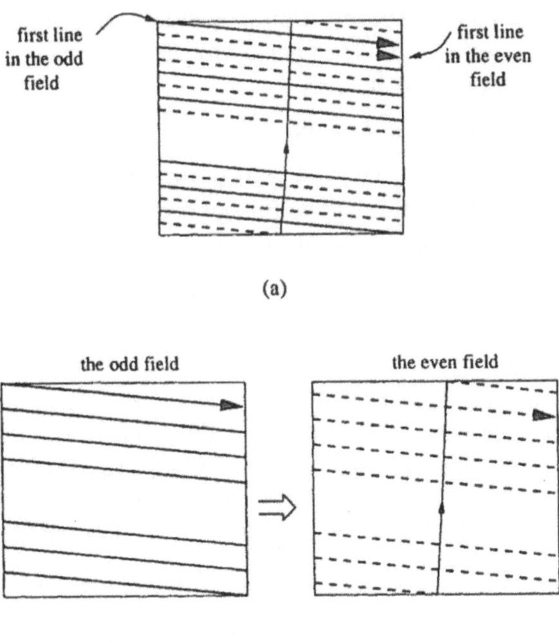

Figure 2.16 (a) One frame in interlaced scanning consists of two fields: the odd field and the even field. (b) The odd field is scanned first, followed by the even field.

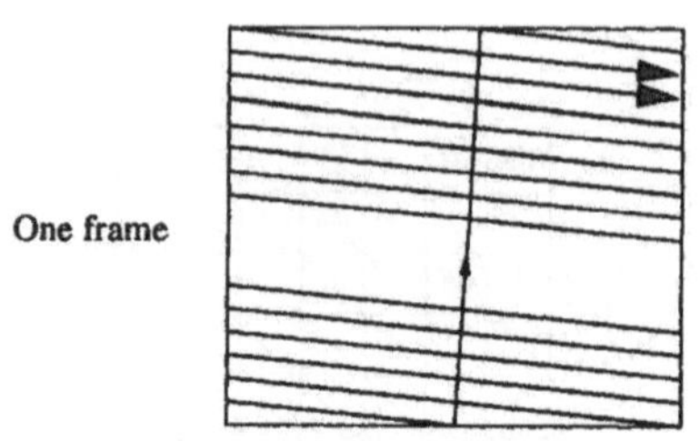

Figure 2.17 One frame in non interlaced progressive scanning.

distinguish the individual flashes from a slowly flashing light. However, as the flashing rate increases, they become indistinguishable at rates above the ***critical fusion frequency***. This frequency generally does not exceed 50 to 60 Hz [13]. Based on this property, images are scanned at a frame rate of 30 frames/sec or 60 fields/sec in interlaced scanning mode; they are scanned at a frame rate of 60 frames/sec in non interlaced (progressive) mode. Although the spatial resolution is somewhat degraded in interlaced scanning since each field is a subsampled image, with an appropriate increase in the scan rate, i.e., lines per frame, interlaced scanning can give about the same subjective quality with smaller bandwidth requirement of the transmitted signals. However, interlacing techniques are unsuitable for the display of high resolution computer generated images which contain sharp edges and transitions. To this end, computer display monitors are refreshed at a rate of 60 frames/sec in non interlaced mode to avoid any flicker perception and to obtained high spatial resolution display images.

2.5.5 MPEG Profiles and Levels

MPEG standards have very generic structure and can support a broad range of applications. Implementation of the full syntax may not be practical for most applications. To this end, MPEG-2 has introduced the "profile" and "level" concepts, which provide means for defining subsets of the syntax and hence the decoder capabilities required to decode a particular video bit stream. The MPEG-2 profiles are listed in Table 2.2, and the upper bound of parameters at each level of a profile are listed in Table 2.3 [10].

Generally, each profile defines a new set of algorithms additional to the algorithms in the lower profile. A Level specifies the range of the parameters such as image size, frame rate, bit rate, etc. The MPEG-2 MAIN profile features nonscalable coding of both progressive and interlaced video sources. A single-chip MPEG2 MP@ML (Main Profile at Main Level) encoder has been presented in [14].

2.6 COMPUTATION DEMANDS IN VIDEO PROCESSING

With compression, the bandwidth of transmitted/stored video sequences can be reduced dramatically. Further improvement of compression rate can be achieved by adopting more complicated compression techniques. These sophisticated compression techniques involve substantial amount of computations at high speed and render new challenges for both hardware and software designers in order to implement these high performance systems in a cost-effective way. For example, the com-

Table 2.2 Algorithms and Functionalities Supported in MPEG-2 Profiles

Profile	Algorithms
HIGH	Supports all functionality provided by the spatial profile plus the provision to support: • 3 layers with the SNR and spatial scalable coding modes • 4:2:2 YUV-representation for improved quality requirements
Spatial Scalable	Supports all functionality provided by the SNR scalable profile plus an algorithm for: • spatial scalable coding (2 layers allowed) • 4:0:0 YUV-representation
SNR Scalable	Supports all functionality provided by the MAIN profile plus an algorithm for: • SNR scalable coding (2 layers allowed) • 4:2:0 YUV-representation
MAIN	Nonscalable coding algorithm, supports functionality for: • coding interlaced video • random access • B-picture prediction modes • 4:2:0 YUV-representation
SIMPLE	Includes all functionality provided by the MAIN profile except: • does not support B-picture prediction modes • 4:2:0 YUV-representation

Table 2.3 Upper Bound of Parameters at Each Level of a Profile

Level	Parameters
HIGH	1920 samples/line, 1152 lines/frame, 60 frames/sec, 80 Mbit/sec
HIGH 1440	1440 samples/line, 1152 lines/frame, 60 frames/sec, 60 Mbit/sec
MAIN	720 samples/line, 576 lines/frame, 30 frames/sec, 15 Mbit/sec
LOW	352 samples/line, 288 lines/frame, 30 frames/sec, 4 Mbit/sec

plexity of a full-search block-matching algorithm is proportional to $3(2p+1)^2 N_h N_v F$ operations/sec, where $N_h \times N_v$ is the frame size, $+/-$ p is the search area and F is the frame rate in frames/sec. For a CIF (Common Intermediate Format) frame with a frame size of 288 × 352 pixels, a frame rate of 30 frame/sec and a search range of +/− 7 pixels, the full-search BMA requires about 2 Giga operations per sec (Gops/s). The required number of operations gets even larger for higher resolution pictures with higher frame rates and larger search range. For high-definition TV (HDTV) video with a frame size of 1920 × 1250 pixels, a frame rate of 50 frame/sec and a search range of +16/ − 15 pixels, the full-search BMA demands a computation rate of about 368.64 Gops/s. The DCT in video communications is also very demanding. The $N \times N$ 2D-DCT requires $2N^3$ multiply-add operations, or $4N^3$ arithmetic operations. For a CIF (Common Intermediate Format) frame with image blocks of size 8 × 8, the computation requirement for 2D-DCT is 97.32 Mops/sec (Mega operations per second). For high-definition TV (HDTV) video with image blocks of size 8 × 8, the computation requirement for 2D-DCT is 3.84 Gops/sec (Giga operations per second). These high processing requirements can only be met using parallel processing techniques with carefully designed hardware and software. [15] Design and implementation of video compression and multimedia signal processing systems are quite challenging!

2.7 CONCLUSIONS

This chapter has presented basic video coding schemes, especially those adopted by MPEG-2 video compression standard. These compression techniques are the keys of realizing real-time high quality digital video processing. These increasingly complex coding schemes render many new challenges for hardware and software designers.

Acknowledgement

The author is thankful to Leilei Song for her help in the preparation of this chapter.

REFERENCES

[1] D. Huffman, "A method for the construction of minimum redundancy codes," *Proc. of IRE*, vol. 40, pp. 1098–1101, 1952.

[2] G. Langdon, "An introduction to arithmetic coding," *IBM J. Research Develop*, vol. 28, pp. 135–149, March 1984.

[3] N. Ahmed, T. Natarajan, and K. R. Rao, "Discrete cosine transform," *IEEE Trans. on Computers*, pp. 90–93, Jan 1974.

[4] P. Pirsch, N. Demassieux, and W. Cehrke, "VLSI architectures for video compression - a survey," *Proceeding of the IEEE*, pp. 220–245, Feb. 1995.

[5] O. Rioul and M. Vetterli, "Wavelets and signal processing," *IEEE Signal Processing Magazine*, pp. 14–38, Oct. 1991.

[6] P. P. Vaidyanathan, *Multirate Digital Signal Processing*, Prentice Hall, Englewood Cliffs, New Jersey, 1993.

[7] R. E. Crochiere and L. R. Rabiner, *Multirate Digital Signal Processing*, Prentice Hall, Englewood Cliffs, New Jersey, 1983.

[8] G. A. Davison, P. R. Cappello, and A. Gersho, "Systolic architectures for vector quantization," *IEEE Trans. on Acoustics Speech*, vol. 36, pp. 1651–1664, Oct. 1988.

[9] S. Y. Kung, *VLSI Array Processors*, Prentice Hall, Englewood Cliffs, NJ, 1988.

[10] T. Sikora, "MPEG digital video-coding standards," *IEEE Signal Processing Magazine*, pp. 82–100, Sep. 1997.

[11] B. Bhatt, D. Birks, and D. Hermreck, "Digital television: Making it work," *IEEE Spectrum*, pp. 19–28, Oct. 1997.

[12] B. Furht, J. Greenberg, and R. Westwater, *Motion Estimation Algorithms for Video Compression*, Kluwer Academic Publishers, 1997.

[13] A. K. Jain, *Fundamental of Digital Image Processing*, Prentice Hall, Englewood Cliffs, NJ, 1989.

[14] M. Mizuno *et al.*, "A 1.5w single-chip mpeg2 MP@ML encoder with low-power motion estimation and clocking," *in Proc. of ISSCC97*, pp. 256–257, Feb. 1997.

[15] K.K. Parhi, *VLSI Digital Signal Processing Systems: Design and Implementation*, John Wiley and Sons, 1999.

Chapter 3

Audio Compression

Akihiko Sugiyama and Masahiro Iwadare
NEC Corporation
Kawasaki, Japan
{sugiyama,iwadare}@dsp.cl.nec.co.jp

3.1 STANDARDIZATION ACTIVITIES OF HIFI AUDIO CODING

This chapter describes the audio coding algorithm for the ISO/IEC international standard. Three standard algorithms have been established for use depending on the number of channels and the sampling frequency: MPEG-1 audio, MPEG-2 MC, and MPEG-2 LSF. Depending on the complexity and quality achieved, each of the above three layers is classified into Layer I/II based on sub-band coding or Layer III based on the combination of the sub-band coding and adaptive transform coding.

Efficient transmission and storage of audio signals are important for several applications such as digital audio, digital satellite broadcasting (DSB), audio signal storage, remote conferencing, and multimedia. ISO/IEC JTC 1/SC 29/WG 11 (the International Organization for Standardization/the International Electrotechnical Commission, 1st Joint Technical Committee, 29th Subcommittee, 11th Working Group) studied a means of designing an international standard for compressing audio signals together with video signals down to 1.5 Mb/s. The outcome of the study for two channel signals of 32 to 48 kHz sampling was published by ISO/IEC on August 1, 1993, after approval by vote of the participating countries [1].

This is referred to as the MPEG/Audio Phase 1, or more simply MPEG-1 Audio. A part of the MPEG-1 Audio has been also established as the ITU-R (the International Telecommunication Union-Radio Sector, the former CCIR: the International Radio Communication Counsel Committee) standard [2] and is of increasing significance.

Studies into the extension of the MPEG-1/Audio to multichannel and multilingual systems, and for further bitrate reduction by adoption of low sampling frequency were then considered. These form Phase 2, and are called MPEG-2/Audio [3].

MPEG-2/Audio was approved, based on the outcome of an international vote at the Singapore Meeting in November, 1994, and was published as the international standard in 1995. This chapter first gives an outline of MPEG/Audio and explains

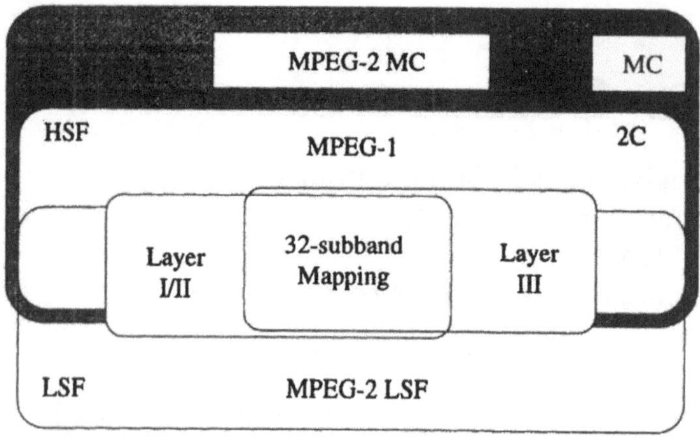

Figure 3.1 Basic structure of MPEG/Audio algorithm.

the technological elements of audio coding, and then describes the actual MPEG-1 Audio and MPEG-2 Audio coding algorithms.

3.2 MPEG AUDIO ALGORITHM STRUCTURE

The MPEG Audio algorithm is comprised of three different algorithms: Layer I, Layer II, and Layer III. The complexity increases from Layer I to Layer III, with corresponding improvement in the sound quality. They can also be classified as MPEG-1, MPEG-2 MC (multichannel), and MPEG-2 LSF (Low Sampling-Frequency), depending respectively on the number of channels coded and the sampling frequency. Because Layer I and Layer II greatly resemble each other, Layers I and II are often treated together.

The commonality between Layer I/II and Layer III is that their algorithms are based on the sub-band coding of 32 bands. The number of channels is two, and the sampling frequency is 32, 44.1, or 48 kHz. These features comprise the MPEG-1 Audio algorithm that forms the basis for all coding. Based on the MPEG-1 Audio algorithm, halving the sampling frequencies, 32, 44.1, and 48 kHz, that is, 16, 22.05, and 24 kHz, creates MPEG-2 LSF. On the other hand, increasing the number of channels to 5 and further adding a low-freqeuncy enhancement channel creates MPEG-2 MC. The number of the channels that MPEG-2 Audio can handle is sometimes called 5.1, by counting the enhancement channel as 0.1. When locating MPEG-2/Audio against MPEG-1/Audio as the standard, MPEG-1/Audio could well be called MPEG-2 2C (2-channels) or MPEG-2 HSF (high sampling-frequency). Fig. 3.1 shows the basic structure of the MPEG Audio algorithm. The figure clearly shows that MPEG-1/Audio is the core of the MPEG/Audio-2 algorithms.

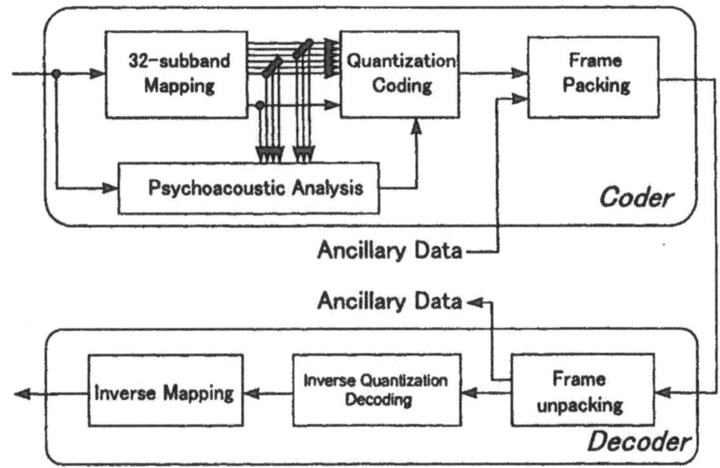

Figure 3.2 Basic structure of MPEG-1/Audio algorithm.

Fig. 3.2 shows a basic block diagram describing the MPEG-1/Audio algorithm. The algorithm is based on the subband coding system, and band splitting is achieved by a polyphase filter bank (PFB) [4] with a quadrature mirror filter (QMF).

A 16-bit linear quantized PCM input signal is mapped from the time domain to 32 frequency bands. At the same time, the masking levels are calculated through psychoaoustic analysis to find the magnitude of the allowed quantization errors. The mapped signal is quantized and coded according to the bit allocation based on a psychoacoustic model, and then taken into the frame, combined with ancillary data. This ancillary data is not used in encoding and decoding processes and users may make use of it for their purposes. To decode, the ancillary data is separated first, and then the frame is disassembled. Decoding and dequantizing are then executed, based on the bit allocation sent as accompanying information. The time domain signal is reconstructed by demapping the dequantizing signal. In practice, the three kinds of algorithm, Layer I, Layer II, and Layer III, will have been specified, based on the basic structure in Fig. 3.2 (see Fig. 3.3). Sub-band coding, psychoacoustic weighting, bit allocation, and intensity stereo are used in all the layers. Layer III further employs adaptive block length transform coding [5, 6], Huffman coding, and MS (middle-sides) stereo for coding quality improvement.

Sound quality depends upon not only the algorithmic layers, but also on the bit rates used. It may be noted that 14 kinds of bit rate have been specified from 32 kb/s to 448 kb/s, 384 kb/s, and 320 kb/s, for Layers I through III respectively. The main target bit rate for each layer is shown in Table 3.3.

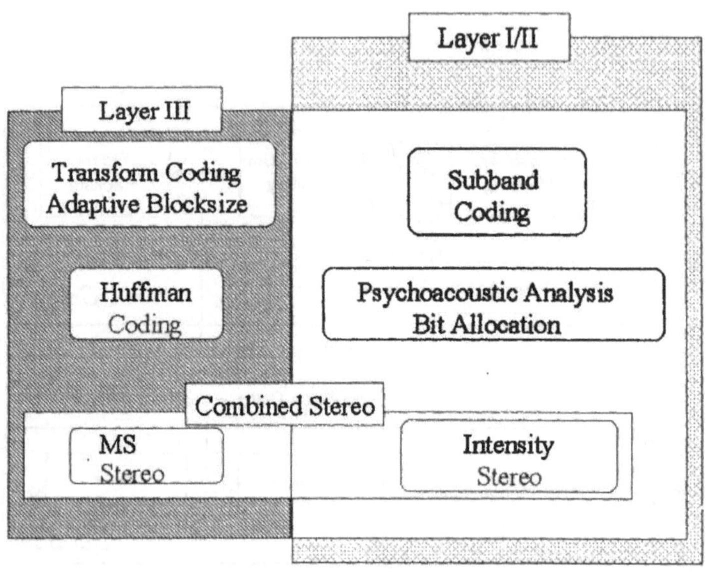

Figure 3.3 Inter-layer correspondence of basic technologies.

Table 3.1 Target Bit Rate

Layer	Target Bitrate (kb/s)
I	128, 192
II	96, 128
III	64, 96, 128

3.2.1 Basic Audio Coding Technologies

(1) Sub-band coding and adaptive transform coding

Typical algorithms for audio coding are sub-band coding (SBC) and adaptive transform coding (ATC) [7]. Both can improve coding efficiency, making use of signal energy maldistribution even though the audio signal has much wider bandwidth than speech signals.

Subband coding divides the input signal into multiple frequency bands, and performs coding indepedently for each of the bands. In this division into subbands signal energy maldistribution is reduced in each sub-band, thus reducing the dynamic range. Bits are then allocated in accordance with the signal energy of each band. Band division can be achieved using a tree structure to repeatedly divide bands into two, using quadrature mirror filters (QMF). The signal samples of the divided low and high bands are decimated by 2, reducing the sampling frequency to 1/2.

The filter bank that performs band division/synthesis by QMF is called the QMF filter bank. The filter bank with a tree structure can be called 'a tree struc-

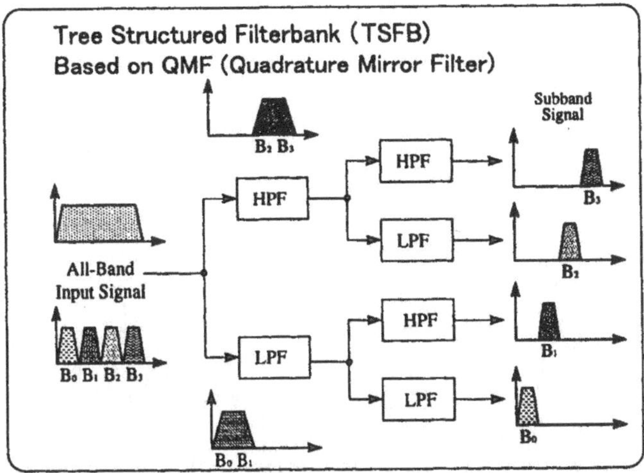

Figure 3.4 Band division by TSBF and PFB.

tured filter bank' (TSFB). The polyphase filter bank (PFB) provides a presentation equivalent to a TSFB. As filters for TSFB and PFB, either a FIR (Finite Impulse Response) filter or an IIR (Infinite Impulse Response) filter can be used. Assuming the adoption of a FIR filter, PFB can reduce the computing complexity by more than TSFB, taking advantage of the filter bank structure and operation of decimation. PFB also has a shorter delay time than TSFB. In practice, therefore, a FIR-based PFB is normally used.

Fig. 3.4 is an example of quad-band-division. The design procedures for QMF filter banks (TSFB/PFB) that can completely reconstruct the input signal by band division and band synthesis as the reverse operation, have been established [4].

Transform Coding improves coding efficiency by concentrating power intensity by applying a linear transform to the input signal before quantization. In particular, the coding algorithm that incorporates adaptive bit allocation is usually called adaptive transform coding [7]. Fourier conversion, cosine conversion [7], etc. are used for the linear transform. It has been pointed out that ATC, which applies a linear transform after multiplying the window function by the overlapped input signal, is equivalent to sub-band coding [8, 9]. Figure 3.5 is an example of the time domain waveform of a piano signal and the frequency domain waveform, obtained by using a cosine transform with a block length of N=1024. In the time domain waveform, the energy is distributed relatively evenly from sample No. 1 to No. 1024. On the other hand, in the frequency domain waveform, the energy is concentrated in the low frequencies, showing that an improvement of coding efficiency is possible.

3.2.2 Adaptive Block Length ATC

Adaptive block length ATC performs linear transform on multiple samples. Usually a larger block length results in a higher resolution, thus improving the coding quality. However, when adopting a large block length in an area where the

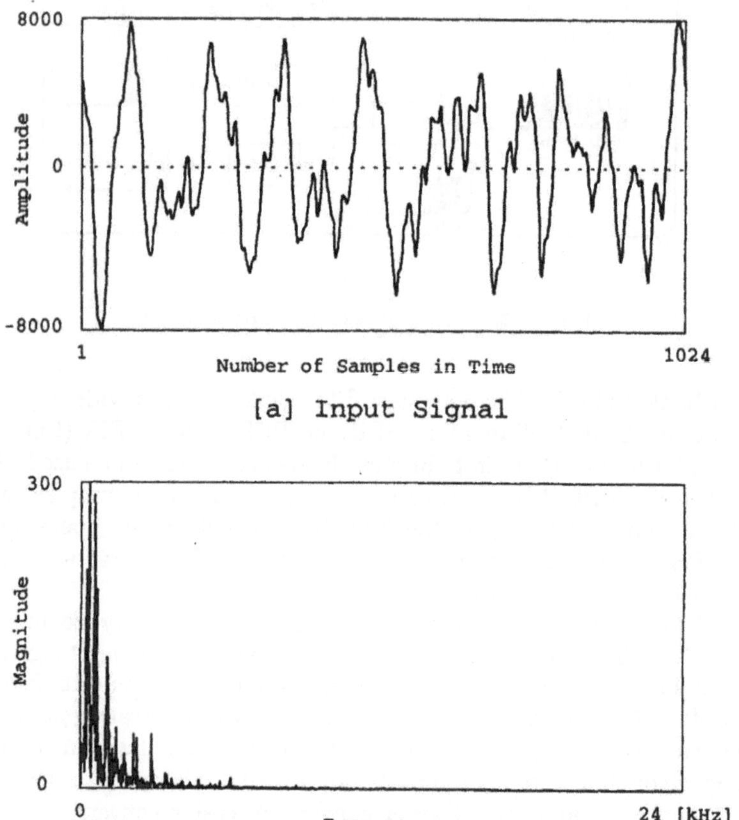

Figure 3.5 Energy concentration by cosine transform.

signal amplitude rises rapidly, a preceding echo, called a pre-echo, is generated. This is because, while the quantizing distortion by coding is distributed evenly in a unit block, the distortion is more clearly perceived where the signal amplitude is small.

Fig. 3.10 shows differences in pre-echo with different block length. Figures 3.6 (a), (b), and (c) represent, respectively, the original sound (drums); coded/decoded signals with a block length of N=256; and coded/decoded signals with a block length of N=1024. In Fig. 3.6 (c), the noise is generated in advance of the part (attack) where the signal amplitude rises steeply. In Fig. 3.6 (b), the time over which the pre-echo is generated is shorter than in Fig. 3.6 (c). By adopting a short block length (size), therefore, pre-echo can be suppressed.

However, when applying a short block size to a relatively static signal, the resolution will fall, as will the coding efficiency. Further, one individual set of supplementary information is required per block, which means that a longer block length will result in better efficiency. These contradictory requirements related to pre-echo can be dealt with by switching the block size in accordance with the input signal properties [5].

3.2.3 Modified Discrete Cosine Transform (MDCT)

Another problem with ATC is block distorbtion. Unfortunately for block coding, two signal samples that are adjacent across the block border are quantized with unequal accuracy, because they belong to different blocks, in spite of the fact that they are continuing on the time coordinate. Therefore, discontinuity in quantizing noise tends to be perceived in the vicinity of block borders. To solve this problem, a method of overlapping windowing has up to now been adopted to reduce the discontinuity [10]. It means, however, that the overlapped section is repeatedly coded in the adjacent two blocks, risking further degradation of coding efficiency due to the longer block size which has a larger effect on reducing block distortion. This problem can be solved by a modified discrete cosine transform (MDCT), which is also called time-domain aliasing cancellation (TDAC) [11].

MDCT first applies a 50% overlap across adjacent blocks, and filters by window functions, and then introduces an offset to the time term for DCT computing the resulting in symmetry in the obtained transform coefficients. The number of transform coefficients to be coded becomes 1/2 of the block length. This cancels the inefficiency generated by the 50% term introduced into the DCT computation. This is often referred as MDCT: modified discrete cosine transform.

3.2.4 Combination of MDCT and Adaptive Block Length

To combine MDCT and adaptive block length, attention is paid to the shape of the window function, since MDCT is originally designed on the assumption that the length of blocks is equal. When the lengths differ between two successive windows, some condition is required on the window shapes to cancel errors (time-domain aliasing) which is caused by overlapping windowing. The detailed neccesary conditions are reported in [12]. One possible solution is to make use of shape windows to connect the different block-length windows where the special window consists of the first half of the previous frame window and the last half of the next frame window.

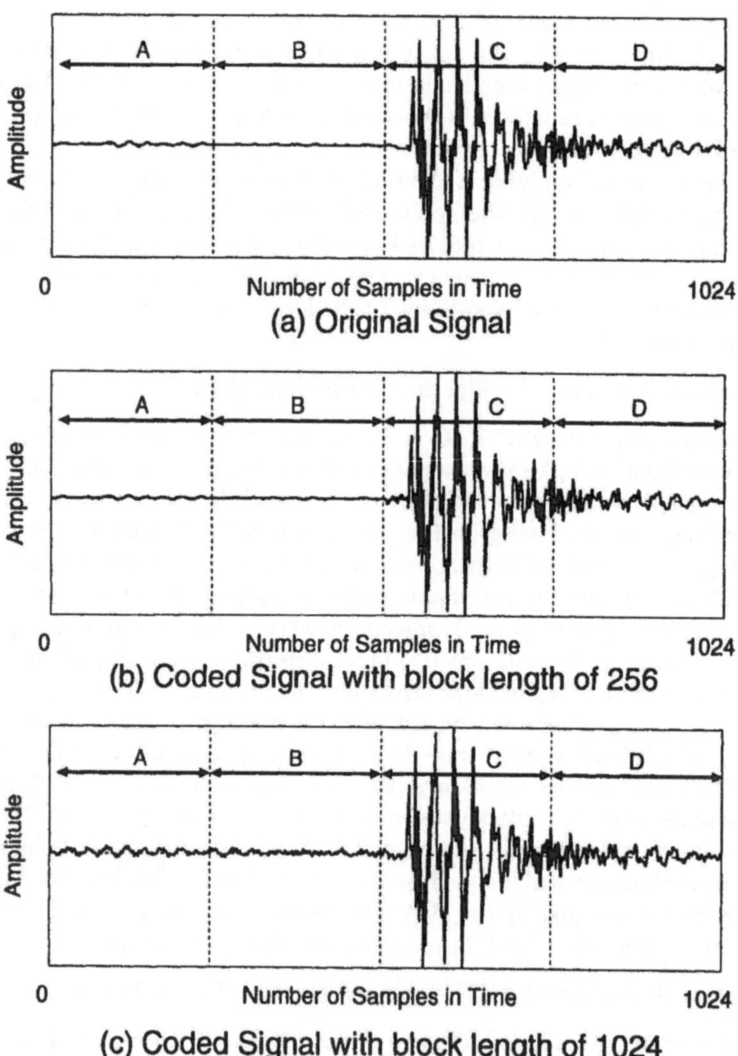

Figure 3.6 Pre-echo (drums) on different block lengths.

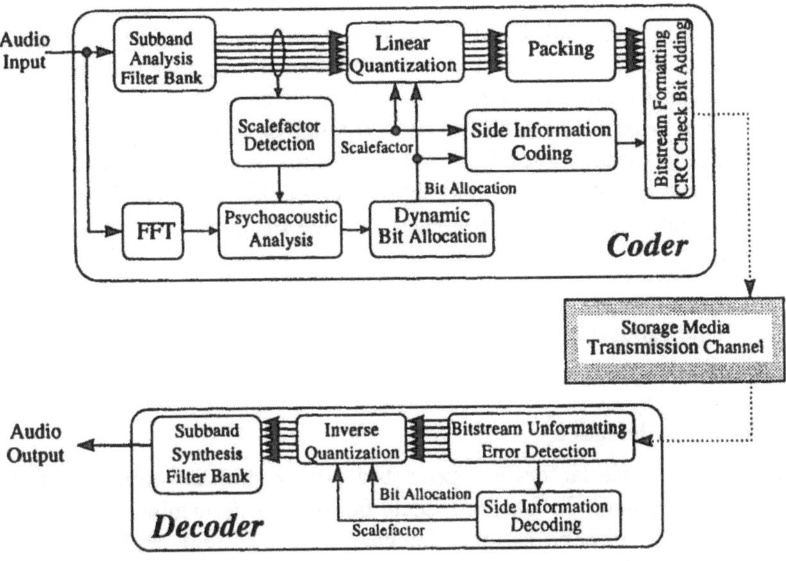

Figure 3.7 Layer I/II algorithm block diagram.

3.2.5 Quantization with Psychoacoustic Weighting

For both sub-band coding and adaptive transform coding, further improvements in the overall coding quality are possible. One technology is human property utilization of psychoacoustic perception where weightings are applied in bit allocation for quantization to minimize the signal degradation over the band area where perceptibility is high. Refer to the next section for details of psychoacoustic weighting.

3.3 MPEG-1 AUDIO ALGORITHM

3.3.1 Layers I/II Encoding

Layers I/II mostly follow the basic structure in Fig. 3.2 and the block diagram shown in Fig. 3.7. The 16-bit linear quantized input signal is divided by the subband analyzing filter into 32-band subband signals. The filter consists of a 512-tap PFB. The system calculates the scale factor for each of the sub-band signals, and aligns the dynamic ranges. Calculation of the scale factor is performed for every 12 subband samples in each band, i.e., for every 384 PCM audio input samples for all at Layer I. For Layer II, the calculation is performed for every 384 subband samples where one frame is composed of one triple block, i.e., 1152 subband samples. In Layer II, the scale factors are further compressed based on the combination of 3 scale factors.

At the same time, the system calculates masking labels, using the result of a fast Fourier transform (FFT) applied to the input signal, and determines the bit allocation for each sub-band. Here, a psychoacoustic weighting approach is used for bit allocation. The sub-band signal that has been quantized according to the obtained bit allocation is formatted into a bitstream, together with the header and the side information, and is output from the encoder.

Decoding is basically achieved by retracing the encoding process. The compressed signal is disassembled from the bitstream into the header, the side information, and the quantized subband signal. The subband signal is dequantized by the allocated number of bits, synthesized through the subband synthesis filters, and then output.

1. Sub-band analysis

   Sub-band analysis is performed using a 512-tap PFB.

2. Scalefactor detection

   In Layer I, scale factors are extracted per 12 subband samples as one block for each subband. In Layer II, scale factors are determined for 3 consecutive blocks of 12 subband samples for each subband, and presented in the form of 2-bit scale factor select information and scale factors which are trasmitted with the selected format.

3. Psychoacoustic analysis

   Model 1 and Model 2 are presented in the standard as examples of psychoacoustic analysis. This article describes the outline of Model 1 only.

   In Model 1, signal-to-mask ratio (SMR) are obtained using the following procedure.

   - Input signal FFT analysis
   - Sound pressure calculation for each subband
   - Classification of tonal and non-tonal components
   - Integration of tonal and non-tonal components
   - Calculation of individual masking threshold
   - Calculation of overall masking threshold
   - Determination of maximum masking level
   - Calculation of signal-to-mask ratio

4. Bit allocation

   The bit allocation are culculated for each sub-band, based on SMR obtained through psychoacoustic analysis.

5. Quantization

   Linear quantization is performed to the subband samples. Quantized values are obtained by A(n)X(n)+B(n), where the value X(n) resprents the magnitude of each subband sample normalized by the scale factor, and A(n) and B(n) are given by the number of bits allocated for each subband. The most siginificant N bits are taken, reversing the most significant one bit.

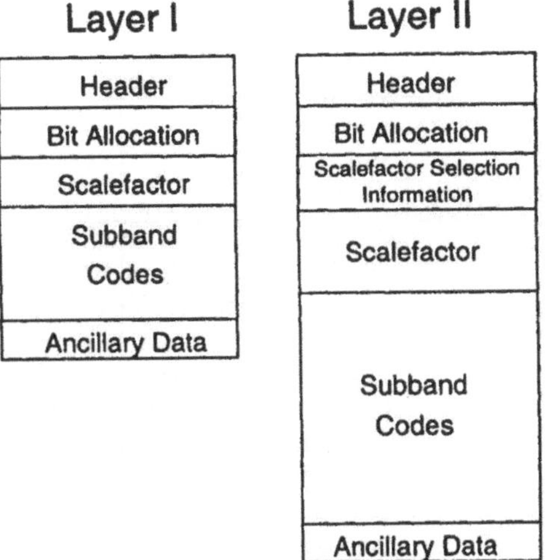

Figure 3.8 Bitstream format of layers I/II.

6. Bitstream formatting

   Quantized data forms the bitstream, together with side information. Fig. 3.8 shows the bitstream format in Layer I and II. The format in Layer I and Layer II differs, mainly in scale factor-related portion. The header shown in Fig. 3.8 includes the synchronization word '1111 1111 1111', followed by the configuration bits as shown in Table 3.2.

3.3.2 Layer I/II Decoding

1. Synchronization

   Synchronization is established by searching for the 12-bit synchronization word '1111 1111 1111'. This is the common step in all the layers. The position of the continuing synchronizing word can be identified, using the 7 bits that follow the protection bit, namely, bit rate, sampling frequency, and padding bit. The length of the current frame between the starting positions of the two consecutive synchronizing word can be calculated using the following formula

$$N = int(\frac{Ni \times (Bit\ rate)}{sampling\ frequency}) + (padding\ bit)[slot] \qquad (1)$$

   where 'slot' is the minimum control unit of the bitstream length, and is equivalent to 4 bytes in Layer I, and 1 byte in Layers II and III. For Layer I, Ni is 12, and for Layer II/III, Ni is 144. When the average number of slots per frame is not an integer, it is truncated to an integer value. The actual slot number is adjusted by 'padding bit'.

   When 'protection bit' is 0, cyclic redundancy codes (CRC) are inserted immediately after the header. Error detection is done using the CRC-16 method, based on the polynomial

Table 3.2 Number of Bits

Contents	Number of Bits	Definition
ID	1	0: MPEG-2/BC, 1: MPEG-1 audio
layer	2	00: reserved, 01: layer III, 10: layer II, 11: layer I
protect-bit	1	0: error detection code added 1: no error detection code added
bitrate	4	index to define bitrate
sampling frequency	2	00: 44.1 kHz, 01: 48 kHz, 10: 32 kHz, 11: reserved
padding bit	1	0: the frame that includes no additional slot 1: the frame that includes one additional slot
private bit	1	private use bit not used bit in coding
mode	2	00: stereo, 01: joint stereo, 10: dual channel 11: single channel
mode extension	2	in Layer I/II the number of subbands for joint stereo in Layer III the intensity and ms stereo configuration
copyright	1	0: no copyright, 1: copyright protected
original/copy	1	0: copy, 1: original
emphasis	2	the type of emphasis to be used

$$G(X) = X^{16} + X^{15} + X^3 + 1. \tag{2}$$

2. Decoding in Layer I

   The basic sequence includes reading the bit allocation information for all subbands, reading the scale factors for all subbands where the bit allocation is not zero, dequantizing the subband samples, and synthesizing the output audio signal with 32 subband samples using the filter bank.

   (a) Inverse Quantization of subband samples

   According to the bit allocation information, the bit series corresponding to each sample is read, and the most significant bit (MSB) is reversed. This operation obtains s"', the complement of 2 that represents -1.0 in MSB. The dequantized value, s", is calculated by

$$s'' = \frac{2^{nb}}{2^{nb} - 1} \times (s''' + 2^{-nb+1}) \tag{3}$$

   using the number of allocated bits, nb. The scale factor is multiplied to the dequantized value s", resulting in s'.

   (b) Synthesizing 32 subband signals by the filter bank

   The audio sample S_i is calculated at the synthesizing filter bank every time 32 subband samples are dequantized per channel. The procedure is as follows:

i. A frequency shift is applied to the 32 subband samples S_i, and V_i are calculated by

$$V_i = \sum_{k=0}^{32} S_k - \frac{cos(2k+1)(i+16)\pi}{64}. \quad (4)$$

ii. The series of 512 sample U_i is obtained by modifing the order of V_i.

$$U_{i\times 64+j} = V_{i\times 128+j} \quad (5)$$

$$U_{i\times 64+32+j} = V_{i\times 128+96-j}. \quad (6)$$

iii. The window function is multiplied tp U_i to calculate W_i

$$W_i = U_i \times D_i. \quad (7)$$

iv. The output signal S_j is calculated by iterative addition

$$S_j = \sum_{t=0}^{15} W_{j+32\times i}. \quad (8)$$

3. Decoding in Layer II

The basic procedure includes decoding the bit allocation information for all subbands, decoding the scale factors for subbands with non-zero bit allocation, inverse quantizing the subband samples, and synthesizing 32 bands using the filter banks. The difference between Layer I and Layer II is that the operation for bit allocation information and scale factor is not "reading" but "decoding".

(a) Decoding of bit allocation information

The bit allocation information is stored in 2 to 4 bits to indicate the quatizaion level. The number of bits are defined by the subband number, bitrate and sampling frequency. It should be noticed that the value may indicate different levels depending on their condition even though the same number of bits are assinged to the bit allocation.

(b) Decoding of scale factor selection information

Coefficients that indicate scale factor selection information, called scfsi (scale factor selection information) are read out from the bitstream. Scfsi is defined as shown in Table 3.3. The scale factors are decoded based on scfsi.

(c) Inverse quantization of subband samples

According to the number of bits identified by the decoded bit allocation information, the bits that corresponds to the three consecutive samples are read. When 3 samples are grouped, they are ungrouped before decoding. The MSB of each sample is reversed to obtain s"', where its MSB means -1.0 in the form of 2's complement. The inverse quantized value s" is calculated by

Table 3.3 Scale Factor Selection Information

SCFSI Value	Scale Factor Coding Method
00	3 scale factors are transmitted independently.
01	two scale factors are transmitted: one is common to the first and the second blocks, and the other is for the 3rd block only.
10	one scale factor that is common to all blocks is transmitted.
11	two scale factors are transmitted: one for the first block only, and the other common to the second and the third blocks.

$$s'' = C \times (s''' + D), \tag{9}$$

using the constants C and D which are decided based on the number of allocated bits. The scale factor is multiplied by the inverse quantized value s", resulting in s'.

(d) Synthesizing of 32 bands by filter banks.
The same sysnthsis filtering is performed as Layer I.

3.3.3 Layer III

More fine ideas have been incorporated into Layer III to improve the coding quality based on Layer I/II. Fig. 3.9 shows a block diagram of Layer III. Compared with Layer I/II, Layer III has recently introduced the adaptive block length modified cosine transform (MDCT), the alias distortion reduction butterfly, nonlinear quantization, and variable length coding (Huffman coding). These contribute to further improvement in frequency resolution and reduction of redundancy. The rest of the basic processes are performed as in Layer I/II.

The 16-bit linear quantized PCM signal is mapped from the time domain to the 32 frequency bands by PFB, and each band is further mapped into narrower-bandwidth spectral lines by the adaptive block length MDCT to reduce pre echos [5]. Either a block length of 18 or 6×3 is used, based on the psychoacoustic analysis. Adoption of a hybrid filter bank increases the frequency resolution from 32 to $32 \times 18 = 576$. The obtained mapping signal is processed by aliasing distortion reduction and then by nonlinear quantization. The mapping with cascade transform of the filter bank, MDCT and aliasing distortion reduction is called the Hybrid Filter Bank (HFB). Quantization is accompanied by an iteration loop for bit allocation. The bitrate of each frame is variable. The quantized signal is Huffman-coded, and then built into the frame. Decoding is achieved by disassembling the frame first, and then decoding the Huffman table index and the scale factors that have been sent as side information. Further, Huffman decoding and dequantization are performed based on the side information. The time domain signal can be reconstructed by demapping the quantized signal through the hybrid filter banks.

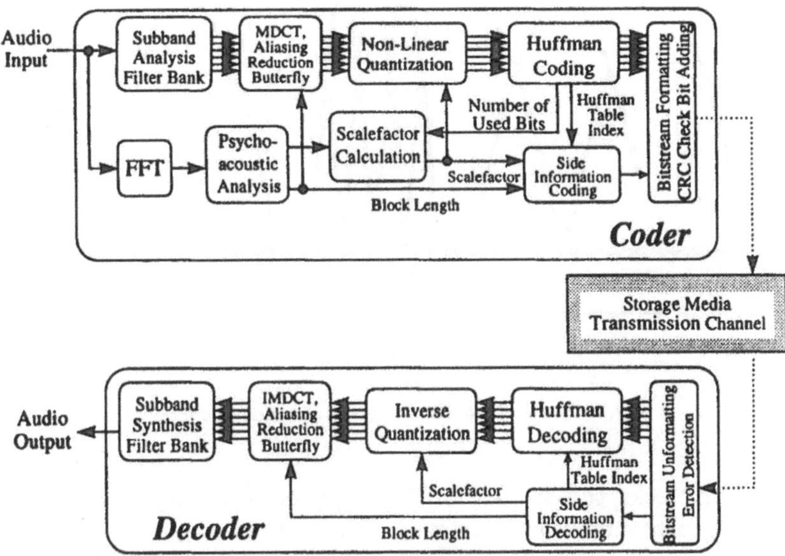

Figure 3.9 Layer III algorithm.

1. Psychoacoustic analysis

   Psychoacoustic analysis is performed to find the masking level of each MDCT component as well as to determine the block length for MDCT. It is recommended to employ a modified version of Psychoacoustic Model II for Layer II.

   The block length is selected based on psychoacoustic entropy, using the unpredictability theory. The unpredictability is measured comparing the spectrum of the current and privious time frames. In the vicinity of attack where pre-echo is generated, the shape of the spectrum differs between the two frames and the psychoacoustic entropy increases. When the entropy exceeds a predetermined value, the system evaluates it as an attack and switches MDCT to short blocks.

   The masking levels is calculated changing the internal parameters depending on the block length. FFT is used to reduce the comuptational complexity, where the FFT block length is 256 for short blocks and 1024 for long blocks.

2. Adaptive block length MDCT and the window shape

   In HFB, 576 samples of input signal comprise 1 granule. A granule is a set of samples, and is one component in the formation of a block. Two granules, i.e., granule 0 and granule 1, are processed as 1 block composed of the total 1152 samples. When subband analysis is performed to PCM samples of one granule, each subband has 18 samples.

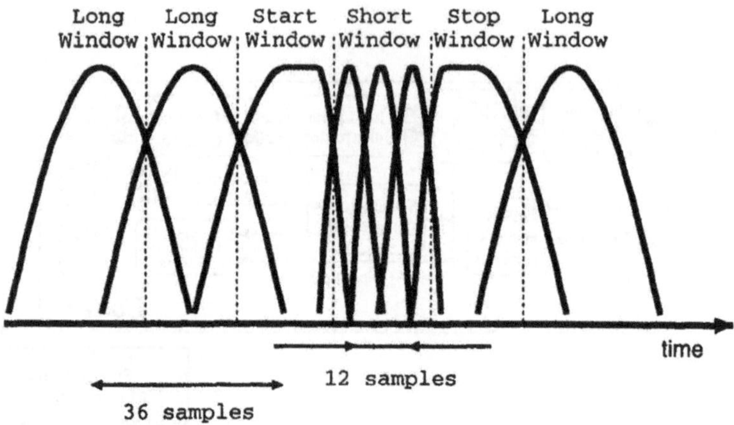

Figure 3.10 Alteration pattern of window functions.

For long blocks, the 36-point MDCT is performed. The 18 subband samples of the current granule are combined with 18 samples in the preceding granule. Because of the coefficient symmetry, independent output of MDCT becomes $36/2 = 18$. For short blocks, the number of MDCT input samples is reduced to 12 and three times of MDCT are applied in one granule. The first 6 samples are combined with the last 6 samples of the previous granule. The number of independent short-block MDCT output is 18, which is same for the long-block MDCT.

Four types of window functions, Normal Window, Start Window, Stop Window, and Short Window are prepared. The 36-point MDCT is applied in the first three windows and the 12-point MDCT is used in the last window. The Start Window has to be placed before the Short Window, and the Stop Window after the Short Window to realize noiseless transform.[12] Fig. 3.10 shows the example shift pattern of window functions.

3. Aliasing distortion reduction in the frequency domain

The MDCT cofficients of long blocks are treated to aliasing distortion reduction via the butterfly circuit as shown in Fig. 3.11. Butterfly operation is performed on the mutually adjacent 32 subbands, using 8 subband samples close to band borders. The butterfly circuit coefficients cs_i and ca_i are given by

$$cs_i = \frac{1}{\sqrt{1+c_i^2}} \tag{10}$$

$$ca_i = \frac{c_i}{\sqrt{1+c_i^2}}. \tag{11}$$

The value for c_i is determined so that it becomes smaller as the frequency distance of applied MDCT coefficients becomes larger [1].

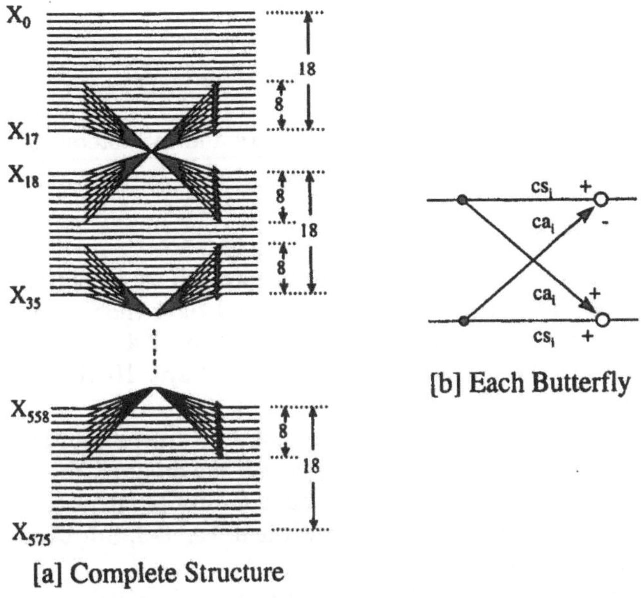

Figure 3.11 Aliasing distortion reduction butterfly.

4. Quantization

   Non-linear quantization is employed in Layer III instead of linear one of Layer I and II. The following equation shows the relation among the inverse quantized MDCT coefficient x, code i and scale factor:

$$x = sign(i) \times |i|^{4/3} \times 2^{scale\ factor}. \tag{12}$$

5. Bitstream formatting and bit buffering

   The bitstream format of Layer III is approximately the same as in Layer II, and the frame size is also the same. Each frame of 1152 samples is divided into two granules of 576 samples. The frame header is followed by the accompanying information that is common to both granules, and the accompanying information of each granule.

   As explained before, the psycho entropy increases in the frame that contains attack(s), and the frame requires larger number of bits. A technology called 'bit reservoir' has been introduced for this purpose. This technology makes use of the skew of the information volume that is generated by each frame. Bit reservoir volume is usually held slightly below the maximum reservoir volume. When entropy increases in a frame that contains an attack, the system uses the reserved bits in addition to the ordinary bits and again starts storing a small number of bits in the next frames, and keeps storing until the volume reaches to slightly below the maximum storage level.

Table 3.4 Joint Stereo Mode

Layer	Available stereo coding mode
Layer I/II	Intensity stereo
Layer III	Combined (intensity and MS) stereo

3.3.4 Stereo Coding

In the standard, stereo coding has been specified as an option. The bitrate reduction utilizing the correlation between left and right channels is performed in the joint stereo mode. This mode is specified as in Table 3.4, corresponding to each layer. Layer I/II has the intensity stereo only, and Layer III the combined stereo comprising intensity and MS.

The intensity stereo uses the same shape, but different-amplitude subband data between the left and right signals in place of the original two-channel signals. Four modes are prepared to change the subband for use as intensity stereo, as 4-31, 8-31, 12-31, and 16-31. The subbands below them, i.e., 0-3, 0-7, 0-11, and 0-15 are coded independently for each channel.

The MS stereo is the simplest available 2-point-orthogonal transform, in which the sum amd the difference of two signals are used instead of original signals. When the correlation between both channels is high, a data compression effect are expected due to the skewed energy distribution. In the combined stereo, the system adds the total sum of each FFT of both channels, and further multiplies by a predetermined constant. If the resultant value is greater than the difference in the power spectra of both channels, then the system selects MS stereo, and if it is not greater, the system selects intensity stereo, and performs coding; i.e., when the ratio between the above sum signal and difference signal is greater than the predetermined threshold value, the system selects MS stereo.

3.3.5 The Performance of MPEG-1/Audio

Subjective evaluation using the hardware of each layer was performed for 128, 96, and 64 kb/s in Stockholm in May 1991, and for re-evaluation of 64 kb/s in Hannover in November 1991 [13, 14]. Figure 3.12 is the result of subjective evaluations. In Fig. 3.12, each score corresponds respectively to equality in Table 3.5 [15]. In practice, there are perception errors of evaluators, etc., and the score for the original sound does not reach 5.0. After these 2 sessions of subjective evaluation, both Layer II and III have been approved as of sufficient quality for distribution purposes of broadcasting stations at 128 kb/s per channel.

3.4 MPEG-2/AUDIO ALGORITHM

The MPEG/Audio Phase 2 algorithm [3], usually called the MPEG-2/Audio, is basically divided into two algorithms for the lower sampling frequency and larger number of channels for the multi-channel/multilingual. For audio, the difference between the MPEG-1 algorithm and MPEG-2 algorithm is smaller than in video systems. It is possible to say that the MPEG-2 algorithm is an extension of the

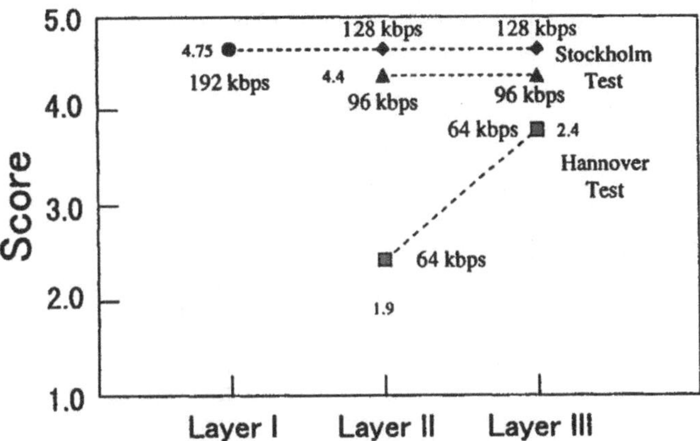

Figure 3.12 MPEG-1/Audio algorithm subjective evaluation.

Table 3.5 Score Criteria for Subjective Evaluation

Score	Quality
5.0	Imperceptible
4.0	Perceptible, but not annoying
3.0	Silightly annoying
2.0	Annoying
1.0	Very annoying

MPEG-1 algorithm. In this section the MPEG-2 algorithm is outlined and the difference between the MPEG-1 and MPEG-2 algorithms is pointed out.

3.4.1 Low Sampling Frequency Algorithm

To achieve high quality at low bit rates below 64 kb/s, three kinds of sampling frequencies are introduced in MPEG-2 algorithm. They are 16 kHz, 22.05 kHz, and 24 kHz, and the target is for quality to exceed the ITU-T Recommendation G.722 [16]. From the viewpoint of the bitstream syntax the supported sampling frequencies and bitrates are changed in comparison with MPEG-1. Also, changes have been made in the bit allocation tables and psychoacoustic models.

3.4.2 Multi-Channel and Multilingual Capability

In MPEG-2, up to 6 channel audio coding is supported for multi-channel and multilingual systems while one or two channel audio coding is possible in MPEG-

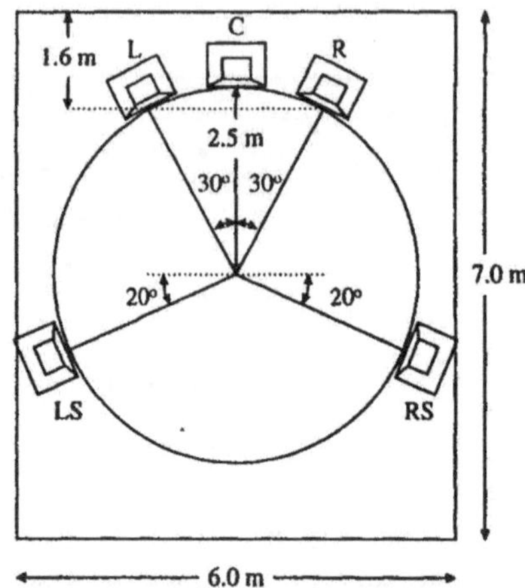

Figure 3.13 Example 3/2 stereo speaker positioning.

1. This system has the remarkable feature of being compatible with the MPEG-1 algorithm. Another aspect worth mentioning is that there was good cooperation with ITU-R in standardization activities.

1. Multichannel format

   The most popular multichannel audio format that is recommended by ITU-T and other specialists is the so-called 3/2 stereo. This system places a center speaker between the left and the right speakers, and places two surround speakers at the left and right of the rear side. Figure 3.13 is a typical speaker positioning for 3/2 stereo. This arrangement was also used at the official subjective evaluation in February 1994 [17]. The MPEG-2 algorithm presumes a multichannel format as described in Table 3.13. Note that the system allows more kinds of format for input than for output. L is the right channel signal, C is center channel signal, LS is the left surround channel signal, L1 and L2 represent the first language left channel signal and the second language left channel signal, respectively, and the right channels are described correspondingly in the same way.

   In addition to these channels, the system allows the addition of a low frequency enhancement (LFE) channel as an option. This has been defined to match the LFE channel in the movie industry. The LFE channel contains information from 15 Hz to 120 Hz, and the sampling frequency is 1/96 of the main left and right channels.

   To reduce redundancy among multi channels, interchannel adaptive prediction is introduced. Three kinds of interchannel prediction signal are calculated

Table 3.6 Audio Input/Output Format Supported in MPEG-2

Format	Input	Output
3/2 stereo (L, R, C, LS, RS)	Yes	Yes
3/1 stereo (L, R, C, S)	Yes	Yes
3/0 stereo (L, R, C)	Yes	Yes
3/0+2/0 stereo (L1, R1, S1, L2, R2)	Yes	No
2/2 stereo (L, R, LS, RS)	Yes	Yes
2/1 stereo (L, R, S)	Yes	Yes
2/0 stereo (L, R; MPEG-1 full comptible)	Yes	Yes
2/0+2/0 stereo (L1, R1, L2, R2)	Yes	Yes
1/0 mono (L or R; MPEG-1 full comptible)	Yes	No
1/0+2/0 stereo (L1 or R1, L2, R2)	Yes	No

Table 3.7 Multichannel Extension Between MPEG-1/2

Base	Extension
MPEG-1 Layer I	MPEG-2 Layer II MC
MPEG-1 Layer II	MPEG-2 Layer II MC
MPEG-1 Layer III	MPEG-2 Layer III MC

within each frequency band, but only the prediction errors in the center channel and surround channel are coded.

2. Compatibility with MPEG-1

   Forward and backward compatibility are assured. Backward compatibility means that an MPEG-1 decoder can decode basic stereo information comprising two (front) left/right channels (L0, R0), from MPEG-2 coded data. These signals consist of down-mix signals given by the following formulae:

$$L0 = L + x \times C + y \times LS \tag{13}$$

$$R0 = R + x \times C + y \times RS. \tag{14}$$

   Four modes are prepared for predetermined values x and y. Forward compatibility means that an MPEG-2 multichannel decoder can correctly decode the bitstream specified by the MPEG-1 algorithm. Combinations are possible as shown in Table 3.7. MC means multi-channel. MPEG-2 information other than the basic stereo information is stored in the ancillary data field of MPEG-1.

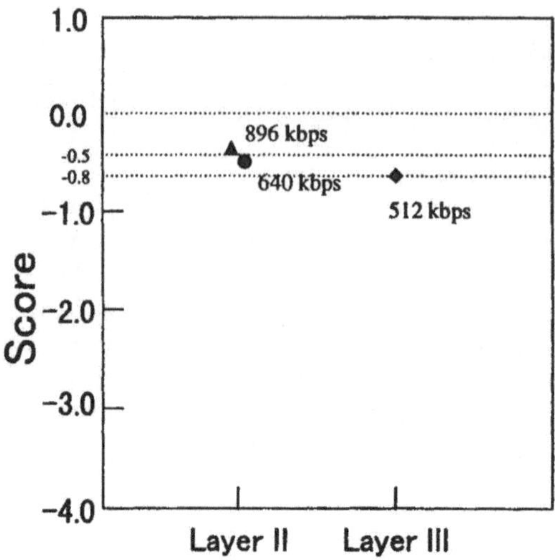

Figure 3.14 Subjective evaluation results for MPEG-2/Audio MC algorithms.

3.4.3 MPEG-2/Audio Performance

Subjective evaluation of MPEG-2/Audio has been performed several times from 1993 to 1996 [17, 18]. Figure 3.14 shows the results of subjective evaluations in 1996 [18]. The evaluation criteria used here are the same as that for MPEG-1 shown in Fig. 3.12, but scoring is different. The original sound quality corresponds to 0.0, not 5.0. The vertical line therefore means the difference in quality between the tested sound and original sound. It was confirmed that 640 kbps Layer II and 512 kbps Layer III achived score of -1.0, or 4.0 by old criteria, which is acceptable (Refer to [17] page 28). Refer to [19] on the evaluation results of MPEG-2/LSF.

3.5 FUTURE WORK

The standardization acitivies at MPEG has realized transparent audio transmission/storage around at 96 to 128 kbps/channel. The number of channels which can be supported now is 6. These technologies are now utilized in the market. For example, it is used in video CD-ROM and audio transmission between the broadcasting centers. Its market continues to grow very rapidly. However, there is no end in the demands for compression algorithms with higher coding efficiency. For this purpose, MPEG is currently exploring MPEG-2/AAC and MPEG-4 activites towards the final goal of transparent coding at 32 kbps/channel.

REFERENCES

[1] ISO/IEC 11172, "Coding of Moving Pictures and Associated Audio for Digital Storage Media at up to about 1.5 Mb/s," Aug. 1993.

[2] Chairman, Task Group 10-2, "Draft New Recommendation, Low Bit-Rate Audio Coding," Document 10-2/TEMP/7(Rev.2)-E, Oct. 1992.

[3] MPEG-Audio Subgroup, "ISO 11172-3 Compatible Low Bit Rate Multi-Channel Audio Coding System and Conventional Stereo Coding at Lower Sampling Frequencies," ISO/IEC JTC1/SC29/WG11 N0803, Nov 1994.

[4] P. P. Vaidyanathan, "Multirate Digital Filters, Filter Banks, Polyphase Networks, and Applications: A Tutorial," *Proc. IEEE* vol. 78, no. 1, pp.56-93, Jan. 1990.

[5] A. Sugiyama *et al.*, "Adaptive Transform Coding with an Adaptive Block Size," *Proc. ICASSP'90*, pp.1093-1096, Apr. 1990.

[6] M. Iwadare *et al.*, "A 128 kb/s Hi-Fi Audio CODEC Based on Adaptive Transform Coding with Adaptive Block Size MDCT," IEEE JSAC pp.138-144, Jan. 1992.

[7] N. S. Jayant and P. Noll, *Digital Coding of Waveforms*, Prentice-Hall, 1984.

[8] M. Vetterli *et al.*, "Perfect Reconstruction FIR Filter Banks: Some Properties and Factorizations," *IEEE Trans. ASSP* vol. 37, pp. 1057-1071, Jul. 1989.

[9] H. G. Musmann, "The ISO Audio Coding Standard," *Proc. Globecom'90*, pp. 0511-0517, Dec. 1990.

[10] J. Tribolet *et al.*, "Frequency Domain Coding of Speech," *IEEE Trans. ASSP* vol. 27, pp. 512-530, Oct. 1979.

[11] J. Princen *et al.*, "Subband/Transform Coding Using Filter Bank Designs Based on Time Domain Aliasing Cancellation," *Proc. ICASSP'87*, pp. 2161-2164, Apr. 1987.

[12] T. Mochizuki, "Perfect Reconstruction Conditions for Adaptive Blocksize MDCT," *Trans. IEICE*, vol. E77-A, no. 5, pp. 894-899, May 1994.

[13] S. Bergman *et al.*, "The SR Report on the MPEG/Audio Subjective Listening Test, Stockholm April/May 1991," ISO/IEC JTC1/SC29/WG11 MPEG91/010, May 1991.

[14] H. Fuchs, "Report on the MPEG/Audio Subjective Listening Tests in Hannover," ISO/IEC JTC1/SC29/WG11 MPEG91/331, Nov. 1991.

[15] CCIR Recommendation BS 562-3, Subjective Assessment of Sound Quality, 1990.

[16] CCITT Rec. G.722, The CCITT Blue Book, Melbourne, 1988.

[17] F. Feige *et al.*, "Report on the MPEG/Audio Multicahnnel Formal Subjective Listening Tests," MPEG94/063, Mar. 1994.

[18] F. Feige *et al.*, "MPEG-2 Backwards compatible codecs Layer II and Layer III: RACE dTTb listening test report," ISO/IEC JTG1/SC29/WG11/N1229, Mar. 1996.

[19] Audio Subgroup, "Report on the Subjective Testing of Coders at Low Sampling Frequencies," ISO/IEC JTC1/SC29/WG11 N0848, Nov. 1994.

Chapter 4

System Synchronization Approaches

Hidenobu Harasaki
C & C Media Research Labs.
NEC Corporation
Kawasaki, Japan
harasaki@ccm.cl.nec.co.jp

4.1 INTRODUCTION

This chapter describes media sampling clock, system clock and inter-media synchronization methods for multimedia communication and storage systems. As a typical example of multimedia communication systems, it focuses on audiovi-

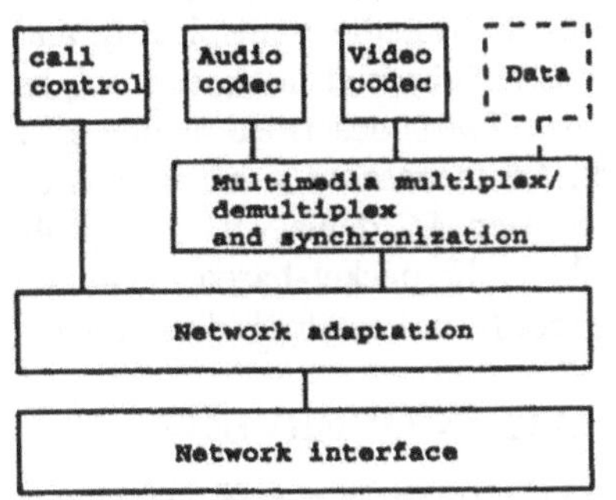

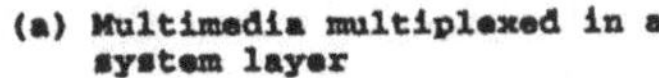
(a) Multimedia multiplexed in a system layer

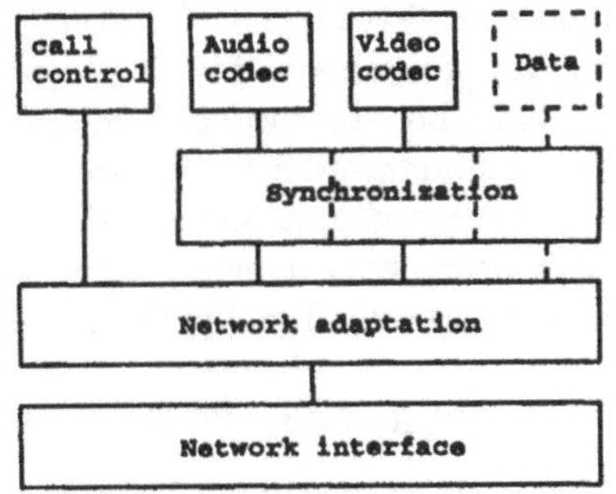

(b) Multimedia multiplexed in a network layer

Figure 4.1 A general protocol stack for audiovidual communication terminal. The ITU-T standardizes audiovisual and multimedia systems in H series recommendations. More specifically, it specifies systems and terminal equipment for audiovisual services in H.31x-H.33x, and call control in Q-series, audio coding in G.71x-G.72x, video coding in H.26x, multimedia multiplexing and network adaptation in H.22x, network interface in I series, and system control which is missing in the figure in H.24x recommendations. Telematic services including data conference and conference control are specified in T-series recommendations.

sual communication systems [1] defined by ITU-T Series H Recommendations. It also describes video-on-demand (VOD) systems [2] as a multimedia storage system example. The difference on the clock synchronization methods between communication and storage systems is derived from real-time vs. non-real-time requirements. Multimedia communication systems always require the clock synchronization between a sender and a receiver, while multimedia storage systems do not. A decoder clock can be independent of an encoder in multimedia storage systems. From a clock synchronization point of view, broadcast such as digital TV broadcast can be included in multimedia communication systems. People may classify a broadcasting system under one-way communication systems, but soon it will be enhanced as an asymmetric two-way communication system, when interactive TV systems are widely spread.

System layer [3] [4] plays an important role for the clock synchronization. Fig. 4.1 shows a general protocol stack for audiovisual communication terminals. The terminal has an audio codec, a video codec, and an optional data interface on the top of the protocol stack. System layer is located just below them, and is in charge of multimedia multiplexing/demultiplexing and system clock synchronization as shown in Fig. 4.1 (a). Network adaptation layer is often included in the system layer. As shown in Fig. 4.1 (b), multimedia multiplex is sometimes achieved in a network layer or below. In ITU-T H.323 [5], audio and video signals are not multiplexed in a system layer, but they are transported in different logical channels provided by the underlying network.

This chapter is organized as follows. Section 4.2 presents system clock synchronization overview, which includes sampling clock and service rate synchronization methods. Time stamp transmission and adaptive clock recovery methods are explained in Section 4.3. Section 4.4 describes multimedia multiplexing and demultiplexing methods. MPEG-2 system layer is enlightened in Section 4.5 as one of the most popular system layers. MPEG over ATM is described as one example of network adaptation in Section 4.6. Section 4.8 focuses on a multipoint extension of multimedia communication system. Section 4.9 deals with the problems in error prune environment, especially in ATM or IP packet-based network (i.e., Internet/Intranet). Finally, Section 4.10 describes future research directions.

4.2 SYSTEM CLOCK SYNCHRONIZATION OVERVIEW

A multimedia system designer has to pay attention to a system clock synchronization method in an early stage of a multimedia system design. Fig. 4.2 shows three clock synchronization models for different system configurations. In general switched telephone network (GSTN) case, a common network clock is available to the both end terminals as shown in Fig.4.2 (a). Sampling clock frequency (**Fs**) for voice signals in an encoder is locked to network clock source (**Fn**), and decoding clock frequency (**Fr**) is also locked to **Fn**. Moreover, service rates at the encoder (**Rs**) and at the decoder (**Rr**) are also locked to **Fn**. Sampling clock, service rate and decoding clock are all locked to the common network clock in GSTN or B-ISDN.

In IP-packet based network or customer premises ATM network (ATM LAN) environment, however, a common network clock is not available to the terminals. ATM Adaptation Layer (AAL) type 1 [6] provides a mechanism to recover service rates at a decoder. But AAL type 1 is not popular in ATM LAN environment.

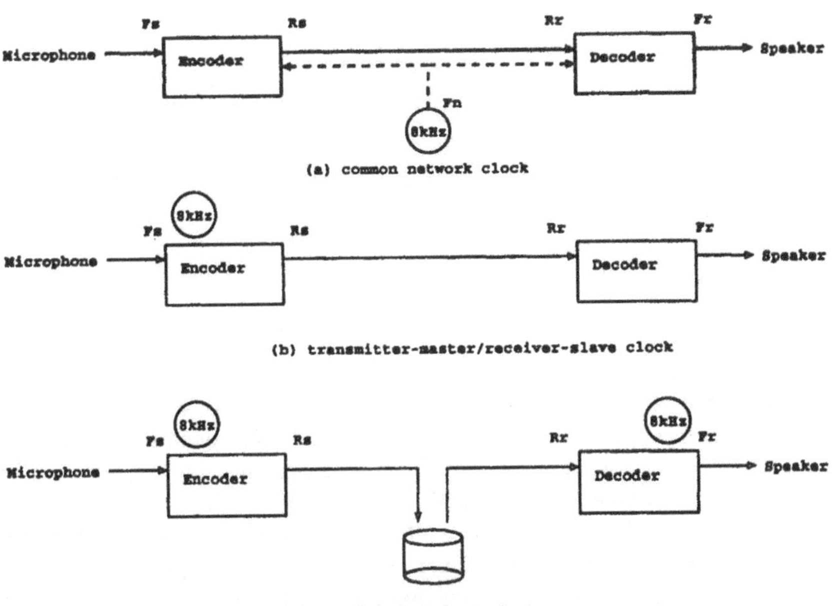

Figure 4.2 Sampling clock synchronization models for different system configurations.

AAL type 5 [7] is widely used instead. In asynchronous network environment, sampling clock frequency and service rates are generated by a local oscillator. Instantaneous service rate is not constant because of IP-packet nature or cell-based bursty transmission. There are two possible implementations for a decoder.

1. Independent: Decoding clock frequency is independently generated by its local oscillator in the decoder.

2. Slave clock generation: Decoding clock frequency is somehow locked to that of the encoder.

Let us consider what occurs when independent clock frequency is used in the decoder. Since the decoding clock **Fr** is different from **Fs**, a sample slip happens periodically. To cope with the sample slips, the decoder should have an adaptation mechanism that repeats a sample when **Fr** > **Fs**, and deletes a sample when **Fr** < **Fs**, when necessary. Mean time between slips is calculated by Eq. (1).

$$Mean\ Time\ Between\ Slips = \frac{1}{Sampling\ Clock\ Frequency \times \frac{PPM}{10^6}}, \quad (1)$$

where *Mean Time Between Slips* is measured in second (sec), *Sampling Clock Frequency* is denoted in Hertz (Hz), and *PPM* is in parts per million (ppm).

Table 4.1 summarizes the mean time between sample slips for typical media types. Sample slip may not be noticeable to users, but sometimes it degrades the

Table 4.1 Mean Time Between Sample Slips

Media type	Mean time between sample slips	
	10ppm difference	100ppm difference
8kHz sampling voice	12.5sec	1.25sec
44.1kHz sampling audio	2.258sec	0.2258sec
30 frame/sec video[1]	55.5min	5.55min

reproduced voice, audio or video quality. Frame skip or repetition is somewhat noticeable, especially when a moving object is in a frame or a camera is panning.

Elastic buffer approach at a decoder should be mentioned here. A decoder on asynchronous network may have a buffer to compensate the frequency difference between sender and receiver. It is achieved by delaying the decode starting time. Suppose that a decoder has 400-sample buffer for 8kHz voice signals and it starts the output to the speaker when buffer is half full. Initial output delay is only 25 msec. A 200-sample elastic buffer can maintain non-slip condition up to 41.6 and 4.16 minutes in the case of 10 ppm and 100 ppm frequency difference, respectively. Note that the elastic buffer approach only works for initial non-slip duration. Once slip happens, the buffer is no longer useful.

Fig. 4.2(c) shows a storage application example. An encoder employs locally generated sampling clock, **Fs**, and stores the encoded data to a media at a service rate of **Rs**. A decoder can retrieve the encoded data from the storage media with a different service rate of **Rr**, and outputs with a different decoding clock frequency **Fr**. In media storage applications (e.g., music CD, DVD, VOD), total playback time may slightly differ from the actual recording time. In VOD applications, a VOD server which has a storage media is usually located far from the decoder and the server and the decoder are connected via a network. The VOD system is characterized as multimedia communication and storage system. There are two senarios for clock master/slave realization. One is a decoder master approach. The decoder can retrieve a stream from VOD server as if its disk is locally attached. Since each decoder has a different system clock, the VOD server will serve a stream per terminal basis. In addition to the local storage type approach, the VOD server reads a multimedia data from the disks and broadcasts or multicasts to a number of clients with its locally determined service rate. In this case, the decoders must recover the service rate to receive all the data. The decoding clock frequency is uniquely determined to coincide with the service rate. When the VOD server sends out the data with 10ppm faster than the nominal service rate, the decoding clock frequency is also 10ppm higher than the nominal one.

System clock synchronization for some multimedia applications can be categorized as follows:

[1] A frame is considered to be a sample, because a slip operation on a video signal usually treats a frame as one unit.

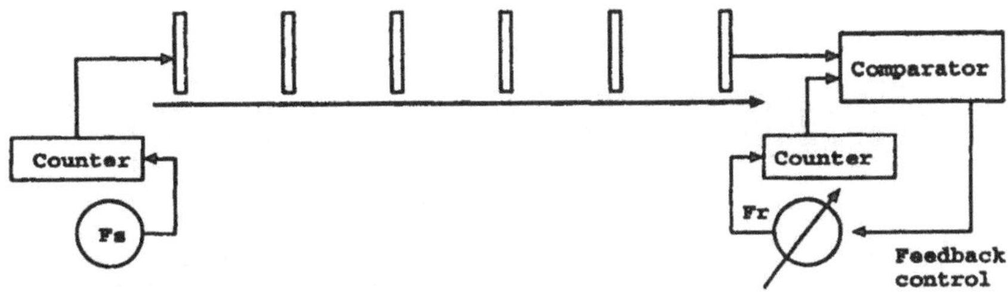

Figure 4.3 Timestamp transmission method.

Network master approach Telephony in GSTN, B-ISDN.

Push approach Digital TV Broadcast, Internet Phone, VOD with multicast or broadcast capability.

Pull approach CD, DVD, VOD without multicast or broadcast capability.

4.3 CLOCK TRANSMISSION METHODS

This section describes two clock synchronization methods. One is a *timestamp transmission method* and the other is an *adaptive clock recovery method.* As described in Section 4.2, if common network clock is available to both the ends and sampling clock frequency can be chosen to be locked to the network clock, clock transmission from a sender to a receiver is not necessary. In video signal transmission applications, however, sampling clock frequency cannot always be chosen to be locked to the network clock. For example, video sampling clock frequency (13.5MHz for component system or 14.3MHz for composite system) is usually generated from the video signal itself. In those cases, the clock transmission is necessary, even in synchronous network environment.

4.3.1 Timestamp Transmission Method

There are two kinds of timestamp transmission methods, one is synchronous timestamp, and the other is non-synchronous timestamp. The synchronous timestamp method can be used where the common network clock is available to both the ends. A source clock frequency is measured by the common network clock, and the measured value is transmitted to the receiver periodically. The better jitter/wander performance is achieved by the synchronous timestamp method. Synchronous Residual Time Stamp (SRTS) method is used in AAL type 1 [6] to recover the service rate which is asynchronous to the network clock. The method is a variation of synchronous timestamp method that only transmits least significant part of the measured values. On the other hand, non-synchronous timestamp can be used in any network environment, because the method does not reply on the common network clock.

Fig. 4.3 shows how a timestamp method works for sample clock frequency transmission. At the sender side, source clock frequency, **Fs**, is fed to a counter. Timestamp which is a copy of a counter value at a certain time is sent out periodi-

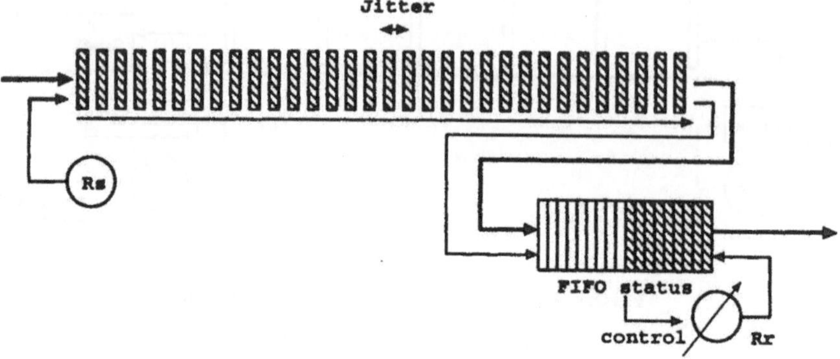

Figure 4.4 Adaptive clock recovery method.

cally. At the receiver side, voltage controlled oscillator (VCO) generates a recovered source clock frequency, **Fr**, and is fed to a counter. Initial timestamp is loaded to the counter. Whenever a succeeding timestamp arrives, a comparator compares the counter value to the timestamp. If the counter value is greater than the received timestamp, **Fr** must be higher than **Fs**. If, on the other hand, the counter value is smaller than the timestamp, **Fr** must be lower than **Fs**. This control loop is called a phase locked loop (PLL).

4.3.2 Adaptive Clock Recovery Method

Fig. 4.4 shows how adaptive clock recovery method works for a service rate recovery. At the sender side, a packet is sent out at a certain rate **Rs**. The packet arrives at the receiving side with some timing fluctuation caused by network jitter. In order to smooth the jitter out, a first-in-first-out (FIFO) buffer is employed. A VCO generates the recovered service clock, **Rr**, and is fed as the read clock to the FIFO. The initial FIFO status is half full. If FIFO fullness decreases, **Rr** is faster then **Rs**. If FIFO fullness increases, **Rr** is slower than **Rs**. Thus, the feedback control loop is created to recover the service rate. Adaptive clock method for service rate recovery is also employed in AAL type 1 [6].

4.4 MULTIPLEXING AND DEMULTIPLEXING

In multimedia communication and/or storage systems, various media, such as voice, audio, video, data are multiplexed in a transmission channel. In digital TV broadcast, several TV channels are multiplexed in a single digital transmission channel.

Multimedia multiplexing methods are categorized as follows:

- Time division multiplex (TDM)
- Packet based multiplex
  - fixed length
  - variable lentgh

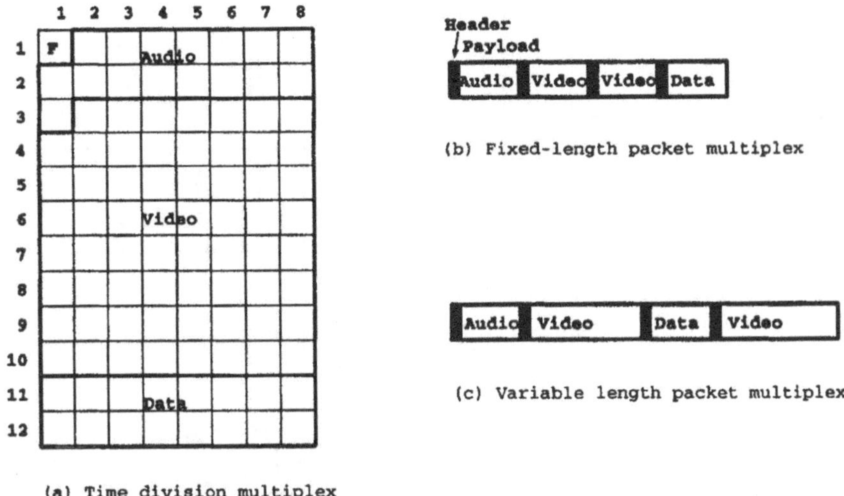

Figure 4.5 A various multiplexing methods.

Fig. 4.5 shows a various multiplexing methods: (a) time division multiplex (TDM), (b) a fixed length packet multiplex, and (c) a variable length packet multiplex. In Fig. 4.5 (a), an eight-column and twelve-row frame is subdivided into a frame start octect (**F**), audio time slots, video time slots and data time slots. Since the audio, video and data boundaries are fixed, TDM is the least flexible among the methods. TDM demultiplexer first hunts a frame by finding several frame start octects in every 96 time slots, and then demultiplexes audio, video and data by its time slot location. ITU-T H.221 [8] multiplex, that is used in H.320 [9] terminal, is based on this TDM multiplex approach. Fig. 4.5 (b) shows a fixed length packet multiplex example. Audio, video and data packets consist of a header and a payload. The header includes a packet start field and a media description field. The fixed length demultiplexer first detects the packet start field, and determines what kind of media is stored in the payload by the media description field. Fig. 4.5 (c) shows a variable length packet multiplex example. A length field is necessary in the packet header. Fixed length packet multiplexing is preferred by hardware multiplex approach, switching and demultiplexing, while variable length packet multiplexing is preferred by CPU based processing approach.

Fig. 4.6 shows a packet based statistical multiplexing example. Let us assume that five data channels are multiplexed to a single transmission channel, and each data channel activity is less than 20% on average over a long period. Since there is a chance that more than two data channels are simultaneously active, buffers are used to delay one of the active data channels. In Fig. 4.6, the first packet of Ch. 2 is collided with the packet of Ch. 5, and then it is delivered to the other end with a short delay. Since the second packet of Ch. 2 is collided with the second packet of Ch. 5 and the packet of Ch. 1, it is delayed significantly. This kind of packet based multiplexer introduces a fluctuation of end-to-end delay. Since continuous media, e.g., voice, speech, audio and video, should be played back without any pause, end-to-end delay must be constant. The system layer designer needs to figure out how

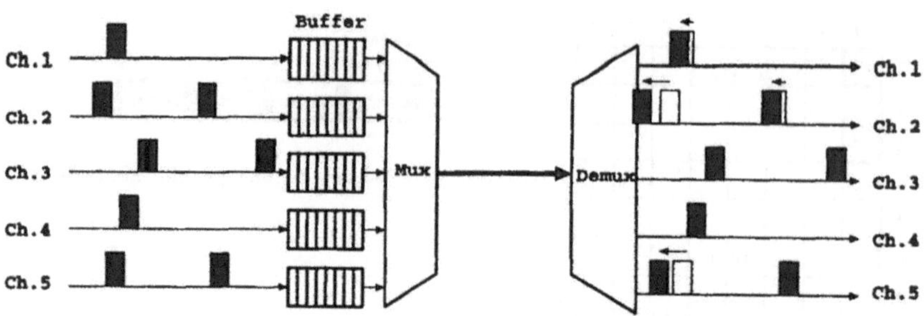

Figure 4.6 A statistical multiplexer.

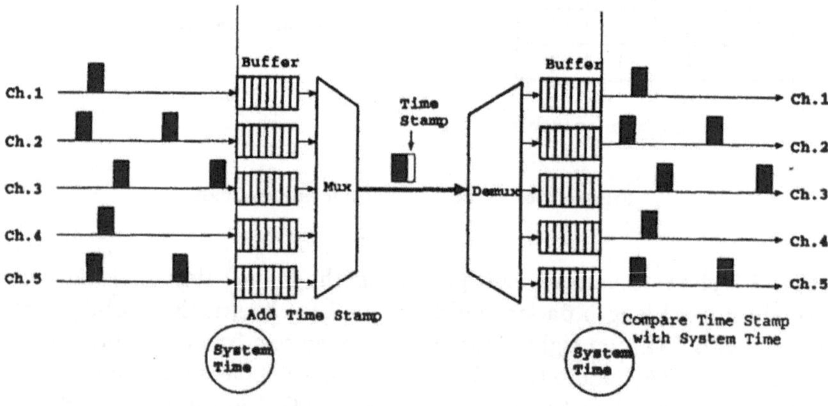

Figure 4.7 A statistical multiplexer with timing recovery.

much delay will be introduced at the multiplexing buffer, and how to compensate the fluctuation of the end-to-end delay.

One method to compensate the end-to-end delay fluctuation is using time stamp as shown in Fig. 4.7. A statistical multiplexer reads the system clock counter when a packet is arrived at the buffer input port, and associates the packet with its arriving time counter value (i.e. timestamp). In the demultiplexer, another buffer per port will delay the packet. The packet is delivered with a predetermined constant end-to-end by comparing the time stamp with the decoder system clock counter. Decoder system clock can be synchronized to the encoder by timestamp method described in Section 4.3.1.

4.4.1 Intermedia Synchronization

Intermedia synchronization, e.g., lip sync, is indispensable for a natural presentation. Fig. 4.8 shows the block diagram of audio and video synchronization. To achieve lip synchronization, end-to-end delay for audio and for video should be the same. The end-to-end delay (Dend2end) is defined as follows:

$$\begin{aligned} Dend2end &= Aenc + Amux + Adem + Adec \\ &= Venc + Vmux + Vdem + Vdec. \end{aligned}$$

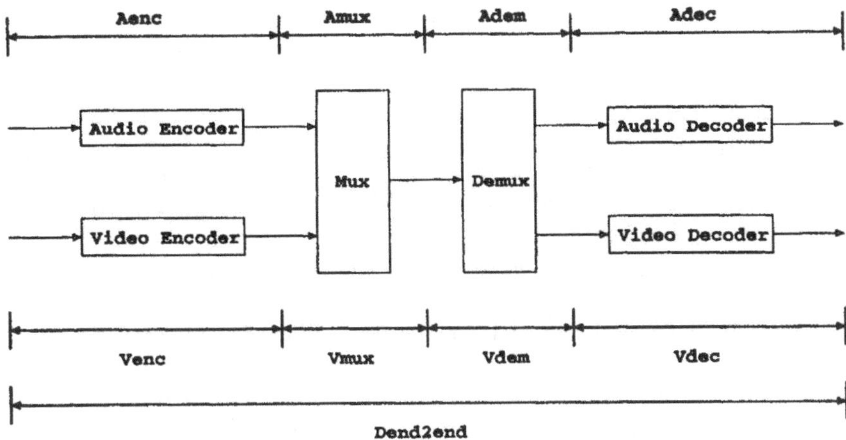

Figure 4.8 Intermedia synchronization.

In general, audio encoding/decoding delay (Aenc, Adec) is smaller than video encoding/decoding delay (Venc, Vdec). Thus audio multiplexing/demultiplexing delay (Amux, Adec) should be larger than that of video to achieve the same end-to-end delay for audio and video. Delaying audio is reasonable because audio bitrate is usually less than video bitrate and the required buffer size is proportional to the bitrate.

4.5 MPEG-2 SYSTEM

The MPEG-2 system (ISO/IEC 13818-1) [3] was standardized in 1995 as an extension of MPEG-1 system (ISO/IEC 11172 standardized in 1992). It is also standardized as a common text in ITU-T, MPEG-2 system and is also referred to as the ITU-T Recommendation H.222.0. The MPEG-2 system is based on a packet based multiplexing, and provides flexible multiplexing mechanisms for storage media, communication, and broadcasting applications. It also provides system clock synchronization and inter-media synchronization.

There are two types of stream format. One is program stream (PS), which is similar to the MPEG-1 system stream. It is classified as a variable length packet based multiplex system. The other is transport stream (TS) which is newly introduced in MPEG-2 system standardization to support multiple program multiplexing for communication and broadcasting. It is designed to be used in error prone network condition. TS is classified as a fixed length packet based multiplex, where the packet length is 188 bytes. Table 4.2 summarizes the differences between PS and TS.

Fig. 4.9 shows an MPEG-2 system encoder configuration example. MPEG-2 video encoder encodes video signals and produces video elementary stream. Audio signals are commonly encoded by MPEG-1 audio encoder. Elementary streams and optional data are fed to MPEG-2 system encoder. The elementary streams are segmented in packetizers. Presentation Time Stamp (PTS) and an optional

Table 4.2 MPEG-2 System: Program Stream vs. Transport Stream.

	Program Stream	Transport Stream
Packet length	variable (e.g. 2KB-4KB)	188 byte fixed
Efficiency	efficient (about 0.1-0.2% PES header)	less efficient (2% TS header)
Application	storage (e.g. video CD, DVD)	transmission (TV conference, Digital TV broadcast)
Error condition	PS is designed for error free condition.	TS is designed for error prone condition.
System clock	one System Clock Reference (SCR)	multiple Program Clock Reference (PCR)

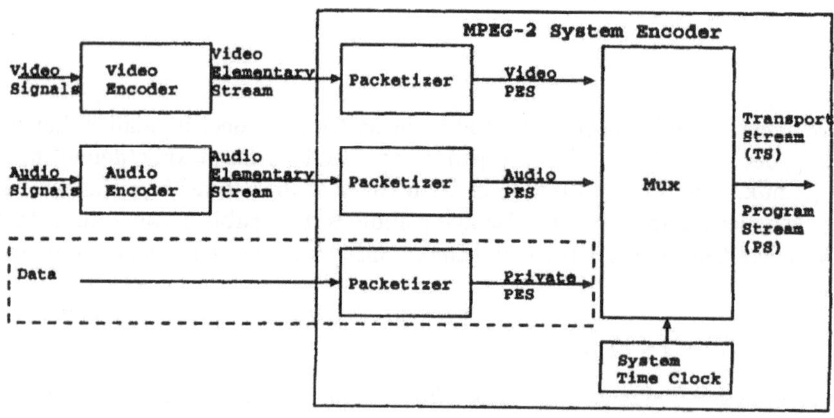

Figure 4.9 MPEG-2 system encoder configuration.

Decoding Time Stamp (DTS) are added to each elementary stream segment to form the packetized elementary stream (PES) header. In the case of transport stream, each PES stream is divided into 184 byte length packets. With 4 byte header, 188-byte fixed length TS packets are generated. Program clock reference (PCR) is multiplexed in the transport stream. PCR carries system time clock (27MHz) counter value and is used to synchronize the system clock frequency in the decoder. In the case of program stream, each PES stream is divided into a few kilo byte length packets. With pack header which includes system clock reference (SCR) and system header, variable length packs can be generated. In video CD and DVD application, pack length is fixed to 2048 byte, because of the storage media access unit size. SCR carries system time clock (27MHz) counter value and is used to synchronize the system clock reference in the decoder.

Fig. 4.10 shows an MPEG-2 system decoder configuration example. MPEG-2 system decoder demultiplexes video, audio and optional data elementary stream,

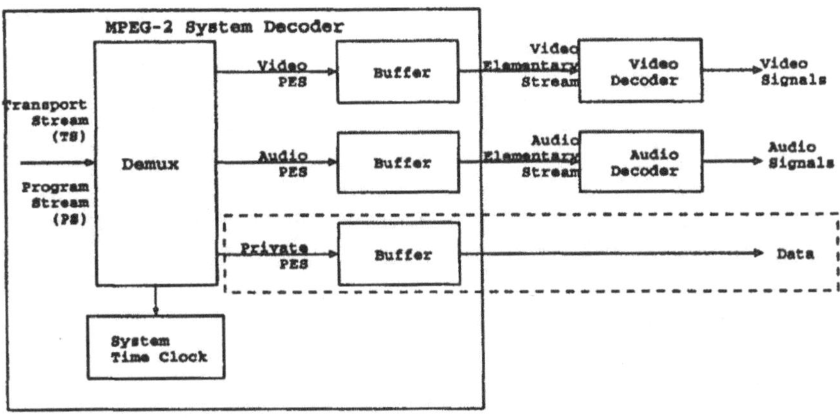

Figure 4.10 MPEG-2 system decoder configuration.

and recovers system time clock as described in Section 4.3.1 by referencing PCR in a transport stream case. In a program stream case (i.e., storage application), decoder can have an independent system clock as described in Section 4.2. The video CD or DVD decoder has a locally maintained 27MHz system time clock (STC) and read program stream from a storage media with comparing SCR in the pack header with own STC counter. Consequently, the decoder can retrieve the stream as it is stored. Elementary streams are buffered and output in time as it is multiplexed by comparing PTS and/or DTS with its own STC clock.

The MPEG-2 system standardizes:

- what are mandatory and optional components for PS and TS,
- how the components are expressed by bit strings,
- what is the maximum rate for PS, TS and each elementary stream,
- what is the maximum allowance for jitter in the encoder side,
- what is the maximum duration between successive timestamps,
- what is the necessary decoder buffer sizes for each elementary stream,
- how a standard target decoder works.

Since the MPEG-2 system does not specify how MPEG-2 system encoder works, there are many implementation alternatives for Parameters such as PES packet size, frequency of timestamp occurrence. For system clock recovery, the more frequently PCRs are transmitted, the shorter transient time is achieved [10].

4.6 NETWORK ADAPTATION

The ITU-T has standardized audiovisual terminal or system recommendation one by one per network interface. Many audiovisual transports are now available, such as telephone (GSTN), dedicated digital line, Narrowband ISDN, IP packet network and ATM network. When we consider the future heterogeneous network

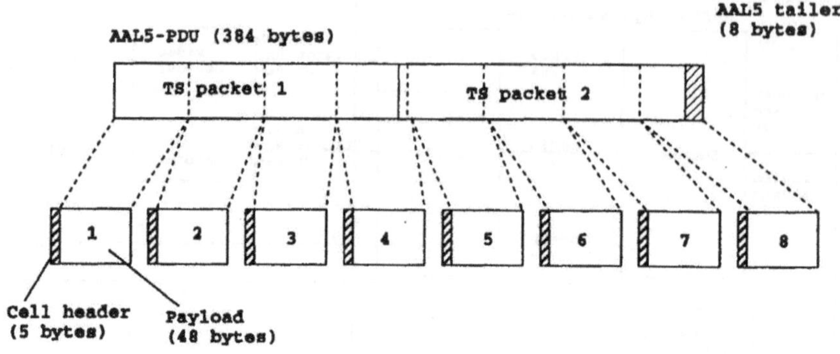

Figure 4.11 MPEG-2 transport stream to ATM cell mapping defined by MPEG over ATM (H.222.1).

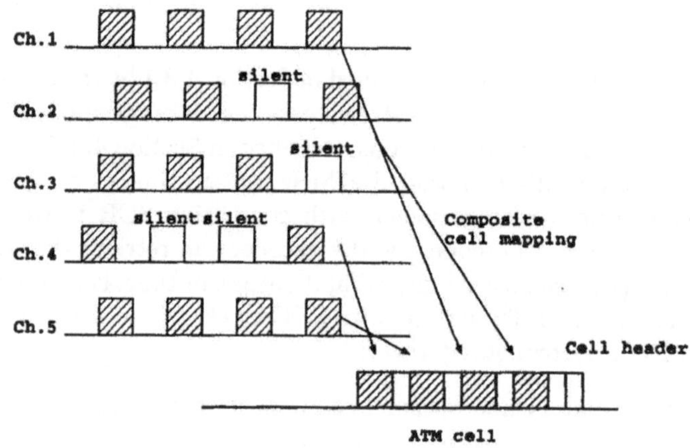

Figure 4.12 Composite ATM cell mapping for low bitrate speech.

environment, audiovisual terminal that can connect any network interface is preferable. The network adaptation layer has been introduced to hide the network transport specific characteristics, and to provide a common transport service interface. Fig. 4.11 shows the MPEG-2 transport stream to ATM cell mapping defined by ITU-T H.222.1 [4] or ATM Forum's MPEG over ATM specification. The network adaptation layer provides a constant bitrate transport service for MPEG-2 transport stream using AAL type 5. Two TS packets are combined to form one AAL5 PDU (protocol data unit) with the 8 bytes AAL5 trailer. The trailer includes length field and cyclic redundancy code (CRC). AAL5 PDU is 384 byte long, thus 8 ATM cells are adequate for the PDU. The network adaptation layer also provides the service rate recovery and jitter removal at the decoder side using either network common clock base or adaptive clock recovery method.

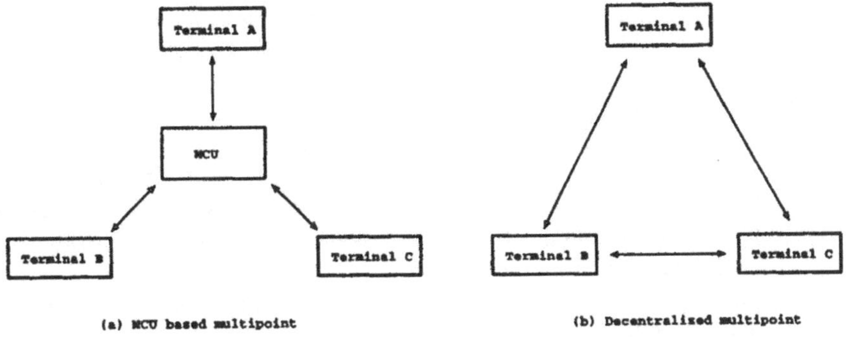

Figure 4.13 A basic multipoint conference configuration.

4.7 ATM ADAPTATION FOR LOW BITRATE SPEECH

Several speech coding algorithms, e.g., 64kbit/s PCM, 32kbit/s ADPCM, 16kbit/s LD-CELP and 8kbit/s CS-ACELP, are used in telephone communications including mobile telelphony. As an ATM cell has 53 byte fixed length, assembly delay for speech signal is inverse proportional to the bitrate. Table 4.3 summarizes the assembly delay for AAL1.

Table 4.3 Speech Signal Assembly Delay for AAL Type 1

Rec.	Coding Algorithm	Bitrate	Delay
G.711	PCM	64kbit/s	5.875 msec
G.721	ADPCM	32kbit/s	11.75 msec
G.728	LD-CELP	16kbit/s	23.5 msec
G.729	CS-ACELP	8kbit/s	47 msec

Large delay is annoying for conversation, and echo cancellation is sometimes required. To minimize the assembly delay for low bitrate speech channel trunking, ATM adaptation layer type 2 [11] is specified in ITU-T. An ATM virtual channel is used to trunk many low bitrate speech channels from a mobile base station to a local telephone switch.

Fig. 4.12 shows a concept for composite cell mapping. In AAL type 2, a cell carries several speech channel chunks. Each chunk has a header that includes channel identifier, length indicator, User-to-user indication and header error control fields. AAL type 2 can multiplex up to 256 speech channels, because the channel identifier is 8 bits long.

4.8 MULTIPOINT COMMUNICATION

This capter, so far, has dealt with several system synchronization issues for storage, point-to-point communication and point-to-multipoint/broadcast applications. But the system synchronization for multi-party multipoint communication, such as multipoint conference, is more difficult than other cases. Fig. 4.13 shows a basic multipoint conference configuration. Fig. 4.13 (a) shows multipoint control unit (MCU) [12, 13] based multipoint. Fig. 4.13 (b) shows decentralized multipoint [5], where each terminal has multiple direct links with other terminals, and there is no MCU in a center of the conference. In the case of MCU based multipoint, MCU selects the audiovisual stream from the current speakers terminal, and distributes it to other terminals. If transmitter-master/receiver-slave system clock is applied to this case, system clock jumps whenever current speaker's terminal switches. In multipoint conference application, audio signals from all end terminals should be mixed. MCU receives audio signals, mixes them and redistributes the result. For video signals, MCU switches a current speaker's video or merges several streams into a single stream like 4 in 1 picture. Since each transmitter has its own system/sampling clock, MCU will have to synchronize before adding audio samples or processing 4 pictures into one picture. A frame synchronizer is used for video clock synchronization, and sample skip or repeat is used for audio clock synchronization. Since both are build on time-domain signal processing, the MCU needs to decode the streams, and add or merge them, and encode the results again. In the decentralized multipoint case as shown in Fig. 4.13 (b), each terminal needs to synchronize audio sample clock frequency before sending a playback device.

4.9 RESILIENCE FOR BIT ERROR AND CELL/PACKET LOSSES

In multimedia communication or broadcast application, transmission error is unavoidable. The error can be categorized as a single bit or octet error, burst error, cell or packet losses, and uninvited cell or packet insertion. Forward Error Correction (FEC) methods are widely used to cope with the transmission errors. But FEC only works for a single or a few bit/octet errors. Burst error might not be correctable by the FEC. FEC with interleaver can correct burst error.

As described in Section 4.3.2, adaptive clock method for service rate recovery expects a constant rate transmission. If a cell or packet is lost in a network, the service rate at the receiver side is no longer constant. When an uninvited cell or packet insertion occurs, the service rate can't also be constant. Adding a sequence number for a cell or packet is useful to detect cell or packet loss and misinsertion, although the transmission efficiency will decrease. A system layer designer needs to know in advance what bit error rate (BER) or cell loss rate (CLR) is expected in the underlying transport service. In an IP packet network, however, there is no guaranteed packet loss rate. Therefore, when the transport service cannot provide the error free transport, the system layer decoder must be equipped with the error resilience mechanism to cope with the transmission errors.

4.9.1 Corrupted Data Delivery

In an IP packet transmission, UDP checksum is optionally encoded in an UDP header. When a receiver detects a checksum error, it will discard the whole UDP packet. In an ATM AAL5 transmission, cyclic redundancy code (CRC) is encoded in AAL5 trailer, as well. A single bit error may result in a whole packet discard. The whole packet discard is not good for multimedia transmission, because it amplifies a single bit error to a burst bit error. Therefore, AAL type 5 has a corrupted data delivery option. If this option is enabled at the decoder, AAL does not discard the whole packet, but delivers the packet with an error indication. In the upper layer, i.e., MPEG-2 system, video and audio decoder, they can try their error concealment action. For example, MPEG-2 system will not refer to the possibly corrupted timestamps.

4.9.2 Cell Loss or Packet Loss Compensation

As described in Section 4.6, MPEG-2 transport stream is transferred by ATM network. In ATM, cell might be lost by network overload or other reasons. Cell losses can be detected by CRC check or length check in the AAL5 at the receiver. But AAL5 does not know what cell in the PDU is lost.

4.10 FUTURE WORK

In a future network, common network clock may not be available. Packet based transmission channels (i.e., ATM and IP packet based) will be widely used in any environment. Constant bitrate transmission will no longer be necessary. Variable bitrate transmission is attractive for multimedia communication. However, variable bitrate transmission has a few drawbacks, e.g., higher probability of cell loss, greater cell delay variation and higher network management cost.

Generally speaking, clock transmission over a variable bitrate channel is more difficult than the one over a constant bitrate channel. A receiver side on variable bitrate channel need to smooth out the jitter with more complex PLLs [10]. System synchronization for variable bitrate transmission must be studied.

REFERENCES

[1] S. Okubo, S. Dunstan, G. Morrison, M. Nilsson, H. Radha, D. Skran, G. Thom, "ITU-T standardization of audiovisual communication systems in ATM and LAN environments," *IEEE Journal on Selected Areas in Communication*, vol. 15, no. 6, pp. 965-982, August 1997.

[2] Audiovisual Multimedia Services: Video on Demand Specification 1.0, af-saa-0049.000, *The ATM Forum*, December 1995.

[3] Information Technology – Generic Coding of Moving Pictures and Associated Audio Information: Systems, ISO/IEC 13818-1—ITU-T Recommendation H.222.0, 1995.

[4] Multimedia Multiplex and Synchronization for Audiovisual Communication in ATM Environments, ITU-T Recommendation H.222.1, 1996.

[5] Packet based multimedia communication systems for local area networks which provide a non-guaranteed quality of service, ITU-T Recommendation H.323 version 2, 1998.

[6] "B-ISDN ATM Adaptation Layer (AAL) Specification, Type 1," ITU-T Recommendation I.363.1, 1996.

[7] "B-ISDN ATM Adaptation Layer (AAL) Specification, Type 5," ITU-T Recommendation I.363.5, 1996.

[8] Frame structure for a 64 to 1920 kbit/s channel in audiovisual teleservices, ITU-T Recommendation H.221 (1995).

[9] Narrow-band ISDN visual telephone sy stems and terminal equipment, ITU-T Recommendation H.320 (1996).

[10] M. Nilsson, "Network adaptation layer support for variable bit rate video services," in *Proc. 7th Int. Workshop Packet Video*, Brisbane, Australia, March 1996.

[11] "B-ISDN ATM Adaptation Layer (AAL) Specification, Type 2," ITU-T Recommendation I.363.2, 1997.

[12] Multipoint Control Units for Audiovi sual Systems Using Digital Channel up to 1920kbps, ITU-T Recommendation H.231, 1997.

[13] Procedures for Establishing Communication Between Three or More Audiovisual Terminals Using Digital Channels up to 1920 kbps, ITU-T Recommendation H.243, 1997.

[14] Information Technology – Generic Coding of Moving Pictures and Associated Audio Information: Video, ISO/IEC 13818-2—ITU-T Recommendation H.262, 1995.

[15] Voice and Telephony Over ATM to the Desktop Specification, af-vtoa-nnn.000, February 1997.

[16] Broadband audio-visual communications systems and terminal equipment, ITU-T Recommendation H.310 version 2, 1998.

[17] Adaptation of H.320 visual telephone terminals equipment to B-ISDN environment, ITU-T Recommendation H.321 version 2, 1998.

[18] Media Stream Packetization and Synchronization for Visual Telephone Systems on Non-Guaranteed Quality of Service LANs, ITU-T Recommendation H.225.0, 1998.

[19] Multimedia terminal for receiving Internet-based H.323 conferences, ITU-T Recommendation I.332, 1998.

[20] Multipoint Extension for Broadband Audiovisual Communication Systems and Terminals, ITU-T Recommendation H.247, 1998.

Chapter 5

Digital Versatile Disk

Shin-ichi Tanaka, Kazuhiro Tsuga, and Masayuki Kozuka
Matsushita Electric Industrial Co., Ltd.
Kyoto/Hyogo/Osaka, Japan
stanaka@drl.mei.co.jp, *tsuga@hdc.mei.co.jp*
mk@isl.mei.co.jp

5.1 INTRODUCTION

A digital versatile disc (DVD) is a new recording medium, replacing the compact disc (CD), for storing digital moving picture data compressed by using MPEG2. The data format of CD is suited to music (audio) which is continuous stream data. It is not always suited to recording of computer data which is often rewritten partly. The DVD is developed "as a completely new package medium suited to both computer applications and AV (audio visual) applications." Recording of movie was taken into consideration as an AV application. As a result, in the DVD specification, the memory capacity is 4.7G bytes single layer disc and 8.5G bytes for dual layer disc on one side of a 12-cm disc (Fig. 5.1). It corresponds, for example, to 135 minutes of MPEG2 data containing the picture, voice in three languages and subtitles in four languages. This capacity is enough to record most films completely.

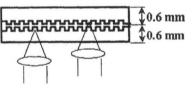

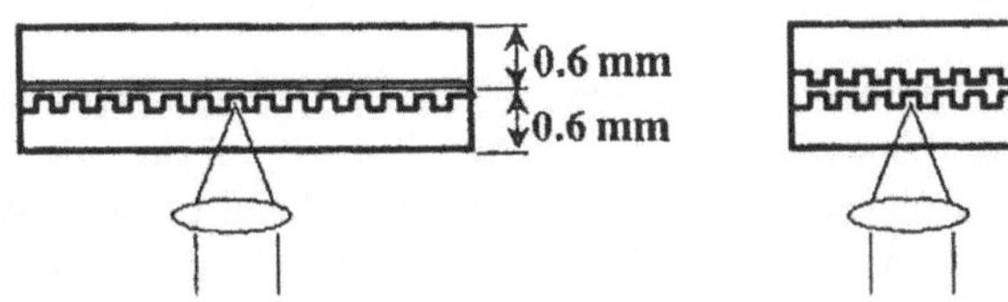

Figure 5.1 Single layer disk and dual layer disk.

5.2 PHYSICAL FORMAT

5.2.1 Physical Recording Density

The physical recording density is enhanced by reducing the diameter of a light spot formed by focusing a radiation light from a laser diode. The light spot diameter is proportional to the wavelength of incident light, and is inversely proportional to the numerical aperture (NA) of an objective lens for converging it (Fig. 5.2).

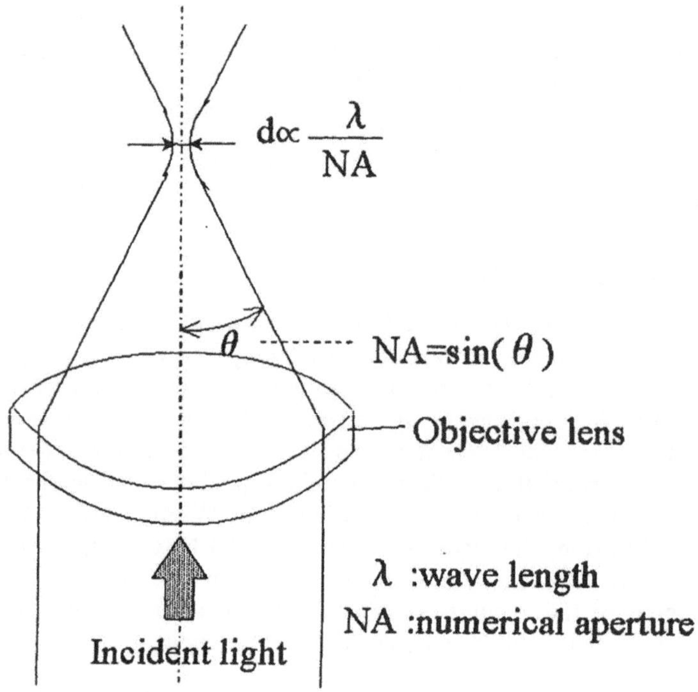

Figure 5.2 Light spot size vs. wave length and NA.

In the DVD, a red laser diode in the wavelength band of 635 to 650 nm and an objective lens with the NA of 0.6 are employed. The combination of this shortening of wavelength and elevation of the NA is effective to enhance the physical recording density by 2.6 times as compared with the CD. When the NA is larger, aberrations due to disc inclination, that is, degradation of focusing performance of the light spot becomes larger (Fig. 5.3). To suppress it, in the DVD, the disk thickness is reduced to 0.6 mm, half of the CD.

Moreover, the linear recording density and the track density are more enhanced than the effect of light spot shrinkage in comparison with the CD. Factors to deteriorate the reproduction signal of digital recording include interference from preceding and succeeding signal waveforms of the same track (inter-symbol interference), and interference from signal waveforms of adjacent tracks (crosstalk). An inter-symbol interference can be suppressed by a waveform equalization filter, but it is difficult to eliminate the crosstalk [1]. The linear recording density and the track density are enhanced in good balance based on the waveform equalization.

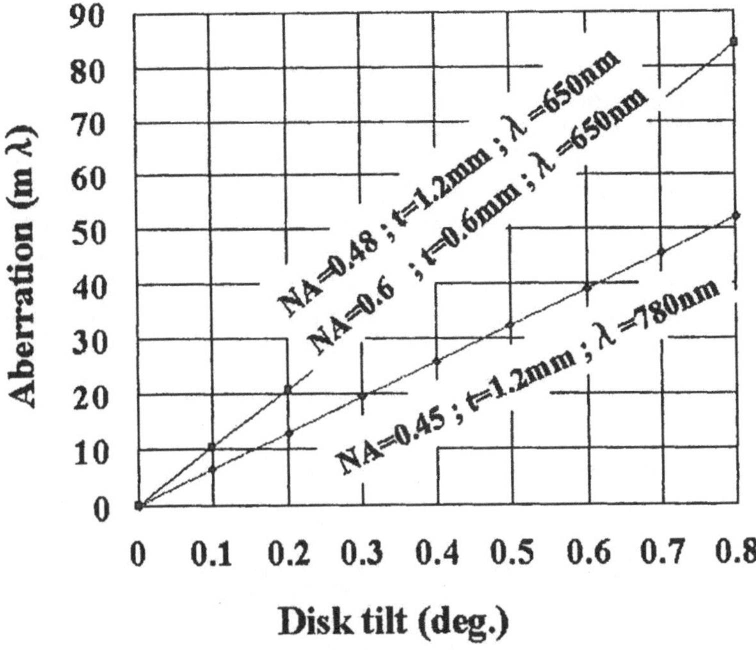

Figure 5.3 Aberration due to disk tilt.

The track pitch of the CD follows the tradition of the video disc of analog recording. In the case of digital recording, however, the allowance to crosstalk is broader than in analog recording. Considering digital recording, in the DVD, the ratio of the track pitch to the light spot diameter is set narrower than in the CD. As a result, the track pitch is defined at 0.74 micron and the minimum transition interval (minimum pit length) is defined at 0.4 micron. It corresponds to 4.2 times the physical recording density of the CD in spite of the fact that the area of a light spot is 1/2.6 of the CD (Fig. 5.2). This difference is absorbed by employing the waveform equalization circuit and curtailing the margin for disk inclination or focus deviation.

5.2.2 Two-Layer Disk

In the DVD standard, four kinds of disks are defined by the combination of one side and both sides, and single layer and two layers.

In a two-layer disk, the focus is adjusted to either layer, and the information of the selected layer is read. The reflection film provided on the closer recording layer is a half-mirror. To suppress inter-layer interference (crosstalk), a transparent adhesive layer of about 40 microns is provided between the two layers. However, due to the thickness of the transparent adhesive layer, the lens focusing performance is lowered slightly. Accordingly, the recording density is lowered by about 10%, and the tolerance to the disk inclination (tilt margin) is set to be nearly the same as in the single-layer disk.

5.2.3 Data Scramble

When sectors recording data of same pattern continue long and sectors of adjacent tracks are similar in pattern, tracking control becomes unstable.

In the DVD, accordingly, the tracking stability is enhanced by scrambling the data to be recorded. That is, the data is scrambled by using a pseudo random number sequence circulating in a longer period than one round of a track.

User data of about 1.2M bits can be recorded in the outermost track of the disk. When realizing a longer pseudo random number sequence, generally, an M sequence (a kind of random number sequence) of 21 bits or more is generated by using a shift register of 21 stages or more. In the DVD standard, however, it is designed to obtain the nearly same effect as as the M sequence of 22 bits in the M sequence of 15 bits in the CD-ROM (Fig. 5.4).

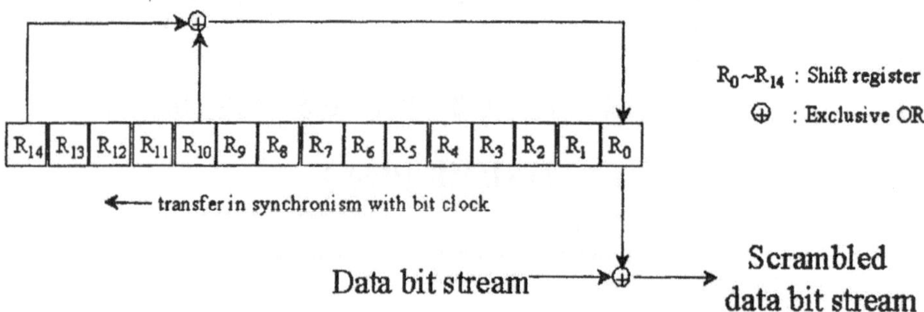

Figure 5.4 Data scrambling circuit.

In the DVD standard, data worth 16 sectors is scrambled by using the same M sequence with the same initial value. Since the number of sectors in one track is 26 or more even in the innermost track, the problem of crosstalk does not occur in the 16 sectors.

In the next 16 sectors, the data is scrambled by using the same M sequence having an initial value different from the first case. Sixteen different patterns of M sequence are prepared so that all these become completely different byte-sequences when these sequences are subdivided into byte units. As a result, the same effects are obtained as when scrambled by using a pseudo number sequence fully circulating in 256 sectors. The number of sectors in a track is smaller than 100 even at the outermost track. That is, the correlation of adjacent tracks can be suppressed.

5.2.4 Modulation Code

The EFM (eight-to-fifteen modulation [2]), which is the modulation code employed in the CD, is a kind of so-called RLL (run length limited) code for limiting the maximum and minimum of the interval of state transition interval in a channel signal. The RLL code is employed in order to facilitate forming and fabrication of master disk even in high density recording. The EFM has its own reputation. Hence, in the DVD, too, a code similar to the EFM is employed. It is the eight-to-sixteen (8/16) modulation that is higher in coding efficiency than in the EFM, while maintaining the performance of the EFM.

In the modulation code used in the read-only optical disk, suppression of low frequency component is an important subject. The noise in the optical disk is high at low frequency, and the low frequency component of recording signal often gets into the control signal as noise.

The 8/16 modulation employed in the DVD is capable of suppressing low frequency components as in the EFM. In addition, the minimum transition interval (shortest mark or space length) that determines the recording density is larger than in the EFM by 6.25 %.

5.2.4.1 Method of 8/16 modulation

In the EFM, an eight-bit data word is converted into a 14-bit codeword, and it is connected through three merging bits. That is, an eight-bit data word is converted into a 17-bit codeword. On the other hand, in the 8/16 modulation employed in the DVD, an eight-bit data word is converted into a 16-bit codeword. No merging bit is used. The recording density of 8/16 modulation is 17/16 times higher than that of the EFM.

Besides, in the EFM, the run length (the number of continuous zeros) is limited to 2 to 10, and the pattern of the merging bits is selected to satisfy the run length limit. That is, the number of continuous zeros including the merging portion is controlled by the merging bits. In the case of DVD, the limitation of the run length is 2 to 10, that is, same as in the EFM. Four kinds of conversion tables are prepared so as to conform to the limitation of the run length including the merging portion of codewords.

The concept of conversion is shown in Fig. 5.5. One of the four states is assigned for each conversion. When starting 8/16 modulation, it begins with state 1. Thereafter, every time converting in the unit of word (8 bits), the next state is assigned depending on the run length at the end of the codeword. The limitation of the run length at the beginning of the codeword differs with the state. By using four types of conversion tables, the limitation of the run length is satisfied also in the merging portion of the codeword.

The codewords with ending run length of 2 to 5 are used twice, that is, these codewords correspond to two different data words. Each of these codewords assigns State 2 for the next conversion in case that the codeword corresponds to one of the two data words and assigns State 3 in the other case. Then these duplicated codewords can be uniquely decoded by checking the next codeword is State 2 or 3. State 2 and 3 can be discriminated to each other by testing the MSB (b15) and the fourth bit from the LSB (b3). If both bits are 0 the state is considered as State 2 and the other State 3.

The conversion tables corresponding to the individual states include the main table of correspondence between data words of 0 to 255 (256 values) and 256 patterns of codewords, and the sub-table of correspondence between data words of 0 to 87 (88 values) and 88 patterns of codewords. That is, as for data words of 0 to 87, main and sub conversion tables are prepared. In a pair of codewords of both conversion tables corresponding to a same data word, the disparities, which are imbalances between 0's and 1's in the codewords after NRZI conversion, have different signs to each other.

Of the main and sub codewords generated by using the conversion tables, the one making the absolute value of the DSV (digital sum variation) smaller is employed. A DSV is an accumulation of disparities of whole codewords converted

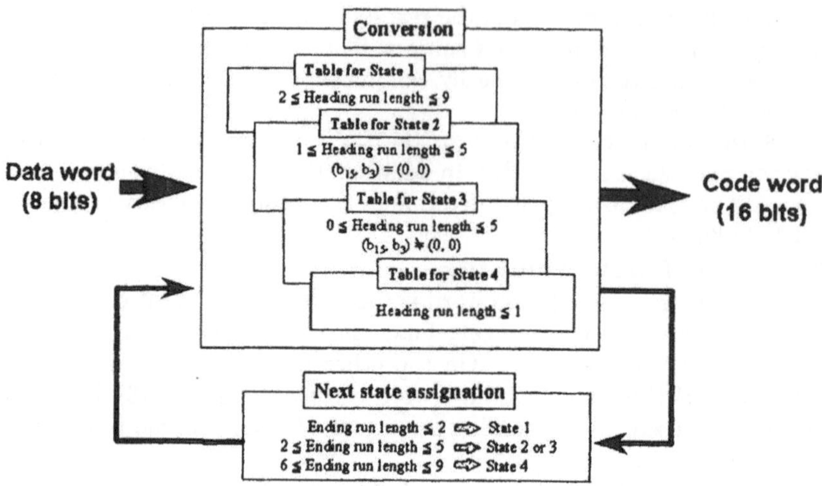

Figure 5.5 Concept of 8/16 modulation.

till then. The smaller absolute value of the DSV becomes the smaller DC component of channel signal suppressed.

Such method of using multiple channel bit patterns selectively is also applied in the EFM. What differs from the EFM is the timing for selecting the bit pattern. In the case of the EFM, the selection is fixed when a data word capable of selecting the bit pattern is entered. In 8/16 modulation too, the selection may be fixed when selective data word is entered. That is EFMPlus [3]. By contrast, in the case of the 8/16 modulation of the DVD, two candidate codewords are stored and put pending. After conversion, when a data word of 0 to 87 is entered again, the better one of two pending codewords is selected.

5.2.4.2 Synchronization code

In the synchronization code positioned at the beginning of each frame, 14T transition interval ("10000000000001" : run length of 13) is defined as violation pattern (non-appearing pattern). In the synchronization code of the EFM, the violation pattern is two repetitions of 11T inversion interval ("1000000000010000000001").

There are two reasons why the violation pattern of the EFM was not employed. That is, (1) by shortening the violation pattern, many types of two-byte synchronization code can be prepared, and (2) if the readout codeword has an erroneous bit "1" of one bit shift, neither a miss detection nor extra detection of synchronization code can occur. In other words, discrimination between the violation pattern and normal patterns cannot be disturbed by any error of one bit shift.

5.2.5 Error Correction Code (ECC)

In the data format of DVD, the correction power is enhanced by employing a product error correction code of large size with long parity. The redundancy is as low as 12.8%. The redundancy of CD (Red Book [1]) is 25.0 %, and that of CD-ROM is 34.7% (Mode 1 in Yellow Book [2]).

As a result of lowering the redundancy, the user data efficiency of the DVD in total is improved by about 20 % (or about 40 %) as compared with the CD (or CD-ROM).

The error correcting code (ECC) employed in the CD is called CIRC (cross interleave Reed-Solomon code) (Fig. 5.6A). The CIRC has the advantage that the memory capacity of the decoding circuit may be small. It is suited to error correction of sequential data, for example, music. However, it is necessary to insert dummy data to disconnect the seamless chain for rewriting a part of data. This causes to lower the data efficiency.

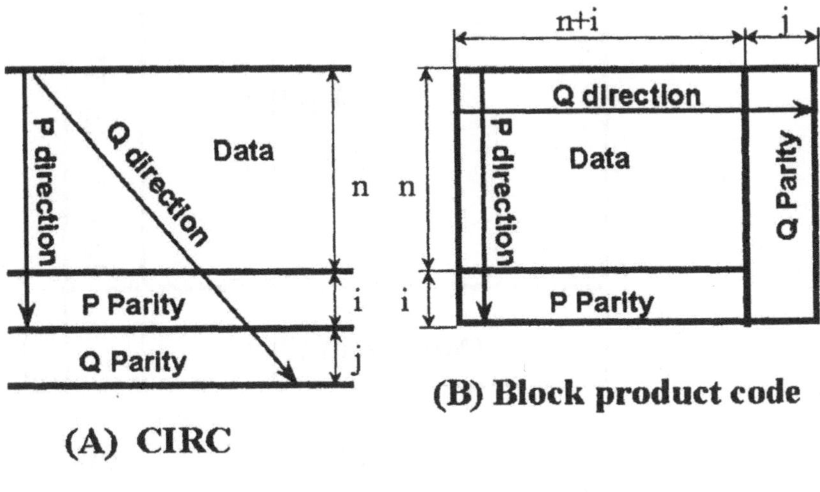

Figure 5.6 Two types of product ECC and their redundancy.

In the DVD, on the other hand, the block product code of larger size, having a higher data efficiency than that of the CIRC, is used (Fig. 5.6B). In the block product code, a specific amount of data is arranged in two dimensions, and the column direction and the row direction are coded respectively. The redundancy of the block product code is exactly same as that of the product code of the type

[1]Red Book is the specification book for basic physical format of CD.

[2]Yellow Book is the specification book for CD-ROM format. Yellow Book is based on Red Book. The user data in Red Book is encoded by the ECC called CIRC. This CIRC was developed for PCM audio recording. In PCM audio few uncorrectable errors can be allowed because the errors can be interpolated to prevent audible noise. But in computer usage the correcting power of CIRC is not sufficient. Then in Yellow Book, user data is encoded by additional ECC to form an ECC-code. The ECC-code is recorded as user data in Red Book. As the result, user data according to Yellow Book is ECC-encoded twice.

such as CIRC when their correcting powers are same as shown in Fig. 5.6. Dummy data is not necessary. In the case of a rewritable medium, since dummy data is not necessary, the coding efficiency is higher than in the CIRC.

Fig. 5.7 shows the ECC format of DVD standard. The error correction codes in the row direction called inner codes are Reed-Solomon codes RS (182,172) having parity symbols of 10 symbols (1 symbol = 1 byte) added to data symbols of 172 symbols. The error correction codes in the column direction called outer codes are RS (208,192) having parity symbols of 16 symbols added to data symbols of 192 symbols. The block of error correction is set lager than that of the CIRC of RS(28,24) and RS (32,28) used in the CD.

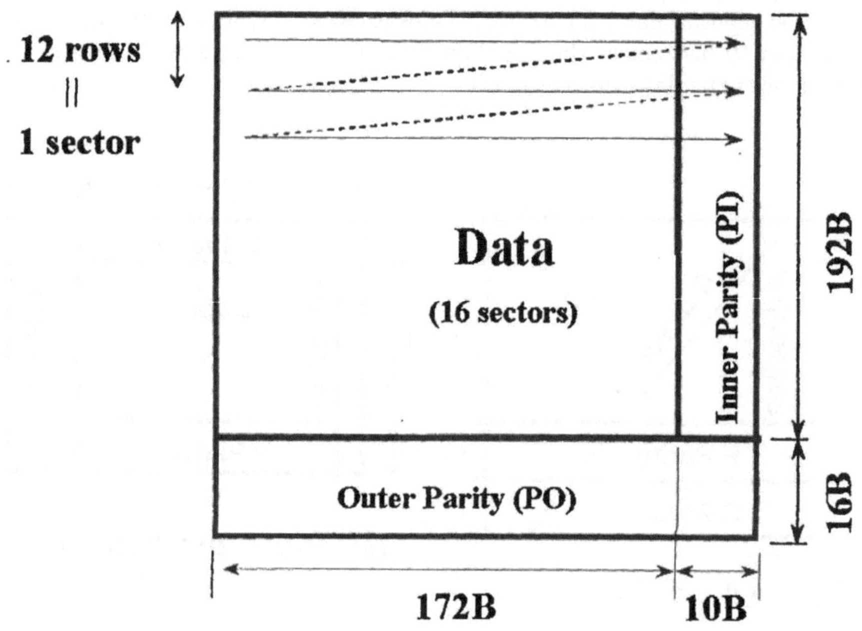

Figure 5.7 Error correction code format.

Fig. 5.8 shows the correction powers against random error for three cases of error correction strategy. Line A shows the error correction power of two corrections in the sequence of inner codes and outer codes, line B shows three corrections in the sequence of inner codes, outer codes and inner codes, and line C shows four corrections in the sequence of inner codes, outer codes, inner codes, and outer codes. As seen from the diagram, the error correcting codes of DVD are extremely high. It is sufficiently practical if the byte error rate before error correction is about 0.01.

5.2.6 Sector Format

The size of the block for error correction is 32k bytes in the DVD. This block is divided into 16 sectors of 2k bytes each. One block contains 192 data rows. It is divided into 16 sectors of 12 rows each.

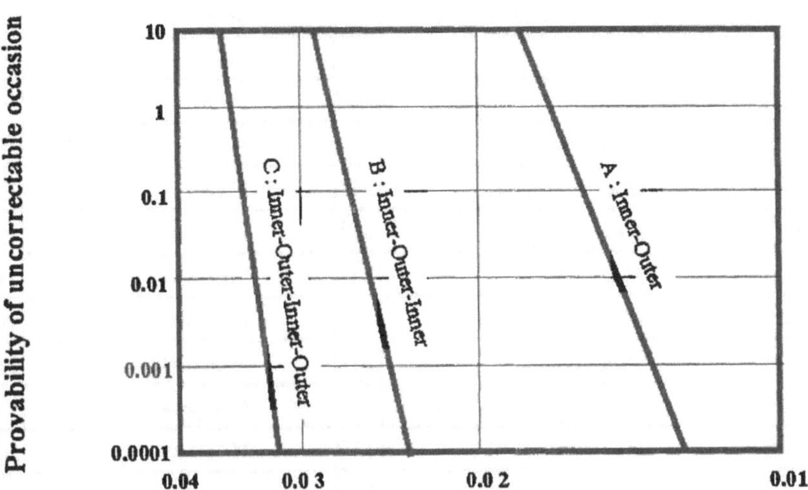

Figure 5.8 Error correcting power for random error.

Parities of outer codes consist of 16 rows, same as the number of sectors. Dividing into each row, it is disposed after each sector. It is composed so that each sector may correspond to 13 rows of the block.

On the disk, data is recorded sequentially from top to bottom, and from left to right. The repeating period of sector is always constant. Therefore, it is possible to access without having consciousness of the boundary of blocks.

Fig. 5.9 shows a sector format. Two-byte synchronization codes are inserted at the beginning and middle of each row. That is, a synchronization code is inserted in every 91 bytes, and the synchronization code and the main data of 91 bytes following it compose one synchronous frame.

As synchronization codes, eight types are prepared from SY0 to SY7. The type of synchronization code varies with the position in the sector. As seen from Fig. 5.9, by detecting a synchronization code between two continuous synchronous frames, the position in the sector of the frame can be identified.

Each data row consists of 172-byte data and 10-byte inner code parity. That is, each sector contains 2064 bytes of data. The 2064-byte data contains ID information of 12 bytes, user data of 2048 bytes, and error detection code (EDC) of 4 bytes.

5.2.7 Address Assigning Method of Two-Layer Disk

In the DVD-ROM standard, whether in one-layer disk or in two-layer disk, one side is handled as one data region (logic volume). In the case of double sided disk, one disk has two logic volumes.

The assignment of logic sector address of DVD-ROM differs in method between the one-layer disk and two-layer disk. In the case of one-layer disk, the

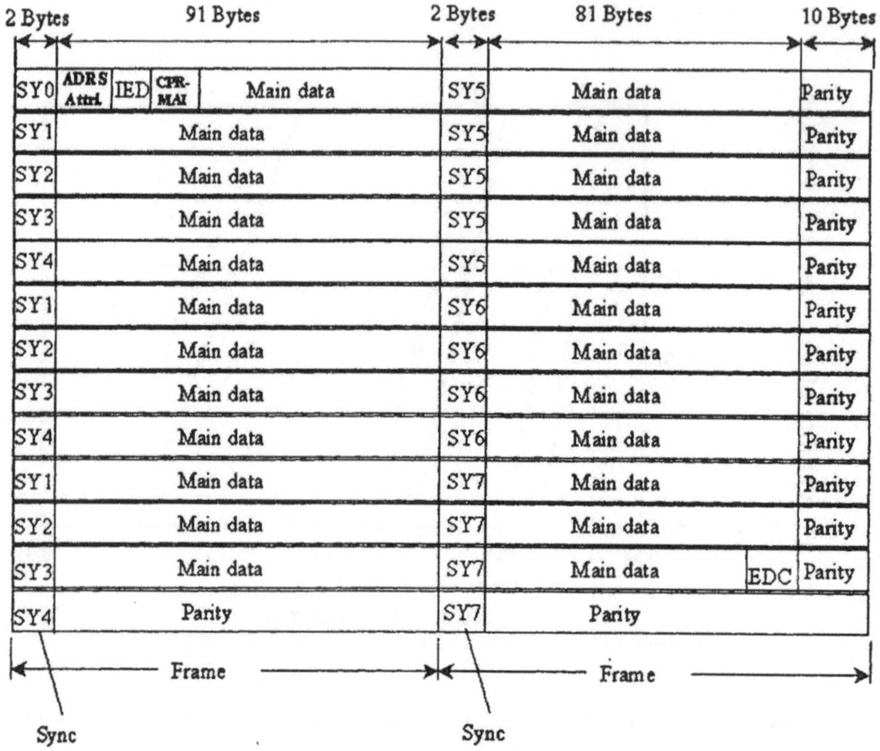

2 Bytes	91 Bytes	2 Bytes	81 Bytes	10 Bytes
SY0	ADRS Attrl. / IED / CPR-MAI / Main data	SY5	Main data	Parity
SY1	Main data	SY5	Main data	Parity
SY2	Main data	SY5	Main data	Parity
SY3	Main data	SY5	Main data	Parity
SY4	Main data	SY5	Main data	Parity
SY1	Main data	SY6	Main data	Parity
SY2	Main data	SY6	Main data	Parity
SY3	Main data	SY6	Main data	Parity
SY4	Main data	SY6	Main data	Parity
SY1	Main data	SY7	Main data	Parity
SY2	Main data	SY7	Main data	Parity
SY3	Main data	SY7	Main data / EDC	Parity
SY4	Parity	SY7	Parity	

Frame — Frame

Sync — Sync

Figure 5.9 Sector format.

address numbers are assigned sequentially from the center to the periphery of the disk.

In the two-layer disk, there are two methods of assigning. That is, (1) parallel track path method, and (2) opposite track path method. In both methods, address numbers in the first layer (layer 0) are assigned sequentially from the center to the periphery of the disk.

In the parallel track path method of (1), in the second layer (layer 1), address numbers are assigned sequentially from the center to the periphery of the disk as well as the layer 0 (Fig. 5.10). That is, sectors of a same address number are located on both layers at same radius. The layer information is recorded in the ID information, and it is detected to judge whether the first layer or second layer.

On the other hand, in the opposite track path method of (2), the address numbers on the layer 1 are assigned from the periphery to the center continuing the first layer (Fig. 5.11). At this time, an address number of a sector on the layer 1 has bit inverted relation (1's complement) to the address number of the sector in the first layer at the same radius. This opposite track path method is effective in case that a long video extending from layer 0 to layer 1 must be reproduced in seamless way.

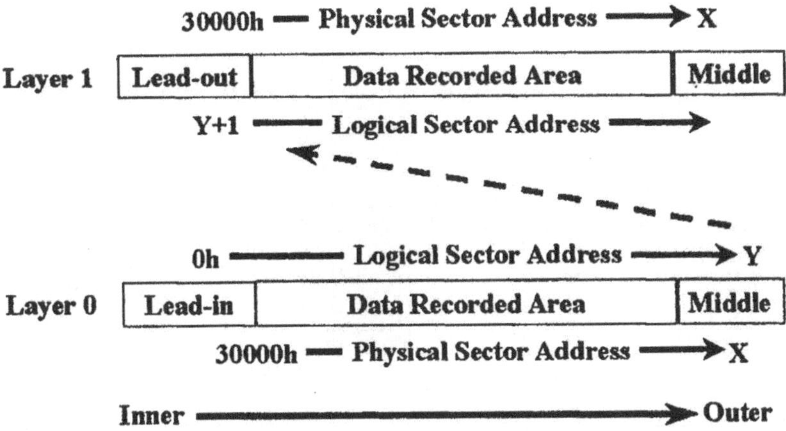

Figure 5.10 Address assignment for the parallel track path.

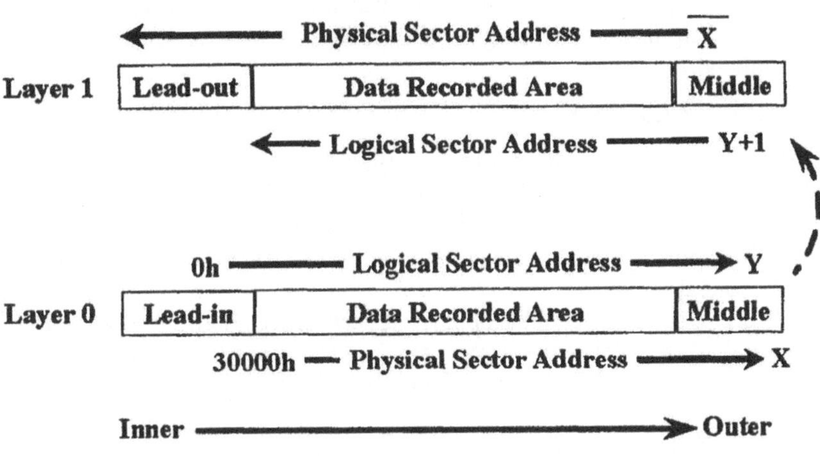

Figure 5.11 Address assignment for the opposite track path.

5.3 FILE SYSTEM LAYER

5.3.1 UDF Bridge

The file system employed in the DVD specification is a newly developed scheme called "UDF Bridge" capable of using both the UDF (universal disk format) usable in combination with the specification of all physical layers, and the ISO 9660 [4] globally distributed among the personal computers as the CD-ROM standard (Fig. 5.12).

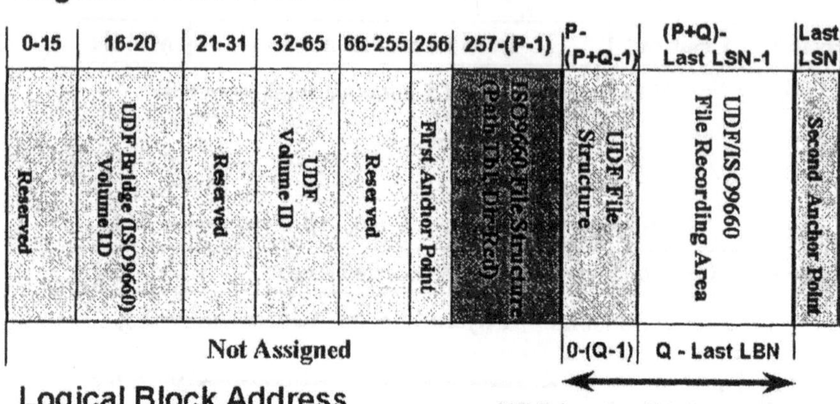

Figure 5.12 UBF Bridge Structure.

The UDF Bridge is a subset of the UDF which is a read-only file system, and is capable of reading files conforming to ISO 9660. Data structure for recording medium is omitted.

5.3.2 AV Data Allocation

The detail of the format of AV data (picture, sound, subtitle and other data) is defined in the application layer. These files can be handled same as files of general computer data. Besides, before and after the AV data, arbitrary computer data can be recorded. However, according to the limitations, the AV data must be recorded in continuous logical sectors.

The DVD player will be capable of playing back the AV data recorded in DVD-R (once recordable) or DVD-RAM. (rewritable). Accordingly, in the file system, it is recommended to use the UDF rather than the ISO 9660.

5.4 APPLICATION LAYER

It is the application layer that formats contents to be stored in the disk, or designates specifications necessary for reproducing. The outline of the already established DVD-video specifications is described below. When designing the specifications of application layer, hitherto, it was important to assume first a specific application, and conform to the requirements of the contents provider.

As applications of DVD, movie, karaoke, music video, and electronic publishing may be considered. In particular, as for the movie of the highest rank of priority, the specification was compiled in cooperation with the Hollywood movie makers.

Special consideration was given to saving of production cost, protection of copyright, and optimization of distribution. Much has been discussed about the future viewing modes of movies. It was also considered so that the contents providers of conventional recording media such as laser disc and video CD could transfer to the DVD smoothly.

As a result, the following demands have been satisfied. That is, (1) the picture quality and continuous playing time satisfied by the movie makers, (2) multi-language service of audio, subtitles and menu (selection screen), (3) compatible with both screens of aspect ratio of 4:3 and 16:9, (4) surround effect of sound, (5) prevention of copying, (6) limitation of reproduction enabled district (reproduction region control), (7) compatible with plural versions, such as original version and edited version, (8) parental control to protect children from violent scenes, etc.

5.4.1 Action Mode

In the DVD-video specification, the title is classified into two types. That is, the "linear AV mode" and the "interactive mode". The former is the title of movie, karaoke, etc., and the latter is the title making use of interactive playback control function such as electronic catalogue and education software.

These two modes are distinguished because the function of trick play of the DVD player and others are different. For example, in the "linear AV mode" title, time search function, displaying of elapse time, repeat function of specific interval and other functions corresponding to the conventional CD and laser disc can be applied, but these functions cannot be used in the "Interactive mode" title.

5.4.2 File Composition

A data structure of application layer is shown in Fig. 5.13. As shown in the explanation of the file system layer, the AV data (including both presentation information and navigation information) is stored in the VIDEO_TS directory.

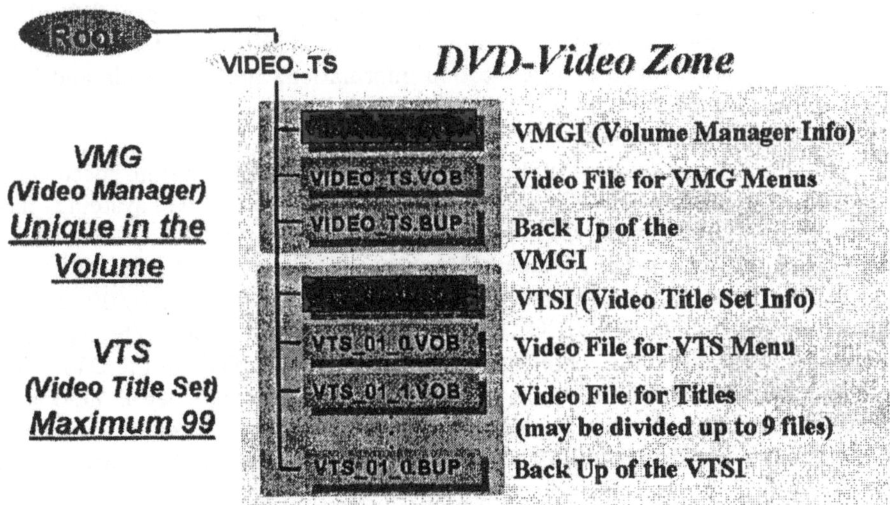

Figure 5.13 File structure of DVD video.

There are two types of file under VIDEO_TS directory. That is, the VMG (video manager) storing the information relating to the entire Volume, and the VTS (video title set) storing the information relating to individual titles. Herein, the title is a unit of content, corresponding to, for example, one movie or one karaoke song.

One volume can store at least one set of VMG, and multiple titles of VTS, and up to 99 sets can be stored. The VTS is merely the control unit for producing the content, and the titles in the VTS can share the video data.

There are three types of files associated with the VMG. These include (1) VMGI (video manager information) file containing the control information of the file, (2) VOBS (video object set) file containing AV data of menu commonly called in the entire disk, and (3) backup file of VMGI file.

On the other hand, there are four types of file associated with the VTS. That is, (4) VTSI (video title set information) file containing the control information of the file, (5) VOBS file containing the AV data of the menu called by the title, (6) VOBS file containing the AV data of the title content, and (7) backup file of VTSI file.

The size of each file is limited to 1G byte. Therefore, a title of a very long time such as a movie cannot be stored in one VOBS file. In such a case, the AV data of the title content is divided into plural files, and disposed in physical continuous regions of the volume.

5.4.3 Separation of Two Types of Information

The information defined in the application layer may be classified into presentation information and navigation information. Although the specifications is not hierarchical in a strict sense of meaning, each may be called presentation layer and navigation layer.

The presentation information is a set of MEPG2 data containing picture stream, sound stream, subtitle stream, etc. On the other hand, the navigation information is reproduction control data (reproduction control program) for designating the playing sequence and branching of individual MPEG2 data. For example, the VBOS file is classified as presentation information, and VMGI file and VTSI file as navigation information.

5.4.3.1 Presentation data

The data of presentation information is based on MPEG2 system specification (ISO/IEC 13818-1 [5]). The size of one pack designated in MPEG2 system is fixed in one sector (2048 bytes) which is a physical recording unit of a disk. This is determined in consideration of the random access performance of pack unit.

As the data to be multiplexed, in addition to MPEG2 video (ISO/IEC 13818-2 [5]) and MPEG2 audio (ISO/IEC 13818-3 [5]), linear PCM data, Dolby digital [6], audio data, and sub-picture (subtitle, etc.) can be also handled. In the DVD-video specification, such data unit is called the VOB (video object). A set of VOB is the VOBS file mentioned above. To satisfy the requirement for multi-language service, up to eight voice streams and up to 32 sub-picture streams can be multiplexed in one VOB.

1. Video data

   Table 5.1 shows an example of specification of video data handling NTSC format television signals. Generally, when the data is compressed, the picture quality degradation is determined by the compression rate. To contain a movie in a limited memory capacity, it is necessary to readout at a lower coding data speed for a determined picture quality. Accordingly, the coding technique of

Table 5.1 An Example of Video Specifications in Case of NTSC

Video	MPEG2 MP@ML
Resolution	Maximum 720 × 480
Frame Rate	29.97
Aspect Ratio	4:3 / 16:9
Display Mode	Normal / Pan-Scan / Letter-Box
Bitrate	Variable Bitrate Maximum 9.8 Mbps

Note: MPEG1 may be used

variable bit rate (VBR) is introduced, and the average data transfer speed is suppressed [7].

The video data of which screen aspect ratio is 16:9 as in the case of a movie is recorded in the DVD by using vertically long pixels (Fig. 5.14). When it is displayed in a wide television, pixels are stretched in the horizontal direction, and the original picture is reproduced. When displaying this video data in a conventional non-wide television receiver, the DVD player converts and delivers the video data into signals of pan/scan or letterbox.

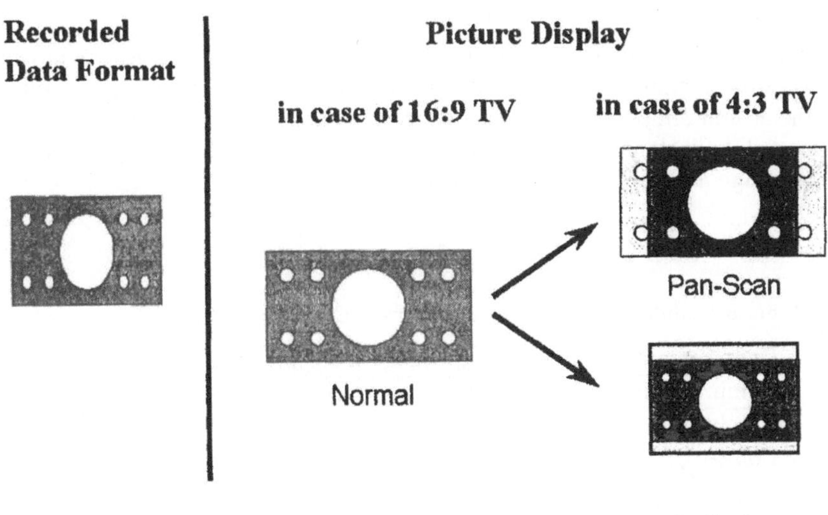

Figure 5.14 Display mode in case of 16:9 video source.

2. Audio data

   Table 5.2 shows an example of audio data to be combined with NTSC format picture. For example, it is applicable to the linear PCM suited to classical music, Dolby digital for emphasizing the feel of presence as in the movie, and MPEG audio of international standard.

Table 5.2 An Example of Audio Specifications in Case of NTSC Video

Linear PCM or Dolby AC-3 is mandatory.

	Linear PCM	Dolby Digital
Sampling Freq.	48, 96 kHz	48 kHz
Quantization	16/20/24 bits	Compressed
No. of CHs	Max. 8	Max. 5.1
Bitrate	Max. 6.75 Mbps	Max. 448 kbps

Note: MPEG audio may be used as an option.

The sampling frequency is 96 kHz at maximum, and the quantizing precision is high, max. 24 bits. There are also settings for multi-channel and karaoke mode.

It is, however, difficult for the DVD player makers to conform to all systems from the beginning. In the DVD-video specification, hence, the range of the mandatory specifications is limited. The linear PCM is required in all DVD players. In addition, Dolby digital is required in the region of NTSC format television broadcast, and MPEG audio in the region of PAL format.

3. Sub-picture data

The specification of sub-picture used when producing subtitles for movie and karaoke is shown in Table 5.3. In addition to characters used in subtitles, menu and simple graphics can be displayed. The sub-picture is not simple text data such as closed caption (teletext for the handicapped), but is coding of image data of four colors (four gradations).

Table 5.3 Sub-Picture Data Specifications in Case of NTSC Video

Data Format	Bitmap Image 2 bit/pixel
Compression	Run Length
Resolution	720 × 480
Colors	4 Colors (extendable up to 16)
Display Commands	Color Palette, Mixture Ratio to Video, Display Area may be dynamically changed

Fade-in/fade-out, scrolling, karooke color change may be realized using Display Commands embedded in the data stream

The sub-picture is composed of image data of run length coding and sequence of commands called DCSQ (display control sequence) for controlling its display method. Using the DCSQ, the display region and color of sub-picture, and the mixing rate of sub-picture and main picture are varied depending on the frame period of picture. It also realizes the fade-in and fade-out display of subtitles, and color changes of verses of karaoke.

The sub-picture is combined with main picture and delivered to the video output. This combination is executed after conversion when converting the main picture into pan/scan or letterbox form. If the screen aspect ratio of main picture is 16:9, the configuration of main picture and sub-picture is changed after changeover of display mode. Accordingly, depending on the case, several patterns of sub-picture data must be prepared. The DVD player is required to have a function to select a proper sub-picture stream among the ones which are prepared corresponding to the display mode variations and multiplexed with the video stream.

4. Hierarchical structure of presentation data

   When reproducing MPEG2 data from disk medium such as DVD, functions for chapter search, time search, and trick play (fast forward, fast rewinding) are indispensable. Moreover, in the title utilizing interactive reproduction control function such as game software, it is also required to jump from an arbitrary position of moving picture to an arbitrary position of other moving picture.

   To be flexible to the reproduction function including such random access, the presentation information is built in a hierarchical structure (Fig. 5.15). That is, it consists of six layers: VOBS, VOB, cell, VOBU (video object unit), pack, and packet.

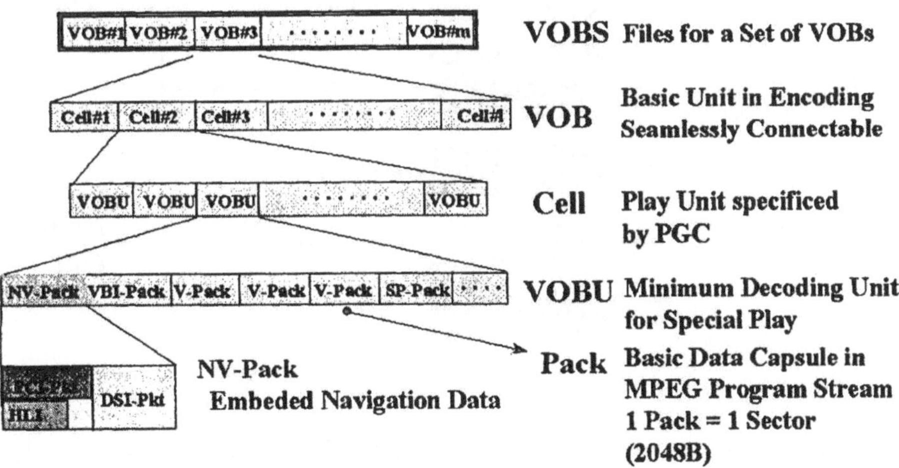

Figure 5.15 Hierarchical data stucture in presentation data.

The VOBS corresponds to, for example, one title. The VOBS is composed of multiple VOBs. One VOB is divided into several cells. In the case of a movie, for example, one cell corresponds to one chapter. The cell is further divided into VOBUs. The VOBU is the minimum unit of time search or random access. It corresponds to about 0.5 second of playback time. The

VOBU is further divided into smaller units called packs and packets. The specification of pack and packet conforms to the Program Stream designated in the MPEG2 standard. Types of pack include the NV pack for stream control, V pack containing video data, and SP pack containing sub-picture.

According to the DVD-video specification, data is recorded by employing the VBR coding technique. Therefore, even if the reproduction time of VOBU is constant, the quantity of data assigned in each VOBU may become variable, and the beginning sector address of the VOBU cannot be calculated simply. Accordingly, when producing, the jump destination address for trick play is recorded in the NV pack. The data is read out while skipping in the VOBU unit by using this information. Besides, the NV pack also contains the reproduction control data (highlight information) relating to remote control operation.

5. Decoder model

Fig. 5.16 shows an action model of a decoder for reproducing the multiplexed presentation information. The presentation information being read out from the disk is separated into picture stream, sound stream, and sub-picture stream according to the stream ID information of each pack and packet, and is fed into individual decoding modules. When several sound data or sub-picture data are multiplexed, unnecessary streams are removed by demultiplexer.

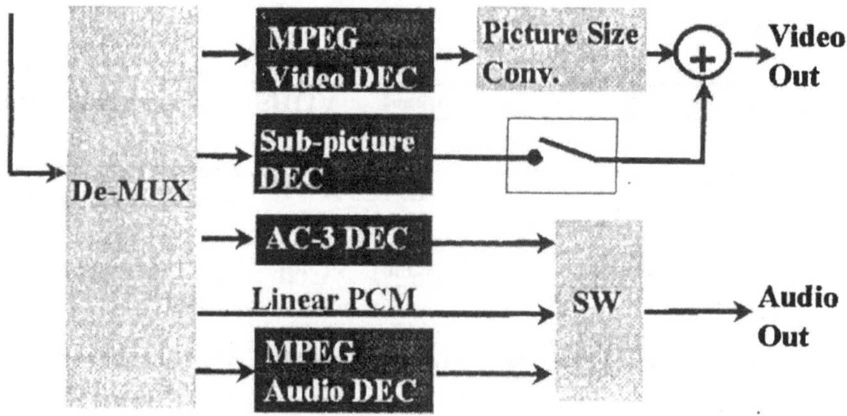

Figure 5.16 DVD decoder model.

The video data recorded in a display model of aspect ratio 16:9 is decoded, and the image size is changed according to the television receiver. The decoded main picture and sub-picture are combined at a specified mixing rate of luminance and chrominance. Then the picture is delivered as an analog signal.

6. Seamless play of multiple MPEG streams

Seamless connection is a function for reproducing both picture and sound without interruption when mutiple video data (MPEG2 data) are connected.

Such concept of continuous play is not present in the MPEG2 standard. It is hence newly defined in the DVD-video specifications.

Seamless connection is available in two types: (1) simple seamless connection, and (2) selected seamless connection.

The simple seamless connection of (1) is a function of connecting several MPEG2 data in cascade, and reproducing as if processing one continuous MPEG2 data. In many contents, they are coded individually, and finally combined together. It enhances the efficiency of contents production.

The selected seamless connection of (2) is a function of reproducing continuously while selecting a desired version if there are several versions in one content. For example, the title credit may be prepared in different languages, or the original version and re-edited version of the movie may be efficiently recorded in one disk. This function is also utilized in the music video containing a live concert taken from different angles, or for realizing the so-called parental control for skipping violent or obscene scenes not recommended to children.

7. Interleaved data allocation

   The simple seamless connected MPEG2 data (VOB) is recorded in contiguous regions on a disk. To realize the selected seamless connection, however, even if reproducing by jumping on the disk, the data to be decoded must be fed continuously into the decoder. Moreover, when connected by selecting one from MPEG2 data differing in playing time, a mechanism for matching the time information is needed.

   Continuous data feed into the decoder in the selected seamless connection is guaranteed by interleaved recording of MPEG2 data. That is, if branching or coupling of MEPG2 data occurs, all presentation information cannot be recorded in continuous regions on a disk. If there is a problem due to the allocation of MEPG2 data, the seek time when skipping unnecessary data becomes too long, and underflow occurs in the buffer in the DVD player while seeking. To avoid such situation, the data is interleaved in consideration of jump performance (seek performance and buffer capacity) of the DVD player (Fig. 5.17).

   While seeking, data cannot be read out from the disk. It requires a mechanism of feeding data into the decoder without interruption while utilizing the buffer memory. Accordingly, a model of buffer occupation capacity while reading and seeking the disk tracks is also prepared (Fig. 5.18). The jump performance required in the DVD player is defined by the parameter of this model.

8. Extended system target decoder model

   On the other hand, in selected seamless connection, when MPEG2 data differing in playing time are connected, the problem is solved by devising an action model of the decoder in consideration of correction of time information.

   The time information (time stamp) is provided in the pack header or packet header of the MPEG2 data composing the presentation information. In the decoder, such time information of MPEG2 data and reference time (system

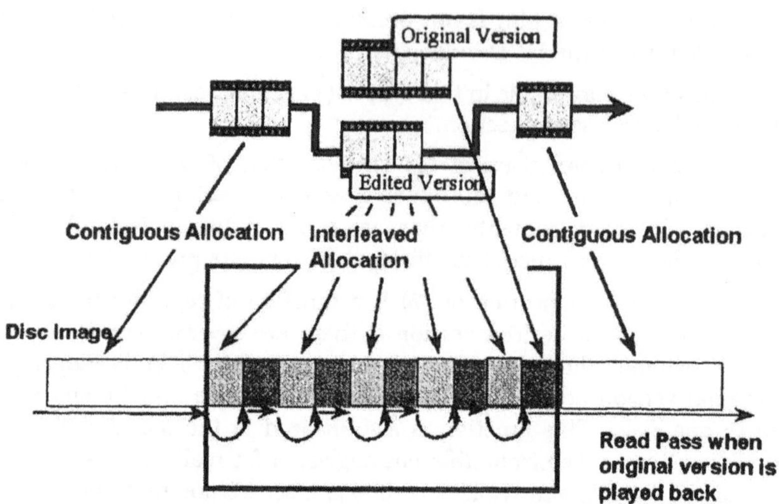

Multiple AV streams are chopped into small pieces and are allocated in interleaved way. When an AV stream is read out, the rest of the streams are to be skipped over.

Figure 5.17 Interleaved allocation of AV streams.

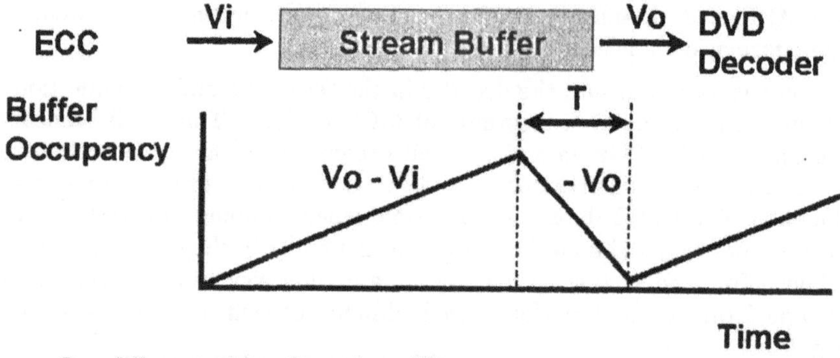

Condition enables Seamless Play:

T = Seek+Latency+ECC Delay < Data Consumption Time

Figure 5.18 Buffering model when seamless play.

time clock) of the decoder are compared, and the timing for input, decoding and display of data is determined. As a result, the video data and audio data are accurately synchronized, and occurrence of underflow or overflow of input buffer is avoided.

In order to cope with different playing time of MPEG2 data, a change-over - switch of reference clock is provided in each decoder of picture, sound, and sub-picture (Fig. 5.19). It is also designed to set the offset amount of the time. By changing over the switch in seamless playing, the time information described in the MPEG2 data and the reference time supplied in the demultiplexer or each decoding module are matched. When changing over all switches, the reference time is set again according to the time information of the next MPEG2 data.

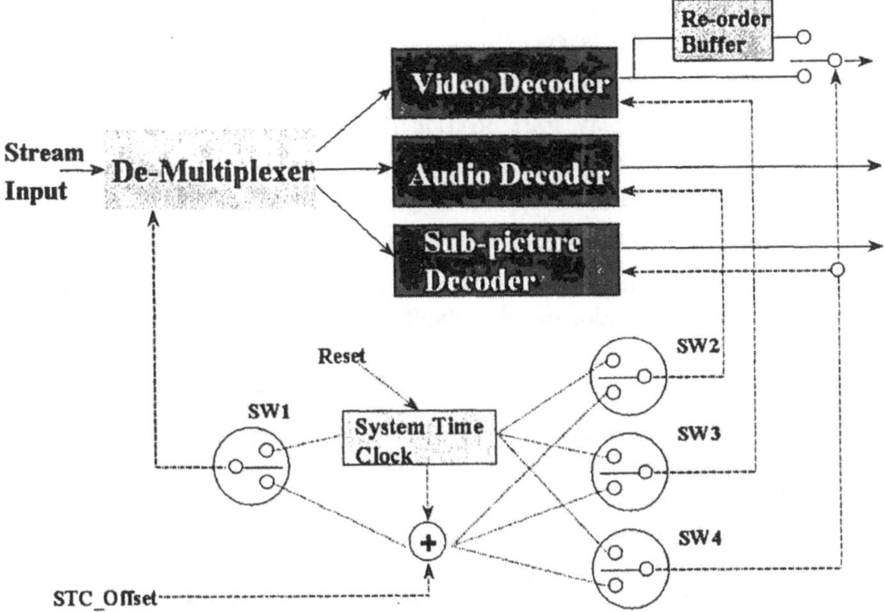

Figure 5.19 MPEG extended STD model.

5.4.3.2 Navigation information

1. Designation conforming to remote control action

   The navigation information (playback control information) is divided into two layers: PG (program) which is a basic unit for playing back MPEG2 data, and PGC (program chain) for describing the playing back sequence of the PG.

   The PG is a skip unit. For example, when the content is a movie, the PG is one chapter. The DVD player sequentially plays back the data of different cells described in the PG. The cells designated by the PG are not always required to be arranged in continuous regions on a volume.

   On the other hand, the PGC is a unit of continuous playback by making use of the function of seamless connection. In the case of a movie, for example, it is usually played back in a seamless manner from start to end. That is, one title is composed in one PGC, and cells are stored in continuous regions.

   The PGC is composed of (1) information for designating the playback sequence of PG, (2) information of pre-processing (pre-command) and post-processing (post-command), and (3) link information for designating the preceding and succeeding PGCs or PGC of upper layer.

2. Navigation API (navigation commands) In the DVD-Video specification, the following actions 2a to 2g are designated as the minimum user operations to be provided in the DVD player.

   (a) Basic AV operation (play start, stop, fast forward, fast rewind, etc.)
   (b) Basic interactive operation (skip to preceding and succeeding PG, jump to PGC of upper layer)
   (c) Changeover of playback stream (changeover of sound, sub-picture, angle)
   (d) Designation of play start position (beginning, designation by chapter, designation by time)
   (e) Call of menu (title selection menu, menus for each titles)
   (f) Highlight button control (move or select highlight)
   (g) Change of setting of DVD player (setting up player configurations such as TV picture ratio, sound, default language selection, parental control level)

   The meaning of operations are designated in (1) through (7), but the player implementation such as, the key in a remote control to be used for operation, is not designated.

   In the DVD-video specification, these user operations correspond to the Navigation commands. For example, when jumping by designating the title number, it corresponds to JumpTitle command. The action to be guaranteed when the DVD player executes a JumpTitle command is stipulated as the specification of navigation command.

3. Highlight function

   In the DVD-video specification, for the ease of production of title utilizing the interactive playback control function, data called highlight is introduced. This is the information relating to the highlight display (emphasis display), and is stored in the NV pack in the presentation information. Each selection item contained in the menu is called a highlight button, and the button selection method, color, position, executing navigation command and others are described in the highlight information (Fig. 5.20). The menu is realized by synthesizing the video data of background, sub-picture data for expressing the button characters and selection frame, and highlight data.

   The user operates a moving of the button by using the cursor key or executes a selection by an activating key of the remote controller. By combining with the sub-picture function, a highlight button of an arbitrary shape can be created.

   It is not always necessary to compose the highlight button on the menu of still picture. A highlight button can be arranged on a moving picture, too. For example, it can be applied in the menu with the changing picture in the background, or it can be applied at each branch for selecting going direction in a labyrinth game.

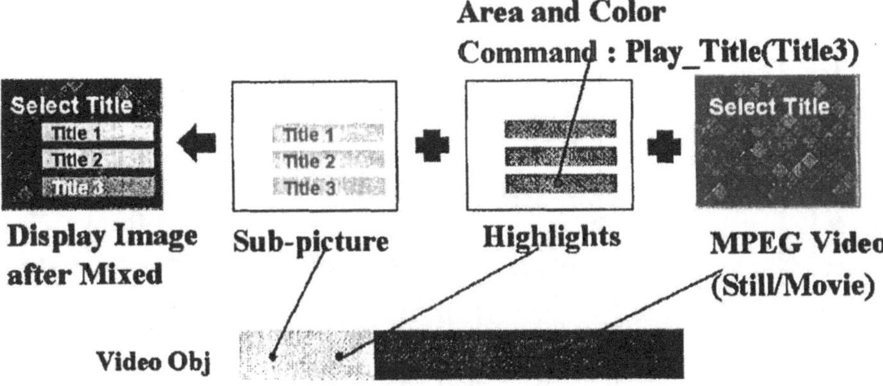

Figure 5.20 Highlight data structure.

4. Three types of interactive functions

   In designing of interactive playback control function, functions used in the existing application such as video-CD and CD-ROM are incorporated as many as possible. What was noted at this time is the viewing environment of television programs (passive viewing). For example, if the user selects nothing, it is designed to play back automatically according to the scenario intended by the title producer. It is also considered to operate by a minimum number of keys.

   Principal interactive playback control functions are (1) basic key operation, (2) operation of highlight button, and (3) index jump.

   The basic key operation of (1) is a function of skipping to a preceding or succeeding PG, or jumping to PGC of higher layer. The link destination of menu or the like is selected by the highlight button control key (up, down, right, left highlight move keys, and select confirm key) (Fig. 5.21).

   The operation of (2) is a function of designating the valid term of highlight button and the behavior upon expiration (presence or absence of automatic execution, etc.), by using highlight data. Moreover, the button of highlight display can be designated before menu display.

   The operation of (3) is a function for jumping to an arbitrary PG (Fig. 5.22). Up to 999 indices can be designated in each title.

5. Navigation Command and Parameter

   The playback control program for interactive playback control is described by using navigation command and navigation parameter.

   The navigation command is a control command for playing back the presentation information. For example, it is executed before and after PGC processing, after cell playback, or after confirming button by user operation.

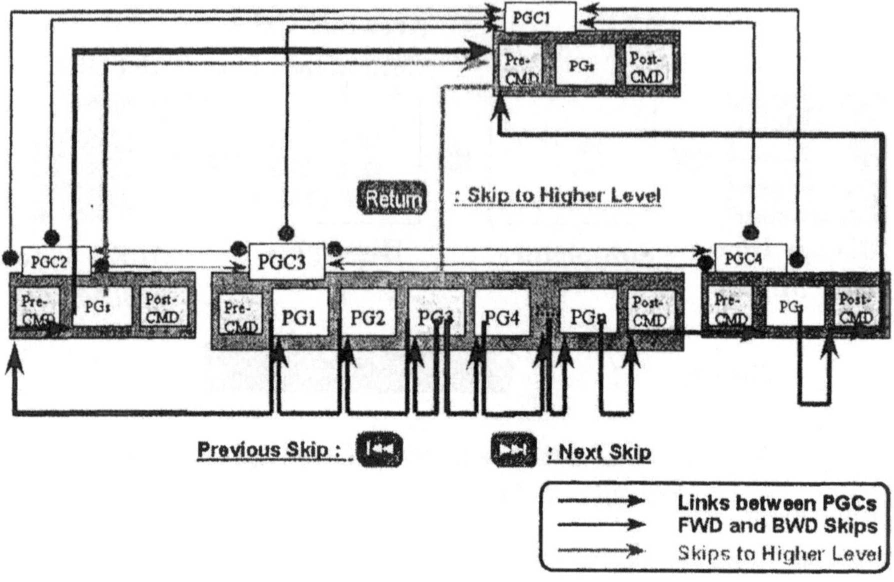

Figure 5.21 Navigation using remote controller.

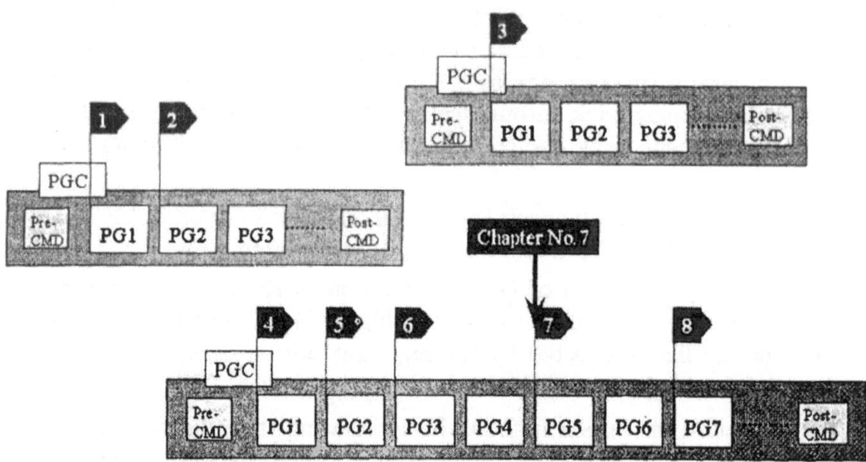

Direct jumping to any arbitrary points in a title may be realized by setting PTT markers shown as flags.

Figure 5.22 Program indices to be jumped to.

The navigation command comprises six groups and 48 commands. The six groups are (1) link group (branching within title), (2) jump group (branching between titles), (3) goto group (branching within command rows), (4) set system group (control of player function), (5) set group (calculation of parameter value), and (6) compare group (comparison of parameter values). These navigation commands are designed to lessen the processing burden of the command interpreter provided in the player.

On the other hand, the navigation parameter corresponds to the register of a computer. There are 16 general parameters capable of referring to and writing, and 21 system parameters (running state of DVD player, timer for playback control, etc.). Each parameter is a 16-bit positive integer.

6. Playback control not depending on hardware (virtual machine approach)

   Usually, in the case of appliance for interactive playback, the OS of the player and the microprocessor are defined in order to keep compatibility among different players. In this system, however, it is hard to change the architecture according to the technical development while keeping compatibility of the players. Besides, the degree of freedom of player design is lowered.

   Accordingly, in the DVD-video specification, a virtual machine for DVD playback control is defined. It does not depend on the hardware or OS. That is, the variations of technique for realization are increased. It is also flexible to get along with new hardware and OS coming out in future.

   Fig. 5.23 shows a configuration of virtual machine for playback control. On the basis of the remote control operation by the user, the information of PGC of disk, etc., the PGC playback machine gives a playback instruction to the presentation engine.

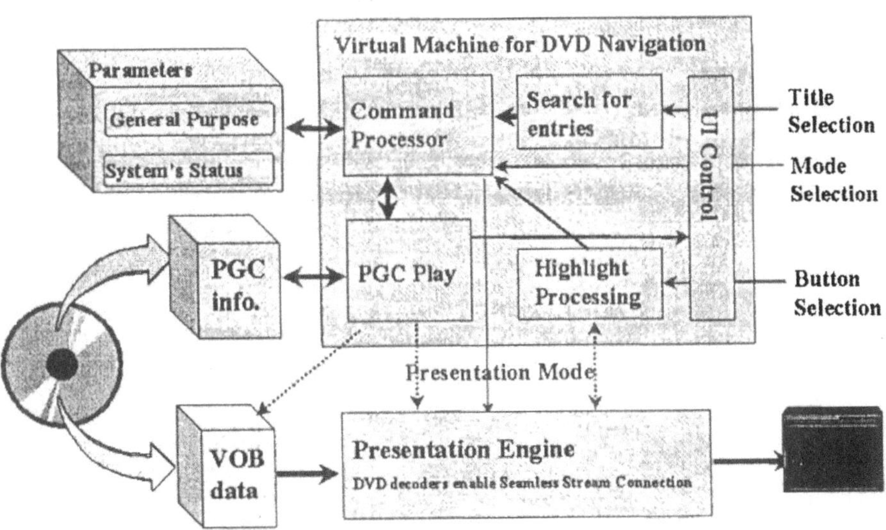

Figure 5.23 Configuration of DVD virtual player.

The user's operation may be temporarily banned depending on the playing position on the disk. The instruction from the user is cut off in the UI control unit. When not cutting off the instruction from the user, the user instruction is transformed into a format of navigation command, and is transmitted to the navigation processor.

The navigation command processor interprets the navigation command transformed from the user instruction in the UI control unit (in the case of operation corresponding to the highlight button, the navigation command transformed in the highlight processing unit), or the navigation command stored in the PGC. Then, the control instruction is transmitted to the PGC playback machine.

Usually, processing is done automatically in the presentation engine unit. Processing of navigation command processor occurs only when instructed by the user, or in the boundary of cell or PGC. Hence, the virtual machine for playback control does not require microprocessor of high performance or memory of large capacity.

REFERENCES

[1] K. Kayanuma, *et. al.*, "High track density magneto-optical recording using a crosstalk canceler," *Optical Data Storage Proceedings*, SPIE vol. 1316, pp. 35-39, 1990.

[2] K. Immink *et. al.*, "Optimization of Low-Frequency Properties of Eight-to-Fourteen Modulation," *Radio Electoron. Eng.*, vol. 53, no. 2, pp. 63-66, 1983.

[3] K.A.S. Immink, "EFMPlus: The Coding Format of the MultiMedia Compact Disc," *IEEE Trans on Consumer Electronics*, vol. CE-41, pp. 491-497, 1995.

[4] ISO 9660: 1988.

[5] MPEG: ISO/IEC 13818.

[6] ATSC standard digital audio compression (AC-3), 1995.

[7] K. Yokouchi, "Development of Variable Bit Rate Disc System," *Symposium on Optical Memory Technical Digest*, pp. 51-52, 1994.

Chapter 6

High-Speed Data Transmission over Twisted-Pair Channels

Naresh R. Shanbhag
Department of Electrical and Computer Engineering
Coordinated Science Laboratory
University of Illinois at Urbana-Champaign
shanbhag@uivlsi.csl.uiuc.edu

6.1 INTRODUCTION

Numerous high-bit rate digital communication technologies are currently being proposed that employ unshielded twisted-pair (UTP) wiring. These include asymmetric digital subscriber loop (ADSL) [1, 2], high-speed digital subscriber loop (HDSL) [3], very high-speed digital subscriber loop (VDSL) [1, 4], asynchronous transfer mode (ATM) LAN [5] and broadband access [6]. While newly installed wiring tends to be fiber, it is anticipated that the data carrying capabilities of UTP will be sufficient to meet consumer needs well into the next century.

The above mentioned transmission technologies are especially challenging from both algorithmic and VLSI viewpoints. This is due to the fact that high data rates (51.84 Mb/s to 155.52 Mb/s) need to be achieved over severely bandlimited (less than 30 MHz) UTP channels which necessitate the use of highly complex digital communications algorithms. Furthermore, the need to reduce costs is driving the industry towards increased levels of integration with stringent requirements on the power dissipation, area, speed and reliability of a silicon implementation. Successful solutions will necessarily require an integrated approach whereby algorithmic concerns such as signal-to-noise ratio (SNR) and bit-error rate (BER) along with VLSI constraints such as power dissipation, area, and speed, are addressed in a joint manner.

One way to integrate algorithmic concerns (such as SNR) and implementation issues such as area, power dissipation and throughput is to employ ***algorithm transformation techniques*** [7] such as pipelining [8, 9, 10], parallel processing [9], retiming [11] etc. This is in contrast to the traditional approach (see Fig. 6.1(a)), which consisted of two major steps: 1.) Algorithm design, and 2.) VLSI implementation. Constraints from the VLSI domain (area, power dissipation and

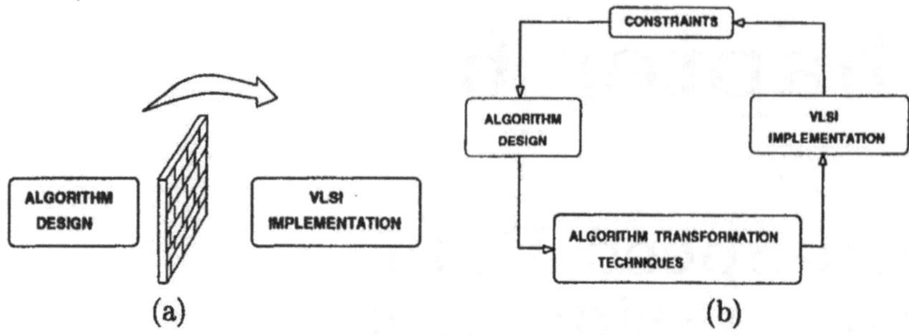

Figure 6.1 VLSI systems design: (a) the traditional and (b) the modern approach.

throughput) were addressed only after the algorithmic performance requirements (SNR and/or BER) were met. The modern approach (see Fig. 6.1(b)) advocated in this chapter incorporates implementation constraints directly into the algorithm design phase thus eliminating expensive design iterations.

In this chapter, we discuss algorithmic and VLSI architectural issues in the design of low-power transceivers for broadband data communications over UTP channels. After presenting certain preliminaries in Section 6.2, we study the UTP-based channel for ATM-LAN and VDSL in Section 6.3 and the commonly employed carrierless amplitude/phase (CAP) modulation scheme in Section 6.4. Next, two algorithmic low-power techniques based upon *Hilbert transformation* (in Section 6.5) and *strength reduction* (in Section 6.6) are introduced along with a high-speed pipelining technique referred to as *relaxed look-ahead.* The application of these techniques to the design of 51.84 Mb/s ATM-LAN, 155.52 Mb/s ATM-LAN and 51.84 Mb/s VDSL is demonstrated via instructive design examples in Section 6.7.

6.2 PRELIMINARIES

In this section, we present the preliminaries for power dissipation [12] in the commonly employed complementary metal-oxide semiconductor (CMOS) technology, the relaxed look-ahead pipelining technique [10], the Hilbert transformation [13], and the strength reduction technique [14, 15, 16].

6.2.1 Power Dissipation in CMOS

The dynamic power dissipation, P_D, in CMOS technology (also the predominant component) is given by

$$P_D = P_{0\to 1} C_L V_{dd}^2 f, \tag{1}$$

where $P_{0\to1}$ is the average '0' to '1' transition probability, C_L is the capacitance being switched, V_{dd} is the supply voltage and f is the frequency of operation. (see also Chapter 24) Most existing power reduction techniques [17] involve reducing one or more of the three quantities C_L, V_{dd} and f. The Hilbert transformation [13] and the strength reduction transformation [14, 16] techniques achieve low-power operation by reduction of arithmetic operations, which corresponds to the reduction of C_L in (1). On the other hand, the relaxed look-ahead pipelining technique [10] permits the reduction of V_{dd} in (1) by trading off power with speed [17].

In order to compare the effectiveness of low-power techniques, we employ the power savings PS measure defined as,

$$PS = \frac{P_{D,old} - P_{D,new}}{P_{D,old}}, \tag{2}$$

where $P_{D,new}$ and $P_{D,old}$ are the power dissipations of the proposed and existing architectures, respectively.

6.2.2 Relaxed Look-Ahead Transformation

The relaxed look-ahead technique was proposed in [10] as a hardware-efficient pipelining technique for adaptive filters. This technique is obtained by approximating the algorithms obtained via the look-ahead pipelining technique [9].

Consider an N-tap serial LMS filter described by the following equations

$$e(n) = d(n) - \mathbf{W}^H(n-1)\mathbf{X}(n), \quad \mathbf{W}(n) = \mathbf{W}(n-1) + \mu e^*(n)\mathbf{X}(n), \tag{3}$$

where $\mathbf{W}(n) = [w_0(n), w_1(n), \ldots, w_{N-1}(n)]^T$ is the weight vector with $\mathbf{W}^H(n)$ being the Hermitian (transpose and complex conjugate), $\mathbf{X}(n) = [x(n), x(n-1), \ldots, x(n-N+1)]^T$ is the input vector, $e^*(n)$ is the complex conjugate of the adaptation error $e(n)$, μ is the step-size, and $d(n)$ is the desired signal.

In this subsection, we assume that the input $\mathbf{X}(n)$ and the weight vector $\mathbf{W}(n)$ are real signals. A direct-mapped architecture for an N-tap serial LMS algorithm is shown in Fig. 6.2(a). Note that the critical path delay for an N-tap serial LMS filter is given by

$$T_{clk,serial} = 2T_m + (N+1)T_a, \tag{4}$$

where T_m is the computation time of multiplier and T_a is the computation time of an adder. For the applications of interest in this chapter, the critical path computation time would prove to be too slow to meet the sample rate requirements. Therefore, pipelining of the serial LMS filter is essential.

The pipelined LMS algorithm (see [18] for details) is given by,

$$e(n) = d(n) - \mathbf{W}^T(n - D_2)\mathbf{X}(n) \tag{5}$$

$$\mathbf{W}(n) = \mathbf{W}(n - D_2) + \mu \sum_{i=0}^{LA-1} e(n - D_1 - i)\mathbf{X}(n - D_1 - i), \tag{6}$$

where D_1 ($D_1 \geq 0$) and D_2 ($D_2 \geq 1$) are algorithmic pipelining delays and LA ($1 \leq LA \leq D_2$) is referred to as the look-ahead factor. Substituting $D_2 = 1$ in (5)-(6) and $LA = 1$ in (6) gives the 'delayed LMS' [19] algorithm. Convergence analysis of the pipelined LMS algorithm in [18] indicates that the upper bound on the step-size μ reduces and the misadjustment $\mathcal{M}$ degrades slightly as the level of pipelining D_1 and D_2 increase. The architecture corresponding to the pipelined LMS algorithm with $N = 5$, $D_1 = 51$ and $D_2 = 4$ is shown in Fig. 6.2(b), where each adder is pipelined with 4 stages and each multiplier is pipelined with 8 stages. Assuming $T_m = 40$ and $T_a = 20$, we find (from (4)) that $T_{clk,serial} = 200$ while the critical path delay of the pipelined architecture in Fig. 6.2(b) is 5. This implies a speed-up of 40.

Note that the relaxed look-ahead technique has been successfully employed to pipeline numerous adaptive algorithms such as the adaptive LMS algorithm

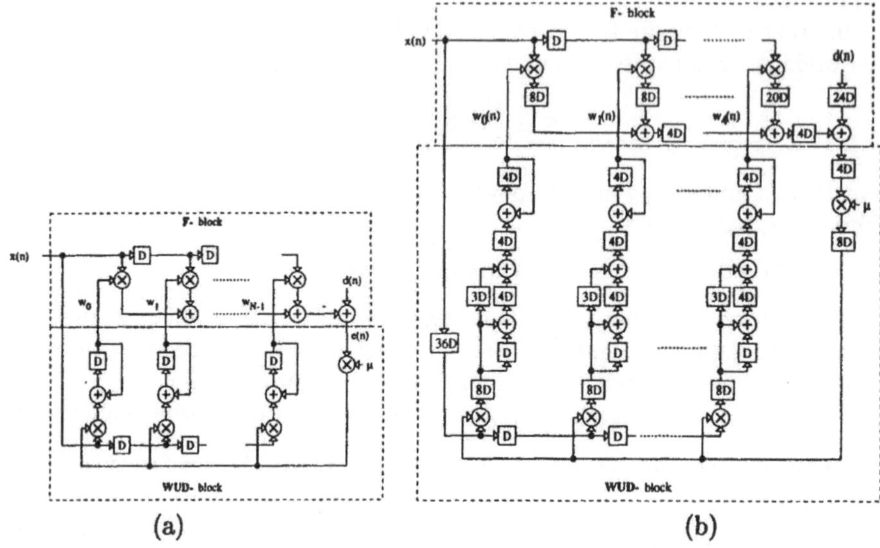

Figure 6.2 Relaxed look-ahead: (a) serial LMS architecture, and (b) pipelined architecture with a speed-up of 40.

[18] and the adaptive differential pulse-code modulation (ADPCM) coder [18]. In both ATM-LAN and VDSL, an adaptive equalizer is employed at the receiver that operates at high sample rates. A pipelined adaptive equalizer architecture based on relaxed look-ahead has proved to be very useful for 51.84 Mb/s ATM-LAN [20] and 51.84 Mb/s VDSL [21, 20].

6.2.3 Hilbert Transformation

Hilbert transform [22] relationships between the real and imaginary parts of a complex sequence are commonly employed in many signal processing and communications applications. In particular, the Hilbert transform of a real sequence $x(n)$ is another real sequence whose amplitude in the frequency domain is identical to that of $x(n)$ but the phase is shifted by 90°. The Hilbert transform of a unit pulse is given by,

$$\begin{aligned} h_I(n) &= \frac{2sin^2(\pi n/2)}{\pi n} \quad \text{for} \quad n \neq 0 \\ &= 0 \quad \text{for} \quad n = 0. \end{aligned} \tag{7}$$

For example, the sine and cosine functions are Hilbert transforms of each other. It will be seen later in Section 6.4 that the coefficients of the in-phase and the quadrature-phase shaping filter in a CAP transmitter are Hilbert transforms of each other. Furthermore, the in-phase and the quadrature-phase equalizer coefficients are also Hilbert transforms of each other. From a low-power perspective, the Hilbert transform relationship between two sequences allows us to compute one from another via a Hilbert filter whose impulse response is given in (7). In Section 6.5, a Hilbert filter is employed to calculate the quadrature-phase equalizer coefficients from those of the in-phase equalizer.

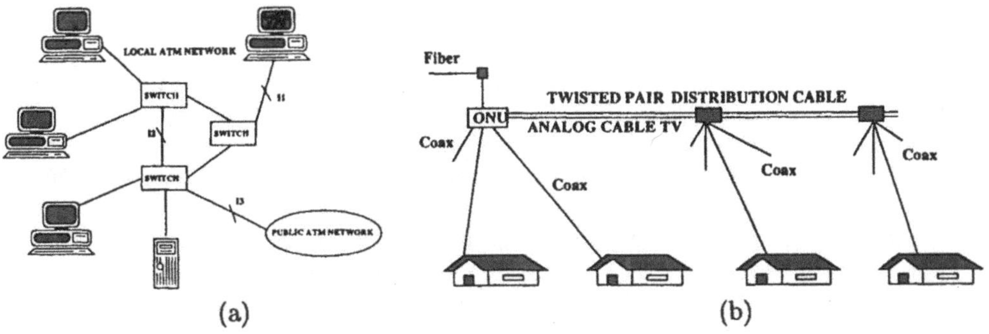

Figure 6.3 Network configurations: (a) ATM-LAN and (b) VDSL.

6.2.4 Strength Reduction Transformation

Consider the problem of computing the product of two complex numbers $(a + jb)$ and $(c + jd)$ as shown below:

$$(a + jb)(c + jd) = (ac - bd) + j(ad + bc). \tag{8}$$

From (8), a direct-mapped architectural implementation would require a total of four real multiplications and two real additions to compute the complex product. Application of strength reduction involves reformulating (8) as follows:

$$(a - b)d + a(c - d) = ac - bd \quad (a - b)d + b(c + d) = ad + bc, \tag{9}$$

where we see that the number of multipliers has been reduced by one at the expense of three additional adders. Typically, multiplications are more expensive than additions and hence we achieve an overall savings in hardware. It can be shown [16] that power savings accrue as long as the average switching capacitance of a multiplier is greater than that of an adder by a factor $K_C > 3$. Furthermore, the power savings asymptotically approach a value of 25% as K_C increases.

6.3 THE CHANNEL

A proper understanding of the physical environment is essential in order to design communications systems that meet the specified performance requirements. In this section, we describe the UTP-based channel for ATM-LAN first and then the VDSL channel.

6.3.1 The LAN Environment

Initially, ATM networks were envisioned to be a wide-area transport technology for delivering integrated services on public networks. However, the potential benefits of this technology has led to the acceptance of ATM technology in a new generation of LANs [23]. Unlike existing LAN technologies such as Ethernet, token-ring, token-bus and fiber distributed data interface (FDDI), data in ATM is transferred between systems via point-to-point links and with switched fixed 53 byte cells.

Fig. 6.3(a) shows a vendor's view of an ATM-based LAN. The environment of interest for the UTP category three (UTP-3) User Network Interface (UNI) consists of the "I1" and "I2" interfaces (see Fig. 6.3(a)). The wiring distribution system runs either from the closet to the desktop or between hubs in the closets. The wiring employed consists mostly of either TIA/EIA-568 UTP-3 4-pair cable or the DIW 10 Base-T 25-pair bundle. Therefore, bandwidth efficient schemes become necessary to support such high data rates over these channels. The CAP transmission scheme is such a scheme and is the standard for 51.84 Mb/s [5] and 155.52 Mb/s [4] ATM-LAN over UTP-3 wiring.

In the LAN environment, the two major causes of performance degradation for transceivers operating over UTP wiring are *propagation loss* and *crosstalk* generated between adjacent wire pairs. The local transmitter produces a signal with amplitude V_t, which propagates on a certain wire pair j and generates spurious signals V_{next} (at the near-end) and V_{fext} (at the far-end) on pair i. The signal V_{next} appears at the end of the cable where the disturbing source V_t is located and is called near-end crosstalk (NEXT). The signal V_{fext} appears at the other end of the cable and is called far-end crosstalk (FEXT). In the LAN environment, NEXT is usually much more severe than FEXT and therefore we will focus on the former. We will see in Section 6.3.2 that the reverse is true for VDSL.

The propagation loss that is assumed in system design is the worst-case loss given in the TIA/EIA-568 draft standard for category 3 cable [24]. This loss can be approximated by the following expression:

$$L_P(f) = 2.320\sqrt{f} + 0.238f, \tag{10}$$

where the propagation loss $L_P(f)$ is expressed in dB per 100 meters and the frequency f is expressed in MHz. The phase characteristics of the loop's transfer function can be computed from $\sqrt{LC}$, where R, L, G, and C are the primary constants of a cable. These constants are available in published literature including [3].

The worst-case NEXT loss model for a single interferer is also given in the TIA/EIA draft standard [24]. The squared magnitude of the NEXT transfer function corresponding to this loss can be expressed as:

$$L_N(f) = 43 - 15\log f, \tag{11}$$

where the frequency f is in megahertz. Measured pair-to-pair NEXT loss characteristics indicate the presence of minima and maxima ocurring at different frequencies. However, the curve of (11) is a worst case envelope of the measured loss and is also referred to as the 15 dB per decade model.

For example, we can derive the average loss in a $100m$ of UTP-3 wiring for a frequency spectrum that extends from d.c. to 25.92 MHz as would be the case for 51.84 Mb/s ATM-LAN. In this case, from (10), we obtain an average propagation loss of 11.4 dB, which is computed as the loss at the center frequency of 12.96 MHz. Similarly, the average NEXT loss from (11) is approximately equal to 26.3 dB. Hence, the signal-to-NEXT ratio at the input to the receiver (SNR_i) would be about 15 dB for ATM-LAN. Note that as the length of the UTP wire increases, the NEXT remains the same while the propagation loss increases resulting in a reduced SNR_i.

6.3.2 The VDSL Environment

In case of VDSL, the community network connects the video server and the set-top equipment. There are two community network architectures being considered to deliver broadband services in the local loop, which are based on hybrid fiber-coax (HFC) and fiber-to-the-curb (FTTC) technologies [25]. The difference between the two being the relative proportion of fiber and coaxial cable in the network.

In an FTTC network architecture shown in Fig. 6.3(b), the optical fiber goes to a curbside pedestal which serves a small number of homes [6]. At the pedestal, the optical signal is converted into an electrical signal and then demultiplexed for delivery to individual homes on copper wiring. These functions are performed in an optical network unit (ONU). The ONU also performs the multiplexing and signal conversion functions required in the opposite direction, i.e. from the homes to the network. The FTTC system considered here makes usage of existing telephone drop wiring or coaxial cable to provide local distribution of VDSL to the home. .

In the VDSL system considered here, the downstream channel (from the ONU to the home) operates at the STS-1 data rate of 51.84 Mb/s, and the upstream channel (from the home to the ONU) operates at a data rate of 1.62 Mb/s. Both channels carry ATM cells and the downstream channel uses SONET framing. The transmission scheme used for the downstream channel is CAP to be described in Section 6.4, while that for the upstream channel is quadrature phase-shift keying (QPSK).

When the VDSL signals propagate on the UTP distribution cable, they interfere with each other by generating FEXT. The downstream CAP signals interfere with each other, and so do the upstream QPSK signals. However, there is minimal interaction between the downstream and upstream signals, because the CAP and QPSK signals use different frequency bands. This is the reason why NEXT is not as significant an issue in broadband applications as compared to FEXT.

In this subsection, we briefly discuss channel and FEXT characteristics of a 600-ft BKMA cable, which is employed for UTP distribution cable in Fig. 6.3(b). The propagation loss characteristics of a BKMA cable are similar to that of a category 5 cable. The worst-case propagation loss for category 5 cable is specified in the TIA/EIA-568A Standard [24], which can also be expressed as follows:

$$L_P(f) = 3.597\sqrt{f} + 0.043f + 0.0914/\sqrt{f}, \tag{12}$$

where the propagation loss $L_P(f)$ is expressed in dB and the frequency f is expressed in MHz.

As far as FEXT is concerned a quantity of interest is the ratio V_r^2/V_{fext}^2, where V_r is the received signal. This ratio (also called equal-level FEXT ($EL-FEXT$) loss or the input signal-to-noise ratio SNR_i in a FEXT dominated environment) can be written as:

$$EL - FEXT = \frac{V_r^2}{V_{fext}^2} = \frac{K}{\Psi f^2 d} \qquad K = (49/N)^{0.6}, \tag{13}$$

where Ψ is the coupling constant which equals 10^{-10} for 1% equal level 49 interferers, d is the distance in kilofeet, f is the frequency in kilohertz and N is the number of interferers. The FEXT impairment can be modeled as a Gaussian source as the FEXT sources are independent of each other.

For example, in the VDSL application, a 600-ft UTP cable has 11 FEXT interferers in the worst case. For this channel, the average SNR_i can be calculated from (13) to be 24 dB. This value is obtained by substituting $\Psi = 10^{-10}$, $N = 11$, $f = 12960$ khz and $d = 0.6$ kft into (13).

There are also several other factors which impair channel function such as splitters, terminated and open-ended stubs, light dimmers, and narrowband interferences [6]. Splitters used in the in-house coaxial cabling system introduce a severe amount of propagation loss and deep notches in the channel transfer function at frequencies below 5 MHz. An open-ended stub connected to an output port of a splitter introduces notches in the channel transfer function corresponding to the other output ports of the splitter. RF interference is generated by AM radio and amateur radio is also one of major impairments for the downstream and upstream channel signals. Light dimmers generate impulse noise which has significant energy up to 1 or 2 MHz.

We conclude this section by noting that the UTP channel has many impairments that necessitate the use of a bandwidth-efficient modulation scheme for high data-rates. Such a scheme is described next.

6.4 THE CARRIERLESS AMPLITUDE/PHASE (CAP) MODULATION SCHEME

In this section, we describe the carrierless amplitude modulation/phase modulation (CAP) scheme and the CAP transceiver structure. The CAP is a bandwidth-efficient two-dimensional passband transmission scheme, which is closely related to the more familiar quadrature amplitude modulation (QAM). At present, 16-CAP modulation scheme is the standard for ATM-LAN over UTP-3 at 51.84 Mb/s [26, 5] and VDSL [25] over copper wiring while 64-CAP is the standard for ATM-LAN [27] over UTP-3 at 155.52 Mb/s. First, the CAP transmitter is described in Section 6.4.1 and then the CAP receiver in Section 6.4.2.

6.4.1 The CAP Transmitter

The block diagram of a digital CAP transmitter is shown in Fig. 6.4(a). The bit stream to be transmitted is first passed through a scrambler in order to randomize the data. The bit-clock which is employed to synchronize the scrambler equals R the desired bit-rate. For the applications of interest in this paper, $R = 51.84$ Mb/s and therefore the bit-clock is equal to 51.84 MHz. The scrambler functionality is defined in terms of a scrambler polynomial $SP(x)$. For example, in case of ATM-LAN, there are two scrambler polynomials defined in the standard: $SPH(x) = 1 + x^{18} + x^{23}$ and $SPW(x) = 1 + x^5 + x^{23}$ for the switch/hub side and the workstation side, respectively. These scramblers can be implemented with 23 1-bit registers and two exclusive-OR logic gates and hence can be operated at these speeds quite easily.

The scrambled bits are then fed into an encoder, which maps blocks of m bits into one of $k = 2^m$ different complex symbols $a(n) = a_r(n) + ja_i(n)$. A CAP line code that uses k different complex symbols is called a k-CAP line code. In this

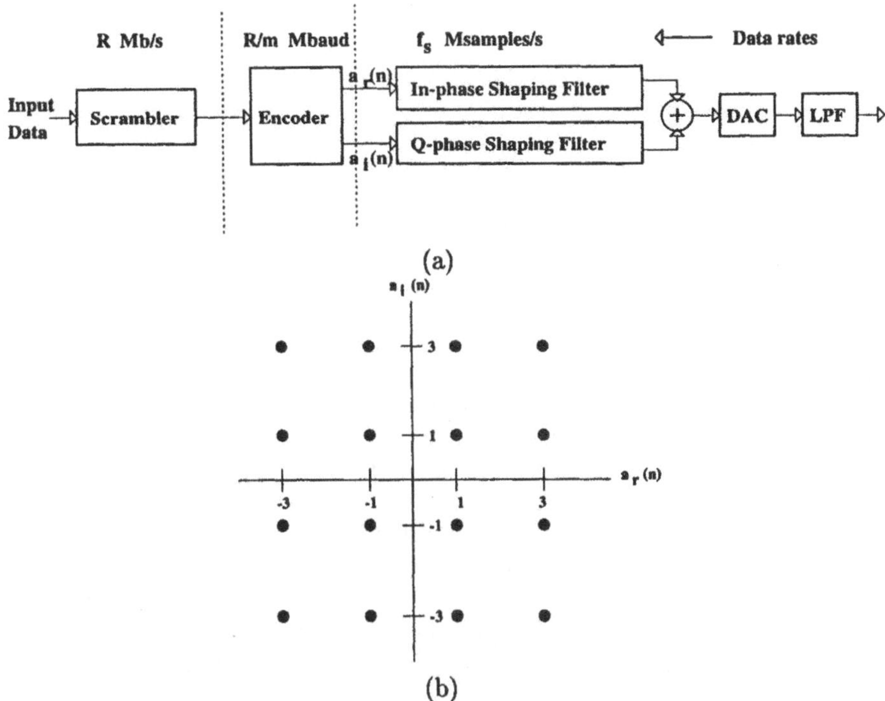

Figure 6.4 The CAP transmitter: (a) block diagram and (b) a 16-CAP signal constellation.

case, the symbol rate $1/T$ given by

$$\frac{1}{T} = \frac{R}{m} = \frac{R}{log_2(k)}, \tag{14}$$

where R is the bit-rate and m is the number of bits per symbol. The encoder block would accept blocks of m-bits and generate symbols $a_r(n)$ and $a_i(n)$ per symbol period. Given that $R = 51.84$ MHz and $m = 4$, then from (14), we have the symbol rate $1/T = 12.96$ Mbaud. Therefore, the symbol clock employed in the encoder block would have a frequency of 12.96 Mhz. The encoder can be implemented as a table look-up. The two-dimensional display of the discrete values assumed by the symbols $a_r(n)$ and $a_i(n)$ is called a signal constellation, an example of which is shown in Fig. 6.4(b).

After the encoder, the symbols $a_r(n)$ and $a_i(n)$ are fed to digital shaping filters. The outputs of the filters are subtracted and the result is passed through a digital-to-analog converter (DAC), which is followed by an interpolating low-pass filter (LPF). The signal at the output of the CAP transmitter in Fig. 6.4(a) can be written as

$$s(t) = \sum_{n=-\infty}^{\infty} [a_r(n)p(t - nT) - a_i(n)\tilde{p}(t - nT)], \tag{15}$$

where T is the symbol period, $a_r(n)$ and $a_i(n)$ are discrete multilevel symbols, which are sent in symbol period nT, and $p(t)$ and $\tilde{p}(t)$ are the impulse responses of in-phase and quadrature-phase passband shaping filters, respectively.

The passband pulses $p(t)$ and $\tilde{p}(t)$ in (15) can be designed in the following manner,

$$p(t) \triangleq g(t)cos(2\pi f_c t) \qquad \tilde{p}(t) \triangleq g(t)sin(2\pi f_c t), \tag{16}$$

where $g(t)$ is a baseband pulse which is usually the square-root raised cosine pulse described below,

$$g(t) = \frac{sin[\pi(1-\alpha)\frac{t}{T}] + 4\alpha\frac{t}{T}cos[\pi(1+\alpha)\frac{t}{T}]}{\pi\frac{t}{T}[1-(4\alpha\frac{t}{T})^2]} \tag{17}$$

and f_c is called the *center frequency* and is larger than the largest frequency component in $g(t)$. The two impulse responses in (16) form a Hilbert pair (see Section 6.2.3), i.e., their Fourier transforms have the same amplitude characteristics, while their phase characteristics differ by 90°.

While the bit-rate R and the choice of the signal constellation determine the symbol rate $1/T$ (see (14)), the transmit spectrum is generated by the shaping filters. It is well-known [28] that the bandwidth of a passband spectrum cannot be smaller than the symbol rate $1/T$. In practice, the transmit bandwidth is made greater than $1/T$ by a fraction α. In that case, the upper and lower edges of the transmit spectrum are given by

$$f_{upper} = f_c + \frac{(1+\alpha)}{2T} \tag{18}$$

$$f_{lower} = f_c - \frac{(1+\alpha)}{2T}, \tag{19}$$

where f_c is the center frequency, f_{upper} is the upper edge and f_{lower} is the lower edge of the transmit spectrum. The fraction α is also referred to as the *excess bandwidth*. The excess bandwidth is 100% ($\alpha = 1.0$) for 51.84 Mb/s ATM-LAN and 20% to 50% ($\alpha = 0.2$ to 0.5) for 51.84 Mb/s VDSL. The sampling frequency f_s is given by

$$f_s = 2f_{upper}, \tag{20}$$

as the spectral shaping is done digitally in the CAP modulation scheme.

Consider an example of the design of CAP shaping filters for 51.84 Mb/s ATM-LAN. As described in Section 6.3.1, this environment has NEXT from multiple sources. It has been shown in [29] that an excess bandwidth of 100% ($\alpha = 1.0$) is necessary for perfect suppression of one NEXT source. With $\alpha = 1.0$ and assuming that the lower edge of the transmit spectrum starts at 0 Hz, we find (from (19)) that $f_c = 1/T$. Substituting this value into (18), we obtain a value of $f_{upper} = 2/T$. The emissions requirements from FCC [24] state that f_{upper} be limited to 30 MHz. This can be achieved if $m \geq 4$, so that from (14) the symbol rate $1/T = 12.96$ Mbaud with $m = 4$ (or 16-CAP). All that remains is now to define the sample rate f_s and determine the coefficients of the shaping filters. From (20), the sampling frequency is given by $f_s = 4/T = 51.84$ MHz. Substituting $t = n/f_s$, $1/T = 12.96$ Mbaud, $\alpha = 1.0$ into (16-17), we can obtain the shaping filter coefficients. The number of taps in the shaping filters is a function of the required stop-band attenuation

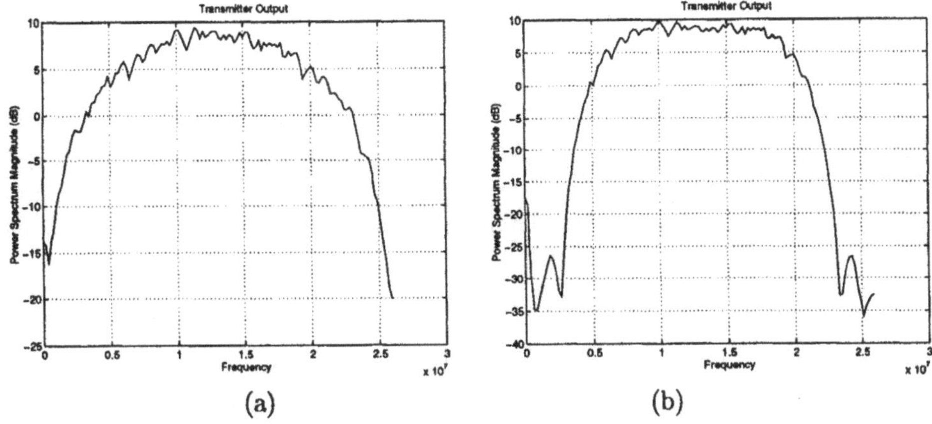

Figure 6.5 Examples of transmit spectrum with excess bandwidths of: (a) 100% for 51.84 Mb/s ATM-LAN and (b) 50% for 51.84 Mb/s VDSL.

and the bandwidth of the transmit spectrum. The resulting transmit spectrum for 51.84 Mb/s ATM-LAN is shown in Fig. 6.5(a), while that for 51.84 Mb/s VDSL with 50% excess bandwidth is shown in Fig. 6.5(b).

The digital shaping filters and the DAC operate at a sampling rate $1/T_s = K/T$, where K is a suitably chosen integer such that the sample rate is greater than $2f_{upper}$ (see (18(a)). In addition to this requirement, the sample rate is also chosen to be an integral multiple of the symbol rate in order to ease the requirements on the clock generation circuitry. As indicated in the above example, the sample rates can be quite high. The shaping filters are usually implemented as finite-impulse response (FIR) filters and hence operating at high sample rates is not difficult. Nevertheless, some degree of pipelining may be required. The transmitter design requires a trade-off which encompasses algorithmic and VLSI domains. In particular, this trade-off balances the roll-off in transmit spectrum band edges and the silicon power-dissipation. It can be seen that most of the signal processing at the transmitter (including transmit shaping) is done in the digital domain.

6.4.2 The CAP Receiver

The structure of a generic digital CAP receiver is shown in Fig. 6.6. It consists of an analog-to-digital converter (ADC) followed by a parallel arrangement of two adaptive digital filters. It has been shown that the optimum coefficients of the receive equalizer are Hilbert transforms of each other. The ADC and the digital filters operate at a sampling rate $1/T_s = M/T$, which is typically the same as the sampling rate employed at the transmitter. The adaptive filters in Fig. 6.6 are referred to as a T/M fractionally spaced linear equalizers (FSLEs) [28]. In addition to the FSLEs, a CAP receiver can have a NEXT canceller and a decision feedback equalizer (DFE). The decision to incorporate a NEXT canceller and/or a DFE depends upon the channel impairments (described in Section 6.3) and the capabilities of an FSLE, which is described next.

The received signal consists of the data signal (desired signal), the ISI, and the NEXT/FEXT signal. The performance of a receiver is a function of the input signal-to-noise ratio SNR_i, which is given by:

$$SNR_i = \frac{\sigma_{ds}^2 + \sigma_{isi}^2}{\sigma_{noise}^2}, \tag{21}$$

where σ_{ds}^2 is the data signal power, σ_{isi}^2 is the intersymbol interference (ISI) power and σ_{noise}^2 is the noise power. Here, $\sigma_{noise}^2 = \sigma_{NEXT}^2$ in case of ATM-LAN and $\sigma_{noise}^2 = \sigma_{FEXT}^2$ in case of VDSL. Similarly, the signal-to-noise ratio at the output of the equalizer SNR_o is defined as:

$$SNR_o = \frac{\sigma_{ds}^2}{\sigma_{r,noise}^2 + \sigma_{r,isi}^2}, \tag{22}$$

where $\sigma_{r,noise}^2$ is the residual noise (NEXT/FEXT) and $\sigma_{r,isi}^2$ is the residual ISI at the equalizer output. Typically, at the input to the receiver (for both ATM-LAN and VDSL), the data signal power σ_{ds}^2 is only 6 dB above the ISI signal power σ_{isi}^2 and hence ISI is the dominant impairment. Therefore, the FSLE first reduces ISI before it can suppress the NEXT/FEXT signal. Thus, the function of the FSLE is to perform *NEXT suppression* (for 51.84 Mb/s ATM-LAN), *FEXT suppression* (for VDSL) and ISI removal (for both). In addition, due to the fractional tap spacing, the FSLE also provides immunity against sampling jitter caused by the timing recovery circuit.

An important quantity for the performance evaluation of transceivers is the *noise margin*, which is defined as the difference between the SNR_o and a reference $SNR_{o,ref}$. Taking 16-CAP as an example, a value of $SNR_{o,ref} = 23.25$ dB corresponds to a $BER = 10^{-10}$. Let SNR_o be the SNR at the slicer for a given experiment, and let $SNR_{o,ref}$ be the SNR required to achieve a given BER. The margin achieved by the transceiver with respect to this BER is then defined as

$$\text{margin} \triangleq SNR_o - SNR_{o,ref}. \tag{23}$$

A positive margin in (23) means that the transceiver operates with a BER that is better than the targeted BER.

While the FSLE is indeed a versatile signal processing block it may become necessary to augument it with a NEXT canceller and/or a DFE in certain situations. For example, in case of the 51.84 Mb/s ATM-LAN, the FSLE eliminates ISI and suppresses NEXT. NEXT suppression is feasible for 51.84 Mb/s ATM-LAN because the excess bandwidth is 100% (see Fig. 6.5(a)) and it has been shown [5, 29] that one cyclostationary NEXT interferer can be suppressed perfectly if the CAP transmitter uses an excess bandwidth of at least 100%. For 155.52 Mb/s ATM-LAN the symbol rate with 64-CAP (from (14)) is 25.92 Mbaud. Hence, it is not possible to have an excess bandwidth of 100% as that would violate the FCC emissions requirements [24]. Therefore, a NEXT canceller is employed as shown in Fig. 6.7. Similarly, in case of 51.84 Mb/s VDSL, the presence of radio frequency interference (RFI) necessitates the use of a DFE as shown in Fig. 6.8.

The two outputs of the FSLE are sampled at the symbol rate $1/T$ and added to the outputs of: 1.) the NEXT canceller for 155.52 Mb/s ATM-LAN, 2.) the DFE for 51.84 Mb/s VDSL or taken as is for 51.84 Mb/s ATM-LAN and the results are fed to a decision device followed by a decoder, which maps the symbols into

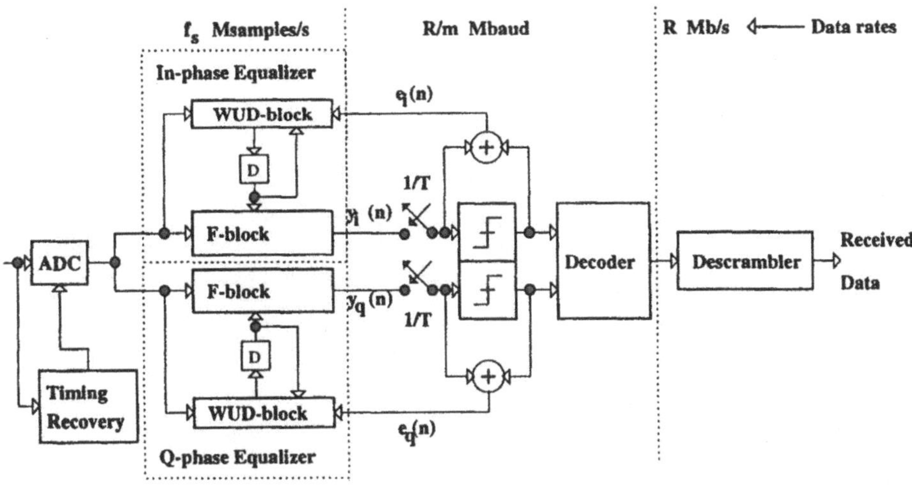

Figure 6.6 The CAP receiver structure for 51.84 Mb/s ATM-LAN.

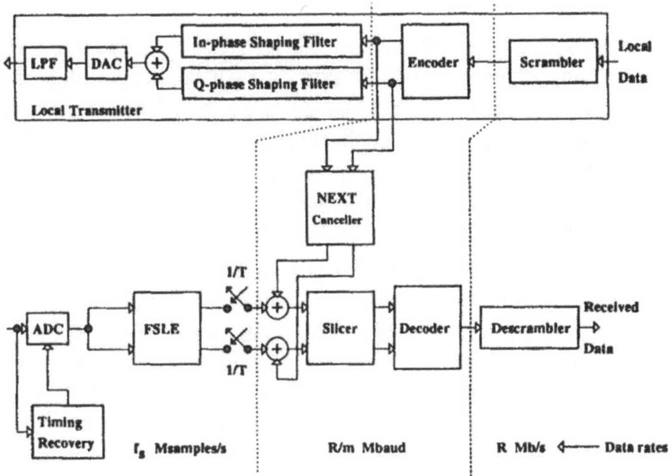

Figure 6.7 The CAP receiver structure for 155.52 Mb/s ATM-LAN.

bits. The output of the decoder is then passed to a descrambler. It must be noted that the decoder and the descrambler perform the inverse operation of the encoder and the scrambler, respectively. Thus, we see that most of the signal processing in a CAP transceiver is done in the digital domain. This minimality of analog processing permits a robust VLSI implementation. Another attractive feature is the fact that it is easy to operate a CAP transceiver at different bit-rates by simply altering the signal constellation generated by the encoder without changing the analog front-end. The interested reader is referred to [30] for more details on the design of CAP receivers.

As seen in Fig. 6.7, the FSLE, the NEXT canceller and the DFE need to be implemented as adaptive filters. The FSLE operates at the sample rate while the

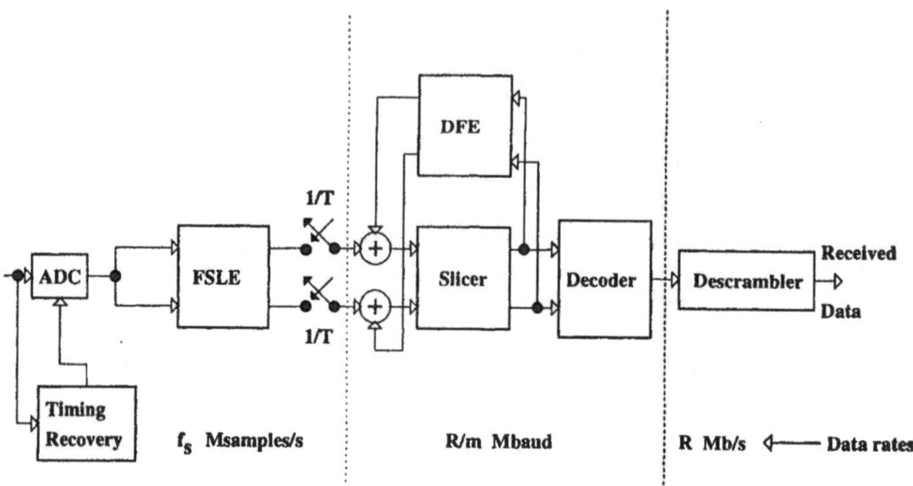

Figure 6.8 The CAP receiver structure for 51.84 Mb/s VDSL applications.

NEXT canceller and the DFE operate at the symbol rate $1/T$. However, from a VLSI perspective, implementing a high sample rate adaptive filter that also consumes low power is a difficult task. We describe two low-power adaptive filter techniques in Sections 6.5 and 6.6 which can be applied to the FSLE, the NEXT canceller and the DFE. In addition, we can employ the relaxed look-ahead (see Section 6.2.2) to develop hardware-efficient high-speed architectures for these blocks.

For example, in case of 51.84 Mb/s ATM-LAN, the two adaptive filters operate on an input sampling rate of 51.84 MHz and produce outputs at the symbol rate of 12.96 Mbaud. The length of the FSLE is usually given in terms of multiples of the symbol period T and is a function of the delay and amplitude distortion introduced by the channel. While the channel characteristics can provide an indication of the required number of equalizer taps, in practice this is determined via simulations. A symbol span of $8T$ has been found to be sufficient for the 51.84 Mb/s ATM-LAN application.

In the past, independent adaptation of the equalizers in Fig. 6.6 has typically been employed at the receiver. In the next section, we show how the filters can be made to adapt in a dependent manner so that low-power operation can be achieved.

6.5 THE HILBERT TRANSFORM BASED FSLE ARCHITECTURE

As described in Section 6.3.1, the LAN environment has ISI and NEXT as the predominant channel impairment, while ISI and FEXT are present in the VDSL environment. In either case, designing adaptive FSLE's with sufficient number of taps to meet a BER requirement of 10^{-10} at these sample rates and with low power dissipation is a challenging problem. In this section, we present a Hilbert transform based low-power FSLE architecture and then pipeline it using the relaxed look-ahead technique [10, 18] to achieve high throughput.

6.5.1 Low-Power FSLE Architecture via Hilbert Transformation

As mentioned in Section 6.4.2, the in-phase and the quadrature-phase equalizers of the CAP receiver in Fig. 6.6 are Hilbert transforms of each other. In this subsection, we show how this relationship can be exploited to obtain a low-power structure. We then compute the power consumed by the CAP equalizer and the low-power equalizer and show that the proposed equalizer can lead to substantial power savings with marginal degradation in performance.

If the in-phase and the quadrature-phase equalizer filter impulse responses are denoted by $f(n)$ and $\tilde{f}(n)$, respectively, then

$$\tilde{f}(n) = h_I(n) * f(n), \tag{24}$$

where the symbol "*" denotes convolution. Let $y_i(n)$ and $y_q(n)$ denote the in-phase and the quadrature-phase components of the receive filter output, respectively, and $x(n)$ denote the input. Employing (24), the equalizer outputs can be expressed as

$$y_i(n) = f(n) * x(n), \quad y_q(n) = \tilde{f}(n) * x(n) = f(n) * [h_I(n) * x(n)]. \tag{25}$$

From (25), we see that $y_q(n)$ can be computed as the output of a filter, which has the same coefficients as that of the in-phase filter with the Hilbert transform of $x(n)$ as the input. Hence, the CAP receiver in Fig. 6.6 can be modified into the form as shown in Fig. 6.9, where **HF** is the Hilbert filter. The structures in Fig. 6.9 and Fig. 6.6 are functionally equivalent as long as the Hilbert filter is of infinite length. However, in practice, an M-tap finite length Hilbert filter is employed whose impulse response $h_F(n) = h_I(n)$ for $n = -(M-1)/2, \ldots, (M-1)/2$ (M is odd and $h_I(n)$ is defined in (7)).

The low-power receiver structure in Fig. 6.9 has several attractive features: 1.) the **WUD** block in the quadrature-phase filter is completely eliminated, and 2.) there is no feedback in the quadrature-phase filter which eases the pipelining of this filter, and 3.) in a blind start-up scheme, the equalizer will converge quicker to the correct solution as as there is only one adaptive equalizer. There is an addition of a Hilbert filter in the feed-forward path of the quadrature-phase arm, which necessitates an additional M sample rate delays in the in-phase path to compensate for the phase delay introduced by the Hilbert filter.

Hence, from the point of view of power consumption and silicon area, the proposed structure would result in power savings as long as the complexity of the Hilbert filter is smaller than that of the **WUD** block. From a power dissipation perspective, the Hilbert filter length should be as small as possible. However, from a performance perspective, the length of the Hilbert filter should be as large as possible. This trade-off is explored in the next two subsections.

6.5.2 Power Savings

The traditional CAP receiver in Fig. 6.6 has a parallel arrangement of two adaptive filters of length N, where the **F** block consists of N multipliers and $N+1$ single precision adders and the **WUD** block contains $N+1$ multipliers and N double precision adders.

Assuming that the switching capacitances of the multiplier and double precision adders are $K_m C_a$ and $K_a C_a$, the average power dissipated by a CAP receiver

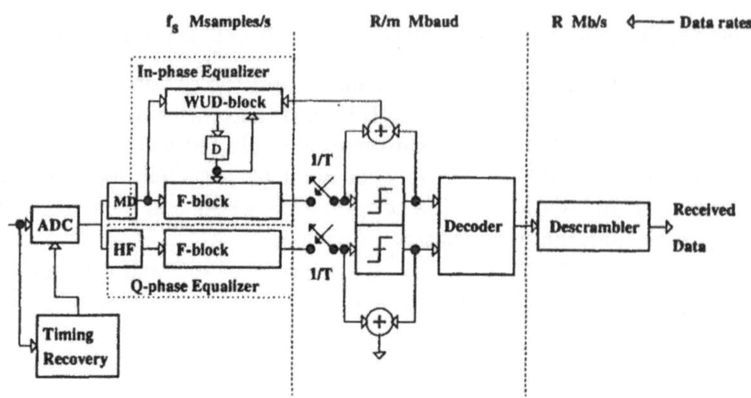

Figure 6.9 The low-power Hilbert transformation based CAP receiver.

of length N ($P_{D,cap}$) is given by (see (1))

$$P_{D,cap} = 2[(2N+1)K_m + NK_a + (N+1)]C_a V_{dd}^2 f_s, \tag{26}$$

where C_a is the switching capacitance of a single precision adder and f_s is the sampling frequency. The low-power structure in Fig. 6.9 has no **WUD** block for the quadrature-phase filter but instead has a Hilbert filter of length M, which requires $\frac{M-1}{4}$ multipliers and $\frac{M-1}{2}$ adders (see (7)). Hence, the average power dissipated by the proposed structure ($P_{D,hilbert}$) is given by

$$P_{D,hilbert} = [(3N + \frac{M}{4} + \frac{3}{4})K_m + NK_a + (2N + \frac{M}{2} - \frac{1}{2})]C_a V_{dd}^2 f. \tag{27}$$

Employing (26) and (27), we can show that the power savings PS (see (2)) is given by,

$$PS = \frac{(1 - \frac{M}{4N} + \frac{5}{4N})K_m + K_a - \frac{M-3}{2N}}{2[(2+\frac{1}{N})K_m + (1+\frac{1}{N})K_a + 1]}. \tag{28}$$

Hence, in order to have positive power savings, we need to choose M such that,

$$M < \frac{2[(N+1)K_m + NK_a + 1}{1 + \frac{K_m}{2}} + 1. \tag{29}$$

Assuming typical values of $K_m = 16$, $K_a = 2$, and $N = 32$, (29) indicates that there is a net saving in power as long as the Hilbert filter length $M < 131$.

6.5.3 Excess MSE

We now derive an expression to compute the decrease in SNR_o due to the use of finite length Hilbert filter. Note that the maximum value of SNR_o achievable via the proposed structure in Fig. 6.9 with an infinite length Hilbert filter is the same as that achieved by the original CAP structure in Fig. 6.6.

It can be shown that [13] the excess error $e_{EX}(n)$ due to the use of an M-tap finite length Hilbert filter is given by.

$$E[e_{EX}^2] = \mathbf{W}_{opt}^T \mathbf{R}_{XX} \mathbf{W}_{opt}, \tag{30}$$

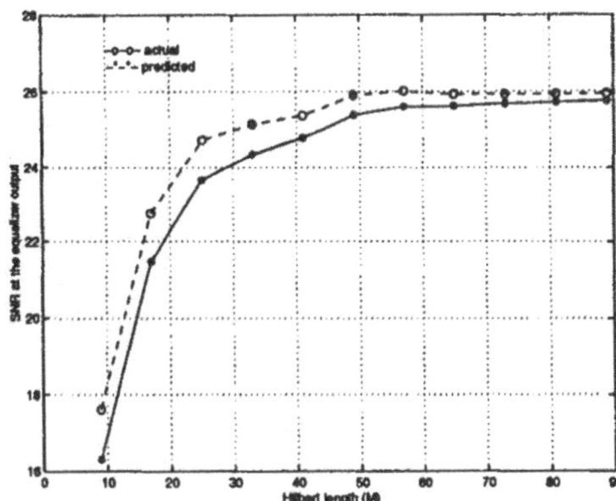

Figure 6.10 The excess MSE due to finite-length Hilbert CAP receiver.

where $E[.]$ represents the statistical expectation operator, $\mathbf{W}_{opt}^{T}$ is the transpose of the optimal equalizer coefficients $\mathbf{W}_{opt}$, $\mathbf{R}_{XX} = E[\mathbf{X}_{err}\mathbf{X}_{err}^{T}]$ is the error correlation matrix, $\mathbf{X}_{err}^{T} = [x_{err}(n), x_{err}(n-1), \ldots, x_{err}(n-N+1)]$ is the vector of outputs of an error filter $h_{err}(n)$, where $x_{err}(k) = \sum_{k=-\infty}^{\infty} h_{err}(k)x(n-k)$ and $h_{err}(n) = h_I(n) - h_F(n)$. Given the channel model such as the ones in Section 6.3, both $\mathbf{R}_{XX}$ and the optimum weight vector $\mathbf{W}_{opt}$ can be easily computed. Note that $\mathbf{R}_{XX}$ is a function of $h_{err}(n)$, which in turn depends on the the length of the Hilbert filter. Hence, we can employ (30) to estimate the increase in the MSE due to the finite length Hilbert filter of length M.

We now verify the results of Sections 6.5.2 and 6.5.3 for the 51.84 Mb/s ATM-LAN application. The spans of the in-phase adaptive filter and the quadrature-phase FIR filter are fixed at $8T$. The length of the Hilbert filter is varied and the SNR_o values are compared against those predicted using the expression for the *excess error* derived in (30). The results shown in Fig. 6.10 indicate that the analysis and simulations match quite well. We obtain SNR_o values of more than 25 dB, when the Hilbert transformer length M is more than 33. This value asymptotically approaches 25.9 dB as the Hilbert transformer length increases. As per our analysis of power consumption in Section 6.5.2, a net power saving is obtained as long as $M < 131$. Therefore, the proposed structure will provide a noise-margin better than 1.75 dB and enable power saving as long as the Hilbert transformer length $33 < M < 131$, when $N = 32$ and the desired $SNR_o = 25.9$ dB.

6.5.4 Pipelined FSLE Architecture

From Fig. 6.9, we see that the Hilbert CAP receiver requires one adaptive FSLE in the in-phase arm with the quadrature arm being non-recursive. Thus, the in-phase equalizer can be pipelined via the relaxed look-ahead technique [10] (see Section 6.2.2) and the quadrature-phase equalizer can be pipelined via feedforward cutset pipelining method [31]. In this subsection, we describe a pipelined FSLE architecture developed using relaxed look-ahead [20].

The architecture of the FSLE (shown in Fig. 6.11) consists of $N1$ hardware taps, where $N1 = N/K$ and $K = Tf_s$. The number of hardware taps $N1$ is less than N due to the K-fold down sampling at the $\mathbf{F}$-block output. It will be shown

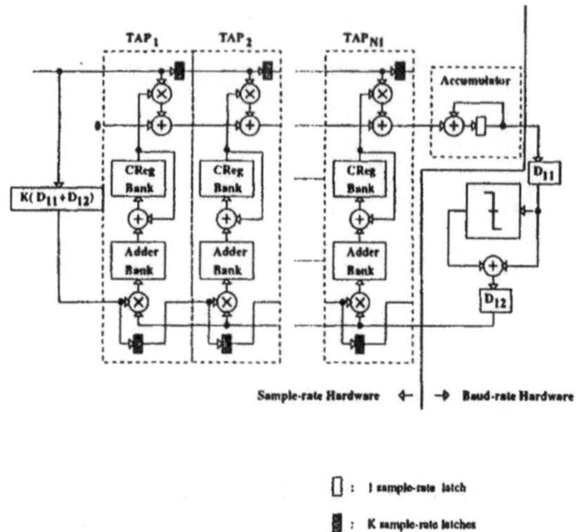

Figure 6.11 The CAP receiver architecture.

in Section 6.7 that $N1 = 8$ is sufficient for 51.84 Mb/s ATM-LAN. However, this value of $N1$ along with the signal precisions (described in the previous paragraph) requires a value of $D_1 = 5$ (baud-rate algorithmic latches). Of these $D_{11} = 2$ and $D_{12} = 3$ latches can be employed to pipeline the **F**-block and the **WUD**-block, respectively. Note that it is only the D_{11} latches that would result in an increased end-to-end delay due to pipelining. Retiming the D_{11} latches resulted in a pipelining latch at the output of every multiplier and adder in the **F**-block.

The **WUD**-block consists of a multiplier (in order to compute the product of the slicer error and data, an adder bank to compute the summation and a coefficient register (CReg) bank to store the coefficients. The CReg bank consists of $D_{22} = KD_2$ latches, where D_2 delayed versions of K algorithmic taps are stored. The product $D_2\gamma$ was chosen to be a power of two such that $D_2\gamma\mathbf{W}_R((n - D_2)T)$ gives the sign bit of $\mathbf{W}_R((n - D_2)T)$. Hence, tap-leakage is implemented by adding the sign of the current weight [28] to the least-significant bit in the **WUD**-block. Thus the **WUD**-block adder shown in Fig. 6.11, is in fact two 23-bit additions, which need to be accomplished within a sample period of $19ns$. This is not difficult to do so as the latches in the CReg bank can be employed to pipeline the additions.

The accumulator at the output of TAP_{N1} is reset after every K sample clocks. This is because the N-tap convolution in Fig. 6.11 is computed $N1 = N/K$ taps at a time by the architecture in Fig. 6.11. The slicer is a tree-search quantizer which is also capable of slicing with two or four levels. This dual-mode capability of the slicer is needed in order to have a blind start-up employing the reduced constellation algorithm [28].

The low-power Hilbert transformation technique presented here applies to a specific but more general class of CAP receivers shown in Fig. 6.6. In the next section, we present another low-power technique that is applicable to any system involving complex adaptive filters. These include the NEXT canceller, the DFE (in CAP systems), the equalizer in a QAM system and many others.

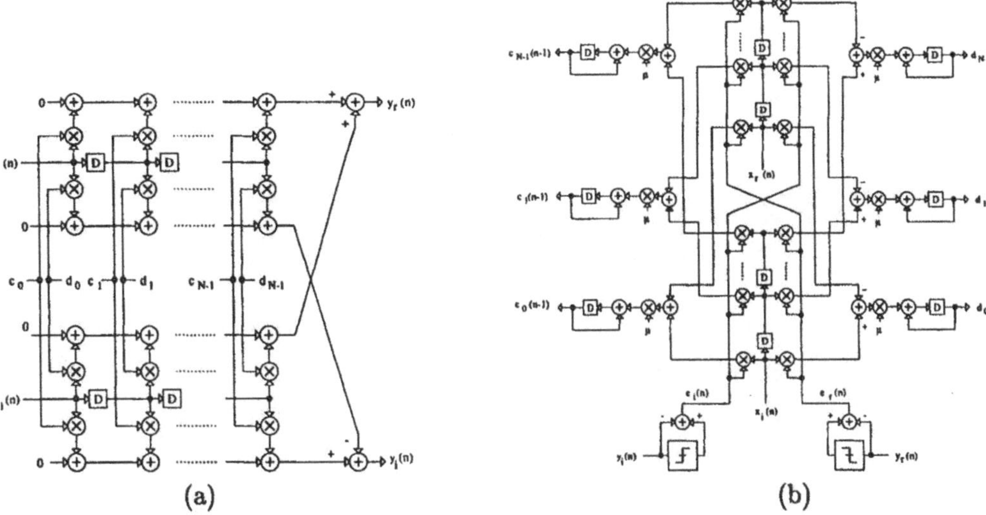

Figure 6.12 The crosscoupled equalizer structure: (a) the **F** block and, the (b) **WUD** block.

6.6 STRENGTH REDUCED ADAPTIVE FILTER

In this section, we present the strength reduced low-power adaptive filter and develop a pipelined version [16] from the traditional cross-coupled (**CC**) architecture. While we present only the final results here, the reader is referred to [16] for more details.

6.6.1 Low-power Complex Adaptive Filter via Strength Reduction

The **SR** architecture [16] is obtained by applying strength reduction transformation at the algorithmic level instead of at the multiply-add level described in Section 6.2.4. Starting with the complex LMS algorithm in (3), we assume that the filter input is a complex signal **X**(n) given by $\mathbf{X}(n) = \mathbf{X}_r(n) + \jmath\mathbf{X}_i(n)$, where $\mathbf{X}_r(n)$ and $\mathbf{X}_i(n)$ are the real and the imaginary parts of the input signal vector $\mathbf{X}(n)$. Furthermore, the filter $\mathbf{W}(n)$ is also complex, i.e., $\mathbf{W}(n) = \mathbf{c}(n) + \jmath\mathbf{d}(n)$.

From (3), we see that there are two complex multiplications/inner-products involved. Traditionally, the complex LMS algorithm is implemented via the cross-coupled **CC** architecture, which is described by the following equations:

$$y_r(n) = \mathbf{c}^T(n-1)\mathbf{X}_r(n) + \mathbf{d}^T(n-1)\mathbf{X}_i(n) \tag{31}$$

$$y_i(n) = \mathbf{c}^T(n-1)\mathbf{X}_i(n) - \mathbf{d}^T(n-1)\mathbf{X}_r(n) \tag{32}$$

$$\mathbf{c}(n) = \mathbf{c}(n-1) + \mu\left[e_r(n)\mathbf{X}_r(n) + e_i(n)\mathbf{X}_i(n)\right] \tag{33}$$

$$\mathbf{d}(n) = \mathbf{d}(n-1) + \mu\left[e_r(n)\mathbf{X}_i(n) - e_i(n)\mathbf{X}_r(n)\right], \tag{34}$$

where $e(n) = e_r(n) + \jmath e_i(n)$ and the **F**-block output is given by $y(n) = y_r(n) + \jmath y_i(n)$. Equations (31-32) and (33-34) define the computations in the **F**-block (see Fig. 6.12(a)) and the **WUD**-block (see Fig. 6.12(b)), respectively. It can be seen from Fig. 6.12 that the cross-coupled **CC** architecture would require $8N$ multipliers and $8N$ adders.

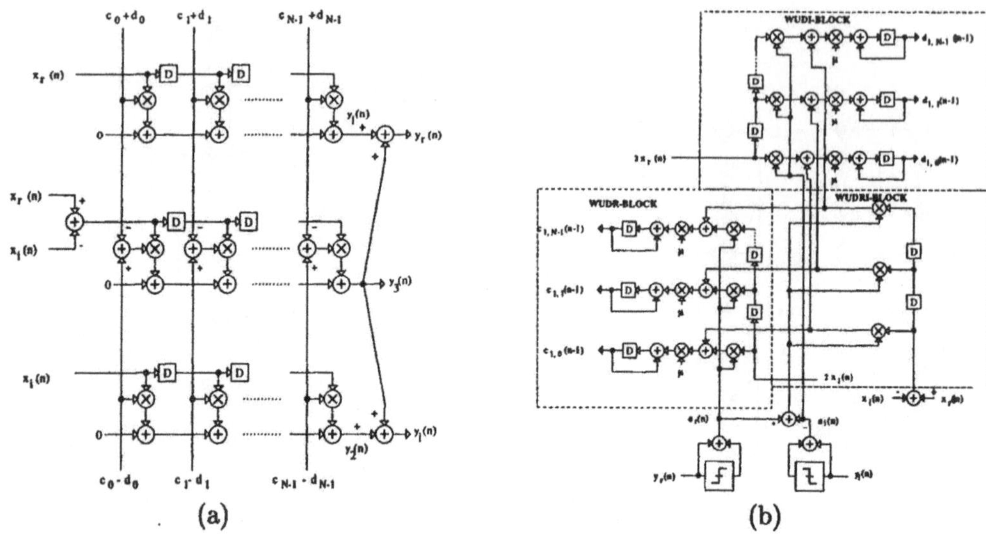

Figure 6.13 The strength reduced equalizer structure: (a) the **F** block and, the (b) **WUD** block.

We see that (3) has two complex multiplications/inner-product and hence can benefit from the application of strength reduction. Doing so results in the following equations, which describe the **F**-block computations of the **SR** architecture [16]:

$$y_1(n) = \mathbf{c}_1^T(n-1)\mathbf{X}_r(n), \quad y_2(n) = \mathbf{d}_1^T(n-1)\mathbf{X}_i(n), \quad (35)$$

$$y_3(n) = -\mathbf{d}^T(n-1)\mathbf{X}_1(n), \quad y_r(n) = y_1(n) + y_3(n), \quad y_i(n) = y_2(n) + y_3(n), \quad (36)$$

where $\mathbf{X}_1(n) = \mathbf{X}_r(n) - \mathbf{X}_i(n)$, $\mathbf{c}_1(n) = \mathbf{c}(n) + \mathbf{d}(n)$, and $\mathbf{d}_1(n) = \mathbf{c}(n) - \mathbf{d}(n)$. Similarly, the **WUD** computation is described by,

$$\mathbf{c}_1(n) = \mathbf{c}_1(n-1) + \mu[\mathbf{eX}_1(n) + \mathbf{eX}_3(n)] \quad (37)$$

$$\mathbf{d}_1(n) = \mathbf{d}_1(n-1) + \mu[\mathbf{eX}_2(n) + \mathbf{eX}_3(n)], \quad (38)$$

where $\mathbf{eX}_1(n) = 2e_r(n)\mathbf{X}_i(n)$, $\mathbf{eX}_2(n) = 2e_i(n)\mathbf{X}_r(n)$, $\mathbf{eX}_3(n) = e_1(n)\mathbf{X}_1(n)$, $e_1(n) = e_r(n) - e_i(n)$, $\mathbf{X}_1(n) = \mathbf{X}_r(n) - \mathbf{X}_i(n)$. It is easy to show that the **SR** architecture (see Fig. 6.13) requires only $6N$ multipliers and $8N + 3$ adders. This is the reason why the **SR** architecture results in $21 - 25\%$ power savings [16] over the **CC** architecture.

6.6.2 Pipelined Strength-reduced Architecture

Combining the **F**-block in Fig. 6.13(a) with the **WUD** block in Fig. 6.13(b), we obtain the **SR** architecture in Fig. 6.14(a), where the dotted line in Fig. 6.14(a) indicates the critical path of the **SR** architecture. As explained in [16], both the **SR** as well as **CC** architectures are bounded by a maximum possible clock rate due the computations in this critical path. This throughput limitation is eliminated via the application of the relaxed look-ahead transformation [18] to the **SR** architecture (see (35-38)). Doing so results in the following equations that describe the **F**-block

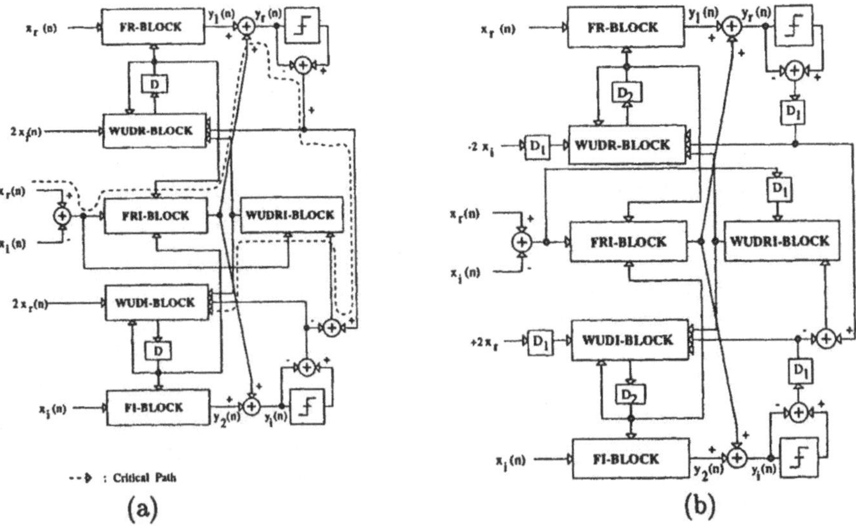

Figure 6.14 The strength reduced equalizer block diagram: (a) serial and (b) pipelined architectures.

computations in the **PIPSR** architecture:

$$y_1(n) = \mathbf{c}_1^T(n - D_2)\mathbf{X}_r(n), \quad y_2(n) = \mathbf{d}_1^T(n - D_2)\mathbf{X}_i(n), \quad (39)$$

$$y_3(n) = -\mathbf{d}^T(n - D_2)\mathbf{X}_1(n), \quad y_r(n) = y_1(n) + y_3(n), y_i(n) = y_2(n) + y_3(n), \quad (40)$$

where D_2 is the number of delays introduced before feeding the filter coefficients into the **F**-block. Similarly, the **WUD** block of the **PIPSR** architecture is computed using

$$\mathbf{c}_1(n) \quad = \quad \mathbf{c}_1(n - D_2) + \mu \sum_{i=0}^{LA-1} [\mathbf{eX}_1(n - D_1 - i) + \mathbf{eX}_3(n - D_1 - i)] \quad (41)$$

$$\mathbf{d}_1(n) \quad = \quad \mathbf{d}_1(n - D_2) + \mu \sum_{i=0}^{LA-1} [\mathbf{eX}_2(n - D_1 - i) + \mathbf{eX}_3(n - D_1 - i)], \quad (42)$$

where $\mathbf{eX}_1(n)$, $\mathbf{eX}_2(n)$ and $\mathbf{eX}_3(n)$ are defined in the previous subsection, $D_1 \geq 0$ are the delays introduced into the error feedback loop and $0 < LA \leq D_2$ indicates the number of terms considered in the sum-relaxation. A block level implementation of the **PIPSR** architecture is shown in Fig. 6.14(b) where D_1 and D_2 delays will be employed to pipeline the various operators such as adders and multipliers at a fine-grain level.

6.6.3 Power Savings

As was seen in this section, relaxed look-ahead pipelining results in an overhead of $2N(LA - 1)$ adders and $5D_1 + 2D_2$ latches (without retiming). Employing the fact that these additional adders are double-precision, we get the power savings

PS with respect to the cross-coupled architecture as follows:

$$PS = \frac{2NK_C(4K_V^2 - 3) + 2N(6K_V^2 - 2LA - 4) - (5D_1 + 2D_2)K_L - 3}{K_V^2(8NK_C + 12N)}, \quad (43)$$

where K_L is the ratio of the effective capacitance of a 1-b latch to that of a 1-b adder, and $K_V > 1$ is the factor by which the power-supply is scaled. Employing typical values of $K_V = 5V/3.3V$, $K_C = 8$, $K_L = 1/3$, $N = 32$, $D_1 = 48$, $D_2 = 2$ and $LA = 3$ in (4.9), we obtain a total power savings of approximately 60% over the traditional cross-coupled architecture. Clearly, 21% of the power savings are obtained from the strength reduction transformation, while the rest (39%) is due to power-supply scaling. Note that, this increased power savings is achieved in spite of the additional $2N(LA-1)$ adders required due to relaxed look-ahead pipelining.

Based upon the transistor threshold voltages, it has been shown in [17] that values of $K_V = 3$ are possible with present CMOS technology. With this value of K_V, (4.9) predicts a power savings of 90%, which is a significant reduction. Thus, a judicious application of algebraic transformations (strength reduction), algorithm transformations (pipelining) and power-supply scaling can result in substantial power reduction.

6.6.4 Finite-Precision Requirements

In this section, we will present a comparison of the precision requirements of the **CC** and **SR** architectures. First, we will consider the **F**-block and then the **WUD**-block.

Consider the **F**-block with N as the tap-length and $SNR_{o,fl}$ as the output SNR in dB of the floating point algorithm. It has been shown [32] that precision of the **F**-block in the **CC** ($B_{\mathbf{F,CC}}$) and the **SR** architecture ($B_{\mathbf{F,SR}}$) is given by,

$$B_{\mathbf{F,CC}} > \frac{1}{2}\log_2\left(\frac{N\sigma_x^2}{6\beta\sigma_d^2}\right) + \frac{SNR_{o,fl}(dB)}{6} \quad (44)$$

$$B_{\mathbf{F,SR}} > \frac{1}{2}\log_2\left(\frac{N\sigma_x^2}{4\beta\sigma_d^2}\right) + \frac{SNR_{o,fl}(dB)}{6}, \quad (45)$$

where σ_x^2 is the input power to the **F**-block, σ_d^2 is the power of symbol constellation (or the desired signal) and $\beta \ll 1$ is the ratio between the quantization noise at the output and floating point MSE.

Thus it can be seen that the coefficient precision of **F**-block for **CC** and **SR** architectures is related by,

$$B_{\mathbf{F,SR}} = B_{\mathbf{F,CC}} + 0.3, \quad (46)$$

implying that the **F**-block in the **SR** architecture requires at the most one bit more than in the **CC** architecture. The **F**-block precision requirements for **PIPSR** architecture (see (39)-(40)) is same as that of the **SR** architecture.

We now consider the precision requirements of the **WUD** blocks. The "stopping criterion" [33] is employed to determine the **WUD**-block coefficient precision, $B_{\mathbf{WUD}}$. For **CC** architecture, The weight correction terms for the **CC** architecture are given in (33-34) while those for the **SR** architecture are defined by (37)-(38).

Employing the stochastic estimates for these terms and applying the stopping criterion we get,

$$B_{\mathbf{WUD,CC}} \geq \frac{1}{2}\log_2\left(\frac{2}{\mu^2\sigma_x^2\sigma_d^2}\right) + \frac{SNR_o(dB)}{6} \quad (47)$$

$$B_{\mathbf{WUD,SR}} \geq \frac{1}{2}\log_2\left(\frac{1}{\mu^2\sigma_x^2\sigma_d^2}\right) + \frac{SNR_o(dB)}{6}, \quad (48)$$

where SNR_o is the desired SNR at the output. Typically, we choose $SNR_o = SNR_{o,fl}$ because we assume that the quantization error at the output is small.

Comparing (47) and (48), we see that the precision requirements for **WUD**-block in the **SR** architecture are 0.5 bits less than that of the **CC** architecture. The precision requirements for **WUD** block of **PIPSR** architecture can be determined by replacing μ in the above analysis by μLA.

We illustrate the use for equations (44), (45), (47) and (48) to determine the precisions of the NEXT canceller in the 155.52 Mb/s ATM-LAN receiver shown in Fig. 6.7. For the floating point algorithm, we choose $SNR_{o,fl} = 36.1$ dB, which is achieved with $N = 32$ taps. With input power $\sigma_x^2 = 42$ (64-CAP signal constellation), we choose $\beta = 0.1$ so that the SNR_o of the fixed point algorithm is within 0.41 dB of that of the floating point algorithm.

The **F**-block precisions can be obtained by substituting the above mentioned values for J_{fl}, N, σ_x^2 and β into (44) for **CC** architecture and (45) for both **SR** and **PIPSR** architectures. In doing so, we obtain $B_{\mathbf{F,CC}} = 8.76$ bits, $B_{\mathbf{F,SR}} = 9.06$ bits and $B_{\mathbf{F,PIPSR}} = 9.06$ bits. These will be supported via simulations in Section 6.7.

The coefficient precision in the **WUD**-block can be determined by employing (47) for **CC**, (48) for **SR** and (48) with μ replaced by μLA for the **PIPSR** architectures. For proper convergence, μ was chosen to be 2^{-12}, 2^{-12} and 2^{-13} for **CC**, **SR** and **PIPSR** algorithms respectively. The precisions are then determined to be $B_{\mathbf{WUD,CC}} = 13.0$ bits, $B_{\mathbf{WUD,SR}} = 12.5$ bits and $B_{\mathbf{WUD,PIPSR}} = 13.5$ bits. These results are confirmed by simulation results in Section 6.7.3, where desired performance is obtained with 14 bit precision for the **SR** architecture.

In the next section, we employ the concepts presented in Sections 6.2-6.6 to design UTP channel data communications receivers.

6.7 DESIGN EXAMPLES

In this section, we provide the rationale behind choosing the system parameters for 51.84 Mb/s ATM-LAN, 155.52 Mb/s ATM-LAN and 51.84 Mb/s VDSL. Employing these system parameters, we then illustrate the design of data transceivers based on the low-power techniques presented in Sections 6.5 and 6.6. We will assume that the implementation technology permits the delay of a 1 bit adder to be $1ns$ under nominal load. The data precision is assumed to be in the range 8 – 10 bits.

6.7.1 Designing A 51.84 Mb/s 16-CAP ATM-LAN

In this subsection, we describe the design of a $51.84Mb/s$ ATM-LAN transceiver operating over unshielded twisted pair (UTP-3) wiring. In particular, the performance of the original receiver (see Fig. 6.6) and the low-power Hilbert transform-based architecture in Fig. 6.9 functioning as a receive requalizer in an

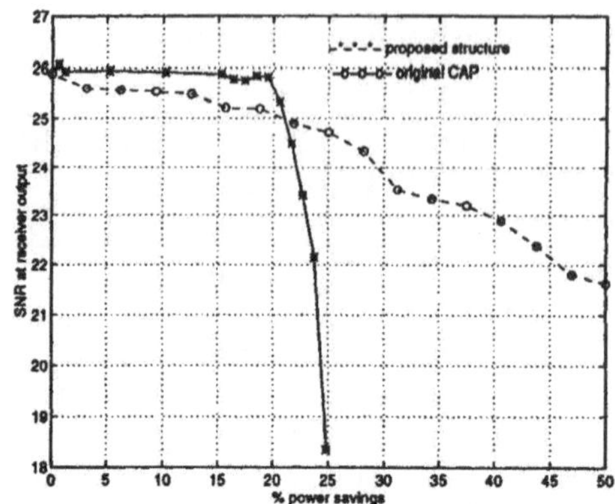

Figure 6.15 The excess MSE due to finite-length Hilbert CAP receiver.

ATM-LAN transceiver are compared. The system parameters for 51.84 Mb/s ATM-LAN are determined by propagation loss (see (10)), near end cross-talk (NEXT) (11) and the FCC restriction [24] on the transmit spectrum to less than 30 MHz. This results in the following parameters [5] which have also been derived in Section 6.4.1:

- Bits per symbol: $m = 4$
- Symbol rate: $\frac{1}{T} = 12.96 Mbaud$
- Excess bandwidth: $\alpha = 1(100\%)$
- Center freq.: $f_c = 12.96$ MHz
- Bandwidth utilization: 25.92 MHz
- Bit-error rate desired: 10^{-10}

Recall that a desired bit-error rate (BER) of 10^{-10} translates to a desired SNR_o of 23.25dB at the output of the equalizer. First, the original CAP receiver (see Fig. 6.6) was simulated with different tap lengths as shown in Fig. 6.15 and it was found that a filter span of $8T$ provided an SNR_o of 25.9dB. The transmit filter has a span of $4T$. This filter will now be employed as a benchmark against which the performance of the proposed structure will be compared.

Fig. 6.15 also shows the corresponding plot for the proposed structure with a receive filter span of $8T$. The following conclusions can be drawn from Fig. 6.15: 1.) for a comfortable noise margin of 3 dB, a desired SNR_o of more than 26.0dB is needed. In this case, the proposed structure offers higher power savings than the original structure even if the number of taps are reduced. 2.) If the desired BER is lower than 10^{-10}, the original structure would then need a higher number of taps, and in that case the proposed structure in Fig. 6.9 would lead to even higher power savings as compared to the original structure in Fig. 6.6.

We are now in a position to determine the parameters of the low-power receiver for 51.84 Mb/s ATM-LAN. From Fig. 6.10 and Fig. 6.15, we fix the length of the equalizer to be $8T$ and that of the Hilbert filter to be $M = 57$ so as to achieve a desired $SNR_o = 25.9$ dB. Next, we employ the analysis in Sections 6.6.4 to determine the precisions of the Hilbert filter ($B_H = 8$ bits), the **F**-block ($B_F = 9$ bits) and the **WUD** block ($B_{WUD} = 16$ bits). In order to achieve a sample rate of 51.84 MHz with these precisions and with a 1ns single-bit adder delay, it can be shown that the pipelined FSLE in Fig. 6.11 with $D_1 = 10$, $D_2 = 2$ and $LA = 1$ will suffice. These conclusions are supported in the convergence plot that is shown in Fig. 6.16 where the pipelined finite-precision receiver is shown to achieve a final

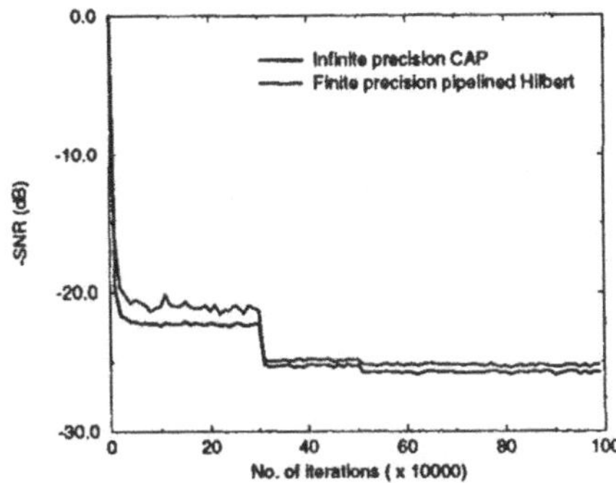

Figure 6.16 The convergence curves for floating and fixed point 51.84 Mb/s ATM-LAN CAP receiver.

$SNR_o = 25.08$ dB. This is about 0.7 dB worse than that of the infinite precision original CAP receiver. This degradation is due to the sum total of Hilbert and relaxed look-ahead transformations and the quantization error.

Similar simulation results and the corresponding laboratory experiments were the basis on which the ATM-LAN transceiver chip [20] was designed.

6.7.2 155.52Mb/s ATM-LAN

Given the data rate of $R = 155.52$ Mb/s and the FCC imposed limit of 30 MHz on the transmit spectrum, we choose a value of $m = 6$ so that from (14) we find the symbol rate $1/T$ to be 25.92 Mbaud. With $m = 6$ (or 64-CAP) the value of $SNR_{o,ref} = 29.75$ dB for a $BER = 10^{-10}$. Choosing $f_{upper} = 30$ MHz and $f_{lower} = 0$ Hz to maximize the excess bandwidth, we obtain from (18)-(19) a value of $\alpha = 0.15$. Clearly, the sample rate f_s needs to be greater than 60 MHz. We choose the next nearest multiple of the symbol rate or 77.76 MHz as the value of f_s. As the excess bandwidth is less than 100%, it implies (see [29] and the discussion in Section 6.4.1) that the equalizers will not be able to suppress NEXT completely. Therefore, a NEXT canceller is required. Hence, the rationale for the receiver architecture in Fig. 6.7.

It has been observed in [4] that the NEXT canceller span is weakly dependent upon the equalizer and that it should be about $1\mu s$ or greater. Given the symbol rate of 25.92 Mbaud, we find that the number of NEXT canceller taps $N_{nextc} \geq 26$. We chose a value of $N_{nextc} = 32$ for our simulations. The next step is to obtain the equalizer symbol span via simulations as shown in Fig. 6.17. We see from Fig. 6.17 that with an equalizer symbol span equal to 40, $SNR_o = 36.1$ dB is obtained. Note that this is also the value of $SNR_{o,fl}$ employed in Section 6.6.4 to determine the precisions. Employing the results of Section 6.6.4, we choose the **F**-block precision in the NEXT canceller $B_{\mathbf{F},nextc} = 10$ bits and the **WUD**-block precision in the NEXT canceller $B_{\mathbf{WUD},nextc} = 14$ bits. A similar analysis as in Section 6.6.4 can be applied to the equalizer. In doing so, we obtained the **F**-block precision $B_{\mathbf{F},eq} = 12$ bits and $B_{\mathbf{WUD},eq} = 18$ bits for the equalizer. With these

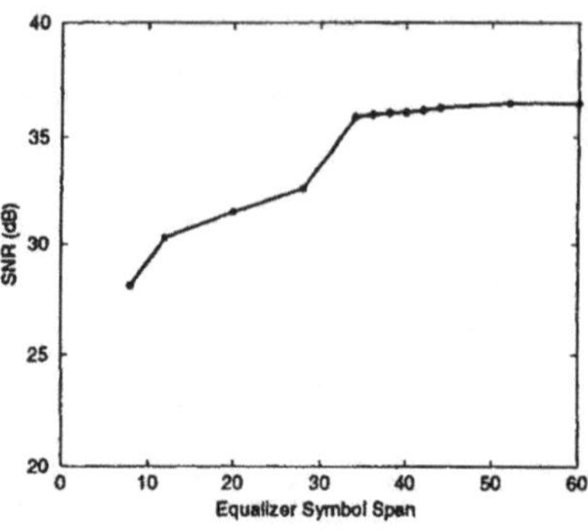

Figure 6.17 SNR_o vs. tap length for 155.52 Mb/s ATM-LAN CAP receiver.

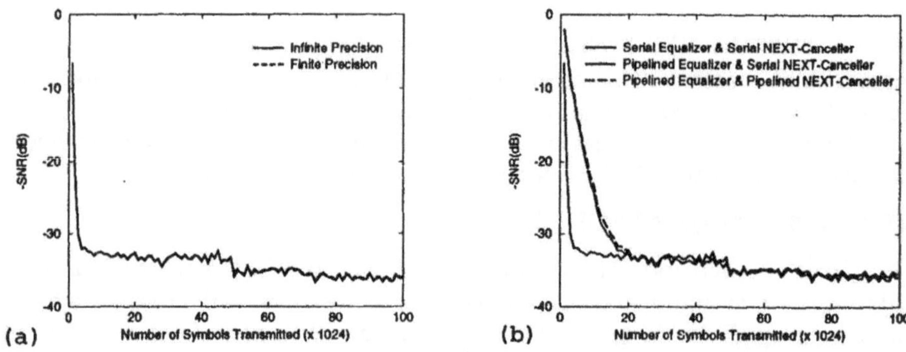

Figure 6.18 Convergence curves for the fixed and floating point receivers for 155.52 Mb/s ATM-LAN CAP receiver: (a) serial receiver and (b) pipelined receiver.

precision values, the finite precision algorithm achieved an $SNR_o = 35.9$ dB (see Fig. 6.18(a)), which is 0.2 dB less than the floating point SNR_o.

Given the tap lengths, the precisions (derived above) and the assumption of a 1 bit adder delay of 1 ns, it can be shown that pipelined architectures for the NEXT canceller and the equalizer becomes necessary. Hence, we employ the pipelined strength reduced adaptive filter architecture (see Section 6.6.2) for the NEXT canceller with $D_1 = 16$, $D_2 = 4$, and $LA = 1$ and the pipelined FSLE architecture (see Section 6.5.4) with $D_1 = 84$, $D_2 = 4$, and $LA = 1$. The simulation results for the finite-precision, pipelined 155.52 Mb/s ATM-LAN receiver is shown in Fig. 6.18(b), where we see that the final $SNR_o = 35.3$ dB has been achieved. This degradation of 0.8 dB (compared to the SNR_o of the floating point serial receiver) is due to pipelining and finite-precision effects. However, this final value of $SNR_o = 35.3$ dB provides a noise margin (see Eq.(23)) of 5.8 dB, which is quite sufficient for practical purposes.

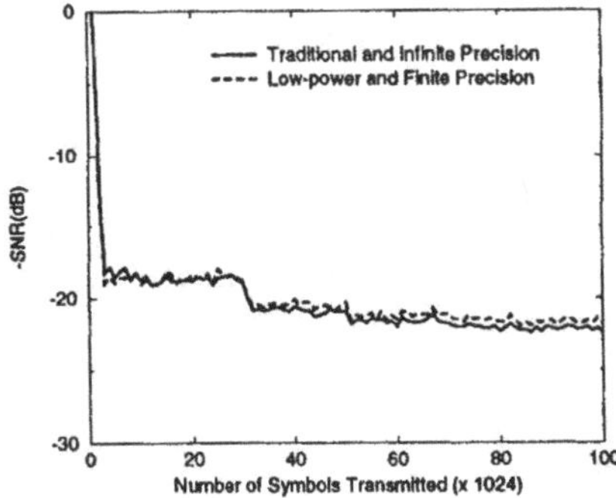

Figure 6.19 Convergence curves for the fixed and floating point receivers for 51.84 Mb/s VDSL CAP receiver.

6.7.3 Results of the 51.84 Mb/s VDSL

In this subsection, we present simulation results for the 51.84 Mb/s VDSL application. An analysis similar to the ones employed in Sections 6.7.1 and 6.7.2 can be employed to determine the parameters of the 51.84 Mb/s VDSL receiver shown in Fig. 6.8. A key point to remember here is that the lower edge of the transmit spectrum cannot start from 0 Hz because of the spectral nulls introduced by splitters below 5 MHz and due to the presence of telephony signals. For the VDSL receiver in Fig. 6.8, we can employ the Hilbert transformation based low-power equalizer (see Section 6.5) for the FSLE and the strength reduced low-power architecture (see Section 6.6) for the DFE. The channel consisted of a 1 Kft UTP distribution cable with 11 FEXT interferers (see Section 6.3.2) and the transmit spectrum had an excess bandwidth of $\alpha = 0.3$.

The final design had an equalizer span of 16 symbol periods and a DFE span of 8 symbol periods. Clearly, other combinations of equalizer and DFE spans are also possible. The precisions of the equalizer were: $B_{\mathbf{F},eq} = 10$ bits and $B_{\mathbf{WUD},eq} = 16$ bits. The low-power strength reduced DFE had precisions: $B_{\mathbf{F},dfe} = 9$ bits and $B_{\mathbf{WUD},dfe} = 13$ bits. A Hilbert filter with length $M = 65$ and coefficient precision of 9 bits was chosen. With these parameters, simulation results (see Fig. 6.19) indicate that the finite-precision VDSL receiver can achieve an $SNR_o = 21.8$ dB. This is 0.3 dB lower than the SNR_o of the floating point receiver model. Note that with $SNR_{o,ref} = 21.5$ dB (corresponding to a $BER = 10^{-7}$ for a 16-CAP transceiver), we have a noise margin of 0.3 dB. Clearly, it is harder to achieve the desired data rates with comfortable noise margins in a VDSL environment as compared to the ATM-LAN environment.

6.8 CONCLUSIONS

Achieving high-bit rates in the range of 50 Mb/s - 155 Mb/s over bandlimited channels such as the UTP-3 with a transmit spectrum limited to 30 MHz is indeed a challenging proposition and currently of great interest. This chapter has discussed various algorithmic low-power techniques for these applications via illustrative examples. An important conclusion that we draw here is the critical importance of developing communications algorithms for VLSI rather than just for the sake of BER performance.

Efforts are underway at various industrial labs to approach data rates far above 100 Mb/s given that the channel capacity for these environments is about three times the present data rates. Our approach to this problem is to jointly optimize power dissipation and algorithmic performance via the use of algorithm transformations. Future work is being directed towards a new class of algorithm transformation referred to as *dynamic algorithm transformations* (DAT) [34], whereby these transformation are applied in real-time to exploit non-stationarities in the signal and user environment.

Acknowledgement

The author would like to acknowledge the work of and the discussions with Keshab Parhi, Jit Kumar, Gi-Hong Im, J. J. Werner, V. B. Lawrence, Manish Goel and Rajmohana Hegde which has led to the development of the subject matter of this chapter. He would also like to thank Manish Goel for his help in reviewing preliminary versions of the manuscript. This work was supported via funds from National Science Foundation CAREER Award MIP-9623737.

REFERENCES

[1] P. S. Chow, J. C. Tu, and J. M. Cioffi, "Performance evaluation of a multichannel transceiver system for ADSL and VHDSL services," *IEEE J. Select. Areas in Commun.*, vol. 9, pp. 909-919, Aug. 1991.

[2] D. W. Lin, C.-T. Chen, and T. R. Hsing, "Video on phone lines: Technology and applications," *Proc. IEEE*, vol. 83(2), pp. 175-193, Feb. 1995.

[3] J. J. Werner, "The HDSL environment," *IEEE J. Select. Areas Comm.*, vol. 9, no. 6, pp. 785-800, August 1991.

[4] G. H. Im and J. J. Werner, "Bandwidth-efficient digital transmission up to 155 Mb/s over unshielded twisted-pair wiring," *IEEE J. Select. Areas Comm.*, vol. 13, no. 9, pp. 1643-1655, Dec. 1995.

[5] G. H. Im, D. D. Harman, G. Huang, A.V. Mandzik, M-H Nguyen, and J.J Werner., "51.84 Mb/s 16-CAP ATM LAN Standard," *IEEE J. Select. Areas Comm.*, vol. 13, no. 4, pp. 620-632, May 1995.

[6] D. D. Harman, G. Huang, G. H. Im, M.-H. Nguyen, J.-J. Werner, and M. K. Wong, "Local distribution for interactive multimedia TV," *IEEE Multimedia Magazine*, pp. 14-23, Fall, 1995.

[7] K. K. Parhi, "Algorithm transformation techniques for concurrent processors," *Proceedings of the IEEE*, vol. 77, pp. 1879-1895, Dec. 1989.

[8] H. H. Loomis and B. Sinha, "High speed recursive digital filter realization," *Circuit System and Signal Processing*, vol. 3, no. 3, pp. 267-294, 1984.

[9] K. K. Parhi and D. G. Messerschmitt, "Pipeline interleaving and parallelism in recursive digital filters - Part I,II," *IEEE Trans. on Acoust. Speech and Signal Proc.*, vol. 37, pp. 1099-1134, July 1989.

[10] N. R. Shanbhag and K. K. Parhi, *Pipelined Adaptive Digital Filters.* Kluwer Academic Publishers, 1994.

[11] C. Leiserson and J. Saxe, "Optimizing synchronous systems," *J. of VLSI and Computer Systems*, vol. 1, pp. 41-67, 1983.

[12] J. M. Rabaey, *Digital Integrated Circuits: A Design Perspective.* Prentice-Hall, New Jersey, 1996.

[13] R. Hegde and N. R. Shanbhag, "A low-power phase-splitting adaptive equalizer for high bit-rate communications systems," *Proc. of IEEE Workshop on Signal Processing Systems*, Nov. 1997, Leicester, U.K.

[14] A. Chandrakasan *et al.*, "Minimizing power using transformations," *IEEE Trans. on Comp.-Aided Design*, vol. 14, no. 1, pp. 12-31, Jan. 1995.

[15] D. A. Parker and K. K. Parhi, "Low-area/power parallel FIR digital filter implementations," *Journal of VLSI Signal Processing*, vol. 17, pp. 75-92, Sept. 1997.

[16] N. R. Shanbhag and M. Goel, "Low-power adaptive filter architectures and their application to 51.84 Mb/s ATM-LAN," *IEEE Trans. on Signal Processing*, vol. 45, no. 5, pp. 1276-1290, May 1997.

[17] A. Chandrakasan and R. W. Brodersen, "Minimizing power consumption in digital CMOS circuits," *Proceedings of the IEEE*, vol. 83, no. 4, pp. 498-523, April 1995.

[18] N. R. Shanbhag and K. K. Parhi, "Relaxed Look-ahead pipelined LMS adaptive filters and their application to ADPCM coder," *IEEE Trans. on Circuits and Systems*, vol. 40, pp. 753-766, Dec. 1993.

[19] G. Long, F. Ling, and J. G. Proakis, "The LMS algorithm with delayed coefficient adaptation," *IEEE Trans. Acoust., Speech, Signal Processing*, vol. 37, no. 9, pp. 1397-1405, Sept. 1989.

[20] N. R. Shanbhag, and G.-H. Im, "VLSI systems design of 51.84 Mb/s transceivers for ATM-LAN and broadband access," *IEEE Trans. on Signal Processing*, vol. 46, no. 5, pp. 1403-1416, May 1998.

[21] L. Goldberg, "Brains and bandwidth: Fiber service at copper price," *Electronic Design*, pp. 51-60, October 2, 1995.

[22] A. V. Oppenhiem and R. W. Schaffer, *Discrete-Time Signal Processing.* Prentice Hall, Englewoods Cliffs, New Jersey, 1989.

[23] *IEEE Journal on Sel. Areas in Commun.*, vol. SAC-13, no. 4, May 1995, Special Issue on ATM LAN.

[24] *Commercial Building Telecommunications Cabling Standard*, TIA/EIA-568-A Standard, 1994.

[25] *DAVIC 1.0 Specification Part 8*; Lower Layer Protocols And Physical Interfaces, Jan. 1996.

[26] *af phy-0018.000*, ATM Forum Technical Committee; Midrange Physical Layer Specification for Category 3 Unshielded Twisted Pair, September 1994.

[27] *af phy-0047.000*, ATM Forum Technical Committee; 155.52 Mb/s Physical Layer Specification for Category 3 Unshielded Twisted-Pair; November, 1995.

[28] R. D. Gitlin, J. F. Hayes and S. B. Weinstien, *Data Communications Principles*. Plenum Press: New York, 1992.

[29] B. R. Petersen and D. D. Falconer, "Minimum mean square equalization in cyclostationary and stationary interference: Analysis and subscriber line calculations," *IEEE Journal on Selected Areas in Communications*, vol. 9, no. 6, pp. 931-940, Aug. 1991.

[30] W. Y. Chen, G.-H. Im and J.-J. Werner, "Design of digital carrierless AM/PM transceivers," AT&T/Bellcore Contribution T1E1.4/92-149, Aug. 19, 1992.

[31] S-Y. Kung, "On supercomputing with systolic/wavefront array processors," Proceedings of the IEEE, vol. 72, pp. 867-884, July 1984.

[32] M. Goel and N. R. Shanbhag, "Finite precision analysis of the pipelined strength-reduced adaptive filter," *IEEE Trans. on Signal Processing*, vol. 46, no. 6, June 1998.

[33] C. Caraiscos and B. Liu, "A roundoff error analysis of the LMS adaptive algorithm," *IEEE Trans. on Acoust., Speech, and Signal Processing*, vol. ASSP-32, no. 1, pp. 34-41, Feb. 1984.

[34] M. Goel and N. R. Shanbhag, "Low-power reconfigurable signal processing via dynamic algorithm transformations (DAT)," *Proc. of ICASSP98*, Seattle, Washington.

Chapter 7

Cable Modems

Alan Gatherer
DSPS R & D Center
Texas Instruments
Dallas, Texas
gatherer@ti.com

7.1 INTRODUCTION

A cable modem is a device that sits on a coaxial cable network and transmits to, and receives from, the network modulated digital data. By coaxial cable network we mean a network that was originally designed for broadcast of analog television signals. In what follows, for clarity we discuss US cable systems only. The user should understand that there are several differences between US and, for example, European cable but the philosophy of cable modem design and most of the problems encountered are essentially the same. As the emphasis of this book is on digital signal processing we also concentrate on the physical layer of the system where the modulation and demodulation occurs.

A cable modem network consists of user modems and one or more headend modems. The headend modems connect the network to the outside world. The headend modems are owned and operated by the cable companies and can be significantly more expensive than the user modems. The standardization of cable modem systems has occurred in several places. At the time of writing standards are either complete or being developed in MCNS (Multimedia Cable Network Systems, a limited partnership of cable companies whose standard is called the Data Over Cable Interface Specification or DOCIS) [1], SCTE (Society of Cable Telecommunications Engineers, an ANSI accredited organization) [2], IEEE802.14 (part of the IEEE LAN/MAN network standardization effort) [3]-[4], ITU (International Telecommunications Union, an international multi-governmental standards organization) [5] and DAVIC (Digital Audio and Video Council, a consortium of manufacturers) [6]. Several vendors already have proprietary solutions. In this chapter our primary goal is to describe the solutions converged on by the above standard organizations. In order to do this in Section 7.2 we first briefly look at cable TV broadcast systems and how they have been designed to deliver analog video of acceptable quality. We show that not all of the choices made to improve the quality of analog TV transmission are helpful to digital data transmission. In Section 7.3 we give a general overview of a cable modem system and describe the media access layer that controls the physical layer operation. In Section 7.4 we outline the channel model for the physical layer transmission that was developed in IEEE802.14. In Section 7.5 and Section 7.6 we discuss the physical layer modulation and demodulation.

7.2 CABLE SYSTEM TOPOLOGIES FOR ANALOG VIDEO DISTRIBUTION

A cable system consists of a group of residences and a single headend connected to a cable network. The "classical" cable system topology is shown in Fig. 7.1. In this case a group of users are connected like the leaves of a tree with the headend modem as the root. The headend is the name given to the cable company office which may receive its NTSC video signals by land line or microwave. Such a network is designed for transmission of analog NTSC video from the headend to the users. We shall call this transmission direction downstream, with the transmission direction towards the headend being upstream. The tree is not necessarily symmetrical with some users much further away from the headend than others. Long trunk lines have amplifiers that boost the signal and also perform a "detilting" operation (which in digital communications literature would be called analog equalization). Detilting is the removal of the effect of the coax preceding the amplifier (a slow rolloff in the frequency domain). In some systems there will also be a module attached to the downstream amplifier that allows upstream transmission. In this case we draw the amplifier with a smaller reversed triangle inside it as in Fig. 7.1. Every so often the signal path splits so it can go to different parts of the served area. Once in the neighborhood the cable runs past the houses and is tapped off in groups of two, four etc. and goes to individual residences. Once inside the residence the cable may split several times to serve connectors in different rooms. Note that a splitter is generally defined as a device that sends half of the power in each of two directions whereas a tap will redirect a small amount of power for each of its residence outputs and send the majority of the power to the next tap.

One of the challenges in designing such a system is to get about the same power level to all residences. To do this there is a wide selection of taps available, each with a different number of output ports and a different attenuation from the input port to a residence output port, leaving a different percentage of the input power passed out of the main output port of the tap. It has been the author's experience that the attenuation occurring from the input to a residence output will be the same attenuation observed if one were to transmit into the residence output port and observe the input of the tap [7]. Therefore, due to the large variation in attenuations required to equalize the power at the residences, there will be a large variation in the attenuation of a signal transmitted from a residence upstream through a series of taps.

The quality of the analog signal is heavily dependent on reflections in the network. A reflection occurs if a cable is not properly terminated with its characteristic impedance, which for television cables is 75 ohms. Whether a reflection causes visible impairment depends on its amplitude relative to the original signal, and its delay. In analog TV the reflection can cause the "ghosting" effect on the image that those of us who have experienced bad cable and terrestrial reception are

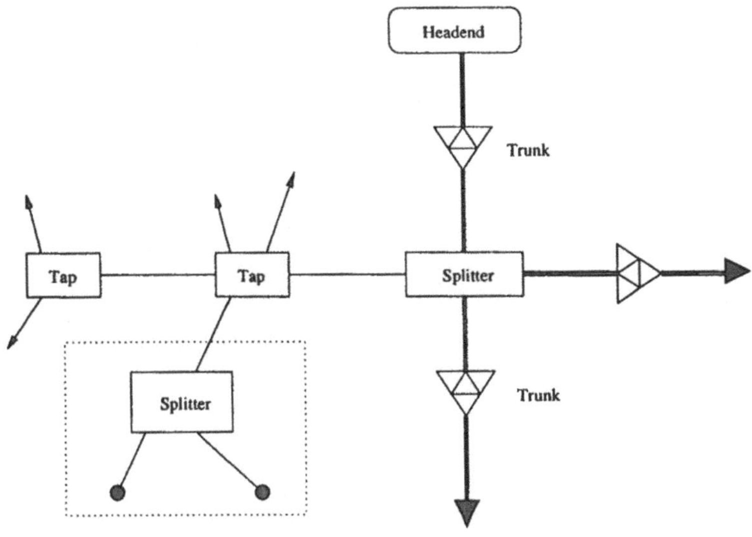

Figure 7.1 Classical cable topology.

familiar with. An image ghost is a faint, displaced version of the original image. In its simplest form a reflection can be represented by the impulse response

$$h(t) = \delta(t) + \alpha\delta(t - \Delta) \tag{1}$$

where the reflection is attenuated by α and delayed by Δ and $\delta(t)$ is the impulse function. Cable TV systems are designed with tolerances on reflection attenuation such that no ghosts are visible. This is a perceptual criteria and controlled studies [8] in the 60s produced tolerance plots similar to the one shown in Fig. 7.2. To a first approximation (as several reflections could occur in practice and we ignore this possibility) for delays less than 1 μs a cable TV system will perform acceptably (with no visible ghosts) provided

$$\alpha(dB) < -20\log_{10}(\Delta(ns)) + 25 \tag{2}$$

and for any given delay the worst case impulse response is

$$h_{\Delta}(t) = \delta(t) + \frac{10^{-7.75}}{\Delta}\delta(t - \Delta). \tag{3}$$

In the frequency domain the reflection impulse response is

$$H(f) = 1 + \alpha\exp^{-j2\pi\Delta f}. \tag{4}$$

Hence a reflection looks like a ripple in the frequency domain with period Hz and amplitude α riding on unity. We use equations (3) and (4) in Section 7.4 when we assess the effect of digital transmissions on an analog video system.

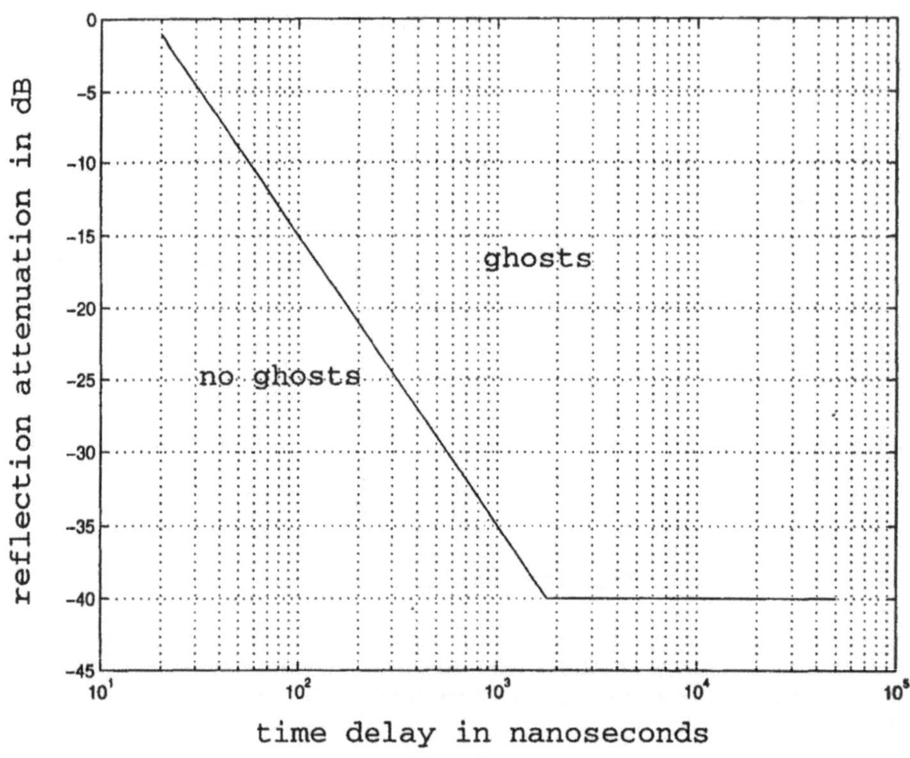

Figure 7.2 Perceptual effect of reflections.

The fraction of a signal reflecting off a bad termination depends on the mismatch between the terminating impedance and the characteristic impedance of the line. To calculate the amount of reflection we note that ideally the coaxial cable used in video transmission is a transmission line with characteristic impedance of $Z_o = 75\Omega$. Therefore from basic transmission line theory [9] there is a reflection coefficient

$$\rho = \frac{Z_L - Z_o}{Z_L + Z_o} \tag{5}$$

equal to the ratio of the voltage wave amplitude incident on the load, Z_L, and the voltage wave amplitude reflected from the load. The situation is shown graphically in Fig. 7.3.

The magnitude of the reflection coefficient plotted as a function of Z_L is shown in Fig. 7.4.

The shape of the impulse response is also affected by the response of the coaxial cable itself. If this response was allowed to get out of hand it would cause the video signal to be low pass filtered, distorting the picture. Because of this the trunk amplifiers have analog equalizer modules that reverse the effect of the coaxial cable on the signal. Because the cable rolls off gently in frequency this does not lead to significant noise enhancement. There will be a trunk amplifier every 1000-2000

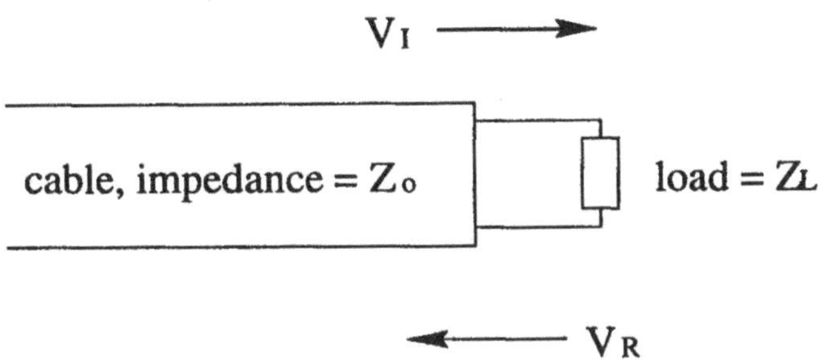

Figure 7.3 Cable incident, V_I, and reflected, V_R, waves.

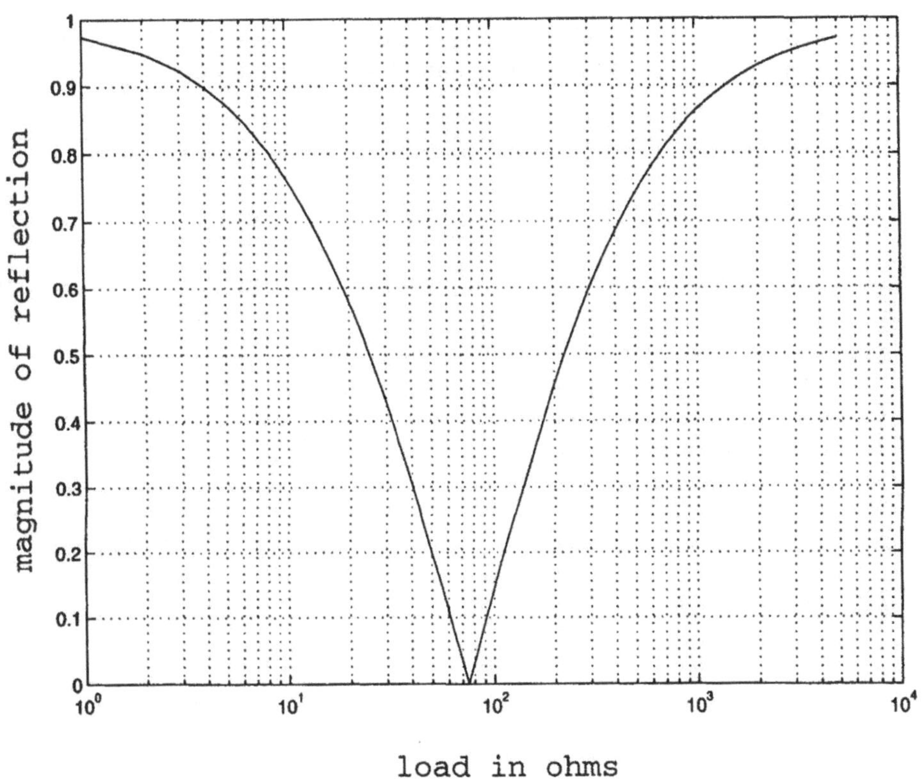

Figure 7.4 Reflection coefficient as a function of load.

feet [10] and for the sort of coax cable used in the trunk section of the system this causes a rolloff towards higher frequencies of a few tenths of a dB every 6MHz. This will be essentially canceled by the analog equalizer. Therefore the only significant rolloff is due to the last few hundred feet after the coax enters the neighborhood. This last section of cable is generally called drop cable and is a different type from the trunk cable. Typical losses for drop cable can be found in [11]-[12]. In Table 7.1 we tabulate loss for two different drop cable types. In Fig. 7.5 the loss is plotted

Table 7.1 Cable Loss per 1000 Feet as a Function of Frequency

Frequency (MHz)	Cable Type RG-6	Cable Type RG-59
55	1.60	2.06
300	3.70	4.72
450	4.58	5.83
550	5.09	6.47

against frequency. There is approximately 0.5dB drop across a 6MHz channel. The coaxial cable itself is therefore not a significant source of distortion.

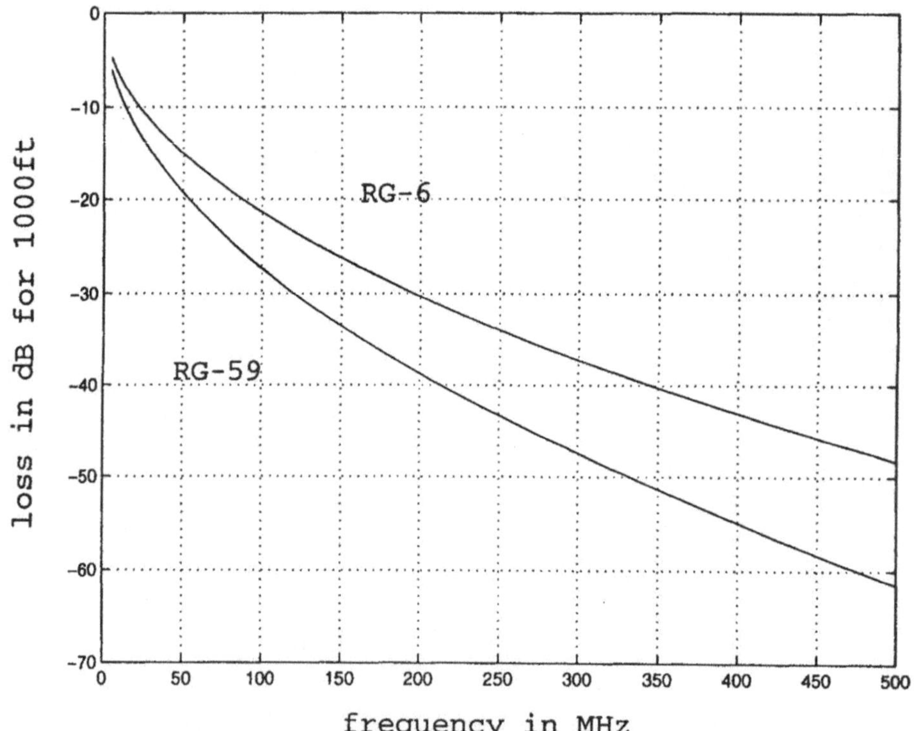

Figure 7.5 Loss versus frequency for RG-6 and RG-59 drop cable.

Referring to Fig. 7.5, a channel sitting at about 50MHz will see a 15dB loss over 1000 feet of RG-6 drop cable. Therefore over 100 feet it will experience 1.5dB loss. For a reflection returning from a bad connection 50 feet away (and therefore travelling 100 feet or about 31m) the delay is about 155ns. Now referring to Fig. 7.2 at 155ns about 20dB of attenuation is required. The amount of signal reflected from the bad connection must be at least 18.5dB down from the incoming signal for there to be no visible ghosts. Therefore, to avoid ghosts the reflection coefficient should be less than 0.12. This corresponds to a load between 60 and 90 ohms. This may be difficult to reliably achieve and the situation becomes a lot worse when lower loss cabling is used [8]. Therefore the cable industry uses directional couplers [8]. Without going into details here, a directional coupler suppresses the return path signal so that reflections are suppressed. This is one reason the taps mentioned earlier suppress so much of the return signal. This is a good idea in a purely broadcast environment but can cause difficulties in a two way environment as we shall see in Section 7.4.

Finally it is important to mention that recently many cable operators have been upgrading their systems to Hybrid Fiber Coax (HFC) where part of the trunk network is replaced by a fiber optic link. These systems are easier to maintain, reduce the need for amplifiers [10] and have a larger bandwidth over the trunk network allowing a single headend to serve a larger area. We will examine the effect of HFC on cable modems in Section 7.4 but for now we note that HFC was originally introduced to improve analog broadcast and not to allow improved digital transmission. The signal is therefore transmitted along the fiber in analog form using a high quality analog laser.

7.3 AN OVERVIEW OF A CABLE MODEM SYSTEM

In this section we concentrate on the system layer issues for cable modems, giving a brief overview of the media access layer (MAC) as it relates to the physical layer (PHY). In all proposed systems known to the author this a strict master/slave relationship with the headend controlling all the traffic on the network. This is done for several reasons. Firstly there is generally only one headend and it is guaranteed to be running all the time, whereas the user modems may be turned off and on, entering and leaving the network at random times. More importantly, the headend hears all the upstream and downstream traffic but each user is only aware of the downstream traffic and its own upstream. The isolation between users is a result of the isolation required to prevent ghosts on the analog signals, as described in Section 7.2. Therefore in the upstream direction we cannot use techniques such as carrier sense multiple access (CSMA) [13] which requires a user to sense the transmission of another user.

In the downstream direction the cable companies wanted to use the same PHY layer as had been developed for digital video transmission. For instance the downstream PHY layer developed by DAVIC was originally designed for transmission of MPEG2 packets and has been modified for transmission of ATM [6]. So in compliance with the requirements for downstream video delivery, each downstream channel will carry up to about 30Mb/s of data (see Section 7.5 for details). Many users share the data on this channel in a time division multiple access (TDMA) manner, selecting the packets addressed to them from the data stream.

The cable companies also want the option of replacing analog channels over time with digital video channels and possibly cable modem channels. To this end each digital video and therefore cable modem downstream channel is 6MHz in width just like an NTSC signal. The downstream digital channel allocation is the same as presently used for NTSC allocation. The spectrum allocation for a subsplit system is shown in Fig. 7.6. A subsplit spectrum allocation has upstream channels at 5-42MHz, downstream above 550MHz and the regular analog TV channels from 45 to 550MHz as usual. There are other possible spectrum allocation schemes that are presently not as common. These include mid-split (upstream extended to 108MHz) and high-split (upstream extended to 174MHz). These other spectrum allocations may become more popular as cable modem use increases and more upstream bandwidth is required at the expense of analog TV channels. The downstream bandwidth extends up to 750MHz and may reach 1GHz as the amount of fiber in the system increases.

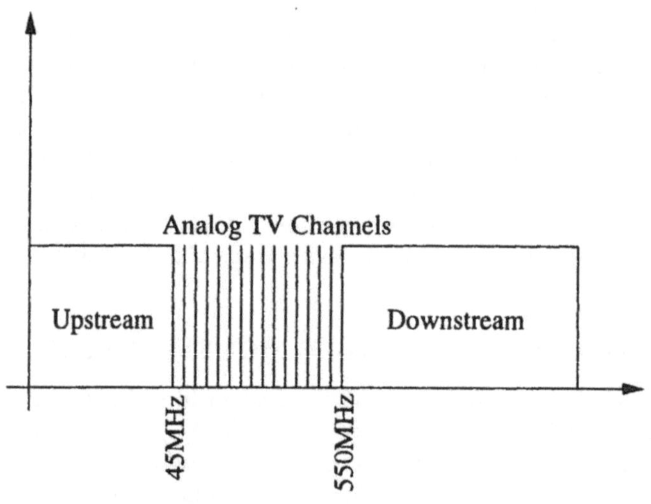

Figure 7.6 Subsplit spectrum allocation.

For upstream transmission we can use any protocol which does not require individual users to monitor each other's calls. One possibility is the well known packet radio access protocol ALOHA [13]. In this technique a user sends data whenever it has data to send. This transmission may collide on the channel with other users' data and not be received. So if a user sends data and does not receive an acknowledgment after a defined amount of time it will wait another random amount of time and re-send. This is a simple protocol but does not support a high throughput. Because upstream bandwidth is a precious resource a more efficient protocol was required. Though there is still a lot of discussion on the details of such a protocol, most developers have agreed on the use of a reservation based protocol for cable modems [13]. This is essentially the approach taken by IEEE802.14 and MCNS. In a reservation protocol, a user requiring upstream bandwidth will request it from the headend who controls all access to the system. The reservation protocol is therefore centralized at the headend. One can think of the headend as having a master clock that wraps back to zero every so often. Once the headend has received the user's request, the user will be given a time at which to start its transmission and a length of time for which it can transmit. Provided all users are synchronized with the master clock the user will be the only one transmitting in that time slot. If a user is presently transmitting then a request for more bandwidth can be attached to the packets it is presently transmitting. This is called "piggybacking". To allow users to request bandwidth if they are not presently transmitting, there must be a way for a user to get the headend's attention. This implies that contention is still required. The headend sets aside a portion of the upstream packets for contention transmission. The optimum ratio of data packets to contention request packets is data type dependent and may have to vary with time to maximize the data throughput of the system. For instance, if there is a lot of large file transfer then most users will be able to piggyback further requests and few contention packets will be required. If there is a lot of interactive gaming occurring on the network then there will be sporadic transmission of keystrokes from the users and a lot of contention requests will have to occur.

The issue of synchronizing the users deserves a detailed description as it can affect the efficiency of the network significantly. The approach followed in IEEE802.14 is to have a master clock that is distributed downstream using a series of time stamp messages [14]. How the users lock to the clock depends on their position in the network. This can be seen by referring to Fig. 7.7. Assuming the headend sends regular timestamps that are used to lock the user clock to the headend clock in frequency then there will still be a phase offset between the clocks due to the propagation delay between the headend and this particular user. It is very important to note that the signals sent by the users combine at the headend so they must be synchronized at the headend. Therefore it is most convenient for time to be reference to the headend. A user does this by adding a phase advance to its clock equal to the propagation delay, T. Then, as shown in Fig. 7.7, if the user sends a packet at time a relative to the user clock it will reach the headend at time a relative to the headend clock. When a user modem switches on it will not know the propagation delay. To determine the propagation delay the user sends a packet at time a by its clock and the headend notes the time of reception of this packet by the headend clock. If the headend receives the packet at time $a + \delta$ then it will tell the user to adjust its clock by $-\delta$ so the next packet will be received at the correct time.

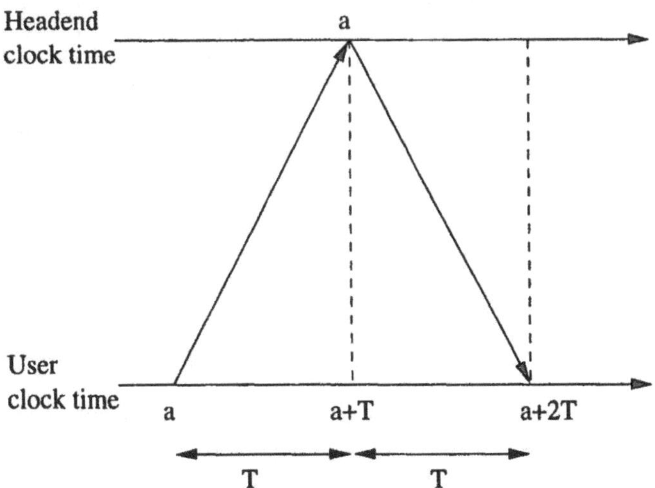

Figure 7.7 Relative timing of headend and user.

Note that the timestamp method is similar to, but a little more flexible than, the more conventional method of assigning time slots where each user is given a slot number in which it can start transmission. This is essentially a coarser grain time assignment than the method described above.

The clock generated as described in the last paragraph is primarily designed to time the start and end of a users transmission. There will be a guard band between user transmission so that there can be a small amount of difference in the clocks of two users without their signals overlapping at the headend. We discuss this in more detail in Section 7.6. The upstream transmitter and receiver also have to be synchronized in sample and symbol rate. The most obvious way to do this is to treat the upstream transmit/receive pair independently of the original system. The transmitter sends data out at a nominal symbol rate T_u. The receiver symbol clock is physically separate from the transmitter clock and therefore is only nominally at the same frequency. A timing recovery algorithm of some sort can be employed in the receiver to determine T_u and adjust the receiver symbol clock to align with this. This method is displayed graphically as method (a) in Fig. 7.8. The downstream symbol period T_d and the upstream symbol period T_u can be unrelated. In method (b) the user receiver recovers the downstream timing, T_d, and then uses it to set the upstream symbol rate, T_u. This is called loop timing. The user modem cannot estimate the downstream symbol rate exactly but this will nevertheless bring the upstream symbol rate very close to a function of the downstream symbol rate known in the headend. Frequency locking in the headend receiver is therefore very simple though this will not help phase locking due to unknown system latency. Because data is transmitted in bursts with phase synchronization between bursts, if the estimate of the frequency is close enough there may be no need for frequency locking at the receiver. The disadvantage of this method is that in a realistic system T_d and T_u will be related by a simple multiple and this constrains the data rate options. A third method proposed in the IEEE802.14 standard body is to use the timing synchronization obtained from the downstream timestamps mentioned earlier. This means the time stamps method must be capable of the increased accuracy required to specify the symbol rate as well as the start of transmission.

The accuracy required for the start of transmission is generally significantly less because a guard band can be added allowing the clock to be in error by more than a symbol period. For it to be of any practical use the timestamp accuracy must also be better than the worst case frequency difference found incommercially available crystal oscillators; about 100ppm. If the timestamp method is capable of enough accuracy then method (c) can be used. In this case a third time period, T_s, is transported in the downstream by the use of timestamps and is used to set the upstream period. The downstream symbol period does not now have to be simply related to the upstream symbol period. In [14] a simulation study implied that timestamps could lock an 8MHz clock frequency to about 25ppm. Whether this is good enough depends on the symbol period in the upstream and is discussed in Section 7.6. Note that even if the timestamps cannot supply enough accuracy to set the clock all by themselves, they can at least narrow the gap between T_u and the nominal value of T_u assumed in the headend, hence making the job of the headend timing recovery easier.

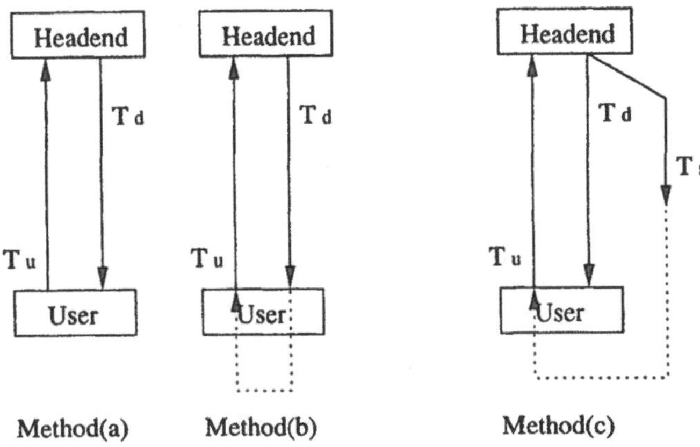

Figure 7.8 Symbol timing methods.

The length of cable separating a user and a headend can vary widely from user to user and the number of splitters and taps may also vary. Also, as described in Section 7.2, the attenuation of a signal passing upstream through a tap depends on the type of tap. Therefore the power level required at the user to achieve a power level within a given range at the headend may vary significantly. When the user first powers up it will have to test various power levels. The method used in IEEE802.14 [15] involves a slow ramp of the signal power until the receiver hears the transmitter. The receiver then responds with the desired transmit level. In this way the transmitter never exceeds the allowed transmit power at the receiver input.

When a user powers on, they must obtain an estimate of the propagation delay and its power level before one can start using regular data packets. This is because incorrect power will cause it to either not be heard or exceed the power limits at the receiver. If the transmitter has an incorrect estimate of the propagation delay it will send its data either too early or too late. In either case its signal will overlap in time with adjacent signals from other users. The headend therefore assigns periods of time in which only users logging onto the system can transmit. In this period enough space is left so that a user with a worst case propagation delay can send an initial packet without overlapping with adjacent signals (see Fig. 7.9). The worst case propagation delay will depend on the particular cable system so it makes sense for the log on period to be set by the headend for that particular system.

To summarize the last few paragraphs we show a simplified version of the most recent version of the IEEE802.14 [15] log on timing diagram [4] in Fig. 7.10. The diagram is divided into MAC and PHY for the headend and user. The PHY is considered to be the owner of the system clock because the delay through the PHY layer is constant and therefore the time from data input to the user PHY and data output at the headend PHY is a constant offset of the true propagation delay. Once in the MAC the data can be delayed by a stochastic amount dependent on the system loading and other factors. Hence in the downstream direction the headend PHY layer sets the timestamps with the correct time and the user PHY layer notes the arrival time of timestamps and tags them for use by the MAC. The

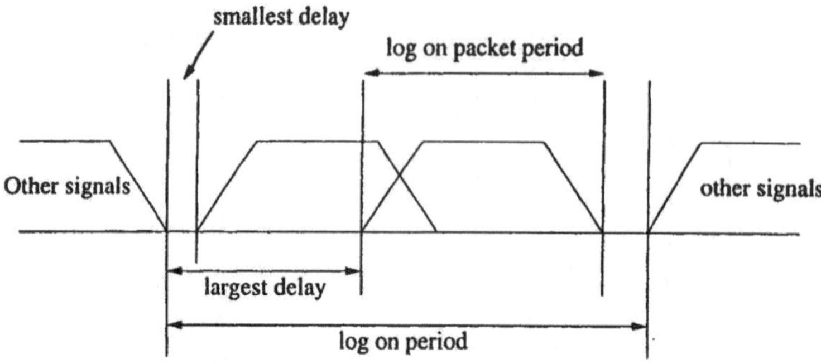

Figure 7.9 Log on slot assigned by the headend.

MAC calculates the correct frequency for the user clock and sends this information back to the PHY. Now the user is locked in frequency. At each notification of a log on period from the headend the user sends a log on packet with information about itself. The user MAC tells the user PHY what power to use for this operation and the power is gradually ramped up until the headend MAC sends an acknowledgment along with accurate power and delay adjustment information. There may be several iterations of modification to the power and delay and then the headend sends a connect packet and the user is logged on. The user is now ready to request access to the data stream proper. During the power up phase a user may achieve a suitable power level but may not successfully communicate with the headend because there is another user trying to log on at the same time. In this case some sort of back off mechanism is required to prevent these two users from blocking each other at every log on period. The back off mechanism will ensure that in the event of no acknowledgment the user will not try to transmit at the very next log on period every time but instead will choose a future log on period at random.

To wrap up the overview of cable modem systems we briefly describe the change over to Hybrid Fiber Coax (HFC) systems. For a cable modem this change has two significant effects. Firstly, the use of fiber in both the downstream and upstream trunk improves the SNR but makes clipping noise the dominant error source in the trunk. This is because the signal is not transmitted digitally on the fiber (because there are still analog channels to transmit) but is transmitted using high quality analog lasers. The analog laser is only linear over a certain range. If the cable system designer were to reduce the signal power to a point that operation within the linear region was guaranteed, the SNR would degrade unacceptably. Instead the signal power is set to the maximum that still produces acceptable performance. Presently acceptable performance is judged in terms of picture quality and the effect of this choice on digital channels is still being assessed [16]-[18].

This is discussed further in Section 7.4. Secondly, the use of fiber in the upstream has increased the capacity of the trunk and therefore reduced the number of trunks required. For this to happen the trunk must be shared over several 5-40MHz bands coming from different coax cable nodes. This situation is shown in Fig. 7.11 [16]. Each coax node delivers a 5-40MHz signal band to the frequency stacker. The frequency stacker then modulates the bands so that they do not over-

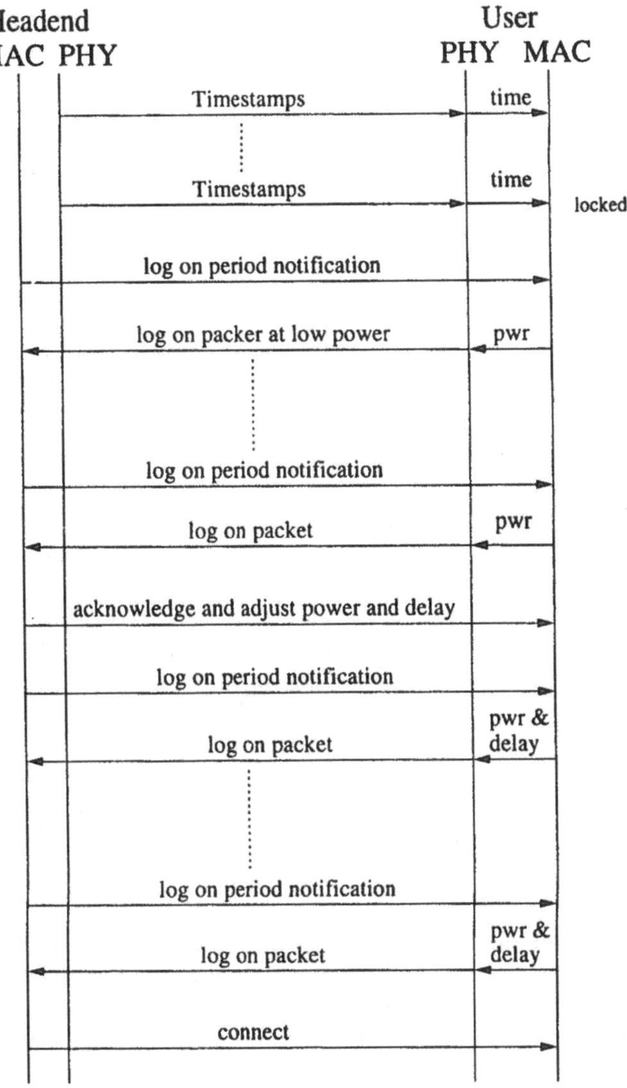

Figure 7.10 Log on timing diagram.

lap in frequency on the trunk fiber. At the headend the signals are separated and demodulated. The frequency stacker adds two new problems to the signal. Firstly the modulation frequency at the stacker and the demodulation frequency at the headend are only nominally the same. In practice the frequency stacking adds a significant amount to the frequency offset in a channel making carrier recovery significantly more difficult. There are techniques by which these frequencies could be locked but they are not generally used as they would cost extra money and do not impact the quality of analog video. Secondly, the frequency stacker modulation waveform may not use the same rule for sine/cosine relationship. Essentially the frequency stacker may use modulation waveforms $cos(\omega t)$ and $sin(\omega t)$ but the headend may assume the frequency stacker used waveforms $cos(\omega t)$ and $-sin(\omega t)$. The effect is to invert the imaginary part of the spectrum and this is called ***spectral inversion***. This is not a rotation and will not be solved by standard differential encoding. Presently spectral inversion occurs only in the upstream but as cable modems increase in popularity and the downstream bandwidth becomes a limitation, cable companies might want to frequency stack the downstream too [19]. In this case spectral inversion will become a downstream problem unless the cable companies are willing to specify their frequency stackers differently.

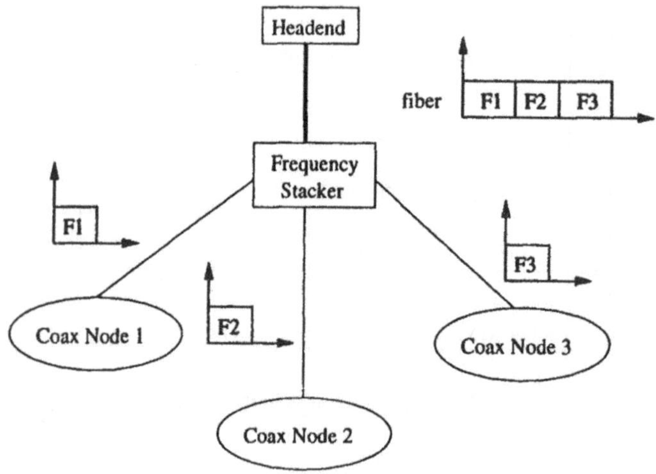

Figure 7.11 HFC system with frequency stacking.

7.4 CHANNEL MODEL

The downstream and the upstream suffer from significantly different impairments and their channel models are defined separately. The development of such channel models has been hampered by a lack of suitable information. There are several reasons for this. Firstly, there is a wide variety of cable network systems in place with widely differing topologies and performance. This is clear from a study made by CableLabs in 1994 [20] [21] in which they studied the noise and microreflection characteristics of 20 different cable systems and transmitted QPSK signals over 5 different systems to study changes in BER. It has not generally been possible to define a small set of canonical impulse responses as is done in the ethernet PHY and ADSL standards. The cable modem channel must instead be characterized statistically. Secondly, it has been difficult to characterize the noise impairments in cable modem systems despite creative attempts to do so [22].

In the following subsections we essentially describe the channel models developed in the IEEE802.14 [15] committee along with some side references to other material.

7.4.1 Downstream Channel Model

The downstream model described here is based on the model developed in IEEE802.14 [15]. Fig. 7.12 is the downstream channel block diagram. The first distortion the signal encounters is the clipping in the fiber. The fiber also adds some Gaussian noise but this will be lumped in with the thermal noise term. There is a considerable amount of literature on the problem of modeling and analyzing the effect of laser clipping [16] [18]. Our description of clipping is based on the results in reference [17]. In order to maximize the SNR the laser will be run at a power level at which clipping will occasionally occur. Therefore in a well designed system clipping will be relatively rare. When it does occur its duration will be much less than the symbol period of the downstream QAM signal. Therefore, to a demodulated QAM signal clipping appears to be impulse noise. For low probability of clipping, the impulses are distributed in time with a Poisson arrival time. Therefore the model for the laser clipping noise is

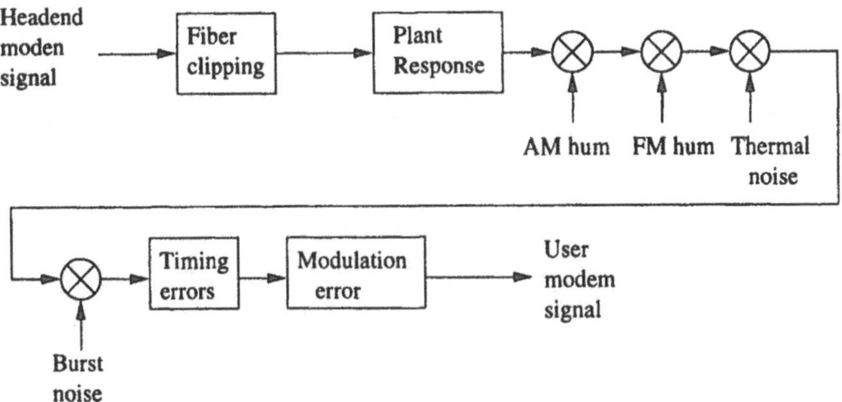

Figure 7.12 Downstream channel block diagram.

$$n_{clip}(t) = \sum_k A_k \delta(t - \tau_k) \tag{6}$$

with τ_k the time of the kth impulse, A_k is the amplitude of the kth impulse and $\delta(t)$ is the delta function. The probability of the amplitude distribution is dependent on the type of fiber and laser used and is consequently difficult to define in a general sense. One easy and conservative assumption is that each impulse wipes out the sample it occurs closest to. We can model this by replacing the sample by white noise spread uniformly over the amplitude range of the sample. The average rate of arrival of the impulses has been observed to be about 360Hz [23].

The coaxial cable adds an impulse response to the channel model and this is the second box in Fig. 7.12. The model used in IEEE802.14 is a "tilt and ripple" specification. This specification was developed from the experience of Tom Koltze, who was at that time a GI engineer working on system level issues for cable modems. He divided the effect of the coax into transmission line response, which causes the frequency response to tilt, and microreflections, which cause ripple in the frequency domain. The tilt is set at 2dB drop per 6MHz channel. From the description of coaxial cable characteristics in Section 7.2 this is a good upper bound when the effect of RF receivers is also taken into account. The microreflections are defined by first picking a group of up to 20 microreflection delays between 0 and $5\mu s$.

The power of each microreflection is obtained by dividing the microreflections into 200ns groups, allocating a certain energy to that group and sharing that energy equally among all microreflections in the group. For the group between $200Kns$ and $200(K+1)ns$ the energy is

$$\alpha(dB) = -11.926 - 2.538K. \tag{7}$$

In Fig. 7.13 we compare this observed microreflection attenuation with the worst case (for no visible ghosts) attenuation given by (2) where for $\Delta = 200Kns$ the worst case becomes

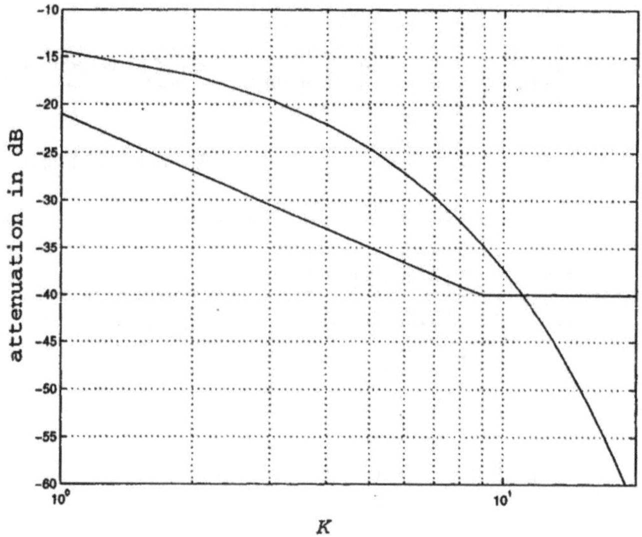

Figure 7.13 Observed attenuation vs. perceptual worst case.

$$\alpha(dB) = -21 - 20\log_1 0(K). \tag{8}$$

The observed reflection attenuation is up to 10dB less than the worst case attenuation for no perceptible ghosts. Therefore the model used in IEEE802.14 for microreflections also applies to a large class of cable systems in which users experience perceptible ghosts.

Referring to equation (7) we have a bound on the depth of the ripple in the frequency domain for a given ripple period. First note that the worst case ripple depth is for $K = 1$ where the attenuation can be $\alpha = 0.2533$. The spectrum will therefore vary in magnitude from 1.2533 to 0.7467 which means the signal strength can vary by about 4.5dB. This reflection will have ripple period less than 5MHz so for any given 6MHz channel it will look like either a constant amplitude change or a single frequency selective fade. Greater reflection delays may produce more fades over a 6MHz band but the fades will be significantly shallower. Therefore we may consider a single 4.5dB fade to be the worst case channel condition and this is not a particularly challenging channel to equalize when compared to the unshielded twisted pair channels used in telephony digital subscriber loops.

There will be slow changes in the channel due to changes in temperature and perhaps humidity. In some cable systems which experience significant temperature changes (such as in Arizona), the trunk amplifiers have thermal compensators for their gain [24]. However, the only sudden change in channel impulse response is due to "channel surfing" where a neighboring modem moves in and out of the channel you are using. If a neighboring user moves into or out of your channel, the impedance of that user seen by your channel will change and hence the microreflection from that user will either appear or disappear. The effect of this on a decision feedback equalizer was studied in [25] but for the purpose of the channel model we use a simple gain change of $\pm 1dB$ occurring over 100 to 200ns. This is a good approximation because the worst microreflection changes occur close to the user and therefore ripple slowly in the frequency domain. At a first approximation over a single 6MHz channel, the slow ripple will appear to be a gain change (this would be called a flat fade in wireless communication terminology). In [26] experimental results for changing from channel 13 to channel 14 show this slow ripple over 750MHz.

As a final note on microreflections, there is now interest in RF receivers that are designed to have 75Ω input impedance across a large band of frequencies no matter what channel they are tuned to. Such devices are starting to appear on the market [27] and may make the microreflection problem less common.

AM hum and FM hum are both due to ingress of the power line into the signal. The power used to run the amplifiers in cable systems is sent along the wire with the signal. This power is supplied using a periodic waveform, usually at about 60Hz. AM hum is an amplitude modulation of the signal by a function of the power waveform. Hence the signal is multiplied by $1 + dm(t)$ where $m(t)$ is a unit amplitude, zero mean periodic waveform and d is called the depth of modulation and is usually significantly less than one. The power supply waveform is not a sinusoid as is used by power utilities but is a quasi-square wave [28]. AM hum occurs when power subtractive effects in the cabling cause the available power to dip at the peak of the power supply voltage. The dip in the power supply voltage below the required operating voltage of the amplifier causes the amplifier to lose gain and hence there is a 60Hz gain variation on the signal [28]. Because the power variation depends on the cabling, the power supply variations, and the make of amplifier determine , it is difficult to define the AM hum modulating waveform in general. The shape of the waveform is at least as important as its depth of modulation because it defines the final maximum rate of change of the signal amplitude which in turn strongly affects the choice of auto gain algorithm and/or equalizer algorithm used to track this change. For instance, if the AM hum modulating waveform was a perfect square wave then it would be impossible to track the instantaneous amplitude change, a burst of errors would be inevitable, and a demodulator would be designed to detect the amplitude change as quickly as possible. If the amplitude changes gradually the demodulator would be designed to track the slow amplitude variation and would not need to cope with a burst of errors every 1/120Hz. In order to provide some sort of model for AM hum we note that from the sheath current study in [28] the depth of modulation appears to be less than 5%. The AM hum effect is on the order of 60Hz and we are dealing with a signal with symbol rate of 5MHz so we do not expect the rate of change of the modulating waveform to be large compared with the symbol rate. A peak to peak slew over $100\mu s$ is probably representative. FM hum is even less well defined. It is again a power supply effect except that this time it is due to the power supply in the cable modem itself. The center frequency of the modulated signal varies at 120Hz with a maximum frequency deviation of up to several kHz [29].

There are several sources of additive noise in downstream cable modem. The laser adds additive noise and there is thermal noise on the coaxial cable. Other channels will also be mirrored into the user channel due to nonlinearities in the RF generating harmonics of the signal. In the cable industry the nonlinearity effects are modeled as composite second order and composite triple beat and are not considered to be an important effect as long as the QAM signal is 8 to 10dB below the analog carriers [30]. To simplify the model, all these additive noise sources are lumped together and modeled as a Gaussian noise source 40dB down from the received QAM signal power.

Burst noise is, as would be expected from its name, a randomly occurring, finite duration burst of noise that tends to wipe out the signal for a short period of time. As the clipping noise is impulsive the burst noise is the only source of noise that requires an interleaver to correct (see Section 7.5). Its source is somewhat mysterious possibly being due to leakage from the upstream RF into the downstream RF in the user's modem. Proof of its existence however is the presence of interleavers that have been added to systems after they were tested. All of the proposed systems can correct bursts of up to $25\mu s$. Therefore we can only assume that bursts of length up to about $25\mu s$ exist and longer bursts are possible. As far as channel modeling is concerned one possible test would be to randomly replace a section of demodulator input data by random noise for $25\mu s$. This test determines if the adaptive systems within the demodulator function correctly after a burst of this length.

Timing errors are not really a channel model problem as, like FM hum, they are dependent on the modems only. However from a digital point of view they become part of the channel model as the digital section of the modem is affected by the sampling rate. Usually, off the shelf crystals are used to generate the same sampling frequency in the transmitter and receiver and crystals that are guaranteed to be within 100 parts per million (ppm) of each other are readily available. So a timing error randomly chosen within 100ppm is suitable.

Modulation error, in the downstream at least, also occurs due to differences in the transmitter and receiver RF sections so is not really a channel model parameter. However a standard must specify the stability of the transmitter modulation frequency and in the DAVIC standard [6] this is specified at $\pm 200ppm$ at the highest modulation frequency. If we take 750MHz as the highest frequency we get a frequency error of 15kHz. Assuming there is the same accuracy in the receiver RF section then the digital portion of the demodulator will observe a frequency error of up to 30kHz.

The 256 QAM constellation is the largest constellation and therefore the most sensitive to phase rotation. The most sensitive point to rotation of the constellation is the (15,15) point which can rotate by 3.7° before an error is made. For at 30kHz frequency error on a 5MHz symbol rate the constellation will rotate by 30/5000 of a full revolution or 2.16° every symbol. Hence the frequency error is quite large and good carrier frequency tracking is required.

7.4.2 Upstream Channel Model

The upstream channel is dominated by RF interference ingressing onto the cable from external sources. The 5 to 40MHz band is a noisy environment which has been called "the electronic sewer" by some in the cable business. So why transmit in this region at all? There are some historical reasons for doing so. This part of the spectrum is useless for analog TV transmission due to noise ingress. With the advent of HFC networks the useful analog TV spectrum started to include spectrum above 550MHz so this region was a potential revenue source. Therefore cable operators with more sophisticated cable system designs that had low bit rate upstream transmission for system performance feedback, would use the 5MHz to 40MHz band for their upstream, safe in the knowledge that they would never want to put a TV channel there. There are also economic reasons for doing this. It is easier to design RF front ends for this band than for the other available regions above 550MHz. The most important reason may be that the use of this region for upstream communications is adequate for the foreseeable future. However, if cable modem use becomes very popular then eventually other parts of the spectrum will have to be utilized. One way to do this would be to use mid-split or high-split spectrum allocations as described in Section 7.3.

The ingress noise in upstream transmission has its source in shortwave broadcast radio, Citizens' Band (CB) and ham radio transmissions. Though the cable is shielded the noise ingresses at points where the cable shielding is damaged (animals may chew through it or temperature variations may crack it), at unshielded connections and at open connections in residences. Most of the ingress noise appears to come from the residences, where the cable company has no control over the quality of the wiring. To determine its effect, the noise should be measured at the headend receiver and the noise at this point is due to the "noise funneling" effect. Because the upstream is multipoint to point all the noise generated at the residences is funneled back up to the headend and adds at the receiver. So, even though the headend is listening to only one user's signal at any point in time, it is experiencing all the users' noise. This problem is worst for large cable systems but is better in HFC systems where the noise funneling stops at the coax to fiber interface. For examples of upstream noise spectrums see [31] [32] and Fig. 7.14.

From a practical point of view the ingress noise does not have to be completely modeled. As will be explained in Section 7.6, the upstream PHY avoids the impulsive noise spikes. The headend has to have a mechanism to detect and locate low SNR regions but this is not part of the upstream PHY. Therefore each upstream channel will be positioned so that it avoids noise spikes and sees a fairly flat noise spectrum.

The channel model is shown in Fig. 7.15. The AM Hum is, as for downstream, due to power effects on the amplifiers. In [23] a modulation depth of 3% and a trapezoidal waveform with 1ms transition times is proposed. The plant response is described in terms of Koltze's tilt and ripple response with the exception that a 0.3dB per MHz tilt and a slightly worse microreflection distribution given by the table above were suggested.

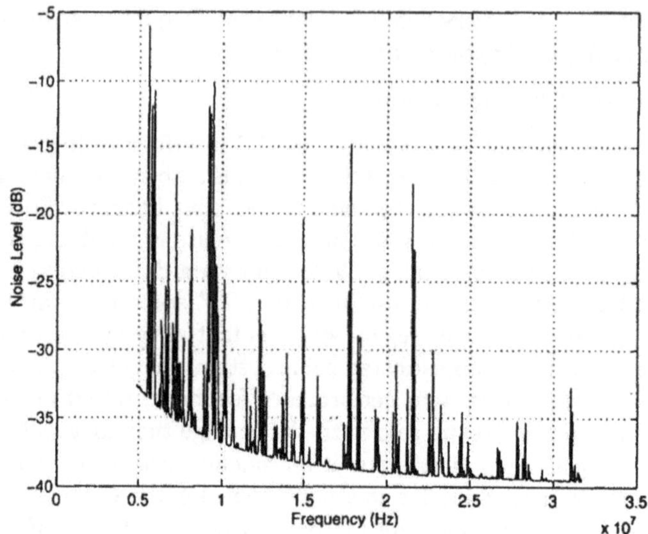

Figure 7.14 An example of the upstream noise spectrum.

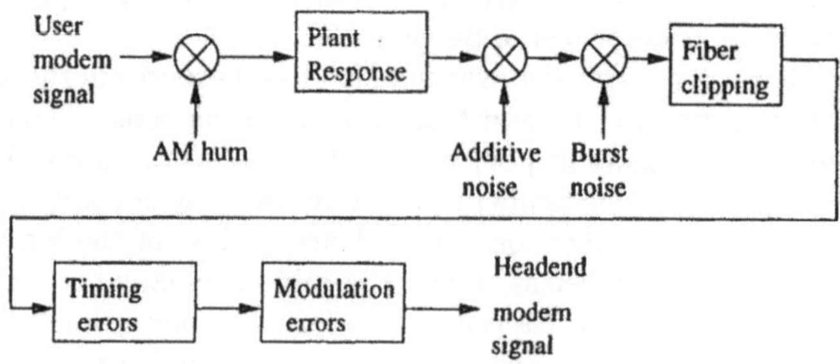

Figure 7.15 Upstream channel medol.

Dealys(ns)	Attenuation(dB)
0-200	-10
200-400	-12
400-800	-14
800-1200	-18
1200-2500	-24
2500-5000	-30
5000-15000	-35

Additive noise includes thermal noise, photon noise, ingress and adjacent channel interference. This noise will vary significantly within the 5-40MHz, from 10dB to 20dB down from the signal. The burst noise is due to other ingress sources that are high power and short lived RF generators. These include the initial transient RF emissions from the start of any electric motor source such as refrigerators, air conditioning and car starter motors. Lightning strikes will also cause burst noise. The burst noise may be Poisson distributed with an average of one every few seconds. The fiber will add impulsive noise to the signal as in the downstream. As in the downstream the noise will essentially destroy one sample at the receiver. In [23] an average impulse rate of 360Hz is suggested. This means that for a 500ksymbol/s system the uncoded symbol error rate will be a little under 1×10^{-3} just due to laser clipping! The timing errors will depend on the accuracy of the lock to the downstream timing and how the downstream timing is used for upstream synchronization (see Section 7.3 for the various options), though the timing error should be less than 50ppm. The modulation error can be large due to frequency stackers that are not locked in modulation frequency to the headend. A modulation error of several hundred kHz would not be uncommon.

The signal attenuation of the channel is very large in upstream communication for the reasons stated in Section 7.3. Therefore any upstream communications system should be tested with a wide range of channel attenuation with about 60dB range [33].

7.5 DOWNSTREAM PHY

Though there are several different standards bodies that are either working on specifying the PHY layer or who have finished specifying it, only two main standards for downstream PHY have emerged. One is favored in Europe and is proposed by DAVIC [6] and in Annex A and C of the ITU recommendation [5], and the other is favored by the US Cable companies and is proposed by MCNS [1], SCTE [2] and Annex B of the ITU recommendation [5]. The ITU and IEEE802.14 approved both and we describe them using their ITU designations of Annex A (for the "DAVIC" standard) and Annex B (for the "MCNS" standard). The reader can obtain copies of these standards from any of the above bodies so we do not attempt to describe them in the detail that can be found in the standards. Rather, we describe the coding systems used (which are the essential differences between the two PHYs) so as to highlight the different choices and what they imply about the performance of the systems.

Both Annex A and Annex B are similar in that they use a subset of 16, 64 and 256QAM constellations with no spectral shaping and a symbol rate of a little over 5MHz. In the calculations that follow we assume a symbol rate of 5MHz for the sake of simplicity. Taking into account the effect of error correction coding this leads to bit rates from a little less than 40Mb/s down to 20Mb/s. The pulse shaping is about 15% and the standard square root raised cosine is used. Because the downstream was initially intended for video delivery it is set up for broadcast transmission with no return channel. Therefore the transmitter blindly transmits and the receiver has to lock to this signal. This actually simplifies the design of the system somewhat because there are no complicated handshaking procedures like those that occur in telephony modems and which tend to complicate the transmitter and receiver controllers. Cable modem designs are therefore mostly made up of simple static data flow and are well suited for ASIC implementation. At the time of writing several companies are already selling cable modem chips information on which can be found in [34]-[39].

The issue of spectral shaping is quite interesting because it potentially could provide low cost dB gain. Spectral shaping is a technique that makes the signal output look more Gaussian. This has the effect of bringing the signal closer to the ideal Shannon theory signal and consequently decreases the required transmit power. The gain in power can be as high as 1.53dB [40] and the complexity of implementation of a shaping system is practically all in the transmitter [41] [42]. It therefore seems ideal for use in a broadcast system where there is one transmitter for many receivers. Unfortunately there is an implementation problem in the receiver with transmitting a Gaussian distributed set of symbols. The most commonly used blind equalization techniques which are used to adaptively determine the filtering required to remove the channel distortion, do their job by extracting information about higher order statistics from the received signal. If the received signal is Gaussian its higher order statistics are deterministically dependent on its first and second order statistics and these algorithms will not work [43]. Therefore spectral shaping has not as yet been used in broadcast systems.

7.5.1 Annex A

The Annex A solution [6][5] allows the user to send either 256QAM (8 bits per symbol), 64QAM (6 bits per symbol) or 16QAM (4 bits per symbol) at 5Msymbols/s rate. It uses Reed-Solomon [44] followed by interleaving to protect against errors. Because it was originally designed to support digital video, and in particular MPEG2 transmission, the unit of transport is the MPEG2 packet. An MPEG packet consists of 188 bytes one of which is a sync byte whose value is always 71 (01000111 binary). This is modified slightly for Annex A with every eighth sync byte being bitwise inverted to give 184 (10111000 binary). This makes up a superframe as shown in Fig. 7.16.

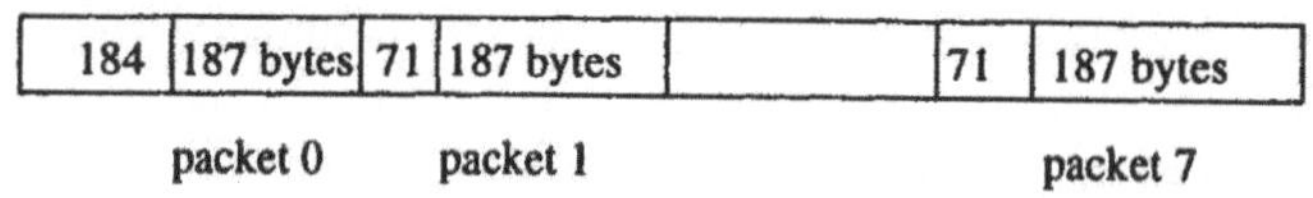

Figure 7.16 Annex A superframe.

There is a scrambler before the Reed-Solomon that ensures the data looks fairly random even if there is long runs of zeros or ones. This is required because the adaptive systems in the demodulator will not operate correctly if the data does not look white (i.e., having a flat power spectrum). A long run of a particular value may cause one or more adaptive systems in the demodulator to lose lock. A self synchronizing scrambler could be used. In this scrambler the input enters into the scrambler state and the output is the final bit of the scrambler state (see Fig. 7.17). By observing the output for a short amount of time, the scrambler state can be determined and the effect of the scrambling logic reversed to give the input bit stream. Therefore the descrambler can synchronize itself by observing its input. However, whenever there is a bit error at the input to the descrambler the descrambler state will be incorrect errored and it will take certain amount of time to resynchronize. During this time the descrambler can make several errors. The self synchronizing scrambler therefore has an error propagation problem. In Annex A a parallel scrambler is used (see Fig. 7.17). In this case the state of the scrambler and therefore the descrambler is unaffected by the input to the scrambler. If the initial state is known to both the scrambler and descrambler then they both know the random sequence being exclusive-ored with the data bit stream. Just as the scrambler exclusive-ors its input with the random sequence to randomize the data, the descrambler exclusive-ors its input with this random sequence to remove the random sequence and recover the data. There will be no error propagation as the input bits of the descrambler do not affect the state of the descrambler. However, the parallel scrambler will not self synchronize. This is why every eighth sync byte is inverted. On observing an inverted sync byte the transmitter and receiver reset their scrambler/descrambler states to a predefined value. Needless to say, the sync bytes are not scrambled so that they can be detected by the receiver.

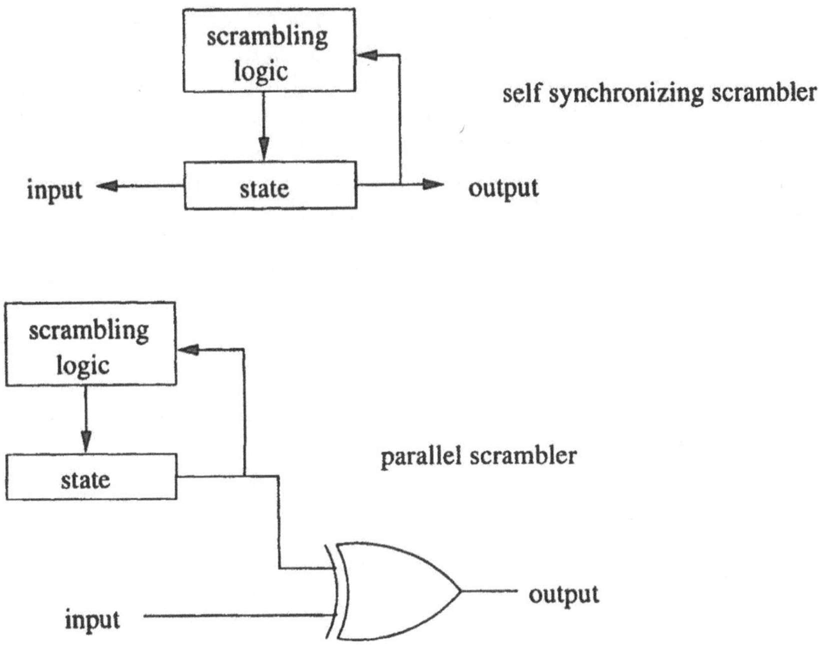

Figure 7.17 Self synchronizing and parallel scrambler.

After scrambling, the Reed-Solomon Coder (R-S) adds 16 parity bytes to each 188 byte packet so that up to 8 byte errors can be corrected in every length 204 block. The R-S in Annex A codes symbols in GF(256) which are 8 bit symbols. These are then passed into a convolutional interleaver. The combination of R-S and convolutional interleaving is optimal for burst error correction [45] and therefore Annex A is optimized for burst errors. For interleaver depth I and interleaver unit delay M, a convolutional deinterleaver is illustrated in Fig. 7.18. We ask the reader to believe that this structure undoes the effect of the interleaver so we can concentrate on what the deinterleaver does to an error burst. The input and output switches change in lock step at the symbol rate performing a scalar to vector and vector to scalar conversion. The delay blocks range from $(I-1)M$ delays to zero and each delay in the system is clocked once every I symbols. So effectively the input is organized into vectors and the vectors are clocked through the delays at the rate at which the vectors are generated and then the output is generated by unpacking the vectors. A burst of error is spread as shown in Fig. 7.19 for $M = 2$ and $I = 6$. In Fig. 7.19 each column is the vector at the output of the delays with the black boxes being the errors at the output caused by a series of completely errored vector inputs. A burst of length IMQ enters the deinterleaver, with in this case $Q = 4$, and is spread out so that when the deinterleaver reads out the vectors columnwise from Fig. 7.19 there is never more than Q errors in each block of length I. In Annex A for 256QAM $I = 204$, $M = 1$ and using the sync bytes we align each R-S block with each column in Fig. 7.19. Therefore the R-S can correct bursts of length 8 in each column so $Q = 8$ is possible. This means 1632 bytes or 0.33ms can be corrected. For the 64 and 16 QAM case $I = 12$ and $M = 17$ and we use the sync bytes to align the R-S block over 17 columns. So the R-S can correct a maximum of 8 errors in every 17 columns. In general if the R-S block is spread over more than P columns then we can only correct a length IP burst. An example error burst distribution at the output of the deinterleaver is shown in Fig. 7.20 for $I = 12$, $M = 3$, and R-S block size equal to 36 with $P = 2$. For Annex A 64QAM and 16QAM the burst length is therefore 96 bytes. For 64QAM this is 128 symbols or 25.6 μs. For 16QAM it is 192 symbols or 38.4 μs. Note that the position of the sync bytes is not changed by interleaving because they are always applied to the top branch of the interleaver and therefore stay in the same position relative to each other. Therefore synchronization can be achieved before deinterleaving, which, of course, has to be the case.

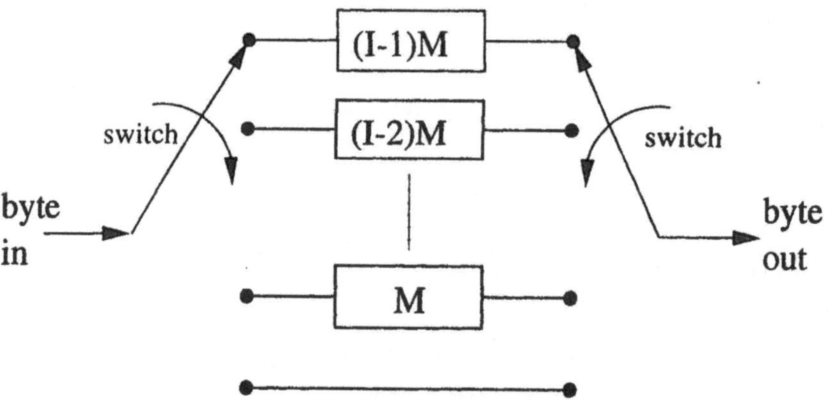

Figure 7.18 Convolutional deinterleave.

If burst noise dominated the error budget then there are many, many different R-S interleaver pairs that can achieve the same performance [46]. For instance, a R-S code with block size 51 that is capable of correcting 2 errors is much simpler to design than a size 204, 8 error correcting code. However, if the size 51 code uses an interleaver of size $I = 51$, $M = 16$, $P = 2$, then the burst error protection is identical. The amount of memory required is $I(I-1)M/2$ which is 20706 bytes for the length 204 code and 20400 for the length 51 code. This looks like a good solution

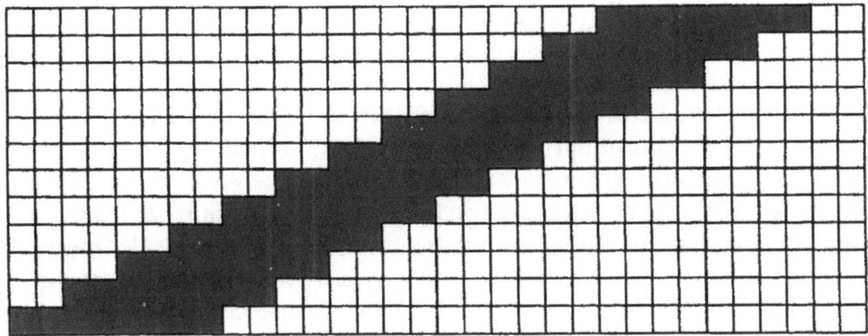

Figure 7.19 Convolutional deinterleave burst noise spreading for M=2 and I=12.

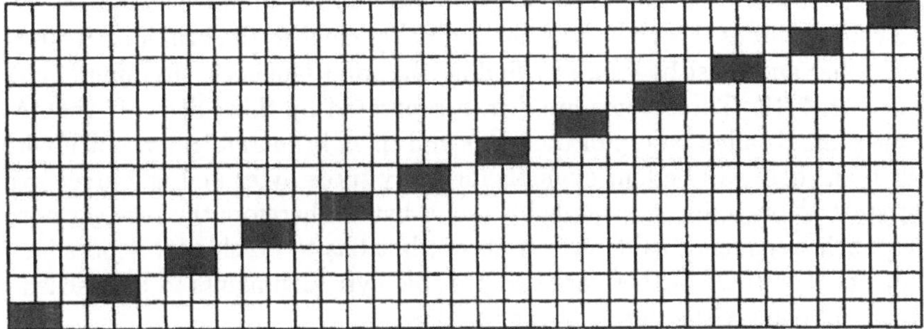

Figure 7.20 Convolutional deinterleave burst noise spreading for M=3 and I=12 and P=2.

from a PHY perspective but the total latency of the interleaver/deinterleaver pair is $(I-1)M$ which is 203 for the length 204 code and 800 for the length 51 code and hence the length 51 code would add 120 μs to the latency in the downstream for 256QAM. As the total latency requirement for the MAC is 200 μs [4] this is an unacceptable increase in latency.

Another reason for requiring a code with a large error correcting capability is to correct random errors. For errors occurring randomly the probability of symbol (with a symbol being a byte in this case) error is approximately [46]

$$P_s = \left(\frac{t+1}{n}\right)\binom{n}{t+1}p^{t+1} \quad for \quad p \ll 1 \tag{9}$$

where the R-S block length is n, the number of correctable errors is t and the probability of symbol error at the input to the R-S. For the Annex A code the output probability of symbol error is $6.22 \times 10^{13} \times p^9$. For "flawless" (less than one corruption a day) video DAVIC requires a bit error rate of 1×10^{-12}[6] which is roughly equal to a 4×10^{-12} symbol error rate. Annex A therefore requires an input symbol error to the decoder of 1.59×10^{-3} to achieve flawless video. For TCP/IP transmission the required bit error rate is better than 1×10^{-8} [47] [48] so Annex A requires an input symbol error rate to the decoder of 4.4×10^{-3}.

The synchronization byte structure in Annex A forces a rather complex synchronization procedure on the receiver. The receiver has to detect a sequence of known bytes (184 followed by seven 71s repeated as in Fig. 7.16) embedded in the data stream. If these bytes were all together then a simple correlation would suffice, but as the bytes are spread out through the system, potential sequence candidates will have to be checked. For instance, if the receiver receives a 71 there is roughly a 50% probability that it is the sync byte (in any R-S block of 204 there will be one sync byte and about one other byte equal to 71). Therefore the receiver will have to note its position and look for other sync bytes occurring at multiples of 204 bytes later. If the receiver keeps track of three sequences at once then the true sequence will probably be found the first time around but it is also possible that the first three 71s or 184s found are data bytes. There must therefore be a mechanism by which the receiver can decide that a sequence it is tracking is not the valid sequence so it can drop that sequence and find another possible sequence. To add to the complexity of this structure, the byte alignment in the symbol stream is not known either and must be found by finding the valid sequence. To understand this look at Fig. 7.21. Three bytes are drawn separated by thick lines. If 256QAM is sent the symbols align on top of the bytes and there is no ambiguity, but if 64QAM is sent then there will be four symbols for every three bytes and any given symbol might start at bit 0, bit 2, bit 4 or bit 6 of a byte. Therefore the receiver not only has to track a sequence as described above, but has to do this for 4 possible byte alignments. For 16QAM there are two possible byte alignments to be tracked with each symbol being either a low or high nibble.

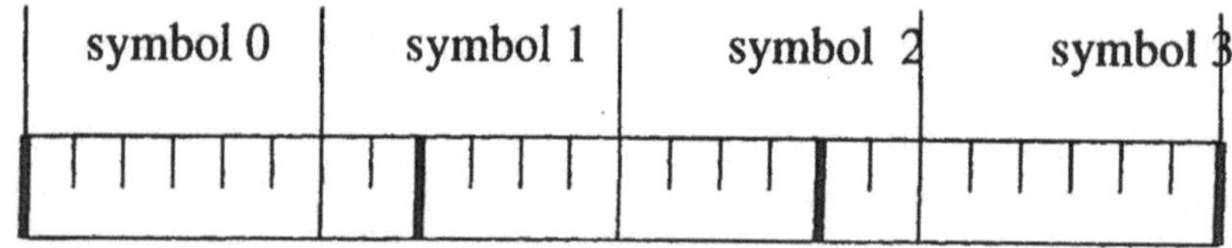

Figure 7.21 Byte alignment in Annex A for 64QAM.

7.5.2 Annex B

The Annex B solution implements 256QAM, 64QAM but not 16QAM. It uses a concatenated code and an interleaver to achieve a higher coding gain than Annex A. Like Annex A it was originally designed to transport MPEG2 frames but there are significant differences in the way this is accomplished. Annex B is less dependent on the MPEG2 frame structure. Firstly, they do not transmit the sync byte for synchronization purposes. Instead they do a checksum on the 187 bytes of data and then put the checksum byte in place of the sync byte. The sync byte is reinserted after checksumming in the demodulator. After the checksum in the transmitter, the transmitter codes the data.

The order of the coding elements in Annex A is scrambler, R-S, interleaver. In Annex B the order is R-S, interleaver, scrambler, trellis code. Apart from the use of a trellis code the significant difference between Annex A and Annex B is the use of the scrambler after the R-S encoding. If a self synchronizing scrambler was used then the error propagation would be amplified by the steep (output error rate)/(input error rate) slope of the R-S decoder, through which small changes in input error rate can cause orders of magnitude changes in output error rate. Therefore the scrambler in Annex B is a parallel scrambler. The R-S in Annex B uses symbols in GF(128) or 7 bit symbols. There is therefore not the same straightforward relationship between the MPEG2 packet and the R-S block that there is in Annex A. The R-S code used has block size 128 symbols and corrects up to 3 errored symbols per block. Using (9) the performance of this code is $P_s = 333375p^4$ where now the symbols are 7 bits long. This is much worse than the Annex A R-S but the trellis code will take up the slack. Between the R-S and trellis code is an interleaver. This is done because the trellis code exhibits error propagation and hence the errors tend to come out of the trellis decoder in bursts. The interleaver can be used to separate the errors due to error propagation as well as error bursts due to the channel.

The interleaver for Annex B is a convolutional interleaver. There is a fixed mode for 64 QAM to be backward compatible with a previous version and a variable mode for both 64QAM and 256QAM. For the fixed mode $I = 128$, $M = 1$ and the maximum length of a correctable burst is $128 \times 1 \times 3 = 394$ 7 bit symbols or 1379/3 64QAM symbols which is about $92\mu s$. This is about 0.28 the burst error protection capability of Annex A, and not all of this will be used on channel burst errors as some of the burst error seen by the R-S will be due to error propagation in the trellis. This disadvantage is due to the concatenated code which though it has a bigger overall gain due to the trellis code has a significantly worse burst error protection because $Q = 3$ instead of 8 in the R-S and the burst error protection is linearly dependent on Q. To achieve the same level of burst protection a much larger interleaver would be required and this would lead to unacceptable latency through the interleaver. For the variable mode, when $I < 128, (I, M) = (64, 2), (32, 4), (16, 8)$ or $(8, 16)$ and when $I = 128, M = 1, 2, 3$ *or* 4. The best burst protection is given by $(I, M) = (128, 4)$ which gives a maximum correctable burst length of $358\mu s$ for 64QAM and $268\mu s$ for 256QAM. This is 0.8 the burst protection of Annex A for 256QAM and about 14 times the burst protection of Annex A for 64QAM. However, these results are obtained at the cost of enormous latency, on the order of 10s of milliseconds, which is much larger than allowed by the MAC layer [47]. This comparison does illustrate how much worse the burst protection of the Annex A 64 QAM is compared to the Annex A 256 QAM.

As the MPEG2 sync sequence is replaced by a checksum sequence it cannot be used for synchronization. Instead the FEC in the receiver is synchronized using a 42 bit sync sequence every 60 R-S blocks for 64QAM and a 40 bit sync sequence every 88 R-S blocks for 256QAM. Each R-S block contains 896 bits so the sync sequence has less than 0.1% overhead in both cases. The transmitter adds this sync sequence after the interleaving so that the sync is not interleaved and the receiver can search for it using a simple correlator. The search will have to be performed over all possible byte alignments but nevertheless this is a much simpler synchronization technique compared to that in Annex A. The main disadvantage with this method is that during initialization you may have to wait about 2ms before finding a sync with 256QAM. In Annex A, assuming you can lock on in two superframes then you have to wait about 0.7ms. Note that in annex B the sync sequence synchronizes the interleaver and R-S but does not synchronize the trellis coder which has to be self synchronizing. The scrambler is synchronized using this sync too because the sync sequence is not scrambled and after the sync sequence is detected the scrambler is reinitialized.

There has been no publications of any analysis of the Annex B trellis code. Even within the various standard bodies that have recommended it, there have only been statements of the overall performance in dB of the concatenated code without much explanation as to why this is so. The analysis of the trellis code that follows is, to the author's knowledge, the first to appear in print. The trellis coder is illustrated in Fig. 7.22. First one must realize that the in-phase and quadrature components of the signal are coded separately. Therefore we can analyze the code for the I rail only as separate PAM constellation, knowing that the same performance will be attained on the Q rail and therefore the overall performance will be the same. Next, note that the least significant bit is all that is coded, with the other bits directly mapped to the constellation point. In the PAM constellation mapping the LSB alternates from point to point so that if the distance between points in the constellation is d then the distance between points with the same LSB is $2d$. Therefore if there was never any mistake on the LSB of the point the minimum distance squared of the code would be $4d^2$.

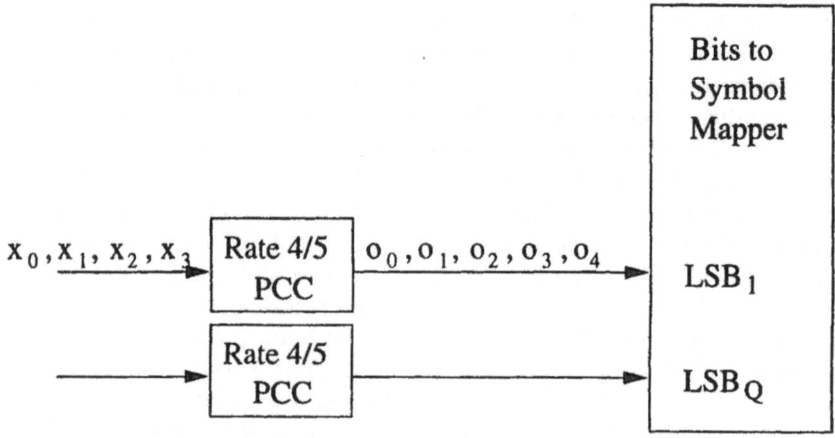

Figure 7.22 Annex B trellis coder.

The LSBs are protected by the convolutional code shown in Fig. 7.23. The top EXOR output produces the first 4 bits as the 4 input bits are being shifted in and the bottom EXOR gives the 5th bit when x_0 reaches the input. Because the bottom output is not used some of the time this is called a punctured convolutional code [49]. The code is nonstationary from cycle to cycle but becomes stationary and linear if the input is considered to be a 4 bit number x and the output a 5 bit number 0. When 4 bits are shifted in, the state of the system becomes those 4 bits and therefore the trellis of this code is fully connected, that is for any two states there is a value of x to take you between those states.

As the code is linear the performance of the code does not have to be calculated for any input to find the overall performance. The overall performance is equal to the performance for an all zero input. The performance of the code is the minimum number of bit errors that must occur before a decoder decides to take a nonzero

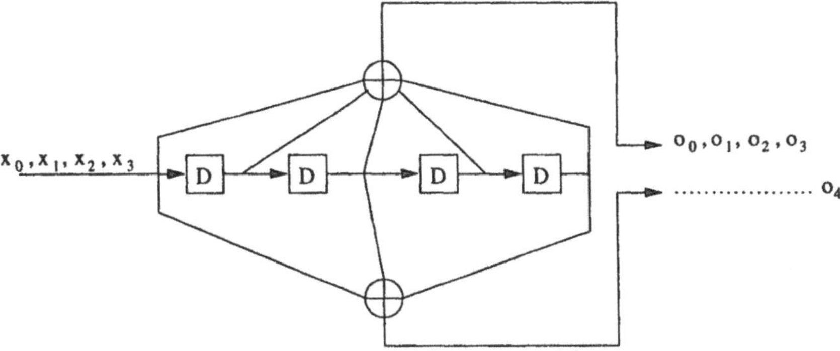

Figure 7.23 Annex B convolutional coder.

path when zero is the correct output. We can show that there are no infinite length inputs to this coder that produce all zero outputs except for the all zero input path. Therefore any nonzero path with finite Hamming distance must by definition start at the zero state and end at the zero state. The minimum Hamming distance of the code can be found by looking at the results of leaving and returning to state zero.

Leaving state zero. When the initial state is zero and you are moving to a nonzero state, the input must be nonzero. When the first nonzero input bit arrives at the shift register input, the top EXOR will generate a one. For there to be no more ones due to the top EXOR there must either be no more bits or the next bit must also be a one and all other bits be zero. This corresponds to x equal to 3,6,8 or 12. In all these cases there are just two ones in the shift register and they must occur together. Therefore the bottom EXOR must produce a one output. This is the purpose of the bottom EXOR; to increase the minimum cost of leaving state zero to 2.

Entering state zero. To enter state zero the input must be zero and therefore the output depends only on the state. When the last bit of the state reaches the end of the shift register the top EXOR will output a one. Transitions from states 1,3 and 6 to state zero produce a single one at the output.

The total number of bits in error when the all zero input is mistaken for any other input must therefore be at least 3. The input 3 surrounded by zeros meets this bound. Hence there must be three LSB errors for an incorrect path to be chosen in the trellis. To choose this path the noise on the signal must be great enough in at least three places to move the received point closer to a neighboring point (which has a different LSB as LSBs alternate in the constellation). The distance between points is d so that the squared distance for three symbols is $3d^2$ because the shortest path from a point (0,0,0) to a point (d,d,d) is of length $\sqrt{3d^2}$. So the squared distance through the trellis is less than the squared distance if the LSBs are correct. Therefore an error in the trellis path is more likely and for high enough SNR will dominate the performance. Though the code gains distance it does so at the expense of power. The power increase is because every 5th symbol the transmitter sends a constellation that is one bit bigger than it has to be. The power in a b bit PAM constellation is [50]

$$\frac{2^{2b}-1}{3} \tag{10}$$

so the power increase ratio is

$$\frac{5\frac{2^{2b}-1}{3}}{4\frac{2^{2b}-1}{3}+\frac{2^{2b-2}-1}{3}} = \frac{5}{4+\frac{0.25-2^{-2b}}{1-2^{-2b}}} \approx \frac{5}{4.25}. \tag{11}$$

The power normalized gain of this code is therefore

$$G(dB) = 10\log_{10}\frac{3}{5/4,25} = 4.07dB. \tag{12}$$

This calculation does not take into account the number of nearest neighbors and the distance and number of the next to nearest neighbors etc., but it does allow us a good upper bound on the code performance and allows us to compare it to other trellis codes in the literature. In particular the famous Wei 4D and 8D codes have coding gains of 4.66dB and 5.41dB respectively when calculated in this manner. The Annex B code is therefore a good but not outstanding code.

To compare Annex A and Annex B we take the waterfall diagrams for the two Reed-Solomon codes used and move the Annex B code up by 4dB. The result for 64QAM is shown in Fig. 7.24. We used (9) to calculate the output symbol error rates of the R-S and the equation

$$p = Q\sqrt{\frac{3SNR}{2^b-1}} \tag{13}$$

to get from SNR to symbol error probability into the R-S, where $Q(x)$ is the Q function that gives the probability of a unit variance Gaussian random variable exceeding x. For convenience we ignore the slight error caused by assuming the 64QAM symbol probability is equal to both the 7bit symbol and 8bit symbol error probability going into the R-S decoders. We also plot for convenience the error probability of an uncoded QAM symbol and note that the results for Annex A and B are only accurate when the uncoded probability is small (less than 0.01) because of the approximations used in (9). Note that in the region that cable modems are most often required to operate, output error probabilities between 10^{-8} and 10^{-12}, the Annex B code is overall about 2dB better than the Annex A code.

7.6 UPSTREAM PHY

In this section we describe the upstream PHY devised by MCNS and adopted by IEEE802.14. It is a simple implementation when understood within the context of the cable modem system as described in Section 7.3. Note that there are some alternative upstream PHY proposals which have the potential of increasing the bandwidth utilization in the upstream and have been grouped as "advanced" PHYs within IEEE802.14 as possible future PHYs to replace the present MCNS PHY [51]-[54].

The headend will first monitor the upstream channel to determine the areas in the spectrum where the SNR is high enough for transmission. These are areas

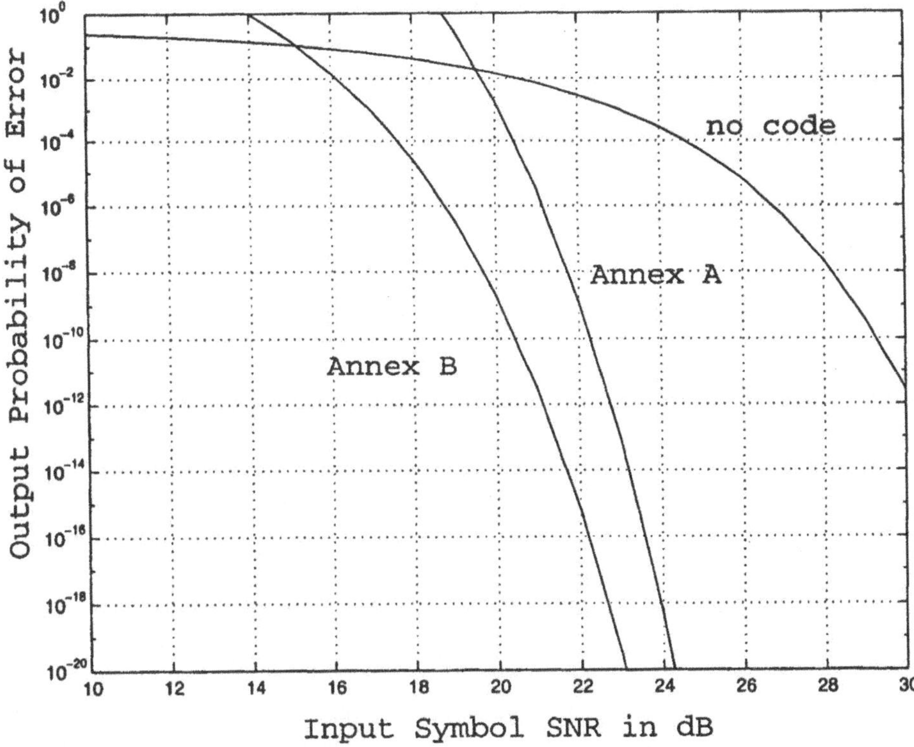

Figure 7.24 Performance of Annex A and Annex B codes.

in which there is little noise ingress. The headend then specifies different channels in terms of their bandwidth and SNR using a set of allowed symbol rates in a way that fills the good spectrum as fully as possible so maximizing the total available bit rate. For each channel the headend specifies the constellation and the error protection used, called the PHY profile for the channel, so that data can be reliably transmitted on this channel. Given this is done, the PHY layer can be described for a single channel in which there is no significant noise ingress.

For a single channel the upstream PHYs proposed in MCNS and IEEE802.14 are quite simple when understood in the context of the cable system overview given in Section 7.3. The MAC layer of the modem will instruct the PHY to send a burst of data starting at a certain time. The PHY layer transmits the data on a headend specified channel using either QPSK or 16QAM depending on a predefined PHY profile set up by the headend for the channel. So that different users' signals do not overlap in time there is a guard band between user transmissions. The guard band is used to let the user ramp up into transmission and ramp down after a transmission as any sudden change in signal energy from a user will cause energy to splatter into other channels. This is shown graphically in Fig. 7.25. Specifically, in the MCNS specification the guard time is specified between 5 and 255 symbols. The size of the guard band is determined by the accuracy of the estimate of packet time obtained by the methods described in Section 7.3. If you can estimate the start of a packet to an accuracy of 3.5 symbols then you have to pad the minimum guard time of 5 symbols by 3.5 symbols on each side to allow for a 3.5 symbol late packet followed by a 3.5 symbol early packet and still achieve 5 symbols in between the last and first symbols. So the guard time in this example would have to be greater than 12 symbols. The guard time of 5 was chosen as enough time for a burst to ramp down and not affect the next or previous burst.

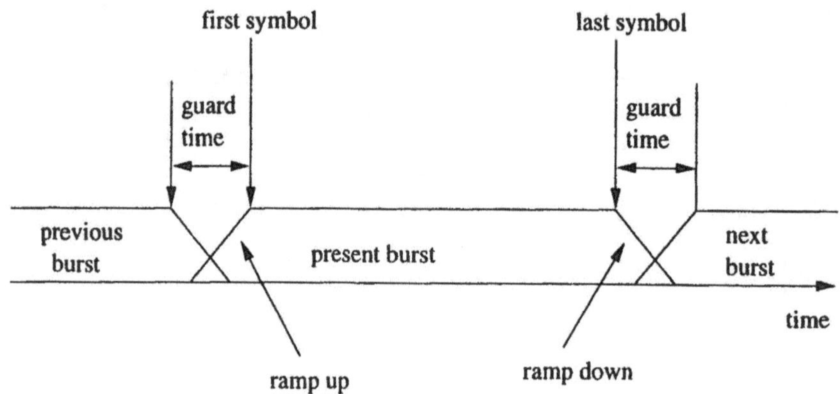

Figure 7.25 Burst profile in upstream.

At the start of a burst there will be a preamble to allow the receiver in the headend to lock to the burst. How big the preamble is depends on how accurate the a priori estimates of the burst's parameters are in the headend. If for example the system has accurate loop timing then the headend will have a very good estimate of the sampling frequency and will only have to learn the sampling phase. Similarly, it is possible for the headend receiver to store estimates of the user's frequency offset obtained from previous bursts transmitted from that user. Then the headend will have an accurate estimate of the carrier frequency too. Another possible approach is for the headend to transmit to the user the carrier error observed during a burst so that on the next burst the user can adjust its carrier frequency to match the frequency desired by the headend receiver.

Typically the more information learned a priori the shorter the preamble will be. This sounds like a win/win situation but in practice the designer has to be careful not to add a lot of hardware complexity to either the receiver or the transmitter only to get a reduction in header size of a couple of symbols on a burst that is several hundred symbols long. In general it is desirable to have a burst header that is variable in length and in MCNS the header is between 0 and 1024 bits. The headend defines the header it wishes to see in a downstream message to the user. The maximum length of a burst affects the length of the preamble in two ways. First, the ratio of the preamble length to the burst length defines the redundancy of the preamble so a longer burst allows a longer preamble for the same redundancy. Second, a longer burst requires either a more accurate initial estimate of timing and carrier frequencies or circuitry in the receiver to track drifts in timing and carrier frequencies. For instance, a carrier phase error of 10° may be acceptable for a QPSK signal. If the initial phase error is about 1° and the carrier frequency estimate is within 100Hz then for a 500ksymbol/s signal the residual frequency error will take at least 125 symbols to move the constellation to an unacceptable phase error. If the burst length is over 125 symbols the receiver will have to track the phase error and provide some correction during the reception of the burst or do more accurate initial frequency estimation.

The SNR of the signal varies significantly depending on the channel used. Therefore the amount of error correction required will vary. In MCNS R-S is used for error correction and the size of the R-S block as well as the number of errors that can be corrected in a block are variable and defined by the headend.

Even though the channels are arranged by the headend to miss significant noise spikes, microreflections will still cause the channel to distort the signal. In fact microreflections are more of a problem at low frequencies because the loss in the cable is less at low frequencies and therefore the microreflections will come back with less attenuation. Equalization can be used to counteract the effect of the channel but it is inconvenient to do this in the receiver. There are two reasons for this. First, the receiver will have to rapidly swap equalizer coefficients from burst to burst because each burst comes from a different user on a different path in the network who therefore sees a different channel. Second, when a user initially asks for bandwidth in the contention period (as explained in Section 7.3), the headend will not know which user this is and not be able to equalize the signal. Therefore it seems advisable that the user pre-equalize its signal before sending it to the headend. This will cause some power increase in the transmitted signal, but there is so much power variation already in the transmitted signal that this will probably not be noticeable. If power increase becomes a problem then Tomlinson precoding [55] could be used. Simulations using the channel model described in Section 7.4.2 [56] predict that no equalization will be required for QPSK and that the number of equalizer taps required for a DFE implementation in a 16QAM system is only a couple for the feedforward filter and a couple for the feedback filter.

7.7 ACKNOWLEDGEMENTS

The author would like to thank Jeyhan Karaoguz of Motorola for taking time out of his busy schedule to review this chapter, and Tom Kolze of ComStar Communications for clarifying many of the more obscure channel modeling issues.

REFERENCES

[1] "Data over cable interface specifications, radio frequency interface specification," *SP-RFIIO1-970326*, MCNS, March 1997.

[2] "Digital video transmission standard for cable television," *SCTE DVS-031*, October 15, 1996.

[3] "Physical layer specification for HFC networks," *Draft Supplement to IEEE Std 802.14*, Version 1.0 May 16, 1997.

[4] "Media access and control," *Draft 2 Revision 1 to IEEE Std 802.14*, June 20, 1997.

[5] "Multi-programme systems for television sound and data services for cable distribution," *ITU-T Recommendation J.83*, October 1995.

[6] DAVIC 1.0 Specification Part 8. *Lower Layer Protocols and Physical Interfaces*, Digital Audio Video Council, Geneva Switzerland, 1996.

[7] A. Gatherer, "Splitter and drop tap measurements," *IEEE Project 802.14 Cable TV Protocol Working Group*, Document # IEEE 802.14-96/84, March 1996.

[8] W. A. Rheinfelder, *CATV System Engineering: How to Plan and Design Modern Cable TV Plants*, 2nd Ed., Tab Books, 1967.

[9] J. Dunlop and D. G. Smith, *Telecommunications Engineering*, Van Nostrand Reinhold, 1984.

[10] C. M. Chamberlain, "Development of a laser diode clipping modem - part 1," *SPIE*, vol. 2609, pp. 194-202.

[11] "Generic requirements for coaxial drop cable," *Bellcore Technical Reference*, TR-NWT-001398, Issue 1, December 1993.

[12] "Generic requirements for coaxial distribution cable," *Bellcore Technical Reference*, TR-NWT-001399, Issue 1, December 1993.

[13] W. Stallings, *Data and Computer Communications*, MacMillan Publishing Company, 3rd Ed., 1991.

[14] J. Karaoguz, "Time synchronization in CATV-HFC systems using time-stamp messages: simulation results," *IEEE Project 802.14 Cable TV Protocol Working Group*, Document # IEEE 802.14-97/043, March 1997.

[15] A. Gatherer, "CATV downstream channel model. rev. 1.2," *IEEE Project 802.14 Cable TV Protocol Working Group*, Document # IEEE 802.14-96/144, July 1996.

[16] D. Raskin, D. Stoneback, J. Chrostowski, R. Menna, *Don't Get Clipped on the Information Highway*, Communications Engineering and Design, pp. 26-36, June 1996.

[17] Q. Shi, "Asymptotic clipping noise distribution and its impact on M-ary QAM transmission over optical fiber," *IEEE Trans. on Communications*, vol. 43, no. 6, pp. 2077-2084, June 1995.

[18] Q. Shi, "Performance limits on M-QAM transmission in Hybrid multichannel AM/QAM fiber optic systems," *IEEE Photonics Technology Letters*, vol. 5, no. 12, pp. 1452-1455, Dec 1993.

[19] A. Gatherer, T. Wright, "Spectral inversion in the downstream," *IEEE Project 802.14 Cable TV Protocol Working Group*, Document # IEEE 802.14-97/036, March 1997.

[20] "Digital transmission characterization of cable television systems," *Cable Television Laboratories Inc.*, Louisville, CO, Nov. 1994.

[21] R. Prodan, M. Chelehmal, T. Williams, C. Chamberlain, "Analysis of cable system digital transmission characteristics," *1994 NCTA Technical Papers*, pp. 254-262.

[22] R. Prodan, M. Chelehmal, T. Williams, C. Chamberlain, "Cable system transient impairment charaterization," *1994 NCTA Technical Papers*, pp. 263-268.

[23] T. Koltze, "Proposed HFC channel model," *IEEE Project 802.14 Cable TV Protocol Working Group*, Document # IEEE 802.14-96/196, July 1996.

[24] R. Russell, *Private Communication*, CSW Inc.

[25] A. Gatherer, "The effect of microreflections on decision feedback equalization," *IEEE Trans. on Signal Processing*, vol. 45, no. 1, pp. 228-230, January 1997.

[26] B. Currivan, "Channel surfing test results," *IEEE Project 802.14 Cable TV Protocol Working Group*, Document # IEEE 802.14-96/133, April 1996.

[27] *Temic Tuner 4737PY5, 3X 1856, Target - Specification Electrical Data(Preliminary).*

[28] T. Osterman, "Grounding, sheath current and power reliability," *Cable Television Engineering*, vol. 15, no. 7, Fourth Quarter 1992.

[29] K. Laudel, E. Tsui, J. Harp, A. Chun, J. Robinson, "Performance of a 256-QAM demodulator/equalizer in a cable environment," *1994 SCTE Technical Papers*, pp. 283-304.

[30] J. Hamilton, D. Stoneback, "The effect of digital carriers on analog CATV distribution systems," *1993 NCTA Technical Papers.*

[31] "Two way cable television system characterization," *Cable Television Laboratories, Inc.*, Louisville, CO, April 1995.

[32] W. Y. Chen, C. F. Valenti, E. Edmon, "HFC upstream noise model," *IEEE Project 802.14 Cable TV Protocol Working Group*, Document # IEEE 802.14-95/110, September 1995.

[33] T. Koltze, "Proposed upstream transmit power specifications," *IEEE Project 802.14 Cable TV Protocol Working Group*, Document # IEEE 802.14-96/191, July 1996.

[34] P. McGoldrick, "Super chip is the first to get the cable modem down to size," *Electronic Design*, June 1997.

[35] L. Goldberg, "IC opens 500-channel frontier to cable systems," *Electronic Design*, November, 1994.

[36] "VES4613 64/256 QAM subsystem," *VLSI Technologies Inc. product bulletin*, May 1995.

[37] H. Samueli, "Update on silicon IC technology for downstream 64-QAM transmission," *IEEE Project 802.14 Cable TV Protocol Working Group*, Document # IEEE 802.14-96/095, March 1996.

[38] K. Laudel *et. al*, "Performance of a 256-QAM demodulator/equalizer in a cable environment," *NCTA Technical papers*, 1994, pp. 283-304.

[39] K. Laudel, "Performance results of a Low-cost alternative equalizer architecture for 64/256 QAM demodulation in a CATV receiver," *NCTA Technical papers*, 1995.

[40] G. D. Forney *et. al*, "Efficient modulation for band-Limited channels," *IEEE Journal on Selected Areas in Communications*, vol. SAC-2, no. 5, Sept. 1984, pp. 632-647.

[41] G. David Fourney Jr., "Trellis shaping," *IEEE Trans. on Information theory*, vol. 38, no. 2, March 1992.

[42] R. Laroia, "Coding for intersymbol interference channels - combined coding and precoding," *IEEE Trans. on Information Theory*, vol. 42, no. 4, July 1996.

[43] S. Haykin, *Blind Deconvolution*, Prentice Hall, 1994.

[44] R. E. Blahut, *Theory and Practice of Error Control Codes*, Addison-Wesley Publishing Co., 1983.

[45] G. D. Forney, "Burst-correcting codes for the classic bursty channel," *IEEE Trans. on Communications Technology*, vol. Com-19, no. 5, pp. 772-781, October 1971.

[46] A. Gatherer, T. Wolf, "Reed-Solomon performance estimates," *IEEE Project 802.14 Cable TV Protocol Working Group*, Document # IEEE 802.14-96/170, June 1996.

[47] J. M. Ulm *et. al*, "IEEE P 802.14 cable TV functional requirements and evaluation criteria," *IEEE Project 802.14 Cable TV Protocol Working Group*, Document # IEEE 802.14-94/002R2, March 1995.

[48] E. Amir *et. al*, "Efficient TCP over networks with wireless links," *Proceedings of the 5th Workshop on Hot Topics in Operating Systems*, IEEE Computer Society Press, May 1995.

[49] L. H. Charles Lee, *Convolutional Coding Fundamentals and Applications*, Artech House, 1997.

[50] John G. Proakis, *Digital Communications*, 2nd Ed., McGraw Hill,1989.

[51] K. S. Jacobsen, J. M. Cioffi, "An efficient digital modulation scheme for multimedia transmission on the cable television network," *NCTA Technical Papers*, 1994, pp. 305-312.

[52] R. Gross, S. Quinn, M. Lauback, G. Edmon, "Graceful evolution to DWMT," *IEEE Project 802.14 Cable TV Protocol Working Group*, Document # IEEE 802.14-96/064, March 1996.

[53] R. Gross, M. Tzannes, S. Sandberg, H. Padir, X. Zhang, "Discrete wavelet multitone (DWMT) system for digital transmission over HFC links," *SPIE*, vol. 2609, pp. 168-175.

[54] J. A. C. Bingham, "Synchronized discrete multitone: a bandwidth solution for the upstream channel of an HFC system," *SPIE*, vol. 2609, pp. 176-181.

[55] "New automatic equalizer employing modulo arithmetic," *Electronics Letters*, vol. 7, pp. 138-139, Mar 1971.

[56] J. Min, "Simulation results of a QPSK/16QAM upstream transceiver in the presence of microreflections," *IEEE Project 802.14 Cable TV Protocol Working Group*, Document # IEEE 802.14-96/096, March 1996.

Chapter 8

Wireless Communication Systems

Elvino S. Sousa
Dept. of Electrical and Computer Engineering
University of Toronto, Toronto, Canada
sousa@comm.utoronto.ca

8.1 INTRODUCTION

Wireless communication networks provide telephone service to mobile users have been rapidly developed since 1980s. Beginning with the first generation analog systems, which provided voice service, systems have developed to the standardization of third generation systems which will have the capability to offer multimedia services. The most fundamental concepts in the development of economically viable wireless telecommunication networks are the concepts of frequency re-use and accurate transmitter power control which allows the use of a relatively small amount of radio spectrum to serve a large population of mobile users in some population area such as a city. The key issue in the development of these systems has been the maximization of the capacity of the system in terms of the number of users that can be served per base station per unit of bandwidth. Due to the limitation in technology the first generation systems were analog and as a result of propagation impairments they used a relatively inefficient modulation technique - wideband FM. As a result of the rapid success of these systems there was a strong push for the design of second generation systems. The main objective for this development in North America was the capacity issue - where the capacity limit of the early systems was quickly reached in cities such as Chicago and Los Angeles. A goal of 10 times the capacity of the analog system was set for the second generation systems in North America. These systems also had to maintain downward compatibility with the analog system. In Europe the second generation systems were motivated by the need to develop a pan-European system which would work with the same hand terminal in all the countries of Europe. The second generation system was also designed to operate in a new frequency band - hence downward compatibility with the older first generation systems was not necessary.

There were various reasons to require that the second generation systems would be digital. One of the key reasons is that digital transmission allows for the transmission of data in what we may call a "native-mode". The other reason is that it was thought that digital transmission could potentially yield a higher capacity. This second characteristic would have been somewhat counter-intuitive in the early days of the development of cellular systems since the common wisdom was that digital transmission always results in an expanded bandwidth. However, the novelty in the approach to second generation cellular systems in comparison to digital transmission over wire-line telephone networks was the high degree of compression of state-of-the-art speech coders in comparison to the classical PCM speech coders of the telephone network developed in the sixties.

Given the requirement for channel compatibility with the analog system, various proposals for a second generation North American cellular system amounted to the utilization of a modulation and multiplexing scheme where a number of users could share an analog channel. Various FDMA and TDMA based proposals were advanced. At a late stage in the process, a proposal based on spread spectrum was also advanced. It was initially rejected since the decision on TDMA had already been made. A high degree of interest by some large cellular operating companies kept the spread-spectrum CDMA proposal alive and the industry eventually standardized two second generation systems for North America which became known as TDMA (IS-54) and CDMA (IS-95).

In this chapter we will discuss the basic characteristics of these systems and the basic issues involved in the design of such systems. We will discuss a series of concepts that are inherent to the field of mobile communications. We will discuss the basic characteristics and design of the systems: AMPS, IS-54/136, GSM, IS-95, and introduce some of the new concepts incorporated in third generation proposals.

8.2 AMPS

The AMPS (Advanced Mobile Phone System) system has defined the basic cellular concept [1] and the concept of frequency re-use. In this concept a telephone service area such as a city is divided into a set of radio cells with radius on the order of 1 Km. A radio base station is placed at the center of each cell. Each base station is connected by a set of voice trunk circuits to a telephone switch that interfaces to the public switched telephone network (PSTN). Mobile terminals in each radio cell communicate with the corresponding base station over a radio link that utilizes a set of frequency bands that are re-used over the network. This system was designed for telephony and for simplicity utilizes analog modulation. Due to the problem of multi-path fading and the limitations on the technology of the time, an FM modulation technique was chosen where the bandwidth of the modulated signal including guard bands is B = 30 KHz. A telephone quality voice signal has a bandwidth of approximately 3 KHz. The minimum bandwidth for analog modulation would therefore be attained with SSB and would be equal to 3 KHz. The actual AMPS FM modulation scheme utilizes a bandwidth equal to approximately ten times the minimum required.

If the allocated bandwidth in an AMPS system is W then there are $C = \frac{W}{B}$ radio frequency (RF) channels. The communication between the base station and mobile terminal is full duplex. Two possibilities exist to implement a full duplex channel: frequency division duplex (FDD) and time division duplex (TDD) which is also called ping-pong TDMA. In AMPS FDD is used due to the delay in the channel. The use of TDD would require a large guard time and would result in greater overhead in the transmission channel. In a typical system two frequency blocks of 12.5 MHz are assigned for base station transmission and mobile station transmission respectively. Each voice circuit consists of two RF channels with a frequency separation of 45 MHz. Each base station defines a coverage area that is referred to as a cell. For analytic purposes the radio cells are considered to be hexagonal in shape. However in practice the actual cell corresponding to a given base station is defined by signal coverage and usually has an irregular shape which depends on the characteristics of buildings and hills in the environment.

The above typical system consists of C = 416 RF channels. In ideal situations we would like to re-use these channels in all cells in the system. However it is not possible to use the same RF channel in two adjacent cells due to interference, and in practice there is a minimum distance that must be maintained between two cells utilizing the same RF channel. This requirement leads to the basic concept of frequency re-use cluster size. In an optimum frequency assignment in an ideal system the set of C RF channels are divided into N channel groups with $S = \frac{C}{N}$ channels per group and these groups are assigned to the cells in such a manner as to maintain the minimum separation between cells using the same channel. The parameter N is referred to as the frequency re-use cluster size.

In a practical network the cluster size is dependent on the modulation scheme and the directivity of the antenna at a base station. Directive antennas minimize the degree of co-channel interference by restricting the amount of radiation for a given transmitter-receiver link, hence their utilization in a system results in a smaller frequency re-use cluster size. A wide band modulation scheme such as spread spectrum can better cope with interference and can result in a practical cluster size as low as $N = 1$. The hard capacity of a system can then be calculated as $K = \frac{W}{BN}$. For example in [1] it is suggested that with 30 KHz analog FM and omni-directional antennas the required cluster size should be $N = 12$, whereas for the same modulation with a cell utilizing three antenna sectors (120 degree radiation pattern) the required cluster size is $N = 7$. Other systems, more recently deployed, utilize antennas with a 60 degree radiation pattern (6 sectors per cell) and have a required frequency re-use cluster size of $N = 4$.

As the mobile terminal travels throughout the cell it eventually leaves the coverage area of the cell. In the AMPS system this is detected by signal level monitoring equipment at the base station. The mobile then switches to a new channel belonging to the new cell that it is entering. This is called the ***hand-off*** process. The AMPS system utilizes what is called a hard hand-off. This means that the terminal switches completely from the old cell to the new cell. The AMPS hand-off is also classified as a base station initiated hand-off. We will discuss other types of hand-off with digital systems. The AMPS system hand-off is also sometimes classified as a "break-before-make" hand-off in the sense that the communication with the old cell is terminated before communication with the new cell is initiated.

There are a number of different versions of first generation cellular systems throughout the world. These are all similar except for the RF channel bandwidth which is 25 KHz instead of the 30 KHz of AMPS.

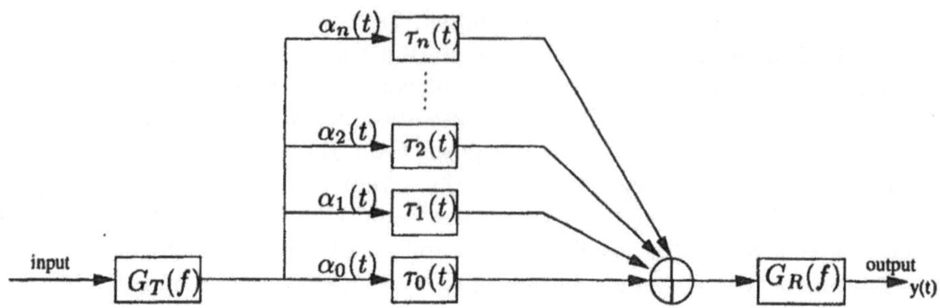

Figure 8.1 Multi-path fading channel model along with transmitter and receiver filters.

8.3 DIGITAL WIRELESS SYSTEMS

Second generation systems are all digital. The principal requirement for a second generation system is an efficient speech coder. Various speech coders have been developed over the years. These coding schemes have achieved bit rates of 8 Kbps, which is eight times lower than the 64 Kbps standard for telephony developed in the sixties. The GSM system was the first of the three second generation systems to be developed. Second generation systems also extensively employ forward error correction techniques.

8.3.1 Mobile Channel Modeling

In order to discuss the system issues involved in the design of digital cellular systems we need to discuss channel propagation characteristics. A wireless communication channel can be modeled as a time varying linear system and is therefore characterized by a time varying impulse response: $h_c(t, \tau)$ where t is a time parameter and τ is the independent variable of the impulse response. If there is no motion, either in transmitter and receiver or in reflectors in the environment, then $h_c(t, \tau)$ has no dependency on t. In general a mobile communications channel can be modeled as shown in Fig. 8.1. The filter $I(f) = G_T(f)G_R(F)$ represents the band-limiting of the channel imposed by the equipment, transmitter and receiver filters, and can be modeled as an ideal low-pass filter in the most common case. The coefficients $\alpha_i(t)$ represent the attenuation in the various multi-path components. If an impulse is applied to the channel the output of the channel is a sequence of $m < n$ pulses where m depends on the bandwidth of the channel (i.e., bandwidth of the filter $H(f)$) and on the relative difference of the various multi-path component delays τ_i; i.e., the multi-path components are "blurred" into a small set of resolvable components. Roughly, two channel components are resolvable in the channel (i.e., appear as two distinct multi-path components) if the difference in their delays is greater than the inverse of the bandwidth of the filter $H(f)$. In the case of one resolvable component we say that the channel behaves as a narrow band channel, otherwise the channel behaves as a wideband channel.

8.3.1.1 Signal variation in time: Coherence time

In general the attenuation and delay parameters in the channel model of Fig. 8.1 depend on the environment and change with time in an unpredictable manner as a result of motion. One way to characterize the channel is to determine the response due to a complex exponential input $Ae^{j\omega t}$. The output $Y(\omega, t)$ can be modeled as a random process of time with parameter ω. We can study this process by fixing $\omega = \omega_0$ and determining the variation in time. The rate of variation, in a statistical sense, is related to the rate of change of the delays τ_i which depends on the velocity v of the receiver (or transmitter) through some environment, or the velocity of surrounding reflectors. For a given velocity the rate of change of the delay of a given multi-path component depends on the angle of arrival of the component relative to the velocity vector. This change results in a Doppler shift of the frequency which is given by $\Delta\omega = \frac{v\omega_0}{c}\cos\theta$ where θ is the angle of arrival relative to the velocity vector and c is the velocity of light. In a standard propagation model we assume that there are a very large number of multi-path components with arrival angles uniformly distributed over 360 degrees. The total received signal, due to the sinusoidal input, has an envelope that varies with time. The degree of variation can be quantified by the auto-correlation function of the process $Y(\omega_0, t)$. The power spectral density of the received scattered signal $Y(\omega_0, t)$ (under uniform distribution of power over arrival angle) is given by

$$S_X(\omega) = \frac{P}{\sqrt{{\omega_m}^2 - (|\omega| - \omega_0)^2}}, \qquad ||\omega| - \omega_0| \le \omega_m \tag{1}$$

where P is the total power of the received signal. The auto-correlation function of $Y(\omega_0, t)$ is then $R(\tau) = Pe^{j\omega_0\tau}J_0(\omega_m\tau)$ and the autocorrelation function for the complex envelope of $Y(\omega_0, t)$ is $PJ_0(\omega_m\tau)$, where $J_0(.)$ is the Bessel function of zero order and $\omega_m = \frac{\omega_0 v}{c}$ is the maximum Doppler shift which occurs for the component that arrives in the direction of motion. One characteristic of the channel is measured by the width of the auto-correlation function of the complex envelope. Generally in modulation design we need to determine the time spacing over which the signal is correlated. This time spacing is generally referred to as the *coherence time*. If we choose the first zero of the above auto-correlation function as the point that we consider the signal to be uncorrelated, then the coherence time is given by

$$\tau_c = \frac{2.4}{\omega_m}. \tag{2}$$

Regardless of the exact definition, the coherence time is inversely proportional to ω_m. The parameter ω_m can be called the *Doppler spread width*. Thus the coherence time is inversely proportional to the Doppler spread width.

8.3.1.2 Signal variation in frequency: Coherence bandwidth

The above concepts of coherence time and Doppler spread width are determined by the velocity of the receiver/transmitter through the environment, and they determine the rate of change of the signal with time. We may consider the dual to the above as the rate of change of the signal versus the frequency ω. For a given set of delays (sometimes referred to as delay spread) we consider the correlation of the complex envelopes of the channel output due to the inputs at the frequencies ω and

$\omega + \upsilon$; i.e., due to the two inputs $e^{j\omega t}$ and $e^{j(\omega+\upsilon)t}$. Under the channel input $e^{j\omega t}$ the output is $r(t) = \sum_{k=1}^{n} \alpha_k e^{j\omega(t-\tau_k)}$, where n is the number of multipath components. To simplify the analysis we may assume a single random output component as follows: $r_1(t) = \alpha e^{j\omega(t-\tau)}$ where α and τ are random amplitude and delay components. At the other frequency the output of the channel is $r_2(t) = \alpha e^{j(\omega+\upsilon)(t-\tau)}$. The correlation of these two components defines what we call the spaced frequency correlation for the channel: $R_\omega(s) = E(r_1(t)r_2^*(t)) = e^{-j\upsilon t}E(\alpha^2)E(e^{j\upsilon\tau})$. The normalized envelope of the spaced-frequency correlation is then $r_\omega(s) = E(e^{j\upsilon\tau})$ where υ is the frequency spacing between the two signals and τ is a random variable which models the delay spread in the channel. A classical model for the delay spread is the exponential distribution where the probability density function is given as follows: $p(\tau) = \frac{1}{\delta_t}e^{-\tau/\delta_t}$. The parameter δ_t is the expected value of the excess delay of the signal. For the exponential delay spread model the normalized spaced-frequency correlation function is $r_\omega(\upsilon) = \frac{1}{1+j\delta_t\upsilon}$. We may now consider the width of the function $r_\omega(\upsilon)$ as a measure of the frequency selectivity of the channel. This measure is referred to as the *coherence bandwidth* of the channel. A typical measure of the width of the spaced-frequency correlation function is the value of υ where the square of the magnitude drops by 3 dB with respect to the value for $\upsilon = 0$. For the exponential delay spread model described above this value is $\upsilon_{3dB} = B_c = \frac{1}{\delta_t}$. We may summarize the above concepts as follows. We transmit a signal $e^{j\omega t}$ in a mobile communications channel. The received signal is a two dimensional random process over time and frequency. To characterize this process we determine the *spaced − time* and *spaced − frequency* correlation functions. These functions will characterize the behavior of the channel in time and frequency. The spaced-time correlation function is determined by the *Dopplerspread* of the various arriving multi-path components which depends on the relative velocity of the receiver with respect to the transmitter and the angle of arrival of the various multipath components. The spaced-frequency correlation function is determined by the distribution of the delays of the multi-path components which we refer to as the *delay spread.*

The Doppler spread and spaced-time correlation function are Fourier transform pairs. The spaced-frequency correlation and delay spread are also Fourier transform pairs. All of these concepts are illustrated in Fig. 8.2.

The delay spread and Doppler spread functions, when normalized, may be viewed as probability density functions. From the general properties of Fourier transforms, for a Fourier transform pair, the widths of the corresponding functions (in the transform pair) are inversely related: if one increases the other decreases.

8.3.1.3 Signal variation in space: Coherence distance

In some systems it may be desirable to utilize multiple antennas at the transmitter or receiver in order to achieve a mitigation of the effect of fading. To consider the use of multiple receiver antennas we consider the correlation of the signals received at two antennas with an antenna spacing of d. In a dense scattering environment the signal arrives over a large number of multi-path components with different angles of arrival. We may model this as a single component with the arrival angle treated as a random variable and do some averaging (expectation) over the arrival angle. To determine the correlation of the signals at the two antennas we utilize the diagram in Fig. 8.3.

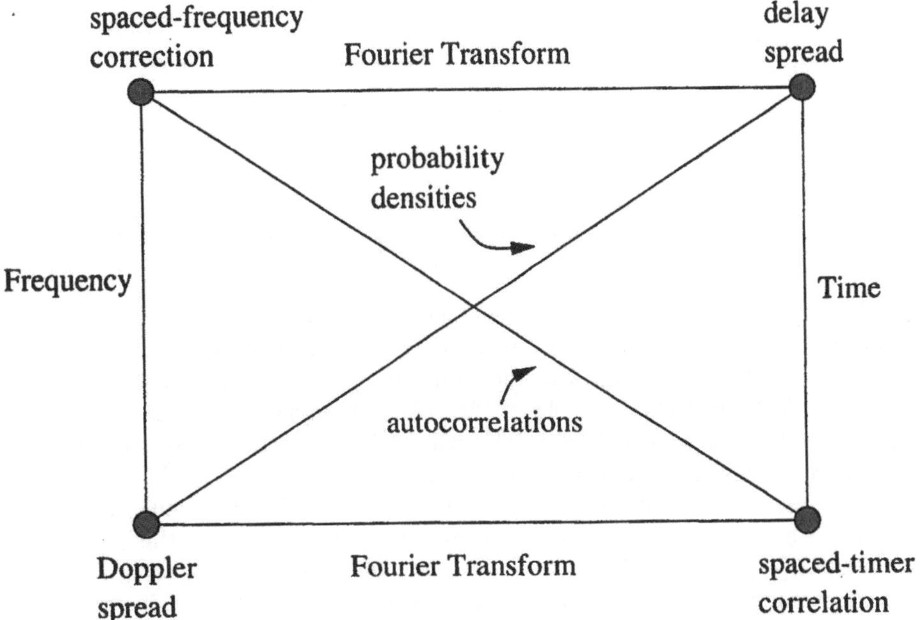

Measures of the width of the above functions	
Spaced-frequency correlation	Coherence bandwidth
Spaced-time correlation	Coherence time
Doppler spread	Doppler spread width
Delay spread	Delay spread width

Figure 8.2 Multipath fading channel concepts.

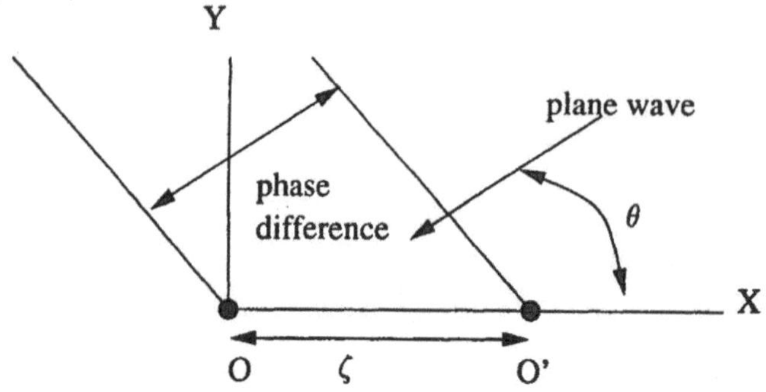

Figure 8.3 Signal correlation in space.

In Fig. 8.3 the two antenna elements have axes perpendicular to the page and are located at the points O and O' respectively. We consider one of the arriving components with arrival angle θ as shown in the figure. Let the signal arriving at the element located at O be $\mu = Ae^{j(\omega t+\phi)}$. Then the signal arriving at the element located at O' is $v(t) = Ae^{j(\omega t+\phi-\beta\zeta cos\theta)}$ where $\beta = \frac{2\pi}{\lambda}$ is the wave number. The correlation of these two signals is then $R_s(\zeta) = E(\mu^*(t)\nu(t)) = A^2E(e^{-j\beta\zeta\cos\theta})$. If we now assume the case where signal components arrive with angle uniformly distributed over the arrival angle θ and take the expectation over this angle then we obtain $R_s(\zeta) = PJ_0(\beta\zeta)$ where $J_0(.)$ is the Bessel function of order 0. In an analogous manner to the case of time and frequency correlations this correlation function may be called the *spaced – distance* correlation function. We may also define a measure of the width of this function and call it the *coherence distance*. If we define this width as that which corresponds to the first zero of the Bessel function (2.4) then we have a coherence distance $d_c \approx \frac{2.4}{\beta} \approx 0.38\lambda$.

8.3.2 Physical Layer Design

The performance of a modulation scheme depends greatly on the bandwidth of the signal in relation to the coherence bandwidth of the channel. Systems where the coherence bandwidth is larger than the signal bandwidth (or RF channel bandwidth) are said to suffer from *flat* or *frequency non-selective fading*. Systems where the signal bandwidth is larger than the coherence bandwidth are said to suffer from frequency selective fading. Multi-path fading is a major consideration in the design of the physical layer of a wireless communication system. The variation of the channel as a result of fading results in a variation of the energy of the received data symbols with time. This further results in the variation of the SNR of the detected symbols. This type of variation is undesirable since it results in errors when the signal power decreases and it results excessive interference at other receivers in the network in the case that the signal power is larger than the minimum required. There are two main approaches to dealing with this variation in received signal power: use of data symbol level diversity, where multiple copies of the same data symbol are transmitted in such a way that the fading for the different symbols is uncorrelated, use of forward error correction coding with soft-decision decoding where the memory of the code is greater than a sequence of transmitted symbols that undergo uncorrelated fading, i.e., a fading burst. In the first case diversity is achieved by transmitting the same symbol over channels with a frequency separation greater than the coherence bandwidth (known as *frequency diversity*), or over the same channel but with a time separation greater than the coherence time of the channel (known as *time diversity*). In either of these cases the utilization of diversity results in a loss of spectral efficiency. An alternative approach is to transmit a single symbol and receive it over two antennas with a distance separation greater than the coherence distance (known as *space* or *antenna diversity*). Receiver antenna diversity does not result in a loss of spectral efficiency but requires the extra antenna hardware which may not be convenient to implement in small portable terminals.

In a wireless communication system the delay spread depends on the environment, the link distance, antenna directivity, and base-station antenna tilt angle. Environments may be classified as dense-urban, urban, sub-urban, and rural. Typically the delay spreads are large in rural areas with a mountainous landscape where values up to $100\mu s$ may be attained in a system with link distances of tens of kilometers and omnidirectional antennas. In a dense urban environment with a link distance of a few kilometers delay spreads are several micro-seconds (typically less than $5\mu s$). Suburban and rural environments with a low density of small buildings and relatively flat terrain have delay spreads less than $1\mu s$, and indoor environments typically have delay spreads with values less than 100 ns. These values correspond to coherence bandwidths approximately equal to 1.6 KHz, 32 KHz, 160 KHz, and 1.6 MHz respectively. Hence the actual coherence bandwidths for typical instances of the above environments are typically greater than the corresponding values given above.

To characterize coherence times we consider the degree of Doppler spread which depends on the velocity of the receiver with respect to the transmitter. In mobile communication systems this value is typically in the range of 0 - 120 Km/H. At zero velocity the Doppler spread equals zero, the space-time correlation is a constant, hence the coherence time equals infinity. At a velocity of 120 Km/H and at a carrier frequency of 1 GHz the maximum Doppler spread is $\omega_m = \frac{\omega_c v}{c} \approx 698$ or $f_m \approx 110Hz$. The coherence time is $\tau_c = 3.4ms$. All of these values depend on our assumptions of uniformly distributed received power over angle and exponentially distributed delay spread. In practice these assumptions do not hold in all cases and we can expect smaller delay spreads (larger coherence bandwidths) and smaller Doppler spreads (larger coherence times).

8.3.3 Approaches to Mitigate Fading

The mitigation of signal fading in a wireless communication system is the most important task in system design. In the design of an efficient wireless communication network there are two fundamental resources which must be efficiently managed. One of these resources is that of radio spectrum, also referred to as frequency management, or frequency allocation. The other traditionally less well known resource management problem which is becoming increasingly important in modern spectrally efficient networks is that of transmitter power management. Proper management of the transmitter power requires the mitigation of the fading problem. Ineffective mitigation of the fading problem will result in a higher required transmitter power to achieve a given SNR for the required quality of service, which will cause higher interference elsewhere in the network.

There are various approaches to system design that mitigate the fading problem. First let us assume that we utilize radio channels with a bandwidth less than the coherence bandwidth as in the case of the AMPS/IS-54 system. In this case when the delay spread is small, as in most cases, fading is flat across the channel, the coherence bandwidth is larger than the channel bandwidth. There is no signal processing method at the receiver (e.g., equalization) which can mitigate the fading problem. The only solution to the fading mitigation problem is to utilize antenna diversity. In the case that the terminal velocity is small or zero the channel will change very slowly with time. At a given point in time some of the channels in the system will be faded (destructive interference of the multipath signals), others will have strong signal (constructive interference of the multipath signals). A solution to the fading problem in the case of a slowly changing channel (low or zero velocity) is to change the utilized channel when the channel fades. This is sometimes known as intra-cell hand-off. This method works well in systems with low or zero terminal velocity such as fixed terminals and especially in indoor radio channels.

Another alternative to mitigate the effect of fading when the rate of change of the channel is too fast to allow for efficient intra-cell hand-off is to utilize periodic frequency hopping. In such a case a change of frequency (i.e., a change of the channel) does not always result in a better channel, but on the average the power of the signal attains an intermediate value and time diversity in the form of error control coding can be effective. This method is used in the GSM system where slow frequency hopping is utilized. In such a system the hopping should occur over a set of channels that has a frequency separation greater than the coherence bandwidth of the radio channel and should occur with a frequency (hopping rate) that is higher than the coding memory of the channel.

The other case is the case where the channel bandwidth is greater than the coherence bandwidth of the channel. In this case the fading is frequency selective. The frequency response across the channel is not constant but has an average value that is sufficiently high to achieve good performance. A transmitted data symbol is typically distorted in time, but maintains a significant level of energy. The approach to mitigate the effect of fading in this case is to use channel equalization. Generally the equalizer performs a function which is somewhat equivalent to the insertion of a filter in cascade with the channel so as to give an overall flat frequency response for the channel. There are various algorithms that can be utilized in the equalization process. Typically the equalizer requires a function which estimates the channel response. The standard method to achieve this estimation is to transmit a sequence of symbols, called a training sequence, periodically. The receiver processes the channel transformed training sequence in order to estimate the channel frequency response which is then used in the receiver equalization algorithm. In wireless communication systems information is typically transmitted in frames. The equalizer training sequence can be sent once or a multiple number of times per frame depending on the coherence time of the channel. Ideally the training sequence should be transmitted once per coherence time period of the channel. If this corresponds to once per frame, the sequence may be transmitted at the beginning or (preferably) in the middle of the frame.

Other approaches utilize equalizers which do not require training sequences. This is advantageous if the coherence time is small where the transmission of a training sequence incurs overhead. These approaches are referred to as blind equalization. In effect the channel is estimated from the received distorted data signal itself.

8.3.4 The Rake Receiver

The rake receiver is a type of equalizer employed in systems that utilize spread spectrum modulation (i.e., systems where the bandwidth of the transmitted signal is much greater than the data rate) and where the channel undergoes frequency selective fading. Let us consider the transmission of one data symbol using the pulse $s(t) = \delta(t)$ as the input to the system shown in Fig. 8.1. Let the output of the channel, as in Fig. 8.1, be $y(t)$ and let its complex envelope (i.e., low-pass representation) be $Y(t)$, where the real and imaginary parts of $Y(t)$ represent the in-phase and quadrature components of the band-pass signal $y(t)$. Then we have

$$y(t) = \sum_{k=0}^{n} \alpha_k g(t - \tau_k) \tag{3}$$

where $g(t)$ is the inverse Fourier transform of $G(f) = G_T(f)G_R(f)$. We can write the low-pass representation (complex envelope) of $y(t)$ as follows

$$Y(t) = \sum_{k=0}^{n} \alpha_k e^{j2\pi f_c \tau_k} \Gamma(t - \tau_k). \tag{4}$$

The above equation may be viewed as the convolution of the functions $\Gamma(t)$ and $\sum_{k=0}^{n} \alpha_k e^{j2\pi f_c \tau_k} \delta(t - \tau_k)$. The latter function is the impulse response of the system assuming an infinite bandwidth. The function $\Gamma(t)$ determines the time resolution of the channel and what we call the resolvability of the channel multipath components. If this function is sufficiently narrow (wide-bandwidth system) then the channel contains $n+1$ resolvable components. As the system bandwidth is made narrower the number of resolvable components decreases and ultimately becomes one. In Fig. 8.4 we show an example of the impulse response that we could say corresponds to a very large bandwidth. This actually shows the peak values of resolvable multi-path from a measurement done with a bandwidth of 10 MHz in a typical urban area with a link distance of the order of 1 Km. In Fig. 8.5 we show the same impulse response (same transmitter and receiver locations) for the case of a system bandlimited to 5 MHz.

The complex signal $Y(t)$ can be represented, using the sampling theorem, by its samples taken at the rate of $R = T^{-1} = 2B$ as follows:

$$Y(t) = \sum_{k=-\infty}^{\infty} Y(kT) sinc(\frac{t}{T} - k). \tag{5}$$

The energy of the signal $Y(t)$ is then given by

$$E_b = T \sum_{k=-\infty}^{\infty} |Y(kT)|^2. \tag{6}$$

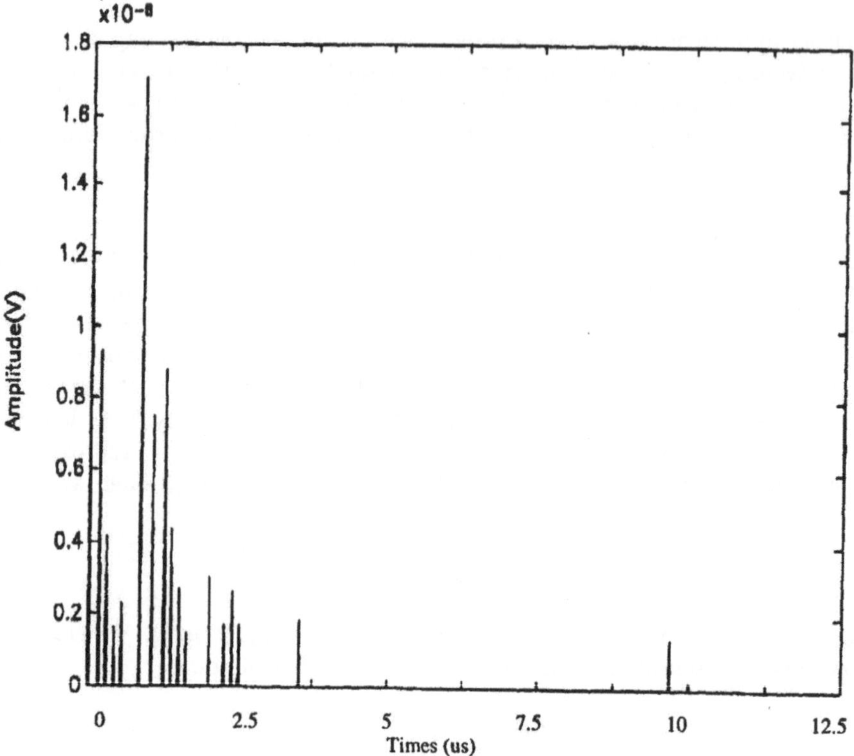

Figure 8.4 Peak values of channel impulse response in an urban location, measured with a bandwidth of 10 MHz.

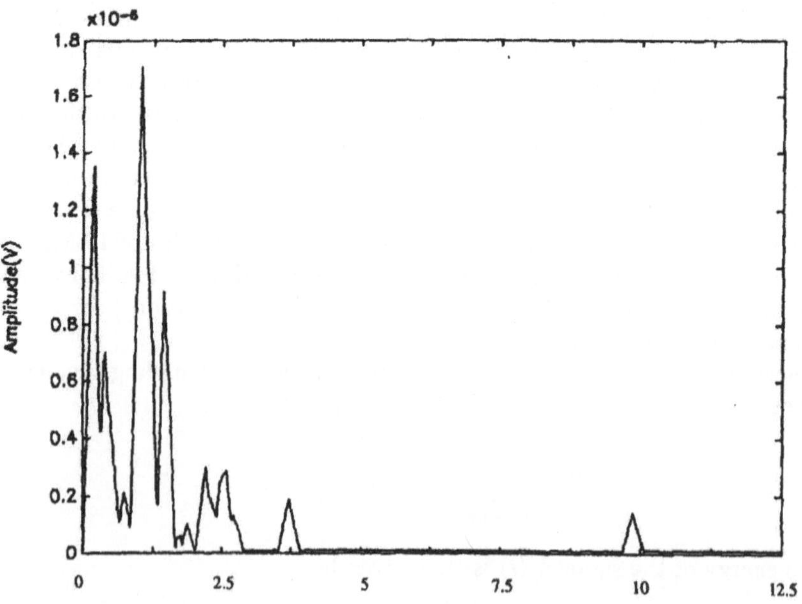

Figure 8.5 Impulse response measured with bandwidth of 5 MHz.

Now, for a given set of delays and a given level of resolvability of the channel (width of $\Gamma(t)$) the values $Y(kT)$ are equal to a sum of shifted versions of $\Gamma(t)$ (corresponding to non-resolvable components) weighted by the complex exponentials as in (4). Typically the argument of these exponential functions (the phases) are independent random variables if the differences in the delays τ_k are significantly larger than the wavelength. In such a case we can show, using the central limit theorem, that with a significant number of non-resolvable multipath components the samples $Y(k)$ are complex Gaussian random variables. The variable $|Y(k)|$ has a Rayleigh distribution, and $|Y(k)|^2$ has the exponential distribution. The assumptions required here are equivalent to the usual assumptions for the Rayleigh fading model of mobile radio channels.

The energy of the received signal corresponding to the transmitted signal $s(t)$ depends on the number of significant components in the sum (6). If this number is one, i.e., the channel consists of one resolvable component, then the received signal energy is an exponential random variable. In typical situations the random variables $Y(kT)$ are independent complex Gaussian random variables with zero mean and variances ${\delta_k}^2$. Let the number of significant values in (6) be m. Then $|Y(kT)|^2$ has a cumulative distribution function (CDF) given by $F_k(y) = 1 - exp(-\frac{y}{2{\delta_k}^2})$. The symbol energy E_b is a random variable with characteristic function given by

$$\Phi_E(\omega) = \prod_{l=1}^{m}(\frac{1}{1 + j2T{\delta_l}^2\omega}). \tag{7}$$

The probability density function for E_b may be obtained from the inverse Fourier transform of (7) using the standard partial fraction expansion to obtain a sum of exponential functions. In a system utilizing an equalizer such as a Rake receiver the performance measure of interest is the cumulative distribution function of the received energy E_b. This can be used to directly compute the channel outage probability and to compute the fading margin of the system which is a very important parameter. The closer to a step function this cumulative distribution function is the better the system performance. In Fig. 8.6 we have shown the above CDF for a different numbers of paths. The perfect case is the case of an infinite number of paths where the CDF becomes a step function. For an outage probability of 1% the fading margin is approximately 18 dB for a 1 path channel and 5 dB for a 6 path channel.

8.3.5 Equalization

One of the main approaches for mitigation of fading is to employ equalization on the channel. A multi-path channel has frequency response characteristics which are non-ideal, i.e., non-constant. There are three main parameters that determine the approach to fading mitigation: the delay spread of the radio environment, the bandwidth of the channel, and the transmission symbol rate. Depending on the values of these different parameters the approach to system design can be determined. In Table'8.1 we show the approach to system design for the different cases in terms of two system parameters: the delay spread normalized to the data symbol period and the bandwidth of the signal normalized by the data symbol rate. Systems with a small normalized bandwidth may be referred to as *narrow band* systems whereas

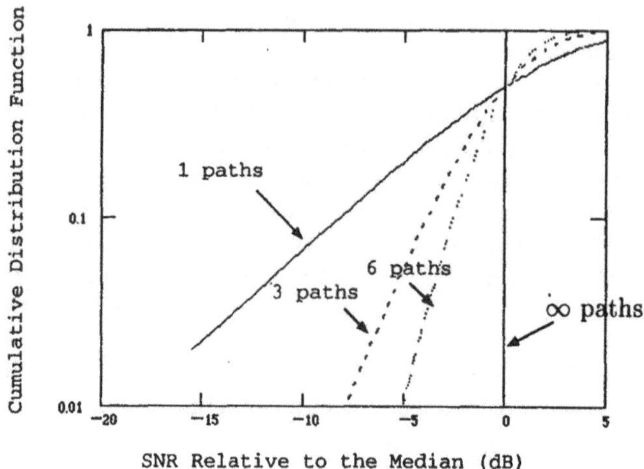

Figure 8.6 Cumulative distribution functions for total received power for a channel with a number of Rayleigh faded multi-path components.

systems with a large normalized bandwidth are referred to as ***spread spectrum*** systems. For narrow band systems a small normalized delay spread means that the system is essentially flat fading and there is no signal processing algorithm that can be used to mitigate the fading. As the normalized delay spread increases the channel becomes frequency selective with a variation in the frequency response within the bandwidth of the transmitted signal. Intersymbol interference results and equalization techniques can be employed to remove the effect of inter-symbol interference. With a large normalized bandwidth (second row of Table 8.1) for a small delay spread a standard Rake receiver may be utilized. In this case the processing is solely on a data symbol by symbol basis. For a large normalized bandwidth and a large normalized delay spread the various delay components of the channel can be resolved and a Rake receiver could be utilized. However we need a complex algorithm to determine the association of the various multi-path components with data symbols before signal combining can be performed. An alternative to signal design for this channel is to utilize orthogonal frequency division multiplex or multi-carrier modulation techniques.

8.4 IS-54/136

The IS-54/136 system is sometimes referred to as the North American TDMA system or the digital AMPS system (DAMPS). The system was designed to be RF channel compatible with the AMPS system. As such it utilizes an RF bandwidth of 30 KHz. Given that a TDMA approach is to be used, the design of this system amounts to the design of a spectrally efficient digital modulation scheme with a bandwidth of 30 KHz and a TDMA frame structure for sharing the channel by multiple users. Highly efficient modulation schemes for band-limited channels have

Table 8.1 Approaches to Equalization for Multi-Path Fading Channels

		Delay spread/symbol period	
		small	large
Bandwidth /symbol rate	small	no signal processing solution	standard equalization
	large	Rake receiver	Rake equalizer, OFDM, other approaches

been developed for the telephone channel and for digital radio point to point links utilizing microwaves. The modulation scheme for these channels typically utilizes a high order signaling alphabet specified by a two-dimensional signal constellation. These constellations have sizes up to 256 points and higher. If (m_c, m_s) is a point on this constellation then the transmitted symbol is given by

$$s(t) = m_c h(t) cos(2\pi f_c t + \theta) + m_s h(t) sin(2\pi f_c t + \theta) \tag{8}$$

where $h(t)$ is the basic signaling pulse whose shape depends on the transmitter and receiver pulse shaping filters.

The bandwidth of this signal depends on the transmission symbol rate and the bandwidth of the pulse $h(t)$. If the size of the signaling alphabet is M and the symbol rate is R then the bit rate is

$$R_b = R \log_2 M. \tag{9}$$

The bandwidth of the signal equals twice the bandwidth of the signaling pulse $h(t)$. This pulse should be designed to satisfy the Nyquist I criterion for zero intersymbol interference, i.e., $h(kT_s) = 1$, for $k = 0$ and 0 for $k \neq 0$, where T_s is the symbol period. The solution that minimizes the bandwidth is the well known sinc pulse which has a bandwidth of $\frac{1}{2T_s}$. For implementation a pulse with a higher bandwidth is used where the bandwidth is $B = \frac{1+\alpha}{2T}$ where α is known as the excess bandwidth (typically $\alpha \leq 1$). Now, for mobile radio channels, as a result of fading, a small signaling constellation is used. In IS-54/136 this constellation has a size of 4, i.e., QPSK, and the excess bandwidth is approximately $\alpha = 0.35$.

Besides the spectral efficiency discussed above, in mobile radio, especially for the mobile transmit channel, it is very important to design a modulation scheme that has a signal envelope that has a low peak to average value ratio. This is required to ensure that the power amplifier for the mobile terminal has a high power efficiency. Ideally we would like the envelope to be constant. However there is typically a trade-off between the envelope peak to average ratio and the spectral sidelobe intensity. To improve the peak to average ratio we typically insert a half symbol delay in the quadrature arm of the transmitter in Fig. 8.7 and obtain the so-called offset QPSK, or OQPSK. The IS-54 scheme also rotates the signalling constellation by $\frac{\pi}{4}$ radians every other symbol and utilizes differential encoding in

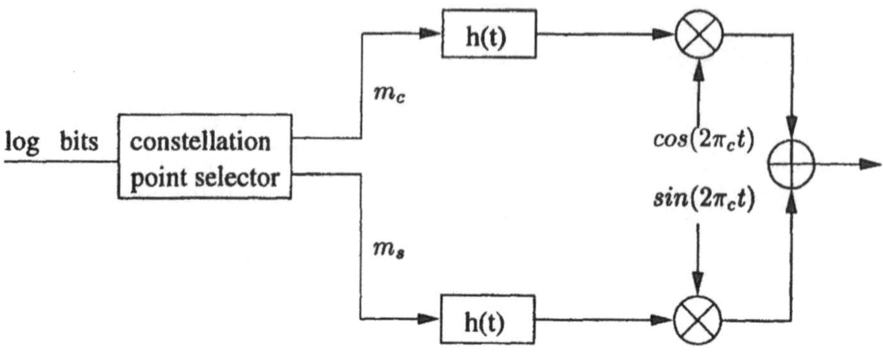

Figure 8.7 QPSK transmitter with high order signaling constellation.

order to facilitate detection of the signal after the carrier undergoes phase changes in the fading channel. The resulting scheme is then known as $\frac{\pi}{4}$-shifted DQPSK. The transmitter symbol rate is 24 Ksymbols/s and the bit rate is 48 kb/s.

With the above bit rate there will be occasional errors in the channel due to signal fading, background thermal noise, and co-channel interference. To reduce the error rate error control coding is used. To determine the number of users that can be supported in one 30-kHz channel we need to consider the speech coder rate.

There has been a vast amount of research since the 1960's on the efficient encoding of speech signals. The first speech coders used in the digital public telephone network employed companded PCM (pulse coded modulation) and achieved a bit rate of 64 kb/s including the synchronization bit. This rate is much larger than the rate required to obtain a higher capacity digital cellular system in comparison to the AMPS system. In fact, with this rate we would not even be able to support one user in the 30 kHz channel of AMPS. Modern speech coders are based on the accurate modeling of the vocal chords and achieve state of the art rates that can go as low as a few kbits/s. The standard full-rate coder used in IS-54 has a rate of 8 kb/s, and the system has been designed for compatibility with future so-called half-rate coders with rates equal to 4 kb/s.

All digital second generation wireless communication systems utilize error control codes. Classically there have been two types of error control codes that have been popular in applications - block codes and convolutional codes. In the case of block codes the Reed-Solomon codes have been the most popular, for example for applications in the audio recording in a compact disc. In the case of mobile communications for telephony the convolutional codes have been the preferred codes up to now. An important parameter of an error control code is the rate of the code; it is equal to the fraction of the transmitted bits that are information carrying bits. In IS-54 the bit rate of the coded speech signal is approximately 13 kbps. If we also consider other factors such as guard times and overhead bits for various functions including training sequences used for equalization, CRC, etc, then the 48 kbps (raw) channel peak rate can support three users.

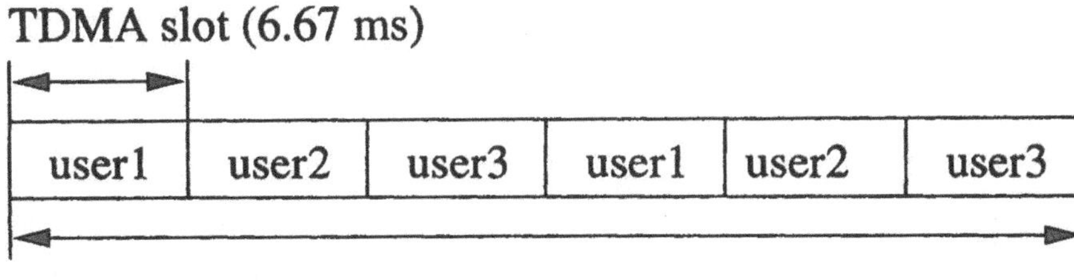

Figure 8.8 TDMA frame structure for the IS-54 standard.

The next issue we need to consider is the multiple access frame structure. The frame size is dictated by the constraint in delay of the speech signal. This constraint also typically determines the speech frame size. In most second generation cellular systems the speech frame is 20 ms long. Typically the speech frame size equals the TDMA multiple access frame size. However, in this case we have a provision for future half-rate speech coders. With these coders we can accommodate 6 users in the 30 kHz channel. As a result the TDMA frame has been designed to consist of 6 multiple access slots. With the present full rate speech coder two slots per multiple access frame are assigned to a single user as shown in Fig. 8.8.

We have seen previously that at a velocity of 120 Km/Hr and at a carrier frequency of 1 GHz, the coherence time for the mobile channel is $\tau_c = 3.4$ms. In order to improve the performance of the error control code it is advantageous to perform interleaving, in order to distribute the errors that occur during a fade, over as large a time span as possible. However we have the constraint on the delay of the speech bits. In IS-54 interleaving is performed over two speech frames and the interleaving depth in time approaches 20 ms, hence there is some gain to be achieved from interleaving. At lower velocities, however, this gain is quickly reduced.

To determines what happens at lower velocities we consider a fixed two path channel where the two paths have equal strength. The transfer function for the channel is $H(f) = 1 + e^{-j2\pi f\tau}$ where τ is the excess delay for the second path relative to the first. The performance of the receiver will depend on the exact equalization algorithm; however, one measure of performance is the energy of the transfer function through a 30 KHz band which we determine as follows:

$$E(d) = \int_{f_c - \frac{W}{2}}^{f_c - \frac{W}{2}} |1 + e^{-j2\pi f \frac{d}{c}}|^2 df \tag{10}$$

where $d = c\tau$ is the excess distance for the reflected path, f_c is the center frequency, and $W = 30$ KHz is the bandwidth of the channel. We have plotted $E(d)$ versus d (the excess distance) in Fig. 8.9. The energy $E(d)$ varies greatly over a distance of one wavelength. In the figure we have plotted the maximum and minimum envelopes. The actual energy varies between these two envelopes with a period equal to the wavelength. For a large excess distance there is no variation since there are many nulls in the bandwidth W. For a small excess distance the received

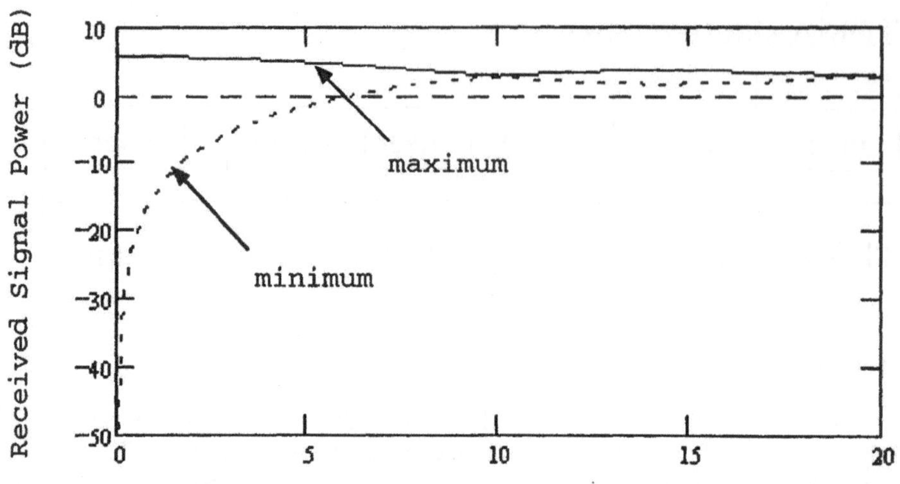

Figure 8.9 Total received signal power for a system with two equal power propagation paths and a system bandwidth of 30 KHz.

energy is very sensitive to the distance and varies greatly. This figure shows that for a small excess distance there is no signal processing algorithm that can be employed at the receiver which mitigates the fading. The fading is very sensitive to the position of the receiver and varies over a distance of a wavelength. The only solution that results in a smaller variation of SNR over distance is the utilization of antenna diversity. This behavior is due to the fading being flat over the frequency band of interest. For large excess distances the channel is frequency selective. There are several nulls and peaks in the frequency band and the received energy approaches a constant which is 3 dB in our example.

As we have discussed above frequency allocation in the AMPS network leads to the concept of frequency re-use cluster size (e.g., see [1]). The standard design for a system utilizing antennas with 120 degree beamwidth yields a frequency re-use cluster size of 7. The capacity of a cellular system in terms of the number of users per cell is inversely proportional to the cluster size. During the initial design of IS-54 it was thought that with digital transmission the system would be more robust with respect to co-channel interference and perhaps allow the operation with a smaller cluster size. In practice this has not been done to this time and the systems that have been deployed employ a similar frequency allocation (i.e., RF channel allocation) and a similar frequency re-use cluster size. As a result the capacity of the IS-54 system is 3 times the capacity of the AMPS system, since it supports 3 users in a 30 KHz channel.

8.5 GSM

The GSM system was designed as a project of the European Community. The acronym initially stood for Groupe Special Mobile, but has since been changed to denote Global System for Mobile. In Europe there was no requirement for backward compatibility with analog systems. The main motivation for the design of this system was to achieve a pan-European system which would be allocated a new frequency band and would allow automatic roaming over the different countries. Without any constraints the designers could opt for one of the main multiple access techniques, FDMA, TDMA, and CDMA. A variation on a TDMA technique was chosen - actually a hybrid of FDMA/TDMA with frequency hopping. The first main parameter to decide on the design of such a system is the RF channel bandwidth. In such a design an RF bandwidth is chosen, a high efficiency modulation scheme is selected to achieve the maximum possible bit rate through the RF channel, and then a TDMA frame structure is introduced. Suppose that we choose a large RF channel bandwidth, then we can support a higher bit rate and will have to share the channel among a large number of users. However there is a maximum time spacing between speech frame transmissions for a given user as a result of the delay constraint e.g., $T_f = 20ms$ speech frame. If we utilize this speech frame as the TDMA frame size, then the TDMA slot time per user is $\frac{T_f}{n}$ where n is the number of users. As the number of users increases the TDMA slot duration decreases. However a guard time is required between TDMA slots, and this time is related to propagation delays and cell sizes rather than to the channel bit rate. As a result the guard time overhead entails an upper limit on the number of users that can efficiently be multiplexed in a given RF channel and consequently results in an upper limit on the RF channel bandwidth.

Let us now look at the considerations of RF channel bandwidth from the other direction, i.e., for small RF bandwidth. The extreme situation is the case where there is only one user per RF channel, hence the FDMA/TDMA hybrid scheme reduces to FDMA. However in this case the RF channel bandwidth is quite narrow and may suffer from flat fading if the delay spread is small and the terminal is stationary (see Fig. 8.9). Under such a condition there is no receiver signal processing scheme that can mitigate the fading. Also if the transmission for a given user is continuous then the receiver is occupied 100 % of the time receiving the signal. If the transmission is bursty (i.e., some form of TDMA with each RF channel) then there are periods in time that the receiver is idle. The receiver can use this time to scan other channels for possible hand-off to neighboring cells. The result is the concept of mobile assisted hand-off that is implemented in GSM systems.

The RF channel bandwidth that was selected is 200 KHz. This is approximately the coherence bandwidth for a typical urban cellular channel. With this wider bandwidth, in comparison to IS-54, equalization will become more effective.

The modulation scheme of GSM is related to a QPSK scheme as follows: Starting with QPSK we introduce a half-symbol delay in the quadrature arm of the transmitter to obtain offset QPSK. Now, if we utilize a half-sine pulse instead of the square signalling pulse we obtain a scheme which could be viewed as frequency modulation and is called minimum shift keying (MSK) [2]. This scheme produces a constant envelope signal. Now if we take the MSK modulator implemented as a frequency modulator (with a square pulse signal as the input) and filter the square pulse with a Gaussian filter we obtain another constant envelope scheme but with a more compact frequency spectrum. The resulting scheme is referred to as Gaussian

Minimum Shift Keying (GMSK). The parameters of this scheme are such that it supports a bit rate of 271 Kb/s in the 200 KHz bandwidth.

At 13 Kb/s the GSM speech coder utilizes a higher bit rate than the IS-95 speech coder. The error control scheme is similar to that of IS-54 in that convolutional codes are used, speech coder bits are divided into two classes in terms of importance, class 1 and class 2. Class 1 bits are coded and the less important class 2 bits are transmitted uncoded. After accounting for the various bit rates, the coding overhead, and various overhead bit fields, we can support a total of 8 users in the 200 kHz channel. As a result if the frequency allocation of the 200 kHz RF channels follows a rule similar to that in analog systems with 25 kHz channels the GSM system would have the same capacity as analog systems utilizing 25 KHz RF channels.

If we consider a transmitted GSM signal (200 KHz bandwidth) and measure the total received power as a function of position of the receiver, there will be a significant variation in the received signal power over different positions with relative distance greater than half of a wavelength. Since the coherence bandwidth is of the order of the channel bandwidth for typical environments this variation in received signal power will differ from channel to channel. One approach to take advantage of this variation and mitigate the effect of fading is to utilize frequency hopping. The carrier frequency of the transmitted signal is hopped over carriers sufficiently separated so as to undergo uncorrelated fading in the different hops. The hopping sequence (of carrier frequencies) should be sufficiently frequent within the transmission of each speech frame so that the symbols transmitted in the different faded hops fall within the memory of the error correcting code. In the GSM system the carrier frequency is hopped over 8 different values within one speech frame. The interleaving and frequency hopping is illustrated in Fig. 8.10. The speech frame encodes a 20 ms segment of speech. After appending error control bits each of these frames contains 456 bits. Two of these coded frames are then distributed over 8 slots (in 8 different TDMA frames) as shown in the Figure. In each slot the even interleaved bits from one frame are combined with the odd interleaved bits from the preceding frame.

Let us assume a deployment of the system consisting of two frequency bands (mobile transmit and base transmit) each with 25 MHz bandwidth. The number of duplex 200 KHz channels is then 124. In a typical deployment these channels are assigned to cells in such a manner as to re-use the channels as frequently as possible but while maintaining a minimum frequency re-use distance between cells utilizing the same channel. An algorithm for the allocation of the RF channels to cells leads to the concept of the frequency re-use cluster size (see [1]). This cluster size may be as low as 3; i.e., each channel in the set of 124 channels is used exactly once every 3-cell cluster. This entails that each cell contains approximately 41 RF channels. In the usual selection of the channels for each of these groups we would select every third channel in the 25 MHz (wide) band so as to minimize adjacent channel interference for the channels allocated to a given cell. In the case of frequency hopping we also need the different channels in one of these groups (i.e., allocated to a given cell) to be as far apart as possible to ensure uncorrelated fading of the signal in different hops.

With the above 41 channels in a cell we introduce frequency hopping. We create a set of frequency hopping sequences so that during each hop no two of the

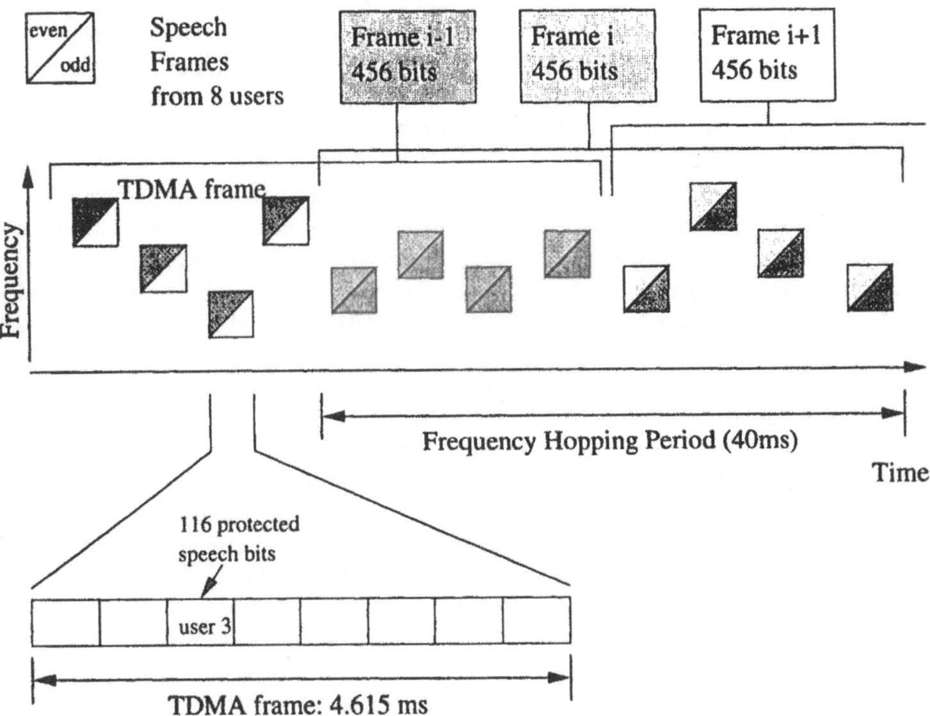

Figure 8.10 Physical layer structure for the GSM system.

41 users in a given cell hop to the same carrier frequency. These sequences are said to be orthogonal hopping sequences.

Let us denote the different channels in the various groups of the frequency re-use cluster as follows: $A = \{f_1, f_4, f_7, ...]$, $B = \{f_2, f_5, f_8, ...]$, and $C = \{f_3, f_6, f_9, ...]$. We could assign these frequencies to the different cells as shown in Fig. 8.11. Now if we introduce frequency hopping we create hopping sequences with the frequencies from each group as follows: For example, for the set A create a matrix $M_1(A)$ with n rows and l columns where n is the number of frequencies (RF channels) in the set A. The elements of this matrix belong to the set A and for each column of $M_1(A)$ each element of A occurs precisely once. A row of the matrix $M_1(A)$ corresponds to a hopping sequence that we will assign to the cell labelled "A_1". We can refer to this matrix as a matrix of orthogonal hopping sequences. With the same set A we can create other orthogonal hopping sequences $M_i(A)$ for $(i = 1, ..., 7)$ such that any two sequences belonging to two different matrices contain at most one element in common; i.e., two users using these hopping sequences would hop to the same frequency and suffer a hit exactly once during a period of the hopping sequence. Such sequences can be designed with the use of Reed-Solomon codes. If we take these matrices and assign them, as hopping sequences, to the cells labelled "A_i" in Fig. 8.11 then we have ensured that co-channel interference has the following peculiar characteristic. The co-channel interference suffered by a given user during a period of the hopping sequence is equal to the cumulative value of bursts of interference from many users in the various co-cells. This is in contrast to a system without frequency hopping where the interference suffered by a user in the center cell ("A_1") for example is equal to the constant interference suffered from the 6 users in co-cells also labelled as A_2 to A_7.

The co-channel interference received from a terminal in a co-cell depends on the position of that terminal in its own cell and the propagation path to the receiver of interest in a co-cell. Some co-channel interferers will be significantly worse than others by virtue of their position. Without frequency hopping some users in the cell A_1 will have constant high co-channel interference, while others will have a constant low co-channel interference. With frequency hopping a given user in the cell A_1 will experience interference from many co-channel interferers at different times from the cells A_2 to A_7. The resulting effect amounts to an averaging of the co-channel interference over all the interferers and is referred to as *interferer diversity*. This concept is fundamental to wireless communications systems; it was first developed in the GSM system but is also applicable to many other systems.

8.6 CDMA

The preceding two digital cellular systems may be considered to utilize narrowband modulation techniques in the sense that the bit rate is of the same order of the RF channel bandwidth. Spread spectrum is an alternative modulation technique which utilizes a much higher bandwidth than these systems in the sense that the bandwidth may be hundreds or even thousands of times greater than the information bit rate. Spread spectrum was initially devised for communications in a military environment [3]. As a modulation technique, and due to its utilization of a wide bandwidth, spread spectrum offers the following benefits: interference rejection, anti-jamming communications, anti-multi-path fading, low probability of

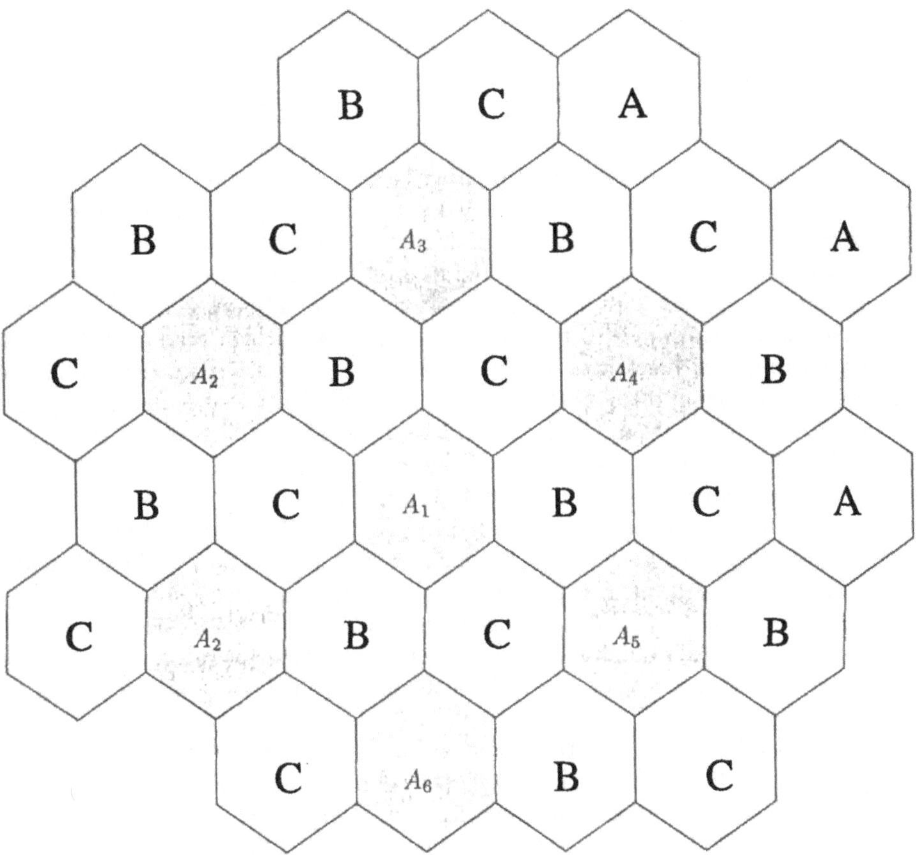

Figure 8.11 Frequency allocation plan for a cellular system with a frequency re-use cluster size of 3.

intercept, modest level of secrecy (encryption), multiple access capability, and good characteristics for frequency re-use in a cellular type of network architecture. The utilization of spread spectrum techniques in a multiple access context is typically referred to as Code Division Multiple Access (CDMA) as a result of the use of different spreading codes to distinguish the different users in the channel.

There are many different types of spread spectrum techniques. The two most important ones are *frequency hopping* (FH) and *direct sequence* (DS). With frequency hopping a classical narrowband modulation technique is used but the carrier frequency is changed (hopped) periodically. Depending on the duration of time for each hop (the dwell time in each carrier frequency) the scheme may be referred to as fast FH or slow FH. The frequency hopping scheme of GSM discussed above is a prime example of slow frequency hopping.

With direct sequence a baseband signal is multiplied by a pseudo random signal. The pseudo random signal has a much higher bandwidth than the information signal and the result is a DS signal with a much wider spectrum than the spectrum of the baseband signal. The requirement for the spreading code (the pseudo random signal) is that it have noise-like properties. In practice it is implemented as a binary pseudo random signal generated by a shift register circuit with feed-back connections. The binary pulses of this signal are commonly referred to as *chips*.

The basic principle of direct sequence utilizing the simplest possible modulation technique is illustrated in Fig. 8.12. The power spectral density of the transmitted DS signal along with the power spectral density of an interfering signal are shown in Fig. 8.13. The bandwidth of the DS signal is given as follows: $W = NB$ where B is the bandwidth of the data signal and N is the ratio of the spreading code clock rate to the input data rate and is called the processing gain in this case. The processing gain determines the capability of the system to reject interference. Fig. 8.14 shows the power spectral densities of the signal components at the input to the receiver low-pass filter. The signal component, upon the second multiplication by the PN code, has been despread back to its original bandwidth whereas the interference that entered in the channel remains spread over the whole band. As a result only a fraction (1/N) of the interference power affects the signal at the detector.

The basic DS spread spectrum scheme that we have described above could be thought of as a BPSK modulation scheme at the chip pulse level since the channel input signal has the spectral characteristics of a BPSK signal with data rate equal to the above chip rate. We refer to this scheme as DS/BPSK.

It is possible to generalize the above scheme to obtain the analogous form of QPSK. However, in the case of spread spectrum there are a few possibilities to generalize BPSK. We may split the data sequence into two sub-streams and create a DS/BPSK signal for each stream where the carrier phases for the resulting two modulated signals have a 90 degree phase difference. An alternative is to transmit each data bit simultaneously on the two DS/BPSK systems as in a diversity scheme. These alternatives are illustrated in Fig. 8.15. In these modulators the in-phase arm is equivalent to the DS/BPSK scheme that we have discussed above with the chip pulse filter included to remove the sidelobes of the spectrum of the SS signal.

The system shown in Fig. 8.12 utilizes a low-pass filter. The optimum approach is to use a matched filter that is usually implemented in discrete time as a chip matched filter followed by a discrete correlator. The above DS/QPSK schemes

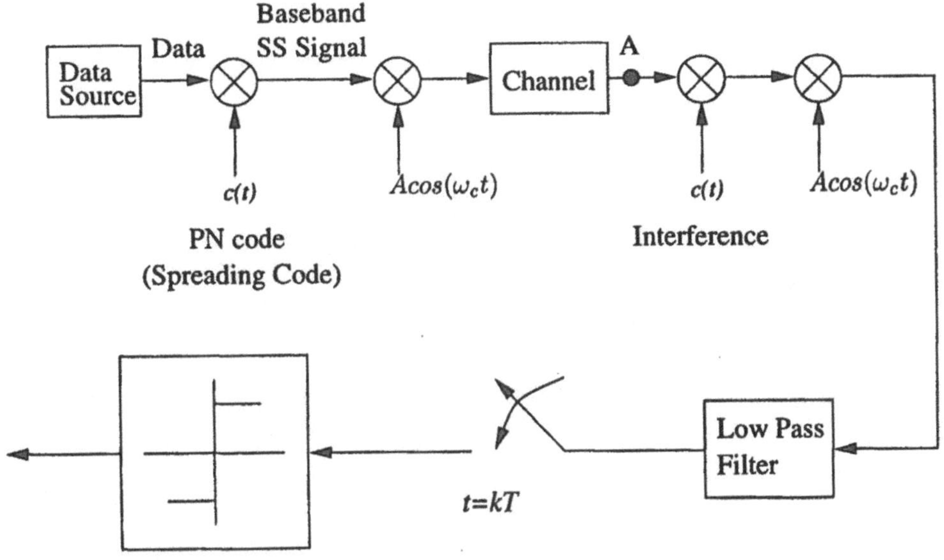

Figure 8.12 Basic direct sequence spread spectrum system with no chip pulse filtering.

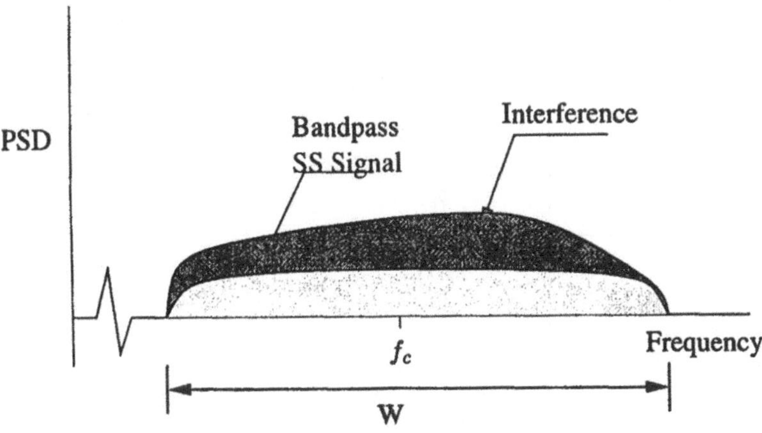

Figure 8.13 Power spectral density of received signal (point A).

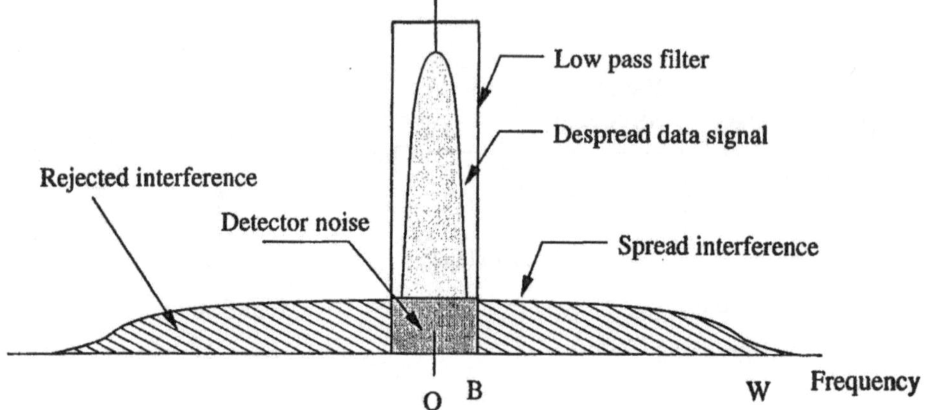

Figure 8.14 Power spectral densities at input to low-pass filter at the receiver.

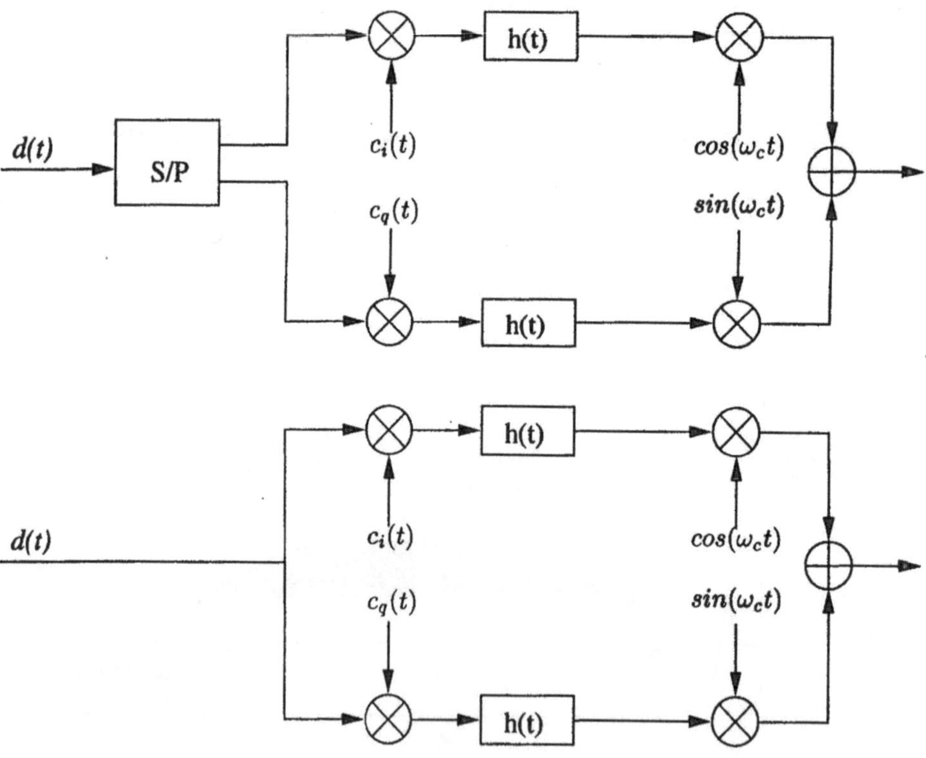

Figure 8.15 Two basic types of DS/QPSK modulators including the chip filter $h(t)$. "S/P" indicates a serial to parallel converter.

can be modified to obtain OQPSK versions by inserting a half-chip delay element on the quadrature channel. It is also possible to obtain DPSK versions of the above, although the relative performance of the DPSK system to the BPSK system tends to be worse than as in the case of narrow band systems.

The advantages of the QPSK versions lie in a greater degree of robustness against a single strong tone interferer or DS/BPSK interferer. In the case of a DS/BPSK system the effect of one of these interferers is dependent on the carrier phase of the interferer with respect to the carrier phase of the signal of interest. In this case if the phase between the carriers of the signal and interferer is zero then depending on the power of the interferer we may have a large bit error probability. On the other hand if the phase between the same two signals is ninety degrees then the bit error probability is zero. In practice the phase between the carriers of two such signals will drift slowly depending on the offset in carrier frequencies and there will be an error pattern that is bursty with the burst length depending on the frequency offset between the two carriers. On the other hand the DS/QPSK scheme of Fig. 8.15(b) is not sensitive to the phase of the carrier of the interferer. This scheme is sometimes referred to as balanced QPSK.

8.6.1 Symbol Error Probability in General Interference

The probability of symbol error of DS spread spectrum scheme depends on the power of the received signal and on the form of the interference, i.e., whether the interference is white Gaussian noise, colored Gaussian noise, tone interference, multiple access interference, etc. For moderate to large processing gains (say greater than 20), and error probabilities higher than 10^{-4}, we may derive a closed form expression for the error probability which is reasonably accurate. Let us consider DS/BPSK (similar results may be obtained for other forms of DS spread spectrum), and a band-pass interference signal $Y(t)$. The received signal is

$$r(t) = \sqrt{2P}\left(\sum_{k=-\infty}^{\infty} d_{\lfloor \frac{k}{N} \rfloor} c_k h(t - kT_c)\right) cos(\omega_c t + \theta) + Y(t) \tag{11}$$

where c_k is the PN code sequence, d_k is the data sequence, $h(t)$ is the chip shaping pulse, T_c is the chip pulse duration, P is the received signal power, and $Y(t)$ is a band-pass interference process. Assuming $Y(t)$ to be a stationary process we may represent it as follows

$$Y(t) = Y_I(t)cos(\omega_c t + \theta) + Y_Q(t)sin(\omega_c t + \theta) \tag{12}$$

where $Y_I(t)$ and $Y_Q(t)$ are low-pass processes with power spectral densities given by

$$S_I(f) = S_Q(f) = Low\ Pass[S_Y(f - f_c) + S_Y(f + f_c)] \tag{13}$$

where $S_Y(f)$ is the power spectral density of $Y(t)$. If we use a correlator receiver, the output of the detector is

$$\pm ANE_h + 2\int_0^T Y(t)c(t)cos(\omega_c t)dt = Signal + Noise \tag{14}$$

where the sign depends on the value of the detected bit. For moderate to large processing gain and symbol error probabilities greater than 10^{-4} the above integral may be modeled as a Gaussian random variable. The probability of error can then be obtained as

$$P_{error} = \frac{1}{2} erfc\left(\sqrt{\frac{E_b}{N_{0eff}}}\right) \tag{15}$$

where $N_{0eff} = \frac{1}{E_h} \int_{-\infty}^{\infty} |H(f)|^2 S_I(f) df$ is an equivalent white noise power spectral density for the interference, and $E_h = \int_{-\infty}^{\infty} |H(f)|^2 df$. As an example, if $Y(t)$ is a tone close to the carrier with power P_Y the $S_I(f) = P_Y \delta(f)$ and $\frac{E_b}{N_{0eff}} = \frac{NP_S}{P_Y}$ where P_S is the power of the signal. We can see from this development that if the interference $Y(t)$ is a signal that can be represented with only an in-phase component (i.e., $Y_Q(t) = 0$) then the symbol error probability depends on the phase of the carrier of the interferer with respect to the phase of the carrier of the signal. On the other hand if $Y(t)$ is a stationary process with equal power on the in-phase and quadrature components then the symbol error probability will not depend on the phase of the carrier.

8.6.2 Spreading Codes

The spreading codes are Pseudo Random (PN) binary sequences. An algorithm is required to generate them at the receiver. If they are to be generated by a finite state machine then they have to be periodic. Hence a major issue is the period of the codes. An efficient technique to generate these codes is by using a shift register circuit such as the six stage circuit shown in Fig. 8.16. If we consider the state of this register as the contents of the n-stage circuit, then there are $2^n - 1$ possible states (excluding the all zeros state), hence this is the maximum possible length of the sequence generated, since the successive bits generated are functions of the state. Sequences of maximum length are always possible for circuits with any number of stages and specific feedback connection patterns. These patterns are usually specified by a polynomial as shown in Fig. 8.16 where we have a circuit that generates a sequence of length $2^5 - 1 = 31$. For a shift register with n stages it is always possible to find a feedback connection pattern that generates a sequence with a period equal to $2^n - 1$.

The choice of the PN code length in a DS system is an important issue. The length of the code can be as short as one data symbol period or extremely long (e.g., months). With short codes the exact structure of the code is a key issue to consider and is typically designed to have specific properties. The code is typically designed to have a "sharp" auto-correlation function. In the case of a long code, the code is typically modeled as a random binary sequence. Different segments of the code are used to spread successive data symbols. With a short code the spectrum of the transmitted signal has a certain line structure with the line spacing inversely proportional to the sequence period. Also in a multiple access situation where other users are also utilizing a short code the interference may have a periodic structure and lack the desirable random properties of true spread spectrum.

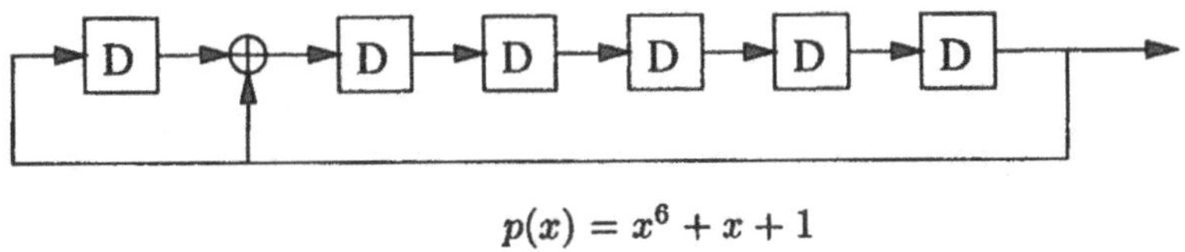

$$p(x) = x^6 + x + 1$$

Figure 8.16 Shift register sequence generator. Generates m-sequence with period 63.

The IS-95 system utilizes two spreading codes of lengths 2^{15} and $2^{42} - 1$. According to our terminology these are both long codes. Short codes with a period equal to the transmitted symbol period are also utilized to differentiate between users.

8.6.3 Synchronization and Pilot Tones

One of the key requirements for a spread spectrum link is synchronization of the PN code at the receiver. The receiver knows the algorithm to generate the PN code but it does not have the phase of the PN code, i.e., the starting time reference for the code sequence in order to align its timing with the timing of the incoming signal. In addition to this, the receiver also requires the usual carrier synchronization for coherent detection, data symbol synchronization, and frame synchronization. Of all of these synchronization requirements, the PN code synchronization is typically the first step in synchronization. Carrier synchronization is performed after PN code synchronization in order to achieve an SNR advantage equal to the processing gain. Data symbol synchronization is typically locked to PN code synchronization. Frame synchronization is similar to the non-spread spectrum case in the case of short PN codes; in the case of long PN codes, frame synchronization is also locked to PN code synchronization.

PN code synchronization typically consists of two steps which are referred to as *PN code acquisition* and *PN code tracking*. PN code acquisition typically brings the timing of the receiver generated PN code to within a chip time of the arriving signal's PN code. PN code tracking typically uses a tracking loop to bring the timing to within a very small fraction of a chip and locks the receiver PN code clock to the transmitter PN code clock.

The acquisition time is an important system performance measure. There are various techniques to perform acquisition. In some cases the PN code is very short and we need to synchronize rapidly. A matched filter type of a circuit is used. The circuit effectively outputs a sharply defined maximum value when the particular code chip pattern enters the matched filter. For the case of a long code there are various acquisition techniques. Typically a correlator with a given programmable code is used. The code is set to various values and correlation is repeatedly performed until one of the code segment values yields a correlation peak. The acquisition time for a system with a long code depends on the following parameters: the PN code period, the uncertainty in the timing that exists at the receiver clock with respect to the transmitter clock, the signal to noise ratio of the incoming signal, and the required operating value for false alarm (i.e., false synchronization).

Typically for transmitter power efficiency it is desirable to derive the PN code clock from the spreading code that is embedded in the received signal, i.e., from the spreading code that spreads the data. However the code acquisition process is quite complicated if there is data (with unknown values) that is modulating the code. A solution may be to transmit periodic short unmodulated bursts of the PN code which the receiver can utilize for quick acquisition and then for the receiver to track the PN code clock from the data modulated PN code during the other times. The easier technique would be to transmit the PN code continuously on a side channel. In the case of a single user system this technique may be inefficient in power usage. However, in a point to multi-point system where a single transmitter transmits to various terminals simultaneously the same side-channel unmodulated PN code signal can be used by all the terminals in the network and it is desirable to utilize such a technique. This scenario is the case in the forward link (base station transmits to mobile) of a CDMA cellular system. The IS-95 CDMA system utilizes this principle of transmission of a continuous unmodulated PN code by each base station for the utilization of all transmissions from that base station. Given that this PN code signal is utilized by all terminals in the given cell, its power can be made larger than the power of a typical information bearing signal transmitted to a mobile terminal, so as to improve the performance of the PN code tracking loop. If the number of users in the cell is large the overhead in pilot power per user is small even if a large pilot power is used.

The value of the pilot signal in a CDMA system such as IS-95 is also important for different reasons. In a typical urban environment, as a result of multipath propagation, the wideband CDMA signal arrives at the receiver over multiple paths. An optimum receiver needs to synchronize to all of these paths in order to maximize the SNR. The pilot transmitted by the base station provides reference signals for the coherent demodulation of the signals of the various users in a fading channel. This technique has been so successful that in third generation systems it has also been adopted for the reverse channel (mobile transmits to the base station) even though in this case, given the multipoint-to-point nature of the channel, the pilot transmitted by a mobile terminal can only be used for demodulating the signal from that terminal. However the gain in SNR obtained from the utilization of coherent detection in the fading channel compensates for the extra power allocated to the pilot signal.

8.6.4 Spread Spectrum Multiple Access

In a networking context spread spectrum modulation is utilized for multiplexing on a channel by assigning a different spreading code to each user. The spreading code acts like a channel. However, all users transmit simultaneously on the same frequency band. The different users can co-exist on the channel as a result of the interference rejection capability of spread spectrum. There is however a limit on the number of users that can be supported with a given error probability. Typically as the number of users increases the error probability increases. The resulting multiplexing scheme is known as spread spectrum multiple access, or more commonly in a commercial context as Code Division Multiple Access (CDMA).

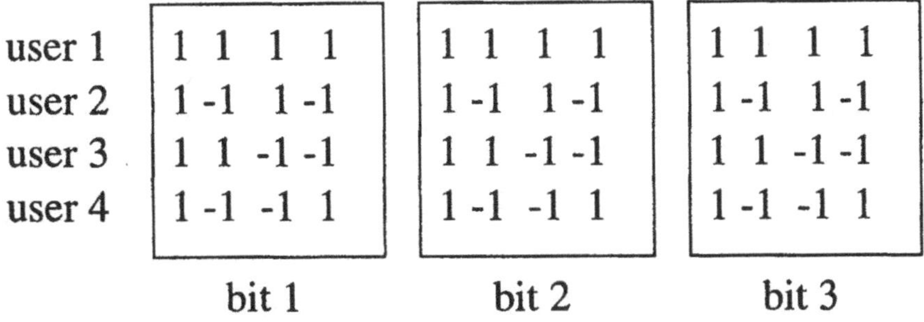

Figure 8.17 4 ×4 Walsh functions for four equal rate users.

8.6.4.1 Orthogonal CDMA

There are many different strategies to assign PN codes to the different users in a CDMA network. The key issue it to choose PN codes in such a manner that the cross-correlation between any two codes is as low as possible. Perhaps the first issue to consider here is whether or not it is possible to synchronize the different transmissions at the chip level. If this is possible then we can actually design a family of orthogonal spreading codes for the different users. In such a case there will be no interference between users and we can support a number of users equal to the processing gain N. Such a set of spreading codes may be represented as the rows of a matrix. The required matrix for orthogonal codes is a Hadamard matrix. A special case of a Hadamard matrix is the matrix generated recursively as follows: $H_0 = 1$, and

$$H_{n+1} = \begin{bmatrix} H_n & H_n \\ H_n & -H_n \end{bmatrix}. \tag{16}$$

As an example, if the processing gain were $N = 4$ then we would have the situation shown in Fig. 8.17. The rows of the matrix defined above are known as Walsh functions. The IS-95 system utilizes the set of Walsh functions of dimension 64, i.e., H_6 in the above.

If we use the above approach to generate orthogonal spreading codes then we end up with a system where the data rate for the different users is the same, this is by virtue of the fact that all of the sequences in the sequence set are of the same length. In third generation systems it is desirable to support different bit rates simultaneously for different services. We can generate a set of orthogonal spreading codes to support different bit rates as shown in Fig. 8.18. In this figure we can support 5 users. Three users have a data rate of say R and two users have a data rate of $R/2$. This method to generate spreading codes of variable rates can be generalized to arbitrary combinations of codes where the rates are equal to a power of 2 times some basic rate. In a system with variable bit rates the transmitter power assigned to the different signals will depend on the desirable SNR $\frac{E_b}{N_0}$ desirable for that service. In the above example if all the services are to operate with the same SNR then the half rate users would utilize half of the transmitter power relative to the full rate users.

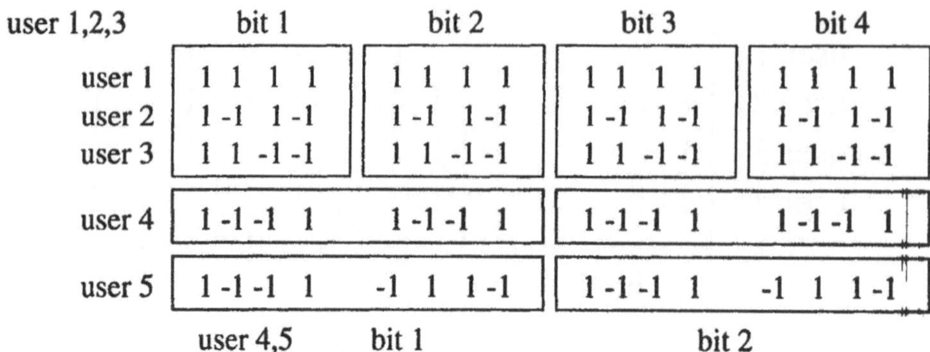

Figure 8.18 Variable rate spreading codes. Three codes to support a data rate R and two codes to support a data rate $R/2$.

The above Hadamard matrices are orthogonal hence the different signals utilizing the different rows as spreading sequences create zero mutual interference. However the rows as spreading sequences lack the essential characteristic of being random. Consider any Hadamard matrix of dimension M. If we take any binary vector ($\pm$ 1 valued) of dimension M and multiply componentwise with all rows of the given matrix then we preserve the orthogonality of the Hadamard matrix; i.e., the resulting matrix is still orthogonal, it is still a Hadamard matrix. In particular the first row of the new matrix will be equal to the multiplying random sequence. However, if we use the same multiplying sequence for the transmission of every data symbol, i.e., if the spreading codes for a given user is the same for successive symbols then we have lost the property of randomness. As a result the above multiplying vector which creates a randomized Hadamard matrix should change for every transmitted data symbol. One method to generate different pseudo random multiplying sequences of this type is to generate them as segments of a long PN sequence.

The above approach to generate randomized orthogonal spreading codes is used in IS-95 to generate spreading codes for the forward link. If we represent the above matrices by 1's and 0's ($1 \rightarrow 0$ and $0 \rightarrow 1$) then the multiplication referred to above becomes addition modulo 2. The random sequence used to randomize the Hadamard matrix is referred to as a masking sequence. In IS-95 the Hadamard matrix has dimension equal to 64, and the masking sequence has period $2^1 5$. This sequence can therefore provide 64-chip PN code segments to randomize the matrix.

8.6.4.2 Non-Orthogonal CDMA

In the above we have considered the design of sets of orthogonal codes which are applicable in synchronous transmission cases. If we consider the reverse channel of a cellular system (mobile transmits to base) then it is not as easy to maintain the different transmissions aligned. There are now two possibilities to proceed here. First we may utilize a family of spreading sequences with short sequences. The cross-correlation between the sequences of different users will not be zero. However, we have an identifiable family of sequences with a possibly relatively small number of sequences in the set. The other alternative is to use very long sequences with each data symbol being spread by a segment of a given sequence.

In the first case we may actually design the family of spreading sequences to have low cross-correlations. Such a family is the well known family of Gold sequences. Along with this we may also utilize a multi-user receiver where all the signals are detected simultaneously. There are various algorithms for multi-user receivers which can be applied here. This approach of using short sequences for multiple access alongwith a multi-user receiver is not utilized in the IS-95 system but is a viable option for future CDMA systems.

In the second case above the different sequence segments are modeled as being random. The different users will act on one another as random interferers and for a given symbol error probability it is desirable to determine the multiple access capability of the system; i.e., to determine how many users the system can support subject to a given operating error probability. The higher the operating error probability the higher the number of users that the system can support.

For a moderate to large processing gain, and if the number of users is not too small, the error probability may be computed using (15) [4][5] where

$$S_I(f) = N_0 + \frac{1}{2T_c}|H(f)|^2 \sum_{i=1}^{K-1} A_i. \tag{17}$$

For BPSK modulation the error probability is then

$$P_{error} = \frac{1}{2} erfc(\sqrt{SNR_0}) \tag{18}$$

where

$$SNR_0 = \frac{E_b}{N0 + \frac{\Psi}{N}\sum_{k=1}^{K-1} E_{b,k}}, \tag{19}$$

K is the number of users, $E_{b,k}$ is the energy per bit for the kth user, E_b is the energy per bit for the user of interest, N is the processing gain, N_0 is the one-sided power spectral density of the background noise, and

$$\Psi = \frac{1}{E_H^2 T_c} \int_{-\infty}^{\infty} |H(f)|^4 df \tag{20}$$

is a factor which depends only on the chip pulse shape. E_h, T_c, and $H(f)$ are the energy, duration, and Fourier transform of the chip pulse. For example, for a rectangular chip pulse (time limited) $\Psi = 2/3$, and for a sinc chip pulse $\Psi = 1$. For different types of modulation, including coded modulation schemes, the bit error probability is given by a similar expression to (18) except that the function $\frac{1}{2}erfc(\sqrt{.})$ is replaced by some other function $f(.)$.

Equation(18) illustrates the necessity for power control in CDMA systems. To support the maximum number of users subject to a maximum probability of error the energy per bit of the various signals at the receiver (or received power) should be constant over all users. Also if the power of the signals is sufficiently large then we can neglect the background noise N_0.

For a given application the important parameter is the multiple access capability. This depends on the error probability requirement which depends on the type of service. A given error probability requirement translates into a nominal SNR requirement SNR_0. The multiple access capability is then

$$K = \frac{N}{\Psi \cdot SNR_0} + 1. \tag{21}$$

As an example, if $SNR_0 = 6$dB, and we use a sinc chip pulse then $K = \frac{N}{4} + 1$. Thus for a processing gain of N (bandwidth expansion factor) we can accommodate approximately 25% of N users. For orthogonal schemes the number of users that we could accommodate for a system with the same bandwidth would be equal to N and the error probability would depend only on the background noise intensity N_0. In general the relationship between SNR_0 and the bit error rate depends on the modulation and error correction coding schemes. With a stronger error correcting code we can achieve the same bit error rate at a lower value of SNR_0, hence we can achieve a larger multiple access capability.

8.6.5 CDMA Cellular Systems

The use of CDMA in a cellular network was initially proposed in [6]. Spread spectrum, with its interference rejection capability, allows for the operation of a cellular system with a frequency re-use cluster size of one. The whole allocated frequency band is used in each cell. The number of users per cell depends on the level of multi-user interference. Multi-user interference arises from in-cell interferers and out-of-cell interferers. The nature of these types of interference on the forward and reverse channels is different. In general, to minimize interference we may attempt to orthogonalize the signals so as to reduce multi-user interference. For the set of signals transmitted from one base station this is possible in practice but the multi-path propagation tends to destroy the orthogonality to some extent. For the signals in the reverse channel (and assuming no multi-path propagation) it is possible to orthogonalize the signals in principle but difficult to achieve in practice. If we consider signals from different cells then it is impossible even in principle and assuming single path propagation to orthogonalize the whole set of signals, over all cells, and simultaneously achieve high spectral efficiency.

To minimize the interference level and to maintain a minimum performance level, power control is a necessity in a CDMA cellular network. Each terminal adjusts its power so that the received power at a base station is constant over all mobiles. In cellular systems, typically the bandwidths assigned to the forward and reverse channels are equal. In principle the capacity of the system is limited by the capacity of the reverse channel due to its multipoint-to-point nature and the difficulty of maintaining the different signals orthogonal, although this may not necessarily be the case in a given implementation. In this section we consider the capacity of the reverse channel.

Assuming a one-cell system, the capacity is given by (21). Let this capacity be denoted by $K1$. For a multiple cell system there will be inter-cell interference in addition to the in-cell (or intra-cell) interference, hence for a given error rate we must reduce the number of users per cell to a value $K_m < K_1$. The exact value of K_m depends on the positions of the users within the cell. If all users are near their respective base stations then there will be little inter-cell interference and $K_m \approx K_1$. The other extreme is where most users are near their respective cell boundaries where the inter-cell interference is the largest and will be significantly lower than K_1. In general the total inter-cell interference will depend on the number of users per cell, their positions within their cells, the propagation power loss law, e.g., r^{-4} over flat ground, and the macro environment characteristics. If we model the terminal positions as random quantities then the inter-cell interference is a random variable. The variance of this random variable decreases as the number of users increases. If the number of users per cell is a few tens then we may approximate the inter-cell interference by its mean. The inter-cell interference is then $\alpha K_m P$ where α is a parameter which depends on the propagation power loss law ($\alpha < 1$), K_m is the number of users per cell and P is the common received power of a user at its base station. For an ideal propagation environment with an r^{-4} power loss law $\alpha \approx 0.4$. The total interference at a base station is

$$Total\ Interference = Inter\text{-}cell\ Interference + Intra\text{-}cell\ Interference. \tag{22}$$

The inter-cell interference is $(K_m - 1)P$, hence the total interference is approximately $((1+\alpha)K_m - 1)P$ and the capacity per cell may be obtained by re-deriving (21) from (19) by replacing the inter-cell interference term in the denominator of (19) with the total interference, which yields

$$K_m = \left(\frac{N}{\Psi \cdot SNR_0} + 1\right) F \tag{23}$$

where $F = 1/(1 + \alpha)$. From a capacity viewpoint the parameter plays the role of a frequency re-use factor. It represents the reduction in cell capacity (i.e., in the number of users allowed in a cell) which is required to reduce inter-cell interference to acceptable levels. For narrow band systems this corresponds to the reciprocal of the cluster size.

From (18), (19), and (23) we can see that the capacity of the CDMA system is limited by interference. Various enhancements to the network can be made which lead directly to a decrease in interference and a corresponding increase in capacity. We now discuss some of these.

8.6.5.1 Sectorized antennas

Let us assume that directional antennas are used and that there are G directional antennas per cell which cover the cell and have non-overlapping beams. If the interferers are uniformly distributed in space then the amount of interference at each antenna at a base station is reduced by the factor G^{-1}. Hence from (19) assuming that N_0 is negligible the SNR_0 is increased by a factor of G. Hence we can increase the number of users by the factor G and still maintain the required constraint for the required quality of service. In practice the antenna radiation patterns are not ideal and there is some sector to sector interference and the interference reduction is somewhat less than the above.

As a mobile terminal moves from sector to sector in a cell a hand-off is performed in the sense that the transmission/reception of the signal switches to a different antenna sector in the same cell. If the antenna beamwidth is very small then we have a large capacity per cell but hand-offs occur very frequently as a terminal travels across the cell. This may lead to a degradation in performance depending on the implementation of the hand-off. Hence there is a practical upper limit to the number of sectors per cell.

Another technique to reduce interference is to use adaptive antennas. These antennas typically contain a number of elements whose inputs/outputs are combined with the amplitudes and phases adjusted dynamically in such a way as to null out the interferers and maximize the signal to interference ratio. These techniques have the potential to offer very large increases in capacity but require a large degree of signal processing capability at the receiver.

There are many different approaches to designing antenna systems that offer greatly reduced interference and hence a greater capacity in a CDMA network. One approach is to utilize a large number of antenna sectors (e.g., 24) with a small sector angle and then utilize an automatic (seamless) technique to implement sector-to-sector hand-off. With this technique the base station consists of a receiver which scans all the sectors for multi-path components of a given user. The standard Rake receiver may be used not only to combine multi-path components from a single sector but also components from different sectors. The resulting receiver may be referred to as a space-time Rake receiver [15].

The other main approach to designing antenna systems for capacity enhancement is to use adaptive beamforming techniques. A type of adaptive beamforming algorithm is run for each demodulated user. The algorithm effectively implements spatial filtering of the interferers based on the arrival angle of the signal. The algorithm adapts to the angular position of the signal of interest. These algorithms are typically implemented in DSP processing of the composite baseband received signal [7][8].

8.6.5.2 Non-continuous traffic sources

Traffic sources for general multimedia networks may be classified into three basic types: circuit switched with continuous transmission at a constant rate; circuit switched with bursty transmission or non-constant rate transmission, and packet switched. For the second type of traffic the throughput of the system utilizing multiple access techniques such as FDMA or TDMA can be increased by using statistical multiplexing techniques where the number of physical channels serving a group of users is less than the number of users. Spread spectrum offers a natural method to implement this type of statistical multiplexing. If the power of the transmitted signal is reduced when the bit rate of the signal decreases then the interference in the network is reduced and the capacity of the network (in terms of the number of traffic sources supported) is increased. In the case of voice signals vocoders with variable bit rates have been designed which take advantage of the variable information content of the speech signal. The vocoder for the IS-95 system utilizes four bit rates: full rate, half rate, quarter rate, and eighth rate. However the bit rate of real signals concentrates on two main values: full rate, corresponding to a talk spurt, and the eighth rate, corresponding to a pause in speech. For a typical conversation approximately 33% of the time is occupied with talk spurts and the remaining time is occupied with pauses. We refer to this value as the *voice activity factor*.

For the purpose of capacity analysis we model the speech coder with the two extreme states which we refer to as *talkspurt* and *pause*. To decrease interference we may turn off the CDMA signal during the speech pauses. The difficulty here is to re-synchronize when the next talk spurt arrives. One technique is to introduce a slot structure to the system and initiate transmissions only at some predetermined times with a slot governed by a pseudo-random algorithm.

For a given number of users in a cell the number of users which are transmitting a talkspurt at a given time is a Binomial distributed random variable with parameter V, the speech activity factor. For a large number of users the standard deviation of the number of active users (transmitting a talkspurt) becomes small relative to the mean. In such a case the interference in the network is reduced to a fraction (slightly greater than V) of the interference in the case of the same number of continuous stream transmissions, depending on the number of interferers and link outage probability.

CDMA can also accommodate constant traffic streams with different rates in a natural way. For a given error probability (performance level) the important parameter is $SNR_0 = E_b/N_{0eff}$, hence if the data rate is lowered then we can achieve a given value of E_b with a lower transmitted power. This property makes it easy to integrate services with different data rates into one transmission scheme with a fixed chip rate and system bandwidth. The transmitted power of a given user is proportional to its transmitted data rate.

8.6.5.3 CDMA cellular system capacity

With the above two methods for interference reduction (sectorization and speech activity factor) the sum in the denominator of (19) is reduced by the factor V/G. This value must be accounted for in rederiving (21). Also, (21) was derived assuming uncoded BPSK modulation where N is the number of chips per data bit, the processing gain. In the case of the use of error correcting codes the relevant parameter is still the processing gain, however it is now more generally defined as the ratio of the system bandwidth to the information data rate, W/R. If we use a high degree of filtering the chip pulses become sinc pulses and the parameter Ψ in (21) approaches unity. Accounting for all of these factors, the capacity equation becomes

$$K_m = \left(\frac{W}{R \cdot SNR_0} + 1\right)\frac{FG}{V}. \tag{24}$$

As an example we compute the capacity for a system such as the IS-95 system. Let $W = 1.25$ MHz, $R = 9600$ bits/s, $G = 3$, $V = 0.4$, $F = 0.71$, and $SNR_0 = 7$ dB. The capacity is then $K_m = 144$ users/cell. As a comparison with the analog cellular system AMPS, in the same bandwidth we can have 41.6 30-KHz channels. If the frequency re-use cluster size is 7 then the number of users per cell is approximately $42/7 = 6$. Thus for the above parameters the capacity advantage of the IS-95 system over AMPS is $130/6 = 24$. This calculation is optimistic since we have not considered a realistic propagation environment. References [9]-[12] contain a more detailed analysis of the capacity taking into account typical propagation environments.

There has been some controversy as to the capacity of CDMA systems relative to narrowband cellular systems. Part of the reason for this is the validity of the various approximations made in calculating the parameters in (24). The calculation of these parameters depends on the positions of the terminals and on the accuracy of various averaging operations. However the most important parameter and the one that causes the greatest uncertainty in the value of the capacity of the system is the operating value of SNR, SNR_0 required to achieve a pre-specified speech frame error probability that is typically assumed to be 10^{-2}. For each 3 dB change in the value of the parameter the capacity calculation yields a result that changes by a factor of 2. For a given level of performance (error rate), the required value of SNR_0 depends on the propagation environment and the effectiveness of the power control scheme. For typical symbol frame error rates (e.g., 10^{-2}) such a value may range anywhere from 3 dB to 15 dB. This corresponds to a range of 12 dB, or a capacity range given by a factor of 1:16. If we assume that a more realistic requirement, given the fading process, is $SNR_0 = 10$ dB, then the above capacity advantage over AMPS would be approximately 12.

8.7 POWER CONTROL

The propagation environment has a large effect on the capacity of a CDMA network. The actual effect of the environment depends on the velocity of the mobile terminal and on the effectiveness of the power control scheme in tracking changes in the power of the received signal. As the mobile travels the signal fades according to two component processes: a slow fading process due to shadowing, and macro characteristics of the environment and a fast fading process due to multi-path propagation. To maximize the number of users, a power control scheme which attempts to maintain a constant received power must be used. The variations in signal strength due to shadow fading are frequency independent and the variations due to multi-path propagation are frequency dependent. Frequency independent variations can be compensated for based on the received signal power on the forward channel. The mobile adjusts its transmitted power based on the received signal strength. This type of compensation is sometimes called $open - loop$ power control. Variations due to multi-path propagation cannot be compensated by the above technique since the channel coherence bandwidth is smaller than the frequency separation between the forward and reverse channels; i.e., the fast fading processes in the forward and reverse channels are independent.

Considering multi-path fading, the channel may be modeled with the impulse response given by (4), where $\Gamma(t)$ is a pulse determined by the bandlimiting filter of the channel.

Over small distances the α_k's in (4) remain relatively constant but the phases $\theta_k = 2\pi f_c \tau_k$ vary unpredictably. If the relative delays between any two paths are greater than the chip time T_c then the signals in the different paths are uncorrelated (resolvable) and if the number of resolvable components is large the received signal power is relatively constant over short distances. If the relative delays are smaller than T_c then the baseband signals are correlated and the RF signal varies drastically over small distances due to constructive/destructive interference.

The optimum receiver for the multi-path channel is the Rake receiver [13]. It consists of a set of correlators (sometimes called the Rake fingers) which demodulate the signals on the various resolvable components. The outputs of the various correlators are then combined to achieve what we may call ***path diversity***. For optimum performance the required number of fingers in the Rake receiver depends on the delay spread of the channel and the bandwidth of the system - for an environment with a large degree of multi-path propagation it is given approximately by DW, where D is the delay spread and W is the system bandwidth.

With a Rake receiver the effect of multi-path propagation on the spectral efficiency of a CDMA network can be practically mitigated if the total received power on all paths can be maintained constant. The relevant parameters here are the system bandwidth (or chip rate) which determines the variability of the power of the received signal with distance, the channel delay spread which determines the number of received signal components, and also the variability of the received signal, the velocity of the mobile terminal, the speed of power control adjustments, and the number of fingers in the Rake receiver. In Fig. 8.19(a) we show an example of the variation of the total received power assuming a perfect Rake receiver over a distance of approximately 6.5m in an urban environment for different system bandwidths. With an infinite bandwidth the number of resolvable components is very large. There is practically no fading in each of the components and the total received power is constant - even if there is fading in the individual components this fading is averaged out at the output of the Rake receiver combiner due to the large number of components. As the bandwidth decreases the variation in signal power increases to a value of approximately 18 dB below the average for the case of a system with a 30 kHz bandwidth and practically one resolvable component. In Fig. 8.19(b) we show the cumulative distribution functions for the total received power. The constant value corresponds to the case of an infinite bandwidth.

The above variation in signal power may be compensated for by power control. The receiver sends power correction commands back to the transmitter at a sufficiently high rate to account for the speed in signal variation which depends on the speed of the receiver relative to the transmitter. This type of power control is referred to as closed loop power control since it relies on a feedback channel between the receiver and transmitter. In the IS-95 system this feedback channel operates at the rate of 800 commands per second (1 bit per command). In other CDMA systems this power control sub-channel may have a rate anywhere between 500 and 2000 bits/s.

After we have done our best, to achieve a constant received power, in system design, taking all of the above factors into account (Rake receiver, power control, etc.), we have a system where the SNR at the receiver is still not constant with time. The SNR may vary with time or distance as shown in Fig. 8.20. To maintain acceptable performance the SNR must be greater than some minimum value SNR_m most of the time. This value is determined by the modulation, error control scheme, and target channel symbol error rate for the service under consideration. On the other hand, in the capacity formula the relevant parameter is the average SNR, i.e., (SNR_0). The difference $SNR_0 - SNR_m$ is the *fading margin*. The larger this value the lower the spectral efficiency of the system. This concept of fading margin can be generalized to systems with any type of modulation. The fading margin is a fundamental parameter in the design of any radio system which suffers from multi-path fading. The reduction of the fading margin is extremely important to achieve a high system capacity.

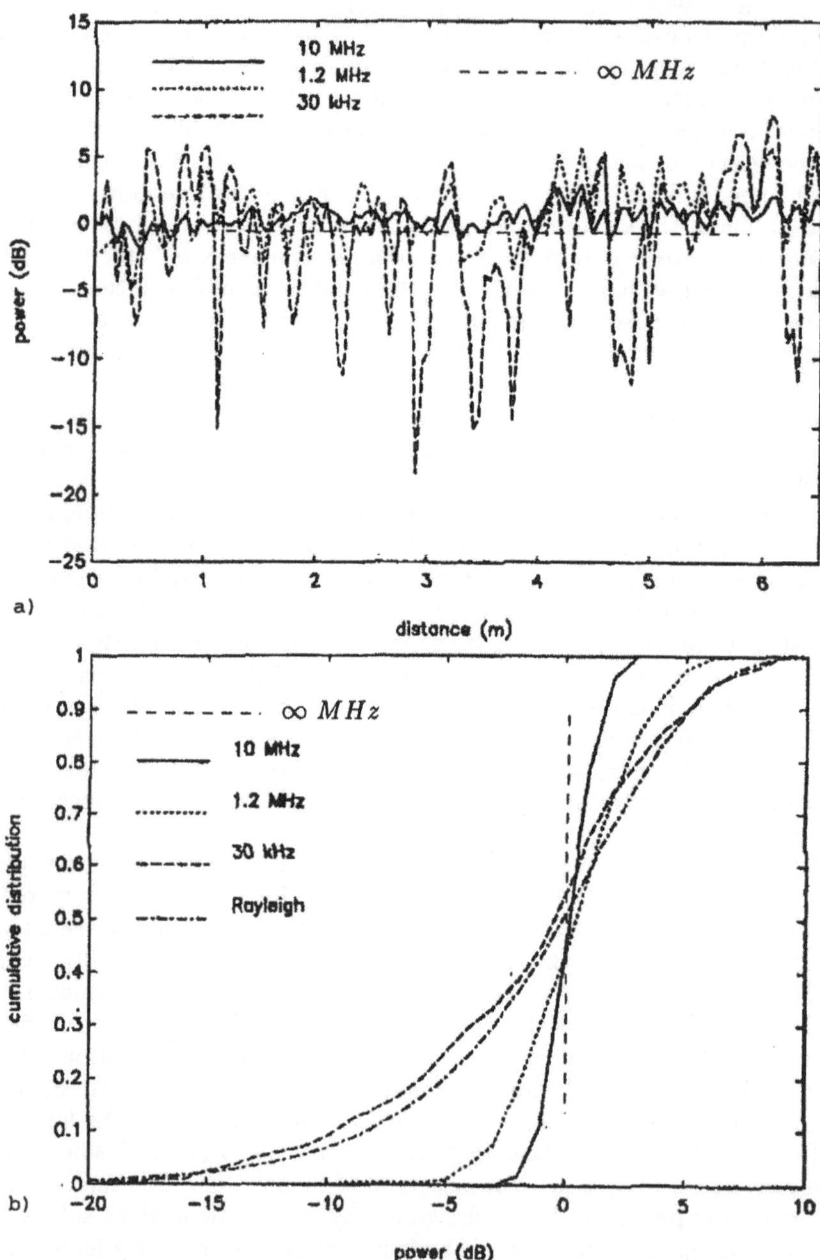

Figure 8.19 a) Variation of total received power versus distance in an urban environment for different system bandwidths- Illustrates requirement for power control. b) Cumulative distribution function of the total received power.

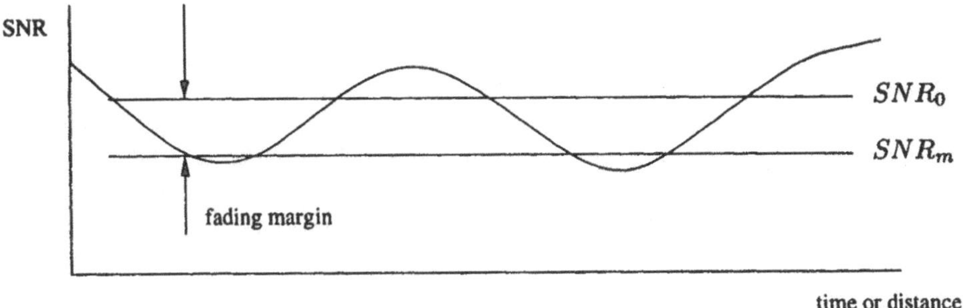

Figure 8.20 Residual SNR variation due to imperfections in power control scheme and implementation of the Rake receiver.

8.8 HAND-OFF PROCESSES

In a cellular network the connection of a terminal to the switch is done through one or more base stations. CDMA, through the much touted soft hand-off mechanism, allows for the simultaneous connection through a multiple number of base stations in order to achieve a type of diversity. This represents a significant evolution in call hand-off techniques since their introduction in the AMPS network. In this section we discuss hand-offs in a cellular system in a more general context.

The simplest way for the mobile to connect to the switch is through a single base station. When the propagation loss on the path from the mobile to the current station increases beyond a threshold, a different base station is found to take up the call. In all cases the communication between a terminal and the network occurs via a single base station. This is the method of operation of systems such as AMPS, IS-54, and GSM.

The above method of operation may evolve to include multiple bases stations. There are three possibilities here. We may utilize multiple base stations only in the forward link, or only in the reverse link, or in both the forward and reverse links. The IS-95 system utilizes multiple base stations in both the forward and reverse links. We may refer to the different approaches with respect to the base stations as transmit diversity, receive diversity, or both transmit and receive diversities, although the issues that we are addressing here are somewhat different than those in the usual cases where one considers these types of diversity. As with other cases of diversity there are different approaches for making use of the different signals, referred to as combining techniques in the usual setting.

Receive diversity can be used with any system whatsoever regardless of the modulation scheme. It always results in a higher system capacity if the fading of the different signals is uncorrelated or if the thermal noise affecting the different signals is uncorrelated. However receive diversity will result in significantly more complex equipment in the infrastructure.

There are different approaches to implement this type of receive diversity at the base stations. The simplest approach is for the signals received at the different base stations to be individually demodulated and transmitted to the switch in the usual packet format. The switch checks the CRC's (cyclic redundancy checks) of the multiple received packets and implements a selection combining rule based on the CRC's. The transmission is successful if at least one of the packets correctly decodes during the CRC computation. This type of processing is used in the IS-95 system. It tends to implement a sort of macro diversity. It improves the capacity of the radio system at the cost of extra trunk circuits in the base station to switch link. If each active communication (each ongoing call) is using this type of diversity of order two; i.e., if each call is being demodulated at two base stations, then the number of trunk circuits from the base stations to the switch is twice the number that would be required in the case where no diversity is being used.

The other possible approach to base station receive diversity is for each base station to do soft decision detection of each received symbol and transmit the resulting symbols to the switch where they will be used alongwith the soft-detected symbols from all the other corresponding base stations in a single decoder [14]. The information transmitted by each base station represents what we refer to in detection theory as a set of sufficient statistics corresponding to the received analog waveform at the corresponding base station. This approach would require higher capacity links between the base stations and the switch. With optimum processing at the switch this approach represents the optimum approach to maximize the capacity of the system for a given set of base stations. Typically we would apply some type of maximal ratio combining for the different symbols at the switch and then feed the resulting soft-decision symbols to some type of maximum likelihood decoder. This approach could be implemented in an enhanced version of the current infrastructure for any digital cellular system including IS-95 and GSM. In such an optimum system the terminal would always be in soft hand-off, in the reverse link, with every base station in the system that is within radio range.

We now consider the case of transmitter diversity. With transmitter diversity, the same information bearing signal is transmitted by a multiple number of base stations. In this case there are two possibilities. In the first case we consider spread spectrum modulation. The signals transmitted by the different base stations are resolvable at the terminal receiver and can be combined if either they utilized different spreading codes or if they utilize the same spreading code with a significantly large shift in time. In IS-95 the same spreading code is utilized with a shift that is a multiple of 64 chips with respect to the spreading code of any of the other cells in the vicinity.

Contrary to the case of base station receiver diversity, utilizing a large number of base stations is not necessarily the optimum strategy. For example, transmitting the signal from a base station that is far from the mobile terminal will create more harm in the network by virtue of the extra multiple access interference than any benefit gained from the use of diversity. In fact, in a network where the terminal is stationary and the links from the base stations to the terminal are time invariant, the optimum approach is to transmit through a single base station and not to utilize transmitter diversity. In this case the transmitted signal should be from the base station whose base-to-terminal link has the lowest path loss taking into account all the fading factors including shadowing and Rayleigh fading. The benefit of transmitting from multiple base stations arises when the links from the different base stations to the terminals undergo a fading that is fast with respect to the time required for the system to determine the fading parameters of the different links and to choose the optimum link for transmission. The classical signal combining method utilized with spread spectrum is the Rake receiver with maximal ratio combining method. The resulting scheme is also referred to as ***path diversity***.

The other case of transmitter diversity to consider is the case where the modulation scheme is not spread spectrum. In this case the multiple transmitted signals will cause interference if they occupy the same carrier frequency. A possible approach is to transmit the same signal on different carrier frequencies, i.e., different RF channels. This however is spectrally inefficient since it reduces the number of users that can be supported in a given cell that has a given number of allocated RF channels.

As we have seen from the above, there are a number of possibilities for the utilization of diversity either in the reverse link or in the forward link. Assuming the use of both transmit and receive diversity, for a given terminal at a given point in time there will be a given set of base stations B involved in the forward link transmission and a number of base stations R involved in the reverse link reception. In the IS-95 system these sets are equal and are referred to as the active set. These two sets will change with time as the terminal moves. The hand-off process determines how these sets are changed; i.e., how we add and drop base stations from the above sets. First we consider the case where there is no diversity - there is only one base station involved both in the forward and reverse links. The hand-off process is initiated by a signal strength measurement which determines that the current link has an excessive path loss and that an alternate link with a significantly lower path loss is likely available. In the AMPS system this measurement is made by receivers at the base station. These receivers continually monitor the received signal strength and alert the mobile when the received signal strength drops below a given threshold. Upon being alerted to the low level of its signal, the terminal scans the set-up channels of the neighboring cells and informs the base as to which one has a sufficiently large signal strength. The switch allocates a channel in the new cell and the terminal is directed by the base station to perform a hand-off to the allocated channel in the new cell. This type of a hand-off is termed base-station initiated hand-off since it is the base station that performs the initial signal strength measurement.

Ideally the signal strength monitoring function should be performed at the terminal since it is in the best position to find a base station with a sufficiently low path loss. The terminal would continuously monitor signals transmitted by the neighboring base stations and initiate a hand-off process by communicating with the current base station when the signal from an alternate base station is significantly stronger than the signal that it is receiving from the current base station. However, a typical terminal contains a single RF front-end. Hence it is not possible for the terminal to continuously monitor signals from neighboring cells without interrupting the main communication channel in FDMA systems where transmission/reception of the traffic channel is continuous. For this reason, the AMPS system utilizes a base-station initiated hand-off procedure.

In TDMA systems, the terminal can perform the signal strength monitoring function during time slots when it is not transmitting or receiving on its traffic channel. In CDMA systems such as IS-95, the terminal receiver contains a searcher correlator that continuously looks for pilot signals with sequence offsets that correspond to offsets expected for the neighboring cells. This information is acquired by the terminal during call set-up and when the active set of base stations changes during a soft hand-off process. This type of hand-off is referred to as *mobile assisted hand-off*.

An alternative to the above process, that would shift the system complexity from the terminal to the infrastructure, would be for the neighboring base stations that are not involved in the active set to be alerted by the switch to search for the signal of a given user that the switch expects should be within their radio range. This entails either measuring the strength of a signal in a given RF channel and/or time slot, in the case of non-spread spectrum signals, or checking for a specific PN code, i.e., correlating with a specific PN code in the case of spread spectrum signals. The base station that finds the user then joins the active set in the forward channel. In the case of non-spread spectrum signaling, the terminal would then be allocated a different channel for transmission in the new cell so as to maintain the minimum distance between cells that use the same RF channel and hence prevent co-channel interference. In the case of spread spectrum with non-synchronous transmission in the reverse link the terminal continues transmission with the same PN code and the hand-off becomes transparent from the point of view of the reverse link. In the case of the forward link if the terminal can maintain the same spreading code it merely detects a time shift in the signal being received which can be similar to the time shift observed with changing multi-path propagation. If the terminal can maintain the same PN code for reception then this hand-off becomes transparent to the terminal. This would be possible with the utilization of non-synchronous CDMA transmission in the forward link [15] since with synchronous and orthogonal CDMA the spreading codes belong to the base station and must be allocated to terminals with the constraint that no two terminals should be allocated the same spreading code. We refer to this type of hand-off as *mobile transparent hand – off*.

8.9 MULTIMEDIA SERVICES

So far we have discussed systems where all the calls have the same average bit rate. Third generation systems are characterized by the requirement to support services of variable bit rates. These services are generally referred to as multimedia services. There are a number of different approaches to designing a modulation scheme that supports variable bit rates. First if we consider a system utilizing a hybrid of FDMA and TDMA as in the GSM system, we can utilize channels with a wider bandwidth to support a bit rate that is at least equal to the bit rate of the highest bit-rate service. A TDMA frame structure is then introduced where different possible slot lengths are utilized. It is also possible to assign a set of different RF channels to a given user. The GSM system for example can be upgraded to support higher bit rate services first by assigning multiple time slots to a given user - up to 8 slots per frame in some cases, and then by assigning multiple 200 KHz RF channels to a single user. A proposed concept to offer higher data rate multimedia services that is compatible with the GSM system is discussed in [16].

In the case of spread spectrum modulation there are a number of different approaches to achieve a variable bit rate transmission. First we can consider a scheme where direct sequence spreading (one of the basic forms) is utilized with a high bit rate R. The signal is then transmitted in a burst manner to achieve rates that are fractions of R. This scheme is a generalization of the IS-95 scheme for the reverse channel where $R = 9600$ bps and the allowed fractional rates are $1, 1/2, 1/4$, and $1/8$. We may refer to this mode of operation as the *bursting approach*. The actual scheduling of the transmitted bursts within one frame is randomized so that the variation in the total interference from all the terminals is reduced. Although utilized in the IS-95 system this scheme is not proposed for third generation systems due to possible problems with interference to hearing aids.

An alternative to the above bursting approach is to utilize a variable received power and processing gain. All the services utilize the same chip rate. This implies that all the services utilize the same bandwidth. The information bit rate is variable, hence the processing gains for the different services are different. The transmitter power is adjusted so that a given service has the required signal to noise ratio E_b/N_0 at the receiver for that service. For services with the same SNR requirement the required transmitter power is proportional to the bit rate for the corresponding service. This type of approach may be referred to as the *dimming approach* and represents a generalization of the scheme employed in the forward link of the IS-95 system.

The other approach to obtaining variable bit rates with CDMA is to utilize a signal that consists of the sum of a number of lower rate CDMA component signals. Each of the component signals utilizes its own spreading code. To prevent multiple access interference between the component signals, the different spreading codes for the component signals can be designed to be orthogonal. This scheme has the disadvantage, compared to the above schemes, that the envelope of the total signal is variable. This results in a loss of efficiency of the power amplifier which may be an important issue at the mobile transmitter, i.e., in the reverse link. We refer to this approach as the multi-code approach.

A proposal for a third generation CDMA system which combines some of the elements of the dimming and multi-code approaches discussed above is presented in [16].

8.10 CONCLUSIONS

In this chapter we have discussed many of the wireless system concepts that have evolved with the development of the first three generations of wireless cellular communication systems. Due to the limited availability of radio spectrum, the main theme that unifies all of these concepts is that of the maximization of system capacity. Various modulation and coding techniques have been developed in the quest to achieve higher spectral efficiencies. The design of a spectrally efficient wireless system entails the joint optimization of many parameters involving the type of modulation, error control coding, multiple access technique, spatial frequency re-use, co-channel interference control, adjacent channel interference control, various levels of power control, call hand-off techniques, and variable rate transmission techniques, and various spatial processing techniques utilizing adaptive and fixed beam antennas. Wireless transmission techniques that will generalize various concepts of radio cell, power control, and call hand-off will evolve from the present concepts with the continuing relaxation of complexity constraints in the system infrastructure.

REFERENCES

[1] V. H. MacDonald, "The cellular concept," *The Bell System Technical Journal*, vol. 58, no. 1, Jan. 1979, pp. 15-41.

[2] S. Pasupathy, "Minimum shift keying: A spectrally efficient modulation," *IEEE Communications Magazine*, July 1979, pp. 14-22.

[3] R. A. Scholtz, "The origins of spread-spectrum communications," *IEEE Trans. on Commun.*, vol. COM-30, no. 5, May 1982, pp. 822-854.

[4] E. S. Sousa, "Interference modeling in a direct sequence spread spectrum packet radio network," *IEEE Trans. on Commun.*, Sept., 1990, pp. 1475-1482.

[5] E. S. Sousa, "The effect of clock and carrier frequency offsets on the performance of a direct sequence spread spectrum multiple-access system," *IEEE J. on Select. Areas in Commun.*, vol. 8, no. 4, May 1990, pp. 580-587.

[6] G.R. Cooper and R.W. Nettleton, "A spread-spectrum technique for high capacity mobile communications," *IEEE Trans. on Veh. Tech.*, vol. VT-7, no. 4, Nov. 1978.

[7] R. Kohno, "Spatial and temporal communication theory using adaptive antenna array," *IEEE Communications Magazine*, Feb. 1998, pp. 28-35.

[8] A. J. Paulraj and B. C. Ng, "Space-time modems for wireless personal communications," *IEEE Communications Magazine*, Feb. 1998, pp. 36-48.

[9] K. S. Gilhousen, I. M. Jacobs, R. Padovani, A. J. Viterbi, L. A. Weaver and C. E. Wheatley, "On the capacity of a cellular CDMA system," *IEEE Trans Veh. Techn.*, vol. VT-40, pp. 303–312, May 1991.

[10] R. Padovani, "Reverse link performance of IS-95 based cellular systems," *IEEE Personal Communications Magazine*, vol. 1, no. 3, 1994, pp. 28-34.

[11] A. J. Viterbi, A. M. Viterbi, and E. Zehavi, "Performance of power-controlled wideband terrestrial digital communication," *IEEE Trans. on Commun.*, vol. 41, April 1993, pp. 559-569.

[12] C. Kchao, and G. L. Stuber, "Analysis of a direct-sequence spread-spectrum cellular radio system," *IEEE Trans. on Commun.*, vol. 41, 1507-1516.

[13] R. Price and P. E. Green, Jr., "A communication technique for multi-path channels," *Proc. IRE*, pp. 555-570, Mar. 1958.

[14] S. Kandala, E S. Sousa, and S. Pasupathy, "Multi-user multi-sensor detectors for CDMA networks," *IEEE Trans. on Commun.*, vol. 43, no. 2/3/4, Feb./Mar./Apr. 1995, pp. 946-957.

[15] J. H. M. Sau, and E. S. Sousa, "Non-orthogonal CDMA forward link offers flexibility without compromising capacity," *Proc. International Symposium on Spread Spectrum Techniques and Applications*, Mainz, Germany, Oct. 1996.

[16] E. Nikula, A. Toskala, E. Dahlman, L. Girard, and A. Klein, "FRAMES multiple access for UMTS and IMT-2000," *IEEE Communications Magazine*, April 1998, pp. 16-24.

[17] M. Mahmoudi and E. S. Sousa, "Sectorized antenna system for CDMA cellular networks," *Proc. IEEE Vehicular Technology Conference*, Phoenix, Arizona, May 1997, pp. 6-10.

Chapter 9

Programmable DSPs

Wanda K. Gass and David H. Bartley
Texas Instruments Incorporated
Dallas, Texas
{gass,bartley}@ti.com

9.1 INTRODUCTION

Programmable Digital Signal Processors (DSPs) offer the flexibility of a programmable processor with much of the efficiency of customized Application Specific Integrated Circuits (ASICs). These specialized embedded processors target low cost, low power consumption, small code size, and real-time response. DSP architectures achieve high computational throughput through the use of multiple data path elements, multiple memory accesses, and pipelined instruction execution. Recent advancements in DSP architectures have brought increased parallelism, lower power, and easier programming.

This chapter is organized as follows. The history of programmable DSPs is briefly reviewed in Section 9.2. Architectural features and real-time processing aspects of DSPs are discussed in Sections 9.3 and 9.4, respectively. Low-cost implementation aspects are presented in Section 9.5. Reductions of code size and power dissipation are addressed in Sections 9.6 and 9.7 , respectively. Specialization of DSPs for domain-specific applications is briefly outlined in Section 9.8.

9.2 HISTORY OF PROGRAMMABLE DSPS

Programmable DSPs evolved out of two DSP trends in the late 1970s. The first trend was the development of dedicated single-chip DSP devices targeted at specific applications. The second trend was the development of multi-chip DSP systems for high-performance military applications which used bit-slice technology to create special-purpose solutions. These two trends and the development of the microprocessor to the invention of the programmable DSP. The features that are common to most programmable DSPs included: a data path with integrated hardware multiplier; on-chip memory, which enables stand-alone system implementations; and an instruction set architecture, which supports a wide range of signal processing applications.

Programmable DSPs helped transform DSP from academic study into a viable commercial business for consumer products that require low-cost, real-time performance, and upgradable solutions. Programmable DSPs moved DSP implementations away from hardware design into software design which greatly increased the number of DSP-based system designs [1].

The first commercially available DSP processor, the NEC 7720, was introduced in 1980. The 7720 had a hardware multiplier, and a Harvard architecture which supported concurrent memory access to data memory and program memory, but the instruction set did not support saturation arithmetic. The 7720 was more like a dedicated single-chip DSP rather than a microprocessor. An emulation version of the chip was used for software development. Then once the software was validated, the program was hard-coded in on-chip ROM for mass production. Therefore it was a one-time programmable special-purpose DSP [2].

The AMI S2811 was announced in 1978, but was not available for several years because of problems with the VMOS technology. The 2811 had a hardware multiplier and concurrent memory accesses like the NEC 7720. It could either execute out of on-chip program ROM, or it could execute instructions that were provided on the I/O pins one instruction at a time. It was designed to be a peripheral part to a microprocessor such as the Motorola 6800 and therefore could not fetch and execute instructions out of program RAM [3].

Intel introduced the 2920 in 1980. It was aimed at signal processing applications because it had A/D and D/A converters on-chip. But its performance for compute intensive tasks was not impressive because it had no hardware multiplier. In addition, it was hard to get parameters into the chip because it lacked a digital interface [4].

In 1980 AT&T developed the DSP1 which was used for internal product developed but never marketed outside of AT&T. It had a hardware multiplier, but was a traditional Von Neumann architecture. Like NEC 7720, program memory was confined to execute out of on-chip ROM and so the DSP1 could not be reprogrammed once it was fabricated [5].

Texas Instruments introduced the TMS32010 in 1982. The 32010 had a hardware multiplier and Harvard architecture with separate on-chip buses for data memory and program memory. This was the first programmable DSP to support executing instructions from off-chip program RAM without any performance penalty. This feature brought programmable DSPs closer to the microprocessor/microcontrollers programming model. In addition, TI's emphasis on development tools to wide spread use.

In 1982 Hitachi introduced the HD61810, the first floating-point format with 12-bit mantissa and 4-bit exponent [6]. In 1984 AT&T introduced their first commercially available DSP to the external market which was the first 32-bit floating-point DSP with 24-bit mantissa and 8-bit exponent [7]. Through the end of the 1980s, AT&T provided the fastest DSPs with the DSP16A running at 30 MHz in 1988.

9.3 ARCHITECTURE OVERVIEW

Programmable DSPs are single-chip systems specialized to the computational requirements of real-time data stream processing. They comprise multiple on-chip

Table 9.1 Contrasts Between Typical Microcontrollers and DSPs

Microcontrollers	DSPs
ALU performs multiplication	additional multiplier unit
ALU generates addresses	additional address generation unit
small data words (8,16 bits)	larger data words (16, 32)
Von Neumann architecture	Harvard architecture
single memory access/cycle	multiple memory accesses/cycle
on-chip data cache	no h/w managed data cache
h/w managed instr. cache	optional s/w control of instr. cache
single cycle instructions	pipelined instructions
hidden branch delays	exposed branch delay
orthogonal instruction set	nonorthogonal instruction set
multi-level memory cache	on-chip DMA support

memories, specialized processing units, and on-chip peripherals dedicated to particular signal processing tasks.

Programmable DSPs cover the portion of the design space that falls between microcontrollers and ASICs. DSPs offer the flexibility of a programmable processor with much of the efficiency of customized ASICs. While programmable DSPs are not able to satisfy the performance requirements of all DSP applications, they have proved to be a viable component of a wide variety of products [8, 9, 10].

Custom ASIC designs typically outperform programmable DSP solutions in terms of cost/performance and power/performance ratios. In fact, this book illustrates this point for several signal processing functions. But these advantages often come with higher time-to-market costs and so usually work best for fixed functions and standard algorithms that do not change appreciably over time. Programmable devices, on the other hand, are better suited for products that support several standards or handle multiple functions and when the processor MIP rate is hundreds of times faster than the sample rate of the signal to be processed.

Programmable DSPs share several features in common with microcontrollers [11, 12, 13]. They target low cost, low power consumption, small code size, and real-time response. Though they share these common goals, their architectures differ significantly because they are specialized to perform different types of operations on data having different characteristics. Microcontrollers are proficient at managing control flow, processing random events, and randomly accessing data. DSPs excel at processing data flow, operating on predictable events, and accessing data in an orderly manner. Table 9.1 illustrates some of the characteristics that distinguish DSPs from microcontrollers.

Where microcontroller architectures favor code with frequent conditional branches and intricate control flow, DSP architectures emphasize higher computational throughput for tight loops. Instruction execution is pipelined to permit faster clock rates and often multiple operations are performed concurrently. For example, DSPs can typically perform all the operations required to do one tap of a

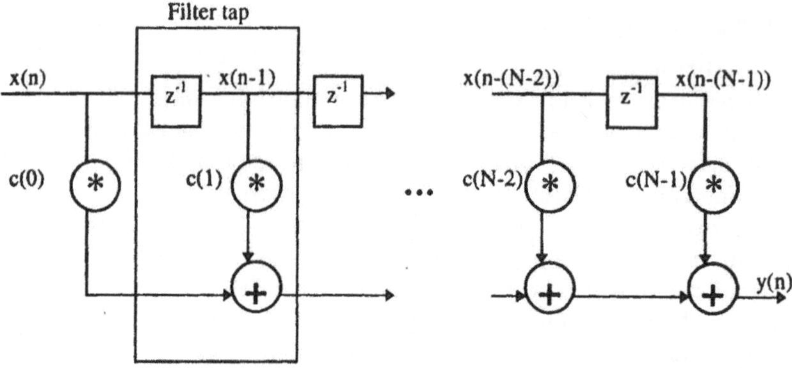

Figure 9.1 An example of the direct form of an FIR filter. One tap of the filter (enclosed by the box) includes a delay element, a multiplier, and an adder to accumulate the sum.

Finite Impulse Response (FIR) filter in a single clock cycle. Fig. 9.1 show the signal flow graph of one tap of an FIR filter. To achieve single-cycle execution per tap of FIR, several operations must occur concurrently: fetching instructions from memory, computing operand addresses, fetching the operands from memory, executing a multiply operation, adding the product to the previous result, and moving data along in memory (simulating a unit time delay). DSPs therefore must comprise several execution units (multiplier, ALU, and address generator), support multiple memory accesses per cycle, and pipeline instruction execution.

This chapter discusses the characteristics of programmable DSPs that enable them to provide real-time response, low cost, small code size, and low power consumption. In addition, it outlines some specialized traits of application-specific programmable DSPs that make them even more cost-effective for targeted DSP applications.

9.4 HARD REAL-TIME PROCESSING

To satisfy the demands of most signal processing applications, a DSP must guarantee low worst-case execution times for time-critical processes operating on input and output data streams. To do so, the programmer must carefully consider the precise timing of the processor as it executes each code segment. The programmer cannot rely on statistical or average behavior, but must quantify the worst case cycle count for all possible paths through each operation performed for each data sample interval.

Because data stream processing is so characteristic of digital signal processing, most compute-intensive loops are statically schedulable. Therefore with proper care in software development, most programs can be implemented without the need for interruption. But this requires that the programmer have both a detailed understanding of the architecture and access to simulators that are timing-accurate.

Minimizing the use of interrupts in a design improves execution predictability. Interrupts entail high overhead cost in most DSP implementations and make precise instruction timing analysis difficult. Therefore, interrupts are most appropriate for events which occur infrequently and unpredictably.

Some DSPs incorporate shadow registers to reduce the amount of processor state that must be preserved during interrupt processing. These shadow registers are usually just one level deep, so interrupts must be disabled during interrupt processing and re-enabled when control is returned to the application process. Programmers often mask interrupts across time-critical loops and when the processor is already servicing an interrupt routine. This reduces the state-saving overhead when interrupts are enabled, because fewer program variables and temporary values need to be saved. Polling statements placed outside key loops can be more effective than interrupts for synchronizing processing with those time-critical events that recur periodically and reasonably predictably. Examples of critical events include reading data from an Analog to Digital Converter (ADC) or passing a block of data from one processing stage to another.

To reduce interrupt frequency for input and output data samples, some DSPs provide auto-buffering on their serial ports to support a form of Direct Memory Access (DMA). Some forms of statically determined multitasking can be efficiently supported by DSPs, but nondeterministic multitasking, whether implemented with interrupts or by polling, involves considerable overhead for most DSPs.

Some DSPs support streamlined interrupt modes that can emulate DMA efficiently in software. For example, when the instruction at the interrupt service location is not a branch, a processor might insert it and one or two following instructions into the instruction pipe and immediately recommence normal operation. With but a two cycle delay, this permits a memory fetch and memory store to be executed asynchronously.

To achieve real-time processing, DSPs must minimize interrupt latency, which is defined as the interval from the time an interrupt signal is asserted until the first instruction of the interrupt routine is executed. But as discussed in the next section, interrupt latency is closely tied to the depth and management of the processor's pipeline.

9.5 LOW COST

The low cost/performance ratio that characterizes a DSP derives principally from architectural features that greatly improve computational throughput while maintaining acceptable cost. Each hardware feature added to a DSP is justified by a worth-while increment in performance [14, 15]. The features which improve computational throughput can be categorized into three classes: multiple data path elements, multiple memory accesses, and pipelined instruction execution.

9.5.1 Data Path

During compute intensive tasks, it is often possible to have several units work in parallel to maximize hardware utilization. Therefore, the data paths of DSPs comprise several functional units which can operate concurrently. While this complexity adds cost to the processor—because each unit is tailored to do a specific

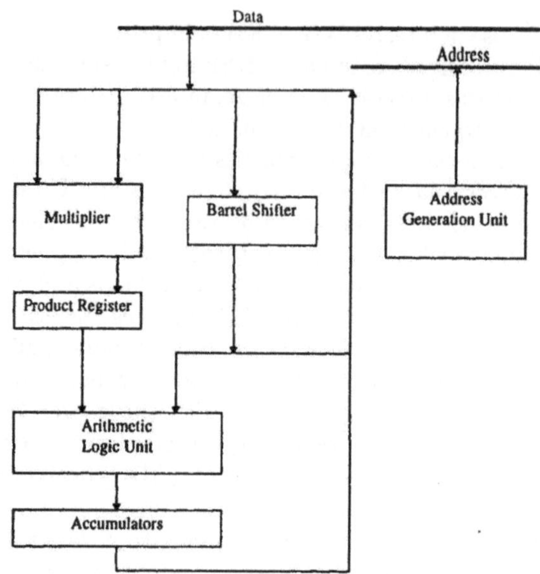

Figure 9.2 A typical DSP data path contains a multiplier, shifter, ALU, and address generator.

task very efficiently—that cost is kept to a minimum. Fig. 9.2 shows a typical DSP data path with multiple function units.

9.5.1.1 Arithmetic

The functional units in low-cost DSPs usually support only integer and fixed-point arithmetic efficiently. Fixed-point arithmetic is adequate for most DSP algorithms if care is taken to scale the intermediate results of computations along the way. Consequently, developing software for fixed-point processors is much more laborious than for floating-point machines [16]. But when the processor is used in a system which will sell in large volumes, the extra amount of time spent in development can be justified by delivering a less costly solution.

Fixed-point arithmetic is generally emulated in software on integer units. The programmer specifies the location within a word of the binary point, separating the integer and fractional components of the number. Before adding two numbers with unaligned binary points, one of the numbers must be scaled up or down to align it with the other. Care must also be taken when multiplying fixed-point numbers to track the location of the binary point in the result. The popular "Q" notation specifies a value's implicit binary point by enumerating its fractional bits. Consider a 16-bit signed value with magnitude less than one. Reserving the most significant bit for the sign leaves 15 bits for the fractional value, scaled by 2^{15}. This is denoted a "Q15" number.

Fixed-point values must be scaled not only prior to arithmetic operations but often after the operations as well. It is important that the programmer analyze average and worst-case data characteristics to preserve adequate precision without risking overflow or excessive saturation. Because excessive scaling surrounding fixed-point arithmetic could degrade performance, floating-point hardware is often justified by a system's overall cost/performance ratio. On the other hand, most DSPs can shift data values as they are fetched or stored to memory, so scaling executes concurrently with other operations and overall performance is not seriously diminished. Higher performance DSPs have larger data paths and sometimes provide extended precision. These too reduce the need for floating-point arithmetic.

Certain DSP operations need the precision or dynamic range of floating-point data and so require hardware-supported floating-point arithmetic. A 32 bit representation with 24 bit mantissa and 8 bit exponent usually suffices. Some floating-point DSPs implement the IEEE floating-point specification. But for embedded applications it is sometimes not necessary or desirable to support the IEEE standard, as compliance with the standard raises the cost without increasing performance. However, standards compliance is often demanded to ensure high-quality, portable results.

Floating-point processors usually support converting data between floating-point and fixed-point representations. If there is a frequent need to convert between the representations, this overhead can reduce performance.

Software developers frequently employ floating-point arithmetic only when prototyping their applications; they then rewrite their code to use fixed-point arithmetic to reduce system costs.

When only a small percentage of a DSP algorithm requires a larger amount of dynamic range and precision than fixed-point representation can provide, it is sometimes possible to use block floating-point representation. Block floating-point arithmetic puts the mantissa in one word and the exponent in a second word. This is an inefficient use of data space unless an exponent word is shared by a "block" of several mantissa words. The relative advantages of block floating-point representation are highly data-specific.

For example, when computing each pass of a Fast Fourier Transform (FFT), it is common to prescale the data by one-half (or shift right one bit) to avoid overflow. Overflow is likely when the input values are large, so scaling the intermediate data between passes is advantageous. On the other hand, overflow will not occur unless the data values are large. Therefore unconditionally shifting the intermediate data to the right after each pass loses valuable precision. The decision to shift the data to the right is a function of the largest magnitude of all the intermediate values. In this situation, a blocked representation leaves the shared exponent holding the total scale amount for the entire FFT computation. This exponent applies to all the output values of the FFT; decrementing the exponent halves every value simultaneously.

9.5.1.2 Multiplier

As illustrated by the FIR filter example in the introduction, many of the most important signal processing algorithms involve the accumulated sum of products. Therefore, the performance of a DSP can be dominated by the speed of its multiply-accumulate (MAC) operation. DSP processors certainly perform many other functions and each application has its own mix of arithmetic and control operations. Still, the MAC is a simple benchmark which best represents a DSP's raw speed. Early DSPs improved materially over contemporary general-purpose microprocessors in this regard because their data paths included hardwired multipliers. Even the earliest DSPs provided single-cycle MAC execution by pipelining operations.

9.5.1.3 Shifter

DSP algorithms usually work with data samples that can be adequately represented as binary fixed-point data values. To maintain both high precision and a low likelihood of overflow, the programmer (or the hardware) must scale the results of integer arithmetic operations. Scaling is also applied when moving the result of a fixed-point multiplication into the ALU or accumulator and when saving the accumulator to memory. Consequently, a shifter is often located between the multiplier and the ALU and at the output of the ALU or accumulator.

Two kinds of shifters are used. A barrel shifter can shift its operand any number of places to the left or right in one cycle. This simplifies programming but requires a significant amount of chip area for signal routing. Less expensive shifters are limited to a small number of specific shift amounts.

9.5.1.4 Arithmetic logic unit

Arithmetic overflow severely degrades signal quality for most DSP applications. While overflow detection is common to most microprocessors, testing a status register and branching on overflow are incompatible with efficiently pipelined inner loops. Many developers prefer arithmetic that saturates on overflow and underflow, since saturation simulates what happens to analog signals when overflow and underflow occur.

Many DSPs implement saturating arithmetic directly in hardware. To simulate positive saturation when an operation generates an overflow, the ALU forces the result to the largest representable positive value. To simulate negative saturation when an operation generates an underflow, the ALU forces the result to the largest negative value. Saturating arithmetic does not eliminate the need to scale intermediate results in order to maintain a proper balance between precision and dynamic range, but it does provide a hardware solution that avoids managing overflow and underflow exceptions in software. This greatly speeds up time-critical code while making its precise timing more predictable.

Sometimes it is known that intermediate values in a summation may overflow or underflow but that the ultimate value will lie within the allowable representative range. In this case a more accurate result is obtained by disabling saturation and allowing overflows to occur undetected. The programmer must have a good understanding of the algorithm and its data to determine the proper arithmetic operations to apply.

Rounding is frequently needed in fixed-point algorithms, and hardware support can be important. For example, more precision is maintained by rounding rather than truncating a 32 bit result when storing to a 16 bit memory word.

Another method for improving performance is to increase parallelism within the ALU itself. Often times an algorithm uniformly applies a single operation to all members of a large set of data. When multiple data samples fit in an ALU word, processing them concurrently can provide considerable speed improvement. This capability is most useful for image processors and multimedia processors which must handle small precision video data as well as large precision audio data. In fact this Single Instruction, Multiple Data (SIMD) execution mode is one of the features adopted by general-purpose processors to improve their DSP performance. For a

SIMD capability to be most useful, the processor must also provide instructions to pack and unpack data samples quickly.

9.5.1.5 Address generation unit

Start address 00101000

End address 01000000

$$\begin{array}{r} 00111111 \\ +\ \underline{00000010} \\ 00101001 \end{array}$$

Figure 9.3 An example of modulo addressing occurs when the pointer is incremented past the end of the circular buffer and wraps around to the start address.

DSPs also increase performance by employing a separate address generation unit. This allows calculation of the next memory address to occur in parallel with data computation[17]. Without the parallel logic for address generation, the calculation would have to be done on the ALU and it would take longer to execute an instruction. DSPs vary widely in the complexity of their address generation units. On the simple end of the spectrum, a register has a minimum amount of adder circuitry attached that can increment and decrement by one. On the other end of the spectrum, the address generation unit is a separate arithmetic unit which supports complex addressing modes and shares full access with the regular ALU to the register file.

Modulo addressing facilitates management of First-In, First-Out (FIFO) data buffers (queues). FIFOs in DSP applications commonly handle real-time buffering and processing of data streams. The size of a FIFO must be carefully adjusted to minimize storage requirements while ensuring against buffer overflow. Because DSPs do not directly support dynamic data management, FIFOs are implemented in software as circular buffers with two pointers to track where data is written to and read from. The *head pointer* indicates the address of the first element in the buffer and is incremented as data is read out of the buffer by the routine that is emptying the FIFO. Likewise, the *tail pointer* specifies the address of the last element in the buffer and is incremented as data is written into the buffer. Circular buffering requires that each pointer value wrap around to the beginning of the buffer whenever it is incremented past the end.

Fig. 9.3 illustrates an example. Suppose the start address of the circular buffer is 28_{16} and the end address is 40_{16}. If the address pointer holds $3F_{16}$ and is incremented by 2, modulo addressing hardware would detect that the pointer had passed the end of the buffer and would subtract the length of the buffer from the pointer. This wraps the address back around to the beginning of the buffer.

Managing FIFO pointers purely in software can be computationally expensive, so DSPs include hardware to help. Some DSPs use special registers for the start and end addresses of the circular buffer. Modulo addressing logic on an increment checks to see if the incremented value falls outside the circular buffer space. If the result points beyond the buffer's end, the processor subtracts the buffer length from the pointer. If the result points before the buffer start (after a decrement), it adds the buffer length to the pointer. DSPs that support modulo addressing typically manage more than one buffer at a time.

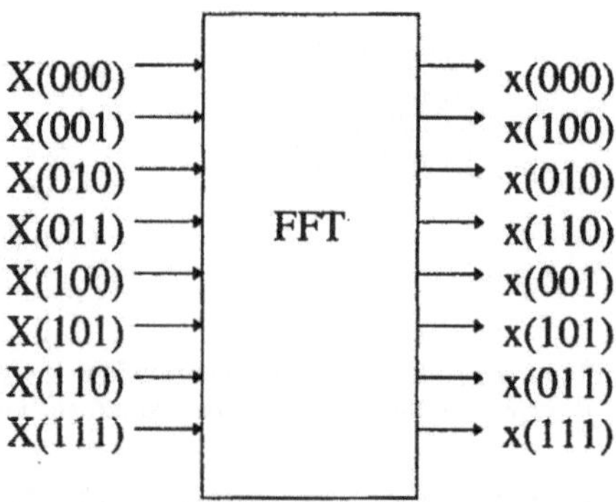

Figure 9.4 A fast Fourier transform generates its results in bit-reversed order relative to its inputs.

A DSP's performance on the fast Fourier transform (FFT) is critical to many DSP applications. Data produced by the FFT are no longer in the original input order, but in bit-reversed order (see Fig. 9.4), and unscrambling the data can take several cycles. To reduce this overhead, most DSPs provide a special addressing mode that generates the bit-reversed address sequence directly. For the example in the figure, computing an eight point FFT, the second element in the output array corresponds to x(4) rather than x(1). Likewise, the fourth element corresponds to x(6) rather than x(3).

9.5.1.6 Very long instruction word processors

Unlike their general-purpose cousins in the personal computer and workstation worlds, DSP families are free to abandon strict compatibility when innovations dramatically improve performance or lower costs. So, while mainstream vendors are boosting performance with complicated superscalar implementations of existing architectures, DSP architects are able to explore alternative means to achieve instruction-level parallelism.

One of the most promising alternatives is the Very Long Instruction Word (VLIW) processor. Like superscalar designs, a VLIW processor issues multiple instructions per cycle. It differs though in leaving all instruction scheduling and parallelizing to the programmer (or compiler). This eliminates a great deal of complicated and expensive on-chip logic to parallelize serially coded operations on the fly while avoiding data and control hazards. VLIW delivers good cost/performance because little control hardware is needed; essentially all the added gates contribute directly to the performance of compute intensive tasks.

A VLIW design incorporates pipelined functional units to permit single-cycle instruction execution at high clock rates. By its very nature, it exposes the complexity of these data paths to the compiler and assembly language programmer. Finding and exploiting instruction-level parallelism in a signal processing algorithm is itself a daunting task. When combined with the difficulty of managing pipelined resources, this makes the architecture extremely hard to program manually. Yet VLIW is inherently highly regular and orthogonal, making it amenable to advanced optimizing compiler technology. Indeed, VLIW architectures have fostered high-quality compilers since they were first introduced as supercomputers in the 1980s. As VLIW DSP compilers and related development tools, such as assembly language optimizers, emerge, VLIW promises the most bang for the buck for the foreseeable future [18].

The chief drawback to VLIW processors are their long instruction words. These typically reach 100 bits or more in length and often must be packed with no-ops when the program code has too little intrinsic parallelism. Thus VLIW programs are generally much larger than those for classical DSPs. Data compression techniques have been used to gain greater code density [19].

9.5.2 Memory

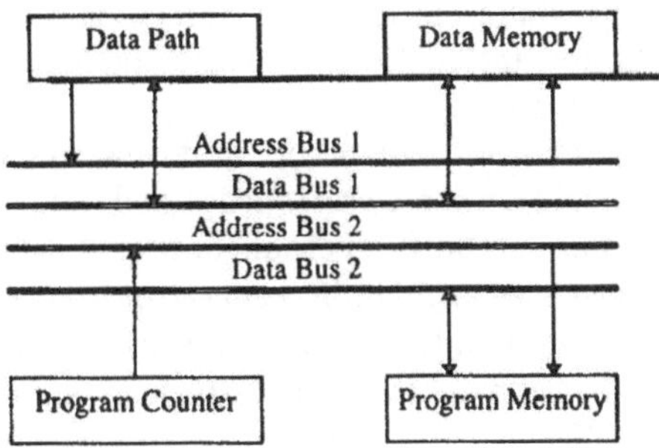

Figure 9.5 A block diagram of the Harvard architecture.

DSP algorithms are compute intensive, so it is important to allow multiple memory accesses per cycle to keep a steady stream of data and instructions feeding the processor [20]. Some early DSPs used a Harvard architecture, which segregates the memory into separate instruction and data banks with separate buses supporting each. This partitioning improves the memory bandwidth, but is awkward to program. Fig. 9.5 shows the split memory systems of a Harvard architecture.

Later generation DSPs migrated to the unified memory of a Von Neumann architecture, in which data and program share one address space (see Fig. 9.6). These retained multiple memory access capability by providing on-chip program

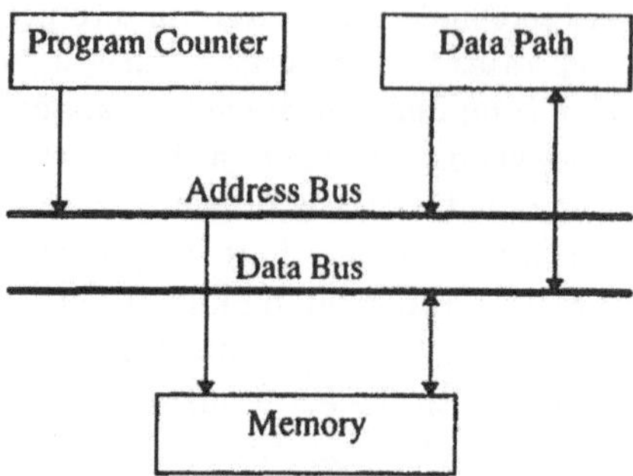

Figure 9.6 A block diagram of the Von Neumann architecture.

memory to feed the instruction decoder on an internal bus separate from the data bus. While this approach resembles the instruction cache that is a standard part of many microprocessors, for most DSPs the burden remains with the programmer to manage which parts of the program are in on-chip program memory, which routines to page in and out of on-chip memory, and when to move code segments. Instruction caches, which automatically bring code segments into on-chip memory on demand, can ease the programming task. But because it is crucial to keep time-critical code resident in on-chip memory, the ability to lock and unlock cache sectors must be under software control. If it is not, instruction caching could significantly degrade the performance of time-critical loops.

DSPs seldom support data caches. In practice, developers want to control memory accesses explicitly to avoid timing uncertainties and data coherency problems. In addition, DSP algorithms exhibit predictable data access patterns; a cache is more suited to the random accesses typified by general-purpose applications.

Even with caching a memory bandwidth problem arises when one memory port is used to fetch instructions while another simultaneously accesses data. Some DSPs provide a small repeat buffer in the processor core to cache a short sequence of instructions that are executed repeatedly. Program memory fetches then occur only on the first of a loop's iterations. This scheme frees the instruction fetch unit to read a second data operand on each cycle. For this sequence of code, the programmer presets a hardware repeat counter and the sequence executes continually until the counter decrements to zero. These *zero-overhead loops* eliminate the need for explicit decrement-and-branch instructions. However, interrupts must generally be disabled while the dedicated repeat register counts down to zero. This can unacceptably lengthen interrupt response time for some applications.

To better exploit two data accesses on each cycle, some DSPs adopt a modified Harvard architecture in which the program memory space contains some data as well as instructions. When executing a zero-overhead loop, the processor is free to access data from program/data memory concurrently with the normal data access from data memory.

Two other techniques increase the number of memory accesses allowed per cycle. One sets the instruction cycle time at twice or more the memory access time. This allows two sequential accesses within one instruction cycle and effectively doubles the memory accesses to one or more memory banks. This approach gives additional bandwidth without increasing the number of internal busses. But as data paths get faster than memories, this technique becomes less applicable.

The second technique makes use of dual-ported memory to perform simultaneous accesses to memory. Dual-ported memories require more transistors than single ported memory, and the cost differential grows exponentially as the memory size increases, but they can improve performance substantially and can be cost effective when the memory size is kept fairly small. Dual-ported memories also require a separate address and data bus for each concurrent access. The additional area devoted to buses also raises the expense of this approach.

9.5.3 Instruction Execution

	Instruction Cycles					
	1	2	3	4	5	6
Instruction Fetch	I1	I2	I3	I4	I5	I6
Instruction Decode		I1	I2	I3	I4	I5
Data Read/Write			I1	I2	I3	I4
Execute				I1	I2	I3

Figure 9.7 An example of the timing of a 4-stage pipelined instruction cycle.

A third way DSPs meet the demands of compute intensive tasks is by pipelining instruction execution to achieve higher throughput. Instruction execution is pipelined into several phases so that during one clock tick, several instructions are executing concurrently, each in a different phase of the pipeline (see Fig. 9.7). During sequential execution of instructions, the pipeline may be ignored if all arithmetic operations execute in one cycle and there are no resource or data hazards. But when program flow is altered by a branch or interrupt, the pipeline must be flushed before execution can continue. Therefore, unstructured control flow in DSPs is inefficient and complicated.

For deeply pipelined architectures, changes to the program's control flow create control hazards (branch delays), which can reduce performance sharply. Two ways to cut branch delay overhead are to allow execution of other instructions to proceed while branches execute and to predicate instructions rather than branch around them. (A third technique, zero-overhead looping, is discussed in the previous section on memory.)

A control hazard occurs when a branch is executed on a pipelined processor. The evaluation of the branch condition and the resolution of the program counter's value take place relatively late in the pipe, after subsequent instructions have already begun to execute. One way to deal with this type of hazard is to stall the pipe when a branch is recognized. This simplifies the programming effort but wastes precious cycles. Another is to expose the pipe to the programmer and allow instructions following the branch to proceed while the branch destination is resolved. These instructions are said to be in the "delay slots" of the branch. Compilers are moderately successful at filling these slots with useful work by moving instructions into them from both before and after the branch and by executing instructions speculatively. This must be done with care, however, to avoid data hazards.

Predicated, or conditionally executed instructions are an effective alternative to short forward branches. These are like normal instructions but take an additional input value, which is typically specified as a register operand. The instruction executes normally when the predicate operand value matches some criterion (typically "non-zero"), but acts as a no-op when it does not. For simple *if-then* and *if-then-else* code patterns, instructions that test a condition and branch on the result can be replaced with a test followed by the conditionally executed *then* and *else* operations. This moves the point at which the condition must be resolved from near the front of the pipeline (for a branch) to the end (where register writes occur), so there is no delay. Many DSPs provide predicated move instructions. VLIWs and other architectures with more parallelism typically predicate all, or nearly all, instructions.

Compilers are effective at exploiting predicated instructions and reordering code to fill delay slots. Another compiler optimization is to apply static branch prediction techniques so as to arrange that the most likely path of a conditional branch is the one that falls through. Some claim an average miss rate as low as 26% for non-loop branches and 20% overall [21].

Interrupts also alter program flow. On an interrupt, the processor must flush the pipeline before beginning executing the interrupt service routine. Three programming models illustrate the spectrum of views of the pipeline.

The first view is a *time-stationary* model. This model provides the most flexibility for controlling the execution timing of operations and shortens the interrupt latency, but is difficult to program. The second is a *data-stationary* model, in which data hazards are not protected against. This model restricts the programmer's control over instruction execution and normally takes longer to process an interrupt, but is easier to program. The third view is a data-stationary model, in which data hazards are prevented. This gives the programmer the least control over instruction timing and takes longest to service interrupts, but provides the most straight-forward programming environment.

9.6 MINIMUM CODE SIZE

Orthogonality is a characteristic of an instruction set architecture that describes its regularity and self-consistency. An instruction set that supports the same addressing modes for all similar operations and does not gratuitously restrict the use of registers or the specification of source and destination operands is considered orthogonal. Orthogonal instruction sets are not only easier to program in assembly language but also greatly facilitate the development of effective optimizing compilers.

Unfortunately, orthogonal instruction sets have somewhat redundant encodings and tend to require wider instruction words. DSP designers want narrow instructions in order to minimize external memory bandwidth and on-chip program storage and thereby reduce cost. Consequently, those features of an orthogonal and flexible design that are least frequently utilized are often eliminated to make the instruction set encoding more efficient.

The first DSP architectures limited the sorts of operations that could execute concurrently as well as much freedom to specify source and destination operands. These constraints ensured compact code but led to irregular instruction sets that poorly accommodate high-level language compilers [22].

9.6.1 Specialized Registers

The simplest architectures have only a few registers and constrain some of them, called accumulators, to data manipulation. Other registers are available only for address generation. Register specialization is often the most cost-effective way to fulfill moderate performance requirements. For example, ***accumulator-based architectures*** are particularly efficient for FIR and other filters that consist of little more than a multiply-accumulate operation in a tight loop. Generally, the ALU operates on the contents of the accumulator and one other operand, typically in memory, and returns the result to the accumulator. Most DSPs provide more than one accumulator, but seldom more than eight. Since there are so few data registers, high data-memory bandwidth through multiple memory accesses per cycle is vital to keeping the execution units busy. Like the accumulator-oriented ALU, the multiplier unit usually has one or two special registers to hold input values and save results. All these registers are distinct from the address registers, which support only loads and stores.

More recent DSP designs are more similar to general-purpose processors, with a unified register file attached to both the data path and address generation units. By and large, registers can hold either data or data memory addresses. These ***register-based architectures*** need more encoding bits to designate the source and destination of the operands, but they are easier to program, are amenable to convenient C-like assembly language syntax, and foster better compiled code quality [23]. On the down side, code size for register-based architectures is normally larger than for accumulator-based designs.

9.6.2 Compressed Addressing Modes

Most DSPs support familiar operand addressing modes found in general-purpose processors, including immediate data, memory-direct, register-direct, register-indirect, register-indirect with pre- and post-increment, and register indirect with indexing. To minimize instruction word width, though, DSPs often restrict the addressing modes available for a given instruction to those that are most commonly used with it. Generally these instructions use implied addressing, short constants, and page registers.

Implied addressing applies when the source or destination of an operand is constrained to be a single resource or to be the same as one of the other operands. For instance, for a typical two-input operation on an accumulator-based DSP, the output and one of the inputs is always the accumulator; in this encoding scheme, only one operand need be specified explicitly. The assembly language syntax may permit both sources and the destination to be explicitly stated to improve readability, but two of the three operand specifiers are fixed.

Short immediate fields encompass a small range of constant data, but are often sufficient in practice. Extended-length instructions allow for larger immediates and address literals when needed.

Implicitly referenced page registers extend the instruction encoding space for memory-direct addressing. The address generator forms a paged operand address by concatenating the page register contents with low-order address bits specified in the instruction encoding. For example, for a 24-bit data space, a page register might specify the upper 8 bits of an address and the instruction the lower 16. Programmers group related data within pages so as to minimize page register set-up overhead in time-critical code regions.

9.7 LOW POWER DISSIPATION

Portable and wireless consumer products are driving DSP specifications towards lower energy consumption in order to extend battery life and reduce battery weight. Multimedia notebook PCs, digital cellular phones, and other wireless communications devices all require low power digital signal processing. Low power operation is also a contributing factor to controlling heat dissipation in high-performance DSPs.

The same architectural features which contribute to a processor's low cost/performance ratio are also a benefit in reducing power consumption [15, 24]. Multiple functional units in the data path and the multiple memories which are common in DSPs can be exploited to lower the operating voltage and selectively enable only the active parts of the circuit to a much greater extent than general-purpose processors. Brodersen, *et. al.*, [25] review techniques and issues in the design of low power VLSI circuits and systems.

One way to reduce power is to lower voltage, because the power consumed is proportional to V_{DD}^2. Therefore, reducing the supply voltage yields a quadratic improvement in the power-delay product. However, reducing the supply voltage also increases gate delays. To regain performance, DSP designers add parallelism to the data path and memory accesses. Though this increases power consumption linearly with the added circuitry, there is still a net reduction overall. To push the state-of-the-art for low power, 1 volt DSPs have been designed which are capable of operating at up to 100MHz [26, 27, 28].

Lower power can be achieved by lowering the frequency of the clock when fewer operations need to be processed. A very simplistic form of managing the clock frequency is to have two different modes of operation: one for compute-intensive periods of time, and another for periods when very little processing is required.

A more effective way to manage clock frequency is to use gated clocks to shut down circuits not currently in use. Because DSPs have multiple memories and multiple units in the data path, it is easy to selectively enable those parts of the processor that are required for execution. These power management techniques can be coarsely adjusted over a long period of time or dynamically adjusted during instruction execution.

Sleep modes are most beneficial when the processor loading is fairly predictable, as it is with most DSP applications. Some DSPs support up to three levels of sleep mode. The first level shuts down the CPU, but leaves the peripherals and Phase Locked Loop (PLL) active. In this mode, the peripherals can be set to wake up the CPU. The second level shuts down both the CPU and the peripherals, but leaves the PLL active. In this mode, an external signal is required to wake up the processor. The third level of sleep mode shuts down all three. This mode saves the most power, but takes thousands of processor cycles to wake up.

Adding functional units to the data path increases the amount of parallelism and enables the clock rate and voltage of the processor to be lowered without affecting throughput and response. For example, some low power DSPs have two multipliers in the data path rather than just one. Each of the functional units can be activated with gated clocks so they need not dissipate energy when they are not in use. During periods of low computational demand, the supply voltage can be lowered to achieve a greater energy shavings.

Multiple memories enable the program and data memory to be separated. The data currently read from data memory is more likely to correlate with the next data sample, thereby reducing the number of bit transitions on the data bus. The data memory addresses are also more likely to be sequential in a DSP application. Multiple memories also allow the individual memory banks to be smaller. In addition each bank can be enabled only when it must be accessed. For some applications, all of the memory can be on-chip, greatly reducing or eliminating input/output operations to access off-chip data.

To reduce the power consumed by on-chip static RAMs, memories can be partitioned so only the memory section that contains data to be accessed is enabled. Program memory and data constants can be stored in ROM rather than RAM to further reduce power. And ROMs can also use diffusion-programming to reduce the ROM area per bit. This reduces power consumption by lowering the bit-line and word-line capacitances.

Because instruction execution is pipelined, several instructions can be operating in parallel. The code size of a DSP is usually small which reduces the number of off-chip accesses of program memory, and sometimes can be stored entirely in on-chip ROM. Some specialized DSPs encode frequently used operations to further reduce code size.

9.8 SPECIALIZATION

The more specialized a programmable DSP, the closer it comes to an ASIC design in performance, cost, and power. Initially DSPs specialized within a family of products, all with the same instruction set, but each with different program and data memory configurations. Because DSP families offer a variety of memory sizes, the memory size closest to the application requirements could be selected to minimize system cost and power consumption. Customers can further customize the DSP by specifying the program memory contents in mask-programmable ROM.

DSPs are also specialized by the sets of peripherals integrated on-chip with the processor core. On-chip peripherals reduce system costs and sometimes enable single-chip solutions. Peripherals range from simple timers to complicated DMA interfaces. The set of peripherals integrated on-chip with a DSP core often tailors the processor to a certain application domain.

More recently, ASIC and DSP cores have been combined in one chip, allowing the customer to add the specific set of peripherals that addresses his particular system requirements. These hybrids integrate the glue logic with the processor and even include hardwired functions to boost performance for application-specific algorithms. Some DSP cores allow the customer to specify the exact program and data memory configuration that meets the application requirements.

The final level of customization entails customized instruction and data path elements targeted to specific application domains and even specific applications. These DSPs are becoming more common for high-volume markets. For example, today several DSPs target wireless communications. Multimedia applications and devices comprise another emerging market targeted by customized designs. Chapters 10 and 11 discuss both of these specializations.

9.9 SUMMARY

The first programmable DSP devices were simple and unpretentious, attaining improved signal processing performance over general-purpose processors simply by adding a multiplier to the data path. These architectures constrained the types of operations that could execute concurrently as well as the selection of source and destination operands. Though these constraints permitted smaller code size—a key requisite for low cost applications—they brought about highly irregular instruction sets that were difficult for compilers to manage. Thus, most software for programmable DSPs has been developed in hand-coded assembly language.

Programmable DSP architectures have improved on all fronts over the two decades of their existence. Lengthened pipelines have permitted faster clocks with more parallelism. Added execution units in the data path have allowed more computation on each clock cycle. Dual-ported memories have doubled data bandwidth. And wider data paths and hardware-supported floating-point arithmetic have simplified programming and improved precision. Recent trends toward more regular and orthogonal instruction sets have also made high-level language compilers more effective, allowing much, if not yet all, of many applications to be written in C, C++, or Ada.

REFERENCES

[1] J. Rabaey, W. Gass, R. Brodersen, and T. Nishitani, "VLSI design and implementation fuels the signal-processing revolution," *IEEE Signal Processing Magazine*, pp. 22–37, Jan. 1998.

[2] T. Nishitani, R. Maruta, Y. Kawakami, and H. Goto, "A single-chip digtial signal processor for telecommunications applications," *IEEE Journal of Solid-State Circuits*, vol. 16, pp. 372–376, 1981.

[3] W. Nicholson, R. Blasco, and K. Reddy, "The S2811 signal processing peripheral," in *Proceedings of WESCON*, pp. 1–12, 1978.

[4] M. Townsend, M. Hoff, and R. Holm, "An NMOS microprocessor for analog signal processing," *IEEE Journal of Solid-State Circuits*, vol. 15, Feb. 1980.

[5] J.R.Boddie, G. Daryanani, I. Eldumiati, R. Gadenz, J. Thompson, and S. Walters, "Digital signal processor: Architecture and performance," *Bell Systems Technical Journal*, vol. 60, pp. 1449–1462, 1981.

[6] Y. Hagiwara, Y. Kita, T. Miyamoto, Y. Toba, H. Hara, and T. Akazawa, "A single chip digital signal processor and its application to real-time speech analysis," *IEEE Transactions on ASSP*, vol. 31, no. 1, 1983.

[7] R. Kershaw, L. Bays, R. Freyman, J. Klinikowsi, C. Miller, K. Mondal, H. Moscovitz, W. Stocker, and L. Tran, "A programmable digital signal processor with 32b floating point arithemtic," in *ISSCC Digest of Technical Papers*, Feb. 1985.

[8] E. A. Lee, "Programmable DSP architectures: Part I," *IEEE ASSP Magazine*, pp. 4–19, Oct. 1988.

[9] E. A. Lee, "Programmable DSP architectures: Part II," *IEEE ASSP Magazine*, pp. 4–14, Jan. 1989.

[10] B. D. T. Inc., "Buyer's guide to DSP processors, 3rd edition." Unpublished technical report, 1997.

[11] J. L. Hennessy and D. Patterson, *Computer Architectures: A Quantitative Approach.* San Francisco, CA: Morgan Kaufmann Publishers, 2nd ed., 1996.

[12] M. Dolle and M. Schlett, "A cost-effective RISC/DSP microprocessor for embedded systems," *IEEE Micro*, pp. 32–40, Oct. 1995.

[13] J. Stankovic, "Real-time and embedded systems," *ACM Computing Surveys*, vol. 28, no. 1, pp. 205–208, 1996.

[14] P. Lapsley and G. Blalock, "How to estimate DSP processor proformance," *IEEE Stectrum*, vol. 33, no. 7, pp. 74–78, 1996.

[15] P. Lapsley, "Low power programmable DSP chips: Features and system design strategies," in *5th International Conference on Signal Processing Applications and Technology*, vol. 1, (Waltham, MA), pp. 501–505, DSP Associates, 1994.

[16] T. Egolf, S. Famorzadeh, and V. Madisetti, "Fixed-point co-design in DSP," in *VLSI Signal Processing, VII*, (New York, NY), pp. 3–13, IEEE, 1994.

[17] G. Araujo, A. Sudarsanam, and S. Malik, "Instruction set design and optimization for address computation in DSP architectures," in *9th International Symposium on Systems Synthesis*, pp. 102–107, 1996.

[18] V. Zivojnovic, "DSPstone: a DSP-oriented benchmarking method," in *5th International Conference on Signal Processing Applications and Technology*, vol. 1, (Waltham, MA), pp. 715–720, DSP Associates, 1994.

[19] S. Liao, S. Devadas, and K. Keutzer, "Code density optimization for embedded dsp processors using data compression techniques," in *16th Conference on Advanced Research in VLSI*, (Los Alamitos, CA), pp. 272–285, IEEE Computer Society, 1995.

[20] M. A. R. Saghir, P. Chow, and C. G. Lee, "Exploiting dual data-memory banks in digital signal processors," in *SIGPLAN Notices*, no. 9, pp. 234–243, Sept. 1996.

[21] T. Ball and J. R. Larus, "Branch prediction for free," in *Conference on Programming Language Design and Implementation*, pp. 300–313, 1993.

[22] M. Saghir, P. Chow, and C. Lee, "Application-driven design of DSP architectures and compiler," in *IEEE International Conference on Acoustics, Speech and Signal Processing*, 1994.

[23] B. Krepp, "A better interface to in-line assembly," in *5th International Conference on Signal Processing Applications and Technology*, vol. 1, (Waltham, MA), pp. 802–811, DSP Associates, 1994.

[24] V. Tiwari, S. Malik, and A. Wolfe, "Power analysis of embedded software: A first step towards software power minimization," *IEEE Transactions on Very Large Scale Integration (VLSI) Systems*, vol. 2, pp. 433–445, Dec. 1994.

[25] R. Brodersen, A. Chandrakasan, and S. Sheng, "Low-power signal processing systems," in *VLSI Signal Processing, V*, (New York, NY), pp. 3–13, IEEE, 1992.

[26] M. Iaumikawa, "A 0.9V 100MHz 4mW $2mm^2$ 16b DSP core," in *ISSCC Digest of Technical Papers*, pp. 84–85, Feb. 1995.

[27] S. Mutoh, "A 1V multi-threshold voltage CMOS DSP with an efficient power management technique for mobile phone applications," in *ISSCC Digest of Technical Papers*, pp. 168–169, Feb. 1996.

[28] W. Lee, "A 1V DSP for wireless communications," in *ISSCC Digest of Technical Papers*, pp. 92–93, Feb. 1997.

Chapter 10

RISC, Video and Media DSPs

Ichiro Kuroda
NEC Corporation
Kawasaki, Kanagawa, Japan
kuroda@dsp.cl.nec.co.jp

10.1 INTRODUCTION

Multimedia processing is expected to be the driving force for the evolution of both microprocessors and DSPs. Introduction of digital audio and video is the starting point of multimedia, because audio and video as well as texts, figures and tables are all in the digital form in a computer, and therefore, they can be handled in the same manner. However, digital audio and video require tremendous amount of information bandwidth, unless compression technology is employed. Audio and video information amount is deeply dependent on their required quality and varies in a wide range. For example, HDTV (1920×1080pixels with 60 fields/sec) is expected to be compressed into around 20 Mbit/sec while the H.263 [1] videophone terminal having sub-QCIF (128x96pixels) with 7.5 frames/sec is expected to require around 10 to 20 kbit/sec. Information rate difference reaches around 1000 times in this case. Compression techniques also require tremendous amount of processing, which is dependent on quality and information rate. The required processing rate has a range from one hundred MOPS (Mega Operations Per Second) to more than TOPS (Tera Operations Per Second). NTSC resolution MPEG-2 (MPEG-2 MP@ML) [2] decoding, for example, requires more than 400 MOPS and 30 GOPS are required for the encoding. This wide variety of processing amount and multimedia quality lead us to employ a variety of programmable implementations.

Recently developed microprocessors and programmable DSP chips provide powerful processing capabilities for realizing real-time video and audio compression/decompression. One of the most important target of these processors is to realize MPEG-2 decoder by software in real-time. Therefore, the performance and functionality for MPEG2 decoding are the key issues in designing these processors architectures. In order to meet such target, these processors have some built-in functions for supporting MPEG processing in addition to the conventional architectures.

Decoding and playback of the MPEG compressed bitstream start with the variable length decoding, followed by the inverse quantization which retrieve DCT coefficients from compressed bit stream. Then, inverse DCT (IDCT) operation produces the prediction error signal. This signal retrieves the decoded frame by the addition of a motion prediction signal. The motion prediction signal is calculated by the pixel interpolation using one or two previously decoded frames. The decoded frame is transformed to the display format and transferred to the video memory (VRAM) and a video output buffer, as there are several display formats such as RGB, YUV and dithering.

This decompression process is carried out by the square image block called macroblock (16 x 16 pixels with color components) or block (8 x 8 pixels). Table 10.1 shows the major parameters for MPEG1 [3] and MPEG2 [2] such as the frame size, the frame rate, and the bitrate of compressed bitstream. The number of macroblocks as well as the number blocks per second are also shown in the table.

Table 10.1 MPEG Parameters

	MPEG1	MPEG2 MP@ML
Horizontal Size (pixels)	352	720
Vertical Size (lines)	240	480
Frame/sec	30	30
Display Byte/sec	3.8M	15.55M
Compressed bits/sec	1.0M	4-15M
Number of Macroblocks/sec	9,900	40,500
Number of Blocks/sec	59,400	243,000
Number of Bytes/frame	126,720	518,400
Number of Bytes/frame to decode	380K	1.55M

MPEG decoding requires five important architectural features listed below and should be considered before selecting each types of processor.

(1) Bit manipulation function which consists of parsing and selecting bit string in serial bit streams. Variable length encoding and decoding are classified in this category.

(2) Arithmetic operations which consist of multiplications and add/subtract operations and other specific arithmetic operations such as the sum of the absolute difference for motion estimation. Different processing precision or word are also preferable for hardware efficiency to handle many different medias such as 8-bit video data and 20-bit audio. Parallel processing units are also very important points for efficient IDCT processing which requires a lot of multiplications, due to the nature of the two-dimensional IDCT algorithms.

(3) Memory access for large memory space enabling a video frame buffer which cannot usually reside in the processor on chip memory. The frequent access to frame buffer in motion compensation requires high bandwidth memory interface.

(4) Stream data I/O for media streams such as video, audio and compressed bitstream. This functionality for video signals, for example, consists of capturing and display as well as a video format conversion such as RGB to YUV. There are

also this type of I/O for compressed bitstream for storage media such as hard disk, compact disk and DVD and the communication network.

(5) Realtime task switching which supports hard realtime. This requires sample by sample or frame by frame time constraint in its nature. The switching is among simultaneous media processing for different types of medias for realizing synchronization of video decoding and audio decoding for example.

10.2 MEDIA MPU

Microprocessors have improved their performance for 26 years by increasing their clock frequency. Pipelining inherent to RISC chips is the key issue in this area. They are used as host CPUs for personal computers (PCs) and workstations with operating systems. They are also used as controllers for game machines and other consumer electronic equipment.

Microprocessors enhanced the capability for multimedia processing by introducing new instruction set. A long word ALU in a microprocessor is divided into several small word ALUs and multiple independent short word data are processed in the same instruction. This approach was first introduced to graphics instructions [4] [5] and after that for RISC processors based for workstations [6] [7] to process MPEG decoding. Recent microprocessors for PCs [8] also employ this approach.

10.2.1 MPU and Multimedia Processing

Today's high-end general purpose microprocessors can issue two to four instructions per cycle by the superscaler control [9]. Superscalar control enables to issue more than one floating point instructions or several multimedia instructions at a time. This control mechanism has two types of issuing mechanisms. One is the in-order-issue which issues instructions in the order stored in the program memory. The other is the out-of-order-issue which does not always issue instruction in the order stored in a memory by keeping the data dependency. This is effective for more than 200Mhz microprocessors which have long pipeline latency instructions, and out-of-order issue can maximize high speed pipelined ALU performance.

Fig. 10.1 shows an example of an out-of-order superscalar microprocessor [10]. Implementation of out-of-order control requires a lot of additional hardware functional units such as a reorder buffer which controls the instruction issue and completion, a reservation station which re-orders the actual instruction issues for execution units and renamed register files as well as control circuits for them. They consume a large part of silicon area as well as power dissipation.

In out-of-order control, the number of registers can actually be increased by register renaming. This is effective for the multimedia processing performance by reducing the stores of intermediate calculation results to the memory. On the other hand, image processing allows the inherent parallelism in pixels and macroblocks. Although large amount of hardware is implemented for superscalar control, two to four parallel instruction issues does not fully make use of the parallelism in image processing. Other aspects in general purpose microprocessor architectures which are related to media processing are considered by using Fig. 10.2, with respect to the five performance features introduced in section 10.1.

Arithmetic operations of microprocessors have a wordlength problem. Microprocessors have enhanced their wordlength into 32-bit or 64-bit. On the other hand,

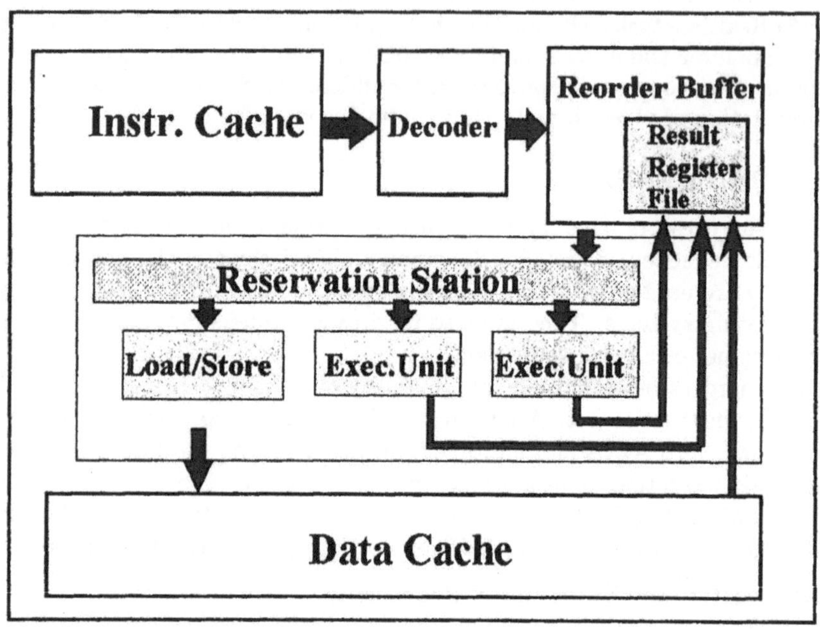

Figure 10.1 Out-of-order superscalar MPU.

required wordlengths for the multimedia processing are 8-bit, 16-bit or 24-bit which are much shorter, compared with the today's microprocessor wordlength. Therefore it has large margins to handle multimedia data by arithmetic instructions on microprocessors. Moreover, the data type available in the high level language such as C is not suitable for the multimedia data. This is really a problem, because general purpose microprocessors are commonly programmed by high level languages.

Memory bandwidth can also be a problem for multimedia processing by microprocessors. In the image processing, there are a number of accesses to a large video frame memory (4M bit/frame or 512Kbyte/frame) which cannot reside in the first-level or second-level cache. Therefore, we cannot expect a high cache hit rate usually assumed in general purpose applications. Moreover, there is a difference in data locality which is related to the performance of memory access. Cache mechanism is designed to make use of one-dimensional locality of consecutive addresses. On the other hand image processing have the two-dimensional locality of access.

The hard real-time nature of multimedia applications is also a problem for microprocessors. In media processing for audio and video, each processing task on an audio sample or a video frame should be completed within a predetermined sample or frame interval time. This requires task switching with less overhead. However predictability of the execution time is decreased in microprocessors by introducing cache mechanism. Internal state introduced by out-of-order superscalar

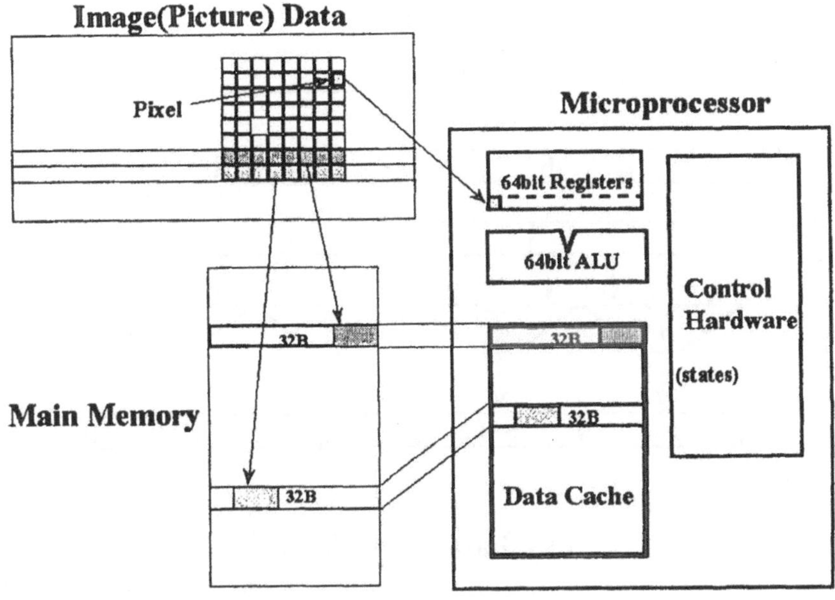

Figure 10.2 Microprocessor architecture.

processors also make it difficult for fast task switching. Moreover, operating systems are not designed for hard real-time processing.

Other features such as the performance for bit manipulation for microprocessors are not so efficient, because they are designed to treat continuous text stream in a fixed wordlength. Stream I/O functions are usually not included in the microprocessor. They are realized by other chips in the system.

10.2.2 Multimedia Enhancements on MPU

Recent microprocessors improve multimedia processing capability by multimedia extensions which can utilize pixel parallelism in video processing. They enhance arithmetic performance by dividing the long word arithmetic unit (ex. 64-bit) to execute 2 to 8 operations in parallel by SIMD type multimedia instructions, such as shown in Fig. 10.3.

For example, split addition/subtraction can perform parallel 8-bit, 16-bit or 32-bit word saturated or wrapped operations. Video signal can be expressed as an unsigned byte. Arithmetic operations for video signal are usually performed in 16-bit units. Audio signal is expressed in 16 bit, arithmetic operations for audio signals are usually performed in 16-bit or 32-bit. If we use 16-bit units for audio, the quality of the signal is degraded due to round-off noise in arithmetic operations.

Multimedia instruction set includes the split addition/subtraction operations, multiplication/multiply-accumulate operations, special operations such as for mo-

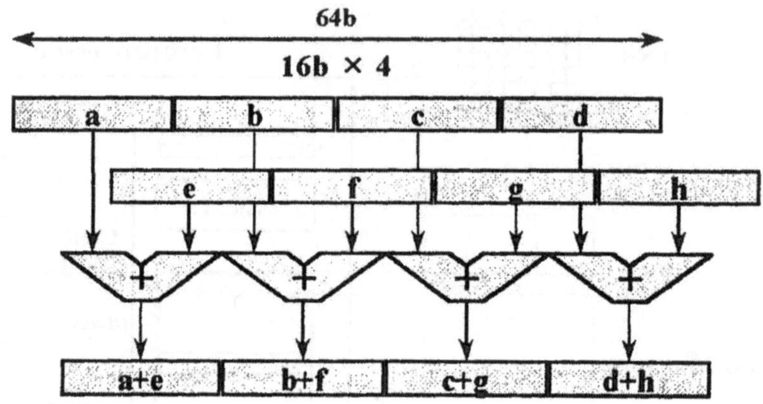

Figure 10.3 Multimedia instruction.

tion estimation, split conditional operations and shuffle operations. The complexity of the multimedia instruction set depends on the processors. Table 10.2 summarizes the number of instructions for multimedia processors.

Table 10.2 List of Media Instruction Set

	UltraSPARC [11]	Pentium [8]	V830R [12]	MIPS [13]	PA-RISC [14]	Alpha [15]
Add/Subt	8	14	10	12	6	-
Mul/Mac	7	3	11	10	-	-
Special	3	-	3	5	3	1
Comparson	12	6	4	6	-	8
Pack/Shuffle	5	9	8	2	2	4
Logical	18	4	4	8	-	-
Shift	0	18	8	5	3	-
Transfer	19 (2)	2	8	-	-	-
etc.	11 (3)	1	-	16 (5)	-	-
Total	83	57	56	64	14	13

These instructions are implemented in either an integer datapath [14] [15] or a floating point (sometimes called coprocessor) datapath [7] [8] in microprocessors. Table 10.3 shows the types of register files (integer registers (INT) or floating point (sometimes called co-processor) registers (FPU)) as well as the sizes of the register files for multimedia instructions and the number of access ports for multimedia enhanced microprocessors . With multimedia instruction set which performs parallel operations, the large register file with many access ports increase the number of operations executed in one processor cycle. Moreover, enough registers make it possible to avoid storing the intermediate value into the memory.

Table 10.3 Register Files

	register file type	# of regs	wordlength	# of multimedia units
UltraSPARC[11]	FPU	32	64	2
Pentium[8]	FPU	8	64	2
V830R[12]	FPU	64	64	1
MIPS[13]	FPU	32+Acc	64 (acc192)	NA
PA-RISC[14]	INT	31	64	2
Alpha[15]	INT	31	64	1

There are several advantages of implementing multimedia instructions on a floating point datapath [7]. One is that floating point registers can be used to store the data for multimedia instructions while integer registers for address and control variables. The second is that the multimedia instructions and floating point instructions are usually not used simultaneously which enables parallel issue of integer and multimedia instructions. Moreover, modifications of the integer datapath to have multimedia instructions might have a significant effect on the critical path of the integer data path pipeline. On the other hand, the advantage of implementing multimedia instructions on integer datapath is that the integer functions such as shift or logical function need not be replicated for the floating point datapath [14]. This results in the relatively smaller number of new instructions and amount of circuits. As shown in Table 10.2, the processors in this approach such as PA-RISC [14] and Alpha [15] do not have the split multiplication/MAC instructions. Especially, Alpha [15] has a very simple multimedia instruction set which does not even have the split addition operation.

10.2.3 Multimedia Instruction Examples

10.2.3.1 Multiply/MAC instruction and IDCT

The design issue in the multiply-accumulate instructions, comes from the fact that the result register requires double wordlength of the input registers. There are several solutions to this. Followings are the list of solutions for the 64-bit architectures.

[Type1] Four parallel 8-bit x 16-bit multiplications with scaling to 16-bit outputs which are shown in Fig. 10.4(a). This is designed for the picture format transformations such as RGB to YUV [11].

[Type2] Four parallel 16-bit x 16-bit multiplications with two 32-bit results. Each result is the sum of the two multiplications as shown in Fig. 10.4(b). Four 16 x 16 multipliers work in parallel with no truncations [8]. However, it suffers overhead of shuffle operations for the four parallel SIMD algorithms, such as the parallel 1D-IDCT.

[Type3] Four parallel 16x16 multiplications with 16-bit accumulations which are shown in Fig. 10.4(c) [16][12]. To get enough accuracy, a rounding and a shift are included in the accumulation. Applications are limited due to the accuracy of the result.

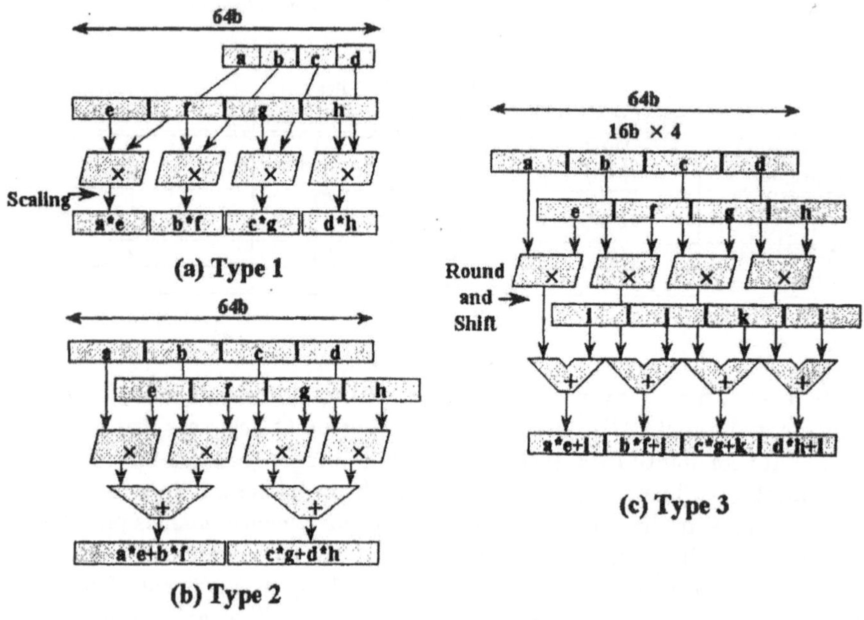

Figure 10.4 Split MAC instructions.

As shown in Fig. 10.5(b), parallel SIMD multimedia instructions realize four parallel 8-point one-dimensional IDCTs for four rows. Straightforward algorithm can be efficiently used for the processors which have fast multiply-accumulate instructions [17]. One-dimensional IDCTs for 8 rows can be realized by repeating this twice. A matrix transposition which is realized by shuffle instructions [14] is needed to start one-dimensional DCT for eight columns.

Consider the implementation of the multimedia instructions described above. Type 1 has a problem in accuracy without the double precision arithmetic. Type 2 requires shuffle operations to realize four parallel one dimensional IDCT. Type 3 can realize four parallel IDCT. The 8-point IDCT algorithm using multiply accumulate operations is shown in Fig. 10.6. This straightforward method requires around 200 cycles to realize 8x8 two-dimensional inverse DCT using Type 3. This is more than 50 times faster than original and 3.5 times faster than fast algorithms for non-media enhanced processors.

Accuracy of the arithmetic operation is important in IDCT calculation. IEEE1180 [18][19] defines the requirement of the accuracy for IDCT, in order not to degrade the quality of the decoded image. The truncation of the 16-bit multipli-

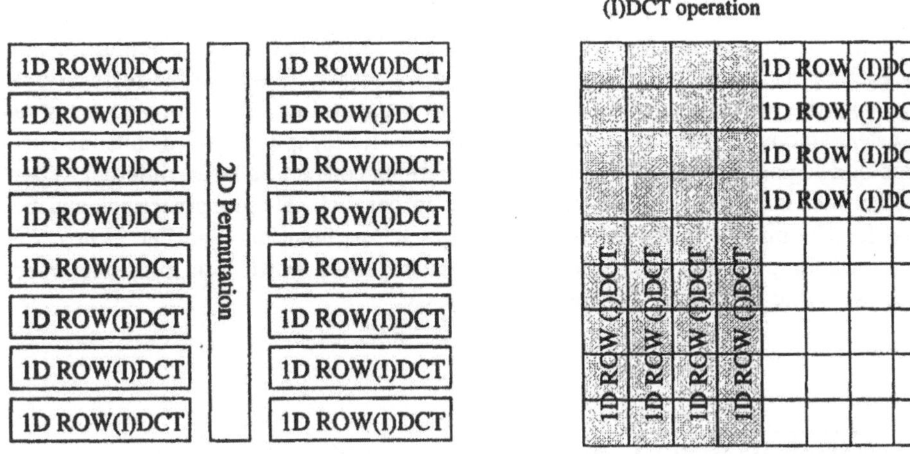

(a) Row-column algorithm

(b) Implementation using SIMD instructions

Figure 10.5 IDCT algorithm.

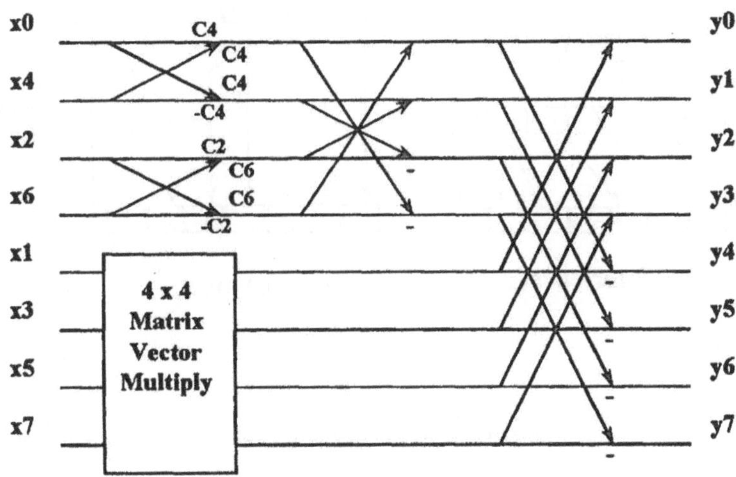

Figure 10.6 IDCT algorithm.

cation result into 16-bit does not provide enough accuracy for IDCT. However, it is reported that enough accuracy can be obtained by shifting and rounding before truncating to 16-bit [12] [16].

10.2.3.2 Shuffle instruction

Shuffle instructions such as the interleave instructions and the precision conversion instructions, shown in Fig. 10.7, have an important role in the multimedia instruction set which divides the long word registers. These shuffle instructions exchange the data among SIMD parallel arithmetic units. They are used for data transfers among splitted words in combination with full/splitted word shift instructions and logical instructions. Shuffle instructions convert between different types of data such as, 8-bit to 16-bit or vice versa in parallel. These instructions can be used for the matrix transpose in (I)DCT which was mentioned earlier. They are also used for precision conversion which convert short word data into long word data as having all 0 as one of the operand.

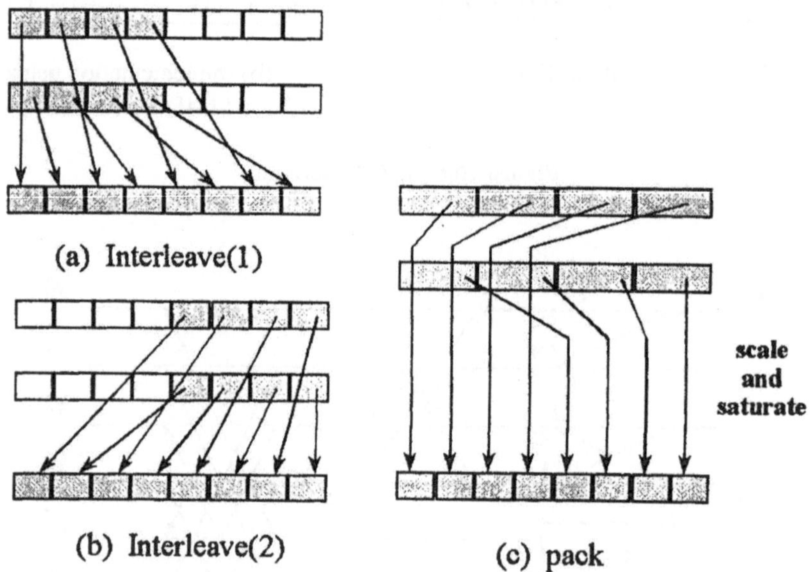

Figure 10.7 Shuffle instructions.

On the other hand, precision conversion instructions which convert long word data into short word data need to realize scaling operations and saturations. Therefore precision conversion instructions (pack instruction) are prepared for this purpose which is shown in Fig. 10.7(c).

10.2.3.3 Instructions for motion compensation

The motion compensation produces prediction signal by using one or two reference (past/future) frame with the pixel interpolation, if necessary. When the motion vector is located between each pixel, the pixel interpolation is required to produce prediction signal by the interpolation of neighbor pixels. Fig. 10.8 shows possible locations of interpolation sample, denoted by pa, pb, pc, pd and pe.

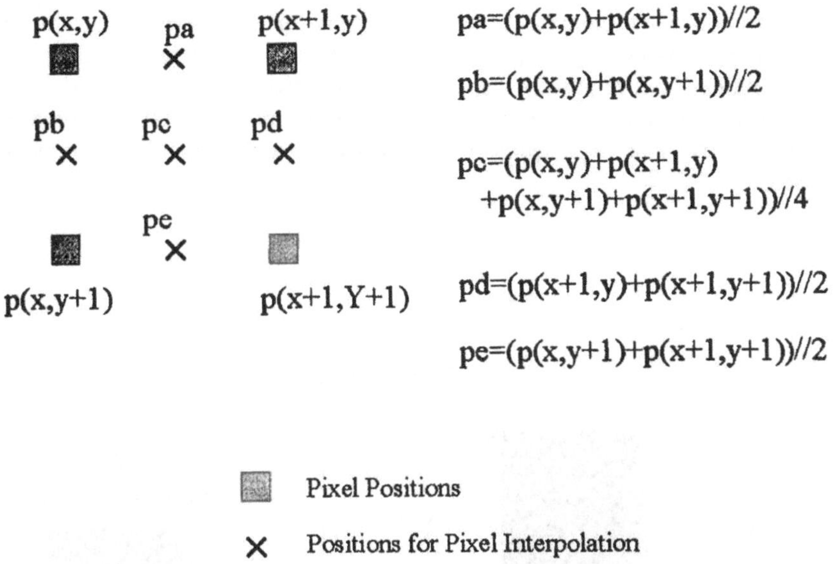

Figure 10.8 Pixel interpolation on motion compensation.

In the addition of two pixel data for interpolation using SIMD type addition operations, more than 8-bit arithmetic is required. Usually 16-bit four parallel operations are used for this purpose. To do this, packed 8-bit data have to be converted into 16-bit data. Unidirectional motion compensation operations for 4 pixel data require two 8-to-16 conversions, four split-additions, one split-shift and one pack operation i.e., 128 operations for 8x8 block which is 3 times faster. Bidirectional motion compensation operations for 4 pixel data require four 8-to-16 conversions, eight split-additions, three split-shifts and one 8-bit clipping. That is 256 operations for 8x8 block which is also 3 times faster. Some multimedia processors [14] [20] have SIMD type multimedia instructions for this pixel average operation. On the other hand it is reported [21][22] that this split-word type operation can be simulated using non-split-word instructions by the special treatment on the carry propagation between each byte in a word.

Another consideration in the motion compensation is unalignment data access. Unaligned memory access is needed depending on the motion vector. Instruction cycles for the alignment are required for the processors which do not have the function of unaligned data access.

10.2.3.4 Instruction for motion estimation

Among each function block of MPEG compression/decompression, the motion vector estimation which predicts input video frame from previous frame requires most computational power. This function is realized by block matchings of the current 16 x 16 macroblock and a 16 x 16 image of the reference block in the specified search area which is shown in Fig. 10.9. This matching operation is performed using the L1 norm (absolute sum of difference) or the L2 norm (squared sum of difference). If we count the absolute sum of difference as three operations, the amount of operations required for block matching for one current macroblock with another 16 x 16 block is 768 operations. If we assume the search area as 32 x 32, 786,300 operation are required to get one motion vector. This results in the required processing performance of 7.8 GOPS for MPEG1 and 31.8 GOPS for MPEG2. However, several fast algorithms to reduce this computational requirements are proposed.

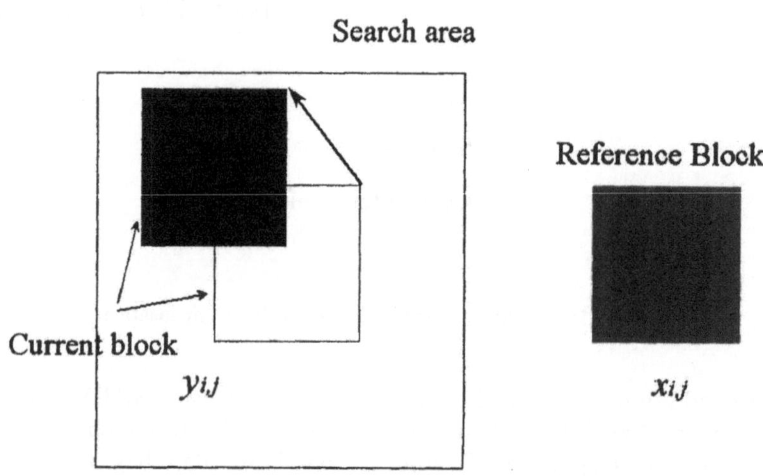

Figure 10.9 Motion estimation.

The media instructions for motion estimation calculate the sum of absolute difference operations. This can be realized using the approaches listed below.

[Type1] Instruction that calculates the sum of the absolute difference for 8 pixel data as shown in Fig. 10.10.

[Type2] Instruction that calculates for sum of the absolute difference for 32 pixel data.

[Type3] Instruction that calculates for sum of absolute difference for 2 pixel data.

Block matching operations in motion estimation can be realized in 325MIPS for MPEG1 and 1325MIPS for MPEG2 by using Type 1 or Type 3 motion estima-

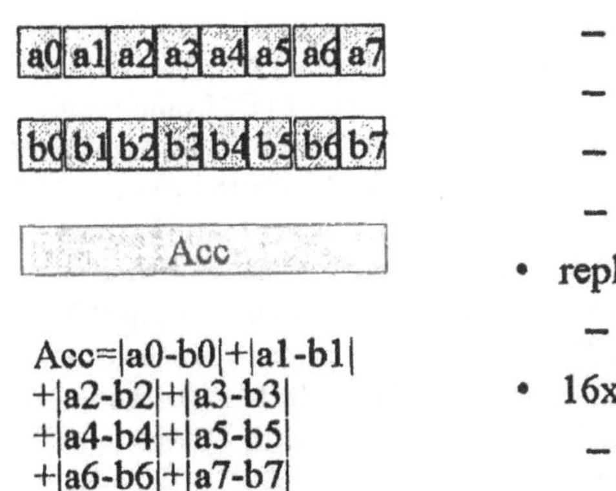

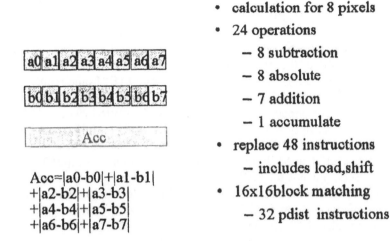

- calculation for 8 pixels
- 24 operations
  - 8 subtraction
  - 8 absolute
  - 7 addition
  - 1 accumulate
- replace 48 instructions
  - includes load,shift
- 16x16block matching
  - 32 pdist instructions

Figure 10.10 Instructions for motion estimation.

tion instructions. For Type 2, it requires 20MIPS and 82.8MIPS respectively. In addition to this, unaligned data loads are also required.

10.2.3.5 Conditional instruction

Conditional instructions are introduced to realize independent conditional operation (such as the comparison of two data) for each data in parallel operations. Followings are the list of conditional operations implemented.

[Type1] Depending on the condition, the result is set to all 0 or all 1 for each data in parallel. Conditional operation can be performed by using the result pattern as the mask pattern. There is some overhead in the mask operation.

[Type2] Prepare the maximum instruction which gets the larger value and the minimum instruction which gets the smaller value. Various conditional operations can be realized by using these instructions. This method is not general purpose, but effective in image processing such as MPEG.

[Type3] Conditional operation is realized by using the saturated addition/subtraction instructions. This method is also effective in MPEG which requires saturate operations.

Most of the conditional operations in MPEG are saturation operations of 16 bit data into 8 bit, they can be covered by the method 2 and 3.

10.2.4 Requirements on Memory Access

The motion compensation in the MPEG video decoder is an extremely memory bandwidth dependent operation. The uni-directional motion compensation for one macroblock requires access to one macroblock of reference frame data, while bi-directional motion compensation requires two macroblocks. The address for these reference macroblocks can be random depending on the motion vectors. Decoding of one P(B) frame requires reading one (two) frame and write one frame of data at most. As shown in Table 10.1, the size of the frame memory is 126.7KB for MPEG1 and 518KB for MPEG2. Motion compensation for one P(B) frame requires access of 253.4KB (380KB) for MPEG1, 1.03MB (1.55MB) for MPEG2. For MPEG2 motion compensation, frame memory read of $8\times8\times6\times45\times30\times(8+20\times2)$=24.8832MB and frame memory write of $8 \times 8 \times 6 \times 45 \times 30 \times 30 = 15.55$MB are at most necessary for each second. Given the second level (L2) cache size of today's PC system which is 256KB to 512KB, it is possible to contain the input and output frames for MPEG1, but not for MPEG2. Therefore frequent access to main memory is required for motion compensation in MPEG2. Table 10.4 shows the cycle time of cache memory access for 200MHz clock CPU with 66MHz CPU bus and L2 bus (today's typical PC). In the case of 64-bit bus with 32byte cache line, cycle times to refill for each types of memories are 45, 36, 27.

Table 10.4 Cache Miss Penalty

MPU Clock Frequency 200MHz
Bus/L2Bus Clock Frequency 66MHz

	Access	Bus Clocks	MPU Clocks
L1 Cache	1-1-1-1	-	4
L2 Cache	3-1-1-1	6	18
FP-DRAM	6-3-3-3	15	45
EDO-DRAM	6-2-2-2	12	36
SDRAM	6-1-1-1	9	27

MPEG2 motion compensation on microprocessors results in 32 to 64 times cache misses in the worst case for macroblock read for uni-directional prediction and bi-directional prediction, respectively. This corresponds to 43.2K to 86.4K times cache miss for one frame, which leads to a total 2.07M times cache miss per second. As shown above, these cache misses for the MPEG2 size frame memory require access to the main memory. An assumption of 45 cycles to refill, leads to around 100M cycles cache miss penalty cycles, in this case.

10.2.4.1 Enhancement for memory access performance

The external (off-chip) memory access performance is enhanced in recent microprocessors for media processing. This is effective, because the memory access latency of microprocessors tends to be long because of the high clock frequency. Moreover, for most of the image processing applications, address for the memory access can be calculated beforehand. Some processors make it possible to get the data into the cache line using software prefetch instructions. Moreover, one processor [7] has block transfer instructions which can transfer a block of data (64 bytes) in one instruction outside of the cache mechanism.

10.2.5 MPU Enhancement and Performance

In the actual implementations of microprocessor based software MPEG decoders, there are performance factors other than the above enhancements. Due to the following factors, cycle time for the actual implementations differ depending on the systems and bitstreams.

(1) Distribution of zero coefficients in DCT block and skipped blocks/macroblocks reduce the number of operations. They depend on the bit stream, especially their bitrate.

(2) Prediction types of the motion compensation change by macroblock. For example, it is possible to have uni-directional prediction macroblocks in B frame. This reduces the number of operations depending on the bitstream.

(3) Performance of memory (access time) and bus (bandwidth) have great impact of total decoding performance. Sometimes, the stall time due to these factors exceeds the core processing time.

(4) Performance and functionality of graphics accelerators also affects the performance in combination with the bus bandwidth.

In general, the media instructions enhance the decoding performance only 1.3 to 2 times. This is because, the other performance such as bit manipulation performance, bus/memory performance and display performance are not so enhanced as the arithmetic performance on the media enhanced microprocessors. Therefore, in addition to the arithmetic enhancements such as the multimedia instruction set, other performance features introduced in section 10.1 should also be enhanced for the media-enhanced microprocessors as well as in the multimedia PC system.

10.2.6 Embedded Microprocessors

Microprocessors for embedded applications (not for PCs or workstations) also enhance the capability of multimedia processing. This class of microprocessors target the applications which originally used DSPs. In addition to these applications, multimedia applications such as internet terminal, STB, car navigator and PDAs are going to be the new targets for these microprocessors.

Embedded RISC processors are low power and low cost microprocessors. This is because they do not employ a complicated control mechanism such as out-of-order controls. Therefore, they can be used in low cost systems such as game machines, consumer electronics and PDAs. Multimedia performance of these processors are enhanced with the requirements from these applications. The arithmetic performance of embedded microprocessors are enhanced by employing a hardware multiplier accumulator as shown in Fig. 10.11(a) [23] as well as SIMD type multimedia instructions.

The requirements for real-time processing and large memory space are achieved by having both caches and internal memories (buffers). An example of cache/RAM memory structure is also seen in V830 [23] shown in Fig. 10.11(b). V830 has 8KB direct mapped/write through caches (4KB for instruction, 4KB for data) and 8KB RAMs (4KB for instruction, 4KB for data). As embedded microprocessors are normally used in low-cost systems which do not have a second-level cache, cache miss penalty is likely to be considerable. As a compensation for this,

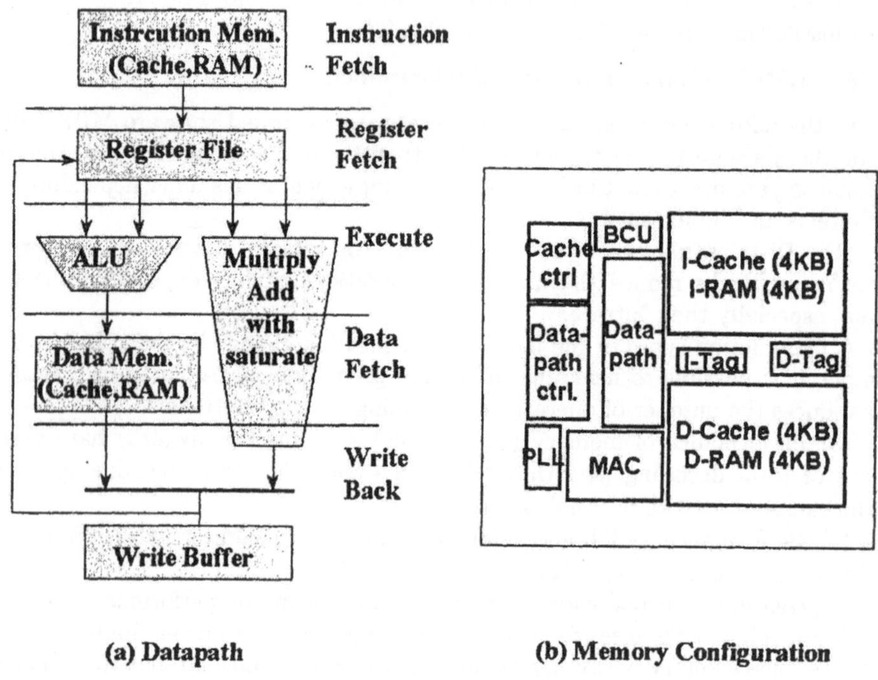

Figure 10.11 Embedded RISC (V830).

internal RAM which is guaranteed not to cause cache miss has been realized. This is effective in realizing high performance program development of MPEG1 [21].

10.3 VIDEO DSP AND MEDIA PROCESSORS

10.3.1 Overview

Programmable DSP chips were developed in 1980. They employ a built-in multiplier in addition to the conventional ALU, and their architectures were based on a pipelined multiply-and-accumulate with parallel control. Therefore, their processing capability was an order of magnitude higher than that of general purpose microprocessors. As a result, they could realize a single-chip modem or a single-chip low bit rate speech codec, while general purpose microprocessors at that time could not realize them.

Fig. 10.12 shows the improvement of DSP processing capability. The DSP processing capability was improved year by year, but it suddenly rose twice in its history, when the DSP applications were shifted to more complex areas. The first jump was to process video signals by higher performance DSPs called video DSP [24]-[26]. However, the video format employed at that time is quarter NTSC for TV conference purposes, called CIF (common intermediate format of 352 × 288 pixels).

The second jump occurred mainly for MPEG-2 applications [20][27] where full NTSC resolution format was required. This type of DSP is referred to as mediaprocessor. Applications of MPEG2 are set-top boxes for digital CATV and VOD (video-on-demand), and DVD (digital versatile disk). These applications highlight the recent multimedia boom.

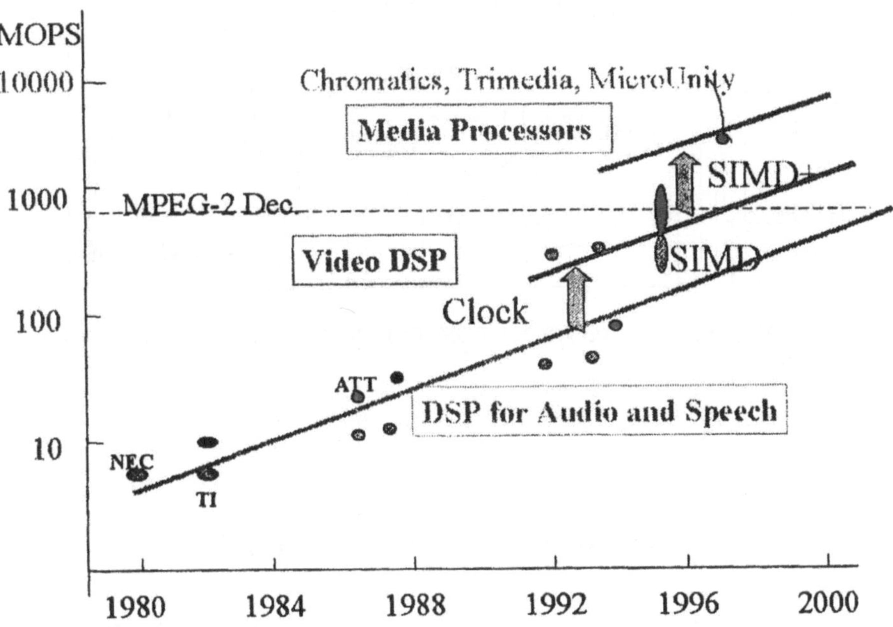

Figure 10.12 Performance improvements of DSPs.

10.3.2 DSP and Video DSP

DSPs have been developed for speech processing and communication processing such as modems as main applications [28] [29]. An example of a DSP architecture shown in Fig. 10.13 is designed to realize high speed multiply-accumulate operations, and is capable of two operand data transfers with two internal memory address modifications and one multiply-accumulate operation for each cycle. They are extensively used in speech compression/decompression for mobile phones. Lower power consumption maintaining high performance are going to be the requirements for these applications. DSP are attractive for these applications due to their high performance/power ratio.

Video DSPs enhance their performance by parallel processing as well as datapath architectures for video processing. Let's check the features listed in section 10.1. Although performance of bit manipulation is enhanced by employing special function units, other performances such as arithmetic operation, external memory access as well as stream I/O are rather moderate for a single processing

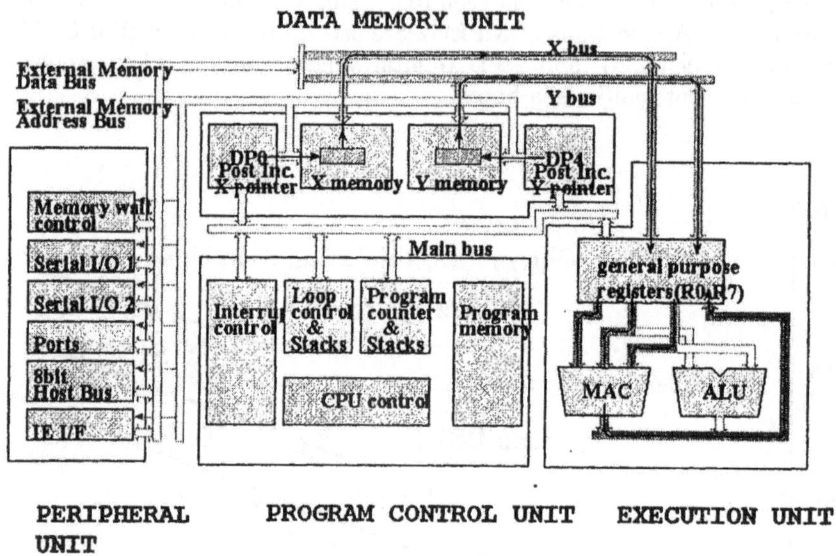

Figure 10.13 DSP architecture (PD7701x).

unit compared with today's high-speed microprocessors. Therefore, several parallel video processing architectures have been proposed to achieve higher performance.

10.3.2.1 Parallel DSPs

One extension form of the DSP architecture is to get higher performance by integrating parallel DSPs on a chip [30] to realize multimedia applications, shown in Fig. 10.14(a). This is a MIMD type parallel processor which integrates several DSPs as well as a microprocessor. DSPs and memory blocks are connected by a crossbar switch and realize a shared memory multiprocessor system. For the concurrent operations of several tasks, a multi-task kernel on the microprocessor controls each task on the parallel DSPs. Each DSP has functionalities for media processing in addition to a multiply accumulate operation. It also has multimedia instructions which realize a parallel operation by dividing the data path into two 16-bit or four 8-bit as well as the instruction for motion estimation which process one pixel in 0.5 cycle. H.324 [31] video conference system which runs multiple tasks such as an audio codec, a video codec and a system control concurrently, is realized using multi-task capability of the parallel processors [32].

Another parallel DSP which is an array processor of twenty DSPs for the PC accelerator application is shown in Fig. 10.14(b) [33]. Each DSP is connected in two-dimensional mesh and is controlled in an SIMD scheme. Some of them have a control capability over other set of DSPs. A memory interface supports block transfer using DMA.

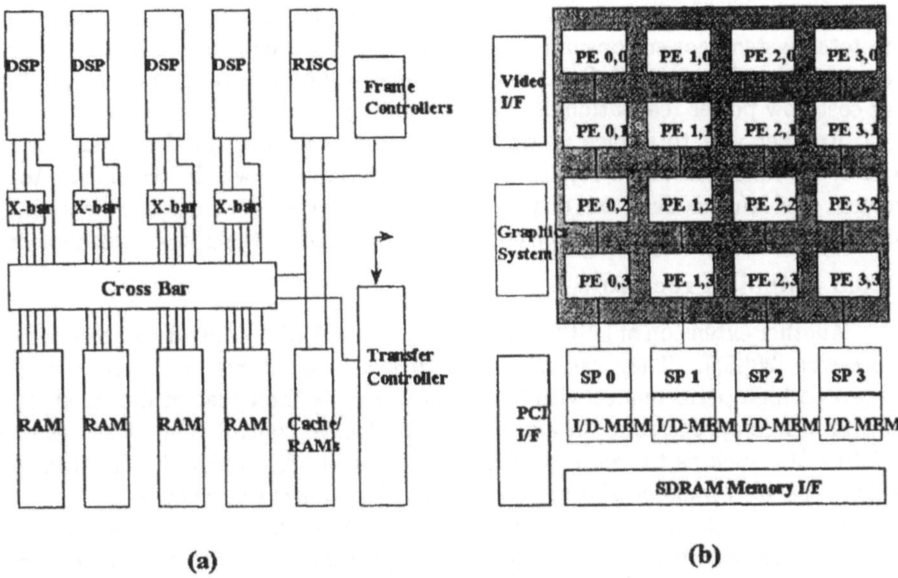

Figure 10.14 Parallel DSPs.

10.3.2.2 DSPs for portable multimedia

Yet another area in which we can expect growth is the mobile multimedia communication. Video communications using wireless phone systems are going to be realized using video codecs using DSPs [34]. Standardization of mobile video compression algorithm such as H.324 for mobile and MPEG4 has accelerated this trend.

A mobile video phone is being developed using the DSP because of this lower power consumption [34]. Limited communication bandwidth makes required performance for this application less comparable with other video applications such as MPEG2.

Currently, DSP are most suitable for mobile phone application (compression and decompression of speech and video). However, when the communication bandwidth is increased in the next generation communication systems such as wide-band CDMA or high speed wireless LAN, higher performance will be required for DSP while maintaining low power consumption.

Multimedia performance required for this application will be within the range of today's media processors. However low power realization is required. Voice codec LSIs which are currently used in mobile phones consume around 0.1W. On the other hand, even the media processors which consume less power than microprocessors, require around 1W which are ten times more.

This low power realization of processors is important as well as the low power realization of the input/output devices such as CCD camera and LCD display. It seems possible to make the power consumption of the media processor into one tenth from now by reducing the supply voltage from 3.3V to 1V. However, to reduce the total system power consumption, frame buffer memory access to the off-chip memory has to be reduced. On-chip frame buffer will be the solution which requires low cost, low power realization of on-chip memory.

Another way to reduce power consumption is having ASIC circuits for the part of the function. ASIC solutions of motion estimation which consume 50 to 90 percent of video encoders, and display processing functions on video decoders enable to lower the clock frequency which contributes to reducing the power consumptions.

10.3.3 Media Processors

Another extension of DSPs which enhanced the VLIW control are called media processors [20][27]. They can issue two to five instructions in parallel. VLIW achieves a high performance with less control circuits than that of the superscalar control in microprocessors. In addition to the VLIW control, some features used in DSPs [35] such as the zero overhead loop control are employed. This function realizes specified number of iterations without introducing overheads.

Let's check the features of multimedia applications as described in section 10.1. To get higher arithmetic performance, media processors have more function units than issue slots. One instruction controls several function units. Some instructions realize SIMD type parallel operations like multimedia instructions for microprocessors.

Media processors inherit many features from traditional DSPs in the datapath architectures such as the direct connection of function units as shown in Fig. 10.15. This feature has been employed in the multiply-accumulate architectures of DSPs. However, there are also new features of media processors which were not usually supported in DSPs. One is operations for parallel packed data type which can be efficiently utilized in image processing like multimedia instructions. Another new feature is a large register file which is effective for the video compression/decompression to store intermediate data.

In addition to the above programmable function units, special purpose function blocks are employed to achieve high multimedia performance with lower clock frequency. Especially, performance for bit manipulation is enhanced by employing a special circuit block for variable length decoding as the VLD coprocessor in Fig. 10.16.

Media processors employ non-cache memories to store program and data which are used in same manner as the internal memories for DSPs. Although they do not have the capability for general purpose processors such as a virtual memory, one processor [20] has functions of CPUs such as an instruction and a data cache, shown in Fig. 10.16. They also employ a high speed memory interface for RDRAM and SDRAM [36] which are effective for image processing. Frame memory access and the data transfer for video output which cause cache miss penalties for microprocessors, can be realized using DMA transfers without introducing overheads.

They have bit stream data I/O interfaces for audio, video and other media with special function block for I/O processing such as a YUV to RGB format

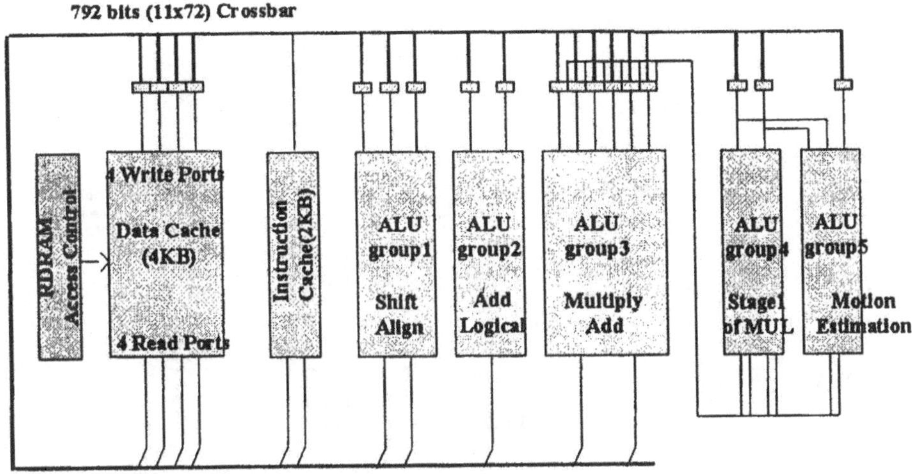

Figure 10.15 Media processor datapath (Mpact).

conversion for display processing which is shown in Fig. 10.16. These video I/O interface as well as a PCI interface to use as a PC accelerator enable the media processor to replace the conventional graphic accelerators.

Media processors have advantages for real-time processing. They realize short interrupt response time by deadline scheduling in its own real-time kernel. No virtual memory support and minimum cache support make hard real-time multimedia processing possible. Unfortunately, use of the compilers is very limited because applications require highly optimized VLIW code. Therefore, extensive hand optimization of assembler or high performance fancy optimizing compilers are needed.

10.3.3.1 Another media processor

Another type of mediaprocessor has special memory interfaces as well as stream I/O interface with accompanying chips. An example is shown in Fig. 10.17 [37]. High bandwidth memory access is realized by special memory interfaces such as SDRAM and RDRAM. This processor also has the capability for the general purpose microprocessor such as virtual memory and memory management for stand alone use. The arithmetic performance of the processor is enhanced by employing SIMD type multimedia instructions with long wordlength [38]. A large register file (128-bit x 32) also contributes to realizing a higher arithmetic performance. Memory mapped I/O avoids coherency and latency problem in DMA based I/O system. Analog interfaces for audio and video are implemented in the separate chip [39]. With advanced features described above as well as a higher

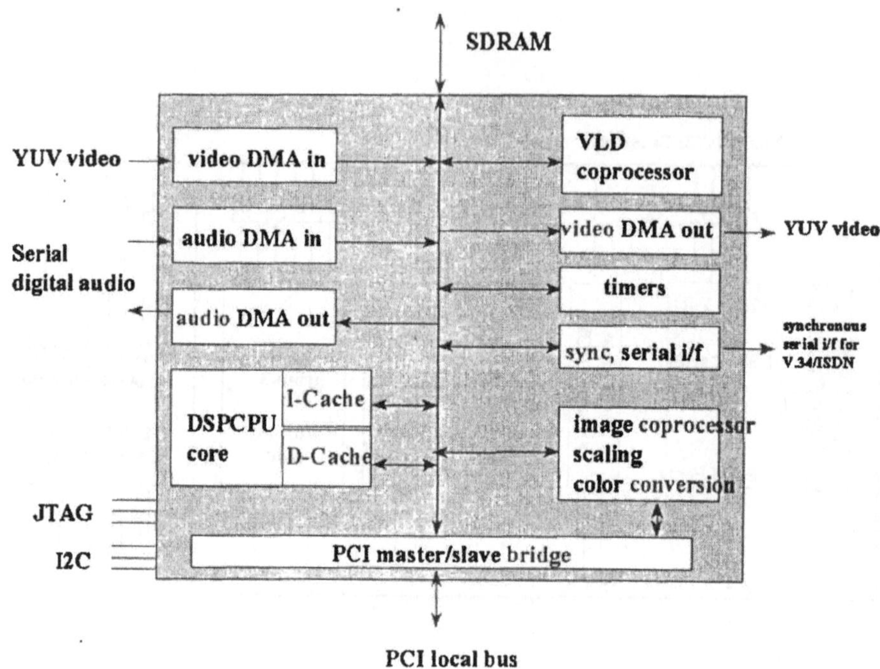

Figure 10.16 Media processor system (Trimedia).

clock frequency (1GHz), this processor is expected to handle broadband media, which results in higher power consumption at the same time.

10.4 COMPARISON OF ARCHITECTURES

Microprocessors for servers and PCs are enhanced to handle multimedia processing by keeping the compatibility to their previous generations. In order to inherit the huge amount of software programs, there are a lot of restrictions in their enhancement. This is why they use very high clock frequency and consume a lot of power. On the other hand, new architectures with aggressive performance enhancements are introduced for embedded microprocessors and DSPs, although some of them have functionality of a stand-alone CPU with high level language support.

Mobile DSPs are not so aggressive right now in terms of their architectural modifications, due to the restriction of low power dissipation for battery operation and limited bandwidth of wireless channels. Video DSPs or media processors have the internal structures which are designed to be tuned for more complex multimedia processing such as MPEG1 encoding. These embedded microprocessors and DSP based chips enable lowering system clock frequency and reducing power dissipation.

Fig. 10.18 shows the clock frequency vs. parallelism with power dissipation, for media enhanced microprocessors and media processors. The measure of the horizontal axis in Fig. 10.18 is parallel 16-bit operations or total bit length of

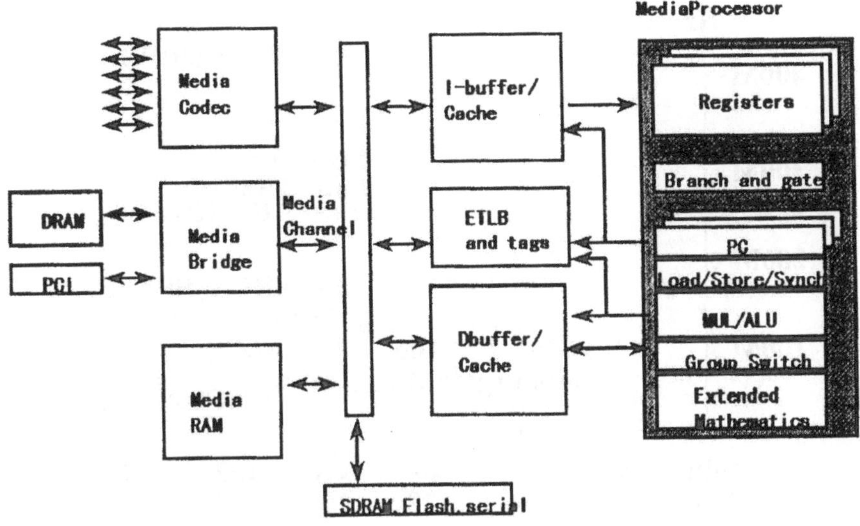

Figure 10.17 Microunity mediaprocessor.

arithmetic units. Microprocessors increase frequency up to 450MHz with maximum 8 level parallelism. On the other hand, media processors increase parallelism by keeping the clock frequency between 50MHz to 100MHz. RISC microprocessors and CISC microprocessors dissipate more than 20 Watts, while power dissipation range of media processors lies around 4 Watts. Embedded RISCs [23] and mobile DSPs power dissipation [35] is less than 1 Watt.

10.4.1 MPU vs Media Processors

For the acceleration of PC multimedia, one solution is the software solution using a highend CPU with multimedia instructions. Another solution is the acceleration by DSPs or media processors with a lower-end CPU. Currently, media processors seem to have advantages in realizing media processing with better quality.

Fig. 10.19 shows the typical data flow for each processing block on MPEG decoding using a media-enhanced microprocessor in (a) as well as a media processor in (b). Multimedia performance for the media-enhanced microprocessor based system depends on the system configurations of the memory system and the bus system and they are summarized in the following. (1).Performance of bit manipulation intensive VLD depends on the integer performance of microprocessors and L1/L2 cache system. (2).Performance of arithmetic intensive IDCT mainly depends on the media unit. (3).Performance of memory access intensive motion compensation de-

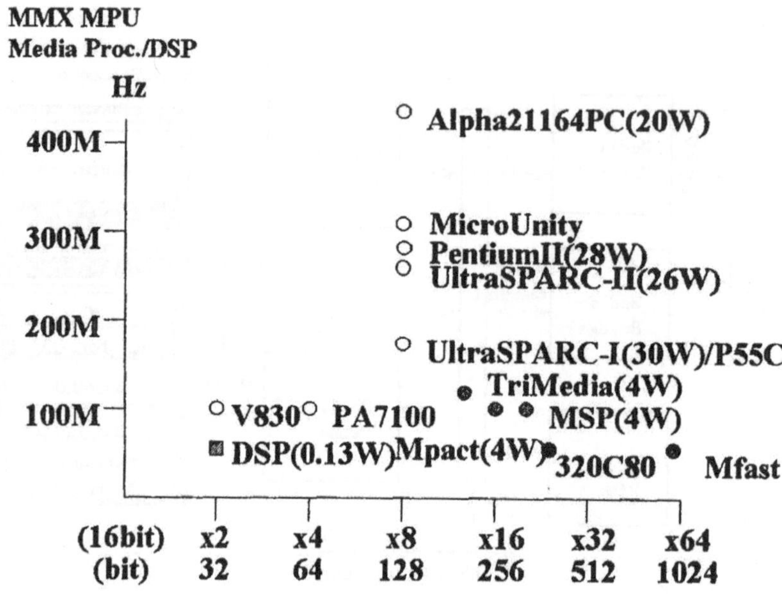

Figure 10.18 Frequency vs. parallelism.

pends on the bus system as well as the memory system. (4).Performance of stream data I/O intensive display operation depends on the bus and the I/O peripherals such as a graphics accelerator. (5.)Performance of real-time task switching intensive A/V synchronized playback depends on the microprocessor control mechanism and the operating system.

On the other hand, for media processors, all performance depends on the mediaprocessor architecture itself and the dedicated memories. Media processors realize the five performance feature enhancements described in the first section mainly by employing special hardware.

The advantage of enhancing the multimedia instructions on microprocessors is that the performance improvement is scalable with clock frequency which continues to increase at a rapid pace. in frequency improvement, compared with the ASIC approach. However there are performance bottlenecks which come from the external bus and the memory access. This approach has other constraints to keep full compatibility with existing operating systems and application softwares. For example, new register set, control registers and other state variables like condition code cannot easily be introduced for the enhanced instruction set. The solution is to share the existing floating point registers for floating point instructions and multimedia instructions, which introduces several cycles of overheads in switching between these two classes of instructions [8].

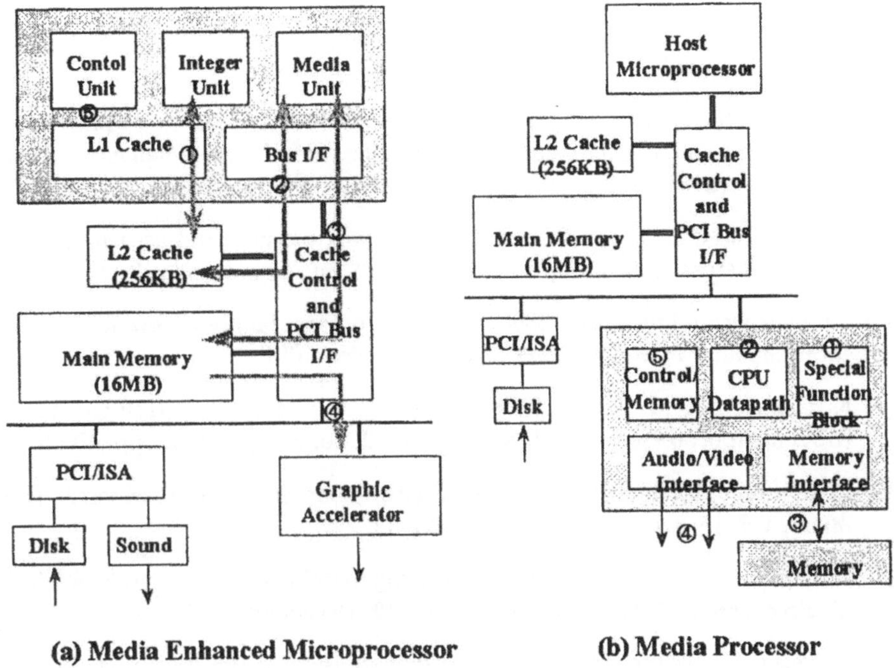

Figure 10.19 Dataflow in multimedia system.

Media processors are introduced as the accelerators for host processors. They are likely to be threatened by the multimedia performance increase of host processors. Moreover, the disadvantages on host processor software processing approach are not necessarily unavoidable problems. They can to be solved if PC systems employ architectures which enhance memory/bus performance for multimedia processing and an advanced OS with real-time support. In this case the media processors can achieve higher cost performance and low power realization.

10.5 CONCLUSIONS

This chapter has summarized the programmable processors for multimedia processing such as media enhanced MPUs, video and media DSPs. Performance and functional requirements of multimedia processing such as the MPEG video have been introduced. Then various multimedia processor architectures were introduced and compared. Data path for multimedia processing and design of the multimedia instruction set with the effect of media operations were described using functions in the MPEG video as examples. Video DSPs and media processors have been described with the functionality for video decoding. Finally, media-enhanced MPUs and media processors were compared in the chip performance for media processing as well as the system performance on PCs.

REFERENCES

[1] ITU-T Recommendation H.263: "Video coding for low Bitrate communication," *ITU*, 1995.

[2] ISO/IEC JTC1/SC29/WG11, "Information technology-generic coding of moving pictures and associate audio information: video," *Recommendation H.262*, ISO/IEC 13818-2, 1994.

[3] ISO/IEC JTC1 CD 11172, "Coding of moving pictures and associated audio for digital storage media up to 1.5 Mbits/s," *Int. Org. Standard (ISO)*, 1992.

[4] Intel corporate literature, "i860^{TM} microprocessor family programmers reference manual," *Intel Corporate Literature Sales*, 1991.

[5] K. Deifendorff, M. Allen, "Organization of the Motorola 88110 superscalar RISC microprocessor," *IEEE Micro,* vol. 12, no. 2, April 1992, pp. 40-63.

[6] R. B. Lee, "Accelerating mutimedia with enhanced microprocessors," *IEEE Micro,* vol. 15, no. 2, April 1995, pp. 22-32.

[7] M. Tremblay, *et. al.*, "The design of the microarchitecture of UltraSPARC^{TM}-I," *Proceedings of the IEEE*, vol. 83, no. 12, December 1995.

[8] A. Peleg, U. Weiser, "MMX technology extension to the Intel architecture," *IEEE Micro,* vol. 16, no. 4, August 1996, pp. 42-50.

[9] M. Johnson, *Superscalar Microprocessor Design*, Prentice Hall Inc., Englewood Cliffs, New Jersey 1991.

[10] M. R. Choudhury, *et. al.*, "A 300MHz CMOS microprocessor with multi-media technology," *Dig. Tech. Papers ISSCC97*, pp. 170-171, Feb. 1997.

[11] M. Tremblay, *et. al.*, "VIS speeds new media processing," *IEEE Micro*, vol.16, no.4, August, 1996, pp. 10-20.

[12] T. Arai, *et. al.*, "Embedded multimedia superscalar RISC processor with rambus interface," *HOTCHIPS IX*, Aug. 1997.

[13] L. Gwennap, "Digital, MIPS add multimedia extensions," *Microprocessor Report*, vol. 10, no. 15, pp. 24-28, Nov. 1996.

[14] R. B. Lee, "Subword parallelism with MAX-2," *IEEE Micro*, vol. 16, no. 4, August 1996, pp. 51-59.

[15] P. Bannon, *et. al.*, "The Alpha 21164PC microprocessor," *Proc. Compcon*, IEEE Computer Science Press, pp. 20-27, 1997.

[16] E. Murata, *et. al.*, "Fast 2D IDCT Implementation via split ALU operations," *IEICE*, NC D-11-95, 1997.

[17] M. Yoshida, *et. al.*, "A new generation 16-bit general purpose programmable DSP and its video rate application," *VLSI Signal Processing VI*, IEEE, pp. 93-101, Oct. 1993.

[18] IEEE Std 1180-1990, "IEEE standard specification for the implementation of 8 by 8 inverse discrete cosine transform," 1990.

[19] ISO/IEC 13818-2:1996/Cor.2 1996(E) in Annex A, "Inverse discrete cosine transform".

[20] S. Rathnam and G. Slavenburg, "An architectutal overview of the programmable multimedia processor, TM-1," *Proc. Compcon*, IEEE Computer Science Press, 1996, pp. 319-326.

[21] K. Nadehara, *et. al.*, "Real-time software MPEG-1 video decoder design for low-cost, low-power applications," *VLSI Signal Processing IX*, IEEE, pp. 438-447, Oct. 1996.

[22] S. Eckart, "High performance software MPEG video player for PCs," in *Proc. SPIE*, San Jose, CA, vol. 2419, pp. 446-454, 1995.

[23] K. Nadehara, *et. al.*, "Low-power multimedia RISC," *IEEE Micro*, vol. 15, no. 6, pp. 20-29, December 1995.

[24] J. Goto, *et. al.*, "250MHz BiCMOS super-high-speed video signal processor (S-VSP)ULSI," *IEEE J.Solid-State Circ.*, vol. 26, no. 12, pp. 1876-1884,1991.

[25] K. Aono, *et. al.*, "A video digital signal processor with a vector-pipeline architecture," *IEEE J.Solid-State Circ.*, vol. 27, no. 12, pp. 1886-1893, 1992.

[26] P. Pirsh, *et. al.*, "VLSI architecture for video compression-A survey," *Proceedings of The IEEE*, vol. 83, no. 2, pp. 220-246, February 1995.

[27] P. Foley, "The mpact media processor redefines the multimedia PC," *Proc. Compcon*, IEEE Computer Science Press, pp. 311-318, 1996.

[28] E. A. Lee, "Programmable DSP architectures: Part I," *IEEE ASSP Magazine*, pp. 4-19, October, 1988.

[29] E. A. Lee, "Programmable DSP architectures: Part II," IEEE ASSP Magazine, pp. 4-14, January, 1989.

[30] K. Guttag, *et. al.*, "A single-chip multiprocessor for multimedia: The MVP," *IEEE Computer Graphics & Applications*, vol. 12, no. 6, pp. 53-64, November 1992.

[31] ITU-T Recommendation H.324: "Terminal for low Bitrate multimedia communication, international telecommunications union," *ITU*, 1995.

[32] J. Golston, "Single-chip H.324 videoconferencing," *IEEE Micro*, vol. 16, no. 4, pp. 21-33, August 1996.

[33] D. Epstein, "IBM Extends DSP Performance with Mfast," *Microprocessor Report*, vol. 9, no. 16, pp. 1,6-8, Dec. 1995.

[34] Y. Naito, *et. al.*, 1997 IEICE, NC A-4-10

[35] NEC Corporation, "μPD77016 User's Manual," 1992.

[36] Y. Oshima, *et. al.*, "High-speed memory architectures for multimedia applications," IEEE Circuits & Devices Magazine, vol., no., January 1997. pp. 8-13.

[37] C. Hansen, "Architecture of a broadband mediaprocessor," *Proc. Compcon*, IEEE Computer Science Press, pp. 334-340, 1996.

[38] C. Abbott, *et. al.*, "Broadband algorithms with the microUnity mediaprocessor," *Proc. Compcon*, IEEE Computer Science Press, pp. 349-354, 1996.

[39] T. Robinson, *et. al.*, "Multi-gigabyte/sec DRAMs with the microunity mediachannel interface," *Proc. Compcon*, IEEE Computer Science Press, pp. 378-381, 1996.

Chapter 11

Wireless Digital Signal Processors

Ingrid Verbauwhede
Dept. of Electrical Engineering
University of California
Los Angeles, California
ingrid@janet.ucla.com

Mihran Touriguian
ATMEL Corporation
Berkeley, California
tourigui@berkeley.atmel.com

11.1 INTRODUCTION

The human need to communicate with peers at any time and in any place around the world is behind the many technological advances in communications systems in the past decade. However this simple need has put many requirements and constraints on engineers designing such systems. This chapter discusses the effects of these technological advances on the device that is the heart of these systems, the digital signal processor, or more commonly known as the DSP. This chapter addresses the design of DSPs for wireless applications.

To illustrate the present state of where we are in terms of communication gadgets, take a marketing or sales person of a big company as an example. Nowadays, this example even applies to people who like to be up to date with the most recent technological advances just for self gratification. This person travelling on the road is most likely to carry with him a briefcase full of different types of wireless products. This is because many of these devices handle different types of communications. For example, a pager is useful for a robust one way and recently two ways transfer of short data messages. Air time is cheap, the device itself is cheap and very convenient for communicating short messages. In the briefcase, we can also find a cellular phone. The cellular phone provides a two-way voice communication with more expensive air time. However, due to the real time voice communication, business can be performed conveniently even when moving in a fast train or idling in highway traffic. A more technically sophisticated business person would carry a computer laptop as well. A laptop would provide two-way data and even multimedia access to the user. Important email can be accessed and responded to while traveling. In addition, the extra processor power of the laptop can be used for other purposes such as large data-base storage, computations, fax and data transfers, etc.

It should be clear from the above example that there are requirements, constraints, and future trends imposed on emerging communications devices. Notably, the need for portability, integration and size, battery lifetime and high computational power.

In the first part of the chapter we explain the applications and algorithms which typically run on a wireless terminal. This is necessary to understand the enhancements which are made to the DSPs. In the second part, we describe the limitations of general purpose DSPs and discuss several alternatives to enhance the performance. In the last part we describe one wireless digital signal processor in detail: the core architecture, the general purpose features and the domain specific features.

11.2 DIGITAL WIRELESS COMMUNICATIONS

11.2.1 Overview

A generalized wireless communication system is illustrated in the block diagram in Fig. 11.1. Different wireless standards have different adaptations of these blocks to match their particular system requirements. On the transmitter side, the function of the source encoder is to modify the structure of the input signal to match the particular communication system being used. In digital communication, most often this module performs the function of removing redundancy from the input source signal to compress the amount of data for transmission through the channel. A speech encoder in a wireless cellular phone is such an example. The output of the source encoder enters the channel encoder. The channel encoder adds redundancy to the signal to make it more robust when it passes through the channel. At a first glance, the channel encoder seems to reverse the effect of the source encoder to reduce the transmission data rate. However, the channel encoder adds controlled redundancy to combat or even correct the transmission errors incurred in the channel. Obviously, a trade-off exists between transmission rate and performance. In wireless channels where multipath fading is one of the prevailing degradations, interleaving and convolutional encoding techniques are mostly used. The output of the channel encoder enters the modulator. The task of the modulator is twofold. It converts the encoded digital message into a suitable analog signal for transmission through a specific channel, and then modulates a high frequency carrier by this signal before putting it on the channel. The modulation method and the carrier frequency chosen depend strongly on the transmission medium. The digital cellular standard in northern America, IS-136, uses a PI/4 shifted QPSK modulation techniques in a frequency band between 824 MHz to 894 MHz. The channel can take many forms from free space to a solid medium such as glass.

On the receiver side, the inverse operations occur. The demodulator down-converts the analog signal to a baseband (original source frequency band) digital signal and then extracts the digital bit stream from it. The recovery of the bit stream from the modulated digital signal can be a complex sub-system. This includes the functions of synchronization, clock and timing recovery, frequency recovery, equalization, and slicing. The channel decoder reverses the transformation performed by the channel encoder on the transmitter side. Operations such as de-interleaving and Viterbi decoding are commonplace. Errors induced by the channel can be detected and corrected in this module. The source decoder recovers the

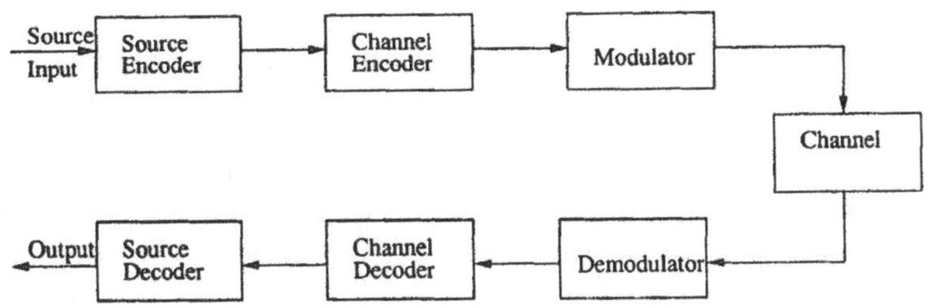

Figure 11.1 Generalized communications system.

message that was transmitted by the source on the transmitter side. In a digital cellular phone, this consists of the speech decoder module that converts the received bit stream into speech signals over the speaker.

Although the system architecture described above is general enough to cover most wireless communications systems, the specifics of each module can differ vastly. One such example is the different speech codecs adopted by different digital cellular standards around the world. Whereas the existing GSM digital cellular standard employs the RPE-LTP (Regular Pulse Excitation - Long Term Prediction) speech codec requiring around 4.6 MOPS (Million Operations Per Second), the Japanese half rate coding standard, known as PSI-CELP (Pitch Synchronous Innovation - Code Excited Linear Prediction), requires around 100 MOPS!

11.2.2 Algorithmic Requirement

It will help us understand better the complexity of wireless communication systems, if we have a closer look at the common signal processing algorithms used in the modules described in the previous section (see also Chapter 8). The most common form of a wireless communication device is today's digital cellular phone. For the purpose of our discussions, we shall concentrate on the digital cellular phone regardless of which standard it implements. The standards we shall borrow examples from will be one the following four: the Northern American Digital Cellular (NADC IS-136) standard, the IS-95 CDMA Standard, the European GSM Standard or the Japanese Digital Cellular Standard.

11.2.2.1 Source encoder/decoder

The most common type of source encoding and decoding used in a digital cellular phone is a speech compression module. The purpose of the coder is to reduce the amount of bits representing speech by removing redundancies in the source bit stream and hence reduce the bandwidth requirements for storage or transmission over the air. A 'toll quality' digital speech requires the sampling of the analog speech waveform at 8 kHz. Each sample requires 8 bits of storage (*μ-law compressed*) for a total data bandwidth of 64 kbits/sec. The IS-136 traffic channel provides a data rate of 13 kbits/sec after channel coding. The raw bit rate at the input of the channel coder is only 7950 bits/sec. Obviously a compression ratio of at least 8 is needed to transport adequate speech information in this case. A half rate IS-136 cellular phone would require a compression ratio of 16!

Speech coders are divided into two main categories: waveform coders and parametric coders. Waveform coders perform sample by sample coding of the analog speech waveform in time domain without taking advantage of the nature of speech or human vocal tract to reduce bandwidth. These are relatively simple coders and can achieve toll quality at rates of 32 kbits/sec. The most common example is the PCM (pulse code modulation) coder and its derivatives the DPCM (differential PCM) and ADPCM (Adaptive DPCM). The DPCM and ADPCM achieve compression by quantizing the difference between consecutive samples instead of the samples themselves. The dynamic range of the amplitude of the difference being smaller than that of the sample itself, requires less information bits to code and hence smaller bandwidth. Increasing compression while keeping the quality of speech acceptable, requires higher and higher complexity. In addition, increasing compression ratios means removing more redundancies in the speech itself. This is possible if coders exploit the physical nature of human speech and the mechanism by which it is created. Parametric coders achieve exactly this. These class of coders take advantage of the high redundancy and correlation properties of speech by modeling the human vocal tract and extracting the few parameters that are necessary in generating speech waveforms. The encoder section extracts these parameters and codes them for transmission. The decoder section uses these parameters to synthesize the original speech. These coders rely on linear predictive coding (LPC) methods for the extraction of the essential parameters. Although the details of LPC and speech codec is outside the scope of this section, it is useful to consider the types of algorithms used in wireless DSPs. The purpose of the LPC analysis is to predict the output sample at t=n+1 given the output sample at t=n and the input samples from t=n-N to t=n-1, where N is the order of the LPC. This translates to solving the coefficients for an autoregressive (all-pole) filter of the form

$$H(z) = \frac{1}{A(z)}. \tag{1}$$

The process involves calculating the autocorrelation or the covariance matrix of the input speech samples over the LPC analysis interval; in most cellular environments this is 160 samples long. The autocorrelation equation takes the form of

$$r(m) = \sum_{n=0}^{L-m} s(n)s(n+m) \tag{2}$$

where L is the analysis interval length and $0 \leq m \leq N$, and N is the LPC order. One of the most common methods known to solve a system of equations in a very efficient iterative way is the Levinson-Durbin algorithm. The iterative core of this algorithm contains a very common butterfly structure also used in implementing other filters having similar lattice structures. This iterative computation is described by

$$a_{n,p} = a_{n,p-1} + K_p a_{p-n,p-1} \qquad a_{p-n,p} = a_{p-n,p-1} + K_p a_{n,p-1} \tag{3}$$

where p=1,2,...,N. A second important process takes place in vocoders employing Code Excited Linear Prediction (CELP). A quantized version of an excitation vector is transmitted to the decoder in addition to the LPC parameters. The excitation vector models the excitation source in human speech generation, the vocal cords. This vector quantization is achieved by a method known as analysis by synthesis. This involves exciting the synthesis filter, which models the human voice tract (throat, mouth, teeth, etc...), with all possible excitation vectors from a codebook which is found in both the encoder and decoder. The vector from the codebook synthesizing a speech section with the closest match to the input speech segment is chosen as the best excitation codeword and its index transmitted to the decoder. Hence the name, analysis by synthesis. This process is called vector quantization and can take the form of exhaustively trying each vector in the codebook or is sometimes optimized in some vocoders to search through subsections of the codebook instead of the whole. Nevertheless, codebook sizes can be above 1000 entries, and in most cases, this process is computationally the most intensive. The basic operation involved here is finding the codeword that has the minimum distance from the input vector. The operation is of the form

$$d = \sum_{n=0}^{N-1} (a_n - b_n)^2 \tag{4}$$

which implements the sum of distance square function.

The vocoder module is by no means limited to the above functions, although in most cases the above consumes around 50% of the available processor horsepower. Vocoders also include many FIR and IIR filtering operations using Multiply/Accumulate type of instructions. The decoder side in CELP type vocoders is usually much more simpler than the encoding side. In fact, for analysis by synthesis type of vocoders, the decoder is a subset of the encoder and requires minimal amount of extra coding effort. For more details on speech vocoders, the reader is encouraged to refer to books with related topics [1].

11.2.2.2 Channel encoder/decoder

The function of the channel codec is to add controlled redundancy to the bit stream generated by the source encoder on the transmitter side, and to decode, detect, and correct transmission errors incurred over the channel on the receiver side. For this reason, channel encoding and decoding is also known as error control coding. The type of error control chosen for a particular application depends heavily on the transmission channel used for that application. In our case, the channel is the air interface between the communicating device and a base station. Some of the most severe impairments the signal encounters in this channel are: the Rayleigh fading due to the destructive interference of the reflected waves arriving at the mobile phone, interference from same and neighboring cells, intersymbol interference due to delayed received signals and low or interrupted signal power to name a few. For enhanced reliability in both speech and data communications in mobile phones, digital cellular standards use several error control coding techniques to combat different types of impairments. Error correcting block codes such as Hamming codes and the more powerful Bose-Chaudhuri-Hocquenghem (BCH) codes are used to detect and correct errors in short message transfers between the mobile and the base; even more powerful convolutional codes, codes that create dependencies over longer bit streams, are used to protect speech and data frames. Although these codes can correct short burst errors in addition to random errors, interleaving across several

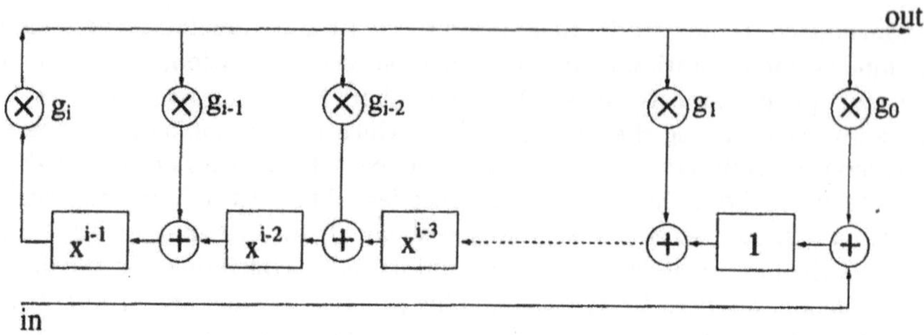

Figure 11.2 GF(2) polynomial division circuitry.

TDMA frames is performed to disperse longer burst errors across frame boundaries and thus enhance the correcting power of these codes. In addition to error correction, CRC checks are added to some data blocks to detect errors and discard the block if one occurs.

The underlying arithmetic operations involved in error control coding belong to the realm of finite field algebra and Galois Fields (GF) arithmetic (see Chapter 21). In addition, functions such as interleaving and de-interleaving involves bit manipulation operations that can be cumbersome for some general purpose processors. Galois field arithmetic can be easily performed by logic circuits, specially when the field size is 2. In fact, the bit streams can be represented as polynomials over GF(2), and the encoding and decoding functions treated as polynomial division or multiplication easily implemented in shift-register circuits. For example, the CRC check involves the division of the data polynomial by the CRC generator polynomial to give a quotient and a remainder. The remainder is the CRC polynomial. The equation takes the form of

$$\frac{D(x)}{G(x)} = Q(x) + \frac{R(x)}{G(x)} \tag{5}$$

where D(x) is the data polynomial, G(x) the generator polynomial, Q(x) the quotient and R(x) is the required CRC polynomial. Although the operation can be easily implemented by the circuit in Fig. 11.2, it can take several DSP cycles per bit to achieve the same function.

Another example is convolutional encoding, which includes polynomial multiplication of the form

$$C(x) = G(x)D(x) \tag{6}$$

where C(x) is the output polynomial, G(x) is the generator polynomial and the D(x) is the input data polynomial. The circuit for multiplication is a feed-forward shift-register implementation that can be easily realized in hardware but can be cycle consuming in software.

The most commonly decoding method used on convolutional codes is the Viterbi decoding algorithm. To summarize, the Viterbi algorithm tries to emulate the encoder's behaviour in creating the transmitted bit sequence. By comparing it to the received bit sequence, the algorithm determines the discrepancy between

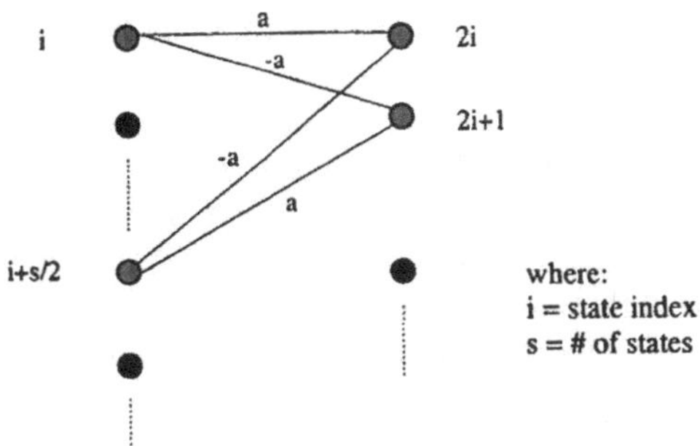

Figure 11.3 Radix 2 viterbi butterfly.

each possible path through the encoder and the received bit sequence. The decoder outputs the bit sequence that has the least discrepancy (also called the minimum distance) compared to the received bit sequence. The Viterbi algorithm becomes impractical for software implementation when the constraint length or the memory of the encoder becomes greater than 8. This is due to the increase in the computations needed to decode and not due to any increase in algorithmic complexity. The basic unit of the Viterbi algorithm is what is known as the Viterbi butterfly. For simplicity reasons, we will discuss convolutional encoders of rate 1/n, where 1 input bit generates n output bits. The butterfly structure of the decoder for such an encoder is called a radix 2 Viterbi butterfly which is shown in Fig. 11.3. For such a decoder each state has only two possible transitions to the next state. The constraint length, m, of the encoder, determines the number of states, 2m present in the decoder. Each transition has a cost a associated with it as shown in the figure. The decision on which transition is chosen is made by computing a set of operations known as the Add-Compare-Select (ACS) operation. The cost for each transition, from the present state, is added to the previously accumulated cost to reach the current state, the two new accumulated costs are compared, and the branch or transition with the minimum (or maximum) cost is selected. This basic ACS operation is performed $2^{m-1}w$ times, where w is the decoding window.

It is clear that for large m and long w, the amount of computation can become prohibitive. For example, for the CDMA IS-95 standard in the US, m=8 and w=192. This will require $2^7 \times 192 \times$ (cycles for ACS) cycles per decode block. Future applications and standards are demanding higher transmission rates and hence will require even larger constraint lengths for improved error correction capabilities. Efficient Viterbi decoding capabilities are essential to current and future designs of wireless DSPs.

To learn more about error control coding, the reader is referred to [2].

11.2.2.3 Modulation/demodulation

The function of the baseband modulator is to convert the digital message into an analog waveform compatible with the channel over which the message is to be transmitted. In wireless systems, and in other bandpass transmissions, this analog waveform is further upconverted by modulating a sinusoid called a carrier. In this section we concentrate on the baseband digital modulation and demodulation functions.

In digital baseband modulation, the message bit stream is mapped into fixed constellation points determined by the chosen modulation scheme, bandlimited with a pulse shaping filter such as a raised cosine filter, interpolated for better approximation and finally fed into a D/A for the analog conversion of the baseband signal. If carrier modulation is desired, this output is fed into an RF (Radio Frequency) subsystem. As an example, the IS-136 employs a $\pi/4$ DQPSK ($\pi/4$ shifted differential quadrature phase shift keying) modulation techniques in which di-bit symbols are mapped into phase shifts, rather than an absolute phase and hence called differential. Each alternating di-bit symbol is mapped onto one out of two 4 point constellation that has a shift of $\pi/4$ relative to the other. The transmit as well as the receive filter is a square root raised cosine filter. Although these two functions can be performed efficiently in a dedicated ASIC, many designers still prefer to perform them in a programmable DSP for increased flexibility. A DSP having bit manipulation, table lookup and fast FIR filtering capabilities is desired.

The digital demodulation functions are much more complex and almost always performed on a programmable DSP. The demodulator receives a highly degraded, noisy, and distorted signal, and extracts the best possible estimate of the transmitted message stream. The demodulator has to deal with recovering the carrier phase for coherent detection schemes, tracking the amplitude and phase of the signal corrupted by the multipath fading wireless channel, acquiring and tracking carrier frequency errors and sampling time errors due to inaccurate crystals, using sophisticated equalizers to compensate for intersymbol interference due to the dispersive nature of the channel, and using MLSE (maximum likelihood sequence estimator) detectors to correct and estimate the most likely transmitted bit stream. Due to the quadrature nature of these modulation schemes the signal processing involves complex arithmetic. To learn more about digital communications, the reader is referred to [3].

To be efficient in these functions, wireless DSPs should be capable of performing fast filtering operations over complex data, adaptive filtering for error tracking and predicting, and fast Viterbi butterfly decoding for the MLSE detector (this is slightly different from the channel decoding Viterbi butterfly structure discussed earlier). Upcoming wireless standards that demand faster and faster transmission bandwidths will put more strain on DSPs to perform these functions more efficiently. Already today the need exists for a multiply/add operation to be performed in 0.5 clock cycles or less.

11.2.3 Advances in Wireless Technology

One of the fastest growing markets in the world is in the area of the wireless cellular communications. Today's consumer is technically aware and demands progressively higher quality and standards. This in turn drives researchers into new signal processing domains that are more and more complex and challenging.

Higher data rates The need for higher data rate transfers is growing in leaps. Soon audio and video transmission over wireless cellular channels will become commonplace. The bit rate requirements for such an application will be in several Mb/s. To maintain performance and robustness at these high rates, more complex channel coding techniques are applied. This in turn requires complex decoders on the receiving side. Channel decoders employing Viterbi algorithms with large number of states will be common.

Higher speech quality First generation speech codecs were met with disappointing reactions from the consumer. By going from analog voice to digital voice transmission, and paying higher prices per unit cellular phone, the consumer expected superior voice quality. Of course this was not true in reality. The main goal for going to digital transmission was to achieve more capacity and to be compatible with the digital world of computers but not necessarily to get higher voice quality. First generation speech codecs include the IS-136 VSELP (Vector Sum Excited Linear Prediction) and the GSM RPE-LTP. However, to meet the need for higher quality, new second generation, more complex speech codecs are under development such as the IS-136 ACELP (Algebraic Code Excited Linear Prediction) and its derivative for the GSM, the Enhanced Full Rate Vocoder.

Higher Capacity Network providers would like to increase the number of subscribers that are active in a cell at a given time. To this end new speech vocoders are being developed that run at half the rate while maintaining acceptable quality and thus doubling the capacity. Reducing the bit rate for a vocoder means increasing the compression ratio of the input speech. This comes at the expense of increased computational complexity. An example of this is the Japanese half rate vocoder, PSI-CELP, which requires in the neighborhood of 100 MOPS.

Portability and lower power consumptions Integration and longer talk- and standby- times contribute to the success of a product. Integrating the functions of a pager, telephone, fax machine, data modem, and video is the goal for cellular phone manufacturers. This device will be more of a portable personal communicator than a simple cellular phone which should fit in a shirt pocket or a purse. Power consumption is an important consideration for these battery operated devices and as more sophisticated functions are included in them, the interval between battery recharge cycles will become the distinguishing factor. Table 11.1 shows the direction cellular phones are headed in the next decade to meet consumer demands [4].

These advances in technology demand the emergence of a new line of digital signal processors. Standard DSP platforms are no longer efficient or, in some case, are not even adequate to handle the high complexity required by these functions. New DSP architectures are evolving that are optimized for digital cellular applications. The focus in these new designs is geared towards creating new specialized complex instructions supplemented with architectural innovations, applying low power design methodologies at all levels of abstraction from silicon to software, and increased parallelism in the datapath to increase the number of operations performed per unit time without necessarily increasing the master clock rate.

11.2.4 Limitations of General Purpose DSPs

Programmable DSPs were introduced in the early 80's for compute intensive applications. These applications run mostly under tight real-time constraints and involve a large number of operations which are in tight repetitive loops involving many memory accesses.

The basic Harvard DSP architecture consists of a datapath, able to perform a multiply-accumulate in an efficient way by employing a separate program bus and data bus to improve the memory bandwidth, as shown in Fig. 11.4. Since then, different generations of DSPs have emerged using variations of the basic Harvard architecture. The main difference between the generations revolve around the bus

Table 11.1 Cellular Phone Requirements [4]

	Now(1997)	Target
Size	100 cc	10 cc
Weight	50 g	10 g
Components	300 - 500	< 50
Standby time	3 days	> 1 week
Talk time	1 to 4 hours	> 10 hours
Chips	8-12	< 3

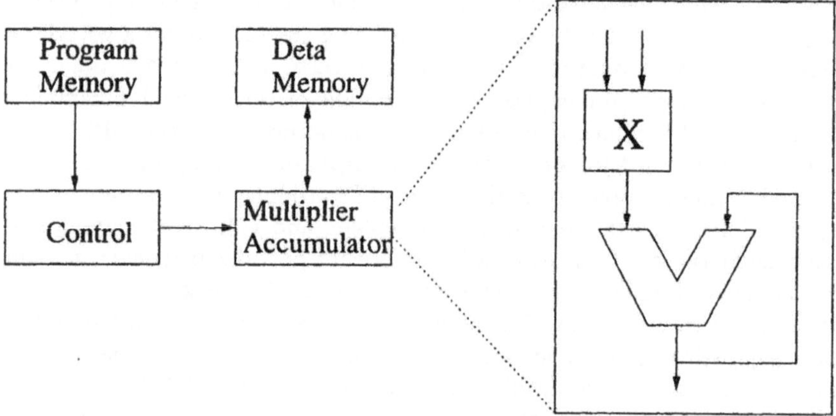

Figure 11.4 Basic Harvard architecture.

structure and the memory bandwidth such that the multiply-accumulate operation

$$z = x + ys \tag{7}$$

can be executed in one cycle or less.

In the 1st generation processors, such as the TMS320C10, z and x are mapped on the same accumulator, y and s come from the data memory. But since the processor has only one data bus and hence one data fetch per cycle, this instruction takes 2 cycles to execute.

In the 2nd generation processors, such as the TMS320C25, part of the program memory can be mapped onto the data memory (RAM). Therefore during a single instruction repeat operation, the program bus is freed and is used to fetch data from the data memory. In this case two data can be fetched in one cycle and hence a multiply-accumulate instruction takes 1 cycle to execute.

Typical for the 3rd generation is the addition of extra addressing features, such as circular buffers, and special functions, such as bit manipulations. These features are introduced to address new and more complex algorithms and support the emerging digital cellular standards, such as the North American Digital Cellular IS-54.

It is clear from the above that the main driving force behind the advent of new generations of DSPs, was the acceleration of the multiply-add operation. Although, this was and still is one of the most basic and integral part of digital signal processing operations, the development of new and complex algorithms is pushing forward other operations that are becoming as basic and commonplace as the multiply-add operation was a decade ago. These new operations either require higher memory bandwidth per unit operation, such as the FFT radix-2 butterfly, or require specialized instructions such as finding the sum of absolute distances for vector quantization in speech compression, or even requiring both features such as the Viterbi butterfly decoding and the add-compare-select operation.

A general purpose DSP can still perform these functions. However, for a wireless digital cellular standard, it would require clock rates in the order of 50 or 60 MHz, depending on the standard chosen to meet the performance requirements. One of the main issues a cellular system designer worries about is how much the system clock can be reduced without adversely affecting performance. A lower clock frequency results in lower power consumption and therefore longer battery life, and minimizes other potential problems such as clock interference in the system. However, reducing clock speed means either not adding extra features which might distinguish the product from others or sacrificing performance.

The two measures of performance in the industry to qualify a DSP for a particular application are MIPS (million instructions per second) and MOPS (million operations per second). As an example, consider a DSP that can fetch one instruction per clock cycle. In a pipelined machine this means that one instruction is effectively performed per clock cycle. In reality, there are instructions requiring effectively two or more clock cycles, such as many program flow instructions and instructions that are more than one word long. For such a DSP, a clock frequency of 20 MHz will yield an instruction execution rate close to 20 MIPS.

In a general purpose DSP, an instruction is defined as one fundamental signal processing operation such as an add or multiply-add. In this case the MOPS figure will almost be equivalent to the MIPS figure. Therefore, a higher clock frequency (with the associated increase in MIPS), is required to increase the MOPS figure. On the other hand, when parallelism is employed, an instruction can incorporate several operations (such as subtract, square and accumulate) in parallel. In this case, the MOPS figure will be much higher than the clock rate or the MIPS figure. Sasaki, [5], shows that the increase in clock frequency (and hence MIPS) exhibited in the last decade will not continue into the next. Instead, application or domain specific DSPs, with specialized functions and parallelism, will become dominant and will allow further increases in the MOPS figure.

It is to this effect that a different class of DSP architectures are playing a crucial role. These are the domain specific architectures that are tailored for efficient implementations of the next generation digital cellular standards. Keeping in mind that in most cases the domain specificity of these DSPs does not undermine their generality to address current signal processing applications. These are referred to as "Wireless DSPs".

11.3 WIRELESS DIGITAL SIGNAL PROCESSORS

To implement the functions described in the previous sections, a wide selection of approaches are possible. Current generation of general purpose DSPs do not have enough processing power to execute the complete baseband signal processing functions on one processor.

11.3.1 Application Specific Solutions

One option is to implement part of the baseband functions in ASICs. This is done for two reasons. In previous generation wireless chipsets, this option is chosen because it reduces the load for the general purpose DSP. Functions such as channel coding and decoding require extensive bit manipulation which cannot be always efficiently implemented on a general purpose DSP. These type of functions are ideal for an ASIC implementation.

An example is the first generation of a GSM Handy Phone of Siemens, [6]. One ASIC chip implements the channel encoder, decoder and the equalizer (PMB 2705). A second chip is a general purpose DSP which implements the GSM full rate speech codec in software. Next generations of GSM implementations of Siemens will implement more functions in software (PMB 2707 and PMB 2708).

In current generation wireless chipsets, this option may be chosen to reduce power. An ASIC implementation has the advantage of having a tuned logic to the application and hence a lower power consumption. It has the disadvantage that once the device is cast, it cannot be modified.

11.3.2 Clock Speed Versus Parallelism

One way to improve the performance is by increasing the clock frequency. This can be done by increasing the pipeline stages. An example is this, is the evolution of the Motorola processors [7]. The previous generation, the 56000 family, has a pipeline depth of three: fetch, decode and execute stage. The new generation, the 56300 and 56600 family, has a pipeline depth of seven! It consists of 2 fetch stages, 1 decode stage, 2 address generation stages and 2 execute stages. The clock frequency has increased to 80 MHz. A deep pipeline creates however pipeline conflicts. For instance, a transfer stall will occur when a value is written to the accumulator registers in a first instruction and read again in a next instruction. The hardware will insert one extra cycle.

The memory accesses also become difficult at high clock frequency. This problem is addressed in [8]. The MAC unit runs at double frequency compared to the memory accesses. The drawback is that memory management becomes difficult: 32 bit words are read from memory and split in even and odd parts in the datapath. This requires the programmer to allign the data items neatly in the data memory. The datapath will operate on the two halfs consecutively.

11.3.3 General Purpose DSPs with Coprocessors

Another option to improve the performance is the addition of specific coprocessors to the general purpose processor. An example of this is Lucent's DSP1618 digital signal processor [9]. This DSP has two coprocessors specific for wireless applications: the error correction coprocessor and the bit manipulation unit. The error correction coprocessor is designed to implement the Viterbi radix

2 butterflies, as shown in Fig. 11.3. The bit manipulation unit is mainly used to perform normalization of vectors of data. These functions can also be integrated in the datapath units. In the processor presented later in this section, the Viterbi butterfly is integrated in a Dual-Mac structure and the normalization and exponent extraction operations are integrated in the ALU module.

A cellular basestation processing requirements are much higher than the handset requirements. This also reflects in the processors designed for it. An example of a DSP processor with multiple co-processors is the Motorola DSP56305 [10], which can perform all signal processing functions within a single radio carrier base tranceiver station for GSM. It contains a filter, a Viterbi, and a cyclic code coprocessors. When too many coprocessors are included, the main DSP is busy handling the data and dispatching the instructions to the coprocessors. One could argue that the signal processing capabilities of the master DSP are underutilized and hence it can be replaced by a smaller general purpose microprocessor.

11.3.4 Microprocessors with DSP Functionality

Another approach is the addition of DSP functionality to general purpose micro-processors. Examples of this are the Hitachi SH-DSP, the ARM-Piccolo processor or the Hyperstone processor [11], [12], [13], [14]. A digital cellular phone usually contains a microcontroller and a digital signal processor. The microcontroller usually performs the control flow tasks, while in parallel the DSP processor performs the signal processing tasks. The motivation for merging both functionalities in one processor is a reduction in processor count.

Designers have added a multiplier or a multiply-accumulate unit to the microprocessor together with the necessary data buffers and communication mechanisms. The integer datapath unit of the RISC processor is now mainly used as an address generation unit. This is at the same time too much and too little. Too much, because a general purpose processor is used for the specific task of address generation. It might also be insufficient because special operations such as modulo arithmetic or bit reverse addressing are not supported. Moreover, the number of address generation units in DSPs is traditionally equal to the number of data busses. For example, the ARM-Piccolo has a multiply-accumulate co-processor. It also contains several input and output buffers to reduce the memory bottleneck. However, now the microcontroller is busy generating addresses and moving data around for the DSP. It does not have enough processing power left to perform the tasks of a microprocessor in a traditional two processor configuration.

11.3.5 Parallel Digital Signal Processors

To increase the MOPS without increasing the clock frequency, several designs use parallel units. There are several ways to organize parallel units. One way is the very long instruction word (VLIW) architecture [15]. Because each unit operates completely independent from one another, this requires a full crossbar network, where each output is connected to each input. Moreover, the memory bandwidth has to be very wide to allow multiple memory accesses at the same time. The instruction word becomes very wide because each of the units is programmed independently and each operand has to be specified. This is an expensive solution both in terms of area and power. An example of this is the TMS320C6x [16].

When all units operate in parallel and the pipeline is complete full, it generates 1600 MOPS. However, the power consumption is 3W (simulated) [17].

This processor and others, such as the Lucent DSP16210 [18] aim at the wireless infrastructure application domain. This requires higher performance as the wireless terminal applications. The disadvantage is the higher power consumption, unacceptable for a mobile terminal. Domain specific DSPs which aim at both high performance and low power are the main topic of the remainder of this chapter.

11.3.6 Domain Specific Digital Signal Processors

In the recent past, domain or application specific processors have been developed for speech processing. An example is the implementation of the Japanese half rate coding standard, PSI-CELP. This is a very compute-intensive speech coder. Special purpose programmable digital signal processors have been developed for it. Three examples are described in [19], [20], and [21]. All three processors need two multiply-accumulator data path units to realize the necessary performance. The processors differ in the bus structure, the memory, and buffer configurations.

To increase the performance of existing DSP architectures, extra computational units are added to operate in parallel. For example, the MAC and the ALU datapaths can be split and operated in parallel. The increase in the number of datapath units necessitates an increase in the memory bandwidth. If each datapath unit operates on two data items, four data items have to be brought to the inputs to keep both datapath units operational every clock cycle. This increase in memory bandwidth resulted in an increase in local and global busses and, in particular, the two separate data busses to the data memory. This is the trend in the 4th generation DSPs.

Today's domain specific processors strife to implement the complete system of Fig. 11.1. Two examples of this are the Texas Instruments TMS320C54x processor [22], [23], and the domain specific processor described in the next section. The TMS320C54x processor is a general purpose DSP containing all the basic features in addition to domain specific enhancements; it has a separate MAC and ALU, three parallel data busses and a separate program bus. The architecture of the TMS320C54x shows the most important trend of the 4th generation DSPs, where we see the separation of the ALU from the MAC unit. Now both units can operate independently and at the same time if needed.

An example of a domain specific enhancement in the TMS320C54x is the implementation of the ACS operation to compute the Viterbi butterflies of Fig. 11.3. Two domain specific features are added, as illustrated in Fig. 11.5. First, the 40 bit ALU is split into two halves, each of 16 bits. In the MSB side, state i is updated with the branch metric, which is stored in register TREG. At the LSB side, the branch metric is subtracted from state i+s/2. The accumulator has also two halves: it stores the result of the addition and the subtraction. Secondly, a special unit, called the compare, select and store unit (CSSU) is added to the architecture. It will compare the MSB and the LSB parts of the accumulator and store the decision bit in a special register, TRN. The decision bit will determine which half of the accumulator will be written back to memory. For a detailed description of the implementation of the Viterbi butterfly operations on the TMS320C54x, refer to [23].

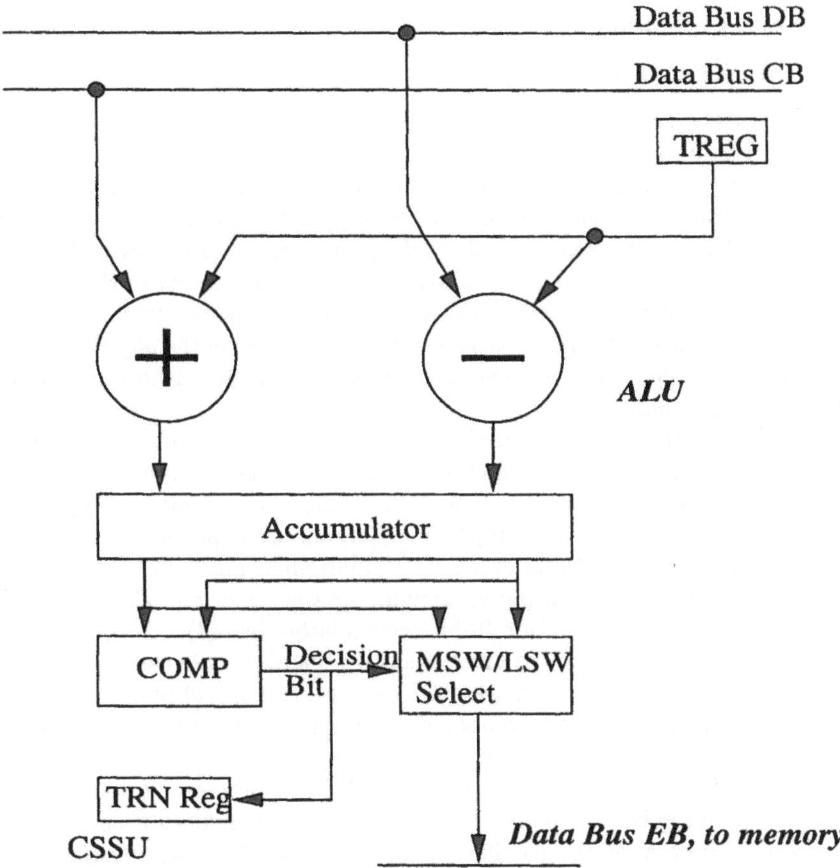

Figure 11.5 ACS operation on TMS320C54x.

11.4 A DOMAIN SPECIFIC DSP CORE: LODE

In this section, we study in detail a domain specific DSP called Lode, designed and implemented by Atmel Corporation, [24], [25]. In addition to all the features which can be found in a general purpose DSP, it contains domain-specific features for wireless communications.

The lode core can be considered a 5th generation processor. Unique to this DSP are its Dual-MAC and ALU which operate in parallel. The internal bus structure, the two data busses which connect to the data RAM, and the instruction set are built such that all three units are kept working in parallel with no bottlenecks in the memory bandwidth.

11.4.1 Architecture

The Lode architecture is shown in Fig. 11.6. Lode is a 16 bit fixed-point DSP core which has been designed to contain specific features for efficient implementation of next generation wireless digital cellular systems and speech compression applications. The 32-bit wide instruction set is architected to optimize the usage of the encoded fields that control the three datapath units and the two address generators. For the most general types of instructions, the flexibility of the choices for the input sources and the destinations is very wide. For more domain specific instructions, the choices are limited to those that make sense in the algorithms for the given domain. These choices were made by analyzing the algorithmic requirements for the targeted domains.

Datapath units: The main features of the DSP core are its Dual-Mac and AMU (arithmetic manipulation unit) datapath units. These three units operate in parallel. The AMU contains a 40-bit ALU (arithmetic logic unit), an input shifter for arithmetic, logical and rotate types of shifts, and an exponent extraction unit. Each of the MAC units consists of a 17×17 multiplier and a 40-bit accumulator. Each unit can execute a multiply-accumulate operation in one clock cycle. The datapath units connect to all four 40-bit accumulators (a0-a3).

Bus network: When three units operate in parallel, care must be taken that all three units have operands to work on. In general this would require a three fold increase in memory bandwidth compared to only one datapath unit, making this solution too expensive in terms of area and power.

Instead of choosing a general VLIW or SIMD architecture, a domain specific DSP has been designed. Therefore, the possible input combinations are restricted while making sure that all three units can operate in parallel. The selection of input combinations and the bus structure has been made such that typical DSP operations and especially operations for wireless communications and speech processing are optimally supported. Internally, the bandwidth is increased by the introduction of a special delay register, lreg. Furthermore, small local connections are provided to route the data from one unit to the next. Local connections are more effective from power consumption viewpoint. The flow of the data through the processor is illustrated for several instructions in the next section.

Data memory busses: Two address busses and two data busses are provided to connect the Lode core with the data memory. Its size and organization depends on the application. Dual or single port memory is supported. For opti-

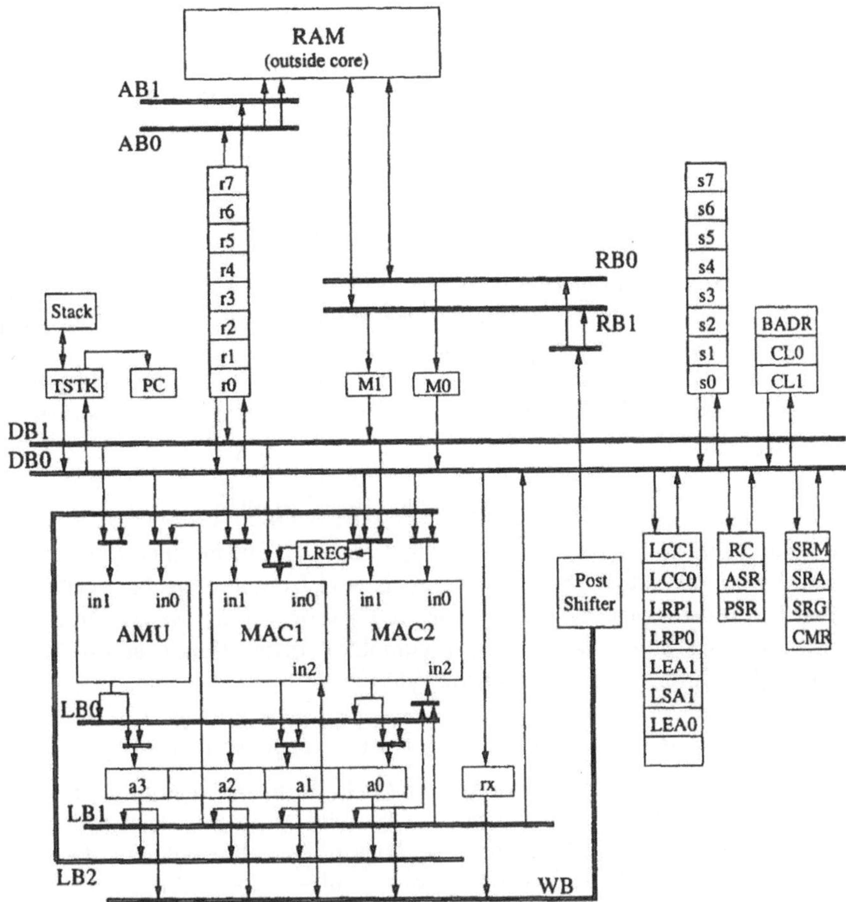

Figure 11.6 Lode DSP core.

mum performance dual port memory is recommended which allows two reads or one read/one write each clock cycle.

Address generation units: The DSP core has two independent address generation units (AGU). In VLIW terminology, these can be considered as two extra datapath units. Hence, five units operate in parallel! The two AGUs share 8 pointer registers and 8 pointer modifier registers. When these registers are not in use by the AGUs, they can be used by the other datapath units as general purpose registers. This is also true for the other available dedicated registers. The AGUs support the following addressing modes: memory direct, register indirect, register direct, short (8-bit) and integer (16-bit) immediate data, and special support for double precision operations. For memory indirect accesses, the following pointer post modifications are allowed: increment, decrement, index, modulo and double fetch.

11.4.2 Instruction Set

The instruction set reflects the features for a domain specific DSP. A complete orthogonal VLIW architecture requires a minimum of three function fields to specify the operations on each datapath unit, and a minimum of six source and destination fields to specify how the inputs and outputs are connected. In addition, it requires the fields for controlling the AGUs. The Lode instruction width is 32 bit. This means that if only one unit is working, the instruction is close to orthogonal. However, when all units work in parallel, certain options for inputs, outputs and modes of operation are predetermined such that the instruction still fits in 32 bit. This selection is determined by evaluating the wireless algorithms. The instruction set will be illustrated by several examples.

Example 1: If only one datapath unit is used, the selection of inputs and outputs is very wide. For example, a single "multiply-accumulate" operation has the following syntax:

$$an = am + op0 \times op1[, ppm][, ppm]; \tag{8}$$

where an and am are any accumulators, op0 and op1 can be data from memory, from another accumulator, from a pointer register or from any special register. ppm describes the address post-modification in case op0 or op1 are reads from memory.

Example 2: The "dual multiply-accumulate" instruction, useful in the computation of FIR filters, has the following syntax:

$$a0 = a0 + op0 \times op1, a1 = a1 + op0 \times Ireg[, ppm][, ppm]; \tag{9}$$

where the first MAC receives two data operands from data memory, the second MAC receives the same operand0, op0, but receives as second input the contents of the delay register lreg. This is used to compute two filter outputs at the same time. As a result, a block FIR computations requires only N/2 instruction cycles and only half of the memory accesses compared to a single MAC implementation. The dual MAC is used in parallel to generate two outputs in one cycle.

A graphical demonstration of the process in presented in Fig. 11.7, where the following FIR filter equation

$$y(n) = \sum_{i=0}^{N-1} c(i)x(n-i) \tag{10}$$

is implemented with the use of the dual MAC.

Example 3: The "square distance and accumulate instruction" is executed in one cycle. This is a basic function of the vector quantization process in speech compression algorithms, It is used to perform the following operation:

$$z = \sum_{i=0}^{N} (x-y)^2. \tag{11}$$

For this instruction, the AMU and one MAC unit are used with the following syntax:

$$a3 = |(op0 - op1 < asr)|, a0 = a0 + (a3)^2[, ppm][, ppm]; \tag{12}$$

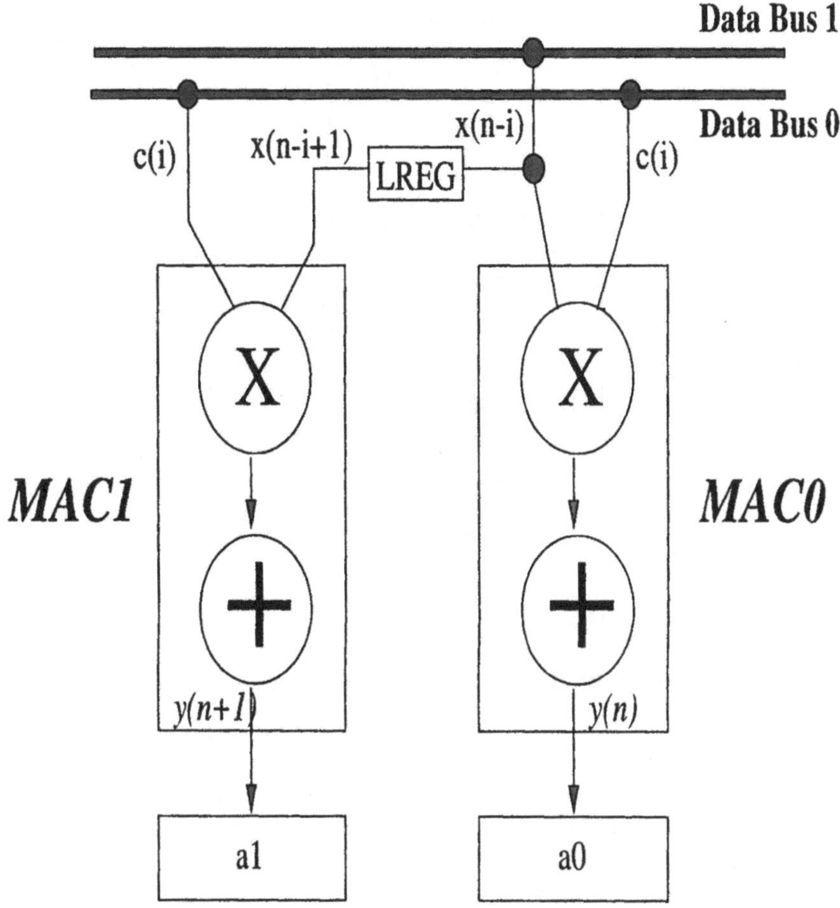

Figure 11.7 Dual MAC operation for an FIR filter.

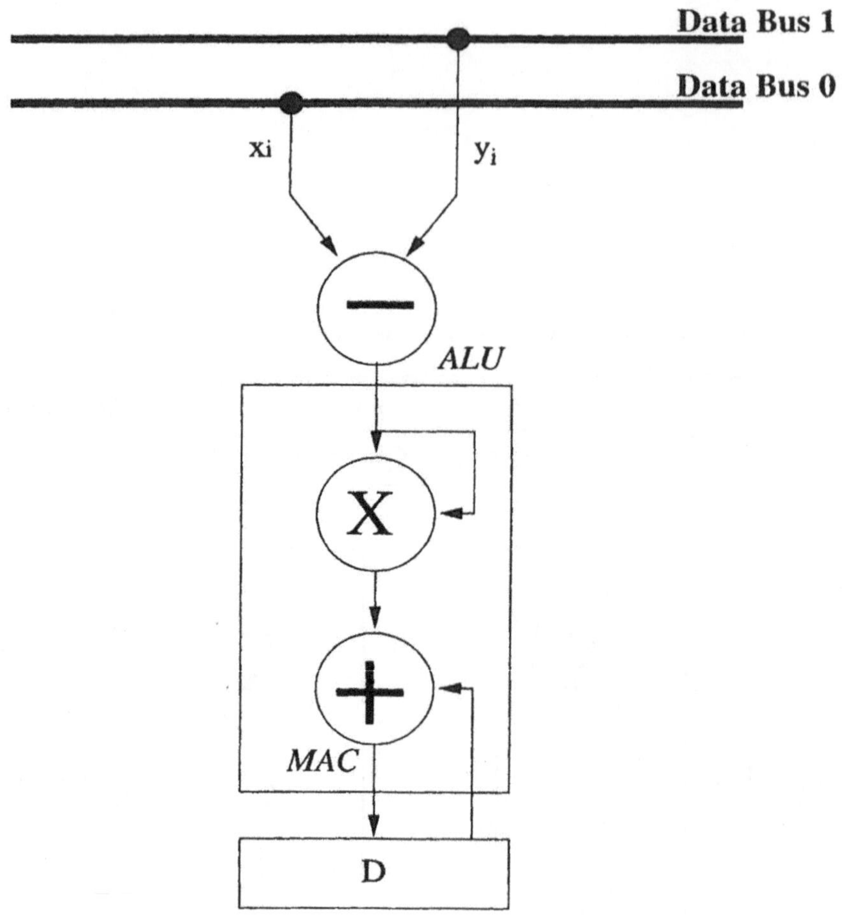

Figure 11.8 ALU II MAC operation for square distance.

The input data operands, op0 and opl, are routed through the AMU, the result is placed in a3. Next, the square of a3 is taken in the MAC unit and the result is accumulated in a0. Asr specifies the input shift value. A graphical demonstration of this process in presented in Fig. 11.8.

Example 4: The "add-compare-select (ACS)" operation for the Viterbi butterfly can be implemented by the dual MAC structure which performs dual additions and dual subtractions combinations by bypassing the multipliers. This, in parallel with the AMU unit performing a maximum or minimum operation, will efficiently compute the ACS butterfly operation of a radix 2 Viterbi core in four clock cycles. For example,

$$sdest = a3 = \max(a0, a1), a0 = a0 + opt < 1, a1 = a1 - op1 < 1, ppm, ppm; \quad (13)$$

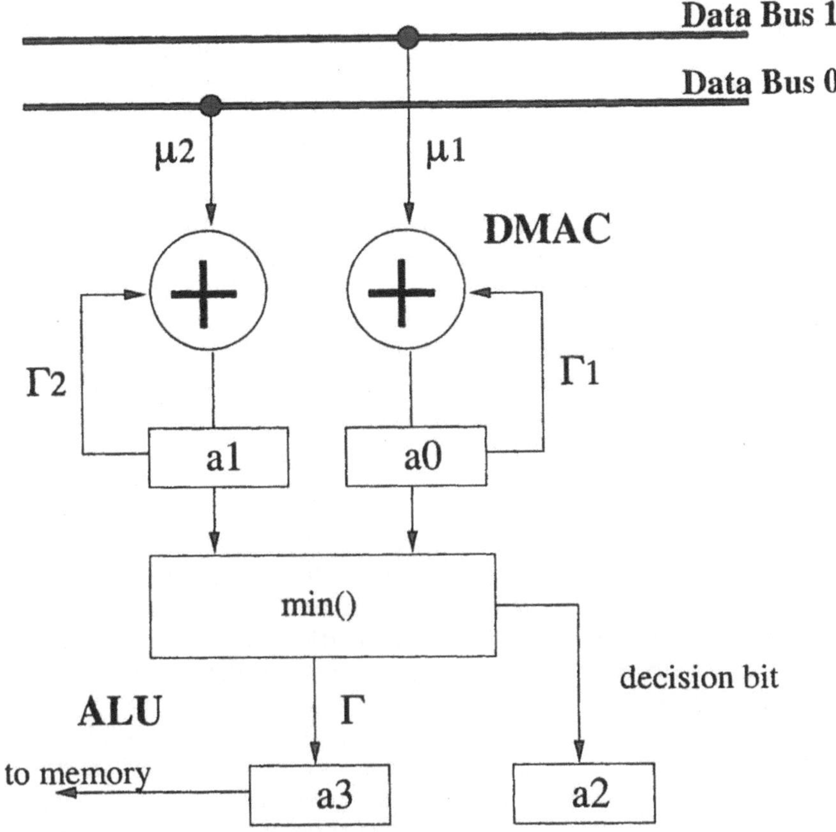

Figure 11.9 ACS implementation on lode.

uses the dual MAC unit to add and subtract op1 from the respective accumulators; simultaneously the AMU finds the maximum of the two accumulators, a0 and a1, and stores the results into memory. The max instruction will implicitly store a decision bit, depending on whether a0 or a1 was a maximum, into the a2 accumulator. Several flavors of this instruction are supported, thus achieving the efficient 4 cycle Viterbi butterfly ACS operation.

A graphical demonstration of this process in presented in Fig. 11.9. The Viterbi butterfly ACS operation that is implemented can be described as follows:

$$\Gamma = \min[(\Gamma_1 + \mu_1), (\Gamma_2 + \mu_2)]. \tag{14}$$

The G represents the accumulated distance of the new state, which is the minimum (or maximum) of the two accumulated distances of the old states, Γ_1 and Γ_2, updated by some metric values, μ_1 and μ_2.

Example 5: "Galois Field (GF) operations." Two specialized instructions in the AMU allow Galois Field operations such as division and multiplication to be performed on polynomials over GF(2). This allows 1 cycle per bit CRC calculations, a basic operation in error control coding for many data fields in the digital cellular standards, and efficient implementation of LFSR (linear feedback shift register) operations which are extensively used in encryption algorithms such as the A5 algorithm in the GSM standard. An example of such an instruction is,

$$a0 = gfdiv(a1 < asr) \tag{15}$$

where the a1 accumulator contains the divisor polynomial over GF(2) and the a0 accumulator contains the dividend. This instruction can be repeated up to 40 times (40-bit accumulator) before saving and reloading a0. At the end of the operation, a2 will have the quotient of the division and a0 will have the remainder. For CRC calculation, the remainder is the result of interest.

Example 6: Bit manipulation instructions such as the bit test instruction, allows simultaneous testing and saving the bit in question in one cycle, thus making interleaving and de-interleaving operations very efficient. These operations are common in channel coding for wireless digital communication.

11.4.3 Design for Low Power

For wireless applications, the measure of importance is the total amount of energy drawn from the battery to finish a task. The peak power is only of secondary importance. For static CMOS implementations, the main energy consumption source is the switching energy, which is proportional to $c \times V_{dd}^2$. Energy will be saved by reducing the amount of switching, which means the amount of operations to finish a given task.

Power or energy can be saved at all levels of abstraction: system level, architectural level, logic and circuit level, and technology level (see also Chapter 24).

At the technology level, the use of smaller geometries and lower voltages will lower the power consumption. The use of two threshold voltages can reduce the power even further while maintaining speed [26].

At circuit level and logic level, registers operate with gated clocks. Clocks are only turned on when new data is available. Also the inputs to functional units are turned off from the main busses such that no unnecessary switching occurs in the units. By making the registers static, the clock frequency can be modified from DC to maximal operating frequency.

The decisions taken at architectural and system level can have a much larger impact. The usage of parallel data path units will increase the MOPS without having to increase the clock frequency. However, only increasing the number of parallel units will not reduce the power consumption of these units. The usage of parallel units will only lower the power consumption if it is combined with reduction in operating voltage [27]. Furthermore, it will reduce overall power savings, if the surrounding overhead is reduced. This technique is applied in the design of the lode DSP core.

The energy savings will be illustrated by the implementation of an N-tap FIR filter. The fundamental operation is the multiply and accumulate function. The overhead sits in the instruction fetching, decoding, memory accesses, and other logic. The main energy savings in this architecture are in the reduction of the memory accesses as summarized in Table 11.2. The total number of multiplications remains the same whether they are executed on one MAC, a DUAL MAC or an N-MAC architecture, because the energy for the multiplications is a fundamental component. Even an ASIC implementation with no programmability will need N multiplications for an N-tap filter. The savings in this architecture is obtained by reducing the memory accesses in half! The energy for one memory access is about the same as for one multiplication. Secondly, only half of the addresses are calculated. This reduces the switching in the address generation units by half resulting in large savings. (Note that for a given clock frequency, the peak power for a dual-MAC instruction will be higher than a single MAC instruction. However, power per instruction is not the correct measure in this case.)

In addition, the number of instruction cycles is reduced by half. Assuming that each instruction includes a fetch from memory and a decode operation, then the number of program memory fetches is reduced in half as well.

In this architecture, the fetch and decode savings are even higher. Instead of N fetches from memory, the N-taps FIR filter is described by a "repeat" instruction. This repeat instruction indicates that the next instruction, which is a dual-multiply-accumulate instruction, is repeated N/2 times. So, the energy for N fetches is replaced by two fetches and a down counter, which counts from N/2 to 0. On top of these savings, there are the secondary savings of reduced overhead in decoding, clocking, and other logic.

Table 11.2 Number of Operation for N-tap FIR Filter

Energy proportional to	Dual MAC	Single MAC
No. of MAC operations	N	N
No. of Date Memory reads	N	2N
No. of Program memory reads (no repeat instruction)	N/2	N
No. of Program memory reads (with instruction)	2	2
No. of Instruction cycles	N/2	N

11.4.4 Performance

The lode core maximizes performance and keeps the clock rate low by using parallelism in its datapath. The main features of the lode core, as described previously, are its dual MAC structure and its complex AMU. At its best all three units can operate in parallel together with the two address generation units.

Table 11.3 shows the performance of the lode core in some basic operations commonly found in the wireless digital cellular applications as well as many speech compression algorithms. The results are compared to general purpose 16-bit fixed point DSPs, such as the Lucent DSP1616, TI C5x, ADI21xx and Motorola DSP56xxx.

Table 11.4 shows the performance achieved on lode for certain applications in the wireless digital cellular area. The performance is given in lode MIPS. The max-

Table 11.3 Lode Core Benchmarks

Operation	Lode Cycles	General Purpose DSP
FIR filtering, N taps, real × real	N/2	N
FIR filtering, N taps, complex × complex	2N	4N
IIR filtering, N taps, direct form	N/2	N
$\sum_{i=0}^{N-1}(x_i - y_i)^2$, Square distance	n	2N
CRC, N bits	N	5N
Block normalize, N entries	2N	3N
Radix 2 Viterbi decoding (ACS)	4	20
Max/Min search in array of N entries	N	3N

imum likelihood sequence estimator (MLSE) and channel Viterbi decoding benchmarks include metric calculations, ACS operation and trace-back.

Table 11.4 Lode Performance

Application	MIPS
GSM MLSE receiver	3.4 MIPS
GSM channel decoder	0.44 MIPS
IS-95 (9.6kbits/s) Viterbi decoding	5.2 MIPS
GSM RPE-LTP (full rate)	2 MIPS
IS-136 VSELP (full rate)	10 MIPS
PSI-CELP (half rate)	22 MIPS
GSM enhanced full rate	12 MIPS
LD-CELP (G.728, 16kbit/s)	20 MIPS

11.5 CONCLUSION

This chapter has described the motivation and the design of wireless digital signal processors. A wireless communication system consists of three main building blocks: a source encoder/decoder, a channel encoder/decoder and a modem. Each of these blocks require high signal processing performance. At the same time the overall system should operate at low power because it is battery operated. This is addressed by enhancing a general purpose DSP with domain specific features. Typical examples are the add/compare/select instructions to support the Viterbi butterfly operation, the instructions to support code book searches and others.

The applications which are run on a mobile terminal are even getting more aggressive. New generations of cellular phones will not only provide better quality voice communication, but also data communication such as fax or e-mail, and even image or video communication. These new applications will determine the next generations of wireless digital processors.

REFERENCES

[1] T. Parsons, *Voice and Speech Processing*, McGraw-Hill Book Company, New York, 1987.

[2] S. Lin and J. Costello Jr, *Error Control Coding: Fundamentals and applications.* Prentice Hall, New Jersey, 1983.

[3] E. A. Lee and D. G. Messerschmitt, *Digital communication*, Kluwer Academic Publishers, 1988.

[4] J. Rapeli, "Technology solutions towards multi-functional and low-voltage /low-power cost effective radio transceivers," presented at *International Symposium on Low-Power Electronics & Design*, Monterey, CA, August 1997.

[5] H.Sasaki, "Multimedia complex on a chip," in *Digest of Technical papers, 1996 ISSCC*, pp. 16–19, Feb. 1996.

[6] Siemens. Gold chip, "PMB 2705," in *http://www.siemens.de/Semiconductor/products/ICs/*.

[7] Motorola 56300, "56600 family," in *http://www.mot.com/SPS/DSP/documentation/*.

[8] I. Tanaka, H. Yasoshima, S. Marui, M. Yamasaki, T. Sugimura, K. Ueda, T. Ishikawa, H. Suzuki, H. Kabuo, M. Okamoto, and, R. Asahi, "An 80 MOPS-peak high-speed and low-power-consumption 16-b digital siganl processor," *IEEE Journal of Solid-State Circuits*, vol. 31, pp. 494–503, Apr. 1996.

[9] G. Ellard, "AT&T's DSP1618 - optimization in action," *DSP & Multimedia Technology*, pp. 48–54, March/April 1995.

[10] E. Engel, *et. al.*, "DSP56305 - Motorola new optimized single-chip DSP for GSM basestation applications," *Proceedings ICSPAT*, pp. 307–311, Oct. 1996.

[11] M. Schlett, "A new way of integrating RISC and DSP," *Proceedings ICSPAT*, pp. 653–657, Oct. 1996.

[12] *ARM Signal Processing Architecture, (Piccolo), Engineering Specification*, Oct. 1996. Document Number, ARM IP0025A.

[13] D. Walsh, "Integrated DSP/RISC Devices target Cellular/PCS phones," *Wireless System Design*, pp. 50–54 July 1997.

[14] *Design Innovations Characterize SH-DSP Single-Core Architecture*, May/June 1996. Hitachi Technology Partner.

[15] D. Patterson and J. Hennessy, *Computer architecture: a quantitative appraoch.* San Mateo, Calif.: Morgan Kaufman Publishers, 1990.

[16] T. Dillon, "The Velociti Architecture of the TMS320c6x," *Proceedings ICSPAT*, pp. 838–842, Sep. 1997.

[17] M. Gold and A. Bindra, "VLIW design takes DSPs to a new high," *EE Times*, Monday Feb. 3, 1997.

[18] A. Bindra, "Lucent reveals details of next-generation DSP," *EE Times*, Monday Sep. 8, 1997.

[19] Y. Miki, Y. Okumura, T. Ohya, and T. Miki, "High performance DSP for half-rate speech codec," *NTT, R&D*, vol. 43, no. 4, 1994.

[20] J. Zingman, S. Wang, E. Wu, J. Gupta, G. Fettweis, S. Kobayashi, T.Kawasaki, Y.Kameshima, H. Konoma, and K. Yamazoe, "Low-power implementation of the PDC 1/2 rate codec on an application specific DSP," *Proceedings ICSPAT'95*, Oct. 1995.

[21] T. Shiraishi, *et. al.*, "A 1.8V 36mW DSP for the half-rate speech codec," *Proceedings IEEE 1996 Custom Integrated Circuits Conference*, pp. 371–374, 1996.

[22] W. Lee, *et. al.*, "A 1V DSP for wireless communications," *Proceedings IEEE International Solid-State Circuits Conference*, pp. 92–93, Feb. 1997.

[23] *TMS320C54x User's Guide.* available from the Texas Instruments Literature Response Center.

[24] I. Verbauwhede, *et. al.*, "A low power DSP engine for wireless communications," *VLSI Signal Processing IX, IEEE Press*, NY, pp. 471–480, 1996.

[25] I. Verbauwhede and M. Touriguian, "Lode DSP engine: high performance at low power," *DSP & Multimedia Technology*, pp. 24–30, Nov/Dec 1996.

[26] M. Izumidawa, *et. al.*, "A 0.25 μm CMOS 0.9V 100 MHZ DSP Core," *IEEE Journal of Solid-State Circuits*, pp. 52-61, vol. 32, no. 1, Jan 1997.

[27] R. Brodersen, A. Chandrakasan, and S. Sheng, "Low-power CMOS digital design," *IEEE Journal of Solid-State Circuits*, pp. 473–483, vol. 27, no 4, Apr 1992.

Chapter 12

Motion Estimation System Design

Yasushi Ooi
C&C Media Research Labs
NEC Corp., Japan
oioi@dsp.cl.nec.co.jp

12.1 INTRODUCTION

This chapter reviews motion estimation algorithms for video coding and their circuit implementations. We also introduce the video encoder LSIs designed for the H.261, MPEG-1, and MPEG-2 standards. The topics of this chapter include cost functions of block-matching, fast search algorithms, decimated full-search algorithms, classification of full-search circuits, circuit architectures for fast search, LSI design examples, and other motion estimation techniques. The emphasis is on reducing computational complexity, data-flow graph and its projection to logic circuits, utilization of processor elements, field/frame prediction, and search range enhancement techniques.

A motion-compensated interframe prediction is an essential technique for video coding in terms of reducing redundancy between successive frames. In several video compression standards such as ITU H.261, ISO MPEG (the Motion Pictures Experts Group) phase 1 (MPEG-1), and phase 2 (MPEG-2), motion between frames is defined on a block-by-block basis, by using a rectangular block of 16×16 pels (macroblock). A difference between a macroblock shifted in a reference frame and a macroblock in a current frame is then coded and transmitted.

In these standards, syntax and semantics of motion-compensated predictive coding guarantee that a decoding process of video bitstreams always produces the same result in any decoder systems. However, there are no rules on how to find motion vectors in the encoding phase. Since motion estimation is a key in coding efficiency, we can exploit a specific algorithm which depends on coding requirements for motion estimation such as accuracy, computational complexity, and real-time processing.

Most video coding schemes apply a block-matching algorithm for the motion estimation (Fig. 12.1). This is a scheme to find the best matching part of each current macroblock in the reference frame. Matching is evaluated in terms of a distance between the macroblocks; the relative position where the distance takes the minimum is selected as the best match. We discuss several cost functions (criteria) of this distance in Section 12.2.1. The block-matching method is widely accepted not only due to its simplicity but also its robustness against noise and correctness even in a complex motion [1].

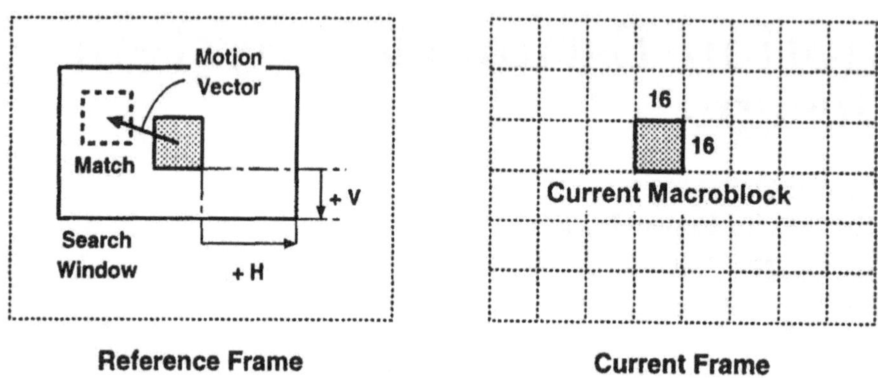

Figure 12.1 Macroblock-based block-matching where a search range in the reference frame is $(\pm H, \pm V)$.

In the 1970's and 1980's, a motion-compensated interframe prediction was mainly adopted to video transmission purposes like video conferencing. The problem for implementation of motion estimation was the complexity of the algorithm. Finding the minimum distance can be guaranteed only by an exhaustive search of candidate points within the search range. However, this full-search scheme needs large computations up to billion operations per seconds (or GOPS: giga operations per second). Hence in the 1980's, several fast search algorithms were exploited as a substitute of the full-search. Most fast algorithms narrow down candidates in a multi-step hierarchical detection to reduce the number of matching distance calculations. Since these algorithms usually have a time complexity of $O(logN)$, where N denotes a search range parameter, they are called logarithmic searches. The 3-step hierarchical search is one of such searches which has been referred many times in related papers.

These algorithms rely on a convex hypothesis; there should be a unique minimum of the matching distance in the search area and from that point the distances monotonously increase for all directions. Because this assumption is not correct in many situations, the search often falls into a local minimum in the distortion space. To test reliability of the fast search algorithms, the search result is to be compared with a full-search motion vector field.

The fast search algorithms contributed to the H.261 transmission applications where the bit rate is around $p \times 64$kbps where p denotes a small integer. Because of the low bit rate, some degree of S/N degradation is unavoidable in these applications. Hence, even if the motion estimation does not work so effectively due to a local minimum convergence, there appear to be no drawbacks for coding.

In the 1990's, when the MPEG standards were discussed, application areas of video coding moved to digital broadcasting and video files. The bidirectional search and the half-pel search were introduced in the MPEG standards as tools of improving image quality.

In this context, the full-search algorithm was regarded as the most important method when high-quality video coding is vital. Realtime operation of the full-search algorithm needs high data throughput, high operating rate and massively concurrent computation. Fortunately, by the rapid progress of semiconductor technology, especially by large-scale circuits integration, the full-search algorithm can be mapped on a silicon die. The full-search algorithm is also suitable for integration because of its regular structure. Several kinds of full-search mapping (projection) to array-type circuits have been investigated. Some of them also have been actually fabricated in a real VLSI.

The motion estimation has a large impact on the computational complexity and coding efficiency. From the viewpoint of cost-performance, finding the best motion estimation algorithm remains still an open problem and several trial-and-errors of algorithm quest have to be repeated to find a reasonable solution.

In this chapter, we review motion estimation algorithms for video coding and their circuit implementations. The emphasis consists of motion estimation by block-matching algorithms, fast search and full-search algorithms, and their VLSI implementations. We describe other motion estimation algorithms in Section 12.6. The surveys on the motion estimation appear in [1], [2], and [3]. Especially [1] shows the superiority of the block-matching to others by several simulation results.

12.2 BLOCK-MATCHING MOTION ESTIMATION

Block matching algorithm is used to find a match of the rectangular region of the current frame in the search area of the reference frame. Though the rectangle can be of the arbitrary size and aspect ratio in general, we here assume only a square of 16×16 pixels (macroblock) because it is adopted in most video coding standards [1].

The search area on the reference frame is larger than the size of macroblock. The width and the height of that area should cover the motion that occurs in the video to be encoded. The search range is normally set symmetric in the vertical and horizontal directions. Assume that the horizontal search range is $\pm H$ and vertical search range is $\pm V$ as shown in Fig. 12.1. In this case total matching operations per macroblock is $(2H+1) \times (2V+1)$.

12.2.1 Cost Functions

For matching criteria, cost function $D(k,l)$ which denotes the distance of the two macroblocks is used. The template of current macroblock is overlaid in the search area with a shift of (k,l) from the original point, to calculate the cost function. The point $(k,l) = (m,n)$ which gives the least distance $D(k,l)$ is considered a match and the argument is the motion vector MV, i.e.,

$$MV = \{(m,n) | arg \min_{-H \leq k \leq H,\, -V \leq l \leq V} D(k,l)\}.$$

[1] Luminance signals are usually used for matching detection. A few experimental reports use both luminance and chrominance signals for matching detection.

One example of $D(k,l)$, which is the most common matching criterion, is the mean absolute difference, (MAD), given by

$$MAD(k,l) = \frac{1}{256}\sum_{i=0}^{15}\sum_{j=0}^{15}|c(i,j) - r(k+i, l+j)|,$$

where $c(i,j)$ and $r(i,j)$ denote the pixel coordinates in the current frame and in the reference frame, respectively. To calculate $MAD(k,l)$, we need 256 absolute-difference operations and 255 additions. To find the minimum value over the search area, we should repeat the calculation $(2H+1) \times (2V+1)$ times.

Another example is the mean square difference (MSD),

$$MSD(k,l) = \frac{1}{256}\sum_{i=0}^{15}\sum_{j=0}^{15}(c(i,j) - r(k+i, l+j))^2,$$

which is closer to the definition of the distance in the natural sense. The distance is regarded as a simplified version of a normalized cross-correlation function [2].

Block matching based on the MAD and/or the MSD gives the best result in that the differences of the macroblocks are encoded eventually in the encoding process. Some reports compare the MAD and the MSD as cost functions and conclude that there are no significant differences in terms of motion vectors detected [4]. But the characteristics are highly dependent on video sequences; accumulation of many small differences is over-evaluated in the MAD, while a few of extreme pixel differences are emphasized in the MSD [1]. The MSD cost function is not used frequently because evaluating multiplications is more time consuming. But the MSD is used software processing on a digital signal processor with enhanced multiply-accumulate operations [5].

To reduce computational complexity, several simplified cost functions have been proposed. One example is the normalized absolute difference, a summation of a value $\{0,1\}$ which is a compared result of absolute difference of pixels and a given threshold value,

$$NAD(k,l) = \frac{1}{256}\sum_{i=0}^{15}\sum_{j=0}^{15}T(i,j,k,l),$$

$$T(i,j,k,l) = \begin{cases} 1 & |c(i,j) - r(k+i, l+j)| > Threshold, \\ 0 & otherwise. \end{cases}$$

Another simplified cost function is simply using the maximum value of absolute difference of pixels (Maximum Pixel Difference),

$$MPD(k,l) = \max_{0 \le i \le 15, 0 \le j \le 15} |c(i,j) - r(k+i, l+j)|.$$

In the MPEG standards, half-pel motion detection is used. It is permitted for k and/or l to have a fraction of 0.5. In this case, $r(k+i, l+j)$ is defined as a mean

[1]Video effects like fading may be over-estimated if we use the MAD. The MSD is effective for a correct match of object boundaries.

value of integer-coordinate points. The search is separated into two levels instead of direct half-pel search, that is, the half-pel search is executed using the result of an integer full-pel search.

In actual hardware implementation, the integer-pel search unit and half-pel search unit are often separated. Simplification of algorithms are mostly for the integer-pel search since the complexity is much higher than the half-pel search. There are, however, some efforts to reduce calculations for half-pel search by approximation of distance cost functions [6].

12.2.2 Cases of Finding Incorrect Motion Vectors

Generally speaking, detection of the true motion is not necessarily the goal if we limit the detection for the coding purpose only. Even if the motion vectors are incorrect, the video coding algorithm works with less fidelity of pictures.

In this sense, motion estimation is used as a means to minimize compressed code which involves prediction error information and motion vector information. Even if the motion vector is near-optimal, it will be justified when the total code size is small enough.

However, minimizing the prediction error is not the best way to obtain the natural flow of moving sequences. To avoid artificial discontinuities, a motion vector field is often desired to be consistent. The quality of a motion estimation algorithm should be eventually evaluated by a subjective test of coded video quality.

In the following discussions, we show several situations where wrong vectors are detected by using the block-matching algorithm [1].

The lack of the search distance is a critical reason of finding wrong vectors. To follow up the real motions, a search range should be large enough.

The search range H and V are highly related to the size of pictures and temporal distance of the video; tracking a horizontal motion which pass over the whole picture per second at a 720-pel width (MPEG-2 MP@ML) must satisfy $H > 720/30 = 24$ (NTSC) or $H > 720/25 = 28.8$ (PAL). Furthermore, if the bidirectional prediction is used and the temporal distance between two core pictures (I-picture and/or P-picture) is 3, the prediction for P-pictures needs a higher H; $H > 720/(30/3) = 72$ (NTSC) or $H > 720/(25/3) = 86.4$ (PAL).

We should also take care of boundary condition of reference pictures. If the search area is outbounded and there is no chance to find a correct match, coding efficiency goes down.

Incorrect vectors are possibly selected if there are plenty of minima of macroblock distance. For examples, a texture with a small repeated pattern leads to many matches between the current and reference macroblocks. The same intensity over a region like a clear sky will be possible as an extreme case of multiple matches.

Several policies are possible on how to choose one among many matches. From the viewpoint of coding, minimizing a code size of the motion vector is the best, which means the smallest magnitude of difference from spatially previous one in video compression standards. But in several hardware implementations, this is not true because of the order of pipeline processing. The selection policy is thus different in every encoding system.

---

[1] Using the simplified cost functions discussed in section 12.2.1 is a typical case which lead to inefficient coding.

12.3 MOTION VECTOR SEARCH ALGORITHMS

12.3.1 Fast Search Algorithms

The fast search algorithms reduce the number of computations by limiting locations searched to find the motion vector. The strategy on how to limit the locations differs in each algorithm. The most simple strategy is to fix the locations as a menu of candidate vectors [7]. It is also proposed in [7] that the vector candidate menu is used with an offset vector at the same macroblock position of the preceding picture to improve prediction efficiency.

The method using a menu cannot track all positions in the search area. A hierarchical search is a way to make stepwise menu-based decision for the next search positions so that all positions can be searched. In general, the number of search positions are limited up to a constant value w in each hierarchy. The total steps usually have the complexity of $O(logN)$ where $N = max(H, V)$. The number of matching calculations in the worst case are around $w\lceil \log_2 N \rceil$ where $\lceil x \rceil$ is a minimum integer which is not less than x.

These techniques assume that a distance $D(k,l)$, a criterion of the motion vector, is always a monotonically increasing (convex) function around the location $(m,n) = (k,l)$, where $D(k,l)$ has a minimum. If the current picture and the reference picture is nearly the same except motion, the highest correlation is observed when the two pictures are overlapped with that motion. The correlation value will decrease according to the skew of the two pictures.

If the convex assumption is correct, motion estimation process is equivalent to finding the minimum as fast as possible along the convex surface. In hierarchical searches, coarse search points are first set up in order to find the sub-minimum distance between two macroblocks. After finding the sub-minimum, smaller, dense search points are recursively set up as the sub-minimum search point is located at the center of the new search.

Some examples of fast search algorithms are described in the following discussion. Koga [8] presented a search that nine points, eight points on a square region and its center, are searched and then the edge of the square is halved as the search step progresses. For the H.261 applications, the search range of ± 7 is said to be enough. The initial size of an edge of the square is 4 and is being changed stepwise such as 4, 2, and 1 (Fig. 12.2(a)). Although this algorithm have been called "3-step search", we can apply this algorithm with more search steps in the case of wider search area.

At the nearly the same time, Jain [9] proposed a similar method using four search points located on vertices of a diamond. This approach can slightly reduce the search calculations. However, as omitting oblique searches, additional search steps are sometimes necessary (Fig. 12.2(b)).

After these proposals, more simplified algorithms have been exploited continually. Since the main purpose of the simplification was reduction of complexity, the correctness of the motion estimation is no longer better than the previous proposals. The improvement of the speed is just the coefficient of the $O(logN)$, or sometimes $O(N)$.

There are some interesting ideas in these modified algorithms. Ghanbari [10] proposed using four search points located on vertices of a search square (Fig. 12.3(a)). Lee [11] added that the size of the square is not halved from the previous

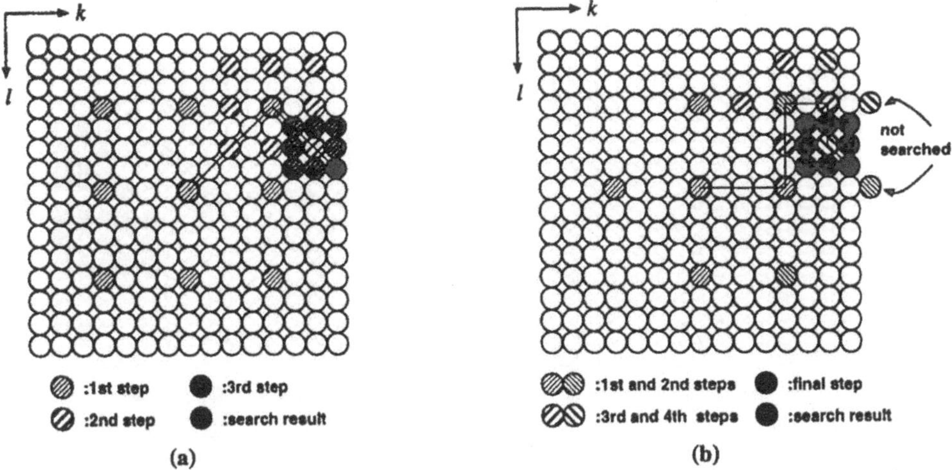

Figure 12.2 Fast search algorithms proposed by Koga (a) and by Jain (b).

step but factor of 1/4 and 3/4 can be adaptively taken according to the convergence speed of distance. Furthermore, Lee suggested that the search points on the square vertices and on the diamond vertices should be selected by turns (Fig. 12.3(b)).

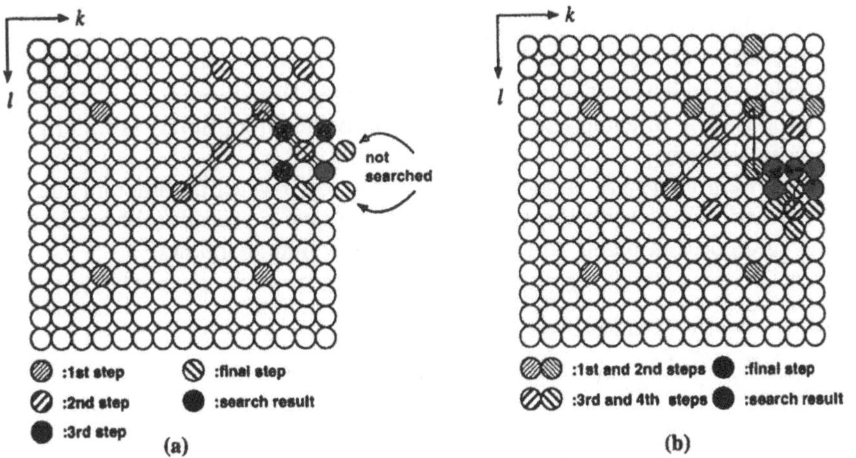

Figure 12.3 Fast search algorithms proposed by Ghanbari (a) and by Lee (b).

Li [12] presented a center-oriented search in addition to the 3-step search (Fig. 12.4(a)). Lin's method [13], described later again, is interesting since multiple candidates are kept during the hierarchical search.

12.3.2 Decimated Full Search

The logarithmic fast searches rely on the convex assumption of $D(k,l)$. However it is not always correct; as we already discussed in 12.2.2, we can easily imagine

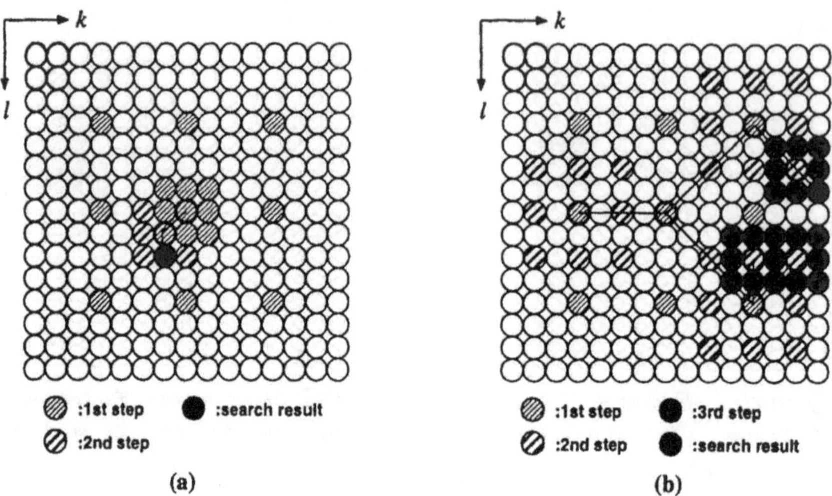

Figure 12.4 Fast search algorithms proposed by Li (a) and by Lin (b).

the picture examples where the assumption is not applicable. They are not rare cases even in actual video sequences.

In general $D(k,l)$ is an uneven function which has many minima, where the logarithmic fast searches often fall into a local minimum depending on the starting point and the intermediate search points.

Instead of $O(logN)$ searches, there are some proposals of simplified $O(N^2)$ (or $O(HV)$) searches which have the same order as the full search. The purpose of these proposals is to avoid the convergence into the suboptimal vectors by taking a nearly full-search.

The main idea here is decimation. There are several parameters that could be decimated. To identify these parameters, we give a procedural notation of the full-search algorithm based on the MAD cost function as follows [1]:

```
m,n = 0; /* initialization of MV */
MAD(m,n) = HUGE;
for k= -H to H
    for l= -V to V
        MAD(k,l) = 0; /* initialization of MAD */
        for i= 1 to 16
            for j= 1 to 16
                MAD(k,l)+=|c(i,j)-r(k+i,l+j)|; /* accumulation */
            end
        end
        if MAD(k,l) < MAD(m,n) then (m,n)=(k,l); /* minimum decision */
    end
end
```

[1]Inner (i,j) loops correspond to matching calculations for every macroblock, while outer (k,l) loops correspond to a series of movement within the search area.

```
return(m,n); /* MV = (m,n) */.
```

We classify here the decimated full-search algorithms into three categories:

- Matching Pixel Decimation

  In matching distance calculation, a part of 16×16 pels will be used to evaluate the distance $D(k,l)$. In the quadruple loop above, this corresponds to reducing (i,j) loop iterations.

- Candidate Vector Decimation

  Search points for a current macroblock are to be reduced. In the quadruple loop above, this correspond to reducing (k,l) loop iterations. Decimated candidates can survive by employing secondary searches. Examples of 4:1 decimation and 2:1 quincunx decimation are shown in Fig. 12.5.

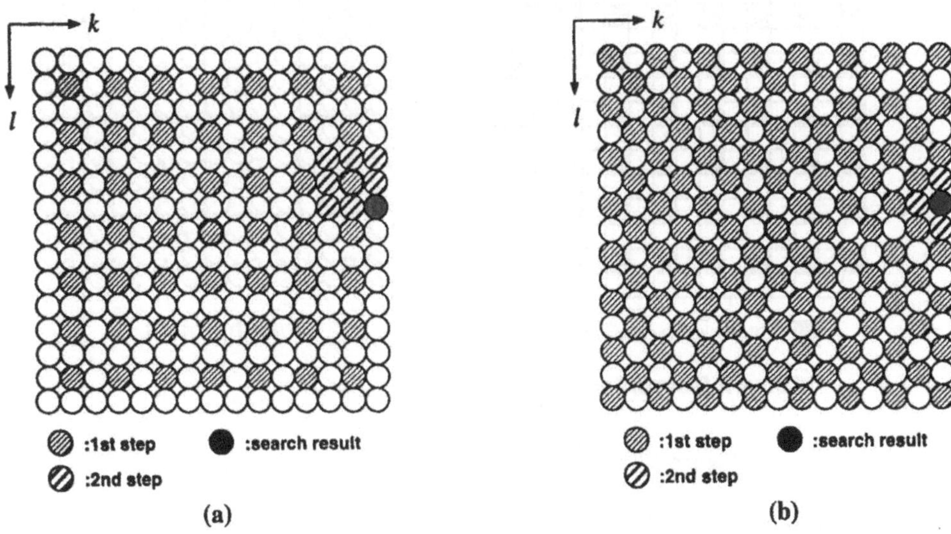

Figure 12.5 Candidate vector decimation: (a) 4:1 subsampling, and (b) 2:1 quincunx subsampling.

- Low-definition Picture Search

  After conversion of the original video into low-definition one, the full search is executed. Low-pass filtering before the conversion is sometimes added to avoid aliasing, which leads to a incorrect of motion vectors. From the view of reducing the number of operations, this scheme is effective because of achieving both of the matching pixel decimation and candidate vector decimation. An additional secondary search is necessary to obtain motion vectors in full-pel resolution.

The matching pixel decimation is a kind of simplification of cost functions mentioned in Section 12.2.1. The logarithmic fast searches in Section 12.3.1 can be regarded as a hierarchical version of the candidate vector decimation.

Note that all these ideas have a possibility of falling into a worse result than the full search.

Liu [14] examined efficiency of the matching pixel decimation for MPEG-1 video compression. Since applying the matching pixel decimation without any change resulted in much degradation of video coding quality, Liu proposed alternating decimation patterns (Fig. 12.6). The access pattern of a frame is changed depending the position of the search point. All pixels in the current macroblock are used eventually, which decreases aliasing effects.

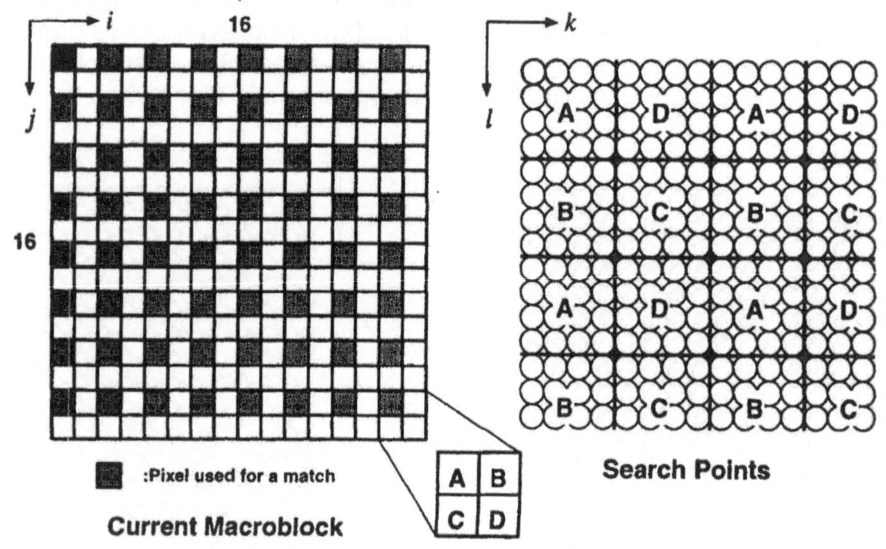

Figure 12.6 Alternating decimation patterns by Liu.

Liu showed by simulation that his algorithms make little degradation compared to the full search as long as the main usage of these algorithms are MPEG-1 applications.

Lin's proposal [13] is a hybrid search where the first step of the hierarchical search uses the candidate vector decimation (Fig. 12.4(b)). The decimation pattern is arbitrary because there is no restriction of the hardware configuration. He also uses the M-algorithm (multi-survival algorithm) to avoid the wrong convergence to the local minima. The value of M, the number of interim candidates of each search hierarchy, depends on the hardware performance.

Ogura [15] presented a projection technique for matching pixel decimation. In both a current macroblock and a reference macroblock, all pixels in a horizontal/vertical line are summed up and then used for matching. For MPEG-2 applications, the low-definition picture search has been often employed. This is discussed in Section 12.5.3.

Liu also proposed an idea of vector field decimation. First, motion estimation is executed at the half of macroblocks, decimated as a 2:1 quincunx pattern, in a frame. Motion vectors of a skipped macroblock is then estimated only at the four search points, that are based on motion vectors in the adjacent four (upper, lower, left, and right) macroblocks (Fig. 12.7).

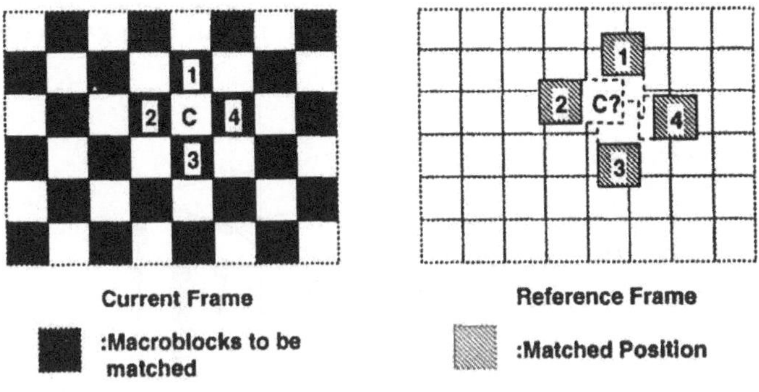

Figure 12.7 Vector field decimation by Liu.

12.4 CIRCUIT ARCHITECTURES FOR MOTION VECTOR SEARCH

In this section, we focus on architectures for the motion vector search circuits. We emphasize classification of the full-search circuits that are derived as combination of the loop variables shown in Section 12.3.2. Note that decimated full-search circuits can be discussed in the same manner as a full-search circuit.

A motion vector search engine is usually designed as a regular array of processor elements (PEs) which consists of an absolute-subtracter and an adder (or an accumulator). The number of PEs determines the concurrency of the operations, which is derived from the trade-off between required performance and maximum switching speed of the circuit primitives.

To obtain a high-performance PE array, we should consider the utilization ratio, which is a spatial and temporal ratio of PEs operating in busy mode. In many cases, data supply of the reference frame data and the current frame data is a key issue for achieving high PE utilization. In some circuit architectures, a lot of data in parallel should be delivered continuously to achieve 100% utilization of all PEs.

12.4.1 Classification of Full Search Circuits

The full-search algorithm is the most simple algorithm in that all of search points are examined as candidates of the motion vector. Even though the computational complexity is so large, this algorithm is often used for VLSI implementation because of its regularity.

In previous discussions, we explained the quadruple loop structure of the full-search algorithm. In a procedural notation of the full-search algorithm, inner (i, j) loops correspond to matching calculations for every macroblock, while outer (k, l) loops correspond to a series of movement within the search area. The circuit architecture is determined such that specific levels of the loop can be executed in parallel. The execution timing is then decided by arrangement of data supply and timing adjustment by inserting pipeline delay registers.

The first discussion of full-search circuit classification appeared in Komarek's paper [16]. Considering the primitive operations like $s_{out} = s_{in} + |x - y|$ as a single node, a data-flow graph of the full-search algorithm can be organized (Fig. 12.8)[1].

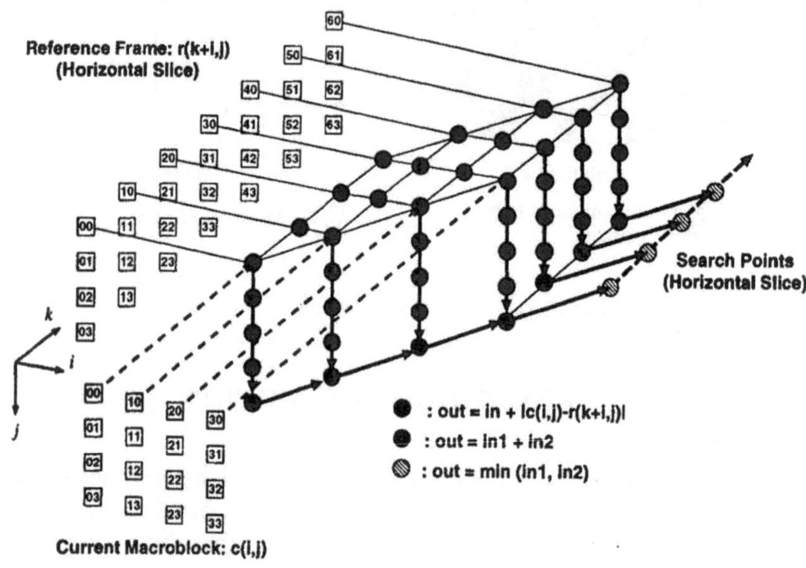

Figure 12.8 Data-flow graph given by Komarek.

Note that the data-flow graph shown by Komarek is an expansion of the (i, j, k) loops of the original quadruple loop shown in Section 12.3.2.

By projecting to a certain plain, the data-flow graph can be mapped onto a specific systolic array architecture. Generally, if a target system is a PE array associated with specific variable combinations from (i, j, k, l), the projection should be made onto that plain (or line) organized by those variables [2]. We can classify the circuit architectures by the result of the projections.

Komarek discussed two-dimensional arrays and one-dimensional arrays generated by reducing the dimension of the data-flow graph. Fig. 12.9 and Fig. 12.10 are examples shown in [16]. These arrays are featured by the degree of concurrency, that is, the number of PEs P. A one-dimensional array associated with variable i (the macroblock width) is defined as AB1 ($P = 16$). Another one-dimensional array associated with k (the horizontal search range) is called AS1 ($P = 2H + 1$). An (i, j) two-dimensional array is named AB2 ($P = 256$), while an (i, k) array is called AS2 ($P = 16 \times (2H + 1)$).

Another important concern is a data input timing of projected circuit. The number of nodes on each path from the primary input to the output in the data-flow graph is considered as a circuit delay. When we use a synchronous scheme for

[1] To simplify illustrations, we sometimes use 4×4-pel notation for a macroblock.

[2] The concept of the projection was introduced in [17].

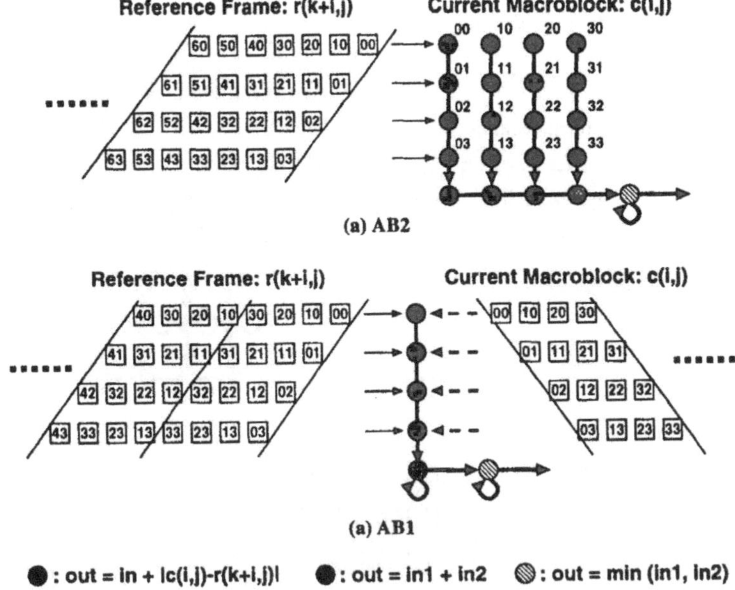

Figure 12.9 Circuit architectures: (a) AB2, and (b) AB1.

circuit design, it corresponds to a clocked register. Since the delays of all paths should take the same value, data at the primary input should be delayed when it is located near the primary output. The input timing of the data $c(i, j)$ and $r(k+i, j)$ should be delayed according to the value of i and/or j [1].

In Komarek's model, the variable l is always fixed. This means projection from a four-dimensional graph to three-dimensional one is assumed in advance. Hence the parallel architecture associated with variables (k, l) cannot be derived.

Vos [18] named a two-dimensional PE array associated with (i, j) (same as AB1) "type-I", and a two-dimensional PE array associated with (k, l) "type-II" $(P = (2H + 1) \times (2V + 1))$. He noted that these two are the most promising schemes. His observation will be discussed later (see the end of this section and also Section 12.5). Hsieh [19] also presented a smart type-I architecture where the reference picture is supplied as a series of one-dimensional data with several delay elements (Fig. 12.11).

Yang [20] showed a realistic solution of an AS1 circuit architecture (Fig. 12.12). One current picture input line and two reference picture input lines are employed to maintain 100% utilization of PEs. This scheme is used in several motion estimation LSIs as shown in Section 12.5.1.

Chang [21] used an extended notation of the data-flow graph, instead of nodes and links, in order to enhance Komarek's model for four-dimensional loop representation. A two-dimensional (i, k) projection (slice) is repeatedly allocated in the (k, l) two-dimensional space (tiling) as shown in Fig. 12.13(a). Combinations of projection onto a two-dimensional array is equivalent to combinations to take two

---

[1] If the projection is a one-dimensional array like the AS1 projection, the priority of the iteration, i or j, on the circuit has an large impact on the data supply scheduling.

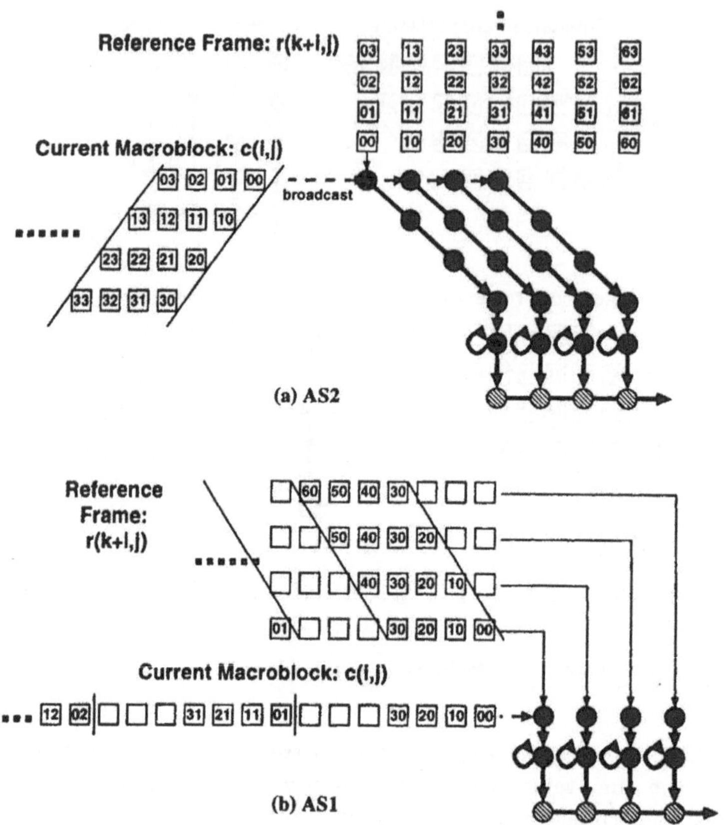

Figure 12.10 Circuit architectures: (a) AS2, and (b) AS1.

of four variables (i, j, k, l). Chang discussed six types of projection that are meaningful.

From these discussions, we know that the full-search circuits are classified by projections of the data-flow graph generated from the quadruple loop. The projection onto a one or two dimensional architecture is suitable for a VLSI implementation. Though the complete discussion is in Chang's model, some important architectures were covered by Komarek and Vos.

Another point of Chang's model is that the reference frame (k, l) is represented as a one-dimensional series to clarify a serial input of the reference frame. The restriction of slice allocation resulted in an introduction of NOP (no operation) nodes (white nodes in Fig. 12.13(a)) that are normally projected onto the shift registers (FIFO) for timing adjustment or actual NOP period which decreases the PE utilization. Chang's discussion concludes that multiple data input lines are necessary to omit NOP nodes.

In many conventional designs, to sustain 100% PE utilization with low data supply rate, the search width $2H + 1$ is set to be equal to the side length of the macroblock. In Fig. 12.13(b)), we obtain 100% utilization with doubled reference picture input lines by the compaction of the graph.

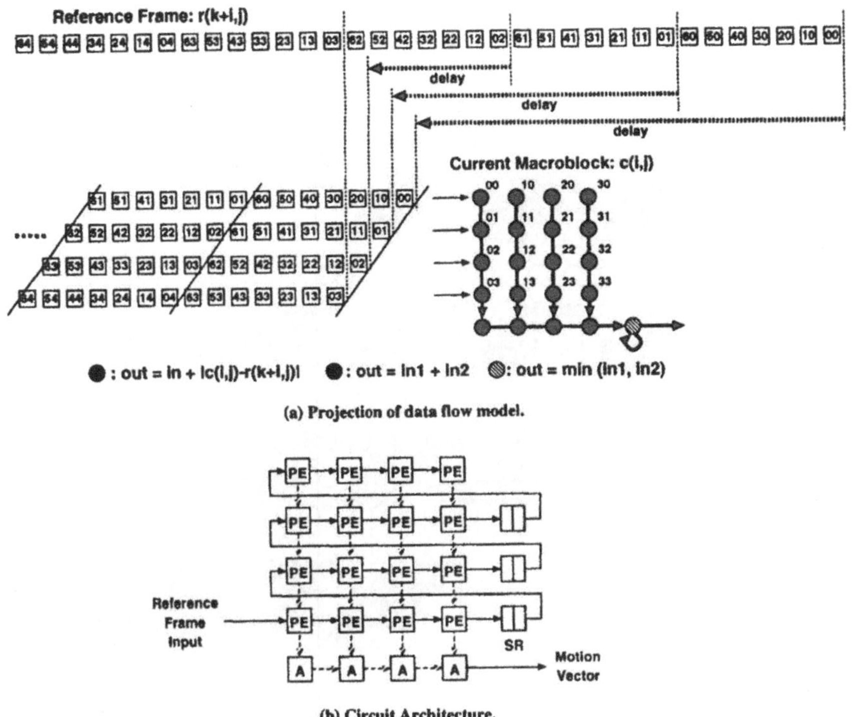

(a) Projection of data flow model.

(b) Circuit Architecture.

Figure 12.11 AB2 (type-I) array where the data supply is achieved by a serial reference input.

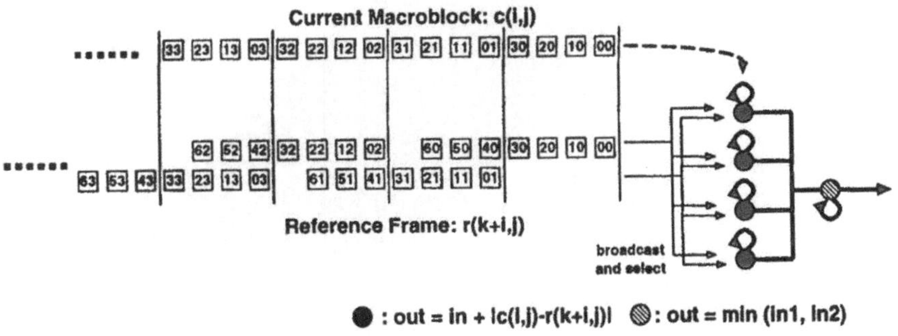

Figure 12.12 Yang's model based on AS1.

If the search range is larger or smaller than the side length of the macroblock, data scheduling after compaction is not managed successfully where input data supply cannot be well organized (Fig. 12.14). To avoid the input line "explosion", we have to add the extra NOP nodes, which eventually lead to degraded throughout or additional shift registers [1]

[1]There are some examples of which the search area is not balanced for plus and minus directions as $-8\cdots+7$ to make the search range be equal to 16.

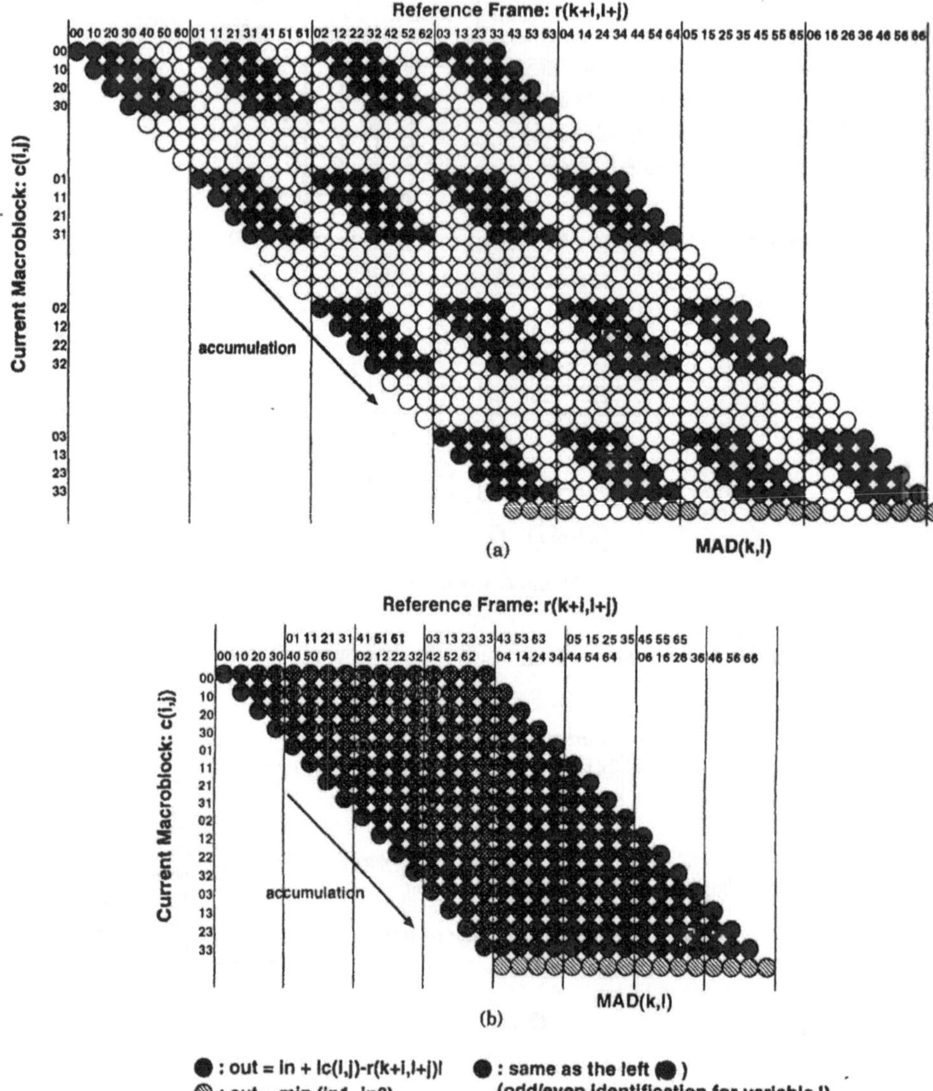

Figure 12.13 Chang's data-flow graph notation: (a) based on one-dimensional reference frame input, and (b) its compressed version. In this illustration, the horizontal search range is equal to 4, that is the same as a side length of the 4×4 block assumed here.

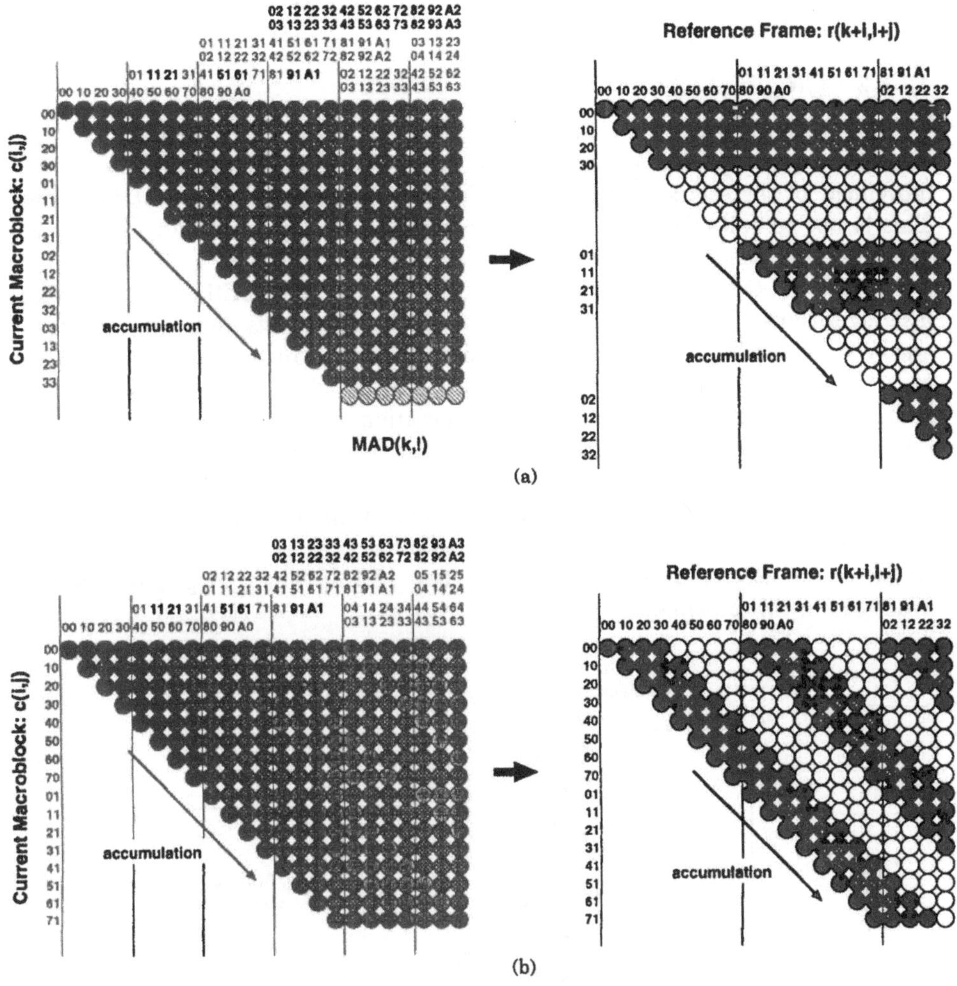

Figure 12.14 Chang's graph: (a) when *search width 8 > macroblock side length 4*, and (b) when *search width 4< macroblock side length 8*. In both cases NOP nodes should be inserted to share the common reference inputs at the same timing, otherwise input supply "explosion" will happen.

Multi-array architectures have been often discussed in order to enlarge the search range, because a single array architecture has difficulty to enhance the search range by above-mentioned reasons.

Let us consider about projection strategy of the data-flow graph with extra NOP nodes. In Fig. 12.14(a), projection on (i, j) plain is the best strategy in terms of minimizing effective PEs though additional timing delays are also necessary. This is true if extremely wide search range is needed. In contrast, in Fig. 12.14(b), projection on (k, l) plain (direction for accumulation) is the best strategy. This is applicable for MPEG half-pel refinement, because the search range in this case is at most ±1-pel range. These projections coincide with Vos's type-I and type-II arrays, which proves his early observation.

12.4.2 Circuit Architecture for Fast Search

In a typical fast search algorithm, the movement among search points ((k, l) loop of the quadruple loop) is simplified in terms of computational complexity. The movement of the search points are decided by evaluating interim cost functions. This is not so suitable for hardware implementation because achieving high PE utilization is not easy in this processing flow.

However, the 3-step search is a considerably regular algorithm among several fast algorithms; it has a fixed number of the search points (nine at each step) and a fixed number of search steps 3 (more generally an integer not more than $\max(\log_2 H, \log_2 V) + 1$ for fixed search ranges H and V) [1]. Only deciding a center position of the next step is a conditional operation. Since the 3-step search has a relatively good performance among the fast searches, there are a number of circuit proposals based on this algorithm.

Jehng's proposal [22] is to make the (i, j) loop the fastest one. The tree-type adders are used to accumulate the absolute differences of (i, j)-pixels. The tree adder can be used in a pipelined fashion (Fig. 12.15). The q loop is successively processed in this pipeline.

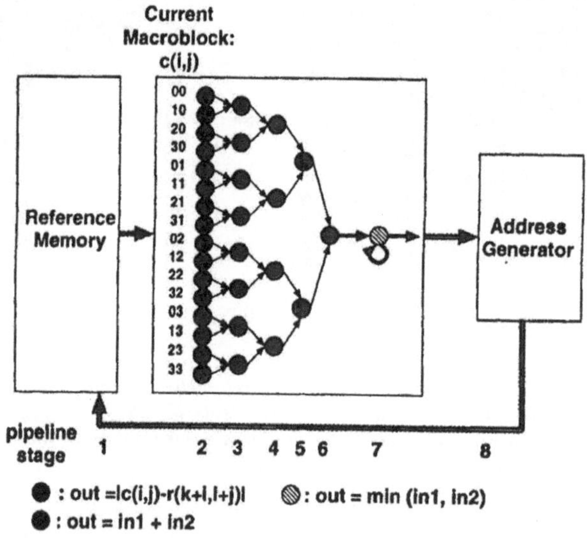

Figure 12.15 Jehng's tree structure for fast calculation of (i, j) loops.

A timing chart of the pipeline is shown in Fig. 12.16. The addition and minimum detection executed in the pipeline need 8 cycles. This pipeline processing is executed 9 times and the total cycles add to $16 = 8 + 9 - 1$.

Since the next search step uses the final result of the previous step, direct concatenation of step pipeline processing is not efficient in terms of PE utilization.

[1]In the following discussion, p and q denotes these numbers respectively and the 3-step search is regarded as the loop execution of variables (p, q).

If the next macroblock of the current frame is processed by turns, the no-operation period is omitted with maximum throughput.

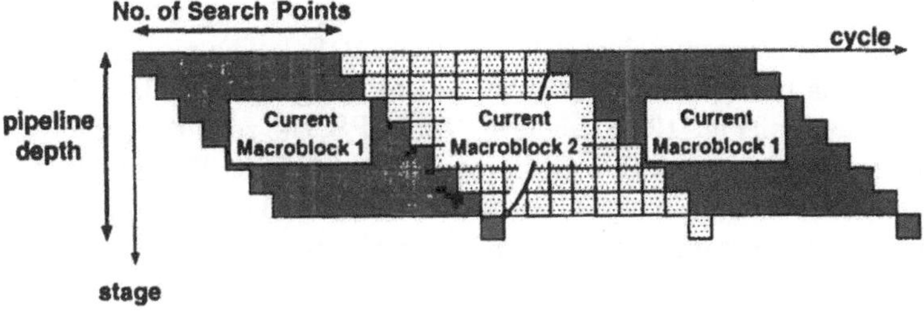

Figure 12.16 Pipeline timing for the tree architecture.

When the pipeline depth is deep, the alternate processing above is different to achieve. Furthermore, parallel delivery of the reference macroblock pixels is another issue to be solved. Jehng noted that a tree for whole 256-pixel addition is not necessary because performance required for H.261-based systems is not so high. He discussed subtree structures of this architecture and concluded that 1/32-cut up to 1/4-cut subtrees are possible solutions for real applications.

In contrast to Jehng's method, Jong [23] presented a circuit of q-loop in parallel instead of (i, j) loop (Fig. 12.17). Similarly in this case it is necessary to deliver the reference picture to 9 PEs in parallel. Jong stated that dynamic exchange of PEs as the values (i, j) change makes it possible to simplify selector circuits between the reference frame memories and PEs.

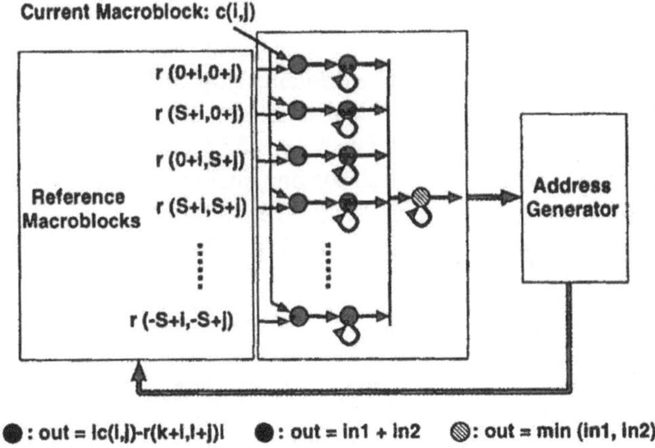

Figure 12.17 Jong's parallel 9-point search architecture.

12.5 VIDEO ENCODER LSI IMPLEMENTATIONS

We have discussed general aspects of motion estimation algorithms and their circuit architectures. It is necessary, however, to consider additional issues for implementation. In MPEG-2 video LSIs, for example, the frame and field prediction of interlaced video is a critical subject. Here, issues on real design are discussed with reference to several examples.

12.5.1 H.261 Encoder LSIs

Ruetz [24] designed a chipset for H.261 CODEC. The motion estimation part of the chipset is based on a parallel version of Yang's proposal. [1] Two sets of 16 internal processors (Fig. 12.18) achieve 778MOPS, the required performance in an H.261 full-search system, with a clock rate below 30MHz. The length of the search range is the same as the macroblock width 16 (32×32 search area) which is the optimum value to sustain 100% utilization as discussed before.

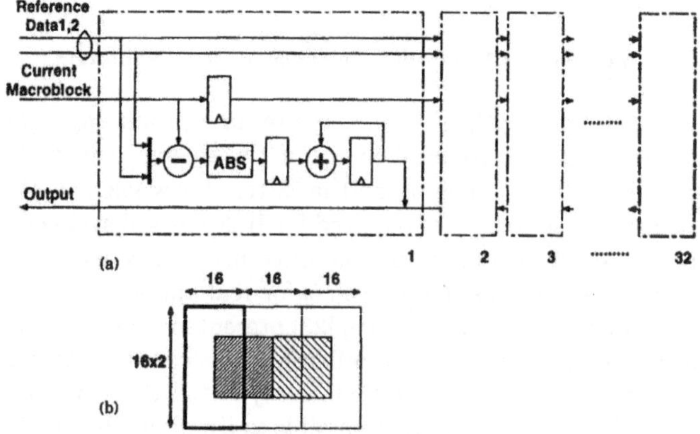

Figure 12.18 (a) Ruetz's circuit architecture, and (b) its prefetch buffers for reference pictures.

To solve a memory bandwidth problem, they incorporated prefetch buffers (cache memories) that form hierarchical memory structure, reducing the number of access directly to the frame memory. A 16×32 rectangular area of the reference picture is prefetched during the search for one current macroblock, which prepares the next current macroblock search. Three rectangles, that are 1.5 times the size of one search window, are enough for buffering. The buffers are filled in a rotated allocation scheme.

12.5.2 MPEG-1 Encoder LSIs

In the MPEG-1 standard, temporal bidirectional motion estimation is introduced. Performance required for motion vector search should be doubled as compared to forward prediction. Furthermore, strict estimation for large motions is required since the applications for video compression is not only the teleconference but general video contents like a TV program where faster motions occur frequently.

Hence, highly-parallel regular PE array architecture for the full-search is of the most importance for MPEG applications. The memory access scheme should be as simple as possible to avoid large increase of the memory bandwidth.

---

[1] As another example of an H.261 encoder, the circuit described in [25] has the same architecture Yang's proposal.

In MPEG-1 examples, however, it is not necessary to parallelize entire (i, j) loop to obtain the performance required. Hence, AB1 architecture or the serial connection of that one is sufficient. In Hayashi's case [26], by using 4:1 subsampling for the low-definition decimation scheme, 16 PEs are enough to execute a real-time bidirectional search of ±16 pixels at 36MHz. [1]

Lin's proposal [27] includes application of the fast search algorithm to MPEG-1 based compression. The strategy is similar to the Jehng's tree architecture in terms of accelerating the (i, j) loop. But he used 8×8 array instead of tree architecture here. An 8-bank memory provides 8×8-bit data to the array. Data in this memory is updated through the other side of the dual memory ports while the search proceeds.

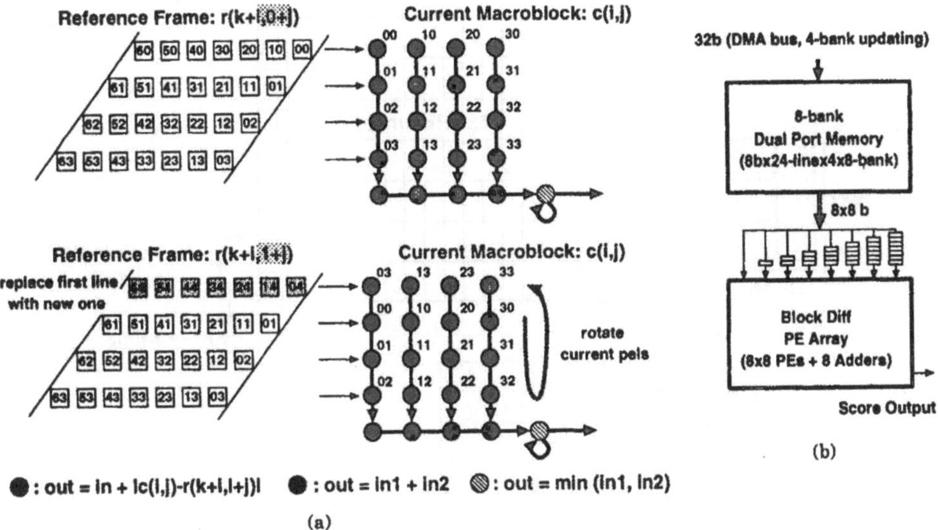

Figure 12.19 Lin's pel-rotation scheme (a) is achieved by a shift register in the PE array (b).

12.5.3 MPEG-2 Encoder LSIs

The common features of MPEG-2 motion estimators are highly parallel architectures, additional circuits for field prediction, and scalability of search range enhancement.

The one-dimensional reference picture data supply model presented by Chang is essential for the realistic implementation of MPEG-2 class motion estimation. Most of the MPEG-2 motion estimators are based on the one-dimensional PE array where entire (i, j) loops are executed serially. They are no longer based on Komarek's AB2 model where the (i, j) loop is hierarchically divided to calculate a grand total by adding each partial sum.

As mentioned before, to sustain 100% PE utilization, the width of the search range is limited to 16. In MPEG-2 applications, the desired range usually exceeds that value. Hence, PE utilization should go down or we should add extra shift

[1] They also used shift registers for a wait to enhance the search range.

registers around PEs. In the following examples several schemes are introduced to overcome this drawback.

In Ohtani's case [28], 16 × 8-PE structure is provided for each picture field. The architecture is a projection onto the (i, j) plain where the MAD is accumulated while marching through the PEs. The width of the search area is 32, which is obtained by executions in the same PE array twice (Fig. 12.20). The lack of the search range is compensated by the telescopic search algorithm, where a shifted origin based on previous searches is used.

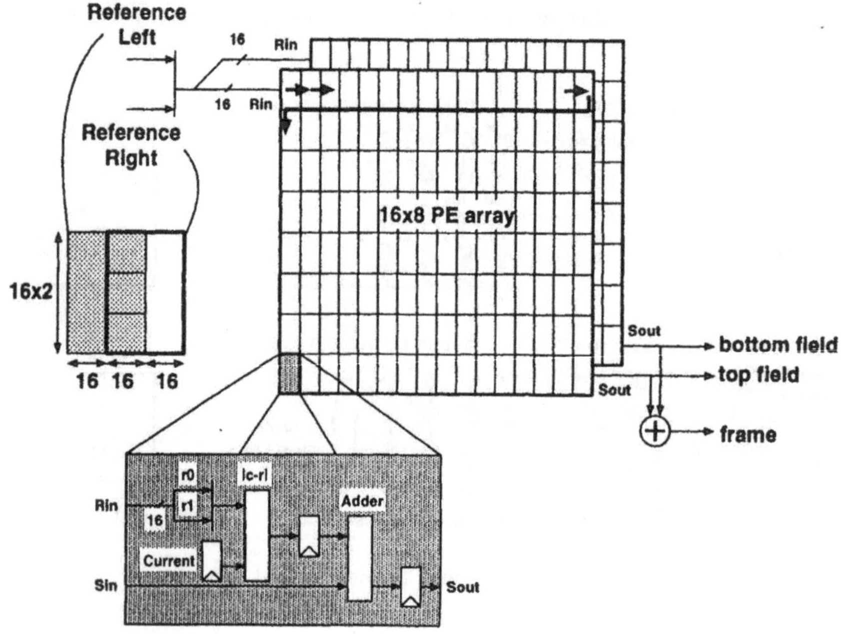

Figure 12.20 Ohtani's 2×128-PE structure uses separate arrays for each picture field.

Ishihara [29] presented a 256-PE architecture which is used for both frame and field prediction (Fig. 12.21). MAD summation required for four predictions (upper-top, upper-bottom, lower-top, and lower-bottom) are executed by four tree adders where 64 8-bit PE's output are summed up by 4-to-2 compressors in the Wallace tree.

The reference image is input in vertical scan order. 16 × 32 shift registers are added to obtain a ±16 vertical search range. The merit of vertical scan is scalability for the horizontal search range; by simply extending the search time, this architecture can achieve a wide search range [1]. With this feature, multi-chip search-range extension without glue logic is possible.

Suguri [30] used compact 8 × 4 PE circuits using a 4:1 subsampling scheme of low-definition search like Hayashi's scheme (Fig. 12.22(a)). The array covers 16 × 8-pel area in full-pel resolution, where each field prediction is executed sequentially

[1]The enhancement of horizontal search range is more important than the vertical one since real video contents contain many horizontal moves like panning.

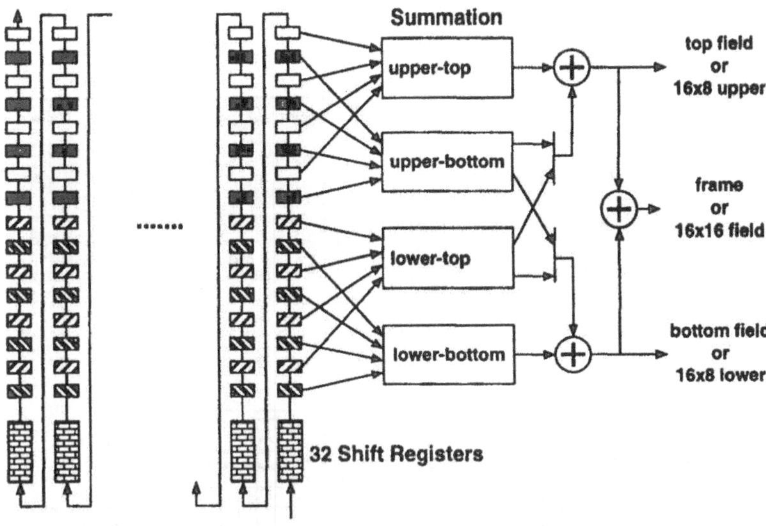

Figure 12.21 Ishihara's 256-PE architecture using vertical scan order.

on the same hardware. To compensate the search range, the telescopic method is also used.

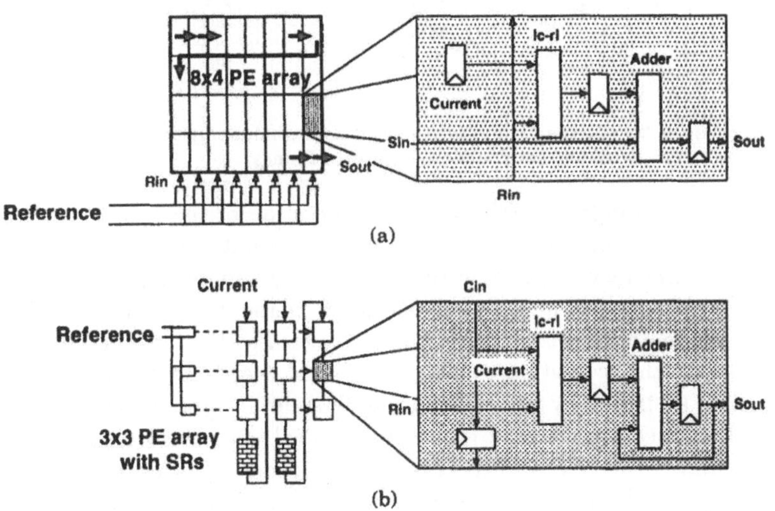

Figure 12.22 Suguri's proposal includes two types of arrays: (a) AB2 (type-I) for full-pel search, and (b) type-II for half-pel search.

As mentioned before, Vos's type-I array is used for the first search, while type-II array is used for one-pel and half-pel refinement (Fig. 12.22(b)).

Mizuno [31] described multi-segment search-window architecture using 16×8 PE array (Fig. 12.23). A horizontal 2:1 subsampling scheme is used (vertical resolution of pictures is the same as original). The four half-field searches are done in the same PE array. Eventually the final results are added to obtain adequate predictions.

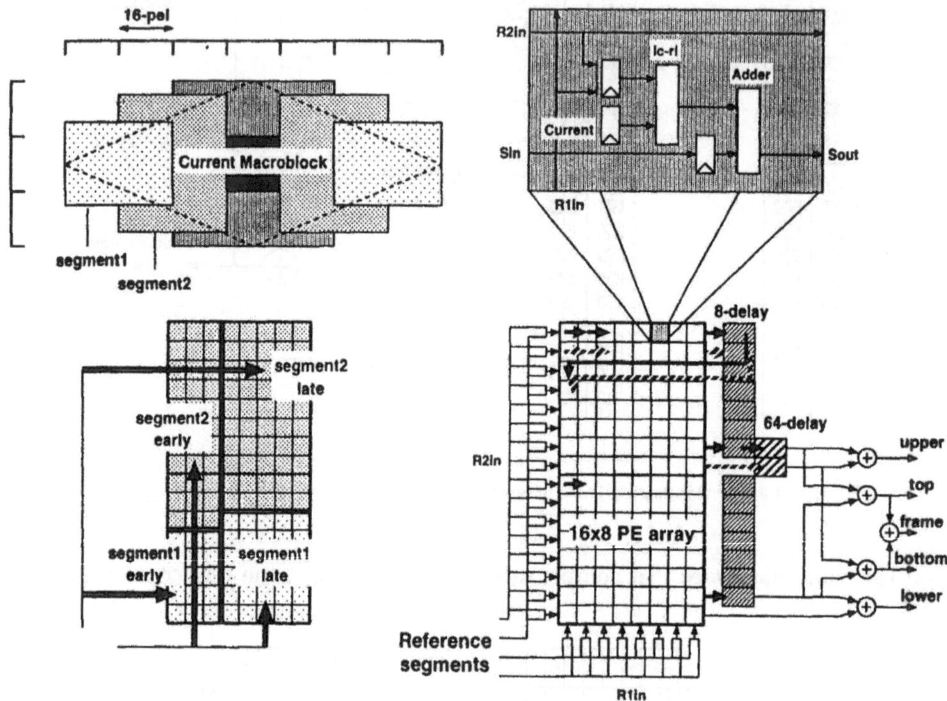

Figure 12.23 Mizuno's multi-segment search-window architecture uses only 4 buses to deal with all combination of reference frames.

The search range is extended by segmentation of the search window, where each segment is defined by a rectangle with constant width (32 pels in original scaling) and arbitrary height and offset. This contributes eliminating restriction of search width. By using a diamond window shape approximated by the search segments, computational complexity can be reduced without loss of coding efficiency [32]. Even at the transition point of continual search segments, the 100% PE utilization is sustained by two dimensional four-data supply architecture.

The architecture adopted four buses to deal with every combination of reference frames. The selectors for the four reference pictures are located at the circumference of the PE array in order to eliminate selectors in each PE circuit. The search window can be shifted, unlike the telescopic search, by one offset per picture to enhance the search range.

Table 12.1 summarizes the architectural features of MPEG-2 encoder LSIs. Though they are classified into the same architecture in terms of the classification we mentioned before (AB2 or Type-I), their detailed specifications are considerably different.

Table 12.1 Architectural Features of MPEG-2 Encoder LSIs

	Ohtani[28]	Ishihara[29]	Suguri[30]	Mizuno[31]
# of PEs	16x8x2	16x16	8x4	16x8
Decimation	none	none	4:1(H and V)	2:1(H)
Length of Delay Lines	none	32	none	8
Search Range Enhancement	telescopic	multichip	telescopic	offset-shift
Reference Scan Order	H	V	H	H
Distance Accumulation in PE	Yes	No	Yes	Yes

12.6 MOTION ESTIMATION - OTHER TECHNIQUES

So far we have reviewed the methods of finding motion vectors by block-matching based on a region of 16 by 16 pixels (macroblock). Here we briefly describe other kinds of motion estimation algorithms that are important in terms of research trends [33]. Some of these methods have been raised through the discussion process of the MPEG-4 standard; however, little discussion about circuit architectures of these motion estimation methods is provided.

12.6.1 Spatial and Temporal Gradient Method

Spatial and temporal gradient methods are a class of motion estimation algorithms which are not based on block-matching [3]. In this algorithm we need the following assumptions: the intensity of the objects and the velocity of the object are both regarded as constant value within a short move.

We define that the intensity of a certain point of a frame $(x(t), y(t))$ and time t is denoted as $f(x(t), y(t), t)$. The first assumption about the intensity implies that the derivative of the intensity with respect to time is zero, that is, $\frac{df}{dt} = 0$, The second assumption about velocity implies that the derivative of the place with respect to time is constant, that is, $\frac{\partial x}{\partial t} = const = v_x$, $\frac{\partial y}{\partial t} = const = v_y$. Using these definitions the spatial and temporal gradient relation can be stated as follows,

$$\begin{aligned}\frac{df}{dt} &= \frac{\partial f}{\partial x}\frac{\partial x}{\partial t} + \frac{\partial f}{\partial y}\frac{\partial y}{\partial t} + \frac{\partial f}{\partial t} \\ &= \frac{\partial f}{\partial x}v_x + \frac{\partial f}{\partial y}v_y + \frac{\partial f}{\partial t} = 0.\end{aligned}$$

In this formula $\frac{\partial f}{\partial x}$ denotes the spatial intensity gradient with respect to the x-direction at time t. It can be approximately estimated from neighborhood pixels around (x, y). This is also the case of $\frac{\partial f}{\partial y}$, the spatial gradient for the y-direction. The temporal gradient $\frac{\partial f}{\partial t}$ can be estimated by comparing the same position (x, y) of video frames at around time t.

The velocities v_x and v_y can be obtained to solve a system of linear equations which are formed by using two or more sets of gradient estimations above. If there are more than two sets of estimations, we should use least squares method since the system of linear equations are over-determined.

It is confirmed by several experiments that the motion estimation based on the spatial and temporal gradient relation does not lead to good results [1]. This means that the assumptions made are not always true. Estimation of gradients is another problem because we have only discrete pixels by sampling and quantization. It is reported that the algorithm has difficulty to find the right vectors especially at the boundary edge of objects as well as in noisy images.

12.6.2 Variable Size Block-Matching

Most of video coding algorithms including MPEG-2 have used a fixed-size macroblock for motion estimation. If we use smaller matching block, the different motions within the macroblock can be detected. Hence we can obtain more consistent motion vector fields. But this is a drawback in terms of video coding since we should encode more information on motion vectors in every picture.

In the H.263 standard, the Advanced Prediction Mode is adopted to compromise the trade-off, where two block sizes are permitted such that 16x16 macroblock is divided into four 8x8 block if necessary for the efficient motion prediction.

An extension of this scheme is non-uniform, inhomogeneous splitting of images, where a part of a frame appears in a larger segment and another appears in a smaller one, by using blocks of different sizes. Considering a set of square blocks with side length of $2^n (n = 0, 1, 2, \cdots)$, optimal motion is estimated using a recursive procedure on the block divided by 4. The division information can be then represented by an unbalanced quad-tree structure. The tree should be encoded as a bit stream if we use this technique for video coding. From a viewpoint of coding efficiency, encoded image quality should be still fair when coding overhead of the tree information is included.

12.6.3 Region Matching

Block matching is not adequate for estimating a complicated motion including rotation, scaling, 3-dimensional motion of real objects, and change of object's shape. Region matching has been proposed to follow complex motion, where the affine transform and the perspective transform are used to describe such a motion.

The affine transform $(X, Y) = (ax + by + c, dx + ey + f)$ is defined by six variables. If these variables are unknown and need to be specified by pairs of transform points $\{(X, Y), (x, y)\}$, we need three pairs to build a well-defined system of linear equations. In other words, two arbitrary triangles (defined by the three points) can be mapped onto each other by the affine transform. Hence the triangular division is used for the region matching based on the affine transform. In the case of perspective transform, the region is defined as quadrilateral.

In a motion estimation process, a current picture is, like a conventional macroblock, divided into uniform triangles. On a reference picture, matching points are searched, based on the distance calculation between a reference triangle region and the transformed current triangle region.

In an advanced region matching, we can modify boundary of the triangles depending on the scene of the picture. In this case we need the modification information to be encoded. If the modification is predicted from the several previous pictures, we do not have to encode it. In other proposals, it is reported that if boundary of the triangles is located to fit the boundary of objects, the coding efficiency is remarkably improved.

12.7 CONCLUDING REMARKS

In this chapter we have discussed the algorithms and VLSI implementations of motion estimation for video coding, focusing on block-matching-based motion vector search techniques.

The full-search algorithm usually provides the best result in terms of coded picture quality and is now easily designed as a part of video coding systems because of algorithm regularity and improvement of VLSI fabrication process. Simplified motion estimation algorithms, however, will be still important and will be discussed continually to achieve further system integration mostly driven by the cost reduction demand from the market, and possibly to achieve wider search range. New motion estimation algorithms may be a mixed version of conventional full-search and fast search algorithms. They make sense only if they are mapped onto a smaller circuit with smaller memory bandwidth than the conventional circuits with comparable coding efficiency.

Another interesting area is software video encoding. In the near future we will see a software implementation of real-time video encoder as we already have an MPEG-2 real-time software decoder on microprocessors with multimedia extension. The objective here is selecting the motion estimation algorithm suitable for the software. We can take advantage of conditional instruction execution for algorithm optimization. One possible solution is a data-dependent algorithm with variable execution time, which is in contrast to the conventional hardware designs.

ACKNOWLEDGMENT

The author expresses his deep appreciation to the members of C&C Media Research Labs. for their valuable comments.

REFERENCES

[1] F. Dufaux and F. Moscheni, "Motion Estimation Techniques for Digital TV: A Review and a New Contribution," *Proc. IEEE*, vol. 83, no. 6, pp. 858-876, June 1995.

[2] H. G. Musmann, P. Pirsch, H-J. Grallert, "Advances in Picture Coding," *Proc. IEEE*, vol. 73, no. 4, pp. 523-548, April 1985.

[3] P. Pirsch, N. Demassieux, and W. Gehrke, "VLSI Architecture for Video Compression - A Survey," *Proc. IEEE*, vol. 83, no. 2, pp. 220-246, Feb. 1995.

[4] B. Furht, J. Greenberg, and R. Westwater, *Motion Estimation Algorithms for Video Compression*, Kluwer Academic Publishers, Norwell, 1997.

[5] Y. Naito, T. Miyazaki, and I. Kuroda, "A fast full-search motion estimation method for programmable processors with a multiply-accumulator," in *Proc. ICASSP*, VLSI1.7, vol. 1, pp. 3222- 3225, 1996.

[6] Y. Senda, H. Harasaki, and M. Yano, "A simplified motion estimation using an approximation for the MPEG-2 real-time encoder," in *Proc. ICASSP*, vol. 4, pp. 2273-2276, 1995.

[7] Y. Ninomiya and Y. Ohtsuka, "A motion-compensated interframe coding scheme for television pictures," *IEEE Trans. Comm.*, vol. COM-30, no. 1, pp. 201-211, Jan. 1982.

[8] T. Koga, K. Iinuma, A. Hirano, Y. Iijima, and T. Ishiguro, "Motion-compensated interframe coding for video conferencing," in *Proc. Nat. Telecomm. Conf.*, New Orleans, LA, pp. G5.3.1-G5.3.5, Nov. 29-Dec. 3, 1981.

[9] J. R. Jain and A. K. Jain, "Displacement measurement and its application in interframe image coding," *IEEE Trans. Comm.*, vol. COM-29, no. 12, pp. 1799-1808, Dec. 1981.

[10] M. Ghanbari, "The cross-search algorithm for motion estimation," *IEEE Trans. Comm.*, vol. COM-38, no. 7, pp. 950-953, July 1990.

[11] L. W. Lee, J. F. Wang, J. Y. Lee, and J. D. Shie, "Dynamic search-window adjustment and interlaced search for block-matching algorithm," *IEEE Trans. Circuits and Syst. for Video Technol.*, vol. 3, no. 1, pp. 85-87, Feb. 1993.

[12] R. Li, B. Zeng, and M. L. Liou, "A new three-step search algorithm for block motion estimation," *IEEE Trans. Circuits and Syst. for Video Technol.*, vol. 4, no. 4, pp. 438-442, August 1994.

[13] H. D. Lin, A. Anesko, B. Petryna, and G. Pavlovic, "A programmable motion estimator for a class of hierarchical algorithms," *VLSI Signal Processing VIII*, pp. 411-420, 1995.

[14] B. Liu and A. Zaccarin, "New fast algorithms for the estimation of block matching vectors," *IEEE Trans. Circuits and Syst. for Video Technol.*, vol. 3, no. 2, pp. 148-157, April 1993.

[15] E. Ogura, *et al.*, "A cost effective motion estimation processor LSI using a simple and efficient algorithm," in *Proc. of ICCE*, pp. 248-249, 1995.

[16] T. Komarek and P. Pirsch, "Array architecture for block matching algorithms," *IEEE Trans. Circuits and Syst.*, vol. 36, no. 10, pp. 1301-1308, Oct. 1989.

[17] S. Y. Kung, *VLSI Array Processors,* Prentice Hall, Englewood Cliffs, 1988.

[18] L. D. Vos and M. Stegherr, "Parameterizable VLSI architectures for the full-search block-matching algorithm," *IEEE Trans. Circuits and Syst.*, vol. 36, no. 10, pp. 1309-1316, Oct. 1989.

[19] C. H. Hsieh and T. P. Lin, "VLSI architecture for block-matching motion estimation algorithm," *IEEE Trans. Circuits and Syst. for Video Technol.*, vol. 2, no. 2, pp. 169-175, August 1992.

[20] K. M. Yang, M. T. Sun, and L. Wu, "A family of VLSI designs for the motion compensation block-matching algorithm," *IEEE Trans. Circuits and Syst.*, vol. 36, no. 10, pp. 1317-1325, Oct. 1989.

[21] S. Chang, J. H. Hwang, and C. W. Jen, "Scalable array architecture design for full search block matching," *IEEE Trans. Circuits and Syst. for Video Technol.*, vol. 5, no. 4, pp. 332-343, August 1995.

[22] Y. S. Jehng, L. G. Chen, and T. D. Chiueh, "An efficient and simple VLSI tree architecture for motion estimation algorithms," *IEEE Trans. Signal Proc.*, vol. 41, no. 2, pp. 889-900, Feb. 1993.

[23] H. M. Jong, L. G. Chen, and T. D. Chiueh, "Parallel architectures for 3-step hierarchical search block-matching algorithm," *IEEE Trans. Circuits and Syst. for Video Technol.*, vol. 4, no. 4, pp. 407-416, Aug. 1994.

[24] P. A. Ruetz, P. Tong, D. Bailey, D. A. Luthi, P. H. Ang, "A high-performance full-motion video compression chip set," *IEEE Trans. Circuits and Syst. for Video Technol.*, vol. 2, no. 2, pp. 111-122, June 1992.

[25] H. Fujiwara, *et al.*, "An all-ASIC implementation of a low bit-rate cideo codec," *IEEE Trans. Circuits and Syst. for Video Technol.*, vol. 2, no. 2, pp. 123-134, June 1992.

[26] N. Hayashi, *et al.*, "A bidirectional motion compensation LSI with a compact motion estimator," *IEICE Trans. Electron.*, vol. E78-C, no. 12, pp. 1682-1690, Dec. 1995.

[27] H. D. Lin, A. Anesko, and B. Petryna, "A 14-GOPS programmable motion estimator for H.26X video coding," *IEEE Journal of Solid-State Circuits*, vol. 31, no. 11, pp. 1742-1750, Nov. 1996.

[28] A. Ohtani, *et al.*, "A motion estimation processor for MPEG-2 video real time encoding at wide search range," *Proc. CICC*, pp. 17.4.1-17.4.4, 1995.

[29] K. Ishihara, *et al.*, "A half-pel precision MPEG2 motion-estimation processor with concurrent three-vector search," *IEEE Journal of Solid-State Circuits*, vol. 30, no. 12, pp. 1502-1509, Dec. 1995.

[30] K. Suguri, *et al.*, "A real-time motion estimation and compensation LSI with wide search range for MPEG2 video encoding," *IEEE Journal of Solid-State Circuits*, vol. 31, no. 11, pp. 1733-1741, Nov. 1996.

[31] M. Mizuno, *et al.*, "A 1.5-W single-chip MPEG-2 MP@ML video encoder with low power motion estimation and clocking," *IEEE Journal of Solid-State Circuits*, vol. 32, no. 11, pp. 1807-1816, Nov. 1997.

[32] Y. Ooi, *et al.*, "An MPEG-2 encoder architecture based on a single-chip dedicated LSI with a control MPU," in *Proc. ICASSP*, Munich, Germany, vol. 1, pp. 599-603, April 21-24, 1997.

[33] G. M. Schuster and A. K. Katsaggelos, *Rate-Distortion Based Video Compression*, Kluwer Academic Publishers, Dordrecht, The Netherlands, 1997.

Chapter 13

Wavelet VLSI Architectures

Tracy C. Denk
Broadcom Corporation
Irvine, California
tdenk@broadcom.com

Keshab K. Parhi
University of Minnesota
Minneapolis, Minnesota
parhi@ece.umn.edu

13.1 INTRODUCTION

Wavelet transforms have proven to be useful tools for signal processing [1] -[7]. Applications of wavelet transforms include signal compression, statistical analysis, adaptive filtering, and solving linear equations, just to name a few. In fact, the United States Federal Bureau of Investigation (FBI) has adopted wavelets as the technology of choice for compressing fingerprint images in order to keep the size of its database manageable [8].

In many applications, the wavelet transform must be computed in real time. Dedicated VLSI implementations are often required for these applications. Many different types of VLSI wavelet architectures have been proposed, and this chapter describes several of these architectures alongwith their advantages and disadvantages.

The chapter begins with an overview of the wavelet transform. This is followed by a description of architectures for computing wavelet transforms of one-dimensional signals (see also Section 2.3.2). Architectures for computing wavelet transforms of two-dimensional signals, such as images, are then described. The chapter concludes with a summary of VLSI wavelet architectures.

13.2 INTRODUCTION TO WAVELET TRANSFORMS

Fourier transforms are useful when dealing with signals with statistical properties that are constant over time (or space). This is because the Fourier transform represents a signal as the sum of sine and cosine functions which have infinite duration in time. In the real world, however, signals are often encountered with statistical properties that vary over time. Rather than using sine and cosine waves to represent these signals, it seems more natural to use localized waves, or "wavelets." Representing non-stationary signals as the sum of basis functions that are localized in time can lead to more compact representations that provide better insight into the properties of the signals than can be provided by using Fourier transforms.

In wavelet analysis, signals are represented using a set of basis functions which are derived from a single prototype function called the "mother wavelet." These basis functions, or wavelets, are formed by translating and dilating the mother wavelet. Another way to state this is that the basis functions are formed by shifting and scaling the mother wavelet in time. As a result, the wavelet transform can be viewed as a decomposition of a signal in the time-scale plane.

The algorithms used to compute the forward and inverse discrete wavelet transforms in this chapter are often referred to as the pyramid algorithm or Mallat's algorithm [3], [4]. Computation of the one-dimensional (1-D) discrete wavelet transform (DWT) begins with an input signal $S_0(n)$. The wavelet transform recursively decomposes this signal into approximation and detail at the next lower resolution. Let $S_i(n)$ and $W_i(n)$ be the approximation and detail, respectively, of the signal at level i. The approximation of the signal at level $i+1$ is given by

$$S_{i+1}(n) = \sum_{k=0}^{L-1} h(k) S_i(2n-k), \tag{1}$$

where $h(k)$ is an $(L-1)$-th order lowpass FIR filter with z-transform $H(z)$. The detail of the signal at level $i+1$ is given by

$$W_{i+1}(n) = \sum_{k=0}^{L-1} g(k) S_i(2n-k), \tag{2}$$

where $g(k)$ is an $(L-1)$-th order highpass FIR filter with z-transform $G(z)$.

The equations (1) and (2) describe the computation of the discrete wavelet transform, and the filters $H(z)$ and $G(z)$ are determined by the continuous-time wavelet. The input signal is $S_0(n)$, and the resolution of the approximations $S_i(n)$ decreases as i increases. Since the signals $S_i(n)$ represent the input signal at various resolutions, the DWT can be viewed as a multiresolution decomposition of the signal. A block diagram of the DWT computation described in (1) and (2) is shown in Fig. 13.1(a). This structure is called the *analysis* DWT filter bank.

The input signal $S_0(n)$ can be synthesized from its wavelet coefficients according to

$$S_{i-1}(n) = \sum_{k} \{ S_i(k) h'(n-2k) + W_i(k) g'(n-2k) \}. \tag{3}$$

This equation describes the *synthesis* DWT filter bank, which uses lowpass filter $H'(z)$ and highpass filter $G'(z)$ (with impulse responses $h'(n)$ and $g'(n)$, respectively) which are determined by the continuous-time wavelet. The synthesis filter bank, which is shown in Fig. 13.1(b), computes the inverse discrete wavelet transform (IDWT). It is interesting to note that the DWT and IDWT require exactly the same number of multiplications and additions per input sample (an explanation of this is provided in [9]).

The filter banks in Fig. 13.1 which are used to compute the DWT and IDWT are *multirate* filter banks due to the presence of decimation by 2 blocks and expansion by 2 blocks in parts (a) and (b), respectively, of Fig. 13.1. Decimation by M is shown in Fig. 13.2(a) and is described by $y_D(n) = x(Mn)$. Expansion by M is shown in Fig. 13.2(b) and is described by

$$y_E(n) = \begin{cases} x(n/M) & , n \text{ is a multiple of } M \\ 0 & , \text{otherwise} \end{cases} .$$

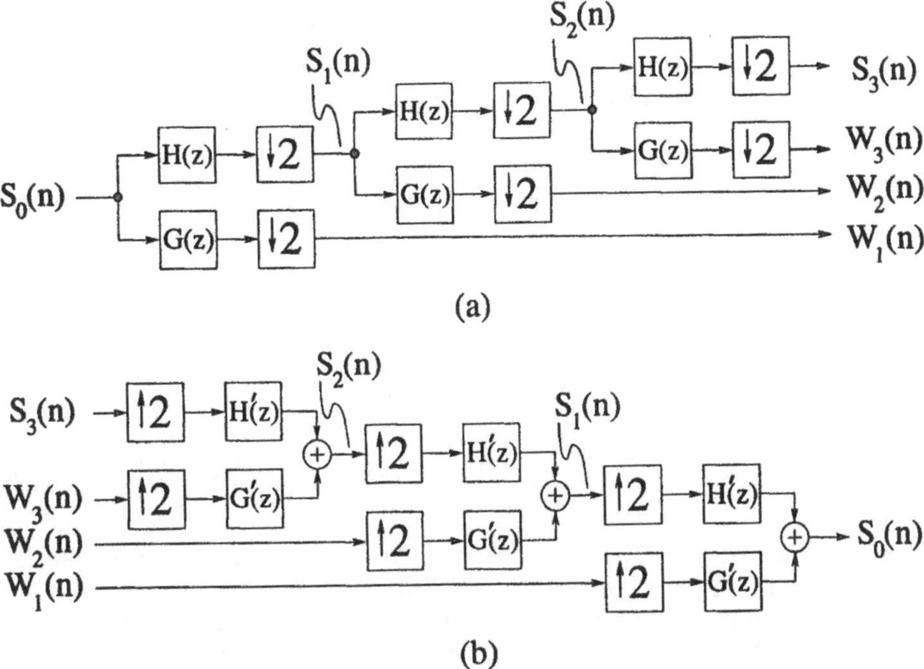

Figure 13.1 Block diagrams of the DWT (a) analysis and (b) synthesis filter banks. The analysis filter bank is used to compute the forward DWT, and the synthesis filter bank is used to compute the inverse DWT.

Further properties of multirate algorithms can be found in [10]. As one might expect from a multirate filter bank, the input signal is decomposed into various frequency bands. Using this description, the DWT can be viewed as a special case of subband coding. The multirate nature of wavelet transforms makes designing VLSI wavelet architectures particularly challenging because traditional hardware design techniques generally do not apply to multirate algorithms.

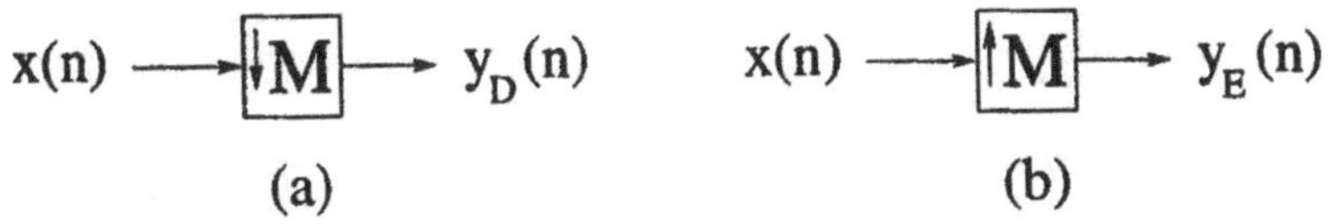

x(n) → ↓M → $y_D(n)$ x(n) → ↑M → $y_E(n)$

(a)   (b)

Figure 13.2 (a) Decimation by M. (b) Expansion by M.

The two-dimensional DWT has applications in multiresolution analysis, computer vision, and image compression [3] [4]. The two-dimensional DWT operates on 2-D signals, such as images. While 1-D filters are used to compute the 1-D DWT, the 2-D DWT uses 2-D filters in its computation. These 2-D filters may be separable or non-separable, where a 2-D filter $f(n_1, n_2)$ is separable if it can be written as $f(n_1, n_2) = f_1(n_1)f_2(n_2)$. In this chapter, we discuss architectures for the separable case. For the interested reader, an architecture for the non-separable

2-D DWT is presented in [11]. For simplicity, we refer to 2-D signals as "images" in spite of the fact that these signals may represent 2-D data other than images.

The separable 2-D DWT decomposes an approximation image $S_i(n_1, n_2)$ into an approximation image and three detail images according to

$$
\begin{aligned}
S_{i+1}(n_1,n_2) &= \sum_{k_1=0}^{L-1}\sum_{k_2=0}^{L-1} h(k_1)h(k_2)S_i(2n_1-k_1, 2n_2-k_2) \\
W_{i+1}^1(n_1,n_2) &= \sum_{k_1=0}^{L-1}\sum_{k_2=0}^{L-1} h(k_1)g(k_2)S_i(2n_1-k_1, 2n_2-k_2) \\
W_{i+1}^2(n_1,n_2) &= \sum_{k_1=0}^{L-1}\sum_{k_2=0}^{L-1} g(k_1)h(k_2)S_i(2n_1-k_1, 2n_2-k_2) \\
W_{i+1}^3(n_1,n_2) &= \sum_{k_1=0}^{L-1}\sum_{k_2=0}^{L-1} g(k_1)g(k_2)S_i(2n_1-k_1, 2n_2-k_2),
\end{aligned}
\tag{4}
$$

where $H(z)$ and $G(z)$ are 1-D wavelet filters. The signal $S_{i+1}(n_1, n_2)$ is an approximation of $S_i(n_1, n_2)$ at a lower resolution. This approximation is computed from $S_i(n_1, n_2)$ by lowpass filtering and decimating by 2 along its rows and columns. The signals $W_{i+1}^1(n_1, n_2)$, $W_{i+1}^2(n_1, n_2)$, and $W_{i+1}^3(n_1, n_2)$ contain the detail of $S_i(n_1, n_2)$. The signal $W_{i+1}^1(n_1, n_2)$ contains the vertical high frequencies (horizontal edges) because this signal is computed by filtering $S_i(n_1, n_2)$ with a lowpass filter in the horizontal direction and with a highpass filter in the vertical direction. Similarly, the signal $W_{i+1}^2(n_1, n_2)$ contains the horizontal high frequencies (vertical edges), and $W_{i+1}^3(n_1, n_2)$ contains the high frequencies in both directions (the corners). The two-level 2-D DWT computation in (4) is shown in Fig. 13.3.

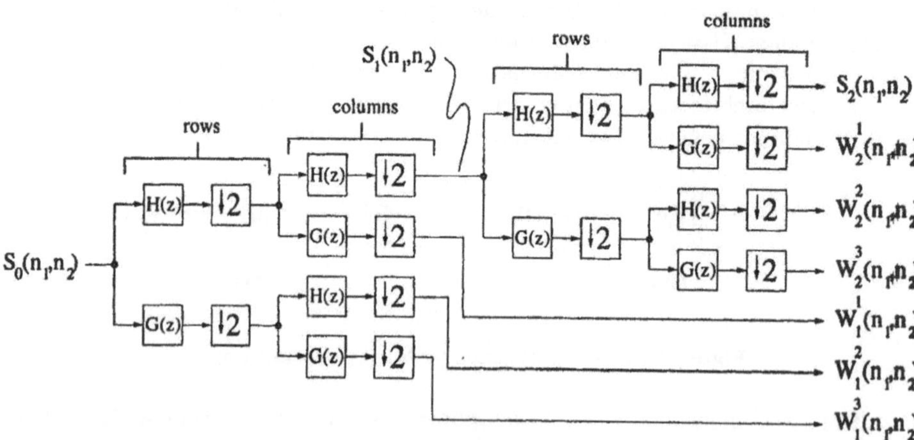

Figure 13.3 Block diagram of the analysis filter bank used to compute the 2-D DWT.

13.3 THE ONE-DIMENSIONAL DWT

This section presents algorithms and architectures for implementing the forward 1-D DWT described by (1) and (2) and shown in Fig. 13.1(a) and the inverse 1-D DWT described by (3) and shown in Fig. 13.1(b). Before discussing specific techniques for computing the DWT, it is instructive to count the number of filtering operations that must be performed per input sample $S_0(n)$ for the analysis DWT filter bank in Fig. 13.1(a) (recall that the DWT and IDWT require exactly the same number of operations per sample). The input signal $S_0(n)$ is input to two filters whose outputs are used one-half of the time (due to the decimation by 2), so the first pair of filters requires $2\left(\frac{1}{2}\right) = 1$ filtering operation per input sample. The signal $S_1(n)$ is half the rate of $S_0(n)$, so the second level requires 1/2 filtering operation per input sample. Extending this argument, we find that the number of filtering operations per input sample is upper bounded by $1 + 1/2 + 1/4 + \ldots = 2$. Therefore, when the analysis and synthesis filters have length L, the forward and inverse DWT each require a maximum of $2L$ multiplications and $2(L-1)$ additions per input sample.

13.3.1 Algorithms Based on Fast FIR Filtering Techniques

The fact that the DWT computation requires a maximum of $2L$ multiplications and $2(L-1)$ additions per input sample suggests that the DWT computation is already efficient; however, further reductions in the number of operations can be obtained using fast Fourier transforms (FFTs) when long filters are used and "fast running FIR filtering" techniques [12] when shorter filters are used [9]. Since short wavelet filters ($L \leq 20$) are commonly used, this section demonstrates how to reduce the number of operations using fast running FIR filtering techniques.

The basic building block of the analysis filter bank in Figure 13.1(a) is shown in Figure 13.4(a). This building block consists of two L-tap filters and two decimators. Considering the fact that one of every two outputs of each filter is discarded, a straightforward implementation of this structure requires L multiplications and $(L-1)$ additions per input sample. The lowpass filter $H(z) = \sum_{k=0}^{L-1} h(k)z^{-k}$ can be rewritten as $H(z) = H_0(z^2) + z^{-1}H_1(z^2)$, where $H_0(z) = \sum_{k=0}^{L/2-1} h(2k)z^{-k}$ and $H_1(z) = \sum_{k=0}^{L/2-1} h(2k+1)z^{-k}$. (Here we assume that the filter length L is even. If the filter length is odd, the filter can be padded with a zero.) In a similar manner, we can write $G(z) = G_0(z^2) + z^{-1}G_1(z^2)$ and $X(z) = X_0(z^2) + z^{-1}X_1(z^2)$ for the highpass filter and the input sequence, respectively. With these definitions, the computations in the building block in Figure 13.4(a) can be reorganized as shown in Figure 13.4(b). This structure, which uses a biphase decomposition of the filters, consists of four $(L/2)$-tap filters and two adders and requires L multiplications and $(L-1)$ additions per input sample. The number of operations required to implement this structure is reduced by using fast running FIR filtering techniques on $H_0(z)$, $H_1(z)$, $G_0(z)$, and $G_1(z)$.

We now describe how to reduce the number of operations required to implement an FIR filter. Consider an N-tap FIR filter $F(z)$ with input signal $U(z)$ and output signal $V(z)$. Writing $U(z) = U_0(z^2)+z^{-1}U_1(z^2)$, $V(z) = V_0(z^2)+z^{-1}V_1(z^2)$, and $F(z) = F_0(z^2)+z^{-1}F_1(z^2)$, an efficient implementation of this filter is shown in Figure 13.5. While a direct implementation of the filter requires N multiplications and $(N-1)$ additions per input sample, the implementation in Figure 13.5 requires

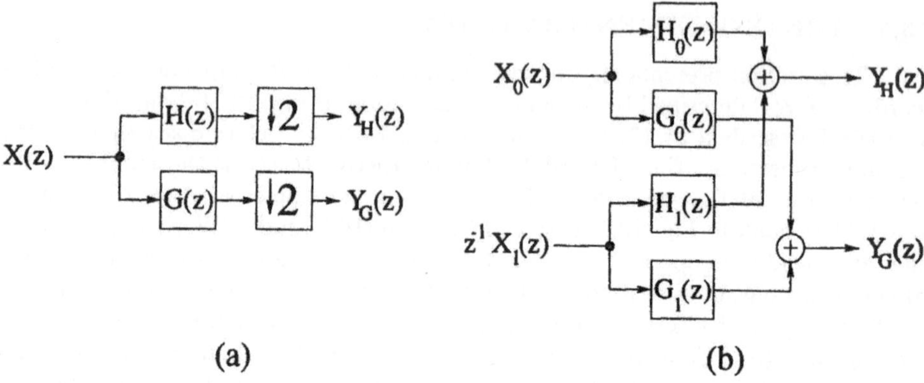

Figure 13.4 (a) The basic building block of the DWT analysis filter bank. (b) A reorganization of the computations in the basic building block. This reorganization uses the biphase decomposition.

only $3(N/2)$ multiplications and $3(N/2-1)+4$ additions for each pair of inputs, for an average of $3N/4$ multiplications and $(3/4)N+1/2$ additions per input sample. Therefore, the structure in Figure 13.5 requires less multiplications and additions per input sample than a direct implementation of the filter.

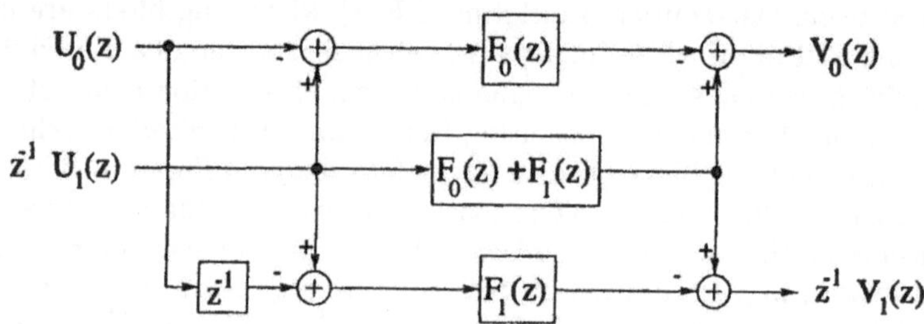

Figure 13.5 An implementation of the FIR filter $F(z)$ using fast running FIR filtering techniques.

The basic building block in Figure 13.4(a) can be implemented by substituting the structure in Figure 13.5 for each of the four $(L/2)$-tap filters in Figure 13.4(b). Using this substitution, the structure in Figure 13.6 implements the basic building block in Figure 13.4(a) for the case when $H(z)$ and $G(z)$ are 8-tap filters ($L=8$). The input and output signals in Figure 13.6 are defined by $X_i(z)=\sum x(4k+i)z^{-k}$ for $i=0,1,2,3$, $Y_{G_i}(z)=\sum y_G(2k+i)z^{-k}$ for $i=0,1$, and $Y_{H_i}(z)=\sum y_H(2k+i)z^{-k}$ for $i=0,1$. Notice that the adders at the inputs in Figure 13.6 are shared between filters. The structure in Figure 13.6 requires 24 multiplications and 28 additions for every four input samples, for an average of 6 multiplications and 7 additions per input sample. This is compared to 8 multiplications and 7 additions required for a straightforward implementation.

In this example, the number of operations required to implement the basic building block in Figure 13.4(a) for $L=8$ is reduced from $8+7=15$ to $6+7=13$ for a 13% reduction. Since the analysis filter bank in Figure 13.1(a) consists

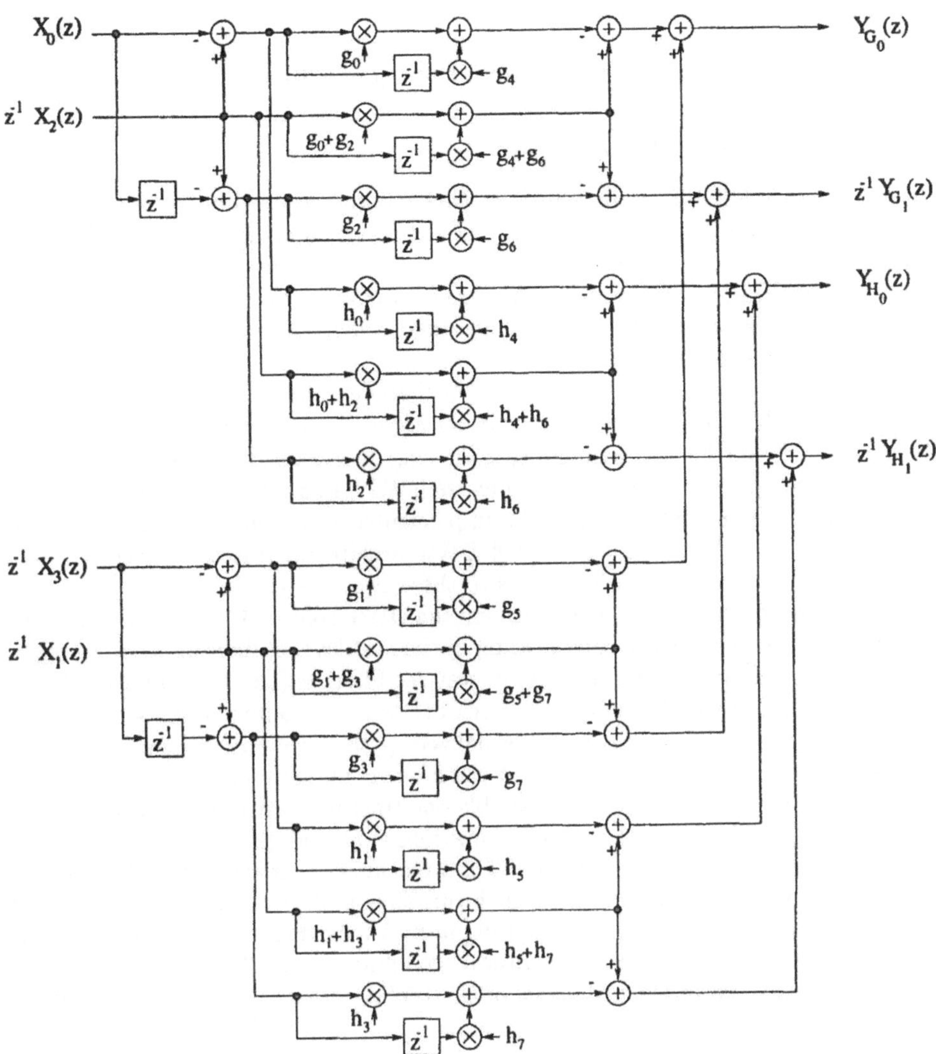

Figure 13.6 An implementation of the basic building block in Figure 13.4(a) constructed by substituting the structure in Figure 13.5 for each of the four $(L/2)$-tap filters in Figure 13.4(b).

of replications of this building block, the total number of operations required to implement the DWT is reduced by 13% using this technique. There are many ways to decompose a filter using the fast running FIR filter techniques, and filters can also be recursively decomposed using these techniques to further reduce the computational burden. These issues are discussed in detail in [9].

The structure in Figure 13.6 reduces the number of operations required to implement the DWT. This can be useful for software implementations of the DWT, but it is not useful for implementing the DWT in VLSI. Its downfalls are its irregularity and the fact that J of these structures running at J different clock rates are required to implement a J-level DWT analysis filter bank. In the following sections, we discuss architectures which require a single clock and reduce the amount of hardware by time-multiplexing several filtering operations to a small number of hardware filters.

13.3.2 Bit-Parallel Architectures

Most of the DWT architectures that have been proposed in the literature are ***bit-parallel architectures***, where all bits of the input sample are processed in one clock cycle. The concepts behind these architectures are useful not only for designing dedicated VLSI architectures, but also for implementations on programmable processors. This section discusses architectures which can be used for single-chip implementations of the 1-D DWT.

The main idea behind bit-parallel DWT architectures is the concept of time-multiplexing all of the filter operations in the algorithm to a single pair of hardware filters [13]. Recall that a maximum of 2 filtering operations are required per input sample for computing the forward or inverse DWT. As a result, the DWT analysis and synthesis filter banks can each be implemented using only 2 hardware filters. The challenge in designing bit-parallel DWT architectures is to efficiently time-multiplex all of the filtering operations to these two filters.

The approach used to time-multiplex the DWT computation in (1) and (2) to two filters is to schedule the computation of the first level of filtering to every other clock cycle to compute the first level of wavelet filtering (i.e., to compute (1) and (2) for $i = 0$) and use the remaining clock cycles to compute the remaining levels of filtering. This concept has been formalized as the ***recursive pyramid algorithm*** [14] and is described here. Assume that the filtering operation in (1) takes one time unit so that the result $S_{i+1}(n)$ is available one time unit after this computation is scheduled. A valid schedule for the computations in (1) and (2) is given in Table 13.1 for a three-level analysis filter bank. This table shows that for a three-level DWT the filters are performing computations during time units $8k+0$ through $8k+6$ and are idle during time units $8k+7$, resulting in 87.5% hardware utilization efficiency. A similar discussion holds for the synthesis computation in (3).

For single-chip implementations of the DWT, it is desirable for the architectures to

- be scalable with filter length and number of octaves.
- be modular.
- have efficient memory utilization.
- have high hardware utilization efficiency (close to 100%).
- have simple routing and control.
- have silicon area which is independent of input signal length.

The first two properties lead to timely and reliable designs, and the last four properties lead to small silicon area. Throughout the remainder of this section, several bit-parallel architectures are discussed for the 1-D DWT analysis and synthesis filter banks, and these architectures are evaluated in terms of the properties listed

Table 13.1 Schedule for computing a three-level analysis filter bank as described by (1) and (2) for $i = 0, 1, 2$. The schedule for one period of computation can be determined by substituting a nonnegative integer k into the table.

Time Unit	$H(z)$ Output	$G(z)$ Output
$8k+0$	$S_1(4k+0)$	$W_1(4k+0)$
$8k+1$	$S_2(2k+0)$	$W_2(2k+0)$
$8k+2$	$S_1(4k+1)$	$W_1(4k+1)$
$8k+3$	$S_3(k)$	$W_3(k)$
$8k+4$	$S_1(4k+2)$	$W_1(4k+2)$
$8k+5$	$S_2(2k+1)$	$W_2(2k+1)$
$8k+6$	$S_1(4k+3)$	$W_1(4k+3)$
$8k+7$	-	-

above. First, architectures based on direct-form FIR filter structures are examined. These architectures are divided into two types based on how the intermediate results $S_i(n)$ are stored. In the first type of architecture, a single memory unit is used to store these results. In the second type of architecture, the memory is distributed throughout the architecture. This is followed by a description of architectures which are based on a lattice filter structure. All of the 1-D DWT architectures discussed in this chapter have silicon area which is independent of input signal length; therefore, this property is not mentioned in this section.

13.3.2.1 Direct-Form Architectures with Single Memory Unit

When a single memory unit is used, the architecture has the general structure shown in Fig. 13.7. This figure shows the architecture for a three-level wavelet decomposition which computes (1) and (2) for $i = 0, 1, 2$ according to the schedule in Table 13.1. The input $S_0(n)$ is input directly into the memory unit. During even clock cycles, the input $S_0(n)$ is fed into the lowpass filter ($H(z)$) to compute $S_1(n)$ and the highpass filter ($G(z)$) to compute $W_1(n)$. The signal $W_1(n)$ is routed to the output, and $S_1(n)$ is routed back to the memory unit. During cycles $4k+1$, $S_2(n)$ and $W_2(n)$ are computed by the lowpass and highpass filters, respectively, and $W_2(n)$ is output while $S_2(n)$ is routed back to the memory unit. During cycles $8k+3$, $S_3(n)$ an $W_3(n)$ are computed and sent to the output. An alternative to the output demultiplexer in this design would be an output multiplexer which multiplexes all wavelet coefficients to the same bus. This is possible because the wavelet transform has the same number of inputs and outputs samples. We now discuss some implementations of the memory unit in this architecture.

One implementation of the memory unit uses a FIFO of length L for each of the signals $S_i(n)$ [13]. For example, the memory unit in Fig. 13.7 would consist of three FIFOs, one each for $S_0(n)$, $S_1(n)$, and $S_2(n)$, as shown in Fig. 13.8. Each FIFO requires its own shift enable signal because the data moves through each FIFO at a different rate, i.e., the FIFO holding $S_i(n)$ is shifted with twice the frequency as the FIFO holding $S_{i+1}(n)$. In addition to the three FIFOs, the memory unit contains L multiplexers to select the correct inputs to the filters. This architecture

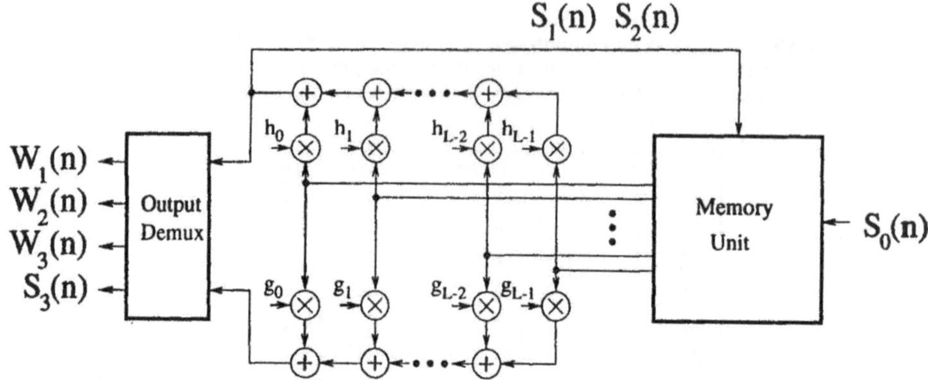

Figure 13.7 The general structure of a direct-form bit-parallel DWT architecture with a single memory unit.

is scalable, modular, and has good memory and hardware utilization. Its major disadvantage is its complex routing.

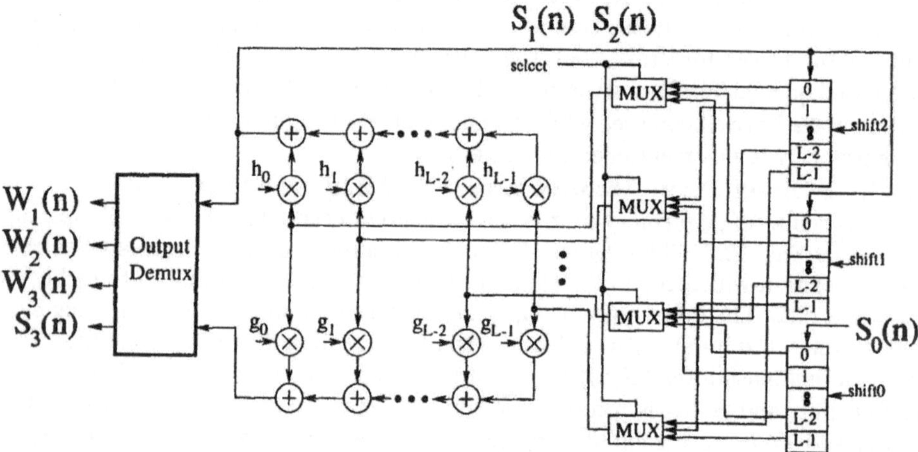

Figure 13.8 A DWT architecture which uses a single memory unit which consists of J FIFOs and L multiplexers.

Another implementation has been proposed which uses the minimum possible number of registers required to implement the memory unit [15]. This minimum number of registers is determined using life-time analysis and the memory unit is designed using forward-backward register allocation [16]. The architecture for the analysis filter bank is shown in Fig. 13.9. The filters in this architecture are pipelined by two stages. As a result, the architecture implements the schedule in Table 13.2 rather than the schedule in Table 13.1. The advantages of this architecture are efficient memory usage and high hardware utilization efficiency. Some disadvantages are that it is not scalable or modular and it requires complex routing of signals.

Table 13.2 Schedule for the three-level analysis filter bank described by (1) and (2) for $i = 0, 1, 2$ where the latency of each filter is two cycles. The schedule for one period of computation can be determined by substituting a nonnegative integer k into the table.

Time Unit	$H(z)$ Output	$G(z)$ Output
$8k+0$	$S_1(4k+0)$	$W_1(4k+0)$
$8k+1$	–	–
$8k+2$	$S_1(4k+1)$	$W_1(4k+1)$
$8k+3$	$S_2(2k+0)$	$W_2(2k+0)$
$8k+4$	$S_1(4k+2)$	$W_1(4k+2)$
$8k+5$	$S_3(k)$	$W_3(k)$
$8k+6$	$S_1(4k+3)$	$W_1(4k+3)$
$8k+7$	$S_2(2k+1)$	$W_2(2k+1)$

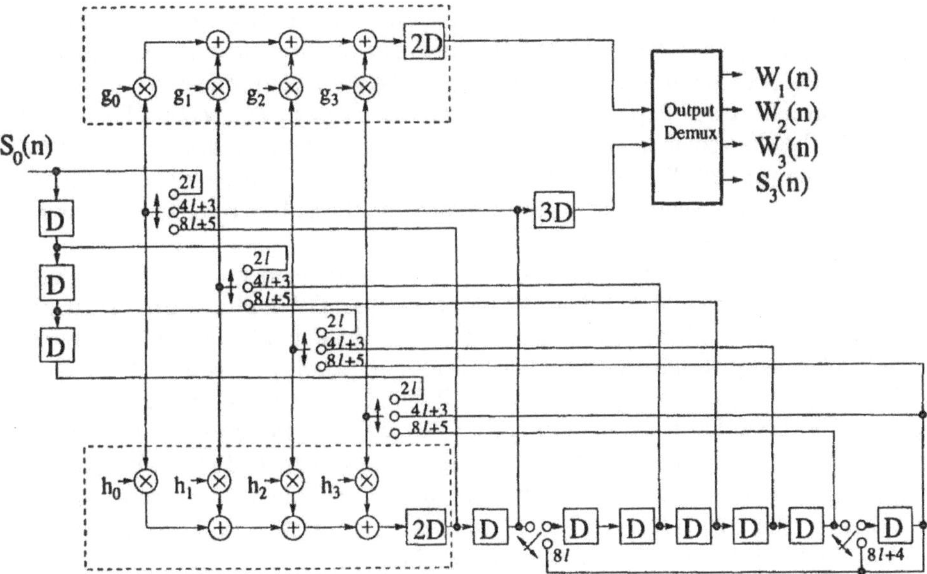

Figure 13.9 Three-level analysis wavelet architecture which uses the minimum number of registers. This architecture makes use of two-stage pipelined filter modules and implements the schedule shown in Table 13.2.

Another possibility is to use a register file or RAM to implement the memory unit. This can reduce switching activity in the memory unit because the intermediate results stay in the same memory location rather than moving through registers as shown in Fig.s 13.8 and 13.9. The reduced switching activity leads to reduced power consumption in the memory unit [17]. Some disadvantages of this architecture are that the register file requires multiple read ports [18] and address signals must be generated for the memory unit.

Other proposals for the memory unit include systolic and semi-systolic routing networks [18]. These have the advantage of a modular hardwired memory unit, but the memory elements and the multipliers and adders are underutilized, and the control of the routing networks is not trivial.

The same concepts that are used in this section for the analysis filter bank architectures can also be used to derive architectures for synthesis filter banks. While some synthesis filter bank architectures have been proposed, past research has focused mainly on analysis filter bank architectures. It turns out, however, that architectures for the synthesis filter banks can be quite similar to the architectures for the analysis filter banks. This is demonstrated in the next section where architectures with distributed memory are presented for both the analysis and synthesis filter banks. As we shall see, these architectures have many nice properties which allow for efficient VLSI implementations.

13.3.2.2 Direct-Form Architectures with Distributed Memory

Architectures with distributed memory are presented for both the analysis and synthesis DWT filter banks. Traditionally, such architectures have been designed using *ad hoc* design techniques; however, it has recently been shown that systematic design techniques can be used to design these architectures [19]. In this section, the methodologies used to design distributed memory architectures are not discussed; however, the architectures are quite simple so the interested reader can investigate how the architectures implement the desired algorithms. The section begins with a simple distributed memory architecture which implements the analysis DWT computation (1) and (2) using the schedule in Table 13.1. This architecture is then transformed into two new architectures. Three synthesis DWT architectures which implement (3) are also presented.

The architecture in Fig. 13.10 is a distributed memory architecture which implements the schedule in Table 13.1 for an analysis DWT filter bank. This architecture is conceptually similar to the architecture in Fig. 13.8. Both architectures consist of a highpass filter $G(z)$, a lowpass filter $H(z)$, and three FIFOs which are used to store the signals $S_i(n)$ for $i = 0, 1, 2$, but the distributed memory architecture in Fig. 13.10 has simpler routing than the architecture in Fig. 13.8 because of the way the memory is organized. The architecture in Fig. 13.10 is scalable, modular, has good hardware utilization efficiency, and has simple routing and control. As it is shown, the architecture in Fig. 13.10 uses more registers than the architecture in Fig. 13.8, but the number of registers can be reduced to the same number as in Fig. 13.8 by replacing each instance of $4D$ and $2D$ by a single register controlled by an appropriate shift signal.

The architecture in Fig. 13.10 executes the schedule in Table 13.1. During clock cycles $2k + 0$, the multiplexer control (MC) signal has the value x and $S_0(n)$ is switched into the filters to compute $S_1(n)$ and $W_1(n)$. The signal $S_1(n)$ is routed back into the circuit, and during clock cycles $4k + 1$, MC has value y and $S_1(n)$ is switched into the filters to compute $S_2(n)$ and $W_2(n)$. The signal $S_2(n)$ is routed back into the circuit, and during clock cycles $8k + 3$, MC has value z and $S_2(n)$ is switched into the filters to compute $S_3(n)$ and $W_3(n)$.

The architecture in Fig. 13.10 has a long combinational path which passes through one multiplier and four adders. This path limits the maximum clock rate of the circuit, which may become a problem when the number of taps in $H(z)$ and $G(z)$ is large. One way to resolve this is to change the locations of some of the

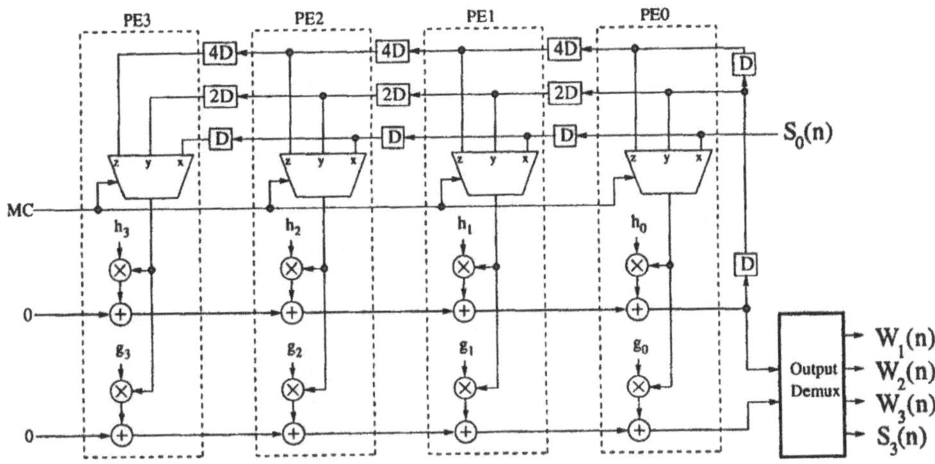

Figure 13.10 A three-level analysis DWT architecture which uses four-tap filters and implements the schedule in Table 13.1. The mux control (MC) signal repeats the sequence $x, y, x, z, x, y, x, \phi$ starting at time $= 0$, where ϕ denotes "don't care."

registers in the circuit so the path through the adders contains some registers. In the following, a simple *retiming* transformation [20] is described which allows us to move registers into the path through the adders in Fig. 13.10, therefore increasing the maximum clock speed of the architecture.

Retiming is a technique used to change the locations of registers in a circuit without affecting the input/output characteristics of the circuit. To retime the architecture in Fig. 13.10, we can simply examine all connections between two adjacent processing elements (PEs) and remove one delay from each connection which goes from right-to-left and add one delay to each connection which goes from left-to-right. This technique is known as *cutset retiming.* For example, consider the connections between PE1 and PE2 in Fig. 13.10. Starting from the top of the figure, there are six connections which contain 4, 2, 1, 0, 0, and 0 sample delays. If we remove one delay from each connection which goes from right-to-left and add one delay to each connection which goes from left-to-right, then the connections contain $4-1=3$, $2-1=1$, $1-1=0$, $0+1=1$, $0+1=1$, and $0+1=1$ delays. Performing cutset retiming on all connections between each adjacent pair of PEs in Fig. 13.10 results in the architecture shown in Fig. 13.11.

It is interesting to compare the architecture in Fig. 13.10 with the architecture in Fig. 13.11. The architecture in Fig. 13.10 has the advantage that none of the inputs to the filters are broadcast to more than one multiplier and the disadvantage of the long combinational path through one multiplier and four adders. On the other hand, the architecture in Fig. 13.11 eliminates the long combinational path through the adder chain but has inputs $S_0(n)$ which are broadcast to all of the PEs in the architecture, which means that the input lines have large routing capacity and must drive many gates.

If the long adder chain in Fig. 13.10 and the broadcast inputs in Fig. 13.11 are both unacceptable, there are two possible alternatives. One possibility is to use a hybrid architecture where cutset retiming is performed between some but not all

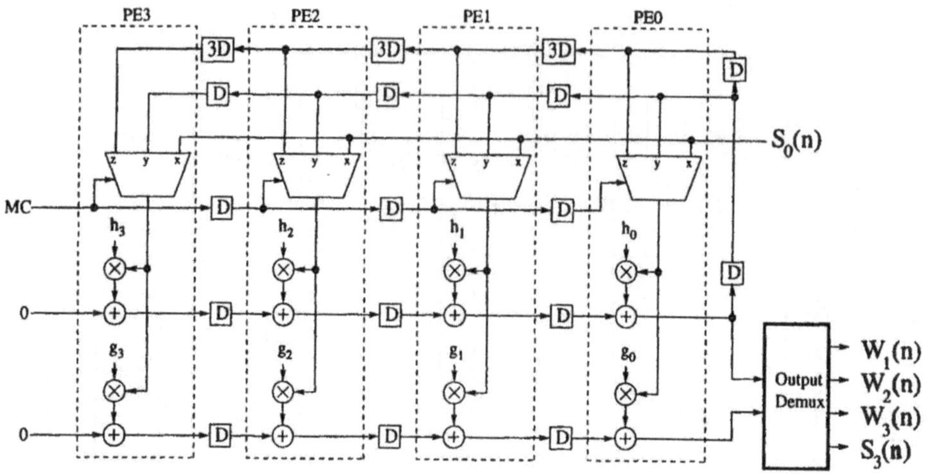

Figure 13.11 A three-level analysis DWT architecture which is derived from the architecture in Fig. 13.10 using cutset retiming. The mux control (MC) signal repeats the sequence $x, y, x, z, x, y, x, \phi$ starting at time = 0, where ϕ denotes "don't care."

adjacent pairs of PEs. For example, if cutset retiming is only performed between PE1 and PE2 in Fig. 13.10, the resulting architecture has a critical path which passes through one multiplier and two adders, and the inputs of the architecture are broadcast to only two PEs. The second solution is to use a fully systolic architecture which can be derived by replacing each delay element in Fig. 13.10 with two delay elements (this is known as creating a *2-slow* circuit) and then performing cutset retiming as before. The resulting architecture is shown in Fig. 13.12. The input is also 2-slow, i.e., input samples $S_0(n)$ are input only during even clock cycles. This architecture has the advantage of being fully systolic, which eliminates both the long combinational path through the adder chain *and* eliminates the broadcast inputs, but it has the disadvantage that its hardware utilization efficiency is only one-half that of the architectures in Fig.s 13.10 and 13.11.

Architectures for the synthesis filter bank shown in Fig. 13.1(b) and described by (3) are now considered. Due to the organization of the computations in (3), the synthesis filter bank uses two filters: an even filter and an odd filter. The even filter consists of the even-indexed filter taps from $H'(z)$ and $G'(z)$, and the odd filter consists of the odd-indexed filter taps. For example, when the synthesis filters $H'(z)$ and $G'(z)$ are third-order FIR filters, the even filter has coefficients $g'(0)$, $h'(0)$, $g'(2)$, and $h'(2)$, and the odd filter has coefficients $g'(1)$, $h'(1)$, $g'(3)$, and $h'(3)$. The computations in (3) can be time-multiplexed to the even and odd filters for all levels of wavelet synthesis.

A distributed memory architecture for the synthesis filter bank is shown in Fig. 13.13. This architecture is similar to the analysis DWT architecture in Fig. 13.10. As a result, this architecture has the same properties as the analysis DWT architecture in Fig. 13.10, including the combinational path through the adders of the even filter and the odd filter. Retiming is used to eliminate this long combinational path in the same manner as for the analysis architecture.

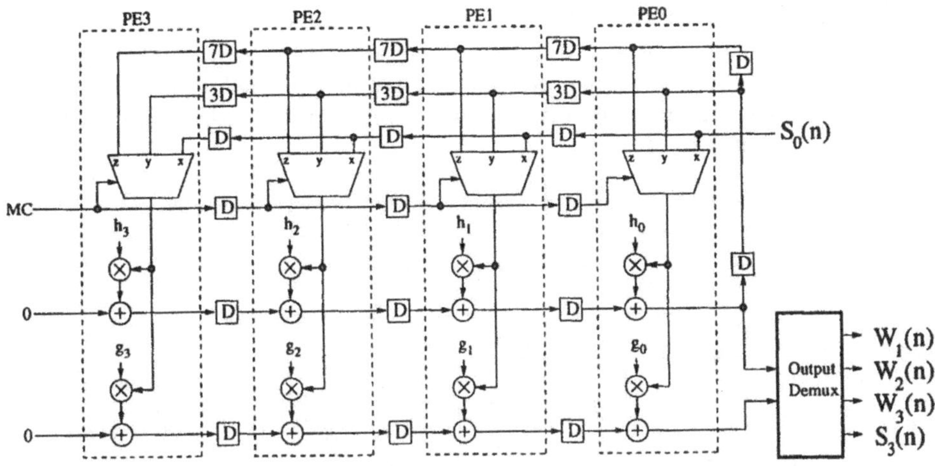

Figure 13.12 A fully systolic three-level analysis DWT architecture which is derived from the architecture in Fig. 13.10 using 2-slow circuit design and cutset retiming. The mux control (MC) signal repeats the sequence $x, \phi, y, \phi, x, \phi, z, \phi, x, \phi, y, \phi, x, \phi, \phi, \phi$ starting at time = 0, where ϕ denotes "don't care."

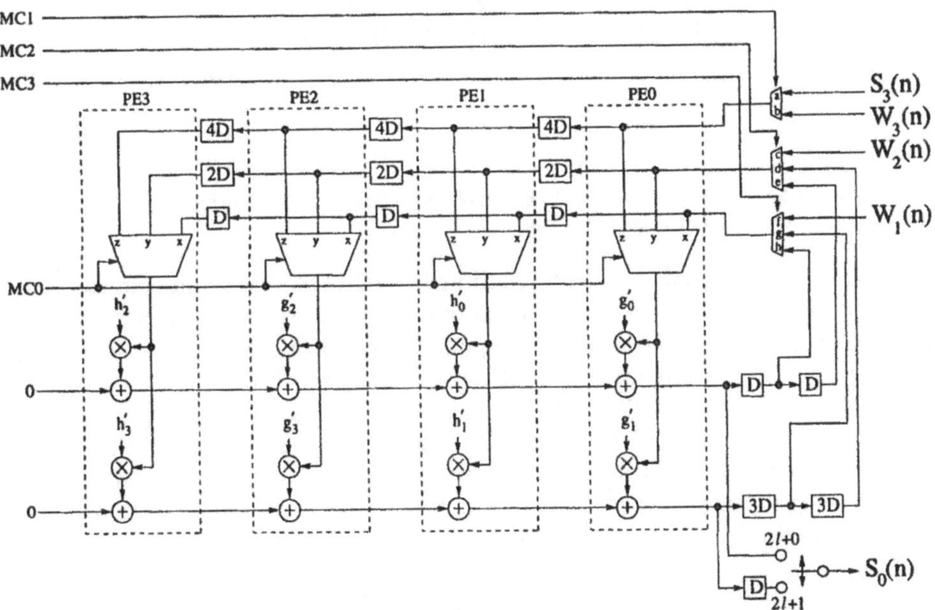

Figure 13.13 A three-level synthesis DWT architecture. The multiplexer control signals are periodic with period 8 and have the values $MC0 = x, \phi, x, y, x, z, x, y$ and $MC1 = \phi, b, \phi, \phi, \phi, a, \phi, \phi$ and $MC2 = \phi, c, \phi, d, \phi, c, \phi, e$ and $MC3 = h, f, g, f, h, f, g, f$.

A second architecture for the synthesis DWT can be designed by performing cutset retiming on the architecture in Fig. 13.13. This architecture is shown in Fig. 13.14, and has the same properties as the analysis DWT architecture in Fig. 13.11, namely, broadcast inputs and short combinational paths which allow for higher clock frequencies.

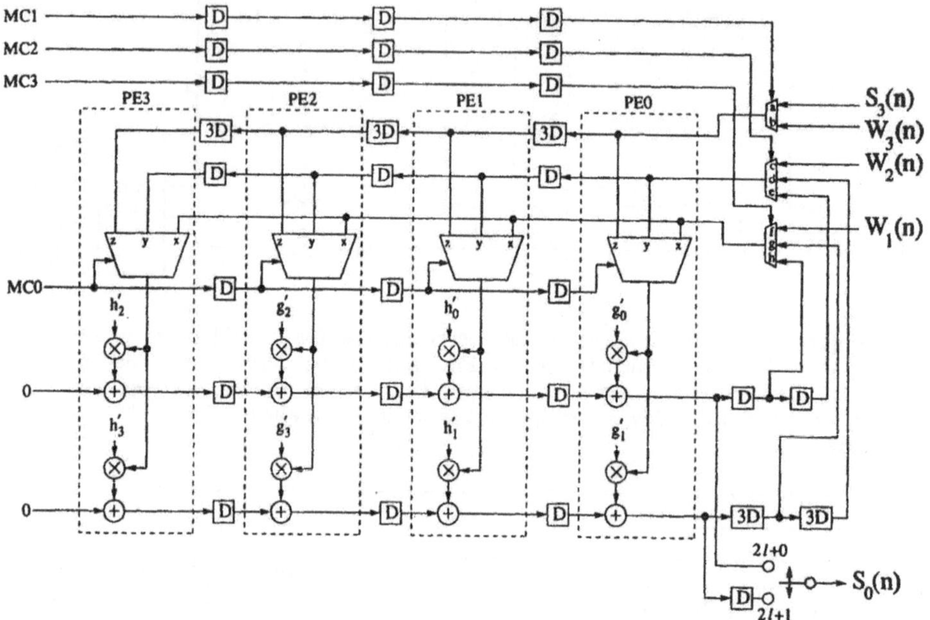

Figure 13.14 A three-level synthesis DWT architecture. The multiplexer control signals are the same as in Fig. 13.13.

If long combinational paths through adders and broadcast inputs in the synthesis DWT architecture are both unacceptable, there are two alternatives. One alternative is to perform cutset retiming at selected locations in the architecture, and the other is to design a fully systolic architecture. The fully systolic architecture for the synthesis DWT can be designed by transforming the architecture in Fig. 13.13 into a 2-slow circuit and then using cutset retiming. The resulting architecture, which is fully systolic, is shown in Fig. 13.15. It has the advantages of no broadcast inputs and short critical path, but its disadvantage is that it has only 50% of the hardware utilization efficiency of the architectures in Fig. 13.13 and 13.14.

In some cases, it may be desirable to have a single architecture which computes both the DWT and IDWT. Such architectures have been explored in the past [21]. The distributed memory architectures discussed in this section are well-suited for such an implementation. For example, the analysis architecture in Fig. 13.10 and the synthesis architecture in Fig. 13.13 are quite similar and could easily be integrated into a single architecture which can compute the DWT or IDWT.

13.3.2.3 Lattice Structure Based Bit-Parallel Architectures

Lattice structure based DWT architectures can be used to implement the analysis and synthesis filter banks shown in Fig. 13.1 [22]. The lattice structure

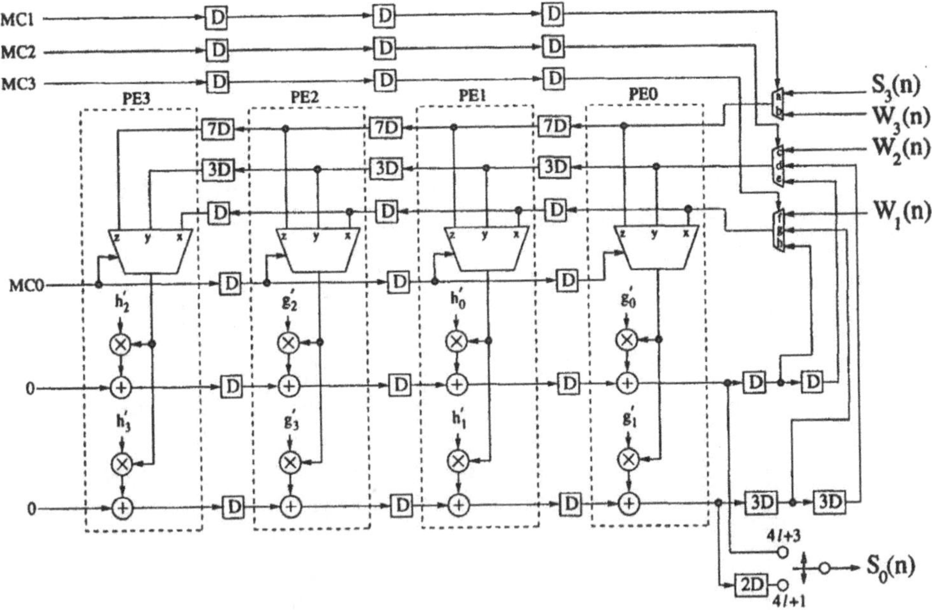

Figure 13.15 A fully systolic three-level synthesis DWT architecture. The multiplexer control signals are periodic with period 16 and have the values $MC0 = x, \phi, \phi, \phi, x, \phi, y, \phi, x, \phi, z, \phi, x, \phi, y, \phi$ and $MC1 = \phi, \phi, b, \phi, \phi, \phi, \phi, \phi, \phi, \phi, a, \phi, \phi, \phi, \phi, \phi$ and $MC2 = e, \phi, c, \phi, d, \phi, c, \phi, e, \phi, c, \phi, d, \phi, c, \phi$ and $MC3 = \phi, \phi, f, \phi, \phi, \phi, g, \phi, \phi, \phi, f, \phi, \phi, \phi, h, \phi$.

based architectures can be more area efficient than direct-form structure based architectures because the lattice structure uses less multipliers and adders. The architectures discussed in this section are based on the quadrature mirror filter (QMF) lattice structure introduced in [23] and discussed in [10]. While the QMF lattice cannot be used to implement the DWT in all cases, it can be used to implement the orthonormal DWT, which includes DWT computation when the popular Daubechies wavelets [2] are used.

The QMF lattice structure equivalent to the three-level analysis filter bank in Fig. 13.1(a) is shown in Fig. 13.16 for the case when $H(z)$ and $G(z)$ are four-tap filters. The parameters α_0, α_1, and S in the figure can be computed from $H(z)$ and $G(z)$ (see [10]). While a direct-form implementation of $H(z)$ and $G(z)$ would require 8 multipliers and 6 adders, the lattice structure implementation of these filters requires only 6 multipliers and 4 adders. This savings in multipliers and adders in the algorithm description translates into fewer functional units in the architecture.

An architecture for implementing the QMF lattice in Fig. 13.16 is shown in Fig. 13.17. For the interested reader, details on the design of this architecture can be found in [24]. In this architecture, the three levels of wavelet decomposition are time-multiplexed to the same hardware. The architecture basically consists of two scaling multipliers, two identical lattice stages, and some memory and control circuits. The lattice based architecture has the advantages of modularity (uses

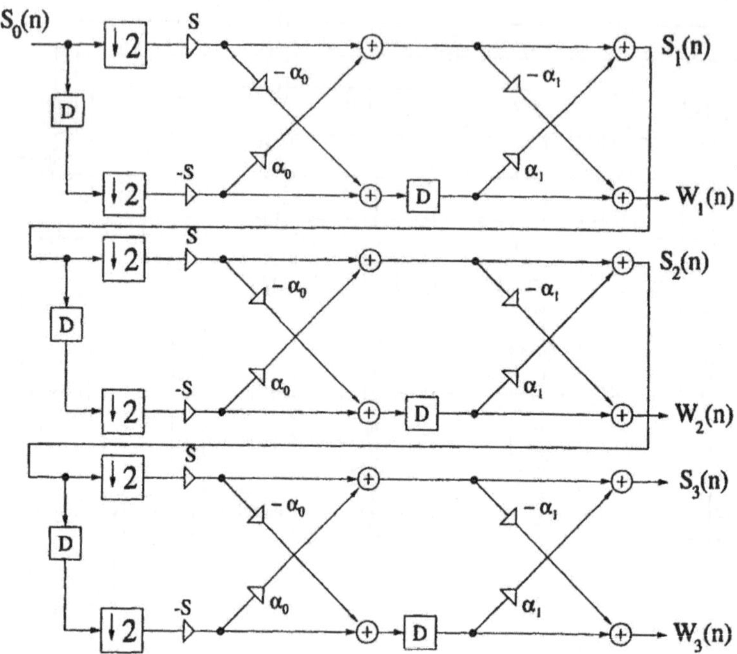

Figure 13.16 The QMF lattice structure which implements a three-level orthonormal DWT analysis filter bank for the case when $H(z)$ and $G(z)$ are four-tap filters.

identical lattice stages), scalability (longer filters need more lattice stages), high hardware utilization, and area which is independent of the length of the input signal. Furthermore, it uses fewer adders and multipliers than a functionally equivalent direct-form based architecture. A disadvantage of the architecture is that long wires are used to feed the approximation signals $S_1(n)$ and $S_2(n)$ back into the filter structure, but this disadvantage can be reduced with proper floorplanning. Further information on using algorithmic properties of the DWT to design efficient architectures can be found in [25].

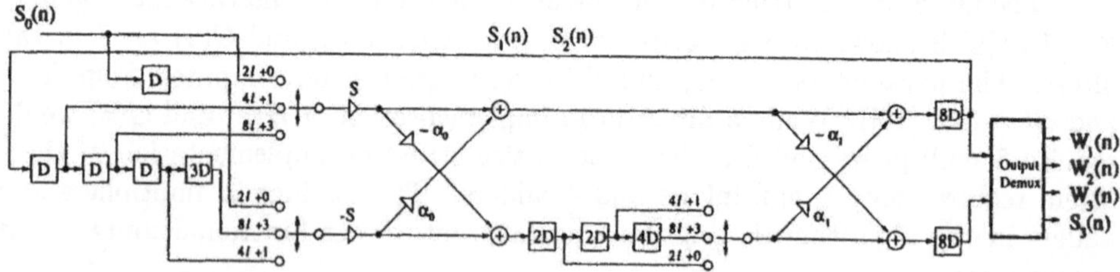

Figure 13.17 A lattice structure based architecture for a three-level orthonormal DWT. The two sets of 8 registers at the right of the figure can be distributed using retiming to increase the clock rate of the circuit.

13.3.3 Digit-Serial Architectures

To improve the hardware utilization efficiency and to reduce the routing and interconnection overhead in the bit-parallel architectures presented thus far, digit-serial architectures can be used to implement analysis and synthesis wavelet filter banks shown in Fig. 13.1. A digit-serial processor processes more-than-one but not-all bits of a word or sample [26]-[28]. The number of bits processed per cycle is referred to as the ***digit-size***. If the digit-size is 1, the architecture is bit-serial, and if the digit-size is the same as the word-length, then the architecture is bit-parallel.

The digit-serial architecture closely resembles the algorithmic structure of the DWT. For the three-level wavelet, the four output signal wires in the analysis wavelet algorithm carry the signals $W_1(n)$, $W_2(n)$, $W_3(n)$, and $W_4(n)$, and these signals generate 4, 2, 1, and 1 output samples every 8 cycles, respectively. Taking this into consideration, it seems natural to use 1/2-word, 1/4-word, 1/8-th word, and 1/8-th word digit-sizes in the processing of these four signal wires. This processing would always require eight clock cycles for transmission of the respective number of outputs. This type of processing can reduce the routing and interconnection cost (since this architecture only involves local interconnection) and can achieve complete hardware utilization. The drawback of this architecture is that the word-length must be multiple of 8 or 16 for 3-level or 4-level wavelet cases, respectively. Another drawback is the increase in the system latency of the architecture. Since the different wavelet levels are realized with different digit-sizes and a single clock is used in the entire architecture, this architecture can be considered a single-clock, non-uniform style architecture.

The i-th level wavelet is implemented using digit-size $W/2^i$ where W is the word-length and $i = 1, 2, 3$ for a 3-level wavelet. Fig. 13.18(a) shows the high-level structure for a three-level DWT analysis filter bank, and Fig. 13.18(b) shows a possible floor-plan. The filters use digit-serial implementation style, and they may be based on the direct-form filter structure [15], or on the QMF lattice structure in Fig. 13.16 when the wavelets are orthonormal [22].

The basic modules in the digit-serial architecture of Fig. 13.18(a) are the converter modules and the digit-serial filtering blocks. Digit-serial design techniques are discussed in [26]. The converter between levels i and $i+1$ in Fig. 13.18(a) converts word-serial bit-parallel data with wordlength $w/2^i$ to two-word-parallel half-word-serial format. This conversion is denoted as $(1, w/2^i) \rightarrow (2, w/2^{i+1})$ in the figure. These converters can be designed using systematic life-time analysis and register allocation schemes. Using the closed-form expressions provided in [16], the minimum number of registers required to design each of the converters in Fig. 13.18(a) can be shown to be W, which is the word-length. More detail on the converter design can be found in [15].

The advantages of using digit-serial DWT architectures are 100% hardware utilization (including register utilization) and local interconnect. These advantages lead to efficient designs in terms of silicon area. A disadvantage is that each level of wavelet filtering is different, which requires longer design time. Also, there is a restriction on the wordlength, i.e., the wordlength must be multiple of 2^J for J-level DWT.

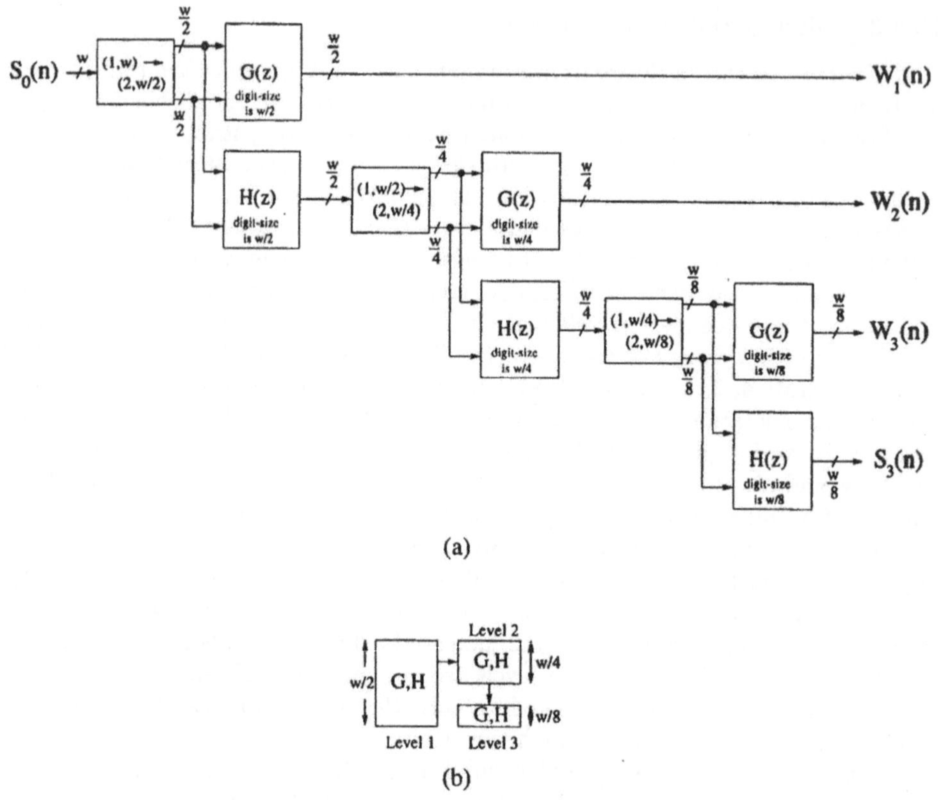

Figure 13.18 (a) The high-level digit-serial architecture block diagram of a three-level analysis wavelet decomposition. (b) The floor-plan of the digit-serial architecture.

13.4 ARCHITECTURES FOR 2-D DWT

This section discusses implementation of the separable 2-D DWT as described by (4) and shown in Fig. 13.3. This section begins with an analysis of the number of functional units (multipliers and adders) required to implement the 2-D DWT. Architectures for the 2-D DWT are then discussed. There have been fewer architectures described in the literature for 2-D DWTs than for 1-D DWTs, and the description of operation of the 2-D DWT architectures is often lengthy. For this reason, we provide a high-level description of 2-D DWT architectures and refer the reader to relevant work for details.

It is instructive to count the number of filtering operations that must be performed per input sample to compute the 2-D DWT. The row filters in the first level of computation in Fig. 13.3 perform $2(1/2) = 1$ filtering operation per input sample, where the factor $(1/2)$ is due to decimation-by-two in the horizontal direction. The column filters in the first level perform $4(1/4) = 1$ filtering operations per input sample, where the factor $(1/4)$ is due to decimation-by-two in both the horizontal and vertical directions. Therefore, the first level of computation performs two filtering operations per input sample. The second level of computation operates on one-fourth as much data as the first level (again, due to decimation in both directions), and therefore performs one-fourth as many filtering operations as the first level per input sample. Extending this argument, the number of filtering operations per input sample for a J-level 2-D DWT is upper bounded by

$$2(1 + 1/4 + 1/16 + \cdots + 1/4^{J-1}) = \frac{8}{3}\left(1 - 4^{-J}\right),$$

which is upper bounded by 8/3. Since the number of lowpass and highpass filtering operations are the same, the 2-D DWT requires a maximum of 4/3 lowpass and 4/3 highpass filtering operations per input sample.

One approach for implementing the 2-D DWT is to use two lowpass and two highpass filters. Using this approach, the filters can be implemented using bit-parallel word-serial design style, and the filter coefficients can be wired into the multipliers using canonic representation to reduce the hardware complexity. One proposed architecture based on this approach is the systolic-parallel architecture in [18]. A downfall of using 4 filters is that at most 8/3 filters are required, so the filters are idle at least

$$1 - (8/3)/4 = 1/3$$

of the time.

Another approach is to use only the number of bit-parallel filters needed and time-multiplex all filtering operations to these filters. For example, for a two-level 2-D DWT, the first level uses 1 lowpass filter and 1 highpass filter, and the second level uses 1/4 lowpass filter and 1/4 highpass filter. If we time-multiplex the second level computations onto 1/2 filter, then the architecture uses the minimum required number of filters, which is 2.5 [29]. One drawback of this technique is that the 1/2 filter which computes the second level wavelet must have programmable coefficients because it implements both highpass and lowpass wavelet filters. This technique can also result in large memory requirements [29].

A third approach is to use a combination of bit-parallel and digit-serial processing styles [15]. For example the architecture can use one bit-parallel and one half-word digit-serial filter for each highpass and lowpass types. The first level of wavelet computation is performed on the two bit-parallel filters and subsequent levels of wavelet computation are performed on the two half-word digit-serial filters. This architecture uses a total of 3 filters, and all multipliers have fixed coefficients.

In another method described in [15], the first level of wavelet computation uses bit-parallel filters, and each subsequent level of wavelet computation uses its own set of digit-serial filters. This method has the advantage of 100% hardware utilization efficiency but imposes restrictions on the wordlength that can be used.

13.5 SUMMARY

The wavelet transform is a useful tool for signal processing. The DWT can be computed efficiently using the pyramid algorithm. One way to improve upon the pyramid algorithm is to use fast FIR filtering techniques. These techniques reduce the number of operations per input sample, but these structures do not lend themselves to efficient VLSI architectures.

The design of VLSI architectures based on the pyramid algorithm is challenging because the pyramid algorithm is multirate in nature. The 1-D DWT architectures have been categorized into those with a single memory unit and those with distributed memory. Two architectures based on the single memory unit design were discussed. The architecture in Fig. 13.8 (commonly referred to as Knowles' architecture) is scalable and modular and makes efficient use of memory and functional units, but its downfall is the complex routing of signals into the multiplexers. The architecture in Fig. 13.9 was designed using a more systematic method, where the DWT computations were scheduled and then the memory unit was designed using life-time analysis and systematic register allocation techniques described in [16] and [30]. This architecture has the advantages of good hardware utilization and is optimal in terms of the number of registers used, but is not scalable and has complex routing. Notice that both architectures which use a single memory unit require complex routing of signals.

Several architectures with distributed memory were discussed. These architectures are scalable, modular, and have simple routing and control. The architecture in Fig. 13.10 operates in a manner similar to Knowles' architecture but has simpler routing because the memory is distributed. This architecture has high hardware utilization efficiency and can be easily modified to have high memory utilization efficiency. Starting with this architecture, retiming can be used to derive the architecture in Fig. 13.11 which has the advantage of a shorter critical path. The fully systolic architecture in Fig. 13.12 is derived using 2-slow circuit techniques and retiming. The advantages of this architecture are that it has a shorter critical path and no broadcast inputs, but its disadvantage is low hardware utilization. Architectures similar to those in Fig. 13.10, 13.11, and 13.12 are given in Fig. 13.13, 13.14, and 13.15, respectively. These synthesis architectures, which are used to compute the IDWT, have similar properties to the corresponding analysis architectures. Other architectures which have similar properties to the distributed memory architectures can be found in [31] [19] [18] [32].

While most DWT architectures are based on direct-form FIR filter structures, the orthonormal DWT can be implemented based on the QMF lattice structure. These architectures have reduced number of multipliers and adders because they take advantage of the structure of the QMF lattice.

Digit-serial DWT architectures provide an alternative to bit-parallel DWT architectures. The digit-serial architectures have simple routing because there is hardware dedicated to each level of wavelet filtering. Like the bit-parallel architectures, the digit-serial architectures use a single clock to control the entire circuit. A disadvantage of digit-serial architectures is that there is a restriction on the wordlength which can be used.

The two-dimensional DWT is typically computed using the pyramid algorithm with separable 2-D wavelet filters. Designing architectures for the 2-D DWT is particularly challenging because the 2-D DWT is a multidimensional and multirate algorithm. Several architectures for computing the 2-D DWT are outlined. These designs use a variety of techniques to reduce the number of functional units and the amount of memory required.

REFERENCES

[1] O. Rioul and M. Vetterli, "Wavelets and signal processing," *IEEE Signal Processing Magazine*, pp. 14–38, October 1991.

[2] I. Daubechies, "Orthonormal bases of compactly supported wavelets," *Comm. in Pure and Applied Math.*, vol. 41, pp. 909–996, November 1988.

[3] S. G. Mallat, "A theory for multiresolution signal decomposition: The wavelet representation," *IEEE Transactions on Pattern Analysis and Machine Intelligence*, vol. 11, pp. 674–693, July 1989.

[4] S. G. Mallat, "Multifrequency channel decompositions of images and wavelet models," *IEEE Transactions on Acoustics, Speech, and Signal Processing*, vol. 37, pp. 2091–2110, December 1989.

[5] G. Strang, "Wavelets and dilation equations: A brief introduction," *SIAM Rev.*, vol. 31, pp. 614–627, December 1989.

[6] I. Daubechies, "The wavelet transform, time-frequency localization and signal analysis," *IEEE Transactions on Information Theory*, vol. IT-36, pp. 961–1005, September 1990.

[7] M. Vetterli and J. Kovačević, *Wavelets and Subband Coding*. Englewood Cliffs, NJ: Prentice Hall PTR, 1995.

[8] J. N. Bradley and C. M. Brislawn, "The wavelet/scalar quantization compression standard for digital fingerprint images," in *Proceedings of IEEE ISCAS*, vol. 3, (London, England), pp. 205–208, May 1994.

[9] O. Rioul and P. Duhamel, "Fast algorithms for discrete and continuous wavelet transforms," *IEEE Transactions on Information Theory*, vol. 38, pp. 569–586, March 1992.

[10] P. P. Vaidyanathan, *Multirate Systems and Filter Banks*. Englewood Cliffs, NJ: Prentice Hall, 1993.

[11] C. Chakrabarti and M. Vishwanath, "Efficient realizations of the discrete and continuous wavelet transforms: From single chip implementations to mappings on SIMD array computers," *IEEE Transactions on Signal Processing*, vol. 43, pp. 759–771, March 1995.

[12] Z.-J. Mou and P. Duhamel, "Short-length FIR filters and their use in fast nonrecursive filtering," *IEEE Transactions on Signal Processing*, vol. 39, pp. 1322–1332, June 1991.

[13] G. Knowles, "VLSI architecture for the discrete wavelet transform," *Electronics Letters*, vol. 26, pp. 1184–1185, July 1990.

[14] M. Vishwanath, "The recursive pyramid algorithm for the discrete wavelet transform," *IEEE Transactions on Signal Processing*, vol. 42, pp. 673–676, March 1994.

[15] K. K. Parhi and T. Nishitani, "VLSI architectures for discrete wavelet transforms," *IEEE Transactions on VLSI Systems*, vol. 1, pp. 191–202, June 1993.

[16] K. K. Parhi, "Systematic synthesis of DSP data format converters using lifetime analysis and forward-backward register allocation," *IEEE Transactions on Circuits and Systems–II: Analog and Digital Signal Processing*, vol. 39, pp. 423–440, July 1992.

[17] A. Chandrakasan, S. Sheng, and R. Brodersen, "Low-power CMOS digital design," *IEEE Journal of Solid-State Circuits*, vol. 27, pp. 473–484, April 1992.

[18] M. Vishwanath, R. M. Owens, and M. J. Irwin, "VLSI architectures for the discrete wavelet transform," *IEEE Transactions on Circuits and Systems–II: Analog and Digital Signal Processing*, vol. 42, pp. 305–316, May 1995.

[19] J. Fridman and E. Manolakos, "Discrete wavelet transform: Data dependence analysis and synthesis of distributed memory and control array architectures," *IEEE Transactions on Signal Processing*, vol. 45, pp. 1291–1308, May 1997.

[20] C. Leiserson, F. Rose, and J. Saxe, "Optimizing synchronous circuitry by retiming," *Third Caltech Conference on VLSI*, pp. 87–116, 1983.

[21] S. Syed and M. Bayoumi, "A scalable architecture for discrete wavelet transform," in *Proceedings of Computer Architecture for Machine Perception*, pp. 44–50, September 1995.

[22] T. C. Denk and K. K. Parhi, "VLSI architectures for lattice structure based orthonormal discrete wavelet transforms," *IEEE Transactions on Circuits and Systems–II: Analog and Digital Signal Processing*, vol. 44, pp. 129–132, February 1997.

[23] P. P. Vaidyanathan and P. Hoang, "Lattice structures for optimal design and robust implementation of two-channel perfect reconstruction QMF banks," *IEEE Transactions on Acoustics, Speech, and Signal Processing*, vol. ASSP-36, pp. 81–94, January 1988.

[24] T.C. Denk and K.K. Parhi, "Synthesis of Folded Pipelined Architectures for Multirate DSP Algorithms," *IEEE Trans. on VLSI Systems*, **6**(4), Dec. 1998

[25] S. Simon, P. Rieder, and J. A. Nossek, "Efficient VLSI suited architectures for discrete wavelet transforms," in *VLSI Signal Processing, IX* (W. Burleson, K. Konstantinides, and T. Meng, eds.), pp. 388–397, IEEE Press, October 1996.

[26] K. K. Parhi, "A systematic approach for design of digit-serial signal processing architectures," *IEEE Transactions on Circuits and Systems*, vol. 38, pp. 358–375, April 1991.

[27] R. Hartley and P. Corbett, "Digit-serial processing techniques," *IEEE Transactions on Circuits and Systems*, vol. 37, pp. 707–719, June 1990.

[28] S. G. Smith and P. B. Denyer, *Serial Data Computation.* Boston, MA: Kluwer Academic, 1988.

[29] J. C. Limqueco and M. A. Bayoumi, "A scalable architecture for 2-D discrete wavelet transform," in *VLSI Signal Processing, IX* (W. Burleson, K. Konstantinides, and T. Meng, eds.), pp. 369–377, IEEE Press, October 1996.

[30] K. K. Parhi, "Video data format converters using minimum number of registers," *IEEE Transactions on Circuits and Systems for Video Technology*, vol. 2, pp. 255–267, June 1992.

[31] A. Grzeszczak, M. Mandal, S. Panchanathan, and T. Yeap, "VLSI implementation of the discrete wavelet transform," *IEEE Transactions on VLSI Systems*, vol. 4, pp. 421–433, December 1996.

[32] H. Chuang, H. Kim, and C.-C. Li, "Systolic architecture for discrete wavelet transforms with orthornormal bases," in *Proceedings SPIE Conf. on Applications of Artificial Intelligence X: Machine Vision and Robotics*, vol. 1708, pp. 157–164, April 1992.

[33] V. Sundararajan and K.K. Parhi, "Synthesis of Folded, Pipelined Architectures for Multi-Dimensional Multirate Systems," *in Proc. of IEEE Int. Conf. on Acoustics, Speech and Signal Processing*, pp. 3089-3092, May 1998, Seattle

Chapter 14

DCT Architectures

Ching-Yu Hung
Texas Instruments
Dallas, Texas
cy-hung@hc.tc.com

14.1 INTRODUCTION

The Discrete Cosine Transform (DCT) first appeared in Ahmed, Natarajan, and Rao's pioneering paper in 1974 [1]. It has become one of the most important transformation in signal processing, second only to the Discrete Fourier Transform (DFT). DCT and its inverse transform, IDCT, are essential operations in audio, image, video coding, and have been included in several image/audio/video standard specifications, including JPEG still image compression [2], MPEG-1 Audio [3], MPEG-1/2 Video [4, 5], Dolby AC-3 Audio [6], and the ATSC Standard for HDTV [7].

There have been many proposals on algorithms and implementations of DCT/IDCT in the literature, probably due to the richness of its computational variations as much as the importance of its practical applications. This chapter provides an overview of DCT algorithms and different types of architectures. Within each type, a few representative papers are briefly described and contrasted, and a handful of interesting design cases are discussed in depth.

Definition and algorithms for DCT/IDCT are covered in Section 14.2. Fundamental techniques in reducing computation complexity are discussed, both for one-dimensional and two-dimensional transforms. DCT/IDCT architectures are reviewed in Section 14.3. Several different styles are described and contrasted. Small numerical examples are shown as appropriate to illustrate the concept in these proposals. Discussion of the future trends of DCT implementation in Section 14.4 concludes this chapter.

14.2 DCT ALGORITHMS

14.2.1 Definitions

There are a few variations of DCT and IDCT in the literature (see also Section 2.3.1). In this report, one-dimensional DCT (1-D DCT) is defined as

$$X(m) = u(m)\sqrt{\frac{2}{N}}\sum_{i=0}^{N-1} x(i)\cos\frac{(2i+1)m\pi}{2N}, \text{ for } m = 0, 1, \ldots, N-1, \quad (1)$$

$$\text{where } u(m) = \begin{cases} 1 & \text{for } m = 0; \\ \frac{1}{\sqrt{2}} & \text{otherwise.} \end{cases}$$

Input vector $x(i)$ represents the time domain (or spatial) signal, and the output vector $X(m)$ represents the frequency-domain signal.

The corresponding inverse transform, 1-D IDCT, is

$$x(i) = \sqrt{\frac{2}{N}}\sum_{m=0}^{N-1} u(m)X(m)\cos\frac{(2i+1)m\pi}{2N}. \quad (2)$$

Most papers on DCT/IDCT adopt the above definitions. However, some use an alternative definition of DCT, sometimes called the type-IV DCT [8]:

$$X(m) = \sqrt{\frac{2}{N}}\sum_{i=0}^{N-1} x(i)\cos\frac{(2i+1)(2m+1)\pi}{4N}, \text{ for } m = 0, 1, \ldots, N-1. \quad (3)$$

The two-dimensional DCT (hereinafter, 2-D DCT) is defined as

$$X(m,n) = u(m)u(n)\frac{2}{N}\sum_{i=0}^{M-1}\sum_{j=0}^{N-1} x(i,j)\cos\frac{(2i+1)m\pi}{2M}\cos\frac{(2j+1)n\pi}{2N}, \quad (4)$$
$$\text{for } m = 0, 1, \ldots, M-1, n = 0, 1, \ldots, N-1.$$

Here we have the input 2-D array $x(i,j)$ representing the spatial signal (e.g., an image), and the output 2-D array $X(m,n)$ representing the frequency-domain signal.

The corresponding inverse transform, 2-D IDCT, is

$$x(i,j) = \frac{2}{N}\sum_{m=0}^{M-1}\sum_{n=0}^{N-1} u(m)u(n)X(m,n)\cos\frac{(2i+1)m\pi}{2M}\cos\frac{(2j+1)n\pi}{2N}, \quad (5)$$
$$\text{for } i = 0, 1, \ldots, M-1, j = 0, 1, \ldots, N-1.$$

Most practical applications of the 2-D transforms are square, i.e., $M = N$. While the input image is not necessarily square, it is partitioned into a number of small square blocks, 8 by 8 or 16 by 16 being the most common, and a 2-D transform is performed on each block.

Among the standard specifications, the MPEG-1 and MPEG-2 Video [4, 5] as well as JPEG [2] specify 8×8 2-D IDCT in (5) with $M = N = 8$. Layers I and

II of MPEG-1 Audio standard [3] specify a variation of 1-D DCT in the matrixing step of audio decoding,

$$V(m) = \sum_{i=0}^{31} S(i) \cos \frac{(m+16)(2i+1)\pi}{64}, \text{ for } m = 0, 1, \ldots, 63,$$

which can be computed with a 32-point 1-D DCT plus some postprocessing. The Dolby AC-3 Audio standard [6] specifies yet another variation of DCT in the encoding process [6],

$$X(m) = \frac{-2}{N} \sum_{i=0}^{N-1} x(i) \cos((2m+1)(2i + \frac{N}{2} + 1)\frac{\pi}{2N}),$$

where N can be either 256 or 512. This is just a minor modification of Eq. (3).

The 2-D transforms, DCT and IDCT, can be applied directly on the N^2 input data items. Algorithms that optimize the 2-D transform will be called *direct 2-D algorithms.* Alternatively, the 2-D transforms can be carried out with multiple passes of 1-D transforms. The *separability* property of 2-D DCT/IDCT allows the transform to be applied on one dimension (for example, on each row) then on the other (on each column). This approach, called the *row-column method* in the literature, requires $2N$ instances of N-point 1-D DCT to implement an $N \times N$ 2-D DCT.

The leading factors, $\sqrt{\frac{2}{N}}$ and $\frac{2}{N}$, in the equations are consistent with DCT and IDCT definitions in the MPEG-1 and MPEG-2 video specifications. Many papers dealing with DCT/IDCT algorithms use different leading factors or omit them altogether. This usually has no impact on computation complexity since in many applications these factors are powers of 2. Some algorithms also omit the DC normalization factor, $u(m)$. These would require one extra multiplication per 1-D transform, or $2N - 2$ extra multiplications per 2-D transform, to complete the transform as defined in (1) and (4).

The most straightforward approach to compute DCT or IDCT is carrying out the computation as full matrix-vector multiplications. An 1-D transform thus implemented requires N^2 multiplications and $N(N-1)$ additions. A 2-D transform requires N^4 multiplications and $N^2(N^2-1)$ additions. Although requiring the most number of operations, this method has the advantage of being very regular. It is most suitable for vector processors or deeply pipelined architectures where computational irregularities may cause under-utilization of processing resource. Most of the fast algorithms to be presented have complexity $O(N \log N)$ for 1-D transforms, and $O(N^2 \log N)$ for 2-D transforms.

14.2.2 Computing with FFT

In the original DCT paper by Ahmed, Natarajan, and Rao [1], an FFT-based method is described. An N-point DCT can be realized with a $2N$-point FFT and some postprocessing (see Section 2.3.1). Starting with the definition of DCT in (1) and ignoring the leading constant and $u(m)$, we have

$$X(m) = \sum_{i=0}^{N-1} x(i) \cos \frac{(2i+1)m\pi}{2N}$$

$$
\begin{aligned}
&= \text{Re}\left[\sum_{i=0}^{N-1} x(i)e^{-j\frac{(2i+1)m\pi}{2N}}\right] \\
&= \text{Re}\left[e^{-j\frac{m\pi}{2N}} \sum_{i=0}^{N-1} x(i)e^{-j\frac{2\pi im}{2N}}\right] \\
&= \text{Re}\left[e^{-j\frac{m\pi}{2N}} \sum_{i=0}^{N-1} F(m)\right], \qquad (6)
\end{aligned}
$$

where $F(m)$ is the $2N$-point discrete Fourier transform of the vector $[x(0), x(1), \ldots, x(N-1), 0, \ldots, 0]$, i.e., $x(i)$ padded with N zeros. The idea is to compute $F(m)$ using FFT so the computation complexity is reduced from $O(N^2)$ (using matrix-vector multiplication) to $O(N \log N)$.

Since the $2N$-point FFT in (6) takes real-value inputs, the well-known method of utilizing N-point complex FFT plus some post processing can be used to further reduce the computation. See, for example, Singleton [9] for details.

14.2.3 Even/Odd Decomposition

Even/odd decomposition is the most obvious way we can reduce the number of multiplications and additions. This algorithm utilizes the symmetry and antisymmetry in the cosine scalers. Take 8-point DCT for example. The computation written in matrix form is

$$
\mathbf{X} = \begin{bmatrix}
c_4 & c_4 & c_4 & c_4 & c_4 & c_4 & c_4 & c_4 \\
c_1 & c_3 & c_5 & c_7 & -c_7 & -c_5 & -c_3 & -c_1 \\
c_2 & c_6 & -c_6 & -c_2 & -c_2 & -c_6 & c_6 & c_2 \\
c_3 & -c_7 & -c_1 & -c_5 & c_5 & c_1 & c_7 & -c_3 \\
c_4 & -c_4 & -c_4 & c_4 & c_4 & -c_4 & -c_4 & c_4 \\
c_5 & -c_1 & c_7 & c_3 & -c_3 & -c_7 & c_1 & -c_5 \\
c_6 & -c_2 & c_2 & -c_6 & -c_6 & c_2 & -c_2 & c_6 \\
c_7 & -c_5 & c_3 & -c_1 & c_1 & -c_3 & c_5 & -c_7
\end{bmatrix} \mathbf{x} = \mathbf{Cx}, \qquad (7)
$$

where c_i denotes $\cos(i\pi/16)$.

Observe that the first and the last columns have the same magnitude with sign changes, so are the second and the second to last columns, and so on. Making use of the symmetry/antisymmetry in the left and right halves of the matrix, we can rewrite (8) as:

$$
\begin{bmatrix} X(0) \\ X(2) \\ X(4) \\ X(6) \\ X(1) \\ X(3) \\ X(5) \\ X(7) \end{bmatrix} = \begin{bmatrix}
c_4 & c_4 & c_4 & c_4 & 0 & 0 & 0 & 0 \\
c_2 & c_6 & -c_6 & -c_2 & 0 & 0 & 0 & 0 \\
c_4 & -c_4 & -c_4 & c_4 & 0 & 0 & 0 & 0 \\
c_6 & -c_2 & c_2 & -c_6 & 0 & 0 & 0 & 0 \\
0 & 0 & 0 & 0 & c_1 & c_3 & c_5 & c_7 \\
0 & 0 & 0 & 0 & c_3 & -c_7 & -c_1 & -c_5 \\
0 & 0 & 0 & 0 & c_5 & -c_1 & c_7 & c_3 \\
0 & 0 & 0 & 0 & c_7 & -c_5 & c_3 & -c_1
\end{bmatrix} \cdot
$$

$$\begin{bmatrix} 1 & 0 & 0 & 0 & 0 & 0 & 0 & 1 \\ 0 & 1 & 0 & 0 & 0 & 0 & 1 & 0 \\ 0 & 0 & 1 & 0 & 0 & 1 & 0 & 0 \\ 0 & 0 & 0 & 1 & 1 & 0 & 0 & 0 \\ 1 & 0 & 0 & 0 & 0 & 0 & 0 & -1 \\ 0 & 1 & 0 & 0 & 0 & 0 & -1 & 0 \\ 0 & 0 & 1 & 0 & 0 & -1 & 0 & 0 \\ 0 & 0 & 0 & 1 & -1 & 0 & 0 & 0 \end{bmatrix} \cdot \begin{bmatrix} x(0) \\ x(1) \\ x(2) \\ x(3) \\ x(4) \\ x(5) \\ x(6) \\ x(7) \end{bmatrix} \quad (8)$$

This expression can be computed with just 32 multiplications and 32 additions, as opposed to 64 multiplications and 56 additions required by full matrix-vector multiplication.

By applying the decomposition process recursively to the upper-left quarter of the cosine matrix in Eq. (8), we can further reduce the number of multiplications. For the 8-point DCT, we end up with

$$\begin{bmatrix} X(0) \\ X(4) \\ X(2) \\ X(6) \\ X(1) \\ X(3) \\ X(5) \\ X(7) \end{bmatrix} = \begin{bmatrix} 1 & 1 & 0 & 0 & 0 & 0 & 0 & 0 \\ 1 & -1 & 0 & 0 & 0 & 0 & 0 & 0 \\ 0 & 0 & 1 & 0 & 0 & 0 & 0 & 0 \\ 0 & 0 & 0 & 1 & 0 & 0 & 0 & 0 \\ 0 & 0 & 0 & 0 & 1 & 0 & 0 & 0 \\ 0 & 0 & 0 & 0 & 0 & 1 & 0 & 0 \\ 0 & 0 & 0 & 0 & 0 & 0 & 1 & 0 \\ 0 & 0 & 0 & 0 & 0 & 0 & 0 & 1 \end{bmatrix} \cdot$$

$$\begin{bmatrix} c_4 & 0 & 0 & 0 & 0 & 0 & 0 & 0 \\ 0 & c_4 & 0 & 0 & 0 & 0 & 0 & 0 \\ 0 & 0 & c_2 & c_6 & 0 & 0 & 0 & 0 \\ 0 & 0 & c_6 & -c_2 & 0 & 0 & 0 & 0 \\ 0 & 0 & 0 & 0 & c_1 & c_3 & c_5 & c_7 \\ 0 & 0 & 0 & 0 & c_3 & -c_7 & -c_1 & -c_5 \\ 0 & 0 & 0 & 0 & c_5 & -c_1 & c_7 & c_3 \\ 0 & 0 & 0 & 0 & c_7 & -c_5 & c_3 & -c_1 \end{bmatrix} \cdot \begin{bmatrix} 1 & 0 & 0 & 1 & 0 & 0 & 0 & 0 \\ 0 & 1 & 1 & 0 & 0 & 0 & 0 & 0 \\ 1 & 0 & 0 & -1 & 0 & 0 & 0 & 0 \\ 0 & 1 & -1 & 0 & 0 & 0 & 0 & 0 \\ 0 & 0 & 0 & 0 & 1 & 0 & 0 & 0 \\ 0 & 0 & 0 & 0 & 0 & 1 & 0 & 0 \\ 0 & 0 & 0 & 0 & 0 & 0 & 1 & 0 \\ 0 & 0 & 0 & 0 & 0 & 0 & 0 & 1 \end{bmatrix} \cdot$$

$$\begin{bmatrix} 1 & 0 & 0 & 0 & 0 & 0 & 0 & 1 \\ 0 & 1 & 0 & 0 & 0 & 0 & 1 & 0 \\ 0 & 0 & 1 & 0 & 0 & 1 & 0 & 0 \\ 0 & 0 & 0 & 1 & 1 & 0 & 0 & 0 \\ 1 & 0 & 0 & 0 & 0 & 0 & 0 & -1 \\ 0 & 1 & 0 & 0 & 0 & 0 & -1 & 0 \\ 0 & 0 & 1 & 0 & 0 & -1 & 0 & 0 \\ 0 & 0 & 0 & 1 & -1 & 0 & 0 & 0 \end{bmatrix} \cdot \begin{bmatrix} x(0) \\ x(1) \\ x(2) \\ x(3) \\ x(4) \\ x(5) \\ x(6) \\ x(7) \end{bmatrix} \quad (9)$$

This formulation requires just 22 multiplications and 28 additions.

Calculating DCT using (8) will be called the ***first-level even/odd decomposition*** algorithm and that using (9) will be called the ***all-level even/odd decomposition***, or simply ***even/odd decomposition*** algorithm.

Observe that, in (8), 2 of the 22 multiplications involve the scaler $c_4 = 1/\sqrt{2}$. When the entire cosine matrix is scaled by $\sqrt{2}$, we eliminate these 2 multiplications while changing the other scalers. This technique is particularly suitable for the row-column method of 2-D DCT/IDCT. As the two passes of the scaled 1-D transform

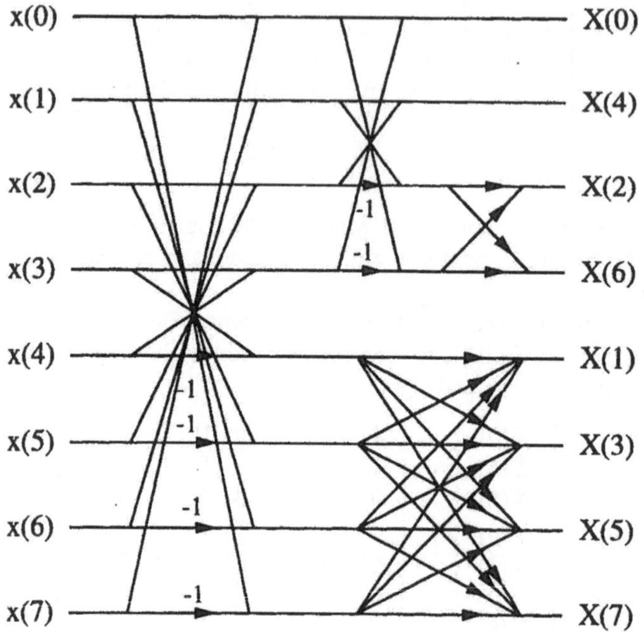

Figure 14.1 8-point DCT signal flow graph for the scaled even/odd decomposition algorithm. Scalers are omitted for readability.

affect the result by $(\sqrt{2})^2 = 2$, the corrective scaling in the end takes just a right shift, and usually does not cost any time or hardware. This method is called the *scaled even/odd decomposition* algorithm. Figure 14.1 shows the signal flow graph for this algorithm.

Compared to full matrix-vector multiplication, the first-level even/odd decomposition algorithm requires fewer operations, but is not as regular, and the all-level even/odd decomposition algorithm and its scaled version require even fewer operations yet are even less regular. Compared to the fast algorithms to be discussed in subsequent subsections, the scaled even/odd decomposition algorithm is not the fastest, yet it has numerical and latency advantages over the faster algorithms, as all the multiplications can be performed concurrently (in parallel or pipelined).

As discussed in Section 14.3, many reported implementations, most notably the ones in the ROM-based distributed arithmetic camp, choose the first-level even/odd decomposition for its balance of regularity and computation complexity. Some other implementations choose the scaled even/odd decomposition algorithm for its low complexity, short latency, and/or better precision.

14.2.4 Chen's Algorithm

Chen, Smith and Fralick's algorithm [10] is among the first proposals for fast IDCT algorithms. Their algorithm has a computation complexity of $O(N \log N)$, and is about 1/6 faster than the previously-proposed FFT method [1]. Many other algorithms improve on Chen's, but the asymptotical complexity remains $O(N \log N)$. For 8-point DCT, Chen's algorithm requires 16 multiplications and 26 additions.

For an N-point transform, $N = 2^L$, the algorithm decomposes the computation into $2L - 2$ sequential stages. Each stage consists of parallel rotation-like operations. The flow graph of Chen's algorithm for 8-point DCT is shown in Fig. 14.2. As the coefficient matrix of the IDCT is the transpose of that of the DCT (of the same size), transposing a DCT flow graph produces an IDCT flow graph with the same number of operations, as shown in Fig. 14.3. Most other non-recursive DCT algorithms can be similarly coverted to perform IDCT.

Chen's algorithm is simple, low-complexity, and has good numerical property. It has been a popular implementation choice, especially for programmable general-purpose or DSP processors.

14.2.5 Loeffler's Algorithms

Loeffler, Ligtenberg, and Moschytz [11] presented a 11-multiplication, 29-addition algorithm that appears to be the most efficient one for 8-point 1-D DCT. This method is best described by an abstracted signal flow graph, as shown in Fig. 14.4. The scale-rotate operator, sCi, takes x_0 and x_1 as inputs and computes

$$\begin{bmatrix} y_0 \\ y_1 \end{bmatrix} = s \cdot \begin{bmatrix} cos\theta & sin\theta \\ -sin\theta & cos\theta \end{bmatrix} \cdot \begin{bmatrix} x_0 \\ x_1 \end{bmatrix},$$

where $\theta = \pi i/16$. Each operator takes 3 multiplications and 3 additions, as opposed 4 multiplications and 2 additions if carried out literally. Here the same technique seen in optimizing complex-number multiplication is used:

$$\begin{aligned} y_0 &= s(\sin\theta - \cos\theta) \cdot x_1 + s\cos\theta \cdot (x_0 + x_1), \\ y_1 &= -s(\cos\theta + \sin\theta) \cdot x_0 + s\cos\theta \cdot (x_0 + x_1). \end{aligned}$$

A scaler of $\sqrt{2}$ is applied to the 1-D transform, like the scaled even/odd decomposition method. The rationale is that in the context of 2-D transform, which is often the case, two passes of the scaled transforms require a correction scaling of 1/2 and is basically free.

Loeffler's algorithm is structurally similar to Chen's. In fact, Chen's algorithm can be easily converted to 11 multiplications using the $\sqrt{2}$ scaling and the complex multiplication technique. The latter technique saves one out of 4 multiplications, but the increase in the depth of operations might imply a larger dynamic range requirement in the internal nodes, and thus inferior numerical performance.

Loeffler proposed many variations of 8-point and 16-point DCT algorithms in his paper. A particularly interesting one is a 12-parallel-multiplication, 30-addition method. The parallel formation of all the multiplications may offer certain implementation advantages.

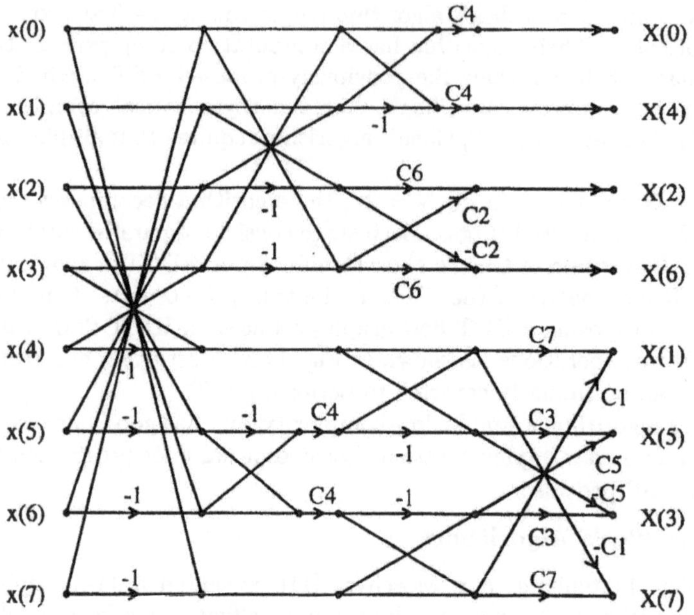

Figure 14.2 DCT flow graph in Chen's algorithm for $N = 8$.

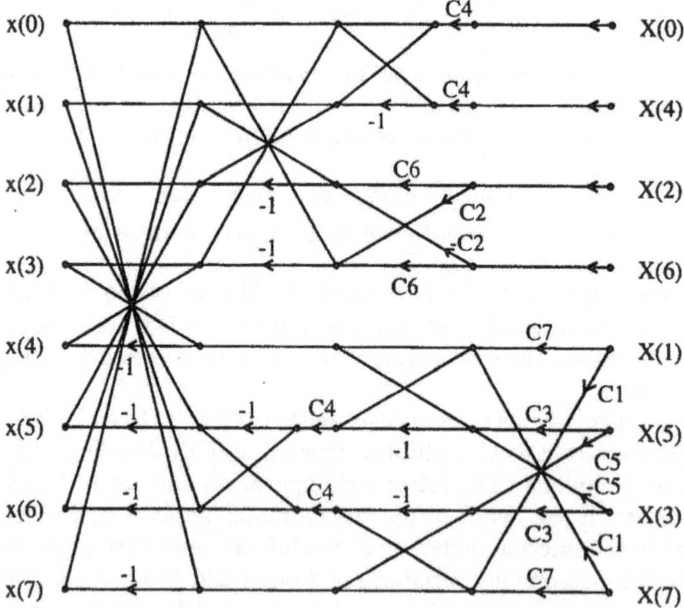

Figure 14.3 IDCT flow graph in Chen's algorithm ($N = 8$).

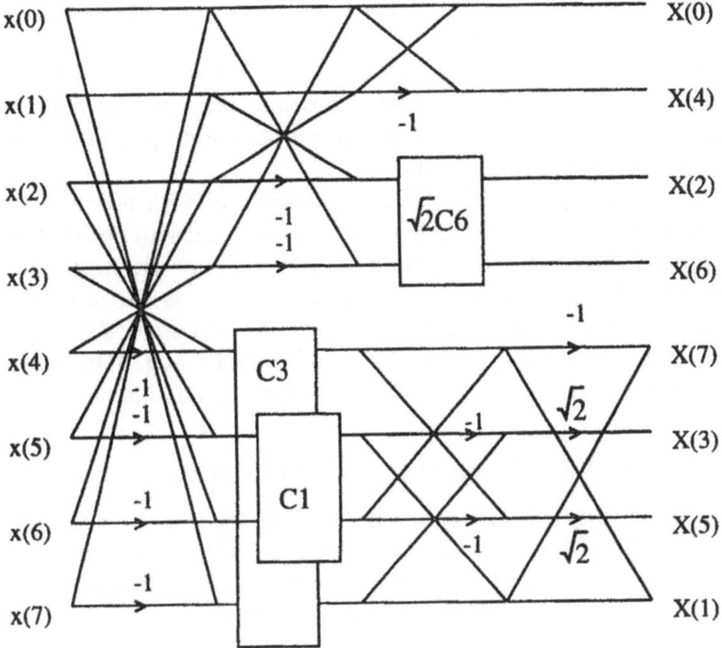

Figure 14.4 Loeffler's algorithm for 8-point DCT in an abstracted signal flow graph. sCi represents a rotation by $\pi i/16$ and scaling by s.

14.2.6 Other 1-D Algorithms

Lee [12], Vetterli and Nussbaumer [13], Wang [8, 14], Suehiro and Hatori [15], Hou [16], and Malvar [17] have also proposed fast algorithms for DCT. Multiplication complexity of these algorithms are close, if not identical, and are between Chen's and Loeffler's.

Arai, Agui and Kakajima [18] showed that with some multiplications absorbed by the quantization step usually performed after the DCT, an 8-point 1-D DCT can be computed with just 5 multiplications. Similar technique was adopted by Feig and Winograd [19] for 2-D DCT. It also applies to IDCT with dequantization performed before the transform, as in the context of MPEG video decoding. Fusing the two steps, however, creates difficulties in the verification of numerical precision in each step. A variation of [18] has been implemented in a hard-wired architecture by Kovac and Ranganathan [20].

Feig and Linzer [21] optimized 1-D and 2-D transforms in terms of number of fused multiply-add operations. These algorithms may execute more efficiently on programmable DSP processors, on which multiply-add operation usually takes as much time as a multiplication alone.

Most of the papers in the DCT literature deal with power-of-2 sized transforms. They are used most often because they are easier to optimize and easier to apply; it takes least work to partition an 1-D signal or an image into power-of-2 blocks to feed the transforms. In the literature, there is also substantial work on efficient algorithms for non-power-of-2 transforms. Recursive filtering-based methods, for example, those proposed by Canaris [22] and Wang, Jullien and Miller [23], are suitable for any transform sizes, so is the convolution method in Chan and Siu [24].

14.2.7 Row-Column Method for 2-D Transform

The *separability* property of 2-D DCT allows the transform to be applied along one dimension then the other. For example, 1-D DCT can be applied on each row, and then on each column of the row-transform result. This approach is called the *row-column method*, and would require $2N$ 1-D transforms of size N to realize an $N \times N$ 2-D transform.

Row-column method is fairly straightforward to implement. Any fast algorithm correctly realizing the 1-D transform can be used as a core or subroutine to perform the 2-D transform. Depending on the flexibility of the hardware platform and the speed and area tradeoffs, the 2-D transform can be carried out using a single, a few, or up to $2N$ parallel 1-D transform units.

14.2.8 Direct 2-D Algorithms

The alternative to the row-column method is the direct 2-D method of computing the transform directly from the N^2 input numbers.

The paper by Kamangar and Rao [25] is the first to work directly on the 2-D transform to reduce the computation complexity. Several other researchers [26, 27, 28, 29, 30] followed up with improvements over prior work. There are also techniques proposed to convert a 2-D $N \times N$ transform into N 1-D transforms plus some pre/post-processing. Cho and Lee's paper [31] was followed by Feig and Winograd's [19] and then by Prado and Duhamel's [32]. All 3 proposals require the same number of multiplications, half of what the row-column method requires, while [19] and [32] have slightly lower addition count than [31].

These 2-D algorithms commonly realize a variant of the DCT; the $u(m)u(n)$ DC-term normalization factor is dropped from (4). Correcting for this normalization factor should take no more than $2N - 2$ post-scalings for 2-D DCT (same number of pre-scalings for IDCT). The multiplication counts reported in the papers generally do not include these extra scalings.

Compared to applying $2N$ instances of 1-D DCT, algorithms in this group generally require fewer multiplications, yet have larger flow graphs, which translate to implementation as more temporary storage and/or larger datapath.

14.2.9 Pruned Transforms

DCT has been a popular technique in image and video compression because it condenses energy to lower frequency components, and therefore the frequency domain output can be represented with fewer bits. A common output pattern is a strong DC and low frequency components, and nearly zero high frequency components. The quantization process afterward takes into account the human visual perception level versus frequency, and usually codes 60% or more of the coefficients as zeros. Pruned transforms make use of this property to reduce computation, by producing fewer output data points for the forward transform, or expecting fewer input data points for the inverse transform.

Pruning the flow graph of a fast 1-D DCT algorithm has been shown by Wang [33] and Skodras [34] to be effective ways to reduce computation. The method proposed by Yang, Shaih, and Bai [35] carries out computation in diagonal groups to facilitate the commonly used zigzag coding order. Yang, Bai, and Hsia in [36] described an IDCT algorithm and its circuit realization that requires $(M - 1)N/2$ multiplications for $N \times N$ 2-D transform, when there are M nonzero input coefficients.

McMillan and Westover [37] presented a table lookup technique that combines the dequantization with the multiplications in IDCT. This approach relies on the fact that usually there are only a small number of nonzero coefficients, and most of them are of small magnitude.

Number of operations for selected 1-D and 2-D algorithms are summarized in Tables 14.1 and 14.2. Note that the efficiency of an implementation depends, to a large extent, how algorithm and architecture match up; choosing an algorithm with the lowest computation complexity may not be the best choice for every architecture.

14.3 DCT ARCHITECTURES

14.3.1 Multiply-Accumulate Architectures

In the literature, multiply-accumulate architectures are designs that are based on the traditional microcode-driven multiply-accumulate (MAC) architectures, usually with multiple multiply-accumulate units to achieve desirable performance levels.

The Programmable Vision Processor (VP) reported by Bailey and others [38] computes 2-D DCT, 2-D IDCT, as well as motion estimation for video compression. For DCT/IDCT, the 64-bit architecture incorporates 4 parallel MAC units in SIMD

Table 14.1 Operation Count for Some 1-D DCT/IDCT Algorithms

	N-point Multiplications	N-point Additions	8-point M	8-point A
Decomp 1 level	$\frac{N^2}{2}$	$\frac{N^2}{2}$	32	32
Decomp all levels	$\frac{N^2+2}{3}$	$\frac{N^2-4}{3}+N$	22	28
Scaled decomp [2]	$\frac{N^2+2}{3}-2$	$\frac{N^2-4}{3}+N$	20	28
Chen [10]	$N\log N-\frac{3N}{2}+4$	$\frac{3N}{2}\log N-\frac{3N}{2}+2$	16	26
Lee [12] [1]	$\frac{N}{2}\log N$	$\frac{3N}{2}\log N-N+1$	12	29
Loeffler [11] [2]	—	—	11	29
Feig [21]	—	—	26 m/a	
Arai [18] [3]	—	—	5	28

Notes:

1. DC-coefficient not normalized.
2. Transform scaled by $\sqrt{2}$.
3. Some multiplications absorbed into quan/dequan so are not counted.

Table 14.2 Operation Count for Some 2-D DCT/IDCT Algorithms

	$N \times N$ point Multiplications	$N \times N$ point Additions	8×8 point M	8×8 point A
Kamangar [25]	—	—	128	430
Vetterli [27]	$\frac{N^2}{2}\log N+\frac{N^2}{3}-2N+\frac{8}{3}$	$\frac{5N^2}{2}\log N+\frac{N^2}{3}-6N+\frac{62}{3}$	104	474
Haque [26]	$\frac{3N^2}{4}\log N$	$3N^2\log N-2N^2+2N$	144	464
Duhamel [29]	—	—	96	484
Feig91 [21]	—	—	480 m/a	
Feig92 [19]	—	—	96	454

mode to simultaneously compute 4 1-D transforms. The full matrix multiplication method is chosen in this implementation, resulting in a processing rate of 4 1-D 8-point transforms in 8 cycles. Running at 33 MHz, the VP delivers 66 M pixels/sec throughput rate for 2-D 8×8 DCT.

Yamauchi and others [39] reported another parallel programmable architecture for video and image processing tasks, including DCT and IDCT. The 4 data processing units communicate through shared memory as well as dedicated buses to allow realization of computation with very different communication needs, including 2-D filtering, motion estimation, fast DCT (Lee's was mentioned as an example), and full-matrix-multiplication DCT. They also incorporated a once-per-task datapath setup mechanism and two-level VLIW control scheme to reduce the width and size of control storage.

The above two designs support a wider range of applications, and are thus much more expensive and complicated than a DCT/IDCT-only design such as the one proposed by Miyazaki and others [40]. This proposal maps the first-level version of even/odd decomposition algorithm into a 4-MAC architecture using a DSP silicon compiler. This design runs at 50 MHz clock rate and delivers 50 M pixels/sec data rate. There are two data processing units, each is a copy of the 4-MAC design and computes one dimension.

Generally speaking, this class of architectures take more silicon area than the other more hard-wired methods. The area disadvantage comes from overheads associated with general multiplier, register file, and microcode control. The advantage of this approach is programmability and adaptation of existing processor architecture.

The DSP processors should be good vehicles for DCT implementation due to their fast (and in some cases, parallel) multiplications and fused multiply-accumulate instructions. Recently, some general-purpose processors are enhanced with SIMD-style parallel processing features to allow more efficient implementation of graphics and video computations. Examples are Texas Instruments' C6200 VLIW DSP architecture, Intel's MMX and Sun's VIS. These architectures, although not dedicated to DCT, can be categorized as multiply-accumulate-based as well.

14.3.2 ROM-based Distributed Arithmetic

There are designs that combine the fixed scalings and summation into table lookups. This approach is called (ROM-based) distributed arithmetic in the literature, attributing to the usage of multiple lookup tables to carry out the computation, and the origin is commonly credited to Liu [41] in the context of fixed-coefficient filtering. The practicality of table size usually dictates the input data to be processed 1-2 bits at a time, so this approach uses some aspect of digit-serial arithmetic as well.

The generic form of this method realizes a fixed-coefficient weighted summation,

$$y = \sum_{i=0}^{K-1} a(i)x(i),$$

where $a(i)'s$ are constants and $x(i)'s$ are the input data. Let $x(i,j)$ represent the individual digits of $x(i)$. In other words, when the digit size is 2^B, for D-digit numbers $x(i)$ we have

$$x(i) = \sum_{j=0}^{D-1} 2^{jB} x(i,j).$$

We can rewrite the weighted summation into a digit-sequential form:

$$\begin{aligned} y &= \sum_{i=0}^{K-1} a(i) \sum_{j=0}^{D-1} 2^{jB} x(i,j) \\ &= \sum_{j=0}^{D-1} 2^{jB} \left(\sum_{i=0}^{K-1} a(i) x(i,j) \right) \\ &= \sum_{j=0}^{D-1} 2^{jB} F(x(0,j), x(1,j), \ldots, x(K-1,j)) \end{aligned} \quad (10)$$

We precompute $F(\cdot)$ for all combinations of the $K \times B$ bits of input, and construct a table with 2^{BK} entries. The weighted summation is then performed by looking up the $F(\cdot)$ value taking each digit slice of input, and shift-accumulating the table outputs. Equation (10) shows a right-to-left evaluation (least-significant digit to most-significant digit). The opposite direction is also possible.

The above technique could be applied directly to the full matrix multiplication form of DCT. However, it is more efficient to use it with the first-level decomposed form. For example, an 8-point transform can be rewritten as 8 weighted summation equations, each having 4 terms. From (8), we have

$$\begin{bmatrix} X(0) \\ X(2) \\ X(4) \\ X(6) \end{bmatrix} = \begin{bmatrix} c_4 & c_4 & c_4 & c_4 \\ c_2 & c_6 & -c_6 & -c_2 \\ c_4 & -c_4 & -c_4 & c_4 \\ c_6 & -c_2 & c_2 & -c_6 \end{bmatrix} \cdot \begin{bmatrix} x(0)+x(7) \\ x(1)+x(6) \\ x(2)+x(5) \\ x(3)+x(4) \end{bmatrix} \quad (11)$$

$$\begin{bmatrix} X(1) \\ X(3) \\ X(5) \\ X(7) \end{bmatrix} = \begin{bmatrix} c_1 & c_3 & c_5 & c_7 \\ c_3 & -c_7 & -c_1 & -c_5 \\ c_5 & -c_1 & c_7 & c_3 \\ c_7 & -c_5 & c_3 & -c_1 \end{bmatrix} \cdot \begin{bmatrix} x(0)-x(7) \\ x(1)-x(6) \\ x(2)-x(5) \\ x(3)-x(4) \end{bmatrix} \quad (12)$$

A basic ROM-based distributed arithmetic structure is shown in Fig. 14.5. It is a bit-serial version, i.e., $B = 1$. Each ROM is 16 words $\times$ 64 bits, assuming 16-bit cosine constant precision. The addition and subtractions on the input elements (explicit $+$ and $-$ in (11) and (12) are performed bit-serially to reduce hardware.

Representatives from this architecture groups are Sun, Chen, and Gottlieb [42], Carlach, Penard, and Sicre [43], Masaki, et al. [44], and Trainor, Heron and Woods [45]. Carlach's design is the most compact (the smallest design surveyed in [46]), taking only 50,000 transistors. It delivers 27 M pixels/sec data rate for 8×8 transforms. It also incorporates a hard-wired network of registers for the transposition operations, as opposed to a regular memory block seen in other reports. Masaki's design supports up to 100 M pixels/sec for 8×8 transforms, and is geared toward HDTV-rate MPEG decoding. Trainor, et al, reported implementing DCT on Xilinx's FPGA (Field-Programmable Gate Array). The circuit takes up 4,800 cells, about 30% of an XC6264 device (rated 64K–100K gates), and delivers 15.36 M pixels/sec performance for 8×8 DCT.

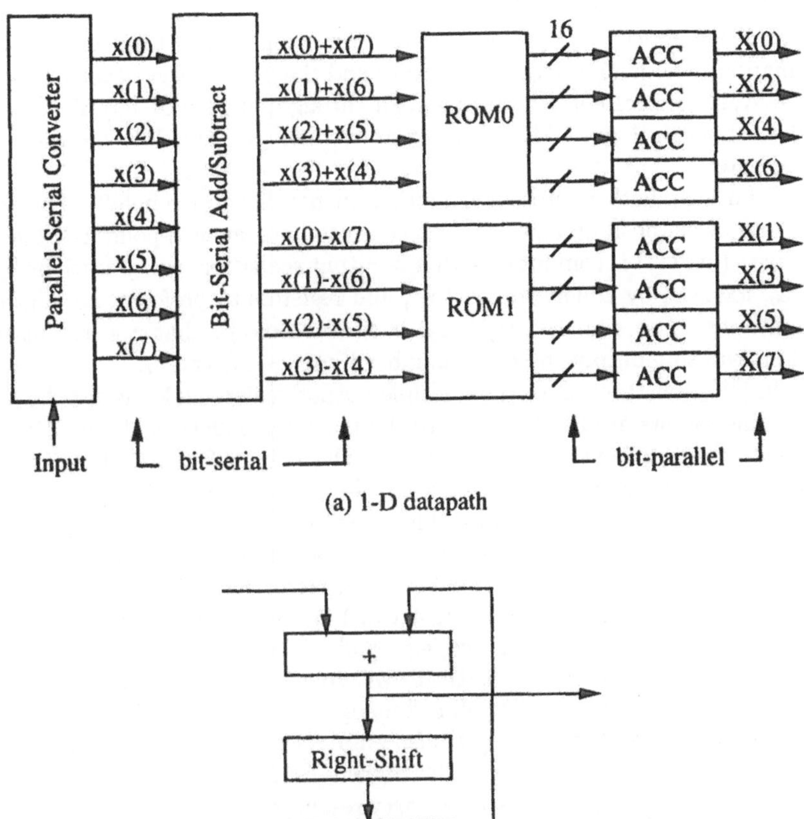

(a) 1-D datapath

(b) The accumulator (ACC) circuit

Figure 14.5 A basic ROM-based distributed arithmetic architecture for 8-point 1-D DCT.

14.3.3 Hard-Wired Multiplier Datapath — Bit Parallel

A number of architecture proposals make use of the constant scalings by hard-wiring the multipliers. Centered on this basic concept, these designs differ on arithmetic-level implementation, and can be roughly categorized into two groups: bit-parallel and digit-serial.

Ruetz and Tong's design [47] realized the first-level even/odd decomposition algorithm of 8-point IDCT, as in (11) and (12). In this design, 8 fixed-coefficient bit-parallel Wallace trees are employed, each computing a fixed 4-term weighted summation. The highly parallel structure and the use of carry-save adders (CSAs) in the Wallace trees contribute to a high throughput rate of 160 M pixels/sec in a relatively early 1-μm CMOS gate array technology. This IDCT design requires 31,000 gates.

Duardo and others presented another bit-parallel design in [48]. The first-level even/odd decomposition algorithm is chosen to implement 8-point 1-D IDCT. The even part (in (11)) is computed with a 3-output scaler for multiplications by c_4, c_2, and c_6, followed by adder, subtractors, and registers to perform the addition and subtractions. The odd part (in (12)) is similar, except that a 4-output scaler is used instead to multiply the input number by c_1, c_3, c_5, and c_7.

In [48], construction of the multiple-output scalers make use of the fact that the scaling factors are fixed and share the partial products in the multiplications. This idea is not unlike the *addition chain* in the Volume 2 of Knuth's "Art of Computer Programming" [49], in the context of exponentiation. A smaller example is presented in Fig. 14.6 to illustrate this technique. In [48], the 4-output scaler for the odd part computation is realized with 12 adders/subtractors. In comparison, separately implementing these 16-bit- scaling-factor scalings should take about 21 adders/ subtractors, based on an average of 1/3 nonzero digit per bit in the canonical signed-digit representation (see, e.g., [50]). This design delivers a data rate of 75 M pixels/sec. The combined IDCT/MC (motion compensation) chip takes up 0.56 mm^2 of area in 0.5 μm CMOS technology.

The design by Tonomura [51] also employs the ideas of hard-wiring scaling factors and sharing of partial products. Signed-digit representation, $-1, 0, 1$, is used in this bit-parallel design to postpone carry resolution until the last stage. The gate count is about 20,000 gates. While the data rate is not shown, the gate depth of the entire circuit is reported to be 37 without pipelining, and it produces 2 pixels per clock cycle.

14.3.4 Hard-Wired Multipliers — Digit-Serial

Digit-serial computation has the advantage of easy area/speed tradeoffs (by adjusting the digit size) and low control overhead. Traditional definition of digit-serial computation is one that utilizes only digit-size processing units [52]. Parallel-serial style (or sometimes serial-parallel) of multiplication, on the other hand, processes one operand digit-serially, while accumulating the other operand (shifted) into a partial product in bit-parallel fashion. This section covers both.

A series of papers and patents by Jutand, Mou, and Demassieux [53, 54, 55, 56, 57] describe a parallel-serial technique to perform multiplications. They apply the fixed operand of the multiplication to the additions in a Booth-encoded shift-and-add process (called hard-wired Booth encoding). Fig. 14.7 shows their

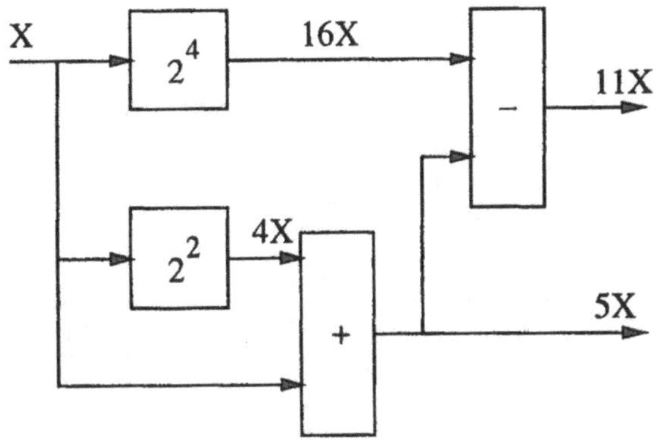

Figure 14.6 A simple example of partial-product sharing in multiple-output scaling. Scaling of 5 and 11 implemented separately would require 3 additions, compared to 2 additions with sharing.

approach and the idea of one-fixed-operand adder. The one-fixed-operand adder they use has half-adder-like cells, so is roughly half the size of a regular two-operand adder. The quasi-serial multiplier idea in Tatsaki, Burekas, and Goutis's paper [58] is similar; except their one-fixed-operand adder is pipelined to the bit level to allow faster clock rate.

Hsu and Wu in [59] showed the fixed-weight weighted summation implemented with the permuted difference coefficient (PDC) algorithm, a method to derive a network of adders and shifters to compute the weighted summation. It is however not clear whether this method results in a lower number of adders compared to separate canonical signed digit recoding of the scaling factors, or to the partial-product sharing in Duarado [48] and in Tonomura [51].

Feher [60] presented a bit-serial design for 2-D DCT. The design implements a direct 2-D transform with a 2-D version of the even/odd decomposition algorithm. Bit-serial adders and flipflops implement the fixed-weight weighted summations of the transform.

Hung and Landman [61] presented a very compact 8×8 IDCT design based on digit-serial arithmetic. The design chooses a digit size of 2 bits and realizes the scaled even/odd decomposition algorithm. The multiple-output scaling form is used instead of weighted summation. (It can be shown that weighted summation and multiple-output scaling are transpose of each other.) Fig. 14.8, 14.9 and 14.10 show the digit-serial building blocks: add, subtract and shift. Multiple-output scalers are constructed by shift, add, and subtract, and like Duardo's bit-parallel architecture, share partial products among the outputs. Fig. 14.11 shows the IDCT datapath made up of these digit-serial components. Note that it is a direct translation of the scaled even/odd decomposition algorithm (transpose of Fig. 14.1) into hardware.

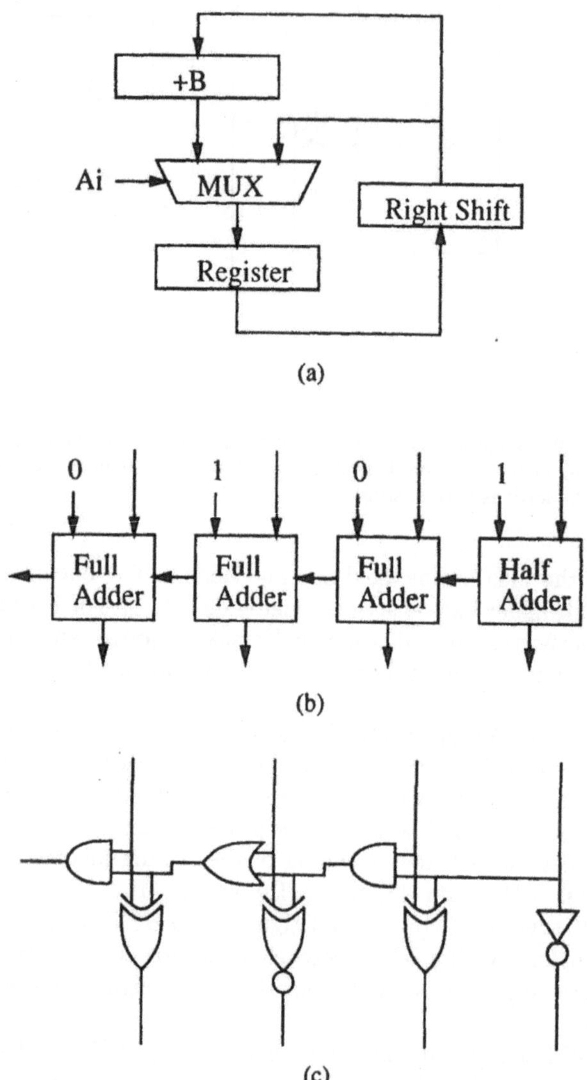

Figure 14.7 A simplified version of the parallel-serial multiplier design by Jutand and others. (a) A shift-and-add multiplier employing a one-fixed-operand adder (the $+B$ block) to perform $P' = P/2 + A_i B$ in each iteration. (b) Functionality of a "5" adder. (c) Logic-minimized "+5" adder.

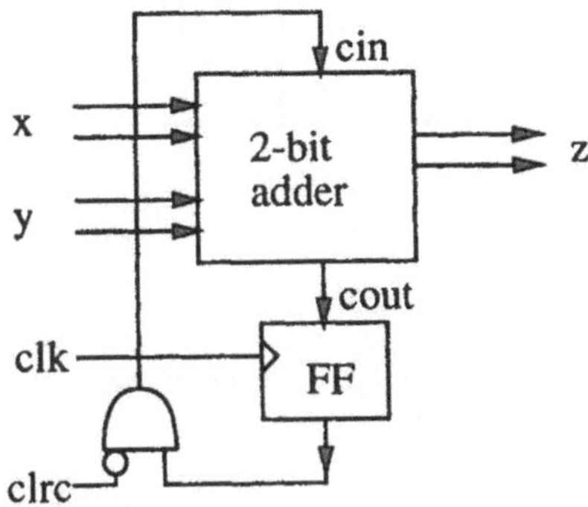

Figure 14.8 The digit-serial adder (DSA) used by Hung and Landman. The clrc signal clears the carry input during addition of the least significant digits.

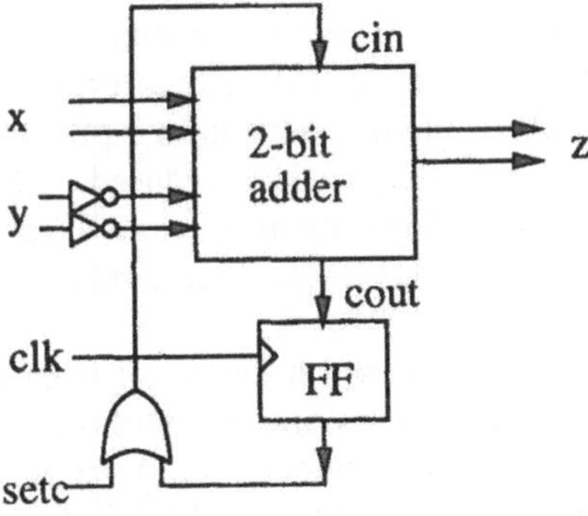

Figure 14.9 The digit-serial subtractor (DSS) used by Hung and Landman. The setc signal asserts the carry input during subtraction of the least significant digits to supply an 1 for two's complement.

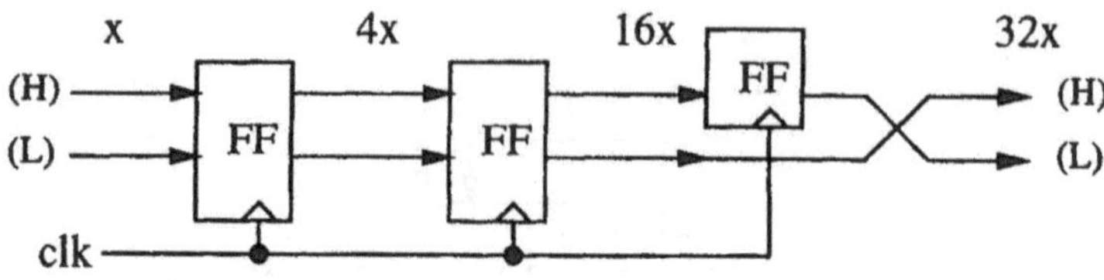

Figure 14.10 Example of shifts (power-of-2 scalings) in digit-serial format. (H)/(L) indicates the more/less significant bit of the 2-bit digit.

Hung and Landman's design requires only 4,600 gates plus 1,024 bits of memory, and achieves 26 M pixels/sec in a 0.35 μm gate array technology for 8×8 IDCT. It should be among the most compact designs reported in the literature. The purely digit-serial construction technique has very low control overhead, as the signal flow graph is mapped one-on-one to hardware. Once the basic arithmetic control signals (set/reset of carry bits, sign extension, and so on) are established, the numbers just flow in and out digit-serially. Parallel-to-serial and serial-to-parallel converters are included so that I/O interface and the transpose memory can work in the conventional bit-parallel format.

14.3.5 Systolic Arrays

Some researchers mapped the signal flow graphs unto systolic arrays. The arrays may be 1-D or 2-D in their connection topology, and the transform they realize may be 1-D transform or direct 2-D transform.

Murthy and Swamy [62] improved the systolic arrays proposed by Chang and Chen [63] and Cho and Lee [64], for 1-D DCT of arbitrary size. Murthy and Swamy's design has a better systolic topology; both data input and output are stream-lined such that the number of boundary cells is reduced. The complexity of the three designs are in the same order; all use about N cells and compute a transform in O(N) clock cycles, for N-point 1-D DCT. Full matrix-vector multiplication algorithm is used.

Chang and Wang [65] proposed a systolic design that computes 2-D DCT. Their design employs N^2 multipliers and completes a transform in N cycles. Although not using a transposition memory, the design in essence performs the row-column method of 2-D transform, and uses the 1-D first-level even/odd decomposition algorithm. This design belongs to the high cost, high performance camp; it requires about 340,000 transistors, and the estimated data rate is 320 M pixels/sec for 8×8 transforms.

Pan and Park [66] recently presented both 1-D and 2-D designs and showed that their new designs have lower latency and/or complexity compared to [63, 64, 65]. Their 1-D array, however, does not appear to have the nice systolic topology in [62].

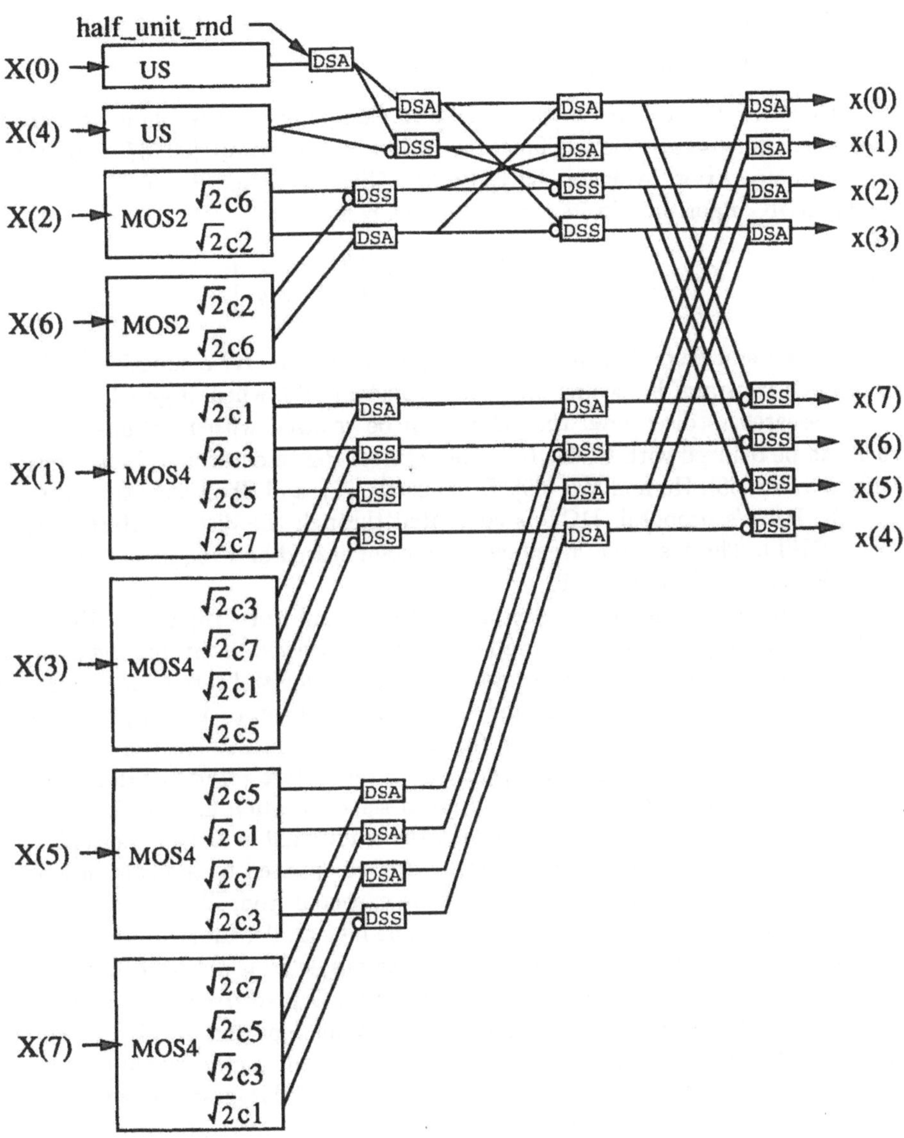

Figure 14.11 The 1-D IDCT datapath in Hung and Landman's design. The half_unit_rnd signal represents the one-half-unit amount added to the DC term and allowed to propagate to all output numbers that need to be rounded. US, MOS2, and MOS4 represent unity-scaler, 2-output scaler, and 4-output scaler.

These systolic array designs, in general, cost more gates to implement for comparable performance level. The idea is that local connectivity and circuit regularity should improve the efficiency of VLSI implementation. By using the simpler symmetry/antisymmetry relation in the transform and not the power-of-2-size-only types of optimizations, they are more flexible in the choice of transform size.

14.3.6 CORDIC-Based Architectures

CORDIC (COrdinate Rotation DIgital Computer) is a technique commonly used to compute trigonometric functions in low-cost hardware [50, 67]. Rotation is a basic computation realizable by CORDIC. Given x, y, and θ, $-\pi/2 < \theta < \pi/2$, the CORDIC algorithm, in its rotation mode, computes

$$\begin{bmatrix} x' \\ y' \end{bmatrix} = K \cdot \begin{bmatrix} cos\theta & -sin\theta \\ sin\theta & cos\theta \end{bmatrix} \cdot \begin{bmatrix} x \\ y \end{bmatrix}, \quad (13)$$

in B iterations. B controls the output precision. K, a function of B, approaches a constant, known as the Volder constant [67], when B gets large.

Researchers discovered that DCT can be realized with rotations, which in turn can be realized with CORDIC architectures. Representatives of such work are Duh and Wu [68], Hu and Wu [69], Hsiao et al. [70], and Zhou and Kornerup [71].

In Hsiao's proposal, DCT is computed through the discrete Hartley transform (DHT). The resulting architecture is complicated and requires 12 CORDIC operations for an 8-point DCT.

Computation complexity is reduced by Zhou and Kornerup to 6 CORDIC operations in the 8-point case. Their approach can be viewed as turning the even/odd decomposed matrix (in (9)) into rotations. The 22 multiplications for an 8-point transform are optimally implemented as 6 rotations, as up to 4 multiplications can be realized by a CORDIC operation (in (13)).

Hekstra and Deprettere [72] proposed a set of fast rotation primitives that generalize CORDIC. Potentially, for a given set of rotation angles, their method may lead to a more efficient solution than the traditional CORDIC method.

By using a pipelined CORDIC structure, as opposed to the conventional iterative structure, very high data rate can be expected from Zhou and Kornerup's design. There is only some routing and and one bit-parallel addition or subtraction in a clock cycle, so the 500 MHz clock rate they projected is not unreasonable with advanced technology. This leads to an amazing 2 G pixels/sec data rate for 2-D 8×8 transforms. Gate count or area estimate is not reported in [71]. There need to be additional scalers to correct the Volder constant factor introduced by the CORDIC algorithm. Including these, the 8-point DCT datapath alone should take about 100 K gates to implement. This appears to be the most promising design for high performance applications.

14.4 CONCLUSION AND FUTURE TRENDS

The design space for DCT/IDCT implementations is quite large, with so many different algorithms and architecture styles to choose from. While there are common metrics in the literature, like number of multiplications, gate count, area and so on, there are more qualitative issues like assumptions, precision, process technology, functionality, and flexibility that complicate a fair comparison. Instead of pushing for numerical comparisons, this chapter described the major architecture categories, a few references therein, and picked a few good ideas to elaborate on. A survey paper by Pirsch [46] tabulates transistor counts, chip sizes, and so on, of 15 DCT/IDCT implementation reports. The table can be used as a starting point for a rigorous evaluation of these architecture proposals.

Since DCT/IDCT functions have become integral parts in many consumer imaging, audio, and video products, in the near future the incentives to develop better schemes will continue to be strong. As practitioners advance the state of the art, it would be interesting to see how some fundamentally different camps compete with each other:

- *Pruned transforms versus full*

  Pruned transform relies on the statistics of the data to reduce computation, and in general requires extra buffering to ensure that extra time taken by occasional dense blocks are evened out. Real-time performance is difficult to guarantee. Full transform is more expensive yet it offers simplicity and robustness.

  An emerging video decompression application is HDTV-rate MPEG decoding, which should have fewer nonzero coefficients, on the average, per 8×8 block compared to the earlier MPEG main level/main profile decoding. While in appearance this may benefit the pruned transform camp, the consumer orientation and increased frame size may favor the full transform approach for robustness and lower memory requirements.

- *Dedicated circuits versus programmable*

  Dedicated circuits generally cost less to deliver the same performance. However, as the circuit technology improves, any given DCT computation task takes up an ever-shrinking percentage workload of a programmable platform. Eventually, the incremental cost of implementing DCT on the platform may become negligible. The VLIW style parallel processing in new DSP processors, e.g., Texas Instruments' C6200, and the SIMD mode in some recent general purpose processors, e.g., Intel's Pentium MMX, make software implementation a very feasible and sometimes desirable choice. Programmable devices allow more end-equipment differentiations and therefore command more value.

  The recent HDTV push also plays an interesting role to these two camps. The higher computation requirement magnifies the cost difference between the two, making the dedicated approach a more reasonable choice. The intense competition of the market, on the other hand, helps to sell the value-adding programmable approach. As in many other applications, we will likely see both approaches thrive; with dedicated implementations in the low end and programmable ones in the high end.

REFERENCES

[1] N. Ahmed, T. Natarajan, and K. R. Rao, "Discrete cosine transform," *IEEE Trans. Computers*, vol. C-23, pp. 90–93, Jan. 1974.

[2] W. B. Pennebaker and J. L. Mitchell, *JPEG Still Image Data Compression Standard.* New York: Van Nostrand Reinhold, 1993.

[3] ISO/IEC 11172-3 Information technology, *Coding of moving pictures and associated audio for digital storage media at up to about 1.5 Mbit/s*, Part 3: Audio, International Organization for Standardization (ISO), 1993.

[4] ISO/IEC 11172-2 Information technology, *Coding of moving pictures and associated audio for digital storage media at up to about 1.5 Mbit/s*, Part 2: Video, International Organization for Standardization (ISO), 1993.ISO/IEC 11172-2 MPEG-1 Video Standard.

[5] ISO/IEC 13818-2 Information Technology, *Generic Coding of Moving Pictures and Associated Audio Information*, Video, International Organization for Standardization (ISO), 1996.

[6] ATSC Document A/52, the Digital Audio Compression (AC-3) Standard.

[7] ATSC Document A/53, the ATSC Digital Television Standard.

[8] Z. Wang, "Fast algorithms for the discrete W tranform and for the discrete Fourier transform," *IEEE Transactions on Acoustics, Speech and Signal Processing*, vol. ASSP-32, no. 4, pp. 803–816, Aug. 1984.

[9] R. C. Singleton, "On computing the fast Fourier transform," *Communication of the ACM*, vol. 10, no. 10, pp. 647–654, Oct. 1967.

[10] W.-H. Chen, C. H. Smith, and S. C. Fralick, "A fast computational algorithm for the discrete cosine transform," *IEEE Transactions on Communications*, vol. COM-25, no. 9, pp. 1004–1009, Sept. 1977.

[11] C. Loeffler, A. Ligtenberg, and G. S. Moschytz, "Pratical fast 1-D DCT algorithms with 11 multiplications," in *ICASSP-89*, pp. 988–991, 1989.

[12] B. G. Lee, "A new algorithm to compute the discrete cosine transform," *IEEE Transactions on Acoustics, Speech and Signal Processing*, vol. ASSP-32, no. 6, pp. 1243–1245, Dec. 1984.

[13] M. Vetterli and H. Nussbaumer, "Simple FFT and DCT algorithms with reduced number of operations," *Signal Processing*, vol. 6, no. 4, pp. 267–78, Aug. 1984.

[14] Z. Wang, "On computing the discrete Fourier and cosine transforms," *IEEE Transactions on Acoustics, Speech and Signal Processing*, vol. ASSP-33, no. 5, pp. 1341–1344, Oct. 1985.

[15] N. Suehiro, "Fast algorithms for the discrete Fourier transform and for other transforms," in *Proceedings, IEEE International Conference on Acoustics, Speech and Signal Processing*, vol. 1, pp. 217–220, 1986.

[16] H. Hou, "A fast recursive algorithm for computing the discrete cosine transform," *IEEE Transactions on Acoustics, Speech and Signal Processing*, vol. ASSP-35, no. 10, pp. 1455–1461, Oct. 1987.

[17] H. Malvar, "Fast computation of the discrete cosine transform and the discrete Hartley transform," *IEEE Transactions on Acoustics, Speech and Signal Processing*, vol. ASSP-35, no. 10, pp. 1484–1485, Oct. 1987.

[18] Y. Arai, T. Agui, and Nakajima, "A fast DCT-SQ scheme for images," *Transactions of the IEICE E*, vol. E71, no. 11, pp. 1095–1097, Nov. 1988.

[19] E. Feig and S. Winograd, "Fast algorithms for the discrete cosine transform," *IEEE Trans. Signal Processing*, vol. 40, no. 9, pp. 2174–2193, Sept. 1992.

[20] M. Kovac, N. Ranganathan, and M. Zagar, "Prototype VLSI chip architecture for JPEG image compression," in *Proceedings of European Design and Test Conference*, pp. 2–6, 1995.

[21] E. Feig and E. Linzer, "Scaled DCT algorithms for JPEG and MPEG implementations on fused multiply/add architectures," in *Image Processing Algorithms and Techniques II, Proceedings of SPIE*, vol. 1452, pp. 458–467, 1991.

[22] J. Canaris, "VLSI architecture for the real time computation of discrete trigonometric transforms," *Journal of VLSI Signal Processing*, vol. 5, no. 1, pp. 95–104, Jan. 1993.

[23] Z. Wang, G. Jullien, and W. Miller, "Recursive algorithms for the forward and inverse discrete cosine transform with arbitrary length," *IEEE Signal Processing Letters*, no. 7, pp. 101–102, July 1994.

[24] Y.-H. Chan and W.-C. Siu, "Mixed-radix discrete cosine transform," *IEEE Transactions on Signal Processing*, vol. 41, no. 11, pp. 3157–316, Nov. 1993.

[25] F. A. Kamangar and K. R. Rao, "Fast algorithms for the 2-D discrete cosine transform," *IEEE Transactions on Computers*, vol. C-31, no. 9, pp. 899–906, Sept. 1982.

[26] M. Haque, "A two-dimensional fast cosine transform," *IEEE Transactions on Acoustics, Speech and Signal Processing*, vol. ASSP-33, no. 6, pp. 1532–1539, Dec. 1985.

[27] M. Vetterli, "Fast 2-d discrete cosine transform," in *Proceedings, IEEE International Conference on Acoustics, Speech and Signal Processing*, vol. 4, pp. 1538–1541, 1985.

[28] C. Ma, "A fast recursive two dimensional cosine transform," in *Proceedings, Intelligent Robots and Computer Vision, SPIE vol.1002*, pp. 541–548, 1988.

[29] P. Duhamel and C. Guillemot, "Polynomial transform computation of the 2-D DCT," in *Proceedings, IEEE International Conference on Acoustics, Speech and Signal Processing*, vol. 3, pp. 1515–1518, 1990.

[30] H. R. Wu and F. J. Paoloni, "A two-dimensional fast cosine transform algorithm based on Hou's approach," *IEEE Transactions on Signal Processing*, vol. 39, no. 2, pp. 544–546, Feb. 1991.

[31] N. I. Cho and S. U. Lee, "Fast algorithm and implementation of 2-D discrete cosine transform," *IEEE Trans. Circuits and Systems*, vol. 38, pp. 297–305, Mar. 1991.

[32] J. Prado and P. Duhamel, "Polynomial-transform based computation of the 2-D DCT with minimum multiplicative complexity," in *Proceedings, IEEE International Conference on Acoustics, Speech and Signal Processing*, vol. 3, pp. 1347–1350, 1996.

[33] Z. Wang, "Pruning the fast discrete cosine transform," *IEEE Transactions on Communications*, vol. 39, no. 5, pp. 640–643, May 1991.

[34] A. N. Skodras, "Fast discrete cosine transform pruning," *IEEE Transactions on Signal Processing*, vol. 42, no. 7, pp. 1833–1837, July 1994.

[35] J.-F. Yang, S.-C. Shain, and B.-L. Bai, "Fast two-dimensional inverse discrete cosine transform for HDTV or videophone systems," *IEEE Transactions on Consumer Electronics*, vol. 39, no. 4, pp. 934– 940, Nov. 1993.

[36] J.-F. Yang, B.-L. Bai, and S.-C. Hsia, "Efficient two-dimensional inverse discrete cosine transform algorithm for HDTV receivers," *IEEE Transactions on Circuits and Systems for Video Technology*, vol. 5, no. 1, pp. 25–30, Feb. 1995.

[37] L. McMillan and L. Westover, "A forward-mapping realization of the inverse discrete cosine transform," in *Proceedings, Data Compression Conference*, pp. 219–228, 1992.

[38] D. Bailey, M. Cressa, J. Fandrianto, D. Neubauer, H. K. J. Rainnie, and C.-S. Wang, "Programmable vision processor/controller for flexible implementation of current and future image compression standards," *IEEE Micro*, vol. 12, no. 5, pp. 33–39, Oct. 1992.

[39] H. Yamauchi, Y. Tashiro, T. Minami, and Y. Suzuki, "Architecture and implementation of a highly parallel single-chip video DSP," *IEEE Transactions on Circuits and Systems for Video Technology*, vol. 2, no. 2, pp. 207–220, June 1992.

[40] T. Miyazaki, T. Nishitani, M. Edahiro, I. Ono, and K. Mitsuhashi, "DCT/IDCT processor for HDTV developed with DSP silicon compiler," *Journal of VLSI Signal Processing*, vol. 5, no. 2–3, pp. 151–158, Apr. 1993.

[41] A. Peled and B. Liu, "A new hardware realization of digital filters," *IEEE Transactions on Acoustics, Speech, and Signal Processing*, vol. ASSP-22, no. 6, pp. 456–462, Dec. 1974.

[42] M.-T. Sun, T.-C. Chen, and A. Gottlieb, "VLSI implementation of a 16*16 discrete cosine transform," *IEEE Transactions on Circuits and Systems*, vol. 36, no. 4, pp. 610–617, Apr. 1989.

[43] J. Carlach, P. Penard, and J. Sicre, "TCAD: a 27 MHz 8*8 discrete cosine transform chip," in *Proceedings, IEEE International Conference on Acoustics, Speech and Signal Processing*, vol. 4, pp. 2429–2432, 1989.

[44] T. Masaki, Y. Morimoto, T. Onoye, and I. Shirakawa, "VLSI implementation of inverse discrete cosine transformer and motion compensator for MPEG-2 HDTV video decoding," *IEEE Transactions on Circuits and Systems for Video Technology*, vol. 5, no. 5, pp. 387–395, Oct. 1995.

[45] D. W. Trainor, J. P. Heron, and R. F. Woods, "Implemention of the 2D DCT using a Xilinx XC6264 FPGA," in *Proceedings, 1997 IEEE Workshop on Signal Processing Systems: Design and Implementation (SiPS97)*, pp. 541–550, Nov. 1997.

[46] P. Pirsch, N. Demassieux, and W. Gehrke, "VLSI architectures for video compression — a survey," *Proceedings of the IEEE*, vol. 83, no. 2, pp. 220–246, Feb. 1995.

[47] P. A. Ruetz, P. Tong, D. Bailey, D. A. Luthi, and P. H. Ang, "A high-performance full-motion video compression chip set," *IEEE Transactions on Circuits and Systems for Video Technology*, vol. 2, no. 2, pp. 111–122, June 1992.

[48] O. Duardo, S. Knauer, and Mailhot, "Architecture and implementation of ics for a DSC-HDTV video decoder system," *EE Micro*, vol. 12, no. 5, pp. 22–27, Oct. 1992.

[49] D. E. Knuth, *The Art of Computer Programming*, Volume 2, *Seminumerical Algorithms*. Reading, Massachusetts: Addison-Wesley, 2nd ed., 1981.

[50] K. Hwang, *Computer Arithmetic, Principles, Architecture, and Design*. New York: John Wiley & Sons, Inc., 1979.

[51] M. Tonomura, "High-speed digital circuit of discrete cosine transform," *Technical Report of IEICE*, pp. 39–46, 1994. SP94-41 DSP94-66 (1994-09).

[52] R. I. Hartley and K. K. Parhi, *Digit-serial computation*. Kluwer Academic Publishers, 1995.

[53] F. Jutand, Z. J. Mou, and N. Demassieux, "DCT architectures for HDTV," in *Proceedings - IEEE International Symposium on Circuits and Systems*, vol. 1, pp. 196–199, 1991.

[54] Z.-J. Mou and F. Jutand, "A high-speed low-cost DCT architecture for HDTV applications," in *Proceedings - IEEE International Conference on Acoustics, Speech and Signal Processing*, vol. 2, pp. 1153–1156, 1991.

[55] N. Demassieux and F. Jutand, "A real-time discrete cosine transform chip," *Digital Signal Processing*, vol. 1, no. 1, pp. 6–14, Jan. 1991.

[56] F. Jutand and Demassieux, "Binary adder having a fixed operand and parallel-serial binary multiplier incorporating such an adder," Aug. 1989. U.S. Patent 4,853,887.

[57] F. Jutand and Demassieux, "Circuit to perform a linear transformation on a digital signal," Feb. 1990. U.S. Patent 4,899,300.

[58] A. Tatsaki and C. Goutis, "Bit-serial VLSI architecture for the 2-D discrete cosine transform," *Microprocessing and Microprogramming*, vol. 40, no. 10–12, pp. 829–832, Dec. 1994.

[59] C.-Y. Hsu and H.-D. Wu, "New architecture for hardware implementation of a 16 multiplied by 16 discrete cosine transform," *International Journal of Electronics*, vol. 72, no. 4, pp. 593–603, Apr. 1992.

[60] B. Feher, "2-d discrete cosine transformation implementation," in *Proceedings, Workshop on Parallel Processing: Technology and Applications*, pp. 112–122, 1994.

[61] C.-Y. Hung and P. Landman, "Compact inverse discrete cosine transform circuit for MPEG video decoding," in *Proceedings, 1997 IEEE Workshop on Signal Processing Systems: Design and Implementation (SiPS97)*, pp. 364–373, Nov. 1997.

[62] N. R. Murthy and M. Swamy, "On the real-time computation of DFT and DCT through systolic architect ures," *IEEE Transactions on Signal Processing*, vol. 42, no. 4, pp. 988–921, Apr. 1994.

[63] L. W. Chang and M. Y. Chen, "A new systolic array for discrete Fourier transform," *IEEE Transactions on Acoustics, Speech, and Signal Processing*, vol. 36, pp. 1665–1666, Oct. 1988.

[64] N. I. Cho and S. U. Lee, "DCT algorithms for VLSI parallel implementations," *IEEE Transactions on Acoustics, Speech, and Signal Processing*, vol. 38, pp. 121–127, Oct. 1990.

[65] S.-L. Chang and T. Ogunfunmi, "Fast DCT (Feig's algorithm) implementation and application in MPEG1 video compression," in *Midwest Symposium on Circuits and Systems*, vol. 2, pp. 961–964, 1995.

[66] S. B. Pan and R.-H. Park, "Unified systolic arrays for computation of the DCT/DST/DHT," *IEEE Transactions on Circuits and Systems for Video Technology*, vol. 7, no. 2, pp. 413–419, Apr. 1997.

[67] J. E. Volder, "The CORDIC trigonometric computing technique," *IRE Transactions on Electronic Computers*, pp. 330–334, Sept. 1959.

[68] W. J. Duh and J. L. Wu, "Two-stage circular-convolution-like algorithm/architecture for the discrete cosine transform," *IEE Proceedings, Part F: Radar and Signal Processing*, vol. 137, no. 6, pp. 465–472, Dec. 1990.

[69] Y. H. Hu and Z. Wu, "Efficient CORDIC array structure for the implementation of discrete cosine transform," *IEEE Transactions on Signal Processing*, vol. 43, no. 1, pp. 331–336, Jan. 1995.

[70] J.-H. Hsiao, L.-G. Chen, T.-D. Chiueh, and C.-T. Chen, "High throughput CORDIC-based systolic array design for the discrete cosine transform," *IEEE Transactions on Circuits and Systems for Video Technology*, vol. 5, no. 3, pp. 218–225, June 1995.

[71] F. Zhou and P. Kornerup, "High speed DCT/IDCT using a pipelined CORDIC algorithm," in *Proceedings, Symposium on Computer Arithmetic*, pp. 180–187, 1995.

[72] G. J. Hekstra and E. Deprettere, "Fast rotations: low-cost arithmetic methods for orthonormal rotation," in *Proceedings, Symposium on Computer Arithmetic*, pp. 116–125, 1997.

Chapter 15

Lossless Coders

Ming-Ting Sun, Sachin G. Deshpande and Jenq-Neng Hwang
Information Processing Laboratory,
Department of Electrical Engineering,
University of Washington
Seattle, Washington
{sun,sachind,hwang}@ee.washington.edu

15.1 INTRODUCTION

Data from various sources often have non-uniform probability distributions. Lossless coding achieves data compression by taking advantage of the non-uniform probability distribution of the data (e.g., using shorter code-words for more frequent symbols and longer code-words for less frequent symbols). It is widely used in text, facsimile, image, and video compression [1]-[2]. It can be used alone (e.g., in text, facsimile, and medical image compression) or used with a lossy coder (e.g., in most image/video compression). Most image/video coders consist of a lossy coder to remove the spatial and temporal redundancies, and a lossless coder to remove the statistical redundancies. With a lossless coder, an image/video coder can achieve an additional compression ratio of about two beyond what can be achieved with a lossy coder alone.

Lossless coding is also often referred to as entropy coding or reversible coding, since it deals with the statistics of the data, and after lossless coding the data can be recovered exactly without any error. For a source with N statistically independent symbols, the entropy is defined as

$$H(s) = -\sum_{i=1}^{N} P_i \log_2 P_i \tag{1}$$

where P_i is the probability of the ith symbol. The unit of the entropy is bits per symbol. The entropy represents a theoretical lower bound on the required average number of bits per symbol to code the source symbols losslessly.

Numerous lossless coding methods have been developed to achieve compression results approaching the entropy [1]-[3]. The choice of a lossless coding method involves tradeoffs among compression ratio, complexity, and delay. For example, some adaptive and high-order entropy coding algorithms [4]-[5] can achieve higher compression ratios, but have much higher complexity. Two-pass algorithms can also achieve higher coding efficiency, but result in longer coding delay. Due to the large amount of information available in the literature on various lossless coders, it is difficult to cover all lossless coding schemes in a chapter. Recently, many image/video coding standards (e.g., JPEG, H.261, H.263, MPEG, etc.) have been established [6]-[11]. The lossless coders used in these standards are based either on Huffman coding or on arithmetic coding. These lossless coders represent best tradeoffs that can be achieved today among the performance, complexity, and coding delay. In this chapter, we focus on the algorithms and implementations of these lossless coders.

The organization of this chapter is as follows. In Section 15.2, we review the Huffman-based lossless coders used in recent image/video coding standards. In Section 15.3, the implementation of the Huffman-based encoders and decoders are discussed. Various architectures discussed for Huffman decoding include bit-serial, constant-output-rate and variable input/output rate decoders. Section 15.4 and Section 15.5 discuss the arithmetic coding and its implementation. Some of the practical implementations of arithmetic coding like QM coder and H.263 syntax-based arithmetic coding (SAC) are explained. Section 15.6 discusses some system issues related to the lossless coders. Finally, a summary is provided in Section 15.7.

15.2 HUFFMAN-BASED LOSSLESS CODING

15.2.1 Huffman Coding

Huffman coding [3] is probably the most widely used lossless coding technique due to its simplicity and good performance. In Huffman coding, shorter code-words are assigned to more frequent symbols so that the average code-length is reduced. The code-words are constructed using the well known Huffman procedure. The construction involves recursive procedures which repeatedly order the symbols according to their probabilities and combine the two symbols with the smallest probabilities. An example of Huffman code and its construction is shown in Fig. 15.1. Using fixed-length coding, the 8 symbols in Fig. 15.1 can be encoded using 3-bits/symbol. With Huffman coding, the average code-length is 1.77 bits/symbol which is close to the entropy of 1.62 bits/symbol (the average code-length can be calculated by: $\sum_{i=1}^{N} P_i l_i$, where P_i is the probability and l_i is the length of the code-word for the ith symbol respectively). The tree structure in Fig. 15.1 is often referred to as a Huffman tree. The highest node is the "root". The end-nodes which correspond to the symbols are called "leaf nodes". The circles represent internal states. The numbers in the rectangles represent the probabilities of the symbols. The numbers in the circles represent the probabilities of the combined symbols. A "0" and "1" is assigned to a branch from each internal state. The bit-pattern from the root to the leaf node gives the code-word for that particular symbol.

The probability of each symbol used in the Huffman code-book construction is usually estimated off-line from training sequences. If the probability model fits the actual input symbol probabilities, then good performance can be achieved. Otherwise, the performance may be degraded. Huffman code-word design is not unique. For example, when there are symbols with the same probability, depending on the order of which symbol is combined first, it will result in different Huffman codes. In the Huffman tree in Fig. 15.1, the assignment of the "0" and "1" to the two branches under any internal node can be reversed which will result in different Huffman codes. If we have a binary representation for the code-words, the complement of this representation is also a valid set of Huffman code-words. However, these different Huffman codes all have the same efficiency.

Symbol	Probability	Fixed-length code	Huffman code
a	0.71	000	0
b	0.10	001	100
c	0.04	010	1010
d	0.04	011	1011
e	0.03	100	1100
f	0.03	101	1101
g	0.03	110	1110
h	0.02	111	1111

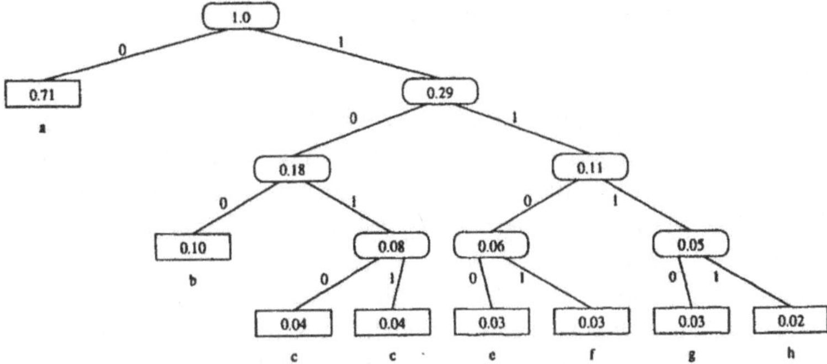

Figure 15.1 A Huffman code and its construction.

To encode a symbol sequence using Huffman coding, the symbols are first mapped into Huffman code-words. The code-words are then concatenated into a continuous bit-stream. Using the codes in Fig. 15.1 to code the sequence "aabad" will result in a bit stream "0010001011". An EOB (End-Of-Block) marker is usually added at the end of the block of symbols to be coded to indicate the end of the sequence.

Huffman code is a "prefix code" which means that no code-word is a prefix of another code-word. Due to this property, the codes are uniquely decodable. Given a coded bit-stream, we can read the bit-stream bit-by-bit and traverse the Huffman tree based on the bit-pattern until a leaf node is reached which is the decoded symbol.

When the number of different symbols in the source is large and the probabilities of the symbols are highly skewed, the Huffman procedure often results in very long code-words. Long code-words are undesirable since it may take more clock cycles to traverse the Huffman tree and may need a large memory for the codebook. In practice, constrained-length design methods are used to limit the maximum code-word length. Some methods for designing constrained length code-words are discussed in [12]-[15].

For practical implementations, it is desirable to also limit the number of entries in the Huffman code-book. A commonly used method to limit the number of entries in the code-book is to use an "escape code". If there are N different symbols in the source and we like to limit the total number of entries in the code-book to be M, where M can be much smaller than N, we can keep the M-1 most frequent symbols and aggregate the remaining N-M+1 symbols into a new symbol which has the probability equal to the sum of probabilities of those N-M+1 symbols. A Huffman code-book is then designed for the new M symbols according to the probabilities. The resulting code for the aggregated symbol is called the escape code. Each member of the aggregated symbol can be represented by the escape code (prefix) followed by a corresponding fixed-length code. In the decoding, when an escape code is decoded, the fixed-length code following the escape code is used to identify the actual symbol from the members of the aggregated symbol. This escape code approach will reduce the coding efficiency slightly. However, as long as the probability of the aggregated symbol is relatively small (which is usually the case in practical applications), the degradation is usually not significant.

15.2.2 Run-Length Coding

Run-length coding [1]-[2] is an efficient way to code a string of repeated symbols. There are many variations of run-length coding. In most image/video coding applications, after source coding the data stream often contains long sequences of "0"s. With the run-length coding used by most image/video coding standards, a stream of symbols consisting of many repeated "0"s is coded with a two dimensional symbol (run, value), where "run" is the number of the consecutive "0"s (i.e., the zero-run-length), and the "value" identifies the non-zero symbol following the sequence of "0"s. For example, an input sequence "0, 0, -3, 5, 1, 0, 0, 0, 0, 0, 0, -2" can be coded as a two-dimensional symbol sequence "(2,-3), (0,5), (0,1), (6,-2)". Run-length coding is a simple technique to reduce the number of symbols to be coded.

The combination of the run-length coding followed by the Huffman coding forms the basis of the lossless coding scheme used in most image/video coding standards. There are various variations in the actual coding scheme. For example, in H.263 very-low bit-rate video coding standard, since it is important to save every bit, the two dimensional symbols were extended to three dimensional symbols (run, value, EOB), where EOB represents the End-Of-Block. When EOB=0, it indicates there are more symbols to be coded. When EOB=1, it indicates the symbol is the last symbol of the block. The three dimensional symbols are then encoded using Huffman coding. In this way, we do not need to have the special symbol EOB at the end of the data block.

In practical image/video coding, the "run" and "value" can result in a large number of combinations. For example, if the maximum zero-run-length is 63, then the "run" may have 64 possibilities (from 0 to 63). If the "value" is in the range of ±255, then the total number of different (run, value) combinations is 64x510 = 32,640 which may be too large for practical Huffman coding implementations. To reduce the number of symbols, we can limit the zero run-length to a smaller number, say 15. When a string of more than 16 zeros occurs, we can break the long string of zeros into segments with each segment having a zero run-length of 16 and the last segment having a zero-run-length of 16 or less. The symbol (0,0) can be used to represent a string of 16 zeros. In this way, the total number of different (run, value) combinations can be reduced to 16x510 + 1 = 8161. To reduce the number of different combinations even further, in JPEG the "value" is represented by a length indicator called "category" and an associated bit-pattern called "magnitude". The "category" represents the number of bits needed to encode the "magnitude". If the "value" is positive, then the code for the "magnitude" is the least significant bit of its direct binary representation. If the "value" is negative, then the code for the "magnitude" is the one's complement of its absolute value. Therefore, code-words for negative "value" always start with a zero bit. For example, -1 and 1 are represented by a 1-bit "magnitude" of "0" and "1" respectively, and a "category" of 1; -3, -2, 2, and 3 are represented by a 2-bit "magnitude" of "00", "01", "10", and "11" respectively, and a "category" of 2, and so on. More "categories" for the "value"s are shown below. With this representation, the two-dimensional symbol (run, value) can be converted into a three-dimensional symbol (run, category)magnitude. For example, the run-length coded sequence in the previous example can be converted into {(2,2)00, (0,3)101, (0,1)1, (6,2)01}. In this way, only the (run, category) part needs to be Huffman encoded. In the encoding, the (run, category) part of a symbol is used to look up the Huffman code-book. The "magnitude" part is then appended to the code-word. In the decoding, once the (run, category) is decoded, from the category we know the code-length for extracting the bit-pattern for the magnitude part. From the bit-pattern, the actual value can be obtained easily. For the previous example which has values in the range ±255 (8 categories), with this method, the Huffman code entries can be reduced to 16x8 + 1 = 129 which is more practical for implementation, although the coding efficiency will be slightly degraded.

Value	Category
-1,+1	1
-3,-2,+2,+3	2
-7,-6,-5,-4,+4,+5,+6,+7	3
-15,-14,...,-9,-8,+8,+9,...,+14,+15	4
-31,-30,...,-17,-16,+16,+17,...,+30,+31	5
-63,-62,...,-33,-32,+32,33,...,62,63	6
-127,-126,...,-65,-64,+64,65,...,126,127	7
-255,-254,...,-129,-128,128,129,...,254,255	8

15.3 IMPLEMENTATION OF HUFFMAN-BASED ENCODERS AND DECODERS

The run-length coding discussed above is relatively easy to implement by using a counter to count the number of repeated symbols. Huffman encoders and decoders, however, are more complicated. The operation of Huffman encoding and decoding involve bit-level manipulation, thus, the implementation on general purpose processors may be relatively slow for real-time video processing. In Huffman encoding, a symbol to be encoded is used to look up a code-book to obtain the variable-length code-word and the corresponding code-length. Based on the code-length information, a concatenation circuit concatenates the variable-length code-words into a continuous bit-stream and segments the concatenated bit-stream into fixed-length words so that they can be stored in a fixed-length memory after the Huffman encoder. After the concatenation of the variable-length code-words, there is no longer clear symbol-boundaries. In Huffman decoding, the bit-stream has to be decoded from the beginning to obtain the decoded symbols one by one. The decoding of the 2nd code-word cannot start before the first code-word and the corresponding code-lengths are decoded. Due to this inherent serial decoding nature, it is difficult to use traditional high-speed implementation techniques such as parallel processing to achieve high-speed Huffman decoding. There are many different approaches proposed for high-speed Huffman decoding [16]-[23]. Some are more suitable for implementations using hardware or special processors. In the following, we only discuss some approaches which are suitable for implementation in both software and hardware.

15.3.1 Bit-Serial Decoder

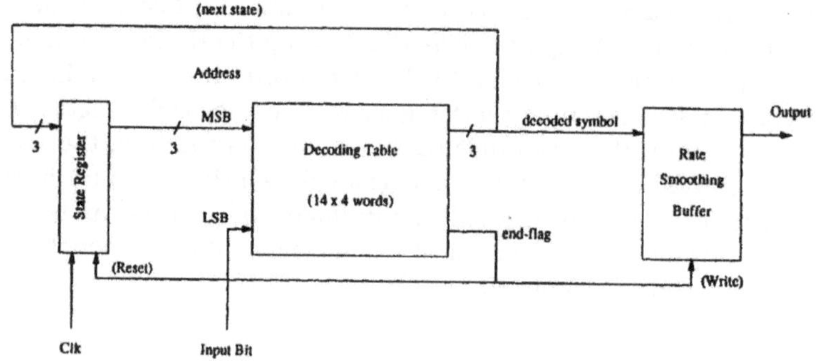

Figure 15.2 A bit-serial Huffman decoder for the codebook in Fig. 15.1.

The basic bit-serial Huffman decoder is shown in Fig. 15.2 and Fig. 15.3. Basically it is a state-machine. Each internal node can be assigned an address to represent its state. The bits above the internal nodes in the Huffman tree in Fig. 15.3 represent the addresses of the states. In the beginning of the operation, the state register is reset to all "0" which corresponds to the root of the Huffman tree. It then accepts the input bit-stream bit-by-bit (can be at a constant bit-rate) and traverses the Huffman tree according to the input bit-patterns. If it does not reach a leaf node, the decoding table will output the address of the state it has reached. The state-address with the next input bit determines the next state. If it reaches a leaf node, a symbol is decoded and the corresponding fixed-length code is output. An end-flag associated with the decoded symbol is used to reset the state register and to write the decoded symbol to the buffer.

Address		Content	
Previous state	Input bit	Decoded symbol/ Next state	End flag
000	0	000 (a)	1
000	1	001	0
001	0	010	0
001	1	011	0
010	0	001 (b)	1
010	1	101	0
011	0	101	0
011	1	110	0
100	0	010 (c)	1
100	1	011 (d)	1
101	0	100 (e)	1
101	1	101 (f)	1
110	0	110 (g)	1
110	1	111 (h)	1

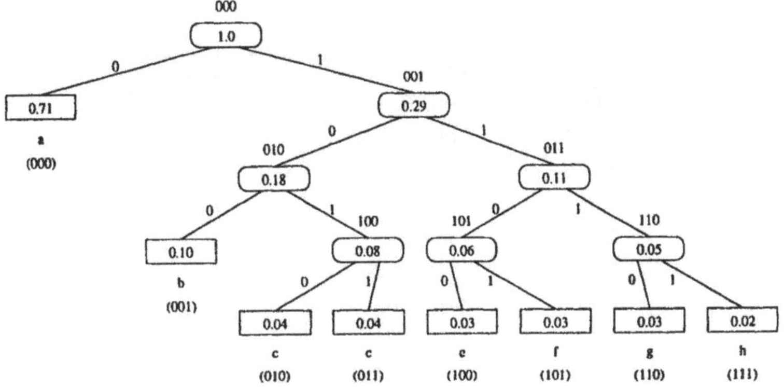

Figure 15.3 The decoding table for the bit-serial decoder in Fig. 15.2 and the code book in Fig. 15.1.

Using this decoder, if a code-word has n bits, it will take n clock cycles to decode the symbol. Since the code-words can have very different lengths, the decoded symbol-rate can be highly variable. Usually a buffer after the Huffman decoder is needed to smooth out the decoded symbol rate.

15.3.2 Constant-Output-Rate Decoder

In the bit-serial decoder, if the maximum code-length is L, then in the worst case, it will take L clock cycles to decode a symbol. For applications which require high throughput, it is desirable to speed up the decoding.

A constant-output-rate decoder decodes a symbol every clock cycle, regardless of its code-length. A straightforward table-lookup Huffman decoder which can achieve this is shown in Fig. 15.4. If the longest code-word in the Huffman codebook is L-bit long, then a look-up-table with 2^L entries is used. The construction of the lookup table for each symbol s_i is as follows: Let c_i be the code-word that corresponds to the symbol s_i. Assume that c_i has l_i bits. We form an L bit address in which the first l_i bits are c_i and the remaining $(L - l_i)$ bits take on all possible combinations of zero and one. Thus, for the symbol s_i , there will be 2^{L-l_i} addresses. At each memory location, we store the two-tuple (s_i, l_i). The decoding is by direct table lookup as follows: From the input, L bits of data are read into a register. The first bit of the register is the beginning of a code-word. The L-bit word in the register is used as an address to read the lookup-table and obtain the decoded symbol s_i, and the corresponding code-length l_i. The first l_i bits from the register are then discarded and we append to the register the next l_i bits from the input, so that the register again has L bits of data which start with the beginning of the next code-word and the operation repeats.

The constant output-rate decoder accepts multiple input bits and decode a code-word every clock cycle. The decoded symbol-rate is constant. Since it decodes variable number of input bits in one clock cycle, the input bit-rate will be variable. This is usually not a problem since in practical systems there is usually a memory before the Huffman decoder which serves as the buffer between the constant bit-rate channel and the video decoder.

Address	Decoded symbol	Code-length
0000	000 (a)	1
0001	000 (a)	1
0010	000 (a)	1
0011	000 (a)	1
0100	000 (a)	1
0101	000 (a)	1
0110	000 (a)	1
0111	000 (a)	1
1000	001 (b)	3
1001	001 (b)	3
1010	010 (c)	4
1011	011 (d)	4
1100	100 (e)	4
1101	101 (f)	4
1110	110 (g)	4
1111	111 (h)	4

The straightforward table look-up method needs a large size memory especially when the maximum word-length is long. For example, if the maximum code-word length is 16-bit, it requires 2^{16} words of memory. A lot of memory spaces are wasted especially for the short code-words. For example, in a Huffman code-book, if there is a 1-bit code and if the maximum code-length is 16, there will be 2^{15} memory locations for this code-word which is very wasteful.

For hardware implementations, it is possible to use a Content-Addressable Memory (CAM) to reduce the required memory size. A CAM performs pattern matching to decode the code-words. For implementation of fixed code-books, a PLA (Programmable Logic Array) can be used for efficient implementation. A content-

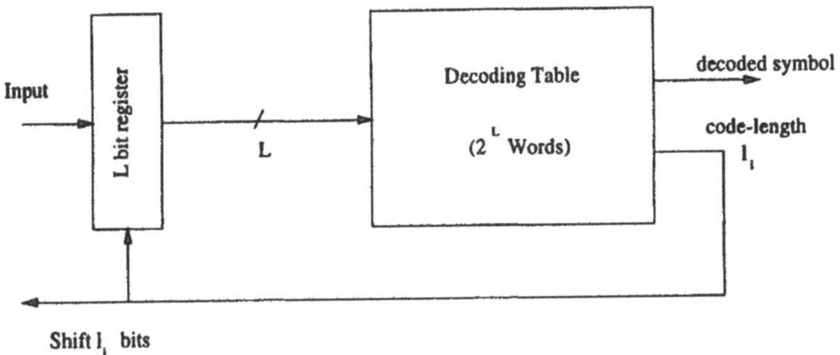

Figure 15.4 A constant output-rate Huffman decoder using table look-up.

addressable memory or PLA does not need to implement those "don't care" terms in the address and thus only needs as many words as the entries in the Huffman code-book. One of the issue of constant output-rate decoder is how to supply the variable number of input bits needed to the decoder from a fixed-length memory in every clock cycle without using a very high-speed clock to shift the data bit-by-bit. An interface circuit between the Huffman encoder/decoder and the fixed-length memory which can achieve this was discussed in [20].

15.3.3 Variable Input/Output-Rate Decoder

For software implementation, a CAM or PLA is not available. To speed up the process of search for a symbol in a Huffman tree and to reduce the memory size, a tree clustering algorithm which is somewhat between the bit-serial and the full table-look-up methods can be used. In the tree clustering algorithm, a Huffman tree is partitioned into clusters form the top. A table is used for each cluster. The algorithm shifts multiple bits in every clock cycle to look up a table containing the end-node and internal states in each cluster. By using this hierarchical memory structure, the memory can be drastically reduced from the full table-look-up method. Although it takes more clock cycles to decode a long code-word compared to the full table-look-up method, since long code-words occur less often, performance degradation may be minor. Using the code-book in Fig. 15.1 as an example, in the first clock cycle, we can shift 4 bits into the register similar to the full table-took-up method. Instead of using all 4-bits to look up a full table, we only use the first bit to look up a memory of 2 words. If the first bit is "0", it will decode the code-word "a", and we shift the register by one bit to start a new decoding. If the first bit is "1", an end-flag indicates that no code-word is matched. We then use the last 3-bits in the register to look up a memory of 8 words to decode the code-word. In this way, a total of 10 words of memory are used, and the code-words are decoded in either one or two clock cycles. The principle can be extended to more general Huffman tree partitions. When the maximum code-length is long and the Huffman tree is sparse, the memory saving can be substantial. This method results in variable input and output rates; so a buffer may be needed before and after the Huffman decoder.

The constant-output-rate decoder discussed above can be modified into another variable-input/output-rate decoder to increase the average decoding throughput further at the expense of more complicated output handling. In the constant-output-rate decoder of Fig. 15.3, since the input register for addressing the memory contains L bits (where L is the maximum code-length), and short cord-words appear quite often in typical situations, the probability that this register contains multiple short code-words is high. Possible combinations of short cord-word patterns can be stored in the memory so that it is possible to decode multiple code-words per clock cycles. Since the number of code-words decoded may not be constant, extra information needs to be provided by the memory to indicate the number of decoded symbols and the decoded symbol patterns.

15.4 ARITHMETIC CODING

Arithmetic Coding [24], [25], [26] is different from Huffman coding in that it does not require integer number of bits to code each specific symbol. Because of this, it can provide better efficiency than Huffman coding. Conventional lossless coding methods map a single symbol to a unique codeword. Arithmetic coding on the other hand codes a sequence of input symbols with a specific code. In its simplest form, a number with a certain precision, in the interval [0,1) is used to represent a sequence of input symbols to be coded. The number of bits of precision to represent this number increases with the number of symbols in the sequence which is encoded.

15.4.1 Introduction

Let the alphabet (Λ) consist of symbols $i \in \{1, 2, \ldots, m\}$. Let S denote the string representing the sequence of symbols coded so far. S will be referred to as context-index in the later discussion. The encoding process consists of following steps:

- **Initialization :** The interval $[0, 1)$ on the number line is partitioned into m non-overlapping subintervals, one for each possible symbol of the alphabet. The ordering of the symbols is agreed upon between the encoder and the decoder.

- **Recursion :** Let at each step, the current interval obtained after encoding a sequence of symbols be denoted by $A(S)$. Then $A(S)$ is subdivided into non-overlapping intervals $A(S1), A(S2), \ldots, A(Sm)$, whose union equals $A(S)$. Each of the subintervals $A(Si), i \in \{1, \ldots, m\}$, is of nonzero length. The recursion continues by choosing the particular i^{th} subinterval $(A(Si))$ as the current interval if i is the new symbol to be coded following the string S.

A typical way to partition the interval on the number line is to use the probability of each symbol. The probabilities must be strictly greater than zero to avoid a zero length subinterval. The recursion step partitions the current interval $A(S)$ in proportion to the symbol probabilities in presence of a new symbol,

$$A(Si) = A(S)P(i|S), \qquad i = 1, \ldots, m. \tag{2}$$

Here $A(Si)$ is the interval corresponding to the symbol i and $P(i|S)$ denotes the conditional probability of the next symbol being i given the earlier coded string of symbols to be S.

After following the recursive partitioning, the entire sequence of symbols will be encoded by sending a bit string representing a point within the final interval $A(S)$. Any point within this interval could be sent. It is important to note that the interval includes only one of its two endpoints. Thus it is represented as a semi-open interval. For example, at the start of the encoding process $[0, 1)$ interval includes starting point "0" but not the end point "1". It is sufficient to send first B bits of the binary expansion of a point within the final interval, for the decoder to be able to uniquely decode it, where

$$B \leq \lceil -\log_2 A(S) \rceil + 1, \tag{3}$$

and the notation $\lceil n \rceil$ represents the smallest integer which is grater than n. More specifically, assuming the sequence of N symbols $s_1, s_2, \ldots, s_N$ to be encoded and the recursion update using the conditional probabilities allows to write the above equation as

$$B \leq \lceil -(\log_2 P(s_1) + \sum_{i=2}^{N} \log_2 P(s_i|s_1, s_2, \ldots, s_{i-1})) \rceil + 1. \tag{4}$$

The arithmetic decoding process traces the following steps,

- **Recursion** This requires at each step to compare the magnitude of the received bit string with the boundaries of each subinterval and to select the one to which it belongs. One symbol is recursively decoded at each step.
- **Termination** There is some termination criterion which is agreed upon between the encoder and the decoder. The decoding is done till this termination condition is encountered.

15.4.2 Arithmetic Coding Issues

15.4.2.1 Decoding termination criterion

There has to be a way for the decoder to know when to stop decoding the received string. A decoding example in the following subsections illustrates how it is possible for the decoder to go on decoding the same string up to any number of symbols. The commonly used decoding termination criteria are:

- The encoder may explicitly send the information about how many symbols the encoded string represents. The decoder then decodes those many symbols and stops.
- A special symbol is encoded at the end of the message. This symbol does not occur in the message. The decoder will go on decoding symbols until it decodes this special symbol which indicates it should terminate the decoding process.

15.4.2.2 Incremental transmission and reception

The basic arithmetic encoding and decoding algorithms described above assume that the encoded output corresponding to the entire message sequence is available at the end of encoding. Similarly the decoder cannot begin decoding until the entire transmitted string is received. In practical situations incremental transmission at the encoder and incremental reception at the decoder is necessary. The various implementations described in the Section 15.5 discuss ways in which this can be achieved. The main idea is to transmit those bits which are not going to change because of the encoding of additional subsequent symbols. Similar analysis is done at the decoder to incrementally decode and discard the high end bits which are no longer needed.

15.4.2.3 Non-adaptive vs. adaptive coding

In non-adaptive coding, the probability for each symbol is fixed at the start. This is known to both the encoder and the decoder. With this the conditional probability is the same as the individual probability, i.e., $P(i|S) = P(i)$.

In the adaptive coding, the probability of the symbols depends upon the symbols encoded till that time. A common choice is for the encoder and the decoder to agree upon some specific rule to update these conditional probabilities dynamically. The encoder and the decoder both have the knowledge of initial probability of each symbol.

15.4.3 Examples

We provide an example of the arithmetic encoding and decoding to make the above discussions more clear.

15.4.3.1 Arithmetic encoding example

Let the alphabet consist of four symbols, $\Lambda = \{a, b, c, !\}$. The respective probabilities are $P(a) = 0.5, P(b) = 0.25, P(c) = 0.125, P(!) = 0.125$. We want to encode a message string "ac!". Fig. 15.5 illustrates the encoding steps.

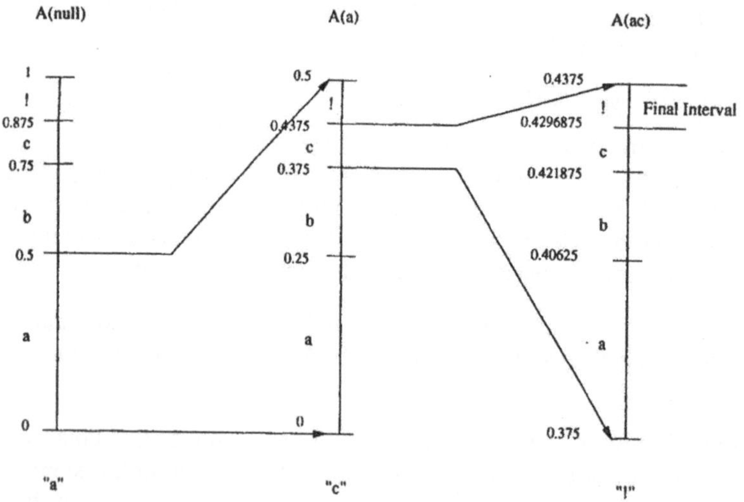

Figure 15.5 An arithmetic encoding example.

1. The interval $[0, 1)$ is divided according to the symbol probabilities. The ordering of the symbols is agreed upon between the encoder and decoder.

2. The first symbol to be encoded is "a" hence the subinterval corresponding to "a" which is [0, 0.5) ($A(a) = A(null)P(a|null)$) is further partitioned into four possible next symbols. Since the second symbol to be encoded is "c", the particular subinterval [0.375, 0.4375) which corresponds to it is chosen and made the current interval for the recursive subdivision.

3. The last symbol to be encoded is "!" and the encoder divides the current interval into appropriate partitions as before and picks the subinterval [0.4296875, 0.4375) which corresponds to "!".

4. The binary representation for this fractional subinterval is given by [0.0110111, 0.0111). It should be noted again that the lower end of the interval is included but the upper end is not included. The encoder can send any number within this interval to represent the sequence of symbols "ac!". One of the widely used conventions is to send the lower end of the interval, which in this case would mean encoding the string as the fractional part given by bit string 0110111. This string will be sent to the decoder which can trace the above steps and map it to the symbols "ac!"

The arithmetic encoding achieves compression because a higher probability symbol occupies larger interval on the number line and a sequence of higher probability symbols will need less number of bits to represent the final subinterval.

15.4.3.2 Arithmetic decoding example

The decoding example shows a typical problem that the decoder may face if it does not know when to stop decoding. In the example the decoder can decode the same received bit string up to any number of symbols. In an actual case, the decoder will just decode till the termination condition is satisfied.

Let us use the same alphabet with four symbols $\{a, b, c, !\}$ and each symbol having same probabilities as that used in the encoder example above. Let the received string at the decoder be "0110111". This bit string is the same as that obtained from the above arithmetic encoding example. Since this string represents the bits after the decimal, this corresponds to 0.4296875 as the decimal fraction value. Fig. 15.6 shows the steps in which this received bit string is decoded up to 3 symbols or up to 5 symbols.

1. The interval [0, 1) is divided according to the symbol probabilities. The ordering of the symbols is the one which is the same as that used at the encoder.

2. Comparing the magnitude of the received string (0.0.4296875), we see that it falls in the rage [0, 0.5), hence the first symbol decoded is "a".

3. Recursively subdividing the [0, 0.5) interval into four subintervals and comparing the magnitude of the received string, it is observed to fall in [0.375, 0.4375) which decodes the symbol "c".

4. Continuing in this manner the interval [0.375, 0.4375) is subdivided and the received string is found to lie in the subinterval [0.4296875, 0.4375). This gives symbol "!" as the decoder output. If the decoder knew that the encoded bit

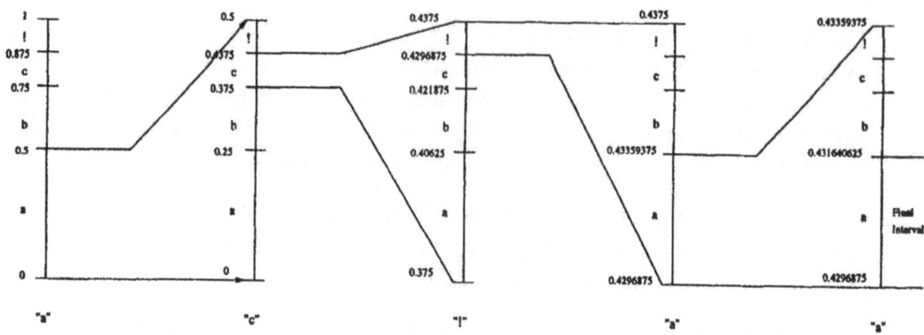

Figure 15.6 An arithmetic decoding example. The decoded output up to 3 symbols is "ac!" while that up to 5 symbols is "ac!aa".

string only contains three symbols (or if "!" was known to be the termination symbol), then it has reached the termination condition and will stop at this point, therefore the decoded output would be three symbol sequence "ac!".

5. However, if the received bit string represents encoding of say five symbols then the decoder can further continue to decode the same bit string. In this case the recursive subdivision process will find that the received string lies in the subinterval [0.4296875, 0.43359375) thus decoding symbol "a". And following the same step, in the subinterval [0.4296875, 0.431640625) at the next recursion, which gives "a" as the decoded output. (Note again that 0.4296875 which is the lower end-point, is included in these subintervals while the upper end point is not a part of these subintervals and instead belongs to the next subintervals.) Now a five symbol sequence ("ac!aa") has been decoded and the decoding process is complete.

15.5 IMPLEMENTATION OF ARITHMETIC CODERS

In the previous section we gave a theoretical overview of the arithmetic coding. This section focuses on various software and hardware implementations of arithmetic coders. Issues that are addressed include:

- Finite precision implementations: Numbers represented in computers do not have infinite precision. Various issues arise when implementing arithmetic coding in software or hardware with a finite precision.
- Multiplication-free implementations: Arithmetic coding requires multiplications which might substantially slow down the encoding/decoding processes. Multiplication-free implementations use some approximations to avoid multiplications.

15.5.1 15.5.1.1 QM Coder

QM coder is an adaptive binary arithmetic coder. It encodes a sequence of binary symbols, 1 and 0. QM coder is used in JPEG (Joint Photographic Experts Group) [28] and JBIG (Joint Bi-level Image Experts Group) [28]. It tries to combine and enhance features of different previous implementations of arithmetic coders [29]-[30]. Different architectures which are suitable for either software (sequential) or hardware (parallel) implementations are discussed.

QM encoder

The QM encoder in JPEG consists of following four procedures for binary arithmetic encoding.

- Initenc : to initialize the encoder.
- Code_0(S) : code a "0" binary decision.
- Code_1(S) : code a "1" binary decision.
- Flush : end of the coded segment.

Here S represents context-index which selects a particular conditional probability estimate to code the binary decision.

15.5.1.2 Symbol ordering

At each stage, the two symbols are denoted as more probable symbol (MPS) and less probable symbol (LPS). Two fixed size registers (or variables) A and C keep track of the width of the current interval and the position of the code-stream (pointer to somewhere within the current interval). Let $Q_e(S)(< 0.5)$ denote the current estimate of LPS probability for context index S. In the following discussion we may drop the context index from our notation to avoid cluttering of various equations. With this, $AQ_e(S)$ and $A(1 - Q_e(S))$ respectively denote the precise probability subintervals for LPS and MPS. Symbol ordering concerns with placing MPS and LPS and C pointer on the number line. C can be pointing to the bottom or top of the interval. However these arrangements result in codes with the same efficiency [28]. In the following, we will assume C points to the bottom of the interval. Fig. 15.7 shows two possible arrangement conventions for the symbol ordering:

- (I) MPS below LPS, and
- (II) LPS below MPS

15.5.1.3 Recursive interval subdivision

To avoid the costly multiplications in the above steps, an approximate algorithm takes advantage of a renormalization strategy which makes $0.75 \leq A \leq 1.5$, i.e., $A \approx 1$. This results in approximating $A \times Q_e$ as Q_e and $A(1-Q_e)$ as $A-Q_e$ so that the sum is still A. Using the symbol ordering convention of (I) and with the above approximations, we have the following algorithm for the recursive interval subdivision.

$A = A - Q_e$
if symbol to code = MPS
$C = C$ /* no operation needed */
if symbol to code = LPS
$C = C + A$
$A = Q_e$

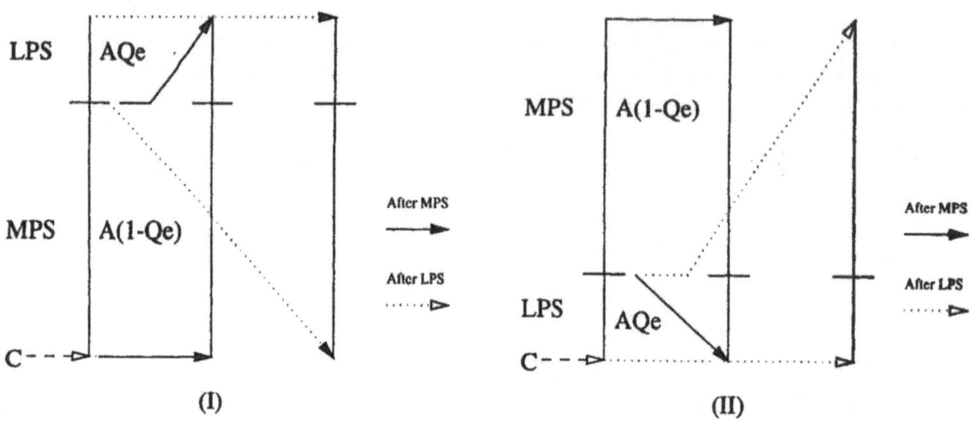

Figure 15.7 Possible MPS, LPS symbol ordering.

If the symbol ordering convention of (II) is used, the algorithm is give by:

if symbol to code = MPS
$A = A - Q_e$
$C = C + Q_e$
if symbol to code = LPS
$A = Q_e$

JPEG Annex D [28] uses convention (I) which places MPS subinterval closer to zero and below LPS subinterval. This symbol ordering is more suitable for software implementation because the more probable path (coding MPS) requires less computation. For hardware implementation, the second ordering (II) is preferred as then MPS path operations of updating A and C can be done in parallel. The similar A and C update operations on LPS path in (I) could not be done in parallel. However the approach (II) requires more computation on more probable path hence reducing efficiency if implemented in software. In the following, we will assume the ordering (I).

15.5.1.4 Renormalization and conditional exchange

Without any normalization, the above ideal algorithm would require an infinite precision for A and C, since A will get smaller and smaller, and C will need more and more bits to represent the end point of the interval. A simple renormalization strategy doubles A every time it is less than a threshold. An identical renormalization of C is also done. After each renormalization, the MSB after the binary point will be shifted into the integer part of the code stream.

With the renormalization, the encoding procedure uses fixed precision integer arithmetic. In this integer representation of fractional values, X'8000' and X'10000' can respectively be regarded as the decimal values 0.75 and 1.5. To prevent overflow of the C register, when the high-order bits of C register contain a data byte which will not change in the later encoding process, they can be output so that the space can be used to accumulate later data bits.

Since A is updated by the approximated value $A-Q_e$, there exists a possibility with the above recursive interval subdivision algorithm with renormalization that can make the updated MPS subinterval smaller than the updated LPS subinterval. The conditional exchange can be used to invert the sizes of MPS and LPS subintervals in this case. The conditional exchange is necessary only when renormalization has been carried out [28].

For each of Code_0(S) and Code_1(S) the binary decision depends upon the current context. Context-index S is determined by prior coding decisions. A renormalization step is followed by probability estimation step. This step determines a new probability estimate for the context which is being encoded currently. Details about the initialization, output procedure and the probability estimation can be found in [28].

15.5.1.5 QM decoder

The decoder retraces the steps followed by the encoder. It decodes a binary decision by searching to find the subinterval pointed by the code stream. This is accomplished by using two registers (A, C) at the decoder. Decoder A register holds the current interval, similar to the encoder A register. Decoder C register can be divided into two parts named: Cx and $Clow$. Part Cx holds the pointer to the current subinterval, where as part $Clow$ holds new data bits. During the renormalization, register A is renormalized and registers Cx and $Clow$ are shifted together in pair, so that the MSB of $Clow$ register gets shifted into the LSB of the Cx register. The decoding procedure consists of conditional exchange, renormalization and probability estimation sub procedures as described above. The flow chart is shown in Fig. 15.8, and is explained as follows:

1. `con_exc` represents the conditional exchange procedure. `renorm` and `prob_est` represent the procedures used for renormalization and probability estimation respectively. D denotes either MPS or LPS.

2. One symbol (MPS/LPS) is decoded at a time. The decision of one symbol leads the decoder to subtract from the code register the amount which the encoder had added. Code register points to the subinterval relative to the base of the current interval.

3. The code register is compared to MPS subinterval size. This can lead to a conditional exchange and renormalization. The conditional exchange is tested by comparing A with $Q_e(S)$. The renormalization consists of shifting both A and C registers till the most significant bit (MSB) of the A register is set.

4. The probability estimation step is similar to that at the encoder.

15.5.2 H.263: Syntax-based Arithmetic Coding Mode

H.263 is a very low bit rate video coding standard [8] from International Telecommunication Union (ITU). The H.263 standard has a number of optional coding modes which can improve the compression efficiency. One of these negotiable optional coding modes is the Syntax-based Arithmetic Coding (SAC) mode.

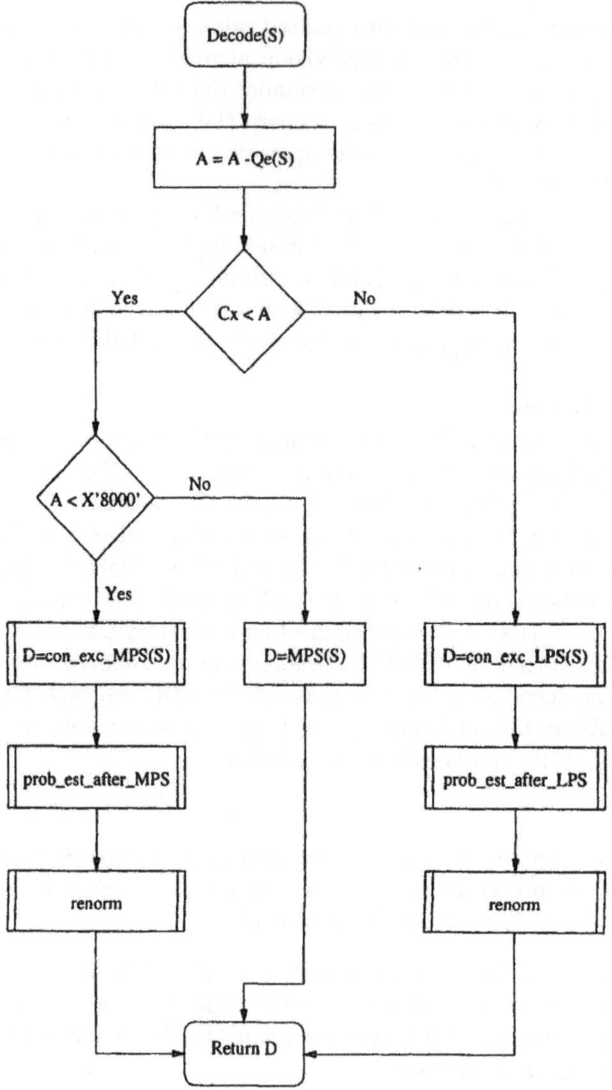

Figure 15.8 The flow chart for a QM decoder procedure.

In this mode, arithmetic coding/decoding replaces corresponding Variable Length Coding/Decoding (VLC/VLD).

Encoder specifications

The H.263 bit stream syntax has various fields which have either fixed or variable length. Examples of the possible fixed length symbol strings include those at Picture Start Code (PSC), Group of Block Start Code (GBSC), and End Of Sequence (EOS). The PSC_FIFO referred to in the flow charts below is a FIFO buffer of size more than 17 bits. It exists at both the encoder and decoder to avoid illegal PSC and GBSC emulations. These will be avoided by stuffing a "1" after every consecutive occurrence of 14 "0"s at the encoder. Similarly the decoder removes the stuffed "1" bits appropriately. To encode the variable length part, the H.263 standard provides SAC models which consist of cumulative frequency tables for the symbols.

The flowchart in Fig. 15.9 illustrates the specifications for the encoder given in the H.263 standard document for the encoding of a single symbol.

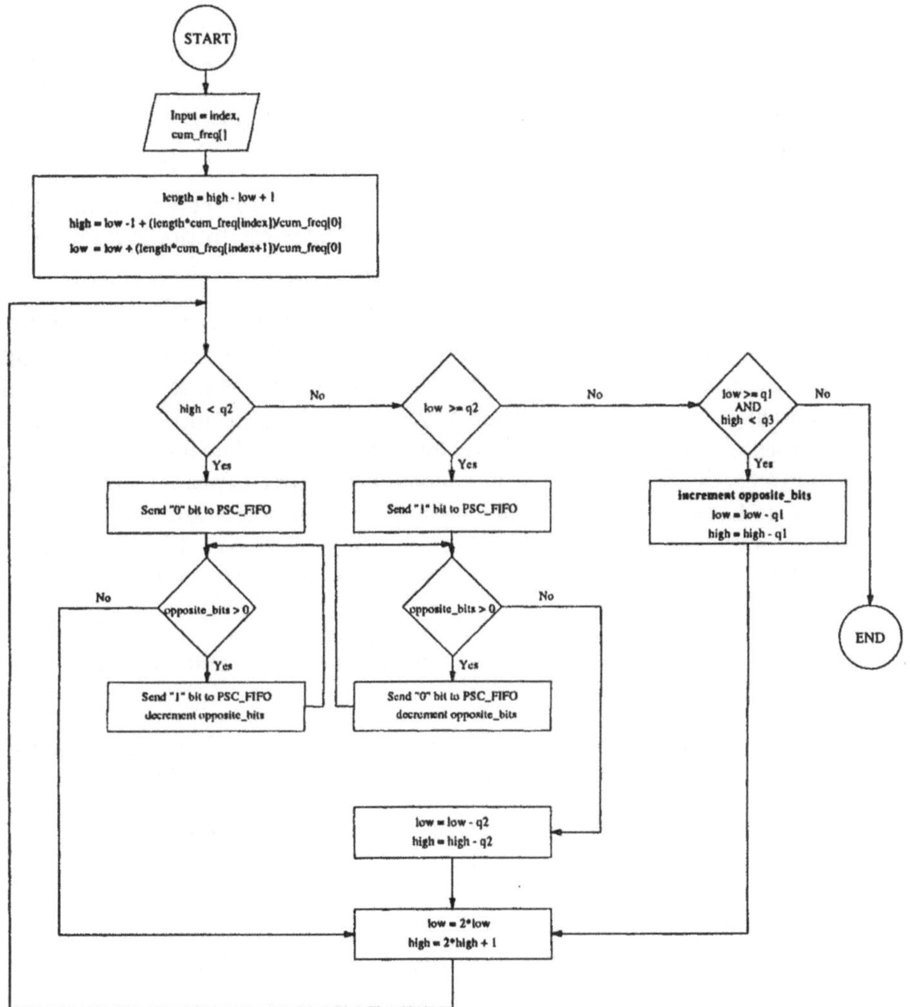

Figure 15.9 The flow chart for encoding a symbol in syntax-based arithmetic coding (SAC) mode of the H.263 video coding standard.

Some of the salient points about the above arithmetic encoding scheme are:

1. It uses integer arithmetic for all the operations.

2. Some of the static constants used are `q1 = 16384, q2 = 32768, q3 = 49512, top = 65535`. q2 is a midpoint of q1 and q3 interval which are at the

quarter positions from the two end points of the [0, 65535] interval. The arithmetic encoding process initialization sets `low`, `high`, and `opposite_bits` to `0`, `top`, and 0 respectively.

3. The input to the single symbol encoder above is an `index` of the symbol to be encoded along with an array containing the cumulative frequencies of the occurrence of the symbols. The array is ordered in opposite direction, i.e., as the array index increases the cumulative frequency decreases. The first entry in the array is the sum of the frequencies of all the individual symbols and hence is used as the normalizing factor. All the symbols have a strictly nonzero positive frequency of occurrence so as to avoid any zero length on the number line for any symbol's interval.

4. The first step in encoding a symbol is to modify the `low` and `high` variables representing the end points of the current subintervals to the end points of the symbol being currently encoded.

5. The next step is a three-way decision to check where the current interval lies with respect to `[q1,q3]` interval's midpoint q2. If it lies in the left or the right half of the midpoint then it indicates the existence of some of the top bits of `low` and `high` being the same. These bits can then be transmitted immediately (incremental transmission) since they cannot get affected by further narrowing of the interval. The number of bits which are output depends upon a count `opposite_bits` which is incremented and the end-points `low` and `high` are shifted, when the arrangement on the number line is `q1`, `low`, `q2`, `high`, q3. This also means the `[low,high]` subinterval lies in the middle half of the number line.

6. After one of the above three steps, the end-points `low` and `high` are scaled and the process is repeated.

7. The termination condition occurs if none of the three possibilities given above are satisfied. This indicates the end of encoding for this particular symbol.

The encoder is flushed after the end of the encoding session or before EOS (End Of Sequence). The details about the encoder flushing in SAC mode of H.263 can be found in [8].

15.5.2.2 Decoder specifications

The decoder traces the steps followed by the encoder. The flowchart in Fig. 15.10 illustrates the operations done while decoding a symbol.

The decoding flow chart retraces the steps of the encoder.

1. The initialization (referred as `decoder_reset` in the standard) sets `low=0`, `high=top` and `code_value` to 16 bits of the received encoded value.

2. The first step searches the cumulative frequencies along with the current `[low,high]` interval to locate the `index` corresponding to the `code_value`.

3. Next step again finds the position of the interval end points `low` and `high` with respect to `q1`, q2 and q3 and gets rid of the top bits which are the same

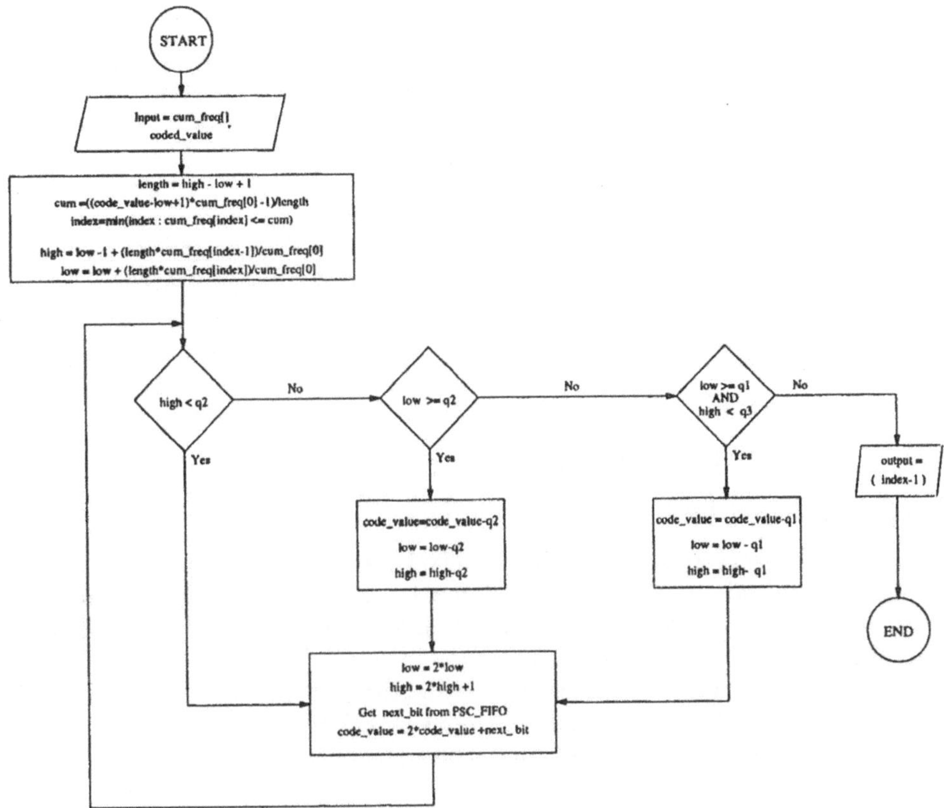

Figure 15.10 The flow chart for decoding a symbol in syntax-based arithmetic coding (SAC) mode of the H.263 video coding standard.

in `low` and `high` and replaces them by next input bits (from `PSC_FIFO`) at the bottom end. These bits are removed because they have already been used and are not useful for decoding the next symbols.

15.5.3 Finite Precision Arithmetic Coder Implementations

In this section we will look at various new finite precision implementation algorithms for arithmetic coding. We will address

- unique decodability: in view of the finite precision and approximations used in the algorithms, and
- performance degradation: arising due to these factors.

The double recursion of arithmetic coding can be expressed with the following equations:

$$A(Si) = A(S) \times P(i|S), \tag{5}$$

$$C(Si) = C(S) + A(S) \times \sum_{j=1}^{i-1} P(j|S), \tag{6}$$

where $P(i|S)$ is the conditional probability of symbol i when S is the previous string. The initial condition is given by $A = 1$ and $C = 0$ which corresponds to interval [0,1) on number line. Let us assume that the alphabet consists of one of m possible symbols which are indexed from 1 to m. Also we will follow the convention in which C register points to the bottom of the current interval and A register holds the size of the interval.

15.5.3.1 Unique decodability

Let S_n denote the entire message string to be coded. The unique decodability conditions guarantee that $C(S_n)$ is different for different message strings.

The necessary and sufficient conditions for unique decodability are given by:

1. $C(S) \leq C(S1)$,
2. $(C(Si) + A(Si)) \leq C(S(i+1))$, where $i = 1, 2, \ldots, m-1$,
3. $C(Sm) + A(Sm) \leq C(S) + A(S)$, and
4. $A(Si) > 0$, for $i = 1, 2, \ldots, m$.

These conditions ensure that there is no interval overlap (conditions 1,2,3), and no zero length intervals are generated (condition 4). Any finite precision and/or approximate arithmetic coding needs to satisfy these unique decodability conditions.

15.5.3.2 Finite precision algorithms

Finite precision representations of probabilities $\{P(i|S)\}$, $A(S)$ and $C(s)$ will cause performance degradation and may result in violation of unique decodability conditions. This section discusses some finite precision algorithms proposed by researchers. These algorithms essentially involve some approximations. The unique decodability condition of Section 15.5.3 is satisfied by these algorithms.

The first two algorithms in [31] use truncation and/or rounding approximation of double recursion to update A and C. The component which has significant impact [31] on the performance concerns the finite precision of $A(S)$ and $C(S)$.

The first algorithm uses truncation. The update equations are given by

$$A(Si) = [[A(S) \times P(i|S)]_T]_N, \tag{7}$$

$$C(Si) = [C(S) + [A(S) \times \sum_{j=1}^{i-1} P(j|S)]_T]_N, \tag{8}$$

with $[\cdot]_T$ and $[\cdot]_N$ representing truncation and normalization operations, respectively. With these update equations for A and C the unique decodability conditions are satisfied [31] provided $P(i|S) \geq 2^{-(L-1)}$ where L is the length of A and C registers and normalized $A \in [1, 2)$. The constraint on $P(i|S)$ is necessary to ensure the unique decodability condition 4.

Let $E(i)$ represent the truncation error in $A(Si)$ update equation above, i.e.,

$$E(i) = A(S) \times P(i|S) - [[A(S) \times P(i|S)]_T]_N. \tag{9}$$

Then the average rate increase due to truncation is given by :

$$D = \sum_{i=1}^{m} P(i|S) \log_2 \frac{A(S)P(i|S)}{[[A(S)P(i|S)]_T]_N} = \sum_{i=1}^{m} P(i|S) \log_2 \frac{A(S)P(i|S)}{A(S)P(i|S) - E(i)}. \tag{10}$$

Assume $E(i)$ is uniformly distributed over $[0, 2^{-(L-1)})$. Further with the normalization operation, $A(S) \in [\alpha, 2\alpha)$ where α is a constant. Assuming $A(S)$ is uniformly distributed and with $\alpha = 1$, the degradation can be approximated by [31]

$$D \approx m2^{-L}. \tag{11}$$

This indicates that the degradation drops by half for each increased bit precision (L) and is proportional to the number of symbols (m).

The second algorithm can use either rounding or truncation. Replacing the truncation operations in the above algorithm by rounding operations can violate unique decodability conditions 2 and 3. The following algorithm which is a modified version can be used either with rounding or truncation operations

$$A(Si) = [[A(S) \times \sum_{j=1}^{i} P(j|S)]_R - [A(S) \times \sum_{j=1}^{i-1} P(j|S)]_R]_N, \tag{12}$$

$$C(Si) = [C(S) + [A(S) \times \sum_{j=1}^{i-1} P(j|S)]_R]_N, \tag{13}$$

with $[\cdot]_R$ representing the rounding operation. With this algorithm the unique decodability condition 2 and 3 are satisfied with an equality. The condition 4 again imposes the same constraint as the above algorithm on $P(i|S)$.

The performance degradation analysis for this algorithm follows the same steps as above. The average rate increase due to rounding equal to

$$D = \sum_{i=1}^{m} P(i|S) \log_2 \frac{A(S)P(i|S)}{A(S)P(i|S) - E(i) + E(i-1)}, \tag{14}$$

with $e(0) = e(m) = 0$. D can be rewritten using a few approximations as,

$$D \approx \frac{2^{-2L}}{6 \ln 2} \left(\frac{1}{2P(1)} + \frac{1}{2P(m)} + \sum_{i=2}^{m-1} \frac{1}{P(i)} \right), \tag{15}$$

after a few approximations and with the uniform distribution assumption similar to the previous algorithm. Hence the rate increase drops by one fourth for every single increased bit, which is better than the previous algorithm (see (11)).

The algorithm is also valid if rounding operation is replaced by truncation and this leads to a performance degradation given by

$$D \approx \frac{2^{-2L}}{6 \ln 2} \left(\frac{2}{2P(1)} + \frac{2}{2P(m)} + \sum_{i=2}^{m-1} \frac{1}{P(i)} \right). \tag{16}$$

This performance is better than the first algorithm with truncation (see (11)). Also this is just slightly inferior to the one using rounding.

15.5.3.3 Improved Algorithm Using Entire Code Space

The algorithm in [32] is an improved version of the first algorithm with truncation given in (7), (8), in which the partitioning of the code space leaves some portions empty so that

$$A(S) \geq A(S1) + A(S2) + \ldots + A(Sm). \tag{17}$$

The improved algorithm makes sure that the above equation is satisfied as an equality at every recursion.

Assuming m possible symbols in the alphabet as above, the symbol having the maximum probability is positioned last on the number line. The ordering of the remaining symbols is arbitrary and does not affect the performance of the algorithm. Let $E(i)$ denote the truncation error for each of the $m-1$ least probable symbols. Then the code space size assigned to the most probable symbol is given by $AP(m) + \sum_{i=1}^{m-1} E(i)$ where $P(m)$ is the probability of the most probable symbol. This makes sure that the entire code space is occupied without any gaps. The recursive update equations for A and C are given by :

$$A(Si) = [A(S) \times P(i|S)]_T, \qquad i \neq m \tag{18}$$

$$A(Sm) = A(S) - [A(S) \times \sum_{i=1}^{m-1} P(i|S)]_T, \tag{19}$$

$$C(Si) = C(S) + [A(S) \times \sum_{j=1}^{i-1} P(j|S)]_T. \tag{20}$$

These update equations satisfy the unique decodability conditions. With this formulation the degradation is given by

$$D = \sum_{i=1}^{m} P(i|S) \log_2 \frac{A(S)P(i|S)}{A(S)P(i|S) - E(i)} - P(m|S) \log_2 \left(1 + \frac{\sum_{i=1}^{m} E(i)}{A(S)P(m|S) - e(m)}\right). \tag{21}$$

Assuming uniform distribution for error $E(i) \in [0, 2^{-(L-1)})$ and $A \in [1, 2)$, the first term is exactly the same as the term in the first algorithm with truncation. The expected value of the second term varies with the probability of the most probable symbol ($P(m)$). However with 16 bit word length, the value converges if $P(m) > 0.1$. Simulations with value of $P(m) > 0.1$ show that the degradation for this algorithm reduces by a quarter for each additional bit for the word length. Also the degradation is smaller as compared to the previous two algorithms proposed in [31].

15.5.3.4 Multiplication Free Algorithm

All the above algorithms except the QM coder involve multiplications which might take up significant time thus hindering a fast implementation. An algorithm proposed by Rissanen and Mohiuddin [33] makes use of the normalization process to make an approximation which eliminates any multiplication. All the operations can be carried out by using addition, subtraction and shifting operations. The algorithm can work either using symbol probabilities or number of symbol occurrences.

Let the alphabet consists of the symbols $i \in \{1, \ldots, m\}$. Let n_i denote the integer valued symbol occurrence count for symbol i. The symbols are arranged such that the last symbol (n_m) is the most probable symbol. The probability of occurrence of a particular symbol i is given by $P(i|S) = n_i/n$, where $n = \sum_{i=1}^{m} n_i$. A normalization of $A(S)$ is done at each step to bring it within a suitable range $[\alpha, 2\alpha)$. Value of $\alpha = 0.75$ is recommended by Rissanen and Mohiuddin [33] based on their simulations. The normalization operation is represented by

$$a(S) = A(S)2^{k(S)}, \tag{22}$$

where $k(S) \geq 0$ is the number of shifts to bring $a(S)$ within the normalization range. The integer occurrence counts n_i are also brought in the normalization range using shifting. Thus

$$\bar{n} = n2^{l(n)}, \qquad \bar{n}_i = n_i 2^{l(n)}. \tag{23}$$

$C(S)$ which represents the code stream and points to within the $A(S)$ interval is also normalized to be in the same range. The update equations are similar to those given by equations (18)-(20) above. Although this update scheme is similar to the previous algorithm the following approximation has been used which makes the algorithm multiplication free.

$$\begin{aligned} a(Si) &= [[a(S)P(i|S)]_T]_N \\ &= [[a(S)\frac{n_i}{n}]_T]_N \\ &= [[a(S)\frac{\bar{n}_i}{\bar{n}}]_T]_N \\ &= [[\bar{n}_i\frac{a(S)}{\bar{n}}]_T]_N \\ &\approx [\bar{n}_i]_N. \end{aligned} \tag{24}$$

The above approximation makes use of the normalization which brings $a(S)$ and $\bar{n}$ both in the same range and their values being of the order of unity. The equation for $C(S)$ is also obtained using the same approximation. The above approximation may violate the unique decodability. Particularly, because of the approximation used, it may turn out that $a(s) < \sum_{j=1}^{m-1} \bar{n}_j$. To avoid this an additional factor 2^{-1} may be used to adjust for the rough approximation. With this, the update equations become:

$$\begin{aligned} a(Si) &= [2^{-\beta}\bar{n}_i]_N, \qquad i \neq m \\ &= \left[a(S) - 2^{-\beta}\sum_{j=1}^{m-1}\bar{n}_j\right]_N, \qquad i = m, \\ C(Si) &= \left[C(S) + 2^{-\beta}\sum_{j=1}^{i-1}\bar{n}_j\right]_N, \end{aligned} \tag{25}$$

where $\beta = 0$ if $a(S) > \sum_{j=1}^{m-1} \bar{n}_j$ and $\beta = 1$ otherwise.

To find the degradation (rate increase), probabilities of symbols are used instead of the occurrence counts. Let L_a and L_i be the actual and ideal lengths obtained after coding N symbols. Then we have

$$\begin{aligned} L_a &= -\sum_{t=0}^{N-1} \log_2 \frac{a(S(t)i(t+1))}{a(S(t))} - \\ &\quad \sum_{t:i(t+1)=m} \log_2 \left(a(S(t)) - \frac{(1-P(m))}{2^{\beta(t)}}\right) - \log_2 \frac{P(m)}{2^{\beta(t)}}, \\ L_i &= \sum_{t=0}^{N-1} \log_2 P(i(t+1)|S(t)). \end{aligned} \tag{26}$$

Here t is used as the time index and the context index $S(t)$ varies with t. Also $i(t+1)$ denotes the symbol $i \in \{1, 2, \ldots, m\}$ being coded at time $(t+1)$. Assume an uniform distribution for $a(S(t))$ in the interval $[0.75, 1.5)$ and define the per symbol code length in ideal (l_i) and actual (l_a) case as the average values after encoding N symbols. Then the excess per symbol rate (difference between actual and ideal values) is given by the following approximation:

$$\begin{aligned}\delta &= l_a - l_i \\ &\approx \frac{(0.25 - P(m))(1 - P(m))}{0.75} + P(m)\log_2 P(m) - \\ &\quad P(m)\left(\frac{1 - P(m)}{1.5}\log_2\frac{1 - P(m)}{2e} + \frac{0.5 + P(m)}{1.5}\log_2\frac{1 + 2P(m)}{e}\right) + \\ &\quad 0.142267, \end{aligned} \tag{27}$$

where e is 2.7182818. The selected normalization range $[0.75, 1.5)$ was chosen because it turns out to be optimal based on simulations [33] for various values of $P(m)$, using derivation for the excess code length for a general scaling range $[\alpha, 2\alpha)$.

15.5.3.5 Generalized multiplication free arithmetic codes

A multiplication free coder is one which has no more than one shift and one add (or subtract) operation for its approximate algorithm. The alphabet size is m as above. The symbol probabilities are $P(1), P(2), \ldots, P(m)$, with $P(m)$ corresponding to the symbol with the maximum probability. For the ease of notation we have dropped the context-index S from the conditional probabilities above. A denotes the exact interval width. Various algorithms approximate this interval width as $\hat{A}$ and they differ in the way the approximation is done. $\hat{A}$ has discrete values which can be represented by simple combination of powers of two so that the multiplication of $\hat{A}P(i)$ can be achieved by shifting. The details of these various approximation schemes can be found in [34],[35]. The multiplication-free codes introduced in [34], [35] were generalized in [36]. The renormalization interval used is $[0.5, 1)$.

The code space is assigned such that least probable $(m-1)$ symbols each occupy width $\hat{A}P(i)$ for $i = 1, \ldots, m-1$ and the remaining width $(A - \hat{A}[1 - P(m)])$ is assigned to the symbol with the maximum probability. Denoting the actually used conditional probabilities by $q(i)$ (again without the context-index) we have

$$\begin{aligned} q(i) &= \frac{\hat{A}}{A}P(i), \qquad i = 1, \ldots, m-1, \\ q(m) &= 1 - \frac{\hat{A}}{A}(1 - P(m)). \end{aligned} \tag{28}$$

The increase in coding rate (excess code length) is then given by [35]

$$D = \sum_{i=1}^{M} p(i)\log_2\left(\frac{P(i)}{q(i)}\right) = \log_2\frac{A}{\hat{A}} - P(m)\log_2\left(1 + \frac{\frac{A}{\hat{A}} - 1}{P(m)}\right). \tag{29}$$

D decreases monotonically with increasing $P(m)$. An upper bound on D is obtained by using the fact that $1/m \leq P(m) \leq 1$,

$$D \leq \log_2\frac{A}{\hat{A}} - \frac{1}{m}\log_2\left(1 + m\left(\frac{A}{\hat{A}} - 1\right)\right). \tag{30}$$

The current interval $[\hat{A}_0, \hat{A}_n)$ is subdivided into n disjoint subintervals $[\hat{A}_0, \hat{A}_1),[\hat{A}_1, \hat{A}_2),\ldots,[\hat{A}_{n-1}, \hat{A}_n)$ so that the values within a particular subinterval $[\hat{A}_{i-1}, \hat{A}_i)$ get mapped to $\hat{A}_{i-1}$ or $\hat{A}_i$ depending upon the truncation or rounding strategy used by the algorithm. Defining the auxiliary functions

$$\begin{aligned} Y &= \frac{A}{\hat{A}} - 1 \\ g(Y) &= \log_2(1+Y) - P(m)\log_2\left(1 + \frac{Y}{P(m)}\right), \end{aligned} \tag{31}$$

and assuming A as uniformly distributed in $[\hat{A}_0, \hat{A}_n)$, leads to the following expression for the expected value of the rate increase [35],[36]

$$E[D] = \frac{1}{\hat{A}_n - \hat{A}_0} \sum_{i=1}^{n} \int_{a_{i-1}}^{a_i} g(Y)dA. \tag{32}$$

In the above equation $a_i = \hat{A}_i$ if truncation is employed. The renormalization intervals commonly used are $[0.5, 1), [0.75, 1.5)$, and $[1, 2)$.

The optimal separation point for truncation and rounding range in a subinterval to achieve compression efficiency was studied in [36]. This will assign a truncation and rounding range in each subinterval. The rounding may not be possible due to code space partition constraint, $A - \hat{A}_{round}(1 - P(m)) \geq 0$ which gives a constraint on $P(m)$. When rounding cannot be employed due to this constraint, then truncation is used. The optimal separation point can thus be determined [36]

$$A_{optimal} = \frac{\hat{A}_{truncate}\hat{A}_{round}(1-P(m))\left[\left(\frac{\hat{A}_{round}}{\hat{A}_{truncate}}\right)^{\frac{1}{P(m)}} - 1\right]}{\hat{A}_{truncate}\left(\frac{\hat{A}_{round}}{\hat{A}_{truncate}}\right)^{\frac{1}{P(m)}} - \hat{A}_{round}}. \tag{33}$$

This can be obtained by solving for the equilibrium point for A in the arbitrary subinterval $(\hat{A}_{truncate}, \hat{A}_{round})$. In this range A is approximated as either $\hat{A}_{truncate}$ or $\hat{A}_{round}$. The actual optimal equilibrium point curve can be approximated by a piecewise constant approximation. This approximation consists of two parts:

- left part for small $P(m)$ values (this is due to code space partition constraint)

and

- right part for large $P(m)$ values (using middle point of the interval as the assigned separation point).

A list of various possible $\hat{A}$ values along with the lookup table size, $P(m)$ values which might cause the rounding constraint, and the maximum alphabet size for which the rounding constraint does not need to be checked is given in [36]. It is observed that rounding algorithm can achieve better coding efficiency than truncation only algorithm. This conclusion is in agreement with the truncation and rounding algorithms discussed in Section 15.5.3 [31].

15.6 SYSTEMS ISSUES

As the data are compressed, the effect of transmission errors becomes more severe. In lossless coding, an erroneous bit from transmission or storage may cause the code-length to be mis-interpreted and result in error propagation and the loss of synchronization [37]-[38]. Using the code-words in Fig. 15.1 as an example, the decoder will decode the bit-stream "00100" which represents the sequence "aab" into "ca", "ae" or "aaaaa" without detecting any illegal code-words if a transmission error occurs at bit 1, 2, or 3 respectively. The effect is more serious for video applications. Even if the decoder can often re-synchronize after an error, the number of the decoded symbols may not be correct as shown in the previous example. It will result in a shift on parts of the reconstructed picture which is very objectionable. Thus, in practical systems, besides using error correction codes to protect the compressed data, special efforts are often needed to provide for recovering from the error propagation and minimizing its effect on the quality of the reconstructed picture. A technique for synchronizing both the code-word and the sample position is the use of synchronizing words after every fixed number of samples. A synchronizing word can be identified by their special bit-patterns (e.g., a long string of "1"s which cannot be generated by any concatenations of other code-words) without decoding other code-words in the bit-stream. If an illegal code-word is detected or the number of decoded symbols between the synchronizing words is incorrect, the decoder can jump to the next synchronizing words to regain synchronization. Synchronizing words also make parallel decoding of the variable-length coded bit stream possible, i.e., the bit-streams separated by the synchronizing words can be decoded in parallel. However, the use of synchronizing words will decrease the coding efficiency.

Lossless coders achieve data compression by utilizing the statistical property of the source symbols. By definition, statistical property is something which cannot be guaranteed definitely. In many situations, those symbols which have longer code-words may actually occur quite often. In these situations, instead of data compression, it may actually result in data expansion. Thus, the system design has to consider the worst case situation. For real-time applications, the decoding speed needs to be able to handle the worst case throughput requirements.

The output of the lossless encoder is inherently variable bit-rate. The output bit-rate will fluctuate according to the statistics of the input source. In many real-time video applications, the combination of lossy coder and lossless coder may produce drastically changing bit-rate. If the channel between the encoder and the decoder is a constant bit-rate channel, rate-smoothing buffers have to be used to interface the variable bit-rate encoders and decoders with the other parts of the system which deal with constant bit-rate data. An overall video codec configuration is shown in Fig. 15.11. The encoder and decoder buffers have to be controlled so that they do not overflow or underflow. To keep the encoder buffer from overflow, when the buffer is getting full, a feedback signal is used to control the lossy coder to produce less symbols to be encoded by the lossless encoder. When the buffer is empty, stuffing bits are generated to fill up the transmission channel. At the decoder side, the buffer also needs to be managed so that it does not overflow or underflow. To guarantee that the decoder buffer will not overflow or underflow, decoding time stamps are usually inserted in the encoder side to help the decoder

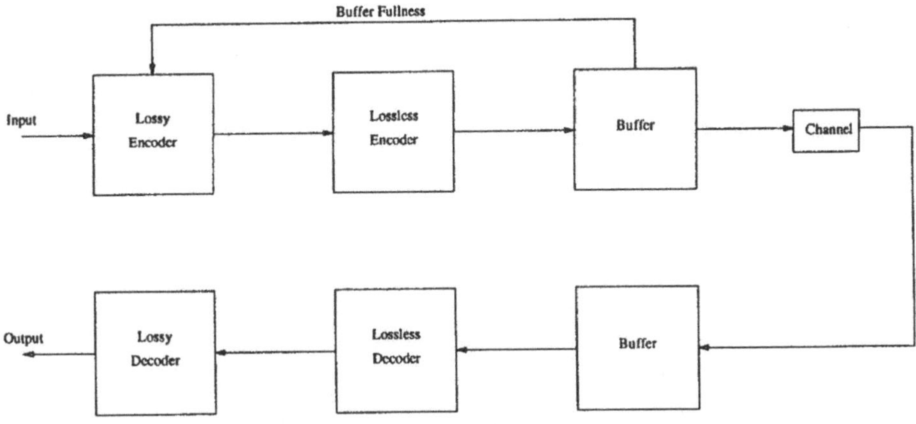

Figure 15.11 Rate-smoothing buffers in a typical video codec.

start decoding the data in the decoder buffer at the right time. This is based on the observation that for real-time video systems, the latency between the time which a piece of data entering the encoder buffer and the time that data leaving the decoder buffer should be a constant. For example, a marker representing the beginning of a video frame should arrive at the input of the encoder buffer and leaving the output of the decoder buffer every 1/30th of a second with a constant delay. Thus, the encoder can attach a decoding time stamp to the marker to tell the decoder when this marker should be leaving the decoder buffer for decoding. With the decoding time stamps, if the channel is a constant bit-rate channel and if the sizes of the transmitter and the receiver buffers are made equal, the fullness of the decoding buffer will be exactly complementary to the fullness of the encoder buffer, thus will not overflow or underflow if the encoder buffer does not underflow or overflow.

15.7 SUMMARY

In this chapter we have looked at various lossless coding techniques and their implementations. The commonly used lossless coding methods in image and video coding are Huffman and arithmetic coding. Huffman coding requires integer number of bits to code each symbol. The arithmetic coding codes a sequence of symbols together which allows possibly fractional number of bits for coding each symbol. Various implementation issues were studied for both the Huffman and arithmetic coding. Various approximations for practical implementation of arithmetic coding were also discussed. The choice of a particular lossless coding algorithm is a tradeoff among coding performance, complexity, and coding delay. System issues should also be carefully considered in the implementation of lossless coders.

REFERENCES

[1] R.C.Gonzalez and R.E.Woods, *Digital image processing*, Addison Wesley, 1992.

[2] V. Bhaskaran and K. Konstantinides, *Image and video compression standards: algorithms and architectures*, Kluwer Academic Publishers, 1995.

[3] D.A. Huffman, "A method for the construction of minimum redundancy codes," *Proc. Of IRE*, vol.40, pp. 1098-1101, Sept. 1952.

[4] L.-Y. Liu, J.-F.Wang, R.-J.Wang, J.-Y.Lee, "Huffman coding," *IEEE Proc. comput. digit. tech.*, vol. 142, no. 6, pp. 411-148, Nov. 1995.

[5] S.M. Lei, M.T. Sun, and K.H. Tzou, "Design and hardware architecture of high-order conditional entropy coding for images," *IEEE Trans. on circuits and systems for video technology*, June 1992, pp. 176-186.

[6] *ISO/IEC JTC1 CD 10918, Digital compression and coding of continuous-tone still images - part 1 : requirements and guidelines*, International Organization for Standardization (ISO), 1993.

[7] *ITU-T Recommendation H.261: video codec for audiovisual services at px64 kbit s/s*, March 1993.

[8] *ITU-T draft recommendation H.263 : video coding for low bitrate communication*, May 1997.

[9] *ISO/IEC JTC1 CD 11172, Coding of moving pictures and associated audio for digital storage media up to 1.5 mbits/s*, International Organization for Standardization (ISO), 1992.

[10] *ISO/IEC JTC1 CD 13818, Generic coding of moving pictures and associated audio*, International Organization for Standardization (ISO), 1994.

[11] R.B. Arps and T.K. Truong, "Comparison of international standards for lossless still image compression," *Proceedings of the IEEE*, June 1994, pp. 889-899.

[12] D.C. Voorhis, "Constructing codes with bounded code-word lengths," *IEEE Trans. on information theory*, March 1974, pp. 288-290.

[13] H.Murakami, S.Matsumoto, and H.Yamamoto, "Algorithms for construction of variable length code with limited maximum word length," *IEEE Trans. on communications*, vol. 32, no. 10, pp. 1157-1159, Oct. 1984.

[14] J.L.P. de-Lameillieure and I.Bruyland, "Comment on 'Algorithm of variable length code with limited maximum word length'," *IEEE Trans. on communications*, vol. 34, no. 12, pp. 1252-1253, Dec. 1986.

[15] C.H. Lu, "Comment on 'Algorithm of variable length code with limited maximum word length'," *IEEE Trans. on communications*, vol. 36, no. 3, pp. 373-375, Mar. 1988.

[16] M.T. Sun, "Design of high-throughput entropy codec," in *VLSI implementations for image communications*, P.Pirsch ed., Elsevier, 1993, pp. 345-364.

[17] S.F. Chang and D.G. Messerschmitt, "Designing high-throughput VLC decoder, Part I-concurrent VLSI architectures," *IEEE Transactions on circuits and systems for video technology*, June 1992, pp. 187-196.

[18] H.D. Lin and D.G. Messerschmitt, "Designing high-throughput VLC decoder, Part II - parallel decoding methods," *IEEE Trans. on circuits and systems for video technology*, June 1992, pp. 197-206.

[19] D.L. Cohn, J.L. Melsa, A.S.Arora, J.M. Kresse, and A.K. Pande, "Practical considerations for variable length source coding," *Proc. IEEE ICASSP*, 1981, pp. 816-819.

[20] S.M. Lei and M.T. Sun, "An entropy coding system for digital HDTV applications ," *IEEE Trans. on circuits and systems for video technology*, vol.1, no.1, March 1991, pp. 147-155.

[21] K.K. Parhi, "High-speed architectures for Huffman and Viterbi decoders," *IEEE Transactions on circuits and systems, Part II: analog and digital signal process ing*, June 1992, pp. 385-391.

[22] K.K. Parhi and G. Shrimali, "A concurrent lossless coder for video compression ," *chapter in VLSI signal processing V*, IEEE press, October 1992 (Proceedings of the 1992 IEEE VLSI signal processing workshop, Napa, Oct. 1992).

[23] R. Hashemian, "Memory efficient and high-speed search Huffman coding," *IEEE Trans. on communications*, vol. 43, no. 10, pp. 2576-2581, October 1995.

[24] G.G. Langdon, "An introduction to arithmetic coding," *IBM J. Res. Develop.*, vol. 28, no. 2, pp. 93–98, Mar. 1984.

[25] J. Rissanen and G.G. Langdon, "Arithmetic coding," *IBM J. Res. Develop.*, vol. 23, no. 2, pp. 149–162, Mar. 1979.

[26] I.H. Witten, R.M. Neal and J.G. Cleary, "Arithmetic coding for data compression," *Commun. ACM*, vol. 30, no. 6, pp. 520–540, Jun. 1987.

[27] P.G. Howard and J.S. Vitter, "Arithmetic coding for data compression", *Proc. IEEE*, vol. 82, no. 6, pp. 857-865, Jun. 1994.

[28] W.B. Pennebaker and J.L. Mitchell, *JPEG Still Image Data Compression Standard.* Van Nostrand, 1993.

[29] W.B. Pennebaker, J.L. Mitchell, G.G. Langdon and R.B. Arps, "An overview of the basic principles of the Q-coder adaptive arithmetic coder," *IBM J. Res. Develop.*, vol. 32, no. 6, pp. 717–726, Nov. 1988.

[30] G.G. Langdon and J. Rissanen, "Compression of black-white images with arithmetic coding," *IEEE Trans. Commun.*, vol. 29, no. 6, pp. 858–867, Jun. 1981.

[31] S.M. Lei, "On the finite-precision implementation of arithmetic codes," *J. of Visual Communication and Image Representation*, vol. 6, no. 1, pp. 80–88, 1995.

[32] B. Fu and K.K. Parhi, "Two VLSI design advances in arithmetic coding," in *IEEE Symposium on Circuits and Systems*, pp. 1440–1443, 28 April-3 May 1995.

[33] J. Rissanen and K. Mohiuddin, "A multiplication-free multialphabet arithmetic code," *IEEE Trans. Commun.*, vol. 37, no. 2, pp. 93–98, Feb. 1989.

[34] D. Chevion, E. Karnin and E. Walach, "High efficiency, multiplication free approximation of arithmetic coding," in *Proc. Data Compression*, pp. 43–52, Apr. 1991.

[35] G. Feygin, P.G. Gulak and P. Chow, "Minimizing excess code length and VLSI complexity in the multiplication free approximation of arithmetic coding," *Information Processing & Management*, vol. 30, no. 6, pp. 805–816, 1994.

[36] B. Fu and K.K. Parhi, "Generalized multiplication-free arithmetic codes," *IEEE Trans. Commun.*, vol. 45, no. 5, pp. 497–501, May 1997.

[37] J.C. Maxted and J.R. Robinson, "Error recovery for variable length codes, *IEEE Trans. inform. theory*, vol. IT-31, Nov. 1985.

[38] T.J. Ferguson and J.H. Rabinowitz, "Self-synchronizing Huffman codes," *IEEE Trans. inform. theory*, vol. IT-30, no. 4, pp. 687-693, July 1984.

Chapter 16

Viterbi Decoders: High Performance Algorithms and Architectures

Herbert Dawid
DSP Solutions Group
Synopsys, Inc.
Herzogenrath, Germany
dawid@synopsys.com

Olaf J. Joeressen
R&D Center
Nokia Mobile Phones
Bochum, Germany
Olaf.Joeressen@nmp.nokia.com

Heinrich Meyr
Aachen University of Technology
Aachen, Germany
meyr@ert.rwth-aachen.de

16.1 INTRODUCTION

Viterbi Decoders (VDs) are widely used as forward error correction (FEC) devices in many digital communications and multimedia products, including mobile (cellular) phones, video and audio broadcasting receivers, and modems. VDs are implementations of the *Viterbi Algorithm* (VA) used for decoding *convolutional* or *trellis codes*[1].

The continuing success of convolutional and trellis codes for FEC applications in almost all modern digital communication and multimedia products is based on three main factors:

- The existence of an optimum *Maximum Likelihood decoding* algorithm – the VA – with limited complexity, which is well suited for implementation.
- The existence of classes of good (convolutional and trellis) codes suited for many different applications.

[1]Other important applications of the VA include equalization for transmission channels with memory like multipath-fading channels and numerous applications apart from digital communications like pattern, text and speech recognition as well as magnetic recording. Due to lack of space, only Viterbi decoding is considered here.

- The advances in digital silicon technology which make the implementation of the VA possible even for sophisticated codes and high bit rate applications.

The coarse system level block diagram shown in Fig. 16.1 illustrates the use of VDs in digital communication systems. The well known discrete time model with the discrete time index k is used here in order to model transmitter, channel and receiver.

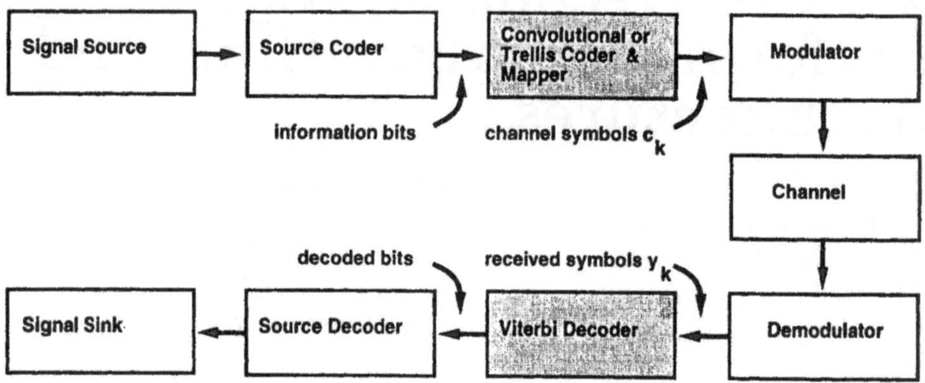

Figure 16.1 Viterbi decoders in digital communication systems.

The signals emitted by the signal source are first compressed in a source encoder (e.g. a speech, audio or video encoder). The compressed information bits then enter a convolutional encoder or trellis encoder, which introduces *channel coding*. While the source encoder removes redundant and irrelevant information from the source signal in order to reduce the transmission rate, the channel encoder deliberately introduces the redundancy necessary to combat transmission impairments by forward error correction (FEC). Coded symbols from a predefined and sometimes multidimensional symbol alphabet are generated by the encoder and mapped onto complex channel symbols which enter the modulator. Here, the signal is modulated according to a chosen modulation scheme and carrier frequency. After transmission over the channel, the received signal is first demodulated. Following, the demodulated received symbols enter the VD. The soft (quantized) channel symbols available in the demodulator can very advantageously be used by the VD. Hence, rather than generating hard symbol decisions, the demodulator delivers a *soft decision* input to the VD[2]. The corrected information bits are finally decompressed in the source decoder.

16.1.1 Viterbi Decoding Applications

Among the numerous applications of Viterbi decoding, we consider three areas to be most important. The completely different characteristics of these applications emphasize the widespread use of Viterbi decoders in almost all modern telecommunication standards.

[2] Soft decision Viterbi decoding leads to an increase in coding gain of about 2dB compared to hard decision Viterbi decoding[1].

1. Mobile (Cellular) Phones:
   For mobile or cellular phone applications, the transmission channels are subject to various impairments like fading. Channel coding is essential in order to obtain the desired transmission quality. Many communication standards like the Global System for Mobile Communications (GSM) standard, the IS–54 digital cellular phone standard and the IS–95 CDMA standard specify the use of convolutional codes.

2. Video and Audio Broadcasting Receivers:
   Channel coding using convolutional codes is essential for almost all satellite communication standards, among them the recent Digital Video Broadcasting (DVB-S) standard for satellite transmission [2], the DSS standard as well as the terrestrial Digital Audio Broadcasting (DAB) standard.

3. Modems:
   Modems represent an application area where very advanced channel coding techniques are used. For example, Viterbi decoding for convolutional and trellis coded modulation codes is employed in modems according to the recent 32kbps and 54 kbps modem standards. Software DSP implementations are commonly used due to the relatively low bit rate.

For low bit rate applications, Viterbi decoding is implemented in software on digital signal processors (DSPs). The bit rate required by modern high quality speech transmission represents the current limit for VD software implementations due to the high computational requirements imposed by the VA. Hybrid DSP architectures were developed with special datapaths supporting the particular VA processing requirements. We focus here on higher bit rate applications, where the VD is implemented in Very Large Scale Integration (VLSI) technology as a separate hardware unit.

16.2 THE VITERBI ALGORITHM

In order to introduce the Viterbi algorithm [3, 4] and the used notation, we discuss the simple convolutional encoder example shown in Fig. 16.2.

The input stream of information bits is mapped onto k-bit *information symbols* u_k, which are input to a finite state machine (FSM) generating $n > k$ coded bits from the information symbols. The ratio k/n (here $1/2$) is called the *code rate.* The larger the code rate, the smaller the amount of redundancy introduced by the coder. With $k = 1$, only code rates $1/n$ are possible. Higher rate codes are known for $k > 1$. Alternatively, higher rate codes can be created by using a $1/n$ *base* or *mother code* and omitting (puncturing) a part of the coded bits after encoding as specified by a given puncturing pattern or puncture mask [5, 6, 7]. It is shown in [5, 6] that the resulting *punctured codes* lead to reduced decoding complexity compared to standard codes with the same code rate and $k > 1$ at negligible performance losses. Today, $k = 1$ holds for virtually all practically relevant base codes [17]; therefore we consider only this case.

The n coded bits $b_{i,k}$ with $i \in \{1, \dots, n\}$ represent the *code symbols* $b_k = \sum_{j=1}^{n} b_{j,k} \cdot 2^{j-1}$ of a given symbol alphabet: $b_k \in \{0, \dots, 2^n - 1\}$. If the encoder FSM has a memory of ν bits, the code symbols are calculated from $K = \nu + 1$ bits, the FSM memory and the current input bit, respectively. K is called the *constraint length* of the code. The kth encoder state can be conveniently written as an integer number:

$$x_k = \sum_{j=0}^{\nu-1} x_{j,k}\, 2^j \quad \text{with} \quad x_k \in \{0,\ldots,2^\nu - 1\}; \quad x_{j,k} \in \{0,1\} \tag{1}$$

Virtually all commonly used convolutional coders exhibit a feedforward shift register structure. Additionally, in contrast to *systematic* codes, where the sequence of input symbols appears unchanged at the output together with the added redundancy, these convolutional codes are *nonsystematic* codes (NSCs). The coder is described by a convolution of the sequence of input bits with polynomials G_i over GF(2)

$$b_{i,k} = \sum_{j=0}^{\nu} g_{i,j} \cdot u_{k-j}; \qquad G_i = \sum_{j=0}^{\nu} g_{i,j} \cdot 2^j. \tag{2}$$

The *generator polynomials* G_i are of degree ν and are usually not written as polynomials, but as numbers in octal notation as shown in (2). Here, $g_{i,j}$ are the binary coefficients of the generator polynomial G_i. For the rate 1/2, $\nu = 2$ coder in Fig. 16.2, the generator polynomials are $G_0 = 7|_{\text{octal}} = 111|_{\text{binary}}$ and $G_1 = 5|_{\text{octal}} = 101|_{\text{binary}}$. Therefore, the structure of the encoder as shown in Fig. 16.2 results[3].

The code symbols generated by the encoder are subsequently mapped onto complex valued *channel symbols* according to a given modulation scheme and a predefined mapping function. In general, the channel symbols $\mathbf{c}_k$ are tuples of complex valued symbols. As an example, in Fig. 16.2, the symbol constellation according to *BPSK* (binary phase shift keying) is shown. Here, each code symbol is mapped onto a tuple of two successive BPSK symbols.

The concatenation of modulator, channel and demodulator as shown in Fig. 16.1 is modeled by adding (complex valued) white noise $\mathbf{n}_k$ to the channel symbols $\mathbf{c}_k$[4]. Hence, for the received symbols $\mathbf{y}_k$

$$\mathbf{y}_k = \mathbf{c}_k + \mathbf{n}_k \tag{3}$$

holds. This model is adequate for a number of transmission channels such as satellite and deep space communication. Even if a given transmission channel cannot be described by additive white noise (e.g., in the case of fading channels), theory [8] shows that the optimum demodulator or *inner receiver* has to be designed in a way that the concatenation of modulator, channel and demodulator appears again as an additive white noise channel. Sometimes, if successive demodulator outputs are correlated (e.g., if equalization is employed in the demodulator or if noise bursts occur), an *interleaver* is introduced in the transmitter at the coder output and the

[3]If not all k bits of the information symbols u_k enter the coder, parallel state transitions occur in the trellis: The parallel transitions are independent of the bypassed bits. Hence, a symbol-by-symbol decision has to be implemented in the receiver for the bypassed bits. This situation can be found in trellis encoders for trellis coded modulation (TCM) [9] codes.

[4]Note that, below, bold letters denote complex valued numbers, and capital letters denote sequences of values in mathematical expressions.

corresponding deinterleaver is introduced prior to Viterbi decoding. Interleaving reduces the correlations between successive demodulator outputs.

The behavior of the encoder is illustrated by drawing the state transitions of the FSM over time, as shown in Fig. 16.2. The resulting structure, the ***trellis diagram*** or just ***trellis***, is used by the Viterbi decoder to find the most likely sequence of information symbols (indicated by the thick lines in Fig. 16.2) given the received symbols y_k. In the trellis, all possible encoder states are drawn as nodes, and the possible state transitions are represented by lines connecting the nodes. Given the initial state of the encoder FSM, there exists a one-to-one correspondence of the FSM state sequence to the sequence of information symbols $U = \{u_k\}$ with $k \in \{0, \ldots, T-1\}$.

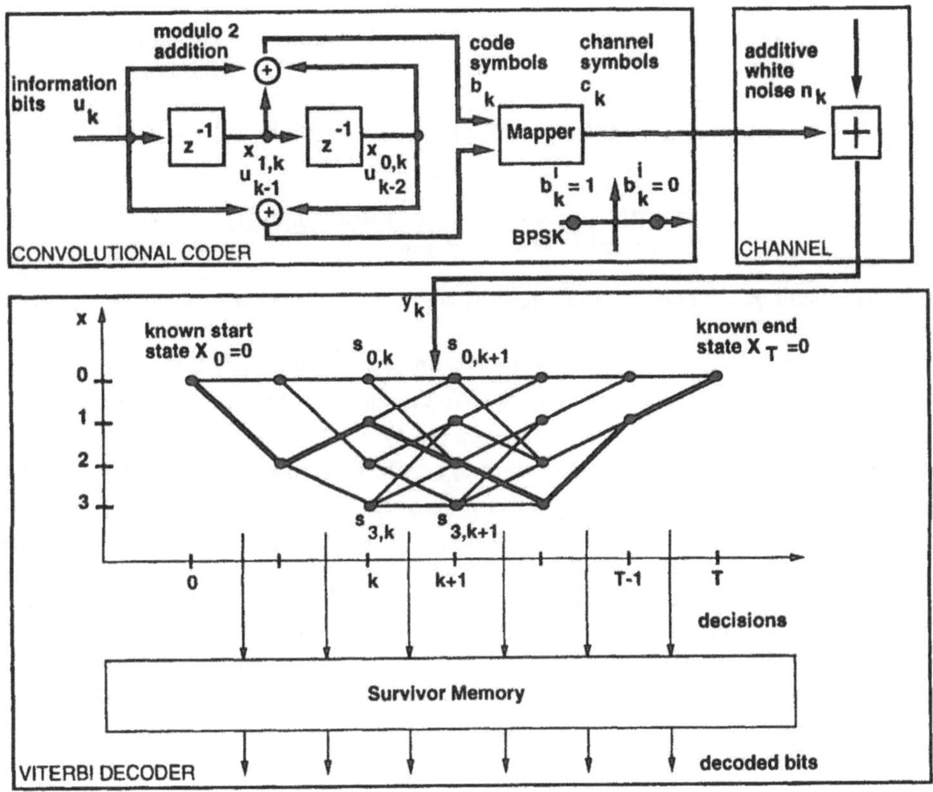

Figure 16.2 Convolutional coder and trellis diagram.

The number of trellis states is $N = 2^\nu = 2^{K-1}$ and the number of branches merging into one state is called M, where $M = 2^k$ is equal to the number of possible information symbols u_k. For binary symbols, $M = 2$ holds as shown in Fig. 16.2. The trellis nodes representing state $x_k = i$ at time k are denoted as $s_{i,k}$.

A possible state transition is a branch in the trellis, and a possible state sequence represents a path through the trellis.

In order to optimally retrieve the transmitted information one searches for the channel symbol sequence $\hat{C}$ which has most likely generated the received symbol sequence $\boldsymbol{Y}$. This approach is called Maximum Likelihood Sequence Estimation (MLSE). Mathematically, this can be stated as follows. Given the received sequence $\boldsymbol{Y}$ one sequence $\hat{C}$ is searched which maximizes the value of the *likelihood function* $P(\boldsymbol{Y}|C)$:

$$\hat{C} = \arg\{\max_{\text{all sequences } C} P(\boldsymbol{Y}|C)\}. \tag{4}$$

Since the noise samples in (4) are statistically independent and the underlying shift register process is a Markov process, the sequence likelihood function can be factorized [4]:

$$P(\boldsymbol{Y}|C) = \prod_{k=0}^{T-1} P(y_k|c_k). \tag{5}$$

Here, $P(y_k|c_k)$ is the conditional *probability density function* (PDF) of one received sample y_k given c_k. In order to express $P(y_k|c_k)$, the PDF of the noise has to be known.

Since the logarithm is a monotonic function, we can equally well maximize:

$$\hat{C} = \arg\{\max_{\text{all sequences } C} \sum_{k=0}^{T-1} \log(P(y_k|c_k))\}. \tag{6}$$

The log–likelihood function $\log(P(y_k|c_k))$ is given the name *branch metric* or *transition metric*[5].

We recall that to every branch in the trellis (see Fig. 16.2) there corresponds exactly one tuple of channel symbols c_k. We therefore assign a branch metric $\lambda_k^{(m,i)}$ to every branch in the trellis. The parameter $\lambda_k^{(m,i)}$ denotes the branch metric of the m-th branch leading to trellis state $s_{i,k}$, which is equal to the encoder state $x_k = i$. Instead of using $\lambda_k^{(m,i)}$, which expresses the branch metric as a function of the branch label m and the current state $x_k = i$, it is sometimes more convenient to use $\lambda_{ij,k}$, which denotes the branch metric of the branch from trellis state $s_{j,k}$ to trellis state $s_{i,k+1}$. The unit calculating all possible branch metrics in a Viterbi decoder is called *transition metric unit* (TMU).

As an important example we consider zero mean complex valued additive white Gaussian noise (AWGN) with uncorrelated inphase and quadrature components and channel symbols c_k consisting of a single complex value. We obtain for the branch metric:

$$P(y_k|c_k) = \frac{1}{\pi\sigma^2} \exp \frac{|y_k - c_k|^2}{\sigma^2} \tag{7}$$

and

$$\log(P(y_k|c_k)) \sim |y_k - c_k|^2 \tag{8}$$

where σ^2 is the variance of the complex valued gaussian random variable n_k. From (8) we observe the important fact that the branch metric is proportional to the *Euclidean distance* between the received symbol y_k and the channel symbol c_k. The sum in (6) represents the accumulation of the branch metrics along a given path through the trellis according to the sequence C. It is called *path metric*. The

[5]The advantage of using the logarithm for the branch metrics will soon become apparent.

path metric for a path leading to state $s_{i,k}$ is called $\gamma_k^{(m,i)}$, where $m \in \{0,\ldots,M-1\}$ denotes the path label of one of the M paths leading to the state $s_{i,k}$.

Conceptually, the most likely sequence $\hat{C}$ can be found by an exhaustive search as follows. We compute the path metric for every possible sequence C, hence for every possible path through the trellis. The maximum likelihood path, which is the path with the smallest Euclidean distance, corresponds to $\hat{C}$:

$$\hat{C} = \arg\{\min_{\text{all sequences } C} \sum_{k=0}^{T-1} |y_k - c_k|^2\}. \tag{9}$$

Hence, maximizing the log–likelihood function as in (6) is equivalent to minimizing the Euclidean distance as in (9).

Since the number of paths increases exponentially as a function of the length of the sequence, the computational effort then also increases exponentially. Fortunately, there exists a much more clever solution to the problem which carries the name of its inventor, the Viterbi algorithm [3]. When using the VA, the computational effort increases only linearly with the length of the trellis, hence the computational effort per transmitted bit is constant.
The VA recursively solves the problem of finding the most likely path by using a fundamental principle of optimality first introduced by Bellman [10] which we cite here for reference:
The Principle of Optimality: An optimal policy has the property that whatever the initial state and initial decision are, the remaining decisions must constitute an optimal policy with regard to the state resulting from the first decision.
In the present context of Viterbi decoding, we make use of this principle as follows. If we start accumulating branch metrics along the paths through the trellis, the following observation holds: Whenever two paths merge in one state, only the most likely path (the best path or the *survivor path*) needs to be retained, since for all possible extensions to these paths, the path which is currently better will always stay better: For any given extension to the paths, both paths are extended by the same branch metrics. This process is described by the *add-compare-select* (ACS) recursion: The path with the best path metric leading to every state is determined recursively for every step in the trellis. The metrics for the survivor paths for state $x_k = i$ at trellis step k are called *state metrics* $\gamma_{i,k}$ below.

In order to determine the state metric $\gamma_{i,k}$, we calculate the path metrics for the paths leading to state $x_k = i$ by adding the state metrics of the predecessor states and the corresponding branch metrics. The predecessor state x_{k-1} for one branch m of the M possible branches $m \in \{0\ldots M-1\}$ leading to state $x_k = i$ is determined by the value resulting from evaluation of the *state transition function* $Z()$: $x_{k-1} = Z(m,i)$.

$$\gamma_k^{(m,i)} = \gamma_{Z(m,i),k-1} + \lambda_k^{(m,i)} \quad , \quad m \in \{0,\ldots,M-1\}. \tag{10}$$

The state metric is then determined by selecting the best path:

$$\gamma_{i,k} = \max\{\gamma_k^{(0,i)},\ldots,\gamma_k^{(M-1,i)}\}. \tag{11}$$

A sample ACS recursion for one state and $M = 2$ is shown in Fig. 16.3. This

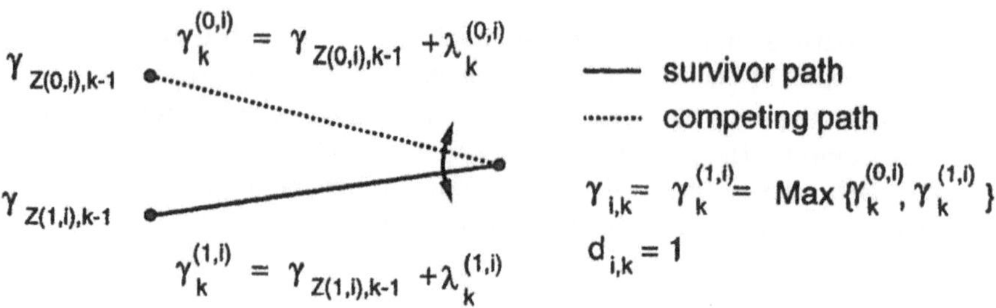

Figure 16.3 ACS recursion for $M = 2$.

ACS recursion is performed for all N states in the trellis. The corresponding unit calculating the ACS recursion for all N states is called ACS unit (ACSU).

Despite the recursive computation, there are still N best paths pursued by the VA. The maximum likelihood path corresponding to the sequence $\hat{C}$ can be finally determined only after reaching the last state in the trellis. In order to finally retrieve this path and the corresponding sequence of information symbols u_k, either the sequences of information symbols or the sequences of ACS decisions corresponding to each of the N survivor paths for all states i and all trellis steps k have to be stored in the *survivor memory unit* (SMU) as shown in Fig. 16.2 while calculating the ACS recursion. The decision for one branch m of $M = 2^k$ possible branches is represented by the *decision bits* $d_{i,k} = m$.

So far, we considered only the case that the trellis diagram is terminated, i.e., the start and end states are known. If the trellis is terminated, a final decision on the overall best path is possible only at the very end of the trellis. The decoding latency for the VA is then proportional to the length of the trellis. Additionally, the size of the SMU grows linearly with the length of the trellis. Finally, in applications like broadcasting, a continuous sequence of information bits has to be decoded rather than a terminated sequence, i.e., no known start and end state exists.

Fortunately, even in this case, certain asymptotic properties allow an approximate maximum likelihood sequence estimation with negligible performance losses and limited implementation effort. These are the acquisition and truncation properties [13] of the VA. Consider Fig. 16.4: the VA is pursuing N survivor paths at time instant k while decoding a certain trellis diagram. These paths merge, when traced back over time, into a single path as shown by the path trajectories in Fig. 16.4. This path is called the *final survivor* below. For trellis steps smaller than $k - D$, the paths have merged into the final survivor with very high probability. The *survivor depth, D*, which guarantees this behavior, depends strongly on the used code. Since *all N* paths at trellis step k merge into the final survivor, it is sufficient to actually consider only *one* path. Hence, it is possible to uniquely determine the final survivor path for the trellis steps with index smaller than $k - D$ already after performing the ACS recursion for trellis step k. This property enables decoding with a fixed latency of D trellis steps even for continuous transmission. Additionally, the survivor memory can be truncated: The SMU has to store only a fixed number of decisions $d_{i,j}$ for $i \in \{0, \ldots, N-1\}$ and $j \in \{k-D, k-D+1, \ldots, k-1, k\}$.

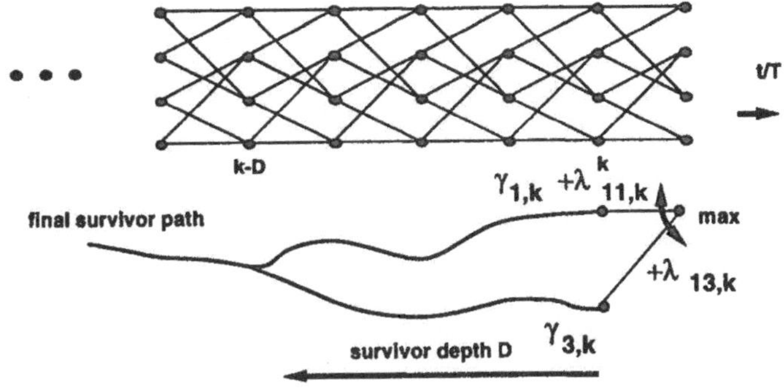

Figure 16.4 Path trajectories for the VA at an intermediate trellis step k.

If the overall best path (the path with the best state metric) at trellis step k is used for determining the final survivor, the value of D guaranteeing that the final survivor is acquired with sufficiently high probability is the survivor depth. This procedure is called *best state decoding* [11, 12]. Sometimes, an arbitrary path is chosen instead, in order to save the computational effort required in order to determine the overall best path, which is called *fixed state decoding.* The properties of these decoding schemes will be discussed in section 16.4.8.

A phenomenon very similar to the just described truncation behavior occurs when the decoding process is started in midstream at trellis step k with an unknown start state. Due to the unknown start state, the ACS recursion is started with equal state metrics for all states. However, the decoding history which is necessary for reliable decoding of the survivor path is not available for the initial trellis steps. What happens if we perform the ACS recursion and try to decode the best path? As indicated in Fig. 16.5, the probability that the final survivor path differs from the correct path is then much larger than for decoding with a known start state. Fortunately, the same decoding quality as for decoding with known start state is achieved after processing a number of initial trellis steps. The number of trellis steps which are required here is called *acquisition depth.* It can be shown that the acquisition depth is equal to the survivor depth D [13, 14, 15]. This is also indicated in Fig. 16.5, where the merging of the paths takes place at trellis step $k + D$.

Summarizing, the three basic units of a VD are depicted in Fig. 16.6. The branch metrics are calculated from the received symbols in the Transition Metric Unit (TMU). These branch metrics are fed into the add-compare-select unit (ACSU), which performs the ACS recursion for all states. The decisions generated in the ACSU are stored and retrieved in the Survivor Memory Unit (SMU) in order to finally decode the source bits along the final survivor path. The ACSU is the only recursive part in a VD, as indicated by the latch. The branch metric computation is the only part which differs significantly if the VA is used for equalization instead of decoding.

Following, we state a computational model for transmitter, AWGN channel and receiver, that will be used in the subsequent sections. The model is shown in Fig. 16.7.

In our model, we assume that the channel symbols have energy normalized to unity after leaving the mapper for reasons of simplicity. Varying transmission

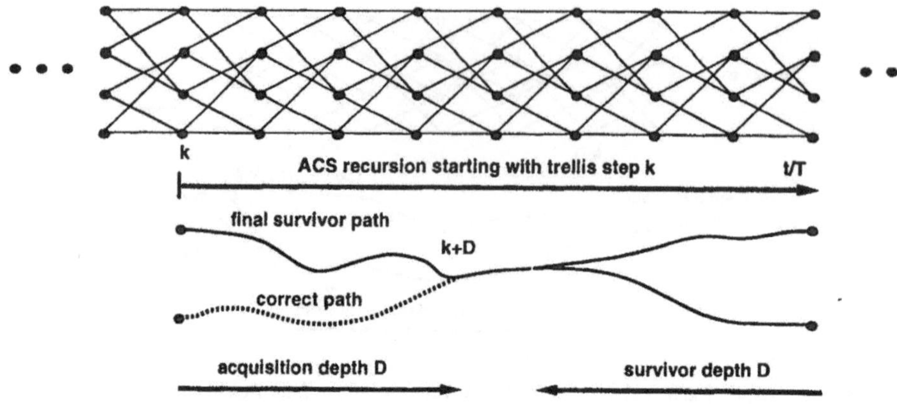

Figure 16.5 Path trajectories for acquisition.

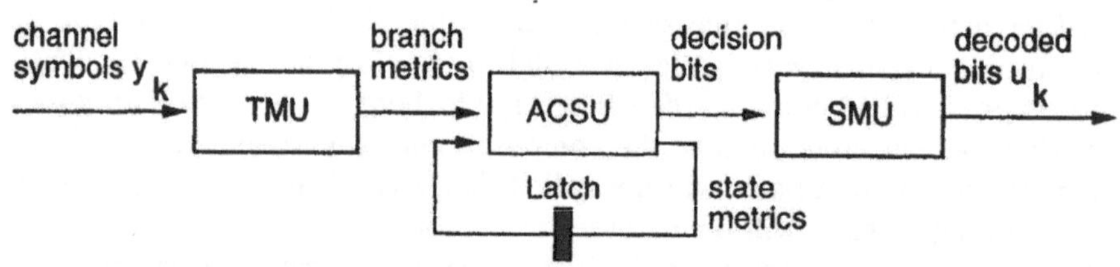

Figure 16.6 Viterbi Decoder block diagram.

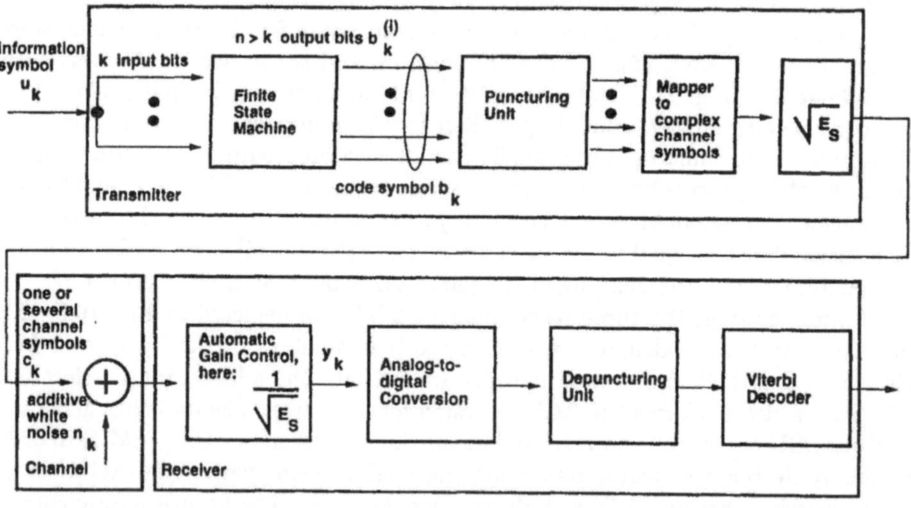

Figure 16.7 Computational model for transmitter, AWGN channel and receiver.

conditions are modeled by changing the signal energy to noise ratio E_s/N_o. E_s is the signal energy, and N_o is the one sided power spectral density of the noise. Since the additive noise is assumed to have a constant variance in the model, changes in

E_s/N_o are modeled by changing the gain in the scaling block at the transmitter output: $\sqrt{E_s}$. In the receiver, a unit implementing *automatic gain control* (AGC) is necessary in front of the analog-to-digital converter (ADC). In our computational model, the AGC just implements a fixed scaling by $\frac{1}{\sqrt{E_s}}$, in order to normalize the energy of the received demodulated symbols y_k to unity again, which is just a matter of mathematical convenience. Therefore, the reference symbols in the decoder have the same magnitude and energy as in the encoder. Several issues related to AGC and ADC are discussed in section 16.3. For actual Viterbi decoder system design and assessment of the performance impact of all parameters and quantization effects, system design and simulation tools like COSSAP^{TM}[16] are indispensable.

16.2.1 Example: $K = 3$ Convolutional Code with BPSK

As an implementation example, we will use the $K = 3$ rate 1/2 code with generator polynomials (7,5) with the trellis shown in Fig. 16.2. For BPSK, the $n = 2$ coded bits for a state transition in the encoder are mapped onto two complex valued BPSK symbols $\mathbf{c}_k = (c_{1,k}, c_{2,k})$ according to the mapping function:

$$\begin{aligned} c_{1,k} &= 1 - 2b_{1,k} = \exp(i\pi b_{1,k}) \\ c_{2,k} &= 1 - 2b_{2,k} = \exp(i\pi b_{2,k}). \end{aligned}$$

If the additive noise is gaussian, the channel is AWGN and the likelihood function $P(\mathbf{y}_k|\mathbf{c}_k)$ for the two successive received complex valued symbols $\mathbf{y}_k = (y_{1,k}, y_{2,k})$ corresponding to a trellis transition is given by:

$$\begin{aligned} P(\mathbf{y}_k|\mathbf{c}_k) &= P(y_{1,k}|c_{1,k}) \cdot P(y_{2,k}|c_{2,k}) \\ &= \sqrt{\frac{E_s}{\pi N_o}} \exp\left(-\frac{E_s}{N_o}|y_{1,k} - c_{1,k}|^2\right) \cdot \sqrt{\frac{E_s}{\pi N_0}} \exp\left(-\frac{E_s}{N_o}|y_{2,k} - c_{2,k}|^2\right). \end{aligned} \quad (12)$$

Hence, the corresponding branch metric is given by:

$$\lambda_k^{(m,i)} = -\frac{E_s}{N_o}\left\{|y_{1,k} - c_{1,k}|^2 + |y_{2,k} - c_{2,k}|^2\right\} + 2ln\left(\sqrt{\frac{E_s}{\pi N_o}}\right). \quad (13)$$

The term $2ln(\ )$ which is common for all branch metrics can be neglected, since this does not affect the path selection. Since the imaginary part of the channel symbols is always zero, the imaginary part $y_{i,k,\mathrm{im}}$ of the received symbols $y_{i,k} = y_{i,k,\mathrm{re}} + iy_{i,k,\mathrm{im}}$ only leads to an additive value which is common for all branch metrics and can be neglected. Furthermore, if the quotient of signal energy and noise power spectral density is constant over time, the factor $\frac{E_s}{N_o}$ can also be neglected:

$$\lambda_k^{(m,i)} \sim -\left\{(y_{1,k,\mathrm{re}} - c_{1,k,\mathrm{re}})^2 + (y_{2,k,\mathrm{re}} - c_{2,k,\mathrm{re}})^2\right\}. \quad (14)$$

This calculation of the branch metrics is performed in the transition metric unit TMU.

In order to calculate the ACS recursion in the ACS unit, we have to define the state transition function for the used $K = 3$ code. For feedforward shift register coders, this function is given by

$$Z(m, x_k) = Z(m, \{x_{K-2,k}, \ldots, x_{0,k}\}) = \{x_{K-3,k}, \ldots, x_{0,k}, m\} \tag{15}$$

if the branch label m is chosen to be equal to the bit shifted out of the encoder for trellis step k.

For the resulting trellis with $N = 2^{3-1} = 4$ states (see Fig. 16.2) the ACS recursion is given by:

$$\begin{aligned}
\gamma_{0,k+1} &= \mathrm{Max}(\gamma_{0,k} + \lambda_{k+1}^{(0,0)}, \gamma_{1,k} + \lambda_{k+1}^{(1,0)}) \\
\gamma_{1,k+1} &= \mathrm{Max}(\gamma_{2,k} + \lambda_{k+1}^{(0,1)}, \gamma_{3,k} + \lambda_{k+1}^{(1,1)}) \\
\gamma_{2,k+1} &= \mathrm{Max}(\gamma_{0,k} + \lambda_{k+1}^{(0,2)}, \gamma_{1,k} + \lambda_{k+1}^{(1,2)}) \\
\gamma_{3,k+1} &= \mathrm{Max}(\gamma_{2,k} + \lambda_{k=1}^{(0,3)}, \gamma_{3,k} + \lambda_{k+1}^{(1,3)}).
\end{aligned} \tag{16}$$

A generalization to more complex codes is obvious.

16.3 THE TRANSITION METRIC UNIT

In the TMU of a Viterbi decoder the branch metrics $\lambda_k^{(m,i)}$ are computed, which are used in the ACSU to update the new state metrics $\gamma_{i,k}$. The number of *different* branch metrics depends on the number of coded bits that are associated with a branch of the trellis. For a code of rate $\frac{k}{n}$, 2^n different branch metrics need to be computed for every trellis step. Since the ACSU uses only differences of path metrics to decide upon survivor selection, arbitrary constants can be added to the branch metrics belonging to a single trellis step without affecting the decisions of the Viterbi decoder. Choosing these constants appropriately can simplify implementations considerably.

Although the TMU can be quite complex if channel symbols of high complexity (e.g., 64-QAM, etc) need to be processed, its complexity is usually small compared to a complete Viterbi decoder. We restrict the discussion here to the case of BPSK modulation, rate 1/2 codes and additive white gaussian noise. We use $y_{i,k}$ instead of $y_{i,k,\mathrm{re}}$ and $c_{i,k}$ instead of $c_{i,k,\mathrm{re}}$ (cf Eq. (14)) in order to simplify the notation[6].

Starting from (14), we write the branch metrics as

$$\lambda_k^{(m,i)} = C_0 \left\{ (y_{1,k}^2 - 2y_{1,k}c_{1,k} + {c_{1,k}}^2) + (y_{2,k}^2 - 2y_{2,k}c_{2,k} + {c_{2,k}}^2) \right\} + C_1 \tag{17}$$

with C_0, C_1 being constants.

Since $c_{1,k}$ and $c_{2,k} \in \{-1, 1\}$ holds and the squared received symbols appear in all different branch metrics independently of the channel symbols that are associated with the branches, the squared terms are constant for a set of branch metrics and

[6]Note that extension to QPSK (quaternary phase shift keying) is obvious. Then, $y_{1,k}$ and $y_{2,k}$ denote the real and imaginary part of a single received complex valued symbol, respectively. $c_{1,k}$ and $c_{2,k}$ denote the real and imaginary part of a single complex valued QPSK channel symbol.

can be removed without affecting the decoding process[7]. Thus we can write the actually computed branch metrics as

$$\lambda_k^{(m,i)'} = C_2 \{-y_{1,k}c_{1,k} - y_{2,k}c_{2,k}\} + C_3 \qquad \text{with constants} \qquad C_2 < 0, C_3. \quad (18)$$

In (8), C_3 can be chosen independently for every trellis step k, while C_2 must be constant for different k to avoid deterioration of the decoding process. For hardware implementations C_3 is advantageously chosen such that $\lambda_k^{(m,i)'}$ is always positive. This enables the use of unsigned arithmetic in the ACSU for path metric computations. For SW implementations it is often advantageous to chose $C_3 = 0$ since then $\lambda_k^{(0,i)'} = -\lambda_k^{(1,i)'}$ holds for all *good* rate 1/2 codes. This can be used to reduce the computational complexity of the ACS computations.

16.3.1 Branch Metric Quantization

While the TMU usually has only a minor impact on the complexity of a Viterbi decoder, the ACSU is a major part. The complexity of the ACSU depends strongly on the wordlength of the branch metrics. It is thus important to reduce the branch metric wordlength to the required minimum.

It is well known for a long time that a wordlength of $w = 3$ bits is almost optimum for the received symbols in the case of BPSK modulation [17]. However, this requires virtually ideal gain control before the Viterbi decoder. Thus larger wordlengths are often used in practice to provide some margin for gain control. For actual determination of the wordlengths in the presence of a given gain control scheme and analog-to-digital conversion, system simulation has to be performed, which can be done easily using tools like COSSAP^{TM}[16].

To compute the branch metrics correctly it must be known how the "original" input value is quantized to the input value of the TMU consisting of w bits. As is pointed out already in [17], the quantization steps do not necessarily have to be equidistantly spaced. However, only such "linear" schemes are considered here.

16.3.1.1 Step at zero quantization

Probably the most widely used quantization characteristic is a symmetrical interpretation of a w-bit 2's complement number, by adding implicitly 0.5. Fig. 16.8 shows the characteristic of such a quantizer for 2 bits output wordlength. Q is the 2's complement output value of the quantizer, on the x-axis the normalized input value is given and the y-axis the interpretation of the interpretation of the output value Q which actually is $Y = Q + 0.5$.

Table 16.1 shows range and interpretation again for a 3-bit integer output value of such a quantizer.

Table 16.1 Step At Zero Quantizer Output Value Interpretation

2's complement quantizer output value	-4	...	-1	0	...	3
interpretation due to quantizer characteristic	-3.5	...	-0.5	0.5	...	3.5

Clearly, the quantizer input value 0 needs to be the decision threshold of the quantizer between the associated normalized integer values −1 and 0, that are

[7]The terms $(c_{x,k})^2$ are not constant for every modulation scheme (e.g., for 16-QAM) and thus cannot be neglected generally.

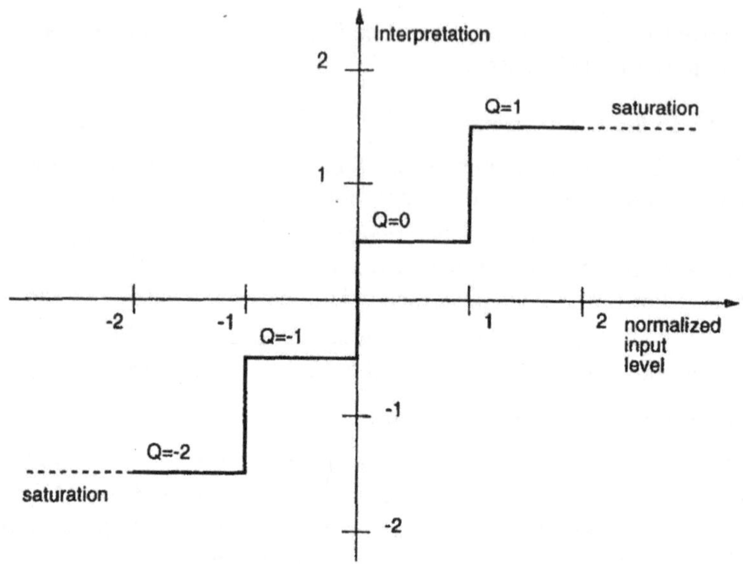

Figure 16.8 Characteristic of a 2-bit step-at-zero quantizer.

interpreted as -0.5 and 0.5, respectively. Thus, the value zero cannot be represented and the actual range of a 2^w-level quantizer is symmetric. Even with a very low average signal level before the quantizer the sign of the input signal is still retained behind the quantizer. Thus the worst case performance using such a quantizer characteristic is equivalent to hard decision decoding. Using this interpretation and chosing $C_2 = 1$ in (18) and $w = 3$, the resulting range of the (integer valued) branch metrics is

$$\begin{aligned} \text{Min}(\lambda_k^{(m,i)}) &= C_3 - 3.5 - 3.5 = C_3 - 7 \\ \text{Max}(\lambda_k^{(m,i)}) &= C_3 + 7 \end{aligned} \tag{19}$$

thus, $w+1$ bits are sufficient for the branch metrics and $C_3 = 2^w - 1$ can be chosen to obtain always positive branch metrics.

16.3.1.2 Dead zone quantizer

A second quantization approach is to take the usual 2's complement value without any offset. Fig. 16.9 shows the characteristic of a 2-bit dead zone quantizer.

In this case the value 0 is output of the quantizer for a certain range around input value 0, nominally for $-0.5 < x \leq 0.5$. In contrast to ***step at zero*** quantization, very low average signal levels before quantization will ultimately result in loosing the information about the input signal completely (even the sign), since the quantizer then outputs zero values only. When using this quantizer characteristic it is advantageous to compute the branch metrics as

$$\lambda_k^{(m,i)'} = C_3 + \frac{1}{2}\left\{(y_{1,k}c_{1,k} + \text{Abs}(y_{1,k})) + (y_{2,k}c_{2,k} + \text{Abs}(y_{2,k}))\right\}. \tag{20}$$

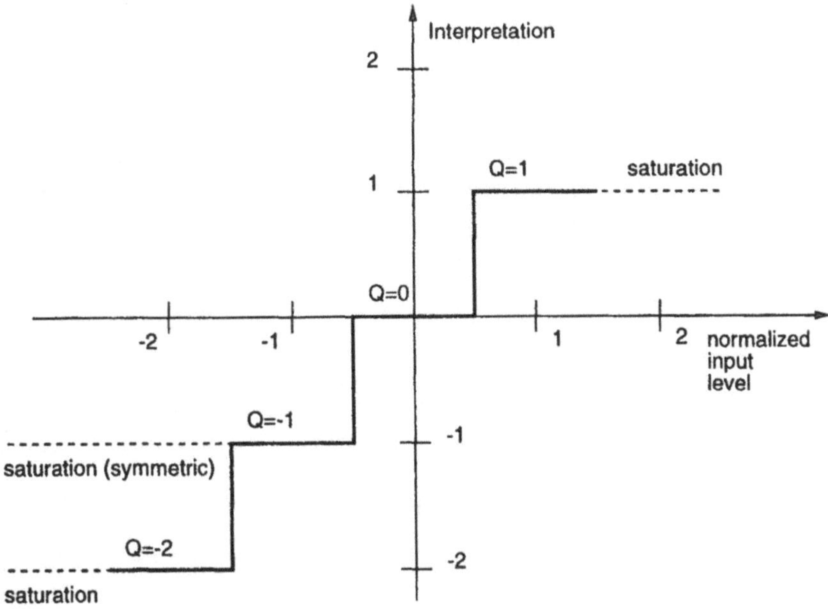

Figure 16.9 Characteristic of a 2-bit dead zone quantizer.

This choice is legal since $\text{Abs}(y_{1,k}) + \text{Abs}(y_{2,k})$ is constant for every trellis step and thus does not influence the selection decisions of the ACSU. By choosing $C_3 = 0$ the range of the branch metrics is described by

$$0 \leq \lambda_k^{(m,i)'} \leq 2\,\text{Max}(\text{Abs}(y_k)). \tag{21}$$

It is easily shown that the branch metrics are still integer values if computed according to (20). For a usual w-bit integer with range $\{-2^{w-1}, \ldots, 2^{w-1} - 1\}$ the resulting branch metric range is $0, \ldots, 2^w$ which requires $(w+1)$-bit branch metrics. However, by making the quantizer output range symmetrical, i.e. constraining $y_{1,k}$ and $y_{2,k}$ to the interval $\{-2^{w-1}+1, \ldots, 2^{w-1}-1\}$ the branch metric range becomes $\{0, \ldots, 2^w - 1\}$ which can be represented with w-bit unsigned branch metrics (cf [18]). Since symmetry is anyway advantageous to avoid a biased decoding process, this is the option of choice.

With this approach we can either reduce the branch metric wordlength by one bit and reduce the quantization levels by one (e.g., from 8 levels to 7 levels for a 3-bit input value) or increase the input wordlength by one bit and thereby provide more margin for non-ideal gain control. Thus the approach either leads to decreased ACSU complexity or better performance at equal complexity since the TMU complexity is in most cases still marginal.

16.3.2 Support of Punctured Codes

Since punctured codes are derived from a *base* code by deleting some code bits prior to transmission, the decoder of the base code can be used if the TMU can compute the branch metrics such that the missing information does not affect the decisions of the remaining decoder. Assuming without loss of generality that the second received value $y_{2,k}$ in the example above is missing, the TMU has to compute the branch metrics such that the terms

$$y_{2,k}c_{2,k} \qquad \text{respectively} \qquad y_{2,k}c_{2,k} + \text{Abs}(y_{2,k}) \tag{22}$$

evaluate to a constant for all different branch metrics. To achieve this it is possible to either replace $y_{2,k}$ with 0, or to manipulate the metric computation such that $c_{2,k}$ is constant for all computed branch metrics, which is equivalent to relabeling part of the branches of the trellis with different code symbols.

Clearly, the first approach is applicable only if one of the quantized values is actually interpreted as 0 (as for the dead-zone quantizer discussed above) since $y_{2,k} = 0$ can easily be chosen. For step at zero quantization, where the quantized values are interpreted with an implicit offset of 0.5, manipulating the branch labels is the better choice since a replacement value of 0 is not straightforwardly available.

16.4 THE ADD–COMPARE–SELECT UNIT

Given the branch metrics, the ACSU calculates the state metrics according to the ACS recursion, which represents a system of nonlinear recurrence equations. Since the ACS operation is the only recursive part of the Viterbi algorithm, the achievable data (and clock) rate of a VLSI implementation is determined by the computation time of the ACS recursion. Due to the repeated accumulation of branch metrics to state metrics, the magnitude of these metrics is potentially unbounded. Hence, *metric normalization schemes* are necessary for a fixed wordlength implementation.

16.4.1 Metric Normalization Schemes

In order to prevent arithmetic overflow situations and in order to keep the register effort and the combinatorial delay for the add and compare operations in the ACSU as small as possible, metric normalization schemes are used.

Several methods for state metric normalization are known, which are based on two facts [19]:

1. The differences $\Delta\gamma_k$ between all state metrics at any trellis step k are bounded in magnitude by a fixed quantity $\Delta\gamma_{Max}$ independent of the number of ACS operations already performed in the trellis.

2. A common value may be subtracted from all state metrics for any trellis step k, since the subtraction of a common value does not have any impact on the results of the following metric comparisons.

Consider all paths starting from a given state $s_{i,k}$ in the trellis, corresponding to the state $x_k = i$ in the encoder. After a certain number n of trellis steps, all other states can be reached starting with $x_k = i$. Since one bit is shifted into the encoder shift register for every trellis step, n is obviously equal to $K - 1$. In other words, after $K - 1$ steps, an arbitrary shift register state is possible independent of the initial state. Hence, the interval n ensures complete connectivity for all trellis states. In the trellis, there are N distinct paths from the starting state to all other states $s_{j,k+n}$, $j \in \{0, \ldots, N-1\}$. An upper bound on the state metric difference $\Delta\gamma_{Max}$ can be found assuming that for one of these paths, the added branch metric $\lambda_k^{(m,i)}$ was minimum for all n transitions, and for another of these paths, the branch metric was always maximum. Hence, an upper bound on the maximum metric difference is given by

$$\Delta\gamma_{Max} = n \cdot (\max(\lambda_k^{(m,i)}) - \min(\lambda_k^{(m,i)}))$$

with $n = K - 1$ and $\max(\lambda_k^{(m,i)})$ and $\min(\lambda_k^{(m,i)})$ being the maximum and minimum metric value possible using the chosen branch metric quantization scheme. The wordlength necessary to represent $\Delta\gamma_{Max}$ is the minimum wordlength required for the state metrics[8]. However, depending on the chosen normalization scheme, a larger wordlength has actually to be used in most cases. We now state two normalization schemes:

16.4.1.1 Subtracting the minimum state metric

After a given number of trellis steps, the minimum state metric is determined and subtracted from all other state metrics. This scheme leads to the minimum state metric wordlength as derived above, if it is performed for every trellis step. The resulting architecture for a single ACS processing element (PE) using this normalization scheme is shown in Fig. 16.10.

If a normalization is performed only after a certain number of trellis steps, an increased wordlength of the state metrics has to be taken into account.

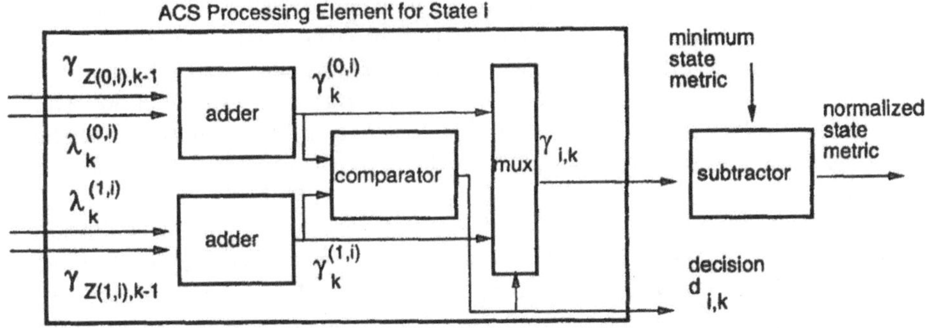

Figure 16.10 ACS processing element and minimum state metric subtraction.

The additional computational effort involved with this scheme is relatively large: first, the minimum state metric has to be determined, second, a subtraction has to be performed in addition to the usual add–compare–select operation. However, it may be suited for low throughput architectures and software applications. The minimum state metric can then be determined sequentially while successively calculating the new state metrics for a trellis transition, and the effort for the additional subtraction does not pose a significant problem.

[8] Even tighter bounds on the state metric differences were derived in [20].

16.4.1.2 "On the fly" normalization schemes

For high throughput applications, the ACS recursion is implemented with a dedicated ACS PE per trellis state. In this case, N new state metrics are calculated in parallel for all states. Determining the minimum of these metrics would require much more processing delay than the ACS calculation itself, hence more efficient ways have to be found for normalization.

A very efficient normalization scheme can be found again exploiting the upper bound on the metric difference $\Delta\gamma_{Max}$. The idea is simply to subtract a fixed value from all state metrics if the state metrics exceed a certain threshold t. Simultaneously, it has to be guaranteed that no overflows or underflows occur for all state metrics. The value of the threshold t can be chosen such that the detection of a threshold excess and the necessary subtraction can be implemented as efficiently as possible, while keeping the state metric wordlength as small as possible. In the following, one of the possible solutions for unsigned branch and state metrics is presented:
The unsigned branch metrics are quantized to b bits, leading to a maximum branch metric value of $2^b - 1 = \max(\lambda_k^{(m,i)})$. The unsigned state metrics are quantized with p bits, corresponding to a maximum value of $2^p - 1$. Of course, $\Delta\gamma_{Max} \leq 2^p - 1$ must hold. If the number of bits p is chosen such that

$$\Delta\gamma_{Max} \leq 2^{p-2}$$

a very efficient normalization without additional subtraction can be derived. It is now sufficient to observe just the value of the most significant bit (MSB). If any state metric value gets equal to or exceeds the value $t = 2^{p-1}$, it is simultaneously known that all other state metrics are equal to or larger than 2^{p-2} because of the limited state metric difference. Hence, it is possible to subtract the value of 2^{p-2} from all state metrics while guaranteeing that all state metrics remain positive.

Both the test of the MSB and the subtraction of 2^{p-2} can be implemented using very simple combinatorial logic involving only the two MSBs and a few combinatorial gates.

The inspection of the MSBs for all state metrics still requires global communication between all ACS PEs. This drawback can be removed by using modulo arithmetic for the state metrics as proposed in [19]. Metric values exceeding the range just wrap around according to the modulo arithmetic scheme, hence no global detection of this situation is necessary. However, the state metric wordlength has also to be increased to a value larger than the minimum given by $\Delta\gamma_{Max}$. Details can be found in [19].

Due to the recursive nature of the ACS processing, the combinatorial delay through the ACS PE determines the clock frequency (and hence the decoded bit rate) of the whole Viterbi decoder. Arithmetic and logic optimization of the ACS PE is therefore essential. Many proposals exist for optimizing the arithmetic in the ACS. Every conventional addition scheme suffers from the fact that the combinatorial delay is some function of the wordlength, since a carry propagation occurs. Redundant number systems allow carry free or limited carry propagation addition [21, 22]. However, the maximum selection cannot be solved straightforwardly in redundant number systems. Nevertheless, a method was proposed allowing to use the redundant carry–save number system for the ACS processing, which can be very beneficial if large wordlengths have to be used [14, 15].

16.4.2 Recursive ACS Architectures

We first consider the case that one step of the ACS recursion has to be calculated in one clock cycle, and later briefly disuss lower throughput architectures. If a dedicated ACS PE is used for every state in the trellis, the resulting node parallel architecture with a throughput of one trellis step per clock cycle is shown in Fig. 16.11. For simplicity, the state metric normalization is not shown in this picture. A complete vector of decisions $d_{i,k}$ is calculated for every clock cycle. These decisions are stored in the SMU in order to facilitate the path reconstruction. Obviously, a large wiring overhead occurs, since the state metrics have to be

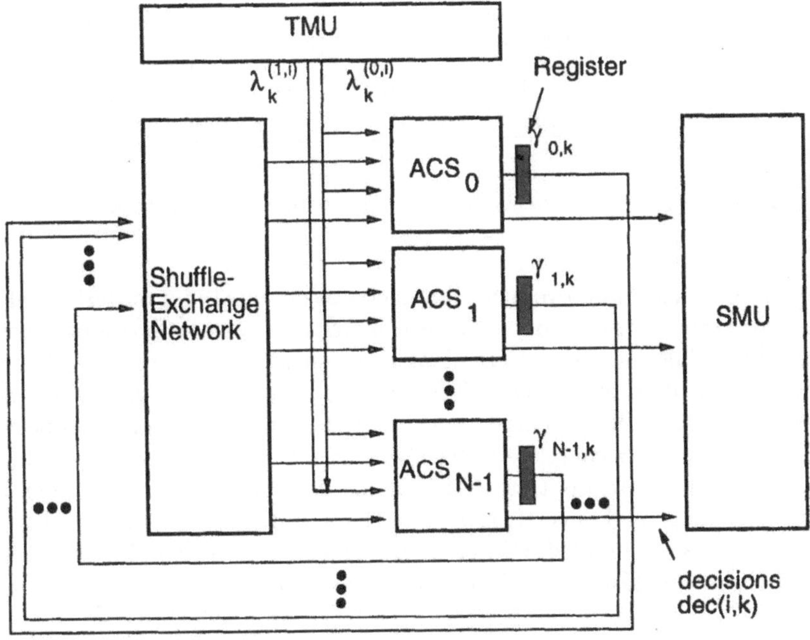

Figure 16.11 Node parallel ACS architecture.

fed back into the ACS PEs. The feedback network is a *shuffle–exchange* network. The possible state transitions for the states $x_k = \sum_{j=0}^{\nu-1} x_{j,k}\, 2^j$ are given by a cyclic shift (perfect shuffle) $x_{0,k}, x_{\nu-2,k}, \ldots, x_{1,k}$ and an exchange $\overline{x_{0,k}}, x_{\nu-2,k}, \ldots, x_{1,k}$. where $\overline{x_{0,k}}$ denotes inversion of $x_{0,k}$. Many proposals exist for optimum placement and routing of this type of interconnection network (see e.g., [23]).

For lower throughput applications, several clock cycles are available for a single ACS recursion. Here, the fact can be exploited that the trellis diagram for nonrecursive rate $1/n$ codes includes butterfly structures known from FFT processing. Since for rate $1/n$ codes, just a single bit is shifted into the encoder FSM, the transition function specifying the two predecessor states for a current state $x_k = \{x_{K-2,k}, \ldots, x_{0,k}\}$ is given by $Z(m, x_k) = \{x_{K-3,k}, \ldots, x_{0,k}, m\}$ as stated in (15). These two predecessor states $Z(0, x_k)$ and $Z(1, x_k)$ have exactly two successor states: $\{x_{K-2,k}, \ldots, x_{0,k}\}$ and $\{\overline{x_{K-2,k}}, \ldots, x_{0,k}\}$ with $\overline{x_{K-2,k}}$ denoting bit inversion. This results in the well known butterfly structure as shown in Fig. 16.12.

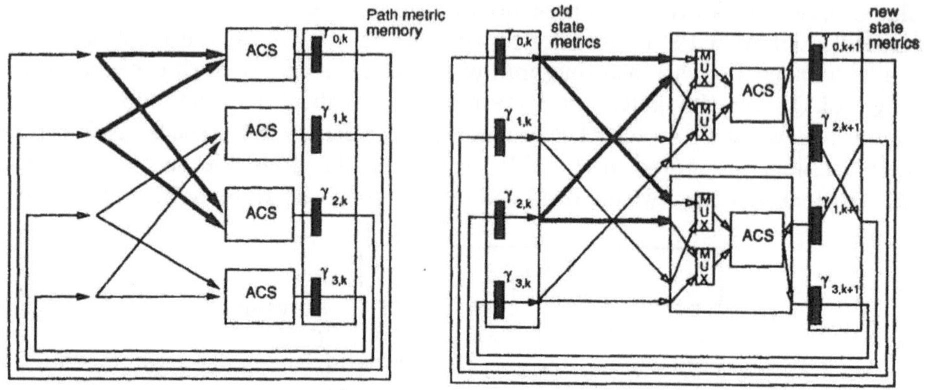

Figure 16.12 Butterfly trellis structure and resource sharing for the $K = 3$, rate 1/2 code.

In order to calculate two new state metrics contained in a butterfly, only two predecessor state metrics and two branch metrics have to be provided. Hence a sequential schedule for calculating all ACS operations in the trellis is given by a sequential calculation of the butterflies. This is shown on the right hand side of Fig. 16.12; two ACS PEs calculate sequentially the two ACS operations belonging to a butterfly, hence, a complete trellis step takes two clock cycles here. The ACS operations according to the thick butterfly are calculated in the first clock cycle, and the remaining ACS operations are calculated in the second clock cycle.

In Fig. 16.12, a parallel ACS architecture with four ACS PEs and a resource shared architecture with two ACS PEs are shown for the $K = 3$ code. As shown in Fig. 16.12, it seems to be necessary to double the memory for the state metrics compared to the parallel ACS architecture, since the state metrics $\gamma_{i,k+1}$ are calculated while the old metrics $\gamma_{i,k}$ are still needed. It was shown in [24], however, that an in–place memory access for the state metrics is possible with a cyclic metric addressing scheme. Here, only the same amount of memory is necessary as for the parallel ACS architecture. Several proposals for resource shared ACSU implementations can be found in [25]-[28].

16.4.3 Parallelized ACS Architectures

The nonlinear data dependent nature of the recursion excludes the application of known parallelization strategies like pipelining or look-ahead processing, which are available for parallelizing linear recursions [29]. It was shown [15, 30, 31] that a linear algebraic formulation of the ACS recursion can be derived, which, together with the use of the acquisition and truncation properties of the Viterbi algorithm, allows to derive purely feedforward architectures [30]. Additionally, the linear algebraic formulation represents a very convenient way to describe a variety of ACS architectures.

Below, the algebraic multiplication $\otimes$ denotes addition and the algebraic addition $\oplus$ denotes maximum selection. The resulting algebraic structure of a semiring defined over the operations $\oplus$ and $\otimes$ contains the following:

- neutral element concerning $\oplus$ (maximum selection): $Q(= -\infty)$
- neutral element concerning $\otimes$ (addition): $\underline{1}(= 0)$

Using the semiring algebra, the ACS recursion for the $K = 3$, rate 1/2 code as stated in (15) can be written as:

$$\begin{pmatrix} \gamma_0 \\ \gamma_1 \\ \gamma_2 \\ \gamma_3 \end{pmatrix}_{k+1} = \begin{pmatrix} \max(\lambda_{00} + \gamma_0; \lambda_{01} + \gamma_1) \\ \max(\lambda_{12} + \gamma_2; \lambda_{13} + \gamma_3) \\ \max(\lambda_{20} + \gamma_0; \lambda_{21} + \gamma_1) \\ \max(\lambda_{32} + \gamma_2; \lambda_{33} + \gamma_3) \end{pmatrix}_k = \begin{pmatrix} (\lambda_{00} \otimes \gamma_0) \oplus (\lambda_{01} \otimes \gamma_1) \\ (\lambda_{12} \otimes \gamma_2) \oplus (\lambda_{13} \otimes \gamma_3) \\ (\lambda_{20} \otimes \gamma_0) \oplus (\lambda_{21} \otimes \gamma_1) \\ (\lambda_{32} \otimes \gamma_2) \oplus (\lambda_{33} \otimes \gamma_3) \end{pmatrix}_k$$

$$= \begin{pmatrix} \lambda_{00} & \lambda_{01} & Q & Q \\ Q & Q & \lambda_{12} & \lambda_{13} \\ \lambda_{20} & \lambda_{21} & Q & Q \\ Q & Q & \lambda_{32} & \lambda_{33} \end{pmatrix}_k \otimes \begin{pmatrix} \gamma_0 \\ \gamma_1 \\ \gamma_2 \\ \gamma_3 \end{pmatrix}_k . \tag{23}$$

Transition metrics not corresponding to allowed state transitions are assigned the metric $Q = -\infty$. Of course, no computational effort is required for terms including the Q–value. Given a state metric vector $\Gamma_k = (\gamma_{0,k}, \ldots, \gamma_{N-1,k})^T$ and an NxN transition matrix Λ_k containing all the transition metrics $\lambda_{ij,k}$, the above equation can be written as a matrix–vector product:

$$\Gamma_{k+1} = \Lambda_k \otimes \Gamma_k.$$

It can be shown, that all rules known from linear algebra are applicable to this linear algebraic formulation of the ACS recursion as well. Hence, this represents much more than just a convenient notation, and allows to derive new algorithms and architectures. It is e.g., possible to arrive at an M-step ACS recursion:

$$\Gamma_{k+M} = {}_M\Lambda_k \otimes \Gamma_k = (\Lambda_{k+M-1} \otimes \Lambda_{k+M-2} \otimes \ldots \otimes \Lambda_k) \otimes \Gamma_k$$

with an M-step transition matrix ${}_M\Lambda_k$ describing the NxN optimum transition metrics from every state at trellis step k to every state at trellis step $k + M$. This approach is just another formulation of the original ACS recursion, i.e., the results are exactly equivalent. Associativity of the $\otimes$ operation allows to reformulate the recursion in this way.

This M step processing already allows a parallelization, since the M matrix–matrix products can be calculated in advance, and the actual recursion now spans M trellis steps, leading to a speedup factor of M. A disadvantage of the M-step approach is the computational effort necessary to calculate matrix-matrix products for the $N \times N$ matrices Λ_k. The matrices for single transitions contain many Q–entries, as shown in the example (22). With successive matrix–matrix multiplications, the number of Q–entries soon becomes zero, leading to an increased effort for matrix multiplications since there is no computation necessary for the Q–entries as stated above. Hence, an implementation for small M and especially $M = 2$ as reported in [32] seems to be particularly attractive. In [32] it is proposed to unfold

the ACS recursion for a number of successive trellis steps, which is equivalent to introducing an M–step recursion. A two–step ACS recursion is also advantageous because there is more room for arithmetic optimization of the recursion equations, since the concatenation of successive additions can be implemented quite advantageously. Since two vectors of decison bits are generated simultaneously for a single clock cycle, the resulting decoded bit rate is two times the clock frequency.
For larger values of M, however, there is a significant increase in the computational effort when using the M–step approach.

However, it was shown by Fettweis [14, 15] that it is even possible to derive an independent processing of blocks of received symbols leading to a purely feedforward solution with an arbitrary degree of parallelism, the so called *minimized method*. The key to this approach is the exploitation of the acquisition and truncation properties of the VA.

We first review conventional Viterbi decoding with regard to the linear algebraic formulation: the M-step transition matrix contains the best paths from every state at trellis step k to every state at trellis step $k+M$:

$$_M\Lambda_k = \begin{pmatrix} _M\lambda_{00} & _M\lambda_{01} & \cdots & _M\lambda_{0(N-1)} \\ _M\lambda_{10} & _M\lambda_{11} & \cdots & _M\lambda_{1(N-1)} \\ _M\lambda_{(N-1)0} & _M\lambda_{(N-1)1} & \cdots & _M\lambda_{(N-1)(N-1)} \end{pmatrix}_k .$$

Each entry $_M\lambda_{ij}$ contains the metric of the best path from state j at trellis step k to state i at trellis step $k+M$.

The conventional VA (for $M=D$) calculates recursively $\Lambda_{k+D}\otimes(\ldots\otimes(\Lambda_k\otimes\Gamma_k))$ which is equal to $_D\Lambda_k\otimes\Gamma_k$. Hence the VA operation can also be interpreted as follows: the VA adds the state metrics at trellis step k to the corresponding matrix entries and then perform a rowwise (concerning $_D\Lambda_k$) maximum selection leading to metrics for the N best paths at time instant $k+D$. If *best state decoding* [11] is applied, the VA finally selects the overall maximum likelihood survivor path with metric $\gamma_{i,k+D} = _D\lambda_{ij}+\gamma_{j,k}$ including decoding the best state $x_k=j$ at time instant k.

The conventional VA with best state decoding for trellis step k can hence also be represented as

$$(\underline{1},\ldots,\underline{1})\otimes_D\Lambda_k\otimes\Gamma_k = \gamma_{i,k+D} = _D\lambda_{ij}+\gamma_{j,k} \tag{24}$$

since the multiplication with $(\underline{1},\ldots,\underline{1})$ in the semiring algebra corresponds to the final overall maximum selection in conventional arithmetic. It is obvious that the best state $x_k=j$ can be immediately accessed via the indices of the overall best metric $\gamma_{i,k+D} = _D\lambda_{ij}+\gamma_{j,k}$.

The state metric vector Γ_k can be calculated by an acquisition iteration. It was already discussed that the acquisition depth is equal to the survivor depth D. Therefore, we start decoding in midstream at $k-D$ with all state metrics equal to zero.

$$\Gamma_k = _D\Lambda_{k-D}\otimes(\underline{1},\ldots,\underline{1})^T.$$

Replacing Γ_k in (24) leads to:

$$\underbrace{((\underline{1},\ldots,\underline{1})\otimes_D\Lambda_k)}_{\text{truncation}}\otimes\underbrace{(_D\Lambda_{k-D}\otimes(\underline{1},\ldots,\underline{1})^T)}_{\text{acquisition}}$$

$$= \underbrace{(({}_D\Lambda_k)^T \otimes (\underline{1}, \ldots, \underline{1})^T)^T)}_{\text{truncation}} \otimes \underbrace{({}_D\Lambda_{k-D} \otimes (\underline{1}, \ldots, \underline{1})^T)}_{\text{acquisition}}$$

$$= (\gamma_{0,k,T}, \ldots, \gamma_{N-1,k,T}) \otimes (\gamma_{0,k,A}, \ldots, \gamma_{N-1,k,A})^T. \quad (25)$$

Both matrix–vector products are equal to the usual ACS operation. Since (24) is purely feedforward, we call the two iterations *ACS acquisition iterations* below. Each of the two iterations leads to a state metric vector as shown in (24), where the index T denotes truncation and the index A denotes acquisition. In a final step, the two state metrics which are resulting per state are added, and the overall maximum metric is determined. This can be verified by writing out the final expression in (24) and replacing semiring arithmetic with conventional arithmetic. The state corresponding to the global maximum is finally decoded. A parallel architecture implementing (24) is shown in Fig. 16.13.

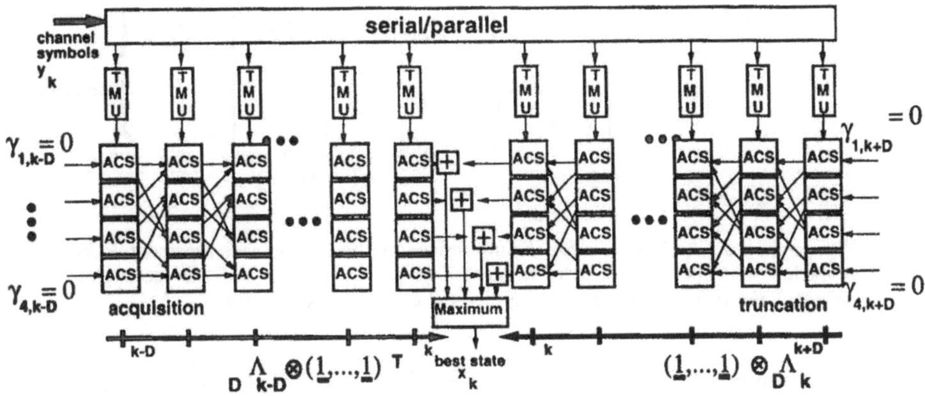

Figure 16.13 Architecture for the ACS acquisition iterations for the $K = 3$, rate 1/2 code. Each node represents a dedicated ACSU for one trellis step.

Obviously, an arbitrary number of ACS acquisition iterations can be performed independently and hence in parallel for an arbitrary number of blocks containing $2M$ symbols with $M \geq D$. It is most efficient to use nonoverlapping contiguous blocks of length $2D$ for this operation. The result is a number of uniquely decoded states with distance $2D$ transitions, i.e., x_k, x_{k+2D}, ...

Using the known states resulting from the ACS acquisition iterations, a second ACS iteration is started. The survivor decisions generated here are used – as for a conventional Viterbi decoder – to finally trace back the decoded paths. For the trace back, best state decoding is performed, since the overall best state x_{k-D} is determined and used as a starting point for the trace back.

The resulting architecture that processes one block at a time is shown in Fig. 16.14. It consumes one block of input symbols for every clock cycle. The latches are necessary to store values which are needed for the following block to be processed in the next clock cycle.

It is possible to extend the architecture shown in Fig. 16.14 by identical modules on the left and right hand side, leading to an even faster architecture that

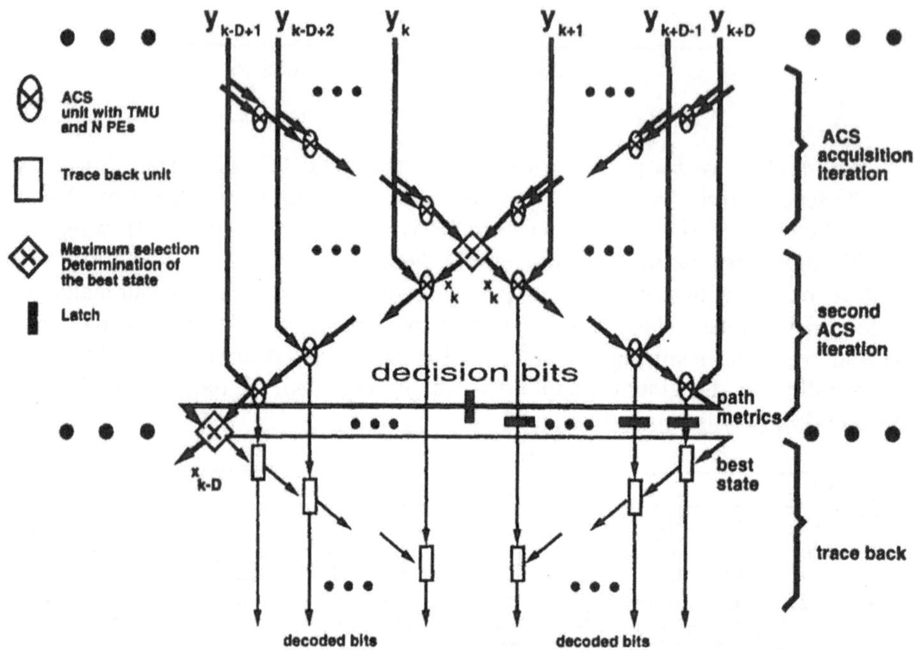

Figure 16.14 Architecture for the minimized method. The ovals represent ACS operation, the diamonds the determination of the best state, and the rectangles represent the trace back operation in forward or backward direction.

consumes a number of blocks at a time. Therefore, in principle, an arbitrary degree of parallelism can be achieved.

A detailed description of the minimized method architecture is given in [33]-[35]. In order to achieve Gbit/s speed, a fully parallel and pipelined implementation of the Minimized Method was developed and realized as a CMOS ASIC for Gbit/s Viterbi decoding [33]. Here, one dedicated ACS unit, with a dedicated ACS processing element (PE) for every state, is implemented for each trellis transition. Bit level pipelining is implemented for the ACS PEs, which is possible since the minimized method is purely feedforward. The fabricated ASIC [33] is one order of magnitude faster than any other known Viterbi decoder implementation.

16.4.4 THE SURVIVOR MEMORY UNIT (SMU)

As was explained earlier, in principle all paths that are associated with the trellis states at a certain time step k have to be reconstructed until they all have merged to find the *final survivor* and thus the decoded information. However, in practice only one path is reconstructed and the associated information at trellis step $k - D$ output (c.f., Fig. 16.4). D must be chosen such that all paths have merged with sufficiently high probablility. If D is chosen too small (taking into account code properties and whether fixed or best state decoding is performed) substantial performance degradations result. The path reconstruction uses stored path decisions from the ACSU. Clearly, the survivor depth D is an important parameter of the SMU since the required memory to store the path decisions is directly proportional to D for a fixed implementation architecture.

Fixed state decoding is usually preferred in parallel HW implementations since finding the largest state metric (for best state decoding) can be both, time critical and costly in HW. However, since a significantly larger D must be chosen for fixed state decoding [11, 12], the involved trade-off should be studied thoroughly for optimum implementation efficiency.

For the actual decoding process that is implemented in the SMU two different algorithms are known: the register exchange algorithm (REA) and the traceback algorithm (TBA) [17, 24]. While register exchange implementations of SMUs are known to be superior in terms of regularity of the design and decoding latency, traceback implementations typically achieve lower power consumption and are more easily adapted to lower speed requirements where more than one clock cycle is available per trellis cycle for processing the data [36, 37]. While we focus here on TBA and REA, it should be noted that the implementation of a hybrid REA/TBA architecture was reported in [38] and further generalizations are treated in [39, 40].

16.4.5 REA

The REA computes the information symbol sequences of the survivor paths associated with all N trellis-states based on the decisions provided by the ACSU.

If we denote the information symbol associated with the reconstructed path belonging to state i at trellis step k as $\hat{u}_k^{[i]}$ and the information symbol associated with the m'th branch merging into state i as $u^{(m,i)}$, we can formally state the algorithm as follows:

```
    Memory:
  (D+1)·N Information Symbols (û_k^[i], ..., û_{k-D}^[i]) ;
Algorithm:
  // Update of the stored symbol sequences according to
  // the current decision bits d_{i,k}
  for t=k-D to k-1 {
    for State=0 to N-1 {
      û_t^[State] := û_{t+1}^[Z(d_{State,k},State)] ;
    }
  }
  // setting the first information symbol of the path
  for State=0 to N-1 {
    û_k^[State] := u^(d_{State,k},State) ;
  }
```

Here, $Z(m, x_k)$ is the state transition function as defined in (15).
The nested loop describes how the stored information sequences corresponding to the best paths at trellis step $k-1$ are copied according to the decision bits di, k obtained at trellis step k in the ACS recursion. In the final loop the information sequences for the N best paths are preceded with the information bits for step k. For example, if at time k and state i, the path according to the branch with label $m = 1$ is selected as the best path, the stored symbol sequence for the state branch

1 emerged from is copied as the new symbol sequence of state i preceded by the information symbol associated with branch 1.
Assuming l-bit information symbols, the algorithm requires $D \cdot N \cdot l$ bits of memory for a code with N states. If we define the decoding latency as the difference between the most recently processed trellis step k and the trellis step, the decoded information is associated with, the decoding latency of the REA is identical to the survivor depth D (neglecting implementation related delays, e.g., pipeline delay). Both figures are the minimum achievable for a given N and D. However, access bandwidth to the memory is very high. Per trellis cycle each survivor symbol sequence is completely overwritten with new survivor symbols resulting in $D \cdot N$ read and write accesses per cycle. Therefore in a fully parallel implementation the cells are usually implemented as flip-flops (registers) and the selection of the possible input symbols is done by means of multiplexors. Fig. 16.15 shows the resulting hardware architecture for the sample $K = 3$, rate 1/2 code with 4 states and binary information symbols. The topology of the connections corresponds to the

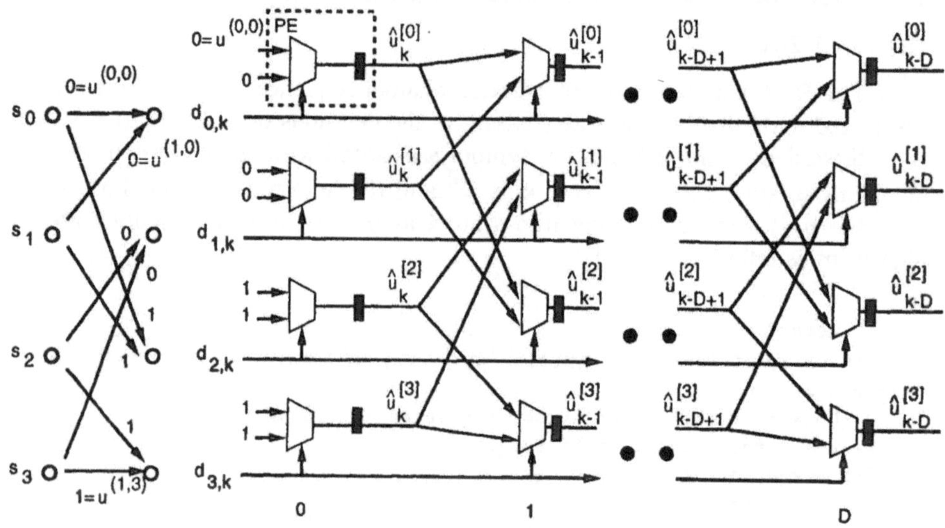

Figure 16.15 REA hardware architecture.

trellis topology, which can be a major drawback of the approach for large number of states. Power consumption can be a problem in VLSI implementations due to the high access bandwidth [41, 42]. As a consequence, the REA is usually not used for low data rate decoders. It is applied if latency, regularity or total memory size are critical parameters.

16.4.6 TBA

In contrast to the REA the TBA does not compute the ***information symbol*** sequence associated with each state. Instead, the ***state*** sequence is computed based on the path decisions $d_{i,k}$. In a second step the associated information symbols $\hat{u}_k$ are computed. In practice, using D reconstruction steps a state of the final survivor is acquired (D : survivor depth). Subsequently, the path is traced back M steps further to obtain M symbols that are associated with the final survivor [36, 37] (M : decoding depth). Fig. 16.16 shows a sample traceback sequence. Formally, we can state the TBA as follows:

```
    Memory:
  (D+M)·N decision bits (d_{i,k}, ..., d_{i,k-(D+M-1)}) ;
Algorithm:
  // every M trellis steps a trace back is started
  if (k-D can be divided by M) then {
    // Initialization
    traceState := startState ;
    // Acquisition
    for t=k downto k-D+1 {
      traceState := Z(d_{traceState,t},traceState) ;
    }
    // Decoding
    for t=k-D downto k-D-M+1 {
      û_t^[d] := u^(d_{traceState,t},traceState) ;
      traceState := Z(d_{traceState,t},traceState) ;
    }
  }
```

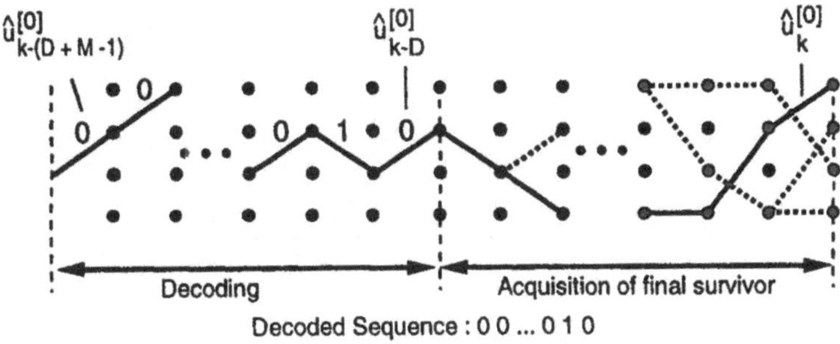

Figure 16.16 Example of the TBA.

The memory size of an implementation of our example must be at least $(D + M) \cdot N$ bits to facilitate performing a data traceback with depth M while maintaining D as survivor depth. Furthermore the decoding latency is increased to at least $D+M$ because tracing back requires $M+D$ ACS iterations to be performed before the first trace back can be started[9]. Blocks of M symbols are decoded in reverse order during the data traceback phase, thus a last-in first-out (LIFO) memory is required for reversing the order before outputting the information. Fast hardware implementations require more memory and exhibit a larger latency.

The algorithm requires write accesses to store the N decision bits. Since a trace back is started only every M trellis steps, on average $(M+D)/M$ decision bits

[9]This minimum figure is only achievable in low rate applications, since the actual computation time for reconstruction is not already included!

are read and trellis steps reconstructed for the computation of a single information symbol. Thus the access bandwidth and computational requirements are greatly reduced compared to register exchange SMUs so RAMs can be used for storage which can be implemented in VLSI with a much higher density than flipflops particularly in semi-custom technologies. Thus TBA implementations are usually more power efficient compared to REA implementations.

Furthermore, the choice of M relative to D (D is usually specified) allows memory requirements to be traded against computational complexity. And the TBA can thus be adapted to constraints of the target technology more easily [36, 37, 43]. We will review the basic trade-offs in the next section.

16.4.7 TBA Trade-offs

The inherent trade-off in the TBA is best understood if visualized. This can be done with a clock-time/trellis-time diagram, where the actual ongoing time measured in clock cycles is given on the x-axis, while the time in the trellis is given on the y-axis. Fig. 16.17 shows such a diagram for the TBA with $M = D$.

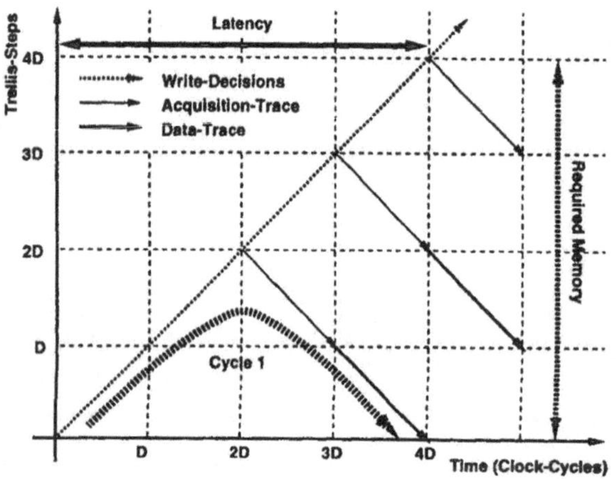

Figure 16.17 Traceback with $M = D$.

Henceforth we assume that a new set of decision bits is generated in every clock cycle as is usually required in fast hardware implementations e.g., for digital video broadcasting applications. Consider the first complete cycle of traceback processing in Fig. 16.17 (Cycle 1). During the first $2 \cdot D$ clock cycles, sets of decision bits are written to the memory. In the subsequent D clock cycles the data is retrieved to facilitate the acquisition of the final survivor (acquisition-trace). Finally the data is decoded by retrieving $M = D$ sets of decision bits and tracing the final survivor further back over time (data-trace), while concurrently a new acquisition-trace is performed starting from trellis step $3 \cdot D$. In fact we obtain a solution where acquisition-trace and data-trace are always performed concurrently, i.e., we obtained a solution which requires two read pointers. The latency of the data-trace is obtained as the difference in clock cycles from data write (at trellis step 0) until the written information is accessed the last time (at clock cycle $4 \cdot D$). The required memory is obtained on the y-axis as the number of kept decision bits before memory can be reused ($4 \cdot D$ trellis steps with a total of $4 \cdot D \cdot N$ bits). We assumed here that during the data-trace the memory is not immediately reused by writing the new data into the memory location just read. Immediate reuse is possible if the used memory technology permits a read and a write cycle to be performed in one clock cycle which is usually not possible in commodity semi-custom technologies for high throughput requirements.

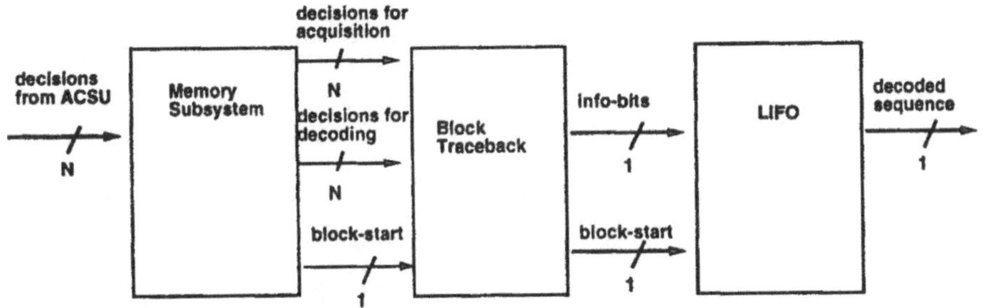

Figure 16.18 Architecture block diagram for traceback with $M = D$.

Fig. 16.18 shows a typical block diagram for a TBA with $M = D$ and one clock cycle per trellis step. Three building blocks are distinguished:
1) The memory subsystem including some control functionality and providing sets of decision bits for acquisition and decoding, as well as a signal indicating the start of a new trace back.
2) The actual path reconstruction which outputs decoded information bits in reverse order and a block start indication.
3) The LIFO required to reverse the decoded information bits blockwise.
Typically, the memory subsystem dominates complexity and is affected strongly by the choice of M and D. By accepting the overhead implied by using dual port RAMs, a very simple architecture can be derived for $M = D$, that uses only 3 RAMs of size $N \cdot (D+1)$ bits. Fig. 16.19 shows the used addressing and multiplexing scheme for the 3 RAMs that repeats after 6 cycles.

Using dual port RAMs, a memory location is overwritten one clock cycle after it was read for the last time. This avoids concurrent read and write access to the same location, which is usually not possible. Consider, e.g., cycle 3. All memories are accessed in ascending order. The first read access for acquisition, as well as data trace, is to address 1 of RAM1 and RAM2, respectively. Concurrently new data is written to address 0 of RAM2. Thus in the subsequent step the read data at address 1 of RAM2 is overwritten with new data. A closer look at the sequence of activity unvails that only a single address generator is needed, that counts up from 1 to D and subsequently down from $D-1$ to 0. The write address equals the read address of the last clock cycle. Fig. 16.20 shows the resulting architecture for the memory subsystem. The inputs *wIdle* and *rIdle* are the access control ports of the RAMs.

A reduction in the memory requirements is possible by choosing a smaller M. Fig. 16.21 shows an example for $M = 0.5 \cdot D$, which reduces latency and memory requirements to $3 \cdot D$ and $3 \cdot D \cdot N$ respectively.

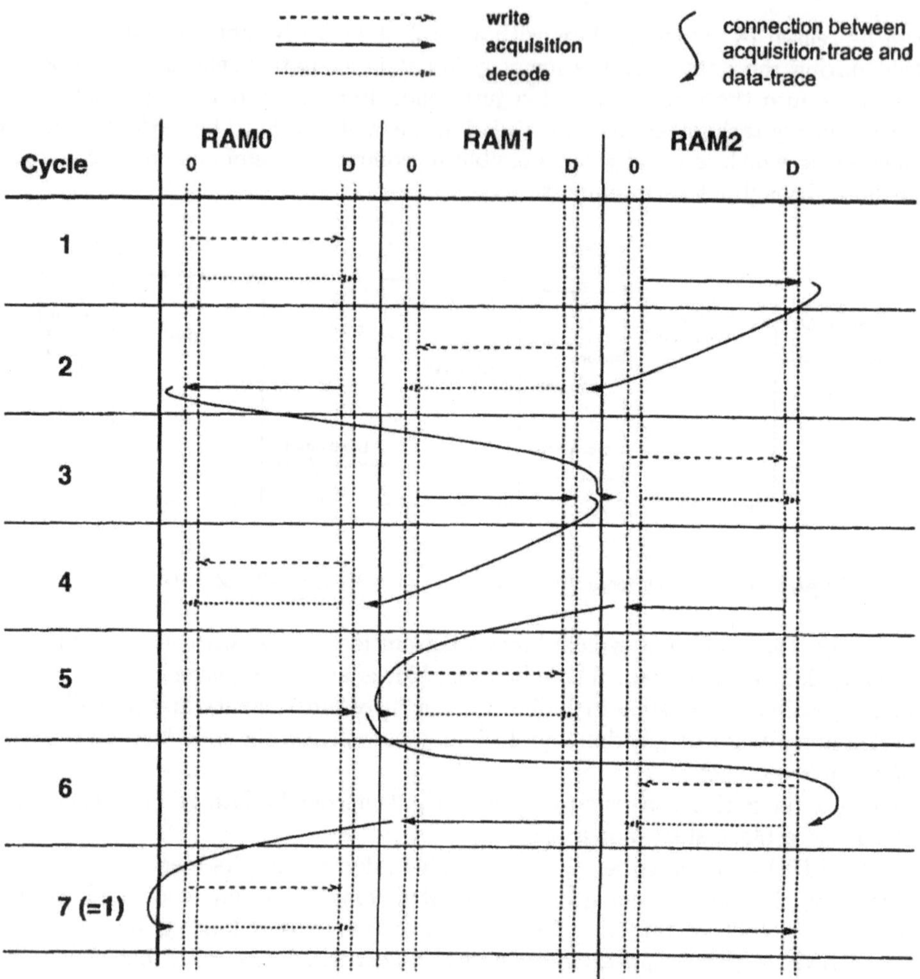

Figure 16.19 Cyclic addressing and multiplexing scheme for 3 dual port RAMs and $M = D$.

However, this amounts to the price of three concurrent pointers [36, 37]. More important, we need to slice the memory into blocks of depth $D/2$ trellis cycles (6 RAMs, denoted M1 ... M6 in the figure) rather than blocks of depth D (as in Fig. 16.17) to be able to access the required data, which complicates the architecture. Clearly, by choosing even smaller M the memory gets sliced more severely and the corresponding architecture soon becomes unattractive.

Tracing more than one trellis cycle per pointer and clock cycle has been considered for the case where more than one trellis step is decoded per clock cycle [32]. This can be done if the decisions from two subsequent trellis steps are stored in a single data word (or parallel memories) and effectively doubles the data wordlength of the RAMs while using half as many addresses. Since we can retrieve the information for two trellis steps in one clock cycle using this approach, the traceback can

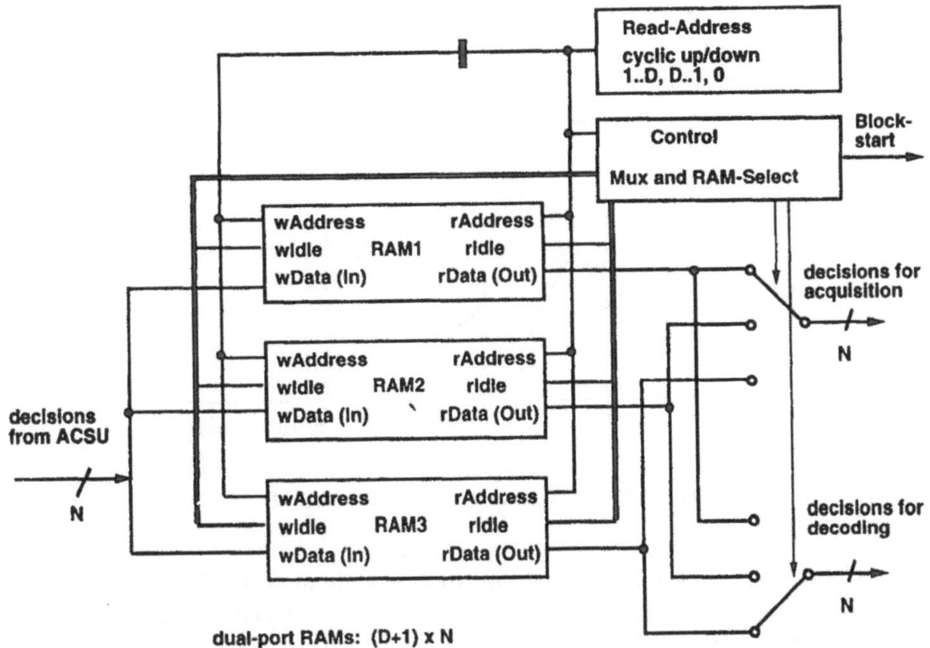

Figure 16.20 Architecture for the memory subsystem.

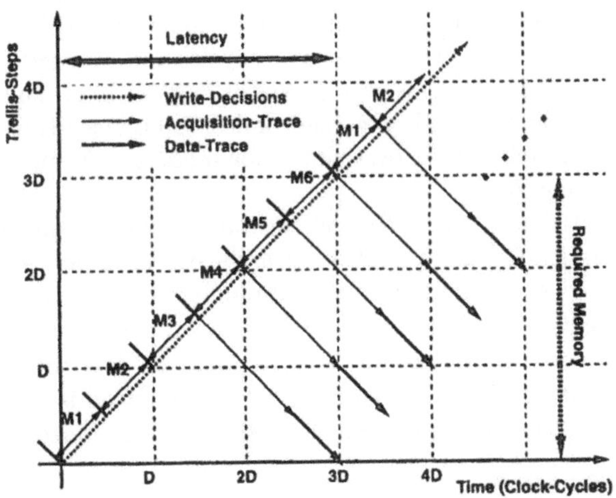

Figure 16.21 Traceback with $M = 0.5 \cdot D$.

evaluate two trellis steps per clock cycle, which leads to architectures with reduced memory size.

Fig. 16.22 shows the resulting clock-time/trellis-time diagram for the scheme proposed in [43] where acquisition-trace and data-trace are performed with even different speeds (c.f., [36]).

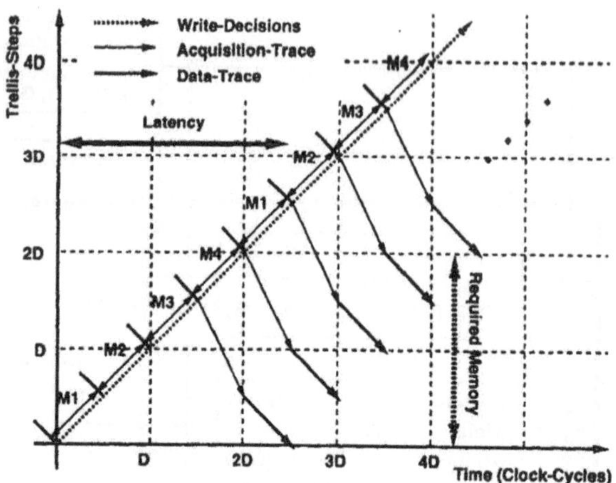

Figure 16.22 Dual timescale traceback with $M = 0.5 \cdot D$.

Consider the first traceback in Fig. 16.22. While in every clock cycle two sets of decision bits ($2 \cdot N$ bits) are retrieved for acquisition-trace, two sets of decision bits are retrieved every other clock cycle for the data-trace. We can thus alternately retrieve the data for the data-trace and write a new set of decision bits to the same memory location, i.e., immediate memory reuse is facilitated. And since we need to read and write to one location within two cycles, this can actually be performed with commodity semi-custom technologies and single port RAMs. The obtained architecture exhibits a latency of $2.5 \cdot D$ clock cycles and we need only $2 \cdot D \cdot N$ bits of memory in four RAM-blocks with $D/4$ words of $2 \cdot N$ bits. The overall required memory is reduced, yet exchanging two sets of decision bits leads to increased wiring overhead. The approach is thus best suited for moderate N and large D, such as for punctured codes and $N = 64$.

As was pointed out, the TBA can be implemented using a wide range of architectures. The choice of the optimum architecture depends on many factors including technology constraints, throughput requirements and in some cases as well latency requirements. Of course software implementations are subject to different trade-offs compared to hardware implementations and low throughput hardware implementations may well use $M = 1$ if power consumption is dominated by other components and the memory requirements need to be minimized.

16.4.8 Survivor Depth

For actual dimensioning of the survivor depth D, $D = 5K$ was stated for rate 1/2 codes as a rule of thumb [17]. However, this figure is applicable only for best state decoding, i.e., if the overall best path is used for determining the final survivor as explained in Section 16.2. For fixed state decoding, D must be chosen larger. In [11, 12], it is reported that if D is doubled for fixed state decoding, even asymptotic performance losses are avoided [10].

[10]This seems to be a pessimistic value, and the authors strongly recommend to run system simulations for a given application in order to determine the required value for the survivor depth.

For punctured codes, D also must be chosen larger than for the non punctured base code. In [44], $D = 96 = 13.5K$ was the reported choice for a $K = 7$ base code punctured to rate 7/8. For codes with rates smaller than 1/2, finally, D can be chosen smaller than $5K$.

Although theoretical results and simulation results available in the literature may serve as a guideline (e.g., [1, 11, 12, 45]), system simulations should be used to determine the required survivor depth for a certain SMU implementation and decoding scheme if figures are not available for the particular code and SMU architecture. System design and simulation tools like COSSAP^{TM}[16] are indispensable here. Simulation can be very useful as well if the performance of the Viterbi decoder (including the choice of D) can be traded against the performance of other (possibly less costly) parts of an overall system (c.f., [46]).

16.5 SYNCHRONIZATION OF CODED STREAMS

As has been pointed out already, a step in the trellis is not generally associated with the transmission of a single channel symbol. Indeed, if punctured codes are used, the number of transmitted channel symbols per trellis step is time variant. Furthermore, channel symbols can exhibit ambiguities that cannot be resolved by synchronizers in the receiver front end. Consider the *simple* case of QPSK channel symbols in conjunction with a punctured code of rate 1/2.

Table 16.2 Puncturing of a Rate 1/2 Base Code to a Rate 2/3 Code

trellis step	k	$k+1$	$k+2$	$k+3$
coded bits	$b_{1,k}, b_{2,k}$	$b_{1,k+1}, b_{2,k+1}$	$b_{1,k+2}, b_{2,k+2}$	$b_{1,k+3}, b_{2,k+3}$
punctured bits	$b_{1,k}, b_{2,k}$	$b_{1,k+1}, -$	$b_{1,k+2}, b_{2,k+2}$	$b_{1,k+3}, -$

Table 16.3 Assignment of the Punctured Rate 3/4 Code Symbols to QPSK Symbols

QPSK Symbol	k	$k+1$	$k+3$
Inphase value	$I_1 = b_{1,k}$	$I_2 = b_{1,k+1}$	$I_3 = b_{2,k+2}$
Quadrature value	$Q_1 = b_{2,k}$	$Q_2 = b_{1,k+2}$	$Q_3 = b_{1,k+3}$

Clearly, 4 trellis cycles and 3 transmitted QPSK symbols are necessary to complete a mapping cycle. It has to be known how the blocks of 3 QPSK symbols are embedded in the received symbols stream to facilitate decoding. Furthermore, phase rotations of 90, 180 and 270 degrees of the QPSK symbols cannot be resolved by the receiver front end and at least the 90 degree rotation needs to be corrected prior to decoding[11]. Thus at least $2 \times 3 = 6$ possible ways of embedding the blocks of symbols in the received symbol stream need to be considered to find the state that is the prerequisite for the actual decoding.

[11]For many codes, including the (177,131) standard code, an inversion of the input symbol corresponds to valid code sequences associated with inverted information symbols. This inversion cannot be resolved without using other properties of the information sequence. Thus resolving the 90 degree ambiguity is sufficient for QPSK modulation in this case.

The detection of the position/phase of the blocks of symbols in the received symbol stream and the required transformation of the symbol stream (i.e., rotating and/or delaying the received channel symbols) is called node synchronization in the case of convolutional coding. We call the different possible ways the blocks can be embedded in the received channel symbol stream ***synchronization states.***

Node synchronization is essential for the reception of infinite streams of coded data, as applied for example in the current digital video satellite broadcasting standards. If data is transferred in frames (as in all cellular phone systems) the frame structure usually provides absolute timing and phase references that can be used to provide correctly aligned streams to the Viterbi decoder.

There are three approaches known to facilitate estimation of the correct synchronization state and thus node synchronization, which will be discussed below.

16.5.1 Metric Growth Based Node Synchronization

This approach was already suggested in the early literature on Viterbi decoding [13]. It is based on the fact that the path metrics in a decoder grow faster in a synchronized decoder compared to a decoder that is out-of-synch. However, the metric growth depends on the signal to noise ratio and the average input magnitude. This effect can substantially perturb the detection of changes in the synchronization state (and thus reacquisition) once the Viterbi decoder is correctly synchronized. Since more reliable approaches are known as described below, we do not consider this method further.

16.5.2 Node Synchronization Based on Bit Error Rate Estimation

This approach is based on the fact that a correctly synchronized Viterbi decoder computes an output data stream that contains much fewer errors than the input data stream. Thus the input data error rate can be estimated by re-encoding the output stream and comparing the generated sequence with the input sequence. Fig. 16.23 shows the resulting implementation architecture.

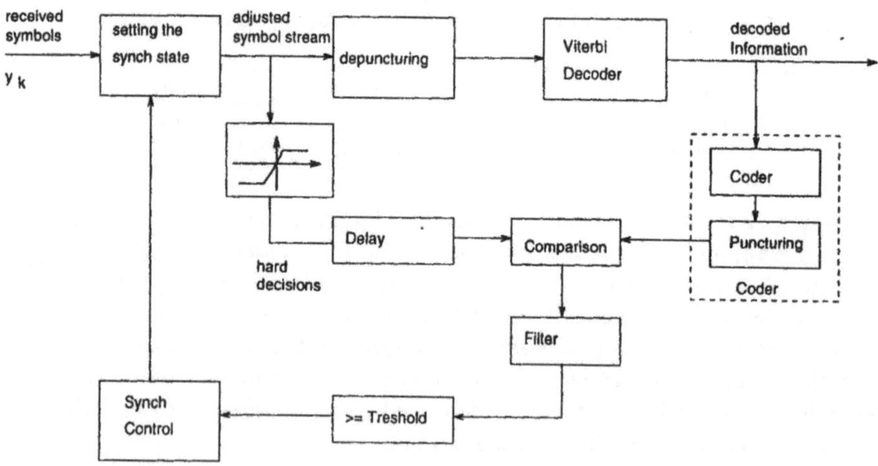

Figure 16.23 Node synchronization based on bit error rate estimation.

The received symbols are processed first in a block that can rotate and delay the input symbols as required for all possible trial synchronization states. The preprocessed stream is depunctured and decoded by a Viterbi decoder. The decoded stream is coded and punctured again. Additionally, the preprocessed symbols are sliced[12] to obtain the underlying hard decision bits which are then delayed according to the delay of the re-encoded data. The comparison result can be filtered to estimate the bit error rate of the input symbol stream. If the estimated error rate exceeds a certain level, a new trial with a new synchronization state is performed steered by some synchronization control functionality.

The approach has been implemented in several commercial designs. In HW implementations, the delay line can be of considerable complexity in particular if the Viterbi decoder exhibits a large decoding delay, i.e., for high rate codes and trace back based SMU architectures.

16.5.3 Syndrome Based Node Synchronization

Syndrome based node synchronization was introduced as a robust high performance alternative to metric growth observation in [48].

The basic idea is depicted in Figures 16.24 and 16.25, assuming a code of rate 1/2 and BPSK transmission. All operations are performed on hard decisions in GF(2), and all signals are represented by transfer functions. The information bits u_k enter the coder, where convolution with the generator polynomials takes place. The code symbols $b_{1,k}$ and $b_{2,k}$ are then corrupted during transmission by adding the error sequences $e1_k$ and $e2_k$, respectively. In the receiver, another convolution of the received sequences with the swapped generator polynomials takes place. In practice, the values of in the receiver are calculated by slicing and de-mapping the quantized received symbols y_k.

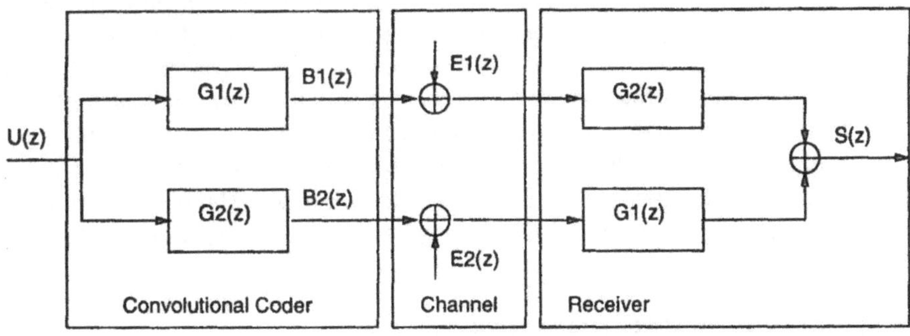

Figure 16.24 Syndrome computation of in-sync state.

From Fig. 16.24 it is easily seen that for the Z-transform of the syndrome

$$S(z) = E1(z) \cdot G2(z) + E2(z) \cdot G1(z) + 2U(z) \cdot G1(z) \cdot G2(z) \tag{26}$$

holds. If the channel error sequences $e1_k$ and $e2_k$ and the corresponding Z-transforms are zero, respectively, the syndrome sequence s_k is also zero, since

[12]Improved performance can be obtained by processing quantized symbols rather than hard decisions [47]. In this case, the required delay line is of course more costly.

$2U(z) = 0$ holds in GF(2). Therefore, the syndrome sequence depends only on the channel error sequences $E1(z), E2(z)$. For reasonable channel error rates, the rate of ones in the syndrome stream s_k is lower than 0.5.

Fig. 16.25 shows the effect of an additional reception delay, i.e., an out-of-synch condition for the Viterbi decoder.

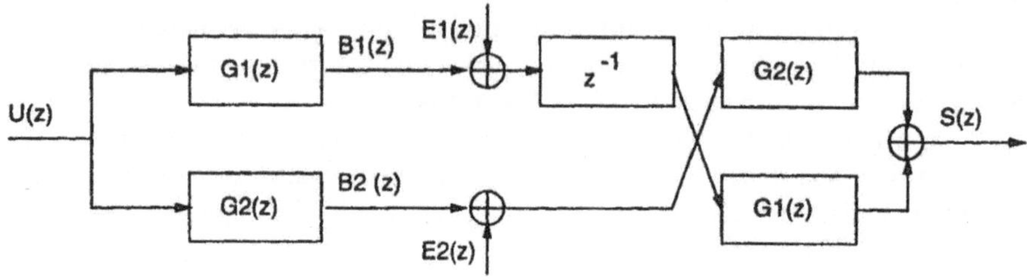

Figure 16.25 Syndrome computation in out-of-sync state.

Now $S(z)$ depends clearly on $U(z)$ as well as on the channel error sequences since

$$S(z) = E1(z) \cdot Z^{-1} \cdot G1(z) + E2(z) \cdot G2(z) + U(z) \cdot (G1^2(z) \cdot Z^{-1} + G2^2(z)) \quad (27)$$

holds. Now, $S(z)$ essentially consists of equiprobable ones and zeros, i.e., the rate of ones in the syndrome stream is 0.5. Thus an estimation of the actual rate of ones in the syndrome serves as a measure to decide whether the actual trial corresponds to an in-synch condition.

A strong advantage of syndrome based node synchronization is the complete independence of the subsequent Viterbi decoder. The involved hardware complexity is sufficiently low to enable the implementation of synchronizers that concurrently investigate all possible synchronization states, which is not economically feasible with other approaches. However, the parameters and syndrome polynomials are more difficult to determine as the parameters of the approach based on bit error rate estimation. In particular, a poor choice of the syndrome polynomials can seriously degrade performance [49]. For a detailed discussion, the reader is referred to [48, 50] for rate 1/2 codes, [51, 52, 53] for rate 1/N codes, and [49] for rate (N-1)/N codes.

16.6 RECENT DEVELOPMENTS

Viterbi decoding as discussed in this chapter is applicable to all current commercial applications of convolutional codes, although the basic algorithms need to be extended in some applications (e.g., if applied to trellis coded modulation with parallel branches in the trellis [9]). In all these applications, the decoder needs to compute an information sequence only (hard output decoding).

However, it has already been shown in [54] that a concatenation of several codes can provide a better overall cost/performance trade-off. In fact serial concatenation (i.e., coding a stream twice subsequently with different codes) has been chosen for deep space communcation and digital video broadcasting. While decoding such codes is possible (and actually done) by decoding the involved (component) codes independently, improved performance can be obtained by passing additional information between the decoders of the component codes [55, 56].

In fact, the most prominent class of recently developed codes, the so called *TURBO codes* [57] can no longer be decoded by decoding the component codes independently. Decoding these codes is an iterative process for which in addition to the information sequences, reliability estimates for each information symbol need to be computed (soft outputs). Thus decoding algorithms for convolutional component codes that provide soft outputs have recently gained considerable attention. Large increases in coding gain are possible for concatenated or iterative (TURBO) decoding systems [58, 59]. The most prominent soft output decoding algorithm is the soft output Viterbi algorithm (SOVA) [60, 61], which can be derived as an approximation to the optimum symbol by symbol detector, the symbol by symbol MAP algorithm (MAP)[13]. The basic structure of the Viterbi algorithm is maintained for the SOVA. Major changes are necessary in the SMU, since now, a soft quantized output has to be calculated rather than decoding an information sequence. Efficient architectures and implementations for the SOVA were presented in [62, 63, 64, 65].

In the MAP algorithm, a posteriori probabilities are calculated for every symbol, which represent the optimum statistical information which can be passed to a subsequent decoding stage as a soft output. Although the MAP was already derived in [4, 66], this was recognized only recently [60]. In its original form, the MAP algorithm is much more computationally intensive than the Viterbi algorithm (VA) or the SOVA. However, simplifications are known that lead to algorithms with reduced implementation complexity [67, 68, 69, 70], It was shown in [71] that acquisition and truncation properties can be exploited for the MAP as for the VA and the SOVA. Thereby, efficient VLSI architectures for the MAP can be derived for recursive [72] and parallelized [73] implementations with implementation complexities roughly comparable to the SOVA and the VA [71].

REFERENCES

[1] J. A. Heller and I. M. Jacobs, "Viterbi Decoding for Satellite and Space Communication," *IEEE Transactions on Communications*, vol. COM-19, no. no. 5, pp. 835–848, Oct. 1971.

[2] M. Vaupel, U. Lambrette, H. Dawid, O. Joeressen, S. Bitterlich, H. Meyr, F. Frieling, and K. Müller, "An All-Digital Single-Chip Symbol Synchronizer and Channel Decoder for DVB," in *VLSI: Integrated Systems on Silicon, R. Reis and L. Claesen*, pp. 79–90, 1997.

[3] A. J. Viterbi, "Error bounds for convolutional coding and an asymptotically optimum decoding algorithm," *IEEE Trans. Information Theory*, vol. IT-13, pp. 260–269, April 1967.

[4] G. Forney, "The Viterbi Algorithm," *Proceedings of the IEEE*, vol. 61, no. 3, pp. 268–278, March 1973.

[13] In fact, it has been shown [62] for the slightly differing algorithms in [60] and [61] that the algorithm from [60] is slightly too optimistic while the algorithm from [61] is slightly too pessimistic compared to a derived approximation on the MAP. However, the performance of either approach seems to be almost identical.

[5] J. Cain, J. G.C. Clark, and J. Geist, "Punctured Convolutional Codes of Rate (n-1)/n and Simplified Maximum Likelihood Decoding," *IEEE Transactions on Information Theory*, vol. IT-25, no. 1, pp. 97–100, Jan. 1979.

[6] Y. Yasuda, K. Kashiki, and Y. Hirata, "High-rate punctured convolutional codes for soft decision," *IEEE Transactions on Communications*, vol. COM-32, no. 3, pp. 315–319, March 1984.

[7] J. Hagenauer, "Rate-compatible punctured convolutional codes (RCPC codes) and their applications," *IEEE Transactions on Communications*, vol. 36, no. 4, pp. 389–400, April 1988.

[8] H. Meyr and R. Subramanian, "Advanced digital receiver principles and technologies for PCS," *IEEE Communications Magazine*, vol. 33, no. 1, pp. 68–78, January 1995.

[9] G. Ungerboek, "Trellis Coded Modulation with Redundant Signal Sets, Parts I+II," *IEEE Communications Magazine*, vol. 25, no. 2, pp. 5–21, 1987.

[10] R. E. Bellman and S. E. Dreyfus, *Applied Dynamic Programming*. Princeton, NJ: Princeton University Press, 1962.

[11] I. Onyszchuk, "Truncation Length for Viterbi Decoding," *IEEE Transactions on Communications*, vol. 39, no. 7, pp. 1023–1026, July 1991.

[12] R. J. McEliece and I. M. Onyszchuk, "Truncation effects in Viterbi decoding," in *Proceedings of the IEEE Conference on Military Communications*, (Boston, MA), pp. 29.3.1–29.3.3, October 1989.

[13] A. J. Viterbi and J. K. Omura, *Principles of Digital Communication and Coding*. New York: McGraw-Hill, 1979.

[14] G. Fettweis and H. Meyr, "A 100 Mbit/s Viterbi decoder chip: Novel architecture and its realisation," in *IEEE International Conference on Communications, ICC'90*, vol. 2, (Atlanta, GA, USA), pp. 463–467, Apr. 1990.

[15] G. Fettweis and H. Meyr, "Parallel Viterbi decoding: Algorithm and VLSI architecture," *IEEE Communications Magazine*, vol. 29, no. 5, pp. 46–55, May 1991.

[16] "COSSAP Overview and User Guide." Synopsys, Inc., 700 East Middlefield Road, Mountain View, CA 94043.

[17] G. Clark and J. Cain, *Error-Correction Coding for Digital Communications*. New York: Plenum, 1981.

[18] O. M. Collins, "The subtleties and intricacies of building a constraint length 15 convolutional decoder," *IEEE Transactions on Communications*, vol. 40, no. 12, pp. 1810–1819, December 1992.

[19] C. Shung, P. Siegel, G. Ungerböck, and H. Thapar, "VLSI architectures for metric normalization in the Viterbi algorithm," in *Proceedings of the IEEE International Conference on Communications*, pp. 1723–1528, IEEE, 1990.

[20] P. Siegel, C. Shung, T. Howell, and H. Thapar, "Exact bounds for Viterbi detector path metric differences," in *Proceedings of the International Conference on Acoustics Speech and Signal Processing*, pp. 1093–1096, IEEE, 1991.

[21] T. Noll, "Carry–Save Architectures for High–Speed Digital Signal Processing," *Journal of VLSI Signal Processing*, vol. 3, no. 1/2, pp. 121–140, June 1991.

[22] T. Noll, "Carry-Save Arithmetic for High-Speed Digital Signal Processing," in *IEEE ISCAS'90*, vol. 2, pp. 982–986, 1990.

[23] J. Sparsø, H. Jorgensen, P. S. Pedersen, and T. R"ubner-Petersen, "An area-efficient topology for vlsi implementation of the viterbi decoders and other shuffle exchange type structures," *IEEE Journal of Solid-State Circuits*, vol. 26, no. 2, pp. 90–97, February 1991.

[24] C. Rader, "Memory Management in a Viterbi Decoder," *IEEE Transactions on Communications*, vol. COM-29, no. 9, pp. 1399–1401, Sept. 1981.

[25] H. Dawid, S. Bitterlich, and H. Meyr, "Trellis Pipeline–Interleaving: A novel method for efficient viterbi decoder implementation," in *Proceedings of the IEEE International Symposium on Circuits and Systems*, (San Diego, CA), pp. 1875–78, IEEE, May 10-13 1992.

[26] S. Bitterlich and H. Meyr, "Efficient scalable architectures for Viterbi decoders," in *International Conference on Application Specific Array Processors (ASAP), Venice, Italy*, October 1993.

[27] S. Bitterlich, H. Dawid, and H. Meyr, "Boosting the implementation efficiency of Viterbi Decoders by novel scheduling schemes," in *Proceedings IEEE Global Communications Conference GLOBECOM 1992*, (Orlando, Florida), pp. 1260–65, December 1992.

[28] C. Shung, H.-D. Lin, P. Siegel, and H. Thapar, "Area-efficient architectures for the viterbi algorithm," in *Proceedings of the IEEE Global Telecommunications Conference GLOBECOM*, 1990.

[29] K. K. Parhi, "Pipeline interleaving and parallelism in recursive digital filters - parts 1&2," *IEEE transactions on Acoustics, Speech and Signal Processing*, vol. 37, no. 7, pp. 1099–1134, July 1989.

[30] G. Fettweis and H. Meyr, "Feedforward architectures for parallel Viterbi decoding," *Journal on VLSI Signal Processing*, vol. 3, no. 1/2, pp. 105–120, June 1991.

[31] G. Fettweis and H. Meyr, "Cascaded feedforward architectures for parallel Viterbi decoding," in *Proceedings of the IEEE International Symposium on Circuits and Systems*, pp. 978–981, May 1990.

[32] P. Black and T. Meng, "A 140MBit/s, 32-State, Radix-4 Viterbi Decoder," *IEEE Journal of Solid-State Circuits*, vol. 27, no. 12, pp. 1877–1885, December 1992.

[33] H. Dawid, G. Fettweis, and H. Meyr, "A CMOS IC for Gbit/s Viterbi Decoding," *IEEE Transactions on VLSI Systems*, no. 3, March 1996.

[34] H. Dawid, G. Fettweis, and H. Meyr, "System Design and VLSI Implementation of a CMOS Viterbi Decoder for the Gbit/s Range," in *Proc. ITG-Conference Mikroelektronik für die Informationstechnik*, (Berlin), pp. 293–296, ITG, VDE-Verlag, Berlin Offenbach, March 1994.

[35] G. Fettweis, H. Dawid, and H. Meyr, "Minimized method Viterbi decoding: 600 Mbit/s per chip," in *IEEE Globecom 90*, (San Diego, USA), pp. 1712–1716, Dec. 1990.

[36] R. Cypher and C. Shung, "Generalized Trace Back Techniques for Survivor Memory Management in the Viterbi Algorithm," in *Proceedings of the IEEE Global Telecommunications Conference GLOBECOM*, (San Diego, California), pp. 707A.1.1–707A.1.5, IEEE, Dec. 1990.

[37] G. Feygin and P. G. Gulak, "Architectural tradeoffs for survivor sequence memory management in Viterbi decoders," *IEEE Transactions on Communications*, vol. 41, no. 3, pp. 425–429, March 1993.

[38] E. Paaske, S. Pedersen, and J. Sparsø, "An area-efficient path memory structure for VLSI implementation of high speed Viterbi decoders," *INTEGRATION, the VLSI Journal*, vol. 12, no. 2, pp. 79–91, November 1991.

[39] P. Black and T. Meng, "Hybrid survivor path architectures for Viterbi decoders," in *Proceedings of the IEEE International Conference on Acoustics, Speech and Signal Processing*, pp. 433–436, IEEE, 1993.

[40] G. Fettweis, "Algebraic survivor memory management for viterbi detectors," in *Proceedings of the IEEE International Conference on Communications*, (Chicago), pp. 339–343, IEEE, June 1992.

[41] T. Ishitani, K. Tansho, N. Miyahara, S. Kubota, and S. Kato, "A Scarce-State-Transition Viterbi-Decoder VLSI for Bit Error Correction," *IEEE Journal of Solid-State Ciruits*, vol. SC-22, no. 4, pp. 575–581, August 1987.

[42] K. Kawazoe, S. Honda, S. Kubota, and S. Kato, "Ultra-high-speed and Universal-coding-rate Viterbi Decoder VLSIC –SNUFEC VLSI–," in *Proceedings of the IEEE International Conference on Communications*, (Geneva, Switzerland), pp. 1434–1438, IEEE, May 1993.

[43] O. J. Joeressen and H. Meyr, "Viterbi decoding with dual timescale trace-back processing," in *Proceedings of the IEEE International Symposium on Personal, Indoor, and Mobile Radio Communications*, (Toronto), pp. 213–217, September 1995.

[44] R. Kerr, H. Dehesh, A. Bar-David, and D. Werner, "A 25 MHz Viterbi FEC Codec," in *Proceeding of the 1990 IEEE Custom Integrated Circuit Conference*, (Boston, MA), pp. 16.6.1–13.6.5, IEEE, May 1990.

[45] F. Hemmati and D. Costello, "Truncation Error Probability in Viterbi Decoding," *IEEE Transactions on Communications*, vol. 25, no. 5, pp. 530–532, May 1977.

[46] O. J. Joeressen, G. Schneider, and H. Meyr, "Systematic Design Optimization of a Competitive Soft-Concatenated Decoding System," in *VLSI Signal Processing VI* (L. D. J. Eggermont, P. Dewilde, E. Deprettere, and J. van Meerbergen, eds.), pp. 105–113, IEEE, 1993.

[47] U. Mengali, R. Pellizoni, and A. Spalvieri, "Phase ambiguity resolution in trellis-codes modulations," *IEEE Transactions on Communications*, vol. 43, no. 9, pp. 2532–2539, September 1995.

[48] G. Lorden, R. McEliece, and L. Swanson, "Node synchronization for the Viterbi decoder," *IEEE Transactions on Communications*, vol. COM-32, no. 5, pp. 524–31, May 1984.

[49] O. J. Joeressen and H. Meyr, "Node synchronization for punctured convolutional codes of rate (n-1)/n," in *Proceedings of the IEEE Global Telecommunications Conference GLOBECOM*, (San Francisco, CA), pp. 1279–1283, IEEE, November 1994.

[50] M. Moeneclaey, "Syndrome-based Viterbi decoder node synchronization and out-of-lock detection," in *Proceedings of the IEEE Global Telecommunications Conference GLOBECOM*, (San Diego, CA), pp. 604–8, Dec 1990.

[51] M.-L. de Mateo, "Node synchronization technique for any 1/n rate convolutional code," in *Proceedings of the IEEE International Conference on Communications*, (Denver, CO), pp. 1681–7, June 1991.

[52] J. Sodha and D. Tait, "Soft-decision syndrome based node synchronisation," *Electronics Letters*, vol. 26, no. 15, pp. 1108–9, July, 19 1990.

[53] J. Sodha and D. Tait, "Node synchronisation for high rate convolutional codes," *Electronics Letters*, vol. 28, no. 9, pp. 810–12, April, 23 1992.

[54] G. D. Forney, Jr., *Concatenated Codes.* Cambridge, MA: MIT Press, 1966. MIT Reasearch Monograph.

[55] E. Paaske, "Improved decoding for a concatenated coding scheme recommended by CCSDS," *IEEE Transactions on Communications*, vol. 38, no. 8, pp. 1138–1144, August 1990.

[56] J. Hagenauer, E. Offer, and L. Papke, "Improving the standard coding system for deep space missions," in *Proceedings of the IEEE International Conference on Communications*, (Geneva, Switzerland), pp. 1092–1097, IEEE, May 1993.

[57] C. Berrou and A. Glavieux, "Near Optimum Error Correcting Coding and Decoding: Turbo-Codes," *IEEE Transactions on Communications*, vol. 44, no. 10, pp. 1261–1271, October 1996.

[58] J. Lodge, R. Young, P. Hoeher, and J. Hagenauer, "Seperable MAP 'Filters' for the decoding of product and concatenated codes," in *Proceedings of the IEEE International Conference on Communications*, (Geneva, Switzerland), pp. 1740–1745, IEEE, May 1993.

[59] C. Berrou, A. Glavieux, and P. Thitimajshima, "Near Shannon Limit Error-Correcting Coding and Decoding: TURBO-Codes," in *Proceedings of the IEEE International Conference on Communications*, (Geneva, Switzerland), pp. 1064–1070, IEEE, May 1993.

[60] J. Hagenauer and P. Höher, "A Viterbi Algorithm with Soft Outputs and It's Application," in *Proceedings of the IEEE Global Telecommunications Conference GLOBECOM*, pp. 47.1.1–47.1.7, Nov. 1989.

[61] J. Huber and A. Rüppel, "Zuverlässigkeitsschätzung für die Ausgangssymbole von Trellis-Decodern," *Archiv für Elektronik und Übertragung (AEÜ)*, vol. 44, no. 1, pp. 8–21, Jan. 1990, (in German.)

[62] O. Joeressen, *VLSI-Implementierung des Soft-Output Viterbi-Algorithmus.* VDI-Fortschritt-Berichte, Reihe 10, Nr. 396, Düsseldorf: VDI-Verlag, 1995. ISBN 3-18-339610-6, (in German).

[63] comatlas sa, Chateubourg, France, *CAS5093, Turbo-Code Codec, Technical Data Sheet*, April 1994.

[64] C. Berrou, P. Adde, E. Angui, and S. Faudeil, "A Low Complexity Soft-Output Viterbi Decoder Architecture," in *Proceedings of the IEEE International Conference on Communications*, (Geneva, Switzerland), pp. 737–740, IEEE, May 1993.

[65] O. J. Joeressen and H. Meyr, "A 40Mbit/s soft output Viterbi decoder," *IEEE Journal of Solid-State Circuits*, vol. 30, no. 7, pp. 812–818, July 1995.

[66] L. Bahl, J. Cocke, F. Jelinek, and J. Raviv, "Optimal decoding of linear codes for minimizing symbol error rate," *IEEE Transactions on Information Theory*, vol. IT-24, no. 3, pp. 284–287, March 1974.

[67] P. Höher, *Kohärenter Empfang trelliscodierter PSK Signale auf frequenzselektiven Mobilfunkkanälen.* VDI-Fortschritt-Berichte, Reihe 10, Nr. 147, Düsseldorf: VDI-Verlag, 1990. ISBN 3-18-144710-2.

[68] G. Ungerböck, "Nonlinear Equalization of binary signals in gaussian noise," *IEEE Trans. Communications*, no. COM-19, pp. 1128–1137, 1971.

[69] J. A. Erfanian, S. Pasupathy, and G. Gulak, "Reduced Complexity Symbol Detectors with Parallel Structures for ISI Channels," *IEEE Transactions Communications*, vol. 42, no. 2,3,4, pp. 1661–1671, Feb./March/April 1994.

[70] P. Robertson, E. Villebrun, and P. Hoeher, "A comparison of optimal and sub-optimal MAP decoding algorithms operating in the log domain," in *Proceedings of the IEEE International Conference on Communications*, (Seattle, WA), 1995.

[71] H. Dawid, *Algorithmen und Schaltungsarchitekturen zur Maximum A Posteriori Faltungsdecodierung.* Aachen: Shaker-Verlag, 1996. ISBN 3-8265-1540-4.

[72] H. Dawid and H. Meyr, "Real-Time Algorithms and VLSI Architectures for Soft Output MAP Convolutional Decoding," in *Proceedings of the IEEE International Symposium on Personal, Indoor, and Mobile Radio Communications*, (Toronto), pp. 193–197, September 1995.

[73] H. Dawid, G. Gehnen, and H. Meyr, "MAP Channel Decoding: Algorithm and VLSI Architecture," in *VLSI Signal Processing VI* (L. D. J. Eggermont, P. Dewilde, E. Deprettere, and J. van Meerbergen, eds.), pp. 141–149, IEEE, 1993.

Chapter 17

A Review of Watermarking Principles and Practices[1]

Ingemar J. Cox
NEC Research Institute
Princeton, New Jersey
ingemar@research.nj.nec.com

Matt L. Miller
Signafy Inc.
Princeton, New Jersey
mlm@signafy.com

Jean-Paul M.G. Linnartz and Ton Kalker
Philips Research
Eindhoven, The Netherlands
{linnartz,kalker}@natlab.research.philips.com

17.1 INTRODUCTION

Digital representation of copyrighted material such as movies, songs, and photographs offer many advantages. However, the fact that an unlimited number of perfect copies can be illegally produced is a serious threat to the rights of content owners. Until recently, the primary tool available to help protect content owners' rights has been encryption. Encryption protects content during the transmission of the data from the sender to receiver. However, after receipt and subsequent decryption, the data is no longer protected and is freely available.

Watermarking complements encryption. A digital watermark is a piece of information that is hidden directly in media content, in such a way that it is imperceptible to observation, but easily detected by a computer. The principal advantage of this is that the content is inseparable from the watermark. This makes watermarks suitable for several applications, including:

[1]Portions of this paper appeared in the Proceedings of SPIE, Human Vision & Electronic Imaging II, V 3016, pp. 92-99, February 1997. Portions are reprinted, with permission, from "Public watermarks and resistance to tampering", I.J. Cox and J.-P. Linnartz, IEEE International Conference on Image Processing, CD-ROM Proc. ©1997 IEEE and from "Some General Methods for Tampering with Watermarks", I. J. Cox and J.-P. Linnartz, IEEE T. of Selected Areas in Communications, 16, 4, 587-593, ©1998 IEEE.

Signatures. The watermark identifies the owner of the content. This information can be used by a potential user to obtain legal rights to copy or publish the content from the contact owner. In the future, it might also be used to help settle ownership disputes [2].

Fingerprinting. Watermarks can also be used to identify the content buyers. This may potentially assist in tracing the source of illegal copies. This idea has been implemented in the DIVX digital video disk players, each of which places a watermark that uniquely identifies the player in every movie that is played.

Broadcast and publication monitoring. As in signaturing, the watermark identifies the owner of the content, but here it is detected by automated systems that monitor television and radio broadcasts, computer networks, and any other distribution channels to keep track of when and where the content appears. This is desired by content owners who wish to ensure that their material is not being illegally distributed, or who wish to determine royalty payments. It is also desired by advertisers who wish to ensure that their commercials are being broadcast at the times and locations they have purchased.

Several commercial systems already exist which make use of this technology. The MusiCode system provides broadcast monitoring of audio, VEIL-II and MediaTrax provide broadcast monitoring of video. Also, in 1997 a European project by the name of VIVA was started to develop watermark technology for broadcast monitioring.

Authentication. Here, the watermark encodes information required to determine that the content is authentic. It must be designed in such a way that any alteration of the content either destroys the watermark, or creates a mismatch between the content and the watermark that can be easily detected. If the watermark is present, and properly matches the content, the user of the content can be assured that it has not been altered since the watermark was inserted. This type of watermark is sometimes referred to as a *vapormark.*

Copy control. The watermark contains information about the rules of usage and copying which the content owner wishes to enforce. These will generally be simple rules such as "this content may not be copied", or "this content may be copied, but no subsequent copies may be made of that copy". Devices which are capable of copying this content can then be required by law or patent license to test for and abide by these watermarks. Furthermore, devices that can play the content might test for the watermarks and compare them with other clues, such as whether the content is on a recordable storage device, to identify illegal copies and refuse to play them. This is the application that is currently envisaged for digital video disks (DVD).

---

[2] In a recent paper [1] it was shown that the use of watermarks for the establishemnt of ownership can be problematic. It was shown that for a large class of watermarking schemes a so called "counterfeit original" attack can be used to confuse ownership establishment. A technical way out may be the use of *one-way watermark* functions, but the mathematical modelling of this approach is still in its infancy. In practical terms the combined use of a copyright office (along the guidelines of WIPO) and a watermark label might provide sufficiently secure fingerprints.

Secret communication. The embedded signal is used to transmit secret information from one person (or computer) to another, without anyone along the way knowing that this information is being sent. This is the classical application of steganography – the hiding of one piece of information within another. There are many interesting examples of this practice from history, e.g., [2]. In fact, Simmons' work [3] was motivated by the Strategic Arms Reduction Treaty verification. Electronic detectors were allowed to transmit the status (loaded or unloaded) of a nuclear missile silo, but not the position of that silo. It appeared that digital signature schemes which were intended to verify the integrity of such status message, could be misused as a "subliminal channel" to pass long espionage information.

There are several public-domain and shareware programs available that employ watermarking for secret communication. Rivest [4] has suggested that the availability of this technology casts serious doubt on the effectiveness of government restrictions on encryption, since these restrictions cannot apply to steganography.

These are some of the major applications for which watermarks are currently being considered or used, but several others are likely to appear as the full implications of this technology are realized.

In the next section, we present the basic principles of watermarking. In Section 17.3 we discuss several properties of watermarking technologies. In section 17.4 we describe a simple watermarking method that then allows for a detailed discussions of robustness (Section 17.5) and tamper-resistance (Section 17.6). Section 17.6.7 gives a brief overview of several watermarking methods.

17.2 FRAMEWORK

Fig. 17.1 shows the basic principle behind watermarking. Watermarking is viewed as a process of combining two pieces of information in such a way that they can be independently detected by two very different detection processes. One piece of information is the media data S_0, such as music, a photograph, or a movie, which will be viewed (detected) by a human observer. The other piece of information is a watermark, comprising an arbitrary sequence of bits, which will be detected by a specially designed watermark detector.

The first step is to encode the watermark bits into a form that will be easily combined with the media data. For example, when watermarking images, watermarks are often encoded as two-dimensional, spatial patterns. The watermark inserter then combines the encoded representation of the watermark with the media data. If the watermark insertion process is designed correctly, the result is media that appears identical to the original when perceived by a human, but which yields the encoded watermark information when processed by a watermark detector.

Watermarking is possible because human perceptual processes discard significant amounts of data when processing media. This redundancy is, of course, central to the field of lossy compression [5]. Watermarking exploits the redundancy by hiding encoded watermarks in them. A simple example of a watermarking method will illustrate how this can be done. It is well known that changes to the least significant bit of an 8-bit gray-scale image cannot be perceived. Turner [6] proposed

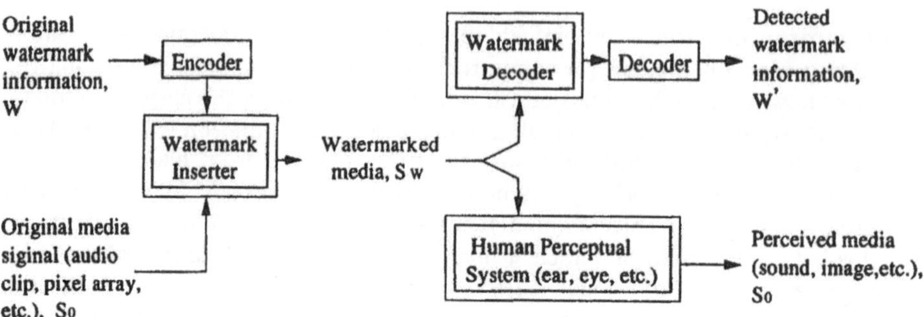

Figure 17.1 Watermarking framework.

hiding a watermark in images by simply replacing the least-significant bit with a binary watermark pattern. The detector looks at only the least-significant bit of each pixel, ignoring the other 7 bits. The human visual system looks at only the 7 most-significant bits, ignoring the least-significant. Thus, the two pieces of information are both perfectly detected from the same data stream, without interfering with one another. The least-significant-bit method of watermarking is simple and effective, but lacks some properties that may be essential for certain applications.

Most watermark detection processes require certain information to insert and extract watermarks. This information can be referred to as a "key" with much the same meaning as is used in cryptography. The level of availability of the key in turn determines who is able to read the watermark. In some applications, it is essential that the keys be widely known. For example, in the context of copy protection for digital video disks (DVD) it is envisaged that detectors will be present in all DVD players and will need to read watermarks placed in all copyrighted video content. In other applications, knowledge of the keys can be more tightly restricted.

In the past, we have referred to these two classes of watermarks as public and private watermarking. However, this could be misleading, given the well known meaning of the term "public" in cryptography. A public-key encryption algorithm involves two secrets; encrypting a message requires knowing one secret, and decrypting a message requires knowing the second. By analogy, a "public watermarking" method should also involve two secrets: inserting a watermark would require knowing one, and extracting would require knowing the second. While watermark messages might be encrypted by a public-key encryption technique before being inserted into media (see, for example, [2]), we know of no watermarking algorithm in which the ability to *extract* a watermark (encrypted or not) requires different knowledge than is required for insertion. In practice, all watermarking algorithms are more analogous to symmetric cryptographic processes in that they employ only one key. They vary only on the level of access to that key. Thus, in this chapter, we refer to the two classes as "restricted-key" and "unrestricted-key" watermarks.

It should be noted that the framework illustrated in Fig. 17.1 is different from the common conceptualization of watermarking as a process of arithmetically adding patterns to media data [7, 8, 9]. When the linear, additive view is employed for public watermarking, the detector is usually conceived of as a signal detector, detecting the watermark pattern in the presence of noise - that "noise" being the original media data. However, viewing the media data as noise does not allow us to consider two important facts: 1) unlike real noise, which is unpredictable, the media data is completely known at the time of insertion, and 2) unlike real noise, which has no commercial value and should be reduced to a minimum, the media data must be preserved. Consideration of these two facts allows the design of more sophisticated inserters.

17.3 PROPERTIES OF WATERMARKS

There are a number of important characteristics that a watermark can exhibit. The watermark can be difficult to notice, survive common distortions, resist malicious attacks, carry many bits of information, coexist with other watermarks, and require little computation to insert or detect. The relative importance of these characteristics depends on the application. The characteristics are discussed in more detail below.

17.3.1 Fidelity

The watermark should not be noticeable to the viewer nor should the watermark degrade the quality of the content. In earlier work [7, 8], we had used the term "imperceptible", and this is certainly the ideal. However, if a signal is truly imperceptible, then perceptually-based lossy compression algorithms either introduce further modifications that jointly exceed the visibility threshold or remove such a signal.

The objective of a lossy compression algorithm is to reduce the representation of data to a minimal stream of bits. This implies that changing any bit of well encoded data should result in a perceptible difference; otherwise, that bit is redundant. But, if a watermark is to be detectible after the data is compressed and decompressed, the compressed unwatermarked data must be different from the compressed watermarked data, and this implies that the two versions of the data will be perceptibly different once they are decompressed and viewed. Thus, as compression technology improves, watermarks that survive compression will cause increasingly perceptible differences in data that has been compressed and decompressed.

Early work on watermarking focused almost exclusively on designing watermarks that were imperceptible and therefore often placed watermark signals in perceptually insignificant regions of the content, such as high frequencies or low-order bits. However, other techniques, such as spread spectrum, can be used to add imperceptible or unnoticeable watermarks in perceptually significant regions. As is pointed out below, placing watermarks in perceptually significant regions can be advantagous for robustness against signal processing.

17.3.2 Robustness

Music, images and video signals may undergo many types of distortions. Lossy compression has already been mentioned, but many other signal transformations are also common. For example, an image might be contrast enhanced and colors might be altered somewhat, or an audio signal might have its bass frequencies amplified. In general, a watermark must be robust to transformations that include common signal distortions as well as digital-to-analog and analog-to-digital conversion and lossy compression. Moreover, for images and video, it is important that the watermark survive geometric distortions such as translation, scaling and cropping.

Note that robustness actually comprises two separate issues: 1) whether or not the watermark is still present in the data after distortion and 2) whether the watermark detector can detect it. For example, watermarks inserted into images by many algorithms remain in the signal after geometric distortions such as scaling, but the corresponding detection algorithms can only detect the watermark if the distortion is first removed. In this case, if the distortion cannot be determined and/or inverted, the detector cannot detect the watermark even though the watermark is still present albeit in a distorted form.

Fig. 17.2 illustrates one way of conceptualizing robustness. Here we imagine all the possible signals (images, audio clips, etc.) arranged in a two-dimensional space. The point S_0 represents a signal without a watermark. The point S_w represents the same signal with a watermark. The dark line shows the range of signals that would all be detected as containing the same watermark as S_w, while the dotted line indicates the range of distorted versions of S_w that are likely to occur with normal processing. This dotted line is best thought of as a contour in a probability distribution over the range of possible distortions of S_w. If the overlap between the watermark detection region and the range of likely distorted data is large, then the watermark will be robust.

Of course, in reality, it would be impossible to arrange the possible signals into a two-dimensional space in which the regions outlined in Fig. 17.2 would be contiguous, but the basic way of visualizing the robustness issue applies to higher dimensional spaces as well.

A more serious problem with Fig. 17.2 is that it is very difficult to determine the range of likely distortions of S_w, and, therefore, difficult to use this visualization as an analytical guide in designing watermarking algorithms. Rather than trying to predetermine the distribution of probable distorted signals, Cox *et al* [7, 8] have argued that robustness can be attained if the watermark is placed in perceptually significant regions of signals. This is because, when a signal is distorted, its fidelity is only preserved if its perceptually significant regions remain intact, while perceptually insignificant regions might be drastically changed with little effect on fidelity. Since we care most about the watermark being detectible when the media signal is a reasonable match with the original, we can assume that distortions which maintain the perceptually significant regions of a signal are likely, and represent the range of distortions outlined by the dotted line in Fig. 17.2. Section 17.5 details particular signal processing operations and their effects on detector performance.

17.3.3 Fragility

In some applications, we want exactly the opposite of robustness. Consider, for example, the use of physical watermarks in bank notes. The point of these watermarks is that they do not survive any kind of copying, and therefore can be used to indicate the bill's authenticity. We call this property of watermarks, fragility. Offhand, it would seem that designing fragile watermarking methods is easier than designing robust ones. This is true when our application calls for a watermark that is destroyed by every method of copying short of perfect digital copies (which can never affect watermarks). However, in some applications, the watermark is required to survive certain transformations and be destroyed by others. For example, a wa-

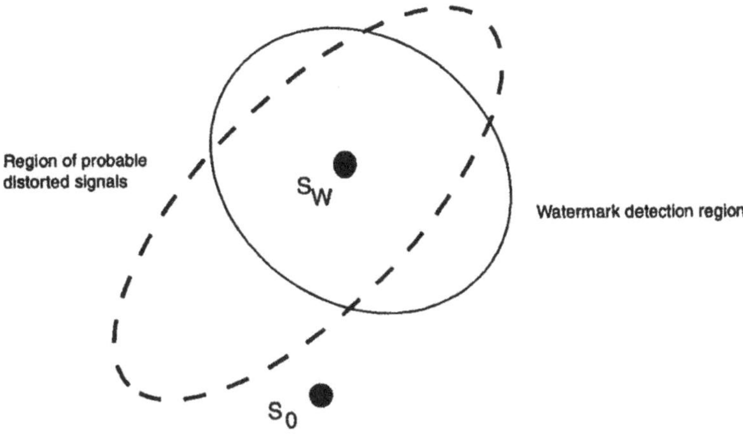

Figure 17.2 Watermark robustness

termark placed on a legal text document should survive any copying that doesn't change the text, but be destroyed if so much as one punctuation mark of the text is moved.

This requirement is not met by digital signatures developed in cryptology, which verify bit-exact integrity but cannot distinguish between various degrees of acceptable modifications.

17.3.4 Tamper-Resistance

Watermarks are often required to be resistant to signal processing that is solely intended to remove them, in addition to being robust against the signal distortions that occur in normal processing. We refer to this property as tamper-resistance. It is desirable to develop an analytical statement about watermark tamper-resistance. However, this is extremely difficult, even more so than in cryptography, because of our limited understanding of human perception. A successful attack on a watermark must remove the watermark from a signal without changing the perceptual quality of the signal. If we had perfect knowledge of how the relevant perceptual process behaved and such models would have tractable computation complexity, we could make precise statements about the computational complexity of tampering with watermarks. However, our present understanding of perception is imperfect, so such precise statements about tamper-resistance cannot yet be made.

We can visualize tamper-resistance in the same way that we visualize robustness, see Fig. 17.3. Here, the dotted line illustrates the range of signals that are perceptually equivalent to S_0. As in Fig. 17.2, this dotted line should be thought of as a contour in a probability distribution, this time the probability that a signal will be perceived as equivalent to S_0 by a randomly chosen observer. In theory, an attacker who precisely knows the range of this dotted line, as well as the range of the black line (the watermark detection region), could choose a new signal which would be perceptually equivalent to S_0 but would not contain the watermark. The

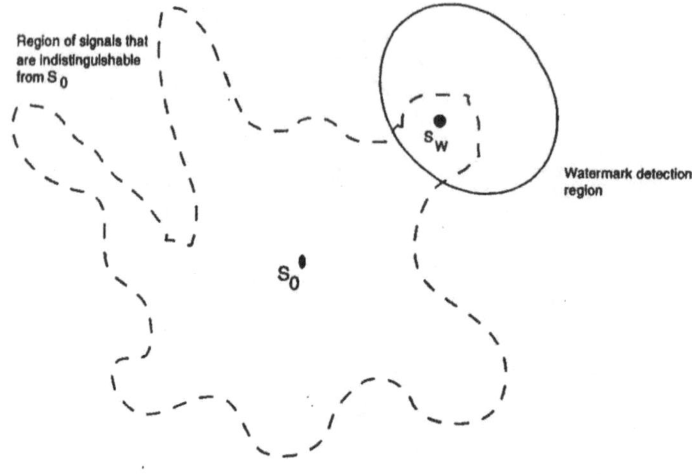

Figure 17.3 Tamper resistance

critical issue here is how well known are these two regions. We can assume an attacker does not have access to S_0 (otherwise, she/he would not need to tamper with S_w) so, even if a perfect perceptual model was available, the tamperer could not have perfect knowledge of the region of perceptually equivalent signals. However, the range of signals which are perceptually equivalent to S_0 has a large overlap with those that are perceptually equivalent to S_w, so, if an attacker finds an unwatermarked signal perceptually equivalent to S_w, it is likely to be equivalent to S_0 as well. The success of this strategy depends on how close S_w is to the dotted line. Tamper resistance will be elaborated upon in Section 17.6.

17.3.5 Key Restrictions

An important distinguishing characteristic is the level of restriction placed on the ability to read a watermark. As explained in earlier sections, we describe watermarks in which the key is available to a very large number of detectors as "unrestricted-key" watermarks, and those in which keys are kept secret by one or a small number of detectors as "restricted-key" watermarks.

While the difference between unrestricted-key and restricted-key is primarily a difference in usage, algorithms differ in their suitability for these two usages. For example, some watermarking methods (e.g., [10]) create a unique key for each piece of data that is watermarked. Such algorithms can be used for restricted-key applications, where the owner of the original data can afford to keep a database of keys for all the data that has been watermarked. But they cannot be used for unrestricted-key applications, since this would require every detector in the world having a complete list of all the keys. Thus, algorithms for use as unrestricted-key systems must employ the same key for every piece of data.

An unrestricted-key algorithm also must be made resistant to a wider variety of tampering attacks than must a restricted-key algorithm. Copy protection applications require that a watermark can be read by anyone, even by potential copyright pirates, but nonetheless only the sender should be able to erase the watermark. The problem is that complete knowledge of the detection algorithm and key can imply knowledge of how to insert watermarks, and, in general a watermark can be erased by using the insertion algorithm to insert the negation of the watermark pattern. The ideal solution would be an algorithm in which knowing how to detect would not imply knowing how to insert, but this would be a true public-key algorithm, and, as pointed out above, we know of no such algorithm.

In the absence of true public watermarking, one alternative for unrestricted-key watermarking is to use an existing algorithm placed in a tamper-resistant box. However, this approach has weaknesses and other disadvantages. An attacker may be able to reverse engineer the tamper resistant box. For the consumer electronics and computer industry, the logistics of the manufacturing process are more complicated and less flexible if secret data has to be handled during design, prototyping, testing, debugging and quality control. Some of the attacks to be described in Section 17.6 exploit the fact that algorithms which are inherently "secret key" in nature, are used in an environment where public detection properties are desired, i.e., access to the key is almost completely unrestricted.

An example of restricted-key watermarking is in the broadcast industry which uses watermarks to automatically monitor and log the radio music that is broadcast. This facilitates the transfer of airplay royalties to the music industry. In a scenario where monitoring receivers are located "in the field", the watermark embedding system as well as any and all receiving monitors can be owned and operated by the royalty collection agency. However, in practice radio stations are more interested in reducing the work load of their studio operators (typically a single disk jockey) than to intentionally evade royalty payments and mostly use watermark readers themselves to create logs. As already mentioned in the introduction, watermarking of television news clips are under research, for instance in the European VIVA project.

A similar scenario is used for a service in which images are watermarked and search robots scan the Internet to find illegally posted copies of these images. In this scenario it is not a fundamental problem that the watermark detector contains sensitive secret data, i.e.,, a detection key, that would reveal how the watermark can be erased. Potential attackers do not, in principle, have access to a watermark detector. However, a security threat occurs if a detector may accidentally fall into the hands of a malicious user. Moreover, the watermark solution provider may offer a service to content publishers to verify on-line whether camera-ready content is subject to copy restriction. Such an on-line service could be misused in an attack to deduce the watermark secrets.

17.3.6 False Positive Rate

In most applications, it is necessary to distinguish between data that contains watermarks and data that doesn't. The false positive rate of a watermark detection system is the probability that it will identify an unwatermarked piece of data as containing a watermark. The seriousness of such an error depends on the application. In some applications, it can be catastrophic.

For example, in the copy control application considered for DVD, a device will refuse to play video from a non-factory-recorded disk if it finds a watermark saying that the data should never be copied. If a couple's wedding video (which would doubtless be unwatermarked and would not be on a factory recorded disk) is incorrectly identified as watermarked, then they will never be able to play the disk. Unless such errors are extremely rare, false positives could give DVD players a bad reputation that would seriously damage the market for them. Most companies competing to design the watermarking method used in DVD place the acceptible false positive rate at one false positive in several tens or hundreds of billions of distinct frames.

17.3.7 Modification and Multiple Watermarks

In some circumstances, it is desirable to alter the watermark after insertion. For example, in the case of digital video discs, a disc may be watermarked to allow only a single copy. Once this copy has been made, it is then necessary to alter the watermark on the original disc to prohibit further copies. Changing a watermark can be accomplished by either (i) removing the first watermark and then adding a new one or (ii) inserting a second watermark such that both are readable, but one overrides the other. The first alternative does not allow a watermark to be tamper resistant since it implies that a watermark is easily removable. Allowing multiple watermarks to co-exist is preferable and this facilitates the tracking of content from manufacturing to distribution to eventual sales, since each point in the distribution chain can insert its own unique watermark.

There is, however, a security problem related with multiple watermarks as explained in Section 17.6.6. If no special measures are taken the availability of a single original with different watermarks will allow a clever pirate to retrieve the unmarked original signal by statistical averaging or more sophisticated methods [7, 10].

17.3.8 Data Payload

Fundamentally, the data payload of a watermark is the amount of information it contains. As with any method of storing data, this can be expressed as a number of bits, which indicates the number of distinct watermarks that might be inserted into a signal. If the watermark carries N bits, then there are 2^N different possible watermarks. It should be noted, however, that there are actually $2^N + 1$ possible values returned by a watermark detector, since there is always the possibility that no watermark is present.

In discussing the data payload of a watermarking method, it is important to distinguish between the number of distinct watermarks that may be inserted, and the number of watermarks that may be detected by a single iteration with a given watermark detector. In many watermarking applications, each detector need not test for all the watermarks that might possibly be present. For example, several companies might want to set up web-crawlers that look for the companies' watermarks in images on the web. The number of distinct possible watermarks would have to be at least equal to the number of companies, but each crawler could test for as few as one single watermark. A watermarking system tailored for such an application might be said to have a payload of many bits, in that many different watermarks are possible, but this does not mean that all the bits are available from any given detector.

17.3.9 Computational Cost

As with any technology intended for commercial use, the computational costs of inserting and detecting watermarks are important. This is particularly true when watermarks need to be inserted or detected in real-time video or audio.

The speed requirements are highly application dependent. In general, there is often an asymmetry between the requirement for speed of insertion and speed of detection. For example, in the DIVX fingerprinting application, watermarks must be inserted in real-time by inexpensive hardware – typically single chips costing only a few dollars each – while they may be detected, in less than real-time, by professional equipment costing tens of thousands of dollars. On the other hand, in the case of copy-control for DVD, it is the detection that must be done in real-time on inexpensive chips, while the insertion may be done on high-cost professional equipment. Note that, in cases like DVD where we can afford expensive inserters, it can actually be desirable to make the inserters expensive, since an inserter is often capable of removing a watermark, and we want them to be difficult for pirates to obtain or reproduce.

Another issue to consider in relation to computational cost is the issue of scalability. It is well known that computer speeds are approximately doubling every eighteen months, so that what looks computationally unreasonable today may very quickly become a reality. It is therefore very desirable to design a watermark whose detector and/or inserter is scalable with each generation of computers. Thus, for example, the first generation of detector might be computationally inexpensive but might not be as reliable as next generation detectors that can afford to expend more computation to deal with issues such as geometric distortions.

17.3.10 Standards

In some application scenarios watermark technology needs to be standardized to allow global usage. An example where standardization is needed is DVD. A copy protection system based on watermarks is under consideration that will require every DVD player to check for a watermark in the same way. However, a standardized detection scheme does not necessarily mean that the watermark insertion method also needs to be standardized. This is very similar to the standardization activities of MPEG, where the syntax and the semantics of the MPEG bitstream is fixed, but not the way in which an MPEG bitstream is derived from baseband video. Thus, companies may try to develop embedding systems which are superior with respect to robustness or visibility.

17.4 EXAMPLE OF A WATERMARKING METHOD

To evaluate watermarking properties and detector performance in more detail, we now present a basic class of watermarking methods.

Mathematically, given an original image S_0 and a watermark W, the watermarked image, S_w, is formed by $S_w = S_0 + f(S_0, W)$ such that the watermarked image S_w is constrained to be visually identical (or very similar) to the original unwatermarked image S_0.

In theory, the function f may be arbitrary, but in practice robustness requirements pose constraints on how f can be chosen. One requirement is that watermarking has to be robust to random noise addition. Therefore many watermark designers opt for a scheme in which image S_0 will result in approximately the same watermark as a slightly altered image $S_0+\epsilon$. In such cases $f(S_0, W) \approx f(S_0+\epsilon, W)$

For an unrestricted-key watermark, detection of the watermark, W, is typically achieved by correlating the watermark with some function, g, of the watermarked image. Thus, the key simply is a pseudo-random number sequence, or a seed for the generator that creates such sequence, that is embedded in all images.

Example: In its basic form, in one half of the pixels the luminance is increased by one unit step while the luminance is kept constant [11] or decreased by one unit step [12] in the other half. Detection by summing luminances in the first subset and subtracting the sum of luminances in the latter subset is a special case of a correlator. One can describe this as $S_w = S_0 + W$, with $W \in R^N$, and where $f(S_0, W) = W$. The detector computes $S_w \cdot W$, where $\cdot$ denotes the scalar product of two vectors.

If W is chosen at random, then the distribution of $S_0 \cdot W$ will tend to be quite small, as the random $\pm$ terms will tend to cancel themselves out, leaving only a residual variance. However, in computing $W \cdot W$ all of the terms are positive, and will thus add up. For this reason, the product $S_w \cdot W = S_0 \cdot W + W \cdot W$ will be close to $W \cdot W$. In particular, for sufficiently large images, it will be large, even if the magnitude of S_0 is much larger than the magnitude of W. It turns out that the probability of making an incorrect detection can be expressed as the complementary error function of the square root of the ratio $W \cdot W$ over the variance in pixel luminance values. This result is very similar to expressions commonly encountered in digital transmission over noisy radio channels. Elaborate analyses of the statistical behavior of $I \cdot W$ and $W \cdot W$are typically found in spread-spectrum oriented papers, such as [7, 8, 13, 14, 15, 16].

17.5 ROBUSTNESS TO SIGNAL TRANSFORMATIONS

Embedding a copy flag in ten seconds of NTSC video may not seem difficult since it only requires the embedding of 4-bits of information in a data stream. The total video data is approximately $720 \times 480 \times 30 \times 10$. This is over 100Mbytes prior to MPEG compression. However, the constraints of (i) maintaining image fidelity and (ii) survive common signal transformations, can be severe. In particular, many signal transformations cannot be modeled as a simple linear additive noise process. Instead, such processes are highly spatially correlated and may interact with the watermark in complex ways.

There are a number of common signal transformations that a watermark should survive, e.g., affine transformations, compression/re-compression, and noise. In some circumstances, it may be possible to design a watermark that is completely invariant to a particular transformation. For example, this is usually the case for translational motions. However, scale changes are often much more difficult to design for and it may be the case that a watermark algorithm is only robust to small perturbations in scale. In this case, a series of attacks may be mounted by identifying the limits of a particular watermarking scheme and subsequently finding a transformation that is outside of these limits but maintains adequate image fidelity.

17.5.1 Affine Transformations

Shifts over a few pixels can cause watermarking detectors to miss the presence of watermark. The problem can be illustrated by our example watermarking scheme. Suppose one shifts S_w by one pixel, obtaining $S_{w,s}$. Let $S_{w,s}$ and W_s denote the similarly shifted versions of S_0 and W. Then $S_{w,s} \cdot W = I_s \cdot W + W_s \cdot W$. As before, the random +/- terms in $S_{w,s} \cdot W$ will tend to cancel themselves out. However, the $W_S \cdot W$ terms will also cancel themselves out, if each +/- value was chosen independently. Hence, $S_{w,s} \cdot W$ will have small magnitude and the watermark will not be detected.

Typical analog VHS recorders cause shifting over a small portion of a line, but enough to cause a shift of several pixels or even a few DCT blocks. Recorder time jitter and tape wear randomly stretch an image. Even if the effects are not disturbing to a viewer, it may completely change the alignment of the watermark with respect to pixels and DCT block boundaries.

There are a number of defenses against such attacks. Ideally, one would like to reverse the affine transformations. Given an original, a reasonable approximation to the distortion can be computed. With unrestricted-key watermarks, and in particular the "do not copy" application, no original is available. A secondary signal, i.e., a registration pattern, may be inserted into the image whose entire purpose is to assist in reversing the transformation. However, one can base attacks on this secondary signal, removing or altering it in order to block detection of the watermark. Another alternative is to place watermark components at key visual features of the image, e.g., in patches whose average luminosity is at a local maximum. Finally, one can insert the watermark into features that are transformation invariant. For example, the magnitudes of Fourier coefficients are translation invariant.

In some applications, it may be assumed that the extent of the affine transformation is minor. Particularly if the watermark predominantly resides in perceptually relevant low-frequency components, the autocorrelation $W_s \cdot W$ can be sufficiently large for sufficiently small translations. A reliability penalty associated with low-pass watermarking is derived in [13].

17.5.2 Noise Addition

A common misunderstanding is that a watermark of small amplitude can be removed by adding random noise of a similar amplitude. On the contrary, correlation detectors appear very robust against addition of a random noise term ϵ. For instance if $f(I, W) = W$ one can describe the attacked image as $S_{w,s} = S_0 + \epsilon + W$. The detector computes $S_w \cdot W$. The product $S_w \cdot W = S_0 \cdot W + \epsilon \cdot W + W \cdot W$. If the watermark was designed with $W \cdot W$ largely exceeding the statistical spreading in $I\cdot$ W, it will mostly also largely exceed the statistical spreading in $\epsilon \cdot W$. In practice, noise mostly is not a serious threat unless (in the frequency components of relevance) the noise is large compared to image I or if the noise is correlated with the watermark.

17.5.3 Spatial Filtering

Most linear filters for image processing create a new image by taking a linear combination of surrounding pixels. Watermark detection can be quite reliable after such filtering, particularly after edge-enhancement type of filters [14]. Such filters typically amplify the luminance of the original image and subtract shifted versions of the surroundings. In effect, redundancy in the image is cancelled and randomness of the watermark is exaggerated. One the other hand, smoothing and low-pass filtering often reduce the reliability of a correlator watermark detector.

17.5.4 Digital Compression

MPEG video compression accurately transfers perceptually important components, but coarsely quantizes high image components with high frequency components. This process may severely reduce the detectability of a watermark, particularly if that resides in high spatial frequencies. Such MPEG compression is widely used in digital television and on DVD discs.

Digital recorders may not always make a bit exact copy. Digital recorders will, at least initially, not contain sophisticated signal processing facilities. For recording of MPEG streams onto media with limited storage capacity, the recorder may have to reduce the bit rate of the content.

For video recorders that re-compress video, image quality usually degrades significantly, as quantization noise is present, typically with large high frequency components. Moreover, at high frequencies, image and watermark components may be lost. In such cases, the watermark may be lost, though the video quality may also be significantly degraded.

17.6 TAMPER RESISTANCE

In this section, we describe a series of attacks that can be mounted against a watermarking system.

17.6.1 Attacks on the Content

Although several commercially available watermarking scheme are robust to many types of transformation (e.g., rotation, scaling etc), these often are not robust to combinations of basic transformations, such as scaling, cropping and a rotation. Several tools have been created by hackers that combine a small non-linear stretching with spatial filtering [17].

17.6.2 Attacks by Statistical Averaging

An attacker may try to estimate the watermark and subtract this from a marked image. Such an attack is particularly dangerous if the attacker can find a generic watermark, for instance one with $W = f(S_0, W)$ not depending significantly on the image S_0. Such an estimate W of the watermark can then be used to remove a watermark from any arbitrary marked image, without any further effort for each new image or frame to be "cleaned".

The attacker may separate the watermark W by adding or averaging multiple images, e.g., multiple successive marked images $S_0 + W, S_1 + W, \ldots, S_N + W$ from a video sequence. The addition of N such images results in $NW + \sum_i S_i$, which tends to NW for large N and sufficiently many and sufficiently independent images $S_0, S_1, \ldots, S_N$.

A countermeasure is to use at least two different watermarks W_1 and W_2 at random, say with probability p_1 and p_2 where $p_2 = 1 - p_1$, respectively. The above attack then only produces $p_1W_1 + (1-p_1)W_2$, without revealing W_1 or W_2. However a refinement of the attack is to compute weighted averages, where the weight factor is determined by a (possibly unreliable but better than random) guess of whether a particular image contains one watermark or the other. For instance, the attacker may put an image in category $i (i \in \{1, 2\})$ if he believes that this image contains watermark W_i. Let P_ϵ denote the probability that an image is put into the wrong category. Then, after averaging a large number (N_1) of images from category 1, the result converges to $x_1 = N_1 p_1 (1 - P_\epsilon) W_1 + N_1 (1 - p_1)(P_\epsilon) W_2$. Similarly the sum of N_2 images in category 2 tends to $x_2 = N_2 p_1 P_\epsilon W_1 + N_2 (1 - p_1)(1 - P_\epsilon) W_2$. Computing the weighted difference gives

$$\frac{x_1}{N_1} - \frac{x_2}{N_2} = p_1 (1 - 2P_\epsilon) W_1 - (1 - p_1)(1 - 2P_\epsilon) W_2.$$

Hence for any $P_\epsilon \neq 1/2$, i.e.,, for any selection criterion better than a random one, the attacker can estimate both the sum and difference of $p_1 W_1$ and $(1 - p_1) W_2$. This reveals W_1 and W_2.

17.6.3 Exploiting the Presence of a Watermark Detector Device

For unrestricted-key watermarks, we must assume that the attacker at least has access to a "black box" watermark detector, which indicates whether the watermark is present in a given signal. Using this detector, the attacker can probably learn enough about the detection region, in a reasonable amount of time, to reliably remove the watermark.

The aim of the attack is to experimentally deduce the behavior of the detector, and to exploit this knowledge to ensure that a particular image does not trigger the detector. For example, if the watermark detector gives a soft decision, e.g., a continuous reliability indication when detecting a watermark, the attacker can learn how minor changes to the image influence the strength of the detected watermark. That is, modifying the image pixel-by-pixel, he can deduce the entire correlation function or other watermark detection rule. Interestingly, such attack can also be applied even when the detector only reveals a binary decision, i.e., present or absent. Basically the attack [18, 19] examines an image that is at the boundary where the detector changes its decision from "absent" to "present". For clarity the reader may consider a watermark detector of the correlator type; but this is not a necessary condition for the attack to work. For a correlator type of detector, our attack reveals the correlation coefficients used in the detector (or at least their sign).

For example:

1. Starting with a watermarked image, the attacker creates a test image that is near the boundary of a watermark being detectable. At this point it does not matter whether the resulting image resembles the original or not. The only criterion is that minor modifications to the test image cause the detector to respond with "watermark" or "no watermark" with a probability that is sufficiently different from zero or one. The attacker can create the test image by modifying a watermarked image step-by-step until the detector responds "no watermark found". A variety of modifications are possible. One method is to gradually reduce the contrast in the image just enough to drop below the threshold where the detector reports the presence of the watermark. An

alternative method is to replace more and more pixels in the image by neutral gray. There must be a point where the detector makes the transition from detecting a watermark to responding that the image contains no watermark. Otherwise this step would eventually result in an evenly gray colored image, and no reasonable watermark detector can claim that such image contains a watermark.

2. The attacker now increases or decreases the luminance of a particular pixel until the detector sees the watermark again. This provides the insight of whether the watermark embedder decreases or increases the luminance of that pixel.

3. This step is repeated for every pixel in the image.

4. Combining the knowledge on how sensitive the detector is to a modification of each pixel, the attacker estimates a combination of pixel values that has the largest influence on the detector for the least disturbance of the image.

5. The attacker uses the original marked image and subtracts (λ times) the estimate, such that the detector reports that no watermark is present. λ is found experimentally, such that λ is as small as possible. Moreover, the attacker may also exploit a perceptual model to minimize the visual effect of his modifications to the image.

The computational effort needed to find the watermark is much less than commonly believed. If an image contains N pixels, conventional wisdom is that an attack that searches the watermark requires an exponential number of attempts of order $O(2^N)$. A brute force exhaustive search checking all combinations with positive and negative sign of the watermark in each pixel results in precisely 2^N attempts. The above method shows that many watermarking methods can be broken much faster, namely in $O(N)$, provided a device is available that outputs a binary (present or absent) decision as to the presence of the watermark.

We can, however, estimate the computation required to learn about the detection region when a black box detector is present, and this opens up the possibility of designing a watermarking method that specifically makes the task impractical.

Linnartz [19] has suggested that a probabilistic detector[3] would be much less useful to an attacker than a deterministic one. If properly designed, a probabilistic detector would teach an attacker so little in each iteration that the task would become impractical.

A variation of the attack above which also works in the case of probabilistic detectors is presented in [20] and [21]. Similar to the attack above the process starts with the construction of a signal S_θ at threshold of detection. The attacker than chooses a random perturbation V and records the decision of the watermark detector for $S_\theta + V$. If the detector sees the watermark, the perturbation V is considered an estimation of the watermark W. If the detector does not see the watermark

[3]A probabilistic detector is one in which two thresholds exist. If the detector output is below the lower threshold then no watermark is detected. Similarly, if the detector output is above the higher threshold then a watermark is detected. However, if the detector output lies between the two thesholds, then then the decision as to whether the watermark is present or absent is random.

the negation $-V$ is considered an estimation of the watermark. By repeating this perturbation process a large number of times and summing all intermediate estimates a good approximation of the watermark W can be obtained. It can be shown that the accuracy of the estimation is $\mathcal{O}(\sqrt{\frac{J}{N}})$ where J is the number of trials and N is the number of samples. In particular it follows that for a fixed accuracy κ the number of trials J is linear with the number of samples N. A more detailed analysis also shows that the number of trials is proportional to the square of the width of the threshold zone (i.e., the zone where the detector takes probabilistic decisions). The designer of a probabilistic watermark detector therefore faces the trade-off between a large threshold zone (i.e., a high security), a small false negative rate (i.e., a small upper bound of the threshold zone) and a small false positive rate (i.e., a large lower bound of the threshold zone).

17.6.4 Attacks Based on the Presence of a Watermark Inserter

If the attacker has access to a watermark inserter, this provides further opportunities to break the security. Attacks of this kind are relevant to copy control in which copy generation management is required, i.e., the user is permitted to make a copy from the original source disc but is not permitted to make a copy of the copied material - only one generation of copying is allowed. The recorder should change the watermark status from "one-copy allowed" to "no more copies allowed". The attacker has access to the content before and after this marking. That is, he can create a difference image, by subtracting the unmarked original from the marked content. This difference image is equal to $f(S_0, W)$. An obvious attack is to pre-distort the original to undo the mark addition in the embedder. That is, the attacker computes $I - f(S_0, W)$ and hopes that after embedding of the watermark, the recorder stores

$$S_0 - f(S_0, W) + f(S_0 - f(S_0, W), W)$$

which is likely to approximate S_0. The reason why most watermarking methods are vulnerable to this attack is that watermarking has to be robust to random noise addition. If, for reasons discussed before,

$$f(S_0, W) \approx f(S_0 + \epsilon, W),$$

and because watermarks are small modifications themselves, $f(S_0, W) \approx f(S_0 - f(S_0, W), W)$. This property enables the above pre-distortion attack.

17.6.5 Attacks on the Copy Protection System

The forgoing discussion of tamper-resistance has concentrated only on the problem of removing a watermark from a given signal. We have not discussed ways of circumventing systems that are based on watermarking. In many applications, it is far easier to thwart the purpose of the watermark than it is to remove the watermark. For example, Craver *et al* [1], discuss ways in which watermarks that are used to identify media ownership might be thwarted by inserting conflicting watermarks into the signal so as to make it impossible to determine which watermark identifies the true owner. Cox and Linnartz [18, 22] discuss several methods of circumventing watermarks used for copy control.

The most trivial attack is to tamper with the output of the watermark detector and modify it in such a way that the copy control mechanism always sees a "no watermark" detection, even if a watermark is present in the content. Since hackers and pirates more easily can modify (their own!) recorders but not their customers' players, playback control is a mechanism that detects watermarks during the playback of discs. The resulting tape or disc can be recognized as an illegal copy if playback control is used.

Copy protection based on watermarking content has a further fundamental weakness. The watermark detection process is designed to detect the watermark when the video is perceptually meaningful. Thus, a user may apply a weak form of scrambling to copy protected video, e.g., inverting the pixel intensities, prior to recording. The scrambled video is unwatchable and the recorder will fail to detect a watermark and consequently allow a copy to be made. Of course, on playback, the video signal will be scrambled, but the user may then simply invert or descramble the video in order to watch a perfect and illegal copy of a video. Simple scrambling and descrambling hardware would be very inexpensive and manufacturers might argue that the devices serve a legitimate purpose in protecting a user's personal video. Similarly, digital MPEG can easily be converted into a file of seemingly random bits. One way to avoid such circumvention for digital recording is to only allow the recording of content in a recognized file format. Of course this would severely limit the functionality of the storage device.

17.6.6 Collusion Attacks

If the attacker has access to several versions of the signal, $S_{w_1}, S_{w_2} \ldots S_{w_N}$, each with a different watermark, but each perceptually equivalent to S_0, then he/she can learn much more about the region of signals that are equivalent to S_0, since it will be well approximated by the intersection of the regions of signals that are equivalent to the watermarked signals. This gives rise to "collusion attacks", in which several watermarked signals are combined to construct an unwatermarked signal. The attacker's knowledge of the detection region is under our direct control. In the case of a restricted-key watermark, she/he has no knowledge of this region at all. This makes it extremely difficult to tamper with restricted-key watermarks. The best an attacker can do is to find a signal that is as far from the watermarked signal as possible, while still likely to be within the range of signals perceptually equivalent to S_0, and to hope that this distant signal is outside the detection range. In the case of a collusion attack, this job is made easier, because the hacker can use the multiple watermarked versions of the signal to obtain closer and closer approximations to S_0, which definitely is not watermarked. However, whether or not the attacker has the advantage of making a collusion attack, he/she can never be sure whether the attack succeeded, since the information required to test for the watermark's presence is not available. This should help make security systems based on restricted-key watermarks more effective. Resistance to collusion attacks is also a function of the structure of the watermark, as discussed in [10].

In the next section, we summarize early work on watermarking and then describe more recent work which attempts to insert a watermark into the perceptually significant regions of an image.

17.6.7 Methods

In this section, we provide a review of watermarking methods that have been proposed. This is unlikely to be a complete list and omissions should not be interpreted as being inferior to those described here. Recent collections of papers can be found in [23, 24].

Early work on watermarking focused on hiding information within a signal but without considering the issues discussed earlier. In an application in which a covert channel between two parties is desired, tamper resistance may not be an issue if only the communicating parties are aware of the channel. Thus, early work can be thought of as steganography [25].

Turner [6] proposed inserting an identification code into the least significant bits of randomly selected words on compact discs. Decoding is accomplished by comparison with the original unwatermarked content. Although the method is straightforward, it is unlikely to be robust or tamper resistant. For example, randomizing the least significant bits of all words would remove the watermark. Oomen *et al.* [26] refined the method exploiting results from the theory of perceptual masking, dithering and noise shaping. Later van Schyndel *et al* [27] proposed a similar method as well as a spread spectrum method that linearly adds a watermark to an image.

Brassil *et al* [28] describe several methods for watermarking text, based on slightly altering the character or line spacings on a page or by adding/deleting serifs from characters. This approach is further refined in [29]. Unfortunately, as the authors note, these approaches are not resistant to tampering. For example, a malicious attacker could randomize the line or character spacing, thereby destroying the watermark. In general, text is particularly difficult to watermark based on adding noise, since optical character technology is, in principle, capable of eliminating it. An alternative approach is to insert the watermark at the symbolic level, by, for example, inserting spelling errors or by replacing words or phrases with alternatives in a predetermined manner, e.g., substituting "that" for "which". However, these approaches also appear susceptible to tampering.

Caronni [30] describes a procedure in which faint geometric patterns are added to an image. The watermark is therefore independent of the image, but because the watermark is graphical in nature, it has a spatial frequency distribution that contains perceptually significant components. However, it is unclear whether such a method is preferable to adding a pre-filtered PN noise sequence.

Tanaka *et al* [31] proposed a method to embed a signal in an image when the image is represented by dithering. Later, Matsui and Tanaka [32] suggested several different methods to encode a watermark, based on whether the image was represented by predictive coding, dithering (monotone printing) or run-lengths (fax). A DCT-based method is also proposed for video sequences. These methods make explicit use of the representation and it is unclear whether such approaches are robust or tamper resistant.

Koch *et al* [33, 34] describe several procedures for watermarking an image based on modifying pairs or triplets of frequency coefficients computed as part of the JPEG compression procedure. The rank ordering of these frequency coefficients is used to represent the binary digits. The authors select mid-range frequencies which typically survive JPEG compression. To avoid creating artifacts, the DC coefficient is not altered. Several similar methods have recently been proposed. Bors and Pitas [35] suggest an alternative linear constraint among selected DCT coefficients, but it is unclear whether this new constraint is superior to that of [33, 34]. Hsu and Wu [36] describe a method in which the watermark is a sequence of binary digits that are inserted into the mid-band frequencies of the 8×8 DCT coefficients.

Swanson *et al* [37] describe linearly adding a PN sequence that is first shaped to approximate the characteristics of the human visual system to the DCT coefficients of 8×8 blocks. In the latter two cases, the decoder requires access to the original image. It is interesting to note that a recently issued patent [38] appears to patent the general principle of extracting a watermark based on comparison of the watermarked and unwatermarked image.

Rhoads [39] describes a method in which N pseudo random (PN) patterns, each pattern having the same dimensions as the image, are added to an image in order to encode an N-bit word. The watermark is extracted by first subtracting a copy of the unwatermarked image and correlating with each of the N known PN sequences. The need for the original image at the decoder was later relaxed. While Rhoads did not explicitly recognize the importance of perceptual modeling, experiments with image compression led him to propose that the PN sequences be spectrally filtered, prior to insertion, such that the filtered noise sequence was within the passband of common image compression algorithms such as JPEG.

Bender *et al* [40] describe several possible watermarking methods. In particular, "Patchwork" encodes a watermark by modifying a statistical property of the image. The authors note that the difference between any pair of randomly chosen pixels is Gaussian distributed with a mean of zero. This mean can be shifted by selecting pairs of points and incrementing the intensity of one of the points while decrementing the intensity of the other. The resulting watermark spectrum is predominantly high frequency. However, the authors recognize the importance of placing the watermark in perceptually significant regions and consequently modify the approach so that pixel patches rather than individual pixels are modified, thereby shaping the watermark noise to significant regions of the human visual system. While the exposition is quite different from Rhoads [39], the two techniques are very similar and it can be shown that the Patchwork decoder is effectively computing the correlation between the image and a binary noise pattern, as covered in our example detector in Section 17.4.

Paatelma and Borland [41] propose a procedure in which commonly occurring patterns in images are located and target pixels in the vicinity of these patterns are modified. Specifically, a pixel is identified as a target if it is preceded by a preset number of pixels along a row that are all different from their immediate neighbors. The target pixel is then set to the value of the pixel a fixed offset away, provided the intensity difference between the two pixels does not exceed a threshold. Although the procedure appears somewhat convoluted, the condition on target pixels assures that the watermark is placed in regions that have high frequency information. Although the procedure does not explicitly discuss perceptual issues, a commercial implementation of this process is claimed to have survived through the printing process.

Holt *et al* [42] describe a watermarking procedure in which the watermark is first nonlinearly combined with an audio signal to spectrally shape it and the resulting signal is then high pass filtered prior to insertion into the original audio signal. Because of the high pass filtering, the method is unlikely to be robust to common signal distortions. However, Preuss *et al* [43] describe an improved procedure that inserts the shaped watermark into the perceptually significant regions of the audio spectrum. The embedded signaling procedure maps an alphabet of signals to a set of binary PN sequences whose temporal frequency response is approximately white. The audio signal is analyzed through a window and the audio spectrum in this window is calculated. The watermark and audio signals are then combined nonlinearly by multiplying the two spectra together. This combined signal will have a shape that is very similar to the original audio spectrum. The resulting signal is then inverse transformed and linearly weighted and added to the original audio signal. This is referred to as spectral shaping. To decode the watermark, the decoder first applies a spectral equalizer that whitens the received audio signal prior to filtering through a bank of matched filters, each one tuned to a particular symbol in the alphabet. While the patent does not describe experimental results, we believe that this is a very sophisticated watermarking procedure that should be capable of surviving many signal distortions.

Cox *et al* [7, 8] describe a somewhat similar system for images in which the perceptually most significant DCT coefficients are modified in a non-linear fashion that effectively shapes the watermark spectrum to that of the underlying image. The decoder requires knowledge of the original unwatermarked image in order to invert the process and extract the watermark. This constraint has been subsequently relaxed. The authors also note that binary watermarks are less resistant to tampering by collusion than watermarks that are based on real valued, continuous pseudo random noise sequences.

Podilchuk and Zeng [44] describe improvements to Cox *et al* by using a more advanced perceptual model and a block based method that is therefore more spatially adapative.

Ruanaidh *et al.*, [45] describe an approach similar to [7, 8] in which the phase of the DFT is modified. The authors note that phase information is perceptually more significant than the magnitude of Fourier coefficients and therefore argue that such an approach should be more robust to tampering as well as to changes in image contrast. The inserted watermark is independent of the image and is recovered using traditional correlation without the use of the original image.

Several authors [7, 8, 13, 14, 33, 34, 43], draw upon work in spread spectrum communications. Smith and Comiskey [15] analyze watermarking from a communications perspective. They propose a spread spectrum based technique that "predistorts" the watermark prior to insertion. However, the embedded signal is not a function of the image, but rather is pre-filtered based on expected compression algorithms such as JPEG. Linnartz *et al.* [13, 14], review models commonly used for detection of spread spectrum radio signals and discuss their suitability in evaluating watermark detector performance. In contrast to typical radio systems in which the signal waveform (e.g., whether it is spread or not) does not affect error performance according to the most commonly accepted channel model,[4] the watermark detector tends to be sensitive to the spectral shape of the watermark signal. A signal-to-noise penalty is derived for placing the watermark in visually important regions, instead of using a spectrally flat (unfiltered) PN-sequence.

17.7 SUMMARY

We have described the basic framework in which to discuss the principle of watermarking, and outlined several characteristics of watermarks that might be desirable for various applications. We covered intentional and unintentional attacks which a watermark system may face. While a watermark may survive many signal transformations that occur in commonly used signal processing operations, resistance to intentional tampering usually is more difficult to achieve. Finally, we surveyed many of the numerous recent proposals for watermarking and attempted to identify their strengths and weaknesses.

REFERENCES

[1] S. Craver, N. Memon, B.-L. Yeo, and M. Yeung, "Resolving rightful ownerships with invisible watermarking techniques: Limitations, attacks and implications," *IEEE Trans. on Selected Areas of Communications*, vol. 16, no. 4, pp. 573–586, 1998.

[2] R. J. Anderson and F. A. P. Petitcolas, "On the limits of steganography," *IEEE Trans. on Selected Areas of Communications*, vol. 16, no. 4, pp. 474–481, 1998.

[3] G. Simmons, "The prisoner's problem and the subliminal channel," in *Proceedings CRYPTO'83*, Advances in Cryptology, pp. 51–67, Plenum Press, 84.

[4] R. L. Rivest, "Chaffing and winnowing: Confidentiality without encryption." http:theory.lcs.mit.edu/ rivest/chaffing.txt, 1998.

[5] N. Jayant, J. Johnston, and R. Safranek, "Signal compression based on models of human perception," *Proc IEEE*, vol. 81, no. 10, 1993.

[6] L. F. Turner, "Digital data security system." Patent IPN WO 89/08915, 1989.

[7] I. Cox, J. Kilian, F. T. Leighton, and T. Shamoon, "Secure spread spectrum watermarking for images, audio and video," in *IEEE Int. Conference on Image Processing*, vol. 3, pp. 243–246, 1996.

[4] The linear time-invariant channel with additive white Gaussian noise.

[8] I. Cox, J. Kilian, F. T. Leighton, and T. Shamoon, "A secure, robust watermark for multimedia," in *Information Hiding: First Int. Workshop Proc.* (R. Anderson, ed.), vol. 1174 of *Lecture Notes in Computer Science*, pp. 185–206, Springer-Verlag, 1996.

[9] I. Cox and M. L. Miller, "A review of watermarking and the importance of perceptual modeling," in *Proceedings of SPIE, Human Vision & Electronic Imaging II*, vol. 3016, pp. 92–99, 1997.

[10] I. Cox, J. Kilian, F. T. Leighton, and T. Shamoon, "Secure spread spectrum watermarking for images, audio and video," *IEEE Trans. on Image Processing*, vol. 6, no. 12, pp. 1673–1687, 1997.

[11] I. Pitas and T. Kaskalis, "Signature casting on digital images," in *Proceedings IEEE Workshop on Nonlinear Signal and Image Processing*, (Neos Marmaras), June 1995.

[12] W. Bender, D. Gruhl, and N. Morimoto, "Techniques for data hiding," in *Proc. of SPIE*, vol. 2420, p. 40, February 1995.

[13] J. Linnartz, A. Kalker, and G. Depovere, "Modelling the false-alarm and missed detection rate for electronic watermarks," in *Workshop on Information Hiding, Portland, OR*, 15-17 April, 1998.

[14] G. Depovere, T. Kalker, and J.-P. Linnartz, "Improved watermark detection using filtering before correlation," in *Proceedings of the ICIP*, (Chicago), Oct. 1998. Submitted.

[15] J. R. Smith and B. O. Comiskey, "Modulation and information hiding in images," in *Information Hiding: First Int. Workshop Proc.* (R. Anderson, ed.), vol. 1174 of *Lecture Notes in Computer Science*, pp. 207–226, Springer-Verlag, 1996.

[16] J. J. Hernandez, F. Perez-Gonzalez, J. M. Rodriguez, and G. Nieto, "Performance analysis of a 2-D multipulse amplitude modulation scheme for data hiding and watermarking still images," *IEEE Trans. on Selected Areas of Communications*, vol. 16, no. 4, pp. 510–524, 1998.

[17] F. Petitcolas, R. Anderson, and M. Kuhn, "Attacks on copyright marking systems," in *Workshop on Information Hiding, Portland, OR*, 15-17 April, 1998.

[18] I. Cox and J.-P. Linnartz, "Public watermarks and resistance to tampering," in *Proceedings of the IEEE International Conference on Image Processing, CDRom*, 1997.

[19] J. Linnartz and M. van Dijk, "Analysis of the sensitivity attack against electronic watermarks in images," in *Workshop on Information Hiding, Portland, OR*, 15-17 April, 1998.

[20] T. Kalker, J. Linnartz, and M. van Dijk, "Watermark estimation through detector analysis," in *Proceedings of the ICIP*, (Chicago), Oct. 1998.

[21] T. Kalker, "Watermark estimation through detector observations," in *Proceedings of the IEEE Benelux Signal Processing Symposium*, (Leuven, Belgium), pp. 119–122, Mar. 1998.

[22] I. J. Cox and J.-P. Linnartz, "Some general methods for tampering with watermarks," *IEEE Trans. on Selected Areas of Communications*, vol. 16, no. 4, pp. 587–593, 1998.

[23] R. Anderson, ed., *Information Hiding*, vol. 1174 of *Lecture Notes in Computer Science*, Springer-Verlag, 1996.

[24] *IEEE Int. Conf. on Image Procesing*, 1996.

[25] D. Kahn, "The history of steganography," in *Information Hiding* (R. Anderson, ed.), vol. 1174 of *Lecture Notes in Computer Science*, pp. 1–5, Springer-Verlag, 1996.

[26] A. Oomen, M. Groenewegen, R. van der Waal, and R. Veldhuis, "A variable-bit-rate buried-data channel for compact disc," in *Proc. 96th AES Convention*, 1994.

[27] R. G. van Schyndel, A. Z. Tirkel, and C. F. Osborne, "A digital watermark," in *Int. Conf. on Image Processing*, vol. 2, pp. 86–90, IEEE, 1994.

[28] J. Brassil, S. Low, N. Maxemchuk, and L. O'Gorman, "Electronic marking and identification techniques to discourage document copying," in *Proc. of Infocom'94*, pp. 1278–1287, 1994.

[29] J. Brassil and L. O'Gorman, "Watermarking document images with bounding box expansion," in *Information Hiding* (R. Anderson, ed.), vol. 1174 of *Lecture Notes in Computer Science*, pp. 227–235, Springer-Verlag, 1996.

[30] G. Caronni, "Assuring ownership rights for digital images," in *Proc. Reliable IT Systems, VIS'95*, Vieweg Publishing Company, 1995.

[31] K. Tanaka, Y. Nakamura, and K. Matsui, "Embedding secret information into a dithered multi-level image," in *Proc, 1990 IEEE Military Communications Conference*, pp. 216–220, 1990.

[32] K. Matsui and K. Tanaka, "Video-steganography," in *IMA Intellectual Property Project Proceedings*, vol. 1, pp. 187–206, 1994.

[33] E. Koch, J. Rindfrey, and J. Zhao, "Copyright protection for multimedia data," in *Proc. of the Int. Conf. on Digital Media and Electronic Publishing*, 1994.

[34] E. Koch and Z. Zhao, "Towards robust and hidden image copyright labeling," in *Proceedings of 1995 IEEE Workshop on Nonlinear Signal and Image Processing*, June 1995.

[35] A. G. Bors and I. Pitas, "Image watermarking using DCT domain constraints," in *IEEE Int. Conf. on Image Processing*, 1996.

[36] C.-T. Hsu and J.-L. Wu, "Hidden signatures in images," in *IEEE Int. Conf. on Image Processing*, 1996.

[37] M. D. Swanson, B. Zhu, and A. H. Tewfik, "Transparent robust image watermarking," in *IEEE Int. Conf. on Image Processing*, 1996.

[38] D. C. Morris, "Encoding of digital information." European Patent EP 0 690 595 A1, 1996.

[39] G. B. Rhoads, "Indentification/authentication coding method and apparatus," *World Intellectual Property Organization*, vol. IPO WO 95/14289, 1995.

[40] W. Bender, D. Gruhl, N. Morimoto, and A. Lu, "Techniques for data hiding," *IBM Systems Journal*, vol. 35, no. 3/4, pp. 313–336, 1996.

[41] O. Paatelma and R. H. Borland, "Method and apparatus for manipulating digital data works." WIPO Patent WO 95/20291, 1995.

[42] L. Holt, B. G. Maufe, and A. Wiener, "Encoded marking of a recording signal." UK Patent GB 2196167A, 1988.

[43] R. D. Preuss, S. E. Roukos, A. W. F. Huggins, H. Gish, M. A. Bergamo, P. M. Peterson, and D. A. G, "Embedded signalling." US Patent 5,319,735, 1994.

[44] C. I. Podilchuk and W. Zeng, "Image-adaptive watermarking using visual models," *IEEE Trans. on Selected Areas of Communications*, vol. 16, no. 4, pp. 525–539, 1998.

[45] J. J. K. O. Ruanaidh, W. J. Dowling, and F. Boland, "Phase watermarking of digital images," in *IEEE Int. Conf. on Image Processing*, 1996.

Chapter 18

Systolic RLS Adaptive Filtering

K. J. Ray Liu
EE Department
University of Maryland
College Park, MD, USA
kjrliu@isr.umd.edu

An-Yeu Wu
EE Department
National Central University
Chung-li, Taiwan, ROC
andywu@ee.ncu.edu.tw

18.1 INTRODUCTION

The least squares (LS) minimization problem constitutes the core of many real-time signal processing problems, such as adaptive filtering, system identification and beamforming [1]. There are two common variations of the LS problem in adaptive signal processing:

1. Solve the minimization problem

$$w(n) = \arg\min_{w(n)} \| \mathcal{B}(n)(X(n)w(n) - y(n)) \|^2, \tag{1}$$

where $X(n)$ is a matrix of size $n \times p$, $w(n)$ is a vector of length p, $y(n)$ is a vector of length n, and $\mathcal{B}(n) = \text{diag}\{\beta^{n-1}, \beta^{n-2}, \cdots, 1\}$, β is the forgetting factor and $0 < \beta < 1$.

2. Solve the minimization problem in (1) subject to the linear constraints

$$c^{iT} w(n) = r^i, i = 1, 2, \cdots, N, \tag{2}$$

where c^i is a vector of length p and r^i is a scalar. Here we consider only the special case of the MVDR (minimum variance distortionless response) beamforming problem [2] for which $y(n) = 0$ for all n, and (1) is solved by subjecting to each linear constraint; *i.e.*, there are N linear-constrained LS problems.

There are two different pieces of information that may be required as the result of this minimization [1]:

1. The optimizing weight vector $w(n)$ and/or

2. The optimal residual at time instant t_n

$$e(t_n) = X(t_n)w(n) - y(t_n), \quad (3)$$

where $X(t_n)$ is the last row of the matrix $X(n)$ and $y(t_n)$ is the last element of the vector $y(n)$.

Recently efficient implementations of the recursive least squares (RLS) algorithm and the constrained recursive least squares (CRLS) algorithm based on the QR-decomposition (QRD) have been of great interest since QRD-based approaches are numerically stable and do not require special initialization scheme [2], [3], [1]. In general, there are two major complexity issues in the VLSI implementation of the QR-decomposition based RLS algorithm (QRD-RLS), and the focus of this chapter is to explore cost-efficient ways to solve complexity problems.

- **Square root and the division operations:** Square root and the division operations, which are the major operations in conventional Givens rotation, are very cost-expensive in practical implementations. To reduce the computational load involved in the original Givens rotation, a number of square-root-free *$\Re$otations algorithms* [1] have been proposed in the literature [4], [5], [6], [7]. Recently, a parametric family of square-root-free $\Re$otation algorithms was proposed [6]. In [6], it was shown that all current known square-root-free $\Re$otation algorithms belong to a family called the *$\mu\nu$-family*. That is, all existing square-root-free $\Re$otation algorithms work in the similar way, except that the settings of the μ and ν values are different. In addition, an algorithm for computing the RLS optimal residual based on the parametric $\mu\nu$ $\Re$otation was also derived in [6].

  In the first part of this chapter, we extend the results in [6] and introduce a parametric family of square-root-free and division-free $\Re$otation algorithms. We refer to this family of algorithms as *parametric $\kappa\lambda$ $\Re$otation*. By employing the $\kappa\lambda$ $\Re$otation as well as the arguments in [2], [3], and [8], we derive novel systolic architectures of the RLS and the CRLS algorithms for both the optimal residual computation and the optimal weight vector extraction. Since the square root and division operations are eliminated, they can save the computation and circuit complexity in practical designs.

- **$O(N^2)$ complexity:** In general, the RLS algorithms do not impose any restrictions on the input data structure. As a consequence of this generality, the computational complexity is $O(N^2)$ per time iteration, where N is the size of the data matrix. This becomes the major drawback for their applications as well as for their cost-effective implementations. To alleviate the computational burden of the RLS, the family of fast RLS algorithms such as fast transversal filters, RLS lattice filters, and QR-decomposition based lattice filters (QRD-LSL), have been proposed [1]. By exploiting the special structure of the input data matrix, they can perform RLS estimation with $O(N)$ complexity. One major disadvantage of the fast RLS algorithms is that they work for data with shifting input only (*e.g.*, Toeplitz or Hankel data

[1] A Givens rotation-based algorithm that can be used as the building block of the QRD algorithm will be called a *$\Re$otation algorithm*.

matrix). However, in many applications such as multichannel adaptive array processing and image processing, the fast RLS algorithms cannot be applied because no special matrix structure can be exploited.

In the second part of the chapter, we introduce an ***approximated*** RLS algorithm based on the ***projection method*** [9], [10], [11], [12]. Through multiple decomposition of the signal space and making suitable approximations, we can perform RLS for non-structured data with only $O(N)$ complexity. Thus, both the complexity problem in the conventional RLS and the data constraint in the fast RLS can be resolved. We shall call such RLS estimation the ***split RLS***. The systolic implementation of the split RLS based on QRD-RLS systolic array in [3] is also proposed. The hardware complexity of the resulting RLS array can be reduced to $O(N)$ and the system latency is only $O(\log_2 N)$.

It is noteworthy that since approximation is made while performing the split RLS, the approximation errors will introduce misadjustment (bias) to the LS errors. Nevertheless, our analyses together with the simulation results indicate that the split RLS works well when they are applied to broad-band/less-correlated signals. Based on this observation, we propose the ***orthogonal preprocessing*** scheme to improve the performance of the split RLS. We also apply the split RLS to the multidimensional adaptive filtering (MDAF) based on the architecture in [13]. Due to the fast convergence rate of the split RLS, the split RLS performs even better than the full-size QRD-RLS in the application of real-time image restoration. This indicates that the split RLS is preferable under non-stationary environment.

The rest of this chapter is organized as follows. Section 18.2 discusses the basic square root and division free operation in Givens rotation. Then the results are applied to the RLS and CRLS systolic architectures in Section 18.3 and 18.4. Section 18.5 discusses the split RLS algorithms and architectures. The performance analysis and simulation results are discussed in Section 18.6. Finally, an improved split RLS algorithm using the orthogonal preprocessing scheme is presented in Section 18.7 followed by the conclusions.

18.2 SQUARE ROOT AND DIVISION FREE GIVENS ROTATION ALGORITHMS

In this section, we introduce a new parametric family of Givens-rotation based algorithms that require neither square root nor division operations. This modification to the Givens rotation provides a better insight on the computational complexity optimization issues of the QR decomposition and makes the VLSI implementation easier.

18.2.1 The Parametric $\kappa\lambda$ $\Re$otation

The standard Givens rotation operates (for real-valued data) as follows:

$$\begin{bmatrix} r_1' & r_2' & \cdots & r_m' \\ 0 & x_2' & \cdots & x_m' \end{bmatrix} = \begin{bmatrix} c & s \\ -s & c \end{bmatrix} \begin{bmatrix} \beta r_1 & \beta r_2 & \cdots & \beta r_m \\ x_1 & x_2 & \cdots & x_m \end{bmatrix}, \tag{4}$$

where

$$c = \frac{\beta r_1}{\sqrt{\beta^2 r_1^2 + x_1^2}}, \quad s = \frac{x_1}{\sqrt{\beta^2 r_1^2 + x_1^2}} \tag{5}$$

$$r'_1 = \sqrt{\beta^2 r_1^2 + x_1^2} \tag{6}$$

$$r'_j = c\beta r_j + s x_j, \quad j = 1, 2, \cdots, m \tag{7}$$

$$x'_j = -s\beta r_j + c x_j, \quad j = 2, 3, \cdots, m\ . \tag{8}$$

We introduce the following data transformation:

$$\begin{array}{ll} r_j = \frac{1}{\sqrt{l_a}} a_j, \quad x_j = \frac{1}{\sqrt{l_b}} b_j, \quad r'_j = \frac{1}{\sqrt{l'_a}} a'_j, & j = 1, 2, \cdots, m \\ x'_j = \frac{1}{\sqrt{l'_b}} b'_j, & j = 2, 3, \cdots, m. \end{array} \tag{9}$$

We seek the square root and division-free expressions for the transformed data $a'_j, j = 1, 2, \cdots, m,\ \ b'_j, j = 2, 3, \cdots, m$, in (6) and solving for a'_1, we get

$$a'_1 = \sqrt{\frac{l'_a}{l_a l_b}(l_b \beta^2 a_1^2 + l_a b_1^2)}. \tag{10}$$

By substituting (5) and (9) in (7) and (8) and solving for a'_j and b'_j, we get

$$a'_j = \frac{l_b \beta^2 a_1 a_j + l_a b_1 b_j}{\sqrt{l_a l_b (l_b \beta^2 a_1^2 + l_a b_1^2)/l'_a}} \quad \text{and} \quad b'_j = \frac{-b_1 \beta a_j + \beta a_1 b_j}{\sqrt{(l_b \beta^2 a_1^2 + l_a b_1^2)/l'_b}} \quad j = 2, 3, \cdots, m\ . \tag{11}$$

We will let l'_a and l'_b be equal to

$$l'_a = l_a l_b (l_b \beta^2 a_1^2 + l_a b_1^2) \kappa^2, \quad l'_b = (l_b \beta^2 a_1^2 + l_a b_1^2) \lambda^2, \tag{12}$$

where κ and λ are two parameters. By substituting (12) in (10)-(11), we obtain the following expressions

$$a'_1 = \kappa (l_b \beta^2 a_1^2 + l_a b_1^2) \tag{13}$$

$$a'_j = \kappa (l_b \beta^2 a_1 a_j + l_a b_1 b_j), \quad j = 2, 3, \cdots, m \quad \text{and} \tag{14}$$

$$b'_j = \lambda \beta (-b_1 a_j + a_1 b_j), \quad j = 2, 3, \cdots, m. \tag{15}$$

If the evaluation of the parameters κ and λ does not involve any square root or division operations, the update equations (12)-(15) will be square root and division-free. In other words, every such choice of the parameters κ and λ specifies a square root and division-free $\Re$otation algorithm. One can easily verify that the only one square root and division-free $\Re$otation in the literature to date [14] is a $\kappa\lambda$ $\Re$otation and can be obtained by choosing $\kappa = \lambda = 1$.

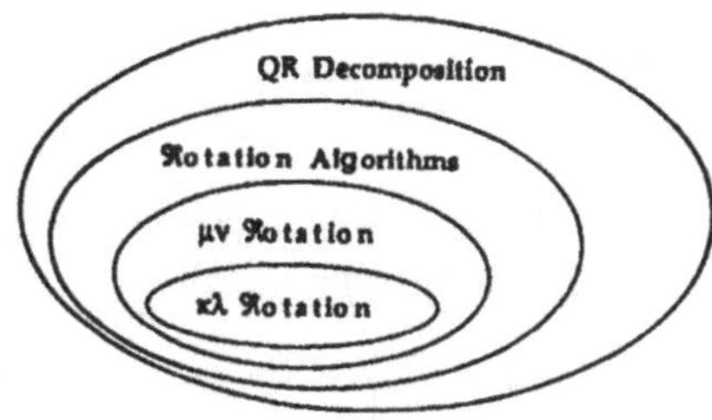

Figure 18.1 The relationship among the classes of algorithms based on QR decomposition, a ℜotation algorithm, a $\mu\nu$ ℜotation , and a $\kappa\lambda$ ℜotation.

18.2.2 Relationship between Parametric $\kappa\lambda$ and Parametric $\mu\nu$ ℜotation

Let

$$k_a = \frac{1}{l_a}, \quad k_b = \frac{1}{l_b}, \quad k'_a = \frac{1}{l'_a}, \quad k'_b = \frac{1}{l'_b}. \tag{16}$$

We can express k'_a and k'_b in terms of k_a and k_b as follows [6]:

$$k'_a = \left(k_a\beta^2 a_1^2 + k_b b_1^2\right)/\mu^2, \quad k'_b = \frac{k_a k_b}{\mu^2\nu^2}\frac{1}{k'_a}. \tag{17}$$

If we substitute (16) and (17) in (12) and solve for μ and ν, we obtain

$$\mu = \frac{\kappa(k_a\beta^2 a_1^2 + k_b b_1^2)}{k_a k_b}, \quad \nu = \lambda. \tag{18}$$

Consequently, the set of $\kappa\lambda$ ℜotation algorithms can be thought of as a subset of the set of the $\mu\nu$ ℜotations. Furthermore, (18) provides a means of mapping a $\kappa\lambda$ ℜotation onto a $\mu\nu$ ℜotation. For example, one can verify that the square root and division-free algorithm in [14] is a $\mu\nu$ ℜotation and is obtained for

$$\mu = \frac{k_a\beta^2 a_1^2 + k_b b_1^2}{k_a k_b}, \quad \nu = 1.$$

In Fig. 18.1, we draw a graph that summarizes the relationship among the classes of algorithms based on QR decomposition, a ℜotation algorithm, a $\mu\nu$ ℜotation, and a $\kappa\lambda$ ℜotation.

18.3 SQUARE ROOT AND DIVISION FREE RLS ALGORITHMS AND ARCHITECTURES

In this section, we consider the $\kappa\lambda$ ℜotation for optimal residual and weight extraction using systolic array implementation. Detailed comparisons with existing approaches are presented.

18.3.1 Algorithm for the RLS Optimal Residual Computation

The QR-decomposition of the data at time instant n is as follows:

$$\begin{bmatrix} R(n) & u(n) \\ 0^T & v(t_n) \end{bmatrix} = T(n) \begin{bmatrix} \beta R(n-1) & \beta u(n-1) \\ X(t_n) & y(t_n) \end{bmatrix}, \tag{19}$$

where $T(n)$ is a unitary matrix of size $(p+1) \times (p+1)$ that performs a sequence of p Givens rotations. This can be written symbolically as

$$\begin{bmatrix} L(n-1)^{-\frac{1}{2}} & 0 \\ 0^T & l_q(n)^{-\frac{1}{2}} \end{bmatrix} \begin{bmatrix} \beta \bar{R}(n-1) & \beta \bar{u}(n-1) \\ \bar{X}(t_n) & \bar{y}(t_n) \end{bmatrix}$$

$$\xrightarrow{T(n)} \begin{bmatrix} L(n)^{-\frac{1}{2}} & 0 \\ 0^T & l_q(n+1)^{-\frac{1}{2}} \end{bmatrix} \begin{bmatrix} \bar{R}(n) & \bar{u}(n) \\ 0^T & b_{p+1}^{(p)} \end{bmatrix}, \tag{20}$$

where

$$\begin{array}{ll} L(\cdot)^{-\frac{1}{2}} \bar{R}(\cdot) = R(\cdot), & L(\cdot)^{-\frac{1}{2}} \bar{u}(\cdot) = u(\cdot), \\ l_q(n)^{-\frac{1}{2}} \bar{X}(t_n) = X(t_n), & l_q(n)^{-\frac{1}{2}} \bar{y}(t_n) = y(t_n) \end{array} \tag{21}$$

and

$$L(n-1) = \text{diag}\{l_1, l_2, \cdots, l_p\} \qquad L(n) = \text{diag}\{l'_1, l'_2, \cdots, l'_p\},$$

$$\bar{R}(n-1) = \begin{bmatrix} a_{11} & a_{12} & \cdots & a_{1p} \\ & a_{22} & \cdots & a_{2p} \\ & & \ddots & \vdots \\ & & & a_{pp} \end{bmatrix} \qquad \bar{R}(n) = \begin{bmatrix} a'_{11} & a'_{12} & \cdots & a'_{1p} \\ & a'_{22} & \cdots & a'_{2p} \\ & & \ddots & \vdots \\ & & & a'_{pp} \end{bmatrix},$$

$$\bar{u}(n-1) = \left[a_{1,p+1}\ a_{2,p+1}\ \cdots\ a_{p,p+1}\right]^T \quad \bar{u}(n) = \left[a'_{1,p+1}\ a'_{2,p+1}\ \cdots\ a'_{p,p+1}\right]^T,$$
$$\left[\bar{X}(t_n)\ \ \bar{y}(t_n)\right] = \left[b_1\ b_2\ \cdots\ b_p\ b_{p+1}\right].$$

Equations (12)-(15) imply that the ith $\Re$otation is specified as follows:

$$l'_i = l_i l_q^{(i-1)} (l_q^{(i-1)} \beta^2 a_{ii}^2 + l_i b_i^{(i-1)^2}) \kappa_i^2 \tag{22}$$

$$l_q^{(i)} = (l_q^{(i-1)} \beta^2 a_{ii}^2 + l_i b_i^{(i-1)^2}) \lambda_i^2, \tag{23}$$

$$a'_{ij} = \kappa_i (l_q^{(i-1)} \beta^2 a_{ii} a_{ij} + l_i b_i^{(i-1)} b_j^{(i-1)}), \quad j = i, i+1, \cdots, p+1 \tag{24}$$

$$b_j^{(i)} = \lambda_i \beta (-b_i^{(i-1)} a_{ij} + a_{ii} b_j^{(i-1)}), \quad j = i+1, i+2, \cdots, p+1, \tag{25}$$

where $i = 1, 2, \cdots, p$, $b_j^{(0)} = b_j$, $j = 1, \cdots, p+1$ and $l_q^{(0)} = l_q$. If the parametric $\kappa\lambda$ $\Re$otation is used in the QRD-RLS algorithm, the optimal residual can be derived as (see Appendix)

$$e_{RLS}(t_n) = -\left(\prod_{i=1}^{p-1} \lambda_i \beta a_{ii}\right) \frac{\kappa_p \beta a_{pp}}{\lambda_p a'_{pp}} b_{p+1}^{(p)} \sqrt{l_q}. \tag{26}$$

Here, l_q is a free variable. If we choose $l_q = 1$, we can avoid the square root operation. We can see that for a recursive computation of (26) only one division operation is needed at the last step of the recursion. This compares very favorably with *the square root free fast algorithms* that require one division for every recursion step, as well as with the original approach, which involves one division and one square root operation for every recursion step.

Note that the division operation in (26) cannot be avoided by proper choice of expressions for the parameters κ and λ. Hence, if a $\kappa\lambda$ $\mathfrak{R}$otation is used, the RLS optimal residual evaluation will require at least one division evaluation.

18.3.2 Systolic Architecture for Optimal RLS Residual Evaluation

McWhirter proposed a systolic architecture for the implementation of the QRD-RLS [3]. We modified the architecture in [3] so that equations (22)-(26) can be evaluated for the special case of $\kappa_i = \lambda_i = 1, i = 1, 2, \cdots, p$ and $l_q = 1$. The systolic array, as well as the memory and the communication links of its components, are depicted in Fig. 18.2 [2]. The boundary cells (cell number 1) are responsible for evaluating (22) and (23), as well as the coefficients $\bar{c}_i = l_q^{(i-1)} a_{ii}$ and $\bar{s}_i = l_i b_i^{(i-1)}$ and the partial products $e_i = \prod_{j=1}^{i} (\beta a_{jj})$. The internal cells (cell number 2) are responsible for evaluating (24) and (25). Finally, the output cell (cell number 3) evaluates (26). The functionality of each one of the cells is described in Fig. 18.2. We will call this systolic array $S1.1$.

On Table 18.1, we collect some features of the systolic structure $S1.1$ and the two structures, $S1.2$ and $S1.3$ in [3], that are pertinent to the circuit complexity. The $S1.2$ implements the square-root-free QRD-RLS algorithm with $\mu = \nu = 1$, while $S1.3$ is the systolic implementation based on the original Givens rotation. In Table 18.1, the complexity per processor cell and the number of required processor cells are indicated for each one of the three different cells [3]. One can easily observe that $S1.1$ requires only one division operator and no square root operator, $S1.2$ requires p division operators and no square root operator, while $S1.3$ requires p division and p square root operators. This reduction of the complexity in terms of division and square root operators is penalized with the increase of the number of the multiplications and the communication links that are required.

Apart from the circuit complexity that is involved in the implementation of the systolic structures, another feature of the computational complexity is the number of *operations-per-cycle*. This number determines the minimum required delay between two consecutive sets of input data. For the structures $S1.2$ and $S1.3$ the boundary cell (cell number 1) constitutes the bottleneck of the computation and therefore it determines the operations-per-cycle that are shown on Table 18.5. For the structure $S1.1$ either the boundary cell or the output cell are the bottleneck of the computation.

18.3.3 Systolic Architecture for Optimal RLS Weight Extraction

Shepherd *et al.* [8] and Tang *et al.* [15] have independently shown that the optimal weight vector can be evaluated in a recursive way. More specifically, one

[2]Note the aliases: $l_q^{(i-1)} \equiv \sigma_{in}, l_q^{(i)} \equiv \sigma_{out}, l_i \equiv l, a_{ij} \equiv r, b_j^{(i-1)} \equiv b_{in}, b_j^{(i)} \equiv b_{out}, e_{i-1} \equiv e_{in}, e_i \equiv e_{out}$.

[3]The multiplications with the constants β and β^2 are not encountered.

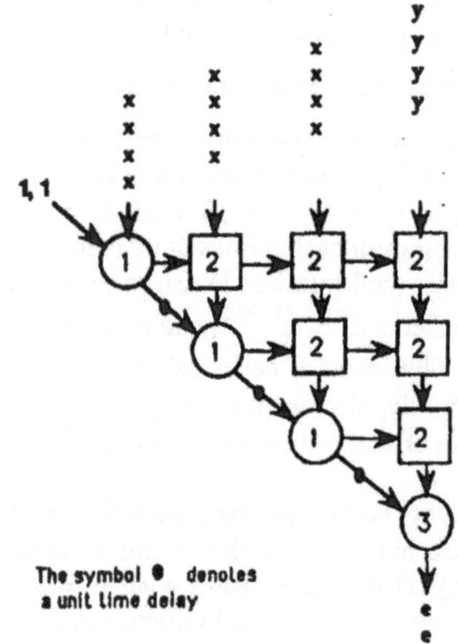

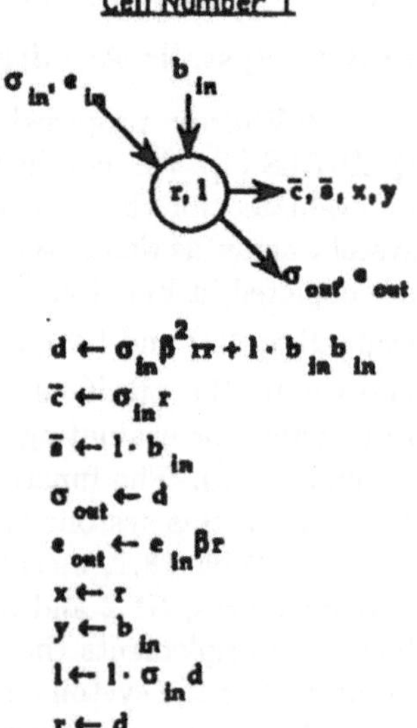

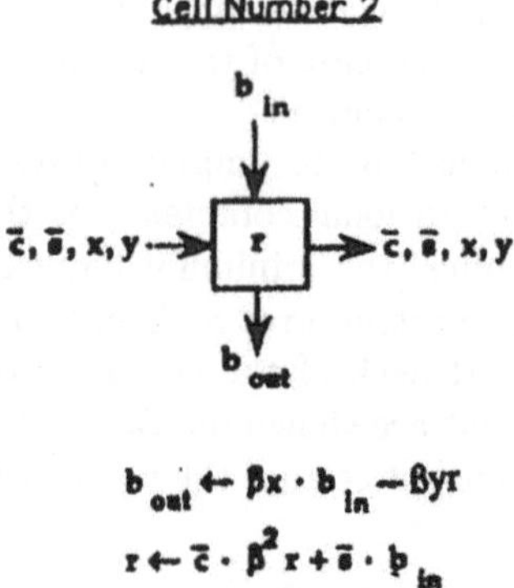

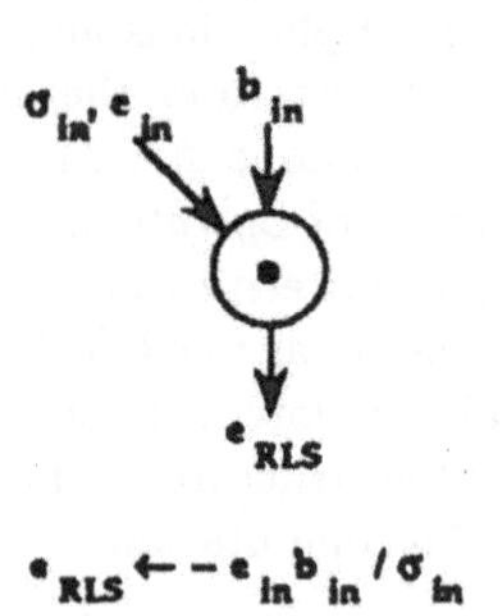

Figure 18.2 $S1.1$: systolic array that computes the RLS optimal residual. It implements the algorithm that is based on the $\kappa\lambda$ $\Re$otation for which $\kappa = \lambda = 1$.

Table 18.1 Computational Complexity for Computing the RLS Residual

	$S1.1:\kappa\lambda$			$S1.2:\mu\nu$			$S1.3$: Givens rotation		
cell type	1	2	3	1	2	3	1	2	3
number of	p	$\frac{p(p+1)}{2}$	1	p	$\frac{p(p+1)}{2}$	1	p	$\frac{p(p+1)}{2}$	1
sq.rt	-	-	-	-	-	-	1	-	-
div.	-	-	1	1	-	-	1	-	-
mult.	9	4	1	5	3	1	4	4	1
i/o	9	10	4	6	8	3	5	6	3

can compute recursively the term $R^{-T}(n)$ by

$$\begin{bmatrix} R^{-T}(n) \\ \# \end{bmatrix} = T(n) \begin{bmatrix} \frac{1}{\beta}R^{-T}(n-1) \\ 0^T \end{bmatrix} \tag{27}$$

and then use parallel multiplication for computing $w^T(n)$ by

$$w^T(n) = u^T(n)R^{-T}(n). \tag{28}$$

The symbol # denotes a term of no interest. The above algorithm can be implemented by a fully pipelined systolic array that can operate in two distinct modes, 0 and 1. The initialization phase consists of $2p$ steps for each processor. During the first p steps the processors operate in mode 0 in order to calculate a full rank matrix R. During the following p steps, the processors operate in mode 1 in order to compute R^{-T}, by performing a task equivalent to forward substitution. After the initialization phase the processors operate in mode 0. In [8] one can find the systolic array implementations based both on the original Givens rotation and the Gentleman's variation of the square-root-free $\Re$otation, that is, the $\mu\nu$ $\Re$otation for $\mu = \nu = 1$. We will call these two structures $S2.3$ and $S2.2$, respectively.

In Fig. 18.3, we present the systolic structure $S2.1$ based on the $\kappa\lambda$ $\Re$otation with $\kappa_i = \lambda_i = 1, i = 1, 2, \cdots, p$. This is a square-root-free and division-free implementation. The boundary cells (cell number 1) are slightly simpler than the corresponding ones of the array $S1.1$. More specifically, they do not compute the partial products e_i. The internal cells (cell number 2), that compute the elements of the matrix R, are identical to the corresponding ones of the array $S1.1$. The cells that are responsible for computing the vector u (cell number 3) differ from the other internal cells only in the fact that they communicate their memory value with their right neighbors. The latter (cell number 4) are responsible for evaluating (28) and (27). The functionality of the processing cells, as well as their communication links and their memory contents, are given in Fig. 18.3. The mode of operation of each cell is controlled by the *mode bit* provided from the input. For a more detailed description of the operation of the mode bit one can see [2] and [8].

On Tables 18.2 and 18.5, we collect some computational complexity metrics for the systolic arrays $S2.1$, $S2.2$ and $S2.3$, when they operate in mode 0[4]. The

[4]The multiplications with the constants $\beta, \beta^2, 1/\beta$ and $1/\beta^2$, as well as the communication links that drive the mode bit, are not encountered.

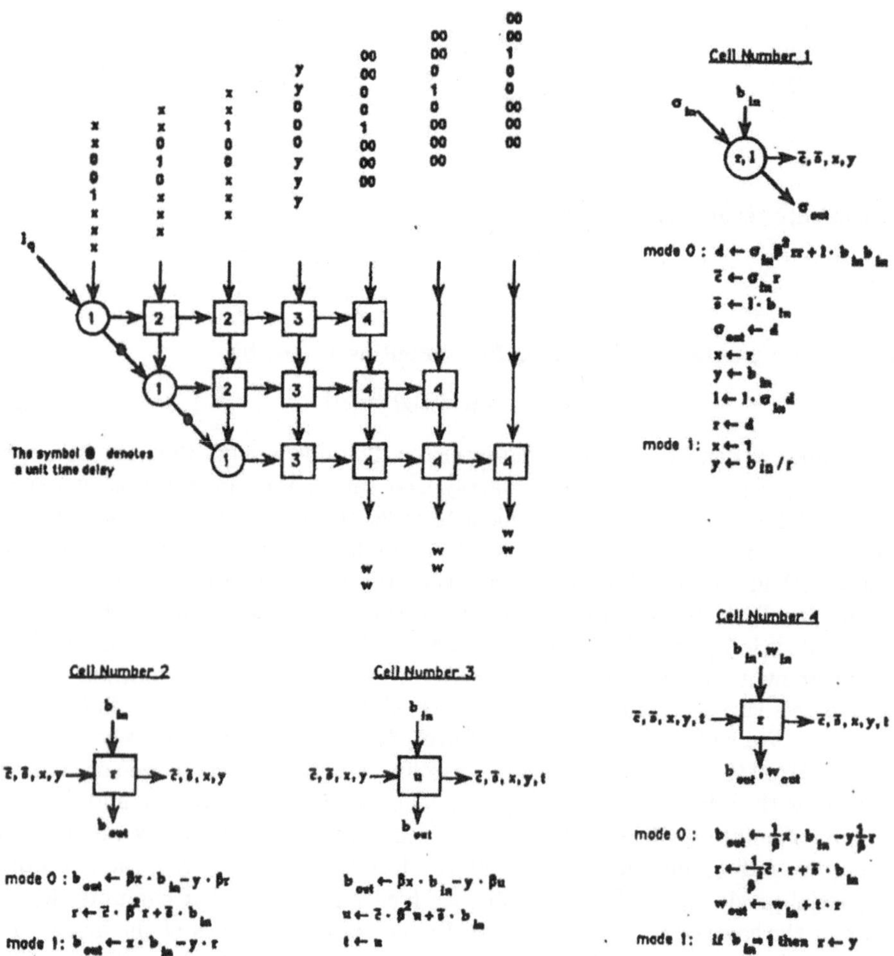

Figure 18.3 $S2.1$: systolic array that computes the RLS optimal weight vector. It implements the algorithm that is based on the $\kappa\lambda$ $\Re$otation for which $\kappa = \lambda = 1$.

conclusions we can draw are similar to the ones we had for the circuits that calculate the optimal residual: the square root operations and the division operations can be eliminated with the cost of an increased number of multiplication o[illegible]ons and communication links. We should also note that $S2.1$ does require t[illegible]lementation of division operators in the boundary cells, since these operators are used during the initialization phase. Nevertheless, after the initialization phase the circuit will not suffer from any time delay caused by division operations. The computational bottleneck of all three structures, $S2.1$, $S2.2$ and $S2.3$, is the boundary cell, thus it determines the operations-per-cycle metric.

Table 18.2 RLS Weight Extraction Computational Complexity (mode 0)

	$S2.1 : \kappa\lambda$				$S2.2 : \mu\nu$			
cell type	1	2	3	4	[illegible]	2	3	4
number of	p	$\frac{(p-1)p}{2}$	p	$\frac{p(p+1)}{2}$	p	$\frac{(p-1)p}{2}$	p	$\frac{p(p+1)}{2}$
sq.rt	-	-	-	-	-	-	-	-
div.	-	-	-	-	1	-	-	-
mult.	8	4	4	5	5	3	3	4
i/o	7	10	11	14	6	8	9	12

	$S2.3$: Givens rotation			
cell type	1	2	3	4
number of	p	$\frac{(p-1)p}{2}$	p	$\frac{p(p+1)}{2}$
sq.rt	1	-	-	-
div.	1	-	-	-
mult.	4	4	4	5
i/o	3	6	7	10

As a conclusion for the RLS architectures, we observe that the figures on Tables 18.1, 18.2 and 18.5 favor the architectures based on the $\kappa\lambda$ $\Re$otation, $\kappa = \lambda = 1$ versus the ones that are based on the $\mu\nu$ rotation with $\mu = \nu = 1$ and the standard Givens rotation. This claim is clearly substantiated by the delay times on Table 18.5, associated to the DSP implementation of the QRD-RLS algorithm. These delay times are calculated on the basis of the manufacturers benchmark speeds for floating point operations [16]. Due to the way of updating R^{-1}, such a weight extraction scheme will have a numerical stability problem if the weight vector at each time instant is required.

18.4 SQUARE ROOT AND DIVISION FREE CRLS ALGORITHMS AND ARCHITECTURES

The optimal weight vector $w^i(n)$ and the optimal residual $e^i_{CRLS}(t_n)$ that correspond to the ith constraint vector c^i are given by the expressions [2]:

$$w^i(n) = \frac{r^i}{\| z^i(n) \|^2} R^{-1}(n) z^i(n) \tag{29}$$

and

$$e^i_{CRLS}(t_n) = \frac{r^i}{\| z^i(n) \|^2} \hat{e}^i_{CRLS}(t_n), \tag{30}$$

where

$$\hat{e}^i_{CRLS}(t_n) = X(t_n)R^{-1}(n)z^i(n). \tag{31}$$

The term $z^i(n)$ is defined as follows

$$z^i(n) = R^{-T}(n)c^i \tag{32}$$

and it is computed with the recursion [2]

$$\begin{bmatrix} z^i(n) \\ \# \end{bmatrix} = T(n) \begin{bmatrix} \frac{1}{\beta} z^i(n-1) \\ 0^T \end{bmatrix}, \tag{33}$$

where the symbol # denotes a term of no interest. In this section, we derive a variation of the recursion that is based on the parametric $\kappa\lambda$ $\Re$otation. Then, we design the systolic arrays that implement this recursion for $\kappa = \lambda = 1$. We also make a comparison of these systolic structures with those based on the Givens rotation and the $\mu\nu$ $\Re$otation introduced by Gentleman [1], [2], [4], [8].

From (32) and (21), we have $z^i(n) = \left(L(n)^{-1/2}\bar{R}(n)\right)^{-T} c^i$, and since $L(n)$ is a diagonal real valued matrix, we get $z^i(n) = L(n)^{1/2}\bar{R}(n)^{-T}c^i$, where c^i is the constraint direction. If we let

$$\bar{z}^i(n) = L(n)\bar{R}(n)^{-T}c^i \tag{34}$$

we obtain

$$z^i(n) = L(n)^{-1/2}\bar{z}^i(n). \tag{35}$$

From (35) we get $\| z^i(n) \|^2 = \bar{z}^{i^T}(n)L^{-1}(n)\bar{z}^i(n)$. Also, from (21) and (35) we get $R^{-1}(n)z^i(n) = \bar{R}^{-1}(n)\bar{z}^i(n)$. Consequently, from (29) (30), and (31), we have

$$e^i_{CRLS}(n) = \frac{r^i}{\bar{z}^{i^T}(n)L^{-1}(n)\bar{z}^i(n)} \hat{e}^i_{CRLS}(n) \tag{36}$$

and

$$w^i(n) = \frac{r^i}{\bar{z}^{i^T}(n)L^{-1}(n)\bar{z}^i(n)} \bar{R}^{-1}(n)\bar{z}^i(n), \tag{37}$$

where

$$\hat{e}^i_{CRLS}(n) = X(n)\bar{R}^{-1}(n)\bar{z}^i(n). \tag{38}$$

Because of the similarity of (31) with (38) and (29) with (37) we are able to use a variation of the systolic arrays that are based on the Givens rotation [2], [8] in order to evaluate (36)-(37).

18.4.1 Systolic Architecture for Optimal CRLS Residual Evaluation

From (26) and (36), if $l_q = 1$, we get the optimal residual

$$e^i_{CRLS}(n) = -\frac{r^i}{\bar{z}^{i^T}(n)L^{-1}(n)\bar{z}^i(n)} \left(\prod_{j=1}^{p-1} \lambda_j a_{jj} \right) \frac{\kappa_p a_{pp}}{\lambda_p a'_{pp}} b^{(p)}_{p+1}. \tag{39}$$

In Fig. 18.4, we present the systolic array $S3.1$, that evaluates the optimal residual for $\kappa_j = \lambda_j = 1, j = 1, 2, \cdots, p$, and the number of constraints is $N = 2$. This systolic array is based on the design proposed by McWhirter [2]. It operates in two modes and is in a way very similar to the operation of the systolic structure $S2.1$ (see Section 18.3). The recursive equations for the data of the matrix $\bar{R}$ are given in (22)-(25). They are evaluated by the boundary cells (cell number 1) and the internal cells (cell number 2). These internal cells are identical to the ones of the array $S2.1$. The boundary cells have a very important difference from the corresponding ones of $S2.1$: while they operate in mode 0, they make use of their division operators in order to evaluate the elements of the diagonal matrix $L^{-1}(n)$, i.e. the quantities $1/l_i, i = 1, 2, \cdots, p$. These quantities are needed for the evaluation of the term $\bar{z}^{i^T}(n)L^{-1}(n)\bar{z}^i(n)$ in (39). The elements of the vectors $\bar{z}^1$ and $\bar{z}^2$ are updated by a variation of (24) and (25), for which the constant β is replaced by $1/\beta$. The two columns of the internal cells (cell number 3) are responsible for these computations. They initialize their memory value during the second phase of the initialization (mode 1) according to (34). While they operate in mode 0, they are responsible for evaluating the partial sums

$$\eta_k = \sum_{j=1}^{k} \parallel \bar{z}_j^i \parallel^2 /l_j. \tag{40}$$

The output cells (cell number 4) are responsible for the final evaluation of the residual[5].

Table 18.3 CRLS Optimal Residual Computational Complexity (mode 0)

	$S3.1 : \kappa\lambda$				$S3.2 : \mu\nu$				$S3.3$: Givens rotation			
cell type	1	2	3	4	1	2	3	4	1	2	3	4
number of	p	$\frac{(p-1)p}{2}$	Np	N	p	$\frac{(p-1)p}{2}$	Np	N	p	$\frac{(p-1)p}{2}$	Np	N
sq.rt	-	-	-	-	-	-	-	-	1	-	-	-
div.	1	-	-	1	1	-	-	1	1	-	-	1
mult.	9	4	6	3	6	3	5	2	5	4	5	2
i/o	10	12	14	7	7	10	12	5	5	6	8	5

McWhirter has designed the systolic arrays that evaluate the optimal residual, based on either the Givens rotation or the square-root-free variation that was introduced by Gentleman [2],[4]. We will call these systolic arrays $S3.3$ and $S3.2$ respectively. In Tables 18.3 and 18.5 we collect some computational complexity metrics for the systolic arrays $S3.1$, $S3.2$ and $S3.3$, when they operate in mode 0 [6]. We observe that the $\mu\nu$ $\Re$otation-based $S3.2$, outperforms the $\kappa\lambda$ $\Re$otation-based $S3.1$. The two structures require the same number of division operators, while $S3.2$ needs less multipliers and has less communication overhead.

[5]Note the alias $r^i \equiv r$.

[6]The multiplications with the constants $\beta, \beta^2, 1/\beta$ and $1/\beta^2$, as well as the communication links that drive the mode bit, are not encountered.

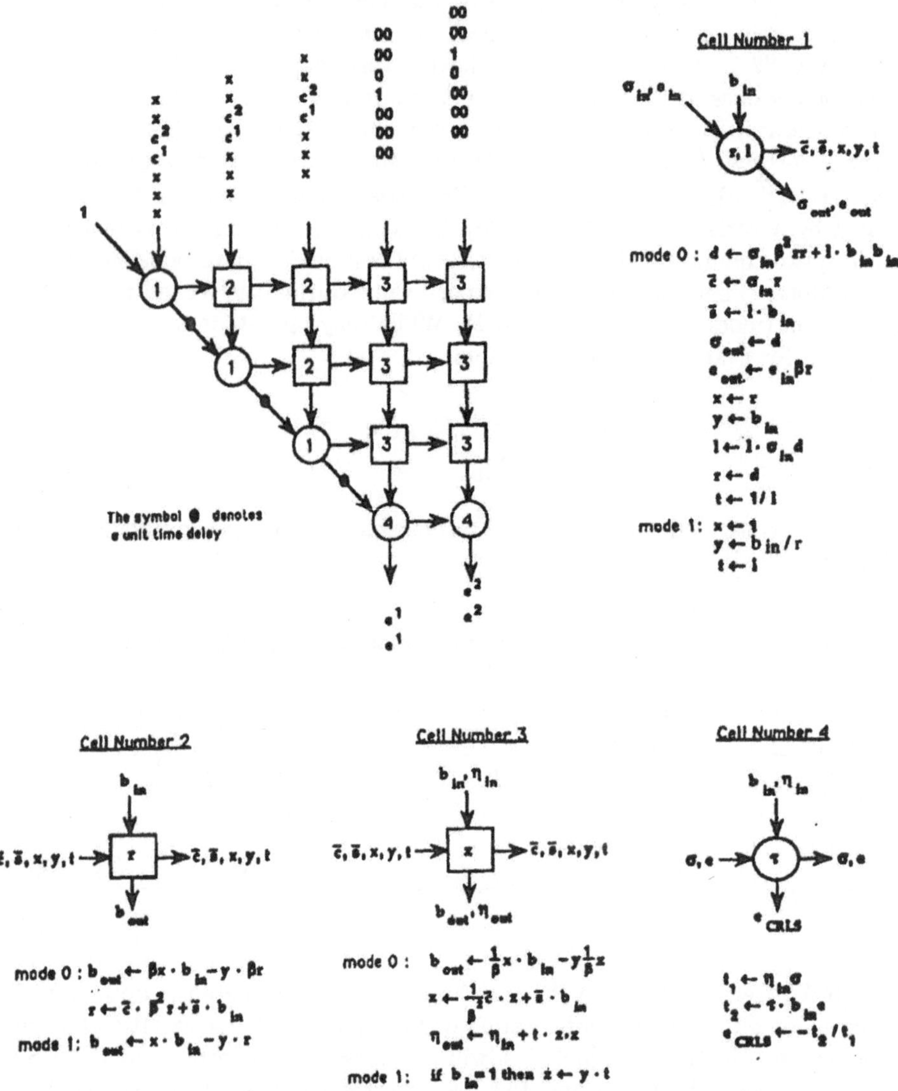

Figure 18.4 *S*3.1 : systolic array that computes the CRLS optimal residual. It implements the algorithm that is based on the $\kappa\lambda$ $\Re$otations for which $\kappa = \lambda = 1$.

18.4.2 Systolic Architecture for Optimal CRLS Weight Extraction

In Fig. 18.5, we present the systolic array that evaluates (37) for $\kappa_j = \lambda_j = 1, j = 1, 2, \cdots, p$ and the number of constraints equal to $N = 2$. This systolic array operates in two modes, just as the arrays $S2.1$ and $S3.1$ do. The boundary cell (cell number 1) is responsible for evaluating the diagonal elements of the matrices R and L, the variable l_q, as well as all the coefficients that will be needed in the computations of the internal cells. In mode 0 its operation is almost identical to the operation of the boundary cell in $S2.1$ (except for t), while in mode 1 it behaves like the corresponding cell of $S3.1$. The internal cells in the left triangular part of the systolic structure (cell number 2) evaluate the non-diagonal elements of the matrix R and they are identical to the corresponding cells of $S3.1$. The remaining part of the systolic structure is a 2-layer array. The cells in the first column of each layer (cell number 3) are responsible for the calculation of the vector z^i and the partial summations (40). They also communicate their memory values to their right neighbors. The latter (cell number 4) evaluate the elements of the matrix R^{-T} and they are identical to the corresponding elements of $S2.1$. The output elements (cell number 5) are responsible for the normalization of the weight vectors and they compute the final result.

Table 18.4 CRLS Weight Vector Extraction Computational Complexity (mode 0)

	$S4.1 : \kappa\lambda$					$S4.2 : \mu\nu$				
cell type	1	2	3	4	5	1	2	3	4	5
number of	p	$\frac{(p-1)p}{2}$	Np	$\frac{Np(p+1)}{2}$	Np	p	$\frac{(p-1)p}{2}$	Np	$\frac{Np(p+1)}{2}$	Np
sq.rt	-	-	-	-	-	-	-	-	-	-
div.	1	-	-	-	1	1	-	-	-	1
mult.	8	4	6	5	1	5	3	5	4	-
i/o	8	12	19	14	4	6	8	14	10	4

	$S4.3$: Givens rotation				
cell type	1	2	3	4	5
number of	p	$\frac{(p-1)p}{2}$	Np	$\frac{Np(p+1)}{2}$	Np
sq.rt	1	-	-	-	-
div.	1	-	-	-	1
mult.	4	4	5	5	-
i/o	4	8	13	10	4

Shepherd *et al.* [8] and Tang *et al.* [15] have designed systolic structures for the weight vector extraction based on the Givens rotation and the square-root-free $\Re$otation of Gentleman [4]. We will call these two arrays $S4.3$ and $S4.2$, respectively. On Tables 18.4 and 18.5, we show the computational complexity metrics for the systolic arrays $S4.1$, $S4.2$ and $S4.3$, when they operate in mode 0. The observations we make are similar to the ones we have for the systolic arrays that evaluate the RLS weight vector (see Section 18.3).

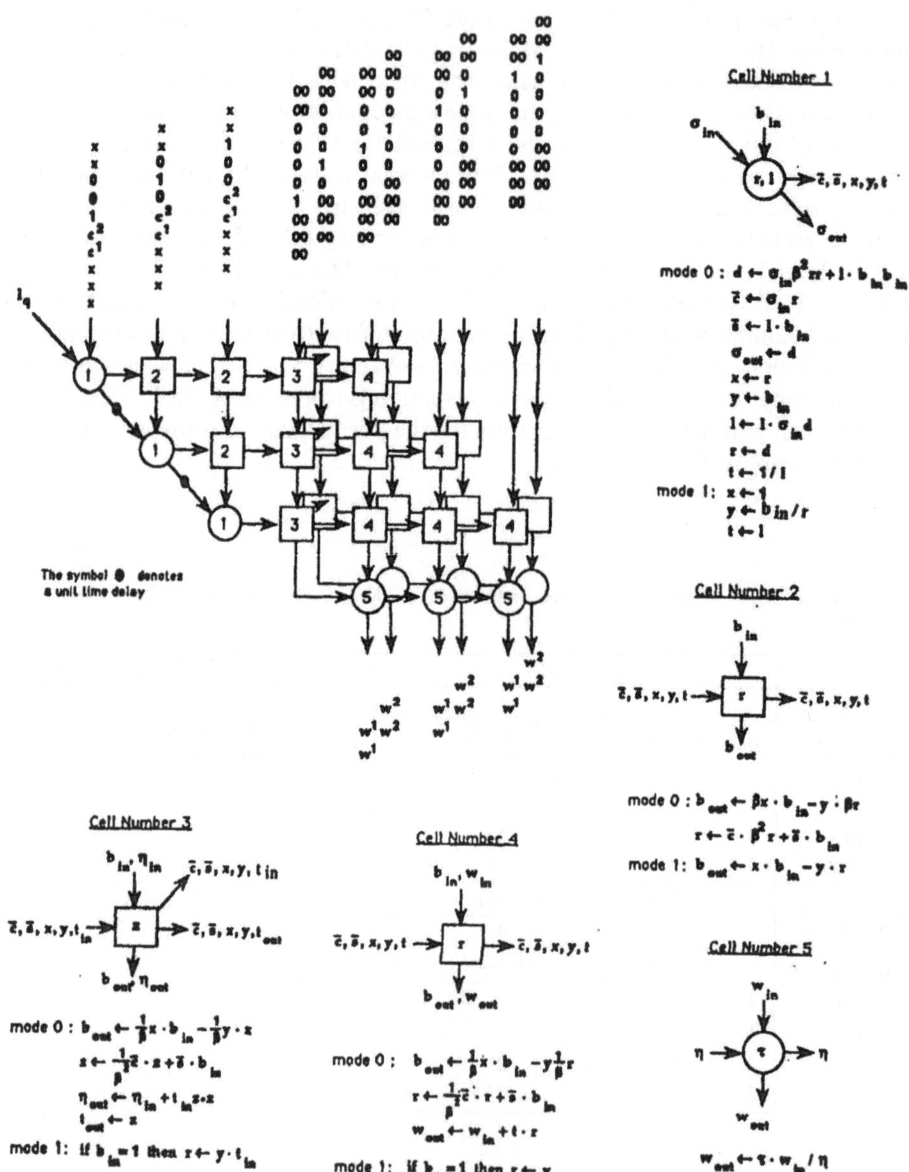

Figure 18.5 *S*4.1 : systolic array that computes the CRLS optimal weight vector. It implements the algorithm that is based on the $\kappa\lambda$ $\Re$otation for which $\kappa = \lambda = 1$.

Note that each part of the 2-layer structure computes the terms relevant to one of the two constraints. In the same way, a problem with N constraints will require an N-layer structure. With this arrangement of the multiple layers we obtain a unit time delay between the evaluation of the weight vectors for the different constraints. The price we have to pay is the global wiring for some of the communication links of cell 3. A different approach can also be considered: we may place the multiple layers side by side, one on the right of the other. In this way, not only the global wiring will be avoided, but also the number of communication links of cell 3, will be considerably reduced. The price we will pay with this approach is a time delay of p units between consequent evaluations of the weight vectors for different constraints.

As a conclusion for the CRLS architectures, we observe that the figures on Tables 18.3, 18.4 and 18.5 favor the architectures based on the $\mu\nu$ $\Re$otation, $\mu = \nu = 1$ versus the ones that are based on the $\kappa\lambda$ rotation with $\kappa = \lambda = 1$.

Table 18.5 Minimum Required Delay Between Two Consequent Sets of Input Data

	operations per cycl	DSP 96000 (ns)	IMS T800 (ns)	WEITEK 3164 (ns)	ADSP 3201/2 (ns)
S1.1	max{1 div. + 1 mult. , 9 mult. }	900	3150	1800	2700
S1.2	1 div. + 5 mult.	1020	2300	2700	3675
S1.3	1 sq.rt. + 1 div. + 4 mult.	1810	4500	5300	7175
S2.1	8 mult.	800	2800	1600	2400
S2.2	1 div. + 5 mult.	1020	2300	2700	3675
S2.3	1 sq.rt. + 1 div. + 4 mult.	1810	4500	5300	7175
S3.1	1 div. + 9 mult.	1420	3700	3500	4875
S3.2	1 div. + 6 mult.	1120	2650	2900	3975
S3.3	1 sq.rt. + 1 div. + 5 mult.	1810	4500	5300	7175
S4.1	1 div. + 8 mult.	1320	3350	3300	4575
S4.2	1 div. + 5 mult.	1020	2300	2700	3675
S4.3	1 sq.rt. + 1 div. + 4 mult.	1810	4500	5300	7175

18.5 SPLIT RLS ALGORITHM AND ARCHITECTURE

In the second part of the chapter, we introduce an approximated RLS algorithm, the *split RLS*, that can perform RLS with only $O(N)$ complexity for nonstructured data. We start with the projection method. Then based on the interpretation of the projection method, the family of split RLS algorithms and systolic architectures are derived.

18.5.1 The Projection Method

Given an observation data matrix $\mathbf{A} = [\mathbf{a}_1, \mathbf{a}_2, \cdots, \mathbf{a}_n] \in \mathcal{R}^{m \times n}$ without any exhibited structure and the desired signal vector $\mathbf{y} \in \mathcal{R}^{m \times 1}$, the LS problem is to find the optimal weight coefficients $\hat{\mathbf{w}}$ which minimize the LS errors

$$||\mathbf{e}||^2 = ||\mathbf{A}\mathbf{w} - \mathbf{y}||^2. \tag{41}$$

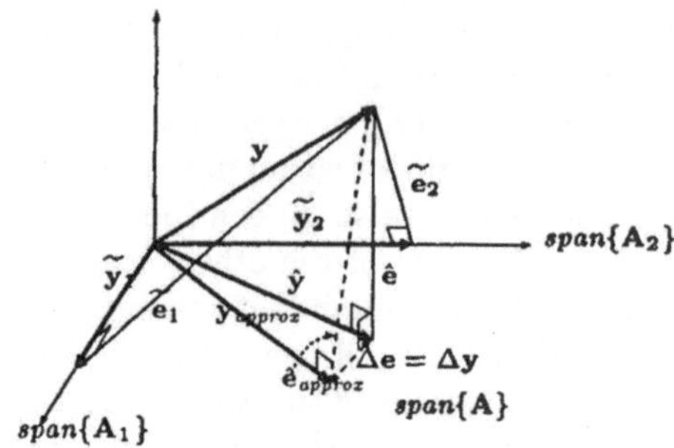

Figure 18.6 Geometric interpretation of the projection method.

In general, $\hat{\mathbf{w}}$ is of the form

$$\hat{\mathbf{w}} = (\mathbf{A}^T\mathbf{A})^{-1}\mathbf{A}^T\mathbf{y}. \tag{42}$$

We also have $\hat{\mathbf{y}} = \mathbf{A}\hat{\mathbf{w}} = \mathbf{P}\mathbf{y}$ and $\hat{\mathbf{e}} = \mathbf{y} - \hat{\mathbf{y}}$, where $\hat{\mathbf{y}}$ is the optimal projection of $\mathbf{y}$ on the column space of $\mathbf{A}$, $\mathbf{P} = \mathbf{A}(\mathbf{A}^T\mathbf{A})^{-1}\mathbf{A}^T$ is the projection matrix, and $\hat{\mathbf{e}}$ is the optimal residual vector. The ***principle of orthogonality*** ensures that $\hat{\mathbf{e}}$ is orthogonal to the column space of $\mathbf{A}$. For RLS algorithms that calculate exact LS solution, such a direct projection to the N-dimensional space takes $O(N^2)$ complexity. Knowing this, in order to reduce the complexity, we shall try to perform projection onto spaces of smaller dimension.

To motivate the idea, let us consider the LS problem with the partition $\mathbf{A} = [\mathbf{A}_1, \mathbf{A}_2]$, where $\mathbf{A}_1, \mathbf{A}_2 \in \mathcal{R}^{n\times(m/2)}$. Now instead of projecting $\mathbf{y}$ directly onto the space spanned by $\mathbf{A}$ (denoted as $span\{\mathbf{A}\}$), we project $\mathbf{y}$ onto the two smaller subspaces, $span\{\mathbf{A}_1\}$ and $span\{\mathbf{A}_2\}$, and obtain the optimal projections $\tilde{\mathbf{y}}_1$ and $\tilde{\mathbf{y}}_2$ on each subspace (see Fig. 18.6). The next step is to find a "good" estimation of the optimal projection $\hat{\mathbf{y}}$, say $\hat{\mathbf{y}}_{approx}$. If we can estimate a 1-D or 2-D subspace from $\tilde{\mathbf{y}}_1$ and $\tilde{\mathbf{y}}_2$ and project the desired signal $\mathbf{y}$ directly on it to obtain $\hat{\mathbf{y}}_{approx}$, the projection spaces become smaller and the computational complexity is reduced as well. In the following, we propose two estimation methods based on their geometric relationship in the Hilbert space.

18.5.2 Estimation Method I (Split RLS I)

The first approach is simply to add the two subspace projections $\tilde{\mathbf{y}}_1$ and $\tilde{\mathbf{y}}_2$ together, *i.e.*,

$$\hat{\mathbf{y}}_{approx} = \tilde{\mathbf{y}}_1 + \tilde{\mathbf{y}}_2. \tag{43}$$

This provides the most intuitive and simplest way to estimate $\hat{\mathbf{y}}_{approx}$. We will show later that as $\tilde{\mathbf{y}}_1$ and $\tilde{\mathbf{y}}_2$ are more orthogonal to each other, $\hat{\mathbf{y}}_{approx}$ will approach to the optimal projection vector $\hat{\mathbf{y}}$. Let Fig.18.7(a) represent one of the existing RLS algorithms that project $\mathbf{y}$ onto the N-dimensional space of $\mathbf{A}$ and compute the optimal projection $\hat{\mathbf{e}}$ (or $\hat{\mathbf{y}}$, depending on the requirements) for the current iteration. The complexity is $O(N^2)$ per time iteration for the data matrix of size N. Now

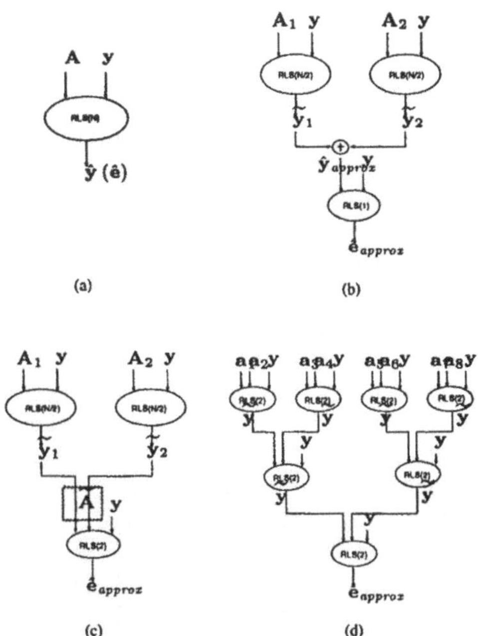

Figure 18.7 Block diagram for (a) an N-input RLS algorithm, (b) the SP-RLS I algorithm, (c) the SP-RLS II algorithm, and (d) the TSP-RLS II algorithm.

using Fig.18.7(a) as a basic building block, we can construct the block diagram for estimation method I as shown in Fig.18.7(b). Because the whole projection space is first split into two equal but smaller subspaces to perform the RLS estimation, we shall call this approach the *split-RLS* (SP-RLS). It can be easily shown that the complexity is reduced by nearly half through such a decomposition.

18.5.3 Estimation Method II (Split RLS II)

In estimation method I, we try to project $\mathbf{y}$ onto the estimated optimal projection vector $\hat{\mathbf{y}}_{approx}$. In this approach, we will project $\mathbf{y}$ directly onto the 2-D subspace $\widetilde{\mathbf{A}} \triangleq span\{\widetilde{\mathbf{y}}_1, \widetilde{\mathbf{y}}_2\}$. As a result, the estimation shall be more accurate with slight increase in complexity.

As with estimation method I, we can construct the block diagram for estimation method II (see Fig.18.7(c)) which is similar to Fig.18.7(b) except for the post-processing part. The projection residual on $span\{\widetilde{\mathbf{y}}_1, \widetilde{\mathbf{y}}_2\}$ is computed through a 2-input RLS block with $\widetilde{\mathbf{y}}_1$ and $\widetilde{\mathbf{y}}_2$ as the inputs.

18.5.4 Tree-Split RLS based on Estimation Method I and II

In estimation method I and II, we try to reduce the complexity by making one approximation at the last stage. Now consider the block diagram in Fig.18.7(c). We can repeatedly expand the two building blocks on the top by applying the same decomposition and approximation to obtain the block diagram in Fig.18.7(d).

We call this new algorithm the *Tree-Split RLS algorithm* (TSP-RLS) due to its resemblance to a binary tree. Likewise, we can derive the TSP-RLS algorithm from estimation method I (TSP-RLS I) by using the block diagram in Fig.18.7(b).

18.5.5 Systolic Implementation

Now we consider the systolic implementation of the above algorithms. First of all, we should note that each RLS building block in Fig.18.7 is independent of choices of RLS algorithms. Because the QRD-RLS array in [3] can compute the RLS estimation in a fully-pipelined way, it is a good candidate for our purpose. However, the original array computes only the optimal residual. In order to obtain the two optimal subspace projections $\widetilde{\mathbf{y}}_1$ and $\widetilde{\mathbf{y}}_2$, we need to modify the QRD-RLS array by keeping the delayed version of $y(n)$ (the desired signal at time n) in the rightmost column of the array. Once the residual is computed, we can use $\widetilde{y}_1(n) = y(n) - \widetilde{e}_1(n)$ and $\widetilde{y}_2(n) = y(n) - \widetilde{e}_2(n)$ to obtain the two subspace projections.

Now based on the block diagram in Fig.18.7, we can implement the Split RLS algorithms in the following way: For those RLS blocks which need to compute the optimal projection, the modified array is used for their implementations, while for those RLS blocks which need to compute the optimal residual (usually in the last stage), the QRD-RLS array in [3] is used. As an example, the resulting systolic implementations of the SP-RLS II and the TSP-RLS II are depicted in Fig.18.8. A comparison of hardware cost for the full-size QRD-RLS in [3] (denoted as FULL-RLS), SP-RLS, TSP-RLS, and QRD-LSL [1], is listed in Table 18.6. As we can see, the complexity of the TSP-RLS is comparable with the QRD-LSL which requires shift data structure.

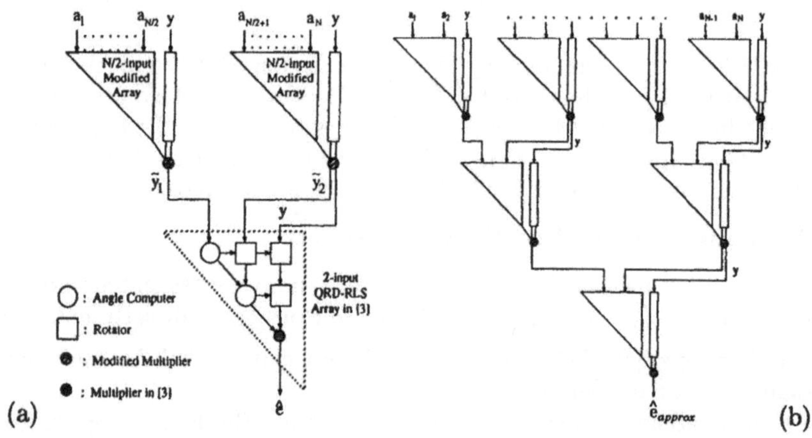

Figure 18.8 Systolic implementations of (a) the SP-RLS II and (b) the TSP-RLS II.

Table 18.6 Comparison of Hardware Cost for the FULL-RLS, SP-RLS, TSP-RLS, and QRD-LSL

	No. of Angle Computers	No. of Rotators	System latency
FULL-RLS	N	$N(N+1)/2$	$N+1$
SP-RLS I	$N+1$	$N^2/4+N/2+1$	$N/2+3$
SP-RLS II	$N+2$	$N^2/4+N/2+3$	$N/2+4$
TSP-RLS I	$2N-1$	$2N-1$	$2(\log_2 N+1)$
TSP-RLS II	$2(N-1)$	$3(N-1)$	$3\log_2 N$
QRD-LSL*	$2N+1$	$3N+1$	$N+1$

*The QRD-LSL requires shift data structure.

18.6 PERFORMANCE ANALYSIS AND SUMULATIONS OF SPLIT RLS

18.6.1 Estimation Error for SP-RLS I

Consider the LS problem in (41) and decompose the column space of $\mathbf{A}$ into two equal-dimensional subspaces, *i.e.*, $\mathbf{A} = [\mathbf{A}_1 \;, \mathbf{A}_2]$. Let $\hat{\mathbf{w}}^T = [\hat{\mathbf{w}}_1^T \;, \hat{\mathbf{w}}_2^T]$, the optimal projection vector $\hat{\mathbf{y}}$ can be represented as

$$\hat{\mathbf{y}} = \mathbf{A}\hat{\mathbf{w}} = \hat{\mathbf{y}}_1 + \hat{\mathbf{y}}_2 \tag{44}$$

where $\hat{\mathbf{y}}_1 = \mathbf{A}_1\hat{\mathbf{w}}_1$ and $\hat{\mathbf{y}}_2 = \mathbf{A}_2\hat{\mathbf{w}}_2$. From the *normal equations*

$$\mathbf{A}^T\mathbf{A}\hat{\mathbf{w}} = \mathbf{A}^T\mathbf{y}, \tag{45}$$

we have

$$\mathbf{A}_1^T\mathbf{A}_1\hat{\mathbf{w}}_1 + \mathbf{A}_1^T\mathbf{A}_2\hat{\mathbf{w}}_2 = \mathbf{A}_1^T\mathbf{y}, \tag{46}$$
$$\mathbf{A}_2^T\mathbf{A}_1\hat{\mathbf{w}}_1 + \mathbf{A}_2^T\mathbf{A}_2\hat{\mathbf{w}}_2 = \mathbf{A}_2^T\mathbf{y}. \tag{47}$$

Let $\widetilde{\mathbf{w}}_i, \widetilde{\mathbf{y}}_i$, $i = 1, 2$, be the optimal weight vectors and the optimal projection vectors when considering two subspaces $span\{\mathbf{A}_1\}$ and $span\{\mathbf{A}_2\}$ separately. From (44) and (45), we have

$$\widetilde{\mathbf{w}}_i = (\mathbf{A}_i^T\mathbf{A}_i)^{-1}\mathbf{A}_i^T\mathbf{y}, \quad \widetilde{\mathbf{y}}_i = \mathbf{A}_i\widetilde{\mathbf{w}}_i, \quad i = 1, 2. \tag{48}$$

Premultiplying $\mathbf{A}_1(\mathbf{A}_1^T\mathbf{A}_1)^{-1}$ on (46) and using the definitions of $\hat{\mathbf{y}}_1, \hat{\mathbf{y}}_2, \widetilde{\mathbf{y}}_1, \widetilde{\mathbf{y}}_2$, (46) can be simplified as

$$\hat{\mathbf{y}}_1 + \mathbf{P}_1\hat{\mathbf{y}}_2 = \widetilde{\mathbf{y}}_1. \tag{49}$$

Similarly, from (47) we can obtain

$$\mathbf{P}_2\hat{\mathbf{y}}_1 + \hat{\mathbf{y}}_2 = \widetilde{\mathbf{y}}_2 \tag{50}$$

where $\mathbf{P}_i = \mathbf{A}_i(\mathbf{A}_i^T\mathbf{A}_i)^{-1}\mathbf{A}_i^T$, $i = 1, 2$ are the projection operators.

In SP-RLS I, we estimate the optimal projection by

$$\hat{\mathbf{y}}_{approx} = \widetilde{\mathbf{y}}_1 + \widetilde{\mathbf{y}}_2, \tag{51}$$

and the estimation error (bias) is given by

$$\| \Delta \mathbf{e}_1 \|^2 = \| \hat{\mathbf{e}}_{approx} - \hat{\mathbf{e}} \|^2 = \| \hat{\mathbf{y}} - \hat{\mathbf{y}}_{approx} \|^2 . \tag{52}$$

Substituting (49)-(51) into (52) yields

$$\| \Delta \mathbf{e}_1 \|^2 = \| \hat{\mathbf{y}} - \widetilde{\mathbf{y}}_1 - \widetilde{\mathbf{y}}_2 \|^2 = \| \mathbf{P}_1 \hat{\mathbf{y}}_2 + \mathbf{P}_2 \hat{\mathbf{y}}_1 \|^2 . \tag{53}$$

In order to lower the bias value, $\mathbf{P}_1\hat{\mathbf{y}}_2$ and $\mathbf{P}_2\hat{\mathbf{y}}_1$ should be as small as possible. Note that

$$\mathbf{P}_1\hat{\mathbf{y}}_2 = \mathbf{A}_1(\mathbf{A}_1^T\mathbf{A}_1)^{-1}\mathbf{A}_1^T\mathbf{A}_2\hat{\mathbf{w}}_2 = \mathbf{A}_1\mathbf{\Phi}_{11}^{-1}\mathbf{\Phi}_{12}\hat{\mathbf{w}}_2, \tag{54}$$

$$\mathbf{P}_2\hat{\mathbf{y}}_1 = \mathbf{A}_2(\mathbf{A}_2^T\mathbf{A}_2)^{-1}\mathbf{A}_2^T\mathbf{A}_1\hat{\mathbf{w}}_1 = \mathbf{A}_2\mathbf{\Phi}_{22}^{-1}\mathbf{\Phi}_{21}\hat{\mathbf{w}}_1 \tag{55}$$

where $\mathbf{\Phi}_{ij} = \mathbf{A}_i^T\mathbf{A}_j$ is the deterministic correlation matrix. When the column vectors of $\mathbf{A}_1$ and $\mathbf{A}_2$ are more orthogonal to each other, $\mathbf{\Phi}_{12}$ and $\mathbf{\Phi}_{21}$ approach to zero and the bias is reduced accordingly.

18.6.2 Estimation Error for SP-RLS II

Consider the block diagram of the SP-RLS II in Fig.18.7(c). The optimal projection of $\mathbf{y}$ onto the space $span\{\widetilde{\mathbf{y}}_1, \widetilde{\mathbf{y}}_2\}$ can be written as

$$\hat{\mathbf{y}}_{approx} = \hat{k}_1\widetilde{\mathbf{y}}_1 + \hat{k}_2\widetilde{\mathbf{y}}_2 \tag{56}$$

where $\hat{\mathbf{k}} = [\hat{k}_1, \hat{k}_2]^T$ is the optimal weight vector. From the *normal equations*, the optimal weight vector can be solved as

$$\hat{\mathbf{k}} = [\hat{k}_1,\ \hat{k}_2]^T = \left[\alpha \frac{\widetilde{\mathbf{y}}_1^T\widetilde{\mathbf{e}}_2}{\| \widetilde{\mathbf{y}}_1 \|^2},\ \alpha \frac{\widetilde{\mathbf{y}}_2^T\widetilde{\mathbf{e}}_1}{\| \widetilde{\mathbf{y}}_2 \|^2}\right]^T \tag{57}$$

where

$$\alpha = \left(1 - \frac{\widetilde{\mathbf{y}}_1^T\widetilde{\mathbf{y}}_2}{\| \widetilde{\mathbf{y}}_1 \|^2}\frac{\widetilde{\mathbf{y}}_2^T\widetilde{\mathbf{y}}_1}{\| \widetilde{\mathbf{y}}_2 \|^2}\right)^{-1} = \csc^2\theta, \tag{58}$$

and θ denotes the angle between $\widetilde{\mathbf{y}}_1$ and $\widetilde{\mathbf{y}}_2$. From Fig.18.6, we have

$$\| \hat{\mathbf{e}}_{approx} \|^2 = \| \mathbf{y} \|^2 - \| \mathbf{y}_{approx} \|^2 = \| \mathbf{y} \|^2 - \mathbf{y}^T\mathbf{y}_{approx} \tag{59}$$

and

$$\| \hat{\mathbf{e}}_{approx} \|^2 = \| \mathbf{y} \|^2 - \hat{k}_1 \| \widetilde{\mathbf{y}}_1 \|^2 - \hat{k}_2 \| \widetilde{\mathbf{y}}_2 \|^2 . \tag{60}$$

Substituting (57) into (60) yields

$$\| \hat{\mathbf{e}}_{approx} \|^2 = \| \mathbf{y} \|^2 - \csc^2\theta \| \widetilde{\mathbf{y}}_1 - \widetilde{\mathbf{y}}_2 \|^2 . \tag{61}$$

Thus, the bias of SP-RLS II is given by

$$\| \Delta \mathbf{e}_2 \|^2 = \| \hat{\mathbf{e}}_{approx} \|^2 - \| \hat{\mathbf{e}} \|^2 = \| \hat{\mathbf{y}} \|^2 - \csc^2\theta \| \widetilde{\mathbf{y}}_1 - \widetilde{\mathbf{y}}_2 \|^2 . \tag{62}$$

For any given θ, it can be shown that $\| \Delta \mathbf{e}_2 \|^2$ is bounded by [17]

$$\| \Delta \mathbf{e}_2 \|^2 \leq \| \Delta \mathbf{e}_1 \|^2 . \tag{63}$$

This implies that the performance of SP-RLS II is better than that of SP-RLS I in terms of estimation error.

18.6.3 Bandwidth, Eigenvalue Spread, and Bias

From (53) and (62) we know that the orthogonality between the two subspaces $span\{\mathbf{A}_1\}$ and $span\{\mathbf{A}_2\}$ will significantly affect the bias value. However, in practice, the evaluation of degree of orthogonality for multidimensional spaces is nontrivial and computationally intensive (e.g., CS-decomposition [18, pp. 75–78]). Without loss of generality, we will only focus our discussion on single-channel case, where the data matrix $\mathbf{A}$ consists of only shifted data and the degree of orthogonality can be easily measured. In such a case, the degree of orthogonality can be measured through two indices: the bandwidth and the eigenvalue spread of the data. If the signal is less correlated (orthogonal), the autocorrelation function has smaller duration and thus larger bandwidth. Noise processes are examples. On the other hand, narrow-band processes such as sinusoidal signals are highly correlated. If the data matrix is completely orthogonal, all the eigenvalues are the same and the condition number is one. This implies that if the data matrix is more orthogonal, it will have less eigenvalue spread. It is clear from our previous discussion that the SP-RLS will render less bias for the broad-band signals than for the narrow-band signals.

As to the TSP-RLS, note that the output optimal projection is a linear combination of the input column vectors. If the inputs to one stage of the TSP-RLS array are less correlated, the outputs of this stage will still be less correlated. Therefore, the signal property at the first stage such as bandwidth plays an important role in the overall performance of the TSP-RLS.

18.6.4 Simulation Results

In the following simulations, we use the autoregressive (AR) process of order p (AR(p)) $u(n) = \sum_{i=1}^{p} w_i u(n-i) + v(n)$, to generate the simulation data. where $v(n)$ is a zero-mean white Gaussian noise with power equal to 0.1. Besides, the pole locations of the AR processes are used to control the bandwidth property. As the poles are approaching the unit circle, we will have narrow-band signals; otherwise, we will obtain broad-band signals. In the first experiment, we try to perform fourth-order linear prediction (LP) with four AR(4) processes using the SP-RLS and TSP-RLS systolic arrays. The simulation results are shown in Fig.18.9, in which the x-axis represents the location of the variable poles, and y-axis represents the average output noise power after convergence. Ideally the output should be the noise process $v(n)$ with power equal to 0.1. As we can see, when the bandwidth of input signal becomes wider, the bias is reduced. This agrees perfectly with what we expected.

Beside the bias values, we also plot the square root of the *spectral dynamic range* D associated with each AR process. It is known that the eigenvalue spread of the data signal is bounded by the spectral dynamic range [19]

$$1 \leq \frac{\lambda_{\text{max}}}{\lambda_{\text{min}}} \leq \frac{\max\{|U(e^{j\omega})|^2\}}{\min\{|U(e^{j\omega})|^2\}} \triangleq D, \tag{64}$$

where $U(e^{j\omega})$ is the spectrum of the signal. From the simulation results, we see the consistency between the bias value and the spectral dynamic range. This indicates that the performance of the split RLS algorithms is also affected by the eigenvalue spread of the input signal. This phenomenon is similar to what we have seen

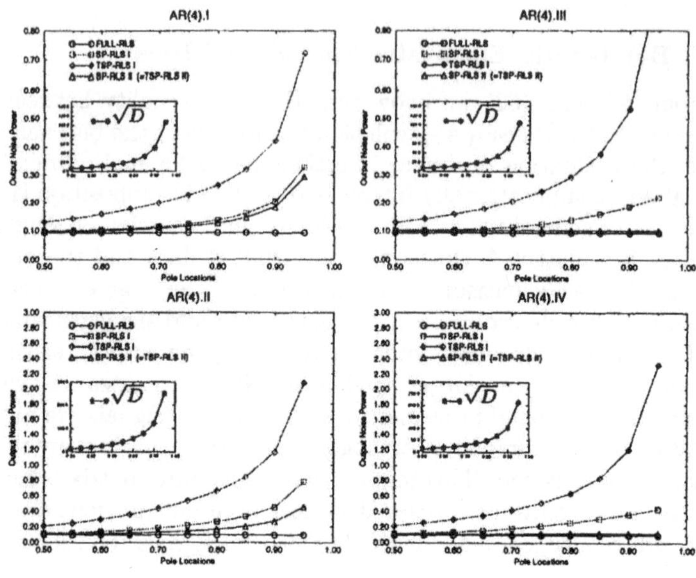

Figure 18.9 Simulation results of AR(4) I-IV, where the square root of the spectral dynamic range (D) is also plotted for comparison.

in the LMS-type algorithms. Besides, two observations can be made from the experimental results: 1) The SP-RLS performs better than the TSP-RLS. This is due to the number of approximation stages in each algorithm. 2) The overall performance of SP-RLS II is better than that of SP-RLS I. This agrees with our analysis in (63).

Next we want to examine the convergence rate of our algorithm. Fig.18.10 shows the convergence curve for the 8-input FULL-RLS and the TSP-RLS II after some initial perturbation. It is interesting to note that although the TSP-RLS II has some bias after it converges, its convergence rate is faster than that of the FULL-RLS. This is due to the fact that the $O(\log_2 N)$ system latency of the TSP-RLS is less than the $O(N)$ latency of the FULL-RLS. Also, to initialize an 8-input full-size array takes more time than to initialize the three small cascaded 2-input arrays. The property of faster convergence rate is especially preferred for the tracking of parameters in non-stationary environments such as the multichannel adaptive filtering discussed below.

We apply the split RLS to the multidimensional adaptive filtering (MDAF) based on the MDAF architecture in [13]. In [13], the McClellan Transformation (MT) [20] was employed to reduce the total parameters in the 2-D filter design, and the QRD-RLS array in [21] was used as the processing kernel to update the weight coefficients. In our approach, we replace the QRD-RLS array with the TSP-RLS array. This results in a more cost-effective ($O(N)$) MDAF architecture. The performance of the proposed MDAF architecture is examined by applying it to a two-dimensional adaptive line enhancer (TDALE) [22], [23] for image restoration. The block diagram is depicted in Fig. 18.11. The primary input is the well-known

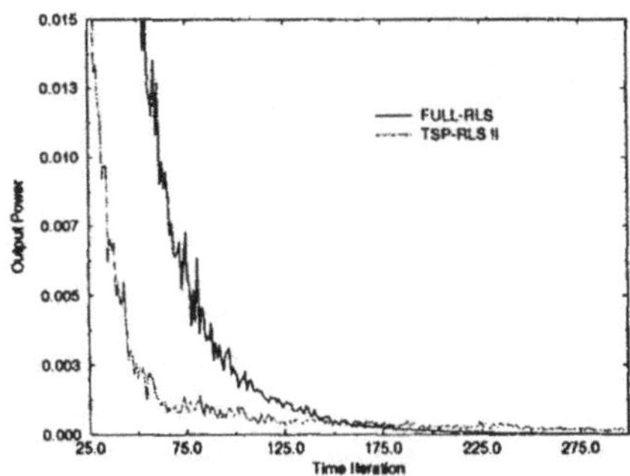

Figure 18.10 Learning curve of the FULL-RLS and TSP-RLS II after some initial perturbation.

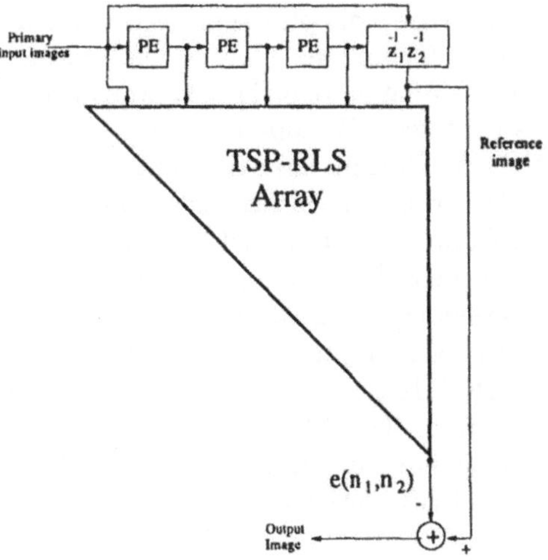

Figure 18.11 Block diagram of the TDALE.

Table 18.7 SNR Results of TDALE in the Application of Restoring Noisy Image

Input SNR (dB)	10.0	3.0	0.0
Output SNR in [23]	12.0	8.0	6.0
Output SNR using FULL-RLS	10.5	9.0	7.6
Output SNR using TSP-RLS II	10.9	9.8	8.7

"LENA" image degraded by a white Gaussian noise. A 2-D unit delay $z_1^{-1}z_2^{-1}$ is used as a decorrelation operator to obtain the reference image. The image signal is fed into the system in the raster scanned format – from left to right, top to bottom. After the input image goes through the TSP-RLS array, the generated estimation error is subtracted from the reference signal to get the filtered image. For comparison, we also repeat this experiment using the FULL-RLS array.

The simulation results are shown in Table 18.7. We can see that the performance of the TSP-RLS is better than the 2-D joint process lattice structure in [23] when the signal-to-noise ratio (SNR) is low. It is also interesting to note that the TSP-RLS outperforms the FULL-RLS. As we can see from Fig. 18.10, although the TSP-RLS has misadjustment after convergence, it converges faster than the FULL-RLS. This fast-tracking property is preferable under non-stationary environments where convergence is very unlikely.

18.7 SPLIT RLS WITH ORTHOGONAL PREPROCESSING

From the analyses in the previous section, we know that the estimated optimal projection will approach to the real optimal projection when all subspaces are more orthogonal to each other. Therefore, if we can preprocess the data matrix such that the column spaces become more orthogonal (less correlated) to each other, a better performance is expected. The operation for the split RLS with orthogonal preprocessing is as follows: First perform the orthogonal transform on the current data vector, then use the transformed data as the inputs of the split RLS. In our approach, the Discrete Cosine Transform (DCT) is used as the preprocessing kernel. As to the hardware implementation, we can employ the time-recursive DCT lattice structure in [24] to continuously generate the transformed data. Fig. 18.12 shows the SP-RLS I array with DCT preprocessing. The transform-domain data are first generated through the DCT lattice structure, then are sent to the SP-RLS I array to perform the RLS filtering. The TSP-RLS array with the preprocessing scheme can be constructed in a similar way. Since both the DCT lattice structure and the TSP-RLS array require $O(N)$ hardware complexity, the total cost for the whole system is still $O(N)$.

In addition to the DCT transform, we also propose a new preprocessing scheme called the *Swapped DCT* (SWAP-DCT). Suppose $\mathbf{Z} = [\mathbf{z}_1, \mathbf{z}_2, \ldots, \mathbf{z}_N]$ is the DCT-domain data. In the DCT preprocessing, the input data is partitioned as

$$\begin{aligned} \mathbf{A}_1 &= [\mathbf{z}_1, \mathbf{z}_2, \ldots, \mathbf{z}_{N/2}], \\ \mathbf{A}_2 &= [\mathbf{z}_{N/2+1}, \mathbf{z}_{N/2+2}, \ldots, \mathbf{z}_N]. \end{aligned} \tag{65}$$

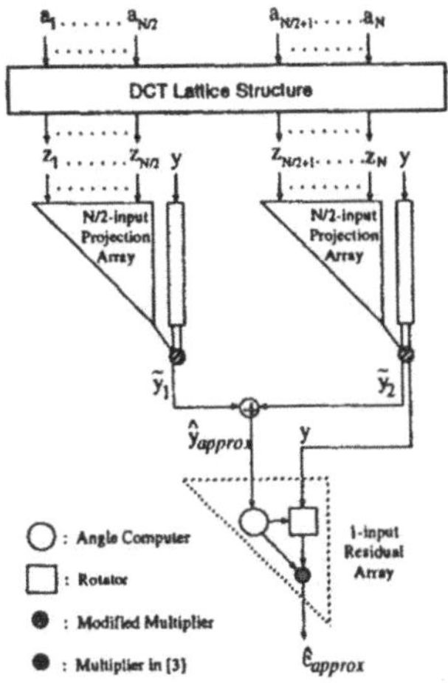

Figure 18.12 SP-RLS I array with orthogonal preprocessing.

To make the input data more uncorrelated, we permute the transformed data column as

$$\begin{aligned} \mathbf{A}_1 &= [\mathbf{z}_1, \mathbf{z}_3, \ldots, \mathbf{z}_{2k-1}, \ldots, \mathbf{z}_{N-1}], \\ \mathbf{A}_2 &= [\mathbf{z}_2, \mathbf{z}_4, \ldots, \mathbf{z}_{2k}, \ldots, \mathbf{z}_N] \end{aligned} \tag{66}$$

in the SWAP-DCT preprocessing scheme. Fig.18.13 shows the spectrum of the normal DCT partitioning and the SWAP-DCT partitioning. Recall that the eigenvalue spread will affect the bias value, and the eigenvalue spread is bounded by the spectral dynamic range. It is obvious that the SWAP-DCT preprocessing scheme will have better performance due to the smaller eigenvalue spread in both $\mathbf{A}_1$ and $\mathbf{A}_2$.

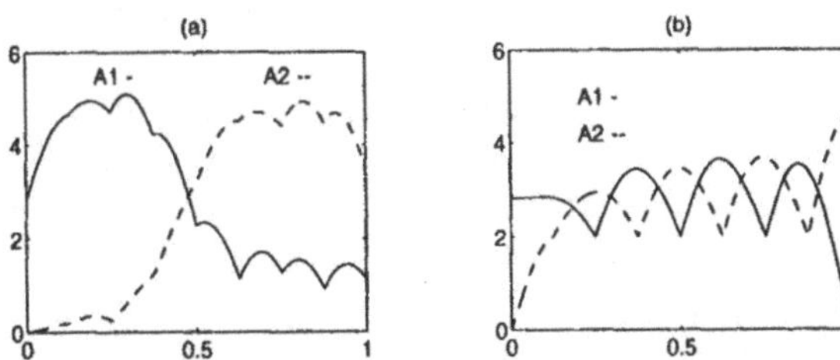

Figure 18.13 Spectrum of (a) the Normal DCT domain and (b) the SWAP-DCT domain.

To validate our arguments for the orthogonal preprocessing, we will repeat the first experiment in the previous section for the TSP-RLS II with different preprocessing schemes (Fig.18.14). In general, the TSP-RLS with DCT preprocessing gives a fairly significant improvement in the bias value over the TSP-RLS without any preprocessing (normal TSP-RLS). Nevertheless, some exceptions can be found in AR(4).III. As expected, the SWAP-DCT performs better than the DCT. This supports our assertion for the effect of the SWAP-DCT.

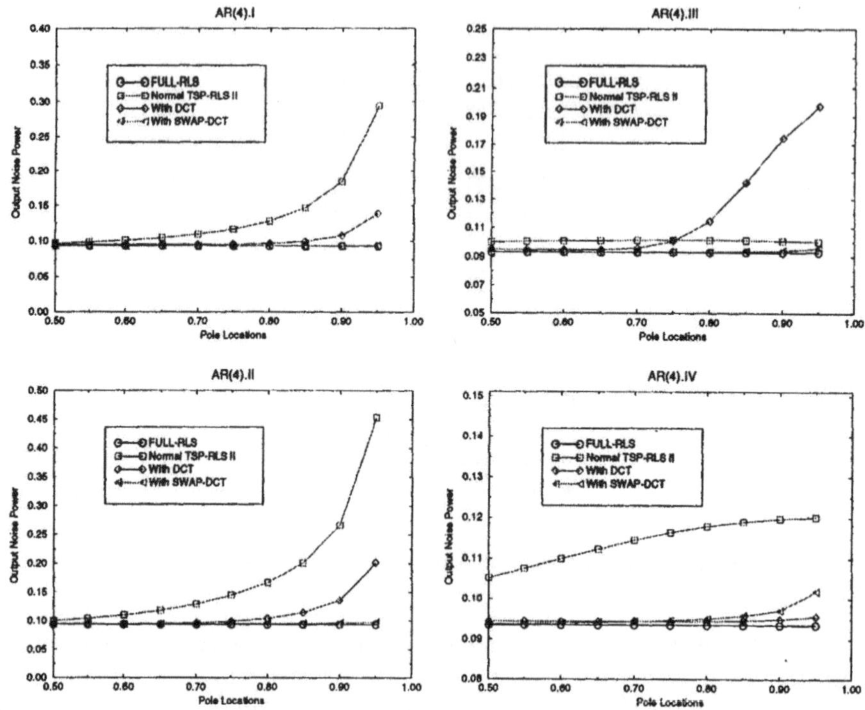

Figure 18.14 Simulation result of AR(4) I-IV with preprocessing schemes.

18.8 CONCLUSIONS

In this chapter, we introduced two novel RLS algorithms and architectures for cost-efficient VLSI implementations. The square root and division free QRD-RLS reduces the computational complexity at the arithmetic level, while the split RLS employs the approximation method to save the total complexity at the algorithmic level.

We first introduced *the parametric $\kappa\lambda$ $\Re$otation*, which is a square-root-free and division-free algorithm, and showed that the parametric $\kappa\lambda$ $\Re$otation describes a subset of the $\mu\nu$ $\Re$otation algorithms [6]. We then derived novel architectures based on the $\kappa\lambda$ $\Re$otation for $\kappa = \lambda = 1$, and made a comparative study with the standard Givens rotation and the $\mu\nu$ $\Re$otation with $\mu = \nu = 1$. Our analysis suggests the following decision rule in choosing between the $\mu\nu$ $\Re$otation-based architectures and the $\kappa\lambda$ $\Re$otation-based architectures: *Use the $\mu\nu$ $\Re$otation-based architectures, with $\mu = \nu = 1$, for the constrained minimization problems, and use the $\kappa\lambda$ $\Re$otation-based architectures, with $\kappa = \lambda = 1$, for the unconstrained minimization problems.* Table 18.5 shows the benchmark comparison of different algorithms using various DSP processors and it confirms our observation. In addition, The dynamic range, numerical stability, and error/wordlength bound of the $\kappa\lambda$ $\Re$otation algorithm can be derived. The readers may refer to [25] for detailed discussion.

We also introduced a new $O(N)$ fast algorithm and architecture for the RLS estimation of nonstructured data. We have shown that the bandwidth and/or the eigenvalue spread of the input signal can be used as a good performance index for these algorithms. Therefore, the users will have small bias when dealing with broad-band/less-correlated signals. For narrow-band signals, we can also employ the orthogonal preprocessing to improve its performance. The low complexity as well as the fast convergence rate of the proposed algorithm makes it suitable for RLS estimation under the non-stationary or fast-changing environments where the data matrix has no structure.

The systolic RLS structures described in this chapter are very promising for cost-effective implementations, since they require less computational complexity (in various aspects) than the structures known to date. Such a property helps to reduce the programmable DSP MIPS count and make VLSI implementation easier.

APPENDIX

Proof of Equation (26): First, we derive some equations that will be used in the course of the optimal residual computation.

If we solve (24), case $i = j = 1$, for $l_q\beta^2 a_{11}^2 + l_1 b_1^2$ and substitute in (22) we get

$$l_1' = l_1 l_q \frac{a_{11}'}{\kappa_1}\kappa_1^2$$

and therefore

$$\frac{l_1'}{l_1} = l_q a_{11}' \kappa_1. \tag{67}$$

If we solve (24), case $j = i$, for $l_q^{(i-1)}\beta^2 a_{ii}^2 + l_i b_i^{(i-1)^2}$ and substitute in (23) we get

$$l_q^{(i)} = \frac{\lambda_i^2 a_{ii}'}{\kappa_i}. \tag{68}$$

If we substitute the same expression in (22) we get

$$l_i' = l_i l_q^{(i-1)} a_{ii}' \kappa_i.$$

In the above expression we substitute $l_q^{(i-1)}$ from (68), and solve for l_i'/l_i to obtain

$$\frac{l_i'}{l_i} = \frac{\lambda_{i-1}^2 \kappa_i}{\kappa_{i-1}} a_{i-1,i-1}' a_{ii}'. \tag{69}$$

If we solve (22) for $l_q^{(i-1)}\beta^2 a_{ii}^2 + l_i b_i^{(i-1)^2}$ and substitute in (23) we get

$$l_q^{(i)} = \frac{\lambda_i^2}{\kappa_i^2}\frac{l_i'}{l_i}\frac{1}{l_q^{(i-1)}}. \tag{70}$$

Also, we note that (4) implies that

$$c_i = \beta r_{ii}/r_{ii}'$$

and by substituting (9) we obtain

$$c_i = \frac{\beta a_{ii}}{a_{ii}'}\sqrt{\frac{l_i'}{l_i}} \quad i = 1, 2, \cdots, p. \tag{71}$$

Similarly, from (4) and (9), we get

$$s_i = \frac{b_i^{(i-1)}}{a'_{ii}} \sqrt{\frac{l'_i}{l_q^{(i-1)}}} \quad i = 1, 2, \cdots, p. \tag{72}$$

The optimal residual for the RLS problem is [1]

$$e_{RLS}(t_n) = -\left(\prod_{k=1}^{p} c_k \right) v(t_n). \tag{73}$$

The expressions in (20) and (19) imply

$$v(t_n) = \frac{1}{\sqrt{l_q^{(p)}}} b_{p+1}^{(p)}.$$

If we substitute the above expressions of $v(t_n)$ and c_i in (73) we obtain

$$e_{RLS}(t_n) = -\prod_{i=1}^{p} \left(\frac{\beta a_{ii}}{a'_{ii}} \sqrt{\frac{l'_i}{l_i}} \right) \frac{1}{\sqrt{l_q^{(p)}}} b_{p+1}^{(p)}. \tag{74}$$

From (70), we get

$$l_q^{(p)} = \frac{\lambda_p^2}{\kappa_p^2} \frac{l'_p}{l_p} \frac{1}{l_q^{(p-1)}} = \frac{\lambda_p^2}{\kappa_p^2} \frac{l'_p}{l_p} \frac{\kappa_{p-1}^2}{\lambda_{p-1}^2} \frac{l_{p-1}}{l'_{p-1}} l_q^{(p-2)}$$

$$= \begin{cases} \prod_{j=1}^{k} \left(\frac{\lambda_{2j}^2}{\kappa_{2j}^2} \frac{l'_{2j}}{l_{2j}} \frac{\kappa_{2j-1}^2}{\lambda_{2j-1}^2} \frac{l_{2j-1}}{l'_{2j-1}} \right) l_q \ , & p = 2k \\ \prod_{j=1}^{k-1} \left(\frac{\kappa_{2j}^2}{\lambda_{2j}^2} \frac{l_{2j}}{l'_{2j}} \frac{\lambda_{2j-1}^2}{\kappa_{2j-1}^2} \frac{l'_{2j-1}}{l_{2j-1}} \right) \frac{\lambda_p^2}{\kappa_p^2} \frac{l'_p}{l_p} \frac{1}{l_q} \ , & p = 2k-1 \end{cases} . \tag{75}$$

Thus, from (74) and (75), for the case of $p = 2k$, we have

$$\begin{aligned} e_{RLS}(t_n) &= -\prod_{j=1}^{k} \left(\frac{\beta a_{2j,2j}}{a'_{2j,2j}} \sqrt{\frac{l'_{2j}}{l_{2j}}} \frac{\beta a_{2j-1,2j-1}}{a'_{2j-1,2j-1}} \sqrt{\frac{l'_{2j-1}}{l_{2j-1}}} \left(\frac{\lambda_{2j}^2}{\kappa_{2j}^2} \frac{l'_{2j}}{l_{2j}} \frac{\kappa_{2j-1}^2}{\lambda_{2j-1}^2} \frac{l_{2j-1}}{l'_{2j-1}} \right)^{-\frac{1}{2}} \right) \\ &\times \frac{1}{\sqrt{l_q}} \cdot b_{p+1}^{(p)}. \end{aligned} \tag{76}$$

By doing the appropriate term cancellations and by substituting the expressions of $l'_i/l_i, i = 1, 2, \cdots, 2k$ from (67) and (69), we obtain the expression (26) for the optimal residual. Similarly, for the case of $p = 2k - 1$, from (74) and (75) we obtain

$$\begin{aligned} e_{RLS}(t_n) &= -\prod_{j=1}^{k-1} \left(\frac{\beta a_{2j,2j}}{a'_{2j,2j}} \sqrt{\frac{l'_{2j}}{l_{2j}}} \frac{\beta a_{2j-1,2j-1}}{a'_{2j-1,2j-1}} \sqrt{\frac{l'_{2j-1}}{l_{2j-1}}} \left(\frac{\lambda_{2j-1}^2}{\kappa_{2j-1}^2} \frac{l'_{2j-1}}{l_{2j-1}} \frac{\kappa_{2j}^2}{\lambda_{2j}^2} \frac{l_{2j}}{l'_{2j}} \right)^{-\frac{1}{2}} \right) \\ &\times \frac{\beta a_{pp}}{a'_{pp}} \sqrt{\frac{l'_p}{l_p}} \sqrt{\frac{\kappa_p^2 l_p l_q}{\lambda_p^2 l'_p}} b_{p+1}^{(p)} \end{aligned} \tag{77}$$

and by substituting (69) we get (26).

REFERENCES

[1] S. Haykin, *Adaptive Filter Theory.* Prentice-Hall, Englewood Cliffs, N.J., 3rd ed., 1996.

[2] J. McWhirter and T. Shepherd, "Systolic array processor for MVDR beamforming," *IEE Prcceedings*, vol. 136, no. 2, pp. 75–80, April 1989. Pt.F.

[3] J. G. McWhirter, "Recursive least-squares minimization using a systolic array," *Proc. SPIE, Real-Time Signal Processing VI*, vol. 431, pp. 105–112, 1983.

[4] W. Gentleman, "Least squares computations by Givens transformations without square roots," *J. Inst. Maths. Applics.*, vol. 12, pp. 329–336, 1973.

[5] S. Hammarling, "A note on modifications to the Givens plane rotation," *J. Inst. Maths. Applics.*, vol. 13, pp. 215–218, 1974.

[6] S. F. Hsieh, K. J. R. Liu, and K. Yao, "A unified approach for QRD-based recursive least-squares estimation without square roots," *IEEE Trans. on Signal Processing*, vol. 41, no. 3, pp. 1405–1409, March 1993.

[7] F. Ling, "Efficient least squares lattice algorithm based on Givens rotations with systolic array implementation," *IEEE Trans. Signal Processing*, vol. 39, pp. 1541–1551, July 1991.

[8] T. Shepherd, J. McWhirter, and J. Hudson, "Parallel weight extraction from a systolic adaptive beamforming," *Mathematics in Signal Processing II*, 1990.

[9] K. Tanabe, "Projection method for solving a singular system of linear equations and its applications," *Numer. Math.*, vol. 17, pp. 203–214, 1971.

[10] A. S. Kydes and R. P. Tewarson, "An iterative methods for solving partitioned linear equations," *Computing*, vol. 15, pp. 357–363, Jan. 1975.

[11] T. Elfving, "Block-iterative methods for consistent and inconsistent linear equations," *Numer. Math.*, vol. 35, pp. 1–12, 1980.

[12] R. Bramley and A. Samem, "Row projection methods for large nonsymmetric linear systems," *SIAM J. Sci. Stat. Comput.*, vol. 13, no. 1, pp. 168–193, Jan. 1992.

[13] J. M. Shapiro and D. H. Staelin, "Algorithms and systolic architecture for multidimensional adaptive filtering via McClellan transformation," *IEEE Trans. Circuits Syst. Video Technol.*, vol. 2, pp. 60–71, Mar 1992.

[14] J. Gotze and U. Schwiegelshohn, "A square root and division free Givens rotation for solving least squares problems on systolic arrays," *SIAM J. Scie. and Stat. Comput.*, vol. 12, no. 4, pp. 800–807, July 1991.

[15] C. F. T. Tang, K. J. R. Liu, and S. Tretter, "Optimal weight extraction for adaptive beamforming using systolic arrays," *IEEE Trans. on Aerospace and Electronic Systems*, vol. 30, pp. 367–385, April 1994.

[16] R. Stewart, R. Chapman, and T. Durrani, "Arithmetic implementation of the Givens QR tiarray," in *Proc. IEEE Int. Conf. Acoust. Speech, Signal Processing*, pp. V-2405-2408, 1989.

[17] A.-Y. Wu and K. J. R. Liu, "Split recursive least-squares: Algorithms, architectures, and applications," *IEEE Trans. Circuits Syst. II*, vol. 43, no. 9, pp. 645-658, Sept. 1996.

[18] G. H. Golub and C. F. Van Loan, *Matrix Computations*. The John Hopkins University Press, Baltimore, MD, 2nd ed., 1989.

[19] J. Makhoul, "Linear Prediction: A tutorial review," *Proc. IEEE*, vol. 63, no. 4, pp. 561-580, April 1975.

[20] R. M. Mersereau, W. F. G. Mecklenbrauker, and T. F. Quatieri, Jr., "McClellan transformations for two-dimensional digital filtering: I - Design," *IEEE Trans. Circuits Syst.*, vol. 23, no. 7, pp. 405-422, July 1976.

[21] W. M. Gentleman and H. T. Kung, "Matrix triangularization by systolic arrays," *Proc. SPIE, Real-Time Signal Processing IV*, vol. 298, pp. 298-303, 1981.

[22] M. M. Hadhoud and D. W. Thomas, "The two-dimensional adaptive LMS (TDLMS) algorithm," *IEEE Trans. Circuits Syst.*, vol. 5, pp. 485-494, May 1988.

[23] H. Youlal, Malika Janati-I, and M. Najim, "Two-dimensional joint proçess lattice for adaptive restoration of images," *IEEE Trans. Image Processing*, vol. 1, pp. 366-378, July 1992.

[24] K. J. R. Liu and C. T. Chiu, "Unified parallel lattice structures for time-recursive Discrete Cosine/Sine/Hartley transforms," *IEEE Trans. Signal Processing*, vol. 41, no. 3, pp. 1357-1377, March 1993.

[25] E. N. Frantzeskakis and K. J. R. Liu, "A class of square root and division free algorithms and architectures for QRD-based adaptive signal processing," *IEEE Trans. Signal Processing*, vol. 42, no. 9, pp. 2455-2469, Sept. 1994.

Chapter 19

Pipelined RLS FOR VLSI: STAR-RLS Filters

K. J. Raghunath
Bell Labs, Lucent Technologies
Warren, New Jersey
raghunath@lucent.com

Keshab K. Parhi
University of Minnesota
Minneapolis, Mnnesota
parhi@ece.umn.edu

19.1 INTRODUCTION

Adaptive filters have wide applications in data communication, system identification, spectrum analysis, adaptive beamforming, magnetic recording, image processing etc. Adaptive filters learn the characteristics of a signal as it is processed and tend to approach the performance of an optimal filter for the given application [1]. The well known adaptation algorithms include the least-mean-square (LMS) [2] algorithm and the recursive least-squares (RLS) algorithm [3],[4]. Traditionally, LMS is the more commonly used algorithm in adaptive filters. The LMS algorithm is an approximation of the steepest descent method [5]. Instead of estimating the cross-correlation and auto-correlation matrices from the data, the instantaneous values of the quantities are used. The LMS algorithm converges to an optimum filter over a period of time. The resulting algorithm is very simple to implement and robust. The LMS algorithm is well understood by the community in the industry. Efficient structures are available for implementation of the LMS algorithm. Blind equalization techniques using LMS algorithm are also well developed [6],[7]. Joint equalization and carrier recovery schemes have been implemented succesfully [8]. These advancements make LMS suitable for practical applications.

On the other hand, the RLS algorithm is an exact approach, i.e., it gives an optimum filter for the given data. The weights of the adaptive filter are chosen to minimize the exponentially weighted average of the estimation error. The RLS algorithm can be considered as a deterministic counterpart of the Kalman filters [9]. The higher computational requirement and concerns about numerical stability discourage potential users of RLS algorithm.

However, the RLS algorithm offers a much faster convergence rate. The RLS algorithm converges within $O(N)$ iterations while the LMS algorithm takes $O(N^2)$ iterations, where N is the number of taps in the adaptive filter. This fast convergence could be used for competitive advantage in many applications. The RLS algorithm using QR decomposition, referred to as QRD-RLS [1], has been proved to be a numerically stable implementation of the RLS algorithm. As will be shown in this chapter, QRD-RLS in fact requires smaller wordlengths in its implementation. The coefficients in the LMS algorithm would typically need about 24 bits while it is possible to implement the QRD-RLS with about 12 to 14 bits. The RLS algorithm allows for graceful trade-off between speed of convergence and hardware requirement. The update rate for QRD-RLS can be reduced so that the computation can be handled within the available hardware. With VLSI technology advancing so fast, the computational requirements of RLS can be met more easily. Another interesting trend in the communication systems area is the move towards using DSP processors instead of ASIC chips. Highly parallel DSP processors are being developed which will be able to meet the demands of any communication chip, even a high data rate cable modem. In a DSP processor environment the design complexity of the RLS algorithm would be less of an issue. It is much easier to code the the RLS algorithm on a DSP processor rather than design ASICs. Also, using RLS algorithm it would also be easier to exploit the sparse nature of the equalizer coefficients.

There are square-root free forms of QRD-RLS which are more computationally efficient than the original algorithm (see [10],[11],[12],[13],[14],[15]). A unified approach to the different square-root free QRD-RLS algorithms, presented in [12], generalizes all existing algorithms. In [13], a low complexity square-root free algorithm is presented and it is shown that this algorithm is stable. In [14], a scaled version of the fast Givens rotation [10] is developed which prevents overflow and underflow. Recently, algorithms which avoid divisions as well as square-roots have been presented [16]. In [17], a fast QRD-RLS algorithm based on Givens rotations was introduced. Recently, other fast QRD-RLS algorithms have been presented in [18],[19]. These new algorithms further encourage the use of RLS algorithms.

With cheaper computational power in the future the focus will be on performance rather than computational complexity. Indeed some researchers believe that the RLS algorithm would practically replace the LMS algorithm in the near future [20]. Hence work is needed to prove in the RLS algorithm for different applications. Development of practical low-cost structures for QRD-RLS, and further analysis of finite-precision performance will be needed.

This chapter is organized as follows. We first explain the notation and introduce the QRD-RLS algorithm. The pipelining bottleneck of the QRD-RLS algorithm is discussed in the next section. Next we introduce the scaled tangent rotations (STAR) and the STAR-RLS algorithm. These lead to pipelined architectures for RLS with performance being similar to QRD-RLS. Finite-precision analysis results for QRD-RLS and STAR-RLS are summarized in the next section. Finally, the VLSI implementation of a 4-tap 100MHz pipelined STAR-RLS filter in 1.2μ technology is described.

19.2 THE QRD-RLS ALGORITHM

In this section we develop the problem formulation and QRD-RLS solution for the same. The notation of [1] is closely followed in this paper wherever possible. We are given a time series of observations $u(1)$, $u(2)$, $\ldots u(n)$ and we want to estimate some desired signal $d(i)$ based on a weighted sum of present sample and a few of the past samples. The data is assumed to be real. The estimation error (denoted by $e(i)$) can be defined as

$$e(i) = d(i) - \mathbf{w}^T(n)\mathbf{u}(i), \tag{1}$$

where $\mathbf{u}(i)$ is a subset of the time series in the form of a M size vector

$$\mathbf{u}^T(i) = [u(i), u(i-1), \ldots, u(i-M+1)], \tag{2}$$

and $\mathbf{w}(n)$ is the weight vector of size M. Some matrix notations can now be defined for simplicity of the equations. The n-by-M data matrix $A(n)$ is defined as

$$A(n) = \begin{bmatrix} \mathbf{u}^T(1) \\ \mathbf{u}^T(2) \\ \vdots \\ \mathbf{u}^T(n) \end{bmatrix}. \tag{3}$$

The error vector $\varepsilon(n)$ and the desired response vector $\mathbf{b}(n)$ are defined as

$$\varepsilon(n) = \begin{bmatrix} e(1) \\ e(2) \\ \vdots \\ e(n) \end{bmatrix} = \mathbf{b}(n) - A(n)\mathbf{w}(n) \quad \textit{where} \quad \mathbf{b}(n) = \begin{bmatrix} d(1) \\ d(2) \\ \vdots \\ d(n) \end{bmatrix}. \tag{4}$$

The weight vector $\mathbf{w}(n)$ is chosen so as to minimize the index of performance $\xi(n)$, given by

$$\xi(n) = \sum_{i=1}^{n} \lambda^{n-i} \mid e(i) \mid^2 = \varepsilon(n)^T \Lambda(n) \varepsilon(n) = \| \Lambda^{1/2}(n)\varepsilon(n) \|^2, \tag{5}$$

where λ is the exponential weighting factor (*forgetting factor*) and $\Lambda(n)$ is a n-by-n diagonal *exponential weighting matrix* given by

$$\Lambda(n) = diag[\lambda^{n-1}, \lambda^{n-2}, \ldots, 1]. \tag{6}$$

Since the norm of a vector is unaffected by premultiplication by a unitary or orthogonal matrix, we can express $\xi(n)$ as

$$\xi(n) = \| Q(n)\Lambda^{1/2}(n)\varepsilon(n) \|^2, \tag{7}$$

where $Q(n)$ is a n-by-n unitary or orthogonal matrix. Now, we can write

$$Q(n)\Lambda^{1/2}(n)\varepsilon(n) = Q(n)\Lambda^{1/2}(n)\mathbf{b}(n) - Q(n)\Lambda^{1/2}(n)A(n)\mathbf{w}(n). \tag{8}$$

The orthogonal matrix $Q(n)$ is chosen so as to triangularize the matrix $\Lambda^{1/2}(n)A(n)$

$$Q(n)\Lambda^{1/2}(n)A(n) = \begin{bmatrix} R(n) \\ \mathbf{0} \end{bmatrix}, \quad n \geq M, \tag{9}$$

where $R(n)$ is an M-by-M upper triangular matrix and $\mathbf{0}$ is a $(n-M)$-by-M null matrix. The desired signal vector is transformed as

$$Q(n)\Lambda^{1/2}(n)\mathbf{b}(n) = \begin{bmatrix} \mathbf{p}(n) \\ \mathbf{v(n)} \end{bmatrix}, \tag{10}$$

where $\mathbf{p}(n)$ is a M dimensional vector and $\mathbf{v}(n)$ is a $(n-M)$ dimensional vector. The least squares solution for the problem (5) is then given by

$$\mathbf{w}_{ls}(n) = R^{-1}(n)\mathbf{p}(n), \qquad n \geq M. \tag{11}$$

The orthogonal matrix $Q(n)$ can be determined as a product of *Givens rotation* matrices. A single Givens rotation matrix can nullify one element in the matrix on which it is applied. Suppose, we want to null the (n,m) element y_{nm} using the (m,m) element y_{mm} of a matrix Y. The Givens rotation matrix G can be defined as

$$\begin{aligned} G(m,m) &= \cos\phi \\ G(m,n) &= \sin\phi \\ G(n,m) &= -\sin\phi \\ G(n,n) &= \cos\phi \\ G(i,i) &= 1, \ i \neq n, m \\ G(i,j) &= 0, \ otherwise, \end{aligned} \tag{12}$$

where $\cos\phi = \frac{y_{mm}}{\sqrt{y_{nm}^2 + y_{mm}^2}}$. A Givens rotation is represented as

$$G = \begin{bmatrix} \cos\phi & \sin\phi \\ -\sin\phi & \cos\phi \end{bmatrix}. \tag{13}$$

The matrix $Q(n)$ is in practice determined recursively. With each incoming data sample $u(n)$ the data matrix $A(n)$ adds a new row $\mathbf{u}(n)$. A set of M Givens rotations are determined to null the last row of $A(n)$. Thus, the triangular matrix $R(n)$ gets updated and the least squares solution can be determined at every time instant. This algorithm is called as the QR decomposition based RLS. This algorithm can be mapped on to a systolic array [1],[21],[10] as shown in Fig. 19.1.

19.3 PIPELINING PROBLEM IN QRD-RLS

The speed (or sample rate or clock rate) of any architecture is limited by the computation time or propagation delay of the longest path between two registers or pipeline stages. Consider the architecture of QRD-RLS algorithm in Fig. 19.1. The paths between the systolic array cells can be pipelined. Hence the critical path will be within the cells. Note that the paths for the computation of $c(n)$ and $s(n)$ in the boundary cell, and $b(n)$ in the internal cell are feed forward paths, and hence can be pipelined using well known cutset pipelining techniques [22],[23]. Hence these are not of concern for speed of processing. However, the cell content $r(n)$, in the boundary cell as well as in the internal cell, is being updated recursively. This feedback loop becomes the critical path or the bottleneck for speed of processing.

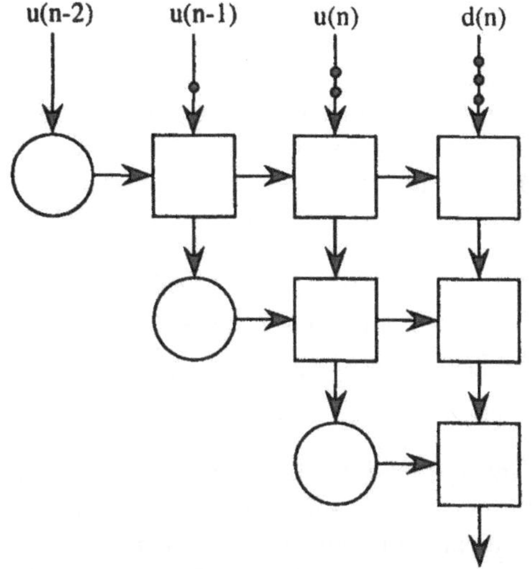

BOUNDARY CELL

INTERNAL CELL

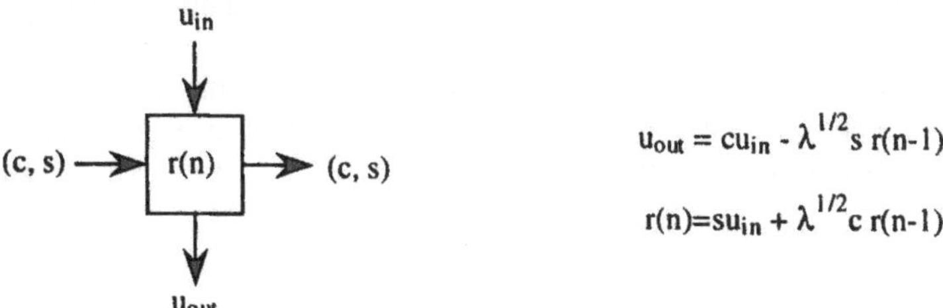

Figure 19.1 Systolic QRD-RLS algorithm.

Thus, the clock speed or proceesing speed is limited by the time required for 2 multiply operations and one square-root operation.

To increase the speed of the QRD-RLS, we could use look-ahead method [24],[25] or block processing techniques [26]. Now consider using look-ahead method to pipeline the recursive loop in the internal cell for $r(n)$. Using the equation for the cell content $r(n)$, and iteratively substituting back into itself, we get the look-ahead equation for $r(n)$ which depends only on $r(n-4)$, as given below

$$\begin{aligned} r(n) &= \lambda^2 c(n)c(n-1)c(n-2)c(n-3)r(n-4) \\ &+ \lambda^{3/2} c(n)c(n-1)c(n-2)s(n-3)u_{in}(n-3) \\ &+ \lambda c(n)c(n-1)s(n-2)u_{in}(n-2) \\ &+ \lambda^{1/2} c(n)s(n-1)u_{in}(n-1) + s(n)u_{in}(n). \end{aligned} \tag{14}$$

This defines the look-ahead of 4 steps for $r(n)$. However, note that the equation tends to grow very fast with increasing levels of look-ahead. In a practical implementation a high order look-ahead will hence not be feasible.

Block processing was used to speedup the QRD-RLS in [27]. However, the hardware will increase linearly with speedup and hence will be expensive. Processing of blocks of data can also be achieved using Householder transformations [28] or vectorized Gram-Schmidt pseudo orthogonalization [29]. The price paid is again a linear increase in hardware.

The square-root free forms of QRD-RLS [10],[11],[12],[13],[14],[15] also have the same problem of pipelining. The overhead for pipelining these algorithms vary, but they grow at least linearly with speedup.

19.4 PIPELINING FOR LOW-POWER DESIGNS

In the above, speed of processing was mentioned as the main motivation for pipelining. However, speed of processing can be traded for reducing power dissipation [30]. In practical designs, fast computational units, such as, fast adders, fast multipliers are used to meet the speed criterion. But the fast adders and multipliers consume much higher power than the standard optimized multiplier. But if the same circuit is instead pipelined, the propagation delay is reduced and hence these fast computational units are no longer required. This leads to power saving.

It is possible to find a simple relationship between pipelining and power savings (see [23] for details), if it is assumed that the Vdd voltage can be scaled.

The power dissipation in a chip can be given as

$$P = CV_{dd}^2 f, \tag{15}$$

where C is the effective switching capacitance, V_{dd} is the supply voltage and f is the frequency of operation. The propagation delay in any part of the circuit is given by

$$Prop.\ Delay = \frac{KV_{dd}}{(V_{dd}-V_t)^2}, \tag{16}$$

where K is a constant and V_t is the threshold voltage. From the above two equations (15),(16), we see that if Vdd is reduced, power dissipation reduces but increases propagation delay, thus reducing speed. We can thus tradeoff speed and power by scaling the supply voltage V_{dd}.

The effect of pipelining is to reduce the propagation delay in a circuit since the pipeline delays decimate the critical path. Since the propagation delay is lower in a pipelined circuit, a lower supply voltage would be sufficient to obtain the same speed. This leads to savings in power. The propagation delay in the critical path of a circuit after pipelining by p levels is

$$Prop.\ Delay\ (pipelined\ circuit) = \frac{KV_{ddp}}{p(V_{ddp} - V_t)^2}. \tag{17}$$

where V_{ddp} is the supply voltage being used in the pipelined circuit. Let us define a power scaling parameter α^2

$$\alpha^2 = \frac{P}{P_p} = \frac{V_{dd}^2}{V_{ddp}^2} \tag{18}$$

where P_p is the power dissipation in the pipelined circuit. Thus, α^2 defines the desired reduction in power. Given α^2, we want to find the amount of pipelining p required and still operate at the same original speed. To do this we equate (16) and (17), and use (18), and obtain the desired number of pipelining levels as

$$p = \frac{(V_{dd} - V_t)^2}{\alpha(V_{dd}/\alpha - V_t)^2}. \tag{19}$$

Of course, the number of levels obtained should be increased to an integral value. For example, to reduce the power dissipation by 4 times (i.e., $\alpha^2 = 4$), we get from (19) a value of $p = 3.55$, assuming $V_{dd} = 5V$ and $V_t = 1V$. Thus a 4 level pipeline would be sufficient.

Voltage scaling is a direct way of reducing power with higher levels of pipelining. However, if voltage scaling cannot be used, other methods such as using slower computational units etc. can be used to reduce power. The relationship (19) shows the extent to which power can be saved.

19.5 STAR-RLS SYSTOLIC ARRAY ALGORITHM

As described in the previous section, pipelining the recursive loops using look-ahead method is not practical because of the growth in the complexity of the equations. The focus of this work was to develop an algorithm which can be pipelined more easily. We make a couple of observations. Firstly, the Givens rotations are at the heart of problem i.e., the look-ahead equations in (14) are growing because of the structure of the Givens rotation matrix. Thus we would like to explore other rotation techniques which could alleviate the problem. Secondly, the goal of RLS is to minimize an exponentially weighted average of estimation errors as shown in (5). Suppose the relative weightage, given to estimation errors at different instants, is changed to a small extent, the resultant RLS solution is going to be a close approximation of the optimum solution. Moreover, if these added weightage reduces with time and settle down to unity at steady state, then the final solution will tend to the exact RLS solution. Thus, we will be solving a weighted least squares solution where the weights settle down to an exponential weighting at steady state.

Such an algorithm is indeed possible and is explained in the following. The derivation of the algorithm is not shown but only the algorithm is described. The Givens rotations are replaced by a new rotation. The sines and cosines in the Givens rotations are replaced by tangents and scaling factors. These rotations are referred to as *scaled tangent rotations (STAR)*. A single STAR rotation, corresponding to the Givens rotation (12), is given as

$$\begin{aligned} T(m,m) &= 1 \\ T(m,n) &= \frac{t(n)}{z(n)} \\ T(n,m) &= -t(n) \\ T(n,n) &= \frac{1}{z(n)} \\ T(i,i) &= 1, \ i \neq n,m \\ T(i,j) &= 0, \ otherwise, \end{aligned} \tag{20}$$

where

$$if\ y_{nm} \leq y_{mm}\ then\ z(n) = 1\ \&\ t(n) = \frac{y_{nm}}{y_{mm}}\ else\ z(n) = \left|\frac{y_{nm}}{y_{mm}}\right|\ \&\ t(n) = sign\left(\frac{y_{nm}}{y_{mm}}\right) \tag{21}$$

A STAR rotation is represented as

$$T = \begin{bmatrix} 1 & \frac{t(n)}{z(n)} \\ -t(n) & \frac{1}{z(n)} \end{bmatrix}. \tag{22}$$

In the STAR rotation $t(n)$ is analogous to a tangent and $z(n)$ is a scaling factor. The scaling factor $z(n)$ ensures the stability of the algorithm by keeping $t(n) < 1$. At the same time $z(n)$ is a weight factor which leads to a weighted least squares solution instead of a uniform weighted or exponentially weighted solution. It can be shown that $z(n)$ goes to unity at steady-state. The main difference as compared to the Givens rotations is that the STAR rotations are not exactly orthogonal, but settle down to an orthogonal solution with time. They can be used to triangularize the data matrix and obtain a solution weight vector as in QRD-RLS. The STAR-RLS algorithm implemented on a systolic array is shown in Fig. 19.2. Simulation results show that the performance of the STAR-RLS algorithm is close to that of the QRD-RLS algorithm (see Fig. 19.5).

19.5.1 Characterization of the Solution

We need to investigate the characteristics of the solution thus obtained. The STAR-RLS algorithm minimizes the performance index $\xi'(n)$ instead of $\xi(n)$ (7), where

$$\xi'(n) = \| B(n)\Lambda^{1/2}(n)\varepsilon'(n) \|^2 . \tag{23}$$

Here, $B(n)$ is the resulting transformation due to all the STAR rotations and $\varepsilon'(n)$ is the error vector for STAR-RLS (defined similar to (4)). $B(n)$ is not exactly an orthogonal matrix but tends to become orthogonal at steady-state. The transformation $\mathbf{B}(n)$ triangularizes the exponentially weighted data matrix $\Lambda^{1/2}(n)\mathbf{A}(n)$

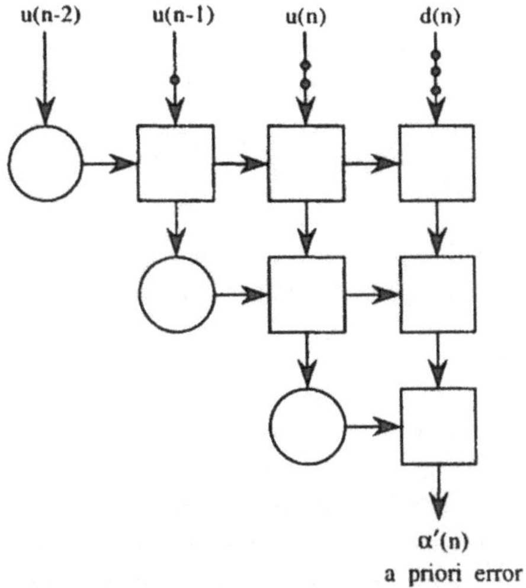

BOUNDARY CELL

u_{in} → r(n) → $(\frac{t}{z}, \rho)$

If $u_{in} \leq \lambda^{1/2} r(n-1)$

$$t = \frac{u_{in}}{\lambda^{1/2} r(n-1)} \;;\; z = 1 \;;\; \rho = 0$$

else

$$t = \mathrm{sign}\left(\frac{u_{in}}{\lambda^{1/2} r(n-1)}\right) \;;\; z = \left|\frac{u_{in}}{\lambda^{1/2} r(n-1)}\right| \;;\; \rho = 1$$

end

$$r(n) = \lambda^{1/2} r(n-1) + \frac{t}{z} u_{in}$$

INTERNAL CELL

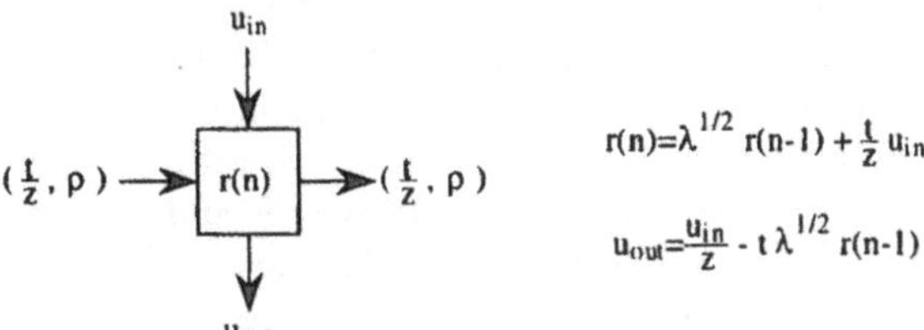

Figure 19.2 Systolic array implementaion of STAR-RLS (the array is same as in Fig. 19.1).

and hence we can write

$$B(n)\Lambda^{1/2}(n)A(n) = \begin{bmatrix} R'(n) \\ \mathbf{0}_{n-M,M} \end{bmatrix}, \quad n \geq M, \tag{24}$$

where $R'(n)$ is an M-by-M upper triangular matrix and $\mathbf{0}$ is a $(n-M)$-by-M null matrix. The desired signal vector is transformed as

$$B(n)\Lambda^{1/2}\mathbf{b}(n) = \begin{bmatrix} \mathbf{p}'(n) \\ \mathbf{v}'(n) \end{bmatrix}, \tag{25}$$

where $\mathbf{p}'(n)$ is a M dimensional vector and $\mathbf{v}'(n)$ is a $(n-M)$ dimensional vector. The least squares solution for the problem (23) is then given by

$$\mathbf{w}'(n) = R'^{-1}(n)\mathbf{p}'(n), \qquad n \geq M. \tag{26}$$

These transformations are carried out recursively. At time $(n-1)$, we have an upper triangular matrix $R'(n-1)$. When a new set of data $\mathbf{u}(n)$ and $d(n)$ is processed, the new row due to the input data is annihilated using a set of M rotations. The transformation matrix $B(n)$ is effectively the product of all these transformations over time. Apart from the rotation being different, the recursive update scheme of STAR-RLS can be explained similar to the QRD-RLS algorithm (see [31],[23],[32]).

In many applications like in beamformers, prediction error filters and adaptive noise cancellers the least-squares weight vector is itself not required. The estimation error is sufficient for such applications. A modified form of systolic array to directly compute the estimation error for the QRD-RLS algorithm was designed in [33]. This avoids the need for backsubstitution to determine the estimation error. It can be shown that the estimation error for the STAR-RLS can also be computed without back-substitution, by modifying its systolic array [23],[31].

19.5.2 Hardware implementation of STAR-RLS

The hardware complexity and the inter-cell communication is reduced as compared to the QRD-RLS algorithm. In the STAR-RLS systolic array implementation (Fig. 19.2), the quantities $\frac{t}{z}$ and ρ are used. The variable ρ is a 1 bit quantity and is zero when z is unity. We can thus exploit the fact that one of the two quantities t and z will have a magnitude of unity at any iteration. The inter cell communication required in STAR is one word and one bit instead of two words that would be required in QRD-RLS. It can be shown that the boundary cells and internal cells in STAR-RLS require three operations (multiplications/divisions/square-roots) each (see [23],[31]). In comparison the QRD-RLS algorithm requires 6 operations in the boundary cells and 5 operations in the internal cells. The computational requirement of the STAR-RLS is even less than that for the square-root free forms of QRD-RLS. When comparing the number of operations required for sqrt-free QRD-RLS the exponentially weighting factor needs to be considered. This is not done in most of the references ([10],[11],[13],[14],[15]).

19.5.3 Look-ahead equations with STAR-RLS

Consider a look-ahead equation for the content of any cell in the STAR-RLS systolic array of Fig. 19.2. Corresponding to the look-ahead equation for QRD-RLS

(14), the equation for STAR-RLS is

$$\begin{aligned} r(n) &= \lambda^2 r(n-4) + \tfrac{\lambda^{3/2} t(n-3) u_{in}(n-3)}{z(n-3)} + \tfrac{\lambda t(n-2) u_{in}(n-2)}{z(n-2)} \\ &+ \tfrac{\lambda^{1/2} t(n-1) u_{in}(n-1)}{z(n-1)} + \tfrac{t(n) u_{in}(n)}{z(n)}. \end{aligned} \quad (27)$$

Note that the equation adds just one extra term for each additional pipelining stage. Also, since the expression has a largely regular structure, it can be implemented more easily. The complexity of this equation does not grow like in QRD-RLS.

19.5.4 Quasi-stable state in STAR-RLS

In [34], it was shown that the boundary cells in the QRD-RLS systolic structure tend to reach a quasi-steady-state. In [23],[31] we show that the boundary cells of STAR-RLS also reach a quasi-steady-state. In such a state the cell content $r(n)$ of the boundary cells tends to reach a constant value and varies very slowly with input data. This property is used in the following sections in developing the pipelined STAR-RLS algorithm and in finite-precision analysis of STAR-RLS.

19.6 PIPELINED STAR-RLS (PSTAR-RLS) ARCHITECTURE

As was shown in the previous section the STAR-RLS has a look-ahead equation which is much more simpler than the QRD-RLS algorithm. However, implementing this look-ahead equation would still require a significant increase in hardware with high levels of pipelining. We would like to design an algorithm such that pipelining would require only a small overhead. To design such an algorithm we make further approximations in the look-ahead equation (27). Since the cell contents of the boundary reach a quasi-stable state (see Section 2.5.4), the STAR rotations defined with respect to a delayed value of cell content $r(n-p)$ would be a good approximation, where p is the number of levels of pipeline. The usage of exponential weighting factor λ is also modified for the look-ahead equation such that exponential weighting is applied with minimum operations (see [23],[31],[35]). The resulting algorithm referred to as pipelined STAR-RLS or PSTAR-RLS. The recursions for the boundary cell can be given as

$$r(n) = \lambda^{p/2} r(n-p) + \sum_{i=0}^{p-1} \frac{t(n-i) u_{in}(n-i)}{z(n-i)} \quad (1)$$

where

$$\begin{aligned} &If \;\; u_{in}(n) \;\leq\; r(n-p) \\ &\quad t(n) \;=\; \frac{u_{in}(n)}{r(n-p)} \;\; and \;\; z(n) = 1; \\ &else \\ &\quad t(n) \;=\; sign\left(\frac{u_{in}(n)}{r(n-p)}\right) \;\; and \;\; z(n) = \left|\frac{u_{in}(n)}{r(n-p)}\right|. \end{aligned} \quad (2)$$

Note that the approximations result in a simplified equation. The complete PSTAR-RLS algorithm is defined in Fig. 19.3 and the hardware for the cells is

Table 19.1 Clock Rate Comparison for Different Algorithms

Algorithm	Clock Period
QRD-RLS	$2T_m + T_{sq} + T_a$
Sq.free QRD-RLS [13]	$T_m + 2T_d + T_a$
STAR-RLS	$2T_m + T_d + T_a + T_s$
PSTAR-RLS	$\frac{T_m+T_d+2T_a+T_s}{p}$

shown in Fig. 19.3. Note that only extra adders and registers are required for PSTAR-RLS as compared to STAR-RLS. The two recursive loops in the boundary cell now have p delay registers which can be used for pipelining. The recursive loop in the internal cell also has p delays. The clock period of a circuit is equal to the computation time of the longest path without any delays. For the PSTAR-RLS algorithm the longest path occurs in the boundary cell (see Fig. 19.2(a)). This path has one multiply, one divide and 2 add operations and has p delays. If these p delays are distributed evenly, the minimum clock period is given by

$$C_{PSTAR} = \frac{T_m + T_d + 2T_a + T_s}{p} \tag{3}$$

where T_m, T_d, T_a, T_s are the computation times for multiplier, divider, adder, and switch respectively. The computation time of multiplier and divider are higher than the time for switch or adder operation. The minimum clock periods for the QRD-RLS, square-root free QRD-RLS [13] and the serial STAR-RLS algorithms are compared with that of the PSTAR-RLS in Table 1 (where T_{sq} is the computation time for a square-root).

The PSTAR-RLS algorithm makes it possible to obtain high speeds with minimum overhead. The level of pipelining should be chosen based on requirements and practical feasibility. The performace of the PSTAR-RLS and STAR-RLS algorithms are compared with the QRD-RLS in Fig. 19.5. Here at iteration 200 the channel changes abruptly. Firstly, note that the performance of STAR-RLS is similar to the QRD-RLS. When started with initial state there seems to be degradation in convergence speed for the PSTAR-RLS with increasing levels of pipeline (p). However, once the RLS filter has settled down, any changes in the channel are tracked by PSTAR-RLS at almost the same rate.

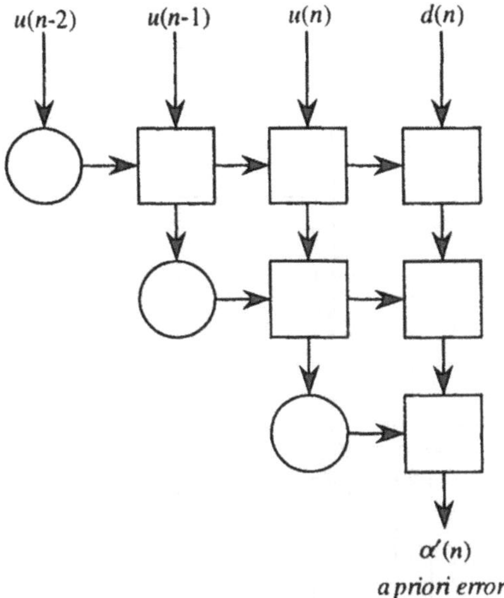

BOUNDARY CELL

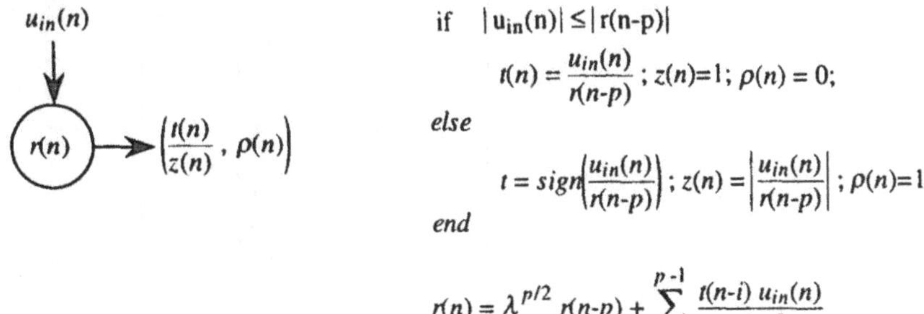

$u_{in}(n)$
$r(n)$
$\left(\frac{t(n)}{z(n)}, \rho(n)\right)$
if $|u_{in}(n)| \leq |r(n-p)|$
$t(n) = \frac{u_{in}(n)}{r(n-p)}$; $z(n)=1$; $\rho(n) = 0$;
else
$t = sign\left(\frac{u_{in}(n)}{r(n-p)}\right)$; $z(n) = \left|\frac{u_{in}(n)}{r(n-p)}\right|$; $\rho(n)=1$;
end
$r(n) = \lambda^{p/2} r(n-p) + \sum_{i=0}^{p-1} \frac{t(n-i)\, u_{in}(n)}{z(n-i)}$

INTERNAL CELL

$u_{in}(n)$

$\left(\frac{t(n)}{z(n)}, \rho(n)\right)$ → $r(n)$ → $\left(\frac{t(n)}{z(n)}, \rho(n)\right)$

$u_{out}(n)$

$$r(n) = \lambda^{p/2} r(n-p) + \sum_{i=0}^{p-1} \frac{t(n-i)\, u_{in}(n)}{z(n-i)}$$

$$u_{out}(n) = \frac{u_{in}(n)}{z(n)} - t(n)\, r(n-p)$$

Figure 19.3 Systolic array implementation of PSTAR-RLS. (the array is same as in Fig. 19.1)

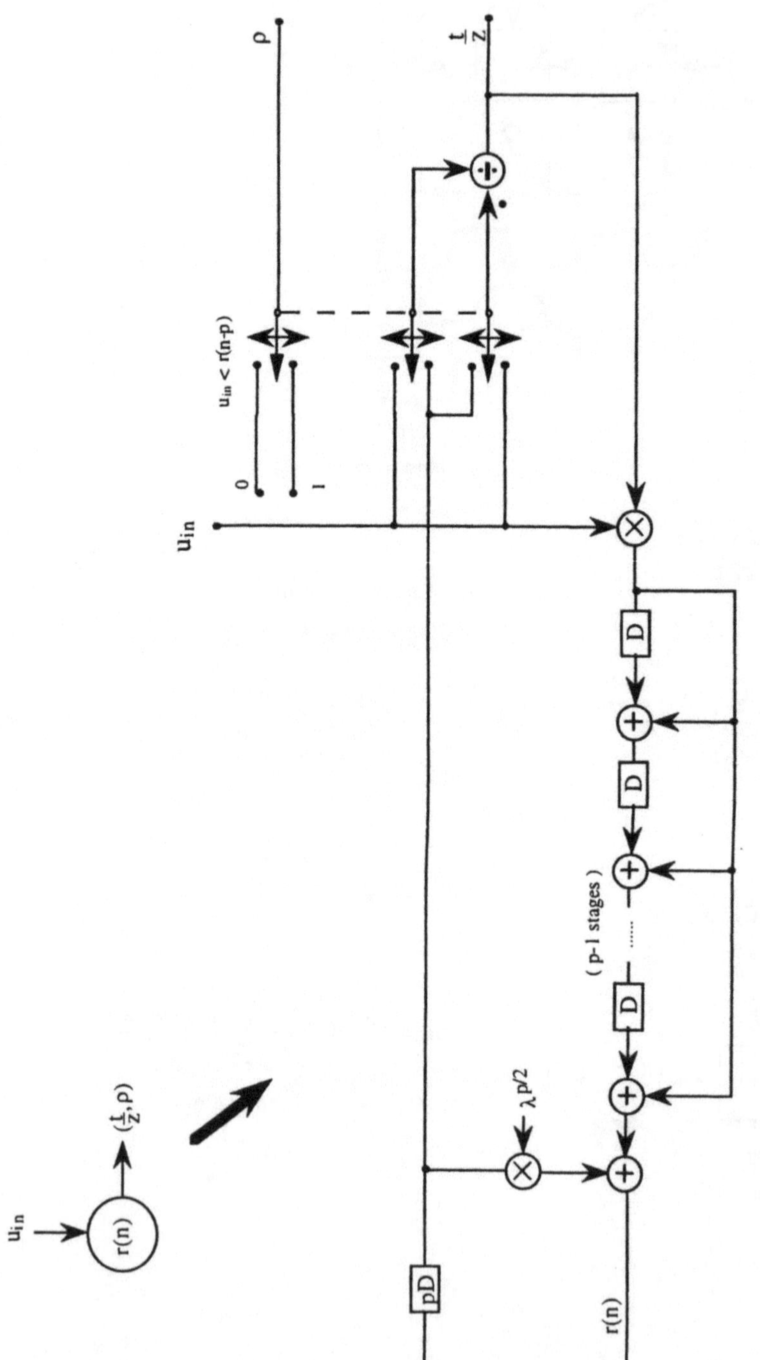

(a) PIPELINED BOUNDARY CELL HARDWARE

Fig. 19.4 The cell hardware for the pipelined STAR-RLS (p is the speedup used).

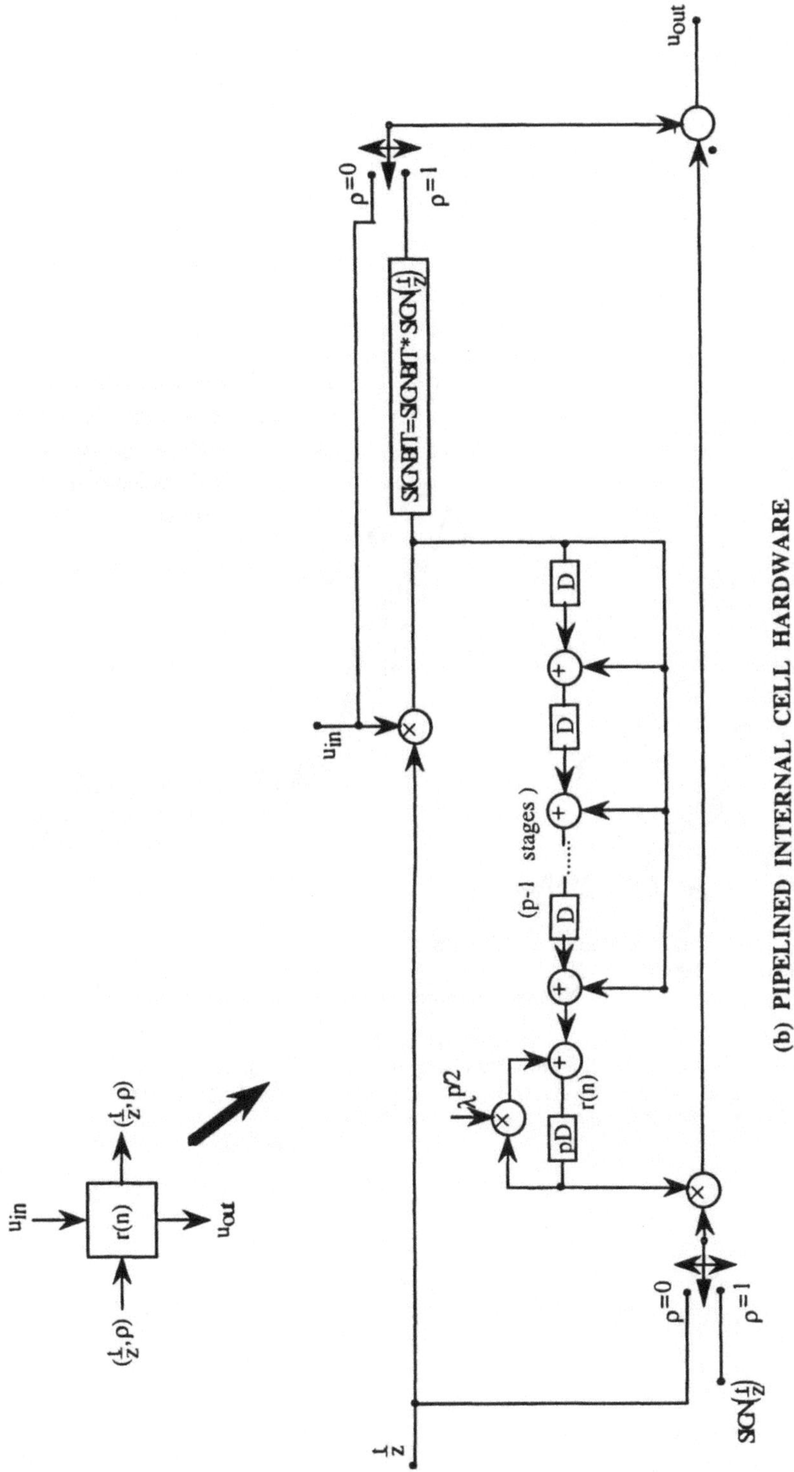

(b) PIPELINED INTERNAL CELL HARDWARE

Fig. 19.4 (continued).

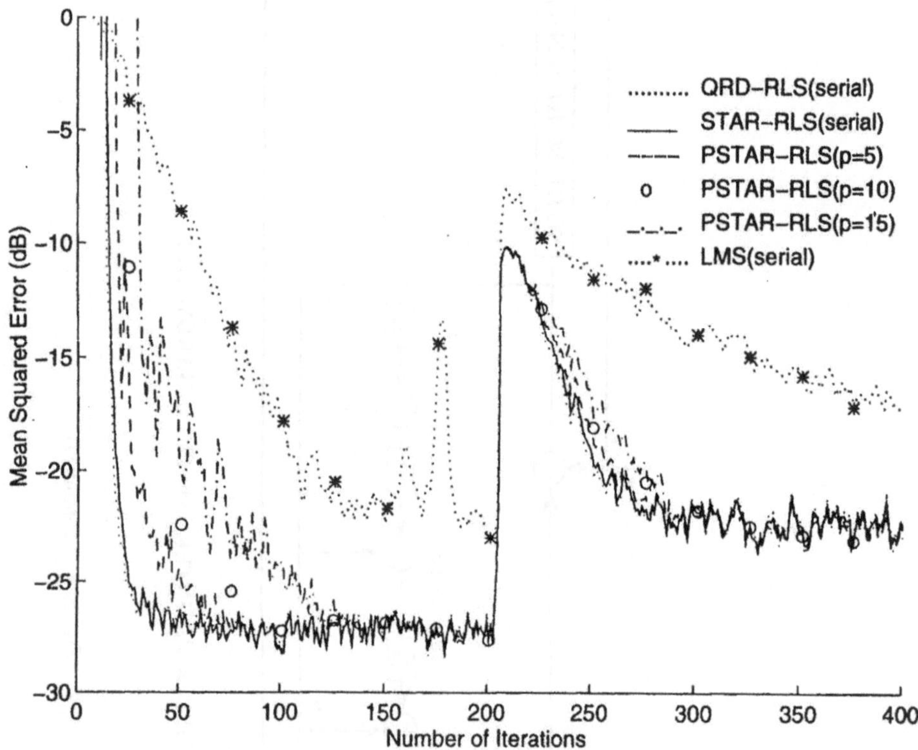

Figure 19.5 Performance with a abruptly changing channel for QRD-RLS ($\lambda = .94$), PSTAR-RLS ($\lambda = .90$), STAR-RLS ($\lambda = .90$) and LMS ($\mu = .065$). The results are averaged over 200 runs.

19.7 NUMERICAL STABILITY ANALYSIS

In this section, we consider the stability and dynamic range properties of the STAR-RLS algorithm.

19.7.1 Stability

The accumulation of quantization errors is an important concern in the implementation of an adaptive filter. An adaptive filter may run over millions of data samples before being reinitialized. Small quantization errors, which appear insignificant over small data sets, can lead to overflows and a consequent breakdown of the filter in the long run. The QRD-RLS algorithm was recently proved to be stable in the bounded input/bounded output (BIBO) sense in [36]. It can be shown that the STAR-RLS algorithm is also stable in the BIBO sense (see [23],[31]).

19.7.2 Numerical benchmark of STAR-RLS

In [37] a numerical error analysis of Givens rotation was carried out. A similar analysis has been done for the different square-root free Givens rotation (see [11],[15]). In this analysis the bound on the finite-precision error for a single Givens rotation implemented in floating-point representation is determined. This bound predicts the finite-precision performance of the algorithms using these rotations. To compare with these algorithms, we determine a corresponding bound for the STAR rotation. The analysis carried out here is similar to that in [37]. The floating point representation is assumed to have a k digit mantissa.

Consider the application of scaled tangent rotations on a vector $\mathbf{a}$. Let T be the scalet tangent rotation which affects the ith and j^{th} elements of the vector $\mathbf{a}$. The rotated vector when computed in floating point arithmetic is

$$\mathbf{b} = fl\{\bar{T}\mathbf{a}\} \tag{1}$$

where $fl\{\}$ represents quantization operation to reduce the quantity to a floating point number and

$$\bar{T} = fl\{T\} \tag{2}$$

where T is a STAR rotation matrix of the form (22). The STAR rotation is computed in different ways depending on whether $z = 1$ or not. The analysis is hence made separately for the two cases. We find that (see derivation in [23])

$$\|fl\{\bar{T}\mathbf{a}\} - T\mathbf{a}\|_2 \leq 4.13(a_i^2 + a_j^2)^{1/2}2^{-k} \quad when\ \ z = 1 \tag{3}$$

and

$$\|fl\{\bar{T}\mathbf{a}\} - T\mathbf{a}\|_2 \leq 6.26(a_i^2 + a_j^2)^{1/2}2^{-k} \quad when\ \ z \neq 1, \tag{4}$$

where a_i and a_j are the ith and j^{th} components of vector $\mathbf{a}$.

We thus see that the quantization error is different for the two cases. The case when $z \neq 1$ holds only for a few initial iterations. Hence, for all practical purposes we can use the bound given by (3).

The bound for the round-off error for the STAR rotations is lower than that of the Givens rotations which has a factor of 6.0 in the place of 4.13 above. The main reason for this is that fewer arithmetic operations are required in computing the STAR rotations. The square-root-free algorithm of [11] leads to a factor of 7.5 (see [15] for a comparison Table).

19.7.3 Dynamic Range

The dynamic range of a variable or a signal in an architecture is the bound on its maximum value. The dynamic range analysis defines the stability of the architecture, and is needed to determine the wordlength needed for the variable under consideration. The dynamic range for the cell contents in the QRD-RLS was determined in [34]. We carried out a similar dynamic range analysis for the STAR-RLS and PSTAR-RLS algorithms ,[23]. The results show that the dynamic range for STAR-RLS and PSTAR-RLS are similar to that of QRD-RLS.

The bound on the cell content for the STAR-RLS is

$$\lim_{n\to\infty} |r(n)| \leq \frac{u_{max}}{\lambda^{1/4}\sqrt{1-\lambda^{1/2}}} = \Re', \tag{5}$$

and the bound on the cell content for the PSTAR-RLS is

$$\lim_{n\to\infty} r(n) < \frac{\sqrt{p}u_{max}}{\sqrt{1-\lambda^{p/2}}} = \Re'_p, \tag{6}$$

It can also be shown that

$$\Re' \approx \Re'_p. \tag{7}$$

19.8 FINITE-PRECISION ANALYSIS

The accumulation of quantization noise is an important concern in the implementation of adaptive digital filters. The stability and dynamic range analysis have been discussed in the previous section. While this ensures that the QRD-RLS and STAR-RLS filters are stable in a finite-precision environment, it still does not give any further insight into the quality of performance. The cost of implementation is strongly dependent on the number of bits used to represent the signals in the filter. Hence one would like to use the fewest number of bits without affecting the performance of the algorithm. This motivates the finite precision error analysis of QRD-RLS and STAR-RLS algorithms.

19.8.1 Previous Work

The effect of finite-precision on adaptive filters has been considered by many authors, for example, for the LMS algorithm [39], for lattice filters [40]-[42] and for recursive least-squares (RLS) [43]-[46]. The RLS algorithms considered in [43]-[46] refer to the original algorithm which does not use any decomposition techniques. [47] also gives a comparitive study of some of previous finite-precision analysis papers. It has been found that using QR decomposition instead of a Cholesky decomposition, for solving a set of normal equations, would reduce the wordlength requirement by about half [48]. Simulation results have been used before to demonstrate that the QRD-RLS algorithm can be implemented with as few as 12 bits in floating-point arithmetic [19]. The only papers that explore a theoretical analysis of the finite-precision behavior of the QRD-RLS, to the best of our knowledge, are [49],[50]. In [50], the classical backward error analysis (see [47]) is used to determine bounds on the errors resulting due to rounding. In [49], the error propagation of quantization noise is investigated assuming a single quantization error has occurred. The analysis in [40] considers only a single stage of QR lattice filter.

19.8.2 Aim of the Analysis

In reality, quatization error occurs at every arithmetic operation depending on the wordlength used, and these propagate from one cell to the other. Thus, the quantization error shown at the final output, say the estimated weight vectors, is an accumulation all such errors in the whole architecture. A comprehensive finite-precision analysis would take all of this into consideration. A statistical average of the deviation due to finite-precision can be a figure of merit for the finite-precision behavior of the algorithm. Such a comprehensive analysis is lacking for QRD-RLS. This analysis is difficult because of the complexity of the RLS algorithm. Even if such an analysis is carried out the final equation may be too complicated to be of any use. In this section we present how this difficulty is overcome by exploiting the steady-state properties of the algorithm. Simple closed form equations are determined for the finite-precision performance of QRD-RLS, STAR-RLS and PSTAR-RLS and its dependence on the wordlength used. Since fixed as well as floating point representations are used, both need to be considered.

19.8.3 Assumptions

The analysis here is based on the assumptions about the roundoff error, such as, being independent of the operands. These are not always satisfied (see [51]). However, we assume that the deviation from the assumed model is not serious enough to affect the results. Such roundoff error models have been successfully used in many papers [39]-[41],[44]-[45] etc.

The QRD-RLS and STAR-RLS (or PSTAR-RLS) are different algorithms and the forgetting parameter λ for the two algorithms can also have different significance. We found that using a value of $\lambda = k$ for QRD-RLS is equivalent to using a value of $\lambda = k^2$ for STAR-RLS (or PSTAR-RLS). For example, the convergence plot for QRD-RLS with $\lambda = .99$ is similar to the convergence plot of STAR-RLS with $\lambda = .98$. This difference arises because of the way the STAR-RLS algorithm has been developed. To make it easy to compare the algorithms, we introduce a common parameter of effective forgetting factor $\underline{\lambda}$. We replace λ by $\underline{\lambda}$ for QRD-RLS and λ by $\underline{\lambda}^2$ for the STAR-RLS (and PSTAR-RLS). Now for a given value of $\underline{\lambda}$, the performance of QRD-RLS and STAR-RLS are similar.

19.8.4 Quasi-Steady State Assumption

A typical phenomenon is seen in the breakdown of the RLS algorithm due to quantization noise. The algorithm initially converges quite nicely but after some time starts degrading gradually and finally breaks down. Apparently, the quantization noise accumulates over a period of time before breaking down. Thus the initial convergence is not dependent on the wordlength assuming it is reasonbly high. This suggests that it is possible to do the finite-precision analysis assuming that convergence has already occured, and a quasi-steady state is reached. The finite-precision analysis is made feasible by exploiting this quasi-stable state of QRD-RLS, STAR-RLS and PSTAR-RLS. At steady state the cell contents of the boundary cells for the three algorithms reaches a near constant state. The steady-state cell content

Table 19.2 Deviation in Estimation Error with Fixed-Point Representation

Algorithm	$E[\{\bar{e}(n)-e(n)\}^2] \approx$
STAR-RLS	$\sigma_u^2\left\{\left(1+\frac{1}{M\sigma_u^2}\right)^M-1\right\}\left(2.5M+\frac{\lambda}{1-\lambda}\right)\sigma_\epsilon{}^2$
PSTAR-RLS	$\sigma_u^2\left\{\left(1+\frac{1}{M\sigma_u^2}\right)^M-1\right\}\left(2M+\frac{p}{1-\lambda^p}\right)\sigma_\epsilon{}^2$
QRD-RLS	$\sigma_u^2\lambda^{M-1}\left\{\lambda^M\left(1+\frac{1}{M\sigma_u^2}\right)^M-1\right\}\left(5M+\frac{1.5\lambda}{1-\lambda}\right)\sigma_\epsilon{}^2$

for the first boundary cell, for example, is determined to be (see [23],[31],[52])

$$r_\infty^2(QRD)=\frac{\sigma_u^2}{1-\lambda};\;\; r_\infty^2(STAR)=\frac{\sigma_u^2}{\lambda(1-\lambda)};\;\; r_\infty^2(PSTAR)=\frac{p\sigma_u^2}{1-\lambda^p}. \qquad (1)$$

19.8.5 Fixed-Point Performance

In this section we analyze the performance of the three algorithms when fixed-point arithmetic is used. All quantities in the algorithm are represented with k_i bits for the integer part and k bits for the fractional part (i.e., k_i bits before the binary point and k bits after binary point). It is assumed that k_i is sufficient to avoid any overflow (for the simulation examples presented in this paper, $k_i = 4$). Thus, the fixed-point representation would require $k_i + k + 1$ bits, with one bit being used for sign. We also assume that rounding is used in the operations. We can model fixed-point errors due to multiplication and division as zero mean additive white noise process with a variance of $\sigma_\epsilon{}^2 = 2^{-2k}/12$ [39],[45]. If $*$ represents a multiply or divide operation, then

$$fix(a*b) = a*b+\epsilon \qquad (2)$$

where ϵ is zero mean white noise process, independent of a and b, with a variance of $\sigma_\epsilon{}^2$. Additions and subtractions do not introduce any error (assuming there is no overflow). The fixed-point deviation expressions are derived separately for the three algorithms below. The approach used here is to follow the arithmetic operations in the different cells and keep track of the errors which arise due to finite-precision quantization, and find the average value of the same (see [53],[31],[52]). All symbols with an overbar represent finite-precision quantities. The results are shown in Table 21.2. Simulation results are compared with the theoretical plots in Fig. 19.6. The results show a good match between the two. For results on the deviation of cell content and optimum weight vector see [52],[23].

19.8.6 Floating-Point Performance

In this section we consider the floating-point representation of the three algorithms. A floating-point representation consists of a $k+1$ bit mantissa and a k_e

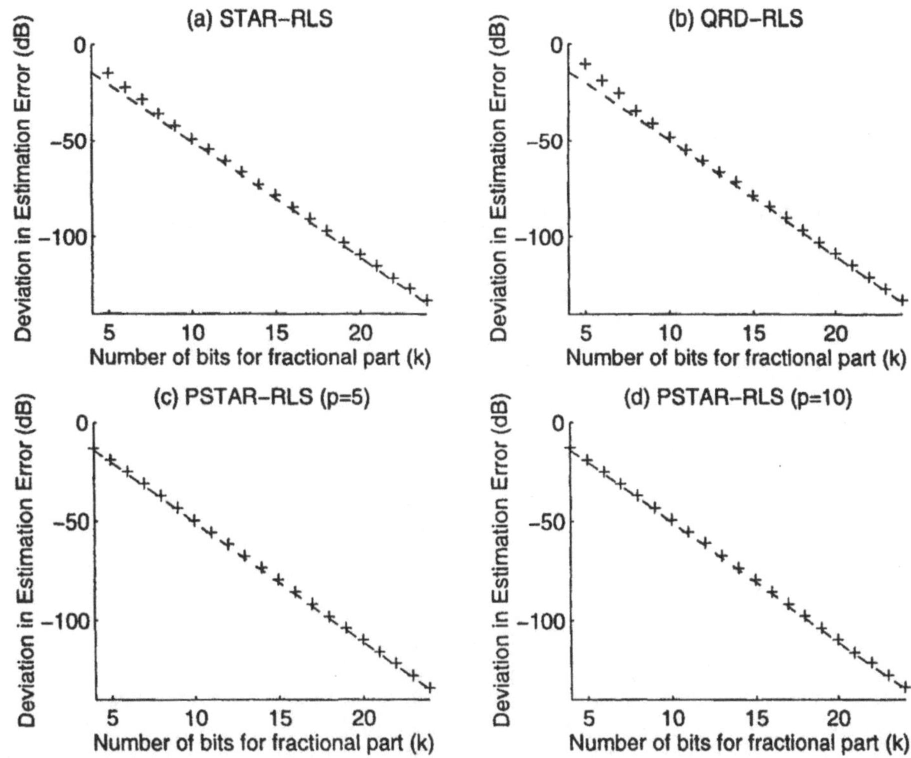

Figure 19.6 The average deviation in estimation error in fixed-point arithmetic implementation (a) STAR-RLS (b) QRD-RLS (c) PSTAR-RLS (p=5) (d) PSTAR-RLS (p=10) ("– – –" theoretical ; "+" Simulation results)

bit exponent. For the mantissa 1 bit is used as sign bit. Rounding is used on the mantissa to bring it down to k bits. The exponent is an integer and it also has a sign bit. Thus the number representation here requires $k_e + k + 1$ bits. In contrast to fixed-point arithmetic, in a floating-point representation any operation including addition or subtraction involves a quantization error. If "#" represents a division, multiplication, addition or subtraction operation, then quantizing the result to a floating-point format would involve an error ϵ, i.e.,

$$float(a\#b) = a\#b(1+\epsilon) \tag{3}$$

where ϵ is assumed to be a zero-mean noise process, independent of a and b, with a variance of ${\sigma_\epsilon}^2$ which is equal to $.18 * 2^{-2k}$ [44],[54].

Using this model for the error due to floating-point representation, we go through the analysis (see [53],[31],[52] for more details). The deviation in the final estimation error is presented in Table 21.3. Simulation and theoretical results are compared in Fig. 19.7. Again there is a good match between the two. For results on the deviation of cell content and optimum weight vector see [52],[23].

Table 19.3 Deviation in Estimation Error with Floating-Point Representation

Algorithm	$E[\{\bar{e}(n) - e(n)\}^2] \approx$
STAR-RLS	$\sigma_u^2 \left\{ \left(1 + \frac{1}{M\sigma_u^2}\right)^M - 1 \right\} \left(1.5M + \frac{2\underline{\lambda}^2}{1-\underline{\lambda}}\right) {\sigma_\epsilon}^2$
PSTAR-RLS	$\sigma_u^2 \left\{ \left(1 + \frac{1}{M\sigma_u^2}\right)^M - 1 \right\} \frac{(M+3)(1-\underline{\lambda}^{2p})+(1+\underline{\lambda}^{2p})^2}{1-\underline{\lambda}^{2p}} {\sigma_\epsilon}^2$
QRD-RLS	$\sigma_u^2 \underline{\lambda}^{M-1} \left\{ \underline{\lambda}^M \left(1 + \frac{1}{M\sigma_u^2}\right)^M - 1 \right\} \left(9M\sigma_u^2 + \frac{8}{1-\underline{\lambda}}\right) {\sigma_\epsilon}^2$

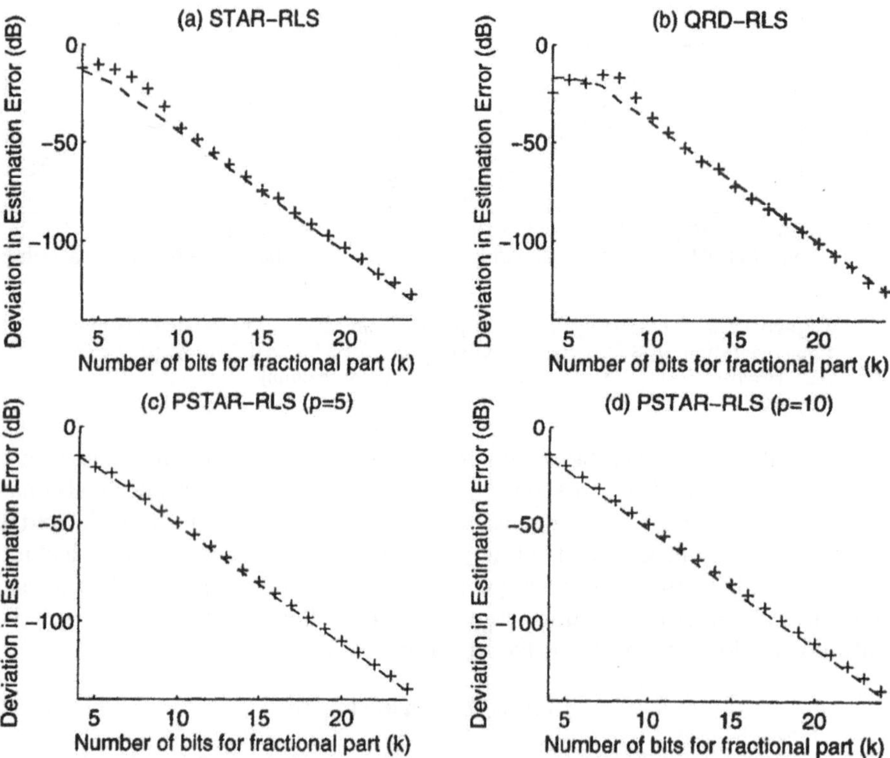

Figure 19.7 The average deviation in estimation error in floating-point implementation (a) STAR-RLS (b) QRD-RLS (c) PSTAR-RLS (p=5) (d) PSTAR-RLS (p=10) ("– – –" theoretical ; "+" Simulation results).

19.8.7 Analysis of the results

The deviation expressions for the three algorithms have a similar structure. The finite-precision performance (fixed and floating-point) degrades dramatically as the forgetting factor approaches 1. Traditionally, the value of forgetting factor decides the tradeoff between tracking property and misadjustment error ([1],page 405). A low value of forgetting factor implies better tracking but a higher misadjustment noise. Our results show that the finite-precision error is also a factor in the decision of the value of forgetting factor, especially when a hardware implementation is required. The theoretical expressions agree well with our simulation results. In general, the PSTAR-RLS and STAR-RLS algorithms perform a little better than the QRD-RLS algorithm, with PSTAR-RLS having the best performance. It is interesting to note that the pipelined algorithm has better performance than the serial version. The algorithms can be implemented with very few bits; for example, for a filter size of 11, about 10 bits for the fractional part would be sufficient for these algorithms. This requirement would vary with the filter size and the forgetting factor used.

19.9 A 100 MHz PIPELINED RLS ADAPTIVE FILTER

It is difficult to implement a RLS adaptive filter on a single chip. There was only one attempt made to design an RLS type of filter chip to the best of our knowledge [55]. But, here a wafer scale integration was needed to implement an RLS type of filter for adaptive nulling. This design used 96 cordic cells which were implemented in a wafer 16 square inches in area. This wafer was implemented in $2\mu m$ technology and the maximum clock speed was 12 MHz.

In this section we present the VLSI design of a 4-tap PSTAR-RLS algorithm on a single chip [56].

19.9.1 Architecture

The aim of this design was to implement a 4-tap PSTAR-RLS with a pipelining level of $p = 5$. A 4-tap PSTAR-RLS systolic array would have 4 boundary cells and 10 internal cells (see [35],[23]). To simplify the implementation we design a *composite* cell which can act as a boundary cell as well as an internal cell. The composite cell requires 4 multiplier/divider operators, while the internal cell and boundary cells, individually, require only 3. This hardware overhead, however, simplifies the design to a great extent.

The 4-tap PSTAR-RLS architecture will not fit in a single chip with 1.2 μ technology. Using the transformation technique of folding, [57], we fold the systolic arcitecture by 14 times to operate on a single composite cell as shown in Fig. 19.8. The cells are folded row-wise from top to bottom. This is the simplest folding set, and it should be noted that other folding sets could be used leading to different architectures. The folding set was so chosen as to minimize the hardware overhead. In Fig. 19.8 note that there is very little extra circuitry added to enable folding.

The folded hardware will process the data 14 times slower. There is hence a speed loss due to folding. However, folding by a factor of 14 scales the number of delays by 14 times. Thus, in the recursive loops we have 70 delays instead of 5 delays. The recursive loops can be pipelined at a finer level now. Now we can have a clock rate that is 14 times higher, which compensates for the speed loss due to folding. Hence, there is no loss in speed due to folding.

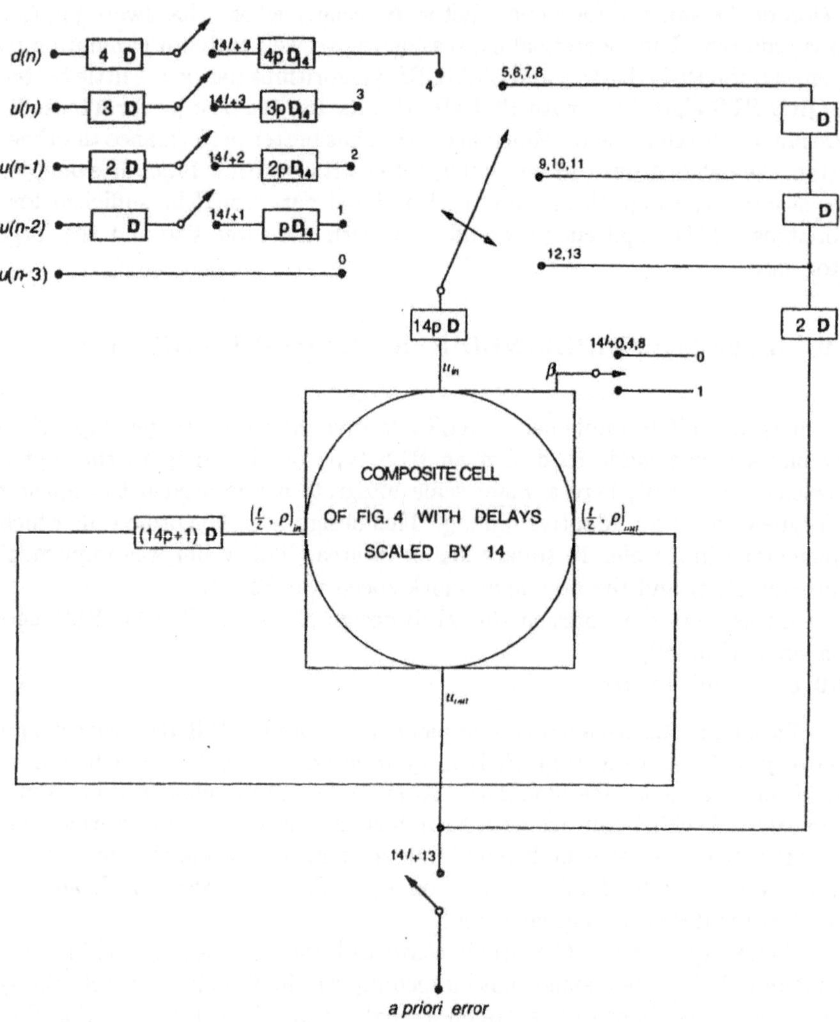

Figure 19.8 Folded systolic array for 4-tap PSTAR-RLS algorithm.

In the final chip circuit the available delays are moved around using retiming [58], so that all operators are bit-level pipelined. The maximum propagation delay is kept below that of a bit adder at all parts in the circuit.

19.9.2 Redundant Number System Arithmetic

We used fixed-point arithmetic since the arithmetic operators would be easier to design. The dynamic range of the quantities is also not so high as to warrant a floating-point representation.

In the previous Section ([52]), a finite-precision analysis was done and that knowledge can be used to determine the wordlengths to be used. In the design of this chip, we decided not to use rounding since that would increase the delay in the arithmetic operators. A direct truncation is instead used. This would mean that more bits would be needed than that is suggested by the analysis in the previous section (see also [53],[23],[52]).

Based on the finite-precision analysis and on simulations with fixed-point arithmetic the wordlength for the the different signals were chosen. It was found that 12 bits for cell contents and outputs, and 8 bits for rotation parameters are sufficient. The performance is almost indistinguishable as compared to the infinite precision results.

The normal arithmetic representation is the binary representation. However, there is growing interest in the use of unconventional arithmetic, such as redundant arithmetic [59]. In a redundant number representation with radix 2, a digit can take the values of $-1, 0, +1$ instead of just $0, 1$ as in binary. Each digit is represented by two bits. Redundant number arithmetic has the property of carry-free addition. This leads to fast addition and multiplication. However, all bus sizes are doubled as compared to a binary representation. We use a hybrid form of representation [60], i.e., we use both redundant and binary representation. The representation at different points is chosen so as to get maximum benefit from the redundant representation, and reduce the extra routing needed due to bus size doubling.

Conventional dividers have a delay proportional to the square of the wordlength. The divider used in this chip is a non-restoring redundant divider [60]. The delay here is $O(n \log n)$, where n is the wordlength. We used a redundant arithmetic based Booth encoded tree multiplier. The partial products here are added in a tree fashion giving a computation delay of $O(\log n)$.

19.9.3 Timing

Since the architecture is pipelined at bit-level, the registers (or delay latches) account for a major part of the chip area. We also need a register that can be useful for high clock rates. We have hence used the true single phase registers that were proposed in [61]. This register is compact needing only 11 transistors and is designed for high-speeds. The two main concerns are clock buffering and race-through due to clock-skew. The clock lines have high loads and hence the load is distributed using multiple-level buffering. The problem of race-through is avoided using reverse direction clock routing [62],[61].

The reverse clock distribution cannot directly be used in cases where there are recursive or feedback loops. In such cases a dummy register is used. The dummy register is an extra register which does not form a part of the logic. The clock line

Table 19.4 STAR Chip Summary

Technology	1.2μ CMOS
Number of Transistors	127,000
Power Supply	5V
Die Size	6.8x8.9mm^2
Active Area	6.0x8.3mm^2
Speed	100 MHz
Architecture	Bit-parallel, Bit level pipelined

is so routed that only the dummy buffer is affected by the skew. This results in a race through but does not affect the logic.

The chip can be operated at a speed of more than 100MHz. The features of the chip are summarized in Table 21.4 and the layout of the chip is shown in Fig. 19.9. This full-custom chip was designed using MAGIC tools.

19.10 CONCLUSIONS

In this chapter we have introduced new results that help the growth in the use of RLS adaptive filtering. We believe the RLS has tremendous potential for the future.

REFERENCES

[1] S. Haykin, *Adaptive Filter Theory.* Englewood Cliffs, NJ: Prentice Hall, 1986.

[2] B. Widrow *et al.*, "Adaptive noise cancelling: Principles and applications," *Proceedings of IEEE*, vol. 63, pp. 1692–1716, December 1975.

[3] R. L. Plackett, "Some theorems in least squares," *Biometrica*, vol. 37, p. 149, 1950.

[4] R. Hastings-James and M. W. Sage, "Recursive generalized-least-squares proceedure for online identification of process parameters," *Proceedings of IEE (London)*, vol. 116, pp. 2057–2062, 1969.

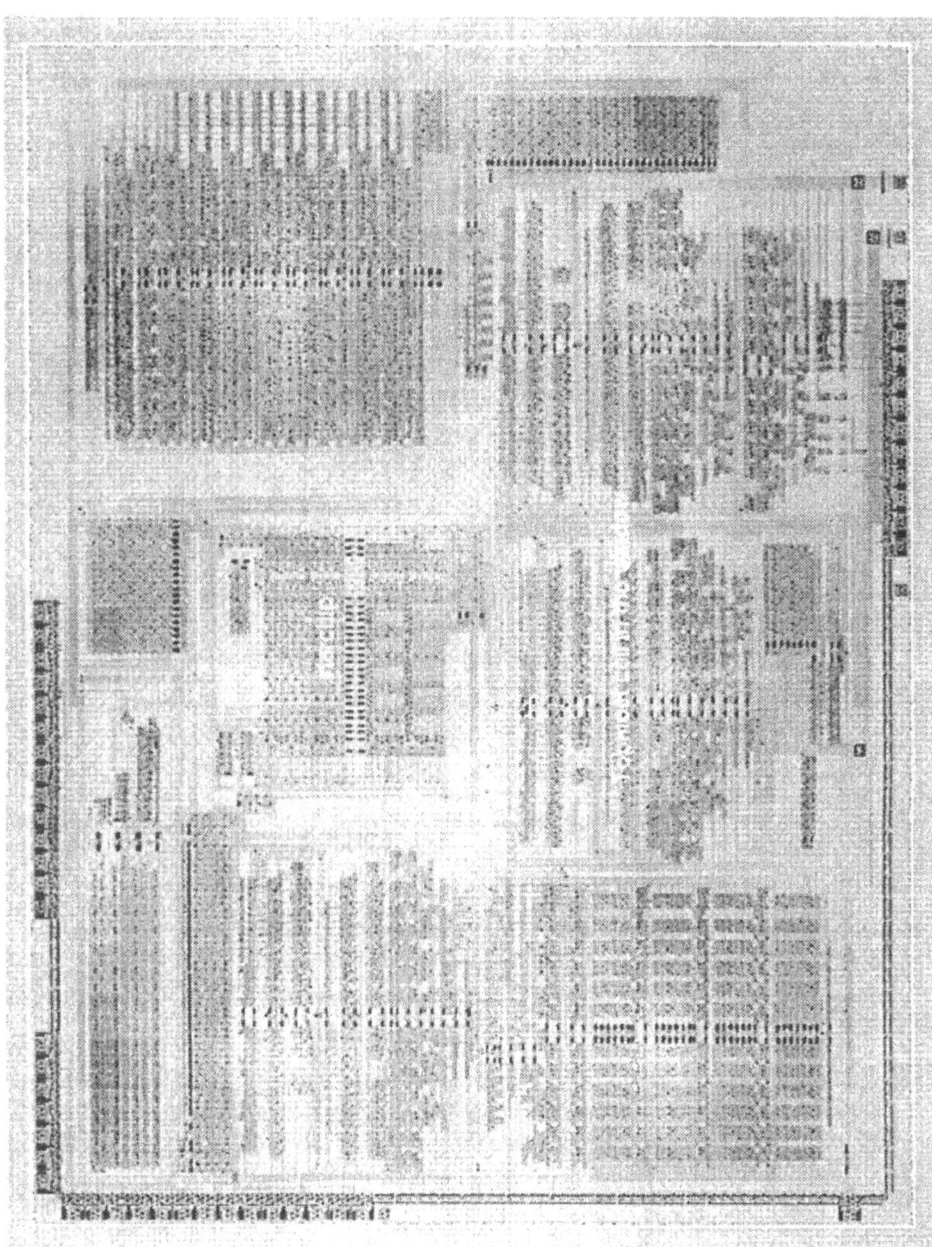

Figure 19.9 The 1.2 μ VLSI layout for a 4-tap PSTAR-RLS algorithm.

[5] D. G. Luenberger, *Optimization by vector space methods.* Wiley, NY, 1969.

[6] Y. Sato, "A method for self-recovering equalization for multilevel amplitude modulated systems," *IEEE Transactions on Communications*, vol. COM-23, pp. 679–682, June 1975.

[7] D. N. Godard, "Self-recovering equalization and carrier tracking in two-dimensional data communication systems," *IEEE Transactions on Communications*, vol. COM-28, pp. 1867–1875, November 1980.

[8] D. D. Falconer, "Jointly adaptive equalization and carrier recovery in two-dimensional digital communication systems," *Bell Syst. Tech. J.*, vol. 55, pp. 317–334, March 1976.

[9] R. E. Kalman, "A new approach to linear filtering and prediction problems," *Trans. ASME J. Basic Engineering*, vol. 82, pp. 35–45, 1960.

[10] G. H. Golub and C. F. V. Loan, *Matrix Computation.* Baltimore, MD: Johns Hopkins, 1989.

[11] W. M. Gentleman, "Least-squares computations by givens transformations without square-roots," *Journal Inst. Maths Applics.*, vol. 12, pp. 329–336, 1973.

[12] S. F. Hsieh, K. J. R. Liu, and K. Yao, "A unified sqrt-free rank-1 up/down dating approach for recursive least-squares problems," *Proc. of IEEE Intl. Conf. on Acoustics, Speech and Signal Processing (ICASSP)*, pp. 1017–1021, 1990.

[13] S. Hammarling, "A note on modifications to Givens plane rotation," *Journal Inst. Maths Applics.*, vol. 13, pp. 215–218, 1974.

[14] J. L. Barlow and I. C. F. Ipsen, "Scaled Givens rotations for solution of linear least-squares problems on systolic arrays," *SIAM J. Sci. Stat. Comput.*, pp. 716–733, September 1987.

[15] J. Götze and U. Schwiegelshohn, "An orthogonal method for solving systems of linear equations without square-roots and with few divisions," *Proc. of IEEE Intl. Conf. on Acoustics, Speech and Signal Processing (ICASSP)*, pp. 1298–1301, 1989.

[16] E. Frantzeskakis and K. J. R. Liu, "A class of square-root and division free algorithms and architectures for QRD-based adaptive signal processing," *IEEE Transactions on Signal Processing*, vol. 42, pp. 2455–2469, September 1994.

[17] J. M. Cioffi, "The fast adaptive ROTORs RLS algorithm," *IEEE Transactions on Acoustics, Speech, and Signal Processing*, pp. 631–653, Aril 1990.

[18] F. Ling, "Givens rotation based least-squares lattice and related algorithms," *IEEE Transactions on Signal Processing*, pp. 1541–1551, July 1991.

[19] I. K. Proudler, J. G. McWhirter, and T. J. Shepherd, "Computationally efficient QR decomposition approach to least squares adaptive filters," *IEE Proceedings-F*, vol. 138, pp. 341–353, August 1991.

[20] M. Bellanger, "A survey of QR based least-squares adaptive filters: From principles to realization," *Proc. of IEEE Intl. Conf. on Acoustics, Speech and Signal Processing (ICASSP)*, pp. 1833–1836, 1991.

[21] W. M. Gentleman and H. T. Kung, "Matrix triangularization by systolic arrays," *Proc. SPIE, Real Time Signal Processing IV*, vol. 298, pp. 298–303, 1981.

[22] K. K. Parhi, "High-level algorithm and architecture transformations for DSP synthesis (invited paper)," *Journal of VLSI Signal Processing*, January 1995.

[23] K. J. Raghunath, "High-Speed RLS adaptive filters," *Ph.D. Thesis, University of Minnesota*, November 1994.

[24] K. K. Parhi and D. G. Messerschmitt, "Pipeline interleaving and parallelism in recusive digital filters- part I: Pipelining using scattered look-ahead and decomposition," *IEEE Transactions on Acoustics, Speech, and Signal Processing*, pp. 1118–1134, July 1989.

[25] K. K. Parhi, "Algorithm transformation techniques for concurrent processors," *Proceedings of IEEE*, pp. 1879–1895, December 1989.

[26] K. K. Parhi and D. G. Messerschmitt, "Pipeline interleaving and parallelism in recusive digital filters- part II: Pipelined incremental block filtering," *IEEE Transactions on Acoustics, Speech, and Signal Processing*, pp. 1099–1117, July 1989.

[27] T. H. Y. Meng, E. A. Lee, and D. G. Messerschmitt, "Least-squares computation at arbitrarily high speeds," *Proc. of IEEE Intl. Conf. on Acoustics, Speech and Signal Processing (ICASSP)*, pp. 1398–1401, 1987.

[28] K. J. R. Liu, S. F. Hsieh, and K. Yao, "Systolic block householder transformation for RLS algorithm with two-level pipelined implementation," *IEEE Transactions on Signal Processing*, pp. 946–958, April 1992.

[29] S. F. Hsieh, K. J. R. Liu, and K. Yao, "Systolic implementation of up/down-dating Cholesky factorization using vectorized Gram-Schmidt pseudo orthogonalization," *Journal of VLSI Signal Processing*, pp. 151–161, 1991.

[30] A. P. Chandrakasan, S. Sheng, and R. W. Broderson, "Low-power CMOS digital desig," *IEEE Journal of Solid-State Circuits*, pp. 473–484, April 1992.

[31] K. J. Raghunath and K. K. Parhi, "Pipelined RLS adaptive filtering using scaled tangent rotations (STAR)," *IEEE Transactions on Signal Processing*, vol. 44, pp. 2591–2604, Oct 1996.

[32] K. J. Raghunath and K. K. Parhi, "High-speed RLS using scaled tangent rotations (STAR)," *Proc. of IEEE Intl. Symp. on Circuits and Systems (ISCAS-93)*, pp. 1959–1962, May 1993.

[33] J. G. McWhirter, "Recursive least-squares minimization using a systolic array," *Proc. SPIE, Real Time Signal Processing IV*, vol. 431, pp. 105–112, 1983.

[34] K. J. R. Liu, K. Yao, and C. T. Chiu, "Dynamic range, stability, and fault-tolerant capability of finite-precision RLS systolic array based on Givens rotation," *IEEE Transactions on Circuits and Systems*, pp. 625–636, June 1991.

[35] K. J. Raghunath and K. K. Parhi, "Pipelined implementation of high-speed STAR-RLS adaptive filters," *Proc. SPIE, Advanced Signal Processing Algorithms, Architectures, and Implementations IV*, vol. 2027, July 1993.

[36] H. Leung and S. Haykin, "Stability of recursive QRD LS algorithms using finite precision systolic array implementation," *IEEE Transactions on Acoustics, Speech, and Signal Processing*, pp. 760–763, May 1989.

[37] J. H. Wilkinson, *The Algebraic Eigenvalue Problem.* London: Oxford University Press, 1965.

[38] K. J. Raghunath and K. K. Parhi, "Parallel adaptive decision feedback equalizers (DFEs)," *IEEE Transactions on Signal Processing*, pp. 1956–1961, May 1993.

[39] C. Caraiscos and B. Liu, "A roundoff error analysis of the LMS adaptive algorithm," *IEEE Transactions on Acoustics, Speech, and Signal Processing*, vol. 32, pp. 34–41, February 1984.

[40] M. A. Syed and V. J. Mathews, "Finite precision error analysis of a QR-decomposition based lattice predictor," *Optical Engineering*, vol. 31, pp. 1170–1180, June 1992.

[41] R. C. North, J. R. Zeidler, W. H. Ku, and T. R. Albert, "A floating-point arithmetic analysis of direct and indirect coefficient updating techniques for adaptive lattice filters," *IEEE Transactions on Signal Processing*, vol. 41, pp. 1809–1823, May 1993.

[42] C. G. Samson and V. U. Reddy, "Fixed point error analysis of the normalized ladder algorithm," *IEEE Transactions on Acoustics, Speech, and Signal Processing*, vol. ASSP-31, pp. 1177–1191, October 1983.

[43] S. Ljung and L. Ljung, "Error propagation properties of recursive least-squares algorithms," *Automatica*, vol. 21, no. 2, pp. 157–167, 1985.

[44] S. H. Ardalan, "Floating-point error analysis of recursive least-squares and least-mean-squares adaptive filters," *IEEE Transactions on Circuits and Systems*, vol. 33, pp. 1192–1208, December 1986.

[45] S. H. Ardalan and S. T. Alexander, "Fixed-point error analysis of the exponential windowed RLS algorithm for time-varying systems," *IEEE Transactions on Acoustics, Speech, and Signal Processing*, vol. 35, pp. 770–783, June 1987.

[46] M. H. Verhaegen, "Round-off error propagation in four generally-applicable, recursive, least-squares estimation schemes," *Automatica*, vol. 25, no. 3, pp. 437–444, 1989.

[47] J. R. Bunch and R. C. LeBorne, "Error accumulation affects for the a posteriori RLSL prediction filter," *IEEE Transactions on Signal Processing*, vol. 43, pp. 150–159, January 1995.

[48] C. L. Lawson and R. J. Hanson, *Solving Least Squares Problems*. Prentice-Hall Inc., 1974.

[49] H. Dedieu and M. Hasler, "Error propagation in recursive QRD LS filter," *Proc. of IEEE Intl. Conf. on Acoustics, Speech and Signal Processing (ICASSP)*, pp. 1841–1844, 1991.

[50] G. W. Stewart, "Error analysis of QR updating with exponential windowing," *Mathematics of Computation*, vol. 59, pp. 135–140, 1992.

[51] C. W. Barnes, B. N. Tran, and S. H. Leung, "On the statistics of fixed-point roundoff error," *IEEE Transactions on Acoustics, Speech, and Signal Processing*, vol. 33, pp. 595–605, June 1985.

[52] K. J. Raghunath and K. K. Parhi, "Finite-precision error analysis of QRD-RLS and STAR-RLS adaptive filters," *IEEE Transactions on Signal Processing*, vol. 45, pp. 1193–1209, May 1997.

[53] K. J. Raghunath and K. K. Parhi, "Fixed and floating point error analysis of QRD-RLS and STAR-RLS adaptive filters," *Proc. of IEEE Intl. Conf. on Acoustics, Speech and Signal Processing (ICASSP)*, pp. 81–84, April 1994.

[54] A. B. Sripad and D. L. Snyder, "Quantization errors in floating-point arithmetic," *IEEE Transactions on Acoustics, Speech, and Signal Processing*, vol. 26, pp. 456–463, October 1978.

[55] C. M. Rader, "Wafer-scale integration of large systolic array for adaptive nulling," *Lincoln Laboratory Journal*, vol. 4, no. 1, pp. 3–28, 1991.

[56] K. J. Raghunath and K. K. Parhi, "A 100-MHz pipelined rls adaptive filter," *Proc. of IEEE Intl. Conf. on Acoustics, Speech and Signal Processing (ICASSP)*, pp. 3187–3190, May 1995.

[57] K. K. Parhi, C. Y. Wang, and A. P. Brown, "Synthesis of control circuits in folded pipelined DSP architectures," *IEEE Journal of Solid-State Circuits*, pp. 29–43, January 1992.

[58] C. E. Leiserson, F. Rose, and J. Saxe, "Optimizing synchronous circuits by retiming," *Proc. of the Third Caltech Conference on VLSI*, pp. 87–116, March 1983.

[59] A. Avizienis, "Signed Digit Number Representation for Fast Parallel Arithmetic," *IRE Transactions on Electronic Computers*, vol. EC-10, pp. 389–400, September 1961.

[60] H. R. Srinivas and K. K. Parhi, "High-speed VLSI arithmetic processor architectures using hybrid number representation," *Journal of VLSI Signal Processing*, vol. 4, pp. 177–198, 1992.

[61] J. Yuan and C. Svensson, "High-speed CMOS circuit technique," *IEEE Journal of Solid-State Circuits*, pp. 62–70, February 1989.

[62] M. Hatamian and K. K. Parhi, "An 85MHz 4th order programmable IIR digital filter chip," *IEEE Journal of Solid-State Circuits*, vol. 27, pp. 175–183, February 1992.

Chapter 20

Division and Square Root

Hosahalli R. Srinivas
Bell Labs, Lucent Technologies
Allentown, Pennsylvania
hsrinivas@lucent.com

Keshab K. Parhi
University of Minnesota
Minneapolis, Minnesota
parhi@ece.umn.edu

20.1 INTRODUCTION

On-chip floating point computational units have become an integral part of the data arithmetic modules of current day microprocessors. This is because the communication overheads and pin input-output bandwidth limitations, with (off-chip) floating point coprocessors, can be totally avoided. In such approaches, a major reduction in cost is possible since implementing computational systems as multi-chip solutions can be avoided.

An important requirement for implementing on-chip computational systems is to employ area efficient architectures. In current day microprocessors much of this area efficiency is achieved through sharing of architectures for performing different computational functions without compromising speed. Low power requirements also form a major factor, driven primarily by battery powered applications, in choosing the right schemes to implement functions.

In this chapter, we briefly study and later compare each of the recently proposed digit-by-digit based radix 2 and radix 4 division and square root algorithms. Unified architectures for performing both division and square root are also compared. In [1] a general comparitive taxonomy of division algorithms, techniques and implementation is presented. A detailed comparison of recently proposed division and square root algorithms has not been included.

High-performance floating point division and square root implementations have focussed on redundant arithmetic [2] (also called signed digit arithmetic) approaches to exploit the carry-free properties of these arithmetic number systems. In such implementations, signed digit numbers are used to represent the partial remainders and the quotient/root digits. These number representations allow the use of carry-free adders to perform carry-free addition/subtraction in the recursion steps. Carry-free addition/subtraction does not involve the overhead of carry/borrow computation and is therefore much faster than conventional carry-propagate addition/subtraction. Furthermore, the addition/subtraction time is independent of the word-length of the operands.

The most prominent of the signed digit number system based division algorithms is the SRT algorithm developed independently by D. Sweeny [3], J. E. Robertson [4], and T. D. Tocher [5] in 1958. This is a division algorithm for normalized operands that allows redundant number representations for the partial remainder and quotient digits, and performs quotient digit selection by observing a constant number n_s (independent of the word-length W) of most significant digits of the partial remainder and by observing the n_d most-significant bits of the divisor. Smaller the values of n_s and n_d, shorter the time required to generate the quotient digit in any division step. A W-digit SRT division algorithm requires $W/\log_2 r$ iterations when radix r arithmetic is used (to produce all the $W/\log_2 r$ radix r quotient digits). Higher the radix, lower the number of cycles required to produce the quotient. The reduction in the number of division cycles (by resorting to higher radix) is accompanied by a corresponding increase in hardware requirement and the cycle time per iteration. A major contributor to the cycle time is the quotient digit selection function. Much of the research efforts in the area of digit-by-digit based signed-digit division have been focussed on reducing the values of n_s and n_d.

Based on the choice of the bit-encoding used to represent the radix-r digits of the partial remainder, the adder/subtractor architecture and the quotient selection logic may vary. Hence the time taken to compute a quotient digit in each iteration step may also vary.

The floating point digit-by-digit based square root algorithms [6] are derived from the SRT division algorithm. In these algorithms the root digits are generated, one digit per iteration step, in most-significant digit first manner.

The division and square root algorithms considered here include those that use regular radix digits and those that use over-redundant radix digits for quotient or root digit representation. If the quotient/root digits are selected from the digit set $\{-a, \ldots, 0, \ldots, +a\}$ for a radix-r number system, the regular radix quotient/root digits would satisfy the relation $a \leq (r-1)$ and the over-redundant radix quotient/root digits would satisfy the relation $a \geq r$.

Certain division schemes prescale the inputs (dividend and divisor) to achieve smaller values for n_s and n_d. These have also been discussed in this chapter.

Finally, the results of comparison of the computation times and the area requirement is presented in a tabular form so as to easily establish the best choice for performing division or square root in a particular application.

This chapter is organized into four major sections. Sections 20.2 and 20.3 explain the division and square root algorithms. Section 20.4 explains the recently proposed unified algorithms for performing division and square root. Finally, section 20.5 compares each of division and square root algorithms discussed and provides the results of comparison in a tabular form.

20.2 DIVISION

A number of division algorithms based on the original SRT division algorithm have been proposed recently. These algorithms have been designed in radix 2 [7]-[13], and radix 4 [14]-[20], etc. In this section, we have restricted our comparison to $r \leq 8$.

As mentioned earlier, the digit-by-digit based division schemes operate on normalized inputs such as the mantissas of the IEEE-754 1985 std., [21] floating point numbers. A normalized floating point number in this standard can be represented as

$$N = m.2^e, \tag{1}$$

where the mantissa m is in the range $[1, 2)$. The word-length of the mantissa m and the exponent e is dependent on the precision of the standard. In division, the digit-by-digit recursive operations are performed only on the mantissas of the dividend and the divisor to generate the mantissa of the quotient. The exponent of the quotient is obtained by subtracting the exponent of the divisor from the exponent of the dividend. The exponent manipulation, being straightforward, is not the main focus of the recently proposed division schemes. All these schemes strive to reduce the computation requirements of the digit-by-digit recursive parts of the algorithms.

The basic division recursion expression can be written as

$$X_i = X_{i-1} - q_i r^{-i} D, \tag{2}$$

and in shifted form as

$$X_i^* = rX_{i-1}^* - q_i D, \tag{3}$$

where i is the index of recursion ($i = 1, 2, \ldots, W$), X_i is the i-th partial remainder, X_i^* is the i-th shifted partial remainder, X_0 (or X) is the dividend (mantissa portion), D is the divisor (mantissa portion), Q_W is the W-digit fractional quotient of X_0/D, and q_i is the i-th quotient digit belonging to the digit set $\{-a, \ldots, 0, \ldots, +a\}$, derived by observing digit string $X_{i-1}^{\prime}$ that is composed of the first n_s digits of the partial remainder and sometimes the first n_d bits of the divisor. The initial value (for $i = 0$) of the quotient, Q_i, is $Q_0 = 0$ and the i-th value is computed as $Q_i = Q_{i-1} + q_i r^{-i}$, for $i \geq 1$. The shifted partial remainder X_i^* is given by $X_i r^i$. In (2), substituting $i = 1$ to W, we get $X_W = X_0 - Q_W D$, where $Q_W < 1$ as X_0 is chosen (or scaled) to be less than D.

The recursion of (3) is convergent (i.e., $X_i \to 0$ for $i \to \infty$) if for all indices i, the remainder X_{i-1}^* and the quotient digit q_i have the same sign, i.e., the shifted partial remainder X_{i-1}^* is bounded for all i ($i > 1$). Such a recursion proceeds in a way that the newly computed partial remainder X_i^* is always bounded by satisfying the relation

$$-\sigma D \leq X_{i-1}^* \leq \sigma D, \tag{4}$$

where σ is called the index of redundancy and is given by

$$\sigma = \frac{a}{(r-1)}. \tag{5}$$

Note that the final remainder should, however, satisfy the relation

$$-D < X_W^* < +D \tag{6}$$

for quotient to be correct to the *ulp* (i.e., the unit in the least-significant digit position) of the quotient.

Fig. 20.1 shows the basic architecture of the recursive portion of the digit-by-digit division algorithms. In this architecture, the hardware for one recursion step is reused for each of the digits of the quotient produced. This type of design is

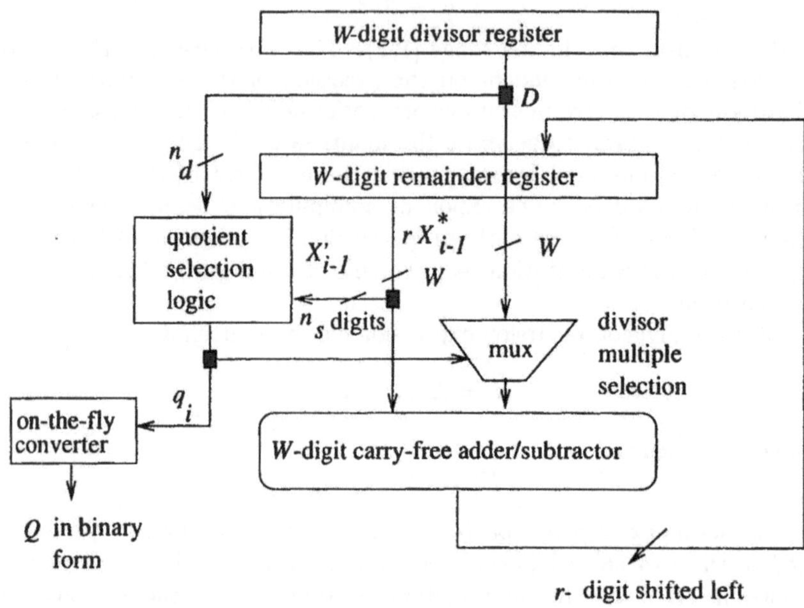

Figure 20.1 Generic architecture of a digit-by-digit based radix-r division algorithm.

typically used in a general purpose microprocessor. Unfolding such an architecture leads directly to a combinational array implementation.

The major constituents of the recursive part include the carry-free controlled adder/subtractor, the quotient selection logic, the storage registers required to hold the partial remainder and the divisor, and the on-the-fly converter [22] required to convert quotient digits (as they are generated in a most-significant digit first manner) to binary form.

The following sections briefly describe some of the recently proposed radix 2, and radix 4 division schemes.

20.2.1 Radix 2 Division

The recursive expression for the radix 2 division algorithm is

$$X_i^* = 2X_{i-1}^* - q_i D, \tag{7}$$

where $Q_0 = 0$ and $Q_i = Q_{i-1} + q_i 2^{-i}$ (the i-th partially accumulated quotient) for $i \geq 1$, and with the symbols defined as before for the generic equation of (3).

We consider each of the radix 2 division algorithms that have been proposed recently and briefly explain and compare them. These algorithms are the SRT radix 2 scheme with regular quotient digits [7, 8, 9], SRT and Svoboda radix 2 division with input prescaling [10, 11], and SRT radix 2 division with over-redundant quotient digits [12, 13].

20.2.1.1 Division with regular radix 2 quotient digits

Radix 2 division algorithm proposed in [7] produces regular radix 2 quotient digits belonging to the digit set $\{-1, 0, +1\}$. This algorithm operates on the dividend and the divisor in the range [1,2). The divisor and the dividend are scaled by 2 and 4, respectively, to bring them into the ranges of [1/2,1) and [1/4,1/2). For this purpose bits are allocated on the least-significant end to retain precision. This scaling allows the first shifted partial residual, i.e., $2X_0$ the scaled dividend, to satisfy the relation

$$-2D < 2X_0 < +2D. \tag{8}$$

Due to this input scaling, the final quotient generated has to be multiplied by 2 to get the correct quotient.

In this division scheme, the partial remainder is retained in redundant form (e.g., carry-sum form or signed digit form) and the divisor in two's-complement binary form. Two different methods of implementing the division algorithm have been suggested in [7], depending on the type of the carry-free adder chosen and the type of digit representation for the partial remainder. The type of the adder chosen has an impact on the number of msds of the partial residual that has to be observed for quotient digit selection. In both the methods the quotient selection logic is independent of the divisor leading to $n_d = 0$. The first method uses a W-bit carry-save adder/subtractor to compute the partial remainder. This scheme requires the first four digits of the shifted partial remainder, $2X_i^*$, to be observed for quotient digit selection, i.e., $n_s = 4$. In other words, X'_{i-1} is composed of the first 4 carry and sum bits. The quotient selection function for this scheme is

$$q_i = \begin{cases} +1 & \text{if } 0 \leq X'_{i-1} \leq 3/2 \\ 0 & \text{if } X'_{i-1} = -1/2 \\ -1 & \text{if } -5/2 \leq X'_{i-1} \leq -1. \end{cases} \tag{9}$$

The above quotient selection logic can be built in two ways. The first way is to convert the 4 msds of the partial remainder into binary form using a fast 4-bit carry propagate adder and then using simple logic to generate the quotient digit. The second method is to directly generate the quotient digits from the 4 msds of the shifted partial remainder.

The critical path delay T_{crit} in each iteration step for this method can be written as

$$T_{crit} = t_{qsel,4} + t_{buffer} + t_{mux,3-to-1} + t_{csa} + t_{reg}, \tag{10}$$

where $t_{qsel,4}$ is the delay of the 4-digit quotient selection logic, t_{buffer} is the delay of the buffers to broadcast the control signals to all the multiplexers in the W adder cells, $t_{mux,3-to-1}$ is the delay of the 3-to-1 multiplexer, t_{csa} the delay of a 3-to-2 bit full adder cell, and finally t_{reg} the delay involved in storing into a register and driving from it the newly computed partial remainder. The 3-to-1 multiplexer is required to select either the divisor D, or 0 or $-D$ (i.e., q_iD term in the division recursion (7)).

The area A required by this division algorithm using a recursive implementation as shown in Fig. 20.1 is

$$A = WCSA + QSEL_4 + ABUFFER + W * AMUX_{3-to-1} + 3 * WREG, \quad (11)$$

where $WCSA$ is the area of a W-bit carry-save adder, $QSEL_4$ is the area required by the 4 digit quotient selection logic, and $ABUFFER$ the area required by the buffers to drive the multiplexer select signals to all the W bit positions, $AMUX_{3-to-1}$ the area of a 3-to-1 multiplexer, and $WREG$ the area of a W-bit register. Note that 3 W-bit registers are required for this scheme as a $2W$-register is used to store the partial remainder and a W-bit register is used to store the divisor.

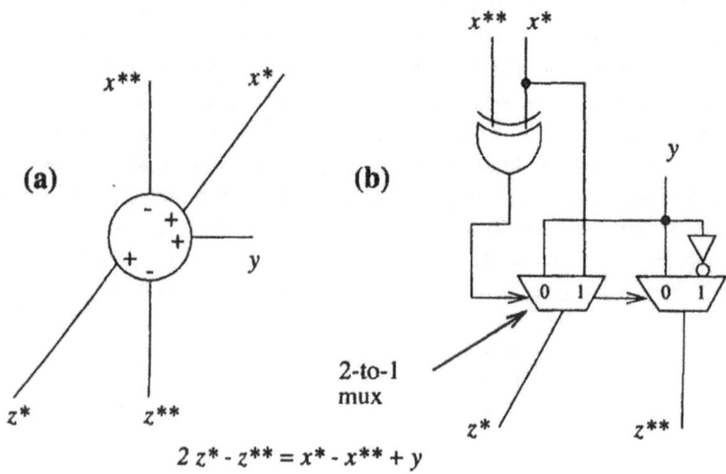

Figure 20.2 A 1-digit carry-free adder/subtractor.

The second method uses a signed-digit radix 2 redundant adder/subtractor to implement the addition/subtraction recursion. Since each digit in this number system can take three possible values, it is therefore encoded using two bits z^* and z^{**} such that the value of the digit is given by $z = z^* - z^{**}$. This digit z is represented by placing z^* to the right of z^{**}, i.e., $z = z^{**}z^*$. Thus, digit 0 is represented by 11 or 00, digit +1 by 01, and digit −1 by 10 [23].

An adder cell used with this digit encoding is shown in Fig. 20.2(a). The logic diagram of this adder cell is shown in Fig. 20.2(b). This adder cell receives 3 bits as inputs x^{**}, x^*, and y. The two bits x^* and y take on values of either a zero or a +1 and the third, x^{**} takes on a value of a 0 or a −1 (i.e., a logic 0 on this input is considered to be of value 0 and a logic 1 is considered to be of value −1). Let $x^{**}x^*$

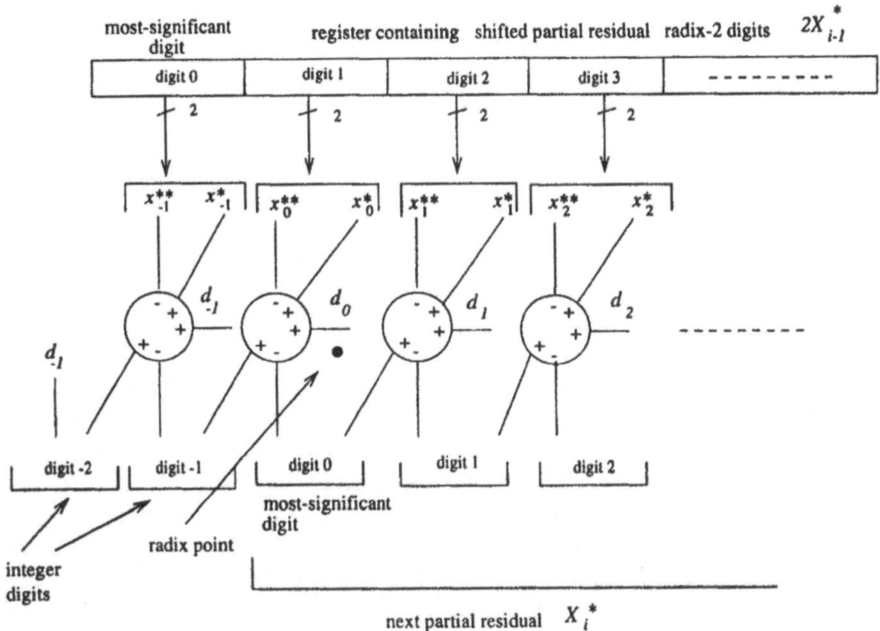

Figure 20.3 A $(W+1)$-digit carry-free adder/subtractor.

represent the encoded representation of signed radix 2 digit x. Let y represent a bit of a binary number (e.g., in two's-complement form). Then, addition is performed by connecting the x and y digits directly to this adder cell. Subtraction can be performed, in this adder cell, by taking two's-complement of y and adding it to x.

A $(W+1)$-digit carry-free adder is designed by placing $(W+1)$ such cells adjacent to each other as shown in Fig. 20.3. The input digits in these operators are represented in such a way that the subscript 0 stands for the most significant bit/digit and the subscript $(W-1)$ stands for the least significant digit. In this adder the radix 2 signed digit number (shifted partial residual $2X^*_{i-1}$) is added to the two's-complement number (i.e., divisor $D = d_{-1}d_0.d_1d_2\ldots$). Controlled subtraction is performed by taking two's-complement of D and adding it to the shifted partial residual $2X^*_{i-1}$. This is performed by logically inverting each and every bit of D and adding a 1 at the carry-input of the least-significant adder cell. This adder cell is logically equivalent to a 3-to-2 full adder used in a carry-save adder arrangement and has a computation time that is equal to the delay of a single binary adder cell (and is independent of word-length).

This method requires a quotient digit selection logic that is a function of the three most-significant digits of the shifted partial remainder. That is X'_{i-1}, is composed of the first 3 signed radix 2 digits of the shifted partial remainder ($n_s = 3$ and $n_d = 0$).

$$q_i = \begin{cases} +1 & \text{if } 1/2 \leq X'_{i-1} \leq 2 \\ 0 & \text{if } X'_{i-1} = 0 \\ -1 & \text{if } -2 \leq X'_{i-1} \leq -1/2. \end{cases} \tag{12}$$

The critical path delay T_{crit} in each iteration step for this method can be written as

$$T_{crit} = t_{qsel,3} + t_{buffer} + t_{mux,3-to-1} + t_{cfa} + t_{reg}, \tag{13}$$

where $t_{qsel,3}$ is the delay of the 3-digit quotient selection logic, and t_{cfa} is the delay of a 1-bit signed-digit adder of Fig. 20.2. The remaining terms are as explained before.

The area required by this design, implemented as a recursive structure of Fig. 20.1, is

$$A = WCFA + QSEL_3 + ABUFFER + W * AMUX_{3-to-1} + 3 * WREG, \tag{14}$$

where $WCFA$ is the area of a W-digit signed digit adder (of Fig. 20.3; $WCFA$ is equal to $WCSA$ as the two adders are logically equivalent), $QSEL_3$ is the area required by the 3 digit quotient selection logic, and $ABUFFER$ the area required by the buffers to drive the multiplexer select signals to all the W bit positions, $AMUX_{3-to-1}$ the area of a 3-to-1 multiplexer, and $WREG$ the area of a W-bit register. Note that 3 W-bit registers are required for this scheme as a $2W$-register is used to store the encoding bits of each of the digits of the partial remainder, and a W-bit register is used to store the divisor.

Unfolding the recursive structure of Fig. 20.1 by a factor of W (the word-length of the input operands) we get a combinatorial array version of the carry-free divider. Such a combinatorial version of the divider has been presented in [8, 9]. For the sake of evaluation, we shall consider only one iteration step (similar to the recursive structure of Fig. 20.1) composed of one row of adder cells with one quotient selection logic, and registers to store the partial remainder and divisor.

The division scheme of [8] operates on mantissas of the dividend and the divisor, which are in the range [1/2,1). This scheme uses signed-digit adder/subtractor to compute the partial radicand in signed-digit form. The adder/subtractor and the encoding used to represent the quotient digits and the digits of the partial remainder are different from those used in the radix 2 scheme of [7]. The quotient selection logic for this design observes the three most-significant digits of the partial remainder. If X'_{i-1} represents a string composed of the 3 most-significant digits of the partial remainder ($n_s = 3$ and $n_d = 0$), then the quotient selection logic for this design can be written as

$$q_i = \begin{cases} +1 & \text{if } X'_{i-1} > 0 \\ 0 & \text{if } X'_{i-1} = 0 \\ -1 & \text{if } X'_{i-1} < 0. \end{cases} \tag{15}$$

The critical path delay for a recursive structure built using this design is

$$T_{crit} = t_{qsel,3} + t_{buffer} + t_{mux,2-to-1} + t_{sda} + t_{reg}, \tag{16}$$

where $t_{qsel,3}$ is the delay of a 3-digit quotient selection logic, t_{buffer} is the delay of the buffer required to drive the control signals for the 2-to-1 multiplexer selecting either $+D$ or $-D$ for recursion, $t_{mux,2-to-1}$ is the delay of the 2-to-1 multiplexer,

t_{sda} is the delay of a signed-digit adder, and t_{reg} is the delay in loading the partial remainder in the registers.

The radix 2 division scheme of [9] operates on the divisor and the dividend in the range [1/2,1). This algorithm uses the same digit representation used by the signed-digit adder of Fig. 20.2 and 20.3 for partial remainder representation. The divisor is retained in two's-complement form. The quotient digit is chosen from the set $\{-1, -0, +0, +1\}$ and from the three most-significant digits of the partial remainder ($n_s = 3$ and $n_d = 0$). The divisor multiple, $0.D$, in the division recursion results in addition of a two's-complement of 0 to the partial remainder. With such a scheme, the representation overflow (i.e., apparent increase in the word-length of a number at the most-significant digit end without increase in the value of the number) is avoided. Fig. 20.4 shows a 1-bit adder/subtractor cell for this division scheme. In this adder cell, two control signals s and v are provided that enable controlled inversion (when $s = 1$) and zeroing (when $v = 1$) of the divisor bit y.

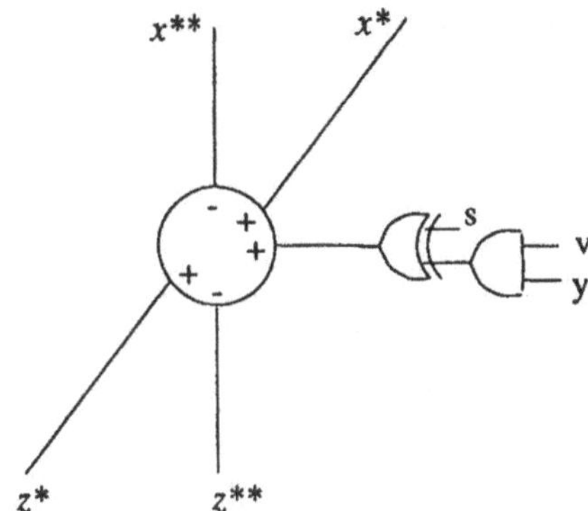

$2\,z^* - z^{**} = x^* - x^{**} + xor(s, and(y,v))$

Figure 20.4 A 1-digit adder/subtractor for division with quotient $\in \{-1, -0, +0, +1\}$.

The critical path delay for a recursive structure built using this design is

$$T_{crit} = t_{qsel,3} + t_{buffer} + t_{xor} + t_{and,2} + t_{cfa} + t_{reg}, \tag{17}$$

where $t_{qsel,3}$ is the delay of the 3 digit quotient selection function ($n_s = 3$), t_{xor} is the delay of a 2-input exclusive-or gate, $t_{and,2}$ is the delay of a 2-input and gate, and the rest of the terms are as defined before for the other designs.

The area required by this design is

$$A = WCFA + QSEL_3 + ABUFFER + W*(AXOR + AAND2) + 3*WREG, \tag{18}$$

where $WCFA$ is the area of a W-digit carry free adder (of Fig. 20.3), $ABUFFER$ is the area of the buffers required to broadcast the sign and the value of the quotient digit to all the W carry-free adder/subtractor cells, $AXOR$ and $AAND2$ are the areas of 2-input exclusive-or gate and a 2-input and gate. The register area requirement is same as before.

20.2.1.2 Division with input prescaling

The division schemes of [10] and [11] employ prescaling of the dividend and the divisor in order to use a simple quotient selection logic. The quotient selection logic in these schemes is a rewrite of the two most-significant digits of the partial remainder (without changing its value) and chosing the quotient digit as the most-significant digit of the rewritten partial remainder (i.e.,$n_s = 2$ and $n_d = 0$). The scheme of [10] is based on Svoboda's division scheme [24, 25]. In this scheme, a divisor in the range [1,2) has to be scaled such that it is transformed into the form $(1 + \beta)$, where $\beta < 2^{-1}$ (i.e., $1 + \beta < 1.5$). Therefore if the divisor is in the range [1.5,2), then both the divisor and the dividend are scaled by 0.75. The scaling by 0.75=(1-0.25) is easily achieved by shifting the divisor by 2 bits to the right and subtracting it from the original divisor. The subtraction is performed in a fast carry-propagate adder. This scheme uses signed-digit adders similar to those shown in Fig. 20.3 to perform the addition/subtraction recursion. Each of W adder cells, however, incorporates two multiplexers at the output to retain the shifted partial remainder if the quotient digit generated is 0. Fig. 20.5 shows a W-digit carry-free controlled adder/subtractor used by this division scheme.

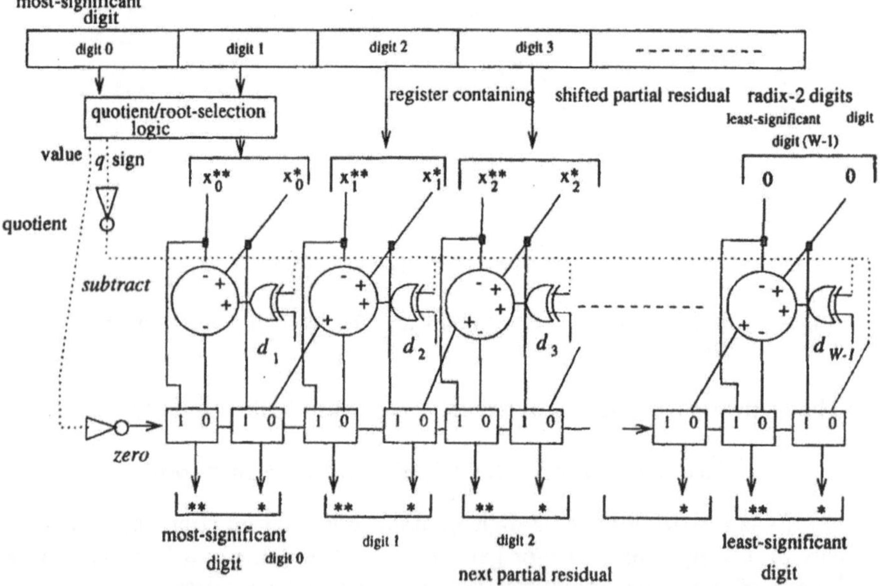

Figure 20.5 A W-digit carry-free controlled adder/subtractor with multiplexers at the output.

The scheme of [11] is derived from the original SRT radix 2 division algorithm [7]. This scheme operates on the divisor and the dividend in the range $1/2 \leq D < 1$ and $1/4 \leq X_0 < 1/2$ and prescales both the divisor and the dividend by a factor of 0.75, if the divisor is in the range $3/4 \leq D < 1$. The prescaling done here is similar to the one used in the previous scheme [10]. The quotient selection process is a rewrite of the two most-significant digits of the partial remainder and selecting the most-significant digit of the rewritten remainder (i.e.,n_s = 2 and $n_d = 0$). The rewrite operation does not change the value of the remainder but avoids representation overflow in the signed-digit adder. The adder/subtractor arrangement is exactly similar to that used in the scheme of [10].

Both these schemes, when implemented as a recursive structure (similar to the Fig. 20.3), have a critical path delay of

$$T_{crit} = t_{qsel,2} + t_{buffer} + t_{xor} + t_{cfa} + t_{mux,2-to-1} + t_{reg}, \tag{19}$$

where $t_{qsel,2}$ is the delay of the quotient selection logic (re-write logic), $t_{mux,2-to-1}$ is the delay of the 2-to-1 multiplexer, and the rest of the terms are as defined before. The area A required for such an implementation of these two designs is

$$\begin{aligned} A &= WCFA + QSEL_2 + ABUFFER + W * AXOR \\ &\quad +2 * W * AMUX_{2-to-1} + 3 * WREG + WCPA, \end{aligned} \tag{20}$$

where $WCFA$ is the area of a W-digit carry-free adder, $W * AXOR$ is the area of a W 2-input exclusive-or gates, $2 * W * AMUX_{2-to-1}$ is the area of $2 * W$ 2-to-1 multiplexers provided to retain the shifted partial remainder as the new remainder when the quotient digit generated is 0, and $WCPA$ is the area of a fast W-bit carry-propagate adder required for input prescaling.

The division schemes of [10, 11] when implemented as an array composed of W rows of signed-digit adders results in computation times that are much shorter than that for the other radix 2 schemes. The only penalty incurred in using this scheme is that a fast carry-propagate adder is required to scale the inputs before performing division.

20.2.1.3 Division with over-redundant radix 2 quotient digits

The algorithm of [13] uses over-redundant radix 2 quotient digits belonging to the digit set $\{-2, -1, 0, +1, +2\}$. The partial remainder digits are, however, maintained in regular radix 2 digit format. The quotient selection logic used by this scheme observes the two most-significant digits of the partial remainder, i.e., $n_s = 2$, and is independent of the divisor, i.e., $n_d = 0$. This division algorithm does not require input prescaling as required by the designs of [10, 11] which also use $n_s = 2$.

From (4), it is obvious that the range of the partial remainder is

$$-2D \le X^*_{i-1} \le +2D. \tag{21}$$

However, restricting this convergence range to

$$-D < X^*_{i-1} < +D, \tag{22}$$

results in a simpler quotient selection function that also causes the i-th partial remainder, X^*_i, to be in the same range as X^*_{i-1}.

Although the quotient digits belong to the over-redundant radix 2 digit set, they satisfy the relation (22) because for any $i \ge 1$,

$$\sum_{j=i}^{\infty} q_j 2^{-j} < 2^{-i+1}. \tag{23}$$

This implies that although the quotient digits belong to the over-redundant radix 2 digit set, a truncated over-redundant radix 2 quotient with W-fractional digits always has acceptable error, i.e., the truncated part that is ignored has a maximum value that is less than 2^{-W}. This quotient is acceptable for a W-fractional digit based division operation since the final remainder also satisfies (22).

The partial remainder for this scheme is generated by a W-digit adder/subtractor similar to the design used in Fig. 20.5. The quotient digit selection logic is a function of the two most-significant integer digit digit-string X'_{i-1} of the shifted partial remainder $2X^*_{i-1}$ (i.e., $n_s = 2$ and $n_d = 0$). The selection function is

$$q_i = \begin{cases} -2 & \text{if } X'_{i-1} = -3 \\ -1 & \text{if } X'_{i-1} = -2 \text{ or } -1 \\ 0 & \text{if } X'_{i-1} = 0 \\ +1 & \text{if } X'_{i-1} = +2 \text{ or } +1 \\ +2 & \text{if } X'_{i-1} = +3. \end{cases} \tag{24}$$

The over-redundant radix 2 quotient digits generated by the division algorithm are used in the division recursion operation. However, the results of division that are presented to the outside world are converted to binary form. The on-the-fly converter originally suggested in [22], for conversion of regular radix 2 digit numbers to binary form, cannot operate on over-redundant radix 2 digits. These digits have to be first transformed into regular radix 2 form and then converted in an on-the-fly converter to binary form.

Conversion of over-redundant radix 2 quotient digits to regular radix 2 digits can be accomplished by a scheme similar to the radix 4 conversion explained in [18]. In this scheme a digit-serial radix 4 adder [26] is used to reduce the over-redundant radix 4 quotient digits to maximally redundant regular radix 4 digits (belonging to the set $\{-3, \ldots, +3\}$). However, it has the disadvantage of producing a $(W+1)$-digit reduced quotient from the W-digit over-redundant quotient resulting in an extra cycle for on-the-fly conversion to binary form. A radix 2 over-redundant quotient digit reduction scheme would involve a digit serial radix-2 adder similar to the one described in [27].

In order to retain the number of digits in the reduced quotient to W and to avoid an extra on-the-fly conversion cycle, a new conversion process called *reduction* is used. The *reduction* uses the fact that the quotient digits satisfy the relation

$$|\sum_{i=j}^{\infty} q_i 2^{-i}| < 2^{-j+1}, \tag{25}$$

for any $j \geq 1$. The quotient digits, generated during the division, which satisfy (25), exhibit the following property. During the j-th iteration, if the quotient digit q_j generated is $+2$, then the remainder X^*_j becomes negative leading to a subsequent non-zero quotient digit that is negative (note that the digits subsequent to $q_j = +2$ may be zeroes but the first non-zero digit encountered will be negative). Similarly, if $q_j = -2$, the partial remainder X^*_j becomes positive and non-zero causing a subsequent non-zero quotient digit to be positive.

Since the W-digit quotient Q_W satisfies the relation $|Q - Q_W| < 2^{-W}$, it is required that the reduced W-digit quotient, Q'_W, composed of regular radix 2 digits also satisfy the relation $|Q - Q'_W| < 2^{-W}$. Furthermore, the reduced quotient Q'_W should also satisfy (25). The reduced quotient Q'_W being composed of regular radix 2 digits obviously satisfies the relation $|\sum_{i=j}^{W} q'_i 2^{-i}| < 2^{-j+1}$, for any $j \geq 1$ ($q_i \in \{-1, 0, +1\}$). It is therefore required of the reduction process to produce Q' from Q such that $Q' = Q$ (note that Q' and Q are assumed to be of infinite precision).

The reduction technique is valid for the division algorithm of [13] as it replaces only the non-overlapping quotient digit strings of $\{+2-1\}$, $\{+2-2\}$, $\{+2\ \ 0\cdots 0-1\}$, $\{+2\ \ 0\cdots 0-2\}$, $\{-2+1\}$, $\{-2+2\}$, $\{-2\ \ 0\cdots 0+1\}$, and $\{-2\ \ 0\cdots 0+2\}$ by the corresponding digit strings of $\{+1+1\}$, $\{+1\ \ 0\}$, $\{+1+1\cdots +1+1\}$, $\{+1+1\cdots +1\ \ 0\}$, $\{-1-1\}$, $\{-1\ \ 0\}$, $\{-1-1\cdots -1-1\}$, and $\{-1-1\cdots -1\ \ 0\}$, without changing their value. All other types of strings in the generated quotient are retained as they are by the reduction process. Therefore the reduction process properly reduces over-redundant quotient digits to regular radix 2 digit form. The reduced digits are input to an on-the-fly converter [22], that produces the final quotient in regular binary form without carry-propagation.

The critical path delay for this design is

$$t_{crit} = t_{qsel,2} + 2 * t_{mux,2} + t_{xor} + t_{cfa} + t_{buffer} + t_{reg}, \tag{26}$$

where $t_{qsel,2}$ is the delay of the two digit quotient selection logic, and $2 * t_{mux,2}$ is the delay of two 2-to-1 multiplexers. The remaining terms are same as those defined for the other radix 2 designs. Of the two multiplexers in a digit position, one is used to select the multiple of the divisor (either D or $2D$ and the other is used to retain the same shifted partial remainder as the new remainder if the quotient digit is zero.

The area requirement for a recursive implementation of this algorithm is

$$A = WCFA + QSEL_2 + ABUFFER + W * AXOR + 3 * W * AMUX_{2-to-1} + 3 * WREG, \tag{27}$$

where $WCFA$ is the area of a W digit carry-free adder as shown in Fig. 20.5. Three multiplexers are required in each digit position as one is used for divisor multiple selection, and the other two are used for retaining the shifted partial remainder if the quotient digit generated is a 0.

20.2.2 Radix 4 Division

The recursive expression for the radix 4 division algorithm is

$$X_i^* = 4X_{i-1}^* - q_i D, \tag{28}$$

where $Q_0 = 0$ and $Q_i = Q_{i-1} + q_i 4^{-i}$ (the i-th partially accumulated quotient) for $i \geq 1$, and with the symbols defined as before for the generic equation of (3).

The main advantage of a radix 4 scheme is that, to produce a W bit quotient, only $W/2$ iteration steps are required. Therefore for a radix 4 division scheme to be faster than any of the radix 2 schemes, the time per iteration stage should be less than twice the time per iteration stage of the fastest radix 2 design.

We consider each of the radix 4 division algorithms that have been proposed recently and briefly explain and compare them. These algorithms are the SRT radix 4 schemes with regular quotient digits [7, 15, 17], radix 4 division with input prescaling [16, 19, 20], and radix 4 SRT division with over-redundant quotient digits [18].

20.2.2.1 Division with regular radix 4 quotient digits

The radix 4 division algorithm proposed in [7] generates minimally redundant quotient digits from the digit set $\{-2, -1, 0, +1, +2\}$. The index of redundancy for this scheme is $\sigma = \frac{2}{3}$. The bound on the remainder is

$$|X_j^*| \leq \frac{2}{3}D. \tag{29}$$

The input dividend X_0 should also satisfy this bound. Therefore any input in the range [1,2) has to be divided by 4 to satisfy this. The final quotient has to be shifted left to accommodate this input scaling. This scheme uses carry-save adder architecture to implement the addition/subtraction of division recursion. The quotient selection logic for this scheme is a function of 3 bits of the divisor (i.e., $n_d = 3$) and 7 msds of the partial remainder (i.e., $n_s = 7$). The quotient selection logic can be implemented in two ways. One way is to reduce the 7 msds of the partial remainder (which are in carry-save form) to 7 bits and then using a table lookup to determine the quotient digit. The other method is to employ a Boolean logic to derive the quotient digit directly from the partial remainder and the divisor.

The critical path delay for this scheme is

$$T_{crit} = t_{qsel,7,3} + t_{buffer} + t_{mux,5-to-1} + t_{csa} + t_{reg}, \tag{30}$$

where $t_{qsel,7,3}$ is the delay of the quotient digit selection logic that is a function of 7 msds of the partial remainder and 3 msds of the divisor, t_{buffer} is the delay of buffer used to broadcast the select signals to the 5-to-1 multiplexer (with a delay of $t_{mux,5-to-1}$), and the remaining terms are as defined before.

The area requirement for this design implemented as a recursive structure (as shown in Fig. 20.1) is

$$A = WCFA + QSEL_{7,3} + ABUFFER + W * AMUX_{5-to-1} + 3 * WREG, \tag{31}$$

where $QSEL_{7,3}$ is the area required by the quotient selection logic, $AMUX_{5-to-1}$ is the area of a 5-to-1 multiplexer, and the remaining terms are same as defined for the other division schemes before.

The total computation time for this design is

$$T_{total-time} = \frac{W}{2}T_{crit}. \tag{32}$$

The radix 4 division scheme of [15] uses minimally redundant quotient digits to perform division. The recursion involved in this scheme is based on the standard SRT division scheme and uses carry-save adders to compute the partial remainders. The quotient digits for this scheme are derived from the 8 most-significant digits of the partial remainder through a $7-bit$ fast carry-lookahead adder and 4 bits of the divisor ($n_s = 8$ and $n_d = 4$). The output bits of this adder are output to a 19-term programmable logic array block. The PLA block then generates the quotient digit that is used to select the divisor multiple. The critical path delay for a recursive stage of this design is

$$\begin{aligned} T_{crit} &= t_{cla,8} + t_{xor} + t_{or} + t_{pla,19} + t_{mux,5-to-1} \\ &\quad + t_{csa} + t_{buf} + t_{reg}, \end{aligned} \tag{33}$$

where $t_{cla,8}$ is the delay of a fast 8-bit carry-lookahead adder, t_{or} is the delay of a or-gate, $t_{pla,19}$ is the delay of a 19-input PLA block, and the remaining terms are as defined before.

The total area requirement for this design is

$$\begin{aligned} A &= ACLA_8 + 5 * AXOR + 4 * AOR + APLA_{19} + ABUFFER \\ &\quad + AMUX_{5-to-1} + WCSA + 3 * WREG, \end{aligned} \tag{34}$$

where $ACLA_8$ is the area of an eight input carry-lookahead adder, $AXOR$ and AOR are the areas of 2-to-1 exclusive-or and logical-or gates, $APLA_{19}$ is the area of the 19-input table lookup PLA for quotient digit selection, and the remaining terms are as defined before for other designs.

The total computation time for this design is

$$T_{total-time} = \frac{W}{2} T_{crit}. \tag{35}$$

The radix 4 division scheme presented in [17] uses a maximally redundant quotient digit set to parallelize addition/subtraction recursion. The recursion expression for this division algorithm is

$$w_i = 4(w_{i-1} - q_i D) = 4^i (X_0 - Q_i D), \tag{36}$$

where w_i is the i-th shifted partial remainder, q_i is the i-th quotient digit, $Q_i = Q_{i-1} + q_i.4^{-i}$. The quotient digit q_i, belonging to the digit set $\{-3, -2, -1, 0, +1, +2, +3\}$, is expressed as a summation of two terms, i.e., $q_i = q_i' + q_i''$, where $q_i' \in \{-2, 0, +2\}$ and $q_i'' \in \{-1, 0, +1\}$. So the computation of the multiple of the divisor for $q_i = 3$ is (i.e., $3D$) is performed as $(q_i' + q_i'')D$, where $q_i' = 2$ and $q_i' = 1$. The problem of generating multiple of divisor is shifted out to the addition/subtraction recursion stages. This operation is performed in two stages of the adder/subtractor. The generation of q_i has been split into two smaller subproblems of generating q_i' and q_i''. Therefore, the original addition/subtraction recursion is split into two parts as shown here.

$$w_i = 4((w_{i-1} - q_i' D) - q_i'' D), \tag{37}$$

where the computation of q_i' and q_i'' are distinct. Since each of these digits belong to a smaller digit set than the original quotient set, their computation complexity is reduced. The recursive division structure required to perform the above dual add/subtract operation is shown in Fig. 20.6.

The quotient digit selection logic for both q_i' and q_i'' is a function of the 7 most-significant digits of the partial remainder (i.e., $n_s = 7$, and $n_d = 3$) and 3 bits of the divisor. Four variations of the generic scheme discussed here have been proposed in [17]. The critical path for all these variants can be written as

$$\begin{aligned} T_{crit} &= t_{cla,7} + max(t_{q'} + t_{mux,3-to-1} + t_{csa}, t_{q''} + t_{mux,3-to-1}) \\ &\quad + t_{csa} + t_{buffer} + t_{reg}, \end{aligned} \tag{38}$$

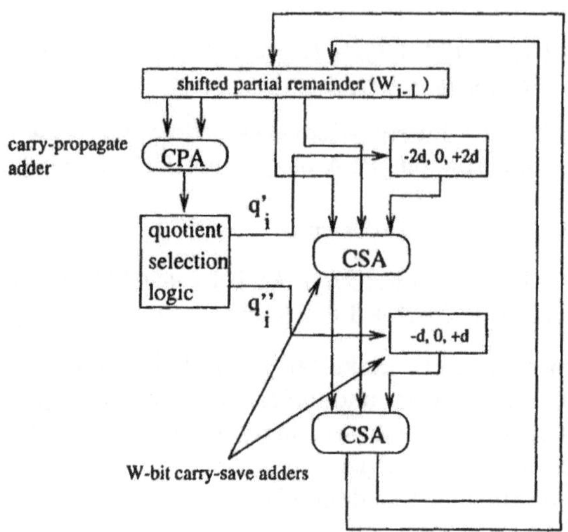

Figure 20.6 A recursive radix 4 division structure with a 2-stage W-digit carry-free adder/subtractor.

where $t_{cla,7}$ is a 7-bit fast carry-propagate adder, $t_{q'}$ and $t_{q''}$ are the delays of the the q' and q'' selection logic, and the rest of the terms are as defined before. The area A required by this design is

$$\begin{aligned} A \;=\; & ACLA_7 + 2 * W * AMUX_{3-to-1} + 2 * WCSA + AQSEL_{q'} \\ & + AQSEL_{q''} + ABUFFER + 3 * WREG, \end{aligned} \quad (39)$$

where $ACLA_7$ is the area of a fast 7-bit carry-lookahead adder, $WCSA$ is the area of a W-bit carry-save adder, $AQSEL_{q'}$ is the area of q' selection logic, $AQSEL_{q''}$ is the area of q'' selection logic, and the remaining terms are as described before.

20.2.2.2 Radix 4 division with input prescaling

The radix 4 division scheme of [16] uses prescaling of the divisor to reduce its range so that only a constant can be used for comparison with the shifted partial remainder during quotient selection. This approach makes the quotient selection logic independent of the divisor. Fig. 20.7 shows a block diagram of the process involved in this scheme. The choice of the scale factor is such that it should move the divisor range to the region that allows maximum overlap of the values of the shifted partial remainder for adjacent values of the minimally redundant quotient digits. This provides a wider range for selection of constants to compare for quotient determination. Furthermore, the scaling should be performed by a simple addition or subtraction operation in the signed digit adder that is also to be used for division recursion. The scaled divisor (signed-digit adder output) is later converted to binary form by using a fast carry-propagate adder so that a 3-to-2 bit carry-save adder or a similar signed-digit adder can be used for division recursion. The scale factor chosen is dependent on the value of the original divisor.

The division recursion used by this scheme is same as (28). The quotient selection logic for this design is dependent on the 6 msds of the partial remainder

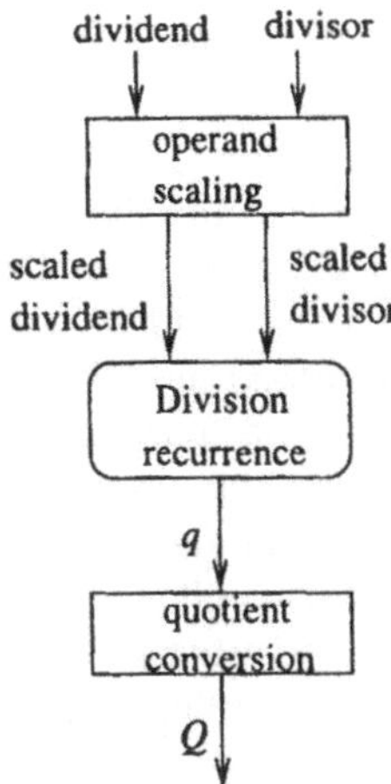

Figure 20.7 Steps in division scheme that use divisor scaling.

and independent of the divisor, i.e., $n_s = 6$ and $n_d = 0$. The critical path delay per recursion stage is

$$T_{crit} = t_{qsel,6} + t_{mux,5-to-1} + t_{csa} + t_{buffer} + t_{reg}, \tag{40}$$

where $t_{qsel,6}$ is the delay of the 6-digit quotient selection logic, and the remaining terms are as defined before for other designs. The 6-digit quotient selection logic can be designed either as a PLA with 12-bits as inputs or as a 6-bit fast carry-lookahead adder to produce a 6-bit sum that can be used as inputs to a simple Boolean function. The total computation time to compute a W-bit quotient is

$$T_{total-time} = \frac{W}{2} T_{crit} + T_{prescale}, \tag{41}$$

where $T_{prescale}$ is the time required for input prescaling. Both the divisor and the dividend are prescaled using the same signed-digit adder and a fast carry-propagate adder. The total area requirement for this scheme is

$$A = WQSEL_6 + WCFA + W * AMUX_{5-to-1} + ABUFFER + 3 * WREG + WCPA, \tag{42}$$

where $WQSEL_6$ is the area of the 6-digit quotient selection logic $AMUX_{5-to-1}$ is the area of the 5-to-1 multiplexer, $WREG$ is the area of the W-bit register, $WCPA$ is the area of a fast W-bit carry-propagate adder.

This division scheme makes the quotient selection logic independent of the divisor by incurring an area penalty equal to that of a fast W-bit carry-propagate adder.

The division scheme of [19] employs maximally redundant radix 4 quotient digits and input prescaling to realize a quotient selection logic that observes only the three most-significant digits of the partial remainder. Note that the number of digits of the partial remainder observed for quotient digit selection is same as the radix 2 SRT schemes of [7, 8, 9]. Thus with a simple quotient selection logic and $W/2$ iteration steps, this radix 4 division scheme presents the shortest division iteration times when compared with the other radix 4 schemes. This algorithm is based on the Svoboda's [24, 25] division scheme and prescales the input divisor (that is in the range [1,2)) to take on a value $(1+\beta)$, where $\beta < 6^{-1}$. The original division scheme by [24, 25] uses input prescaling such that the prescaled divisor has a value $(1+\beta)$, where $\beta < r^{-1}$, where r is the radix. The tighter constraint chosen for β in the scheme of [19] allows a much simpler quotient digit selection to be used. The original division scheme proposed in [24] was modified in [25] such that the two most-significant digits of the partial remainder were required for quotient selection, if $\beta < r^{-1}$. The same scheme was used in the division scheme of [19] where each radix 4 digit of the partial remainder was represented using two radix 2 digits. Instead of observing the four most-significant radix 2 digits (i.e., the 2 most-significant radix 4 digits as in the division scheme of [25]) of the partial remainder for quotient selection, only the 3 most-significant radix 2 digits are observed by constraining the value of β to $\beta < 6^{-1}$.

The scaling operation in this algorithm is performed by a shift-add method where three shifted values of the divisor are added in a carry-free adder arrangement and later converted to binary form using a fast carry-propagate adder. The same prescaling is also performed on the dividend. The dividend is not converted to binary form. The computation of 3β is also performed by the same arrangement using three shifted values. The amount of shift required for each of the three shift values, for both β and 3β computation, is determined by looking at the 5 most-significant fractional bits of the divisor (original divisor is in the range [1,2)).

The adder/subtractor arrangement needed for this algorithm is same as that shown in Fig. 20.5. The critical path for such a recursive implementation is

$$T_{crit} = t_{qsel,3} + t_{xor} + t_{mux,3-to-1} + t_{cfa} + t_{mux,2-to-1} + t_{buffer} + t_{reg}, \tag{43}$$

where $t_{qsel,3}$ is the delay of the 3-digit quotient selection function (i.e., $n_s = 3$ and $n_d = 0$), the remaining terms are as described before for other designs.

The area required for this design is

$$\begin{aligned} A = \ & AQSEL_2 + 2*W*AXOR + W*AMUX_{3-to-1} + WCFA \\ & + W*AMUX_{2-to-1} + ABUFFER + 4*WREG + WCPA + WP, \end{aligned} \tag{44}$$

where $AQSEL_2$ is the area of a 2-digit quotient selection logic, and $4*WREG$ is the area of four W-bit registers. Two of the four registers are used for storing the partial remainder, produced every iteration, in signed-digit form. The other two are used to store β and 3β in binary form. $WCPA + WP$ is the area required by a fast carry-propagate adder and the prescaling hardware (in the form of control circuits for selecting appropriate shifting).

The total computation time to compute a W-bit quotient is

$$T_{total-time} = \frac{W}{2} T_{crit} + T_{prescale}, \tag{45}$$

where $T_{prescale}$ is the time required for input prescaling.

The radix 4 division scheme of [20] is a modified version of the Svoboda division scheme [24, 25]. This scheme uses input prescaling of the divisor and dividend and employs a minimally redundant quotient digit set. The partial remainder digits are also maintained in minimally redundant radix 4 digits. The input divisor D, originally in the range [1,2), is scaled such that it is in the range in the range [1,9/8). The dividend is also scaled by the same factor and right shifted by 1 digit. This brings it into the division convergence bound of $|X^*_{i-1}| \leq 2/3$. The prescaling constants are chosen by observing the first 5 bits of the divisor. The prescaling operation involves addition/subtraction of 3 right shifted values of the divisor. The dividend is also prescaled by the same factor used for the divisor. This addition/subtraction operation is performed in a single row of carry-free adders. The scaled values of the divisor available in signed digit form, as the output of this carry-free adder, is converted to binary form by a fast carry-propagate adder.

A recursive implementation of this radix 4 divider architecture similar to the arrangement shown in Fig. 20.1 uses a row of 3-to-2 bit signed-digit adder cells along with input multiplexers for performing the addition/subtraction portion of the division recursion. The multiplexers select either the divisor multiple or the complement of the divisor multiple to be added or subtracted from the previous shifted partial remainder. The quotient digit is determined by observing the two most-significant minimally redundant radix 4 partial remainder digits and independent of the range of the divisor (i.e., $n_s = 2$ and $n_d = 0$). The Boolean logic for the this selection unit has 6 input bits. The critical path per iteration stage is given by

$$T_{crit} = T_{QSEL,2} + 2 * t_{mux,2-to-1} + 2 * t_{cfa} + t_{buffer} + t_{reg}, \tag{46}$$

where $T_{QSEL,2}$ is the delay of the two radix 4 digit quotient selection logic, and the rest of the terms are as explained before. The total computation time to compute a W-bit quotient is

$$T_{total-time} = \frac{W}{2} T_{crit} + T_{prescale}, \tag{47}$$

where $T_{prescale}$ is the time required for input prescaling.

The total area requirement for this design is

$$\begin{aligned} A \; = \; & AQSEL_2 + 4 * W * AMUX_{2-to-1} + WCFA + ABUFFER \\ & +5/2 * WREG + WCPA + WP, \end{aligned} \tag{48}$$

where $AQSEL_2$ is the area of the two digit quotient selection logic, the number of register required to represent the radix 4 digits of the partial remainder is $3/2 * WREG$ as each radix 4 digit of the partial remainder is represented by 3 bits and a W-bit register is required for storing the scaled divisor. Finally, a fast carry propagate adder is required for input prescaling and some amount of area is taken by the control to implement prescaling on the same row of carry-free adders to be used by division recurrence.

20.2.2.3 Radix 4 division with input prescaling and over-redundant quotient

The division algorithm proposed in [18] operates on divisor in the range [1/2,1) and the dividend in the range [1/4,1/2). Prescaling of the divisor and the dividend is used similar to that used by [16]. The division recursion for this scheme is same as in (36). The quotient digits for this radix 4 division algorithm are derived from the radix digit set $\{-4,-3,-2,-1,0,+1,+2,+3,+4\}$. The index of redundancy for this scheme is $\sigma = 4/3$ (this condition is referred to as over-redundancy). The quotient digit for this algorithm, in any iteration step, is determined from the 4 most-significant digits of the partial remainder (i.e., $n_s = 4$ and $n_d = 0$). Since $q_i = \pm 3$ is allowed, the divisor multiple $\pm 3D$ is required. Two designs have been presented in [18] to realize this. The first scheme uses a 9-to-1 multiplexer to select the divisor multiple. In this scheme, if X'_{i-1} represents the first 4 digits of the partial remainder then the quotient selection, for the case where $\pm 3D$ is assumed available, is

$$q_i = \begin{cases} -4 & \text{if } -6 \le X'_{i-1} \le -4 \\ +4 & \text{if } 4 \le X'_{i-1} \le 6 \\ \pm 3 & \text{if } X'_{i-1} = \pm 3 \\ \pm 2 & \text{if } X'_{i-1} = \pm 2 \\ \pm 1 & \text{if } X'_{i-1} = \pm 1 \\ 0 & \text{if } X'_{i-1} = 0 \end{cases} \tag{49}$$

The critical path delay for a recursive implementation similar to the Fig. 20.1 is

$$T_{crit} = t_{qsel,4} + t_{mux,9-to-1} + t_{csa} + t_{buffer} + t_{reg}, \tag{50}$$

where $t_{qsel,4}$ is a 4 digit quotient selection logic, $t_{mux,9-to-1}$ is the delay of a 9-to-1 multiplexer, and the rest of the terms are same as explained before.

The area required for this design is

$$A = AQSEL_4 + W * AMUX_{9-to-1} + WCSA + ABUFFER + 4 * WREG + WCPA, \tag{51}$$

where $AQSEL_4$ is the area of 4-digit quotient selection logic, $AMUX_{9-to-1}$ is the area of a 9-to-1 multiplexer, $4 * WREG$ is the area of 4 W-bit registers required to store D, $3D$, and the partial remainder in redundant form, $WCPA$ is the area of a fast W-bit carry-lookahead adder, and the remaining terms are as defined before. The fast W-bit carry-lookahead adder is required to generate $3D$ for recursion and for prescaling the inputs.

The design in [18] also proposes another implementation of the above described scheme. In this scheme, the need for computing $3D$ is avoided by splitting the division addition/subtraction recursion into two stages. Each stage composed of a row of carry-free adders. In each of these rows, either a 0 or $\pm D$ or $\pm 2D$ is allowed so that the need to generate qD where $q \in \{-4, -3, -2, -1, 0, +1, +2, +3, +4\}$ can be easily handled. This is similar to the division scheme of [17] that uses maximally redundant quotient digits. The recursion stage is same as Fig. 20.6 except that a 5-to-1 multiplexer is required in the second stage than a 3-to-1 multiplexer, as shown.

The quotient digit q_i in the i-th iteration of this radix 4 over-redundant division is split into two terms $q'_i (\in \{\pm 2, 0\})$ and $q''_i (\in \{-2, -1, 0, +1, +2\})$ such that $q_i = q'_i + q''_i$. These two terms are derived from the expressions

$$q'_i = \begin{cases} +2 & \text{if } 2 \le X'_{i-1} \le 6 \\ -2 & \text{if } -6 \le X'_{i-1} \le -3 \\ 0 & \text{if } -2 \le X'_{i-1} \le +1 \end{cases} \tag{52}$$

and

$$q_i'' = \begin{cases} +2 & \text{if } 4 \leq X_{i-1}' \leq 6 \\ +1 & \text{if } X_{i-1}' = 3 \text{ or } 1 \\ +0 & \text{if } X_{i-1}' = 2 \text{ or } 0 \\ -1 & \text{if } X_{i-1}' = \text{-3 or -1} \\ -2 & \text{if } -6 \leq X_{i-1}' \leq -4 \end{cases} \tag{53}$$

The critical path delay for recursive implementation, similar to the one shown in Fig. 20.6 is

$$T_{crit} = max(t_{q'} + t_{mux,3-to-1} + t_{csa}, t_{q''} + t_{mux,5-to-1}) + t_{csa} + t_{buffer} + t_{reg}, \tag{54}$$

where $t_{q'}$ is the time taken to compute the partial quotient digit q' from the 4 msds (X_{i-1}') of the partial remainder, $t_{q''}$ is the time taken to compute the partial quotient digit q'', and the remaining terms are as defined before for other designs.

The area A required by this design is given by

$$\begin{aligned} A = \; & AQSEL_4 + W * AMUX_{3-to-1} + W * AMUX_{5-to-1} \\ & +2 * WCSA + ABUFFER + 3 * WREG + WCPA, \end{aligned} \tag{55}$$

where $AQSEL_4$ is the area required by the logic to generate the partial quotient digits q' and q'', the remaining terms are as described before for other designs. A fast carry-propagate adder is required here for this design to assist in input prescaling.

The total computation time to compute a W-bit quotient is

$$T_{total-time} = \frac{W}{2} T_{crit} + T_{prescale}, \tag{56}$$

where $T_{prescale}$ is the time required for input prescaling. Both the divisor and the dividend are prescaled using the same signed-digit adder, used for division recursion, and a fast carry-propagate adder.

Conversion of over-redundant radix 4 quotient digits to bits can be accomplished by using a digit-serial radix 4 adder [26] to reduce the over-redundant radix 4 quotient digits to maximally redundant regular radix 4 digits (belonging to the set $\{-3, \ldots, +3\}$) and then converting them to bits using a radix 4 on-the-fly converter. This scheme although does not increase the critical path delay has the disadvantage of producing a $(W+1)$-digit reduced quotient from the W-digit over-redundant quotient resulting in an extra cycle for on-the-fly conversion to binary form.

20.3 SQUARE ROOT

Square rooting techniques are of two types, namely direct techniques (digit-recurrence or digit-by-digit) and successive approximation. The direct techniques produce one digit of the root per iteration [6],[27]-[33]. The second technique uses an approximation method which converges to the result [27, 30, 34, 35].

Most of the high-speed digit recurrence methods for square root are extensions of the redundant arithmetic based non-restoring digit-by-digit SRT division algorithm [4, 14].

In this section, we briefly discuss and later provide a comparative evaluation of the square root algorithms (that operate on a bit-parallel mantissa in the range [1,2)) for radix 2 [6, 7, 11, 12, 36] and radix 4 [32, 33].

The basic square root recursion used in a digit-by-digit scheme is

$$X_i = X_{i-1} - (2Q_{i-1} + q_i r^{-i})q_i r^{-i}, \tag{57}$$

and in the shifted form

$$X_i^* = r.X_{i-1}^* - (2Q_{i-1} + q_i r^{-i})q_i, \tag{58}$$

or

$$X_i^* = r.X_{i-1}^* - Yq_i, \tag{59}$$

where $Q_i = Q_{i-1} + q_i r^{-i}$, $X_i^* = X_i 2^i$, and the symbols are defined as $i \rightarrow$ index of recursion, $i = 1, 2, \ldots, W$; $X_i \rightarrow$ i-th partial remainder; $X_i^* \rightarrow$ i-th shifted partial remainder; $X_0, X \rightarrow$ radicand (mantissa); $Q_i \rightarrow$ i-th partially generated square root, with $Q_0 = 0$; $Q_W \rightarrow$ W-digit square root of X_0; $q_i \rightarrow$ i-th square root digit, and $Y = (2Q_{i-1} + q_i r^{-i}) \rightarrow$ i-th partial radicand (PRD). As is seen from above, an updated Y (PRD) is required in each iteration step i.

This recursion is convergent only if $X_i^* \rightarrow 0$ for $i \rightarrow \infty$, if for all indices i the partial remainder X_{i-1}^* and q_i have the same sign. That is the shifted remainders X_{i-1}^* for all i ($i > 1$) are bounded. The convergence bound (see [32]) for X_{i-1}^* is given by

$$-2\sigma Q_{i-1} + r.\sigma^2.r^{-i} \leq X_{i-1}^* \leq +2\sigma Q_{i-1} + r.\sigma^2.r^{-i}, \tag{60}$$

where σ is the index of redundancy as defined in the section on division.

Fig. 20.8 shows the generic architecture of a recursive portion of a square rooting scheme that reuses the same hardware to generate each digit of the root.

The major constituents of the recursive part include the carry-free controlled adder/subtractor, the root digit selection logic (a function of n_s msds of the partial remainder and n_d most-significant bits of an estimated root), the storage registers required to hold the partial remainder, the PRD (partial radicand) generation block, and the on-the-fly converter needed to permit generation of the PRD in binary form. The use of PRD in binary form enables use of 3-to-2 bit adders for addition/subtraction operation.

The following sections briefly describe some of the recently proposed radix 2, and radix 4 square root schemes.

20.3.1 Radix 2 Square Root

The recursive expression for the radix 2 square root algorithm is

$$X_i^* = 2.X_{i-1}^* - (2Q_{i-1} + q_i 2^{-i})q_i, \tag{61}$$

where $Q_i = Q_{i-1} + q_i 2^{-i}$, $X_i^* = X_i 2^i$, and the symbols are defined as $i \rightarrow$ index of recursion, $i = 1, 2, \ldots, W$; $X_i \rightarrow$ i-th partial remainder; $X_i^* \rightarrow$ i-th shifted partial remainder; $X_0, X \rightarrow$ radicand (mantissa); $Q_i \rightarrow$ i-th partially generated square root, with $Q_0 = 0$; $Q_W \rightarrow$ W-digit square root of X_0; $q_i \rightarrow$ i-th square root digit, and $Y = (2Q_{i-1} + q_i 2^{-i}) \rightarrow$ i-th partial radicand (PRD).

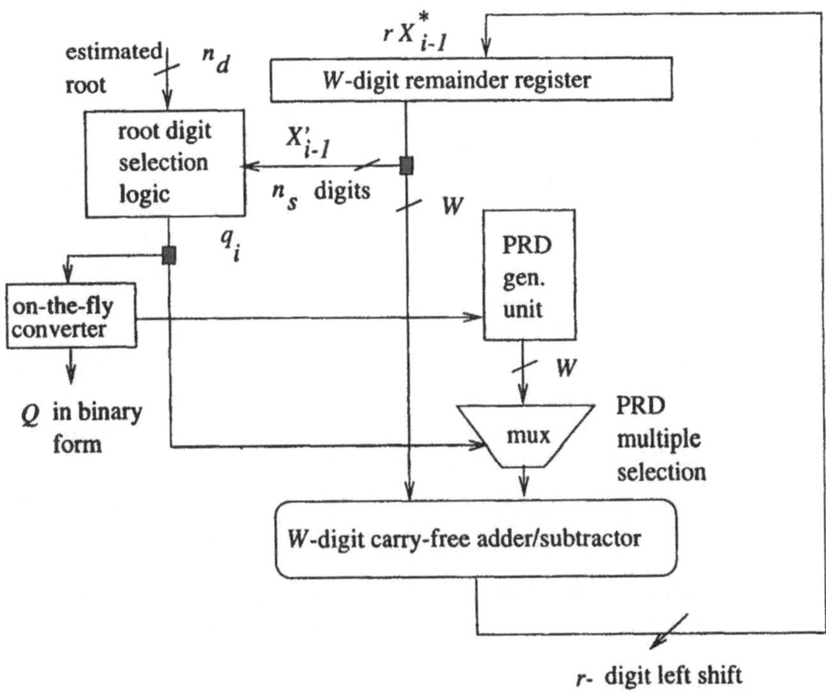

Figure 20.8 Generic architecture of a digit-by-digit based radix-r square root algorithm.

The radix 2 square root algorithm of [6] operates on the input in the range [1/4,1). This algorithm generates root digits from the set $\{-1, 0, +1\}$. The partial remainder in this algorithm satisfies the relation (in the i-th step)

$$-2(Q_{i-1} - 2^{-i}) \leq X^*_{i-1} \leq 2(Q_{i-1} + 2^{-i}). \tag{62}$$

Carry-free adders are used to generate the partial remainder in redundant form. The root digits, in any iteration step, are determined only from the three most-significant digits of the partial remainder ($n_s = 3$ and $n_d = 0$). This scheme uses a complex adder (at the most-significant digit end) that adds the 4 most-significant partial remainder digits (i.e., two rows of 4 bits each since each digit is encoded as 2 bits) with the 3 most-significant bits of the PRD derived from the PRD generation unit. This complex adder allows for a simple root digit selection function to be used.

The PRD is generated in binary form by concatenating the i-th root digit to the partially accumulated root Q_{i-1} (available in binary form from the on-the-fly converter), after appropriate shifting. The critical path delay for a recursive implementation of this design (similar to the one shown in Fig. 20.8) is

$$\begin{aligned} T_{crit} &= t_{qsel,3} + t_{PRD} + t_{mux,2-to-1} \\ &\quad + t_{complex\_add} + t_{buffer} + t_{reg}, \end{aligned} \tag{63}$$

where $t_{qsel,3}$ is the delay of the 3-digit root selection logic, t_{PRD} is the delay incurred in the PRD generation unit, $t_{complex\_add}$ is the delay of the complex adder cell at the msd end of the carry-free adder/subtractor, and the remaining terms are as explained before.

The area required by this design is

$$\begin{aligned} A &= AQSEL_3 + APRD + AOFC + W * AMUX_{2-to-1} + WCFA \\ &\quad + ABUFFER + 2 * WREG, \end{aligned} \tag{64}$$

where $AQSEL_3$ is the area required by the 3-digit root digit selection function, $APRD$ is the area of the PRD generation unit, $AOFC$ is the area of the on-the-fly converter unit, $WCFA$ is the area of the W-digit carry-free adder, and the remaining terms are as explained before.

The square root scheme of [7] is similar to the scheme of [6] except that it does not use a complex adder at the most-significant digit end. It uses a carry-save adder to produce the partial remainder in carry-save form. The root selection logic for this scheme is a function of the 4 most-significant digits of the partial remainder (i.e., $n_s = 4$ and $n_d = 0$). The critical path delay for this design can be written as

$$\begin{aligned} T_{crit} &= t_{qsel,4} + t_{PRD} + t_{mux,3-to-1} \\ &\quad + t_{csa} + t_{buffer} + t_{reg}, \end{aligned} \tag{65}$$

where $t_{qsel,4}$ is the delay of the 4-digit root digit selection logic, and the remaining terms are as defined before. The area required by this design is same as that given in (64).

The square root scheme of [11] operates on input radicand in the range [1/4,1/2). It employs signed-binary representation (i.e., digits belonging to the set $\{-1, 0, +1\}$) for the partial remainder and generates root digits from the two most-significant digits of the partial remainder (i.e., $n_s = 2$ and $n_d = 0$). The root selection function is given by the expression

$$q_i = \begin{cases} -1 & \text{if } -1 \le X'_{i-1} \le -1/2 \\ 0 & \text{if } -1/4 \le X'_{i-1} \le 1/4 \\ +1 & \text{if } 1/2 \le X'_{i-1} \le +1 \end{cases} \tag{66}$$

where X'_{i-1} is as shown in 20.8. The critical path for this design is given by

$$T_{crit} = t_{qsel,2} + t_{xor} + t_{csa} + t_{mux,2-to-1} + t_{buffer} + t_{reg}, \tag{67}$$

where $t_{qsel,2}$ is the delay of the 2-digit root generation function. The area required by this design is

$$\begin{aligned} A &= AQSEL_2 + W * AXOR + 2 * W * AMUX_{2-to-1} + WCSA \\ &\quad + ABUFFER + 3 * WREG + APRD, \end{aligned} \tag{68}$$

where $AQSEL_2$ is the area of the two digit root selection function, $APRD$ is the area of the PRD generation circuitry and the root digit to bit conversion circuitry, and the remaining terms are as explained before. This scheme, although faster than other radix 2 schemes, operates only on inputs in the range [1/4,1/2). If the exponent accompanying a mantissa in the range [1/4,1/2) is odd than this algorithm would require a post-scaling by $\sqrt{2}$ to produce the correct root. This is the only drawback of the design.

20.3.1.1 Radix 2 square root using over-redundant quotient digits

The radix 2 square root algorithm of [12, 36] accommodates both even and odd exponents, uses 3-to-2 digit adders for partial remainder computation, and requires a root selection function that observes only the two most-significant digits of the partial remainder. This square root technique uses two different algorithms which share the same hardware. Of the two algorithms, the one chosen, for any given input, is determined by whether the exponent of the input is even or odd, i.e., whether the least-significant bit of the exponent is 0 or 1. In the case of an even exponent the square-root algorithm [12], based on the Svoboda's division technique [24, 25], is used. This algorithm operates on the mantissa in the range [1,2) (mantissa conforming to the IEEE-754, 1985 standard). This is referred to as the ESQRT algorithm (for even exponents). In the case of an odd exponent another square-root algorithm [12], that uses over-redundant radix 2 root digits belonging to the set $\{-2,-1,0,+1,+2\}$, is used. This algorithm is derived from the over-redundant radix 2 division algorithm briefly discussed in the previous section and uses the same hardware as required by that division algorithm. This algorithm operates on inputs in the range [1/2,1). It is called the OSQRT (for odd exponents) algorithm. An IEEE-754 standard mantissa in the range [1,2), with an odd exponent, can be divided by 2 (and a 1 added to its exponent, to make it even) to bring it to the range [1/2,1) with an even exponent. This mantissa scaling is a simple right shift (by 1-bit) operation.

This square root scheme also employs a modified form of the on-the-fly converter to convert unnormalized root digits produced by the ESQRT algorithm and over-redundant radix 2 digits produced by the OSQRT algorithm to conventional binary form. These two algoriths are explained below.

OSQRT Algorithm: Suppose the floating point number (input radicand), represented as $m.2^e$, has an odd value for the exponent e, then it is prepared as $(m/2).2.2^e = G.2^{(e+1)}$, and G is used as the input to the algorithm.

The OSQRT algorithm uses *over-redundant* radix 2 digits belonging to set $\{-2,-1,0,+1,+2\}$ for representing the root and the regular radix 2 digits belonging to the set $\{-1,0,+1\}$ for the partial remainder digits. This is different from the classical radix 2 square root algorithm that uses digit from the set $\{-1,0,+1\}$ to represent both the root digits and the digits of the partial remainder.

The recursion for the OSQRT algorithm, in the i-the iteration step, is

$$X_i^* = 2X_{i-1}^* - Yq_i, \tag{69}$$

where Y the PRD is given by $(2Q_{i-1} + q_i 2^{-i})$, and $Q_i = Q_{i-1} + q_i 2^{-i}$, the i-th partially accumulated root.

The root digit selection function used by the OSQRT algorithm is

$$q_i = \begin{cases} -2 & \text{if } X_{i-1}' = -3 \\ -1 & \text{if } X_{i-1}' = -2 \text{ or } -1 \\ 0 & \text{if } X_{i-1}' = 0 \\ +1 & \text{if } X_{i-1}' = +2 \text{ or } +1 \\ +2 & \text{if } X_{i-1}' = +3, \end{cases} \tag{70}$$

where term X_{i-1}' is the digit string composed of the two most-significant digits of the shifted partial remainder $2X^*_{i-1}$. The square root of the input radicand m is given by Q_W.

ESQRT Algorithm: The ESQRT algorithm is used to generate the square root of a mantissa (in the range [1,2)) with an even exponent. The ESQRT algorithm is different from the classical radix 2 square root algorithm in that it is derived not from the SRT type of division but from the *Svoboda* type of division [24] .

The mantissa (input) for this algorithm is of the form $m = 1 + M$, where $0 \leq M < 1$. In this algorithm, a partial root, Q', is defined, that is given by

$$Q' = q'_0 + q'_1 2^{-1} + q'_2 2^{-2} + q'_3 2^{-3} + \cdots + q'_{W-1} 2^{-(W-1)}, \tag{71}$$

where $q'_0 = 0$. The quantity q'_i in the above expression represents the partial root digit, generated in the i-th iteration, and is selected from the regular radix 2 digit set $\{-1, 0, +1\}$.

The basic recursion for ESQRT is

$$R^*_i = 2R^*_{i-1} - \frac{1}{2}(2Q'_{i-2} + q'_{i-1} 2^{-(i-1)})q'_{i-1}, \tag{72}$$

where $R^*_i = 2^i y_i = 2^i(Q'_{i-1} + q'_i 2^{-i} + X_i)$, Q'_{i-1} is the $(i-1)$-th accumulated partial root, and $q'_i = s_i y_{i,i}$ the msd of the partial remainder $2^i(Q'_{i-1} + s_i y_{i,i} 2^{-i} + X_i)$.

The term $\frac{1}{2}(2Q'_{i-2} + q'_{i-1} 2^{-(i-1)})$ is the PRD that is used by the ESQRT algorithm in every iteration. The square root of the input radicand $(1 + M)$, for the ESQRT algorithm, is

$$Q = \sqrt{m} = \sqrt{1 + M} = (1 + \frac{Q'}{2}). \tag{73}$$

The root digits selection function is a rewrite operation of the two most-significant digits of the partial remainder and then selecting the most-significant digit of the rewritten partial remainder as the the root digit.

Both OSQRT and ESQRT algorithms use the same adder/subtractor for performing the recursion step. Both the algorithms observe only the two most-significant digits of the partial remainder.

The critical path for both OSQRT and ESQRT square root schemes is

$$T_{crit} = t_{QSEL,2} + t_{PRD} + t_{mux,2-to-1} + t_{csa} + t_{buffer} + t_{reg}, \tag{74}$$

where $t_{QSEL,2}$ is the delay incurred in the two-digit root selection function (i.e., $n_s = 2$ and $n_d = 0$), and the remaining terms are as defined before.

The area required by these designs is

$$\begin{aligned} A \;=\; & AQSEL_2 + APRD + 2 * W * AMUX_{2-to-1} + WCSA \\ & + ABUFFER + 2 * WREG + AMOFC, \end{aligned} \tag{75}$$

where $AQSEL_2$ is the area of the two digit root digit selection function, $AMOFC$ is the delay of the modified on-the-fly converter. The modified on-the-fly converter requires twice the area required by the original on-the-fly converter [22].

20.3.1.2 Radix 4 Square Root

A generalized square root algorithm for higher radix $r \geq 4$ has been proposed in [32]. In this scheme the square root operation is divided into two parts. The first part includes estimation of the square root of the the input radicand by observing only a few most-significant bits of the input radicand. After this step, the estimated root bits along with the most-significant digits of the partial remainder, in every iteration step, are used to determine the root digit.

For a radix 4 square root scheme (using minimally redundant radix 4 root digits) proposed in [32] the initial phase requires a 7-bit input look-up table to determine the first 5 estimated bits of the root. After this step, the correct second root digit is determined by using a root selection logic that is a function of the 7 most-significant digits of the partial remainder and the 5 bits of the estimated root (i.e., $n_s = 7$ and $n_d = 5$). The recursive expression of (58) is used with $r = 4$ for the second phase. The second phase starts from the index $i = 3$ and uses a carry-save adder array to implement the recursive addition/subtraction operation. In order to use a carry-save adder array to implement the recursive addition/subtraction, the PRD is needed in binary form, in every iteration. This is generated from a radix 4 on-the-fly converter.

The critical path delay of a recursive stage of this scheme is given by

$$\begin{aligned} T_{crit} &= t_{root\_sel,7,5} + t_{mux,5-to-1} + t_{PRD} \\ &\quad + t_{csa} + t_{buffer} + t_{reg}, \end{aligned} \tag{76}$$

where $t_{root\_sel,7,5}$ is the delay of the 7-digit and 5-bit root digit selection logic or the delay in reading a (19-bit input) lookup table, and the remaining terms are as explained in other designs.

The area required by this design is

$$\begin{aligned} A &= ARSEL_{7,5} + AISEL_5 + W * AMUX_{5-to-1} + AOFC + WCFA \\ &\quad + ABUFFER + 2 * WREG, \end{aligned} \tag{77}$$

where $ARSEL_{7,5}$ is the area of a 19-bit root selection logic, $AISEL_5$ is the area of the initial PLA (or table look-up) for determining the 5-bit estimated root, $AOFC$ is the area of the radix 4 on-the-fly conversion unit, and the remaining terms are as defined before.

The radix 4 square root scheme of [33] does not employ input table lookup or PLA for determining the estimated root. This saves time and area by avoiding the table lookup requirement. It uses a simple Boolean function to determine a 4-bit initial estimate of the root. The recursive portion for this scheme is similar to that of the scheme of [32]. The partial remainder in each step is generated in carry-save form as the recursive part uses carry-save adders to implement the addition/subtraction. The root digit, in each iteration, is determined from 8 most-significant carry-save digits or 16-bits and the 3-bits of the initial estimated root (i.e., $n_s = 8$ and $n_d = 3$). The critical path delay for this design is same as the design of [32] but its area required by this design is a little less due to the fact that the initial table look-up PLA is reduced to a few gates.

The radix 4 scheme of [15] employs an initial table look-up phase that observes the first 6 bits of the input radicand and the least-significant bit of the exponent to determine a 5-bit estimate of the root. An initial set of 3 square root recursions are employed to determine the partial radicand from which the remaining root digits can be determined. The root digits used for these 3 iterations are derived from the initial estimate. After these three iterations, the square root digit (belonging to

Table 20.1 Y for Unified Division/Square Root Recursion

Operation	Y
Division	$= q_i D$
Square root	$= PRD$

$D \to$ divisor, $PRD \to$ partial radicand
$q_i \to i$-th quotient/root digit, $PRD = q_i(2Q_{i-1} + q_i r^{-i})$, and
$Q_i \to i$-th partial (accumulated) quotient/root.

the minimally redundant radix 4 digit set) in every iteration is derived from the root selection logic that is exactly similar to the quotient selection logic used by a shared division/square root architecture that it implements. A scheme similar to the on-the-fly conversion is used to generate the partial radicand in binary form. The root digit multiples of the partial radicand are selected using a multiplexer. The critical path delay for a recursive structure of this design is slightly more than the division algorithm this scheme implements. The increase in the delay is due to the PRD generation and it's muliple selection.

The area required by a recursive implementation for this design is exactly similar to the division algorithm it implements except that an on-the-fly converter arrangement and partial radicand generation hardware is required.

20.4 UNIFIED DIVISION SQUARE ROOT ALGORITHM

Division and square root operations employ similar controlled addition or subtraction of the divisor/partial radicand to or from the shifted partial remainder, in each and every iteration. An unified expression for the recursion step can be written as

$$X_i^* = r.X_{i-1}^* - Y, \tag{78}$$

where X_i^* is the i-th partial remainder and Y is as given in Table 20.1.

Fig. 20.9 shows a generic architecture of a recursive structure that can perform either digit-by-digit division or digit-by-digit square root. The only difference that this architecture has with that of the Fig. 20.1 and 20.8 is that an on-the-fly converter is required to generate the PRD in binary form. It also incorporates a multiplexer to select either the divisor or the PRD depending upon whether a division or a square root operation is performed. The shared division/square root architectures employ the same digit selection logic to do both the operations. This results in savings in area as the need to implement two separate modules is avoided. The main reason that such sharing of hardware is possible is because both division and square root operations are very similar in nature.

A number of shared division/square root architectures have been proposed recently. They are briefly discussed in this section.

In order to use the same digit selection function for both division and square root, the architecture of [7] scales the input radicand (for square root) such that it

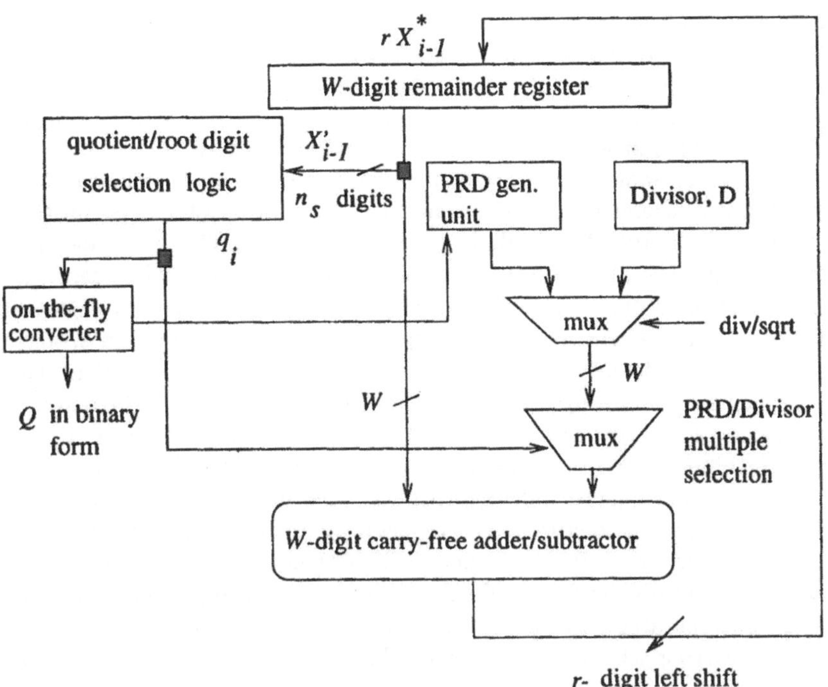

Figure 20.9 Generic architecture of a digit-by-digit based shared radix-r division/square root algorithm.

is divided by 2. The scaling of the mantissa is accounted by an adjustment to the exponent. This scheme uses carry-save adders to generate the partial remainder in each iteration. The quotient or root digit selection function observes only the 4 most-significant digits of the partial remainder (i.e., $n_s = 4$ and $n_d = 0$). An on-the-fly converter is used for conversion of the quotient and the root digits to binary form and to also provide the partial radicand required for square root, in binary form. The critical path delay for this scheme is same as the delay for the square root algorithm it implements. This is because the square root operation requires more circuitry for generation of PRD in the critical path than division.

The radix 2 shared multiplication/division/square root scheme of [37] implements division and square root that requires a 3 digit quotient/root selection function. The critical path delay for this scheme can be written as

$$T_{crit} = t_{qsel,3} + t_{buffer} + t_{mux,3-to-1} + t_{or} + t_{cfa} + t_{reg}, \tag{79}$$

where $t_{qsel,3}$ is the delay of the 3-digit quotient selection function, and the remaining terms are as explained before. The area A required by this design can be expressed as

$$\begin{aligned} A \;=\; & AQSEL_3 + AOFC + W * AMUX_{3-to-1} + W * AOR + WCSA \\ & + ABUFFER + 2 * WREG, \end{aligned} \tag{80}$$

where $AOFC$ is the area required by the on-the-fly converter and the rest of the terms are as defined before. This scheme is the most-efficient of all the shared radix 2 division/square root designs as it offers fast operation with nominal hardware requirement. Furthermore it implements most-significant digit first multiplication thus utlizing the hardware required in the most efficient manner.

The radix 2 shared division/square root scheme of [38] is similar to that of the shared radix 2 scheme of [7], except that it uses multiplexers to retain the partial radicand if the quotient/root digit generated is zero. The main idea here is to skip the addition/subtraction recursion if the quotient/root digit is zero. By use of variable clocks, to handle skipping of addition/subtraction for zero quotient/root digits and to handle addition/subtraction for non-zero quotient digits, this scheme demonstrates a shorter computation times on the average. The critical path delay for this scheme is more by a 2-to-1 multiplexer delay over the scheme of [7]. This design requires an extra area for the $2 * W$ 2-to-1 multiplexers needed to retain the partial remainder when the quotient/root digit turns out to be a zero.

The division/square root algorithm of [11] also implements a shared division/square root architecture. The division algorithm it implements is fast and requires extra hardware (in the form of a W-bit fast carry-propagate adder) to perform divisor and dividend prescaling. The square root algorithm it implements is also fast but operates only on the mantissas with even exponents. This shared division/square root scheme is not practicable because of the limitation of the square root algorithm.

The shared radix 2 division/square root scheme of [36] employs over-redundant radix 2 quotient digits for division and OSQRT algorithm (square root algorithm for mantissas with odd exponents), and regular radix 2 quotient digits for ESQRT algorithm (square root algorithm for mantissas with even exponents). The quotient digits are different for division/OSQRT and ESQRT algorithms primarily to allow the use of carry-free adders with no representation overflow (an increase in word-length of a signed digit number at the msd end without change in its value) and to allow a 2-digit quotient/root selection function. The architecture for this scheme is same as the square root scheme it implements except that a multiplexer is need to select the divisor, instead of the PRD, if division is desired. The critical path delay for this scheme is the delay of the square root scheme it implements (see section 20.3 for time per recursion stage). This scheme is faster than the other radix 2 shared division/square root schemes discussed. The area required for this scheme is similar to its square root scheme it implements (see section 20.3 for area requiremnts) except that an extra W-bit register is required to store the divisor for performing division. The area requirement of this scheme is much more than that of the other schemes primarily because of the use of the modified on-the-fly converter.

The radix 4 shared division square root scheme of [15] is also similar to the square root scheme it proposes. The only difference is that a multiplexer is required to select either the divisor or the partial radicand if division or square root operation is to be performed. The critical path delay for this structure is dictated by the delay of the square root unit.

Table 20.2 Radix 2 Division Schemes

Design	i/p scale	$\lvert q_j \rvert \leq$	n_s	n_d	Time required for a recursion stage Area for a recursion stage
[7]	no	1	3	0	$T_{crit} = t_{qsel,3} + t_{buffer} + t_{mux,3-to-1} + t_{csa} + t_{reg}$ $A = WCSA + QSEL_4 + WBUF + W * MUX_{3-to-1} + 3 * WREG$
[8, 9]	no	1	3	0	$T_{crit} = t_{qsel,3} + t_{buffer} + t_{xor} + t_{and,2} + t_{cfa} + t_{reg}$ $A = WCFA + QSEL_3 + WBUF + W * (AXOR + AAND2)$ $+3 * WREG$
[10, 11]	yes	1	2	0	$T_{crit} = t_{qsel,2} + t_{buffer} + t_{xor} + t_{cfa} + t_{mux,2-to-1} + t_{reg}$ $A = WCFA + QSEL_2 + WBUF + W * AXOR$ $+2 * W * MUX_{2-to-1} + 3 * WREG + WCPA$
[13]	no	2	2	0	$T_{crit} = t_{qsel,2} + 2.t_{mux,2} + t_{xor} + t_{cfa} + t_{buffer} + t_{reg}$ $A = WCFA + QSEL_2 + WBUF + W * AXOR$ $+3 * W * MUX_{2-to-1} + 3 * WREG$

20.5 COMPARISON

Tables 20.2 and 20.3 list the cycle time and area required for a recursion stage of the radix 2 and radix 4 division algorithms that have been discussed. The major factor driving the need for higher radix schemes is to reduce the total computation time required to perform division. An immediate requirement for higher radix division schemes to satisfy this is to reduce the time per iteration step. Recently proposed higher radix division schemes strive to reduce this iteration time. An important effect in resorting to higher radix schemes is that the contribution of negative factors such as register loading and buffer delays are reduced. This is true because these delays now appear only in $W/\log_2 r$ iterations (for a radix $r > 2$) instead of W for radix 2. Using higher radix results in a wider range for the partial remainder causing the quotient selection logic to be a function of more number of digits. Furthermore, divisor multiple selection for a larger quotient digit set adds extra delay.

Some of the recently proposed radix 2 division schemes [10] [11] use very simple quotient selection functions ($n_s = 2$). The delay involved in these schemes is small enough to ignore the need for higher radix division schemes. These schemes incur area penalty as they require a fast carry-propagate adder to perform operand scaling. The scheme of [13] also uses a simple quotient digit selection function with $n_s = 2$ but does not require operand scaling. However, it is not as fast as the schemes of [10] [11]. In fact, the intrinsic radix 2 division recursion has been simplified to the extent that speed improvement, if possible, over already existing schemes is going to be very minimal. If a fast carry-propagate adder is already available (e.g., as in the datapath of a microprocessor) the best choice for a radix 2 division scheme appears to be that of [10] and [11]. The adder, apart from being used for prescaling, can also be used to perform remainder correction and quotient rounding if the results have to conform to a standard such as the IEEE 754 1985 std.

Table 20.3 Radix 4 Division Schemes

Design	i/p scale	$\|q_j\| \le$	n_s	n_d	Timer required for a recursion stage Area for a recursion stage
[7]	no	2	7	3	$T_{crit} = t_{qsel,7,3} + t_{buffer} + t_{mux,5-to-1} + t_{csa} + t_{reg}$ $A = WCFA + QSEL_{7,3} + WBUF + W * MUX_{5-to-1} + 3 * WREG$
[15]	no	2	8	4	$T_{crit} = t_{cla,8} + t_{xor} + t_{or} + t_{pla,19} + t_{mux,5-to-1} + t_{csa} + t_{buf} + t_{reg}$ $A = ACLA_8 + 5 * AXOR + 4 * AOR + APLA_{19} + ABUFFER + AMUX_{5-to-1} + WCSA + 3 * WREG$
[16]	yes	2	6	0	$T_{crit} = t_{qsel,6} + t_{mux,5-to-1} + t_{csa} + t_{buffer} + t_{reg}$ $A = WQSEL_6 + WCFA + W * MUX_{5-to-1} + ABUFFER + 3 * WREG + WCPA$
[17]	no	3	7	3	$T_{crit} = max(t_{q'} + t_{mux,3-to-1} + t_{csa}, t_{q''} + t_{mux,3-to-1}) + t_{cla,7} + t_{csa} + t_{buffer} + t_{reg}$ $A = ACLA_7 + 2 * W * MUX_{3-to-1} + 2 * WCSA + AQSEL_{q'} + AQSEL_{q''} + ABUFFER + 3 * WREG$
[18] arch-1	yes	4	4	0	$T_{crit} = t_{qsel,4} + t_{mux,9-to-1} + t_{csa} + t_{buffer} + t_{reg}$ $A = AQSEL_4 + W * AMUX_{9-to-1} + WCSA + ABUFFER + 4 * WREG + WCPA$
[18] arch-2	yes	4	4	0	$T_{crit} = max(t_{q'} + t_{mux,3-to-1} + t_{csa}, t_{q''} + t_{mux,5-to-1}) + t_{csa} + t_{buf} + t_{reg}$ $A = AQSEL_4 + W * AMUX_{3-to-1} + W * AMUX_{5-to-1} + 2 * WCSA + WBUF + 3 * WREG + WCPA$
[19]	yes	3	3	0	$T_{crit} = t_{qsel,3} + t_{xor} + t_{mux,3-to-1} + t_{cfa} + t_{mux,2-to-1} + t_{buffer} + t_{reg}$ $A = AQSEL_2 + 2 * W * AXOR + W * AMUX_{3-to-1} + WCFA + W * AMUX_{2-to-1} + ABUFFER + 4 * WREG + WCPA + WP$
[20]	yes	2	3	0	$T_{crit} = T_{QSEL,2} + 2 * t_{mux,2-to-1} + 2 * t_{cfa} + t_{buffer} + t_{reg}$ $A = AQSEL_2 + 4 * W * AMUX_{2-to-1} + WCFA + ABUFFER + 5/2 * WREG + WCPA + WP$

Note:

1. arch-1 and arch-2 represent the two different architectures presented in [18].

Of the radix 4 division schemes discussed, the ones that perform input prescaling [16], [18], [19], and [20] are the most promising in allowing shorter recursion times. This is primarily because of the simpler quotient selection functions they use. These schemes incur prescaling time overhead in the form of extra cycles. The overhead is in the form of performing a carry-free addition followed by a fast

Table 20.4 Radix 2 Square Root Schemes

Design	$\lvert q_j \rvert \leq$	n_s	n_d	Timer required for a recursion stage Area for a recursion stage
[6]	1	3	0	$T_{crit} = t_{qsel,3} + t_{PRD} + t_{mux,2-to-1} + t_{complex\_add} + t_{buffer} + t_{reg}$ $A = AQSEL_3 + APRD + AOFC + W * AMUX_{2-to-1} + WCFA + ABUFFER + 2 * WREG$
[7]	1	4	0	$T_{crit} = t_{qsel,4} + t_{PRD} + t_{mux,3-to-1} + t_{csa} + t_{buffer} + t_{reg}$ $A = AQSEL_3 + APRD + AOFC + W * AMUX_{2-to-1} + WCFA + ABUFFER + 2 * WREG$
[11]	1	2	0	$T_{crit} = t_{qsel,2} + t_{xor} + t_{csa} + t_{mux,2-to-1} + t_{buffer} + t_{reg}$ $A = AQSEL_2 + W * AXOR + 2 * W * AMUX_{2-to-1} + WCSA + ABUFFER + 3 * WREG + APRD$
[12, 36]	2	2	0	$T_{Crit} = t_{QSEL,2} + t_{PRD} + t_{mux,2-to-1} + t_{csa} + t_{buffer} + t_{reg}$ $A = AQSEL_2 + APRD + 2 * W * AMUX_{2-to-1} + WCSA + ABUFFER + 2 * WREG + AMOFC$

Table 20.5 Radix 4 Square Root Schemes

Design	$\lvert q_j \rvert \leq$	n_s	n_d	Timer required for a recursion stage Area for a recursion stage
[15]	2	6	0	$T_{crit} = t_{cla,8} + t_{xor} + t_{or} + t_{pla,19} + t_{mux,5-to-1} + t_{csa} + t_{buf} + t_{reg} + t_{PRD}$ $A = ACLA_8 + 5 * AXOR + 4 * AOR + APLA_{19} + ABUFFER + AMUX_{5-to-1} + WCSA + 3 * WREG + AOFC + APRD$
[32]	2	7	5	$T_{crit} = t_{root\_sel,7,5} + t_{mux,5-to-1} + t_{PRD} + t_{csa} + t_{buffer} + t_{reg}$ $A = ARSEL_{7,5} + AISEL_5 + W * AMUX_{5-to-1} + AOFC + WCFA + ABUFFER + 2 * WREG$
[33]	2	8	3	$T_{crit} = t_{root\_sel,7,5} + t_{mux,5-to-1} + t_{PRD} + t_{csa} + t_{buffer} + t_{reg}$ $A = ARSEL_{7,5} + AISEL_5 W * AMUX_{5-to-1} + AOFC + WCFA + ABUFFER + 2 * WREG - AIPLA$

carry-propagate addition. In most of the general microprocessors, there is no area penalty involved as a fast carry-propagate adder is always available in the datapath. If so, the schemes of [19] and [20] offer the shortest critical paths for a recursion operation. The scheme of [19] uses maximally redundant radix 4 quotient digits and therefore requires more complicated prescaling due to the burden of generating ±3 times the divisor. The scheme of [20], however, uses a minimally redundant radix 4 quotient digit set (thereby simplifying the prescaling operation) and is therefore the most attractive choice.

Tables 20.4 and 20.5 list the various features of the different radix 2 and radix 4 square root algorithms that have been discussed. Here again the argument for using higher radix schemes over radix 2 designs is to lower the negative contributions of fanout and register load delays.

The radix 2 square root schemes of [6] and [7] incur almost the same critical path delay in any recursion step. The scheme of [36] presents the shortest critical path delay for a radix 2 scheme. But it incurs hardware penalty in the form of a larger modified on-the-fly converter required. Therefore, the selection of a high performance radix 2 square root scheme is a tradeoff between area and time.

The radix 4 square root schemes of [32] [33] incur the similar critical path delay in a recursion stage. This is slightly better than that of [15]. The area requirement for these designs is almost similar. The scheme of [33] saves some area, over the other two, by reducing the initial PLA table lookup to a simple Boolean function composed of a few logic gates.

Shared division/square root architectures employ the arrangement used to perform square root also for division. This is possible by multiplexing the divisor instead of the PRD whenever division operation has to be performed. Of the shared radix 2 division/square root schemes, the scheme proposed in [37] is the best choice, if area is the main concern. Also, this scheme is capable of performing multiplication using the same hardware. If high speed of operation (shorter critical path delay in a recursive stage), is the main requirement then the architecture of [36] is a better choice. Note, however, that the scheme of [36] uses a modifed on-the-fly converter that requires almost twice the area required by the original on-the-fly converter.

The discussion in this chapter was concerned only with synchronous implementations, i.e., implementations using a fixed clock period. Various implementations have been developed that exploit the fact that with a zero quotient/root digit an addition/subtraction step can be skipped. These include the zero skipping scheme of [38], and the self-timed schemes of [39] and [40]. These schemes are particularly useful if the division or square root module can have variable completion times. In most of the current day processors there is no need to have a synchronous operation. These implementations can then incur variable computation times and indicate to the control unit the finishing of the operations through a completion signal. An advantage that such schemes can also offer is that on the average they consume only as much power as is required to complete the required operation.

This discussion and comparative evaluation has focussed on only recursive implementations of the division and square root algorithms. There are other implementations of these operations that allow better performance. These include the radix 8 schemes with overlapped digit selection ([41],[42]) and non-overlapped digit selection, and digit prediction [43]. On-line arithmetic implementations have have also not been considered primarily as these form a variation in implementation of the same algorithms. Very high radix versions of these algorithms have also not been considered.

REFERENCES

[1] S. F. Oberman and M. J. Flynn, "Division algorithms and implementations," *IEEE Transactions on Computers*, vol. 46, pp. 833–854, August 1997.

[2] A. Avizienis, "Signed digit number representation for fast parallel arithmetic," *IRE Transactions on Electronic Computers*, vol. EC-10, pp. 389–400, September 1961.

[3] O. L. MacSorley, "High-speed arithmetic in binary computers," *Proc. of the IRE*, pp. 67–91, January 1961.

[4] J. E. Robertson, "A new class of digital division methods," *IRE Transactions on Electronic Computers*, vol. EC-7, pp. 218–222, September 1958.

[5] T. D. Tocher, "Techniques of multiplication and division for automatic binary computers," *Quarter. J. Mech. App. Math.*, vol. 2, pt. 3, pp. 364–384, 1958.

[6] S. Majerski, "Square-rooting algorithms for high-speed digital circuits," *IEEE Transactions on Computers*, vol. 34, pp. 724–733, 1985.

[7] M. D. Ercegovac and T. Lang, *Division and Square Root.* Norwell, Massachusetts 02061: Kluwer Academic Publishers, 1994.

[8] S. Kuninobu, T. Nishiyama, H. Edamatsu, T. Tanaguchi, and N. Takagi, "Design of high speed MOS multiplier and divider using redundant binary representation," in *Proc. of 8th Symposium on Computer Arithmetic*, (Como, Italy), pp. 80–86, 1987.

[9] A. Vandemeulebroecke, E. Vanzieleghem, T. Denayer, and P. G. A. Jespers, "A new carry-free division algorithm and its application to a single-chip 1024-b RSA processor," *IEEE Journal of Solid State Circuits*, vol. 25, pp. 748–756, June 1990.

[10] N. Burgess, "A fast division algorithm for VLSI," in *Proc. of IEEE International Conference on Computer Design: VLSI in Computers and Processors*, (Boston, MA), pp. 560–563, October 1991.

[11] S. E. McQuillan, J. V. McCanny, and R. Hamill, "New algorithms and VLSI architectures for SRT division and square root," in *Proc. of 11th Symposium on Computer Arithmetic*, (Windsor, Ontario), pp. 80–86, June 29-July 2 1993.

[12] H. R. Srinivas, *High Speed Computer Arithmetic Architectures.* PhD thesis, Univ. of Minnesota, Dept. of Elect. Engg., Minneapolis, September 1994.

[13] H. R. Srinivas, K. K. Parhi, and L. A. Montalvo, "Radix 2 division with over-redundant quotient selection," *IEEE Transactions on Computers*, vol. 46, pp. 85–92, January 1997.

[14] D. E. Atkins, "Higher-radix division using estimates of the divisor and partial remainders," *IEEE Transactions on Computers*, vol. C-17, pp. 925–934, October 1968.

[15] J. Fandrianto, "Algorithm for high-speed shared radix-4 division algorithm," in *Proc. of 8th IEEE Symposium on Computer Arithmetic*, (Como, Italy), pp. 73–79, May 1987.

[16] M. D. Ercegovac and T. Lang, "Simple radix-4 division with operands scaling," *IEEE Transactions on Computers*, vol. C-39, pp. 1204–1208, September 1990.

[17] P. Montuschi and L. Ciminiera, "Design of a radix-4 division unit with simple selection table," *IEEE Transactions on Computers*, vol. 41, pp. 1606–1611, December 1992.

[18] P. Montuschi and L. Ciminiera, "Over-redundant digit sets and the design of digit-by-digit division units," *IEEE Transactions on Computers*, vol. 43, pp. 269–277, March 1994.

[19] H. R. Srinivas and K. K. Parhi, "A fast radix 4 division algorithm and its architecture," *IEEE Transactions on Computers*, vol. 44, pp. 826–831, June 1995.

[20] L. A. Montalvo, *Number Systems for High Performance Dividers*. PhD thesis, de l'Institut National Polytechnique De Grenoble, Grenoble, France, March 1995.

[21] "IEEE Standard for Binary Floating Point Arithmetic." IEEE Standard 754, IEEE Computer Society, 1985.

[22] M. D. Ercegovac and T. Lang, "On-the-fly conversion of redundant into conventional representations," *IEEE Transactions on Computers*, vol. C-36, pp. 895–897, July 1987.

[23] H. R. Srinivas and K. K. Parhi, "High-speed VLSI arithmetic processors using hybrid number systems," *Journal of VLSI Signal Processing*, vol. 4, pp. 177–198, April 1992.

[24] A. Svoboda, "An algorithm for division," *Information Processing Machines*, vol. 9, pp. 183–190, March 1963.

[25] C. Tung, "A division algorithm for signed-digit arithmetic," *IEEE Transactions on Computers*, vol. 17, pp. 887–889, September 1968.

[26] M. J. Irwin and R. M. Owens, "Design issues in digit serial signal processors ," *in Proc. of IEEE International Symposium on Circuits and Systems*, pp. 441–444, Portland, Oregon 1989.

[27] I. Koren, *Computer Arithmetic Algorithms*. NJ 07632: Prentice Hall, Englewood Cliffs, 1993.

[28] G. Metze, "Minimal square rooting," *IEEE Transactions on Electronic Computers*, vol. EC-14, pp. 181–185, April 1965.

[29] J. C. Majithia, "Cellular array for extraction of squares and square roots of binary numbers," *IEEE Transactions on Computers*, vol. C-21, pp. 1023–1024, September 1972.

[30] K. Hwang, *Computer Arithmetic: Principles, Architecture, and Design.* NY: Wiley, 1979.

[31] B. G. DeLugish, *A Class of Algorithms for Automatic Evaluation of Certain Elementary Functions in a Binary Computer.* PhD thesis, Dep. of Computer Science, Univ. of Illinois at Urbana, 1970.

[32] L. Ciminiera and P. Montuschi, "Higher radix square rooting," *IEEE Transactions on Computers*, vol. 39, pp. 1220–1231, October 1990.

[33] M. D. Ercegovac and T. Lang, "Radix-4 square root without initial PLA," *IEEE Transactions on Computers*, vol. 39, pp. 1016–1024, August 1990.

[34] H. Peng, "Algorithms for extracting square roots and cube roots," in *Proc. 5th IEEE Symposium on Computer Arithmetic*, (Ann Arbor, MI), pp. 121–126, May 1981.

[35] C. V. Ramamoorthy, J. R. Goodman, and K. H. Kim, "Some properties of iterative square rooting methods using high-speed multiplication," *IEEE Transactions on Computers*, vol. C-21, pp. 837–847, August 1972.

[36] H. R. Srinivas and K. K. Parhi, "A floating point radix 2 shared division/square root chip," in *Proc. of IEEE International Conference on Computer Design:VLSI in Computers and Processors*, (Austin, TX), pp. 472–478, October 1995.

[37] M. D. Ercegovac and T. Lang, "Implementation of module combining multiplication, division, and square root," in *Proc. of the International Symposium on Circuits and Systems*, pp. 150–153, 1989.

[38] P. Montuschi and L. Ciminiera, "Reducing iteration time when result digit is zero for radix 2 SRT division and square root with redundant remainders," *IEEE Transactions on Computers*, vol. 42, pp. 239–246, February 1993.

[39] T. Williams *et al.*, "A self-timed chip for division," in *Proc. of Stanford VLSI Conference*, (Cambridge, MA: MIT Press), pp. 75–95, 1987.

[40] G. Matsubara, N. Ide, H. Tago, S. Suzuki, and N. Goto, "30-ns 55-b shared radix 2 division and square root using a self-timed circuit," in *Proc. of the 12-th International Symposium on Computer Arithmetic*, pp. 98–105, July 1995.

[41] G. Taylor, "Radix-16 SRT dividers with overlapped quotient selection stages," in *Proc. of 7th IEEE Symposium on Computer Arithmetic*, (Urbana, IL), pp. 64–71, June 1985.

[42] J. A. Prabhu and G. B. Zyner, "167 MHz Radix 8 floating point divide and square root using overlapped radix 2 stages," in *Proc. of the 12-th International Symposium on Computer Arithmetic*, pp. 155–162, IEEE, IEEE, July 1995.

[43] M. D. Ercegovac, "A division algorithm with prediction of quotient digits," *Seventh IEEE Symposium on Computer Arithmetic*, pp. 51–56, Urbana, IL, 1985.

Chapter 21

Finite Field Arithmetic Architecture

Leilei Song and Keshab K. Parhi
Department of Electrical and Computer Engineering
University of Minnesota, Minneapolis, Minnesota
{llsong,parhi}@ece.umn.edu

21.1 INTRODUCTION

Finite fields have received a lot of attention because of their important and practical applications in cryptography, coding theory, switching theory and digital signal processing. For example, one of the major application of finite fields is algebraic coding theory - the theory of error-correcting and error-detecting codes. Typically, the message transmitted over a discrete communication channel consists of a finite sequence of symbols that are elements of some finite alphabet. In general, the alphabet is assumed to be a finite field. For instance, if the alphabet consists of 0 and 1, it is considered as the finite field $GF(2)$. Since noise is unavoidable in a communication channel, error-control codes have been extensively used in digital communication and computer systems to achieve efficient and reliable digital transmission and storage, and have become an essential part of digital communication and recording system. Due to the inherent properties of digital transmission over a communication channel, most error-control encoding and decoding algorithms are based on finite field arithmetic operations, such as Reed-Solomon codes, Golay codes, *etc.* [1] [4]. In practice, the finite fields of characteristic 2, $GF(2^m)$, are generally used and the multiplications over $GF(2^m)$ are the major building blocks in many error-control coders.

The finite field $GF(2)$ has 2 elements, $0, 1$. The addition and multiplication are performed modulo 2 as in one's complement arithmetic. $GF(2^m)$ is the extension field of $GF(2)$ and has 2^m elements. Each of these elements is represented as a polynomial of degree less than or equal to $m-1$ with coefficients coming from the ground field $GF(2)$. For such representation, addition and subtraction are bit-independent and straightforward. However, multiplication and division involve polynomial multiplication and division modulo some primitive polynomial $p(x)$, which are much more complicated. Hence the design of efficient architectures to perform these arithmetic operations is of great practical concern.

Many approaches and architectures have been proposed to perform finite field multiplication, exponentiation and division efficiently. Different polynomial representations including standard basis, dual basis and normal basis, and power representation have been used to obtain some interesting realizations. In what follows, the systolic arrays with broadcast signals are called ***semi-systolic*** arrays, distinguished from those systolic arrays with only local communications. In terms of standard polynomial basis representation, two semi-systolic array multipliers were presented in [5], one is for fixed finite field and the other is a generalized architecture programmable with respect to the field order m. In [6] and [7], two systolic multipliers were presented. A systolic power-sum circuit to compute $AB^2 + C$ was presented in [8]. A low latency standard basis semi-systolic multiplier and a squarer was presented in [9]. Another systolic squarer architecture was presented in [10]. In [11] a systolic exponentiator was proposed which required $2^m - 1$ parallel multipliers. An exponentiator architecture utilizing ***square-and-multiply*** algorithm and dedicated systolic multipliers was proposed in [12]. The first bit-serial dual basis multiplier was presented in [13]. In [14], a parallel dual basis multiplier was presented. Normal basis was used to compute multiplication and inversion over $GF(2^m)$ in [15] [16]. A normal basis exponentiator was presented in [17]. Efficient architectures to compute multiplication, exponentiation and division using power representation were designed in [18], , and [21]. Finite field division can be performed using exponentiations based on Fermat's Theorem, or utilizing Euclid's algorithm to find the greatest common divisor (GCD) of two polynomials [22] [23], or solving system of equations by Gauss-Jordan eliminations [24][25].

Standard basis polynomial representation is generally used for finite field arithmetic operations. In this chapter, we address the standard basis based array-type algorithms and architectures for finite field arithmetic operations, where one operand is processed in parallel and the other operand is processed one bit at each step. The corresponding algorithms using standard polynomial basis are classified as ***LSB-first*** scheme and ***MSB-first*** scheme. LSB-first scheme processes the least significant bit of the second operand first, while MSB-first scheme processes its most significant bit first. For finite field multiplication and exponentiation, internal computations at each step can be performed concurrently using LSB-first algorithms; while they are computed sequentially in MSB-first algorithms. Hence, multipliers and exponentiators implemented using LSB-first scheme have much less computation delay time compared with their counterparts based on MSB-first scheme, with the same hardware complexity. The LSB-first multiplication algorithm is mapped to a semi-systolic multiplier. Semi-systolic array differs from systolic array due to the presence of broadcast signals. Total number of latches and computation latency of a fully pipelined semi-systolic array is much less than those of their systolic counterpart with the same critical path delay time. As a result, the LSB-first semi-systolic multiplier has smaller area, latency and cycle time compared with previous systolic architectures [7] and MSB-first semi-systolic architectures [5]. Both LSB-first and MSB-first finite field squarers are designed and it turns out that the MSB-first squarer architecture has less hardware complexity. A LSB-first exponentiator is then designed utilizing LSB-first semi-systolic multipliers and MSB-first squarers. This standard basis exponentiator pipelined at bit-level has lower latency and smaller area compared with the systolic implementation in [12]. Another advantage of LSB-first scheme over MSB-first scheme is its capability of achieving substructure sharing among several arithmetic operations. The idea of substructure sharing has been introduced and used to design area efficient Reed-Solomon encoder in [26]. In this chapter, this issue is addressed separately in each subsection devoted to multipliers, squarers and exponentiators.

Dual and normal basis representations are efficient for certain types of operations, i.e., dual basis representation for bit-serial multiplications [13] and normal basis representation for exponentiations [16]. Dual basis multipliers are hybrid architectures, where the multiplicand and product are represented in dual basis and the multiplier is in standard basis. Such hybrid representation leads to efficient implementations of dual basis multipliers and dividers; however, it necessitates the use of basis converters between dual and standard basis. In this chapter, the bit-serial dual basis multiplier is introduced. Schemes to simplify basis conversion circuitry is briefly discussed. In normal basis representation, the squaring of an element in $GF(2^m)$ is a simple cyclic shift of its binary bits. Hence exponentiations using normal basis require only half the number of multipliers required by standard basis exponentiators. However, the multiplications in normal basis are substantially more difficult. The hardware complexity of a parallel normal basis multiplier over $GF(2^m)$ is in the order of $O(m^3)$, while those of the standard basis or dual basis multipliers are in the order of $O(m^2)$. The concept of normal basis representation is to be briefly introduced in this chapter. Readers may refer to [27] [16] for detailed explanations regarding to normal basis arithmetic architectures and their complexity analysis.

21.2 MATHEMATICAL BACKGROUND

This section gives a brief introduction to finite field fundamental concepts and properties. Three types of polynomial representations, standard basis, dual basis and normal basis are introduced. The readers may refer to [1], [28], [29] for more detailed descriptions. In what follows, the symbols in $GF(2^m)$ and the bits of symbols, which are elements of $GF(2)$, are represented using upper and lower case variables, respectively. Unless specified explicitly, "+" and "·" denote logic XOR and AND operations, respectively.

21.2.1 Finite Field Fundamentals

21.2.2.1 Groups and fields

Let's start at the very beginning.

Definition 21.2.1 *A set G and a binary operation * define a group such that the following are true*

- *Set is closed for the operation *;*
- *Operation * is associative;*
- *G contains an identity element e such that for every element* $a \in G, a * e = e * a = a$;
- *For every element* $a \in G$, *there exists an inverse element* a' *such that* $a * a' = a' * a = e$.

In addition, if the operation * is commutative, the group is called an *Abelian Group*. A group is thus completely characterized by the set of elements, the operation * and the identity element e.

Example 21.2.1 *The set of all integers, with the operation of integer addition define a group. The element 0 is the identity element for this group.*

Example 21.2.2 *The set of all rational numbers excluding zero and the operation of multiplication define a group. The element 1 is the identity element for this group.*

Roughly speaking, a group is a set which is closed under one type of operation, i.e., either addition or multiplication but not both. A *field*, however, is a set of elements closed under both addition and multiplication.

Definition 21.2.2 *A set F and two operations +, * define a field such that the following are true*

- *F is an Abelian Group under the operation + with the identity element 0;*
- *The non-zero elements of F form an Abelian group under the operation * with the identity element 1;*
- *The distributive law $a * (b + c) = a * b + a * c$ holds.*

In a field, the additive inverse of an element a is denoted by $-a$ and the multiplicative inverse is denoted by a^{-1}. A *field* is called *finite* or *infinite* according to whether the underlying set is finite or infinite. In this chapter, we consider only finite field. Finite fields are also referred to as *Galois field* in honor of the French mathematician who did pioneering work in this area.

A finite field with q elements is called GF(q). It exists only for $q = p^m$, where p is a prime number and m is a non-negative integer. When m=1, $GF(q) = GF(p)$ is called the *ground field*, where + and * are integer addition and multiplication modulo p, respectively. The *extension field* $GF(p^m)$ for $m \geq 2$ is defined using polynomials with coefficients from $GF(p)$ and the + and * operations are defined as polynomial additions and multiplications over $GF(p)$. In practice, finite field $GF(2)$ and $GF(2^m)$ are generally used. We will therefore lay special emphasis on $p = 2$ in the rest of this chapter.

21.2.2.2 Construction and properties of finite fields

Definition 21.2.3 *A polynomial $p(x)$ of degree m, $p(x) = x^m + p_{m-1}x^{m-1} + \cdots + p_1x + p_0$, is said to be irreducible over $GF(2)$ if $p(x)$ is not divisible by any other polynomial of degree k over $GF(2)$, where $0 < k < m$.*

The concept of irreducible polynomial is analogous to prime numbers and implies that there are no factors for a given polynomial. Considering the fact that an irreducible polynomial of degree m over GF(2) is a factor of the polynomial

$x^{2^m-1}+1$, it is, therefore, possible to obtain the irreducible polynomial of degree m by considering all the factors of the polynomial $x^{2^m-1}+1$.

Definition 21.2.4 *An irreducible polynomial $p(x)$ of degree n is said to be primitive if the smallest integer k for which $p(x)$ divides x^k+1 is $k=2^m-1$.*

It can be proved that there exists at least one primitive polynomial $p(x)$ of degree m for any $m \geq 1$. With the primitive polynomial defined, we are now in a position to state how to construct a finite field GF(2^m).

Definition 21.2.5 *The set of all possible polynomials of degree $(m-1)$ over GF(2) (there are exactly 2^m such polynomials) define the finite field GF(2^m). The additions in $GF(2^m)$ are bit-independent polynomial additions; the multiplications are polynomial multiplications modulo $p(x)$, the primitive polynomial.*

Some important properties of finite field $GF(2^m)$ are listed as follows:

- The *order* of an element A of finite field is defined as the smallest integer n such that $A^n=1$. It can be shown that the order of any element A of finite field $GF(2^m)$ must be a divisor of 2^m-1.
- A non-zero element is called the *primitive element* of $GF(2^m)$ if its order is equal to 2^m-1. The primitive element is denoted as α, which is the root of the *primitive polynomial* $p(x)$ used to generate the finite field.
- A finite field always has one zero element (the additive identity), one unit element (the multiplicative identity) and one primitive element α.
- The *characteristic* of a field F is defined as the smallest integer λ such that $\Sigma_{i=1}^{\lambda} 1 = 0$ where the computation is carried out in F. It can be shown that the characteristic of a finite field GF(p^m) is always p, a prime number. For finite field of characteristic p, the following equation holds true

$$(A_t + \cdots + A_1 + A_0)^p = A_t^p + \cdots + A_1^p + A_0^p. \quad (1)$$

  This implies that for GF(2^m), squaring is a linear operation.

$$(\alpha+\beta)^2 = \alpha^2+\beta^2. \quad (2)$$

- The *trace* of a finite field element over GF(2^m) is defined as follows:

$$Tr(\beta) = \sum_{k=0}^{m-1} \beta^{2^k}. \quad (3)$$

  The trace function has the following properties

  - Trace is in the ground field, i.e., $Tr(A) \in GF(2)$ for $A \in GF(2)$;
  - $Tr(\beta)^2 = \beta + \beta^p + \ldots + \beta^{p^{m-1}} = Tr(\beta)$;
  - Trace is a linear function over $GF(2^m)$, i.e.,

$$Tr(c_1A_1 + c_2A_2) = c_1Tr(A_1) + c_2Tr(A_2), \quad (4)$$

  where $c_1, c_2 \in GF(2)$ and $A_1, A_2 \in GF(2^m)$.

21.2.2.3 Representation of finite field elements

The non-zero elements of $GF(2^m)$ can be represented in two forms, exponential form and polynomial form. In *exponential form*, they are represented as powers of the primitive element α, i.e.,

$$GF(2^m) = \{0, \alpha^0, \alpha^1, \cdots, \alpha^{2^m-2}\}, \tag{5}$$

which is also referred to as *power representation*. The elements of $GF(2^m)$ can also be expressed as polynomials of degree less than m as

$$\begin{aligned} GF(2^m) &= \{A|A = a_{m-1}\alpha^{m-1} + a_{m-2}\alpha^{m-2} + \cdots + a_1\alpha + a_0, \\ & \quad \textit{where } a_i \in GF(2),\ 0 \le i \le m-1\}, \end{aligned} \tag{6}$$

which is referred to as *standard polynomial form*.

Let the primitive polynomial $p(x)$ be written as $p(x) = x^m + P(x)$, where $P(x) = p_{m-1}x^{m-1} + \cdots + p_1x + p_0$. Since α is a root of the primitive polynomial $p(x)$, we have

$$\alpha^m = p_{m-1}\alpha^{m-1} + p_{m-2}\alpha^{m-2} + \cdots + p_1\alpha + p_0, \tag{7}$$

which is equivalent to $\alpha^m = P(\alpha)$. Conversion between power representation and polynomial representation can be accomplished by substituting α^m by $P(\alpha)$, or vice versa.

Power representation is efficient for finite field multiplication, division and exponentiation, where these operations can be carried out by adding, subtracting or multiplying exponents modulo $2^m - 1$. However, addition in power representation requires the use of 2-way *log* and *anti-log* conversion tables, or conversion circuitry to convert operands from power representation to polynomial representation, and convert the sum from polynomial representation to power representation. The conversion circuitry could become very complicated as m increases. Therefore, the polynomial representation is generally used for finite field arithmetic operation, where addition is carried out using bit-independent XOR operations, and multiplication is carried out using polynomial multiplication and modulo operations.

If considering $\{1, \alpha, \cdots, \alpha^{m-1}\}$ as a *standard basis* and expressing every finite field element using its coefficient vector $[a_{m-1}, \cdots, a_1, a_0]$, the finite field $GF(2^m)$ can be viewed as a m-dimensional vector space. Hence it is possible to represent elements of $GF(2^m)$ with respect to other bases. Additions over $GF(2^m)$ in polynomial representations turns out to be trivial and bit-independent, no matter what basis is used. In practice, basis can be selected so as to simplify the complicated arithmetic operations such as multiplication, division or exponentiations. Three types of bases are considered, namely standard basis (also referred to as conventional basis), dual basis and normal basis.

21.2.2.4 Standard basis representation

In this representation, elements of finite field GF(p^m) are represented as polynomials of α (the primitive element) with degree less than m, as shown in (6). Multiplication in standard basis involves multiplication of two polynomials of degree $m-1$ over $GF(2)$, and a polynomial modulo operation (*mod* $p(x)$) to reduce the degree of the product to be less than or equal to $m-1$.

Example 21.2.3 *Consider $\alpha^5 * \alpha^9$ over $GF(2^4)$ with $p(x) = x^4 + x + 1$. Note that α^5 and α^9 can be represented as $\alpha^2 + \alpha$ and $\alpha^3 + \alpha$ in standard basis. Therefore,*

$$\begin{aligned} \alpha^5 * \alpha^9 &= ((\alpha^2 + \alpha) * (\alpha^3 + \alpha)) \ mod \ p(\alpha) \\ &= (\alpha^5 + \alpha^4 + \alpha^3 + \alpha^2) mod(\alpha^4 + \alpha + 1) \\ &= (\alpha^2 + \alpha) + (\alpha + 1) + \alpha^3 + \alpha^2 \\ &= \alpha^3 + 1 = \alpha^{14}. \end{aligned} \tag{8}$$

It has been shown that finite field architectures based on standard basis have the desirable properties of regularity, simplicity [30] and these architectures are usually programmable with respect to the primitive polynomial $p(x)$ and can be easily extended to higher order finite fields.

21.2.2.5 Dual basis representation

Definition 21.2.6 *Let $\{\alpha_0, \alpha_1, \cdots, \alpha_{m-1}\}$ be a basis for $GF(2^m)$, i.e., a set of m linearly independent elements of $GF(2^m)$. The corresponding dual basis is defined to be the unique set of elements $\{\beta_0, \beta_1, \cdots, \beta_{m-1}\} \in GF(2^m)$ such that*

$$Tr(\alpha_i \beta_j) = \left[\begin{array}{ll} 1, & if i = j \\ 0, & if i \neq j. \end{array} \right., \tag{9}$$

where $\{\alpha_i\}$ is called the primal basis.

$\forall Z \in GF(2^m)$, Z can be expressed in the dual basis $\{\beta_j\}$ by the expansion:

$$Z = \sum_{j=0}^{m-1} z_j \beta_j, \tag{10}$$

where $z_j = Tr(Z\alpha_j)$ is the jth coordinate of Z in dual basis. It is true because

$$Tr(Z\alpha_j) = Tr((\sum_{i=0}^{m-1} z_i \beta_i)\alpha_j) = \sum_{i=0}^{m-1} z_i Tr(\beta_i \alpha_j) = z_j. \tag{11}$$

In the rest of this chapter, we consider only the standard or primal basis $\{1, \alpha, \alpha^2, \cdots, \alpha^{m-1}\}$. Its dual basis can be computed using the following theorem (see Exercise 2.40 in [31]).

Theorem 21.2.1 *Suppose that α is the primitive element in $GF(2^m)$ and let $p(x)$ be the primitive polynomial. Factor $p(x)$ as $p(x) = (x + \alpha)(\sum_{i=0}^{m-1} g_i x^i)$. Then the dual basis to $\{1, \alpha, \cdots, \alpha^{m-1}\}$ is $\{(g_i/p'(\alpha)), o \le i \le m-1\}$, where $p'(x)$ is the derivative of $p(x)$.*

Example 21.2.4 *Consider the standard basis $\{1, \alpha, \alpha^2, \alpha^3\}$ over $GF(2^4)$ with $p(x) = x^4 + x + 1$. $p(x)$ can be factored as*

$$p(x) = (x + \alpha)(x^3 + \alpha x^2 + \alpha^2 x + \alpha^{14}), \tag{12}$$

hence we have $g_3 = 1, g_2 = \alpha, g_1 = \alpha^2, g_0 = \alpha^{14}$. Note that $p'(x) = 1$. Therefore, the dual basis is $\{g_0, g_1, g_2, g_3\}$, i.e., $\{\alpha^{14}, \alpha^2, \alpha, 1\}$. Polynomial representation of elements in $GF(2^4)$ in dual basis can be found in Appendix I.

Dual basis representation is efficient for bit-serial multiplications [13] [29]. However, dual basis multiplication is in fact a hybrid operation, where the multiplicand and the product are in dual basis while the multiplier is in standard basis. This hybrid nature makes basis conversion an inherrent operation in dual basis multipliers. The complexity of basis conversion circuitry has to be taken into consideration when designing dual basis multipliers. To simplify the conversion circuitry, a generalized dual basis definition has been proposed.

Definition 21.2.7 *Define the new trace function with respect to certain $\beta \in GF(2^m)$ as*

$$Tr_\beta(A) = Tr(\beta A), \ \forall\, A \in GF(2^m). \tag{13}$$

The dual basis of the primal basis $\{1, \alpha, \alpha^2, \cdots, \alpha^{m-1}\}$ with respect to certain $\beta \in GF(2^m)$ is defined to be the unique set of elements $\{\beta_0, \beta_1, \cdots, \beta_{m-1}\}$ such that

$$Tr_\beta(\alpha^i \beta_j) = \left[\begin{array}{ll} 1, & when\ i = j \\ 0, & when\ i \neq j, \end{array} \right.$$

where β can be selected appropriately to simplify the conversion between standard (primal) and dual basis. In dual basis representation, The coordinates of $Z \in GF(2^m)$, z_i, can be computed as

$$z_i = Tr_\beta(Z\alpha^i), \tag{14}$$

for $0 \leq i \leq m-1$.

In the rest of the chapter, we denote the new trace function with respect to β as Tr, and treat the original definition of trace function as a special case where $\beta = 1$, since the presence of β in the new trace function affects only the basis conversion circuits, not the arithmetic computation itself.

21.2.2.6 Normal basis

Definition 21.2.8 *A normal basis of $GF(2^m)$ over $GF(2)$ is a basis of the form $\{\beta,, \beta^2, \beta^4, \cdots, \beta^{2^{(m-1)}}\}$, $\beta \in GF(2^m)$.*

Notice that this element β need not necessarily be the primitive element of the finite field. However, for most finite fields, primitive polynomials exist such that the set $\{\alpha, \alpha^2, \alpha^4, \cdots, \alpha^{2^{m-1}}\}$ does form a basis for the finite field. In Appendix II, we list the elements of GF(2^4) generated by the primitive polynomial $p(x) = x^4 + x^3 + 1$. For this primitive polynomial, the set $\{\alpha, \alpha^2, \alpha^4, \alpha^8\}$ does form a basis. The reader can verify that for GF(2^4) given in Appendix I, the set $\{\alpha, \alpha^2, \alpha^4, \alpha^8\}$ does not form a basis as they are not linearly independent. For this finite field, the set $\{\alpha^3, \alpha^6, \alpha^{12}, \alpha^{24} = \alpha^9\}$ forms a basis and could be used for normal basis representation.

The salient feature of normal basis representation is that the squaring of an element in $GF(2^m)$ can be obtained by a simple cyclic shift of its binary bits, which simplifies the finite field exponentiation operations; however, the multiplications in normal basis are substantially more difficult.

21.3 FINITE FIELD ARITHMETIC ARCHITECTURES USING STANDARD BASIS

21.3.1 Standard Basis Parallel Array-Type Multipliers Over $GF(2^m)$

Let $p(x)$ be the primitive irreducible polynomial of degree m for $GF(2^m)$ and let α be a root of $p(x)$. Let $A = \sum_{i=0}^{m-1} a_i\alpha^i$ be the multiplicand. Let $B = \sum_{i=0}^{m-1} b_i\alpha^i$ be the multiplier and b_{m-1} and b_0 are the most and least significant bits, respectively. Let $W = \sum_{i=0}^{m-1} w_i\alpha^i$ be the product. The operations involved in finite field multiplication include polynomial multiplication and modulo operation using one's complement arithmetic. Multiplication can be performed in various styles based on different arrangements of internal computations. In this section, parallel *array-type* multiplication are considered, where multiplication is performed in such a way that polynomial multiplication and polynomial modulo operations are interleaved with each other.

21.3.1.1 Multiplication algorithms

Let W be the product of A and B, $A, B, W \in GF(2^m)$. Array-type multiplication can be performed in one of the following ways:

$$\begin{aligned} W &= AB \; mod \; p(x) \\ &= b_0A + b_1(A\alpha \; mod \; p(x)) \\ &\quad +b_2(A\alpha^2 \; mod \; p(x)) + \cdots \\ &\quad +b_{m-1}(A\alpha^{m-1} \; mod \; p(x)) \end{aligned} \tag{15}$$

or

$$\begin{aligned} W &= AB \; mod \; p(x) \\ &= (\cdots(Ab_{m-1}\alpha \; mod \; p(x) + Ab_{m-2})\alpha \; mod \; p(x) + \cdots \\ &\quad +Ab_1)\alpha \; mod \; p(x) + Ab_0. \end{aligned} \tag{16}$$

Depending on the order in which multiplier bits are processed, two schemes can be used to realize array-type multiplication, i.e., LSB-first (least significant bit first) scheme where multiplication starts with the LSB of multiplier B, as shown in (15); or MSB-first (most significant bit first) scheme where multiplication starts with the MSB of multiplier B, as shown in (16) [32]. In LSB-first implementation, basic cell column in step k, $1 \le k \le m$, performs the following computations in parallel:

$$\begin{aligned} A^{(k)} &= (A^{(k-1)})\alpha \; mod \; p(x) \\ W^{(k)} &= A^{(k-1)}b_{k-1} + W^{(k-1)}, \end{aligned} \tag{17}$$

where $W^{(k)} = \sum_{i=0}^{k-1} Ab_i\alpha^i$, and $W^{(0)}$=0, $A^{(0)} = A$. Note that by assigning $W^{(0)} = C$, where $C = \sum_{i=0}^{m-1} c_i\alpha^i \in GF(2^m)$, the LSB-first array multiplier can be easily used to perform multiply-accumulation operation $W = AB + C$. In MSB-first implementation, basic cell column in step k, $1 \le k \le m$, completes the following computation:

$$W^{(k)} = W^{(k-1)}\alpha \; mod \; p(x) + Ab_{m-k}, \tag{18}$$

where $W^{(k)} = \sum_{i=1}^{k} Ab_{m-i}\alpha^{k-i}$, and $W^{(0)}=0$. Note that another row of XOR gates is required if MSB-first array multiplier is used to perform multiply-accumulation operation, i.e., $W = W^{(m)} + C$, where $W^{(m)} = AB \; mod \; p(x)$ is the output of MSB-first array multiplier.

A common computation in both types of multiplication is ***multiply-by-α*** operation, which can be done by following the rules in (20). Assume $A^{(1)} = A\alpha \; mod \; p(x)$. A straightforward *multiply-by-α* operation without polynomial modulo operation is equivalent to ***shift-left-by-one-bit*** operation. Therefore, we have the intermediate result

$$A^{(1)} = a_{m-1}\alpha^m + a_{m-2}\alpha^{m-1} + \cdots + a_1\alpha^2 + a_0\alpha. \tag{19}$$

Substituting (7) into (19) for degree reduction, we have

$$a_i^{(1)} = \begin{cases} a_{i-1} + a_{m-1}p_i, \; 1 \le i \le m-1 \\ a_{m-1}p_0, \; i = 0 \end{cases}, \tag{20}$$

where a_i and $a_i^{(1)}$ are the coordinates of A and $A^{(1)}$, respectively, with respect to standard basis.

Based on (20), operations in each basic cell in LSB-first or MSB-first multiplier array can be derived and the corresponding LSB-first and MSB-first multiplication algorithms are given as follows.

Algorithm 21.3.1 LSB-first Multiplication Algorithm

1. *Initially, set* $A^{(0)} = A$ *and* $W^{(0)} = 0$.

2. *Each basic cell at step* k $(1 \le k \le m-1)$ *computes:*

$$\begin{aligned} a_i^{(k)} &= \begin{cases} a_{i-1}^{(k-1)} + a_{m-1}^{(k-1)}p_i, \; 1 \le i \le m-1 \\ a_{m-1}^{(k-1)}p_0, \; i = 0 \end{cases} \\ w_i^{(k)} &= a_i^{(k-1)}b_{k-1} + w_i^{(k-1)}. \end{aligned} \tag{21}$$

3. *The basic cells at step* $k = m$ *compute:*

$$w_i^{(m)} = a_i^{(m-1)}b_{m-1} + w_i^{(m-1)}. \tag{22}$$

4. *Finally, the product is equal to* $W^{(m)}$. ■

Algorithm 21.3.2 MSB-first Multiplication Algorithm

1. *Initially, set* $W^{(0)} = 0$.

2. *Each basic cell at 1st step computes:*

$$w_i^{(1)} = a_ib_{m-1}, \; 0 \le i \le m-1. \tag{23}$$

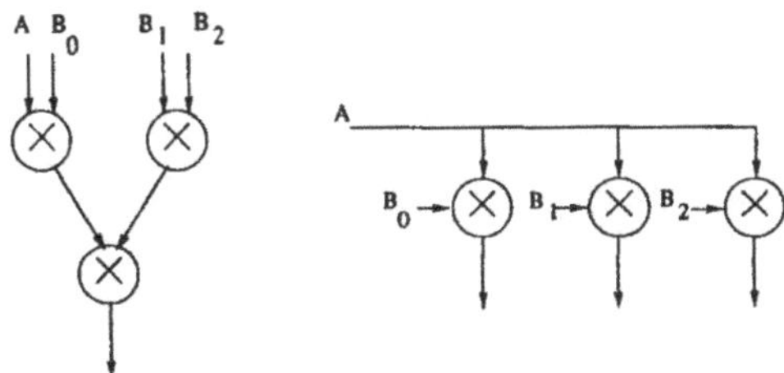

Figure 21.1 Multiplier block in a large system.

3. The basic cells at step k ($2 \le k \le m$) compute:

$$w_i^{(k)} = \begin{cases} w_{i-1}^{(k-1)} + w_{m-1}^{(k-1)} p_i + a_i b_{m-k}, & 1 \le i \le m-1 \\ w_{m-1}^{(k-1)} p_0 + a_0 b_{m-k}, & i = 0 \end{cases} \quad (24)$$

4. Finally, the product is equal to $W^{(m)}$. ■

In both algorithms 21.3.1 and 21.3.2, $a_i^{(k)}$, $w_i^{(k)}$ denote the i-th coefficients in $A^{(k)}$ and $W^{(k)}$, respectively; a_k and b_k denotes the k-th coefficient of A and B, respectively; p_i represents the i-th coefficient of $p(x)$.

Notice that the computations involved in both LSB-first and MSB-first basic cells are *multiply-by-α*, *generate-current-partial-product* and *accumulate-to-previous-result*. These operations are performed concurrently in LSB-first scheme, but sequentially in MSB-first scheme. As a result, the parallelism in LSB-first basic cell leads to a considerable reduction in multiplication computation delay without any increase in hardware complexity.

Another advantage of LSB-first scheme over MSB-first scheme is its capability of achieving substructure sharing among several multiplication operations. Finite field multipliers are used as building blocks for large systems, such as Reed-Solomon encoder and decoder, which contains structures illustrated in Fig. 21.1. For a broadcast structure like the second one in Fig. 21.1, a substructure sharing technique can be used to achieve savings in hardware [26]. Assume A is multiplied by $B_0, B_1, \cdots, B_z$ at the same time. Since the LSB-first multiplier computes *multiply-by-α*, and *generate-current-partial-product* and *accumulate-to-previous-result* in parallel as illustrated in (21), computation of $A\alpha, \cdots, A\alpha^{m-1}$ can be shared among all B_i's provided that these multiplications use the same primitive polynomial $p(x)$.

21.3.1.2 Semi-systolic array-type multipliers

The LSB-first and MSB-first algorithms can be mapped directly to VLSI implementations. This subsection addresses algorithm-architecture mapping for designing efficient finite field semi-systolic multipliers.

A MSB-first semi-systolic cellular-array multiplier based on (16) and (24) was introduced in [5]. A similar LSB-first semi-systolic multiplier can be designed based on (15) and (21). The gate level circuit diagram of each basic cell and the system level block diagram for LSB-first multiplier over $GF(2^4)$ are shown in Fig. 21.2 and 21.3, respectively. This multiplier can be pipelined by placing delay elements at

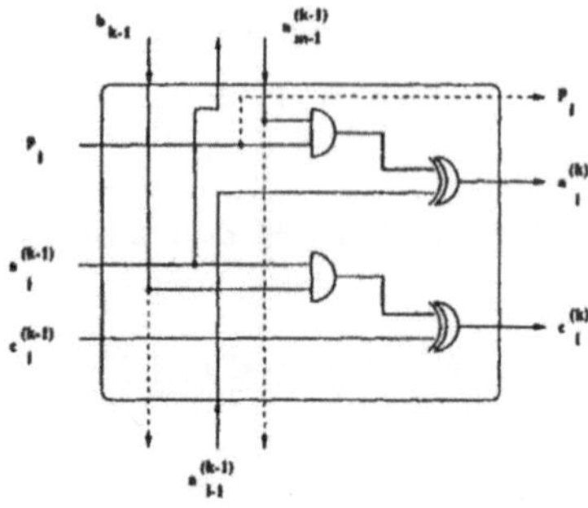

Figure 21.2 Basic cells in the LSB-first semi-systolic array multiplier.

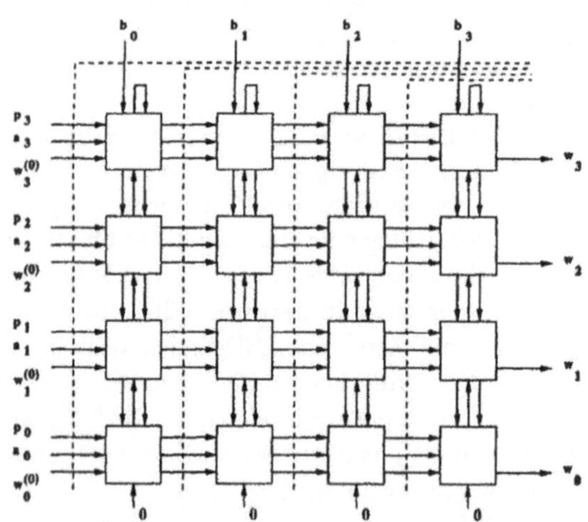

Figure 21.3 System level diagram of the LSB-first semi-systolic array multiplier over $GF(2^4)$.

Table 21.1 Performance Characteristics for Systolic and Bit-Level Pipelined Semi-Systolic Multipliers Based on LSB-first Scheme

Item	Systolic[6]	Proposed
Number of basic cells	m^2	m^2
Basic Cell	2 2-input AND , 2 2-input XOR, 7 1-bit latches	2 2-input AND, 2 2-input XOR 3 1-bit latches
Latency	3m	m+1
Throughput	1	1
Critical path	1 D_{AND} +1 D_{XOR}	1 D_{AND} +1 D_{XOR}

feedforward cutsets shown in dashed lines in Fig. 21.3 to reduce critical path delay time.

Multiplication based on LSB-first scheme has less computation delay time than its counterpart based on MSB-first scheme due to the increase of parallelism among internal computations. The basic cell computation time in LSB-first semi-systolic cellular-array multiplier has been reduced from 1 D_{AND} + 2 D_{XOR} to 1 D_{AND} + 1 D_{XOR}, where D_{AND} and D_{XOR} denote AND and XOR gate delay time, respectively. As a result, 30% reduction in total computation delay of one multiplication operation can be achieved by utilizing LSB-first scheme as opposed to MSB-first scheme. Note that the bidirectional edges in the vertical datapath in Fig. 21.3 do not introduce any feedback since they only carry broadcast signals. This can be observed clearly by replacing the basic cells in Fig. 21.3 by the detailed circuit in Fig. 21.2.

The systolic multiplier in [6] was also based on LSB-first scheme. However, this fully systolic architecture requires twice the number of latches as the proposed bit-level pipelined LSB-first semi-systolic multiplier and has a latency which is three times as that of the proposed multiplier. Their performance characteristics are compared in Table 21.1.

21.3.1.3 Generalized cellular-array multiplier

Although the semi-systolic multipliers obtained by direct algorithm-architecture mapping are programmable with respect to the primitive polynomial $p(x)$, they can only operate over fixed-order finite field. This subsection addresses generalized finite field multiplication algorithms and their corresponding semi-systolic cellular-array architectures which are programmable with respect to not only the primitive polynomial $p(x)$, but also the field order m. Let M be the largest possible field order. The multiplier is designed for multiplication over $GF(2^M)$, where the primitive polynomial $p(x)$ is of degree M and multiplicand and multiplier have wordlength of M bits. Elements in smaller field $GF(2^m)$, $m \leq M$, have wordlength of m bits and can be extended to M-bit words by filling higher order bits with zeros. Besides zero-padding, the multiplication algorithms have to be extended to incorporate the programmability of field order m, and extra control circuitry is required when actual field order is smaller than the hard-wired field order M.

21.3.1.4 MSB-first cellular-array multiplier

A semi-systolic cellular array multiplier was presented in [5], which is programmable with respect to both primitive polynomials and field order m. This array-type multiplier is based on the MSB-first scheme described by (16) and (18). In order to achieve programmability on field order m, extra control circuits, drawn in the shaded regions in Fig. 21.4, are introduced. First, a column of *bit-locator* cells, namely $\boxed{\text{LCELL}}$, are used to detect the actual field order based on the coefficient vector of primitive polynomial

$$\mathbf{P} = \{0, \cdots, p_m, p_{m-1}, \cdots, p_0\}. \tag{25}$$

Note that all $m+1$ coefficients in $p(x)$ must be used with zeros padded to higher order positions and vector $\mathbf{P}$ is of wordlength $M+1$. For example, $p(x) = x^4 + x + 1$ over $GF(2^4)$ will be written as $0 \cdots 010011$ with wordlength equal to $M+1$. The bit-locator column generates a $\mathbf{H}$ vector of length M, in which one and only one bit is equal to 1 indicating the actual field order, based on the M higher order bits in vector $\mathbf{P}$. The truth table and circuit of one bit-locator cell is shown in Fig. 21.4(a). Correspondingly, each basic cell in the array-multiplier is modified such that it is capable of locating the appropriate most significant bit, $w_{m-1}^{(k-1)}$, in the intermediate result $W^{(k-1)}$ and feed this value to all the basic cells in k-th column. The computations in each basic cell in k-th multiplier column are modified from (24) to

$$\left[\begin{array}{lcl} y_{out} & = & w_{i-1}^{(k-1)} \cdot h_i \boxed{\text{OR}}\, y_{in} \\ w_i^{(k)} & = & (w_{i-1}^{(k-1)} + y_{out} \cdot p_i + a_i \cdot b_{m-k}) \cdot \overline{h_i},\ 1 \le i \le m-1 \\ w_0^{(k)} & = & y_{in} \cdot p_0 + a_0 \cdot b_{m-k} \end{array}\right. , \tag{26}$$

where for multiplier cells in the same column, y_{out} of previous cell at higher order position is y_{in} for cell at one bit lower position; y_{in} of the highest order cell is set to $w_{M-1}^{(k-1)}$. Detailed gate level circuits of each basic cell, $\boxed{\text{LCELL}}$ and $\boxed{\text{MCELL}}$, and the system level block diagram of multiplier over $GF(2^4)$ are shown in Fig. 21.4 and 21.5, respectively.

21.3.1.5 LSB-first Cellular-Array Multiplier

The cellular array multiplier in [5] is based on MSB-first scheme. In this section, we propose the generalized LSB-first cellular array multiplier based on (15), (17) and (21).

The LSB-first semi-systolic cellular-array multiplier uses the same bit-locator cells to generate $\mathbf{H}$ vector which indicates the actual field order m. However, unlike in the MSB-first multiplier where M bits in lower order position in vector $\mathbf{H}$, $\{h_{M-1}, h_{M-2}, \cdots, h_1, 0\}$, are used as inputs to the multiplier cells, the M bit in higher order position in vector $\mathbf{H}$, $\{h_M, h_{M-1}, \cdots, h_1\}$, are used in the LSB-first multiplier cells. The computations in each basic cell in k-th multiplier column are

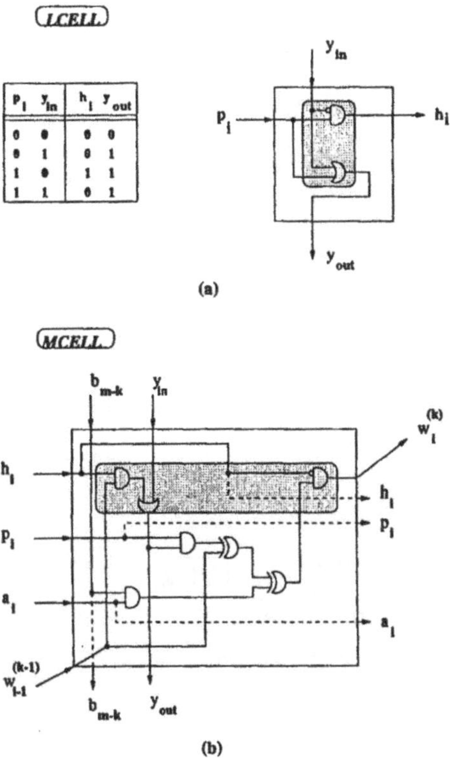

Figure 21.4 Basic cells in the programmable MSB-first semi-systolic cellular-array multiplier.

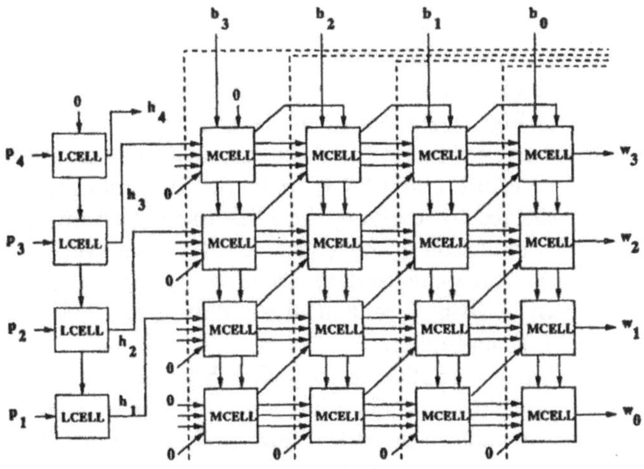

Figure 21.5 System level diagram of programmable MSB-first semi-systolic cellular-array multiplier over $GF(2^m)$ for $m \leq 4$.

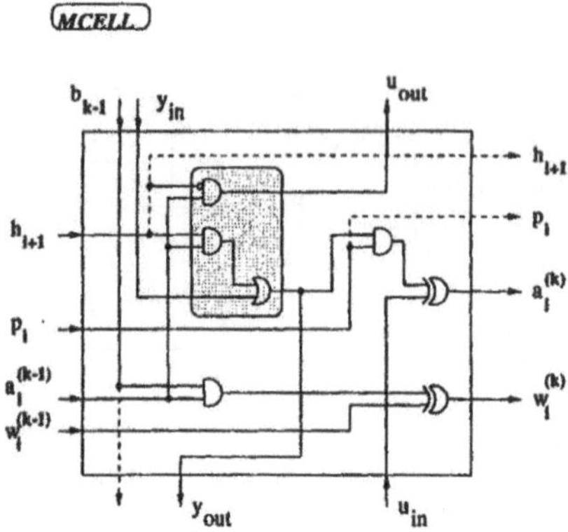

Figure 21.6 Basic cells in the programmable LSB-first semi-systolic cellular-array multiplier.

described by:

$$\left[\begin{array}{lcl} y_{out} & = & a_i^{(k-1)} \cdot h_{i+1} \boxed{\text{OR}}\, y_{in} \\ u_{out} & = & a_i^{(k-1)} \cdot \overline{h_{i+1}} \\ a_i^{(k)} & = & u_{in} + y_{out} \cdot p_i, \; 1 \le i \le m-1 \\ a_0^{(k)} & = & y_{out} \cdot p_0 \\ w_i^{(k)} & = & a_i^{(k-1)} \cdot b_{k-1} + w_i^{(k-1)} \end{array}\right. , \tag{27}$$

where y_{out} of cell at higher order position is y_{in} for cell at one bit lower order position, u_{out} of cell at lower order position is u_{in} for cell at one bit higher order position, and y_{in} of the highest order cell and u_{in} at the lowest order cell are both set to 0. Detailed circuit of basic multiplier cell and system level block diagram of LSB-first cellular-array multiplier over $GF(2^4)$ are shown in Fig. 21.6 and 21.7, respectively.

21.3.1.6 Comparison and comments

Comparing the basic cell of LSB-first multiplier in Fig. 21.6 with that of MSB-first multiplier in Fig. 21.4(b), one can observe that with the same hardware complexity, LSB-first basic cell has reduced the critical path computation time from 6 to 4 gate delays. As a result, about 28.3% reduction in total critical path computation time can be achieved by applying the LSB-first algorithm to cellular-array multiplier. Meanwhile, reduction in the length of critical path reduces propagation of switching activities, which can lead to dramatic savings in total energy consumption and make the LSB-first cellular-array multiplier a low-power structure compared with MSB-first multiplier.

Note that pipelining is applicable to both MSB-first and LSB-first cellular-array multipliers to reduce the critical path computation time. Feedforward cutsets for bit-level pipelining are shown in dashed lines in Fig. 21.5 and 21.7. Due to the presence of bidirectional vertical datapaths, cutsets cannot be drawn across vertical datapaths. Hence, the minimum critical path computation time in both multipliers are lower-bounded by the computation delay along the vertical datapaths, which is equal to $(m-1)\ D_{OR}$, where D_{OR} denotes one OR gate delay time.

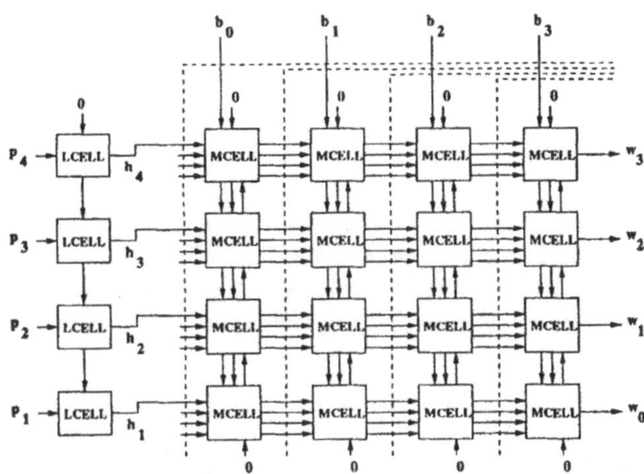

Figure 21.7 System level diagram of programmable LSB-first semi-systolic Cellular-array multiplier over $GF(2^m)$ for $m \leq 4$.

21.3.2 Standard Basis Squarer over $GF(2^m)$

21.3.2.1 Squaring algorithms

In a finite field of characteristic 2, the following equation holds:

$$(\alpha+\beta)^2 = \alpha^2 + \beta^2 \tag{28}$$

where $\alpha, \beta \in GF(2^m)$. Using this property, a hardware efficient squarer for finite field $GF(2^m)$ can be derived.

Let $B = b_{m-1}\alpha^{m-1} + \cdots + b_1\alpha + b_0 \in GF(2^m)$. Then, $W = B^2$ can be computed as

$$\begin{aligned} W &= B^2 = b_{m-1}\alpha^{2m-2} + \cdots + b_1\alpha^2 + b_0 \\ &= (b_{\lfloor (m-1)/2 \rfloor}\alpha^{2\lfloor (m-1)/2 \rfloor} + \cdots + b_1\alpha^2 + b_0) \\ &\quad + b_{\lfloor (m-1)/2 \rfloor + 1}(\alpha^{2\lfloor (m-1)/2 \rfloor} \cdot \alpha^2) + \cdots + b_{m-1}(\alpha^{2m-4} \cdot \alpha^2), \end{aligned} \tag{29}$$

which is referred to as LSB-first scheme where the squaring operation starts with the least significant bits of operand B; or

$$\begin{aligned} W &= B^2 \\ &= ((\cdots((b_{m-1}\alpha^{2(\lceil m/2 \rceil - 1)} + \cdots + b_{\lfloor m/2 \rfloor + 1}\alpha^2 + b_{\lfloor m/2 \rfloor})\alpha^2 \\ &\quad + \cdots)\alpha^2 + b_1)\alpha^2 + b_0, \end{aligned} \tag{30}$$

which is referred to as MSB-first scheme where the squaring operation starts with the most significant bits of operand B. In LSB-first implementation, the basic cells in step k perform the following computations in parallel:

$$\begin{aligned} B^{(k)} &= B^{(k-1)} \cdot \alpha^2 \; mod \; p(x) \\ W^{(k)} &= W^{(k-1)} + B^{(k-1)} b_{\lceil m/2 \rceil + k - 1}, \end{aligned} \tag{31}$$

where $W^{(0)} = b_{\lfloor (m-1)/2 \rfloor}\alpha^{2\lfloor (m-1)/2 \rfloor} + \cdots + b_1\alpha^2 + b_0$, $B^{(0)} = \alpha^{2\lfloor (m-1)/2 \rfloor} \cdot \alpha^2 = \alpha^{2\lceil m/2 \rceil}$ and $k = 1, 2, \cdots, \lfloor m/2 \rfloor$. In MSB-first implementation, the basic cells in step k perform the following computations:

$$W^{(k)} = (W^{(k-1)})\alpha^2 \; mod \; p(x) + b_{\lfloor m/2 \rfloor - k}, \tag{32}$$

where $W^{(0)} = b_{m-1}\alpha^{2(\lceil m/2 \rceil - 1)} + \cdots + b_{\lfloor m/2 \rfloor + 1}\alpha^2 + b_{\lfloor m/2 \rfloor}$ and $k = 1, 2, \cdots, \lfloor m/2 \rfloor$.

The essential computation, *multiply-by-*α^2, can be performed by following the rules in (37). Assume $A^{(2)} = A\alpha^2 \; mod \; p(x)$. A straightforward *multiply-by-*α^2 operation is equivalent to *shift-left-by-two-bit* operation. Therefore, we have the intermediate result

$$A^{(2)} = a_{m-1}\alpha^{m+1} + a_{m-2}\alpha^m + \cdots + a_1\alpha^3 + a_0\alpha^2. \tag{33}$$

Polynomial modulo operations are required to reduce the degree of $A^{(2)}$ from $m+1$ to less than or equal to $m-1$. To simplify the operations in each squarer basic cell, a new input polynomial, $P'(\alpha)$, is introduced and defined as follows [8]. Let

$$\begin{aligned} P(\alpha) &= p_{m-1}\alpha^{m-1} + \cdots + p_1\alpha + p_0 \\ P'(\alpha) &= P(\alpha)\alpha, \end{aligned} \tag{34}$$

where coefficients of $P'(\alpha)$, p'_i, can be pre-computed from coefficients of $P(\alpha)$, p_i, as follows:

$$p'_i = \begin{cases} p_{i-1} + p_{m-1}p_i, & 1 \le i \le m-1 \\ p_{m-1}p_0, & i = 0 \end{cases}. \tag{35}$$

From (7), we have

$$\begin{aligned} \alpha^m &= P(\alpha) \\ \alpha^{m+1} &= P'(\alpha). \end{aligned} \tag{36}$$

Substituting (36) into (33) for degree reduction, we have

$$a_i^{(2)} = \begin{cases} a_{i-2} + a_{m-1}p'_i + a_{m-2}p_i, & 2 \le i \le m-1 \\ a_{m-1}p'_1 + a_{m-2}p_1, & i = 1 \\ a_{m-1}p'_0 + a_{m-2}p_0, & i = 0 \end{cases}, \tag{37}$$

where a_i and $a_i^{(2)}$ are coordinates of A and $A^{(2)}$, respectively.

Therefore, we have the following algorithms for finite field squaring operation.

Algorithm 21.3.3 LSB-first Squaring Algorithm

1. *Initially,*

$$\begin{aligned} W^{(0)} &= b_{\lfloor (m-1)/2 \rfloor}\alpha^{2\lfloor (m-1)/2 \rfloor} + \cdots + b_1\alpha^2 + b_0 \\ B^{(0)} &= \alpha^{2\lceil m/2 \rceil} = \begin{cases} P(\alpha), & for\ even\ m \\ P'(\alpha), & for\ odd\ m \end{cases}. \end{aligned} \tag{38}$$

2. At k-th step, $1 \leq k \leq \lfloor m/2 \rfloor - 1$, each basic cell computes

$$b_i^{(k)} = \begin{cases} b_{i-2}^{(k-1)} + b_{m-1}^{(k-1)} p_i' + b_{m-2}^{(k-1)} p_i, \ 2 \leq i \leq m-1 \\ b_{m-1}^{(k-1)} p_1' + b_{m-2}^{(k-1)} p_1, \ i = 1 \\ b_{m-1}^{(k-1)} p_0' + b_{m-2}^{(k-1)} p_0, \ i = 0 \end{cases}$$
$$w_i^{(k)} = w_i^{(k-1)} + b_i^{(k-1)} b_{\lceil m/2 \rceil + k - 1}. \quad (39)$$

3. The basic cells at $k = \lfloor m/2 \rfloor$ step compute:

$$w_i^{(k)} = w_i^{(k-1)} + b_i^{(k-1)} b_{m-1}. \quad (40)$$

4. Finally, the result $W = W^{(\lfloor m/2 \rfloor)}$. For even field order m, $m/2$ steps are required; for odd field order m, $(m-1)/2$ steps are required. ■

Algorithm 21.3.4 MSB-first Squaring Algorithm

1. Initially,

$$W^{(0)} = b_{m-1}\alpha^{2(\lceil m/2 \rceil - 1)} + \cdots + b_{\lfloor m/2 \rfloor + 1}\alpha^2 + b_{\lfloor m/2 \rfloor}. \quad (41)$$

2. At k-th step, $1 \leq k \leq \lfloor m/2 \rfloor$, each basic cell computes

$$w_i^{(k)} = \begin{cases} w_{i-2}^{(k-1)} + w_{m-1}^{(k-1)} p_i' + w_{m-2}^{(k-1)} p_i, \ 2 \leq i \leq m-1 \\ w_{m-1}^{(k-1)} p_1' + w_{m-2}^{(k-1)} p_1, \ i = 1 \\ b_{\lfloor m/2 \rfloor - k} + w_{m-1}^{(k-1)} p_0' + w_{m-2}^{(k-1)} p_0, \ i = 0 \end{cases}. \quad (42)$$

3. Finally, the result $W = W^{(\lfloor m/2 \rfloor)}$. For even field order m, $m/2$ steps are required; for odd field order m, $(m-1)/2$ steps are required. ■

21.3.2.2 Semi-systolic squarer over $GF(2^m)$

Mapping the operations in (39) to one basic cell and assigning m basic cells to each computation step, a LSB-first semi-systolic squarer can be designed based on the squaring algorithm 21.3.3. The gate level diagram of each basic cell and the system level diagram of the LSB-first squarer over $GF(2^4)$ are shown in Fig. 21.8 and 21.9, respectively. In general, $\lfloor m/2 \rfloor$ basic cell columns are required and the squarer contains $m\lfloor m/2 \rfloor$ basic cells, each of which has 3 2-input AND gates, 1 2-input XOR gate and 1 3-input XOR gate.

MSB-first squarer can be designed based on algorithm 21.3.4 in a similar way. The basic cell and system level diagram of MSB-first squarer are shown in Fig. 21.10 and 21.11, respectively. The MSB-first squarer contains $m\lfloor m/2 \rfloor$ basic cells, each of which has 2 2-input AND gates and 1 3-input XOR gate.

Bit-level pipelining is applicable to both semi-systolic squarers, where pipelining latches are placed on the cutsets (drawn in dashed lines in Fig. 21.9 and 21.11) along the horizontal uni-directional datapaths between inputs and outputs. Note

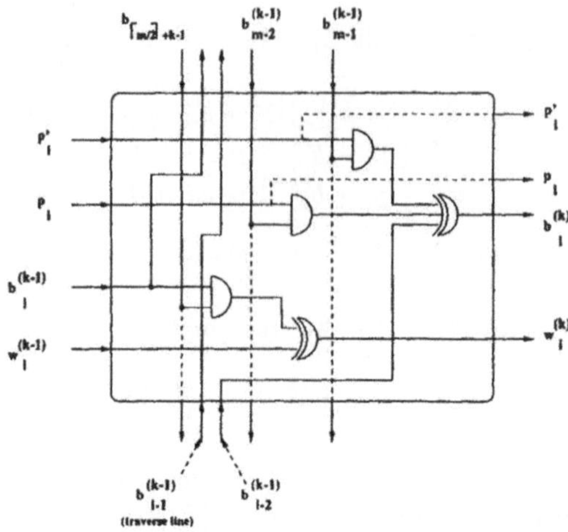

Figure 21.8 Basic cell of the LSB-first squarer.

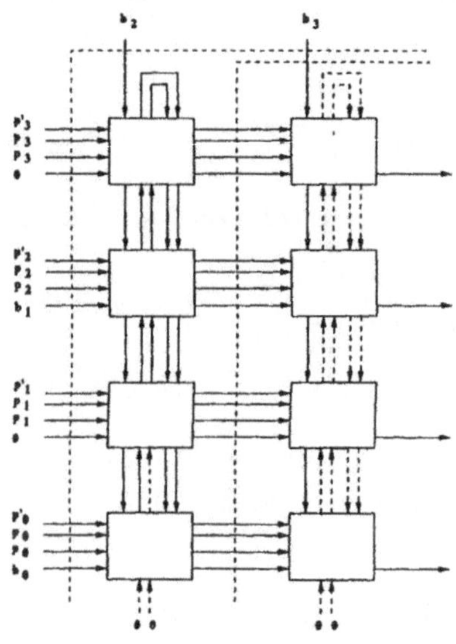

Figure 21.9 LSB-first semi-systolic squarer over $GF(2^4)$.

that the basic cells in the rightmost column and the bottom rows of the LSB-first squarer are degenerate versions of regular basic cell due to some unnecessary computations.

The bit-level pipelined semi-systolic squarers mapped from the LSB-first and MSB-first algorithms are compared with squarer using a dedicated bit-level pipelined semi-systolic multiplier and the systolic squarer circuit in [8]. The comparison results are summarized in Table 21.2, where only 2-input AND and XOR gates are used, and D_{AND} and D_{XOR} denote the gate delay time for one 2-input AND gate and one 2-input XOR gate, respectively.

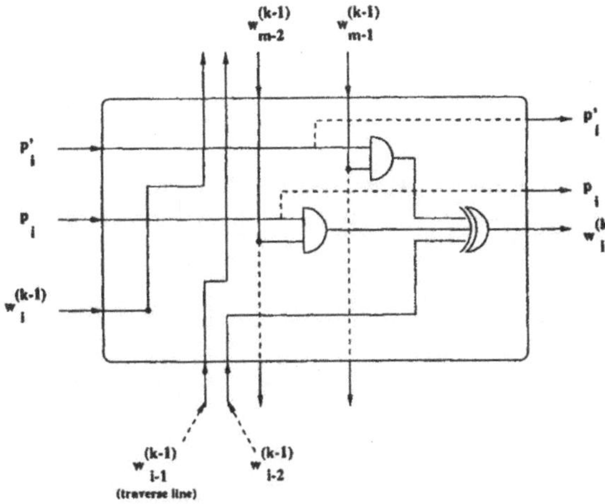

Figure 21.10 Basic cell of the MSB-first squarer.

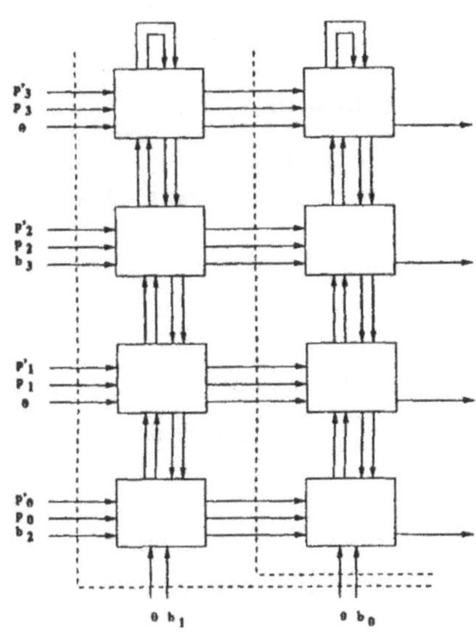

Figure 21.11 MSB-first semi-systolic squarer over $GF(2^4)$.

From the comparison table, we conclude that the semi-systolic squarer based on MSB-first algorithm has the best performance. Compared with the dedicated multiplier approach, it has reduced the total computation time by about 25% (with half area) by using the pre-computed polynomial $P'(\alpha)$. Compared with the systolic power-sum circuit in [8], it leads to more than 50% savings in hardware and reduces the total computation time by about 6 times. Compared with the semi-systolic squarer based on LSB-first approach, its basic cell contains less number of logical gates and latches, which leads to simpler VLSI architecture.

Table 21.2 Performance Characteristics of Different Squarers

Item	Multiplier (semi-systolic)	Power-sum [8] (systolic)	LSB-first (semi-systolic)	MSB-first (semi-systolic)
Number of basic cells	m^2	m^2	$m\lfloor m/2\rfloor$	$m\lfloor m/2\rfloor$
Basic Cell	2 AND, 2 XOR, 3 latches	3 AND, 3 XOR 10 latches	3 AND, 3 XOR, 4 latches	2 AND, 2 XOR, 3 latches
Latency	m	3m	$\lfloor m/2\rfloor$	$\lfloor m/2\rfloor$
Critical path	1 D_{AND} +1 D_{XOR}	1 D_{AND} +1 D_{XOR}	1 D_{AND} +1 D_{XOR}	1 D_{AND} +1 D_{XOR}
Pre-computed $P'(\alpha)$	no	yes	yes	yes

21.3.3 Finite Field Exponentiator

21.3.3.1 Exponentiation algorithms

The exponentiation algorithms introduced in this section are applicable to all basis representations.

Let β be an arbitrary element in GF(2^m). Define the exponentiation of β as

$$W = \beta^N. \tag{43}$$

Since $\forall \beta \in GF(2^m)$, $\beta^{2^m-1} = 1$. We have

$$\beta^N = \beta^{N \ mod \ 2^m-1}. \tag{44}$$

Therefore, it is sufficient to consider the range $1 \leq N \leq 2^m - 1$ for exponent N. Expressing N $(1 \leq N \leq 2^m - 1)$ as a binary number

$$N = n_{m-1} \times 2^{m-1} + n_{m-2} \times 2^{m-2} + \cdots + n_1 \times 2 + n_0, \tag{45}$$

or representing it using the binary vector $[n_{m-1}, n_{m-2}, \cdots, n_0]$ where $n_i \in GF(2)$, we can then compute the exponentiation of β using continuous ***square and multiply*** operations as follows:

$$\begin{aligned} \beta^N &= \beta^{n_{m-1}\times 2^{m-1}+n_{m-2}\times 2^{m-2}+\cdots+n_1\times 2+n_0} \\ &= (\beta)^{n_0}(\beta^2)^{n_1}(\beta^4)^{n_2}\cdots(\beta^{2^{m-1}})^{n_{m-1}}, \end{aligned} \tag{46}$$

or

$$\beta^N = ((\cdots((\beta^{n_{m-1}})^2(\beta^{n_{m-2}})^2)\cdots)^2\beta^{n_1})^2\beta^{n_0}. \tag{47}$$

These lead to the LSB-first (46) and MSB-first (47) implementations of finite field exponentiation depending on the order in which the exponent bits are processed. Here the LSB and MSB denote the least significant bit n_0 and most significant bit n_{m-1} of the exponent N, respectively.

Algorithm 21.3.5 LSB-first Exponentiation Algorithm

1. *Initially,* $B^{(0)} = \beta$ *and* $W^{(0)} = 1$.
2. *The 1-st step computes*

$$\begin{aligned} B^{(1)} &= (B^{(0)})^2 \\ W^{(1)} &= (B^{(0)})^{n_0}. \end{aligned} \tag{48}$$

3. *At k-th step,* $2 \leq k \leq m-1$*, the following operations are performed:*

$$\begin{aligned} B^{(k)} &= (B^{(k-1)})^2 \\ W^{(k)} &= (B^{(k-1)})^{n_{k-1}} W^{(k-1)}. \end{aligned} \tag{49}$$

4. *The* $k = m$*-th step computes*

$$W^{(m)} = (B^{(m-1)})^{n_{m-1}} W^{(m-1)}. \tag{50}$$

5. *The result* $W = W^{(m)} = B^2$. ∎

Algorithm 21.3.6 MSB-first Exponentiation Algorithm

1. *Initially,* $W^{(0)} = 1$.
2. *The 1-st step computes*

$$W^{(1)} = \beta^{n_{m-1}}. \tag{51}$$

3. *At k-th step,* $2 \leq k \leq m$*, the following operations are performed:*

$$W^{(k)} = (W^{(k-1)})^2 \beta^{n_{m-k}}, \tag{52}$$

4. *The result* $W = W^{(m)} = B^2$. ∎

Comparing (49) and (52), we see that by operating the square and multiply operation in parallel, the critical path computation time in each step in LSB-first scheme is equal to the maximum computation delay in one squarer or one multiplier; however, the computation time in each step in MSB-first scheme is equal to the sum of the computation delay in one squarer and one multiplier. As a result, the total computation delay in LSB-first exponentiator is much less than the MSB-first exponentiator provided that same squarers and multipliers are used for both schemes.

Substructure sharing can be achieved using LSB-first scheme when one finite field element is raised to several exponents simultaneously, i.e., to compute β^{N_i}, $1 \leq i \leq z$ at the same time. In this case, computation of $B^{(k)} = (B^{(k-1)})^2$, $1 \leq k \leq m-1$ can be shared among all the exponentiators, which leads to savings of $(m-1)(z-1)$ squarers with little routing overhead.

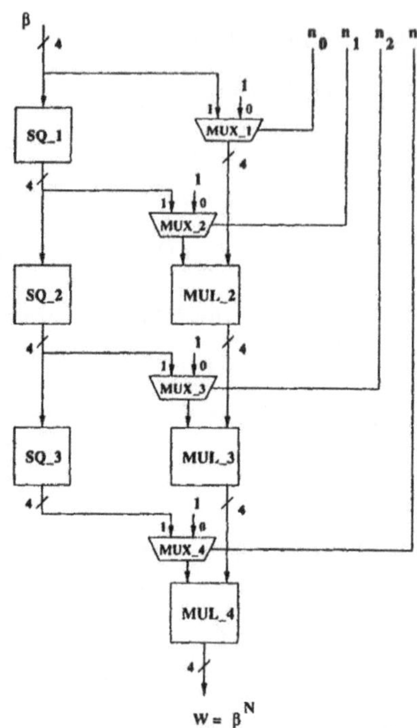

Figure 21.12 LSB-first semi-systolic exponentiator over $GF(2^4)$.

21.3.3.2 Standard basis semi-systolic exponentiator over $GF(2^m)$

The LSB-first squaring algorithm described in (49) can be implemented using the semi-systolic multipliers and squarers developed in the previous sections. The architecture of a LSB-first exponentiator over $GF(2^4)$ is shown in Fig. 21.12.

This architecture consists of $(m-1)$ GF(2^m) multipliers, $(m-1)$ GF(2^m) squarers and m m-bit MUXes. The squarer SQ_k evaluates β^{2^k} for k $=1, 2, \ldots, m-1$. The multiplexer MUX_k selects either $\beta^{2^{k-1}}$ or 1 depending on the value of n_{k-1}. The multiplier MUL_k computes the intermediate result $W^{(k)}$, for k $= 2, 3 \cdots, m$.

Notice that bit-level pipelined exponentiator can be obtained using bit-level pipelined squarers and multipliers. Hence this architecture can accept one new input every clock cycle where the clock period is determined by the delay of a 2-input AND gate followed by a 3-input XOR gate.

The properties of the bit-level pipelined LSB-first exponentiator constructed using LSB-first multipliers and MSB-first squarers are compared with the exponentiator in [7] and the results are summarized in Table 21.3, where only 2-input AND and XOR gates are used, and D_{AND} and D_{XOR} denote the gate delay time for one 2-input AND gate and one 2-input XOR gate, respectively. From this table, we see that the proposed exponentiator results in hardware savings of 50% over the exponentiator in [7] which contains systolic multipliers only. We have also reduced the system latency to $m(m-1) + \lfloor m/2 \rfloor + 1$ from $2m^2 + m$ without any change in the critical path.

Table 21.3 Comparison of exponentiators

Item	Exponentiator [7]	Proposed
Number of multipliers	2(m-1)	m-1
Number of cells in multiplier	m^2	m^2
Basic cell in multiplier	2 AND, 2 XOR, 7 latches	2 AND, 2 XOR 3 latches
Number of squarers	-	m-1
Number of cells in squarer	-	$m\lfloor m/2 \rfloor$
Basic cell in squarer	-	2 AND, 2 XOR, 3 latches
Latency	$2m^2 + m$	$m(m-1) + \lfloor m/2 \rfloor + 1$
Throughput	1	1
Critical path	1 D_{AND} +2 D_{XOR}	1 D_{AND} +2 D_{XOR}

21.4 FINITE FIELD DIVISION ALGORITHMS

In this subsection, we will give a brief overview of algorithms for finite field division.

Division over $GF(2^m)$ involves dividing a polynomial $C(x)$ by $A(x)$ (both of *degree* $< m$) modulo the primitive polynomial $p(x)$. This operation is usually performed in two steps:

1). Find the multiplicative inverse $A^{-1}(x)$;

2). Compute $C(x)/A(x) = C(x) \cdot A^{-1}(x)$.

In other words, division is implemented as multiplication by the multiplicative inverse. Inversion operations are much more involved than multiplication operations. For smaller finite field $GF(2^m)$ where $m < 8$, a table-lookup technique for finding an inverse element has better performance in terms of chip area and computation time [22]. For large fields, however, exponential growth of memory requirements makes this lookup-table based scheme unattractive.

There are three well-known algorithms for finding the multiplicative inverse.

1. *Fermat's Theorem*: For every $\beta \in GF(2^m)$, $\beta^{2^m} = \beta$ and therefore, $\beta^{-1} = \beta^{2^m-2}$. Notice that $2^m - 2 = 2 + 2^2 + 2^3 + \cdots + 2^{m-1}$. Hence β^{-1} can be found by successive *square-and-multiply* operations as

$$\beta^{-1} = \beta^2 \cdot \beta^{2^2} \cdot \beta^{2^3} \cdots \beta 2^{m-1}, \tag{53}$$

   which can be implemented using finite field exponentiators.

2. *Euclid's algorithm*: $GCD(p(x), B(x)) = W(x) \cdot p(x) + U(x) \cdot B(x)$. Suppose that $B(x)$ is the polynomial representation of finite field element $B \in GF(2^m)$. Note that the greatest common divisor of the primitive polynomial $p(x)$ (of degree m) and $B(x)$ (of degree $m-1$) is 1 since $p(x)$ is irreducible. Therefore,

there exist $W(x)$ and $U(x)$ over $GF(2)$ such that $1 = W(x)\cdot p(x)+U(x)\cdot B(x)$, i.e., $U(x)\cdot B(x) = 1\ mod\ p(x)$. Hence $(U(x)\ mod\ p(x))$ is the multiplicative inverse of $B(x)$ in $GF(2^m)$. Two efficient implementations for computing finite field inversion based on Euclid's algorithm were presented in [22] and [23].

3. *Solve simultaneous linear equations over* $GF(2)$: Hasan *et. al.* proposed another scheme for inverting a finite field element by solving simultaneous linear equations over $GF(2)$ in [24]. Assume $AB = C$, where $A, B, C \in GF(2^m)$. This multiplication can be carried out in the following way:

$$[\mathbf{A}\ \mathbf{A}\alpha\ \cdots\ \mathbf{A}\alpha^{m-1}]\mathbf{B} = \mathbf{C}, \tag{54}$$

where $\mathbf{A}$, $\mathbf{B}$ and $\mathbf{C}$ are the column vectors of coordinates of A, B and C, respectively and $\mathbf{A}\ \alpha^i$ is the column vector of coordinates of $A\alpha^i\ mod\ p(x)$. When the coordinates of A and C are known, $B{=}C/A$ can be computed by solving the system of m linear equations for m unknowns. This leads to the division algorithm as follows:

   (a) Form the coefficient matrix $\mathbf{M}$ from A, one column at a time, and the successive column is computed by multiplying its immediate previous column by α (Matrix Generation).

   (b) Solve (54) $\mathbf{M}\ \mathbf{B} = \mathbf{C}$ to obtain the coordinates of B by performing elementary row operations on matrix $[\ \mathbf{M}\ \mathbf{C}\ ]$, where $\mathbf{C}$ is appended to $\mathbf{M}$ as the last column (Gauss-Jordan Elimination).

21.5 FINITE FIELD ARITHMETIC USING DUAL BASIS REPRESENTATION

Dual basis representation is efficient for bit-serial multiplications in finite field since *multiply-by-α* operation is easier in dual coordinate systems [13] [29]. Denote the dual basis of the primal basis $\{1, \alpha, \cdots, \alpha^{m-1}\}$ as $\{\beta_0, \beta_1, \cdots, \beta_{m-1}\}$. Let $A \in GF(2^m)$ be expressed in dual basis as

$$A = a_{m-1}\beta_{m-1} + \cdots + a_1\beta_1 + a_0\beta_0. \tag{55}$$

According to (14), we have

$$a_j = Tr(A\cdot\alpha^j),\ j = 0, 1, \cdots, m-1, \tag{56}$$

and the coordinates of $A^{(1)} = A\cdot\alpha$ in dual basis can be computed as

$$a_j^{(1)} = Tr(A\cdot\alpha^{j+1}) = \left[\begin{array}{l} a_{j+1},\ j = 0, 1, \cdots, m-2 \\ Tr(A\cdot\alpha^m),\ j = m-1 \end{array}\right., \tag{57}$$

where

$$\begin{aligned} a_{m-1}^{(1)} &= Tr(A\cdot\sum_{i=0}^{m-1} p_i\alpha^i) \\ &= \sum_{i=0}^{m-1} p_i Tr(A\cdot\alpha^i) \end{aligned}$$

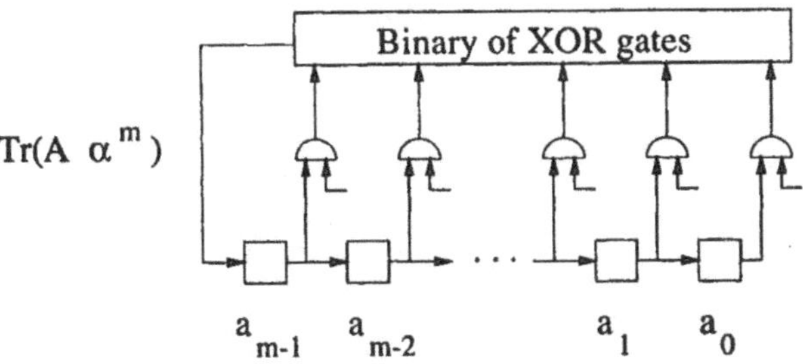

Figure 21.13 Generic *multiply-by-α* circuit.

$$= \sum_{i=0}^{m-1} p_i a_i. \tag{58}$$

A genetic circuit for *multiply-by-α* operation is shown in Fig. 21.13. It is essentially a shift-register of length m and at each clock cycle, only the most significant bit is updated.

As mentioned before, the finite field multiplication $C = AB$ can be written in matrix-vector form as follows, following the LSB-first multiplication algorithm in (15):

$$[\mathbf{A}\ \mathbf{A}\alpha\ \cdots\ \mathbf{A}\alpha^{m-1}]\mathbf{B} = \mathbf{C}, \tag{59}$$

where $\mathbf{A}$, $\mathbf{B}$ and $\mathbf{C}$ are the column vectors of coordinates of A, B and C, respectively and $\mathbf{A}\ \alpha^i$ is the column vector of coordinates of $A\alpha^i \ mod\ f(x)$. More explicitly, we have

$$\begin{bmatrix} a_0^{(0)} & a_0^{(1)} & \cdots & a_0^{(m-1)} \\ a_1^{(0)} & a_1^{(1)} & \cdots & a_1^{(m-1)} \\ & \cdots & \cdots & \\ a_{m-1}^{(0)} & a_{m-1}^{(1)} & \cdots & a_{m-1}^{(m-1)} \end{bmatrix} \begin{bmatrix} b_0 \\ b_1 \\ \cdots \\ b_{m-1} \end{bmatrix} = \begin{bmatrix} c_0 \\ c_1 \\ \cdots \\ c_{m-1} \end{bmatrix} \tag{60}$$

where $a_i^{(k)}$ denotes the i-th coordinate of $A\alpha^k \ mod\ f(x)$, b_i denotes the coordinate of B and c_i denotes the coordinate of C, $0 \leq i \leq m-1$ and $0 \leq k \leq m-1$. In standard basis multiplication, every bit in $\mathbf{A}\ \alpha^i$ is updated from the coordinates of $\mathbf{A}\ \alpha^{i-1}$, as can be seen from (21) and multiplication has to be performed column-wise, starting with the first column $\mathbf{A}$, and the product vector $\mathbf{C}$ is available after m steps. In dual basis representation, define $a_{m+i} = Tr(A\alpha^{m+i})$ for $0 \leq i \leq m-1$. The finite field multiplication in dual basis can be written as

$$\begin{bmatrix} a_0 & a_1 & \cdots & a_{m-1} \\ a_1 & a_2 & \cdots & a_m \\ & \cdots & \cdots & \\ a_{m-1} & a_m & \cdots & a_{2m-2} \end{bmatrix} \begin{bmatrix} b_0 \\ b_1 \\ \cdots \\ b_{m-1} \end{bmatrix} = \begin{bmatrix} c_0 \\ c_1 \\ \cdots \\ c_{m-1} \end{bmatrix}. \tag{61}$$

Due to the properties of dual basis, $\mathbf{MB}$=$\mathbf{C}$ becomes a discrete-time Wiener-Hopf equation ($\mathbf{M}$ is the multiplication matrix)! Multiplication that has to be performed column-wise in standard basis as

$$\mathbf{C} = \mathbf{A} \cdot b_0 + \mathbf{A}\alpha b_1 + \cdots + \mathbf{A}\alpha^{m-1} b_{m-1} \tag{62}$$

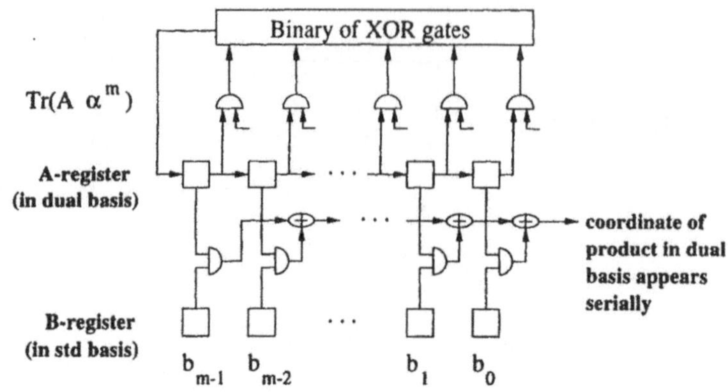

Figure 21.14 Dual basis bit-serial multiplier over $GF(2^m)$.

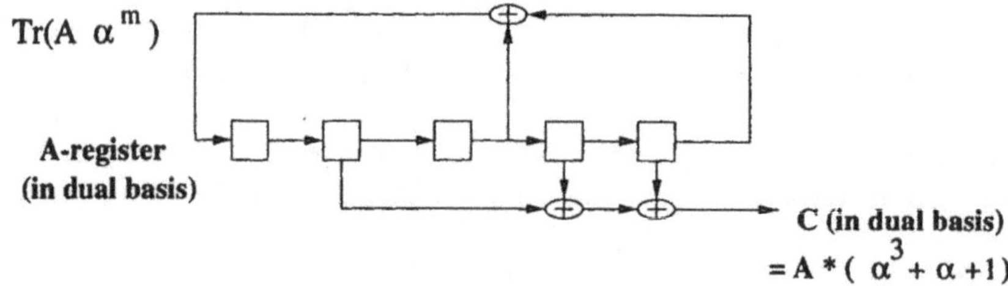

Figure 21.15 Hard-wired variable-by-constant dual basis bit-serial multiplier over $GF(2^5)$ with $p(x) = x^5 + x^2 + 1$.

are now the same as row-wise multiplication since the transpose of ith column in matrix M is the same as its ith row. Therefore, finite field multiplication in dual basis can be performed in serial as follows:

$$a_{m+i} = Tr(A\alpha^{m+i});$$
$$c_i = \left[\ a_i a_{i+1} \cdots a_{i+m-1}\ \right] \begin{bmatrix} b_0 \\ b_1 \\ \cdots \\ b_{m-1} \end{bmatrix}. \tag{63}$$

The architecture of bit-serial dual basis multiplier is shown in Fig. 21.14. Note that the hardware complexity of this variable-by-variable dual basis bit-serial multiplier is the same as a standard basis bit-serial multiplier. However, dual basis multiplier is more efficient when the multiplier is a fixed constant, in which case the multiplier bits can be hard-wired into the multiplier and the total number of gates can be further minimized. Fig. 21.15 shows a hard-wired dual basis variable-by-constant bit-serial multiplier over $GF(2^5)$ with $p(x) = x^5 + x^2 + 1$ and the multiplier is equal to $B = \alpha^3 + \alpha + 1$.

The fact that the finite field multiplication in dual basis can be described by a discrete-time Wiener-Hopf equation, can also be used to design efficient low-area finite field dividers, as shown in [33].

Table 21.4 Polynomial Representation of Elements in $GF(2^m)$ Using Standard Basis and its Dual Basis

Power j	Elements in Standard Basis	Elements in Dual Basis
	$\alpha^3\alpha^2\alpha^1\alpha^0$	$\beta_3\beta_2\beta_1\beta_0$
*	0000	0000
0	0001	1000
1	0010	0100
2	0100	0010
3	1000	1001
4	0011	1100
5	0110	0110
6	1100	1011
7	1011	0101
8	0101	1010
9	1010	1101
10	0111	1110
11	1110	1111
12	1111	0111
13	1101	0011
14	1001	0001

The hybrid nature of dual basis arithmetic operations simplifies their architectures. However, the hybrid representation imposes the necessity of basis conversion between the standard basis coordinates and dual coordinates, whose complexity plays an important role in dual basis arithmetic operations. Using the generalized trace function in (13), it has been shown in [34] [35] [36] [14] that the number of gates in basis conversion circuitry can be minimized by appropriately selecting β in the generalized trace function. Especially, for finite field $GF(2^m)$ in which a primitive polynomial in the trinomial form $p(x) = x^m + x^k + 1$ exists, dual basis is merely a permutation of the standard basis.

21.6 CONCLUSIONS

Theory and implementations of finite field arithmetic operations using different basis representations, standard basis, dual basis and normal basis, have been addressed in this chapter. Algorithms and architectures for finite field multiplication, squaring and exponentiation using standard basis representation has been discussed extensively. We have explored and classified algorithms for standard basis finite field multiplication, squaring and exponentiation into LSB-first scheme and MSB-first scheme. Efficient VLSI implementation of semi-systolic multipliers, squarers and exponentiators have been designed and compared with other architectures in the literature. The use of dual basis representations for finite field bit-serial multiplications were discussed. Various algorithms for finite field division have also been summarized in this chapter.

Acknowledgement

The authors are grateful to Surendra Jain for his early contributions to this work.

Appendix I

The representation of elements of GF(2^4) generated by the primitive polynomial $p(x) = x^4 + x + 1$ in both standard basis $\{1, \alpha, \alpha^2, \alpha^3\}$ and its dual basis $\{\alpha^{14}, \alpha^2, \alpha, 1\}$ are given in Table 21.4. In Table 21.4, the first column is the index or power of an element in exponential form. The index of the zero element is indicated by *. The 4-tuples coordinates of the elements expressed as polynomials in the standard basis and the dual basis representations are given in the second and the third columns, respectively.

Appendix II

The representation of elements of GF(2^4) generated by the primitive polynomial $p(x) = x^4 + x^3 + 1$ in both standard basis $\{1, \alpha, \alpha^2, \alpha^3\}$ and normal basis

Table 21.5 Polynomial Representation of Elements in $GF(2^m)$ Using Standard Basis and Normal Basis

Power j	Elements in Standard Basis	Elements in Normal Basis
	$\alpha^3\alpha^2\alpha^1\alpha^0$	$\alpha^8\alpha^4\alpha^2\alpha$
*	0000	0000
0	0001	1111
1	0010	0001
2	0100	0010
3	1000	1011
4	1001	0100
5	1011	0101
6	1111	0111
7	0111	1100
8	1110	1000
9	0101	1101
10	1010	1010
11	1101	0110
12	0011	1110
13	0110	0011
14	1100	1001

$\{\alpha, \alpha^2, \alpha^4, \alpha^8\}$ are given in Table 21.5, where the first column is the index or power of an element in exponential form. The index of the zero element is indicated by *. The 4-tuples coordinates of the elements expressed as polynomials in the standard basis and the normal basis representations are given in the second and the third columns, respectively.

REFERENCES

[1] R. E. Blahut, *Theory and Practice of Error Control Codes*, Addison Wesley, 1984.

[2] R. J. McEliece, *The Theory of Information and Coding*, Addison-Wesley, 1977.

[3] F. J. MacWilliams and N. J. A. Sloane, *The Theory of Error Correcting Codes*, North-Holland Pub. Co., Amsterdam, 1977.

[4] S. Lin and D.J. Costello, *Error Control Coding : Fundamentals and Applications*, Prentice-Hall, Englewood Cliffs, N.J., 1983.

[5] B. A. Laws and C. K. Rushforth, "A cellular-array multiplier for GF(2^m)", *IEEE Trans. on Computers*, vol. , pp. 1573–1578, Dec. 1971.

[6] C.-S. Yeh, I. S. Reed, and T. K. Truong, "Systolic multipliers for finite fields GF(2^m)", *IEEE Trans. on Computers*, vol. c-33, pp. 357–360, April 1984.

[7] C. L. Wang and J. L. Lin, "Systolic array implementation of multipliers for finite field GF(2^m)", *IEEE Trans. on Circuits and Systems*, vol. 38, pp. 796–800, July 1991.

[8] S. W. Wei, "A systolic power-sum circuit for GF(2^m)", *IEEE Trans. on Computers*, vol. 43, pp. 226–229, Feb 1994.

[9] S. Jain, L. Song and K. K. Parhi, "Efficient semi-systolic VLSI architectures for finite field arithmetic", *IEEE Trans. on VLSI Systems*, **6**(1), pp. 101-113, March 1998.

[10] J. H. Guo and C. L. Wang, "A new systolic squarer and its application to compute exponentiations in $GF(2^m)$", *in Proc. of ISCAS*, pp. 2044–2047, Hong Kong, June 9-12 1997.

[11] A. Ghafoor and A. Singh, "Systolic architecture for finite field exponentiation", *IEE Proc.-E*, vol. 136, pp. 465–470, Nov 1989.

[12] C. L. Wang, "Bit-level systolic array for fast exponentiation in GF(2^m)", *IEEE Trans. on Computers*, vol. 43, pp. 838–841, July 1994.

[13] E. R. Berlekamp, "Bit serial Reed-Solomon encoders", *IEEE Trans on information Theory*, vol. IT-28, pp. 869–874, Nov. 1982.

[14] S. T.J. Fenn, M. Benaissa, and D. Taylor, "$GF(2^m)$ Multiplication and division over the dual basis", *IEEE Trans. on Computers*, vol. 45, pp. 319–327, March 1996.

[15] J. L. Massey and J. K. Omura, "Computational method and apparatus for finite field arithmetic", Technical report, US Patent application, submitted 1981, 1981.

[16] C. C. Wang et al, "VLSI Architectures for Computing Multiplications and Inverses in GF(2^m)", *IEEE Trans. on Computers*, vol. c-34, pp. 709–716, August 1985.

[17] C.C. Wang and D. Pei, "A VLSI design for computing exponentiations in $GF(2^m)$ and its application to generate pseudorandom number sequences", *IEEE Trans on Computers*, vol. 39, pp. 258–262, Feb 1990.

[18] M. Kovac, N. Ranganathan, and M. Varanasi, "SIGMA: a systolic array implementation of a Galois field $GF(2^m)$ based multiplication and division algorithm", *IEEE Trans. on VLSI Systems*, vol. 1, pp. 22–30, March 1993.

[19] M. Kovac and N. Ranganathan, "ACE: A VLSI chip for Galois Field $GF(2^m)$ based exponentiation", *in Proc. of 7th International Conference on VLSI Design*, pp. 291–295, Calcutta, Jan 1994.

[20] S. K. Jain and K. K. Parhi, "Efficient power based Galois field arithmetic architectures", *in IEEE Workshop on VLSI Signal Processing*, pp. 306–316, San Diego, Oct. 1994.

[21] M. Kovac and N. Ranganathan, "ACE: A VLSI chip for Galois field GF(2^m) based exponentiation", *IEEE Trans. on Circuits and Systems-II: Analog and Digital Signal Processing*, vol. 43, pp. 289–297, April 1996.

[22] H. Brunner, A. Curiger, and M. Hofstetter, "On computing multiplicative inverse in $GF(2^m)$", *IEEE Trans. on Computers*, vol. 42, pp. 1010–1015, Aug. 1993.

[23] J. H. Guo and C. L. Wang, "Systolic array implementation of Euclid's algorithm for inversion and division in $GF(2^m)$", *in Proc. of IEEE ICASSP-96*, pp. 481–484, Atlanta, USA, May 1996.

[24] M. A. Hasan and V. K. Bhargava, "Bit-serial systolic divider and multiplier for finite field $GF(2^m)$", *IEEE Trans. on Computers*, vol. 40, pp. 972–980, Aug. 1992.

[25] C. L. Wang and J. L. Lin, "A systolic architecture for computing inverses and divisions in finite field $GF(2^m)$", *IEEE Trans. on Computers*, vol. 42, pp. 1141–1146, Sep. 1993.

[26] S. K. Jain and K. K. Parhi, "Efficient standard basis Reed-Solomon encoder", *in Proc. of IEEE ICASSP*, Atlanta, May 1996.

[27] D. Jungnickel, *Finite Fields: Structure and Arithmetic*, Wissenschaftsverlag, Mannheim, Germany, 1993.

[28] W. W. Peterson and E. J. Weldon, *Error-Correcting Codes*, The MIT Press, 1972.

[29] R. J. McEliece, *Finite Fields for Computer Scientists and Engineers*, Kluwer Academic, 1987.

[30] I. S. Hsu, T. K. Truong, L. J. Deutsch, and I. S. Reed, "A comparison of VLSI architecture of finite field multipliers using dual, normal, or standard bases", *IEEE Trans on Computers*, vol. 37, pp. 735–739, June 1988.

[31] R. Lidl and H. Niederreiter, *Introduction to Finite Fields and Their Applications*, Cambridge University Press, 1986.

[32] L. Song and K. K. Parhi, "Efficient finite field serial/parallel multiplication", *in Proc. of International Conf. on Application Specific Systems, Architectures and Processors*, pp. 72–82, Chicago, Aug 1996.

[33] L. Song and K. K. Parhi, "Low area finite field dual basis divider", *in Proc. of IEEE ICASSP-97*, Germany, April 1997.

[34] M. Morii, M. Kasahara, and D. L. Whiting, "Efficient bit-serial multiplication and the discrete-time Wiener-Hopf equation over finite field", *IEEE Trans. on Information Theory*, vol. 35, pp. 1177–1183, Nov 1989.

[35] M. Wang and I.F. Blake, "Bit-serial multiplication in finite field", *SIAM Journal on Discrete Mathmematics*, vol. 3, pp. 140–148, 1990.

[36] D. R. Stinson, "On bit-serial multiplication and dual basis in GF(2^m)", *IEEE Trans. on Information Theory*, vol. 37, pp. 1733–1736, Nov 1991.

Chapter 22

CORDIC Algorithms and Architectures

Herbert Dawid	Heinrich Meyr
Synopsys, Inc.	Aachen University of Technology
DSP Solutions Group	Integrated Systems in Signal Processing
Herzogenrath, Germany	Aachen, Germany
dawid@synopsys.com	*meyr@ert.rwth-aachen.de*

22.1 INTRODUCTION

Digital signal processing (DSP) algorithms exhibit an increasing need for the efficient implementation of complex arithmetic operations. The computation of trigonometric functions, coordinate transformations or rotations of complex valued phasors is almost naturally involved with modern DSP algorithms. Popular application examples are algorithms used in digital communication technology and in adaptive signal processing. While in digital communications, the straightforward evaluation of the cited functions is important, numerous matrix based adaptive signal processing algorithms require the solution of systems of linear equations, QR factorization or the computation of eigenvalues, eigenvectors or singular values. All these tasks can be efficiently implemented using processing elements performing vector rotations. The COordinate Rotation DIgital Computer algorithm (CORDIC) offers the opportunity to calculate all the desired functions in a rather simple and elegant way.

The CORDIC algorithm was first introduced by Volder [1] for the computation of trigonometric functions, multiplication, division and datatype conversion, and later on generalized to hyperbolic functions by Walther [2]. Two basic CORDIC modes are known leading to the computation of different functions, the *rotation mode* and the *vectoring mode.*

For both modes the algorithm can be realized as an iterative sequence of additions/subtractions and shift operations, which are rotations by a fixed rotation angle (sometimes called *microrotations*) but with variable rotation direction. Due to the simplicity of the involved operations the CORDIC algorithm is very well suited for VLSI implementation. However, the CORDIC iteration is not a perfect rotation which would involve multiplications with sine and cosine. The rotated vector is also scaled making a *scale factor correction* necessary.

We first give an introduction into the CORDIC algorithm. Then we discuss methods for scale factor correction and accuracy issues with respect to a fixed wordlength implementation. In the second part of the chapter different architectural realizations for the CORDIC are presented for different applications:

1. Programmable CORDIC processing element
2. High throughput CORDIC processing element
3. CORDIC Architectures for Vector Rotation
4. CORDIC Architectures using redundant number systems.

22.2 THE CORDIC ALGORITHM

In this section we first present the basic CORDIC iteration before discussing the full algorithm, which consists of a sequence of these iterations.

In the most general form one CORDIC iteration can be written as [2, 3]

$$\begin{aligned} x_{i+1} &= x_i - m \cdot \mu_i \cdot y_i \cdot \delta_{m,i} \\ y_{i+1} &= y_i + \mu_i \cdot x_i \cdot \delta_{m,i} \\ z_{i+1} &= z_i - \mu_i \cdot \alpha_{m,i} \end{aligned} \tag{1}$$

Although it may not be immediately obvious this basic CORDIC iteration describes a rotation (together with a scaling) of an intermediate plane vector $v_i = (x_i, y_i)^T$ to $v_{i+1} = (x_{i+1}, y_{i+1})^T$. The third iteration variable z_i keeps track of the rotation angle $\alpha_{m,i}$. The variable $m \in \{1, 0, -1\}$ specifies a circular, linear or hyperbolic coordinate system, respectively. The rotation direction is steered by the variable $\mu_i \in \{1, -1\}$. Trajectories for the vectors v_i for successive CORDIC iterations are shown in Fig. 22.1 for $m = 1$, in Fig. 22.2 for $m = 0$, and in Fig. 22.3 for $m = -1$, respectively. In order to avoid multiplications $\delta_{m,i}$ is defined to be

$$\begin{aligned} \delta_{m,i} &= d^{-s_{m,i}} \quad ; d: \text{ Radix of employed number system, } s_{m,i}: \text{ integer number} \\ &= 2^{-s_{m,i}} \quad ; \text{ Radix 2 number system} \end{aligned} \tag{2}$$

For obvious reasons we restrict consideration to radix 2 number systems below: $\delta_{m,i} = 2^{-s_{m,i}}$. It will be shown later that the *shift sequence* $s_{m,i}$ is generally a nondecreasing integer sequence. Hence, a CORDIC iteration can be implemented using only shift and add/subtract operations.

The first two equations of the system of equations given in (1) can be written as a matrix-vector product

$$v_{i+1} = \begin{pmatrix} 1 & -m \cdot \mu_i \cdot \delta_{m,i} \\ \mu_i \cdot \delta_{m,i} & 1 \end{pmatrix} \cdot v_i = \boldsymbol{C}_{m,i} \cdot v_i \tag{3}$$

In order to verify that the matrix-vector product in (3) describes indeed a vector rotation and to quantify the involved scaling we consider now a general normalized plane rotation matrix for the three coordinate systems. For $m = 1, 0, -1$ and an angle $\mu_i \cdot \alpha_{m,i}$ with μ_i determining the rotation direction and $\alpha_{m,i}$ representing an unsigned angle, this matrix is given by

$$\boldsymbol{R}_{m,i} = \begin{pmatrix} \cos(\sqrt{m} \cdot \alpha_{m,i}) & -\mu_i \cdot \sqrt{m} \cdot \sin(\sqrt{m} \cdot \alpha_{m,i}) \\ \frac{\mu_i}{\sqrt{m}} \cdot \sin(\sqrt{m} \cdot \alpha_{m,i}) & \cos(\sqrt{m} \cdot \alpha_{m,i}) \end{pmatrix} \tag{4}$$

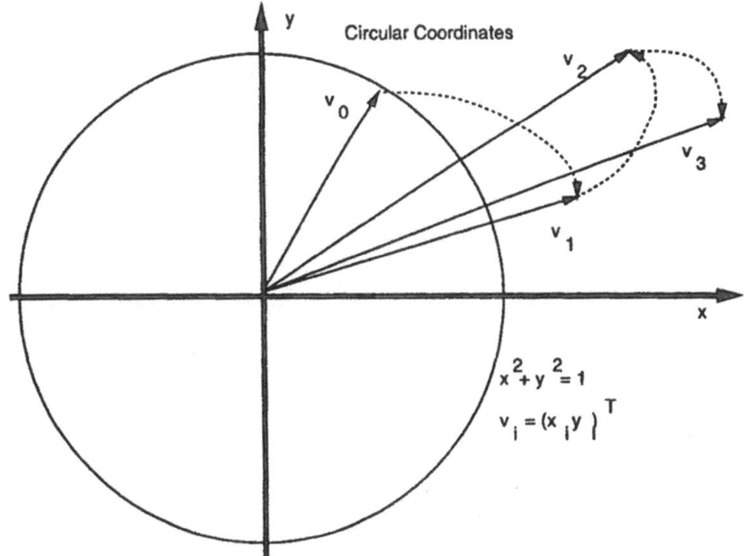

Figure 22.1 Rotation trajectory for the circular coordinate system ($m = 1$).

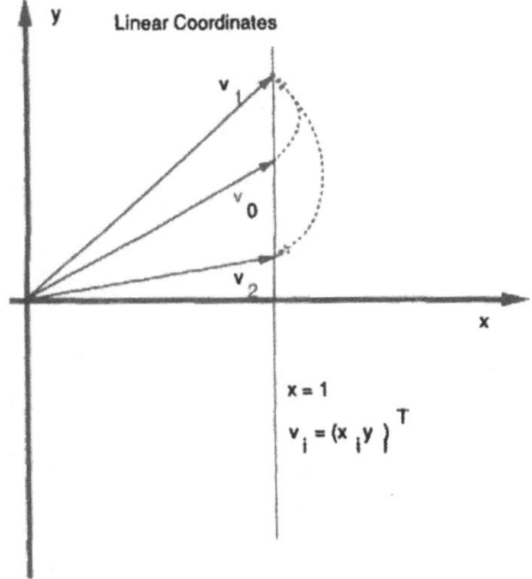

Figure 22.2 Rotation trajectory for the linear coordinate system ($m = 0$).

which can be easily verified by setting m to $1, 0, -1$, repectively, and using (for the hyperbolic coordinate system) the identities: $\sinh(z) = -i sin(iz)$, $\cosh(z) = \cos(iz)$, and $tanh(z) = -i \cdot tan(iz)$ with $i = \sqrt{-1}$. The norm of a vector $(x, y)^T$ in these coordinate systems is defined as $\sqrt{x^2 + m \cdot y^2}$. Correspondingly, the norm preserving rotation trajectory is a circle defined by $x^2 + y^2 = 1$ in the circular coordinate system, while in the hyperbolic coordinate system the "rotation" trajectory is a hyperbolic function defined by $x^2 - y^2 = 1$ and in the linear coordinate system

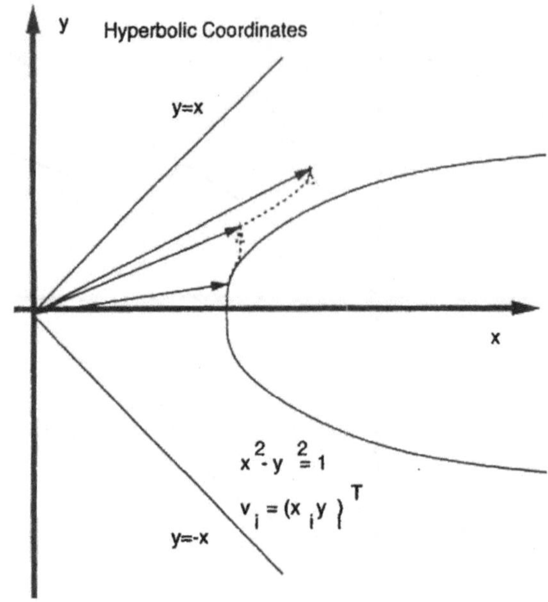

Figure 22.3 Rotation trajectory for the hyperbolic coordinate system ($m = -1$).

the trajectory is a simple line $x = 1$. Hence, the common meaning of a rotation holds only for the circular coordinate system. Clearly

$$\frac{1}{\cos(\sqrt{m} \cdot \alpha_{m,i})} \cdot \boldsymbol{R}_{m,i} = \begin{pmatrix} 1 & -\mu_i \cdot \sqrt{m} \cdot \tan(\sqrt{m} \cdot \alpha_{m,i}) \\ \frac{\mu_i}{\sqrt{m}} \cdot \tan(\sqrt{m} \cdot \alpha_{m,i}) & 1 \end{pmatrix}$$

holds, hence

$$\frac{1}{\cos(\sqrt{m} \cdot \alpha_{m,i})} \cdot \boldsymbol{R}_{m,i} = \boldsymbol{C}_{m,i} \quad \text{holds for} \quad \delta_{m,i} = \frac{1}{\sqrt{m}} \cdot \tan(\sqrt{m} \cdot \alpha_{m,i}).$$

This proves that $\boldsymbol{C}_{m,i}$ is an unnormalized rotation matrix for $m \in \{-1, 1\}$ due to the scaling factor $\frac{1}{\cos(\sqrt{m} \cdot \alpha_{m,i})}$. $\boldsymbol{C}_{m,i}$ describes a rotation with scaling rather than a pure rotation. For $m = 0$, $\boldsymbol{R}_{m,i} = \boldsymbol{C}_{m,i}$ holds, hence $\boldsymbol{C}_{m,i}$ is a normalized rotation matrix in this case and no scaling is involved. The scale factor is given by

$$\begin{aligned} K_{m,i} &= \frac{1}{\cos(\sqrt{m} \cdot \alpha_{m,i})} = \frac{\sqrt{\cos^2(\sqrt{m} \cdot \alpha_{m,i}) + \sin^2(\sqrt{m} \cdot \alpha_{m,i})}}{\cos(\sqrt{m} \cdot \alpha_{m,i})} \\ &= \sqrt{1 + \tan^2(\sqrt{m} \cdot \alpha_{m,i})} \end{aligned} \tag{5}$$

For n successive iterations, obviously

$$v_n = \prod_{i=0}^{n-1} \boldsymbol{C}_{m,i} \cdot v_0 = \prod_{i=0}^{n-1} K_{m,i} \cdot \prod_{i=0}^{n-1} \boldsymbol{R}_{m,i} \cdot v_0$$

holds, i.e., a rotation by an angle

$$\theta = \sum_{i=0}^{n-1} \mu_i \cdot \alpha_{m,i}$$

and is performed with an overall scaling factor of

$$K_m(n) = \prod_{i=0}^{n-1} K_{m,i}. \tag{6}$$

The third iteration component z_i simply keeps track of the overall rotation angle accumulated during successive microrotations

$$z_{i+1} = z_i - \mu_i \cdot \alpha_{m,i}.$$

After n iterations

$$z_n = z_0 - \sum_{i=0}^{n-1} \mu_i \cdot \alpha_{m,i}$$

holds, hence z_n is equal to the difference of the start value z_0 and the total accumulated rotation angle.

In Fig. 22.4, the structure of a processing element implementing one CORDIC iteration is shown. All internal variables are represented by a fixed number of digits, including the precalculated angle $\alpha_{m,i}$ which is taken from a register. Due to the limited wordlength some rounding or truncation following the shifts $2^{-s_{m,i}}$ is necessary. The adders/subtractors are steered with $-m\mu_i$, μ_i and $-\mu_i$, respectively.

A rotation by any (within some convergence range) desired rotation angle A_0 can be achieved by defining a converging sequence of n single rotations. The CORDIC Algorithm is formulated given

1. A shift sequence $s_{m,i}$ defining an angle sequence

$$\alpha_{m,i} = \frac{1}{\sqrt{m}} \cdot \tan^{-1}(\sqrt{m} \cdot 2^{-s_{m,i}}) \quad \text{with } i \in \{0, \ldots, n-1\} \tag{7}$$

which guarantees convergence. Shift sequences will be discussed in Section 22.2.3.

2. A control scheme generating a sign sequence μ_i with $i \in \{0, \ldots, n-1\}$ which steers the direction of the rotations in this iteration sequence and guarantees convergence.

In order to explain the control schemes used for the CORDIC Algorithm the angle A_i is introduced specifying the remaining rotation angle after rotation i. The direction of the following rotation has to be chosen such that the absolute value of the remaining angle eventually becomes smaller during successive iterations [2]

$$|A_{i+1}| = ||A_i| - \alpha_{m,i}|. \tag{8}$$

Two control schemes fulfilling (8) are known for the CORDIC algorithm, the *Rotation Mode* and the *vectoring mode.*

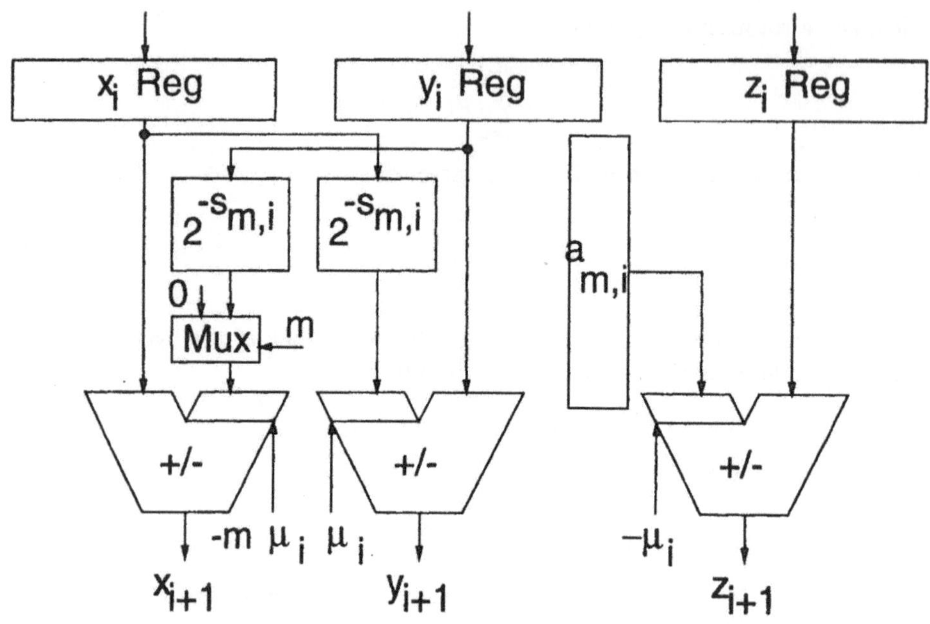

Figure 22.4 Basic structure of a processing element for one CORDIC iteration.

22.2.1 Rotation Mode

In rotation mode the desired rotation angle $A_0 = \theta$ is given for an input vector $(x, y)^T$. We set $x_0 = x$, $y_0 = y$, and $z_0 = \theta = A_0$. After n iterations

$$z_n = \theta - \sum_{i=0}^{n-1} \mu_i \cdot \alpha_{m,i}.$$

If $z_n = 0$ holds, then $\theta = \sum_{i=0}^{n-1} \mu_i \cdot \alpha_{m,i}$, i.e., the total accumulated rotation angle is equal to θ. In order to drive z_n to zero, $\mu_i = \text{sign}(z_i)$ is used leading to

$$\begin{aligned} x_{i+1} &= x_i - m \cdot \text{sign}(z_i) \cdot y_i \cdot 2^{-s_{m,i}} \\ y_{i+1} &= y_i + \text{sign}(z_i) \cdot x_i \cdot 2^{-s_{m,i}} \\ z_{i+1} &= z_i - \text{sign}(z_i) \cdot \alpha_{m,i}. \end{aligned} \tag{9}$$

Obviously, for $z_0 = A_0$ and $z_i = A_i$

$$
\begin{aligned}
z_{i+1} &= z_i - \mathrm{sign}(z_i) \cdot \alpha_{m,i} \\
\text{hence} & \\
\mathrm{sign}(z_i) \cdot z_{i+1} &= \mathrm{sign}(z_i) \cdot z_i - \alpha_{m,i} \\
&= |z_i| - \alpha_{m,i}. \\
\text{Taking absolute values} & \\
|z_{i+1}| &= ||z_i| - \alpha_{m,i}| \\
\text{follows, satisfying (8) for } z_i &= A_i. \qquad (10)
\end{aligned}
$$

The finally computed scaled rotated vector is given by $(x_n, y_n)^T$. In Fig. 22.5, the trajectory for the rotation mode in the circular coordinate system is shown. It becomes clear that the vector is iteratively rotated towards the desired final position. The scaling involved with the successive iterations is also shown in Fig. 22.5.

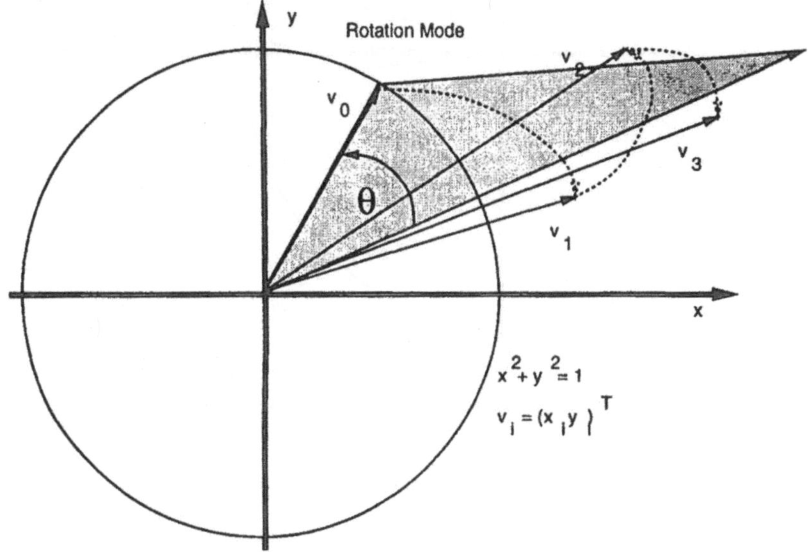

Figure 22.5 Rotation trajectory for the rotation mode in the circular coordinate system.

22.2.2 Vectoring Mode

In vectoring mode the objective is to rotate the given input vector $(x, y)^T$ with magnitude $\sqrt{x^2 + m \cdot y^2}$ and angle $\phi = -A_0 = \frac{1}{\sqrt{m}} \cdot \tan^{-1}(\sqrt{m} \cdot \frac{y}{x})$ towards the x–axis. We set $x_0 = x$, $y_0 = y$, and $z_0 = 0$. The control scheme is such that during the n iterations y_n is driven to zero: $\mu_i = -\mathrm{sign}(x_i) \cdot \mathrm{sign}(y_i)$. Depending on the sign of x_0 the vector is then rotated towards the positive ($x_0 \geq 0$) or negative ($x_0 < 0$) x–axis. If $y_n = 0$ holds, z_n contains the negative total accumulated rotation angle after n iterations which is equal to ϕ

$$z_n = \phi = -\sum_{i=0}^{n-1} \mu_i \cdot \alpha_{m,i}$$

and x_n contains the scaled and eventually (for $x_0 < 0$) signed magnitude of the input vector as shown in Fig. 22.6. The CORDIC iteration driving the y_i variable

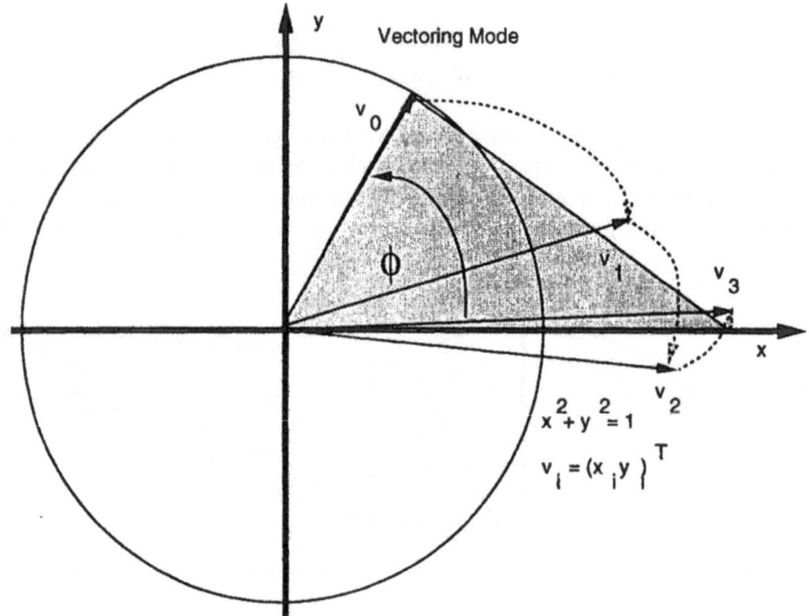

Figure 22.6 Rotation trajectory for the vectoring mode in the circular coordinate system.

to zero is given by

$$\begin{aligned} x_{i+1} &= x_i + m \cdot \text{sign}(x_i) \cdot \text{sign}(y_i) \cdot y_i \cdot 2^{-s_{m,i}} \\ y_{i+1} &= y_i - \text{sign}(x_i) \cdot \text{sign}(y_i) \cdot x_i \cdot 2^{-s_{m,i}} \\ z_{i+1} &= z_i + \text{sign}(x_i) \cdot \text{sign}(y_i) \cdot \alpha_{m,i} \end{aligned} \quad (11)$$

with $x_n = K_m(n) \cdot \text{sign}(x_0) \cdot \sqrt{x^2 + m \cdot y^2}$. Obviously, the remaining rotation angle after iteration i is given by $A_i = -\frac{1}{\sqrt{m}} \cdot \tan^{-1}(\sqrt{m} \cdot \frac{y_i}{x_i})$. Clearly, $\text{sign}(A_i) = -\text{sign}(x_i) \cdot \text{sign}(y_i)$ holds. With $z_0 = 0$ and $A_0 = -\phi$, $A_i = z_i - \phi$ holds. Using $\mu_i = -\text{sign}(A_i)$

$$\begin{aligned} A_{i+1} &= z_{i+1} - \phi \\ &= z_i + \text{sign}(x_i) \cdot \text{sign}(y_i) \cdot \alpha_{m,i} - \phi \quad \text{using Eq. (11)} \\ &= z_i - \text{sign}(A_i) \cdot \alpha_{m,i} - \phi \\ &= A_i - \text{sign}(A_i) \cdot \alpha_{m,i} \end{aligned}$$

holds. Thus, (8) is again satisfied as was shown in (10).

22.2.3 Shift Sequences and Convergence Issues

Given the two iteration control schemes, shift sequences $s_{m,i}$ have to be introduced which guarantee convergence.

First, the question arises how to define convergence in this case. Since there are only n fixed rotation angles with variable sign, the desired rotation angle A_0 can only be approximated resulting in an angle approximation error $\Delta\phi$ [4]

$$\Delta\phi = A_0 - \sum_{i=0}^{n-1} \mu_i \cdot \alpha_{m,i}.$$

$\Delta\phi$ does not include errors due to finite quantization of the rotation angles $\alpha_{m,i}$. In rotation mode, since $z_0 = A_0$, $\Delta\phi = z_n = A_n$ holds, i.e., z_n cannot be made exactly equal to zero. In vectoring mode the angle approximation error is given by the remaining angle of the vector $(x_n, y_n)^T$

$$\Delta\phi = A_n = \frac{1}{\sqrt{m}} \cdot \tan^{-1}(\sqrt{m} \cdot \frac{y_n}{x_n}) = \begin{cases} \tan^{-1}(\frac{y_n}{x_n}) & \text{m = 1} \\ \frac{y_n}{x_n} = \frac{y_n}{x_0} & \text{m = 0} \\ \tanh^{-1}(\frac{y_n}{x_n}) & \text{m =-1.} \end{cases}$$

Hence, convergence can only be defined taking into account the angle approximation error. Two convergence criteria can be derived (for a detailed discussion see [1, 2]): 1.) First, the chosen set of rotation angles has to satisfy

$$\alpha_{m,i} - \sum_{j=i+1}^{n-1} \alpha_{m,j} \leq \alpha_{m,n-1}. \tag{12}$$

The reason for this requirement can be sketched as follows: if at any iteration stage i the current remaining rotation angle A_i is zero, it will be changed by $\pm\alpha_{m,i}$ in the next iteration. Then, the sum of rotation angles for the remaining iterations $\sum_{j=i+1}^{n-1} \alpha_{m,j}$ has to be large enough to bring the remaining angle after the last iteration A_n to zero within the accuracy given by $\alpha_{m,n-1}$. 2.) Second, the given rotation angle A_0 must not exceed the convergence range of the iteration which is given by the sum of all rotation angles plus the final angle

$$|A_0| \leq \sum_{i=0}^{n-1} \alpha_{m,i} + \alpha_{m,n-1}$$

Walther [2] has proposed shift sequences for each of the three coordinate systems for radix 2 implementation. It was shown by Walther [2] that for $m = -1$ the convergence criterion (12) is not satisfied for $\alpha_{-1,i} = \tanh^{-1}(2^{-i})$, but it is satisfied if the integers $(4, 13, 40, \ldots, k, 3k+1, \ldots)$ are repeated in the shift sequence. For any implementation, the angle sequence $\alpha_{m,i}$ resulting from the chosen shift sequence can be calculated in advance, quantized according to a chosen quantization scheme and retrieved from storage during execution of the CORDIC algorithm as shown in Fig. 22.4.

22.2.4 CORDIC Operation Modes and Functions

Using the CORDIC Algorithm and the shift sequences stated above, a number of different functions can be calculated in Rotation Mode and Vectoring Mode as shown in Table 22.2.

Table 22.1 CORDIC Shift Sequences

coordinate system	shift sequence	convergence	scale factor
m	$s_{m,i}$	$\|A_0\|$	$K_m(n \to \infty)$
1	$0, 1, 2, 3, 4, \ldots, i, \ldots$	~ 1.74	~ 1.64676
0	$1, 2, 3, 4, 5. \ldots, i+1, \ldots$	1.0	1.0
− 1	$1, 2, 3, 4, 4, 5, \ldots$	~ 1.13	~ 0.82816

Table 22.2 Functions Calculated by the CORDIC Algorithm

m	mode	initialization	output
1	rotation	$x_0 = x$	$x_n = K_1(n) \cdot (x \cos\theta - y \sin\theta)$
		$y_0 = y$	$y_n = K_1(n) \cdot (y \cos\theta + x \sin\theta)$
		$z_0 = \theta$	$z_n = 0$
		$x_0 = \frac{1}{K_1(n)}$	$x_n = \cos\theta$
		$y_0 = 0$	$y_n = \sin\theta$
		$z_0 = \theta$	$z_n = 0$
1	vectoring	$x_0 = x$	$x_n = K_1(n) \cdot \mathrm{sign}(x_0) \cdot \sqrt{x^2 + y^2}$
		$y_0 = y$	$y_n = 0$
		$z_0 = \theta$	$z_n = \theta + \tan^{-1}(\frac{y}{x})$
0	rotation	$x_0 = x$	$x_n = x$
		$y_0 = y$	$y_n = y + x \cdot z$
		$z_0 = z$	$z_n = 0$
0	vectoring	$x_0 = x$	$x_n = x$
		$y_0 = y$	$y_n = 0$
		$z_0 = z$	$z_n = z + \frac{y}{x}$
- 1	rotation	$x_0 = x$	$x_n = K_{-1}(n) \cdot (x \cosh\theta + y \sinh\theta)$
		$y_0 = y$	$y_n = K_{-1}(n) \cdot (y \cosh\theta + x \sinh\theta)$
		$z_0 = \theta$	$z_n = 0$
		$x_0 = \frac{1}{K_{-1}(n)}$	$x_n = \cosh\theta$
		$y_0 = 0$	$y_n = \sinh\theta$
		$z_0 = \theta$	$z_n = 0$
- 1	vectoring	$x_0 = x$	$x_n = K_{-1}(n) \cdot \mathrm{sign}(x_0) \cdot \sqrt{x^2 - y^2}$
		$y_0 = y$	$y_n = 0$
		$z_0 = \theta$	$z_n = \theta + \tanh^{-1}(\frac{y}{x})$

In addition, the following functions can be calculated from the immediate CORDIC outputs

$$\tan z = \frac{\sin z}{\cos z}$$

$$
\begin{aligned}
\tanh z &= \frac{\sinh z}{\cosh z} \\
\exp z &= \sinh z + \cosh z \\
\ln z &= 2\tanh^{-1}(\frac{y}{x}) \quad \text{with } x = z + 1 \text{ and } y = z - 1 \\
\sqrt{z} &= \sqrt{x^2 - y^2} \quad \text{with } x = z + \frac{1}{4} \text{ and } y = z - \frac{1}{4}.
\end{aligned}
$$

22.3 COMPUTATIONAL ACCURACY

It was already mentioned that due to the n fixed rotation angles a given rotation angle A_0 can only be approximated, resulting in an angle approximation error $\Delta\phi \leq \alpha_{m,n-1}$. Even if all other error sources are neglected, the accuracy of the outputs of the nth iteration is hence principally limited by the magnitude of the last rotation angle $\alpha_{m,n-1}$. For large n, approximately $s_{m,n-1}$ accurate digits of the result are obtained since $s_{m,n-1}$ specifies the last right shift of the shift sequence. As a first guess the number of iterations should hence be chosen such that $s_{m,n-1} = W$ for a desired output accuracy of W bits. This leads to $n = W+1$ iterations if the shift sequence given in Table 22.1 for $m = 1$ is used.

A second error source is given by the finite precision of the involved variables which has to be taken into account for fixed point as well as floating point implementations. The CORDIC algorithm as stated so far is suited for fixed point number representations. Some facts concerning the extension to floating point numbers will be presented in Section 22.3.4. The format of the internal CORDIC variables is shown in Fig. 22.7. The internal wordlength is given by the wordlength W of the

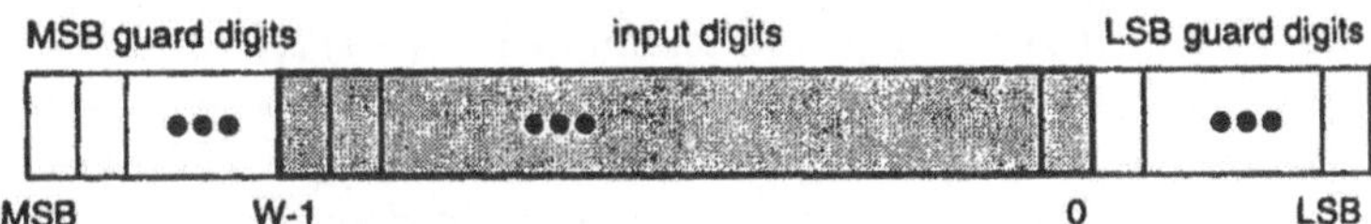

Figure 22.7 Format of the internal CORDIC variables.

fixed point input, enhanced by G guard bits on the most significant bit (MSB) and C guard bits at the least significant bit (LSB) side[1]. The MSB guard bits are necessary since for $m = 1$ the scale factor is larger than one, and since range extensions are introduced by the CORDIC rotation as obvious from the rotation trajectories given in Fig. 22.1, Fig. 22.2 and Fig. 22.3, respectively[2]. The successive right shifts employed in the CORDIC algorithm for the x_i and y_i variables, together with the limited number of bits in the internal fixed point datapath, require careful analysis of the involved computational accuracy and the inevitable rounding errors[3]. LSB

[1] In the CORDIC Rotation Mode and Vectoring Mode the iteration variables z_i and y_i are driven to zero during the n iterations. This fact can be exploited by either increasing the resolution for these variables during the iterations or by reducing their wordlength. Detailed descriptions for these techniques can be found in [5, 6] for the rotation mode and in [7] for the vectoring mode.

[2] For $m = 1$ two guard bits are necessary at the MSB side. First, the scale factor is larger than one: $K_1(n) \sim 1.64676$. Second, a range extension by the factor $\sqrt{2}$ occurs for the x_n and y_n components.

[3] In order to avoid any rounding the number of LSB guard digits must be equal to the sum of all elements in the shift sequence, which is of course not economically feasible.

guard digits are necessary in order to provide the desired output accuracy as will be discussed in Section 22.3.3.

22.3.1 Number Representation

Below, we assume that the intermediate CORDIC iteration variables x_i and y_i are quantized according to a normalized (fractional) fixed point two's complement number representation. The value V of a number n with N binary digits $(n_{N-1}, \ldots, n_0)$ is given by

$$V = (-n_{N-1} + \sum_{j=0}^{N-2} n_j \cdot 2^{-(N-1)+j}) \cdot F \quad \text{with } F = 1. \tag{13}$$

This convenient notation can be changed to a two's complement integer representation simply by using $F = 2^{N-1}$ or to any other fixed point representation in a similar way, hence it does not pose any restriction on the input quantization of the CORDIC.

For $m = 1$ the different common formats for rotation angles have to be taken into account. If the angle is given in radians the format given in (13) may be chosen. F has to be adapted to $F = 2$ if the input range is $(-\pi/2, \pi/2) = (-1.57, 1.57)$. However, in many applications for the circular coordinate system it is desirable to represent the angle in fractions of π and to include the wrap around property which holds for the restricted range of possible angles $(-\pi, +\pi)$. If $F = \pi$ is chosen, the well known wrap around property of two's complement numbers ensures that all angles undergoing any addition/subtraction stay in the allowed range. This format was proposed by Daggett [8] and is sometimes referred to as "Daggett angle representation". The input angle range is often limited to $(-\pi/2, \pi/2)$ which guarantees convergence. The total domain of convergence can be easily expanded by including some "pre-rotations" for input vectors in the ranges $(\pi/2, \pi)$ and $(-\pi, -\pi/2)$. If the Daggett angle format is used it is very easy to determine the quadrant of a given rotation angle A_0 since only the upper two bits have to be inspected. Pre-rotations by $\pm\pi$ or $\pm\pi/2$ are very easy to implement. More sophisticated proposals for expanding the range of convergence of the CORDIC Algorithm for all coordinate systems were given in [9].

22.3.2 Angle Approximation Error

It was shown already that the last rotation angle $\alpha_{m,n-1}$ determines the accuracy achievable by the CORDIC rotation. A straightforward conclusion is to increase the number of iterations n in order to improve accuracy since $\alpha_{m,n-1} = \frac{1}{\sqrt{m}} \cdot \tan^{-1}(\sqrt{m} \cdot 2^{-s_{m,n-1}})$ with $s_{m,i}$ being a nondecreasing integer shift sequence. However, the finite word length available for representing the intermediate variables poses some restrictions. If the angle is quantized according to (13) the value of the least significant digit is given by $2^{-W-C+1} \cdot F$. Therefore, $\alpha_{m,n-1} \geq 2^{-W-C+1} \cdot F$ must hold in order to represent this value. Additionally the rounding error, which increases with the number of iterations, has to be taken into account.

22.3.3 Rounding Error

Rounding is preferred over truncation in CORDIC implementations since the truncation of two's complement numbers creates a positive offset which is highly undesirable. Additionally, the maximum error due to rounding is only half as large as the error involved with truncation. The effort for the rounding procedure can be kept very small since the eventual addition of a binary one at the LSB position can be incorporated with the additions/subtractions which are necessary anyway.

Analysis of the rounding error for the z_i variable is easy since no shifts occur and the rounding error is just due to the quantization of the rotation angles. Hence, the rounding error is bounded by the accumulation of the absolute values of the rounding errors for the single quantized angles $\alpha_{m,i}$.

In contrast the propagation of initial rounding errors for x_i and y_i through the multiple iteration stages of the algorithm and the scaling involved with the single iterations make an analytical treatment very difficult. However, a thorough mathematical analysis taking into account this error propagation is given in [4] and extended in [10]. Due to limited space we present only a simplified treatment as can be found in [2]. Here the assumption is made that the maximum rounding errors associated with every iteration accumulate to an overall maximum rounding error for n iterations. While for z_i this is a valid assumption it is a coarse simplification for x_i and y_i. As shown in Fig. 22.7, $W + G + C$ bits are used for the internal variables with C additional fractional guard digits. Using the format given in (13) the maximum accumulated rounding error for n iterations is given by $e(n) = n \cdot F \cdot 2^{-(W+C-1)-1}$. If W accurate fractional digits of the result are to be obtained the resulting output resolution is $2^{-(W-1)} \cdot F$. Therefore, if C is chosen such that $e(n) \leq F \cdot 2^{-W}$, the rounding error can be considered to be of minor impact. From $n \cdot F \cdot 2^{-(W+C)} < F \cdot 2^{-W}$ it follows that $C \geq \log_2(n)$. Hence, at least $C = \lceil \log_2(n) \rceil$ additional fractional digits have to be provided.

22.3.4 Floating Point Implementations

The shift–and-add structure of the CORDIC algorithm is well suited for a fixed point implementation. The use of expensive floating point re–normalizations and floating point additions and subtractions in the CORDIC iterations does not lead to any advantage since the accuracy is still limited to the accuracy imposed by the fixed wordlength additions and subtractions which also have to be implemented in floating point adders/subtractors. Therefore, floating point CORDIC implementations usually contain an initial floating to fixed conversion. We consider here only the case $m = 1$ (detailed discussions of the floating point CORDIC can be found in [7, 11, 12, 13]). Below, it is assumed that the input floating point format contains a normalized signed mantissa m and an exponent e. Hence, the mantissa is a two's complement number[4] quantized according to (13) with $F = 1$. The components of the floating point input vector v are given by $v = (x, y)^T = (m_x \cdot 2^{e_x}, m_y \cdot 2^{e_y})^T$. The input conversion includes an alignment of the normalized signed floating point mantissas according to their exponents. Two approaches are known for the floating point CORDIC:

1. The CORDIC algorithm is started using the same shift sequences, rotation angles and number of iterations as for the fixed point CORDIC. We consider a floating point implementation of the vectoring mode.
   The CORDIC Vectoring Mode iteration written for floating point values x_i

[4]If the mantissa is given in sign–magnitude format it can be easily converted to a two's complement representation.

and y_i and $m = 1$ is given by

$$
\begin{aligned}
m_{x,i+1} \cdot 2^{e_x} &= m_{x,i} \cdot 2^{e_x} + \mu_i \cdot m_{y,i} \cdot 2^{e_y} \cdot 2^{-s_{m,i}} \\
m_{y,i+1} \cdot 2^{e_y} &= m_{y,i} \cdot 2^{e_y} - \mu_i \cdot m_{x,i} \cdot 2^{e_x} \cdot 2^{-s_{m,i}}.
\end{aligned} \tag{14}
$$

Depending on the difference of the exponents $E = e_x - e_y$ two different approaches are used. If $E < 0$ we divide both equations in (14) by 2^{e_y} to obtain

$$
\begin{aligned}
m_{x,i+1} \cdot 2^{e_x - e_y} &= m_{x,i} \cdot 2^{e_x - e_y} + \mu_i \cdot m_{y,i} \cdot 2^{-s_{m,i}} \\
m_{y,i+1} &= m_{y,i} - \mu_i \cdot m_{x,i} \cdot 2^{e_x - e_y} \cdot 2^{-s_{m,i}}.
\end{aligned} \tag{15}
$$

Hence, we can simply set $y_0 = m_{y,0}$ and $x_0 = m_{x,0} \cdot 2^{e_x - e_y}$ and then perform the usual CORDIC iteration for the resulting fixed point two's complement inputs x_0 and y_0. Of course, some accuracy for the x_i variable is lost due to the right shift $2^{e_x - e_y}$. For $E \geq 0$ we could proceed completely accordingly and divide the equations by 2^{e_x}. Then the initial conversion represents just an alignment of the two floating point inputs according to their exponents. This approach was proposed in [11].

Alternatively, if $E \geq 0$ the two equations in (14) are divided by 2^{e_x} and 2^{e_y}, respectively, as described in [2] to obtain

$$
\begin{aligned}
m_{x,i+1} &= m_{x,i} + m \cdot \mu_i \cdot m_{y,i} \cdot 2^{-(s_{m,i}+E)} \\
m_{y,i+1} &= m_{y,i} - \mu_i \cdot m_{x,i} \cdot 2^{-(s_{m,i}-E)}.
\end{aligned} \tag{16}
$$

Then, the usual CORDIC iteration is performed with two's complement fixed point inputs, but starting with an advanced iteration k: $x_k = m_{x,0}$ and $y_k = m_{y,0}$. Usually, only right shifts occur in the CORDIC algorithm. A sequence of left shifts would lead to an exploding number of digits in the internal representation since all significant digits have to be taken into account on the MSB side in order to avoid overflow. Therefore, the iteration k with $s_{m,k} = E$ is taken as the starting iteration. With $s_{m,k} = E$ all actually applied shifts remain right shifts for the n iterations: $i \in \{k, \ldots, k+n-1\}$. Hence, the index k is chosen such that optimum use is made of the inherent fixed point resolution of the y_i variable whose value is driven to zero. Unfortunately, the varying index of the start iteration leads to a variable scale factor as will be discussed later.

The CORDIC angle sequence $\alpha_{m,i}$ has also to be represented in a fixed point format in order to be used in a fixed point implementation. If the algorithm is always started with iteration $i = 0$ the angle sequence can be quantized such that optimum use is made of the range given by the fixed wordlength. If we start with an advanced iteration with variable index k the angle sequence has to be represented such that for the starting angle $\alpha_{m,k}$ no leading zeroes occur. Consequently, the angle sequence has to be stored with an increased resolution (i.e., with an increased number of bits) and to be shifted according to the value of k in order to provide the full resolution for the following n iterations.

So far we discussed the first approach to floating point CORDIC only for the vectoring mode. For the rotation mode similar results can be obtained (c.f., [2]). To summarize, several drawbacks are involved for $E \geq 0$:

(a) The scale factor is dependent on the starting iteration k

$$K_m(n,k) = \prod_{j=k}^{n-1+k} K_{m,j}. \tag{17}$$

Therefore the inverse scale factor as necessary for final scale factor correction has to be calculated in parallel to the iterations or storage has to be provided for a number of precalculated different inverse scale factors.

(b) The accuracy of the stored angle sequence has to be such that sufficient resolution is given for all possible values of k. Hence, the number of digits necessary for representing the angle sequence becomes quite large.

(c) If the algorithm is started with an advanced iteration k with $s_{m,k} = E$, the resulting right shifts given by $s_{m,i} + E$ for the x_i variable lead to increased rounding errors for a given fixed wordlength.

Nevertheless, this approach was proposed for a number of applications (c.f., [11, 14, 15]).

2. For full IEEE 754 floating point accuracy a floating point CORDIC Algorithm was derived in [12, 13]. Depending on the difference of the exponents of the components of the input vector and on the desired angle resolution an optimized shift sequence is selected here from a set of predefined shift sequences. For a detailed discussion the reader is referred to [12, 13].

Following the fixed point CORDIC iterations, the output numbers are re-converted to floating-point format. The whole approach with input-output conversions and internal fixed point datapath is called block floating-point [15, 12, 13].

22.4 SCALE FACTOR CORRECTION

At first glance, the vector scaling introduced by the CORDIC algorithm does not seem to pose a significant problem. However, the correction of a possibly variable scale factor for the output vector generally requires two divisions or at least two multiplications with the corresponding reciprocal values. Using a fixed-point number representation a multiplication can be realized by W shift and add operations where W denotes the wordlength. Now, the CORDIC algorithm itself requires on the order of W iterations in order to generate a fixed-point result with W bits accuracy as discussed in Section 22.3. Therefore, correction of a variable scale factor requires an effort comparable to the whole CORDIC algorithm itself. Fortunately, the restriction of the possible values for μ_i to $(-1, 1)$ ($\mu_i \neq 0$) leads to a constant scale factor $K_m(n)$ for each of the three coordinate systems m and a fixed number of iterations n as given in (6).

A constant scale factor which can be interpreted as a fixed (hence not data dependent) gain can be tolerated in many digital signal processing applications[5]. Hence it should be carefully investigated whether it is necessary to compensate for the scaling at all.

[5]The drawback is that a certain unused headroom is introduced for the output values since the scale factor is not a power of two.

If scale factor correction cannot be avoided, two possibilities are known: performing a constant factor multiplication with $\frac{1}{K_m(n)}$ or extending the CORDIC iteration in a way that the resulting inverse of the scale factor takes a value such that the multiplication can be performed using a few simple shift and add operations.

22.4.1 Constant Factor Multiplication

Since $\frac{1}{K_m(n)}$ can be computed in advance the well known *multiplier recoding* methods [16] can be applied. The effort for a constant factor multiplication is dependent on the number of nonzero digits necessary to represent the constant factor, resulting in a corresponding number of shift and add operations[6]. Hence, the goal is to reduce the number of nonzero digits in $\frac{1}{K_m(n)}$ by introducing a *canonical signed digit* [17] representation with digits $s_j \in \{-1, 0, 1\}$ and recoding the resulting number

$$\frac{1}{K_m(n)} = \sum_{j=0}^{W-1} s_j 2^{-j}$$

On the average, the number of nonzero digits can be reduced to $\frac{W}{3}$ [16], hence the effort for the constant multiplication should be counted as approximately one-third the effort for a general purpose multiplication only.

22.4.2 Extended CORDIC Iterations

By extending the sequence of CORDIC iterations the inverse of the scale factor may eventually become a "simple" number (i.e., a power of two, the sum of powers of two or the sum of terms like (1 ± 2^{-j})), so that the scale factor correction can be implemented using a few simple shift and add operations. The important fact is that when changing the CORDIC sequence, convergence still has to be guaranteed and the shift sequence as well as the set of elementary angles $\alpha_{m,i}$ both change. Four approaches are known for extending the CORDIC algorithm:

22.4.2.1 Repeated iterations

Single iterations may be repeated without destroying the convergence properties of the CORDIC algorithm [18] which is obvious from (12). Hence, a set of repeated iterations can be defined which lead to a simple scale factor. However, using this simple scheme the number of additional iterations is quite large reducing the overall savings due to the simple scale factor correction [19].

22.4.2.2 Normalization steps

In [20] the inverse of the scale factor is described as

$$\frac{1}{K_m(n)} = \prod_{i=0}^{n-1} (1 - m \cdot \gamma_{m,i} \cdot 2^{-s_{m,i}})$$

[6]In parallel multiplier architectures the shifts are hardwired, while in a serial multiplier the multiplication is realized by a number of successive shift and add operations.

with $\gamma_{m,i} \in \{0,1\}$. The single factors in this product can be implemented by introducing normalization steps into the CORDIC sequence

$$\begin{aligned} x_{i+1,norm} &= x_{i+1} - m \cdot x_{i+1} \cdot \gamma_{m,i} \cdot 2^{-s_{m,i}} \\ y_{i+1,norm} &= y_{i+1} - m \cdot y_{i+1} \cdot \gamma_{m,i} \cdot 2^{-s_{m,i}}. \end{aligned} \tag{18}$$

The important fact is that these normalization steps can be implemented with essentially the same hardware as the usual iterations since the same shifts $s_{m,i}$ are required and steered adders/subtractors are necessary anyway. No change in the convergence properties takes place since the normalization steps are pure scaling operations and do not involve any rotation.

22.4.2.3 Double shift iterations

A different way to achieve a simple scale factor is to modify the sequence of elementary rotation angles by introducing *double shift iterations* as proposed in [21]

$$\begin{aligned} x_{i+1} &= x_i - m \cdot \mu_i \cdot y_i \cdot 2^{-s_{m,i}} - m \cdot \mu_i \cdot y_i \cdot \eta_{m,i} \cdot 2^{-s'_{m,i}} \\ y_{i+1} &= y_i + \mu_i \cdot x_i \cdot 2^{-s_{m,i}} + \mu_i \cdot x_i \cdot \eta_{m,i} \cdot 2^{-s'_{m,i}} \end{aligned} \tag{19}$$

where $\eta_{m,i} \in \{-1,0,1\}$ and $s'_{m,i} > s_{m,i}$. For $\eta_{m,i} = 0$ the usual iteration equations are obtained. The set of elementary rotation angles is now given by

$$\alpha_{m,i} = \frac{1}{\sqrt{m}} \cdot \tan^{-1}(\sqrt{m} \cdot (2^{-s_{m,i}} + \eta_{m,i} \cdot 2^{-s'_{m,i}})).$$

The problem of finding shift sequences $s'_{m,i}$ and $s_{m,i}$, which guarantee convergence and lead to a simple scale factor and simultaneously represent minimum extra hardware effort was solved in [22, 15, 23].

22.4.2.4 Compensated CORDIC iteration

A third solution leading to a simple scale factor was proposed in [19] based on [24]

$$\begin{aligned} x_{i+1} &= x_i - m \cdot \mu_i \cdot y_i \cdot 2^{-s_{m,i}} + x_i \cdot \eta_{m,i} \cdot 2^{-s_{m,i}} \\ y_{i+1} &= y_i + \mu_i \cdot x_i \cdot 2^{-s_{m,i}} + y_i \cdot \eta_{m,i} \cdot 2^{-s_{m,i}}. \end{aligned}$$

The advantage is that the complete subexpressions $x_i \cdot 2^{-s_{m,i}}$ and $y_i \cdot 2^{-s_{m,i}}$ occur twice in the iteration equations and hence need to be calculated only once. The set of elementary angles is here described by

$$\alpha_{m,i} = \frac{1}{\sqrt{m}} \cdot \tan^{-1}(\frac{\sqrt{m}}{2^{s_{m,i}} + \eta_{m,i}}).$$

A comparison of the schemes 1.-4. in terms of hardware efficiency is outside the scope of this chapter. It should be mentioned, however, that the impact of the extra operations depends on the given application, the desired accuracy and the given wordlength. Additionally, recursive CORDIC architectures pose different constraints on the implementation of the extended iterations than pipelined unfolded architectures. A comparison of the schemes for a recursive implementation with an output accuracy of 16 bits can be found in [19].

22.5 CORDIC ARCHITECTURES

In this section several CORDIC architectures are presented. We start with the dependence graph for the CORDIC which shows the operational flow in the algorithm. Note that we restrict ourselves to the conventional CORDIC iteration scheme. The dependence graph for extended CORDIC iterations can be easily derived based on the results. The nodes in the dependence graph represent operations (here: steered additions/subtractions and shifts $2^{-s_{m,i}}$) and the arcs represent the flow of intermediate variables. Note that the dependence graph does not include any timing information, it is just a graphical representation of the algorithmic flow. The dependence graph is transformed into a signal flow graph by introducing a suitable projection and a time axis (c.f., [25]). The timed signal flow graph represents a register-transfer level (RTL) architecture. Recursive and pipelined architectures will be derived from the CORDIC dependence graph in the following.

The dependence graph for a merged implementation of rotation mode and vectoring mode is shown in Fig. 22.8. The only difference for the two CORDIC modes is the way the control flags are generated for steering the adders/subtractors. The signs of all three intermediate variables are fed into a control unit which generates the control flags for the steered adders/subtractors given the used coordinate system m and a flag indicating which mode is to be applied.

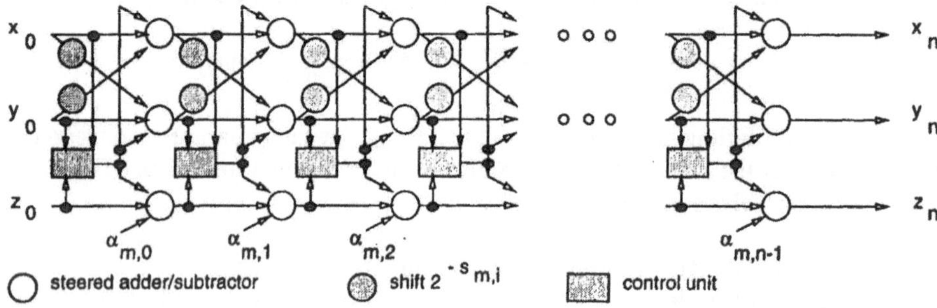

Figure 22.8 CORDIC dependence graph for rotation mode and vectoring mode.

In a one-to-one projection of the dependence graph every node is implemented by a dedicated unit in the resulting signal flow graph. In Fig. 22.9, the signal flow graph for this projection is shown together with the timing for the cascaded additions/subtractions (the fixed shifts are assumed to be hard-wired, hence they do not represent any propagation delay). Besides having a purely combinatorial implementation, pipeline registers can be introduced between successive stages as indicated in Fig. 22.9.

In the following we characterize three different CORDIC architectures by their clock period T_{Clock}, throughput in rotations per second and latency in clock cycles. The delay for calculating the rotation direction μ_i is neglected due to the simplicity of this operation, as well as flipflop setup and hold times. As shown in Fig. 22.9 every addition/subtraction involves a carry propagation from least significant bit (LSB) to most significant bit (MSB) if conventional number systems are used. The length of this ripple path is a function of the wordlength W, e.g., $T_{Add} \sim W$ holds for a carry-ripple addition. The sign of the calculated sum or difference is known

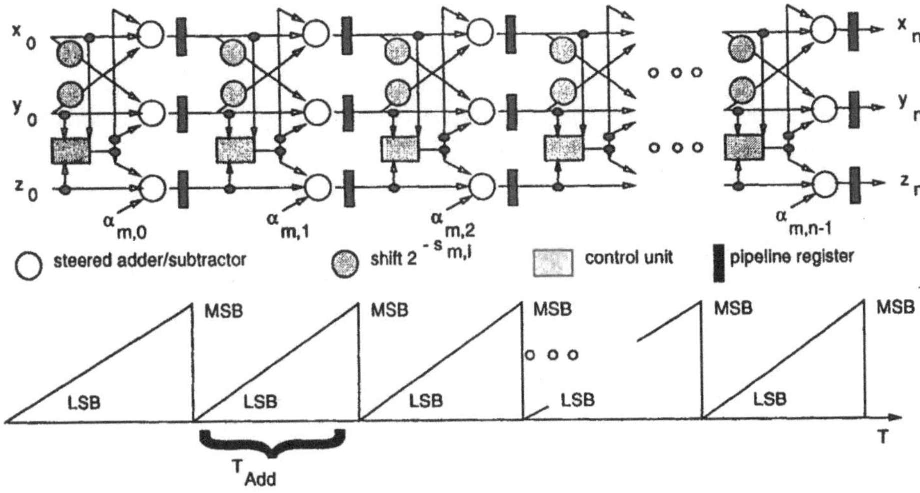

Figure 22.9 Unfolded (pipelined) CORDIC signal flow graph.

only after computation of the MSB. Therefore, the clock period for the unfolded architecture without pipelining is given by $n \cdot T_{Add}$ as shown in Fig. 22.9. The throughput is equal to $\frac{1}{n \cdot T_{add}}$ rotations/s. The pipelined version has a latency of n clock cycles and a clock period $T_{Clock} = T_{add}$. The throughput is $\frac{1}{T_{add}}$ rotations/s.

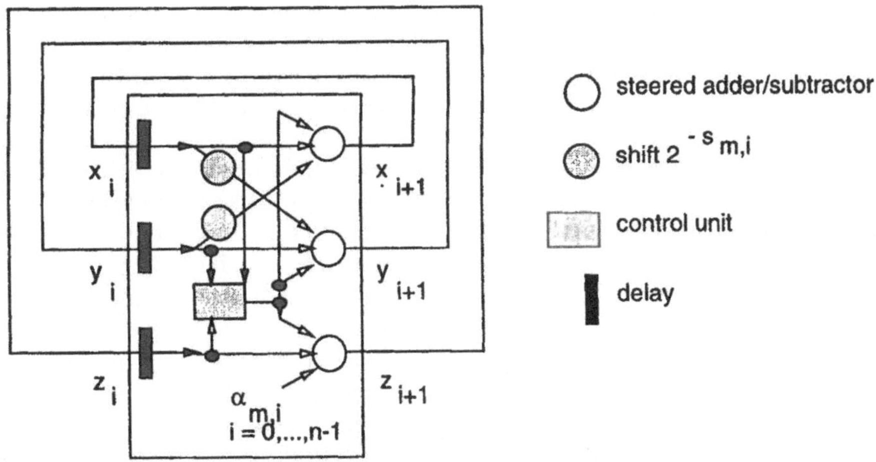

Figure 22.10 Folded (recursive) CORDIC signal flow graph.

It is obvious that the dependence graph in Fig. 22.8 can alternatively be projected in horizontal direction onto a recursive signal flow graph. Here, the successive operations are implemented sequentially on a recursive shared processing element as shown in Fig. 22.10.

Note that due to the necessity to implement a number of different shifts according to the chosen shift sequence, variable shifters (i.e., so called barrel shifters) have to be used in the recursive processing element. The propagation delay associated with the variable shifters is comparable to the adders, hence the clock period is given by $T_{Clock} = T_{Add} + T_{Shift}$. The total latency for n recursive iterations is given by n clock cycles and the throughput is given by $\frac{1}{n \cdot (T_{Add}+T_{Shift})}$ since new input data can be processed only every n clock cycles.

The properties of the three architectures are summarized in Table 22.3.

Table 22.3 Architectural Properties for Three CORDIC Architectures

Architecture	Clock period	Throughput rotations/s	Latency cycles	Area
unfolded	$n \cdot T_{add}$	$\frac{1}{n \cdot T_{add}}$	1	$3n$add, 1reg
unfolded pipelined	T_{add}	$\frac{1}{T_{add}}$	n	$3n$add, $3n$regs
folded recursive	$T_{add} + T_{shift}$	$\frac{1}{n \cdot (T_{add}+T_{Shift})}$	n	3add, $3+n$regs 2shifters

22.5.1 Programmable CORDIC Processing Element

The variety of functions calculated by the CORDIC Algorithm leads to the idea of proposing a programmable CORDIC processing element (PE) for digital signal processing applications, e.g., as an extension to existing arithmetic units in digital signal processors (DSPs) [18, 19]. The folded sequential architecture presented in Fig. 22.10 is the most attractive architecture for a CORDIC PE due to its low complexity. In this section, we give an overview of the structure and features of such a PE.

Standard DSPs contain MAC (Multiply–Accumulate) units which enable single cycle parallel multiply–accumulate operations. Functions can be evaluated using table–lookup methods or using iterative algorithms (e.g., Newton–Raphson iterations [17]) which can efficiently be executed using the standard MAC unit. Since the MAC unit performs single cycle multiplication and addition the multiplication realized with the linear CORDIC mode ($m = 0$) cannot compete due to the sign–directed, sequential nature of the CORDIC algorithm which requires a number of clock cycles for multiplication. In contrast all functions calculated in the circular and hyperbolic modes compare favorably to the respective implementations on DSPs as shown in [19]. Therefore, a CORDIC PE extension for $m = 1$ and $m = -1$ to standard DSPs seems to be the most attractive possibility. It is desirable that the scale factor correction takes place inside the CORDIC unit since otherwise additional multiplications or divisions are necessary in order to correct for the scaling. As was already pointed out in Section 22.4, several methods for scale factor correction are known. As an example, a CORDIC PE using the double shift iteration method is shown in Fig. 22.11. Here, the basic CORDIC iteration structure as shown in Fig. 22.4 was enhanced in order to facilitate the double shift iterations. The double shift iterations (19) with $s'_{m,i} \neq 0$ are implemented in two

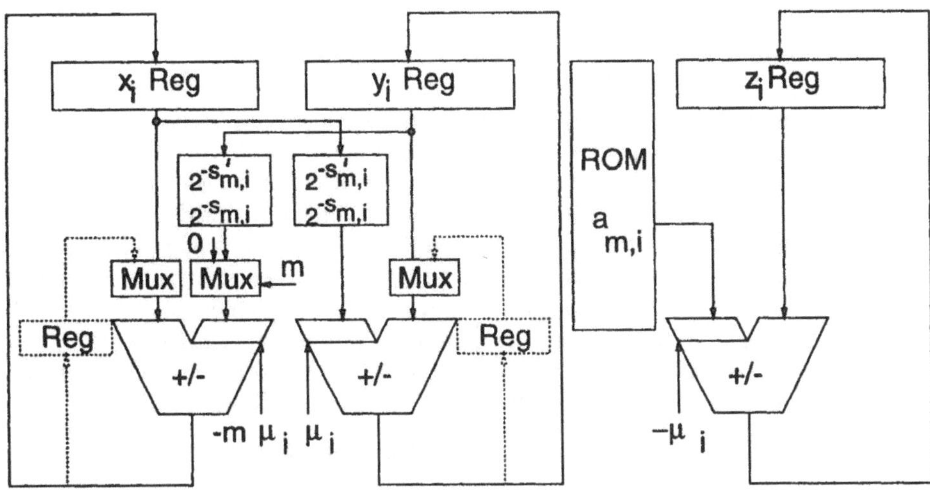

Figure 22.11 Programmable CORDIC processing element.

successive clock cycles[7]. For iterations with $s'_{m,i} = 0$ the shifters are steered to shift by $s_{m,i}$ and the part of the datapath drawn in dashed lines is not used. For double shift iterations with $s'_{m,i} \neq 0$ the result obtained after the first clock cycle is registered in the dashed registers and multiplexed into the datapath during the second clock cycle while the other registers are stalled. During this second clock cycle the shifters are steered to shift by $s'_{m,i}$ and the final result is obtained as given in (19).

The sets of elementary rotation angles to be used can be stored in a ROM as shown in Fig. 22.11 or in register files (arrays of registers), if the number of angles is reasonably small. A control unit steers the operations in the datapath according to the chosen coordinate system m and the chosen shift sequence. In addition, the CORDIC unit may be enhanced by a floating to fixed and fixed to floating conversion if the floating point data format is used. Implementations of programmable recursive CORDIC processing units were reported in [26, 19, 20].

22.5.2 Pipelined CORDIC Architectures

In contrast to a universal CORDIC processing element the dominating motivation for a pipelined architecture is a high throughput constraint. Additionally, it is advantagous if relatively long streams of data have to be processed in the same manner since it takes a number of clock cycles to fill the pipeline and also to flush the pipeline if for example, the control flow changes (a different function is to be calculated).

Although pipeline registers are usually inserted in between the single CORDIC iterations as shown in Fig. 22.9 they can principally be placed everywhere since the unfolded algorithm is purely feedforward. A formalism to introduce pipelining is

[7] Alternatively, an architecture executing one double shift iteration per clock cycle is possible, however, this requires additional hardware. If the number of iterations with $s'_{m,i} \neq 0$ is small, the utilization of the additional hardware is poor.

given by the well known cut-set retiming method [27, 25].
The main advantage of pipelined CORDIC architectures compared to recursive implementations is the possibility to implement hard-wired shifts rather than area and time consuming barrel shifters. However, the shifts can be hard-wired only for a single fixed shift sequence. Nevertheless, a small number of different shifts can be implemented using multiplexors which are still much faster and less area consuming than barrel shifters as necessary for the folded recursive architecture.
A similar consideration holds for the rotation angles. If only a single shift sequence is implemented the angles can be hard-wired into the adders/subtractors. A small number of alternative rotation angles per stage can be implemented using a small combinatorial logic steering the selection of a particular rotation angle. ROMs or register files as necessary for the recursive CORDIC architecture are not necessary. Implementations of pipelined CORDICs are described in [21, 22, 15].

22.5.3 CORDIC Architectures for Vector Rotation

It was already noted that the CORDIC implementation of multiplication and division ($m = 0$) is not competitive. We further restrict consideration here to the circular mode $m = 1$ since much more applications exist than for the hyperbolic mode ($m = -1$).

Traditionally, vector rotations are realized as shown by the dependence graph given in Fig. 22.12. The sine and cosine values are generated by some table-lookup method (or another function evaluation approach) and the multiplications and additions are implemented using the corresponding arithmetic units as shown in Fig. 22.12. Below, we consider high throughput applications with one rotation per clock cycle, and low throughput applications, where several clock cycles are available per rotation.

22.5.3.1 High throughput applications:

For high throughput applications, a one-to-one mapping of the dependence graph in Fig. 22.12 to a possibly pipelined signal flow graph is used. While only requiring a few multiplications and additions, the main drawback of this approach is the necessity to provide the sine and cosine values. A table-lookup may be implemented using ROMs or combinatorial logic. Since one ROM access is necessary per rotation, the throughput is limited by the access time of the ROMs. The throughput cannot be increased beyond that point by pipelining. If even higher throughputs are needed, the ROMs have to be doubled and accessed alternatingly every other clock cycle. If on the other hand combinatorial logic is used for calculation of the sine and cosine values, pipelining is possible in principle. However, the cost for the pipelining can be very high due to the low regularity of the combinatorial logic which typically leads to a very high pipeline register count. As shown in Section 22.5.2, it is easily and efficiently possible to pipeline the CORDIC in rotation mode. The resulting architectures provide very high throughput vector rotations. Additionally, the effort for a CORDIC pipeline grows only linearly with the wordlength W and the number of stages n, hence about quadratically with the wordlength if $n = W + 1$ is used. In contrast the effort to implement the sine and cosine tables as necessary for the classical method grows exponentially with the required angle resolution or wordlength. Hence there is a distinct advantage in terms of throughput and implementation complexity for the CORDIC at least for relatively large

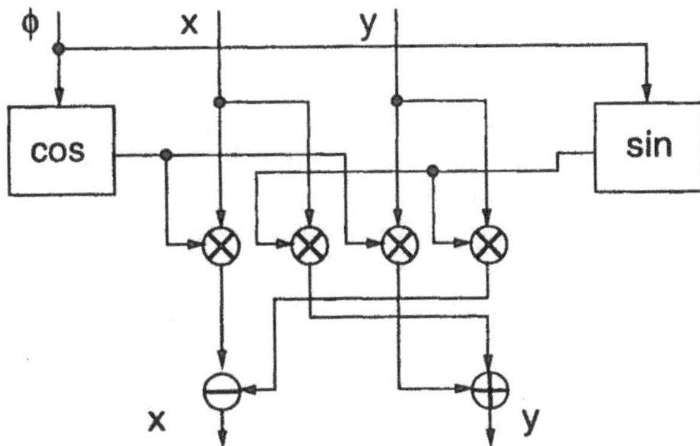

Figure 22.12 Dependence graph for the classical vector rotation.

wordlengths. Due to the n pipelining stages in the CORDIC the classical solution can be advantageous in terms of latency.

22.5.3.2 Low throughput applications:

A single resource shared multiplier and adder is sufficient to implement the classical method in several clock cycles as given for low throughput applications. However, at least one table shared for sine and cosine calculation is still necessary, occupying in the order of $(2^W - 1) * W$ bits of memory for a required wordlength of W bits for the sine and cosine values and the angle Φ[8]. In contrast, a folded sequential CORDIC architecture can be implemented using three adders, two barrel shifters and three registers as shown in Table 22.3. If $n = W + 1$ iterations are used, the storage for the n rotation angles amounts to $(W + 1) * W$ bits only. Therefore, the CORDIC algorithm is highly competetive in terms of area consumption for low throughput applications.

The CORDIC vectoring mode can be used for fast and efficient computation of magnitude and phase of a given input vector. In many cases, only the phase of a given input vector is required, which can of course be implemented using a table–lookup solution. However, the same drawbacks as already mentioned for the sine and cosine tables hold in terms of area consumption and throughput, hence the CORDIC vectoring mode represents an attractive alternative.

Other interesting examples for dedicated CORDIC architectures include sine and cosine calculation [6, 28], Fourier Transform processing [24], Chirp Z–transform implementation [29] and adaptive lattice filtering [30].

22.6 CORDIC ARCHITECTURES USING REDUNDANT NUMBER SYSTEMS

In conventional number systems, every addition or subtraction involves a carry propagation. Independent of the adder architecture the delay of the resulting carry ripple path is always a function of the wordlength. Redundant number systems

[8]Symmetry of the sine and cosine functions can be exploited in order to reduce the table input wordlength by two or three bits.

offer the opportunity to implement carry-free or limited carry propagation addition and subtraction with a small delay independent of the used wordlength. Therefore they are very attractive for VLSI implementation. Redundant number systems have been in use for a long time for example in advanced parallel multiplier architectures (Booth, Carry-Save array and Wallace tree multipliers [16]). However, redundant number systems offer implementation advantages for many applications containing cascaded arithmetic computations. Recent applications for dedicated VLSI architectures employing redundant number systems include finite impulse response filter (FIR) architectures [31], cryptography [32] and the CORDIC algorithm. Since the CORDIC Algorithm consists of a sequence of additions/subtractions the use of redundant number systems seems to be highly attractive. The main obstacle is given by the sign directed nature of the CORDIC algorithm. As will be shown below, the calculation of the sign of a redundant number is quite complicated in absolute contrast to conventional number systems where only the most significant bit has to be inspected. Nevertheless, several approaches were derived recently for the CORDIC algorithm. A brief overview of the basic ideas is given.

22.6.1 Redundant Number Systems

A unified description for redundant number systems was given by Parhami [33] who defined *Generalized Signed Digit* (GSD) number systems. A GSD number system contains the digit set $\{-\alpha, -\alpha+1, \ldots, \beta-1, \beta\}$ with $\alpha, \beta \geq 0$, and $\alpha+\beta+1 > r$ with r being the radix of the number system. Every suitable definition of α and β leads to a different redundant number system. The value X of a W digit integer GSD number is given by:

$$X = \sum_{k=0}^{W-1} r^k x_k \quad , \quad x_k \in \{-\alpha, -\alpha+1, \ldots, \beta-1, \beta\} \qquad (20)$$

An important subclass are number systems with $\alpha + \beta = r$, which are called "minimally redundant", since $\alpha + \beta = r - 1$ corresponds already to a conventional number system. The well known *Carry-Save* (CS) number system is defined by $\alpha = 0, \beta = 2, r = 2$. CS numbers are very attractive for VLSI implementation since the basic building block for arithmetic operations is a simple full adder[9].

In order to represent the CS digits two bits are necessary which are called c_i and s_i. The two vectors C and S given by c_i and s_i can be considered to be two two's complement numbers (or binary numbers, if only unsigned values occur). All rules for two's complement arithmetic (e.g., sign extension) apply to the C and the S number.

An important advantage of CS numbers is the very simple and fast implementation of the addition operation. In Fig. 22.13, a two's complement carry ripple addition and a CS addition is shown. Both architectures consist of W full adders for a wordlength of W digits. The carry ripple adder exhibits a delay corresponding to W full adder carry propagations while the delay of the CS adder is equal to a single full adder propagation delay and independent of the wordlength. The CS adder is

[9]With $\alpha = 1, \beta = 1, r = 2$ the well known *Binary Signed Digit* (BSD) number system results. BSD operations can be implemented using the same basic structures as for CS operations. The full adders used for CS implementation are replaced with "Generalized Full Adders" [32] which are full adders with inverting inputs and outputs.

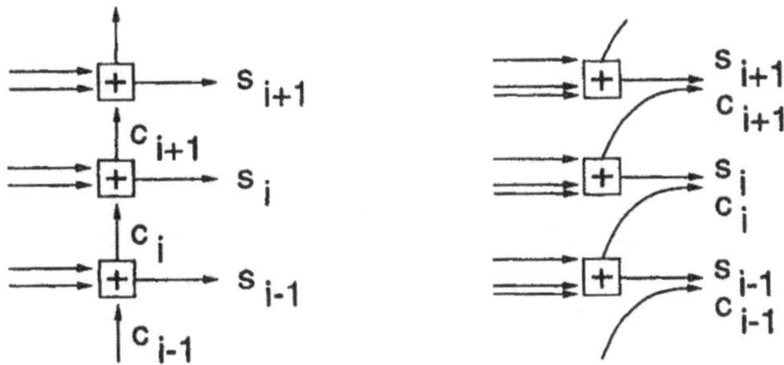

Figure 22.13 Carry ripple addition (left hand side) and 3–2 carry-save addition (right hand side).

called 3–2 adder since 3 input bits are compressed to 2 output bits for every digit position. This adder can be used to add a CS number represented by two input bits for every digit position and a usual two's complement number. Addition of two CS numbers is implemented using a 4–2 adder as shown in Fig. 22.14. CS subtraction is implemented by negation of the two two's complement numbers C and S in the minuend and addition as for two's complement numbers. It is well known that due to the redundant number representation ***pseudo overflows*** [34, 35] can occur. A correction of these pseudo overflows can be implemented using a modified full adder cell in the most significant digit (MSD) position. For a detailed explanation the reader is referred to [34, 35].

Conversion from CS to two's complement numbers is achieved using a so called Vector–Merging adder (VMA) [35]. This is a conventional adder adding the C and the S part of the CS number and generating a two's complement number. Since this conversion is very time consuming compared to the fast CS additions it is highly desirable to concatenate as many CS additions as possible before converting to two's complement representation.

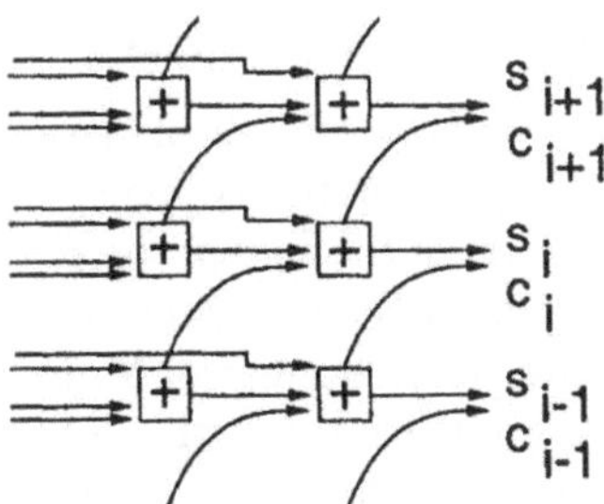

Figure 22.14 Addition of two carry-save numbers (4–2 carry-save addition).

The CORDIC algorithm consists of a sequence of additions/subtractions and sign calculations. In Fig. 22.15, three possibilities for an addition followed by a sign calculation are shown. On the left hand side a ripple adder with sign calculation is depicted. Determination of the sign of a CS number can be solved by converting the

CS number to two's complement representation and taking the sign from the MSD, which is shown in the middle of Fig. 22.15. For this conversion a conventional adder with some kind of carry propagation from least significant digit (LSD) to MSD is necessary. Alternatively, the sign of a CS number can be determined starting with

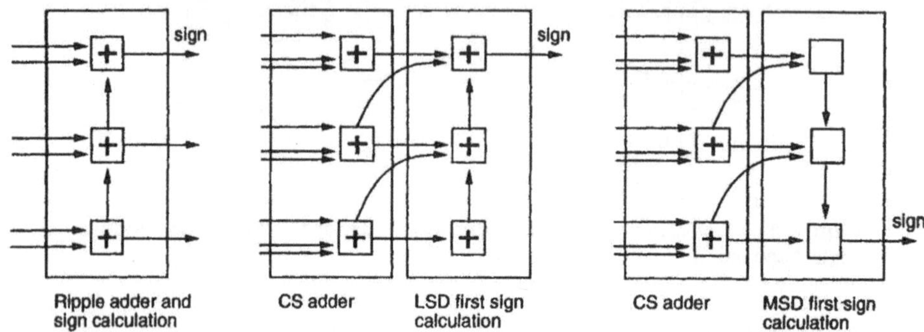

Figure 22.15 Addition and sign calculation using a ripple adder (left hand side), CS adder and LSD first (middle), as well as MSD first (right hand side) sign calculation.

the MSD. If the C and the S numbers have the same sign, this sign represents the sign of the CS number. Otherwise succssive significant digits have to be inspected. The number of digits which have to be inspected until a definite decision on the sign is possible is dependent on the difference in magnitude of the C and the S number. The corresponding circuit structure is shown on the right hand side of Fig. 22.15. Since in the worst case all digits have to be inspected a combinatorial path exists from MSD to LSD.

To summarize, a LSD first and a MSD first solution exists for sign calculation for CS numbers. Both solutions lead to a ripple path whose length is dependent on the wordlength. Addition and sign calculation using the two's complement representation requires less delay and less area. Therefore it seems that redundant arithmetic can not be applied advantageously to the CORDIC iteration.

22.6.2 Redundant CORDIC Architectures

In order to overcome the problem of determining the sign of the intermediate variables in CORDIC for redundant number systems, several authors have proposed techniques based on an estimation of the sign of the redundant intermediate results from a number of MSDs using a particular selection function [7, 36, 28, 5] for the circular coordinate system. If a small number of MSDs is used for sign estimation, the selection function can be implemented with a very small combinatorial delay. It was shown in the last section that in some cases it is possible to even *determine* the sign from a number of MSDs but in other cases not. The proposed algorithms differ in the treatment of the case that the sign cannot be determined.

In [7] a redundant method for the vectoring mode is described. It is proposed not to perform the subsequent microrotation at all if the sign and hence the rotation direction cannot be determined from the given number of MSDs. This is equivalent to expanding the range for the sign sequence μ_i from $\mu_i \in \{-1, 1\}$ to $\mu_i \in \{-1, 0, 1\}$. It is proved in [7] that convergence is still guaranteed. Recall that the total scale factor is given by the product of the scale factors involved with the single iterations. With $\mu_i \in \{-1, 0, 1\}$ the scale factor

$$K_m(n) = \prod_{i=0,\mu_i \neq 0}^{n-1} K_{m,i} \tag{21}$$

is variable. The variable scale factor has to be calculated in parallel to the usual CORDIC iteration. Additionally, a division by the variable scaleing factor has to be implemented following the CORDIC iteration.

A number of recent publications dealing with *constant scale factor redundant* (CFR) CORDIC implementations for the rotation mode ([5] –[35]) describe techniques to avoid a variable scale factor. As proven in [6] the position of the significant digits of z_i in rotation mode changes by one digit per iteration since the magnitude of z_i decreases during the CORDIC iterations. In all following figures this is taken into account by a left shift of one digit for the intermediate variables z_i following each iteration. Then, the MSDs can always be taken for sign estimation.

Using the *double rotation method* proposed in [6, 5] every iteration or (micro–) rotation is represented by two subrotations. A negative, positive or non–rotation is implemented by combining two negative, positive or a positive and a negative subrotation, respectively. The sign estimation is based on the three MSDs of the redundant intermediate variable z_i. The range for μ_i is still given by $\{-1, 0, 1\}$. Since nevertheless exactly two subrotations are performed per iteration the scale factor is constant. An increase of about 50 percent in the arithmetic complexity of the single iterations has to be taken into account due to the double rotations. In the *correcting iteration method* presented in [6, 5] the case $\mu_i = 0$ is not allowed. Even if the sign cannot be determined from the MSDs, a rotation is performed. Correcting iterations have to be introduced every m iterations. A worst case number of $m + 2$ MSDs (c.f., [6]) have to be inspected for sign estimation in the single iterations. In Fig. 22.16 an implementation for the rotation mode with sign estimation is shown with an input wordlength W and a number of M iterations with $M > n$ and n being the usual number of iterations. This method was extended to the vectoring mode in [37]. In [38], a different CFR algorithm is proposed for the rotation mode. Using this "branching CORDIC", two iterations are performed in parallel if the sign cannot be estimated reliably, each assuming one of the possible choices for rotation direction. It is shown in [38] that at most two parallel branches can occur. However, this is equivalent to an almost twofold effort in terms of implementation complexity of the CORDIC rotation engine.

In contrast to the abovementioned approaches a constant scale factor redundant implementation without additional or branching iterations was presented in [39, 40, 41], the *Differential CORDIC Algorithm* (DCORDIC). It was already mentioned (see (8)) that the rotation direction in CORDIC is chosen always such that the remaining rotation angle $|A_{i+1}| = ||A_i| - \alpha_{m,i}|$ eventually decreases. This equation can directly be implemented for the z_i variable in rotation mode and the y_i variable in vectoring mode. The important observation is that using redundant number systems an MSD first implementation of the involved operations addition/subtraction and absolute value calculation is possible without any kind of word–level carry propagation. Hence, successive iterations run concurrently with only a small propagation delay. As an example, the algorithm for the DCORDIC

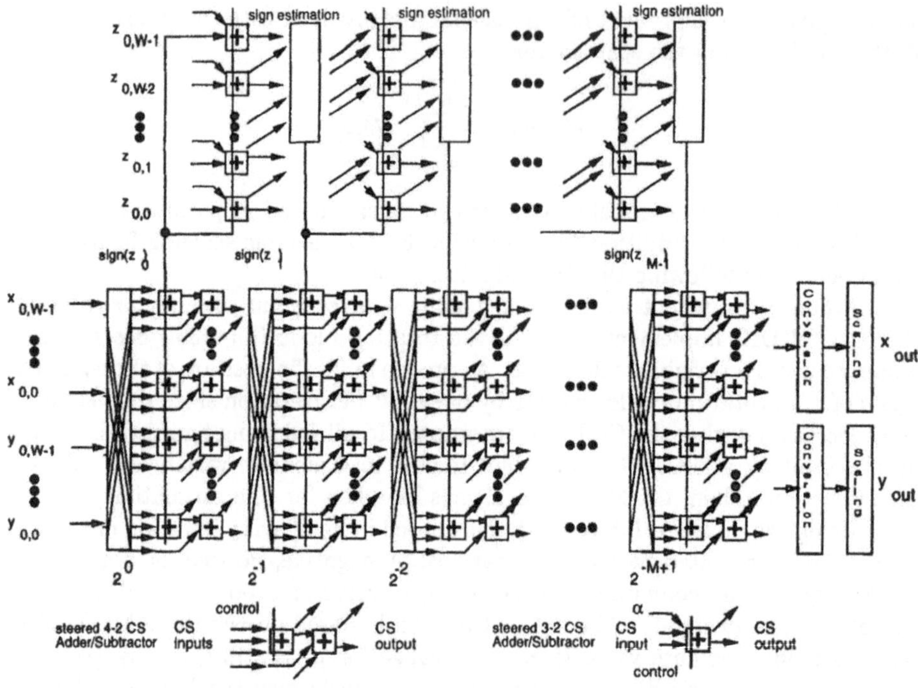

Figure 22.16 Parallel architecture for the CORDIC rotation mode with sign estimation and $M > n$ iterations.

rotation mode is stated below. Since only absolute values are considered in (8) the iteration variable is called $\hat{z}_i$ with $|\hat{z}_i| = |z_i|$.

$$
\begin{aligned}
|\hat{z}_{i+1}| &= ||\hat{z}_i| - \alpha_i| \\
\text{sign}(z_{i+1}) &= \text{sign}(z_i) \cdot \text{sign}(\hat{z}_{i+1}) \\
x_{i+1} &= x_i - \text{sign}(z_i) \cdot y_i \cdot 2^{-i} \\
y_{i+1} &= y_i + \text{sign}(z_i) \cdot x_i \cdot 2^{-i}.
\end{aligned} \tag{22}
$$

As shown in (22) the sign of the iteration variable z_i is achieved by differential decoding the sign of $\hat{z}_i$ given the initial sign sign(z_0). A negative sign of $\hat{z}_i$ corresponds to a sign change for z_i. The iteration equations for x_i and y_i are equal to the usual algorithm. In order to obtain sign($\hat{z}_1$) a single initial ripple propagation from MSD to LSD has to be taken into account for the MSD first absolute value calculation. The successive signs are calculated with a small bit–level propagation delay. The resulting parallel architecture for the DCORDIC rotation mode is shown in Fig. 22.17. Compared to the sign estimation approaches a clear advantage is given by the fact that no additional iterations are required for the DCORDIC.

22.6.3 Recent Developments

Below, some recent research results for the CORDIC algorithm are briefly mentioned which cannot be treated in detail due to lack of space.

In [42] it is proposed to reduce the number of CORDIC iterations by replacing

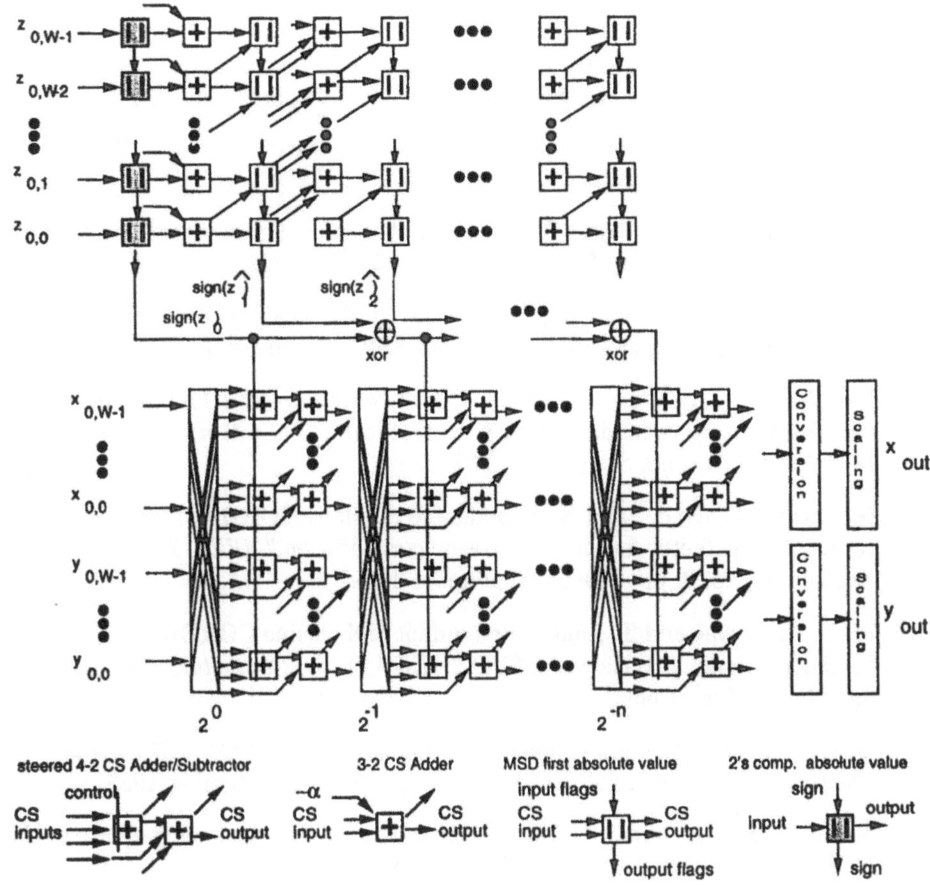

Figure 22.17 Parallel architecture for the DCORDIC rotation mode with n iterations.

the second half of the iterations with a final multiplication. Further low latency CORDIC algorithms were derived in [43, 44] for parallel implementation of the rotation mode and for word-serial recursive implementations of both rotation mode and vectoring mode in [45]. The computation of $\sin^{-1}$ and $\cos^{-1}$ using CORDIC was proposed in [46, 47].

Recently, a family of generalized multi-dimensional CORDIC algorithms, so called ***Householder*** CORDIC algorithms, was derived in [48, 49]. Here, a modified iteration leads to scaling factors which are rational functions instead of square roots of rational functions as in conventional CORDIC. This attractive feature can be exploited for the derivation of new architectures [48, 49].

REFERENCES

[1] J. E. Volder, "The CORDIC trigonometric computing technique," *IRE Trans. Electronic Computers*, vol. EC–8, no. 3, pp. 330–34, September 1959.

[2] J. S. Walther, "A unified algorithm for elementary functions," in *AFIPS Spring Joint Computer Conference*, vol. 38, pp. 379–85, 1971.

[3] Y. H. Hu, "CORDIC–based VLSI architectures for digital signal processing," *IEEE Signal Processing Magazine*, pp. 16–35, July 1992.

[4] Y. H. Hu, "The quantization effects of the CORDIC algorithm," *IEEE Transactions on Signal Processing*, vol. 40, pp. 834–844, July 1992.

[5] N. Takagi, T. Asada, and S. Yajima, "A hardware algorithm for computing sine and cosine using redundant binary representation," *Systems and Computers in Japan*, vol. 18, no. 8, pp. 1-9, 1987.

[6] N. Takagi, T. Asada, and S. Yajima, "Redundant CORDIC methods with a constant scale factor for sine and cosine computation," *IEEE Trans. Computers*, vol. 40, no. 9, pp. 989–95, September 1991.

[7] M. D. Ercegovac and T. Lang, "Redundant and on-line CORDIC: Application to matrix triangularisation and SVD," *IEEE Trans. Computers*, vol. 38, no. 6, pp. 725–40, June 1990.

[8] D. H. Daggett, "Decimal–binary conversions in CORDIC," *IEEE Trans. on Electronic Computers*, vol. EC–8, no. 3, pp. 335–39, September 1959.

[9] X. Hu, R. Harber, and S. C. Bass, "Expanding the range of convergence of the CORDIC algorithm," vol. 40, pp. 13–20, 1991.

[10] X. Hu and S. C. Bass, "A neglected error source in the CORDIC algorithm," in *Proceedings IEEE ISCAS'93*, pp. 766–769, 1993.

[11] J. R. Cavallaro and F. T. Luk, "Floating point CORDIC for matrix computations," in *IEEE International Conference on Computer Design*, pp. 40–42, 1988.

[12] G. J. Hekstra and E. F. Deprettere, "Floating–point CORDIC," *Technical report: ET/NT 93.15, Delft University*, 1992.

[13] G. J. Hekstra and E. F. Deprettere, "Floating–point CORDIC," in *Proc. 11th Symp. Computer Arithmetic*, (Windsor, Ontario), pp. 130–137, June 1993.

[14] J. R. Cavallaro and F. T. Luk, "CORDIC arithmetic for a SVD processor," *Journal of Parallel and Distributed Computing*, vol. 5, pp. 271–90, 1988.

[15] A. A. de Lange, A. J. van der Hoeven, E. F. Deprettere, and J. Bu, "An optimal floating-point pipeline cmos CORDIC processor," in *IEEE ISCAS'88*, pp. 2043–47, 1988.

[16] K. Hwang, *Computer arithmetic: principles, architectures, and design.* New York: John Wiley & Sons, 1979.

[17] N. R. Scott, *Computer number systems and arithmetic.* Englewood Cliffs: Prentice Hall, 1988.

[18] H. M. Ahmed, *Signal processing algorithms and architectures.* 1981. Ph.D. Thesis, Dept. Elec. Eng., Stanford (CA).

[19] R. Mehling and R. Meyer, "CORDIC–AU, a suitable supplementary unit to a general–purpose signal processor," *AEÜ*, vol. 43, no. 6, pp. 394–97, 1989.

[20] G. L. Haviland and A. A. Tuszynski, "A CORDIC arithmetic processor chip," *IEEE Transactions on Computers*, vol. C–29, no. 2, pp. 68–79, Feb. 1980.

[21] E. F. Deprettere, P. Dewilde, and R. Udo, "Pipelined CORDIC architectures for fast VLSI filtering and array processing," in *Proceedings IEEE ICASSP*, pp. 41 A6.1 - 41 A6.4, March 1984.

[22] J. Bu, E. F. Deprettere, and F. du Lange, "On the optimization of pipelined silicon CORDIC Algorithm," in *Proceedings EUSIPCO 88*, pp. 1227–30, 1988.

[23] G. Schmidt, D. Timmermann, J. F. Böhme, and H. Hahn, "Parameter optimization of the CORDIC algorithm and implementation in a CMOS chip," in *Proc. EUSIPCO'86*, pp. 1219–22, 1986.

[24] A. M. Despain, "Fourier transform computers using CORDIC iterations," *IEEE Transactions on Computers*, vol. C–23, pp. 993–1001, Oct. 1974.

[25] S. Y. Kung, *VLSI Array Processors.* Englewood Cliffs: Prentice–Hall, 1988.

[26] D. Timmermann, H. Hahn, B. J. Hosticka, and G. Schmidt, "A programmable CORDIC chip for digital signal processing applications," *IEEE Transactions on Solid–State Circuits*, vol. 26, no. 9, pp. 1317–1321, 1991.

[27] C. E. Leiserson, F. Rose, and J. Saxe, "Optimizing synchronous circuitry for retiming," in *Proc. of the 3rd Caltech Conf. on VLSI*, (Pasadena), pp. 87–116, March 1983.

[28] N. Takagi, T. Asada, and S. Yajima, "A hardware algorithm for computing sine and cosine using redundant binary representation," *Trans. IEICE Japan (in Japanese)*, vol. J69–D, no. 6, pp. 841–47, 1986.

[29] Y. H. Hu and S. Naganathan, "Efficient implementation of the chirp Z–transform using a CORDIC processor," *IEEE Transactions on Signal Processing*, vol. 38, pp. 352–354, Feb. 1990.

[30] Y. H. Hu and S. Liao, "CALF: A CORDIC adaptive lattice filter," *IEEE Transactions on Signal Processing*, vol. 40, pp. 990–993, April 1992.

[31] T. Noll et al, "A pipelined 330 MHz multiplier," *IEEE Journal Solid State Circuits*, vol. SC–21, pp. 411–16, 1986.

[32] A. Vandemeulebroecke, E. Vanzieleghem, T. Denayer, and P. G. A. Jespers, "A new carry–free division algorithm and its application to a single–chip 1024–b RSA processor," *IEEE Journal Solid State Circuits*, vol. 25, no. 3, pp. 748–65, 1990.

[33] B. Parhami, "Generalized signed-digit number systems: A unifying framework for redundant number representations," *IEEE Trans. on Computers*, vol. 39, no. 1, pp. 89–98, 1990.

[34] T. Noll, "Carry-save arithmetic for high-speed digital signal processing," in *IEEE ISCAS'90*, vol. 2, pp. 982–86, 1990.

[35] T. Noll, "Carry–save architectures for high–speed digital signal processing," *Journal of VLSI Signal Processing*, vol. 3, no. 1/2, pp. 121–140, June 1991.

[36] M. D. Ercegovac and T. Lang, "Implementation of fast angle calculation and rotation using on-line CORDIC," in *IEEE ISCAS'88*, pp. 2703–06, 1988.

[37] J. Lee and T. Lang, "Constant–factor redundant CORDIC for angle calculation and rotation," *IEEE Trans. Computers*, vol. 41, pp. 1016–1035, August 1992.

[38] J. Duprat and J.-M. Muller, "The CORDIC algorithm: new results for fast VLSI implementation," *IEEE Transactions on Computers*, vol. 42, no. 2, pp. 168–178, 1993.

[39] H. Dawid and H. Meyr, "The differential CORDIC algorithm: constant scale factor redundant implementation without correcting iterations," *IEEE Transactions on Computers*, vol. 45, no. 3, pp. 307–318, March 1996.

[40] H. Dawid and H. Meyr, "Very high speed CORDIC implementation: algorithm transformation and novel carry–save architecture," in *Proceedings of the European Signal Processing Conference EUSIPCO '92*, (Brussels), pp. 358–372, Elsevier Science Publications, August 1992.

[41] H. Dawid and H. Meyr, "High speed bit–level pipelined architectures for redundant CORDIC implementation," in *Proceedings of the Int. Conf. on Application Specific Array Processors*, (Oakland), pp. 358–372, IEEE Computer Society Press, August 1992.

[42] D. Timmermann, H. Hahn, and B. J. Hosticka, "Modified CORDIC algorithm with reduced iterations," *Electronics Letters*, vol. 25, no. 15, pp. 950–951, 1989.

[43] D. Timmermann and I. Sundsbo, "Area and latency efficient CORDIC architectures," in *Proc. ISCAS'92*, pp. 1093–1096, 1992.

[44] D. Timmermann, H. Hahn, and B. Hosticka, "Low latency time CORDIC algorithms," *IEEE Trans. Computers*, vol. 41, no. 8, pp. 1010–1015, 1992.

[45] J. Villalba and T. Lang, "Low latency word serial CORDIC," in *Proceedings IEEE Conf. Application specific Systems, Architectures and Processors ASAP*, (Zurich), pp. 124–131, July 1997.

[46] C. Mazenc, X. Merrheim, and J. M. Muller, "Computing functions $\cos^{-1}$ and $\sin^{-1}$ using CORDIC," *IEEE Trans. Computers*, vol. 42, no. 1, pp. 118–122, 1993.

[47] T. Lang and E. Antelo, "CORDIC–based computation of ArcCos and ArcSin," in *Proceedings IEEE Conf. Application specific Systems, Architectures and Processors ASAP*, (Zurich), pp. 132–143, July 1997.

[48] S.F.Hsiao and J.M.Delosme, "Householder CORDIC algorithms," *IEEE Transactions on Computers*, vol. C–44, no. 8, pp. 990–1001, Aug. 1995.

[49] S.F.Hsiao and J.M.Delosme, "Parallel singular value decomposition of complex matrices usijng multi–dimensional CORDIC algorithms," *IEEE Transactions on Signal Processing*, vol. 44, no. 3, pp. 685–697, March 1996.

[50] J. Lee and T. Lang, "On-line CORDIC for generalized singular value decomposition," in *SPIE Vol. 1058 High Speed Computing II*, pp. 235–47, 1989.

[51] S. Note, J. van Meerbergen, F. Catthoor, and H. de Man, "Automated synthesis of a high speed CORDIC algorithm with the Cathedral-III compilation system," in *Proceedings IEEE ISCAS'88*, pp. 581–84, 1988.

[52] R. Künemund, H. Söldner, S. Wohlleben, and T. Noll, "CORDIC Processor with carry-save architecture," in *Proc. ESSCIRC'90*, pp. 193–96, 1990.

[53] H. X. Lin and H. J. Sips, "On-line CORDIC algorithms," *IEEE Trans. Computers*, no. 8, pp. 1038–52, August 1990.

[54] H. Yoshimura, T. Nakanishi, and H. Yamauchi, "A 50 MHz CMOS geometrical mapping processor," *IEEE Transactions on Circuits and Systems*, vol. 36, no. 10, pp. 1360–63, 1989.

Chapter 23

Advanced Systolic Design

Dominique Lavenier, Patrice Quinton, and Sanjay Rajopadhye
IRISA, Campus de Beaulieu, France
{lavenier, quinton, rajopadhye} @irisa.fr

23.1 INTRODUCTION

The term systolic arrays was coined by Kung and Leiserson in 1978 to describe application specific VLSI architectures that were regular, locally connected and massively parallel with simple processing elements (PEs). The idea of using such regular circuits was even present in von Neuman's cellular automata in the fifties, Hennie's iterative logic arrays in the sixties, and also in specialized arithmetic circuits (Lyon's bit-serial multiplier [1] is clearly a linear systolic array). However, the emergence of VLSI technology in the late seventies and early eighties made the time ripe for introducing such architectures in order to highlight the characteristics appropriate to the technology.

Systolic arrays immediately caught on, since they involved a fascinating interplay between algorithm and architecture design. When researchers started investigating automatic synthesis, a third stream joined this confluence, namely the analysis, manipulation and transformation of programs. Some of the early designs are classics. The Guibas et al. array for optimal string parenthesization [2] is one of our all time favorites.

The early eighties was a period of intense activity in this area. A large number of "paper designs" were proposed for a wide variety of algorithms from linear algebra, graph theory, searching, sorting, etc. There was also much work on automatic synthesis methods, using *dependency analysis*, *space-time* transformations of inner loops, and also other formalisms such as recurrence equations.

Then, the technological evolutions in the late eighties seemed to invalidate the assumptions of the systolic model, namely that (i) locality, regularity and simplicity were primordial for VLSI and (ii) an elementary computation could be performed in the same time that it took to perform an elementary communication. The first one meant that with the improved CAD tools and increasing levels of integration, circuit designers were not obliged to *always* follow that dictates of the systolic model (which was itself applicable to only a part — albeit, a computationally significant one — of the complete application). The second one meant that the systolic space-time mapping methods were not directly applicable for general purpose parallel machines where the communication latency was an order of magnitude higher than computation speed.

A number of recent developments now lead us to believe that it is time to revive the field. First of all, technology has evolved. With the growing levels of integration, it is feasible to actually implement a number of the old "paper designs", particularly if the elementary operations are not floating point, but come from a simpler algebraic structure (such as semi-rings with additions and comparisons, etc.) Second, the circuits of today are much more complex, and we need to again question the arguments against the regularity, locality and simplicity of systolic arrays. Can an irregular circuits achieve *better* performance? Will they do so *reliably*? What will be the *design time*? A third and very important reason is the emergence of FPGAs, programmable logic and reconfigurable computing. The basic technology underlying FPGAs is regularity, locality and simplicity. It is therefore not surprising that some of the impressive successes in this domain are systolic designs. Fourth, there is the growing understanding that the techniques of systolic synthesis have a close bearing on automatic loop parallelization for general purpose parallel machines. These techniques can thus be seen as very high level synthesis methods, applicable for software *as well as* hardware, and thus form a foundation for codesign, for a certain class of problems. Coupled with the fact that recent advances in integration permit one or more programmable processor cores to be implemented "on-chip", this makes such techniques well suited for *provably correct* codesign. Finally, we believe that the increasing complexity of VLSI system design naturally leads towards formal design methods involving correctness-preserving transformation of high-level specifications.

This chapter presents two aspects of advanced systolic design, namely advanced methods for systolic array design (Secs 23.2–23.2.5), but also design techniques and case studies of advanced, state of the art systolic arrays (Sec 23.3).

In the first part we use a formalism called systems of recurrence equations (SREs) to describe both the initial specification as well as the final array. The entire process of systolic design is viewed as the application of a series of transformations to the initial specification, until one obtains a description of the final array. We desire that all the details, including control, I/O, interface etc. be specified in the same formalism, and that a simple translation step should yield a description that is suitable for a conventional CAD tool (say VHDL). For each transformation we are concerned with two aspects: (i) the manipulation of the SRE to obtain a provably equivalent SRE (la 'correctness-preserving' program transformations), and (ii) the choice of the transformation, usually with a view to optimizing certain cost criteria. In Section 23.2 we present the basic notations of recurrence equations and their domains. In Section 23.2.2 we describe the foundations of the first aspect, namely the formal manipulation of recurrence equations. In Sections 23.2.3 to 23.2.5, we explain how these various transformations can be used to transform a specification into an abstract architecture: Section 23.2.3 details scheduling techniques, Section 23.2.4 deals with allocation of computations on processors, and finally, Section 23.2.5 is concerned with localization and serialization transformations.

In the second part, we illustrate four different ways to implement systolic arrays: general-purpose programmable architectures (Section 23.3.1), application oriented programmable architectures (Section 23.3.2), reconfigurable architectures

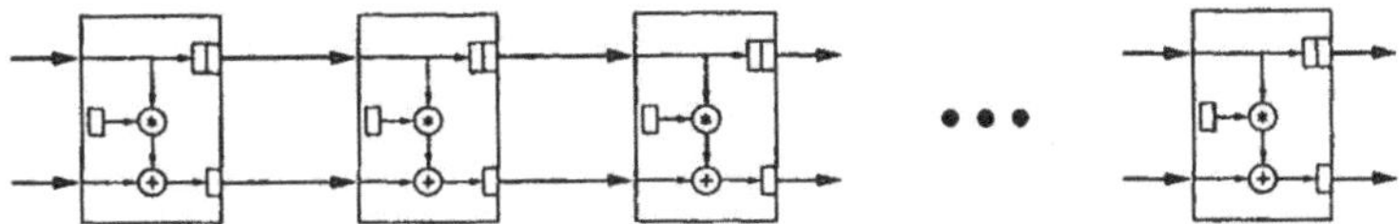

Figure 23.1 The systolic array for Eqns. (1–3).

(Section 23.3.3), and special-purpose architectures (Section 23.3.4), and we illustrate each one with one case study. Although some of the designs presented are not new, a close study of their design is instructive. As the technology of integrated circuits is still in full motion, we believe that lessons have to be learned from these examples, in order to take the best advantage of future architectural opportunities.

23.2 SYSTOLIC DESIGN BY RECURRENCE TRANSFORMATIONS

Systems of Recurrence Equations (SREs) play an important part in the design process. They are useful as (i) behavioral descriptions of the final arrays, and also (ii) high level specifications of the initial algorithm. For example, consider the system of equations given below.

$$X(t,p) = \begin{cases} \{p=0, -n+1 \leq t < 0\} & : \; 0 \\ \{p=0, t \geq 0\} & : \; x_t \\ \{0 < p < n, 2p \leq t\} & : \; X(t-2, p-1) \end{cases} \quad (1)$$

$$W(t,p) = \begin{cases} \{t=0\} & : \; w_p \\ \{0 < t\} & : \; W(t-1,p) \end{cases} \quad (2)$$

$$Y(t,p) = \begin{cases} p=0, t \geq 0 & : \; W(t,p) * X(t,p) \\ n > p > 0, t \geq p & : \; Y(t-1,p-1) + W(t,p) * X(t,p) \end{cases} \quad (3)$$

If we interpret the index t as the time and the index p as the processor, this system can be viewed as a specification of the values that will appear in the 'registers' `X`, `W` and `Y` of a linear array with n-processors as shown in Fig. 23.1. From (1) we see that (other than in the boundary processor $p = 0$), the value in the `X` register of processor p at time instant t is the same as that in the `X` register of processor $p-1$ at time $t-2$ (this corresponds to an interconnection between adjacent processors with a delay of 2 cycles.) Similarly, the `W` values stay in the processors after initialization, and the `Y` values propagate between adjacent processors with a 1-cycle delay (after being incremented by the `W*X` product in each processor). Note that the equations do not clearly specify how the `W` registers are 'loaded' at $t = 0$, nor do they describe the initial values in the `X` registers of the processors other than the first one. These details and the control signals to ensure correct loading and propagation of initial values can be included, but have been omitted in the interests of simplicity. It is well known that this array computes the convolution of the x stream with the coefficients w.

The same SRE formalism, augmented with *reduction operations*, can serve as a very high level, mathematical description of the algorithm for which a systolic array is to be designed. For example, the n-point convolution of a sequence of samples $x_0, x_1, \ldots$ with the weights $w_0 \ldots w_{n-1}$ produces the sequence $y_0, y_1, \ldots$ given by the following equation (assuming that $x_k = 0$, for $-n < k < 0$).

$$y_i = \sum_j w_j x_{i-j} \tag{4}$$

23.2.1 Recurrence Equations: Definitions and Notation

We present the fundamental definitions of recurrence equations, and the domains over which they are defined. In what follows, **Z** denotes the set of integers, and **N** the set of natural numbers.

Definition 23.2.1 *A* **Recurrence Equation** *defining a function (variable) X at all points, z, in a domain, D, is an equation of the form*

$$X(z) = D^X \quad : \quad g(\ldots X(f(z))\ldots) \tag{5}$$

where

- *z is an n-dimensional* **index variable.**
- *X is a* **data variable,** *denoting a function of n integer arguments; it is said to be an n-dimensional variable.*
- *$f(z)$ is a* **dependency function** *(also called an* **index** *or* **access** *function), $f: \mathbf{Z}^n \to \mathbf{Z}^n$;*
- *the "..." indicate that g may have other arguments, each with the same syntax;*
- *g is a strict, single-valued function; it is often written implicitly as an* expression *involving operands of the form $X(f(z))$ combined with basic operators and parentheses.*
- *D^X is a set of points in $\mathbf{Z}^n$ and is called the* **domain of the equation. Often, the domains are parametrized with one or more (say, l) size parameters. In this case, we represent the parameter as a vector, $p \in \mathbf{Z}^l$, and use p as an additional superscript on D.**

A variable may be defined by more than one equation. In this case, we use the syntax shown below:

$$X(z) = \begin{cases} & \vdots \\ D_i & : \quad g_i(\ldots X(f(z))\ldots) \\ & \vdots \end{cases} \tag{6}$$

Each line is called a **case**, *and the domain of X is the union of the domains of all the cases, $D^X = \bigcup_i D_i$ (actually, the convex hull of the union, for analysis purposes). The D_i's must be disjoint. Indeed, the domain appearing in such an equation must be subscripted (annotated) with (i) the name of the variable being* *defined, (ii) the branch of the case and also (iii) the parameters of the domain. When there is no ambiguity, we may drop one or more of these subscripts.*

Finally, the expression defining g, may contain **reduction operators, i.e., associative and commutative binary operators applied to a collection of values such as addition ($\sum$), multiplication (Π), minimum (min), maximum (max), Boolean or ($\bigvee$), Boolean and ($\bigwedge$), etc. These operators are subscripted with one or more auxiliary index variables, z', whose scope is local to the reduction. A domain for the auxiliary indices may also be given.**

Definition 23.2.2 *A recurrence equation (5) as defined above, is called an* **Affine Recurrence Equation (ARE) if every dependence function is of the form, $f(z) = Az + Bp + a$, where A (respectively B) is a constant $n \times n$ (respectively, $n \times l$) matrix and a is a constant n-vector. It is said to be a Uniform Recurrence Equation (URE) if it is of the form, $f(z) = z + a$, where a is a constant n-dimensional vector, called the dependence vector. UREs are a proper subset of AREs, where A is the identity matrix and $B = 0$.**

Definition 23.2.3 *A system of recurrence equations (SRE) is a set of m such equations, defining the data variables $X_1 \dots X_m$. Each variable, X_i is of dimension n_i, and since the equations may now be mutually recursive, the dependence functions f must now have the appropriate (not necessarily square) matrices. We are interested in systems of AREs (SAREs) where all dependence functions are affine, and also a proper subset, systems of UREs (SUREs). Note that in a SURE, the domains of all variables must have the same number of dimensions, since A has to be the identity matrix.*

Domains

An important part of the SRE formalism is the notion of domain, i.e., the set of indices where a particular computation is defined. The domain of the *variables* of an SRE are usually specified explicitly. The domains that we use are *polyhedra* (i.e., the set of integral indices that satisfy a finite number of linear (in)-equality constraints), or a union of finitely many polyhedra. The domains may have one or more *size parameters*. Often, other families of domains such as lattices, sparse polyhedra (the intersections of lattices and polyhedra) and linearly bounded lattices (the image of polyhedra by affine functions) are used, but will not be discussed here for the sake of simplicity. The domains are often *parametrized* by a number of size parameters. An important goal of the research in this field is to elaborate parameter-independent analysis techniques and architectures.

A subtle point to note is that in an SRE each (sub) expression itself can be assigned a domain, and this can be deduced using simple rules. For example, let us do this for each sub-expression on the right-hand side (rhs) of 4. First, note that the expression within the summation has a 2-dimensional domain, since it involves two indices i and j. The j index is local, and not visible outside the summation, the domain of the expression $\sum_j w_j x_{i-j}$ is thus the *projection* along j of the domain of the expression $w_j x_{i-j}$. We now reason as follows:

- The domain of w is $\{i|0 \leq i < n\}$.
- The domain of the expression w_j is the set of (2-dimensional) index points $[i, j]$ for which w_j is defined, i.e., points such that w_j belongs to the domain of w. This imposes the constraint that j must be between 0 and n (and no constraint on i), and gives us the 'infinite strip', $\{i, j|0 \leq j < n\}$.
- The domain of x is $\{i| - n < i\}$ (remember it is padded by $n - 1$ zeros).
- The domain of x_{i-j} is the set of points $[i, j]$ such that $i - j$ belongs to the domain of x, yielding $\{i, j| - n < i - j\}$.
- The expression $w_j x_{i-j}$ is the product of the two sub-expressions, w_j, and x_{i-j}. It is thus defined only where both operands are present, and hence its domain is the *intersection* of the respective domains of w_j and x_{i-j}, namely the domain $\{i, j|0 \leq j < n, j - n < i\}$.
- Finally, the domain of the entire rhs is the *projection* of the domain of $w_j x_{i-j}$ on the i axis, and is given as $\{i|-n < i\}$. Observe that this includes the domain of the variable y, namely the half line $\{i|0 \leq i\}$, and hence the equation is 'well formed', i.e., there exists a valid definition for every point in the domain of y.

Finally, we precisely define which SREs correspond to a systolic array. Actually, it is not easy to give a formal definition without being either too restrictive or overly general. The following definition corresponds to a restrictive version. Some of the restrictions below can be relaxed (yielding, what may not be an efficient or practically viable systolic array).

Definition 23.2.4 *A* ***systolic array*** *is defined as a SRE where*

- *All variables, except possibly inputs and outputs, have the same number of dimensions.*
- *All dependency functions (except possibly those involving input and output variables) are uniform.*
- *Each instance of each input variable is used exactly once (typically, but not exclusively at boundary processors).*
- *One index variable, can be interpreted as the* time, *i.e., all dependency vectors are strictly negative in this dimension.*
- *The other indices are interpreted as processor coordinates.*
- *The spatial components of all dependency vectors are interpreted as interconnections between processors.*

23.2.2 Synthesis by Formal Manipulation of SREs: A Walk-through

We now describe the various transformations that can be applied to an SRE. We will focus primarily on describing the basic formal manipulations and on ensuring that the resulting SRE is computationally equivalent to the original one. Another subject, that we will not consider here, is the choice of the particular transformation to apply. As a running example, we will use the convolution SRE of (4), and we will illustrate how the SRE of (1)–(3) are systematically derived from it.

Serialization of Reductions

In this transformation we replace the reduction operations by a sequence of binary operations. This is needed because the hardware elements that one uses to implement the algorithm have bounded 'fan-in'. The choice of the transformation involves the 'direction of accumulation': either the *increasing* or *decreasing* order of j, for the convolution example (recall that j is the local index introduced by the reduction), and also the name of the temporary variable in which to accumulate the partial results. Suppose we accumulate in the increasing j direction, and use the name Y for our new variable, we will obtain the following SRE:

$$y_i = Y(i, n-1) \quad (7)$$

$$Y(i,j) = \begin{cases} \{j = 0, 0 \le i\} & : \; w_j * x_{i-j} \\ \{0 < j < n, 0 \le i\} & : \; Y(i, j-1) + w_j * x_{i-j} \end{cases} \quad (8)$$

To understand why this SRE is equivalent to the reduction, and how it can be derived automatically, consider the expression within the reduction of (4), namely $w_j x_{i-j}$. At each point $[i, j]$ in its domain, we associate, in addition to the product of the w and x terms, a *partial sum*, which is the accumulation of the terms 'up to j'. The domain of the auxiliary variable is thus the same as that of $w_j x_{i-j}$. Moreover, we see that the accumulation corresponds to a recurrence defining $Y(i, j)$ with the (uniform) dependency $Y(i, j-1)$, and this follows from our choice of the accumulation direction. The initialization of the Y variable (the clause in its recurrence where there is no dependency on Y) corresponds to the (sub) domain where $j = 0$, and the result is available in the $j = n$ subdomain (this is reflected in the $Y(i, n-1)$ dependency in (7). These 'boundaries' of the domain of Y can be deduced by translating it by ± 1 in the j direction and determining the difference.

Alignment

In the SRE (7)–(8), variables x and w, defined over 1-dimensional domains, are used in (8) to define Y, a 2-dimensional variable. We therefore *align* x and w with the 2-dimensional domain of Y. In particular, let us choose to align w with the $i = 0$ line, and x with the $j = 0$ line. Since x and w are inputs of the SRE, we will introduce two auxiliary variables, say W and X, yielding the following SRE.

$$y_i = Y(i, n-1) \quad (9)$$

$$Y(i,j) = \begin{cases} \{j = 0, 0 \le i\} & : \; W(0,j) * X(i-j, 0) \\ \{0 < j < n, 0 \le i\} & : \; Y(i, j-1) + W(0,j) * X(i-j, 0) \end{cases} \quad (10)$$

$$W(i,j) = \{ \{i = 0, 0 \le j < n\} \; : \; w_j \quad (11)$$

$$X(i,j) = \{ \{j = 0, 0 \le i\} \; : \; x_i \quad (12)$$

We will see later that alignment is a particular case of a general transformation called *change of basis.*

Localization

This SRE does not have uniform dependencies and localization enables us to derive an equivalent SURE. To illustrate this transformation, note that although the dependency of Y on X is not uniform, it is *many to one.* Indeed, *all* points in the domain of Y, lying on the straight line $i - j = c$ depend on $X(c, 0)$. Moreover, the point $[c, 0]$ also lies on this line. As a result, we can introduce an auxiliary variable, X' in our SRE whose domain is $\{i, j | 0 \leq j < n, 0 \leq i\}$ and 'pipeline the X value to all the points where it is needed'. Using a similar argument for W, we obtain the following SRE.

$$y_i = Y(i, n-1) \tag{13}$$

$$Y(i,j) = \begin{cases} \{j = 0, 0 \leq i\} & : W'(i,j) * X'(i,j) \\ \{0 < j < n, 0 \leq i\} & : Y(i,j-1) + W'(i,j) * X'(i,j) \end{cases} \tag{14}$$

$$W(i,j) = \{ \{i = 0, 0 \leq j < n\} : w_j \tag{15}$$

$$X(i,j) = \{ \{j = 0, 0 \leq i\} : x_i \tag{16}$$

$$W'(i,j) = \begin{cases} \{i = 0, 0 \leq j < n\} & : W(i,j) \\ \{i > 0, 0 \leq j < n\} & : W'(i-1,j) \end{cases} \tag{17}$$

$$X'(i,j) = \begin{cases} \{j = 0, 0 \leq i\} & : X(i,j) \\ \{0 < j < n, 0 \leq i\} & : X'(i-1,j-1) \end{cases} \tag{18}$$

The legality of this transformation can be shown by an inductive argument, provided that the new pipelining and *initialization* dependency 'spans' the set of points that depend on a given value - any point that needs a given value can be reached by composing one copy of the initialization dependency and a positive number of copies of the pipelining dependency.

SREs can be viewed as *functional programs.* In particular, they are *referentially transparent* and one can substitute equals for equals. Hence, manipulations that one performs on mathematical equations (epitomized by phrases such as "substituting for x and simplifying" that one finds in mathematical discourse) can equally well be performed on SREs. For example, we can substitute the uses of $W(i, j)$ and $X(i, j)$ (on the rhs of (17) and (18) by their definitions, and then (15)-(16)) can be dropped since they are not used in any other equation. This (followed by dropping the primes on the remaining variables) yields the SRE given below. Although this seems utterly obvious since we are manipulating SREs by hand, such transformations can be performed automatically on *any* SRE.

$$y_i = Y(i, n-1) \tag{19}$$

$$Y(i,j) = \begin{cases} \{j = 0, 0 \leq i\} & : W(i,j) * X(i,j) \\ \{0 < j < n, 0 \leq i\} & : Y(i,j-1) + W(i,j) * X(i,j) \end{cases} \tag{20}$$

$$W(i,j) = \begin{cases} \{i = 0, 0 \leq j < n\} & : w_j \\ \{i > 0, 0 \leq j < n\} & : W(i-1,j) \end{cases} \tag{21}$$

$$X(i,j) = \begin{cases} \{j = 0, 0 \leq i\} & : x_i \\ \{0 < j < n, 0 \leq i\} & : X(i-1,j-1) \end{cases} \tag{22}$$

Change of Basis

Finally, we may apply a *reindexing transformation* (also called a *change of basis* or a *space-time mapping*) to the variables of an SRE. The transformation, $\mathcal{T}$, must admit a *left inverse* for all points in the domain of the variable. When applied to the variable X of an SRE defined as follows:

$$X(z) = \begin{cases} & \vdots \\ D_i^X & : \; g_i(\ldots Y(f(z))\ldots) \\ & \vdots \end{cases}$$

we obtain an equivalent SRE as follows:

- Replace each D_i^X by $\mathcal{T}(D_i^X)$, its image by $\mathcal{T}$.
- In the occurrences of a variable on the rhs of the equation for X, replace the dependency f by $f \circ \mathcal{T}^{-1}$, the composition[1] of f and $\mathcal{T}^{-1}$.
- In all occurrences of the variable X on the rhs of *any* equation, replace the dependency f by $\mathcal{T} \circ f$.

The occurrences of X on the rhs of the equation for X itself constitute a special case where the last two rules are *both* applicable, and we replace the dependency f by $\mathcal{T} \circ f \circ \mathcal{T}^{-1}$. Similar transformation rules can be developed for SREs with reduction operations, but are beyond the scope of this overview.

For our example, let us choose a transformation $\mathcal{T}(i,j) = i+j, j$, which maps the point $[i,j]$ to $[i+j, j]$. Here, $\mathcal{T}^{-1}(j,k) = (i-j, j)$ is the required inverse. Let us apply it to *all* the variables, X, W and Y. This gives us the following SRE:

$$y_i = Y(i+n-1, n-1) \tag{23}$$

$$Y(i,j) = \begin{cases} \{j=0, j \le i\} & : \; W(i,j) * X(i,j) \\ \{0 < j < n, j \le i\} & : \; Y(i-1, j-1) + W(i,j) * X(i,j) \end{cases} \tag{24}$$

$$W(i,j) = \begin{cases} \{i=j, 0 \le j < n\} & : \; w_j \\ \{i > j, 0 \le j < n\} & : \; W(i-1, j) \end{cases} \tag{25}$$

$$X(i,j) = \begin{cases} \{j=0, j \le i\} & : \; x_i \\ \{0 < j < n, j \le i\} & : \; X(i-2, j-1) \end{cases} \tag{26}$$

We now see that if we simply rename the new indices as t and p, we obtain the SRE of (1)–(3) which describes the array of Fig. 23.1. As mentioned above, the transformation, $\mathcal{T}$, must admit an integral left inverse (for all points in the domain), and hence it is restricted to what are called *unimodular* transformations[2]. Often, one may desire to apply other transformations. A particular example is non-unimodular non-singular integer transformations. This leads to SREs whose domains are "sparse" (e.g., all even points from 1 to n), but a detailed discussion is beyond the scope of this chapter.

---

[1] Recall that function composition is right associative, i.e., $(g \circ h)(z) = g(h(z))$.

[2] A matrix is said to be unimodular if its determinant is ± 1. Hence an integral unimodular matrix admits an integral inverse.

We emphasize that what we have presented for the convolution example is a particular series of transformations. In general, the transformations can be applied in different orders, and indeed, infinitely many equivalent SREs can be derived from the original specification.

23.2.3 Scheduling

We have so far discussed the formal manipulations for transforming SREs and obtaining provably equivalent SREs. We now address the second question, what transformation to apply. In general, the answer to this question consists of two interrelated parts: the choice of a scheduling, and the choice of a processor allocation.

Scheduling techniques are based on representing each system of recurrence equations with *dependence graphs*, defined as follows. The *dependence graph* (or *reduced dependence graph*) $G = (V_G, E_G)$ of a SRE is the graph whose vertices V_G are the variables of the system, and there exists an arc $e \in E_G$ between two vertices, Y and X if and only if there exists an equation where X is the left-hand side, and Y is an argument of the right-hand side. The arcs of this graph are labeled with information describing the dependencies. For example, in the system of (23)–(26), the nodes of the dependence graph (shown in Fig. 23.2) are y, x, w, Y, X, and W. There exists an arc from Y to X, since the definition of Y depends on that of X.

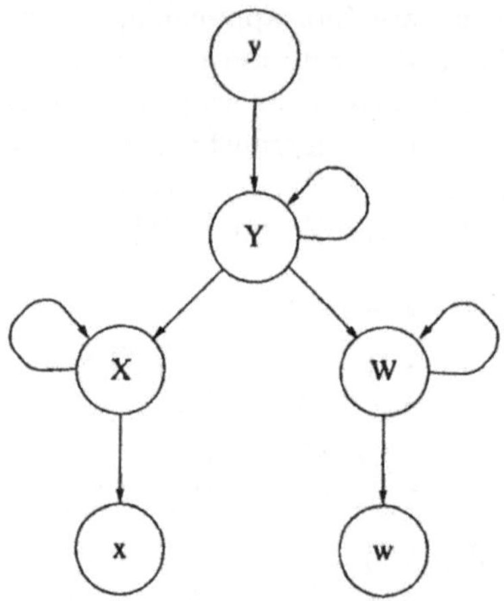

Figure 23.2 Dependence graph associated with the system of (23–26).

Scheduling SUREs

SUREs (Systems of Uniform Recurrence Equations, see Def. 23.2.3) are of particular interest in order to study scheduling. In a SURE, the dependencies can be represented by n-dimensional vectors giving the translation between the index points of dependent instances of two variables. Thus, dependencies can be summarized in a *dependence matrix* D whose columns are the dependence vectors of the SURE.

In the example of (23)–(26), if we consider only variables which are neither inputs (x, and w) nor output (i.e., y), we can see that the system is uniform: $Y[i,j]$

depends on $X[i,j]$, $W[i,j]$ and $Y[i,j-1]$, $X[i,j]$ depends on $X[i-1,j-1]$, and finally, $W[i,j]$ depends on $W[i-1,j]$. We notice also that the dependency between $Y[i,j]$ and $X[i,j]$ is somehow "artificial", in the sense that $Y[i,j]$ depends in fact on $X[i-1,j-1]$, through the definition of X. A similar remark shows that $Y[i,k]$ depends, in fact, on $W[i-1,k]$.

Thus, the following dependence matrix summarizes all the information which is needed to schedule this system of equations:

$$D = \begin{pmatrix} 1 & 0 & 1 \\ 0 & 1 & 1 \end{pmatrix} .$$

The dependence graph can be analyzed via the so-called *dependence relation*. Consider the set $\mathcal{I}$ of pairs (X,x), where X is a variable, and $x \in D^X$. Such a pair is called an *instance* in the following. Given two instances (X,x) and (Y,y) we say that (X,x) depends directly on (Y,y), denoted by $(X,x) \overset{1}{\leftarrow} (Y,y)$, if and only if there exists an instance of equation $X(x) = f[\ldots, Y(y), \ldots]$. The relations "depend in t steps on" $(X,x) \overset{t}{\leftarrow} (Y,y)$, and "depend on" $(X,x) \leftarrow (Y,y)$ are defined in the usual way, by transitivity. The *extended dependence graph* of the SURE, denoted Γ, is the graph of the dependence relation (see for example Fig. 23.3). It describes exactly the set of variable instances to be calculated in order to compute the SURE.

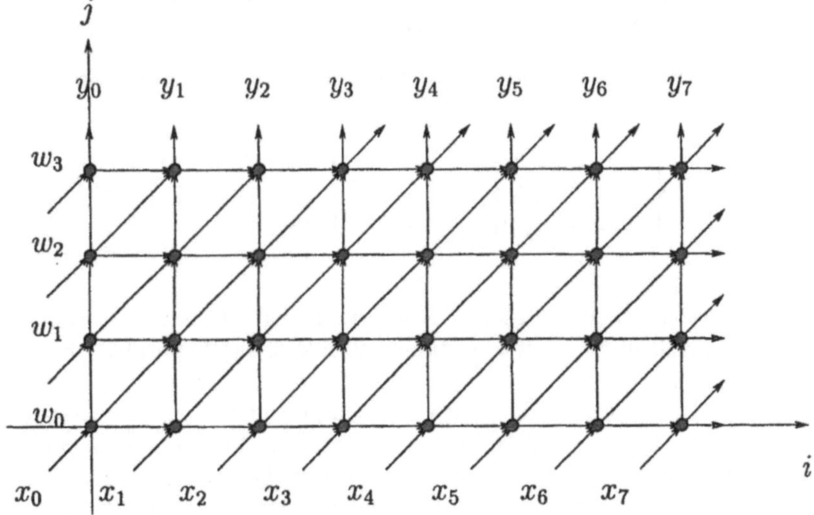

Figure 23.3 Extended dependence graph for the SURE of the convolution algorithm, (23–26), for $n = 3$.

A *schedule* is a mapping $t : \mathcal{I} \to \mathbf{Z}$ such that

1. $(X,x) \leftarrow (Y,y) \Rightarrow t(X,x) > t(Y,y)$ (causality condition).
2. $t(X,x) \geq 0$, $\forall (X,x) \in \mathcal{I}$ (positivity condition).

In this definition, the causality condition guarantees that dependent computations are performed in a consistent order, whereas the positivity condition ensures that we can set a starting time for the computations.

Observe that a scheduling function may not exist. This is the case, for example, when the extended dependence graph contains a cycle. This motivates the *computability analysis* of recurrence equations, which is directly related to scheduling. A variable X is said to be *computable* – or, *explicitly defined* – if and only if for all $x \in D$, there exists a scheduling function t such that $t(X, x) < +\infty$.

The following theorem (due to Karp, Miller and Winograd [3]), sets a framework for the study of scheduling, in the case of a single recurrence equation:

Theorem 23.2.1 *Given a uniform recurrence equation*

$$X(z) = f[X(z - d_1), \ldots, X(z - d_s)], \tag{27}$$

defined on the domain $D = \mathbf{N}^n$, the following conditions are equivalent:

1. *$X(z)$ is computable.*
2. *There exists no semi-positive vector*[3] *$(u_1, \ldots, u_s)$ such that*

$$\sum_{j=1}^{s} -u_j d_j \geq 0 \quad .$$

3. *If $\tau = (\tau_1, \ldots, \tau_n)$ is a vector of $\mathbf{Z}^n$, the system of inequalities $d_j.\tau \geq 1$, $1 \leq j \leq s$, and $\tau_i \geq 0$, $1 \leq i \leq n$ has a solution.*
4. *For all $z \in \mathbf{N}^n$, the following linear programs have an optimal common value $m(z)$:*

$$I \begin{cases} z - \sum_{j=1}^{s} u_j d_j \geq 0, \\ u_j \geq 0, 1 \leq j \leq s, \\ \max \sum_{i=1}^{s} u_j \end{cases} \qquad II \begin{cases} \tau_i \geq 0, 1 \leq i \leq n, \\ d_j.\tau \geq 1, 1 \leq j \leq s, \\ \min z.\tau \end{cases}$$

Condition *3* has a simple geometric interpretation: the extremities of vectors d_i, $1 \leq i \leq s$, must all be strictly on the same side of the hyperplane $\tau.z = 0$, as shown in Fig. 23.4.

Moreover, the τ vector for which the solution of the linear program *II* is obtained defines a *linear schedule.* In other words, we can compute $X(z)$ at time $\tau.z$. Fig. 23.4 depicts a two dimensional case, where the dashed lines (often called *iso-temporal lines*) represent calculations done at successive instants t and $t + \tau.d_i$.

In practice, one is often interested in linear (or affine) schedules, as then, the velocity of the data is constant, and the implementation of the corresponding parallel program is easier.

Let us illustrate the application of this technique to the convolution example of (23)–(26). As far as scheduling is concerned, we may consider the dependency information of this system as being equivalent to that of the following single equation.

$$\begin{pmatrix} Y(i,k) \\ W(i,k) \\ X(i,k) \end{pmatrix} = g(Y(i, j-1), W(i-1, j), X(i-1, j-1)) \tag{28}$$

[3] A vector is said to be semi-positive if it is non-null and all its components are non-negative.

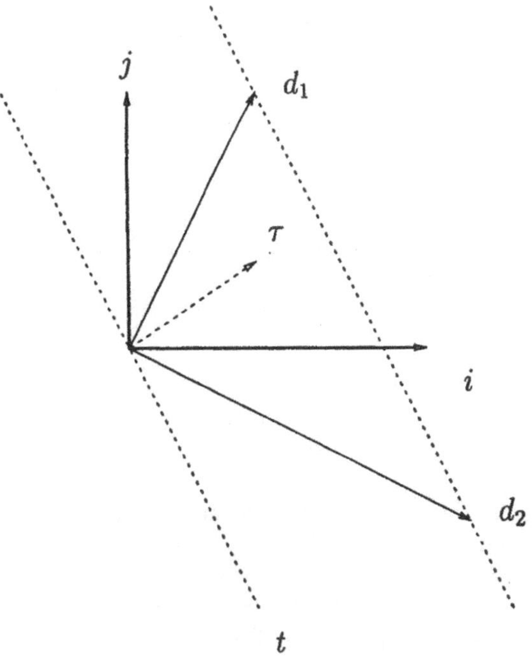

Figure 23.4 Illustration of Theorem 23.2.1. Points are indexed by i and j, and dependence vectors are d_1 and d_2. The vector τ is normal to the iso-temporal lines.

Applying condition 3 of Theorem 23.2.1 leads to the following inequalities:

$$\tau_1 \geq 1$$
$$\tau_2 \geq 1$$
$$\tau_1 + \tau_2 \geq 1$$

A valid schedule is therefore $t(i,j) = i + j$.

Linear Schedules

In practice, systems of recurrence equations are seldom SUREs, and moreover, there is no reason why equations are aligned in such a way that a common schedule exists, e.g., one may have to align two variables in opposite directions in order to be able to compute them. For these reasons, one has to associate different linear schedules to different variables.

Consider our standard SRE (Def. 23.2.1) parametrized by an integer p

$$X(z,p) \quad = \quad \begin{cases} & \vdots \\ D_i^X & : \quad g_i(\ldots Y(f(z,p))\ldots) \\ & \vdots \end{cases}$$

where $f(z) = Az + Bp + C$. To each edge e of the reduced dependence graph G describing the dependence between variables Y and X of this equation, let us associate the domain D_i^X, and the index function f. The edge e can therefore be represented by a 4-tuple (Y, X, D_i^X, f), where D_i^X is a sub-domain of D^X where the dependence effectively occurs.

Definition 23.2.5 *[Affine by variable parametrized schedule] An affine by variable parametrized schedule has the form*

$$t_X(z) = \tau_X z + \sigma_X p + \alpha_X \tag{29}$$

where τ_X and σ_X are fixed vectors with rational coordinates, and α_X is a rational number.

For the sake of simplicity, we will consider only rational timing functions. All the following results can be extended to the case where $t_X(z) = \lfloor \tau_X z + \sigma_X p + \alpha_X \rfloor$ without difficulty.

Def. 23.2.5 assumes that the schedule depends linearly on the parameter. The problem is to characterize the values of τ_X, σ_X and α_X such that the causality and positivity conditions are satisfied, and then to select one particular solution so as to optimize some quality criterion like maximal throughput or minimum latency.

In order to satisfy the causality condition for each edge (Y, X, D_i, f) of E_G one must satisfy the condition:

$$\forall z \in D_i^X : t_X(z) \geq t_Y(I(z)) + 1 \quad . \tag{30}$$

The above definition of timing-function maps directly to the structure of the equations: it is implicitly assumed that the evaluation of the system is to be performed on an architecture where the combinational operators are separated by at least one delay. However, this hypothesis excludes practical cases when one may wish to cascade several combinational elements. Changing the way delay will be mapped is not a problem provided the information is known statically. The real problem is to ensure that condition (30) is satisfied on and only on domain D_i^X.

When one substitutes appropriate values for z in (30), one obtains systems of linear inequalities in the coefficients of t_X and t_Y. This situation is similar to the positivity constraints, $t_X(z) \geq 0$. The problem is that there is a very large number (perhaps an infinity) of values of z and $I(z)$ which result in a large number of inequalities. We need to express all these inequalities using a finite description.

There are two basic methods for that. One uses a fundamental result of the theory of linear inequalities, the affine form of Farkas' lemma [4], and is called *the Farkas method*. The other one, called the *vertex method*, uses the dual representation of polyhedra, and is summarized as follows.

There are two ways of describing a polyhedron. One is as the set of points which satisfy a finite set of linear inequalities, and the other one is as the convex hull of a finite set of points. For each polyhedron, there exists a finite set of points, called its *vertices*, which together generate the polyhedron, none of them being a convex combination of the other. Such a set of points is a minimal generating system. If the polyhedron is unbounded, some of the vertices will be at infinity, they are called *rays* (a ray is an infinite direction in the polyhedron). There are standard algorithms for going from one representation to the other one [5].

The key idea behind the vertex method is the fact that a linear inequality is satisfied at *all* points of a convex polyhedron if and only if it is satisfied at all points of the generating system (which is finite). Hence the method: find generating systems for the polyhedra D_i^X, write (30) at each of these points, and solve for the unknown coefficients. The method is summarized as follows:

- Compute a generating system for the polyhedron D of any edge (Y, X, D, f) of the dependence graph, and for all polyhedra D^X, $X \in \mathcal{V}$.
- Write instances of (30) at all vertices of polyhedra D of each edge.
- Write instances of $t_X(z) \geq 0$ at all vertices and rays of D^X.
- Solve the resulting finite system of linear inequalities by optimizing an appropriate criterion. For example, minimizing the total computation time for a system of equations can be achieved by writing the total time as a linear function of the parameters, and expressing that this function is greater than the schedule of all instances of all variables.

It should be noted that using such a method leads quickly to rather complex linear problems. For example, scheduling the convolution given by the SRE (23)–(26) using the vertex method leads to a linear program of 37 constraints and 15 variables. The schedule obtained is

$$t_Y(i,j,n) = j + 1$$
$$t_y(i,n) = n + 1$$

The results presented in this section are based on work in the eighties, and more recently in the early nineties, when the work was picked up by the loop parallelization community. Undoubtedly, the most important work on this subject is that of Karp, Miller and Winograd [3], which predates systolic arrays by well over a decade! They studied (in fact, defined) SUREs over a particular class of domains (the domain of all variables is the same and is the entire positive orthant), and gave techniques for determining (one- and multi-dimensional) affine schedules[4]. This work is only recently being matched and improved.

Quinton, Rao, Roychowdhury, Delosme-Ipsen [6, 7, 8, 9] and a number of others worked on schedules for UREs and SUREs. Rajopadhye et al. and Quinton-van Dongen [10, 11, 12] developed the basic results of the vertex method for AREs. Mauras et al. [13] extended this result to *systems* of AREs by using *variable dependent* (also called affine-by-variable) schedules. Rajopadhye et al. [14] proposed *piecewise* affine schedules for SAREs. Feautrier [15] gave the alternative formulation using Farkas' lemma, for determining (one-dimensional, variable dependent) affine schedules for a SARE. He further extended the method to multidimensional schedules [16].

As for the optimality of schedules, basic results (for SUREs) were of course obtained by Karp et al. [3] and later by Rao, Shang-Fortes, Darte et al. and Darte-Robert [7, 17, 18, 19], among others. For SAREs, the problem was addressed by Feautrier, Darte-Robert and Darte-Vivien [15, 16, 20, 21].

[4]Today we know them to be multi-dimensional schedules, but this understanding was slow in coming.

Finally, some theoretical results about the undecidability of scheduling are also available. Joinnault [22] showed that scheduling a SURE whose variables are defined over arbitrary domains is undecidable. Quinton-Saouter showed [23] that scheduling even parametrized families of SAREs (each of whose domains is bounded) is also undecidable.

23.2.4 Allocation Functions

The second important aspect in the choice of the space time transformation is the *allocation function*, namely, the function specifying the mapping of the computation, i.e., an index point z, in the domain, D of the SRE to a processor index. In this section we will discuss the nature and choice of the allocation function and how it can be combined with the schedule to obtain an appropriate change of basis transformation. We assume that we have a SURE (variables have been aligned, and non-local dependencies are uniformized, see Section 23.2.2), and a (1-dimensional, affine) schedule specified by the pair (τ, α) has been chosen. For a k-dimensional SRE, the allocation function is $a : \mathbf{Z}^k \to \mathbf{Z}^{k-1}$, i.e., it maps a k-dimensional index point to a $k-1$-dimensional processor. We consider linear (or affine) transformations, and hence the allocation function is represented by a $(k-1) \times k$ matrix and a $(k-1)$-vector.

Consider the allocation function $a(i,j) = j$ that we used for the convolution example in Section 23.2.2. Note that it is not unique — the function $a(i,j) = 1 - j$ gives us *the same architecture*, except that the processors are now labeled $-1 \ldots -n$. Thus, although the allocation function is specified by an $(k-1) \times k$ matrix and a $k-1$-vector (i.e., by k^2-1 integers), they are not really independent. In fact, the allocation function is *completely* specified by means of what is called the *projection vector*, $\vec{u}$ (by convention, $\vec{u}$ is a *reduced* vector: the gcd of its components is 1). The intuition is that any two points in D are mapped to the same processor if and only if their difference is a scalar multiple of $\vec{u}$. Since $\vec{u}$ is reduced, we also see that for any $z \in D$, the *next* point mapped to the same processor is one of $z \pm \vec{u}$. Now, the *time* between two such successive points mapped to the same processor is $\tau^T \vec{u}$. This is true for all processors: they are all active exactly once every $\tau^T \vec{u}$ clock cycles, and $\frac{1}{\tau^T \vec{u}}$ is called the *efficiency* of the array.

An important constraint that must be satisfied is that no two index points are mapped to the same processor at the same time instant, which holds if and only if $\tau^T \vec{u} \neq 0$. Before we discuss the factors influencing the choice of $\vec{u}$, we will summarize how a change of basis transformation, T is constructed, given τ and $\vec{u}$, such that $\tau^T \vec{u} \neq 0$. We need to construct an $n \times n$ matrix (this may or may not be unimodular, and in the latter case, the domains of the SREs we obtain on applying this change of basis will be what are called "sparse polyhedra") such that its first row is τ^T, and whose remaining $n-1$ rows constitute a matrix π such that $\pi \vec{u} = 0$. This can be done by first *completing* $\vec{u}$ into a unimodular matrix S (say by using the right Hermite normal form [4]), so that $\vec{u} = S\vec{e_1}$ (here, $\vec{e_i}$ is the i-th unit vector). Then the matrix consisting of the last $n-1$ rows of S^{-1} is a valid choice for π, and hence $\begin{bmatrix} \tau^T \\ \pi \end{bmatrix}$ is the required transformation, and it is easy to see that it is non-singular, and that its determinant is $\tau^T \vec{u}$.

Choosing the Allocation Function

Unlike the schedule vector, τ, which must belong to the a polyhedral region specified by a set of linear inequalities, there are no such linear constraints that we can impose on the projection vector $\vec{u}$. Indeed, the space of permissible values of $\vec{u}$ is the entire set of primitive vectors in $\mathbf{Z}^n$, except *only* for the normals to τ. In general, we seek the allocation function that yields the "best" array, and the natural choice for the cost function is the number of processors — the number of integer points in the projection of D. Unfortunately, this is not a linear function. Moreover, it is only an approximate measure of the number of processors — in arrays with efficiency $\frac{1}{k}$, one can always cluster k processors together so that the final array is fully efficient, at the cost of some control overhead. Thus, the number of processors can be systematically reduced (albeit at the cost of additional control and temporary registers). In practice, however, a number of other constraints can be brought into play, rendering the space manageable, and indeed, it is often possible to use an exhaustive search.

First of all, for many real time signal processing applications, the domain of computation is infinitely large in one direction, and immediately, since we desire a finite sized array, (and we insist on linear transformations) the projection *has* to be this direction.

A second constraint comes from the fact that the I/O to the array must be restricted to only boundary processors, otherwise we lose many of the advantages of regularity and locality.

In addition, we could also impose that $\tau^T \vec{u} = \pm 1$, which would ensure arrays with 100% efficiency. As mentioned above, arrays that do not achieve this can be "post-processed" by clustering together $\tau^T \vec{u}$ neighboring processors, but this increases the complexity of the control. Moreover, it is also difficult to describe such arrays using only the formalism of linear transformations (the floor and ceiling operations involved in such transformations are inherently non-linear).

Finally, we could impose the constraint that the interconnections in the derived array must be of a fixed type (such as 4, 6 or 8 nearest neighbors, or we may even permit "one-hop" interconnections). It turns out that this constraint is surprisingly effective. When deriving 1-dimensional arrays (from 2-dimensional recurrence equations), if we allow only nearest neighbor interconnections, there are no more than 4 possible arrays that one can derive. Similarly there are no more than 9 2-dimensional arrays (with only north, south east and west connections); the number goes up to 13 if one set of diagonal interconnections is allowed, and to 25 if 8 nearest neighbors are allowed). Moreover, these allocation functions can be efficiently and systematically generated.

23.2.5 Localization and Serialization

We now address one of the classic problems of systolic design, namely localization (and its dual problem, serialization). The key idea is to render uniform all the dependencies of a given SRE. Clearly, this is closely related to the problem of alignment (i.e., defining the SRE such that all variables have the same number of dimensions). For example, consider the SRE defined below (we have deliberately chosen distinct index names for each variable)

$$X(i,j) \quad = \quad \{0 \leq i,j < n\} \; : \; Y(j-1, i-1) \tag{31}$$

$$Y(p,q) \quad = \quad \{0 \le p,q < n\} \; : \; A_{p,q} \tag{32}$$

At first glance, the $X \to Y$ dependency is not uniform. However, this can be made uniform if the variable Y (and the input A) is aligned so that the p dimension is aligned with j and q with i, i.e., we perform a change of basis on Y with transformation, $\mathcal{T}(p,q) = (q,p)$, and then *rename* the indices (p,q) to (i,j). Alignment does not help with ***self dependencies***, and in the rest of this section, we will consider only such dependencies.

Consider a (self) dependency, $X \to X(f(z))$ in an ARE, thus $f(z) = Az+a$ (for simplicity, we ignore the parameters). Applying a change of basis transformation $\mathcal{T}$ will yield a new dependency $\mathcal{T} \circ f \circ \mathcal{T}^{-1}$. This will be uniform, i.e., its linear part will be Id if and only if $A = \mathrm{Id}$, i.e., the original dependency is itself uniform. Hence, change of basis transformations cannot serve to localize a dependency, and we need to develop other transformations.

The fundamental such transformation is (null space) ***pipelining.*** Recall the convolution example and its localization in Sec. 23.2.2. We could view the transformation as consisting of two steps; the first consists of defining a new variable, say X' whose domain is also D (the same as that of X), and whose value at any point z is that of $X(f(z))$ then we can simply replace the $X(f(z))$ on the rhs of the equation for X by $X'(z)$ and get a semantically equivalent SRE. Let us focus on the equation for X', which does not do any "computation" but will be only used for propagating the values appropriately. Now, if we find a constant vector ρ such that $\forall z \in D$, $X'(z) = X'(z+\rho)$, and another constant vector ρ' such that, adding "enough" scalar multiples of ρ to any point $z \in D$ eventually yields a point z' such that $f(z) = z' + \rho'$, then we can replace the affine dependency f by a uniform dependency ρ. This can be proved by a simple inductive argument that we omit. Hence, our problem can be reduced to the following:

- Determining the pipelining vector(s) ρ.
- Resolving the problem of "initializing" the pipelining, i.e., determining the vector(s) ρ' and also *where* it is to be used.

The first problem is resolved by observing that the pipelining vectors must "connect up" all the points that depend on the *same* value; in other words, two points z and z' could be (potentially) connected by a dependency $\rho = z - z'$ if $f(z) = f(z')$, i.e., $A(z - z') = 0$, i.e., $z - z'$ must belong to the null space A. Indeed, any ***basis vector*** of the null space of A is a valid candidate for the pipelining vector.

Now, to initialize the pipelines, once we have chosen the pipelining vectors ρ we need to ensure that for all points $z \in D$, there exists a scalar, k (which could be a function of z), such that $z - k\rho$ is a *constant* distance, ρ' from $f(z)$. In other words, $A(z + k\rho) + a - z + k\rho = \rho'$. Multiplying both sides by A and simplifying shows that this is possible if and only if $A^2 = A$, i.e., A is a *projection.* For illustration, consider the two recurrences below.

$$X(i,j) \quad = \quad \{0 < i,j < n\} \; : \; X(i,0) \tag{33}$$
$$Y(i,j) \quad = \quad \{0 < i,j < n\} \; : \; Y(0,i) \tag{34}$$

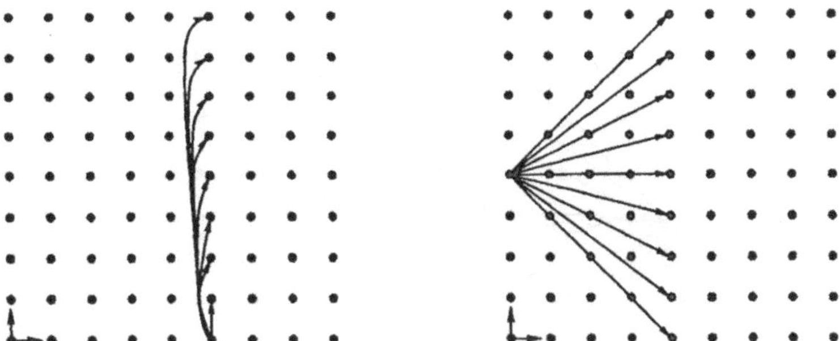

Figure 23.5 Illustration of the conditions for null space pipelining. The diagrams show the dependencies of (33) (left) and (34) (right). In both cases, the pipelining vector is $\pm[0,1]$, since all the points in a given column depend on the *same* point. For (33), these *producer* points are "close" to the *consumers* (because the dependency is idempotent), and the pipeline can be correctly initialized, while for Eqn. 34, this is not the case, so null space pipelining is not possible.

In the first one, we see that all the index points in the i-th column require the same argument. The dependency functions are, respectively, $A_1 = \begin{bmatrix} 1 & 0 \\ 0 & 0 \end{bmatrix}$ and $A_2 = \begin{bmatrix} 0 & 0 \\ 1 & 0 \end{bmatrix}$. For both of them, the null space of A is 1-dimensional, and spanned by $\pm[0,1]$; this is thus the pipelining vector. However, while $A_1^2 = A_1$, we see that $A_2^2 = \begin{bmatrix} 0 & 0 \\ 0 & 0 \end{bmatrix} \neq A_2$. Hence the first recurrence can be pipelined (using null space pipelining as outlined above), while the second one cannot (see Fig. 23.5). For (33), we obtain the following, provably equivalent, SRE.

$$X(i,j) = \{0 < i,j < n\} : P_X(i,i) \quad (35)$$

$$P_X(i,j) = \begin{cases} \{0 < i,j < n\} & : P_X(i,j-1) \\ \{0 < i < n; j = 0\} & : X(i,j) \end{cases} \quad (36)$$

It was derived by

- defining a new equation for a pipelining variable, P_X, whose domain is that of the (sub) expression where the dependency occurs, *plus* a "boundary" $\{j = 0\}$, for the initialization. P_X depends on itself with the uniform dependency ρ everywhere, except for the boundary, where it depends on X with the dependency ρ'.
- replacing, in the original equation for X, the $X(i,0)$ by $P_X(i,j)$. Thus all affine dependencies have been replaced by uniform ones.

The natural question to ask is what happens when null space pipelining is not possible. In this case, a number of solutions exist. First, we could try *multi-stage* pipelining, where we initialize a pipelining variable, not with the original variable,

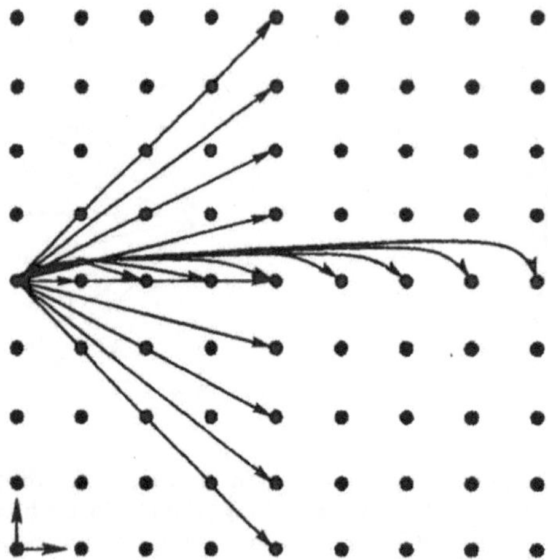

Figure 23.6 Illustration of multi-stage pipelining. The first dependency $[i,j] \rightarrow [0,j]$ can be pipelined (propagation vector $[1,0]$), but the second one $[i,j] \rightarrow [0,i]$ cannot (the propagation vector is $\pm[0,1]$, but the pipelined cannot be initialized). In multistage pipelining, the second pipeline is initialized with the *first* one at the *intersection* of the two pipelines.

but with *another* pipelining variable. Consider the following example.

$$X(i,j) \quad = \quad \{0 < i,j < n\} \ : \ X(0,i) + X(0,j) \tag{37}$$

Observe that the first dependency is the same as that in Eqn. 34, for which, null space pipelining is impossible. The second dependency, however, is similar to the one in (33) (the pipelining is along the row rather than along the columns) as illustrated in Fig. 23.6, and can be pipelined by introducing a local variable P_2 defined as follows.

$$P_2(i,j) \quad = \quad \left\{ \begin{array}{ll} \{0 < i,j < n\} & : P_2(i-1,j) \\ \{0 < j < n; i = 0\} & : X(i,j) \end{array} \right. \tag{38}$$

But now, we may also pipeline the second dependency by going "piggyback" on this pipeline. In particular, we note that the (potential) pipeline for the first dependency *intersects* the second pipeline on the $i = j$ line. Hence, we can initialize the first pipeline at this "boundary" but, the variable used for initialization is P_2, and not X. On the other side of this line, we pipeline in the opposite directions, as follows.

$$P_1(i,j) \quad = \quad \left\{ \begin{array}{l} \{0 < i < j < n\} : P_1(i,j-1) \\ \{0 < j < i < n\} : P_1(i,j+1) \\ \{0 < i = j < n\} : P_2(i,j) \end{array} \right. \tag{39}$$

All that remains is to replace the two dependencies in (37) by $P_1(i,j)$ and $P_2(i,j)$, and we have a SURE.

The next question concerns what to do when such multi-stage pipelining is not possible. At this point, we may still try to pipeline, but by sacrificing the primary advantage of null space pipeline — data values are propagated *only through* those points that use them. If we relax this constraint, we could localize (34) by defining *two* variables, P_Y^1 and P_Y^2. The first one corresponds to propagation along the pipelining vector [0, 1], and is initialized (at the $j = 0$ boundary), but with the P_Y^2 variable. This is similar to multi-stage pipelining, except that we do not have the P_Y^2 variable pipelined "for free", but we must actually construct its pipeline (also called its routing). The following is (one of many possible) solutions.

$$P_Y^1(i,j) = \begin{cases} \{0 < i,j < n\} & : P_Y^1(i,j-1) \\ \{0 < i < n; j = 0\} : P_Y^2(i,j) \end{cases} \tag{40}$$

$$P_Y^2(i,j) = \begin{cases} \{0 < i,j < n\} & : P_Y^2(i-1,j+1) \\ \{0 < j < n; i = 0\} : Y(i,j) \end{cases} \tag{41}$$

What these two equations achieve can be explained as follows. The variable P_Y^2 is used to propagate the values of Y from the first column in the south-east direction (with the $[-1,1]$ dependency vector.) This propagation passes through half the index points in the domain. Once the values reach the bottom row, they are used to initialize the standard pipeline (of P_Y^1) which propagates them up the column to the index points that will use these values. There are many choices for the propagation/routing vector. Indeed, it suffices to choose it so that it does not belong to the *range space* of A. Thus, any SAREs where the dependencies are singular can always be rendered uniform. However, the choice of the propagation/routing vectors can affect the schedule (optimality and/or even the existence of a schedule).

The final question concerns how to treat non-singular dependencies. One simple, but naive approach is to extend the dimension of the index space, and in this extended index space, one can always make the dependency non-singular. However, a much more interesting and elegant solution is possible in certain cases. If the non-singular dependency matrix A is such that, for some integer k, $A^k = I$, then we can "cut" the domain into k pieces by means of $k - 1$ hyperplanes, and then "superpose" each of these pieces so that the dependency become uniform. To illustrate, consider the RE below.

$$X(i,j) = \begin{cases} 0 \le i = j < n : X(i-1,j-1) \\ 0 \le j < i < n : X(j,i) \\ 0 \le i < j < n : X(j-1,i) \quad . \end{cases} \tag{42}$$

Here, the dependency corresponds to the transpose, i.e., $A = \begin{bmatrix} 0 & 1 \\ 1 & 0 \end{bmatrix}$, and since $A^2 = I$, there must be a line (namely the diagonal, $i = j$) such that folding with respect to it will render the dependencies uniform. Indeed, we obtain the following SRE (see Fig. 23.7).

$$X(i,j) = \begin{cases} \{i = j\} : & \Rightarrow & X(i-1,j-1) \\ \{i > j\} : & \Rightarrow & Y(i,j) \end{cases} \tag{43}$$

$$Y(i,j) = X(i-1,j) \tag{44}$$

The results in this section are based on the work in the late eighties, notably by Rajopadhye *et al.* Quinton-Van Dongen, Wong-Delosme, Roychowdhury, and Yaacoby-Cappello [10, 24, 12, 25, 9, 26]. It is also interesting to note that similar ideas underlie much of the recent work on communication optimization in parallelizing loop nests and in data distribution.

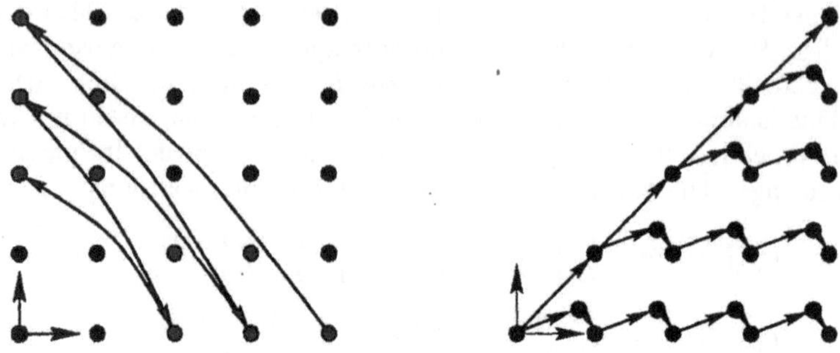

Figure 23.7 Illustration of localization by folding.

23.3 ADVANCED SYSTOLIC ARCHITECTURES

In the second part of this chapter, we describe through several examples, specific techniques used in advanced systolic arrays. In the methodology described in the first part, we have viewed systolic arrays as a dedicated implementation of a single, given algorithm. In practice, the design is (should be) ***application-specific***, and not ***algorithm-specific***, and this introduces a number of problems. A "naive" or abstract systolic array, such as one derived by our transformations, can be efficiently implemented in many different ways, depending on the granularity of the elementary calculations, and/or the versatility of the architecture which is targeted. In particular, the target technology may be one of the following:

- general purpose programmable parallel architectures,
- application oriented programmable architectures,
- reconfigurable FPGA based architectures,
- dedicated (custom) architectures.

23.3.1 General-Purpose Systolic Programmable Architectures

General-purpose systolic programmable architectures, are designed to support a wide range of systolic and parallel algorithms. Their main characteristic is that processors are only connected to their neighbors, and only systolic communications are allowed. Most often, these architectures have a special instruction set for communication. There is no general interconnection network: distant communications are implemented by the routers of the intervening processors. The communication/computation ratio is in general close to 1, and such architectures are ideal to implement fine grain parallel algorithms (indeed, it is this feature that would distinguish them from general-purpose parallel machines based on a mesh topology such as the Intel Paragon, the Cray T3D, T3E, etc.)

The most famous example of such an architecture is the iWARP [27], a circuit designed by Intel Corporation in 1988 in cooperation with Carnegie Mellon University (CMU). Its architecture is derived from the original WARP project [28]

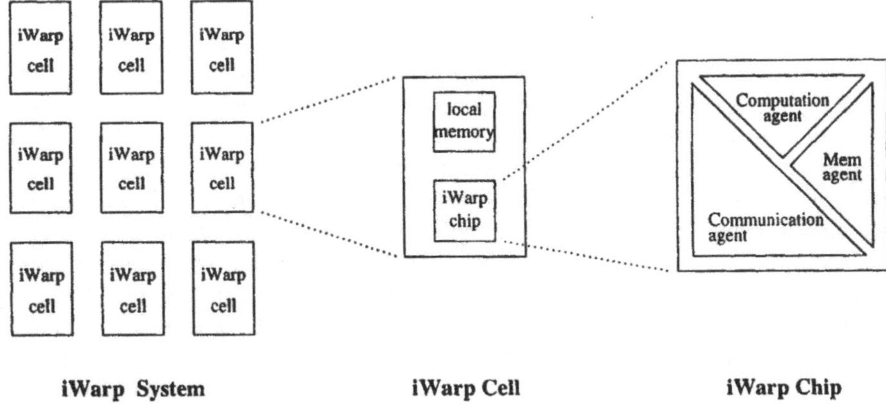

Figure 23.8 An iWARP system is composed of identical iWARP cells. A cell contains an iWARP chip with a local memory. A chip is divided into three sub-systems: a computation agent, a memory agent and a communication agent.

started in 1984 at CMU. The iWARP chip was designed as a building block for developing powerful and programmable parallel systems for high speed signal, image and scientific computing. An iWARP system is simply an assembly of iWARP cells (iWARP chip and memory) connected by means of a dedicated high throughput communication channel. Fig. 23.8 illustrates the iWARP system concept.

The iWARP chip consists of three autonomous sub-systems. The *computation agent* executes programs and contains floating point units with a peak performance of 20 and 10 MFLOPS for single or double precision operations respectively, as well as an integer ALU that performs 20 millions integer or logical operations per second. The *communication agent* implements the iWARP's communication system and operates independently of the computation agent. Finally, the *memory agent* provides an interface to the local memory.

The iWARP processor has many of the more traditional architectural features found in distributed memory machine, such as support for non-neighbor communication, message routing hardware, word-level-flow control between neighboring cells and spooling (a DMA-like mechanism). But what makes the iWARP architecture innovative and very well suited to systolic computations, are the two following additional features: program access to communication and logical channels.

In order to implement systolic communications, an application program can directly access the input and output of the message queues located in the communication agent. These queues are bound to special registers (the *gates*) in the register space of the instruction set. Reading from a gate corresponds to receiving a data from the queue. Similarly, writing to the gate corresponds to sending a data to the queue. Typically, this mechanism is used to smooth the flow of data between cells and to delay one stream of data with respect to the other ones. Such direct access to the message queues is also a way of optimizing the communication protocol. The data can be transmitted using an application-specific protocol which directly analyzes the header of the messages and explicitly tells the communication system where to deposit the messages. This protocol management can be directly used by parallel program generators such as Apply [29], AL [30, 31] or C-stolic [32].

The iWARP's second interesting communication innovation is the implementation of logical channels between processors. Logical channels provide both a higher degree of connectivity than physical links, and a mechanism for delivering guaranteed communication bandwidth for certain classes of messages. A high degree of connectivity is useful for systolic communication: a cell may need to have simultaneous links with several cells. Without logical channels, algorithms that require more physical connections than those provided by hardware cannot be implemented. The iWARP logical channels overcome this problem by proposing multiple *logical* connections on the same physical support. The ability to deliver guaranteed communication bandwidth makes the implementation of systolic algorithms, often characterized by their high throughput and word-level transfer, very efficient. By reserving a channel exclusively to dedicated communications — no other message can pass through this channel — one may guarantee a minimum bandwidth in order to ensure effective low-cost fine-grain communication.

Above all, an iWARP system is a MIMD machine, and no specific language is fundamentally attached to this machine. The W2 language, previously designed for the Warp machine, can be used, as well as many other parallel languages. But imperative languages, such as C, used in conjunction with a communication routine library remain a very simple and effective approach. In that case, the system runs in SPMD mode: each cell executes the same program independently, and synchronizations are ensured through systolic communications.

Other general-purpose systolic architectures are MGAP [33, 34], and Saxpy-1M [35], although they could be justifiably be classified as application oriented programmable. Johnson et al. [36] present an excellent survey of general purpose systolic arrays.

23.3.2 Application Oriented Programmable Architectures

Intermediate between fully programmable architectures and dedicated systolic arrays are *application oriented programmable architectures*. Such architectures target a particular domain of applications, such as image processing, cryptography, or neural network simulation. However, the great variety of algorithms pertaining to this domain calls for a programmable approach.

The SYSTOLA 1024 architecture illustrates this concept. SYSTOLA 1024 [37, 38] is a 32×32 systolic programmable architecture oriented towards image processing applications. It is an add-on board for personal computers with PCI slots. SYSTOLA 1024 is based on the concept of Instruction Systolic Array (ISA).

The structure of SYSTOLA 1024 is shown in Fig. 23.9. It consists of a square array of processors, each processor being controlled by instructions, row and column selectors. The whole array is controlled by a global clock. Instructions are input in the upper left corner, move rightwards and downwards step by step so that all processors of a diagonal of the array execute the same instruction. Selectors move in the same way as instructions, and are used to disable processors selectively: a processor executes the instruction it receives only if both selector bits are 1.

Each processor has read and write access to its own memory. One communication register is accessible by four neighbors, thus allowing neighbor to neighbor communications. In addition, broadcast of information on a column of processors is allowed.

A board contains 4×4 chips. Each chip is a 8×8 array of bit-serial RISC processors. Input and output to the array are done by a cascaded memory. Intelligent memory units, called interface processors are at the North and the West of the array. They access on-board RAM (northern board RAM, or western board RAM), which can communicate with the PC through a PCI interface.

The data transfer between the units of this memory hierarchy is supervised by a controller which either receives instructions from the PC, or operates autonomously. In the second case, it receives controller instruction from an instruction queue on the board, loaded from the PC before the application is started. The ISA gets its instruction from the ISA program memory, also loaded in advance.

The sequence of activities of the parallel computer is determined by so-called configurations, that is to say, controller programs and ISA programs. A programming environment called ISATOOLS allows the application programmer to develop parallel programs for this architecture, using an adaptation of the Pascal language.

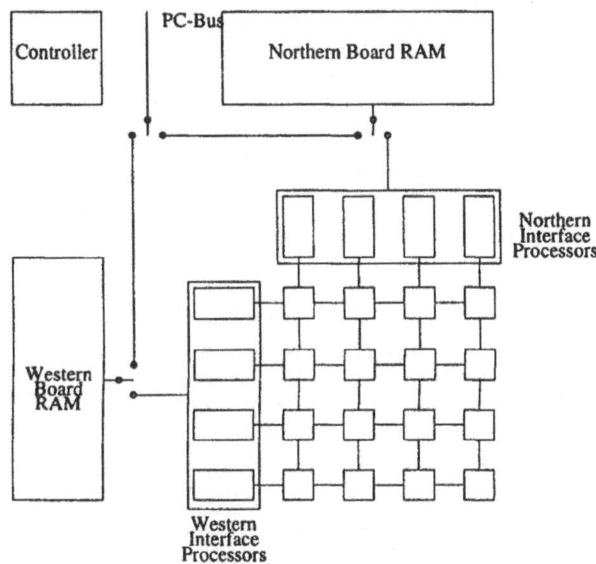

Figure 23.9 Architecture of the Systola 1024 architecture.

Each processor is capable of 3 Mips (16 bit instructions), which leads to a peak performance of 3125 Mips for the whole array. A typical application performs between 1000 and 3000 Mips. The array can reach 142 Mflops on 64 bit IEEE floating point operations, and 237 Mflops on 32 bit numbers. Typical acceleration rates of 60 are reached in numerical applications.

Although SYSTOLA 1024 was targeted for image processing, it can be used to accelerate a much broader class of algorithms: solution of large linear equation systems (for finite element methods), pattern matching, graph algorithms, neural networks, computer graphics, image reconstruction for computer tomography, speech recognition, and microbiology.

Other architectures in this class of application oriented programmable architectures include: (i) Adaptive Solutions' CNAPS [39], designed for accelerating neural networks; (ii) Synapse-1, designed by Siemens [40], another systolic architecture for neural networks; (iii) Kestrel [41], a programmable systolic architecture dedicated to DNA processing; (iv) MicMacs [42, 43] a programmable systolic array for string matching algorithms; (v) MOVIE [44] and Sympathy2 [45], designed for image processing applications.

23.3.3 Reconfigurable Architectures

Another popular approach for the implementation of systolic arrays is to use reconfigurable architectures, using FPGA chips. Examples of such architectures are PeRLe-1 [46], SPLASH-2 [47] or SPACE-2 [48]. Although these systems have not specially been designed to support systolic computations, their architecture make them particularly suitable for this purpose: they are made of a regular arrangement of FPGA chips on which systolic arrays/algorithms can be efficiently implemented.

Example of systolic arrays designed using FPGA boards are:

- systolic filter for fast DNA bank scanning [49],
- 2D convolution [50],
- text searching [51],
- numerical algorithms [52], division and other arithmetic operations [53].

Fitting a systolic array on such architectures requires the following problems to be solved:

- Good partitioning strategies: the boards have limited resources as compared to the size of the systolic array one wishes to implement.
- Efficient synthesis software: the challenge is to be able to rapidly map systolic algorithms to the reconfigurable platform. To this end, high clock speed and quick synthesis may be more important to space optimized designs. New methods need to be invented, as since current, design methodology mimics ASIC design, which is not appropriate.

We illustrate this class using the SPLASH-2 architecture, which is an attached processor system. It has been designed to connect up to 16 array boards to a SPARCstation-2 through a specific interface board. Each array board contains 17 Xilinx XC4010 FPGA chips. 16 of them form a linear array, and the 17th provides a broadcast capability to the other 16 chips. These 17 FPGA chips can access 512 Kbytes of memory (see Fig. 23.10). In addition, a reconfigurable crossbar switch links all the FPGA chips together. At run time, five crossbar configurations can be simultaneously stored and activated instantaneously.

The primary models of computation that were intended to be supported by the SPLASH-2 architecture were a SIMD and a linear (not necessarily systolic) architecture. Viewed as a SIMD machine, instructions can be broadcast through the X0 chip to the other FPGA components. Viewed as a linear array, data can circulate through the linear data-path of one board, then reach the first FPGA chip of the

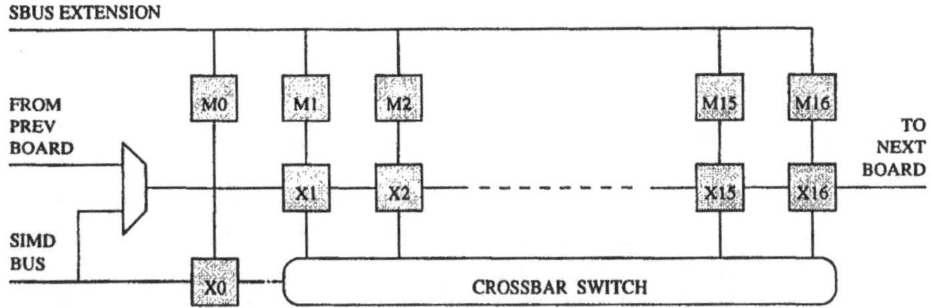

Figure 23.10 SPLASH-2 array board architecture: up to 16 such boards can be connected together to form a larger array. One board contains 17 Xilinx XC4010 FPGA chip connected linearly and through a crossbar switch.

next board, and so on. The presence of the crossbar switch allows the programmer to get beyond the rigid structure of a linear data-path.

The SPLASH-2 machine, as well as the PAM[5] family boards (PERLE-0 and PERLE-1) are the first well-known FPGA structures on which many applications have been tested. Applications of image processing, scientific calculation, data compression and encryption, and molecular biology have been experimented. These experiments have shown that supercomputer performance can be achieved, especially on computations which exhibit high regularity. In most cases, the time-consuming part of these applications can be implemented on systolic architectures.

This is not really surprising since the FPGA and the systolic concepts share some basic features: they are both composed of identical and locally connected cells. Systolic arrays can then be considered as adapted structures for FPGA arrays, even if the cell granularity differs. A systolic processor is made up of several elementary FPGA cells, and thus the same pattern of FPGA cells can be replicated inside a component, and the same configuration applied to every FPGA chip of the array. This last remark has a major consequence on the design process: it shortens the place-and-route stage which appears as the main limitation of the FPGA technology. In the case of a systolic architecture, the place-and-route operation of a single processor can be performed rapidly as only a few FPGA cells have to be configured, and the result is replicated immediately across the array. In practice, defining a placement is sufficient, since experience has shown that a good placement leads to a good and fast routing.

Recently, a number of researchers have explored incorporating an FPGA or similar programmable hardware on general-purpose programmable processors. Although they are not systolic architectures per se, these architectures use regularity and locality within the reconfigurable part. For example, RaPiD (Reconfigurable Pipelined Datapath) [54], is a configurable computing solution for regular computation-intensive applications. It fills the gap between static systolic arrays, which are too inflexible, and programmable systolic arrays like the WARP, which more closely resemble multiprocessors with too much control overhead.

[5] Programmable Activity Memory.

23.3.4 Special-Purpose (Custom) Architectures

Implementing a systolic algorithm on a special-purpose architecture is surely the best way to reach high performance. This comes however at the price of a longer design time. We illustrate this technique by summarizing the design of the SAMBA machine [55].

SAMBA is a full-custom parallel hardware accelerator dedicated to the comparison of biological sequences such as the scan of DNA or protein data-bases. It implements a parametrized version of a well-known biological sequence comparison algorithm — the Smith and Waterman algorithm — which is based on dynamic programming. This algorithm is known to provide good results but not often used when large data sets such as genetic data banks are involved. Other, heuristic algorithms are generally preferred for their speed, even though they are less reliable. The primary goal of SAMBA was to develop a systolic architecture for implementing the full Smith and Waterman algorithm in a very efficient way.

The second objective was to design the machine in the most systematic way in order to **rapidly** get a **correct** hardware. A few decisions — that were never challenged later — were made at the beginning of the project. For example, no attempts to improve the basic comparison algorithm was made. In addition, it was decided that the architecture would be designed using available CAD tools without any attempt to adapt these tools to the particular problem. The general architecture of SAMBA and the global design methodology were thus fixed and remained unchanged during the full duration of the project.

This, it was soon decided that the basic architecture of SAMBA would be a linear systolic array of hardwired full-custom VLSI chips connected to a standard workstation through a reconfigurable interface. This linear structure was imposed by the length of the biological sequences. The hardwired full-custom solution was dictated by performance. Finally, the reconfigurable interface was found to be the easiest way to fill the gap between a programmable von Neuman machine (the workstation) and a fully dedicated hardware (the array). Furthermore, FPGA boards already existed and could be efficiently used for this purpose. For SAMBA, the PERLE-1 FPGA board was used for housing the workstation/array interface.

The main part of the SAMBA project was then the design of the systolic array and, more precisely, of the full-custom VLSI chip. The design methodology that was followed can be summarized as follows:

Functional specification: the specification of the algorithm, including its parallel version was written using the C-stolic language [56] (an extension of C allowing parallel systolic algorithms to be simulated on Unix workstations). The C-stolic compiler produces three outputs: a C program emulating one processor of the systolic array, a C program emulating the interface between the host and the systolic array, and a C program for the host workstation.

Functional simulation: the functional specification was simulated on test data obtained from various data sets.

Architectural specification: the program run by the processors of the systolic array was translated manually from C-stolic code into VHDL code. Only the subset of VHDL accepted by the COMPASS VLSI synthesis software was used, in order to be able to obtain a hardware specification.

Validation of the architecture: the architectural specification was run on the same data as the functional simulation, and validated by comparison.

Synthesis of the VLSI chip: the VLSI chip was synthesized using the COMPASS hardware synthesis software, and sent to fabrication. The chips were fabricated by the ES2 company through the French multi-chip project.

Board design: the systolic array was designed by assembling the chips in printed boards.

Interface design: Due to the flexibility of the PERLE-1 board, the interface was designed in two steps. First, the PERLE-1 board was configured as a very simple link between the host and the systolic array, in order to test the full functionality of the chips and of the architecture. In this configuration, the interface program produced by the C-STOLIC compiler was run by the host. Then, in order to reach the full speed of the systolic array, the PERLE-1 board was configured to implement the final interface.

The total design effort amounted to 11 person/months: 1 month for functional specification and simulation, 1 month for architectural specification and simulation, 2 months for architectural synthesis, 2 months for chip testing, 0.5 month for board design, 4 months for interface design, and 0.5 month for final integration. The chip, as well as the complete machine, was correct at the first fabrication. It must be pointed out that the most difficult part (and the longest one) was the design of the interface.

Like SAMBA, a number of custom architectures have been proposed and built for a number of applications, and it is impossible to enumerate them all. Image compression is an interesting area where systolic algorithms can be applied. For example, Baglietto et al. [57] present a chip for block-matching. The MA7170 Systolic correlator [58] is a single chip architecture designed by InMos in the sixties. Demassieux [59] describes the implementation of a Median filter on VLSI.

To conclude our discussion of custom systolic architectures, we may list a few rules[6], taken from the SAMBA experience, which can help in design of such architectures rapidly and correctly.

- Never work simultaneously on algorithm and architecture: choosing an existing stable application is the best way to converge rapidly to an appropriate architecture.
- Do not design simultaneously tools and architecture: CAD tools are complex and need extensive verification before being operational for real designs.
- Make sure that the anticipated speed-up beats the current general purpose technology by *at least* one order of magnitude: designing a special-purpose architecture takes time. The architecture must not become obsolete due to the technology progress (doubling every 18 months or less) during the time it takes for the design.

[6]Most of these rules also hold for designing other (non-systolic) custom architectures.

- Design a balanced architecture: remember that the speed of a computer system is dictated by its slowest element! Within a complete systolic system, a high speed array is inefficient if it cannot be fed with input data.
- Get the help of the end-user: this is probably the most important rule. The success of a special-purpose machine depends essentially on the quality of the service it provides. And only the future users are able to specify exactly what they expect the machine to do.

23.4 CONCLUSION

In this chapter, we have presented two aspects of advanced systolic design, namely advanced design methods for systolic design and also the design and implementation of advanced systolic arrays.

The theory summarized in the first part is more or less the basis of several transformation systems and tools for regular array synthesis. Without any pretension to be exhaustive, let us cite a few of these systems. A number of classic systems such as SDEF, VACS, etc., were developed as research prototypes at various universities. Among the earliest software based on such transformations was Diastol [60]. This program accepted a few recurrence equations, and produced the sketch of a systolic architectures using the space-time transformation technique. More recent attempts such as the MMAlpha system [61, 62] are based on applicative languages, and correspond closely to the theory that we have presented. The Cathedral IV software [63] is an attempt to produce regular designs from algorithms specified in Silage. (Silage is a single assignment language used as the starting point for synthesis of signal processing architectures. A commercial version of this language, called DFL, is used by Mentor Graphics.) Presage [64] was a prototype tool written in Prolog, also meant for time-space transformations. One characteristic of this software was its ability to explore in a systematic way the solution space, using the backtracking mechanism of Prolog. Another attempt in this direction was presented by Baltus [65], where the solution space was explored using branch and bound techniques. HiFi [66] is developed in Delft, and covers a large spectrum of synthesis techniques, including clustering and partitioning.

One important limitation of the theoretical model is its inability to model partitioning (the problem of designing systolic arrays under resource constraints). There has been considerable work on the *choice* of the partitioning (to satisfy certain resource constraints and to optimize certain cost functions). However, these results are still not satisfactorily incorporated into the framework of design by correctness-preserving transformations of (very) high level specifications in the form of SREs. The key difficulty is that after partitioning, the resulting domains are no longer polyhedral and the dependencies are no longer affine.

There is a wide range of target technologies for implementing systolic architectures. In the future, there will be many more choices and, given the growing complexity of designs, it is important to be able to learn from the lessons of previous designs. For this reason we have described a number of implementation examples in the second part.

The next generations of chips will integrate hundred of millions of transistors. How will microprocessors and other VLSI implementations of DSP and multimedia systems use these resources? One possibility, which is now moving beyond the research prototype to commercial availability, is processors with a reconfigurable area on which custom (co)processing will take place. The success of this concept will mainly depend on two points: the efficiency of the co-processor, and the programming ease and availability of tools.

The efficiency will be provided by the parallelization of some parts of the programs and, as already mentioned, systolic arrays are largely used for this purpose. Experiments carried on the SPLASH-2 and PAM projects are good illustrations of this idea. They have shown that systolic structures fit well onto FPGA components and that they deliver high performance.

On the other hand, programming a reconfigurable architecture is still delicate. It requires knowledge of both hardware, computer architecture and digital design. For example, the translation of a few lines of imperative code into its equivalent in hardware is not obvious for usual programmers. Undoubtedly the challenge is then to make this translation as automatic as possible, even if the efficiency (from a resource allocation point of view) cannot reach that of a hand-crafted solution. The main problem to solve for reconfigurable co-processors to be successful is therefore to design compilers which map sections of code to FPGA in a reasonable amount of time without sacrificing efficiency too much. The long research on automatic synthesis of VLSI systolic arrays can find a new field of application in compilation and optimization for reconfigurable co-processors.

Acknowledgments The authors gratefully acknowledge their colleagues in Irisa and elsewhere who have contributed to the development of the concepts presented in this chapter.

REFERENCES

[1] R.F. Lyon, "Two's complement pipeline multipliers," *IEEE Transactions on Computers*, vol. COM-24, pp. 418–425, April 1976.

[2] L. Guibas, H. T. Kung, and C. D. Thompson, "Direct VLSI implementation of combinatorial algorithms," in *Proc. Conference on Very Large Scale Integration: Architecture, Design and Fabrication*, pp. 509–525, January 1979.

[3] R. Karp, R. Miller, and S. Winograd, "The organization of computations for uniform recurrence equations," *Journal of the Association for Computing Machinery*, vol. 14, no. 3, pp. 563–590, July 1967.

[4] A. Schrijver, *Theory of Integer and Linear Programming*. John Wiley and Sons, 1988.

[5] H. Le Verge, "A note on Chernikova's algorithm," RR 635, IRISA, IRISA, Campus de Beaulieu, 35042 Rennes Cedex, France, Feb 1992.

[6] P. Quinton, "Automatic synthesis of systolic arrays from recurrent uniform equations," in *11th Annual International Symposium on Computer Architecture, Ann Arbor*, pp. 208–214, June 1984.

[7] S. Rao, *Regular Iterative Algorithms and their Implementations on Processor Arrays*. PhD thesis, Standford University, U.S.A., Oct. 1985.

[8] J. Delosme and I. Ipsen, "Systolic array synthesis : computability and time cones," in *Parallel Algorithms and Architectures*, pp. 295–312, North-Holland, 1986.

[9] V. P. Roychowdhury, *Derivation, Extensions and Parallel Implementation of Regular Iterative Algorithms.* PhD thesis, Stanford University, Department of Electrical Engineering, Stanford, CA, December 1988.

[10] S. V. Rajopadhye, S. Purushothaman, and R. M. Fujimoto, "On synthesizing systolic arrays from recurrence equations with linear dependencies," in *Proceedings, Sixth Conference on Foundations of Software Technology and Theoretical Computer Science,* (New Delhi, India), pp. 488–503, Springer Verlag, LNCS 241, December 1986.

[11] S. V. Rajopadhye and R. M. Fujimoto, "Synthesizing systolic arrays from recurrence equations," *Parallel Computing*, vol. 14, pp. 163–189, June 1990.

[12] P. Quinton and V. Van Dongen, "The mapping of linear recurrence equations on regular arrays," *Journal of VLSI Signal Processing*, vol. 1, no. 2, pp. 95–113, 1989.

[13] C. Mauras, P. Quinton, S. Rajopadhye, and Y. Saouter, "Scheduling affine parameterized recurrences by means of variable dependent timing functions," in *Application Specific Array Processors* (S. Kung, E. Schwartzlander, J. Fortes, and K. Przytula, eds.), (Princeton University), pp. 100–110, IEEE Computer Society Press, Sept. 1990.

[14] S. V. Rajopadhye, L. Mui, and S. Kiaei, "Piecewise linear schedules for recurrence equations," in *VLSI Signal Processing, V* (K. Yao, R. Jain, W. Przytula, and J. Rabbaey, eds.), (Napa), pp. 375–384, IEEE Signal Processing Society, Oct. 1992.

[15] P. Feautrier, "Some efficient solutions to the affine scheduling problem. I. one-dimensional time," *International Journal of Parallel Programming*, vol. 21, no. 5, pp. 313–347, Oct. 1992.

[16] P. Feautrier, "Some efficient solutions to the affine scheduling problem. part II. multidimensional time," *International Journal of Parallel Programming,* vol. 21, no. 6, pp. 389–420, Dec. 1992.

[17] W. Shang and A. Fortes, "Time optimal linear schedules for algorithms with uniform dependencies," *IEEE Trans. on Computers*, vol. 40, no. 6, pp. 723–742, June 1991.

[18] A. Darte, L. Khachiyan, and Y. Robert, "Linear scheduling is nearly optimal," *Parallel Processing Letters*, vol. 1, no. 2, pp. 73–82, December 1991.

[19] A. Darte and Y. Robert, "Constructive methods for scheduling uniform loop nests," *IEEE Transactions on Parallel and Distributed Systems*, vol. 5, no. 8, pp. 814–822, Aug. 1994.

[20] A. Darte and Y. Robert, "Affine-by statement scheduling of uniform and affine loop nests over parametric domains," *Journal of Parallel and Distributed Computing*, vol. 29, no. 1, pp. 43–59, February 1995.

[21] A. Darte and F. Vivien, "Revisiting the decomposition of karp, miller and winograd," in *International Conference on Application-Specific Array Processors*, (Strasbourg, France), pp. 12–25, IEEE, July 1995.

[22] P. Gachet and B. Joinnault, *Conception d'algorithmes et d'architectures systoliques.* PhD thesis, Université de Rennes I, September 1987. (In French).

[23] Y. Saouter and P. Quinton, "Computability of recurrence equations," *Theoretical Computer Science*, vol. 114, 1993.

[24] S. V. Rajopadhye, "Synthesizing systolic arrays with control signals from recurrence equations," *Distributed Computing*, vol. 3, pp. 88–105, May 1989.

[25] F. C. Wong and J.-M. Delosme, "Broadcast removal in systolic algorithms," in *International Conference on Systolic Arrays*, (San Diego, CA), pp. 403–412, May 1988.

[26] Y. Yaacoby and P. R. Cappello, "Converting affine recurrence equations to quasi-uniform recurrence equations," in *AWOC 1988: Third International Workshop on Parallel Computation and VLSI Theory*, Springer Verlag, June 1988. See also, UCSB Technical Report TRCS87-18, February 1988.

[27] S. Borkar, R. Cohn, G. Cox, S. Gleason, T. Gross, H. Kung, M. Lam, B. Moore, C. Peterson, J. Pieper, L. Rankind, P. Tseng, J. Sutton, J. Urbanski, and J. Webb, "iWarp: An Integrated Solution to High-speed parallel computing," in *Proceedings of Supercomputing '88*, pp. 330–339, IEEE Computer Society and ACM SIGARCH, Nov 1988.

[28] M. Annaratone et al., "The Warp computer: architecture, implementation, and performance," *IEEE Transactions on Computers*, vol. C-36(12), pp. 1523–38, 1987.

[29] L. G. Hamey, J. A. Webb, and I. C. Wu, "An architecture independent programming language for low-level vision," *Computer Vision, Graphics and Image Processing*, vol. 48, pp. 246-264, 1989.

[30] P. S. Tseng, *A Parallelizing Compiler for Distributed Memory Parallel Computers.* PhD thesis, Carnegie Mellon University, 1988.

[31] H. Printz, H. Kung, T. Mummert, and P. Scherer, "Automatic mapping of large signal processing systems to a parallel machine," in *Proc. of SPIE, Real-Time Signal Processing XII*, (San Diego CA (USA)), Aug. 1989.

[32] D. Lavenier, B. Pottier, F. Raimbault, and S. Rubini, "Fine grain parallelism on a MIMD machine using FPGAs," in *IEEE Workshop on FPGAs for Custom Computing Machines*, (Napa Valley), pp. 2–8, Apr. 1993.

[33] M. Borah, R. Bajwa, S. Hannenhalli, and M. Irwin, "A SIMD solution to the sequence comparison problem on the MGAP," in *ASAP 94, the International Conference on Application Specific Array Processors*, pp. 336–345, IEEE Computer Society Press, 1994.

[34] R. Bajwa, R. Owens, and M. Irwin, "The MGAP's programming environment and the *C++ language," in *ASAP'95* (P. Cappello, C. Mongenet, G.-R. Perrin, P. Quinton, and Y. Robert, eds.), pp. 121–124, IEEE Computer Society Press, July 1995.

[35] D. Foulser and R. Schreiber, "The Saxpy Matrix-1: A general purpose systolic computer," *Computer*, vol. 20, no. 7, pp. 35–43, July 1987.

[36] K. Johnson, A. Hurson, and B. Shirazi, "General-purpose systolic arrays," *Computer*, pp. 20–31, Nov. 1993.

[37] ISATEC Soft - & Hardware GmbH, "The ISATEC Systola 1024 Parallel Computer." Technical Note, 1997. http://www.netzservice.de/Home/isatec.

[38] B. Schmidt, M. Schimmler, and H. Schröder, "Morphological Hough transform on the instruction systolic array," in *Euro-Par'97*, vol. LNCS 1300, pp. 788–806, Springer Verlag, 1997.

[39] D. Hammerstrom, "A VLSI architecture for high-performance, low-cost, on-chip learning," in *International Joint Conference on Neural Networks*, vol. 2, pp. 537–544, 1990.

[40] U. Ramacher et al., "Synapse-1: A high-speed general purpose parallel neurocomputer system," in *IPPS95*, pp. 774–773, IEEE Computer Society Press, Apr. 1995.

[41] J. Hirschberg, R. Hughey, and K. Karplus, "KESTREL: A programmable array for sequence analysis," in *ASAP'96* (I. C. S. Press, ed.), (Chicago, Illinois), pp. 25–34, Aug. 1996.

[42] P. Frison, D. Lavenier, H. Leverge, and P. Quinton, "A VLSI programmable systolic architecture," in *International Conference on Systolic Array*, (Killarney, Irland), IEEE Computer Society Press, Jun. 1989.

[43] P. Frison and D. Lavenier, "Experience in the design of parallel processor arrays," in *International Workshop on Algorithms and Parallel VLSI Architectures II* (P. Quinton and Y. Robert, eds.), (Bonas - France), pp. 233–242, Elsevier Science Publishers B.V., Jun. 1991.

[44] R. Barzic and C. Bouville and F. Charot and G. Le Fol and P. Lemonnier and C. Wagner, "MOVIE: a building block for the desing of real time simulator of moving pictures compression algorithms," in *ASAP'95* (P. Cappello, C. Mongenet, G.-R. Perrin, P. Quinton, and Y. Robert, eds.), pp. 193–203, IEEE Computer Society Press, July 1995.

[45] D. Juvin, J. Basille, H. Essafi, and J. Latil, "Sympathi2 : a 1.5D processor array for image applications," in *EUSIPCO Signal Processing IV : theories and application*, pp. 311–314, North Holland, 1988.

[46] J. Vuillemin, P. Bertin, D. Roncin, M. Shand, H. Touati, and P. Boucard, "Programmable active memories: reconfigurable systems come of age," *IEEE Transactions on VLSI Systemes*, vol. 4, no. 1, Mar. 1996.

[47] D. Buell, J. Arnold, and J. Walter, *Splash 2: FPGAs in a custom computing machine*. IEEE computer society press, 1996.

[48] B. Gunther, "SPACE 2 as a reconfigurable Stream Processor," Tech. Rep. CIS-97-008, School of Computer and Information Science, University of South Australia, 1997.

[49] P. Guerdoux-Jamet and D. Lavenier, "Systolic filter for fast DNA similarity search," in *ASAP'95*, (Strasbourg), July 1995.

[50] J. Peterson and P. Athanas, "High-speed 2-D convolution with a custom computing machine," *Journal of VLSI Signal Processing*, vol. 12, pp. 7–19, 1996.

[51] D. Pryor, M. Thistle, and N. Shirazi, "Text searching on Splash 2," in *IEEE Workshop on FPGAs for Custom Computing Machines*, (Napa, California), 1993.

[52] T. Jebelean, "Implementing GCD systolic arrays on FPGA," in *International Workshop on Field-Programmable Logic and Applications*, vol. LNCS 849, (Prague, Czech Republic), 1994.

[53] J. Andersen, A. Nielsen, and O. Olsen, "A systolic on-line non restoring division scheme," in *Proceedings of the Twenty-Seventh Annual Hawaii International Conference on System Sciences*, 1994.

[54] C. Ebeling, D. Cronquists, and P. Franlin, "Configurable computing: the catalyst for high performance architectures," in *IEEE International Conference on Application Specific, Systems, Architectures, and Processors*, (Zurich, Switzerland), July 1997.

[55] P. Guerdoux–Jamet, D. Lavenier, C. Wagner, and P. Quinton, "Design and implementation of a parallel architecture for biological sequence comparison," in *Euro-Par'96* (L. Boug, P. Fraigniaud, A. Mignotte, and Y. Robert, eds.), no. 1123 in LNCS, (Lyon, France), pp. 11–24, Springer, Aug. 1996.

[56] F. Raimbault and D. Lavenier, "ReLaCS for systolic programming," in *ASAP'93*, Oct. 1993.

[57] P. Baglietto, M. Maresca, A. Migliaro, and M. Migliardi, "Parallel implementation of the full search block matching algorithm for motion estimation," in *ASAP'95* (P. Cappello, C. Mongenet, G.-R. Perrin, P. Quinton, and Y. Robert, eds.), pp. 182–192, IEEE Computer Society Press, July 1995.

[58] B. Christie, "Working demonstration of the MA7170 systolic correlator," in *International Conference on Systolic Arrays* (W. Moore, A. McCabe, and R. Urquhart, eds.), (University of Oxford, UK), pp. 113–122, Adam Hilger, July 2-4 1986.

[59] N. Demassieux, F. Jutand, M. Saint-Paul, and M. Dana, "VLSI architecture for a one chip video median filter," in *ICASSP 85*, pp. 1001–1004, 1985.

[60] P. Frison, P. Gachet, and P. Quinton, "Designing systolic arrays with DIASTOL," in *VLSI Signal Processing II* (S.-Y. Kung, R. E. Owen, and J. G. Nash, eds.), pp. 93–105, IEEE Press, Nov. 1986.

[61] C. Mauras, "Alpha : un langage équationnel pour la conception et la programmation d'architectures parallèles synchrones." Thèse de l'Université de Rennes 1, IFSIC, France, Dec. 1989.

[62] J. Rosseel and M. van Swaaij and F. Catthoor and H. De Man and H. Le Verge and P. Quinton, *Regular Array Synthesis for Image and Video Applications*, ch. 6, pp. 119–142, Kluwer Academic Publisher, 1993.

[63] M. VanSwaaij, "Automating high level control flow transformations for dsp memory management," in *IEEE workshop on VLSI Signal Processing* (IEEE, ed.), 1992.

[64] V. Van Dongen, "PRESAGE, a tool for the design of low-cost systolic circuits," in *IEEE International Symposium on Circuits and Systems*, (Espoo, Finland), June 1988.

[65] J. A. D. Baltus, "Efficient exploration of nonuniform space-time transformations for optimal systolic array synthesis," in *ASAP'93* (L. Dadda and B. Wah, eds.), (Venice), IEEE Computer Society Press, Oct. 1993.

[66] P. Held, P. Dewilde, E. Deprettere, and P. Wielage, *HiFi: from parallel algorithm to fixed-size VLSI processor array*, ch. 2, pp. 71–94. Kluwer Academic Publisher, 1993.

Chapter 24

Low Power CMOS VLSI Design

Tadahiro Kuroda
Microelectronics Engineering Lab.
Toshiba Corporation
Aiwai-ku, Japan
kuroda@sdel.eec.toshiba.co.jp

Takayasu Sakurai
Institute of Industrial Science
University of Tokyo
Tokyo, Japan
tsakurai@iis.u-tokyo.ac.jp

24.1 INTRODUCTION

Multimedia VLSIs demand high-performance as well as low-power consumption [1]. Fig. 24.1 shows the performance requirements for multimedia applications. Higher performance can cover more applications. For example, the MPEG2 decoder, which is the key multimedia function, needs more than 1000 million operations per second (MOPS). Thus, low-power design for multimedia VLSIs means low-power yet high-performance, which is different from the established low-power design techniques for calculators and watches.

The reason why low-power is needed is threefold. It is summarized in Table 24.1. Battery life of portable multimedia products can be lengthened by low-power design. This reason is dominated by VLSIs with power dissipation of less than 0.5 watts. In systems with a power range of up to 5 watts, however, the compelling reason for low-power is to mount the VLSIs in inexpensive plastic packages. Multimedia VLSIs are often used in the consumer market, which is very cost-sensitive. In VLSIs with a power dissipation above 5 watts, on the other hand, low-power design is needed to eliminate a noisy fan and/or to keep the power from exceeding the ceramic package limit.

Fig. 24.2 shows power dissipation of MPUs and DSPs presented in the ISSCC for the past 17 years. The power dissipation has increased fourfold every three years [2], which is the same pace as bit density of state-of-the-art DRAMs. As shown in the figure, the increase in power dissipation comes from an increase in power density.

This trend is understood by the scaling theory which is summarized in Table 24.2. A constant field scaling theory [3] assumes that device voltages as well as device dimensions are scaled by a scaling factor $k(> 1)$, resulting in a constant electric field in a device. This scaling brings a desirable effect that, while power density remains constant, circuit performance can be improved in terms of den-

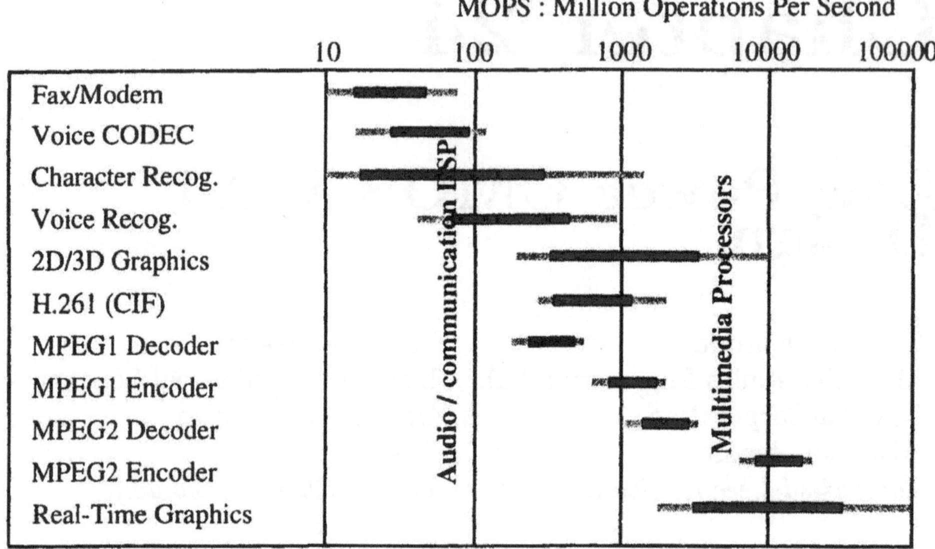

Figure 24.1 Performance requirements for multimedia applications.

Table 24.1 Low-Power Design

Power range	Concerns	Typical applications (All need high-performance)
< 0.5W	• Battery life	Portable • PDA • Communications
0.5W~ 5W	• Plastic package limit • System heat (10W / box)	Consumer • Set-Top-Box • Audio-Visual
> 5W	• Ceramic package limit • IR drop of power lines	Processor • High-end MPUs • Multimedia DSPs

sity (k^2), speed (k), and power ($1/k^2$). But in practice, neither a supply voltage, V_{DD}, nor a threshold voltage of a transistor, V_{TH}, had long been scaled till 1990. The resultant effect is better explained by a constant voltage scaling theory where the circuit speed is further improved (k^2), while the power density increases very rapidly as k^3. For more precise analysis for submicron devices [4], the power density

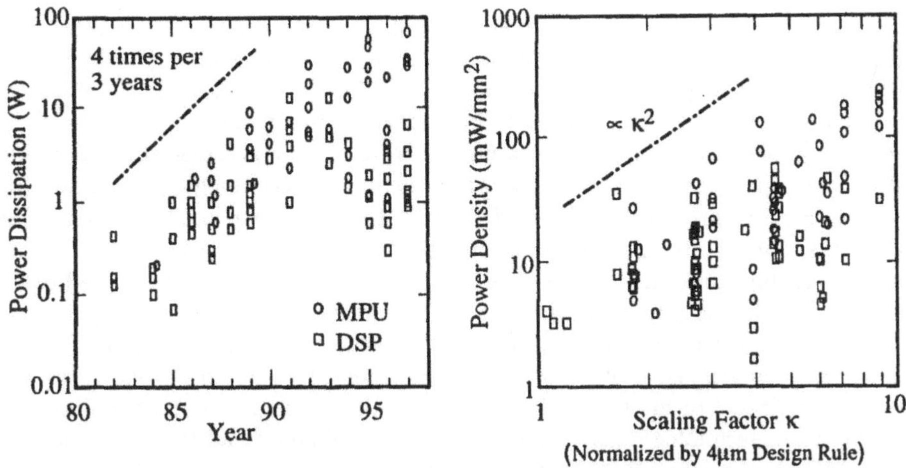

Figure 24.2 Power dissipation for the past 17 years.

increases as $k^{\alpha+1}$, where α represents the velocity saturation effect and is typically 1.3 for 0.5μm MOSFETs. Thus, it is clear that this constant voltage scaling causes the rapid increase in CMOS power dissipation.

Table 24.2 Scaling Theory

Parameters		Scaling scenario		
		Const. field	Const. voltage	$\propto 1/\kappa^{0.5}$ voltage
Device size		$1/\kappa$	$1/\kappa$	$1/\kappa$
Gate thickness	tox	$1/\kappa$	$1/\kappa$	$1/\kappa^{0.5}$
Substrate doping		κ	κ^2	$\kappa^{1.5}$
Supply voltage	V	$1/\kappa$	1	$1/\kappa^{0.5}$
Electric field	E	1	κ	1
Current	I	$1/\kappa$	$\kappa \quad (\kappa^{\alpha-1})$	$1/\kappa$
Area	A	$1/\kappa^2$	$1/\kappa^2$	$1/\kappa^2$
Capacitance	$C=\varepsilon A/tox$	$1/\kappa$	$1/\kappa$	$1/\kappa^{1.5}$
Gate delay	VC/I	$1/\kappa$	$1/\kappa^2 (1/\kappa^{\alpha})$	$1/\kappa$
Power dissipation	VI	$1/\kappa^2$	$\kappa \quad (\kappa^{\alpha-1})$	$1/\kappa^{1.5}$
Power density	VI/A	1	$\kappa^3 \quad (\kappa^{\alpha+1})$	$\kappa^{0.5}$

Since 1990 V_{DD} has been reduced, which is making the power increase slower. However with a plausible scaling scenario where V_{DD} is reduced as $k^{0.5}$ [5] the power density is still increasing as $k^{0.5}$. This means that power dissipation of CMOS VLSIs is steadily increasing as a natural result of device scaling.

Thus, power dissipation will continue to be an important constraint in the future in designing not only portable applications but also for desktop use. Designers should always think of power budget together with delay and area. In this sense, a low-power technology is becoming a must-have technology rather than a preferable technology.

A CMOS technology is reaching a turning point because of the increasing power dissipation. Integrated circuits have been constantly evolving by increasing integration level (e.g., the Moore's law). When they face a limitation of the integration a mainstream technology is to be changed. Historically the limitation was caused by power dissipation. When a bipolar technology was replaced by an nMOS technology in the early 60's, and also when the nMOS technology was replaced by the CMOS technology in the middle of the 80's, the reason was the limitation of the integration due to their high power dissipation. Even with penalties in cost, speed, and turn-around-time the CMOS took the place of the nMOS because the CMOS dissipated negligible power and for the next 15 years no limitation was perceived in terms of integrating circuits. However, as shown in Fig. 24.2, the CMOS power dissipation was increased by 1000 times during this 15 years, and is exceeding 100 watts. The CMOS technology is facing the same situation as the nMOS. Unfortunately no new technology has been discovered yet, while we had the CMOS for the post nMOS technology. In this sense the destiny of the integrated circuits in the era of "system on a chip" will rest upon a low-power CMOS technology.

In Section 24.2, CMOS circuit power dissipation is analyzed and general guidelines for power reduction are described. The power dissipation can be reduced by 1) lowering the supply voltage, 2) reducing load capacitance, 3) reducing switching activity. In theses, using low supply voltages is the most effective way to reduce the power, since the CMOS power dissipation depends quadratically on the supply voltage. Low voltage circuits are discussed in Section 24.3. Low voltage designs, however, essentially suffers from speed degradation. Some of the solutions for this problem are discussed in this section, together with the recent circuit developments including threshold-voltage control circuits and supply-voltage control circuits. In Section 24.4 capacitance reduction is discussed. In some cases pass-transistor logic requires fewer transistors and exhibits smaller stray capacitance than the conventional CMOS static circuits. Variations in its circuit topology as well as a logic synthesis method are presented and discussed. Section 24.5 is dedicated for summary.

24.2 ANALYSIS OF CMOS POWER DISSIPATION

24.2.1 Expression for CMOS Power Dissipation

CMOS power dissipation is given by [2]

$$P = p_t f_{CLK} \cdot (C_L V_S + \bar{I_{SC}} \cdot \Delta t_{SC}) \cdot V_{DD} + (I_{DC} + I_{LEAK}) \cdot V_{DD} \qquad (1)$$

The first term, $p_t f_{CLK} C_L V_S V_{DD}$, represents dynamic dissipation due to charging and discharging of the load capacitance, where p_t is the switching probability, f_{CLK} is the clock frequency, C_L is the load capacitance, V_S is the voltage swing of a signal, and V_{DD} is the power supply voltage. In most cases, V_S is the same as V_{DD}, but in some logic circuits V_S may be smaller than V_{DD} for high-speed and/or low-power operations. The second term, $p_t f_{CLK} \bar{I_{SC}} \Delta t_{SC} V_{DD}$, is dynamic dissipation due to switching transient current (so-called crowbar current), where $I_{\bar{SC}}$ is the mean value of the switching transient current, and Δt_{SC} is time while the switching transient current flows. This dissipation can be held down to from 10% to15% of the first term through careful design [6]. The third term, $I_{DC}V_{DD}$, is static dissipation in such a circuit as a current mirror sense amplifier where current is designed to draw continuously from the power supply. The last term, $I_{LEAK}V_{DD}$, is due to the subthreshold current and the reverse bias leakage between the source/drain diffusions and the substrate.

With respect to the first term, it is understood that, in a CMOS gate, charge of the amount of $C_L V_S$ loses V_{DD} of potential in one operation so that the energy of $C_L V_S V_{DD}$ is consumed per operation. Consequently, this expression is exact even with spurious resistors associated with MOS circuits and non-linear nature of MOS transistors. The power changes only when C_L is non-linear or time-variant. Normally an average of C_L over the voltage swing is used.

24.2.2 General Guidelines for Low-Power Design

The dominant term in a well-designed logic circuit is the charging and discharging term which is given by

$$P = p_t f_{CLK} \cdot C_L \cdot {V_{DD}}^2. \tag{2}$$

As is seen from this expression, general guidelines for CMOS circuit power reduction is basically threefold: to reduce switching probability, to lower operation voltage, and to reduce load capacitance. In these, reducing switching probability is mostly a task of CAD tools, which will not be discussed in this chapter. Other two approaches are discussed more in detail in the following sections.

We now have general understanding of the power dependence on design parameters but there is other standpoint to attack low-power design problems. Location dependence of power dissipation in a VLSI is also of interest. Fig. 24.3 shows power distribution of CMOS VLSIs. It should be noted that the power distribution pattern changes from design to design so that designers should be well aware of the effective points to attack before tackling the low-power design problems of their specific VLSIs.

Table 24.3 summarizes reported low-power circuit design techniques in the two-dimensional fashion, one dimension being the design parameters and the other dimension being the target location of the low-power techniques.

24.3 LOW VOLTAGE CIRCUITS

Lowering the supply voltage is the most attractive choice due to the quadratic dependence. However, as the supply voltage becomes lower, circuit delay increases and chip throughput degrades. There are three means to maintain throughput: 1) to lower V_{TH} to recover circuit speed [7]; 2) to employ parallel and/or pipeline architecture to compensate for the degraded circuit speed [8]; and 3) to make use of multiple supply voltages and provide lower supply voltages only to those circuits which can be slow [9].

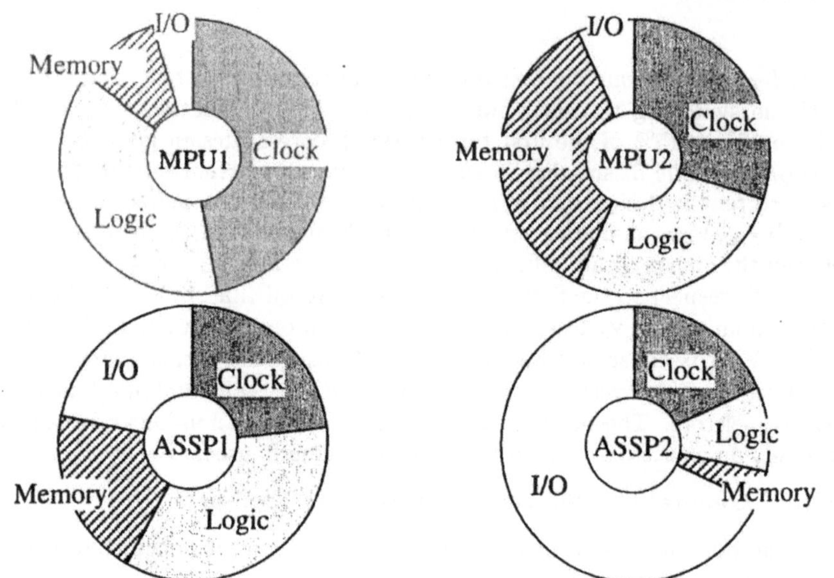

Figure 24.3 Power distribution in CMOS VLSIs.

The idea of the second approach is to utilize the increasing transistor density to trade off silicon area against power dissipation. This approach will be mandatory, but currently the area penalty is considerable. This can be understood from the fact that chip size is still increasing even though the transistor density is increasing by 60% per year.

In the third approach a good voltage level converter is essential for signal interface between circuit clusters in different supply voltages. Otherwise all the efforts are for nothing due to speed and power penalties induced by the level converter. How efficiently can circuits be clustered with the minimum number of the signal interface is to be investigated.

In this section, circuit design techniques for lowering both V_{DD} and V_{TH} are discussed.

24.3.1 $V_{DD} - V_{TH}$ Design Space

In low V_{TH} power dissipation due to the subthreshold current cannot be ignored. Taking the first term and the last term in (1), CMOS power dissipation is given by

$$P = p_t f_{CLK} C_L {V_{DD}}^2 + I_0 10^{-\frac{V_{TH}}{S}} V_{DD} \tag{3}$$

where S is the subthreshold slope and is typically about 100mV/decade.

CMOS circuit propagation delay, on the other hand, is approximately given in [4] by

$$D = \gamma \frac{C_L V_{DD}}{(V_{DD} - V_{TH})^{\alpha}} \tag{4}$$

Table 24.3 Low-Power CMOS VLSI Circuit Design Techniques

	P_t	C_L	V_S	V_{DD}	I_{LEAK}	f_{CLK}	I_{SC}	I_{DC}
General		• device scaling	Small Signal	Low V_{DD} • DC-DC [22] • VS scheme [23,25] • Multiple V_{DD} [9]	V_{TH} Control • VTCMOS [11-15, 18, 20] • EVTCMOS[19]		Careful Design • design verification by CAD	
Clock	• gated clock • data-transition look-ahead DFF [48]	• floorplan to reduce wire length • F/F sizing Charge Recycling • C stacking [36]	• RCSFF [26] • 1/2 swing [36]					
Bus	Glitch Suppress • 3-state-buffer activated after data fix	• C stacking [37] • exclusive bus	• 1/4 swing [37]					
Data Path	• latch insertion to deskew data-in [38]	Tr. Reduction • pass-transistor (CPL[29], DCVSPG[30], SRPL[31],)	• pass-tr. (SAPL) [21]			• parallelism [8]		
Random Logic	CAD • permutation of series-connected tr. order [39]	• library & CAD for pass-tr. logic [35] • Tr. sizing [28][40]	• current switch logic (MCML) [41]		Sleep Mode • MTCMOS [16, 17]			
Memory		• memory hierarchy	• reduced swing WL, BL		• switched source-impedance [43]			Cut Current • latch S/A [42]
I / O		• MCM [44] • area pad [44]	• reduced swing I/O GTL[45], LVDS			• phase modulation [46]		• dynamic termination [47]

where α represents the velocity saturation effect and is typically 1.3 for submicron MOSFETs.

Fig. 24.4 shows the power and the delay dependence on V_{DD} and V_{TH} calculated from (3) and (4). Obviously tradeoffs between the power and the delay exists. However if V_{DD} and V_{TH} are carefully chosen better tradeoffs can be found. Lines on the V_{DD} - V_{TH} plane in the right figure represent equi-delay lines. When moving on the same equi-delay line, for example from A to B, circuit speed is maintained, while power dissipation is reduced by about 50% as shown in the left figure. In this way equi-speed lines and equi-power lines on the V_{DD}-V_{TH} plane are to be investigated.

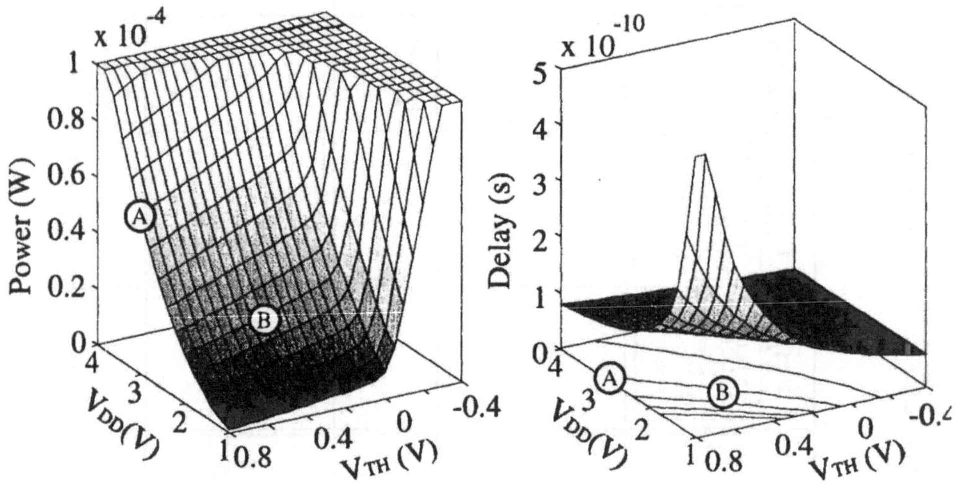

Figure 24.4 Power and delay dependence on V_{DD} and V_{Th}.

Fig. 24.5 depicts equi-speed lines (broken lines) and equi-power lines (solid lines) on the V_{DD}-V_{TH} plane calculated from (3) and (4). Typically circuits are designed at V_{DD} =3.3V ± 10% and V_{TH} =0.55V ± 0.1V as shown as a rectangle in Fig. 24.5. This rectangle is a design window because all the circuit specifications should be satisfied within the rectangle for yield conscious design. In the design window circuit speed becomes the slowest at the corner A while at the corner B power dissipation becomes the highest. Therefore better tradeoffs between the speed and the power can be found by reducing fluctuations of V_{DD} and V_{TH}. The equi-speed lines and the equi-power lines are normalized at the corner A and B, respectively, so that it can be figured out how much speed and power dissipation are improved or degraded compared to the typical condition by sliding and sizing the design window on the V_{DD}-V_{TH} plane. For example at V_{DD}=2.2V ± 10% and V_{TH} =0.19V ± 0.05V the power dissipation can be reduced to about 50% while maintaining the circuit speed. A dotted line in Fig. 24.5 depicts a equi-power-ratio line where power dissipation due to the subthreshold leakage current makes up 10% of the total power dissipation. The power dissipation becomes a minimum around on this line under each speed condition. The results of exploring the low-V_{DD}, low-V_{TH} design space are summarized in Fig. 24.6. Optimum V_{DD} and V_{TH} can save the waste of power and speed caused by the constant V_{DD} and V_{TH}. The waste is depicted as shadow in the figure. In this way optimizing V_{DD} and V_{TH} is essential in low-power high-speed CMOS design while they are treated constant and common parameters in the conventional CMOS design.

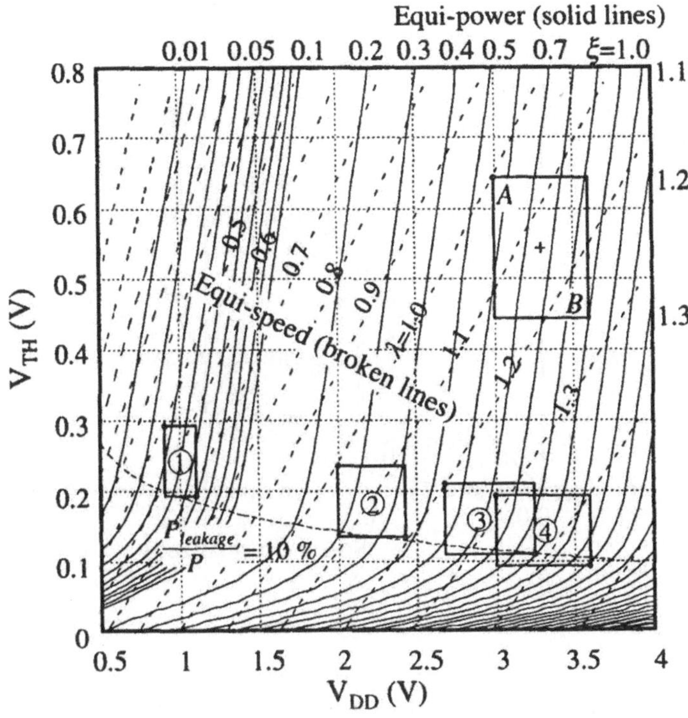

Figure 24.5 Exploring low-V_{DD}, low-V_{TH} design space.

This approach, however, raises three problems : 1) degradation of worst-case speed due to V_{TH} fluctuation in low V_{DD} [10, 11], 2) increase in standby power dissipation in low V_{TH}, [12]-[17], and 3) an inability to sort out defective chips by monitoring the quiescent power supply current (I_{DDQ}) [15]. Delay variations due to V_{TH} fluctuation is increased in low VDD. It can be understood in Fig. 24.5 that the design window should be reduced in size to keep the delay variation percentage constant in low V_{DD} [18]. The second and the third problems come from the increased subthreshold leakage current in low V_{TH}. The design window should be returned to a high V_{TH} region in a standby mode and the I_{DDQ} testing.

In the next section V_{TH} control circuits to solve these problems are discussed.

24.3.2 V_{TH} Control Circuits

V_{TH} control schemes

To solve the three problems in low-V_{DD} and/or low-V_{TH} several circuit schemes have been proposed. These include a multi threshold-voltage CMOS (MTCMOS) scheme [16], a variable threshold-voltage CMOS (VTCMOS) scheme [18], and an elastic- V_t CMOS (EVTCMOS) scheme [19].

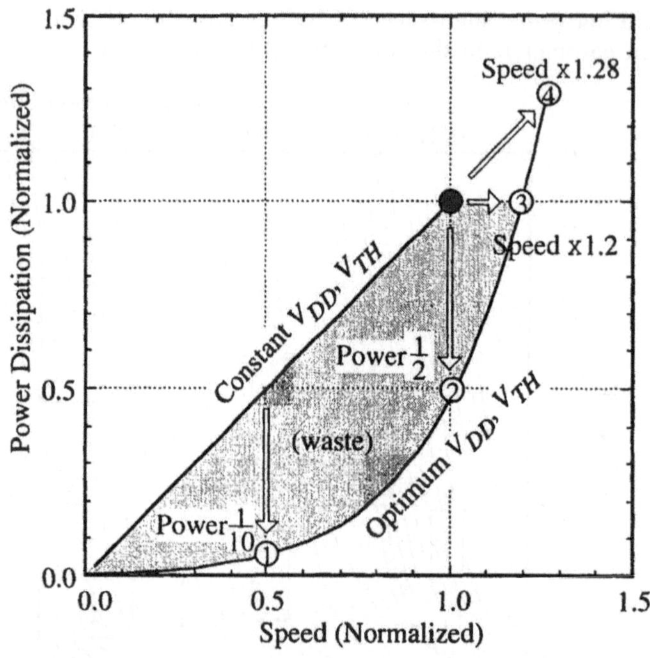

Figure 24.6 Speed and power saving by optimum V_{DD} and V_{TH}.

The three circuit schemes are sketched and compared in Fig. 24.7. Currently in a system such as a PC, power supply is turned off by a power management controller when a chip is inactive, but this idea can even be applied at the chip level. The MTCMOS [16] employs two V_{TH}; low V_{TH} for fast circuit operation in logic circuits and high V_{TH} for cutting off the subthreshold leakage current in the standby mode. Since parasitic capacitance is much smaller on a chip than on a board the on-off control of the power supply can be performed much faster on a chip, in less than 0.5 μs, which in turn enables frequent power management [17]. This scheme is straightforward and easy to be employed in a current design method. However, it requires very large transistors for the power supply control, and hence imposes area and yield penalties, otherwise circuit speed degrades. These penalties become extreme below 0.9V V_{DD} because (V_{GS} - V_{TH}) of the high-V_{TH} transistors becomes too small in the active mode. One potential problem is that the MTCMOS cannot store data in the standby mode. Special latches have been developed to solve this problem [17].

While the MTCMOS can solve only the standby leakage problem, the VTC-MOS [18] can solve all three problems. It dynamically varies V_{TH} through substrate-bias, V_{BB}. Typically, V_{BB} is controlled so as to compensate V_{TH} fluctuations in the active mode, while in the standby mode and in the I_{DDQ} testing, deep V_{BB} is applied to increase V_{TH} and cut off the subthreshold leakage current.

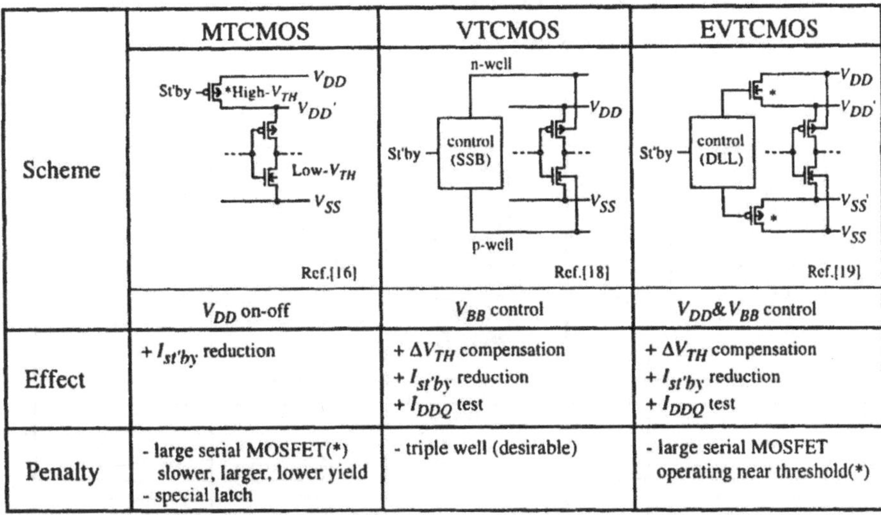

	MTCMOS	VTCMOS	EVTCMOS
Scheme	Ref.[16]	Ref.[18]	Ref.[19]
	V_{DD} on-off	V_{BB} control	V_{DD}& V_{BB} control
Effect	+ $I_{st'by}$ reduction	+ ΔV_{TH} compensation + $I_{st'by}$ reduction + I_{DDQ} test	+ ΔV_{TH} compensation + $I_{st'by}$ reduction + I_{DDQ} test
Penalty	- large serial MOSFET(*) slower, larger, lower yield - special latch	- triple well (desirable)	- large serial MOSFET operating near threshold(*)

Figure 24.7 MTCMOS, VTCMOS, and EVTCMOS.

The EVTCMOS [19] controls both V_{DD} and V_{BB} such that when V_{DD} is lowered V_{BB} becomes that much deeper to raise V_{TH} and further reduce power dissipation. Note that internal V_{DD} and V_{SS} are provided by source-follower nMOS and pMOS transistors, respectively, whose gate voltages are controlled. In order to control the internal power supply voltage independent from the power current, the source-follower transistors should operate near the threshold. This requires very large transistors.

The essential difference among the three schemes is that the VTCMOS controls the substrate-bias while the others control the power lines. Since much smaller (almost no) current flows in the substrate than in the power lines, much smaller circuit can control the substrate-bias. This leads to negligible penalties in area and speed in the VTCMOS. Global routing of substrate-contacts, however, may impose area penalty. It has been experimentally evaluated that the number of substrate (well) contacts can be greatly reduced in low voltage environments [15, 18]. Using a phase-locked loop and an SRAM in a VTCMOS gate-array [15], the substrate noise influence has been shown to be negligible even with 1/400 of the contact frequency compared with the conventional gate-array. A discrete cosine transform (DCT) macro made with the VTCMOS [18] has also been manufactured with substrate- and well-contacts only at the periphery of the macro and it worked without problems realizing more than one order of magnitude smaller power dissipation than a DCT macro in the conventional CMOS design. The VTCMOS will be discussed in more detail in the next section.

Variable threshold-voltage CMOS scheme
VTCMOS variants

Several variants have been developed [11]-[15],[18] for the VTCMOS, whose salient features are summarized in Fig. 24.8.

	SATS	SPR			SATS + SPR
Circuit	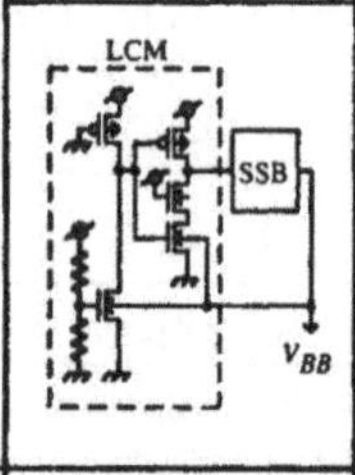 LCM, SSB, V_{BB}	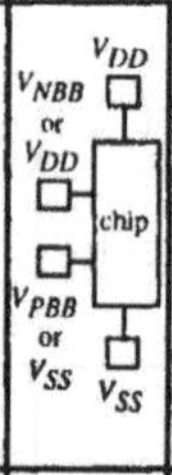 V_{DD}, V_{NBB} or V_{DD}, chip, V_{PBB} or V_{SS}, V_{SS}	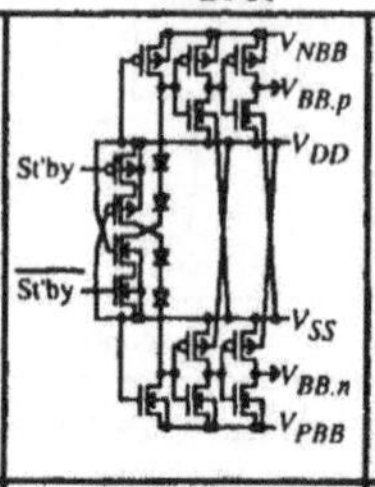 St'by, V_{NBB}, $V_{BB.p}$, V_{DD}, V_{SS}, $V_{BB.n}$, V_{PBB}	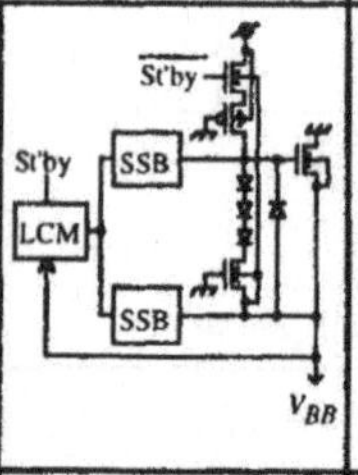 St'by, SSB, LCM, SSB, V_{BB}	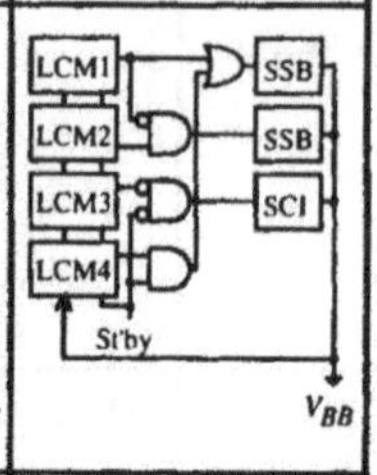 LCM1, LCM2, LCM3, LCM4, SSB, SSB, SCI, St'by, V_{BB}
Active	SSB	—	short to V_{DD},V_{SS}	short to V_{DD},V_{SS}	SSB
St'by	no consideration	—	short to V_{NBB},V_{PBB}	SSB	SSB
Transition time	— —	—	0.1μs (St'by → Active) 0.1μs (Active → St'by)	0.1μs (St'by → Active) 100μs (Active → St'by)	0.1μs (St'by → Active) 100μs (Active → St'by)
Effect	+ ΔV_{TH} compensation	+ I_{DDQ} test	+ $I_{st'by}$ reduction + I_{DDQ} test	+ $I_{st'by}$ reduction + I_{DDQ} test	+ ΔV_{TH} compensation + $I_{st'by}$ reduction + I_{DDQ} test
Penalty	- <5% area + 300μm - triple well (desirable)	- <5% area	- <5% area + 70μm - 3 supplies required - triple well (desirable)	- <5% area + 500μm	- <5% area + 500μm - triple well (desirable)

LCM: Leakage Current Monitor, SSB: Self Substrate Bias, SCI: Substrate Charge Injector

Figure 24.8 VTCMOS variants.

A self-adjusting threshold-voltage scheme (SATS) [11] reduces the V_{TH} fluctuation. When V_{TH} is lower than a target value, larger leakage current flows through a leakage current monitor (LCM) and turns on a self-substrate bias (SSB) circuit. As a result, V_{BB} goes deeper and causes V_{TH} to increase. Thus the substrate bias is controlled such that the transistor leakage current is adjusted to be a constant value. This means that the V_{TH} process fluctuation can be canceled by the SATS. The measured overall V_{TH} controllability including static and dynamic effects is $\pm$ 0.05V while using bare process V_{TH} fluctuation of $\pm$ 0.15V. The same idea is also presented in [20].

Three standby power reduction (SPR) schemes are developed. One is to just take out substrate (well) contacts to Pads. In the I_{DDQ} testing deep V_{BB} is applied from external power supplies through the Pads to raise V_{TH} and cut off the substreshold leakage, while in shipping the chip the Pads are connected to V_{DD} and GND. The other two schemes reported in [12],[14] are to lengthen the battery life in mobile applications. One scheme in [12] switches V_{BB} between the power supply and an additional supply for substrate bias. It requires three external power supplies but takes less than 0.1μs for the substrate bias switching. Triple well technology is a must. The other scheme in [14] employs the SSB for the substrate bias. No additional external power supply or additional steps in process are required. An active to standby mode transition is performed by the SSB, and hence takes about 100μs. On the other hand, a standby to active mode transition is carried out by a MOS switch, and therefore is completed in 0.1 μ s. This "slow falling asleep but quick awakening" feature is acceptable for most applications.

The latest scheme in [18] is realized by combining the SATS [11] and the SPR [14]. It achieves both the SATS and the SPR capability at the same time. Operation principles are the same, and hence the same circuit performance. This scheme will be discussed more in detail as a representative of the VTCMOS scheme.

Control scheme

Fig. 24.9 depicts the VTCMOS scheme block diagram. The VTCMOS scheme consists of four Leakage Current Monitors (LCMs), the Self Substrate Bias circuit (SSB), and a Substrate Charge Injector (SCI). The SSB draws current from the substrate to lower V_{BB}. The SCI, on the other hand, injects current into the substrate to raise V_{BB}. The SSB and the SCI are controlled by monitoring where V_{BB} sits in four ranges. Their criteria are specified in the four LCMs: for example, $V_{active(+)}$=-0.3 volts, V_{active}=-0.5 volts, $V_{active(-)}$=-0.7 volts, and $V_{standby}$=-3.3 volts. The substrate bias is monitored by transistor leakage current, because the leakage current reflects V_{BB} very sensitively.

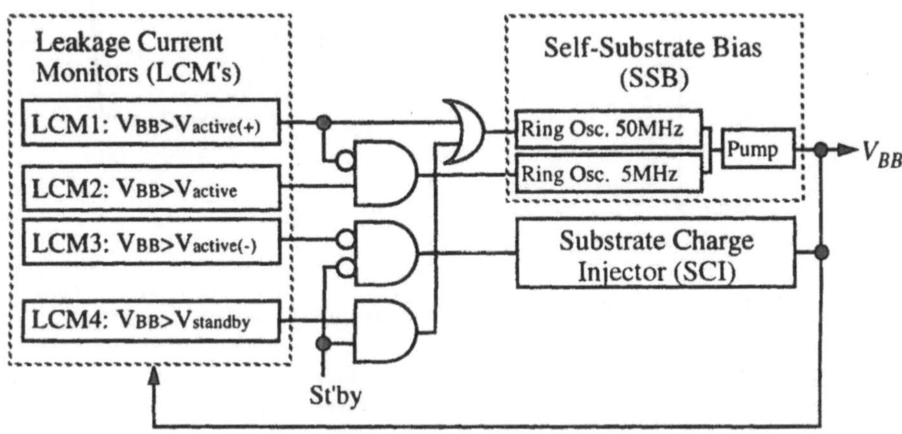

Figure 24.9 VTCMOS block diagram.

Fig. 24.10 illustrates the substrate bias control. After a power-on, V_{BB} is higher than $V_{active(+)}$, and the SSB begins to draw 100μA from the substrate to lower V_{BB} using a 50MHz ring oscillator. This current is large enough for V_{BB} to settle down within 10μs after a power-on. When V_{BB} goes lower than $V_{active(+)}$, the pump driving frequency drops to 5MHz and the SSB draws 10μA to control V_{BB} more precisely. The SSB stops when V_{BB} drops below V_{active}. V_{BB}, however, rises gradually due to device leakage current through MOS transistors and junctions, and reaches V_{active} to activate the SSB again. In this way, V_{BB} is set at V_{active} by the on-off control of the SSB. When V_{BB} goes deeper than $V_{active(-)}$, the SCI turns on to inject 30mA into the substrate. Therefore, even if V_{BB} jumps beyond $V_{active(+)}$ or $V_{active(-)}$ due to a power line bump for example, V_{BB} is quickly recovered to V_{active} by the SSB and the SCI. When "SLEEP" signal is asserted ("1") to go to the standby mode, the SCI is disabled and the SSB is activated again and 100μA current is drawn from the substrate until V_{BB} reaches $V_{standby}$. V_{BB} is set at $V_{standby}$ in same way by the on-off control of the SSB. When "SLEEP" signal becomes "0" to go back to the active mode, the SSB is disabled and the SCI is activated. The SCI injects 30mA current into the substrate until V_{BB} reaches $V_{active(-)}$. V_{BB} is finally set at V_{active}. In this way, the SSB is mainly used for a transition from the active mode to the standby mode, while the SCI is used for a transition from the standby mode to the active mode. An active to standby mode transition takes about 100μs, while a standby to active mode transition is completed in 0.1μs. This "slow falling asleep but fast awakening" feature is acceptable for most of the applications.

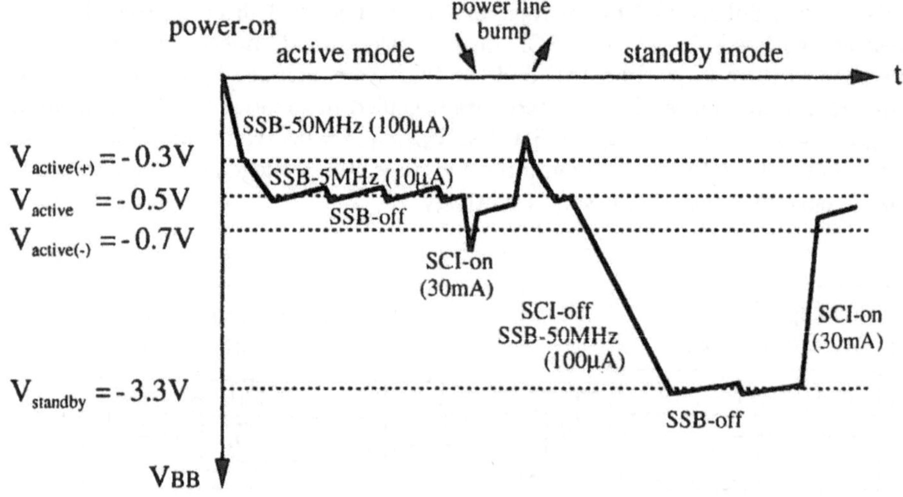

Figure 24.10 Substrate bias control in VTCMOS.

The SSB operates intermittently to compensate for the voltage change in the substrate due to the substrate current in the active and the standby modes. It therefore consumes several microamperes in the active mode and less than one nanoamperes in the standby mode, both much lower than the chip power dissipation. Energy required to charge and discharge the substrate for switching between the active and the standby modes is less than 10nJ. Even when the mode is switched 1000 times in a second, the power dissipation becomes only 10μW. The leakage current monitor should be designed to dissipate less than 1nA because it is always active even in the standby mode.

In the VTCMOS scheme care should be taken so that no transistor sees high-voltage stress of gate oxide and junctions. The maximum voltage that assures sufficient reliability of the gate oxide is about V_{DD}+20%. All transistors in the VTCMOS scheme receive (V_{DD}-V_{TH}) on their gate oxide when the channel is formed in the depletion and the inversion mode, and less than $|V_{standby}|$ in the accumulation mode. These considerations lead to a general guideline that $V_{standby}$ should be limited to -(V_{DD}+20%). $V_{standby}$ of -(V_{DD}+20%), however, can shift V_{TH} big enough to reduce the leakage current in the standby mode. The body effect coefficient, g, can be adjusted independently to V_{TH} by controlling the doping concentration density in the channel-substrate depletion layer.

VTCMOS circuit implementations
Leakage current monitor (LCM)

The substrate bias is generated by the SSB which is controlled by the Leakage Current Monitor (LCM). The LCM is therefore a key to the accurate control in the VTCMOS scheme. Fig. 24.11 depicts a circuit schematic of the LCM. The circuit works with 3.3-volt V_{DD} which is usually available on a chip for standard interfaces with other chips. The LCM monitors leakage current of a chip, $I_{leak.CHIP}$, with a transistor M4 that shares the same substrate with the chip. The gate of M4 is biased to Vb to amplify the monitored leakage current, $I_{leak.LCM}$, so that the circuit response can be shortened and the dynamic error of the LCM can be reduced. If $I_{leak.LCM}$ is larger than a target reflecting shallower V_{BB} and lower V_{TH}, the node N1 goes "*low*" and the output node N_{out} goes "*high*" to activate the SSB. As a result, V_{BB} goes deeper and V_{TH} becomes higher, and consequently, $I_{leak.LCM}$ and $I_{leak.CHIP}$ become smaller. When $I_{leak.LCM}$ becomes smaller than the target, the SSB stops. Then $I_{leak.LCM}$ and $I_{leak.CHIP}$ increase as V_{BB} gradually rises due to device leakage current through MOS transistors and junctions, and finally reaches the target to activate the SSB again. In this way $I_{leak.CHIP}$ is set to the target by the on-off control of the SSB with the LCM.

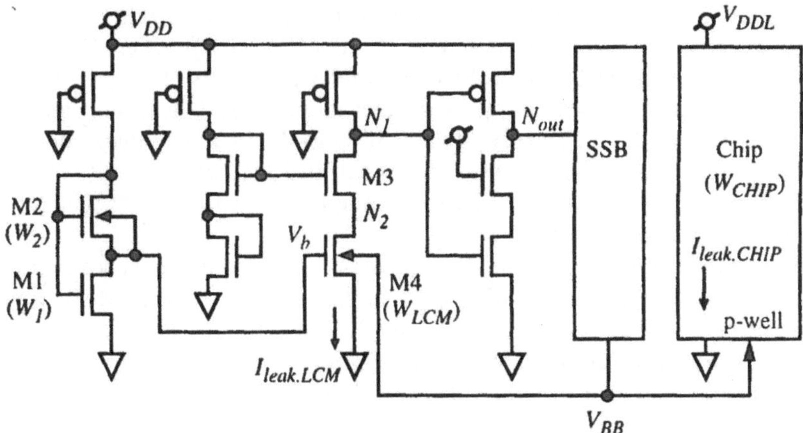

Figure 24.11 Leakage current monitor (LCM).

In order to make this feedback control accurately, the current ratio of $I_{leak.LCM}$ to $I_{leak.CHIP}$, or the current magnification factor of the LCM, X_{LCM}, should be constant. When a MOS transistor is in subthreshold its drain current is expressed as

$$I_{DS} = \frac{I_0}{W_0} W \cdot 10^{\frac{V_{GS}-V_{TH}}{S}} \tag{5}$$

where S is the subthreshold swing, V_{TH} is the threshold voltage, I_0/W_0 is the current density to define V_{TH}, and W is the channel width. By applying (5), X_{LCM} is given by

$$X_{LCM} = \frac{I_{leak.LCM}}{I_{leak.DCT}} = \frac{W_{LCM}}{W_{CHIP}} 10^{\frac{V_b}{S}} \tag{6}$$

where W_{CHIP} is the total channel width in the chip and WLCM is the channel width of M4. Since two transistors M1 and M2 in a bias generator are designed to operate in subthreshold region, the output voltage of the bias generator, V_b, is also given from (5) by

$$V_b = S \log W_2 W_1 \tag{7}$$

where W_1 and W_2 are the channel widths of M1 and M2, respectively. X_{LCM} is therefore expressed as

$$X_{LCM} = \frac{W_2}{W_1} \frac{W_{LCM}}{W_{CHIP}}. \tag{8}$$

This implies that X_{LCM} is determined only by the transistor size ratio and independent of the power supply voltage, temperature, and process fluctuation. In the conventional circuit in [11], on the other hand, where V_b is generated by dividing the V_{DD}-GND voltage with high impedance resistors, V_b becomes a function of V_{DD}, and therefore, X_{LCM} becomes a function of V_{DD} and S, where S is a function of temperature. Fig. 24.12 shows SPICE simulation results of X_{LCM} dependence on circuit condition changes and process fluctuation. X_{LCM} exhibits small dependence on DVTH and temperature. This is because M4 is not in deep subthreshold region. The variation of X_{LCM}, however, is within 15 %, which results in less than 1 % error in V_{TH} controllability. This is negligible compared to 20 % error in the conventional implementation.

The four criteria used in the substrate-bias control, corresponding to $V_{active(+)}$, V_{active}, $V_{active(-)}$, and $V_{standby}$ can be set in the four LCMs by adjusting the transistor size W_1, W_2, and W_{LCM} in the bias circuit. For the active mode, with W_1=10μm, W_2=100μm, and W_{LCM}=100μm, the magnification factor X_{LCM} of 0.001 is obtained when W_{CHIP} =1m. $I_{leak.CHIP}$ of 0.1mA can be monitored as $I_{leak.LCM}$ of 0.1mA in the active mode. For the standby mode, with W_1=10μm, W_2=1000μm, and W_{LCM}=1000μm, X_{LCM} becomes 0.1. Therefore, $I_{leak.CHIP}$ of 10nA can be monitored as $I_{leak.CHIP}$ of 1nA in the standby mode. The overhead in power by the monitor circuit is about 0.1% and 10% of the total power dissipation in the active and the standby mode, respectively.

The parasitic capacitance at the node N_2 is large because M4 is large. This may degrade response speed of the circuit. The transistor M3, however, isolates the N1 node from the N_2 node and keeps the signal swing on N2 very small. This reduces the response delay and improves dynamic V_{TH} controllability.

Compared with the conventional LCM where V_b is generated by dividing the V_{DD}-GND voltage with high impedance resistors, the V_{TH} controllability including

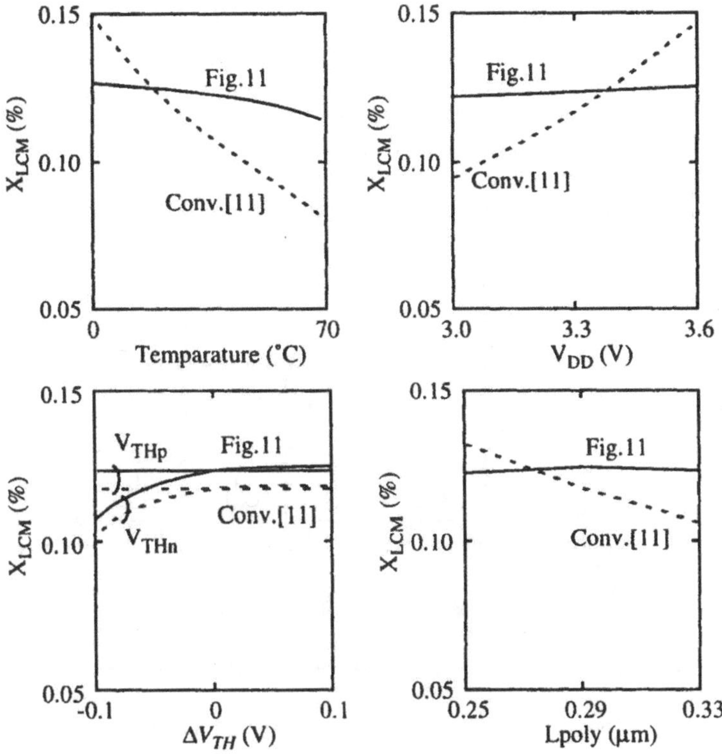

Figure 24.12 Current magnification factor of the LAC, X_LCM, dependence on circuit condition changes and process deviations simulated by SPICE.

the static and dynamic effects is improved from ±0.05 volts to less than ±0.01 volts, response delay is shortened from $0.6\mu s$ to $0.1\mu s$, and pattern area is reduced from $33250\mu m^2$ to $670\mu m^2$. This layout area reduction is brought by the elimination of the high impedance resistors made by polysilicon.

Self substrate bias circuit (SSB)

Fig. 24.13 depicts a schematic diagram of a pump circuit in the Self Substrate Bias circuit (SSB). PMOS transistors of the diode configuration are connected in series whose intermediate nodes are driven by two signals, F1 and F2, in 180 phase shift. Every other transistor therefore sends current alternately from p-well to GND, resulting in lower p-well bias than GND. The SSB can pump as low as -4.5 volts. SSB circuits are widely used in DRAMs and E^2PROMs, but two orders of magnitude smaller circuit can be used in the VTCMOS scheme. The driving current of the SSB is $100\mu A$, while it is usually several milliamperes in DRAMs. This is because substrate current generation due to the impact ionization is a strong function of the supply voltage. Substrate current in a 0.9-volt design is considerably smaller than that in a 3.3-volt design. Substrate current introduced from I/O Pads does not affect the internal circuits if they are separated from peripheral circuits by a triple-well structure. Eventually no substrate current is generated in the standby mode. From these reasons the pumping current in the SSB can be as small as several percent of that in DRAMs. Silicon area is also reduced considerably. Another concern about the SSB is an initialization time after a power-on. Even in a 10mm square chip, V_{BB} settles down within 200μs after a power-on which is acceptable in real use.

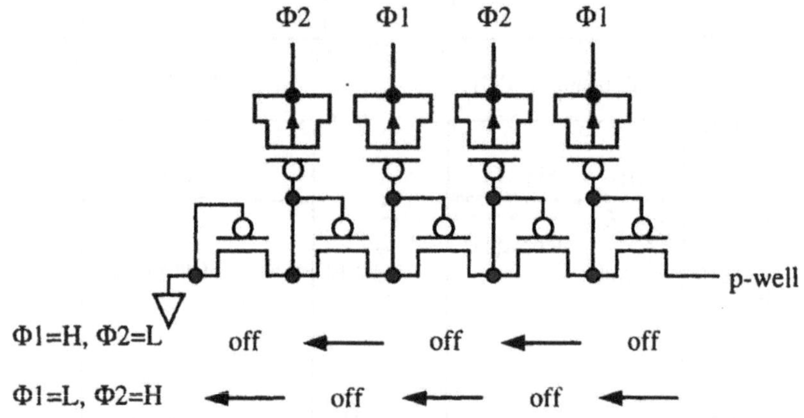

Figure 24.13 Pump circuit in self substrate bias (SSB).

Substrate charge injector (SCI)

The Substrate Charge Injector (SCI) in Fig. 24.14 receives a control signal that swings between V_{DD} and GND at node N_1 to drive the substrate from $V_{standby}$ to V_{active}. In the standby-to-active transition, $V_{DD} + | |V_{standby}| |$ that is about 6.6 volts at maximum can be applied between N_1 and N_2. However, as shown in SPICE simulated waveforms in Fig. 24.14, $| |V_{GS}| |$ and $| |V_{GD}| |$ of M1 and M2 never exceed the larger of V_{DD} and $| |V_{standby}| |$ in this circuit implementation to ensure sufficient reliability of transistor gate oxide.

DCT macro design in VTCMOS

The VTCMOS scheme is employed in a two-dimensional 8 by 8 discrete cosine transform (DCT) core processor for portable HDTV-resolution video compression / decompression. This DCT core processor executes two-dimensional 8 by 8 DCT and inverse DCT. A block diagram is illustrated in Fig. 24.15. The DCT is composed of two one-dimensional DCT and inverse DCT processing units and a transposition RAM. Rounding circuits and clipping circuits which prevent overflow and underflow are also implemented. The DCT has a concurrent architecture based on distributed arithmetic and a fast DCT algorithm, which enables high through-put DCT processing of one pixel per clock. It also has fully pipelined structure. The 64 input data sampled in every clock cycles are output after 112 clock cycle latency.

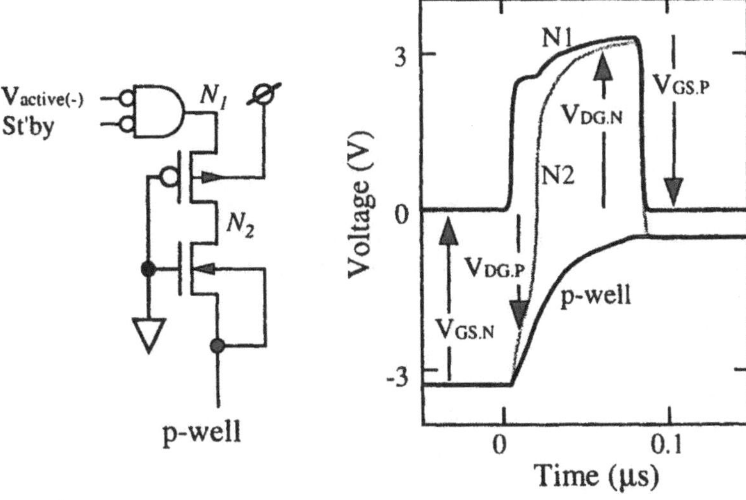

Figure 24.14 Substrate charge injector (SCI) and its waveforms simulated by SPICE.

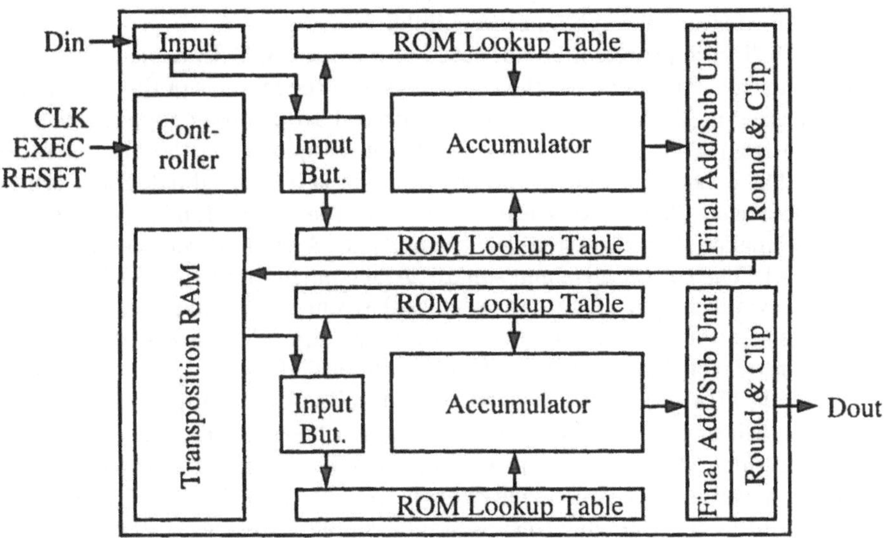

Figure 24.15 DCT block diagram.

Various memories which use the same low V_{TH} transistors as logic gates are employed in the DCT. Table lookup ROMs (16bits by 32words by 16banks) employ contact programming and an inverter-type sense-amplifier. Single-port SRAMs (16bits by 64words by 2banks) and dual-port SRAMs (16bits by 8words by 2banks) employ a 6-transistor cell and a latch sense-amplifier. They all exhibit wide operational margin in low V_{DD} and low V_{TH} and almost behave like logic gates in terms

Table 24.4 Features of DCT Macro

Technology	0.3μm CMOS, tripple-well double-metal, V_{TH}=0.15V+-0.1V
Power supply voltage	1.0V+-0.1V
Power dissipation	10mW @ 150MHz
Standby current	< 10nA @ 70C
Transistor count	120k Tr
Area	2.0 x 2.0 mm2
Function	8 x 8 DCT and inverse DCT
Data format	9-b signed (pixel), 12-b signed (DCT)
Latency	112 clock
Throughput	64 clocks / block
Accuracy	CCITT H.261 compatible

of circuit speed dependence on V_{DD} and V_{TH}. No special care is necessary such as word-line boosted-up or a special sense-amplifier.

The DCT core processor is fabricated in a 0.3 μm CMOS triple-well double-metal technology. Parameters of the technology and the features of the DCT macro are summarized in Table 24.4. It operates with a 0.9-volt power supply which can be supplied from a single battery source. Power dissipation at 150MHz operation is 10mW. The leakage current in the active mode is 0.1mA, about 1% of the total power current. The standby leakage current is less than 10nA, four orders of magnitude smaller than the active leakage current. A chip micrograph appears in Fig. 24.16(a). The core size is 2mm square. A magnified picture of the VT control circuit is shown in Fig. 24.16(b). It occupies 0.37mm by 0.52mm, less than 5% of the macro size. If additional circuits for testability are removed and the layout is optimized, the layout size is estimated to be 0.3mm by 0.3mm.

Figs. 24.17(a)-(c) show measured p-well voltage waveforms. Due to large parasitic capacitance in a probe card the transition takes longer time than SPICE simulation results. Just after the power-on, the VT circuits are not activated yet because the power supply is not high enough. As shown in Fig. 24.17(a) p-well is biased forward by 0.2 volts due to capacitance coupling between p-well and power lines. Then the VT circuits are activated and p-well is to be biased at -0.5 volts. It takes about 8μs to be ready for the active mode after the power-on. The active-to-standby mode transition takes about 120μs as shown in Fig. 24.17(b), while the standby-to-active mode transition is completed within 0.2μs as presented in Fig. 24.17(c).

Compared to the DCT in the conventional CMOS design [21], power dissipation at 150MHz operation is reduced from 500mW to 10mW, that is only 2%. Most of the power reduction, however, is brought by capacitance reduction and voltage reduction by technology scaling. Technology scaling from 0.8 μm to 0.3 μm reduces power dissipation from 500mW to 100mW at 3.3V and 150MHz operation.

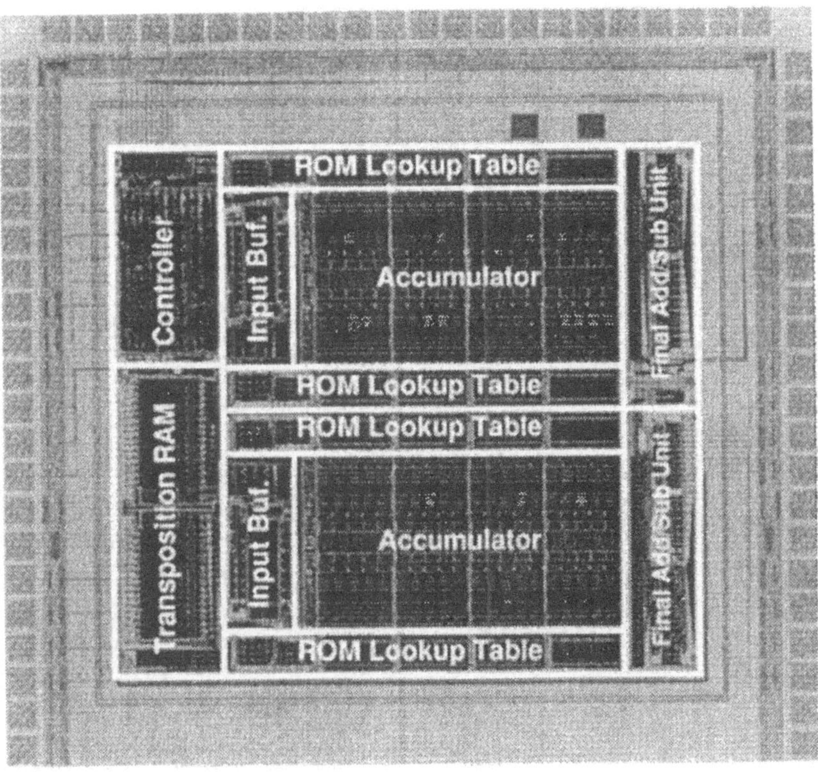

(a) DCT macro

(b) VT macro

Figure 24.16 Chip micrograph: (a) DCT macro and (b) VT macro.

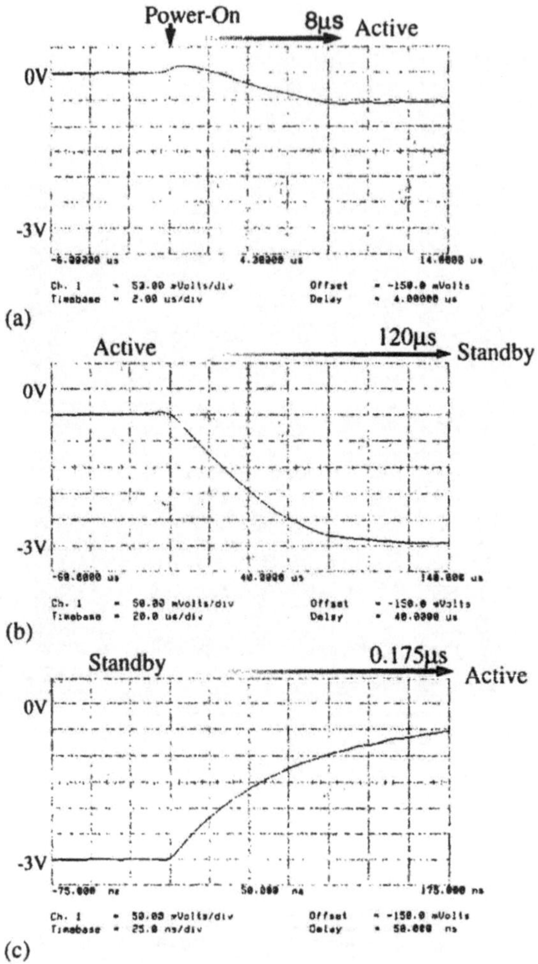

Figure 24.17 Measured p-well bias V_{BB}: (a) after power-on, (b) active-to-standby, and (c) standby-to-active.

Without the VTCMOS scheme, V_{DD} and V_{TH} cannot be lowered under 1.7V and 0.5V, respectively, and the active power dissipation is to be 40mW. It is fair to claim that the VTCMOS scheme reduces the active power dissipation from 40mW to 10mW.

24.3.3 V_{DD} Control Circuits

Variable supply-voltage (VS) scheme

A circuit scheme to control V_{DD} on a chip, namely variable supply-voltage scheme (VS scheme) is discussed in this section. In the VS scheme a DC-DC converter [22] generates an internal supply voltage, V_{DDL}, very efficiently from an external power supply, V_{DD}. V_{DDL} is controlled by monitoring propagation delay of a critical path in a chip such that it is set to the minimum of voltages in which the chip can operate at a given clock frequency, f_{ext}. This control also reduces V_{DDL}

fluctuations, which is essential in low-voltage design. A 32-bit RISC core processor is designed with the VS scheme in the VTCMOS [23], and achieves more than twice improvement in MIPS/W performance compared with the previous CMOS design [24] in the same technology.

The VS scheme is illustrated in Fig. 24.18. It consists of three parts: 1) a buck converter, 2) a timing controller, and 3) a speed detector. The buck converter generates for the internal supply voltage, V_{DDL} . N is an integer from 0 to 64 which is provided from the timing controller. Therefore the resolution of V_{DDL} is about 50mV. A duty control circuit generates rectangular waveforms with duty cycle of N/64 whose average voltage is produced by the second order low-pass filter configured by external inductance, L , and capacitance, C. The lower limit of V_{DDL} can be set in the duty control circuit to assure the minimum operating voltage of a chip. The upper limit can also be set to prevent N from transiting spuriously from 63 to 0 due to noise.

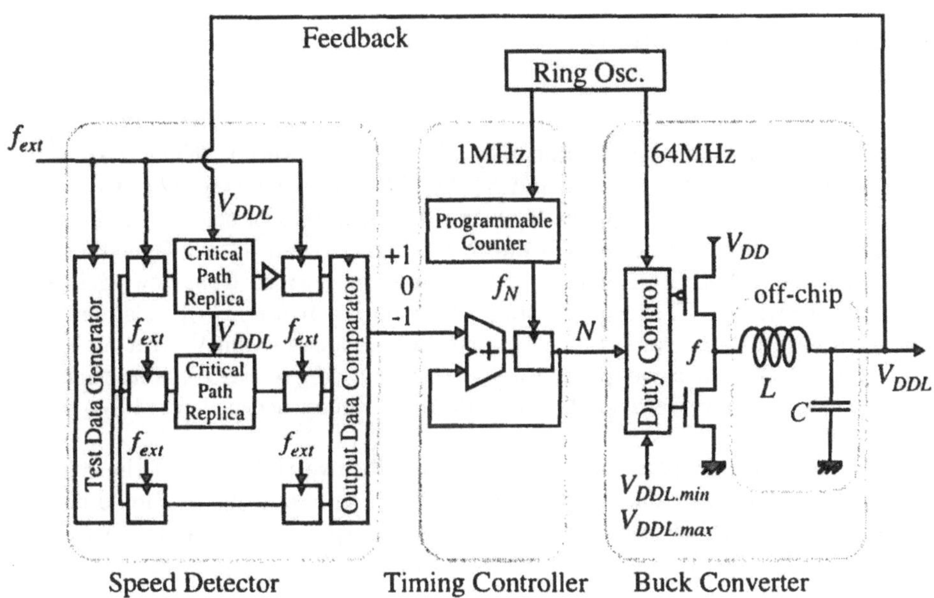

Figure 24.18 Variable supply-voltage (VS) scheme.

The timing controller calculates N by accumulating numbers provided from the speed detector, +1 to raise V_{DDL}, and -1 to lower V_{DDL}. The accumulation is carried out by a clock whose frequency is controlled by a 10-bit programmable counter.

The speed detector monitors critical path delay in the chip by its replicas under V_{DDL}. When V_{DDL} is too low for the circuit operation in f_{ext}, the speed detector outputs +1 to raise V_{DDL}. On the other hand when V_{DDL} is too high the speed detector outputs -1 to lower V_{DDL}. By this feedback control, the VS scheme can automatically generate the minimum V_{DDL} which meets the demand on its operation frequency. For fail-safe control small delay is to be added to the critical path replicas.

Since the speed detection cycle based on f_{ext} (e.g., 25ns) is much faster than the time constant of the low-pass filter (e.g., 16ms) the feedback control may fall into oscillation. The programmable counter in the timing controller adjusts the accumulation frequency, fN, to assure fast and stable response of the feedback control.

There is no interference between the VS scheme and the VTCMOS scheme. The VTCMOS scheme controls V_{TH} by referring to leakage current of a chip, while the VS scheme controls V_{DDL} by referring to f_{ext}. V_{DDL} is also affected by V_{TH} because circuit speed is dependent on V_{TH}. Therefore, V_{TH} is determined by the VTCMOS scheme, and under the condition, V_{DDL} is determined by the VS scheme. Since VTCMOS scheme is immune to V_{DDL} noise (See Sec. 24.3.2), there is no feedback from the VS scheme to the VTCMOS scheme, resulting in no oscillation problem between them.

VS circuit implementations
Buck Converter

Fig. 24.19 depicts a circuit schematic of the buck converter. When the output of a 6-bit counter, n, is between 0 and N, the pMOS device of the output inverter is turned on. When n is between N+1 and 63, an nMOS of the output inverter is turned on. When n is between N and N+1, and between 63 and 0, neither the pMOS nor the nMOS is turned on to prevent short current from flowing in the large output inverter. The output voltage of the buck converter, V_{DDL}, is therefore controlled with 64-step resolution. This resolution causes +50mV error at V_{DDL} from V_{DD} =3.3V, which yields +3.3% V_{DDL} error at V_{DDL} =1.5V. Note that the error is always positive because the speed detector cannot accept lower V_{DDL} than a target voltage.

The external low-pass filter, L and C, an effective resistance of the output inverter, R, and its switching period, DT (or switching frequency, f), should be designed considering DC-DC conversion efficiency, h, output voltage ripple, DV/Vout, time constant of the filter as an index of the response, T0, and pattern area, S.

The efficiency, η, can be expressed as

$$\eta = \frac{V_{out} I_{out}}{V_{out} I_{out} + I_{out}^2 R + P_{VX} + P_{control}} \tag{9}$$

where P_{VX} is power dissipation at the output inverter caused by overshoot and undershoot at VX from V_{DD} and ground potential due to inductance current, and Pcontrol is power dissipation of control circuits. Fig. 24.20 shows simulated waveforms at VX. As shown in the figure inappropriate L increases P_{VX}. Its analytical model can be derived from an equivalent LCR circuit in Fig. 24.21 with the following two assumptions.

1) Duty ratio, D, is assumed to be 0.5 for calculation simplicity.

2) Dumping factor of the low-pass filter is assumed to be 1 for fast and stable response.

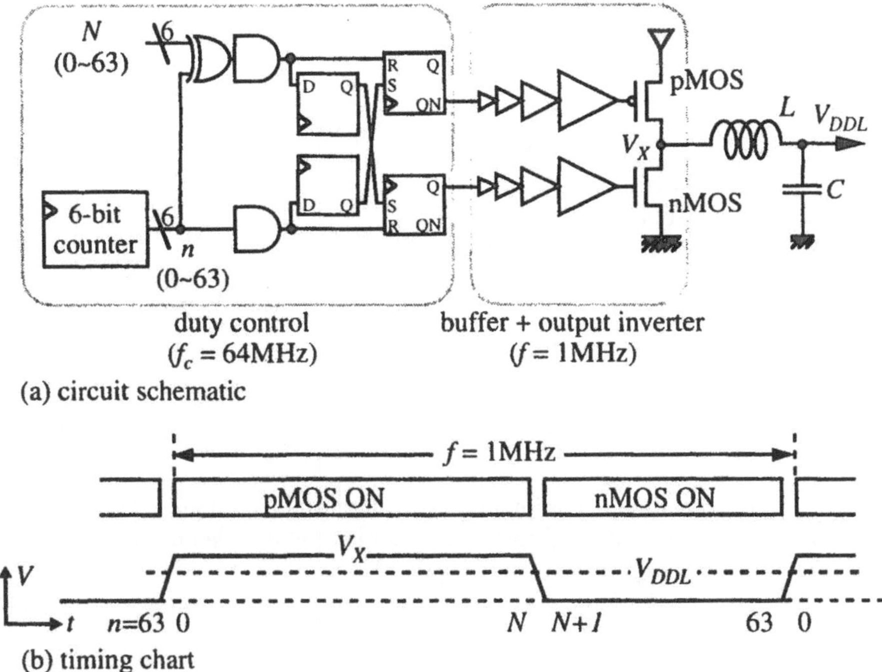

Figure 24.19 Buck converter: (a) circuit schematics, and (b) timing chart.

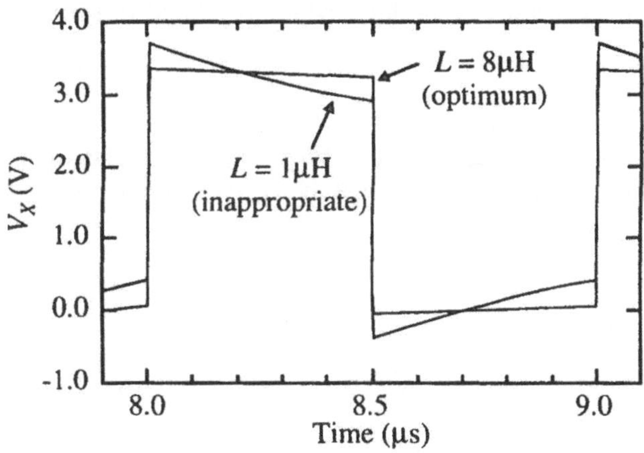

Figure 24.20 Simulated waveforms at V_X.

$$\xi = \frac{R}{2}\sqrt{\frac{C}{L}} = 1 \tag{10}$$

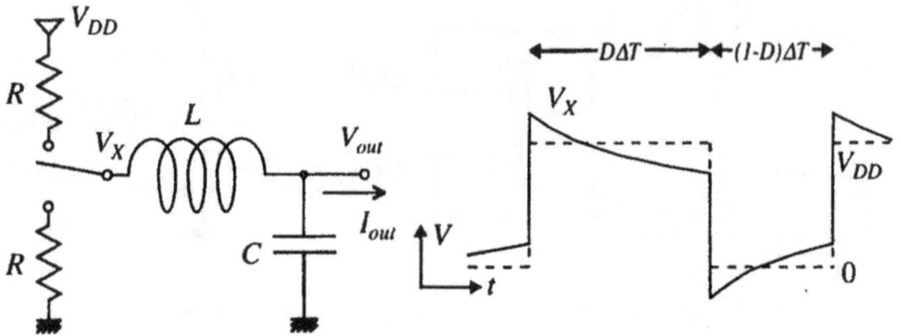

Figure 24.21 Equivalent LCR circuit.

After the conventional manipulation of differential equations of the equivalent circuit, P_{VX} is approximately given as (see [25] for the detailed derivation)

$$P_{VX} = \frac{V_{DD}^2}{R} \cdot \frac{\beta^2}{24} \tag{11}$$

where

$$\beta \equiv \frac{\Delta T}{T_0}. \tag{12}$$

To is the time constant of the filter which is related to settling time, and is given by

$$T_0 = \sqrt{LC}. \tag{13}$$

The output voltage ripple, DV/Vout, can also be derived from the differential equations, and expressed approximately as (see [25] for the detailed derivation)

$$\frac{\Delta V_{out}}{V_{out}} = \frac{\beta^2}{16}. \tag{14}$$

$P_{control}$, on the other hand, is written as

$$P_{control} = \alpha_c N_{max} f C_c V_{DD}^2 + f C_{buffer} V_{DD}^2 + \alpha_{replica} f_{ext} C_{replica} V_{out}^2. \tag{15}$$

where

$$f = \frac{1}{\Delta T}. \tag{16}$$

The first term is power dissipation of the duty control circuits where operating frequency is $N_{max} \cdot f$. N_{max} is the output voltage resolution which is 64 in this design. The second term is power dissipation of the buffer circuit in the buck converter, and the third term is power dissipation of the replica circuits in the speed detector. α is switching probability and C is capacitance.

Since most of the layout pattern is occupied by the large inverter and buffer circuits, pattern area can be expressed as

$$S = \frac{S_1}{R} + S_2 \tag{17}$$

where S_1 and S_2 are constants.

From these equations, the smaller b, the smaller P_{VX} and the smaller output voltage ripple. On the other hand for the smaller settling time, the smaller T_0 is preferable. Therefore ΔT should be reduced, which in turn increases $P_{control}$. In this way there are tradeoffs among these parameters.

For example under the following constraints,
Output voltage: V_{out} =2.1V
Output current: I_{out} =67mA (P_{out}=140mW)
Output voltage ripple: DV/V_{out} ¡ 0.1%
Filter time constant (related to settling time): T_0 ¡ 100 μ s
Pattern area: S ¡ 500μm-square
DC-DC efficiency: η = maximum
L , C, R, and f can be numerically solved as follows.
Low-pass filter inductance: L =8 μ H ,
Low-pass filter capacitance: C =32 μ F ,
Output inverter effective resistance: R =1 Ω ,
Output inverter switching frequency: f =1MHz .

For the equivalent R=1 Ω in the output inverter, transistor size of the pMOS and the nMOS is as large as 7.6mm and 3.8mm, respectively. Cascaded inverters are necessary to drive the output inverter with a typical inverter whose pMOS and nMOS transistor size is about 8 μm and 4 μm, respectively. The optimum scale-up factor, x, and the optimum number of stages, n, to minimize the power dissipation of the cascaded inverters are given by (see [25] for the detailed derivation)

$$x = 1 + \sqrt{1 + K} \tag{18}$$

$$n = \frac{\log \frac{W_n}{W_0}}{\log x} \tag{19}$$

where K is the ratio of power dissipation due to capacitance charging and discharging to power dissipation due to crowbar current when x=1. From simulation study depicted in Fig. 24.22 the above equations hold very accurately with K=8. The optimum scale-up factor, x, becomes 4, and the optimum number of stages, n, becomes 5 in this design.

Speed detector

A circuit schematic of the speed detector is shown in Fig. 24.23(a). It has three paths under V_{DDL} : 1) a critical path replica of the chip, "CPR", 2) the same critical path replica with inverter gates equivalent to 3% additional delay, "CPR+", and 3) direct connection between flip-flops, "REF". Since the direct connection can always transmit the test data correctly within the cycle time of f_{ext} even in low V_{DDL}, it can be referred to as a correct data. Other paths may output wrong data when the delay time becomes longer than the cycle time of the given f_{ext} at the given V_{DDL}. By comparing the outputs of these paths with that of the direct connection, it can be figured out whether or not the chip operates correctly in f_{ext}

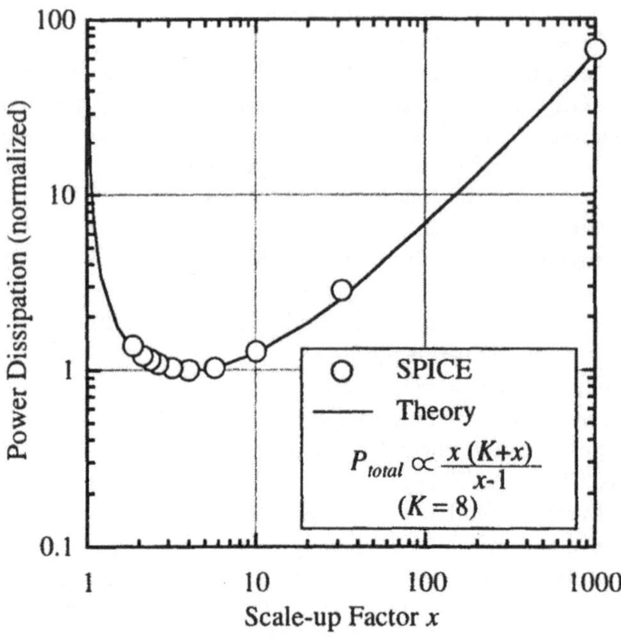

Figure 24.22 Power dissipation dependence on scale-up factor in cascaded inverters.

at V_{DDL}. When V_{DDL} is not high enough, the outputs of the two paths, "CPR" and "CPR+", are both wrong, and the speed detector outputs +1 to raise V_{DDL}. When V_{DDL} is higher equivalent to more than 3% delay in the critical path than the given f_{ext}, the outputs of the two paths are both correct, and the speed detector outputs -1 to lower V_{DDL}. When V_{DDL} is in between, the output of the critical path, "CPR", is correct and that of the longer path, "CPR+", is wrong, and the speed detector outputs 0 to maintain V_{DDL}. This non-detecting voltage gap is necessary to stabilize V_{DDL} but yields an offset error. The offset error should be minimized but no smaller than the minimum resolution of the V_{DDL}. It is because if the gap is smaller than the resolution, no V_{DDL} level may exist in the voltage gap. This may cause the output voltage ripple as large as the resolution. The 3% additional delay corresponds to 80mV in V_{DDL}, which is larger than the resolution of 50mV. In total, V_{DDL} may have 130mV offset error.

A timing chart of the speed detector is illustrated in Fig. 24.23(b). The test data in this figure is an example when the critical path becomes critical in propagating a low-to-high signal. The test is performed every 8 clock cycles. The rest 7 clock cycles are necessary in low V_{DDL} for not evaluating test data provided before. V_{DDL} can be set at very low voltages where the propagation delay becomes eight multiples of the cycle time of f_{ext}. This mislocking, however, can be avoided by setting the lower limit of V_{DDL} in the timing controller. The compared results are registered by flip-flops which are hold by a held signal as shown in Fig. 24.23(a) until the next evaluation.

Since the critical path replicas operate at V_{DDL}, the signals need to be level-shifted to V_{DD}. A sense-amplifier flip-flop [21] is employed to perform level-shifting and registering simultaneously.

Timing controller

A timing controller adjusts the control frequency of N, f_N, to realize the fast and stable response of the feedback control. The higher f_N the faster response, but the lower stability. Conventional stability analysis and compensation techniques, however, are rather difficult to be applied because of several reasons. In the speed detector, circuit speed is a nonlinear function of V_{DDL}. Its output is +1 or -1 regardless of the magnitude of the error in V_{DDL}. Most of the control is performed in digital while the low-pass filter is analog. With these difficulties a programmable counter is introduced as a practical way to control f_N. Based upon experimental evaluation the optimum f_N can be found and set to the programmable counter.

Fig. 24.23 depicts simulation results of V_{DDL} after power-on. When f_N is 1MHz, much faster than the roll-off frequency of the low-pass filter, 10kHz, oscillation appears in V_{DDL}. When f_N is 62.5kHz, on the other hand, the response of V_{DDL} is fast and stable. V_{DDL} can reach the target voltage in 100μs after power-on.

RISC core design with VS scheme in VTCMOS

A 32-bit RISC core processor, R3900, is implemented by about 440k transistors, including 32-bit MAC (Multiplier Accumulator), 4-kB direct mapped instruction cache, and 1-kB 2-way set-associative data cache [24]. Layout is slightly modified for the VS scheme and the VTCMOS. A VS macro and a VT macro are added at the corners of the chip. Many of the substrate contacts are removed [15] and the rest are connected to the VT macro. The chip is fabricated in a 0.4 μm CMOS n-well/p-sub double-metal technology. A chip micrograph appears in Fig. 24.25. Main features are summarized in Table 24.5. The VS and the VT macros occupies 0.45 x 0.59mm2, and 0.49 x 0.72mm2, respectively. The total area penalty of the two macros is less than 1 % of the chip size.

Fig. 24.26 is a shmoo plot of the RISC processor. The RISC core operates at 40MHz at 1.9V, and at 10MHz at 1.3V. In this figure, measured V_{DDL} versus f_{ext} are also plotted. The VS scheme can generate the minimum V_{DDL} of the voltages where the circuit can operate at f_{ext}. Practically fail-free operation should be guaranteed. The VS scheme should be designed such that V_{DDL} is controlled to sit sufficiently inside of the pass region in the shmoo plot by adding supplementary gates to the critical path replicas.

Fig. 24.27 shows a measured power dissipation of the RISC core without I/O. White circles and black squares in this figure represent power dissipation at 3.3V and V_{DDL} determined by the VS scheme, respectively. The VS scheme can reduce power dissipation more than proportionally to its operating frequency. The power dissipation at $f_{ext} = 0$ in the VS scheme is about 20mW which comes from the DC-DC converter. This power loss is mainly due to circuits for experimental purposes and can be reduced to lower than 10mW. The DC-DC efficiency, η, is measured and plotted in Fig. 24.28. The left side of the peak is degraded by the power dissipation in DC-DC itself, while the right side of the peak is degraded by parasitic resistance. Due to the power dissipation of the experimental circuits and

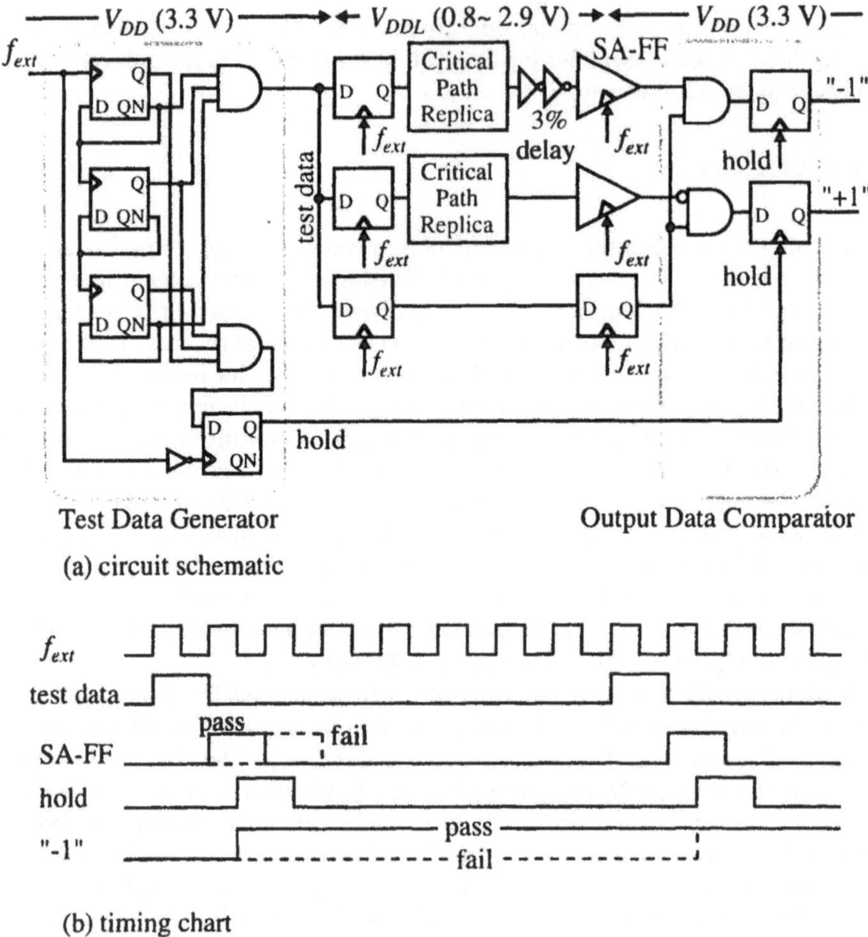

Figure 24.23 Speed detector: (a) circuit schmatics, and (b) timing chart.

due to high contact resistance of about 6W in a probe card the maximum efficiency is lower than anticipated. If the experimental circuits are removed and the chip is bond-wired in a package the maximum efficiency is estimated to be higher than 85% .

Measure performance in MIPS/W are 320MIPS/W at 33MHz, and 480MIPS/W at 20MHz, which are improved by a factor of more than 2 compared with that of the previous design, 150MIPS/W [24].

Fig. 24.29 shows measured V_{DDL} voltage regulated by the VS scheme when V_{DD} is varied by about 50% . The robustness to the supply-voltage fluctuation is clearly demonstrated. V_{DDL} is regulated at a target voltage as long as V_{DD} is higher than the target.

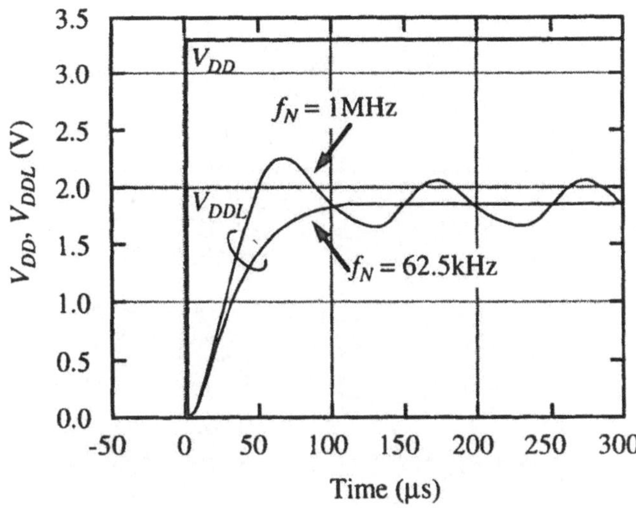

Figure 24.24 Simulated V_{DDL} response after power-on.

24.3.4 Low-Swing Circuits

One interesting observation of the power distribution in Fig. 24.3 is that a clock system and a logic part itself consume almost the same power in various chips, and the clock system consumes 20 % to 45 % of the total chip power. One of the reasons for this large power dissipation of the clock system is that the transition ratio of the clock net is one while that of ordinary logic is about one third on average.

In order to reduce the clock system power, it is effective to reduce a clock voltage swing. Such idea is embodied in the Reduced Clock Swing Flip-Flop (RCSFF) [26]. Fig. 24.30 shows circuit diagrams of the RCSFF. The RCSFF is composed of a current-latch sense amplifier and cross-coupled NAND gates which act as a slave latch. This type of flip-flop was first introduced in 1994 [21] and extensively used in a microprocessor design [27]. The sense-amplifying F/F is often used with low-swing circuits because there is no DC leakage path even if the input is not full swing being different from the conventional gates or F/Fs.

The salient feature of the RCSFF is to accept a reduced voltage swing clock. The voltage swing, V_{CLK}, can be as low as 1V. When a clock driver Type A in Fig. 24.31 is used, power improvement is proportional to V_{CLK}, while it is V_{CLK}^2 if Type B driver is used. Type A is easy to implement but is less efficient. Type B needs either an external V_{CLK} supply or a DC-DC converter.

The issue of the RCSFF is that when a clock is set high to V_{CLK}, P1 and P2 do not switch off completely, leaving leak current flowing through either P1 or P2. The power dissipation by this leak current turns out to be permissible for some cases, but further power improvement is possible by reducing the leak current. One way is to apply backgate bias to P1 and P2 and increase the threshold voltage. The

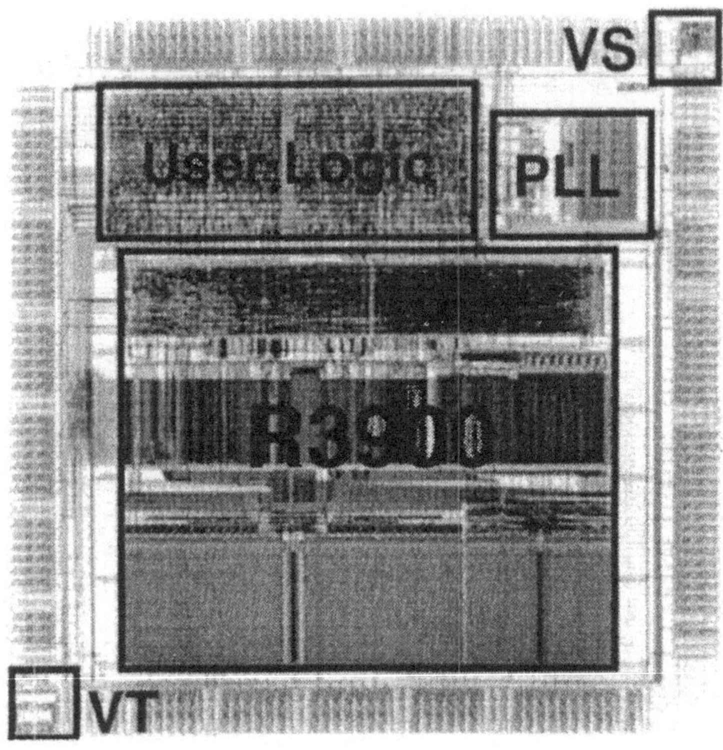

Figure 24.25 Chip micrograph of R3900 with VS scheme in VTCMOS.

other way is to increase the V_{TH} of P1 and P2 by ion-implant, which needs process modification and is usually prohibitive. When the clock is to be stopped, it should be stopped at V_{SS}. Then there is no leak current.

The area of the RCSFF is about 20% smaller than the conventional F/F as seen from Fig. 24.32 even when the well for the precharge pMOS is separated.

As for delay, SPICE analysis is carried out assuming typical parameters of a generic 0.5m double metal CMOS process. The delay depends on WCLK (WCLK is defined in Fig. 24.30). Since delay improvement is saturated at WCLK = 10m, this value of WCLK is used in the area and power estimation. Clock-to-Q delay is improved by a factor of 20% over the conventional F/F even when V_{CLK} = 2.2V, which can be easily realized by a clock driver of the Type A1. Data setup time and hold time in reference to clock are 0.04ns and 0ns, respectively being independent from V_{CLK}, compared to 0.1ns and 0ns for the conventional F/F.

The power in Fig. 24.33 includes clock system power per F/F and the power of a F/F itself. The power dissipation is reduced to about 1/2 to 1/3 compared to the conventional F/F depending on the type of the clock driver and VWELL. In the best case studied here, a 63% power reduction was observed. Table 24.6 summarizes typical performance improvement.

Table 24.5 Features of R3900 Macro

Technology	0.4μm CMOS, double-well, double-metal
Process V_{TH}	0.05V+-0.1V
Compensated V_{TH}	0.2V+-0.05V
External V_{DD}	3.3V+-10%
Internal V_{DDL}	0.8V~2.9V +-5%
Power dissipation	140mW @ 40MHz
Chip size	8.0 x 8.0 mm2
VS macro size	0.45 x 0.59 mm2
VT macro size	0.49 x 0.72 mm2

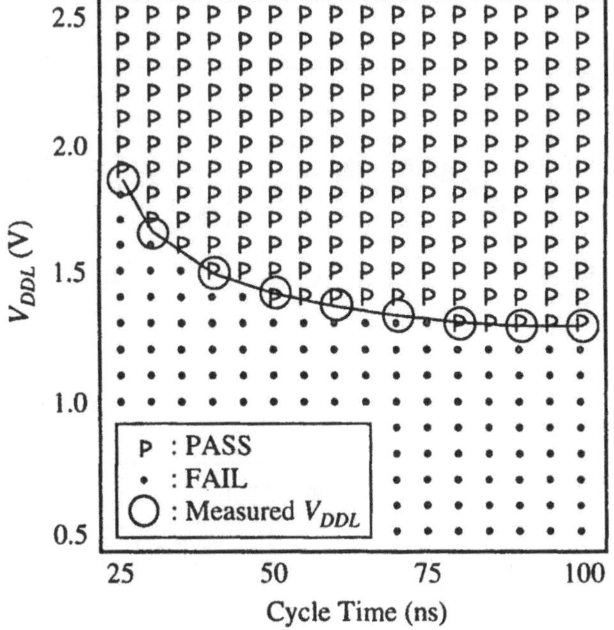

Figure 24.26 Shmoo plot and measured V_{DDL}.

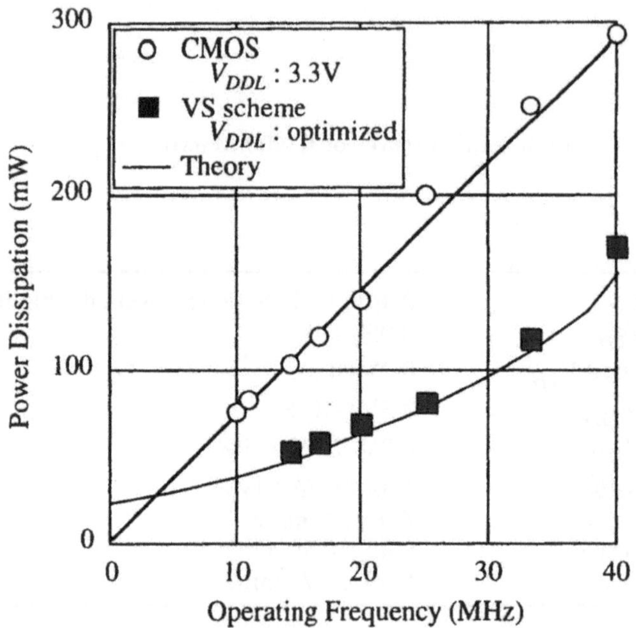

Figure 24.27 Measured power dissipation vs. operating frequency.

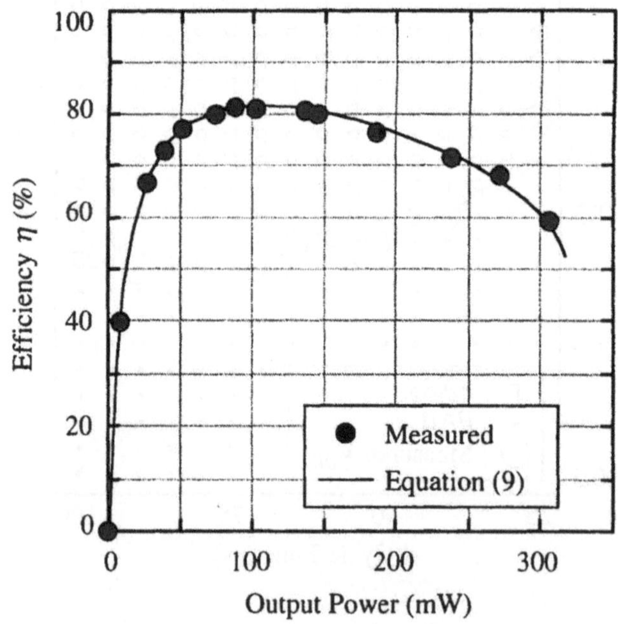

Figure 24.28 Measured DC-DC efficiency.

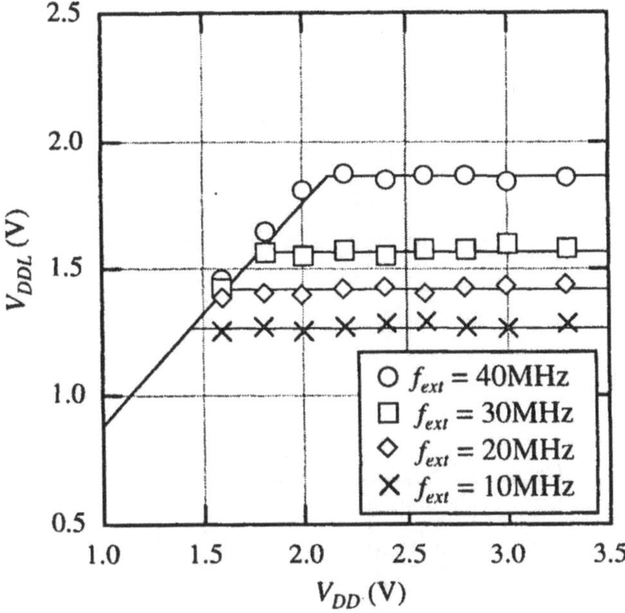

Figure 24.29 Measured V_{DDL} vs. V_{DD}.

24.4 CAPACITANCE REDUCTION

Reducing transistor size reduces the gate capacitance and the diffusion capacitance. In [28] it was reported that the total size of one million transistors in a gate array design was reduced to 1/8 of original design through transistor size optimization while maintaining the circuit speed. Consequently, the total load capacitance was reduced to 1/3, which saved 55 % of the power dissipation on average. It is often seen that bigger transistors are used in macrocells in a cell library so that they can drive even a long wire within an acceptable delay time.

Using small number of transistors contributes to reduce overall capacitance. Pass-transistor logic may have this advantage because it requires fewer transistors than the conventional CMOS static logic. In this section pass-transistor logic is discussed which is expected as a post CMOS logic for low power design.

24.4.1 Pass-Transistor Logic Circuits

The conventional CMOS and the pass-transistor logic are compared in Fig. 24.34. The pass-transistor logic can be constructed with fewer transistor count, which achieves lower overall capacitance. The salient feature of the pass-transistor logic is the existence of pass variables which come through the source of nMOSs.

Various pass-transistor logic circuits are compared in Fig. 24.35. A Complementary Pass-transistor Logic (CPL) [29] uses nMOS pass-transistor circuits where "H" level drops by V_{TH}n. CMOS inverters are provided in the output stage to compensate for the dropped signal level as well as to increase output drive capability. However, the lowered "H" level increases leak current in the CMOS inverters. Therefore the cross-coupled pMOS loads can be added to recover the "H" level and enlarge operation margin of the CMOS inverters in low V_{DD}. In this case, the

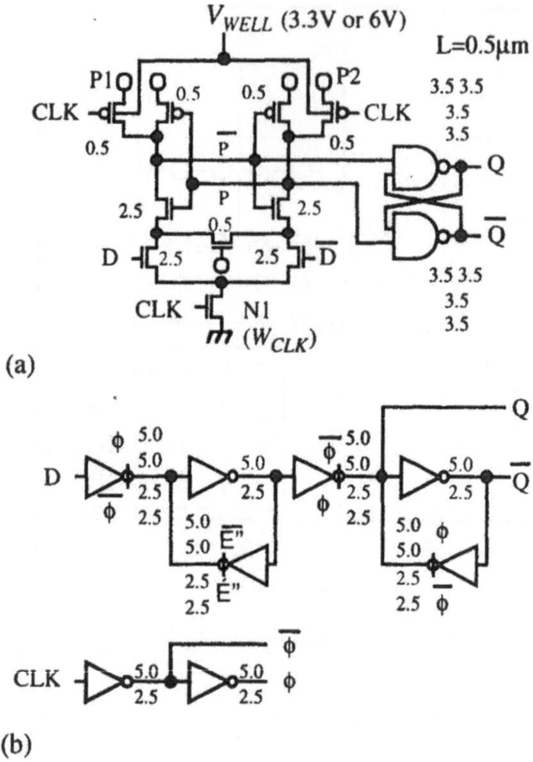

Figure 24.30 Circuit diagram of (a) reduced clock swing flip-flop (RCSFF), and (b) conventional F/F. Numbers in the figure signify MOSFET gate width. W_{CLK} is the gate width of N1.

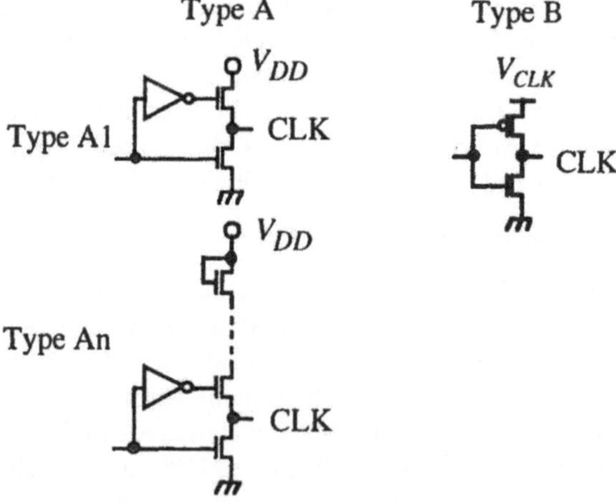

Figure 24.31 Types of clock drivers. In type B. V_{CLK} is supplied externally.

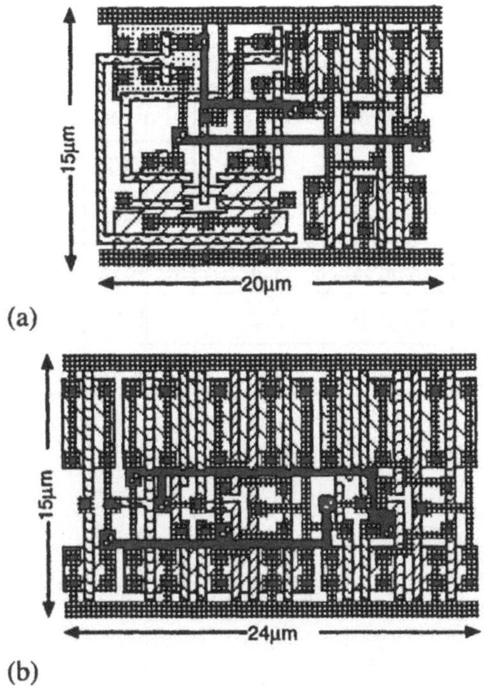

Figure 24.32 Layout of (a) RCSFF and (b) conventional F/F.

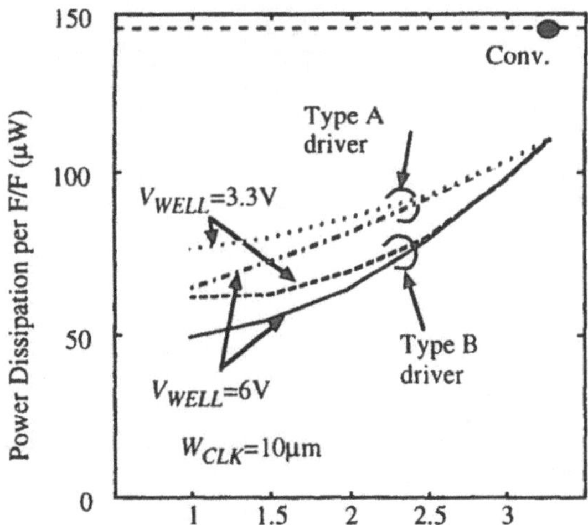

Figure 24.33 Power of dissipation for one F/F. Clock interconnection length per one F/F is assumed to be 200 μm and data activation ratio is assumed to be 30 %. f_{CLK} is 100 MHz.

Table 24.6 Performance Comparison of RCSFF and Conventional F/F

	Driver	V_{CLK} (V)	Power	Delay	Area
Conventional		3.3	100%	100%	100%
RCSFF	Type A1	2.2	59%	82%	83%
V_{WELL}=6.6V	Type A2	1.3	48%	123%	83%
W_{CLK}=10μm	Type B	2.2	48%	82%	83%
f_{CLK}=100MHz	Type B	1.3	37%	123%	83%

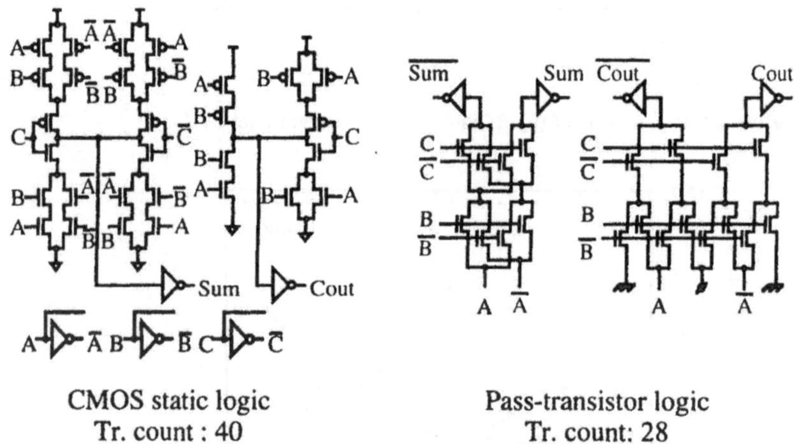

Figure 24.34 CMOS static vs. pass-transistor logic.

cross-coupled pMOS loads are used only for the level correction so that they do not require large drive capability. Therefore, small pMOS's can be used to prevent from degradation in switching speed. A Differential Cascade Voltage Switch with the Pass-Gate (DCVSPG) [30] also uses nMOS pass-transistor logic with the cross-coupled pMOS loads.

A Swing Restored Pass-transistor Logic (SRPL) [31] uses nMOS pass-transistor logic with a CMOS latch. Since the CMOS latch flips in a push-pull manner, it exhibits larger operation margin, less static current, and faster speed, compared to the cross-coupled pMOS loads. The SRPL is suitable for circuits with

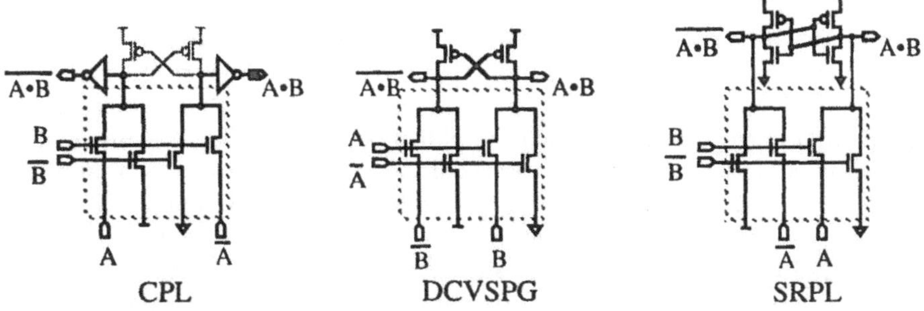

0.4μm device (full adder)

Items / Circuit	Tr. Count	Delay (ns)	Power (mW/100MHz)	P•D (normalized)	E•D (normalized)
CMOS static	40	0.82	0.52	1.00	1.00
CPL	28	0.44	0.42	0.43	0.23
DCVSPG	24	0.53	0.30	0.37	0.24
SRPL	28	0.48	0.19	0.21	0.13

Figure 24.35 Various pass-transistor logic circuits.

light load capacitance. Fig. 24.36 depicts a full adder and its delay dependence on the transistor sizes in the pass-transistor logic and the CMOS latch. The figure shows substantial design margin in the SRPL which means that SRPL circuits are quite robust against process variations. As shown in Fig. 24.35, the CPL is the fastest while the SRPL shows the smallest power dissipation.

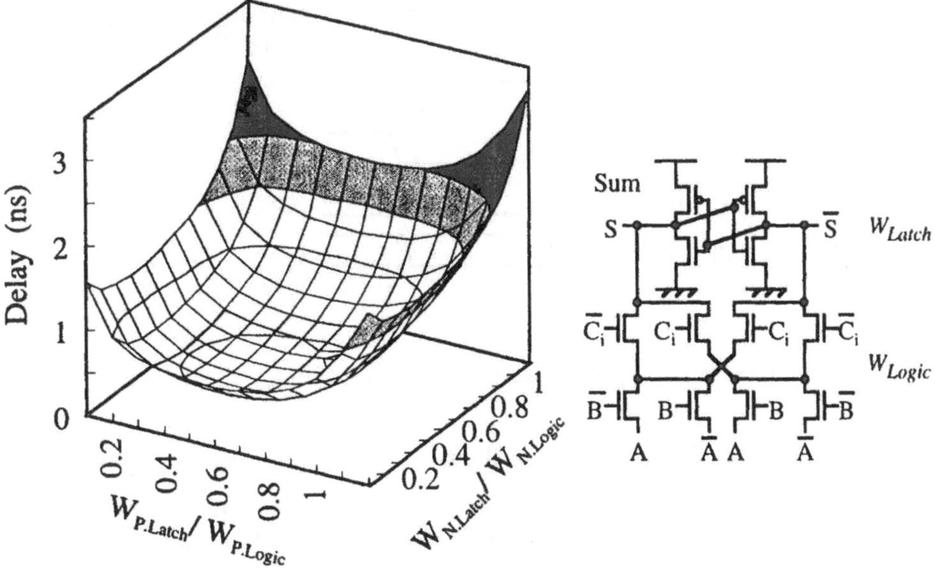

Figure 24.36 SRPL full adder and its delay dependence on transistor size.

An attempt has been made to further reduce the power dissipation by reducing the signal voltage swing. A Sense-Amplifying Pass-transistor Logic (SAPL) [21] is such a circuit. Fig. 24.37 depicts the circuit diagram. In the SAPL a reduced output signal of nMOS pass-transistor logic is amplified by a current latch sense-amplifier to gain speed and save power dissipation. All the nodes in the pass-transistor logic are first discharged to the GND level and then evaluated by inputs. The pass-transistor logic generates complement outputs with small signals of around 100mV just above the GND level. The small signals are sensed by the sense-amplifier in about 1.6ns. Since the signal swings are small just above the GND level, the circuit runs very fast with small power dissipation, even when the load capacitance is large. The SAPL therefore is suitable for circuits with large load capacitance. By adding a cross-coupled NOR latch, the sensed data can be latched so that the SAPL circuit can be used as a pipeline register. Application examples are a carry skip adder and a barrel shifter where multi-stage logic can be constructed by concatenating the pass-transistors without inserting an amplification stage.

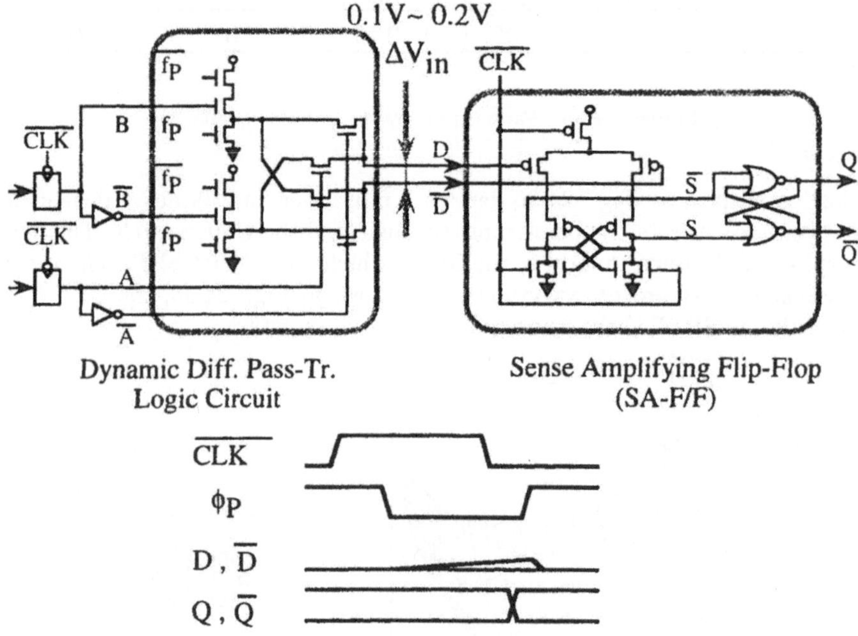

Figure 24.37 Sense-amplifying pass-transistor logic (SAPL).

24.4.2 Pass-Transistor Logic Synthesis

Although pass-transistor logic achieves low power, it is difficult to construct the pass transistor network manually by inspection. A synthesis method of pass-transistor network is studied [32]. It is based on the Binary Decision Diagram (BDD) [33]. The synthesis begins by generating logic binary trees for separate

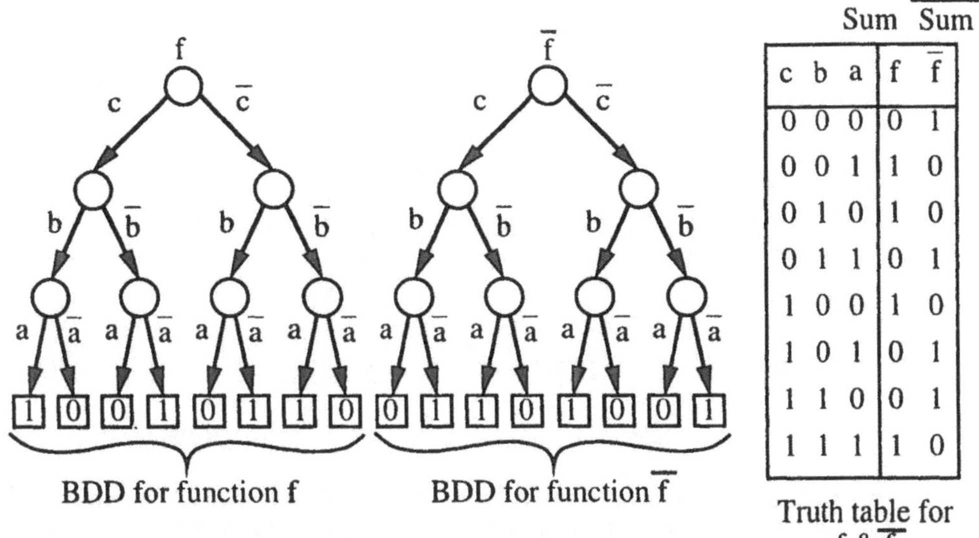

c	b	a	f	$\bar{f}$
0	0	0	0	1
0	0	1	1	0
0	1	0	1	0
0	1	1	0	1
1	0	0	1	0
1	0	1	0	1
1	1	0	0	1
1	1	1	1	0

Figure 24.38 Pass-transistor logic synthesis with BDD.

logic functions which are then merged and reduced to a smaller graph. Lastly the graph is mapped to transistor circuits.

Consider a carry generation function in an adder. The function is expressed as

$$f = abc + a\bar{b}c + \bar{a}\bar{b}\bar{c} + \bar{a}bc. \tag{20}$$

The logic binary trees are instantly generated as shown in Fig. 24.38 from a truth table of the function f. For example, the path from the source node (f) through edges "c", "b", and "a" to the sink node $f = 1$ corresponds to the case with c=b=0 and a=1.

The trees can be reduced by applying in sequence two operations illustrated in Fig. 24.39 from the sink node. Operation A merges two nodes whose corresponding outgoing complement edges reach the same node. Operation B removes from the graph a node with two outgoing complement edges to the same node. In this particular example, a case where the second operation can be applied is not found. Fig. 24.40 illustrates the reduction procedure of the logic binary trees in Fig. 24.38. The reduced graph is mapped to transistor circuits as shown in Fig. 24.41. All the edges are replaced with n-transistors whose gates are provided with the variables marked on the edges. The sink nodes $f = 0$ and $f = 1$ are replaced with V_{SS} and V_{DD}. If a edge "x" reaches the sink node $f = 1$ and the compliment edge "$\bar{x}$" reaches the sink node $f = 0$, "x" can be fed to the node as a pass variable. In this example two transistors are reduced by this rule. Lastly, appropriate buffer circuits should be connected to the output nodes (f) and ($\bar{f}$).

This BDD based method does not always give the optimum circuit in terms of transistor counts but does always give a correct network, which is a desirable characteristic when used in CAD environments. More detailed discussion on how to further reduce the transistor count can be found in [32].

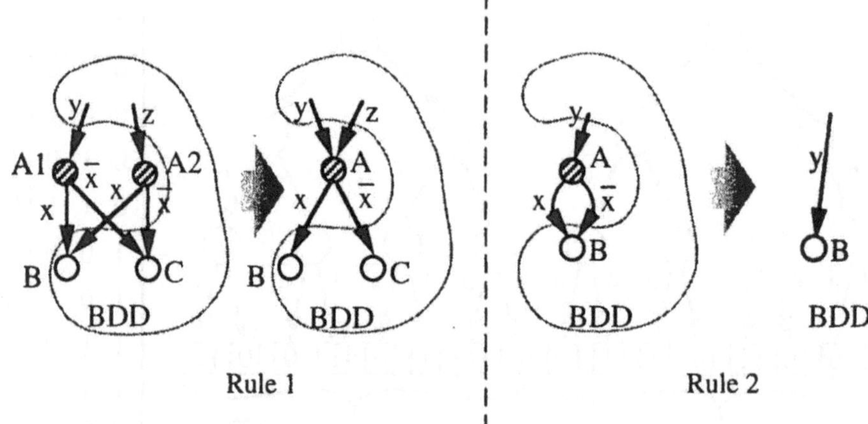

A1
A2
B
C
BDD
A
B
C
BDD
Rule 1
Collapse two nodes A1 and A2 whose right and left branch each point to the same node.
A
B
BDD
B
BDD
Rule 2
Eliminate a node A whose right and left branch point to the same node.

Figure 24.39 BDD reduction rules.

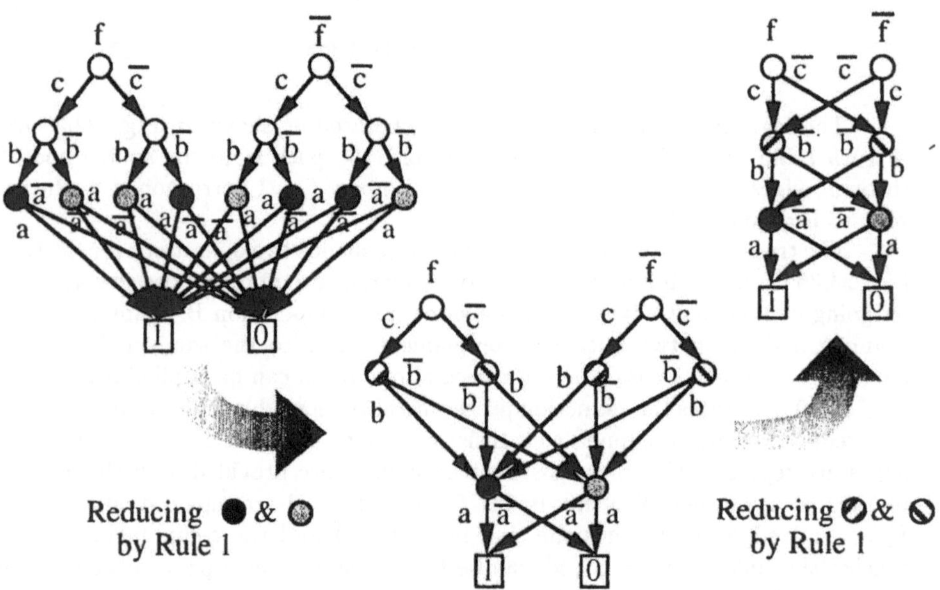

Reducing ● & ◎ by Rule 1
Reducing & by Rule 1

Figure 24.40 BDD reduction procedure.

It should also be noted that test patterns can be generated automatically by using D-algorithm [34] for pass-transistor logic as well as for conventional CMOS static logic.

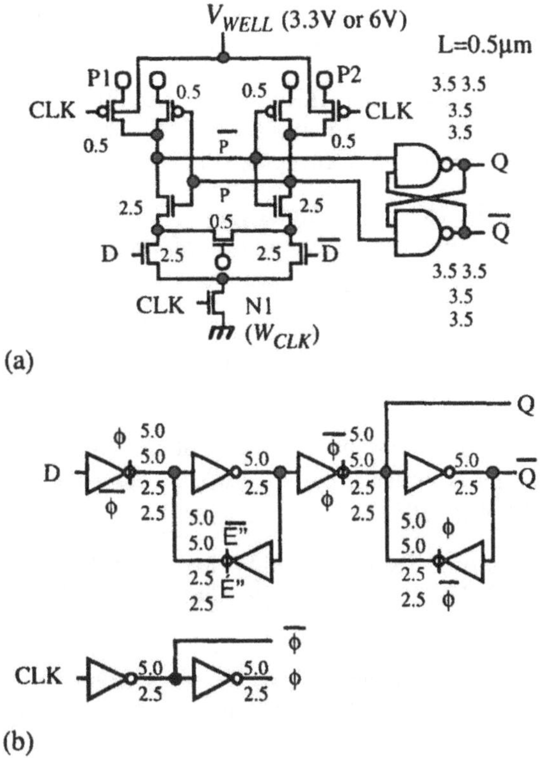

Figure 24.41 Mapping BDD to nMOS circuit.

24.5 SUMMARY

Circuit design techniques for low power CMOS VLSIs are presented from general discussion to detailed description on the VTCMOS, the VS scheme, and pass-transistor logic. Various techniques are to be employed in each design domain from architectural level to algorithmic, logical, circuit, layout, and device levels. Furthermore, circuit design is not sufficient and a broad range of research and development activities is required in areas such as system design, circuit design, CAD tools, and device/process design.

Acknowledgments

The authors would like to acknowledge the encouragement of T. Furuyama, M. Saitoh, and Y. Unno throughout the work. Discussions with T. Fujita, K. Suzuki, S. Mita, and F. Hatori were inspiring and are appreciated. Test chips were designed and evaluated with assistance by K. Matsuda, Y. Watanabe, F. Sano, A. Chiba, and S. Kitabayashi, and their efforts are acknowledged.

REFERENCES

[1] T. Sakurai and T. Kuroda, "Low-power circuit design for multimedia CMOS VLSIs," in *Proc. of SASIMI'96*, pp. 3-10, Nov. 1996.

[2] T. Kuroda and T. Sakurai, "Overview of low-power ULSI circuit techniques," *IEICE Trans. on Electronics*, vol. E78-C, no. 4, pp. 334-344, Apr. 1995.

[3] R. H. Dennard, F. H. Gaensslen, H.-N. Yu, V. Leo Rideout, E. Bassous, and A. R. Leblank, "Design of ion-implanted MOSFETs with very small physical dimensions," *IEEE J. Solid-State Circuits*, vol. 9, no. 5, pp. 256-268, Oct. 1974.

[4] T. Sakurai and A. R. Newton, "Alpha-power law MOSFET model and its applications to CMOS inverter delay and other formulas," *IEEE J. Solid-State Circuits*, vol. 25, no. 2, pp. 584-594, Apr. 1990.

[5] M. Kakumu, "Process and device technologies of CMOS devices for low-voltage operation," *IEICE Trans. on Electronics.*, vol. E76-C, no. 5, pp. 672-680, May 1993.

[6] H. J. M. Veendrick, "Short-circuit dissipation of static CMOS circuitry and its impact on the design of buffer circuits," *IEEE J. Solid-State Circuits*, vol. 19, no. 4, pp. 468-473, Aug. 1984.

[7] D. Liu and C. Svensson, "Trading speed for low power by choice of supply and threshold voltages," *IEEE J. Solid-State Circuits*, vol. 28, no. 1, pp. 10-17, Jan. 1993.

[8] A. P. Chandrakasan, S. Sheng, and R. W. Brodersen, "Low-power CMOS digital design," *IEEE J. Solid-State Circuits*, vol. 27, no. 4, pp. 473-484, Apr. 1992.

[9] K. Usami and M. Horowitz, "Clustered voltage scaling technique for low-power design," in *Proc. of ISLPD'95*, pp. 3-8, Apr. 1995.

[10] S-W. Sun and P. G. Y. Tsui, "Limitation of CMOS supply-voltage scaling by MOSFET threshold-voltage variation," *IEEE J. Solid-State Circuits*, vol. 30, no. 8, pp. 947-949, Aug. 1995.

[11] T. Kobayashi and T. Sakurai, "Self-adjusting threshold-voltage scheme (SATS) for low-voltage high-speed operation," in *Proc. of CICC'94*, pp. 271-274, May 1994.

[12] K. Seta, H. Hara, T. Kuroda, M. Kakumu, and T. Sakurai, "50% active-power saving without speed degradation using standby power reduction (SPR) circuit," in *ISSCC Dig. Tech. Papers*, pp. 318-319, Feb. 1995.

[13] T. Kuroda and T. Sakurai, "Threshold-voltage control schemes through substrate-bias for low-power high-speed CMOS LSI design," *J. VLSI Signal Processing Systems*, Kluwer Academic Publishers, vol.13, no. 2/3, pp. 191-201, Aug./Sep. 1996.

[14] T. Kuroda, T. Fujita, T. Nagamatu, S. Yoshioka, T. Sei, K. Matsuo, Y. Hamura, T. Mori, M. Murota, M. Kakumu, and T. Sakurai, "A high-speed low-power 0.3 μm CMOS gate array with variable threshold voltage (VT) scheme," in *Proc. of CICC'96*, pp. 53-56, May 1996.

[15] T. Kuroda, T. Fujita, S. Mita, T. Mori, K. Matsuo, M. Kakumu, and T. Sakurai, "Substrate noise influence on circuit performance in variable threshold-voltage scheme," in *Proc. of ISLPED'96*, pp. 309-312, Aug. 1996.

[16] S. Mutoh, T. Douseki, Y. Matsuya, T. Aoki, S. Shigematsu, and J. Yamada, "1-V power supply high-speed digital circuit technology with multithreshold-voltage CMOS," *IEEE J. Solid-State Circuits*, vol. 30, no. 8, pp. 847-854, Aug. 1995.

[17] S. Mutoh, S. Shigematsu, Y. Matsuya, H. Fukuda, and J. Yamada, "A 1V multi-threshold voltage CMOS DSP with an efficient power management technique for mobile phone application" in *ISSCC Dig. Tech. Papers*, pp. 168-169, Feb. 1996.

[18] T. Kuroda, T. Fujita, S. Mita, T. Nagamatu, S. Yoshioka, K. Suzuki, F. Sano, M. Norishima, M. Murota, M. Kako, M. Kinugawa, M. Kakumu, and T. Sakurai, "A 0.9V 150MHz 10mW 4mm2 2-D discrete cosine transform core processor with variable-threshold-voltage scheme," *IEEE J. Solid-State Circuits*, vol. 31, no. 11, pp. 1770-1779, Nov. 1996.

[19] M. Mizuno, K. Furuta, S. Narita, H. Abiko, I. Sakai, and M. Yamashina, "Elasti-Vt CMOS circuits for multiple on-chip power control," in *ISSCC Dig. Tech. Papers*, pp. 300-301, Feb. 1996.

[20] V. Kaenel, M. Pardoen, E. Dijkstra, and E. Vittoz, "Automatic adjustment of threshold & supply voltages for minimum power consumption in CMOS digital circuits", in *Proc. of SLPE'94*, pp. 78-79, 1994.

[21] M. Matsui, H. Hara, K. Seta, Y. Uetani, L.-S. Kim, T. Nagamatsu, T. Shimazawa, S. Mita, G. Otomo, T. Ohto, Y. Watanabe, F. Sano, A. Chiba, K. Matsuda, and T. Sakurai, "200MHz video compression macrocells using low-swing differential logic," in *ISSCC Dig. Tech. Papers*, pp. 76-77, Feb. 1994.

[22] A. J. Stratakos, S. R. Sanders, and R. W. Brodersen, "A low-voltage CMOS dc-dc converter for a portable battery-operated system," in *Proc. of IEEE Power Electronics Specialists Conference*, vol. 1, pp. 619-626, Jun. 1994.

[23] K. Suzuki, S. Mita, T. Fujita, F. Yamane, F. Sano, A. Chiba, Y. Watanabe, K. Matsuda, T. Maeda, and T. Kuroda, "A 300MIPS/W RISC core processor with variable supply-voltage scheme in variable threshold-voltage CMOS," in *Proc. of CICC'97*, pp. 587-590, May 1997.

[24] M. Nagamatsu, H. Tago, T. Miyamori, M. Kamata, H. Murakami, Y. Ootaguro, *et al.*, "A 150MIPS/W CMOS RISC processor for PDA applications," in *ISSCC Dig. of Tech. Papers*, pp.114-115, Feb. 1995.

[25] T. Kuroda, K. Suzuki, S. Mita, T. Fujita, F. Yamane, F. Sano, A. Chiba, Y. Watanabe, K. Matsuda, T. Maeda, T. Sakurai, and T. Furuyama, "Variable supply-voltage scheme for low-power high-speed CMOS digital design ," *IEEE J. Solid-State Circuits*, vol. 33, no. 3, pp. 454-462, Mar. 1998.

[26] H. Kawaguchi and T. Sakurai, "A reduced clock-swing flip-flop (RCSFF) for 63% clock power reduction," in *Proc. of Symp. on VLSI Circuits.*, pp. 97-98, June 1997.

[27] J. Montanaro, *et. al.*, "A 160-MHz, 32-b, 0.5-W CMOS RISC Microprocessor ," *IEEE J. Solid-State Circuits,* vol. 31, no. 11, pp. 1703-1714, Nov. 1996.

[28] M. Yamada, S. Kurosawa, R. Nojima, N. Kojima, T. Mitsuhashi, and N. Goto, "Synergistic power/area optimization with transistor sizing and write length minimization," in *Proc. of SLPE'94*, pp. 50-51, Oct. 1994.

[29] K. Yano, T. Yamanaka, T. Nishida, M. Saito, K. Shimohigashi, and A. Shimizu, "A 3.8-ns CMOS 16x16-b multiplier using complementary pass-transistor logic," *IEEE J. Solid-State Circuits,* vol. 25, no. 2, pp. 388-395, Apr. 1990.

[30] F. S. Lai, and W. Hwang, "Differential cascade voltage switch with the pass-gate (DCVSPG) logic tree for high performance CMOS digital systems," in *Proc. of VLSITSA'93,* pp. 358-362, 1993.

[31] A. Parameswar, H. Hara, and T. Sakurai, "A high speed, low power, swing restored pass-transistor logic based multiply and accumulate circuit for multimedia applications," in *Proc. of CICC'94*, pp. 278-281, May 1994.

[32] T. Sakurai, B. Lin and A. R. Newton, "Multiple-output shared transistor logic (MOSTL) family synthesized using binary decision diagram," *Dept. EECS, Univ. of Calif., Berkeley, ERL Memo* M90/21, Mar. 1990.

[33] R. E. Bryant, "Graph-based algorithms for Boolean function manipulation," *IEEE Trans. on Computers,* vol. C-35, no. 8, pp. 677-691, Aug. 1986.

[34] J. P. Roth, V. G. Oklobdzija, and J. F. Beetem, "Test generation for FET switching circuits," in *Proc. of Int'l Test Conference,* pp. 59-62, Oct. 1984.

[35] K. Yano, Y. Sasaki, K. Rikino, and K. Seki, "Lean integration: achieving a quantumleap in performance and cost of logic LSIs," in *Proc. CICC'94*, pp. 603-606, May 1994.

[36] H. Kojima, S. Tanaka, and K. Sasaki, "Half-swing clocking scheme for 75% power saving in clocking circuitry," in *Proc. of Symp. on VLSI Circuits,* pp. 23-24, June 1994.

[37] H. Yamauchi, H. Akamatsu, and T. Fujita, "A low power complete charge-recycling bus architecture for ultra-high data rate ULSI's," in *Proc. of Symp. on VLSI Circuits,* pp. 21-22, June 1994.

[38] C. Lemonds, and S. S. M. Shetti, "A low power 16 by 16 multiplier using transition reduction circuitry," in *Proc. of IWLPD'94*, pp. 139-142, Apr. 1994.

[39] S. C. Prasad, and K. Roy, "Circuit optimization for minimization of power consumption under delay constraint," in *Proc. of IWLPD'94*, pp. 15-20, Apr. 1994.

[40] C. H. Tan, and J. Allen, "Minimization of power in VLSI circuits using transistor sizing, input ordering, and statistical power estimation," in *Proc. of IWLPD'94*, pp. 75-80, Apr. 1994.

[41] M. Mizuno, M. Yamashina, K. Furuta, H. Igura, H. Abiko, K. Okabe, A. Ono, and H. Yamada, "A GHz MOS adaptive pipeline technique using variable delay circuits," in *Proc. of Symp. on VLSI Circuits*, pp. 27-28, May 1994.

[42] T. Sakurai, "High-speed circuit design with scaled-down MOSFETs and low supply voltage," in *Proc. of ISCAS'93*, pp. 1487-1490, May 1993.

[43] M. Horiguchi, T. Sakata, and K. Itoh, "Switched-source-impedance CMOS circuit for low standby subthreshold current giga-scale LSI's," in *Proc. of Symp. on VLSI Circuits*, pp. 47-48, June 1993.

[44] Q. Zhu, J.G. Xi, W.W.-M. Dai, and R. Shukla, "Low power clock distribution based on area pad interconnect for multichip modules," in *Proc. of IWLPD'94*, pp. 87-92, Apr. 1994.

[45] B. Gunning, L. Yuan, T. Nguyen, and T. Wong, "A CMOS low-voltage-swing transmission-line transceiver," in *ISSCC Dig. Tech. Papers*, pp. 58-59, Feb. 1992.

[46] K. Nogami, and A. E. Gamal, "A CMOS 160Mb/s phase modulation I/O interface circuit," in *ISSCC Dig. Tech. Papers*, pp. 108-109, Feb. 1994.

[47] T. Kawahara, M. Horiguchi, J. Etoh, T. Sekiguchi, and M. Aoki, "Low power chip interconnection by dynamic termination," in *Proc. of Symp. on VLSI Circuits*, pp. 45-46, June 1994.

[48] M. Nogawa and Y. Ohtomo, "A data-transition look-ahead DFF circuit for statistical reduction in power consumption," in *Proc. of Symp. on VLSI Circuits*, pp. 101-102, June 1997.

Chapter 25

Power Estimation Approaches

Janardhan H. Satyanarayana
Bell Laboratories
Lucent Technologies
Holmdel, New Jersey
jana@lucent.com

Keshab K. Parhi
Department of Elect. and Comp. Eng.
University of Minnesota
Minneapolis, Minnesota
parhi@ece.umn.edu

25.1 INTRODUCTION

In the past, the major concern of VLSI designers were performance, area, reliability, and cost, with power being only a secondary issue. However, in recent years this has changed and power, area, and speed have become equally important. There are many reasons for this new trend. Primarily, the rapid advancement in semiconductor technology in the last decade has made possible the integration of a large number of digital CMOS circuits on a single chip. Moreover, the desirability of portable operations of these circuits has necessitated the development of low power technology. Portable applications could be anywhere from desk-tops to audio/video based multimedia products to personal digital assistants and personal communicators. These systems demand both complex functionality and low power at the same time, thereby making their design challenging. The power consumption of portable circuits has a direct bearing on the life-time of the batteries. For example, a portable multimedia terminal designed using off-the-shelf components (not optimized for low power), could consume about 30-40 watts of power. If this system were to use the state-of-the-art nickel-metal-hydride battery technology [1], it would require 4.5-6 kilograms of batteries for 10 hours of operation. Therefore, this would mean that portable systems will experience either heavy battery packs or a very short battery life. Reduction in power consumption also plays an important role for producers of non-portable systems. The state of the art microprocessors optimized for performance consume around 20-30 watts of power for operating frequencies of 150-200 MHz. With rapid advancement in technology, the speeds could reach 500-600 MHz with extraordinarily high power consumption values. This would mean that the packaging cost for such devices would be very high and expensive cooling and packaging strategies would be required. Therefore, reduction in power consumption could greatly cut cooling costs. Finally, the issue of reliability is also a major concern for consumer system designers. Systems which consume more power often run hot and acerbate failure mechanisms. In fact the failure rate increases rapidly for a small increase in operating temperature. Therefore, the maximum power consumption of the system is a crucial design factor as it could have an impact on the system cost, battery type, heat sinks, etc. Therefore, reduction in peak power is also an important issue. It is clear that the motivations for reduction in power consumption vary from application to application. In portable applications such as cellular phones and personal digital assistants, the goal is to keep the battery lifetime and weight reasonable. For high performance portable computers such as laptops, the goal is to reduce the power dissipation of the electronics portion of the system. Finally, for non-portable systems such as workstations and communication systems the goal is to reduce packaging, cooling cost and ensure long-term reliability.

In digital CMOS circuits, there are four major sources of power dissipation. They are due to:

- the *leakage current*, which is primarily determined by the fabrication technology, caused by the 1) reverse bias current in the parasitic diodes formed between source and drain diffusions and the bulk region in a MOS transistor, and 2) the sub-threshold current that arises from the inversion that exists at the gate voltages below the threshold voltage,
- the *standby current* which is the DC current drawn continuously from V_{dd} to ground,
- the *short-circuit* current which is due to the DC path between the supply rails during output transitions, and
- the *capacitance current* which flows to charge and discharge capacitive loads during logic changes.

The diode *leakage current* is proportional to the area of the source or drain diffusion and the leakage current density and is typically in the order of 1 *picoA* for a 1 *micron* minimum feature size. The sub-threshold leakage current for long channel devices decreases exponentially with $V_{GS} - V_t$ where V_{GS} is the gate bias and V_t is the transistor threshold voltage, and increases linearly with the ratio of the channel width over channel length. This is negligible at normal supply and threshold voltages but its effect can become pronounced at reduced power supply and device threshold voltages. Moreover, at short channel lengths, the sub-threshold current becomes exponentially dependent on drain voltage instead of being dependent on V_{DS} [2] which is the difference between the drain and the source voltages.

The *standby power consumption* occurs when both the nMOS and the pMOS transistor are continuously on. This could happen, for example, in a pseudo-nMOS inverter, when the drain of an nMOS transistor is driving the gate of another nMOS transistor in a pass-transistor logic, or when the tri-stated input of a CMOS gate leaks away to a value between power supply and ground. The standby power is equal to the product of V_{dd} and the DC current drawn from the power supply to ground. The term *static power dissipation* refers to the sum of the leakage and standby power dissipations. Leakage currents in digital CMOS circuits can be made small with proper choice of device technology. Standby currents play an important role in design styles like pseudo-nMOS and nMOS pass transistor logic and in memory cores.

The *short-circuit power consumption* of a logic gate is proportional to the input rise-time, the load, and the transistor sizes of the gates. The maximum short circuit current flows when there is no load and it decreases as the load increases. Depending on the approximations used to model the currents and to estimate the input signal dependencies, different techniques [3][4] have been derived for the evaluation of the short circuit power. A useful formulae was also recently derived in [5] that shows the explicit dependence of the short circuit power dissipation on the design parameters. The idea is to adopt an alternative definition of the short circuit power dissipation through an equivalent short circuit capacitance C_{SC}. If the gate sizes are selected so that the input and output rise- and fall-times are about equal, the short-circuit power consumption will be less than 15% of the dynamic power consumption [4]. However, if very high performance is desired and large gates are used to drive relatively small loads and if the input rise time is long, then the short-circuit power consumption cannot be ignored.

The dominant source of power consumption in digital CMOS circuits is due to the charging and discharging of the node capacitances (referred to as the *capacitive power dissipation*) and is computed as

$$P_{dyn} = \frac{1}{2}.\alpha.C_l.V_{dd}^2.f_{clk}, \tag{1}$$

where α (referred to as the switching activity) is the average number of output transitions, C_l is the load capacitance at the output node, V_{dd} is the power supply voltage, and f_{clk} is the clock frequency. The product of the switching activity and the clock frequency is also referred to to as the *transition density* [6]. The term *dynamic power consumption* refers to the sum of the short-circuit and capacitive power dissipations. Using the concept of equivalent short-circuit capacitance described above, the dynamic power dissipation can be calculated using (1) if C_{SC} is added to C_l.

Power estimation refers to the problem of estimating *average power dissipation* of digital circuits. Ideally the average power should include both the static and the dynamic power dissipations. However, for well-designed CMOS circuits, the capacitive power is dominant and therefore the average power generally refers to the capacitive power dissipation. It should be noted that this is much different from estimating the instantaneous or the worst case power which is modeled as a voltage drop problem [7][8]. The most straight-forward method of power estimation is to perform a circuit simulation of the design and monitor the power supply current waveform. Then, the average of the current waveform is computed and multiplied by the power supply voltage to calculate the average power. This technique is very accurate and can be applied to any general logic network regardless of technology, functionality, design style, architecture, etc. The simulation results, however, are directly related to the types of input signals used to drive the simulator. Therefore, this technique is strongly *pattern dependent* and this problem could be serious. For example, in many applications the power of a functional block needs to be estimated even when the rest of the chip has not yet been designed. In this case, very little may be known about the inputs to this functional block and complete information about its inputs would be impossible to obtain. As a result, a large number of input patterns may have to simulated and averaged and this could become computationally very expensive; even impossible for large circuits.

Other power estimation techniques start out by simplifying the problem in three ways. First, it is assumed that the logic circuit is assumed to be built of logic gates and latches, and has the popular and well-structured design style of a synchronous sequential circuit as shown in Fig. 25.1. Here, the circuit consists of a combinational block and a set of flip-flops such that the inputs (outputs) of the

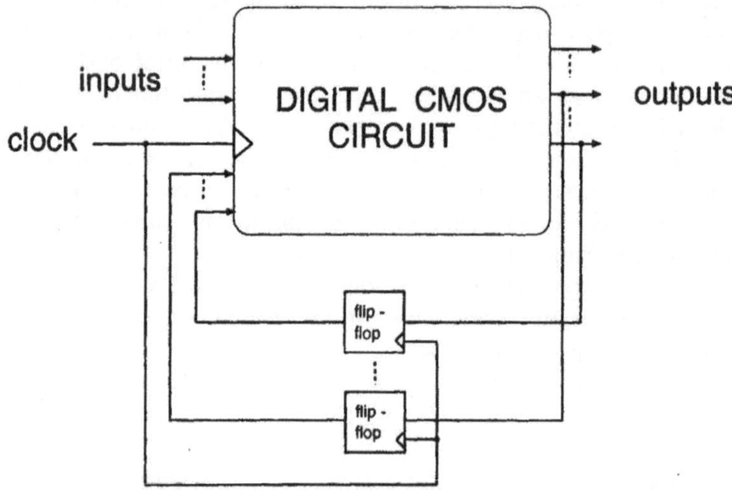

Figure 25.1 A typical synchronous sequential circuit.

combinational block are latch outputs (inputs). It is also assumed that the latches are edge-triggered. Therefore, the average power consumed by the digital circuit is computed as the sum of the power consumed by the latches and the power consumed by the combinational logic blocks. Second, it is assumed that the power supply and the ground voltage levels are fixed through put the chip so that it becomes easier to compute the power by estimating the current drawn by every sub-circuit assuming a given fixed power supply voltage. Finally, it is commonly accepted [4] that it is enough to consider only the charging/discharging current drawn by the logic gate and therefore the short-circuit current during switching is neglected.

The latches are essentially controlled by their clock signals and therefore whenever they make a transition they consume some power. Thus latch power is drawn in synchrony with the clock. However, this is not true with the gates inside the combinational block as they may make several transitions before settling to their steady state value for that clock period. These spurious transitions are referred to as *glitches* and they tend to dissipate additional power. It is observed [9] that this additional power is typically 20% of the total power. However, for functional units such as adders and multipliers thus could be as high as 60% of the total power. This component of the power dissipation is computationally expensive to estimate because it depends on the timing relationships between various signals in the circuit. Only few approaches [10]-[12] have considered this elusive component of power referred to as the *toggle power*.

The problem of getting a pattern-independent power estimate is another challenge and researchers have resorted to probabilistic techniques to solve this problem [6], [11]-[18]. The motivation behind this approach is to compute, from the input pattern set, the fraction of the clock cycles in which an input signal makes a transition (***probability***) and use that information to estimate how often transitions occur at internal nodes, and consequently the power consumed by the circuit. This can be thought of as performing the averaging before, instead of after, running the circuit simulation. This approach is efficient as it replaces a large number of circuit simulation runs with a single run of a probabilistic tool at the expense of some loss in accuracy. Of course, the results of the analysis will still depend on the supplied probabilities. Thus, to some extent the process is still pattern dependent and the user must supply some information about the typical behavior of the input signals in terms of probabilities.

In this chapter, based on [11][18], we present a stochastic approach for power estimation of digital circuits and a tool referred to as HEAT (Hierarchical Energy Analysis Tool) which is based on the proposed approach. The salient feature of this approach is that it can be used to estimate the power of large digital circuits including multipliers, dividers etc. in a short time. A typical approach to estimate power consumption of large digital circuits using stochastic methods would be to model them using state-transition diagrams (***stds***). However, this would be a formidable task as the number of states would increase exponentially with increase in the number of nodes. Therefore, we propose to decompose the digital circuit into sub-circuits, and then model each sub-circuit using stds. This greatly reduces the number of states in the std, thereby reducing the computation time by orders of magnitude. For example, a typical Booth multiplier (designed using full-adders, encoders, and multiplexors) is broken up into three sub-classes, with the first sub-class containing full-adders, the second sub-class containing encoders, and the third containing multiplexors. The circuit belonging to each sub-class is then modeled with the help of a std, facilitated through the development of analytic expressions for the state-update of each node in the circuit. Then, the energy associated with each edge in the state transition diagram is computed using SPICE, and the total energy of the circuit belonging to a given sub-class is computed by summing the energies of all the constituent edges in the state transition diagram. This procedure is repeated for all the sub-classes, and the final energy of the digital circuit is computed by summing the energies of the constituent sub-classes. An estimate of the average power is then obtained by finding the ratio of the total energy to the total time over which it was consumed.

The organization of this chapter is as follows. Section 25.2 discusses some previous work done in the field of power estimation. Section 25.3 is concerned with the basic definitions and terminologies used throughout the chapter. An algorithm for the proposed hierarchical approach to power estimation of combinational circuits is presented in Section 25.4. Here, a technique for modeling a given static CMOS digital circuit using an std is first discussed with the help of a simple example. An approximation based on irreducible Markov chains [19][20] is then used to compute the steady-state probabilities associated with the various states in the std. A technique for the computation of energies associated with the various edges is also presented in this section. A CAD tool called *HEAT* has been developed based on the proposed hierarchical approach and tested on various digital circuits. The proposed approach is extended to handle sequential circuits in Section 25.5. Here, the modeling of an edge-triggered flip-flop facilitated through the development of state-update equations is presented. The experimental results of the HEAT tool are presented in Section 25.6. Finally, the main conclusions of the chapter and future work are summarized in Section 25.7.

25.2 PREVIOUS WORK

The design of low power digital CMOS circuits cannot be achieved without accurate power prediction and optimization tools. Therefore, there is a critical need for CAD tools to estimate power dissipation during the design process to meet the power constraint without having to go through a costly redesign effort. The techniques for power estimation can be broadly classified into two categories: simulation based and non-simulation based.

25.2.1 Simulation Based Approaches

The main advantages of these techniques are that issues such as hazard generation, spatial/temporal correlation, etc. are automatically taken into account. The approaches under this category can be further classified into direct simulation and statistical simulation.

25.2.1.1 Direct Simulation

The approaches in this category basically simulate a large set of random vectors using a circuit simulator like SPICE [21] and then measure the average power dissipated. They are capable of handling various device models, different circuit design styles, tristate drivers, single and multi-phase clocking methodologies, etc. The main disadvantage of these techniques is that they eat up too much memory and have very long execution times. As a result they cannot be used for large, cell-based designs. Moreover, it is difficult to generate a compact vector set to calculate activity factors at various nodes.

Direct simulation can also be carried out using a transistor-level power simulator [22] which is based on an event-driven timing simulation algorithm. This uses simplified table-driven device models, circuit partitioning to increase the speed by two to three orders of magnitude over SPICE while maintaining the accuracy within 10% for a wide range of circuits. It also gives detailed information like instantaneous, average current, short-circuit power, capacitive power, etc.

Other techniques like Verilog-based gate-level simulation programs can be adapted to determine the power dissipation of digital circuits under user-specified input sequences. These techniques rely heavily on the accuracy of the macro-models built for the gates in the ASIC library as well as on the detailed gate-level timing analysis tools. The execution time is 3-4 orders of magnitude shorter than SPICE. Switch-level simulators like IRSIM [23] can be easily modified to report the switched capacitance (and thus dynamic power dissipation) during circuit simulations. This is much faster than the circuit-level simulation techniques but is not as versatile or accurate.

25.2.1.2 Statistical Simulation

Techniques under this category are based on a *Monte Carlo simulation (MCS)* approach which alleviate the pattern-dependence problem by a proper choice of input vectors [24]. This approach consists of applying randomly generated input patterns at the circuit inputs and monitoring the power dissipation for T clock cycles using a simulator. Each such measurement gives a power sample which is regarded as a random variable. By applying the *central limit theorem*, it is found that as

T approaches infinity, the sample density tends to a normal curve. Typically, a sample size of 30-50 ensures normal sample density for most combinatorial circuits. For a desired percentage error in the power estimate, ϵ, a given confidence level, ϑ, the sample mean, μ, and the sample standard deviation, σ, the number of required samples, N, is estimated as

$$N > \left(\frac{t_{\vartheta/2}\sigma}{\epsilon\mu}\right)^2 \tag{2}$$

where $t_{\vartheta/2}$ is defined so that the area to its right under the standard normal distribution curve is equal to $\vartheta/2$. In estimating the average power consumption of the digital circuit, the convergence time of the MCS approach is short when the error bound is loose or the confidence level is low. It should be noted that this method may converge prematurely to a wrong power estimate value if the sample density does not follow a normal distribution. Moreover, this approach cannot handle spatial correlations at the circuit inputs.

25.2.2 Non-Simulative Approaches

These approaches are based on library models, stochastic models, and information theoretic models. They can be broadly classified into those that work at the behavioral level and those that work at the logic level.

25.2.2.1 Behavioral Level Approaches

Here, power estimates for functional units such as adders, multipliers, registers, memories are directly obtained from the design library where each functional unit has been simulated using white noise data and the average switched capacitance per clock cycle has been calculated and stored in the library. The power model for a functional unit may be parameterized in terms of its input bit width. For example, the power dissipation of an adder (or a multiplier) is linearly (or quadratically) dependent on its input bit width. Although this approach is not accurate, it is useful in comparing different adder and multiplier architectures for their switching activity. The library can thus contain interface descriptions of each module, description of its parameters, its area, delay, and internal power dissipation (assuming white noise data inputs). The latter is determined by extracting a circuit or logic level model from an actual layout of the module by simulating it using a long stream of randomly generated input patterns. These characteristics are stored in the form of equations or tables. The power model thus generated and stored for each module in the library has to be modulated by the real input switching activities in order to provide power estimates which are sensitive to the input activities.

Word-level behavior of data input can be captured by its probability density function (pdf). In a similar manner, spatial correlation between data inputs can be captured by their joint pdf. This idea is used in [25] to develop a probabilistic technique for behavioral level power estimation. The approach can be summarized in four steps: 1) building the joint pdf of the input variables of a data flow graph (DFG) based on the given input vectors, 2) computing the joint pdf for some combination of internal arcs in the DFG, 3) calculation of the switching activity at the inputs of each register in the DFG using the joint pdf of the inputs, 4) power estimation of each functional block using input statistics obtained in step 3.

This method is robust but suffers from the worst-case complexity of the joint pdf computation and inaccuracies associated with the library characterization data.

An information theoretic approach is described in [26][27] where activity measure like entropy are used to derive fast and accurate power estimates at the algorithmic and structural behavioral levels. Entropy characterizes the uncertainty of a sequence of applied vectors and thus this measure is related to the switching activity. It is shown in [26] that under a temporal independence assumption the average switching activity of a bit is upper bounded by one half of its entropy. For control circuits and random logic, given the statistics of the input stream and having some information about the structure and functionality of the circuit,the output entropy bit is calculated as a function of the input entropy bit and a structure- and function-dependent information scaling factor. For DFGs, the output entropy is calculated using a compositional technique which has linear complexity in terms of its circuit size. A major advantage of this technique is that it is not simulative and is thus fast and provides accurate power estimates.

Most of the above techniques are well suited for data-paths. Behavioral level power estimates for the controller circuitry is outlined in [28]. This technique provides a quick estimation of the power dissipation in a control circuit based on the knowledge of its target implementation style, i.e., dynamic, precharged pseudo-nMOS, etc.

25.2.2.2 Logic-Level Approaches

It is clear from the discussion in the previous section that most of the power in digital CMOS circuits is consumed during the charging/discharging of load capacitance. Therefore, in order to estimate the power consumption one has to determine the switching activity α of various nodes in the digital circuit. If temporal independence among input signals is assumed then it can be easily shown that the switching activity of a node with probability p_n is found to be

$$\alpha = 2.p_n.(1 - p_n). \tag{3}$$

If two successive values of a node are correlated in time then the switching activity is expressed as [29]

$$\alpha = 2.p_n.(1 - p_n).(1 - \rho_n) \tag{4}$$

where ρ_n denotes the temporal correlation parameter of the signal. Computing the signal probabilities has therefore attracted a lot of attention from the researchers in the past. In [13], some of the earliest work in computing the signal probabilities in a combinational network is presented. Here, variable names are assigned with each of the circuit inputs to represent the signal probabilities of these inputs. Then, for each internal circuit line, algebraic expressions involving these variables are computed. These expressions represent the signal probabilities for these lines. While the algorithm is simple and general, its worst case complexity is exponential. Therefore, approximate signal probability calculation techniques are presented in [14][15][30]. An exact procedure based on ordered binary-decision diagrams (OBDDs) [31] can also be used to compute signal probabilities. This procedure is linear in the size of the corresponding function graph, however, may be exponential in the number of circuit inputs. Here, the signal probability of the output node is calculated by fist building an OBDD corresponding to the global function of the node (i.e., function of the node in terms of the circuit inputs) and then performing a postorder traversal of the OBDD using the equation:

$$prob(z) \;=\; prob(x)prob(f_x) + prob(\overline{x})prob(f_{\overline{x}}). \tag{5}$$

This leads to a very efficient computational procedure for signal probability estimation. For example, if x_1, x_2, x_3 , and x_4 are the inputs of a 4-input XOR-gate, then the probability of the output is computed using the OBDD shown in Fig. 25.2 and is expressed as

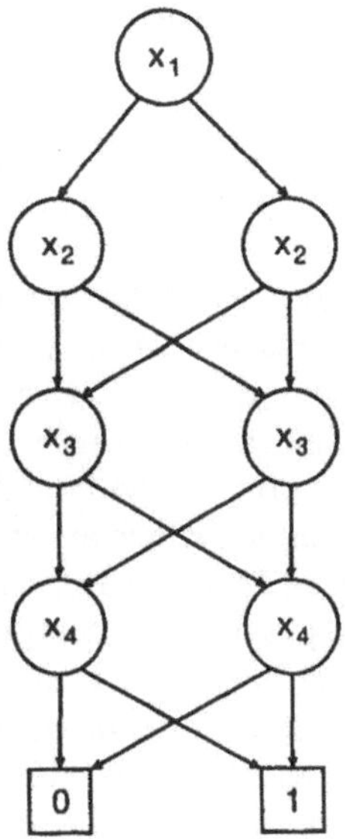

Figure 25.2 Computing signal probabilities using OBDDs.

$$\begin{aligned} p_z \;=\; & \overline{p_{x_1}}\,\overline{p_{x_2}}\,\overline{p_{x_3}}\,p_{x_4} + p_{x_1}p_{x_2}\,\overline{p_{x_3}}\,p_{x_4} + \overline{p_{x_1}}\,p_{x_2}p_{x_3}p_{x_4} + p_{x_1}\,\overline{p_{x_2}}\,p_{x_3}p_{x_4} + \\ & \overline{p_{x_1}}\,\overline{p_{x_2}}\,p_{x_3}\,\overline{p_{x_4}} + p_{x_1}p_{x_2}p_{x_3}\,\overline{p_{x_4}} + \overline{p_{x_1}}\,p_{x_2}\,\overline{p_{x_3}}\,\overline{p_{x_4}} + p_{x_1}\,\overline{p_{x_2}}\,\overline{p_{x_3}}\,\overline{p_{x_4}}. \end{aligned} \tag{6}$$

where p_{x_i} represents the probabilities of the input signals.

If the temporal correlation of a signal between two successive clock cycles is modeled by a time-homogeneous Markov chain, then the activity factor can also be computed as the sum of the transition probabilities. For example, for a signal s the activity factor is computed as

$$\alpha \;=\; p(s_{0->1}) + p(s_{1->0}) \tag{7}$$

where $p(s_{0->1})$ and $p(s_{1->0})$ represent the transition probabilities from 0 to 1 and 1 to 0, respectively. The various transition probabilities can be computed exactly using the OBDD representation of the signal in terms of its circuit inputs.

All the above techniques account for steady-state behavior of the circuit and thus ignore hazards and glitches and are therefore defined to be *zero-delay* model based techniques. There has been some previous work done in the area of estimation under a real delay model. In [32], the exact power estimation of a given combinational logic is carried out by creating a set of symbolic functions that represent Boolean conditions for all values that a node in a circuit can assume at different time instances under a pair of input vectors. The concept of a probability waveform is introduced in [33]. This waveform consists of an *event list* which is nothing but a sequence of transition edges over time from the initial steady state to the final steady state where each event is annotated with a probability. The probability waveform of s node is a compact representation of the set of all possible logical waveforms at that node. In [6], an efficient algorithm based on Boolean difference equation is proposed to propagate the transition densities from circuit inputs throughout the circuit. The transition density $D(y)$ of each node in the circuit is calculated in accordance with

$$D(y) = \sum_{i=1}^{n} P\left(\frac{\partial y}{\partial x_i}\right) D(x_i) \tag{8}$$

where y is the output of a node and x_i's are the inputs of the node and the Boolean difference of the function y with respect to x_i gives all combinations for which y depends on x_i. Although this is quite effective it assumes that the x_i's are independent. This assumption is incorrect because x_i's tend to become correlated due to re-convergent fanout structures in the circuit. The problem is solved by describing y in terms of the circuit inputs which are still assumed to be independent. Although the accuracy is improved in this case, the calculation of the Boolean difference terms becomes very expensive. A compromise between accuracy and efficiency can be reached by describing y in terms of some set of intermediate variables in the circuit. This chapter presents an algorithm which is both non-simulative and real-delay model based. Before going to the actual details of the algorithm, a brief discussion of some theoretical background is given.

25.3 THEORETICAL BACKGROUND

Let a signal $\hat{x}(t)$, $t \in (-\infty, +\infty)$, be a stochastic process [19] which makes transitions between logic zero and logic one at random times. A logic signal x(t) can then be thought of as a sample of the stochastic process $\hat{x}(t)$, i.e., x(t) is one of an infinity of possible signals that make up the family $\hat{x}(t)$. In this chapter, it is also assumed that the input processes are ***strict-sense stationary*** [19] implying that its statistical properties are invariant to a shift in the time origin.

In this section we redefine some discrete-time probabilistic measures, which will be used throughout the chapter. The digital CMOS circuits under consideration are assumed to be operating in a synchronous environment, i.e., they are being controlled by a global clock. Let T_{clk} denote the clock period, and T_{gd} denote the smallest gate delay in the circuit. To capture the glitches in the circuit, the clock period is assumed to be divided into S slots [10] as shown in Fig. 25.3, where

$$S \triangleq \frac{T_{clk}}{T_{gd}}. \tag{9}$$

The duration of a time-slot is determined by performing SPICE simulations with detailed device level parameters. Then, the probability of a signal x_i being one at

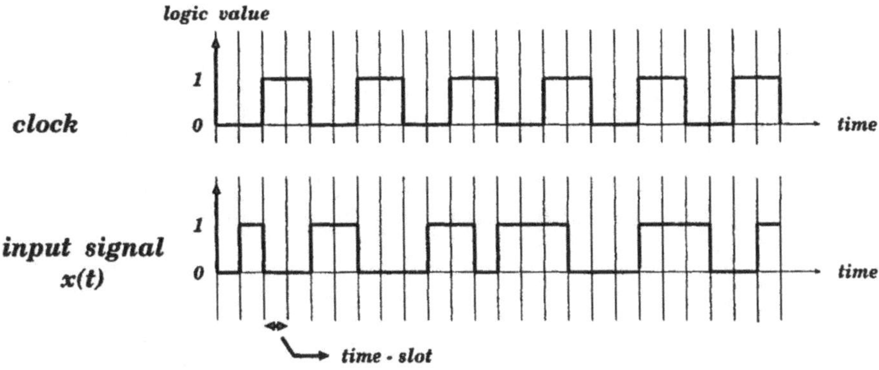

Figure 25.3 Notion of a time-slot.

a given time is defined as

$$p_{x_i}^1 = \lim_{N \to \infty} \frac{\sum_{n=1}^{N \times S} x_i(n)}{N \times S} \tag{10}$$

where, N represents the total number of clock cycles, and $x_i(n)$ is the value of the input signal x_i between the time instances n and $n+1$. Then, the probability that the signal x_i is zero at a given time is defined as

$$p_{x_i}^0 = 1 - p_{x_i}^1. \tag{11}$$

Let us assume that the signal x_i makes a transition from zero to one. Then, the probability associated with this transition is defined as

$$p_{x_i}^{0 \to 1} = \lim_{N \to \infty} \frac{\sum_{n=1}^{N \times S} \overline{x_i(n)} x_i(n+1)}{N \times S}. \tag{12}$$

The other transition probabilities can be obtained in a similar manner. It is easy to verify that

$$p_{x_i}^{0 \to 1} + p_{x_i}^{1 \to 0} + p_{x_i}^{1 \to 1} + p_{x_i}^{0 \to 0} = 1, \tag{13}$$

and

$$p_{x_i}^{0 \to 0} + p_{x_i}^{1 \to 0} = p_{x_i}^0 \tag{14}$$

$$p_{x_i}^{0 \to 1} + p_{x_i}^{1 \to 1} = p_{x_i}^1. \tag{15}$$

The conditional probabilities can be easily derived from the transition probabilities [34], where for example

$$p_{x_i}^{1/0} = \frac{p_{x_i}^{0 \to 1}}{p_{x_i}^{0 \to 1} + p_{x_i}^{0 \to 0}} \tag{16}$$

represents the probability that $x_i(n+1) = 1$ given that $x_i(n) = 0$.

The signal characteristics can be completely determined once the conditional or transition probabilities are known.

25.4 HIERARCHICAL APPROACH TO POWER ESTIMATION OF COMBINATIONAL CIRCUITS

This section presents a hierarchical approach for power estimation of combinational digital circuits. The salient feature of this approach is that it can be used to estimate the power of large digital circuits including multipliers, dividers etc. in a short time. Consider a typical digital circuit consisting of a regular array of cells as shown in Fig. 25.4. The array is treated as an interconnection of sub-circuits

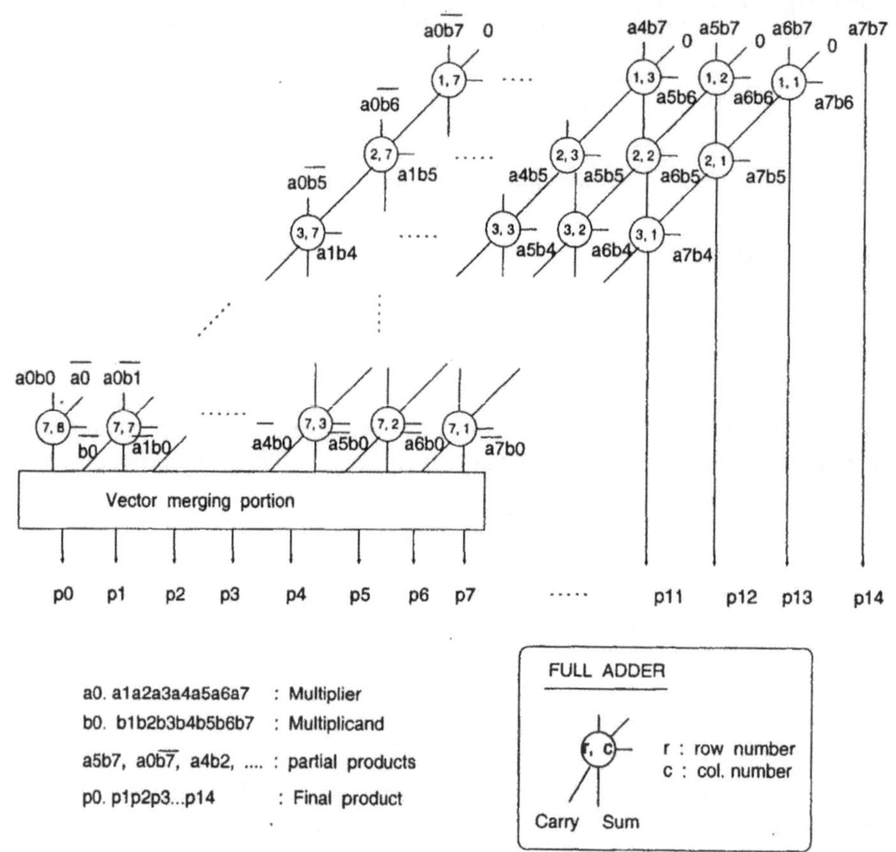

Figure 25.4 8×8-b Baugh-Wooley multiplier.

arranged in rows and columns. The energy of the entire circuit is then computed by summing the energies of the individual sub-circuits. The steps in the proposed approach are summarized in the following algorithm.

Algorithm

INPUT: # of rows, cols. in the circuit, type of sub-circuits, parameters, i.e., signal, conditional probabilities of all input signals

OUTPUT: Estimated average power

est_power () {
total_energy = 0;
for r = 1 to rows

```
  for c = 1 to cols.
     model the sub_circuit(r,c) by using a std;
     compute steady-state probabilities using MATLAB from the input signal;
     parameters sub_circuit(r,c), by treating the std as an irreducible Markov
     chain;
     compute edge activities of the edges in the stds using steady-state
     probabilities and MATLAB;
     compute energy(r,c) associated with the edges of the std using SPICE;
      /* this step has to be executed only once */
     total_energy = total_energy + energy(r,c);
     compute the output signal parameters of sub_circuit(r,c);
  end;
end;
```

average power $= \frac{\text{total energy}}{\text{time over which the energy was spent}}$ }

■

The remainder of this section is concerned with the implementation of each step in the above algorithm.

25.4.1 State-Transition Diagram Modeling

Here, a systematic approach is presented to model digital circuits using state transition diagrams. The modeling is done by deriving analytic expressions for the state-update of all nodes in the corresponding digital circuits.

A) Static CMOS NOR gate

Consider a typical static CMOS NOR gate shown in Fig. 25.5, where x_1 and x_2, respectively, represent the two input signals and x_3 represents the output signal. It is clear from Fig. 25.5 that there are basically two nodes $node_2$ and $node_3$, which

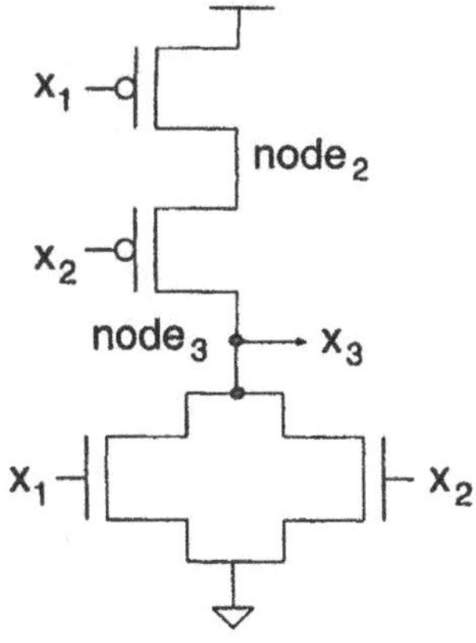

Figure 25.5 A Static CMOS NOR gate.

have their values changing between 1 and 0. The presence of charging/discharging capacitances at these nodes enables us to develop the state-update *arithmetic* equations for the nodes in accordance with

$$node_2(n+1) = (1 - x_1(n)) + x_1(n) * x_2(n) * node_2(n) \tag{17}$$

$$node_3(n+1) = (1 - x_1(n)) * (1 - x_2(n)). \tag{18}$$

The above equations can be used to derive the std for the NOR gate as shown in Fig 25.6, where for example, S_1 represents the state with node values $node_2$ =

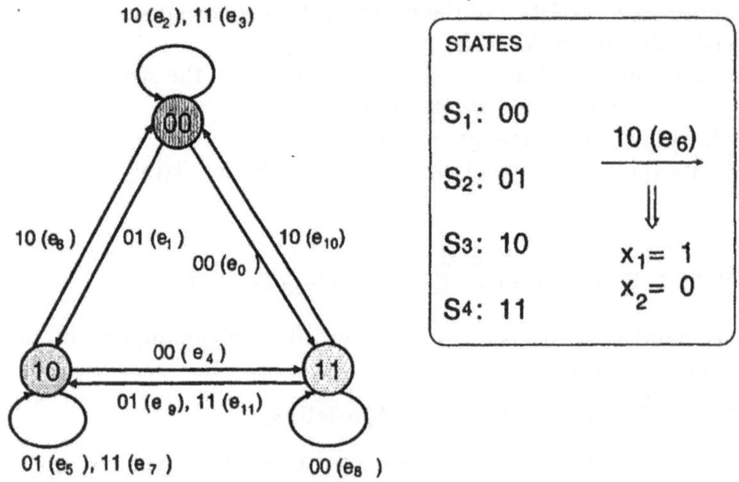

Figure 25.6 State transition diagram for a static CMOS NOR gate.

$node_3$ = 0, and the edge e_1 represents a transition (switching activity) from state S_1 to S_3.

B) Static CMOS NAND gate

A static CMOS NAND fate is shown in Fig. 25.7, where as before x_1 and x_2 represent the two input signals and x_3 represents the output signal. The state

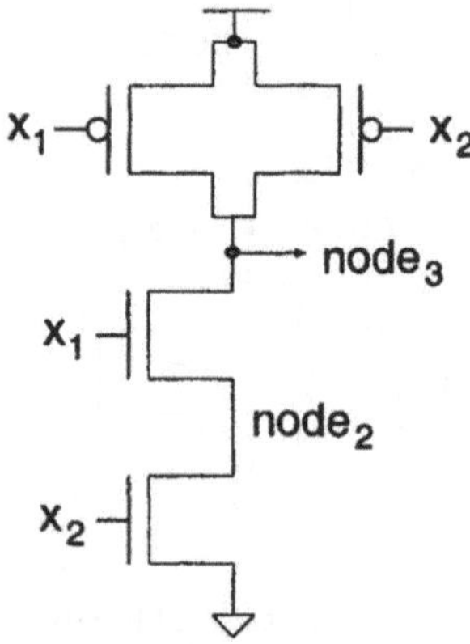

Figure 25.7 A static CMOS NAND gate.

update equations for the static CMOS NAND gate are expressed as

$$node_2(n+1) = (1 - x_2(n)) * (x_1(n) + (1 - x_1(n)) * node_2(n)) \tag{19}$$

$$node_3(n+1) = 1 - x_1(n) * x_2(n). \tag{20}$$

The above equations can be used to derive the std for the NAND gate as shown in Fig 25.8.

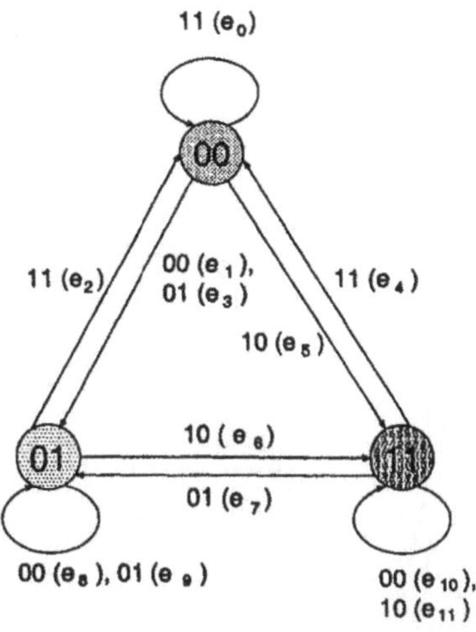

Figure 25.8 State transition diagram for a static CMOS NAND gate.

It turns out that the state-transition diagram thus obtained is identical to the one given in [10].

C) A Static CMOS full adder

Consider the architecture of a static CMOS full adder as shown in Fig. 25.9. It is clear from the figure that the architecture is comprised of a carry generation

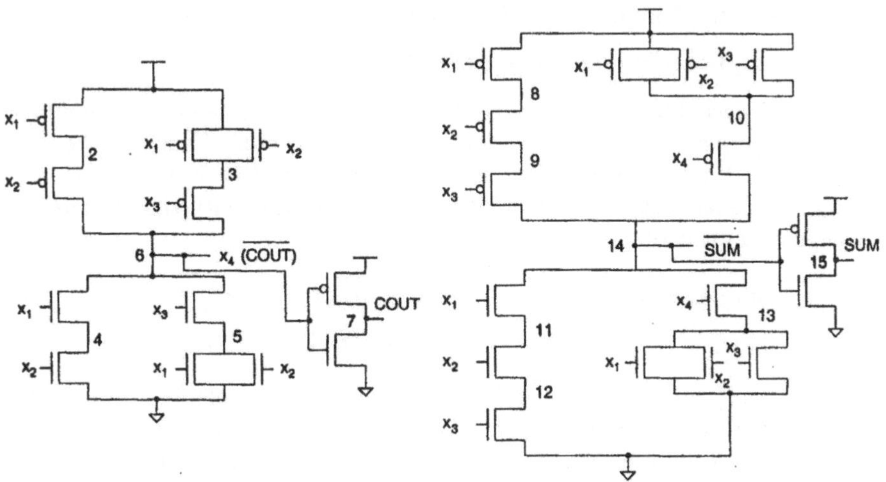

Figure 25.9 Static CMOS full-adder.

portion and a sum generation portion.

The state-update equations can be determined in a similar manner for the carry and the sum portion of the full adder. Then, independent state transition diagrams are constructed for both the portions. For example, the state transition diagram for the carry portion of the full-adder is shown in Fig. 25.10. Here, for the

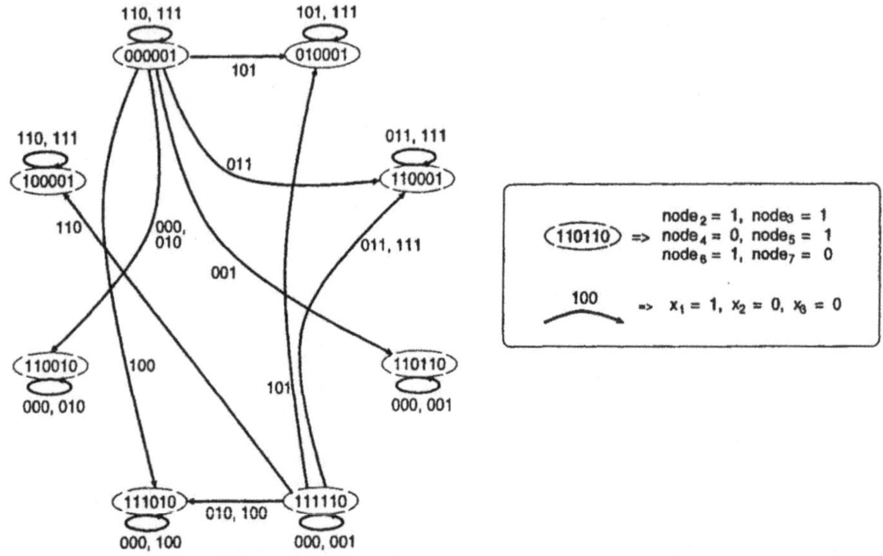

Figure 25.10 State transition diagram for the carry portion of the static CMOS full-adder.

sake of brevity only few edges have been shown. It is clear from Fig. 25.10 that the state transition diagram is comprised of eight states. Each state is associated with the six nodes present in the carry portion of the full-adder, and each edge is associated with the 3 inputs to the full-adder.

25.4.2 Computation of Steady-State Probabilities

An approach based on irreducible Markov chains is used to compute the steady-state probabilities of the various states in the std. Consider the std of the CMOS NOR gate shown in Fig. 25.6. Here, assuming that the input signals x_1 and x_2 are independent, the probabilities $p(e_j)$ associated with the various edges are computed in accordance with

$$p(e_j|e_j : x_m = q, x_n = r) = (q * p^1_{x_m} + (1-q) * p^0_{x_m}) \times (r * p^1_{x_n} + (1-r) * p^0_{x_n}) \tag{21}$$

where $j \in \{0, 11\}$, $m, n \in \{1, 2\}$, and $q, r \in \{0, 1\}$. For example, the probability of edge e_1 in the state transition diagram is $p(e_1) = p^0_{x_1} * p^1_{x_2}$. These edge probabilities are then used to compute the state transition matrix Π_{nor} in accordance with

$$\Pi_{nor} = \begin{bmatrix} p(e_2) + p(e_3) & p(e_1) & p(e_0) \\ p(e_6) & p(e_5) + p(e_7) & p(e_4) \\ p(e_{10}) & p(e_9) + p(e_{11}) & p(e_8) \end{bmatrix}. \tag{22}$$

Here, the $\Pi_{nor_{ij}}$-th element represents the transition probability from state S_i to state S_j, where $i, j \in \{1, 4\}$ and $i, j \neq 2$. Having modeled the transition diagram as an irreducible Markov chain, the steady-state probabilities are then computed by solving

$$P_S = P_S * \Pi_{nor} \tag{23}$$

where

$$P_S = \left[\begin{array}{ccc} P_{S_1} & P_{S_3} & P_{S_4} \end{array} \right] \tag{24}$$

represents the steady-state probabilities of the different states. A simple approach to solve (23) is to first compute the eigenvalues associated with Π^T_{nor}. Then the normalized eigenvector corresponding to an eigenvalue of 1 would be the steady-state probability vector P_S. It may be noted that P_S is computed using MATLAB.

The steady-state probabilities computed by using the above Markov model are then used to compute the edge-activities EA_j (for $j \in \{0, \#ofedges - 1\}$) as proposed in [10]. For example, the edge-activity numbers for the NOR gate are computed using MATLAB in accordance with

$$EA_0 = P_{S_1} * N * S * P(00/10) + EA_3 * (P(00/11) - P(00/10)) \tag{25}$$

$$EA_1 = P_{S_1} * N * S * P(01/10) + EA_3 * (P(01/11) - P(00/10)) \tag{26}$$

$$\vdots$$

$$EA_7 = P_{S_3} * N * S * P(11/01) + (EA_{11} + EA_7) * (P(11/11) - P(11/01)) \tag{32}$$

$$\vdots$$

$$EA_{11} = P_{S_4} * N * S * P(11/00) \tag{36}$$

where, P(11/00) for example, represents the probability that $x_1(n+1) = x_2(n+1) = 1$ given that $x_1(n) = x_2(n) = 0$. The error in the edge-activity numbers using the proposed approach was found to be less than 1.5%.

25.4.3 Energy Computation of Each Edge in the Std

This section presents an algorithm for the computation of energy associated each edge in the std using SPICE. The first step in the algorithm is the identification of the initial state, and the sequence of inputs leading to that state. Two flag vectors; one for the state, and another for each edge in the std are defined. The state flag vector is set whenever that state is first encountered. The edge flag vector on the other hand is set whenever the corresponding edge is traversed. The variable i stores the state number, while the variable k stores the number of the input sequence. For example in Fig. 25.6, i can vary from 1 to 4 (corresponding to states S_1 to S_4), and k can vary from 1 to 4 (corresponding to the sequence of inputs 00, 01, 10, 11). A matrix called edge_mat is formed, the rows of which store the sequence of inputs leading to the traversal of an edge in the std. The steps in the algorithm are summarized below.

Algorithm

INPUT: std of the sub-circuit; initial state (*init_state*), number of inputs to the sub-circuit (*num_inputs*), initialized edge-matrix (*edge_mat*).

```
OUTUT: energy of each edge in the std.
energy_edge() {
reset state flags and edge flags to zero;
i = init_state;k = 1;
while(all edge flags have not been set)
   m = new state;
   if(edge flag vector corresponding to input k not set)
      set edge flag vector;
      update edge_mat;
      if(flag corresponding to state m is not set)
         set flag corresponding to state m;
         update edge_mat;
         prev_state(i) = i;
         i = m;k = 0;
      else
         update edge_mat;
      end;
   end;
   k = k+1;
   if(k > 2^num_inputs)
      k = 1;
      i = prev_state(i);
   end;
end; }
/* a matrix edge_mat with rows containing the sequence of inputs leading to the
traversal of edges in the std has been formed */
rows = number of rows in edge_mat;
cols. = number of columns in edge_mat;
for j = 1 to rows;
   run SPICE for input sequence edge_mat(j,cols-1);
   energy1 = resulting energy;
   run SPICE for input sequence edge_mat(j,cols.);
   energy2 = resulting energy;
   W_j = energy2 - energy1;
end;
```

■

Using the above algorithm, the initial state for the NOR gate shown in Fig. 25.6 was found to be 11, and the edge_matrix was found to be

$$edge\_mat = \begin{bmatrix} 00 & 01 & 10 & 00 \\ 00 & 01 & 10 & 01 \\ 00 & 01 & 10 & 10 \\ 00 & 01 & 10 & 11 \\ 00 & 01 & 00 & \\ 00 & 01 & 01 & \\ 00 & 01 & 10 & \\ 00 & 01 & 11 & \\ 00 & 00 & & \\ 00 & 01 & & \\ 00 & 10 & & \\ 00 & 11 & & \end{bmatrix} \begin{matrix} \rightarrow e_0 \\ \rightarrow e_1 \\ \rightarrow e_2 \\ \rightarrow e_3 \\ \rightarrow e_4 \\ \rightarrow e_5 \\ \rightarrow e_6 \\ \rightarrow e_7 \\ \rightarrow e_8 \\ \rightarrow e_9 \\ \rightarrow e_{10} \\ \rightarrow e_{11} \end{matrix} \tag{37}$$

The energy associated with the sub_circuit is then computed by taking a weighted sum of the energies associated with the various edges in the std representing the sub_circuit, in accordance with

$$energy = \sum_{j=0}^{\#\,of\,edges\,-\,1} W_j * EA_j. \tag{38}$$

25.4.4 Computation of Output Signal Parameters

The final step in the hierarchical approach to power estimation is concerned with the computation of the signal parameters at the output of the sub-circuits. This is best illustrated with the help of a simple example. Consider two NOR gates connected in cascade as shown in Fig. 25.11. Let $x_1(n)$ and $x_2(n)$ represent,

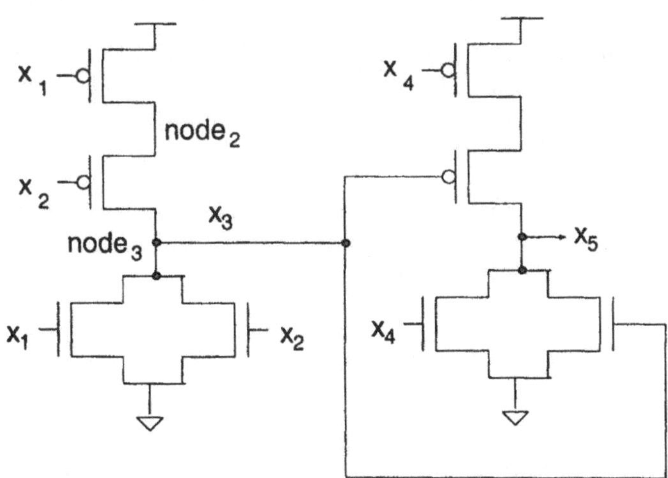

Figure 25.11 Two NOR gates connected in cascade.

respectively, the binary values of the input signals x_1 and x_2 between time instances

n and $n+1$. Then, from (18) one can compute the values of the signal x_3 for all $N \times S$ time-slots. Therefore, once $x_3(n)$, and $x_4(n)$ are well-defined for all time-slots, the signal characteristics for the second NOR gate can be computed. To compute the energy of the second NOR gate using (38), we use the W_j values calculated previously for the first NOR gate and the new EA_j values obtained for the second NOR gate. This enables the computation of the energy values in a very short time.

The above method is easily generalized to multipliers (dividers) which are designed using type-0 or type-1 adders cascaded in a specific manner.

25.4.5 Loading and Routing Considerations

One of the disadvantages of the proposed approach is that it does not take into account the effect of loading and routing capacitances directly. In this section, we propose an approach which enables these effects to be taken into account.

Consider the CMOS digital circuit shown in Fig. 25.12, where an effective load/routing capacitance has been added. The proposed method involves re-

Figure 25.12 Circuit with loading effects.

computation of the edge energies in the state transition diagram of the CMOS circuit with the load capacitance in place. The idea is to simulate the effect of loading when computing the edge energies. Therefore, we see that by a slight modification in the computation of the edge energies, the effect of loading can be taken into account.

One of the main advantages of this approach is that SPICE is used to characterize the effect of loading. Therefore, accurate device models can be used to incorporate the effect of loading. The steps in the proposed approach are summarized in the following algorithm.

Algorithm

INPUT: # of rows, cols. in the circuit, type of sub-circuits, parameters, i.e., signal, conditional probabilities of all input signals

OUTPUT: Estimated average power

```
est_power () {
total_energy = 0;
for r = 1 to rows
   for c = 1 to cols.
      model the sub_circuit(r,c) by using a std;
      compute steady-state probabilities from the input signal parameters of
      sub_circuit(r,c), by treating the std as an irreducible Markov chain;
      compute edge activities of the edges in the stds using steady-state
      probabilities and MATLAB;
      estimate the load/routing capacitance of sub_circuit(r,c);
```

compute energy(r,c) associated with the edges of the std using SPICE (with the load capacitance in place);
total_energy = total_energy + energy(r,c);
compute the output signal parameters of sub_circuit(r,c);
end;
end;
average power = $\frac{\text{total energy}}{\text{time over which the energy was spent}}$ }

25.5 POWER ESTIMATION OF SEQUENTIAL CIRCUITS

In this section, the algorithm presented for combinational circuits is extended to handle sequential circuits as well. A sequential circuit has a combinational block and some storage elements like flip-flops. In the previous section, a method was proposed to model any arbitrary combinational block using state transition diagrams. The method is extended to model flip-flops which are basically designed by cascading latches.

Consider an edge-triggered D flip-flop as shown in Fig. 25.13. Here, D represents the input signal, Q represents the output signal, and $\phi_{1,2}$ represent the

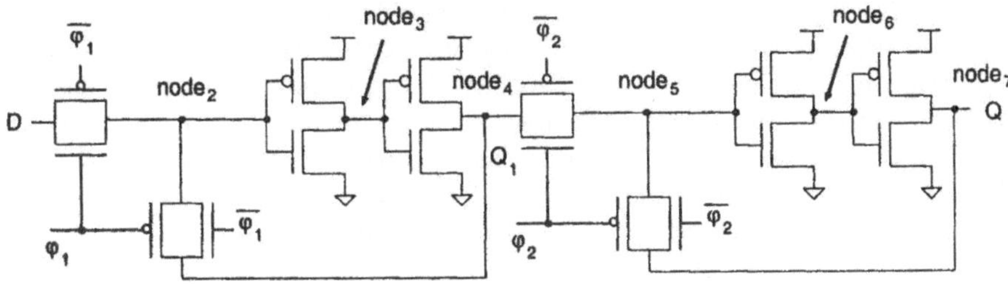

Figure 25.13 An edge-triggered D flip-flop.

non-overlapping two-phase clock signals. It is clear from Fig. 25.13 that the D flip-flop can be viewed as a cascade of two identical latches controlled by different clocks. Therefore, for power estimation it is sufficient to model a single latch with the help of a std. The state-update *arithmetic* equations for the first latch are

$$node_2(n+1) \ = \ D(n) * \phi_1(n) \ + \ (1 \ - \ \phi_1(n)) * node_4(n) \tag{39}$$

$$node_3(n+1) \ = \ 1 \ - \ node_2(n+1) \tag{40}$$

$$node_4(n+1) \ = \ node_2(n+1). \tag{41}$$

Using the above equations, the std for the latch is derived and is shown in Fig. 25.14. Here, the states represent the values of the nodes $node_2$, $node_3$, and $node_4$ at some time instant. For example, S_1 represents the state with node values $node_2 \ = \ 0$, $node_3 \ = \ 1$, and $node_4 \ = \ 0$. The numbers associated with the edges represent the sequence D, phi_1. It is interesting to note that although there are three nodes in the latch, there are only two states. Intuitively, this means that the presence of a latch reduces the glitching activity. The std for the second latch can then be easily obtained by replacing phi_1 with phi_2, and D with Q_1.

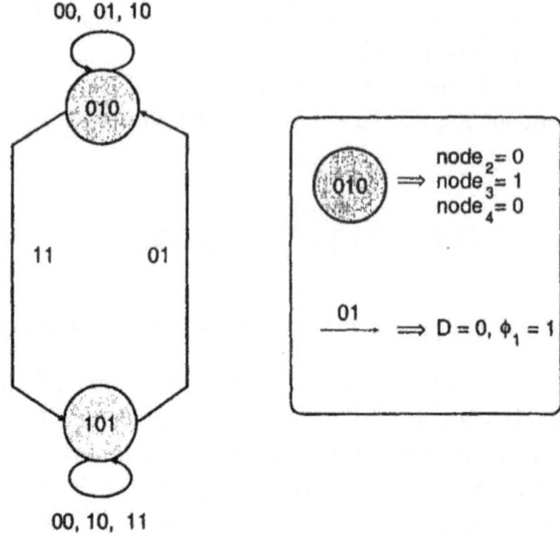

Figure 25.14 State-transition diagram for a latch.

Once the flip-flops have been modeled, the next step is to just simulate the entire sequential circuit without computing any energy values. This enables the computation of both the direct and feedback input signal values at all possible time-slots, and the transition probabilities can then be determined by considering these values. As a result the temporal correlation between the input signal values is taken into account. Then, the algorithm for power estimation of combinational circuits is used to estimate the power of sequential circuits as well.

25.6 EXPERIMENTAL RESULTS

A CAD tool called *HEAT* (*Hierarchical Energy Analysis Tool*) has been developed based on the proposed approach, and power has been estimated for many benchmark circuits. The first experiment on power consumption was conducted on some basic cells and multipliers. The second set of power estimation experiments was performed on some Galois field architectures which form the basis of error control coding. The third set of experiments was performed on different DCT designs which form the basis of video architectures. Finally, the last set of experiments were conducted on fast binary adders.

25.6.1 Power Estimation of Basic Cells and Multipliers

The power estimation results of some basic cells and multipliers designed using these basic cells are presented in Tables 25.1 and 25.2. The experiments were performed on a SUN SPARC 20 workstation. The entries in the first column in Table 25.1 represent the various kinds of basic static CMOS digital circuits for which power has been estimated. The entries in the second column represent the average power consumption computed by using both SPICE and *HEAT*, while those in the third column represent the corresponding run times. The reduction in the number of states in the state transition diagram, obtained by using the proposed algorithm is elucidated in column four. Finally, the entries in the fifth column

Table 25.1 Average Power of Some Basic Static CMOS Digital Circuits

CMOS circuit	avg. power (μW)		run time (sec)		max. # of states possible		error
	SPICE	*HEAT*	SPICE	*HEAT*	existing	*HEAT*	
nor	0.9939	0.9845	188	6	4	4	-0.94%
nand	1.4490	1.4107	219	6	4	4	-2.64%
type-0 adder	33.218	32.12	1289	34	2^{14}	2^{14}	-3.31%
type-1 adder	33.029	31.94	1295	36	2^{15}	2^{15}	-3.29%
D flip-flop	-	4.235	-	8	8	8	-

represent the error in power estimation using the proposed approach. It is clear from these entries that the values obtained by *HEAT* are in close agreement with the actual values obtained by performing exhaustive SPICE simulations. However, the run time of *HEAT* is orders of magnitude less than that of SPICE.

The hierarchical approach for power estimation is exploited to obtain the results in Table 25.2. Here, the subscripts for the basic gates (e.g., NAND, NOR)

Table 25.2 Average Power for Larger Circuits Obtained by Using the Hierarchical Approach

CMOS circuit	avg. power (μW)		run time (sec)		max. # of states possible		error
	SPICE	*HEAT*	SPICE	*HEAT*	existing	*HEAT*	
nor_2	3.4029	3.2567	383	10	16	4	-4.30%
nor_3	5.7795	5.5429	571	15	64	4	-4.09%
nor_6	13.2548	13.8794	1266	34	4096	4	4.71%
$nand_2$	3.7608	3.8205	379	10	16	4	1.56%
$nand_3$	6.4374	6.1255	629	16	64	4	-4.85%
$nand_6$	13.7415	13.0554	1400	35	4096	4	-4.99%
$mult_{BW_4}$[35]	980.28	900.34	4959	200	$2^{14*11}+2^{7*9}$	$2^{14}+2^{7}$	-8.15%
$mult_{BW_6}$	2928.3	3102.2	19007	223	$2^{14*27}+2^{7*20}$	$2^{14}+2^{7}$	5.94%
$mult_{BW_8}$	8847.4	8412.3	43438	318	$2^{14*51}+2^{7*35}$	$2^{14}+2^{7}$	-5.17%
$mult_{HY_4}$[35]	672.49	627.79	4065	189	2^{15*12}	2^{15}	-6.64%
$mult_{HY_6}$	1297.6	1367.5	10011	215	2^{15*30}	2^{15}	5.11%
$mult_{HY_8}$	2802.7	3011.5	29052	289	2^{15*56}	2^{15}	7.45%

represent the number of cells connected in cascade. For example, $nand_6$ represents 6 nand-gates connected in cascade. The subscripts for the Baugh-Wooley (BW) and the redundant hybrid (HY) multiplier architectures proposed in [35] represent the word length. The BW multiplier is designed by cascading type-0 adders in the form of an array, while the HY multiplier is designed by cascading type-1 adders.

The results show that the power consumed by the HY multiplier is much less than that consumed by the BW multiplier.

25.6.2 Power Estimation of Galois Field (GF) Architectures

In recent years, finite fields have received a lot of attention because of their application in error control coding [36] [37]. They have also been used in digital signal processing, pseudo-random number generation, encryption and decryption protocols in cryptography. Well-designed finite field arithmetic units and a powerful decoding algorithm are important factors for designing high speed and low complexity decoders for many error control codes [38].

Addition in GF(2^m), where m denotes the field order, is bit independent and is a relatively straightforward operation. However, multiplication, inversion and exponentiation are more complicated. Hence, design of circuits for these operations with low circuit complexity, short computation delay and high throughput rate is of great practical concern. Reed-Solomon (RS) codes can correct both random and burst errors and have found many applications in space, spread spectrum and data communications [38]. RS codes, as a special class of BCH (Bose-Chaudhuri-Hocquenghem) codes, have both their codeword symbols and the error-locator symbols coming from the same field $GF(2^m)$, which leads to it optimum error correcting capabilities. RS codes are capable of correcting both burst and random errors.

The HEAT tool has been used to estimate the power of various multipliers constituting the encoder/decoder architectures in order to decide which one is best in terms of power consumption. The results are shown in Figs. 25.15 and 25.16

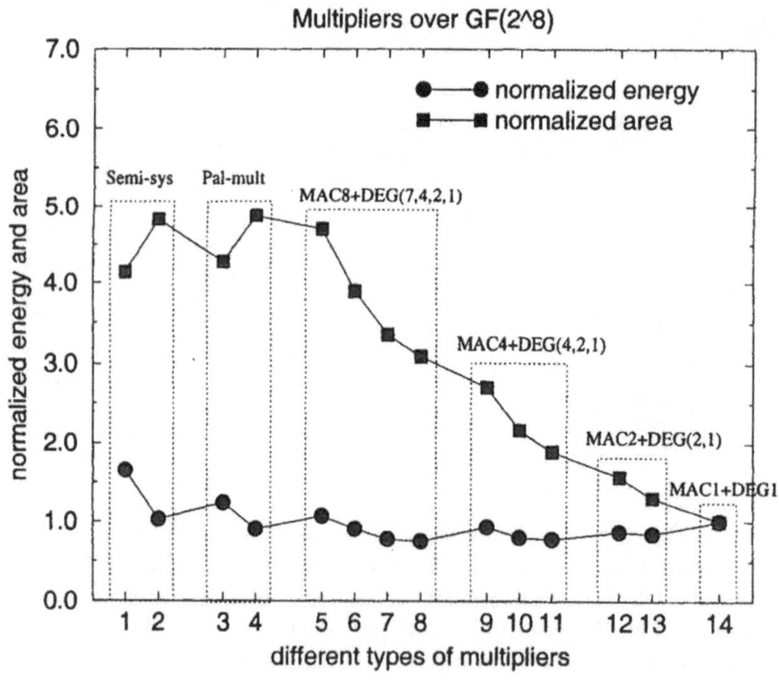

Figure 25.15 Energy consumption of programmable finite-field multipliers.

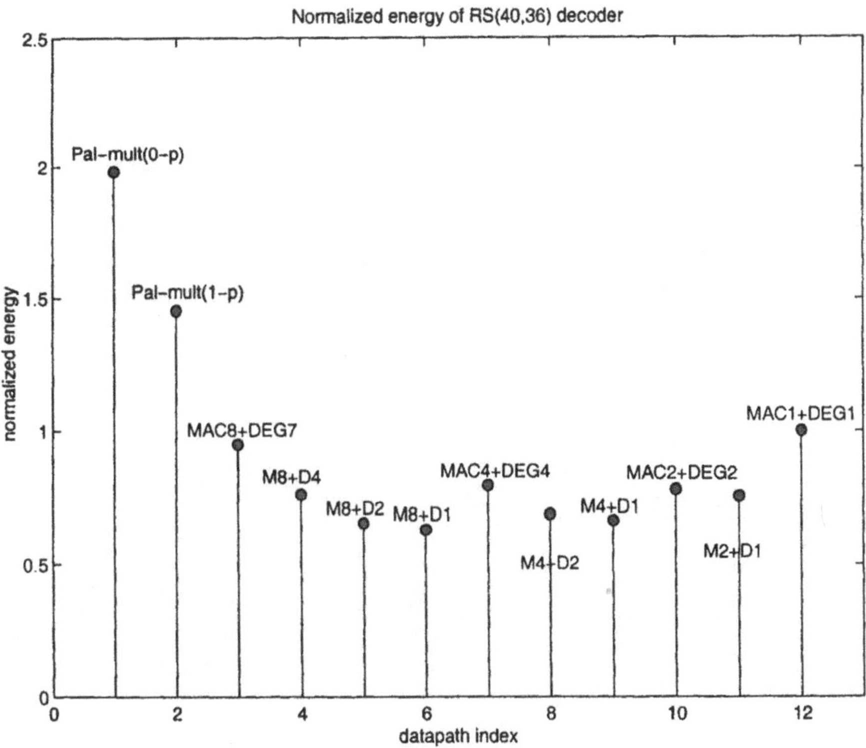

Figure 25.16 Energy consumption of a RS(36,32) encoder.

[39]. Fig. 25.15 shows the power consumption (using 0.3μ technology parameters) of various finite-field multipliers which can be programmable with respect to the primitive polynomial as well as the field order (up to 8). This figure shows 14 types of finite-field multipliers including semi-systolic [40], parallel, and various forms of heterogeneous digit-serial multipliers [41] where different digit sizes are used for the MAC array and the degree-reduction (DEG) array. Here, $MACx$ denotes polynomial multiplication operation with digit-size x and $DEGy$ denotes polynomial modulo operation with digit-size y. The results show that digit-serial multipliers consume less energy and occupy less area. However, the latency required for multiplication is much higher than their bit-parallel counterparts. Thus it is important to match the data-path for a certain application to achieve the least energy consumption. The energy consumption of a Reed-Solomon encoder for RS(36,32) code for these finite-field multipliers is shown in Fig. 25.16. It is seen that the designs MAC8+DEGi, where i can be 1 or 2 or 4, consume the least energy.

25.6.3 Power Estimation of DCT (Discrete Cosine Transform) Architectures

Video compression is one of the most difficult tasks in design of multimedia systems. DCT is an important component in the implementation of JPEG standards used for compression of still images and MPEG standards used for compression of moving images. DCT is simple to implement and provides near-optimal performance as compared with other approaches. Two popular approaches for implementation of DCT algorithms include use of distributed arithmetic and flow graphs based on fast algorithms. The distributed arithmetic architecture (DAA) is more popular than the flow-graph architecture (FGA) due to its reduced area requirements. While the area advantages of the DAA are well known, which architecture is best suited for power-dominated applications has not been addressed so far. The power consumption of the two DCT architectures for HDTV (high-definition television) video compression applications are compared using the HEAT tool and the results are presented in Table 25.3 [42]. The results show that the

Table 25.3 Comparison Between DAA and FGA

	Distributed Arithmetic	Flow Graph
Latency	215.5 ns	808 ns
Frequency	55.68 MHz	18.56 MHz
Sample rate	74.25 Msample/s	74.25 Msample/s
Power @ 3.3 V	39 mW	19.8 mW
Power @ 2.05 V	6.8 mW	4.1 mW
Area	1 mm^2	1.2 mm^2

power consumed by the flow graph architecture is about five times lower than the architecture designed using distributed arithmetic for he same sample rate. However, the DCT architecture designed using distributed arithmetic is much smaller than the flow graph architecture. In terms of latency, it is observed that the DAA is much faster than the FGA. Therefore, if power is an issue the FGA is more suitable and if area and latency are issues, then the DAA architecture is more suitable.

25.6.4 Power Estimation of Fast Binary Adders

Fast binary addition is carried out using a redundant-to-binary converter in [43]. Here, the HEAT tool is used to compare different architectures for power consumption and to decide the best architecture for low-power. A family of fast converter architectures is developed based on *tree-type* (obtained using *look-ahead* techniques) and *carry-select* approaches. These adders are implemented using static CMOS multiplexers only and power estimation is done at a clock frequency of 50 MHz. Table 25.4 shows the power consumption of 16-bit carry-select adders (which include input rewriting circuitry, carry generators, and output sum generation). From this table, we conclude that, for a given latency, the design with more number of carry-select blocks and larger length for the block operating on the least significant bits leads to less switching activity and power consumption.

In order to further understand the effect of block sizes in the carry-select adder, the ordering of the blocks is changed and the experimental results are presented in Table 25.5. All the designs shown in the table have the same latency ($10t_{mux}$) and the same number of multiplexers, i.e., 61. We can observe from the results that including more number of smaller blocks (size 2) between the longer

Table 25.4 Power Consumption of Various Types of Carry-Select Adders.

Design	Power (μW)	Area (# mux)	Latency
CSEL(5, 4, 3, 2, 2)	809.99	67	7
CSEL(3, 3, 3, 3, 2, 2)	762.24	65	8
CSEL(4, 3, 3, 3, 3)	784.60	65	8
CSEL(4, 4, 4, 4)	806.16	65	8
CSEL(5, 4, 3, 4)	812.98	65	8
CSEL(6, 5, 5)	830.71	65	8
CSEL(3, 2, 5, 2, 4)	759.46	63	9
CSEL(3, 3, 3, 3, 4)	759.48	63	9
CSEL(4, 4, 3, 5)	776.41	63	9
CSEL(4, 6, 6)	794.97	63	9
CSEL(5, 5, 6)	796.93	63	9
CSEL(2, 2, 2, 2, 2, 2, 2, 2)	686.37	61	10
CSEL(2, 2, 2, 5, 5)	738.86	61	10
CSEL(8, 8)	788.03	61	10

Table 25.5 Estimated Power of CSEL Adders*

Design	Power (μW)
CSEL(2, 2, 2, 5, 5)	738.86
CSEL(2, 2, 5, 2, 5)	733.89
CSEL(2, 5, 2, 2, 5)	731.90
CSEL(5, 2, 2, 2, 5)	727.75

*:Block size chosen from the set (5, 5, 2, 2, 2) for latency of $10t_{mux}$.

blocks (size 5) leads to lower power consumption. This is because the smaller blocks absorb the glitching introduced by the larger block. Therefore, we conclude that within a family of designs with constant latency and equal number of multiplexers, the design with smaller blocks in between the larger blocks consumes the least power.

Other adders including the Manchester based redundant binary adder, the tree based redundant binary adder, and a combination of tree and carry select adders have also been investigated for power consumption and the results are summarized in Table 25.6. The results show that the RB adder consumes the least power and the tree based RB adder consumes the most power. This is because there is high

Table 25.6 Comparison of Power Consumption (μW) of Binary Adders Implemented Using Different Implementation Styles

Imp. style	Power	Lat.	Power*Lat.
RB	586.01	17	9962
CS RB	810	7	5670
Tree RB	904.08	6	5424
Manchester	685.38	18	12337

capacitive loading in the tree based carry generation circuit. However, the latency is the least for the tree structure and therefore it has the minimum power-latency product.

The power consumption results for the hybrid adders are summarized in Table 25.7, where the letter t is used to denote the block based on a modified tree implementation, and the entire hybrid adder is denoted as the CST adder. The results show a very interesting trend in the number of tree blocks. It is observed that

Table 25.7 Power Dissipated by CSEL (4, 4, 4, 4) Adder*

Design	Power (μW)	# MUX	Latency
CSEL(4,4,4,4)	806.2	65	8
CST(4,4,4,4t)	809.7	67	8
CST(4,4,4t,4)	774.3	65	8
CST(4,4t,4,4)	772.6	65	8
CST(4,4,4t,4t)	778.0	67	7
CST(4,4t,4,4t)	782.0	67	8
CST(4,4t,4t,4)	754.6	65	8
CST(4,4t,4t,4t)	754.6	67	7

*:Some of the carry select blocks are replaced by 4-bit tree blocks.

as the number of tree blocks increases, the power consumption decreases. Moreover, for a fixed number of tree blocks, the power consumption is less if the tree blocks are in the more significant blocks of the architecture. This is because the number of multiplexers is reduced by two for a slight increase in latency. It is interesting to note that among all 16-bit adders, the design of CST(4, 4t, 4t, 4t) has the least power-latency product of 5282.2 as compared with 5424 for the tree adder. Therefore, this adder consumes the least energy.

25.7 CONCLUSIONS

A CAD tool called HEAT has been presented in this chapter. The tool is very versatile and is applied to estimate the power of various architectures. The power estimation results show that the hybrid multiplier consumes much less power than the Baugh-Wooley multiplier. In the area of error control coding it is found that the optimum Reed-Solomon encoder architecture in terms of power consumption is the one that uses a multiplier architecture with digit-size 8 and a degree reduction operation with digit-sizes 1, 2, or 4. The tool is used to estimate the power of various DCT architectures used in image compression and it is found that the flow-graph architecture consumes the least power. The tool is also used to estimate the power of fast binary adders and it is found that a hybrid adder consumes the least power. Future work includes application of the tool for power estimation of various finite impulse response (FIR) and infinite impulse response (IIR) filters.

REFERENCES

[1] R. Powers, "Batteries for low power electronics," *Proceedings of the IEEE*, vol. 38, pp. 687–693, April 1995.

[2] T. A. Fjeldly and M. Shur, "Threshold voltage modeling and the subthreshod regime of operation of short-channel MOSFETs," *IEEE Trans. on Electron Devices*, vol. 1, pp. 137–145, Jan. 1993.

[3] N. Hedenstierna and K. Jeppson, "CMOS circuit speed and buffer optimization," *IEEE Trans. on Computer-Aided Design of Integrated Circuits and Systems*, vol. 3, pp. 270–281, Mar. 1987.

[4] H. M. Veendrick, "Short-circuit dissipation of static CMOS circuitry and its impact on the design of buffer circuits," *IEEE Journal of Solid-State Circuits*, vol. SC-19, pp. 468–473, Aug. 1984.

[5] S. Turgis, N. Azemard, and D. Auvergne, "Explicit evaluation of short circuit power dissipation for CMOS logic structures," in *Proc. IEEE Int. Symp. on Low Power Design*, pp. 129–134, Apr. 1995.

[6] F. N. Najm, "Transition density: A new measure of activity in digital circuits," *IEEE Trans. on Computer-Aided Design of Integrated Circuits and Systems*, vol. 12, pp. 310–323, Feb. 1993.

[7] S. Chowdhury and J. S. Barkatullah, "Estimation of maximum currents in MOS IC logic circuits," *IEEE Trans. on Computer-Aided Design*, vol. 9, pp. 642–654, June 1990.

[8] S. Devadas, K. Keutzer, and J. White, "Estimation of power dissipation in CMOS combinatorial circuits using Boolean function manipulation," *IEEE Trans. on Computer-Aided Design*, vol. 11, pp. 373–383, Mar. 1992.

[9] A. Shen, A. Ghosh, S. Devadas, and K. Keutzer, "On average power dissipation and random pattern testability of CMOS combinational logic networks," in *Proc. IEEE Int. Conf. Computer Aided Design (ICCAD)*, pp. 402–407, 1992.

[10] J.-Y. Lin, T.-C. Liu, and W.-Z. Shen, "A cell-based power estimation in CMOS combinational circuits," in *Proc. IEEE Int. Conf. Computer Aided Design (ICCAD)*, pp. 304–309, 1994.

[11] J. H. Satyanarayana and K. K. Parhi, "HEAT: Hierarchical Energy Analysis Tool," in *Proc. IEEE/ACM Design Automation Conference (DAC)*, (Las Vegas, NV), pp. 9–14, June 1996.

[12] J. H. Satyanarayana and K. K. Parhi, "A theoretical approach to estimation of bounds on power consumption in digital multipliers," *IEEE Trans. Circuits and Systems-II: Special Issue on Low Power Wireless Communications*, vol. 44, pp. 473–381, June 1997.

[13] K. P. Parker and E. J. McCluskey, "Probabilistic treatment of general combinatorial networks," *IEEE Trans. on Computers*, vol. C-24, pp. 668–670, June 1975.

[14] J. Savir, G. Ditlow, and P. Bardell, "Random pattern testability," *IEEE Trans. on Computers*, vol. C-33, pp. 79–90, Jan. 1984.

[15] B. Krishnamurthy and G. Tollis, "Improved techniques for estimating signal probabilities," *IEEE Trans. on Computers*, vol. C-38, pp. 1245–1251, July 1989.

[16] H.-F. Jyu, S. Malik, S. Devadas, and K. W. Keutzer, "Statistical timing analysis of combinatorial logic circuits," *IEEE Trans. VLSI Systems*, vol. 1, pp. 126–135, June 1993.

[17] T.-L. Chou, K. Roy, and S. Prasad, "Estimation of circuit activity considering signal correlations and simultaneous switching," in *Proc. IEEE Int. Conf. Computer Aided Design (ICCAD)*, pp. 300–303, 1994.

[18] J. H. Satyanarayana and K. K. Parhi, "A hierarchical approach to transistor-level power estimation of arithmetic units," in *Proc. IEEE International Conf. Accoustic Speech and Signal Processing (ICASSP)*, (Atlanta, GA), pp. 3339–3342, May 1996.

[19] A. Papoulis, *Probability, Random Variables, and Stochastic Processes.* New York: McGraw-Hill, 2nd ed., 1984.

[20] C. W. Therrien, *Discrete random signals and statistical signal processing*, ch. 1. Englewood Cliffs, New Jersey: Prentice Hall International, Inc., 1992.

[21] T. Quarles, "The SPICE3 implementation guide," Tech. Rep. M89-44, Electronics Research Laboratory, University of California, Berkeley, California, Apr. 1989.

[22] C. X. Huang, B. Zhang, A.-C. Deng, and B. Swirski, "The design and implementation of PowerMill," in *Proc. IEEE Int. Symp. on Low Power Design*, pp. 105–110, Apr. 1995.

[23] A. Salz and M. A. Horowitz, "IRSIM: An incremental MOS switch-level simulator," in *Proc. 26th IEEE/ACM Design Automation Conference (DAC)*, pp. 173–178, June 1989.

[24] R. Burch, F. N. Najm, P. Yang, and T. Trick, "A Monte Carlo approach for power estimation," *IEEE Trans. VLSI Systems*, vol. 1, pp. 63–71, Mar. 1993.

[25] J.-M. Chang and M. Pedram, "Low power register allocation and binding," in *Proc. 32nd IEEE/ACM Design Automation Conference (DAC)*, pp. 29–35, June 1995.

[26] D. Marculescu, R. Marculescu, and M. Pedram, "Information theoretic measures for energy consumption at register transfer level," in *Proc. IEEE Int. Symp. on Low Power Design*, pp. 81–86, Apr. 1995.

[27] F. N. Najm, "Towards a high-level power estimation capability," in *Proc. IEEE Int. Symp. on Low Power Design*, pp. 87–92, Apr. 1995.

[28] P. E. Landman and J. Rabaey, "Activity-sensitive architectural power analysis for control path," in *Proc. IEEE Int. Symp. on Low Power Design*, pp. 93–98, Apr. 1995.

[29] S. Ramprasad, N. R. Shanbhag, and I. N. Hajj, "Analytical estimation of transition activity from word-level signal statistics," in *Proc. 34nd IEEE/ACM Design Automation Conference (DAC)*, pp. 582–587, June 1997.

[30] H. Goldstein, "Controllability/observability of digital circuits," *IEEE Trans. on Circuits and Systems*, vol. 26, pp. 685–693, Sep. 1979.

[31] R. Bryant, "Graph-based algorithms for Boolean function manipulation," *IEEE Trans. on Computers*, vol. C-35, pp. 677–691, Aug. 1986.

[32] A. Ghosh, S. Devadas, K. Keutzer, and J. White, "Estimation of average switching activity in combinational and sequential circuits," in *Proc. 29th IEEE/ACM Design Automation Conference (DAC)*, pp. 253–259, June 1992.

[33] R. Burch, F. Najm, P. Yang, and D. Hocevar, "Pattern independent current estimation for reliability analysis of CMOS circuits," in *Proc. 25th IEEE/ACM Design Automation Conference (DAC)*, pp. 294–299, June 1988.

[34] R. Marculescu, D. Marculescu, and M. Pedram, "Switching activity analysis considering spatiotemporal correlations," in *Proc. IEEE Int. Conf. Computer Aided Design (ICCAD)*, pp. 292–297, 1994.

[35] H. R. Srinivas and K. K. Parhi, "High-speed VLSI arithmetic processsor architectures using hybrid number representation," *Journal of VLSI Signal Processing*, no. 4, pp. 177–198, 1992.

[36] R. E. Blahut, *Theory and Practice of Error Control Codes*. Addison Wesley, 1984.

[37] F. J. MacWilliams and N. J. A. Sloane, *The Theory of Error Correcting Codes*. Amsterdam: North-Holland Pub. Co., 1977.

[38] S. B. Wicker and V. K. Bhargava, *Reed-Solomon Codes and Their Applications.* New York, NY: IEEE Press, 1994.

[39] L. Song, K. K. Parhi, I. Kuroda, and T. Nishitani, "Low-energy heterogeneous digit-serial Reed-Solomon codecs," *Proc. of 1998 IEEE ICASSP,* pp. 3049-3052, May 1998.

[40] S. K. Jain, L. Song, and K. K. Parhi, "Efficient semi-systolic architectures for finite field arithmetic," *IEEE Trans. on VLSI Systems*, vol. 6, pp. 101–113, Mar. 1998.

[41] L. Song and K. K. Parhi, "Efficient finite field serial/parallel multiplication," in *Proc. of International Conf. on Application Specific Systems, Architectures and Processors*, (Chicago), pp. 72–82, Aug 1996.

[42] M. Kuhlmann, "Power comparison of DCT architectures with computation graphs and distributed arithmetic," *Proc. of 1998 Asilomar Conf. on Signals, Systems and Computers*, Pacific Grove, CA, Nov. 1998.

[43] K. K. Parhi, "Fast low-energy VLSI binary addition," in *Proc. IEEE International Conf. on Computer Desgn (ICCD)*, (Austin, TX), pp. 676–684, Oct. 1997.

Chapter 26

System Exploration for Custom Low Power Data Storage and Transfer

Francky Catthoor, Sven Wuytack, Eddy De Greef, Florin Balasa † and Peter Slock ‡
IMEC, Leuven, Belgium
{catthoor,wuytack,degreef}@imec.be
florin.balasa@rss.rockwell.com, peter.slock@student.kuleuven.ac.be
† *Currently at Rockwell Intnl. Corp., Newport Beach, California*
‡ *Currently at KB, Brussels, Belgium*

26.1 INTRODUCTION

For most real-time signal processing applications there are many ways to realize them in terms of a specific algorithm. As reported by system designers, in practice this choice is mainly based on "cost" measures such as the number of components, performance, pin count, power consumption, and the area of the custom components. Currently, due to design time restrictions, the system designer has to select – on an ad-hoc basis – a single promising path in the huge decision tree from abstract specification to more refined specification (Fig. 26.1). To alleviate this situation, there is a need for fast and early feedback at the algorithm level *without* going all the way to assembly code or hardware layout. Only when the design space has been sufficiently explored at a high level and when a limited number of promising candidates have been identified, a more thorough and accurate evaluation is required for the final choice (Fig. 26.1).

In this chapter key parts of our system level power exploration methodology are presented for mapping data-dominated multi-media applications to custom processor architectures. This formalized methodology is based on the observation that for this type of applications the power consumption is dominated by the data transfer and storage organization. Hence, the first exploration phase should be to come up with an optimized data transfer and storage organization. In this chapter, the focus lies on the lower stages in our proposed script, dealing with system-level memory organization and cycle budget distribution. For the most critical tasks in the methodology, prototype tools have been and are further being developed. The methodology is first illustrated in-depth on a typical test-vehicle namely a 2D

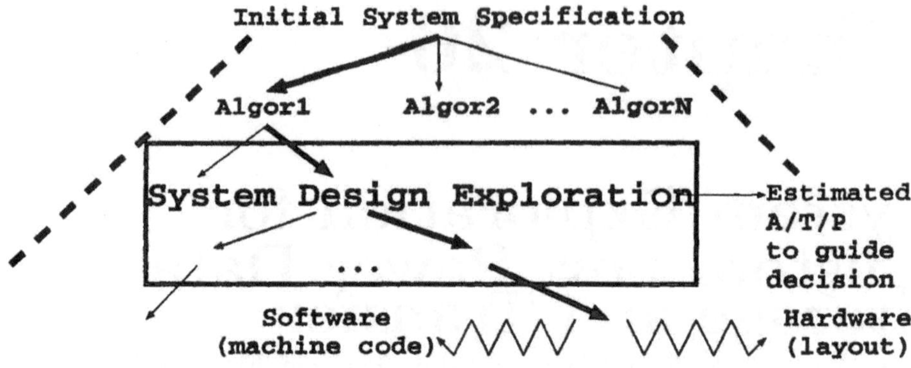

Figure 26.1 System exploration environment: envisioned situation.

motion estimation kernel. The quite general applicability and effectiveness is then substantiated for a number of industrial data-dominated applications.

This chapter is organized as follows. Section 26.2 describes target application domain and architectural styles. Section 26.3 describes the related work. Section 26.4 introduces our methodology. Next, in Section 26.5 the methodology is illustrated in-depth on a small but realistic test-vehicle. We concentrate mostly on the lower stages, related to system-level memory organization and cycle budget distribution. Section 26.6 discusses other experiments on power and/or storage size exploration for real-life applications. Section 26.7 summarizes the conclusions of the chapter.

26.2 TARGET APPLICATION DOMAIN AND ARCHITECTURE STYLE

We cannot achieve this ambitious goal for general applications and target architectural styles. So a clear focus has been selected, together with a number of reasonable assumptions. Our target domain consists of real-time signal and data processing systems which deal with large amounts of data. This happens both in real-time multi-dimensional signal processing (RSMP) applications like video and image processing, which handle indexed array signals (usually in the context of loops), and in sophisticated communication network protocols, which handle large sets of records organized in tables and pointers. Both classes of applications contain many important applications like video coding, medical image archival, multi-media terminals, artificial vision, ATM networks, and LAN/WAN technology.

The top-level view of a typical heterogeneous system architecture in our target application domain is illustrated in Fig. 26.2. Architecture experiments have shown that 50-80% of the area cost in (application-specific) architectures for real-time multi-dimensional signal processing is due to *memory units*, i.e. single or multi-port RAMs, pointer-addressed memories, and register files [1, 2, 3, 4]. Also the power cost is heavily dominated by storage and transfers [5]. This has been demonstrated both for custom hardware [6] and for processors [7] (see Fig.26.3).

Hence, we believe that the organization of the global communication and data storage, together with the related algorithmic transformations, form the dominating

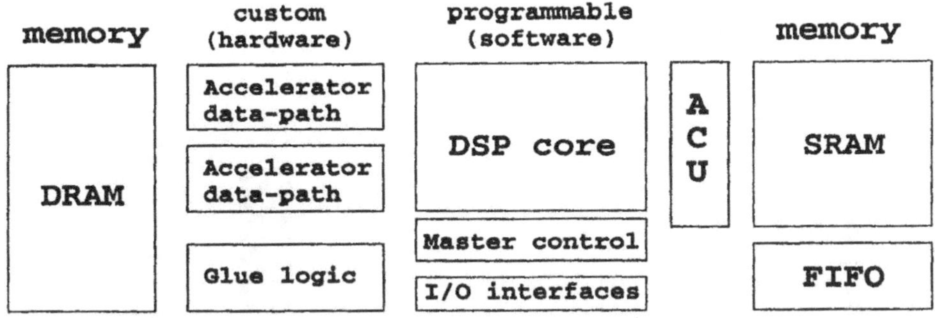

Figure 26.2 Typical heterogeneous VLSI system architecture with custom hardware (application-specific accelerator data-paths and logic), programmable hardware (DSP core and controller), and a distributed memory organization which is usually expensive in terms of area and power cost.

factors (both for area and power) in the system-level design decisions. Therefore, the key focus lies mainly on the effect of system-level decisions on the access to large (background) memories which requires separate cycles, and on the transfer of data over long "distances" (over long-term main storage). In order to assist the system designer in this, a formalized system-level data transfer and storage exploration (DTSE) methodology has been developed for custom processor architectures, partly supported in our prototype tool environment ATOMIUM [8, 9, 10, 3, 11]. We have also demonstrated that for our target application domain, it is best to optimize the memory/communication related issues **before** the data-path and control related issues are tackled [1, 12]. Even within the constraints resulting from the memory decisions, it is then still possible to obtain a feasible solution for the data-path organization, and even a near-optimal one if the appropriate transformations are applied at that stage also [12].

Up till recently, most of our activity has been aimed at application-specific architecture styles, but since 1995 also predefined processors (e.g., DSP cores) are envisioned [13, 14, 15]. Moreover, extensions to our methods and prototype tools are also useful in the context of global communication of complex data types between multiple (heterogeneous) processors [16], like the current generation of (parallel) multi-media processors [17, 18]. All this is the focus of our new predefined processor oriented DTSE methodology which is the topic of our new ACROPOLIS compiler project. Also in a software/hardware codesign context a variant of our approach provides much better results than conventional design practice [19]. In this chapter, the focus lies on custom realizations however.

The cost functions which we currently incorporate for the storage and communication resources are both area and power oriented [5, 20]. Due to the real-time nature of the targeted applications, the throughput is normally a constraint.

26.3 RELATED WORK

Up to now, little design automation development has been done to help designers with this problem. Commercial EDA tools, such as SPW/HDS (Alta/Cadence

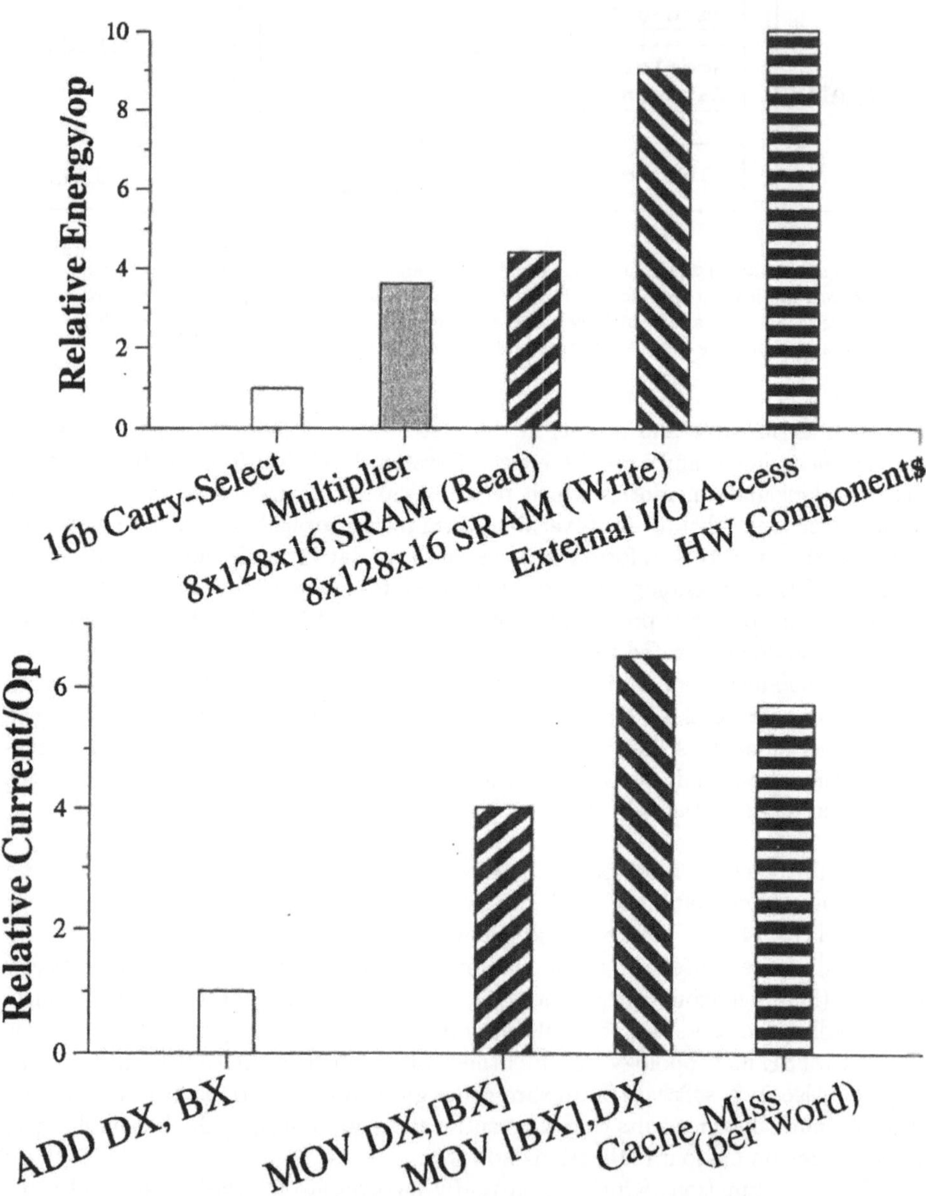

Figure 26.3 Demonstration of dominance of storage and transfer over data-path operations: both in hardware (Meng et al.) and software (Tiwari et al.).

Design), System Design Station (Mentor Graphics) and the COSSAP environment (CADIS/Synopsys), support system-level specification and simulation, but are not geared towards design exploration and optimization of memory or communication-oriented designs. Indeed, all of these tools start from a procedural interpretation of the loops where the memory organization is largely fixed. Moreover, the actual memory organization has to be indicated by means of user directives or by a partial netlist. In the CASE area, represented by, e.g., Statemate (I-Logix), Matrix-X (ISI) and Workbench (SES), these same issues are not addressed either. In the parallel compiler community, much research has been performed on loop transformations for parallelism improvement (see e.g., [21, 22, 23, 24]). In the scope of our multimedia target application domain, the effect on memory size and bandwidth has however been largely neglected or solved with a too simple model in terms of power or area consequences, even in recent work.

Within the system-level/high-level synthesis research community, the first results on memory management support for multi-dimensional signals in a hardware context have been obtained at Philips (Phideo environment [25]) and IMEC (prototype ATOMIUM environment [8, 11]), as discussed in this chapter. Phideo is mainly oriented to stream-based video applications and focuses on memory allocation and address generation. A few recent initiatives in other research groups have been started [26, 27, 28, 29], but they focus on point tools, mostly complementary to our work (see section 26.4.2).

Most designs which have already been published for the main test-vehicle used in this chapter, namely 2D motion estimation, are related to MPEG video coders [30, 31, 32, 33, 34]. These are based on a systolic array type approach because of the relatively large frame sizes involved, leading to a large computational requirement on the DCT. However, in the video conferencing case where the computational requirements are lower, this is not needed. An example of this is discussed in [35]. As a result, a power and area optimized architecture is not so parallel. Hence, also the multi-dimensional signals should be stored in a more centralized way and not fully distributed over a huge amount of local registers. This storage organization then becomes the bottle-neck.[1]

Most research on power oriented methodologies has focussed on data-path or control logic, clocking and I/O [20]. As shown earlier by us [36], in principle, for data-dominated applications much (more) power can be gained however by reducing the number of accesses to large frame memories or buffers. Also other groups have made similar observations [6] for video applications, however, no systematic global approach has been published to target this important field. Indeed, most effort up to now has been spent, either on data-path oriented work (e.g., [37]), on control-dominated logic, or on programmable processors (see [20] for a good overview).

26.4 CUSTOM DATA TRANSFER AND STORAGE EXPLORATION METHODOLOGY

The current starting point of the ATOMIUM methodology is a system specification with accesses on multi-dimensional (M-D) signals which can be statically

[1] Note that the transfer between the required frame memories and the systolic array is also quite power hungry and usually not incorporated in the analysis in previous work.

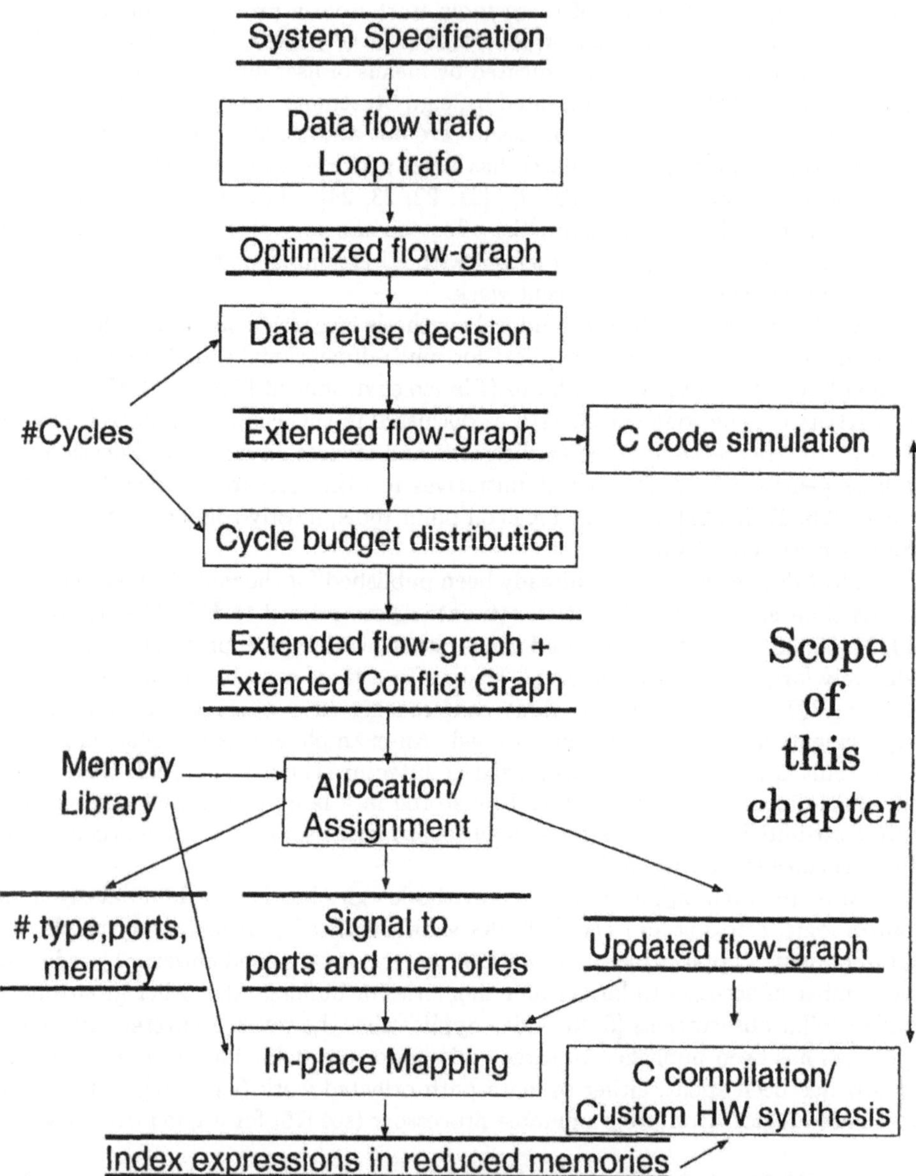

Figure 26.4 ATOMIUM script for data transfer and storage exploration of the specification, to be used for simulation and hardware/software synthesis. This methodology is partly supported with prototype tools.

ordered[2]. The output is a netlist of memories and address generators (see Fig. 26.5), combined with a transformed specification which is the input for the architecture (high-level) synthesis when custom realizations are envisioned, or for the software compilation stage (with a variant of our methodology, not addressed here) in the case of predefined processors. The address generators are produced by a separate address optimization and generation methodology, partly supported with our ADOPT prototype tool environment (see below).

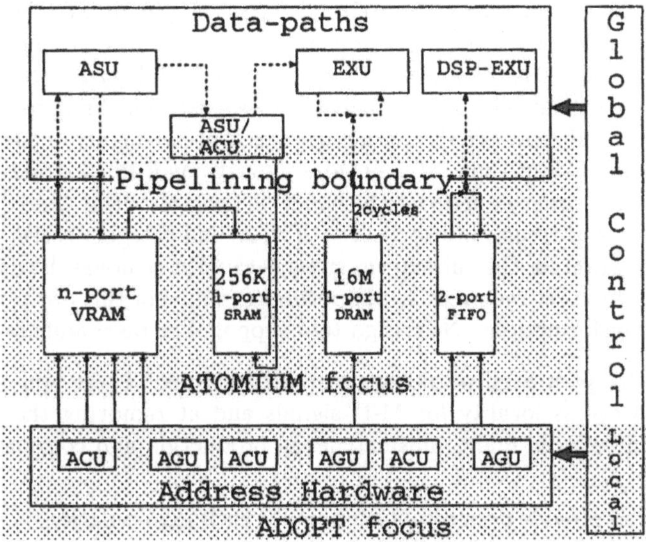

Figure 26.5 Current target architecture model for ATOMIUM and ADOPT: memory organization and address hardware embedded in global heterogeneous VLSI system architecture.

26.4.1 Global Script

The research results on techniques and prototype tools [3] which have been obtained within the ATOMIUM project are briefly discussed now (see also Fig. 26.4). More details are available in the cited references and will be partly provided also in Section 26.5 during the discussion of the test-vehicle results. We concentrate mostly on the lower stages, related to system-level memory organization and cycle budget distribution. The upper stages, focused on system-level transformations have been described elsewhere [39].

1. **Memory oriented data flow analysis and model extraction:** a novel data/control flow model [40, 10] has been developed, aimed at memory ori-

[2]Currently, this specification is written in a Data Flow oriented Language (called DFL) [38] which is applicative in nature (single definition rule) as opposed to procedural languages like C. If a procedural input is desired, either use has to be made of array data-flow analysis tools developed at several other research institutes but which are not yet operational for fully general C code, or the designer has to perform the translation for the relevant code.

[3]All of these prototype tools operate on models which allow run-time complexities which are dependent in a limited way on system parameters like the size of the loop iterators, as opposed to the scalar-based methods published in conventional high-level synthesis literature.

ented algorithmic reindexing transformations, including efficient counting of points in polytopes [41] to steer the cost functions. Originally it was developed to support irregular nested loops with manifest, affine iterator bounds and index expressions. Extensions are however possible towards WHILE loops and to data-dependent and regular piece-wise linear (modulo) indices [42, 43, 10]. A synthesis backbone with generic kernels and shared software routines is under implementation.

2. **Global data-flow transformations:** the set of system-level data-flow transformations that have the most crucial effect on the system exploration decisions has been classified and a formalized methodology has been developed for applying them [44]. Two main categories exist. The first one directly optimizes the important DTSE cost factors and consists mainly of advanced signal substitution (which especially includes moving conditional scopes), modifying computation order in associative chains, shifting of "delay lines" through the algorithm, and recomputation issues. The second category servers as enabling transformation for the subsequent steps because it removes the data-flow bottlenecks wherever required. An important example of this are advanced look-ahead transformations. No design tool support has been addressed as yet.

3. **Global loop and reindexing transformations:** These aim at improving the data access locality for M-D signals and at removing the system-level buffers introduced due to mismatches in production and consumption ordering.

   In order to provide design tool support for such manipulations, an interactive loop transformation engine (SYNGUIDE) has been developed that allows both interactive and automated (script based) steering of language-coupled source code transformations [45]. It includes a syntax-based check which captures most simple specification errors, and a user-friendly graphical interface. The transformations are applied by identifying a piece of code and by entering the appropriate parameters for a selected transformation. The main emphasis lies on loop manipulations including both affine (loop interchange, reversal and skewing) and non-affine (e.g., loop splitting and merging) cases. In addition, research has been performed on loop transformation steering methodologies. For power, a script has been developed, oriented to removing the global buffers which are typically present between subsystems and on creating more data locality [14]. This can be applied manually. Also an automatable CAD technique has been developed, partly demonstrated with a prototype tool called MASAI, aiming at total background memory cost reduction with emphasis on transfers and size. An abstract measure for the number of transfers is used as an estimate of the power cost and a measure for the number of locations as estimate for the final area cost [9]. This tool is based on an earlier prototype [8]. The current status of this automation is however still immature and real-life applications cannot yet be handled. Research is going on to remedy this in specific contexts but much future research effort will be required to solve this in a general context.

4. **Data reuse decision in a hierarchical memory context:** in this step, the exploitation of the memory hierarchy has to be decided, including bypasses

wherever they are useful [46, 47]. Important considerations here are the distribution of the data (copies) over the hierarchy levels as these determine the access frequency and the size of each resulting memories. Obviously, the most frequently accessed memories should be the smallest ones. This can be fully optimized only if a memory hierarchy is introduced. We have proposed a formalized methodology to steer this, which is driven by estimates on bandwidth and high-level in-place cost [48, 47]. Based on this, the background transfers are partitioned over several hierarchical memory levels to reduce the power and/or area cost.

At this stage of the script, the transformed behavioral description can already be used for more efficient simulation or it can be further optimized in the next steps.

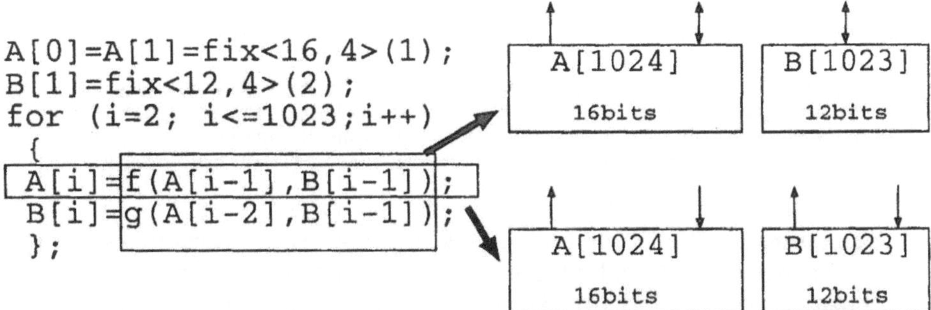
A[0]=A[1]=fix<16,4>(1);
B[1]=fix<12,4>(2);
for (i=2; i<=1023;i++)
{
A[i]=f(A[i-1],B[i-1]);
B[i]=g(A[i-2],B[i-1]);
};
A[1024]
16bits
B[1023]
12bits
A[1024]
16bits
B[1023]
12bits

Figure 26.6 Illustration of two memory allocation alternatives for a simple C specification where it is assumed that 2 cycles are available to execute the inner loop body (and that everything must be stored in background memory). Either the two statements are each executed in a separate cycle (bottom) or the right-hand sides are evaluated in cycle 1, followed by the left-hand side stores in cycle 2 (top). Clearly the top solution is more efficient as it requires a single-port memory for B instead of a dual-port one.

5. **Memory organization issues:** this task is illustrated in Fig. 26.6. Given the cycle budget, the goal is to allocate memory units and ports (including their types) from a memory library and to assign the data to the best suited memory units. This happens in several steps. In a first step, the required bandwidth is minimized within the given cycle budget, by adding ordering constraints to the flow graph. This allows the subsequent memory allocation/assignment tasks to come up with a memory architecture with a small number of memories and memory ports. Also the balancing of the available cycle budget over the different loops and memory accesses in the algorithmic specification is achieved (storage cycle budget distribution CBD) [67, 49]. Then, the necessary number of memory units and their type are determined matching this balanced flow-graph and the required parallelism in the memory access (memory allocation) [50]. In a last step, the multi-dimensional signals are assigned to these background memory units and ports (memory assignment) [50, 51]. Again, this results in an updated flow-graph specification. The cost function for each of these steps is based on a combination of area and power related criteria. A prototype tool, called HIMALAIA, supporting part of this functionality has been developed to illustrate the feasibility of our

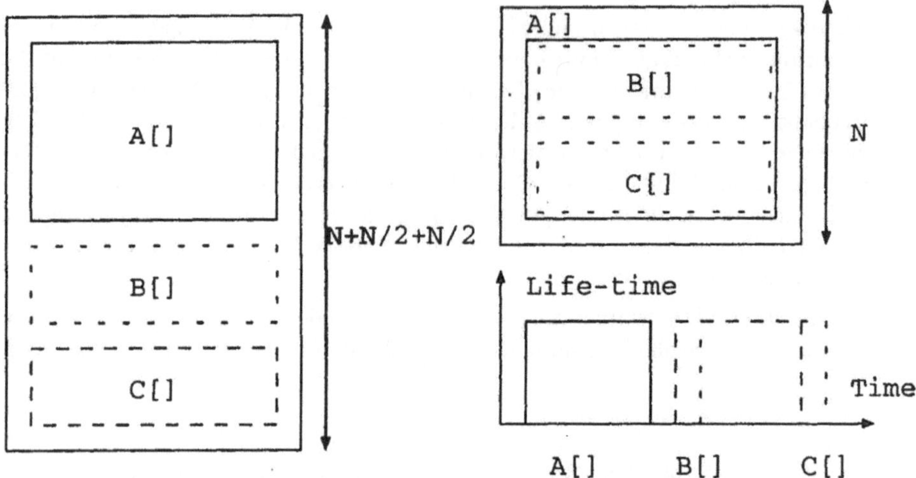

Figure 26.7 Illustration of in-place mapping task. In the worst-case, all signals require separate storage locations (left-hand side solution). However, if the life-times of signals $B[]$ and $C[]$ are not overlapping with the life-time of signal $A[]$, the space reserved in the memory for these groups can be shared (right-hand side, top).

approach [50, 52, 53]. It also includes a high-level but accurate estimation of the memory cost, under the assumption of an applicative (non-procedural) input specification [52]. An extended version of the allocation and assignment prototype tool has been developed recently which now incorporates conflict information from the cycle budget distribution step and which includes several powerful memory model extensions [51].

6. **In-place mapping:** this task is illustrated in Fig. 26.7. Initial methods have been investigated for deciding on in-place storage of multi-dimensional signals [1, 2, 54]. Recently, further experiments have been performed to allow even more reduced storage requirements and to identify extensions for predefined memory organizations as in programmable DSP's [13, 14, 55]. This has lead to a solid theoretical foundation for the in-place mapping task and the development of promising heuristics to solve this very complex problem in a reasonable CPU time in a 2-stage approach [17, 56, 57]: intra-signal windowing [58], followed by inter-signal placement [59]. These techniques have been implemented in a prototype tool for evaluating in-place storage of multi-dimensional signals. Very promising results have been obtained [56, 57, 58, 59].

A number of tasks are related to the above methodology:

1. **System-level validation by formal verification of the global data-flow and loop transformations:** a methodology to tackle this crucial validation task has been developed for a very general context including procedural (e.g., WHILE) loops, general affine indices (including the so-called "lattices" of the form $[a.i + b]$, and data-dependent signal indices. Such a formal verification stage avoids very CPU-time and design time costly resimulation. A prototype verification tool is operational, dealing with affine loop transformations like loop interchange, reversal and skewing on manifest loop nests with affine

signal indices [60]. It demonstrates the feasibility of our approach. Recently, heuristics have been added to deal with very large applications [61] and a methodology has been developed to deal with data-flow transformations and more general cases of data dependencies. We will continue developing the necessary (prototype) tool support.

2. **High-level address optimization and address unit synthesis:** These tasks are very crucial for data-dominated applications. They are error-prone and tedious if done manually for RMSP applications. Therefore, a methodology has been developed which is partly supported in IMEC's ADOPT environment [62, 63]. It is embedded in a separate prototype tool-box however and will not be treated in this chapter. The subtasks addressed in this toolbox are complementary to the traditional high-level synthesis techniques as employed in Synopsys BC; for example so ADOPT can be seen as a high-level preprocessing stage.

The need for design support of each of these different mapping and optimization tasks is motivated in Fig. 26.8.

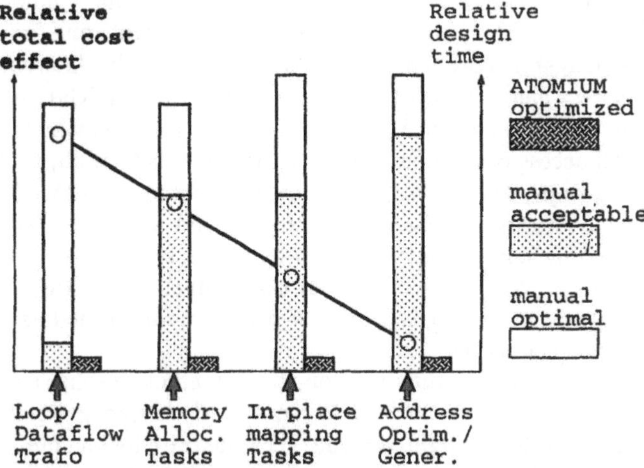

Figure 26.8 Motivation of need for design support of each of the 4 main mapping and optimization tasks in ATOMIUM. Only a qualitative analysis is provided here so the above data only provides a trend. The line connecting the circles indicates the relative total cost effect of each of the tasks. The bars indicate the design time typically spent on each of them using different approaches. It can be seen that a combination of a large design-time spent to arrive at an acceptable solution and/or the need to come up with heavily optimized critical design decisions require the support of all four tasks.

In addition to RMSP applications, also extensions of the methodology towards network applications, as occurring in Layer 3-4 of ATM, have been investigated [64, 65, 51, 66]. The methodology for the steps related to the bandwidth requirements and balancing the available cycle budget over the different memory accesses in the algorithmic specification (flow-graph balancing for non-hierarchical graphs within a single loop body) during background memory allocation, has been worked out in detail [67]. Also the combination with memory allocation and assignment has been explored with very promising results [51].

26.4.2 More Information on Memory Organization Related Subtasks

Here, we will only deal with (storage) cycle budget distribution, memory allocation and assignment, and in-place mapping.

Cycle budget distribution The CBD step orders the memory accesses within the given cycle budget. In several data-dominated contexts like ATM applications, nested loops almost never occur. The existing (single) loops are usually data-dependent (WHILE loop) spanning an entire task in a concurrent specification. In general, the DTSE steps can be performed for the concurrent tasks separately, but also optimizations are feasible between tasks. The latter is however a topic of further research. So here, it will be assumed that CBD has to operate on a single task, possibly generated after combining a number of initially concurrent tasks so that they are statically schedulable. Due to the very limited loop oriented characteristics, our current technique and prototype realization can be restricted to handle a "flat" graph operating on the body of a single loop where typically complex (dynamic) data structures are accessed in a very complicated condition hierarchy. In this case, the most important substep is called "flow-graph balancing" (FGB) [67, 49]. Extensions to handle nested loops are under study but will not be discussed here. It will be assumed for now that in the case of multiple and/or nested loops, the global cycle budget has been manually distributed over the different loop bodies.

Whenever two memory accesses to two basic groups (BGs) within the data structures occur in the same cycle, an access conflict is present between these two basic groups. All access conflicts are collected in a conflict graph, where the nodes represent basic groups, and the edges indicate a related conflict. These conflicts have to be resolved during the memory allocation and assignment steps. This can be done by assigning conflicting basic groups either to different memories or to a multiport memory. When all conflicts in the conflict graph are resolved during the memory assignment step, it is guaranteed that a valid schedule exists for the obtained memory architecture.

We have defined a cost function for these conflict graphs, such that more costly conflict graphs are likely to lead to more costly memory architectures. The cost function includes three weighted terms: 1) a measure for the minimum number of memories needed (the chromatic number[4] of the CG), 2) a term related to the number of conflicts in the conflict graph (each conflict is weighted with its importance), and 3) a term to minimize the number of self-conflicts (based on the notion of forces similar to IFDS [68]). More details about the cost function can be found in [67, 49]. The idea of flow graph balancing is then to come up with a partial ordering of the memory accesses that leads to a conflict graph with minimal cost.

CBD is executed prior to the more conventional memory allocation/assignment tasks [25, 26, 28, 50]. The main difference between our flow graph balancing and the related work is that we try to minimize the required memory bandwidth in advance by optimizing the access conflict graph for groups of scalars within a given cycle budget. We do this by putting ordering constraints on the flow

[4]A *c-coloring* of a graph G is a partitioning of G's nodes in c partition classes $V = X_1 + X_2 + \cdots + X_c$ such that every two adjacent nodes belong to a different partition class. In this case, when the members of partition X_i are colored with color i, adjacent nodes will receive different colors. The chromatic number $\chi(G)$ is the smallest number c for which there exists a c-coloring of G.

graph, taking into account *which* memory accesses are being put in parallel (i.e., will show up as a conflict in the access conflict graph).

Memory allocation and assignment The memory assignment technique discussed below is significantly different compared to existing techniques (including our own early work [43]) because it has to take into account the (extended) conflict graphs produced by the CBD step. Compared to most other approaches, it also works on groups of scalars, thereby improving the accuracy design space exploration and still reducing the complexity significantly.

Once the (extended) conflict graph is available, all the inputs required to do a *valid and cost efficient* allocation and assignment, are present. This is done by minimizing a cost function, containing weighted memory area and power terms, while taking into account all memory access constraints expressed by the conflict graph. During the *allocation* phase, the user decides on the *number of one-port memories* to be allocated. This should be at least the chromatic number of the conflict graph. Remark that this lower bound on the number of memories only holds for a library with one-port memories, which is always assumed further on in this chapter[5]. The main reason for allocating more than the minimal number of memories is to reduce the power, as demonstrated in [51]. The practical upper bound on this number is the number of signal partitions (which we also call "basic groups" [50]).

During the *assignment* phase, each basic group (BG) is assigned *as a whole* to one of the allocated memories. This yields an assignment scheme. Usually a BG can only be assigned to *some* of the allocated memories, because of memory access conflicts with some of the basic groups assigned already. In order to find a minimum cost assignment scheme, the entire assignment search space must be explored, because classical global optimization approaches like (Mixed) Integer Linear Programming solvers do not work for our non-linear problem. The assignment search space can be represented as a tree, as shown in Fig. 26.9. This tree has M^N leaves, where M denotes the number of memories allocated and N the number of basic groups to be assigned, so there is a huge amount of assignment possibilities. We use a *branch-and-bound* algorithm (called *B&B* in the sequel) with an effective bounding to search the complete assignment tree.

In our specific context, a very effective B&B strategy has been implemented. Large parts of the tree can be cut away ('bound', i.e., pruning of subtrees) because of several reasons, evaluated in this order:

1. Paths which give rise to assignment schemes which are fully symmetric with already generated schemes can be discarded.
2. Access conflicts between basic groups can be effectively checked due to the explicit ECG information, which allows to remove the corresponding subtrees.
3. Paths which have too high a cost from a certain basic group on can be pruned from that node. Currently we propose a simple but safe estimate of the minimum remaining cost which assumes storage in a common 1-port memory

[5]We also have a methodology for dealing with multi-port memories, but that feature is not implemented yet in our prototype software environment.

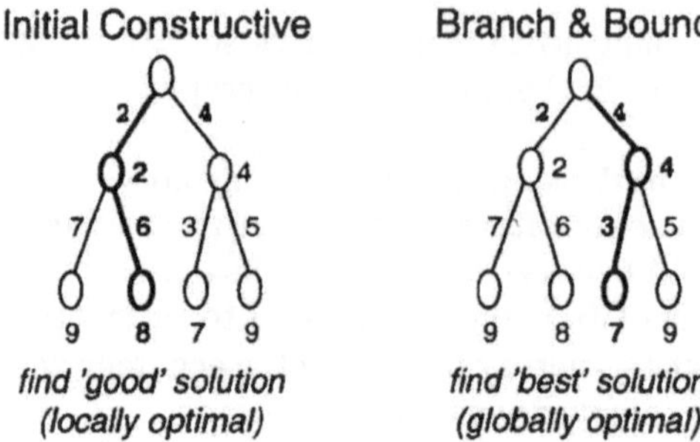

Figure 26.9 Different assignment algorithms for traversing the search tree: each tree level corresponds to the assignment of one basic group, each branch from a node corresponds to one memory.

unit with minimum bitwidth for area and storage in separate 1-port memory units for power, and this for all remaining BGs.

Moreover, in order to have a low cost threshold (for pruning) at an early stage in the B&B process, first involves an initial constructive assignment algorithm. This finds a local optimum in the search tree by iteratively assigning at each level in the search tree the corresponding basic group to the locally cheapest memory.

The order in which the basic groups are assigned also has an enormous effect on the run-time of the B&B algorithm. Experiments have shown that in most cases, ordering based on cost (i.e., assigning basic groups in the order from costly to cheap in terms of area/power consumption) leads to shorter run-times than ordering based on constraints (i.e., assigning BGs in the order from difficult-to-assign to easy-to-assign).

All the pruning criteria mentioned above have been implemented in our prototype tool. As a result, experiments have confirmed that the run times remain very acceptable for all practical examples [51].

In-place mapping Our intention is to optimize the storage order for each multidimensional signal (array), such that the required size of each memory is minimal. However, as shown in [58], the exact constraints in the general formulation for in-place mapping can be very hard to evaluate in practice. Therefore, an optimization strategy that takes them explicitly into account would not be feasible for realistic problems. So a more pragmatic approach is used, that avoids the evaluation of these constraints, but that may lead to suboptimal results. Experiments have confirmed however that the memory sizes remain near-optimal in practice.

Without loss of generality, the focus will be on the size reduction of only one (shared) memory. In general, multiple memories can be present, but our techniques can be applied to each memory separately, as there is no interference between the optimizations for different memories for a given data-distribution (as imposed by the signal-to-memory assignment substep) and background access execution order (imposed by all previous steps in our script).

We can identify two components in the storage order of arrays:

- the *intra*-array storage order, which refers to the internal organization of an array in memory (e.g., row-major or column-major layout);
- the *inter*-array storage order, which refers to the relative position of different arrays in memory (e.g., the offsets, possible interleaving, ...).

This observation has stimulated us to come up with a two-phase approach. In a first phase, an optimal intra-array storage order is searched for each array separately. This storage order then translates into a partially fixed address equation, which is referred to as the *abstract address equation* (AAE). In the second phase, an optimal inter-array storage order is looked for, resulting in a fully fixed address equation for each array. This equation will be referred to as the *real address equation* (RAE).

We will now first outline a few strategies that can be used to obtain a RAE for each array, given their AAE's and corresponding occupied address-time domain (OATD)'s [58] (i.e., the second phase).

In general, for the inter-array storage order, the exact shape of the OATDs is not known, as only implicit descriptions are available. It is however possible to extract certain properties (e.g., the width or the height of an OATD) which allow us to approximate the shape of the OATDs. These approximations can then be used in several ways, depending on their accuracy, as indicated next.

In Fig. 26.10a, the OATDs of five different arrays are shown. The simplest way to allocate memory for these arrays is to assign a certain address range to each array in such a way that the different address ranges do not overlap. The RAE then equals the AAE, shifted by a constant offset, as illustrated in Fig. 26.10b. We will refer to this first strategy as *static allocation*. Note that this approach results in no memory savings at all. The required memory size actually equals the sum of the sizes of the arrays. This is also the approach taken by traditional compilers.

A potentially better strategy is illustrated in Fig. 26.10c. Here a certain address range is allocated for each array, but only during the time that the array is in use. This allows sharing of certain address ranges by more than one array. We will refer to this strategy as *dynamic allocation*[6]. In general, a dynamic strategy requires less memory. Also note that this strategy actually approximates the OATDs of the arrays by rectangles[7].

These first two strategies have in common that the size of the address range assigned to an array equals the size of the array, and that the intra-array storage order (which influences the shape of the OATDs) has no effect on the total memory size.

However, in general an array does not use its complete address range all the time. This has lead to the definition of an *(address reference) window* [54] as the maximum distance between two addresses being occupied by the array during its life-cycle. This address reference window is indicated for each of the arrays in Fig. 26.10a by means of a vertical arrow. If the size of the window is known, the OATDs of the arrays can be "folded" by applying a modulo operation to the AAEs. After the folding, either the static or the dynamic allocation strategy can be applied, as indicated in Fig. 26.10d and 26.10e. Note that the windowed strategies

[6]Note that this allocation strategy can be performed at *compile* time, in contrast to the traditional run-time dynamic allocation provided by certain languages such as C.

[7]In [59], some extensions are discussed .

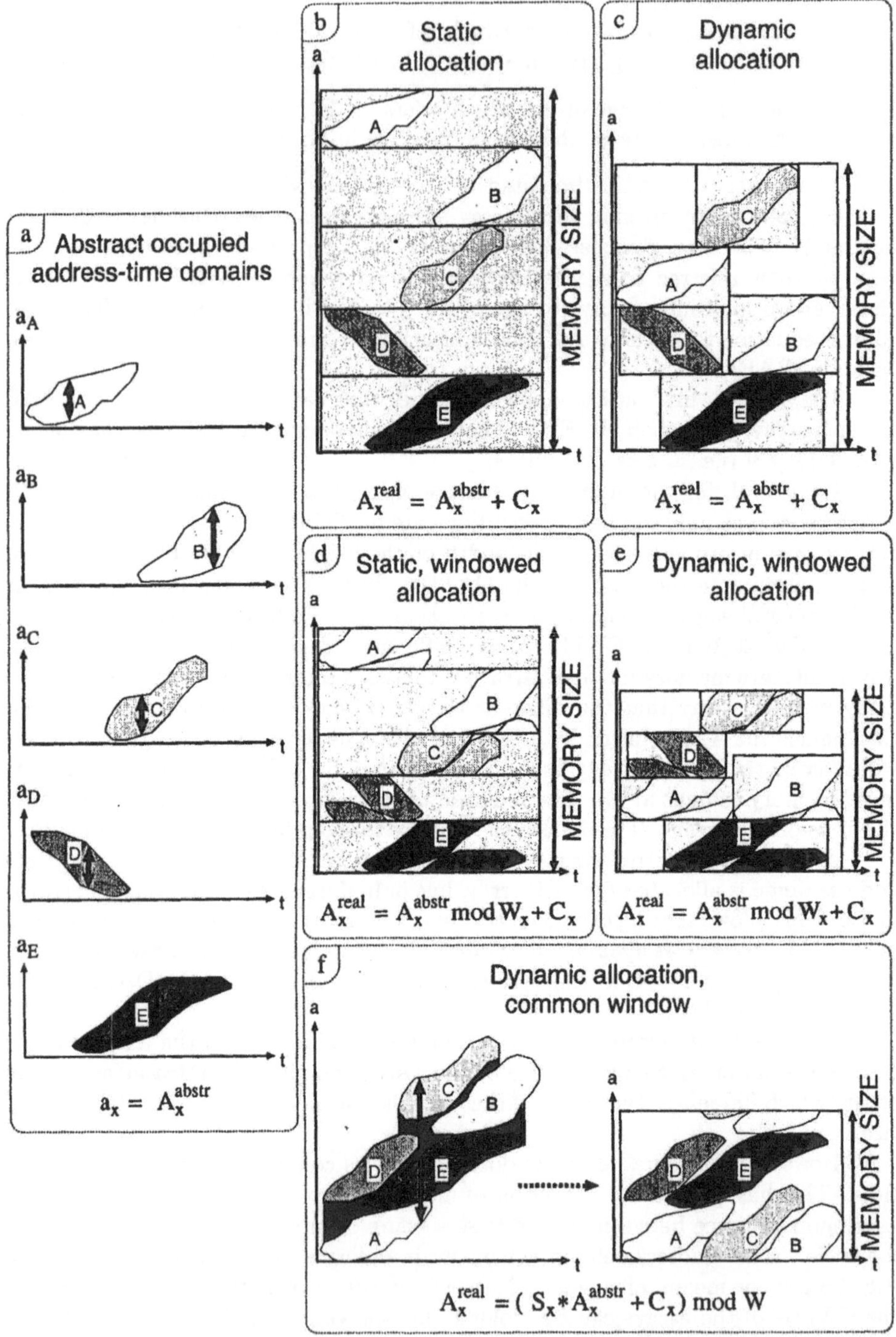

Figure 26.10 Different allocation strategies.

in general require less memory than the non-windowed ones. They require the calculation of the address reference window though. Moreover, the internal storage order of the arrays influences the shape of the OATDs and hence the size of the address reference window. In [58] techniques are described for exact evaluation of the window and optimization of the intra-array storage order to obtain minimal window sizes.

A possibly even better strategy is depicted in Fig. 26.10f. In a first step, the OATDs are shifted (and possibly even vertically scaled or flipped) such that their common window is minimal. After that, the complete address range is folded by this one window. Note that for this example, the last strategy is the best one, but this is *not* true in general. The strategy with separate windows can sometimes yield better results (e.g., when the OATDs don't "fit" together very well). Moreover, the common window strategy requires the evaluation of the abovementioned very complex constraints, i.e. one has to make sure that the OATDs do not overlap, otherwise different arrays would simultaneously use the same memory locations, which is obviously illegal.

Therefore, the strategy with separate windows for each array has been selected, as depicted in Fig. 26.10e, as it offers the best compromise between optimality and complexity. Our detailed placement strategy is discussed in [59], and some results we obtained are also presented.

From the discussion of the inter-array storage order strategies, it can be derived that the intra-array storage orders have to be searched for which result in OATDs that are as "thin" as possible, i.e., that have the smallest address reference window. The number of possible storage orders is huge however, even if we restrict ourselves to the affine ones. Moreover, checking whether a storage order is valid generally requires the evaluation of the abovementioned complex constraints and to our knowledge no practically feasible strategy exist for choosing the best order.

Therefore, the number of possibilities is restricted drastically. First of all, we require that each element of an array is mapped onto an abstract address that is unique w.r.t. the address of the other elements of the same array. In that way, checking for intra-array memory occupation conflicts can be avoided. Another requirement imposed is that the storage order should be dense, i.e., the set of abstract addresses occupied by a rectangular array should be a closed interval. A row-major order as in C for instance, satisfies this requirement. However, for multi-dimensional arrays, *all* possible orders of the dimensions will be considered, and also both directions (i.e., positive and negative) for each dimension. Consequently each AAE will have the following format and properties:

$$AAE_x = \sum_{i=1}^{D} N_{ix} \left(\mp B_{n_i x} \pm a_{n_i x}\right)$$

$$\text{where} \quad N_{ix} = \max\left(\sum_{j=i+1}^{D} N_{jx} \left(\mp B_{n_j x} \pm a_{n_j x}\right)\right) \quad \text{and} \quad N_{Dx} = 1$$

Here, $a_{n_i x}$ and $B_{n_i x}$ represent the index and the upper or lower bound of dimension n_i of array x respectively. The N_{ix} coefficients are constants, obeying certain criteria to obtain dense storage orders. The signs depend on the chosen dimension directions and D equals the total number of dimensions of the array. An example of the possible orders considered for a 2x3 2-D array is given in Fig. 26.11.

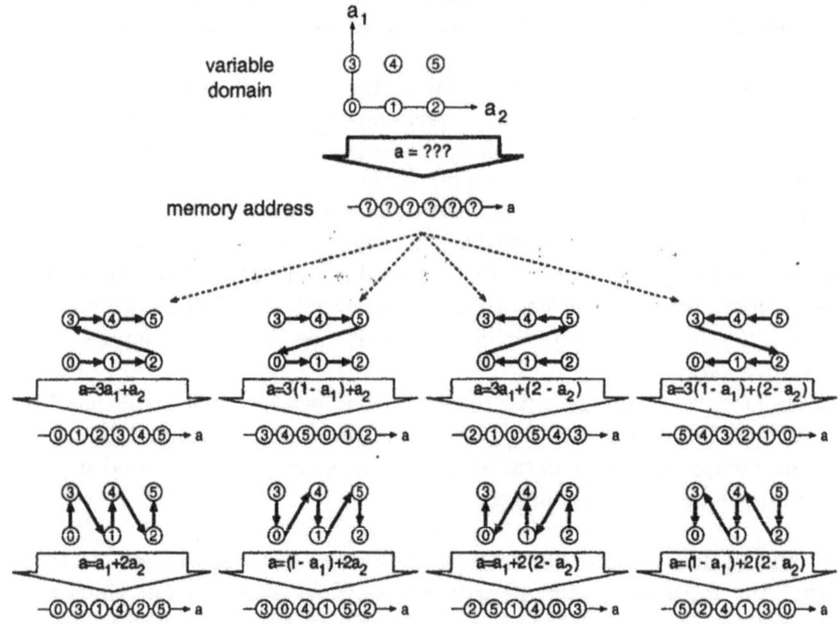

Figure 26.11 The possible dimension orders for a 2x3 2-D array.

In general, for a N-dimensional array, $2^N N!$ possibilities are considered. For a 6-D array[8] for instance, there are no less than 46080 possibilities! It must be clear that even though this is only a very limited subset of all possible storage orders, evaluating this set will be infeasible for arrays with many dimensions. So a more intelligent search strategy has been developed which is discussed in [58].

26.5 DEMONSTRATOR APPLICATION FOR ILLUSTRATING THE METHODOLOGY

26.5.1 Motion Estimation Algorithm

The 2D motion estimation algorithm [69] is used in moving image compression algorithms. It allows to estimate the motion vector of small blocks of successive image frames. We will assume that the images are gray-scaled (for color images, in practice, only the luminance is considered). The version considered here is the kernel of what is commonly referred to as the "full-search full-pixel" implementation [70].

Each frame is divided into small blocks. For each of these blocks (called *current blocks* or CB's in the sequel), a region in the previous image is defined around the same center coordinates (called the *reference window* or RW in the sequel). This is shown in Fig. 26.12. Every CB is matched with every possible

[8] A 6-D array is not unusual for single-assignment code.

region (of the same size as the CB) in the RW corresponding to the CB. The matching is done by accumulating the absolute pixel differences between the CB and the considered regions of the RW. The position of the region that results in the smallest difference is assumed to be the previous location of the CB. In this way, a motion vector can be calculated for every CB in a frame.

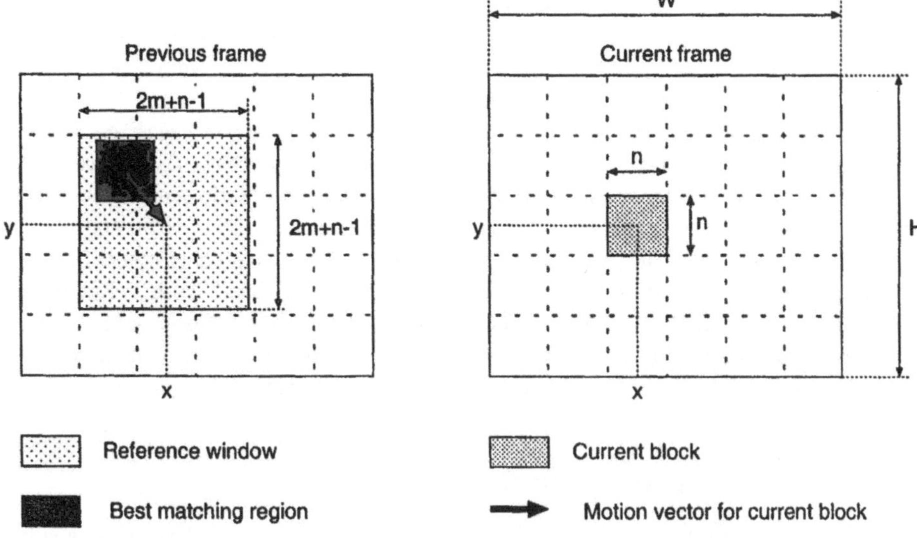

Figure 26.12 The motion estimation algorithm and its parameters.

The algorithm is typically executed in 6 nested loops, next to the implicit frame loop. The choice of the nesting for these loops is partially open, and there is quite a lot of room for parallelization and (loop) reordering.

A C-like description of the algorithm template is given next. The parameters are indicated in Fig. 26.12.

```
for (g=0;g<H/n;g++)          /* vertical CB counter */
{ for (h=0;h<W/n;h++)        /* horizontal CB counter */
  { for (i=-m;i<m;i++)       /* vertical searching of RW */
    { for (j=-m;j<m;j++)     /* horizontal searching of RW */
      { for (k=0;k<n;k++)    /* vertical traversal of CB */
        { for (l=0;l<n;l++)  /* horizontal traversal of CB */
          { [Accumulate absolute pixel difference]
          } }
        [Calculate new minimum and its location]
      } } } }
```

In our experiments, the following typical parameters (QCIF standard) are used: W=176 pixels, H=144 pixels, blocks of $n = 8 \times 8$ pixels with a search range of $2m = 16$ pixels (resulting in a 23×23 search window). The pixels are 8-bit gray scale values.

26.5.2 Power Models

The libraries used in the power models have been uniformly adapted for a CMOS technology operating at 5 V. If a lower supply voltage can be allowed by the process technology, the appropriate scaling has to be taken into account. It will however be (realistically) assumed that V_{dd} is fixed in advance as low as possible within the process constraints and noise tolerance, and that it cannot be lowered any further by architectural considerations.

For the data-paths and address generation units (which were realized as custom data-paths), a standard cell technology was assumed where the cells were *NOT* adapted to low power operation. As a result, the power figures for these data-paths are high compared to a macro-cell design with power-optimized custom cells. The power estimation itself however has been accurately done with the PowerMill tool of EPIC, based on gate-level circuits which have been obtained from behavioural specifications using IMEC's Cathedral-3 custom data-path synthesis environment [71] followed by the Synopsys RT-synthesis Design Compiler. The resulting VHDL standard cell net-list was supplied with reasonable input stimuli to measure average power.

For the memories two power models are used:

- For the embedded background RAMs, power figures are used which are supplied with VLSI Technology's SRAM compilers for a 0.6 μm CMOS technology at 5V. The power figures for the memories (expressed in mW/MHz) depend on two parameters: the bitwidth of the memory (ranging from 8 to 128 bits) and the number of words in the memory (ranging from 16 to 8192). In general, the larger the memory, the more power it consumes. These power figures have to be multiplied with the real access frequency F_{real} of the memories to get the power consumption for each of the memories. Both single and dual port memories are available in VLSI Technology's library.
- For the large off-the-shelf units on a separate chip, SRAMs have been assumed because of the fast random access without the need for a power-hungry additional cache in between [72]. For the SRAMs, the model of a very recent Fujitsu low-power memory is used [73]. It leads to 0.26 W for a 1 Mbit SRAM operating at 100 MHz at 5V [9]. Because this low-power RAM is however internally partitioned, the power will not really be significantly reduced by considering a smaller memory (as required further on) as long as they remain larger than the partitions. The power budget [73] clearly shows that about 50% of the power in this low power version is consumed anyhow in the peripheral circuitry, which is much less dependent on the size. Moreover, no accurate figures are available on the power consumed in the chip-to-chip communication but it is considered as less dominant so that contribution will be ignored. This contribution would more than compensate for the potential

[9] Currently, vendors do not supply much open information, so there are no better power consumption models available to us for off-chip memories.

power gains by having smaller off-chip memories available. This means that, in practice, the power consumption of the off-chip memories will be higher than the values obtained with this model, even when having smaller memories available. Still, we will use a power budget of 0.26 W for 100 MHz operation in all off-chip RAMs further on. For lower access frequencies, this value will be scaled linearly, which is a reasonable assumption.

Note that the real access rate F_{real} should be provided and not the maximum frequency F_{cl} at which the RAM can be accessed. The maximal rate is only needed to determine whether enough bandwidth is available for the investigated array signal access. This maximal frequency will be assumed to be 100 MHz [10]. If the background memory is not accessed, it will be in power-down mode[11].

A similar reasoning also applies for the data-paths, if we carefully investigate the power formula. Also here the maximal clock frequency is not needed in most cases. Instead, the actual number of activations F_{real} should be applied, in contrast with common belief which is based on an oversimplification of the power model. During the cycles for which the data-path is idle, all power consumption can then be easily avoided by any power-down strategy. A simple way to achieve this is the cheap gated-clock approach for which several realizations exist (see e.g., [74]). In order to obtain a good power estimate, it is crucial however to obtain a good estimate of the average energy per activation by taking into account the accurately modeled weights between the occurrence of the different modes on the components. For instance, when a data-path can operate in two different modes, the relative occurrence and the order in which these modes are applied should be taken into account, especially to incorporate correlation effects. Once this is done, also here the maximal F_{cl} frequency is only needed afterwards to compute the minimal number of parallel data-paths of a certain type (given that V_{dd} is fixed initially).

26.5.3 Target Architecture Assumptions

In the experiments it has been assumed that the application works with parameterized frames of $W \times H$ pixels, processed at F frames/s. For our test-vehicle, i.e., the 2D motion estimation kernel for the QCIF standard, this means 176×144 pixel frames in a video sequence of 30 frames/s. This results in an incoming pixel frequency of about 0.76 MHz.

We consider a target architecture template as shown in Fig. 26.13. Depending on the parameters, a number of parallel data-paths are needed. In particular, for the 2D motion estimation this is $2m \times 2m \times W \times H \times F/F_{cl}$ processors for a given clock rate F_{cl}. However, this number is not really important for us because an architecture is considered in which the parallel data-paths with their local buffers are combined into one large data-path which communicates with the distributed frame memory. This is only allowed if the parallelism is not too large (as is the case for the motion estimator for the QCIF format). Otherwise, more systolic organizations, with memory architectures tuned to that approach, would lead to better results [13]. In practice, it will be assumed that a maximal F_{cl} of 50 MHz[12] is feasible for the on-chip components, which means that 4 parallel data-path processors

[10] Most commercial RAMs have a maximal operating frequency between 50 and 100 MHz.

[11] This statement is true for any modern low-power RAM [72].

[12] 48.66 MHz is actually needed as a minimum in this case.

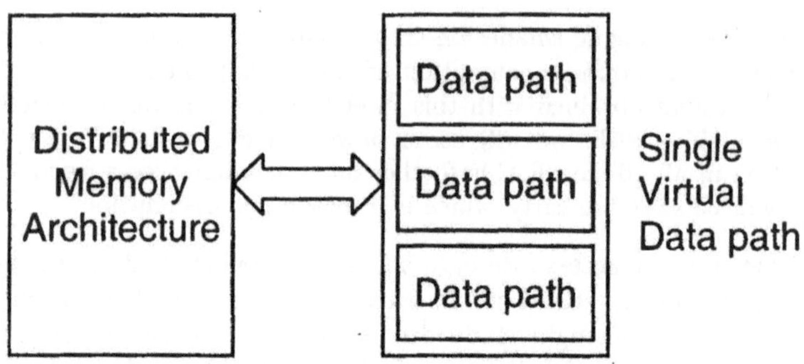

Figure 26.13 Architecture consisting of a distributed memory architecture that communicates with a data-path consisting of a number of parallel data-paths.

are needed. In many applications, the memory organization can be assumed to be identical for each of the parallel processors (data-paths) because the parallelism is usually created by "unrolling" one or more of the loops and letting them operate at different parts of the image data [14].

We will now discuss a power optimized architecture exploration for the motion estimation kernel, as a more detailed illustration of the more general data transfer and storage exploration methodology described in Section 26.4 and Fig. 26.4.

26.5.4 Application of Low-power Exploration Methodology

For the background memory organization, experiments have been performed to go from a non-optimized applicative description of the kernel in Fig. 26.12 to an optimized one for power, tuned to an optimized allocation and internal storage organization. In the latter case, the accesses to the large frame memories are heavily reduced. These accesses take up the majority of the power as we will see later.

Based on our script an optimized memory organization will be derived for the frame memories and for the local memories in the different data-path processors for 2D motion estimation.

STEP 1: Data and control-flow optimization.

The first optimization step in our methodology, is related to data-flow and loop transformations. For the 2D motion estimation, mainly the effect of loop transformations has been very significant. The results are discussed in detail in [39]. It is clear that reordering of the loops in the kernel will affect the order of accesses and hence the regularity and locality of the frame accesses. In order to improve this, it is vital to group related accesses in the same loop scope. This means that all important accesses have to be collected in one inner loop in the 2D motion estimation example. The latter is usually done if one starts from a C specification for one mode of the motion estimation, but it is usually not the case if several modes are present. Indeed, most descriptions will then partition the quite distinct functionality over different functions which are not easily combined. Here is a first option to improve the access locality by reorganizing the loop nest order and function hierarchy amongst the different modes.

The most optimal case is depicted in Fig. 26.14. If a direct mapping of this organization on a memory architecture is compared with a direct mapping of the most promising alternative organization, it consumes only 1015 mW compared to 1200 mW, using the power models of Subsection 26.5.2).

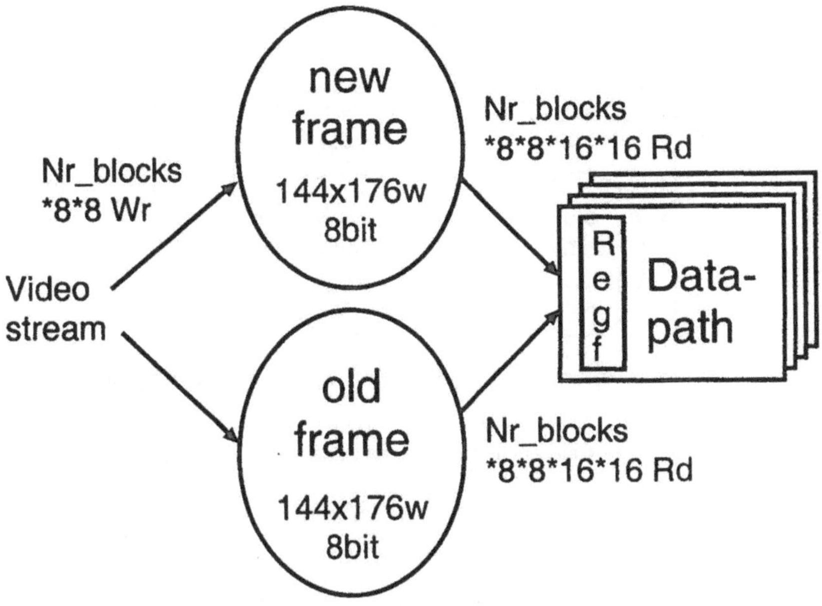

Figure 26.14 Required signal storage and data transfers when the traversal over the current block is done in the inner loops.

STEP 2: Data reuse decision in a hierarchical memory context.

In a second step, we have to decide on the exploitation of the available data reuse possibilities to maximally benefit from a customized memory hierarchy. Also this step is detailed elsewhere [39]. Important considerations here are the distribution of the data (copies) over the hierarchy levels as these determine the access frequency and the size of the resulting memories [48, 47].

After the introduction of one extra layer of buffers, both for the current block and the reference window accesses, and after exploiting "inter-copy reuse" [48], the memory hierarchy shown in Fig. 26.15 is derived. A direct implementation of this organization leads to a memory architecture that consumes about 560 mW [13].

STEP 3: Storage cycle budget distribution.

At this stage, the data has been partitioned over different "levels" in the memory hierarchy and all transfers between the different memory partitions are known. We are now ready to optimize the organization of every memory partition. But before doing the actual allocation of memory modules and the assignment of the signals to the memory modules, it has to be decided for which signals simultaneous access capability should be provided to meet the real time constraints. This storage cycle budget distribution task [67, 49] with as most important substep the flow-graph balancing or FGB (see section 26.4.2) tries to minimize the required memory

[13]Note that due to the parallel data-path architecture target, in the end some extra issues have to be taken into account (see [14]) which are not discussed here either.

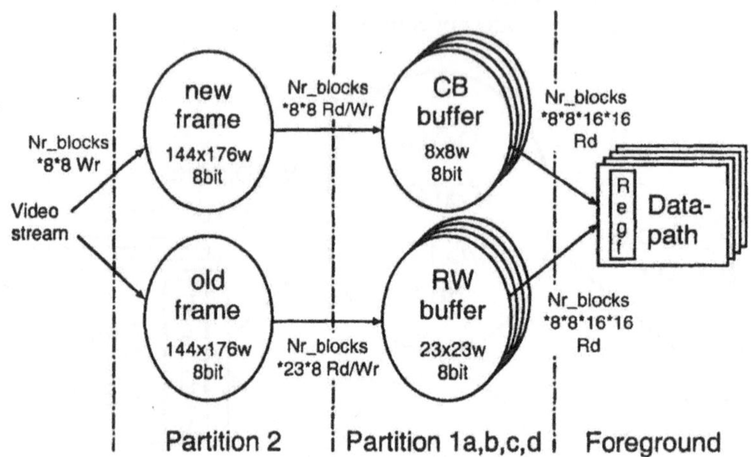

Figure 26.15 Data transfers between stored signals for the fully optimized memory hierarchy.

bandwidth (i.e., # parallel ports) of every memory partition given the flow graph and the cycle budget in which the algorithm has to be scheduled. The result is a conflict graph that indicates for which signals simultaneous access capabilities should be provided in order to meet the cycle budget. The potential parallelism in the execution order of the accesses directly implies the position of the conflicts. This execution order is heavily affected by the final loop organization and the cycle budgets assigned to the different loop scopes. Because there is much freedom in the decision of the latter parameters, an optimization process has to take place. The optimized conflict graphs for the memory partitions obtained in the previous step are shown in Fig. 26.16. In general, obtaining the optimized conflict graphs requires a tool. In this case, however, because there is only one simple loop nest, it is possible to derive it by hand. Given the clock frequency of 50 MHz and a frame rate of 30 Hz, the cycle budget for one iteration of the algorithm is $Nr\_blocks * 8 * 8 * 16 * 16$ cycles. Given the number of transfers from/to signals indicated in Fig. 26.15, it can be seen that for instance the CB signal has to be read *every* clock cycle. But from time to time, the CB signal has to be updated as well. This means that in some cycles there will be 1 read access and 1 write access to the CB signal. This is indicated in the graph as a self conflict for the CB signal annotated with 1/1/2, meaning that there is at most 1 simultaneous read operation, at most 1 simultaneous write operation, and at most 2 simultaneous memory accesses (1 read and 1 write in this case).

The conflict graph for partition 2 doesn't contain any conflicts. This means that both signals can be stored in the same memory, because they never have to be accessed at the same time to meet to cycle budget. The conflict graphs for partitions 1a, b, c and d (they are all the same) show conflicts between the current buffer and the reference window buffer because at every cycle, each memory partition has to

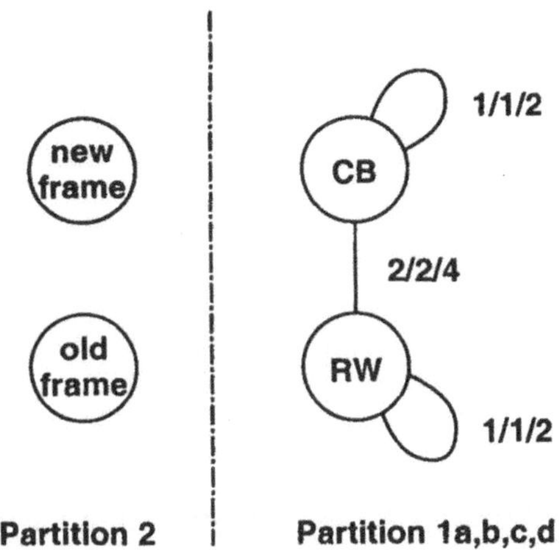

Figure 26.16 Conflict graphs for memory partitions resulting from the memory hierarchy decision step. The labels next to the conflict edges are of the form R/W/RW, where R and W equal the total number of simultaneous read, respectively write, accesses, and RW equals the total number of simultaneous memory accesses (read or write) of the signals that are in conflict.

supply one pixel from the current buffer and one pixel from the reference window to the data-path. Self conflicts, as for CB, will inevitably lead to the allocation of a multiport memory because the same signal has to be accessed twice (1 read for the data path + 1 write for updating the buffer) in the same clock cycle to meet the cycle budget. In our example, allocating two-port memories for the buffers would increase the power consumption from 560 mW to 960 mW. This is of course a bad solution both in terms of area and power. However, because the updating of the buffers is done at a much lower rate than supplying the data to the data-paths, it is better to increase the cycle budget per data-path a little bit, such that cycles become available for sequentializing the parallel accesses. The best way to achieve this is by allowing the minimal clock frequency to be slightly larger, i.e., 48.86 MHz, which still fits in the 50 MHz range. The updating of the memories can then be done in the spare clock cycles, avoiding conflicts with the read accesses to the buffer memories. A more costly alternative would be to use 1 more parallel data-path. This new cycle budget leads after flow graph balancing to the conflict graphs of Fig. 26.17. This time there are no self conflicts and therefore a solution consisting of only single port memories becomes possible. The power consumption is then again 560 mW.

STEP 4: Memory allocation and assignment.

The next step is to allocate the memory modules for every memory partition, and to assign all (intermediate) signals to their memory module. The main input for this task are the conflict graphs obtained during the storage cycle budget distribution step. The memory allocation/assignment step tries to find the cheapest

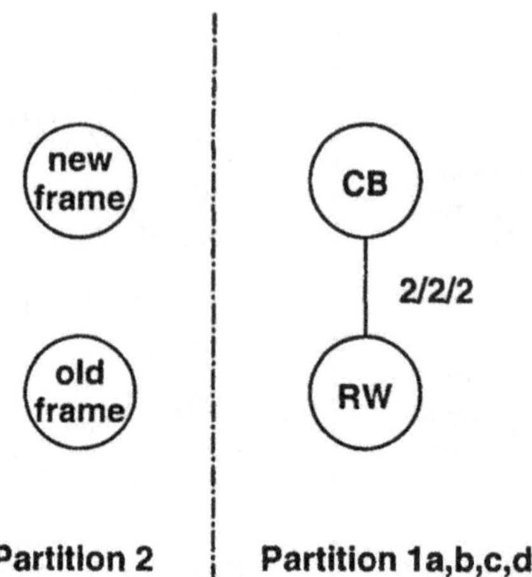

Figure 26.17 Conflict graphs for the memory partitions obtained for a slightly larger cycle budget.

memory organization that satisfies all constraints expressed in the conflict graphs. If all constraints are satisfied, it is guaranteed that there is a valid schedule that meets the cycle budget. Usually the search space for possible memory configurations meeting the cycle budget is very large. In this case ,however, the conflict graphs are very simple and therefore the optimal allocation and assignment is quite obvious: for memory partition 2 where there are no conflicts, 1 memory that stores both the old and the new frame is the best solution, and for the other partitions, which contain two signals that are in conflict, the best solution is to assign both of them to a separate memory. This then results in the memory organization of Fig. 26.18. The power consumption of this memory organization is 560 mW.

STEP 5: In-place mapping optimization.

In a final step, each of the memories – with the corresponding M-D signals assigned to it – should be optimized in terms of storage size applying so-called in-place mapping for the M-D signals. This will directly reduce the area and indirectly it can also reduce power further if the memory size of frequently accessed memories is reduced. Instead of the two frames *oldframe* and *newframe* used in the initial architecture, it is possible to overlap their storage based on detailed "life-time analysis" but then extended to array signals where the concept becomes much more complex. Life-times do not just overlap anymore when the part of the array is still in use! Instead, a polyhedral analysis is required to identify the part of the M-D storage domains which can be reused "in-place" for the different signals [58]. The results of this analysis are depicted in Fig. 26.19.

Because of the operation on a 8×8 block basis and assuming the maximal span of the motion vectors to be 8, the overhead in terms of extra rows in the combined frame buffer is then only $8 + 8 = 16$ lines, by using a careful in-place compaction. This leads to a common frame memory of about $(H + 16) \times W \times 8$

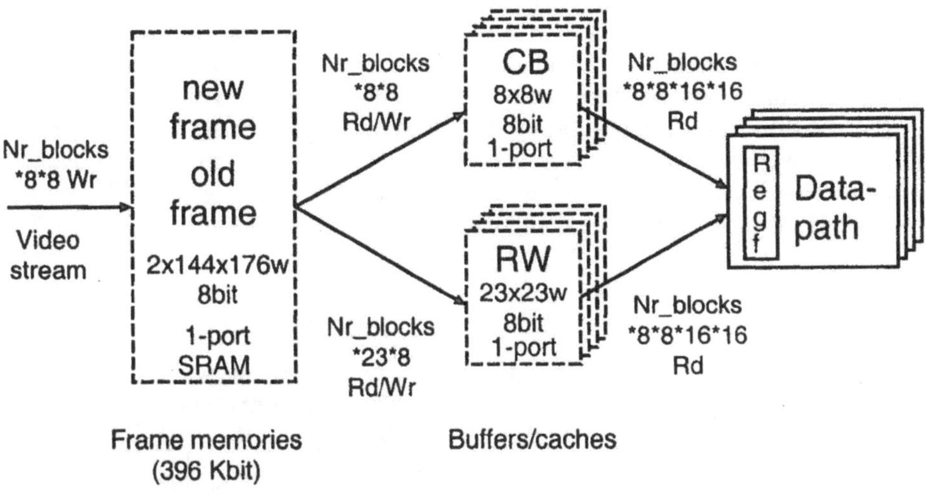

Figure 26.18 Memory organization after allocation and assignment.

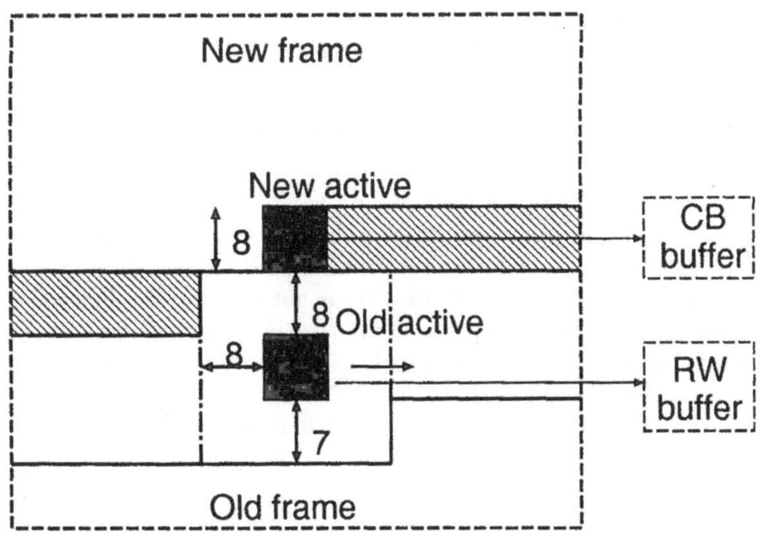

Figure 26.19 In-place storage scheme for the optimized frame organization.

bits, in addition to the already minimal window buffer of $(2m + n - 1) \times n \times 8$ bits and a block buffer of $n \times n \times 8$ bit. For the parameters used in the example, the frame memory becomes about 0.225 Mbit.

In practice, however, the window around the block position in the "old active" frame is buffered already in the window buffer so the mostly unused line of blocks on the boundary between "new" and "active old" (indicated with hashed shading in Fig. 26.19) can be removed also. This leads to an overhead of only 8 lines in the combined *new/oldframe* (the maximal span of the motion vectors), namely 1408 words, with a total of 26752 instead of 2 × 25344 = 50688 words (47% storage reduction). The corresponding final memory organization is shown in Fig. 26.20.

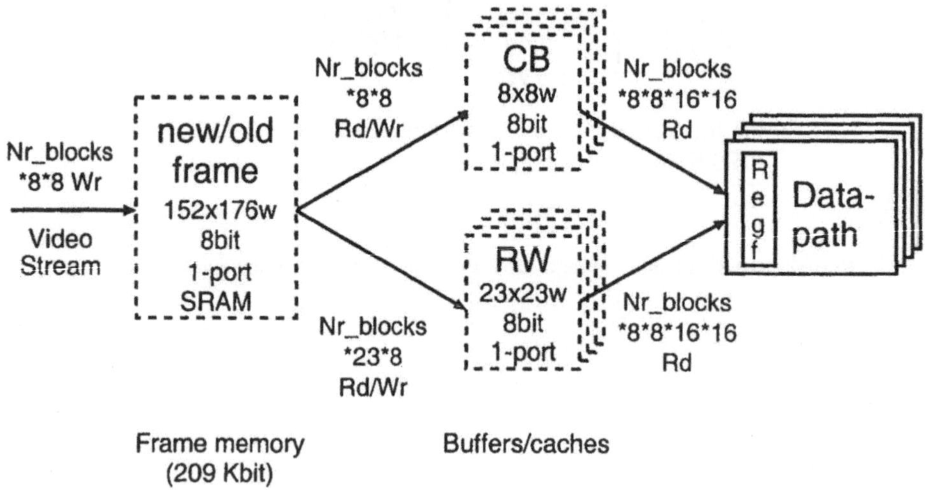

Figure 26.20 Final memory organization after in-place optimization.

The effect of this in-place optimization on the power consumption is small in this case, because the power consumption of off-chip RAMs is not very sensitive to the memory size and the amount of transfers is not changed by this optimization. So, finally we have arrived at a power optimized memory organization that consumes about 560 mW. However, because of the relatively small frame size in QCIF, it can be considered now to put this combined frame memory of 209 Kbit on chip. If low-power embedded RAMs are used[14], this will reduce the power budget further because the expensive off-chip communication is totally avoided.

26.5.5 Global Summary of Results

Taking into account all the memory transfers to the frame memory, from the combined frame memory to the two buffers, and from these two buffers to the datapaths, leads to a total power budget for the memory architecture of about 560 mW

[14]Due to the low access speed, very low-power circuitry should be feasible.

using the memory power models discussed in Section 26.5.2. This corresponds to a substantial saving compared to the initial situation (almost a factor 2).

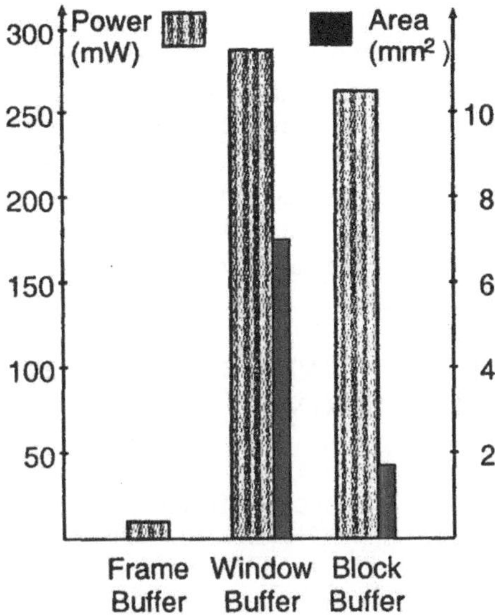

Figure 26.21 A breakdown of the power consumption for the three memories in the optimized memory architecture.

The effect of the different optimizations in our exploration methodology is shown in Table 26.1. It is assumed that the frame memories are stored in-off chip memories, while all intermediate buffers are considered to be on-chip memories. A breakdown of the final power figure over the different buffers is shown in Fig. 26.21.

Table 26.1 Results of Methodology Exploration*

Step	*Resulting Power* (mW)	*Off-chip (# words)*	*On-chip (# words)*
Control-flow optimization	1015	50688	0
Data reuse decision	560	50688	2372
Flow Graph Balancing + Memory Allocation/Assignment	(960 →) 560	50688	2372
In-place optimization	560	26752	2372

*:Note that the initial 1015 and 560 mW figures were based on lower bound estimates at that stage, so in reality the data reuse decision stage solution would have required 960 mW as indicated between parentheses.

In addition to the memory architecture optimization, also the address generation unit and data-path/controller realizations have been explored for the original architecture. It can be concluded from these experiments that after optimization of the memories, data-paths and address generation units, the power which goes into the memory accesses (560 mW) dominates the other contributions, which are both comparable (less than 90 mW for all the data-paths and about 140 mW for the optimized address generators). This is true even when low-power circuits are used in the memories and when power hungry standard cells are used in the data-paths and address units. Moreover, the current figures do not yet include the power of the data transfers themselves which will also consume much power (especially the off-chip ones). These transfers are also within the focus of ATOMIUM and they are equally optimized by reducing the background accesses. These figures clearly justify our approach of first optimizing the memory architecture for this type of data-dominated video applications.

26.6 INDUSTRIAL APPLICATION DEMONSTRATORS FOR CUSTOM REALIZATIONS

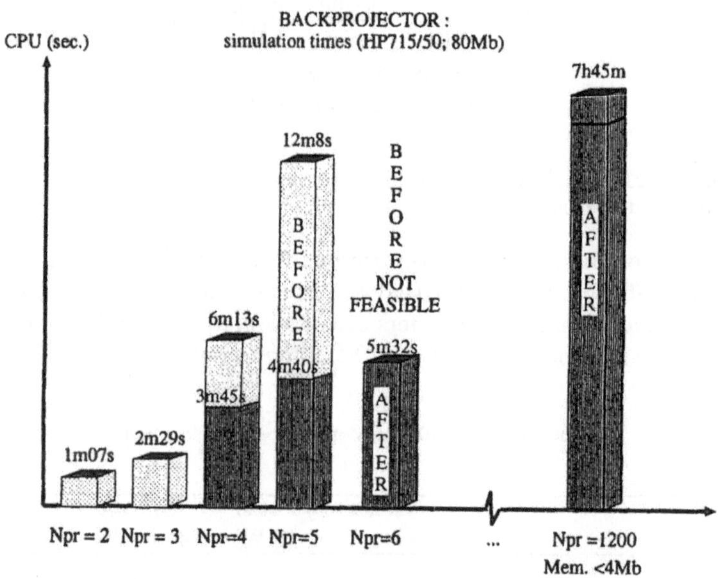

Figure 26.22 Execution times for system-level simulation (run of 1 image frame of 256 pixels) of medical computer tomography (CT) back-projection subsystem both before and after automatic optimization. For NrProj larger than 5, the 80 Mb swap space of the mid-size HP station was not sufficient any longer to start the actual simulation originally. Realistic modern scanners require a range of NrProj from 300 to 1200 which is feasible only after optimization of the specification, as shown at the right.

We have coupled the memory transformation prototype tools to our simulation engine for the high-level applicative Silage language, with excellent results for the virtual memory usage and run-times on mid-size workstations [9]. An example of the effect for a realistic medical image processing application is shown in Fig. 26.22. For relatively small parameter values of NrProj (the number of pro-

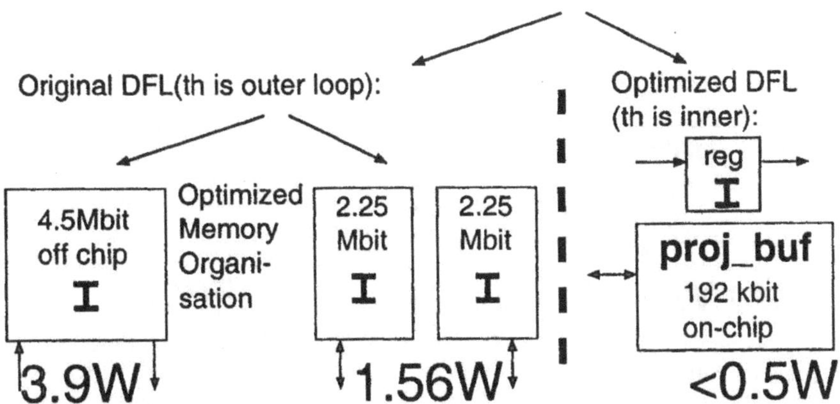

Figure 26.23 Estimated power cost for several system exploration options in the low resolution alternative of the back-projector (with $NrProj = 12$). Notice that both system-level loop transformations and memory organization decisions are needed to support these steps. The loop transformations have been automatically obtained with our prototype MASAI tool.

jections to be used for reconstruction), namely NrProj=6, the simulation of the original description (BEFORE) becomes infeasible due to memory shortage even with a workstation swap space of 80 Megabyte. Also for lower NrProj values, the CPU times are becoming rapidly dominated by the memory swapping. After automatic optimization of the description (AFTER), the simulation run-time (on a HP715/50) and the memory requirements are drastically reduced as shown in the figure, even for large NrProj values. The number of scalar signals which would have to be analyzed when the loops would be expanded as in conventional simulation paradigms is then up to several billion words.

For this same application with realistic parameters (image size 512×512), both for small ($NrProj = 12$) and high ($NrProj = 1200$) resolution versions, also the results on off-chip memory power cost for a customized hardware realization have been obtained [5].

For the high-resolution version, they lead to a reduction from 77 Mbit to 2 times 2.25 Mbit for the size and a reduction from 20W to 1.56W for the system power budget. These 2 versions are illustrated in Fig. 26.23 and Fig. 26.24.

Also for another demanding test-vehicle, namely the entire video decoder algorithm in a H.263 video conferencing application [75], large reductions of the power consumption associated with the background memory accesses have been obtained [76]. The starting point for our exploration has been a C specification of the video decoder, available in the public domain from Telenor Research. We have transformed the data transfer scheme in the initial system specification and have optimized the distributed memory organization. This results in a memory architecture with significantly reduced power consumption. For the worst-case mode using Predicted (P) frames, memory power consumption is reduced by a factor of 7 when compared to a good direct implementation of the reference specification. For the worst-case mode using Predicted and Bi-directional (PB) frames, the maximum memory power consumption is even reduced by a factor of 9. To achieve

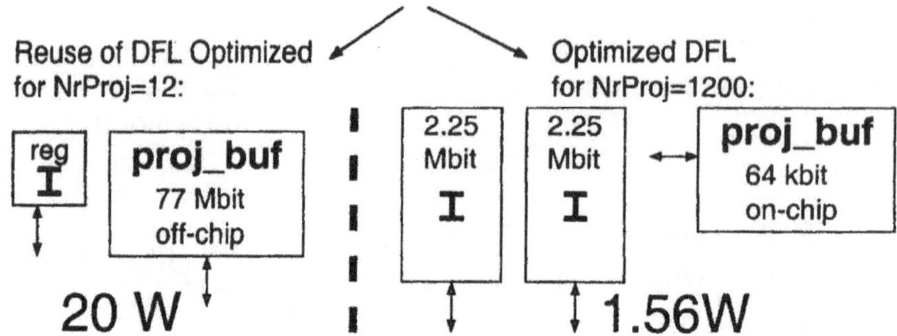

Figure 26.24 Estimated power cost for high resolution alternative of the back-projector with $NrProj = 1200$. At the left, reuse has been made of the DFL code which was optimal for the $NrProj = 12$ case. At the right a new exploration for the optimal choice has been performed with a different outcome. This shows that no simple reuse from a "code library" should be attempted for such data-dominated applications.

these results, we have used our formalized data transfer and storage exploration methodology, partly supported in the ATOMIUM environment and script.

For the same design, also the global power associated with the most demanding arithmetic module, namely the IDCT component, has been determined. Here, the combined power of the unoptimized data-path, including its local controller and the local memories was more than a factor 150 lower than the unoptimized background memory contribution. This clearly motivates our strategy to initially ignore the data-path effects in our system-level exploration.

We believe these results are very convincing for the effect on the global system power of the proposed approach.

We will now also show the effect of a few steps in our approach on memory size/area for another version of the 2D motion estimation test-vehicle (the basic algorithm was already described above for power exploration). Here it is assumed that different clock cycle budgets are evaluated and that both background and small intermediate signals are considered, leading to a larger bandwidth requirement.

The result of the memory allocation (automatically obtained with our prototype HIMALAIA tool) for several cycle budgets in the inner loop body is provided in Fig. 26.25. Note that the frame memory which is needed to store the delayed frames is not included. This large frame memory needs to be stored off-chip. What is included here are the smaller background (and partly also foreground) memories for storing the intermediate data on-chip, relatively close to the data-paths. The area figures have been obtained for an accurate RAM layout model, instantiated for a fast 1- or 2-port embedded SRAM generator in a 1.2 μm CMOS technology. Notice the large range of acceptable/feasible solutions for the 2 cycle case. From these alternatives, the optimal solution has been identified by the tool, which (surprisingly if one is used to the "normal" A-T trade-off curves) has comparable area as the 8 cycle solution. Clearly, less optimal solutions result in very significant area losses for such applications, much larger than the cost of the other chip components like data-path or control logic.

Other major driver vehicles during 1993-98 have included a voice coder in a mobile terminal application (auto-correlation, Schur decomposition and Viterbi

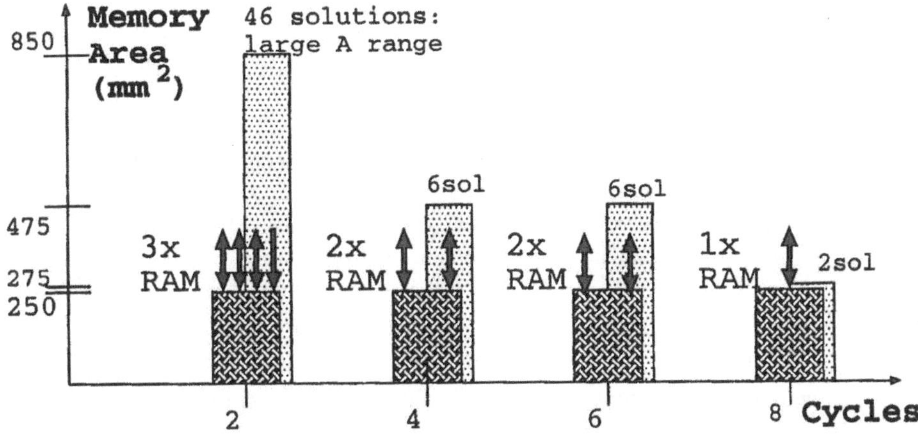

Figure 26.25 Results for automated memory allocation for several cycle budgets in inner loop of a video motion estimation kernel. The cross-hatched boxes show the area required for the optimal memory organization found, including the number of RAMs needed and their Read or Read/Write port requirements (arrows on top of boxes). The dashed boxes represent the area range for the feasible solution space.

coder), a regularity detector kernel for vision [9], a quadrature-structured differential PCM kernel (QSDPCM) [77] for image processing, a non-linear diffusion algorithm for image enhancement [78, 79], updating singular value decomposition for beam-forming [53], memory-dominated submodules of the MPEG2 video decoder/encoder [53, 13, 11, 33], and a 2D wavelet coder (forward and backward) for image coding [80]. Recent results on predefined and parallel processors have been obtained also for the full QSDPCM application [19] and the complete video motion estimation module in the new MPEG4 standard [81].

26.7 CONCLUSIONS

In this chapter we have introduced our system level data transfer and storage exploration methodology for data-dominated video applications, oriented to custom processor architectures. It tackles the power and area reduction of the dominant cost components in the architecture for this target domain, namely the system-level busses and the background memories. The focus has been on the lower stages in our proposed script, targeting system-level memory organization and cycle budget distribution. For the most critical tasks in the methodology, prototype tools were implemented and are being developed further to reduce the design time. The general applicability and effectiveness has been substantiated for several industrial data-dominated applications, including H.263 video conferencing decoding and medical computer tomography (CT) back-projection.

Acknowledgements

We wish to thank our colleagues at the VSDM Division of IMEC Leuven for the stimulating discussions and the fruitful co-operation. We especially acknowledge the contributions of our former colleagues in the memory management research domain, who have performed the many application driver studies and co-operated on the software backbone without which little of the presented work would have been feasible.

Note: more information on the ATOMIUM and ADOPT projects can be obtained at our web site:
`http://www.imec.be/vsdm/domains/designtechno/index.html`
A report version of many of the publications listed below can be also obtained there.

REFERENCES

[1] I. Verbauwhede, F. Catthoor, J. Vandewalle, H. De Man, "Background memory management for the synthesis of algebraic algorithms on multi-processor DSP chips", *Proc. VLSI'89, Int. Conf. on VLSI,* Munich, Germany, pp. 209-218, Aug. 1989.

[2] I. Verbauwhede, F. Catthoor, J. Vandewalle, H. De Man, "In-place memory management of algebraic algorithms on application-specific IC's", *Journal of VLSI signal processing,* vol. 3, Kluwer, Boston, pp. 193-200, 1991.

[3] W. Geurts, F. Franssen, M. van Swaaij, F. Catthoor, H. De Man, M. Moonen, "Memory and data-path mapping for image and video applications", in *Application-driven architecture synthesis,* F. Catthoor, L. Svensson (eds.), Kluwer, Boston, pp. 143-166, 1993.

[4] T. Gijbels, F. Catthoor, L. Van Eycken, A. Oosterlinck, H. De Man, "An application-specific architecture for the RBN-coder with efficient memory organization", *Journal of VLSI signal processing,* special issue on "Video/image signal processing", T. Nishitani, P. Ang, F. Catthoor (eds.), vol. 5, no. 2-3, Kluwer, Boston, pp. 221-236, April 1993.

[5] F. Catthoor, F. Franssen, S. Wuytack, L. Nachtergaele, H. De Man, "Global communication and memory optimizing transformations for low power signal processing systems", *IEEE workshop on VLSI signal processing,* La Jolla CA, Oct. 1994. Also in *VLSI Signal Processing VII,* J. Rabaey, P. Chau, J. Eldon (eds.), IEEE Press, New York, pp. 178-187, 1994.

[6] T. H. Meng, B. Gordon, E. Tsern, A. Hung, "Portable video-on-demand in wireless communication", special issue on "Low power design" of the *Proceedings of the IEEE,* vol. 83, no. 4, pp. 659-680, April 1995.

[7] V. Tiwari, S. Malik, A. Wolfe, "Power analysis of embedded software: a first step towards software power minimization", *IEEE Trans. on VLSI Systems,* vol. 2, no. 4, pp. 437-445, Dec. 1994.

[8] M. van Swaaij, F. Franssen, F. Catthoor, H. De Man, "Automating high-level control flow transformations for DSP memory management", *Proc. IEEE workshop on VLSI signal processing,* Napa Valley CA, Oct. 1992. Also in *VLSI Signal Processing V,* K. Yao, R. Jain, W. Przytula (eds.), IEEE Press, New York, pp. 397-406, 1992.

[9] F. Franssen, L. Nachtergaele, H. Samsom, F. Catthoor, H. De Man, "Control flow optimization for fast system simulation and storage minimization", *Proc. 5th ACM/IEEE Europ. Design and Test Conf.*, Paris, France, pp. 20-24, Feb. 1994.

[10] M. van Swaaij, F. Franssen, F. Catthoor, H. De Man, "High-level modelling of data and control flow for signal processing systems", in *Design Methodologies for VLSI DSP Architectures and Applications*, M.Bayoumi (ed.), Kluwer, Boston, pp. 219-259, 1994.

[11] L. Nachtergaele, F. Catthoor, F. Balasa, F. Franssen, E. De Greef, H. Samsom, H. De Man, "Optimisation of memory organization and hierarchy for decreased size and power in video and image processing systems", *Intnl. Workshop on Memory Technology, Design and Testing*, San Jose CA, pp. 82-87, Aug. 1995.

[12] F. Catthoor, W. Geurts, H. De Man, "Loop transformation methodology for fixed-rate video, image and telecom processing applications", *Proc. Intnl. Conf. on Applic.-Spec. Array Processors*, San Francisco, CA, pp. 427-438, Aug. 1994.

[13] E. De Greef, F. Catthoor, H. De Man, "Mapping real-time motion estimation type algorithms to memory-efficient, programmable multi-processor architectures", *Microprocessors and Microprogramming*, special issue on "Parallel Programmable Architectures and Compilation for Multi-dimensional Processing" (eds. F.Catthoor, M.Moonen), Elsevier, pp. 409-423, Oct. 1995.

[14] E. De Greef, F. Catthoor, H. De Man, "Memory organization for video algorithms on programmable signal processors", *Proc. IEEE Int. Conf. on Computer Design*, Austin TX, pp. 552-557, Oct. 1995.

[15] L. Nachtergaele, D. Moolenaar, B. Vanhoof, F. Catthoor, H. De Man, "System-level power optimization of video codecs on embedded cores : a systematic approach", special issue on *Future directions in the design and implementation of DSP systems* (eds. Wayne Burleson, Konstantinos Konstantinides) of *Journal of VLSI Signal Processing*, vol. 18, no. 2, Kluwer, Boston, pp. 89-110, Feb. 1998.

[16] G. Goossens, I. Bolsens, B. Lin, F. Catthoor, "Design of heterogeneous ICs for mobile and personal communication systems", *Proc. IEEE Int. Conf. Comp. Aided Design*, San Jose CA, pp. 524-531, Nov. 1994.

[17] E. De Greef, F. Catthoor, H. De Man, "Reducing storage size for static control programs mapped onto parallel architectures", presented at *Dagstuhl Seminar on Loop Parallelisation*, Schloss Dagstuhl, Germany, April 1996.

[18] K. Danckaert, F. Catthoor, H. De Man, "System-level memory management for weakly parallel image processing", *Proc. EuroPar Conference*, Lyon, France, August 1996. "Lecture notes in computer science" series, Springer Verlag, pp. 217-225, 1996.

[19] K. Danckaert, F. Catthoor, H. De Man, "System level memory optimization for hardware-software co-design", *Proc. IFIP Intnl. Workshop on Hardware/Software Co-design*, Braunschweig, Germany, pp. 55-59, March 1997.

[20] D. Singh, J. Rabaey, M. Pedram, F. Catthoor, S. Rajgopal, N. Sehgal, T. Mozdzen, "Power conscious CAD tools and methodologies: a perspective", special issue on "Low power design" of the *Proceedings of the IEEE*, vol. 83, no. 4, pp. 570-594, April 1995.

[21] S. Amarasinghe, J. Anderson, M. Lam, and C. Tseng, "The SUIF compiler for scalable parallel machines", in *Proc. of the 7th SIAM Conf. on Parallel Proc. for Scientific Computing*, 1995.

[22] U. Banerjee, R. Eigenmann, A. Nicolau, D. Padua, "Automatic program parallelisation", *Proc. of the IEEE*, invited paper, vol. 81, no. 2, Feb. 1993.

[23] W. Pugh, "The Omega Test: a fast and practical integer programming algorithm for dependence analysis", *Communications of the. ACM*, vol. 35, no. 8, Aug. 1992.

[24] M. Wolfe, "The Tiny loop restructuring tool", *Proc. of Intnl. Conf. on Parallel Processing*, pp. II.46-II.53, 1991.

[25] P. Lippens, J. Van Meerbergen, W. Verhaegh, A. van der Werf, "Allocation of multiport memories for hierarchical data streams", *Proc. IEEE Int. Conf. Comp. Aided Design*, Santa Clara CA, Nov. 1993.

[26] L. Ramachandran, D. Gajski, and V. Chaiyakul, "An algorithm for array variable clustering", *Proc. 5th ACM/IEEE Europ. Design and Test Conf.*, Paris, France, pp. 262-266, Feb. 1994.

[27] D. Kolson, A. Nicolau, N. Dutt, "Minimization of memory traffic in high-level synthesis", *Proc. 31st ACM/IEEE Design Automation Conf.*, San Diego, CA, pp. 149-154, June 1994.

[28] O. Sentieys, D. Chillet, J. P. Diguet, J. Philippe, "Memory module selection for high-level synthesis", *Proc. IEEE workshop on VLSI signal processing*, Monterey CA, Oct. 1996.

[29] H. Schmidt and D. Thomas. "Address generation for memories containing multiple arrays", *Proc. IEEE Int. Conf. Comp. Aided Design*, San Jose CA, pp. 510–514, Nov. 1995.

[30] T. Akiyama, H. Aono, K. Aoki, K. Ler, B. Wilson, T. Araki, T. Morishige, H. Takeno, A. Sato, S. Nakatani, T. Senoh, "MPEG2 video codec using image compression DSP", *IEEE Trans. on Consumer Electronics*, vol. CE-40, no. 3, pp. 466-472, August 1994.

[31] K. Ishihara *et. al.*, "A half-pel precision MPEG2 motion-estimation processor with concurrent three-vector search", *Proc. IEEE Int. Solid-State Circ. Conf.*, pp. 288-289, Feb. 1995.

[32] T. Miyazaki, I. Kuroda, and M. Imanishi, "A low-cost MPEG1 video encoder based on a single chip DSP", *Proc. DSPX*, pp. 136-142, 1994.

[33] J. Rosseel, F. Catthoor, H. De Man, "The systematic design of a motion estimation array architecture", *Proc. Intnl. Conf. on Applic.-Spec. Array Processors*, Barcelona, Spain, pp. 40-54, Sep. 91.

[34] M. Toyokura *et. al.*, "A video DSP with a macroblock-level-pipeline and a SIMD type vector-pipeline architecture for MPEG2 codec", *IEEE J. Solid-state Circ.*, vol. SC-29, pp. 1474-1480, Dec. 1994.

[35] M. Harrand, M. Henry, P. Chaisemartin, P. Mougeat, Y. Durand, A. Tournier, R. Wilson, J. Herluison, J. Langchambon, J. Bauer, M. Runtz and J. Bulone, "A single chip videophone encoder/decoder", *Proc. IEEE Int. Solid-State Circ. Conf.*, pp. 292-293, Feb. 1995.

[36] S. Wuytack, F. Catthoor, F. Franssen, L. Nachtergaele, H. De Man, "Global communication and memory optimizing transformations for low power systems", *IEEE Intnl. Workshop on Low Power Design*, Napa CA, pp. 203-208, April 1994.

[37] A. Chandrakasan, M. Potkonjak, R. Mehra, J. Rabaey, R. W. Brodersen, "Optimizing power using transformations," *IEEE Trans. on Comp.-aided Design*, vol. CAD-14, no. 1, pp. 12-30, Jan. 1995.

[38] P. N. Hilfinger, J. Rabaey, D. Genin, C. Scheers, H. De Man, "DSP specification using the Silage language", *Proc. Int. Conf. on Acoustics, Speech and Signal Processing*, Albuquerque, NM, pp. 1057-1060, April 1990.

[39] F. Catthoor, S. Wuytack, E. De Greef, F. Franssen, L. Nachtergaele. H. De Man, "System-level transformations for low power data transfer and storage", in paper collection on "Low power CMOS design" (eds. A.Chandrakasan, R.Brodersen), IEEE Press, pp. 609-618, 1998.

[40] M. van Swaaij, F. Franssen, F. Catthoor, H. De Man, "Modelling data and control flow for high-level memory management", *Proc. 3rd ACM/IEEE Europ. Design Automation Conf.*, Brussels, Belgium, pp. 8-13, March 1992.

[41] F. Balasa, F. Catthoor, H. De Man, "Practical solutions for counting scalars and dependences in ATOMIUM – a memory management system for multi-dimensional signal processing", *IEEE Trans. on Comp.-aided Design*, vol. CAD-16, no. 2, pp. 133-145, Feb. 1997.

[42] F. Franssen, F. Balasa, M. van Swaaij, F. Catthoor, H. De Man, "Modeling multi-dimensional data and control flow", *IEEE Trans. on VLSI systems*, vol. 1, no. 3, pp. 319-327, Sep. 1993.

[43] F. Balasa, F. Franssen, F. Catthoor, H. De Man, "Transformation of nested loops with modulo indexing to affine recurrences", in special issue of *Parallel Processing Letters* on "Parallelization techniques for uniform algorithms", C. Lengauer, P. Quinton, Y. Robert, L. Thiele (eds.), World Scientific Pub., pp. 271-280, 1994.

[44] F. Catthoor, M. Janssen, L. Nachtergaele, H. De Man, "System-level data-flow transformation exploration and power-area trade-offs demonstrated on video codecs", special issue on "Systematic trade-off analysis in signal processing systems design" (eds. M.Ibrahim, W.Wolf) in *Journal of VLSI Signal Processing*, vol. 18, no. 1, Kluwer, Boston, pp. 39-50, 1998.

[45] H. Samsom, L. Claesen, H. De Man, "SynGuide: an environment for doing interactive correctness preserving transformations", *IEEE workshop on VLSI signal processing*, Veldhoven, The Netherlands, Oct. 1993. Also in *VLSI Signal Processing VI*, L.Eggermont, P.Dewilde, E.Deprettere, J.van Meerbergen (eds.), IEEE Press, New York, pp. 269-277, 1993.

[46] S. Wuytack, F. Catthoor, L. Nachtergaele, H. De Man, "Power exploration for data dominated video applications", *Proc. IEEE Intnl. Symp. on Low Power Design*, Monterey, pp. 359-364, Aug. 1996.

[47] S. Wuytack, J. P. Diguet, F. Catthoor, H. De Man, "Formalized methodology for data reuse exploration for low-power hierarchical memory mappings", *IEEE Trans. on VLSI Systems*, vol. 6, 1998.

[48] J. P. Diguet, S. Wuytack, F. Catthoor, H. De Man, "Formalized methodology for data reuse exploration in hierarchical memory mappings", *Proc. IEEE Intnl. Symp. on Low Power Design*, Monterey, pp. 30-35, Aug. 1997.

[49] S. Wuytack, F. Catthoor, G. De Jong, H. De Man, "Minimizing the required memory bandwidth in VLSI system realizations", *IEEE Trans. on VLSI Systems*, vol. 6, 1998.

[50] F. Balasa, F. Catthoor, H. De Man, "Background memory area estimation for multi-dimensional signal processing Systems", *IEEE Trans. on VLSI Systems*, vol. 3, no. 2, pp. 157-172, June 1995.

[51] P. Slock, S. Wuytack, F. Catthoor, G. De Jong, "Fast and extensive system-level memory exploration for ATM applications", *Proc. 10th ACM/IEEE Intnl. Symp. on System-Level Synthesis*, Antwerp, Belgium, pp. 74-81, Sep. 1997.

[52] F. Balasa, F. Catthoor, H. De Man, "Exact evaluation of memory area for multi-dimensional processing systems", *Proc. IEEE Int. Conf. Comp. Aided Design*, Santa Clara CA, pp. 669–672, Nov. 1993.

[53] F. Balasa, F. Catthoor, H. De Man, "Dataflow-driven memory allocation for multi-dimensional signal processing systems", *Proc. IEEE Int. Conf. Comp. Aided Design*, Santa Jose CA, pp. 31-34, Nov. 1994.

[54] J. Vanhoof, I. Bolsens, H. De Man, "Compiling multi-dimensional data streams into distributed DSP ASIC memory", *Proc. IEEE Int. Conf. Comp. Aided Design*, Santa Clara CA, pp. 272-275, Nov. 1991.

[55] E. De Greef, F. Catthoor, H. De Man, "Program transformation strategies for reduced power and memory size in pseudo-regular multimedia applications", publication in *IEEE Trans. on Circuits and Systems for Video Technology*, 1998.

[56] E. De Greef, F. Catthoor, H. De Man, "Memory size reduction through storage order optimization for embedded parallel multimedia applications", special issue on "Parallel Processing and Multi-media" (ed. A.Krikelis), in *Parallel Computing* Elsevier, vol. 23, no. 12, Dec. 1997.

[57] E. De Greef, "Storage size reduction for multimedia applications", *Doctoral dissertation*, ESAT/EE Dept., K.U.Leuven, Belgium, Jan. 1998.

[58] E. De Greef, F. Catthoor, H. De Man, "Memory size reduction through storage order optimization for embedded parallel multimedia applications", *Intnl. Parallel Proc. Symp.(IPPS)* in Proc. Workshop on "Parallel Processing and Multimedia", Geneva, Switzerland, pp. 84-98, April 1997.

[59] E. De Greef, F. Catthoor, H. De Man, "Array placement for storage size reduction in embedded multimedia systems", *Proc. Intnl. Conf. on Applic.-Spec. Array Processors*, Zurich, Switzerland, pp. 66-75, July 1997.

[60] H. Samsom, F. Franssen, F. Catthoor, H. De Man, "Verification of loop transformations for real time signal processing applications", *IEEE workshop on VLSI signal processing*, La Jolla CA, Oct. 1994. Also in *VLSI Signal Processing VII*, J.Rabaey, P.Chau, J.Eldon (eds.), IEEE Press, New York, pp. 269-277, 1994.

[61] M. Cupak, F. Catthoor, "Efficient functional validation of system-level loop transformations for multi-media applications", *Proc. Electronic Circuits and Systems Conference*, Bratislava, Slovakia, pp. 39-43, Sep. 1997.

[62] M. Miranda, F. Catthoor, M. Janssen, H. De Man, "ADOPT: Efficient hardware address generation in distributed memory architectures", *Proc. 9th ACM/IEEE Intnl. Symp. on System-Level Synthesis*, La Jolla CA, pp. 20-25, Nov. 1996.

[63] M. Miranda, F. Catthoor, M. Janssen, H. De Man, "High-level address optimisation and synthesis techniques for data-transfer intensive applications", *IEEE Trans. on VLSI Systems*, 1998.

[64] J. L. da Silva Jr, C. Ykman-Couvreur, M. Miranda, K. Croes, S. Wuytack, G. de Jong, F. Catthoor, D. Verkest, P. Six, H. De Man, "Efficient system exploration and synthesis of applications with dynamic data storage and intensive data transfer", *Proc. 35th ACM/IEEE Design Automation Conf.*, San Francisco CA, pp. 76-81, June 1998.

[65] G. de Jong, B. Lin, C.V erdonck, S. Wuytack, F. Catthoor, "Background memory management for dynamic data structure intensive processing systems", *Proc. IEEE Int. Conf. Comp. Aided Design*, San Jose CA, pp. 515-520, Nov. 1995.

[66] S. Wuytack, F. Catthoor, H. De Man, "Transforming set data types to power optimal data structures", *IEEE Trans. on Comp.-aided Design*, vol. CAD-15, no. 6, pp. 619-629, June 1996.

[67] S. Wuytack, F. Catthoor, G. De Jong, B. Lin, H. De Man, "Flow graph balancing for minimizing the required memory bandwidth", *Proc. 9th ACM/IEEE Intnl. Symp. on System-Level Synthesis*, La Jolla CA, pp. 127-132, Nov. 1996.

[68] W. Verhaegh, P. Lippens, E. Aarts, J. Korst, J. Van Meerbergen, A. van der Werf, "Improved force-directed scheduling in high-throughput digital signal processing", *IEEE Transactions on CAD and Systems*, vol. 14, no 8, Aug. 1995.

[69] C. Lin and S. Kwatra, "An adaptive algorithm for motion compensated colour image coding", *Proc. IEEE Globecom*, pp. 47.1.1-4, 1984.

[70] T. Komarek, P. Pirsch, "Array architectures for block matching algorithms", *IEEE Trans. on Circuits and Systems*, vol. 36, no. 10, Oct. 1989.

[71] W. Geurts, F. Catthoor, S. Vernalde, H. De Man, "Accelerator data-paths synthesis for high-throughput signal processing applications", Kluwer Academic Publishers, Boston, 1996.

[72] K. Itoh, K. Sasaki, Y. Nakagome, "Trends in low-power RAM circuit technologies", special issue on "Low power electronics" of the *Proceedings of the IEEE*, vol. 83, no. 4, pp. 524-543, April 1995.

[73] T. Seki, E. Itoh, C. Furukawa, I. Maeno, T. Ozawa, H. Sano, N. Suzuki, "A 6-ns 1-Mb CMOS SRAM with latched sense amplifier", *IEEE J. of Solid-state Circuits*, vol. SC-28, no. 4, pp. 478-483, Apr. 1993.

[74] P. Van Oostende, G. Van Wauwe, "Low power design: a gated-clock strategy", *Low Power Workshop*, Ulm, Germany, Sep. 1994.

[75] ITU-H.263, "Video coding for narrow telecommunications channels at less than 64 kbits/s," *http://www.nta.no/brukere/DVC/h263_wht/* .

[76] L. Nachtergaele, F. Catthoor, B. Kapoor, D. Moolenaar, S. Janssens, "Low power data transfer and storage exploration for H.263 video decoder system", special issue on *Very low-bit rate video coding* (eds. Argy Krikelis et al.) of *IEEE Journal on Selected Areas in Communications*, vol. 15/16, no. 12/1, Dec. 1997 - Jan. 1998.

[77] J. M. Janssen, F. Catthoor, H. De Man, "Memory management aspects in the architecture design of a QSDPCM coder for video-phone", *Proc. Picture Coding Symposium*, Lausanne, Switzerland, March 1993.

[78] T. Gijbels, P. Six, L. Van Gool, F. Catthoor, H. De Man, A. Oosterlinck, "A VLSI architecture for parallel non-linear diffusion with applications in vision", *IEEE workshop on VLSI signal processing*, La Jolla CA, Oct. 1994. Also in *VLSI Signal Processing VII*, J. Rabaey, P. Chau, J. Eldon (eds.), IEEE Press, New York, pp. 398-407, 1994.

[79] J. Rosseel, F. Catthoor, T. Gijbels, P. Six, L. Van Gool, H. De Man, "An optimization methodology for mapping a diffusion algorithm for vision into a modular and flexible array architecture", *Proc. Intnl. Workshop on Algorithms*

and Parallel VLSI Architectures, Leuven, Belgium, August 1994. Also in "Algorithms and Parallel VLSI Architectures III" (eds. M.Moonen, F.Catthoor), Elsevier, pp. 131-142, 1995.

[80] G. Lafruit, F. Catthoor, J. Cornelis, H. De Man, "An efficient VLSI architecture for the 2-D wavelet transform with novel image scan", *IEEE Trans. on VLSI Systems*, 1998.

[81] E. Brockmeyer, F. Catthoor, J. Bormans, H. De Man, "Code transformations for reduced data transfer and storage in low power realization of MPEG-4 full-pel motion estimation", *Proc. IEEE Int. Conf. on Image Proc.*, Chicago, IL, pp. III.985-989, Oct. 1998.

Chapter 27

Hardware Description and Synthesis of DSP Systems

Lori E. Lucke
Minnetronix, Inc.
St. Paul, Minnesota
lelucke@minnetronix.com

Junsoo Lee
University of Minnesota
Minneapolis, Minnesota
jlee@ece.umn.edu

27.1 INTRODUCTION

The design process for an application specific-architecture of a digital signal processing algorithm is shown in Fig. 27.1. This process consists of the following steps. Beginning with a known algorithm, such as a finite impulse response (FIR) filter, the design is mapped onto functional units. This process can be automated using high level synthesis tools [1] - [11]. Once the functional blocks and the connection topology have been identified, it is necessary to begin mapping the architecture onto a physical design. This process can be automated by using hardware description languages such as Verilog or VHDL to first describe the functional blocks. Finally the hardware description modules can be mapped to a particular physical implementation using logic synthesis.

In this chapter we demonstrate the design flow necessary to implement an application specific architecture for a signal processing algorithm. Several examples are used to illustrate this process. Using high level synthesis, a functional implementation is generated for a common signal processing problem. The remainder of the chapter focuses on the steps necessary to transform the functional implementation into a physical design. We demonstrate both standard cell and gate array implementations since these are the most automated.

The design process consists of several major steps: description, verification, synthesis, analysis, and mapping to physical blocks. During design description, the design is formulated using a hardware description language or schematic capture. In this chapter we demonstrate the use of VHDL for implementing functional blocks. Throughout the design process, the design must be verified to ensure that it is operating properly using logic simulation. Designs implemented using VHDL must be synthesized to a specific physical set of gates. In this chapter we will synthesize designs to the 1.2 μm CMOSN standard cell library and to the Xilinx gate array library. The performance of the design is analyzed to make sure it meets the initial design goals. Speed, area, and power consumption are used to gauge the performance of the design. The last step is to map the design to its final physical form. For example, layout of the standard cell circuit is performed. Of course, the design process is an iterative process. Should verification fail, or should the performance of the design, not meet the design goals, then the design must be specified, synthesized, simulated, and analyzed again.

In this chapter we demonstrate the design flow necessary to implement an application specific architecture for a signal processing algorithm. We follow a specific design example throughout the entire process while introducing the necessary steps. We begin with high level synthesis in Section 27.2. Section 27.3 details the steps necessary to transform the high level description to a physical implementation. Design entry including a discussion on synthesizable VHDL is detailed in Section 27.4. The design is verified in Section 27.5. The design is synthesized in Section 27.6. More detailed design verification is performed in Section 27.7. Design analysis including presentation of a power estimation tool is performed in Sections 27.8 and 27.9. Finally layout and post layout verification are detailed in Sections 27.10 and 27.11.

27.2 HIGH LEVEL SYNTHESIS

A digital signal processing algorithm is described using a synchronous data-flow graph (DFG) [12, 13]. The data-flow graph describes an iterative, deterministic, multiprocessing algorithm. An example of a synchronous data-flow graph is shown in Fig. 27.2. Nodes represent operations or tasks with predetermined execution times. No preemption of tasks is allowed. Arcs represent the data flow within the graph, which is predetermined and does not depend on the data. The arcs between tasks with no delays represent precedence constraints between tasks within an iteration. Delays on arcs represent precedence constraints between tasks executed in different iterations. The tasks in the data-flow graph are periodically executed. When all the operations have been executed once, one iteration is completed. Digital signal processing algorithms do not have the same amount of overhead as the data paths typically synthesized by high level synthesis systems. They have no loop overhead and no branches and execute repetitively. Therefore, there are many transformations which can be used during high level synthesis of these algorithms to develop highly concurrent architectures.

High level synthesis [1] - [11] consists of many subtasks including algorithmic transformations, scheduling, and allocation. Algorithmic transformations [4] - [17] attempt to improve the speed or reduce the resource usage of an algorithm without changing the underlying function of the algorithm. The scheduling task assigns operations within the data-flow graph to control steps within the architectural description. Scheduling may be either time-constrained where the goal is to reduce the number of processors necessary to execute the data-flow graph within a certain time period, or processor-constrained where the goal is to reduce the number of time steps necessary to execute the data-flow graph on some predetermined number of processors. There are several methods for scheduling the DFG [13] - [27]. The allocation subtask consists of assigning resources such as processors, memory, communication busses, muxes, and controllers to develop an architecture which will implement the schedule determined during the scheduling step.

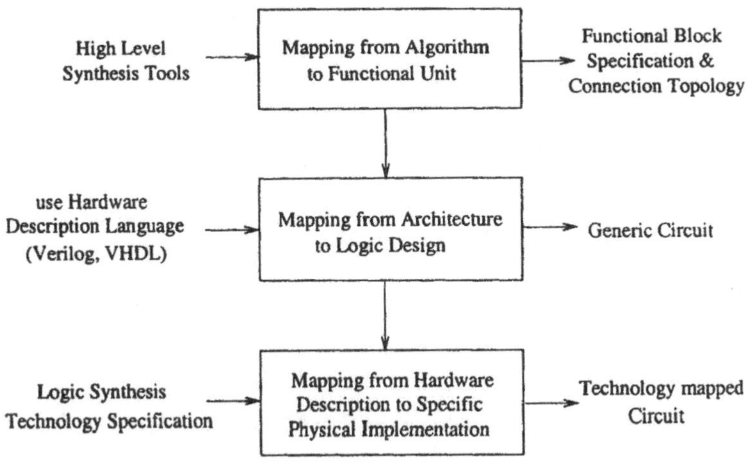

Figure 27.1 Design flow using hardware description.

There exist many high level synthesis systems. In the following we demonstrate how to use high level synthesis to map the FIR (finite impulse response) filter algorithm shown in Fig. 27.2 to a processor-constrained architecture. Fig. 27.2 shows the DFG for a 4 tap FIR filter implementing $y(n) = x(n)h3 + x(n-1)h2 + x(n-2)h1 + x(n-3)h0$. In Fig. 27.2 (a) the original 4 tap FIR filter is shown and in Fig. 27.2 (b) one multiplier and one adder is merged to form a MA (multiplier and adder) unit. We will apply the design process to the 4 tap FIR filter shown in Fig. 27.2.

We begin by scheduling the DFG to a fixed set of processors. Assume there are two processors (P1, P2) available to compute the MA operation. Each processor is internally pipelined so that a MA operation takes 2 clock cycles but a new operator can be initiated each clock cycle. A schedule using these processors is shown below.

TIME	0	1	2	3	4	5	6	7	8	9
IN	X_0	X_0	X_1	X_1	X_2	X_2	X_3	X_3	X_4	X_4
P1	$MA1_0$	$MA2_0$	$MA1_1$	$MA2_1$	$MA1_2$	$MA2_2$	$MA1_3$	$MA2_3$	$MA1_4$	$MA2_4$
P2		$MA3_0$	$MA4_0$	$MA3_1$	$MA4_1$	$MA3_2$	$MA4_2$	$MA3_3$	$MA4_3$	$MA3_4$
OUT		$y{-4}$		$y{-3}$		$y{-2}$		$y{-1}$		yn_0

Here the subscript references a particular iteration of the algorithm. By fixing the number of processors, the throughput of the circuit is limited to two clock cycles,

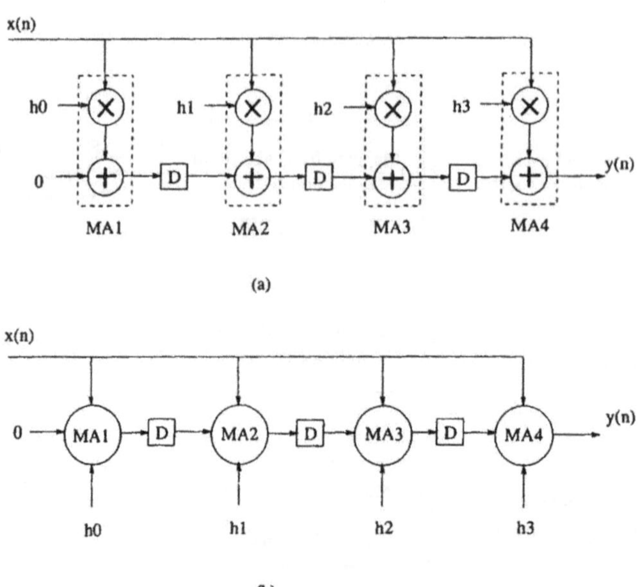

Figure 27.2 Four tap FIR filter.

while the latency (time from X_i to y_i) is ten clock cycles. Many other schedules could also be generated.

The resulting functional description for the FIR filter is shown in Fig. 27.3. In Fig. 27.3, each MA (multiplier adder) unit has two levels of pipelining. By using only two MA units for the four MA operations in the original algorithm, it is necessary to multiplex the data as shown by the switches in figure 27.3. The switches or multiplexers require control inputs that must be generated by an additional control unit. The complete functional block diagram for the FIR filter is shown in Fig. 27.4. Although not shown in the Fig., the FIR filter also requires a simple controller to control the data flow through the muxes.

27.3 TOP DOWN DESIGN

Once an architecture has been specified it is necessary to implement the physical design. However it is not possible or desirable to jump straight to transistors or even logic gates from the architectural description. In this section we focus on the use of top-down design for implementation.

The top down design process consists of the following steps.

1. **Design specification.**
2. **Design entry.** - The design is described using a hardware description language such as VHDL or Verilog.
3. **Functional simulation.** - The design is verified using functional logic simulation.

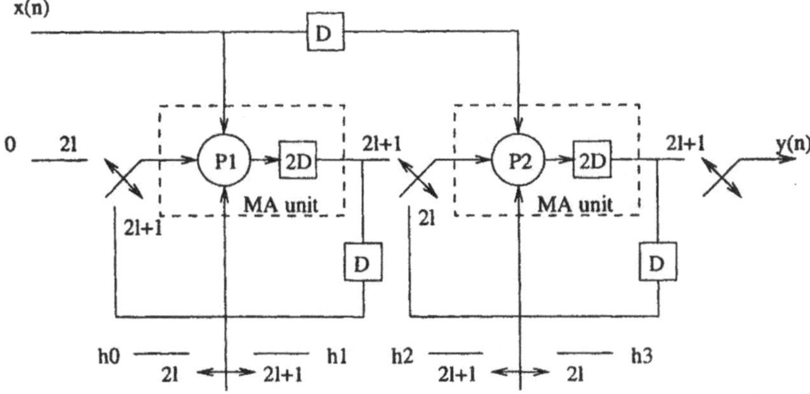

Figure 27.3 FIR filter functional specification.

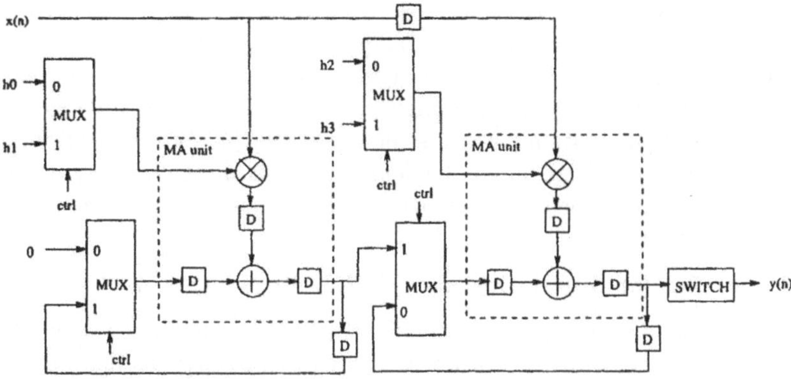

Figure 27.4 Completely specified FIR filter.

4. **Logic synthesis.** - The design is synthesized to a set of logic gates. It is at this point that we must specify the technology that will be used, such as standard cells or an FPGA (Field Programmable Gate Array).

5. **Structural simulation.** - The logic design is verified using structural simulation.

6. **Design analysis.** - The actual area, speed, and power consumption of the design are computed.

7. **Physical layout.** - The design is mapped to the actual VLSI layout.

8. **Post layout verification.** - Prior to sending the design out for fabrication, it should be verified at the low level. This includes a schematic versus layout check, logic simulation with actual physical timing parameters, design rules check, and perhaps a circuit level simulation using SPICE.

9. **Fabrication.** - Finally the device can be made.

10. **Physical part testing.** - The actual physical device should be tested to see if it meets the design specification.

In the following sections we will demonstrate the top down design process using the FIR filter architecture specified in the previous section. We utilize the Mentor Graphics design automation tools including Autologic for logic synthesis, QuickSim for logic simulation, QuickPath for critical path analysis, and ICStation for layout and backannotation.

27.4 DESIGN ENTRY

Each of the functional blocks in the architectural description must be specified using either a hardware description language or schematic entry. The use of a hardware description language gives us many benefits including: the ability to describe the design at a higher level, better support for design changes, and support of logic synthesis. In this chapter, we concentrate on developing the design using VHDL [28, 29]. Several examples are given to emphasize the design process. A working knowledge of VHDL is assumed.

VHDL is a hardware description language. It was designed to support both logic emulation and logic synthesis. However it must be noted that logic emulation is distinctly different from logic synthesis. A VHDL model designed to emulate a specific logic system must only provide the correct outputs given a set of inputs at the correct time. A VHDL model for logic synthesis must correctly describe the underlying hardware. The difference between emulation and synthesis is described in the following sections.

27.4.1 VHDL for Simulation and VHDL for Synthesis

VHDL has many language constructs designed to support the emulation of a logic model. Many of these cannot be used when writing logic models for synthesis. The differences between VHDL for simulation and VHDL for synthesis are explained using examples in this section. For correct logic synthesis of VHDL it is necessary to remember exactly what is being synthesized. Basically the VHDL must be used to specifically describe either combinational logic or sequential logic rather than emulating logic. Therefore only a restricted set of VHDL constructs can be used to describe the logic for synthesis. The following example demonstrates this difference.

In VHDL, we can assign specific delays to an operation.

```
delay_proc:  PROCESS (a, b)
             BEGIN
                     x <= a AND b AFTER 5ns;
             END PROCESS delay_proc;
```

Here, we used an assignment statement "x <= a and b after 5ns". During simulation x will actually change state 5 ns (nanoseconds) after a or b changes. However this *after* clause cannot be synthesized since a time delay is not a logical construct. Remember the synthesizer is attempting to synthesize combinational or sequential logic. It is not possible to construct digital logic capable of an exact 5 ns delay.

In this section we review various VHDL language constructs and discuss how they are mapped to combinational or sequential logic.

27.4.1.1 Synthesizing combinational logic

To synthesize combinational logic properly it is necessary to remember a few key rules. (1) A straightforward signal assignment statement in a concurrent area will always generate combinational logic. For example,

```
s <= in1 XOR in2;
```

will always generate a connection of gates to implement an XOR function. (2) Conditional assignments must include all possible conditions to generate combinational logic. (3) If a signal is present on both the left hand side and the right hand side of an assignment statement, then a latch will be synthesized. (4) Process statements must include all combinational inputs (all signals on the right hand side of a statement or in condition clauses) in the sensitivity list. A few examples will demonstrate these rules.

The following demonstrates the difference between a correctly specified conditional statement and one which does not contain all possible conditions.

```
out1 <= i0 WHEN sel = "00" ELSE
        i1 WHEN sel = "11" ELSE
        i2 WHEN sel = "10" ELSE
        i3;

out2 <= i0 WHEN sel = "00" ELSE
        i2 WHEN sel = "10";
```

In the first conditional assignment, all possible conditions for *sel* (including "XX") are specified. When synthesized, this will generate the combinational logic to implement a mux. In the second conditional assignment, every condition for *sel* is not specified. Therefore *out*2 will be connected to the output of a latch. To see why this is true, imagine what happens when simulating *out*2 and *sel* changes from "00" to "11". Any change in the right hand side of *out*2 will cause it to be executed. There is no clause for *sel* = "11", thus *out*2 must retain its previous value. This implies there is memory in the circuit. During synthesis this statement will generate a latch.

The following demonstrates what happens when a signal is present on both the right- and left-hand side of a assignment statement.

```
WITH control SELECT
    out3 <= i1 WHEN "00";
            i2 WHEN "11";
            out3 WHEN OTHERS;
```

Again imagine what happens when the VHDL code is simulated and *control* changes from "00" to "01". Although all possible conditions are specified, for the "01" condition, *out*3 must retain its last value. Again this implies there is memory in the circuit. When synthesized, this statement will connect *out*3 to the output of a latch.

The following demonstrates what happens when a process statement does not contain the correct sensitivity list. In VHDL we use a process sensitivity list to control when a statement is executed. When a change occurs on one of the inputs in the sensitivity list, the corresponding statements in the process will be executed. To see the effect of the sensitivity list, let's take a look at the following example which attempts to implement a simple mux.

```
sens_procl:  PROCCESS (i0,i1)
             BEGIN
                     IF sel = "0" THEN
                           x <= i0;
                     ELSE
                           x <= i1;
             END PROCESS sens_procl;
```

Although all conditional clauses are specified here, when we simulate sens_procl, x changes when there is a change on $i0$ or $i1$ but there is no change on the output when *sel* changes since it is not in the sensitivity list. This is not the proper operation for a mux. During synthesis unforeseen results may occur. The proper process statement would include *sel* in the sensitivity list.

27.4.1.2 Synthesizing clocked logic

Clocked circuits consist of latches, registers or flip-flops, or state machines. To design a clocked circuit in VHDL requires the use of processes or guarded blocks. A latch can be designed as follows.

```
latchp:  PROCESS (enable, d)
BEGIN
              IF enable = '1' THEN
                    q <= d;
              END IF;
END PROCESS latchp;
```

Here it is clear that not all the conditions are specified for the IF statement. Thus a latch will be synthesized.

For a register a clock edge must be specified. A process statement may be used to specify a register as follows.

```
PROCESS (clock)
BEGIN
        IF (clock = '1' AND clock'LAST_VALUE = '0') THEN
              q <= d;
        END IF;
END PROCESS;
```

Signal attributes are used to clearly specify the clock. The process statement will synthesize a rising edge triggered flip-flop. Here it is clear that the process will only be entered when the clock changes and the value of q will only change on the rising clock edge. A guarded block may also be used as follows.

```
BLOCK (NOT clock'stable AND clock = '1' AND clock'LAST_VALUE = '0')
```

```
BEGIN
        q <= d;
END BLOCK;
```

This block statement will also synthesize a rising edge triggered flip-flop.

Typically it is required to synthesize a flip-flop with a reset. For a flip-flop with a synchronous reset, the following code can be used.

```
sync_reset:   PROCESS (clock)
BEGIN
              IF (clock = '1' AND clock'LAST_VALUE = 0) THEN
                        IF (reset = '1') THEN
                              q <= '0';
                        ELSE
                              q <= d;
                        END IF;
              END IF;
END PROCESS sync_reset;
```

For a flip-flop with an asynchronous reset, the following code can be used.

```
async_res:   PROCESS (clock, reset)
BEGIN
             IF (reset = '1') THEN
                       q <= '0';
             ELSIF (clock'EVENT AND clock = '1' AND
             clock'LAST_VALUE = '0') THEN
                       q <= d;
             END IF;
END PROCESS async_res;
```

The difference between the synchronous and asynchronous reset is clear. For the synchronous reset, the output can only change when the clock changes whereas for the asynchronous reset, the output changes either when the clock changes or when the reset changes.

A state machine consists of sequential registers to hold the present state and combinational logic to calculate the next state. The classic state machine is shown in Fig. 27.5. Remember there are two types of state machines, a Moore machine where the output is a function of only the present state; and a Mealy machine where the output is a function of both the present state and the present input. To write a synthesizable state machine in VHDL simply requires a process statement to specify the next state registers, and a process statement to specify the combinational logic for the output and the next state.

An example of a 2-bit counter with a count enable is shown in Fig. 27.6. This is a Mealy machine implementation. The synthesizable VHDL code for this machine is as follows.

```
ENTITY counter IS
PORT (clock,count_en,reset: IN STD_LOGIC;
      ctrl: OUT STD_LOGIC);
```

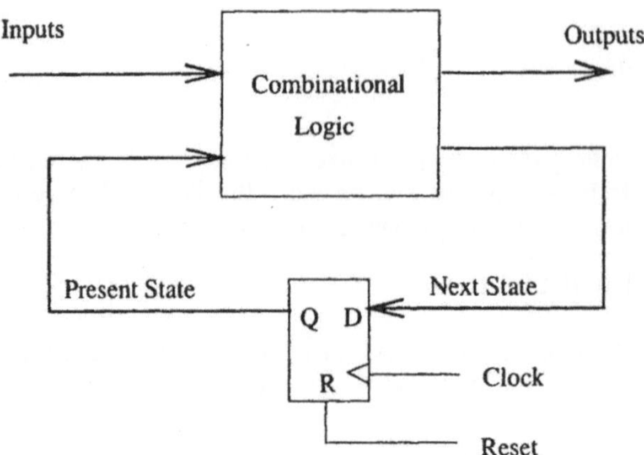

Figure 27.5 State machine.

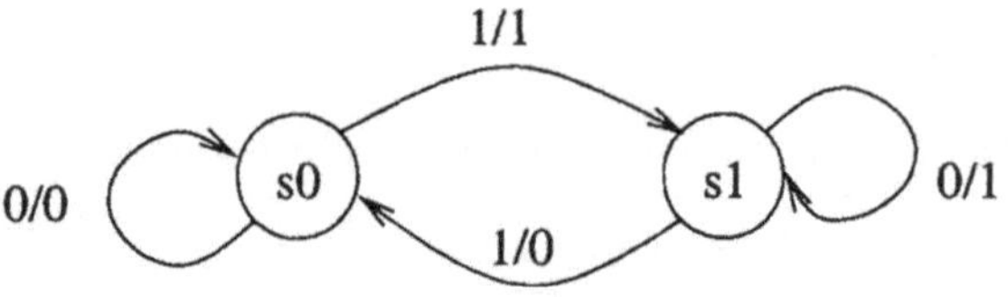

(a) State transition diagram

Present_State	Next_State/Output(ctrl)	
	count_en = 0	count_en = 1
s0	s0/0	s1/1
s1	s1/1	s0/0

(b) State transition table

Figure 27.6 Two bit counter. (a) State transition diagram; (b) State transition table.

```
END counter;
ARCHITECTURE mealy OF counter IS
    TYPE state_enum IS (s0,s1);
    SIGNAL present_state: state_enum := s0;
    SIGNAL next_state: state_enum := s1;
BEGIN
    registerp: PROCESS (clock, reset)
```

```
    BEGIN
        IF (reset = '0') THEN
           present_state <= s0;
        ELSIF (clock'EVENT AND clock = '1' AND
        clock'LAST_VALUE = '0') THEN
           present_state <= next_state;
        END IF;
    END PROCESS registerp;
    comblogicp: PROCESS (present_state,count_en)
    BEGIN
        next_state <= present_state;
        ctrl <= '0';
        CASE present_state IS
           WHEN s0 => IF count_en = '0' THEN
                          next_state <= s0;
                          ctrl <= '0';
                      ELSE
                          next_state <= s1;
                          ctrl <= '1';
                      END IF;
           WHEN s1 => IF count_en = '0' THEN
                          next_state <= s1;
                          ctrl <= '1';
                      ELSE
                          next_state <= s0;
                          ctrl <= '0';
                      END IF;
        END CASE;
    END PROCESS comblogicp;
END mealy;
```

In the state machine the clock process is identical to that of the register process described earlier. The combinational logic process contains all the inputs to the combinational block in the sensitivity list. In addition there is an initial statement setting each output to a value. This ensures that all possible conditions are satisfied so that no latches are generated. An example of a Moore state machine is shown in the appendix for the controller for the FIR filter example.

27.4.2 VHDL for the FIR Filter

For the FIR filter shown in Fig. 27.4, the VHDL code for each functional unit is shown in the appendix. There is a separate module for the controller, multiplexers, switches, and pipelined multiplier adder unit. These units are connected using structural VHDL. The complete structural VHDL representation of the FIR filter is omitted for space reasons.

27.5 FUNCTIONAL SIMULATION

Functional logic simulation is used to verify correctness of the design. The correct outputs must be generated for a given set of inputs. Typically we generate a set of simulation vectors that are used to determine if the design functions as it should. During functional simulation, we are simulating a VHDL model which has yet to be synthesized. Thus it is not possible to verify the physical timing of the underlying system.

A set of functional simulation test vectors is generated to verify the design. It is necessary to provide enough test vectors to verify the design, not to exhaustively test the design. For example, for a 4 bit adder which is implementing the function $A + B = C$, there are $2^8 = 256$ possible inputs. However, to verify the correctness of the design, a test vector set for the 4 bit adder need only contain a test of the carry, a test of the extreme ranges of both the inputs and outputs, and a short test of some nominal inputs. Typically such a test is enough to provide the designer with a high degree of confidence in the correctness of the design.

The VHDL code for the 4 bit multiplier used in the FIR filter in Fig. 27.4 is given in the appendix. Fig. 27.7 shows a portion of the corresponding functional simulation results. All the signal names are the same as in the VHDL code in the appendix. Here it is clear that the designed multiplier is working as expected.

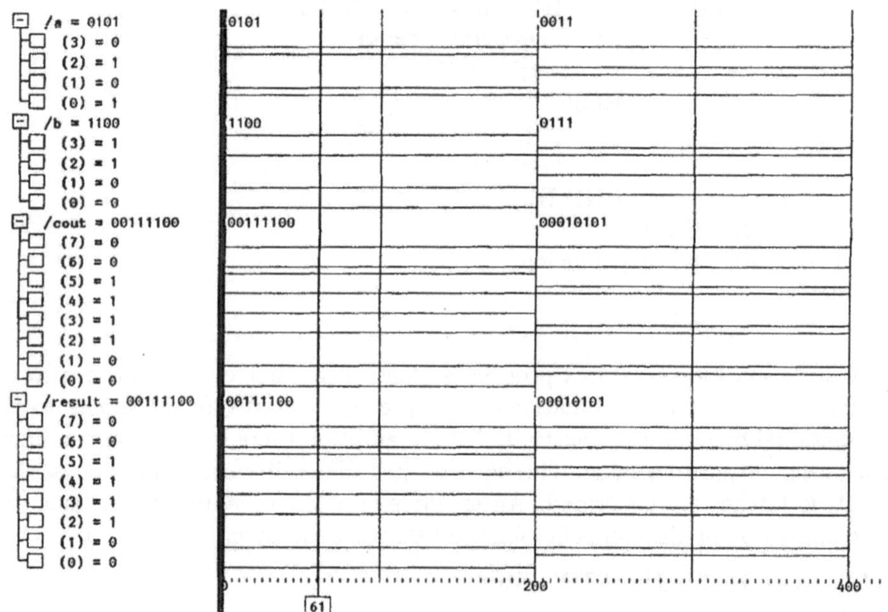

Figure 27.7 Four bit multiplier functional simulation.

27.6 LOGIC SYNTHESIS

After functional simulation of the design using VHDL code, logic synthesis is performed using synthesis tools to map the design to a specific technology. Logic synthesis provides an automated way to generate a gate-level circuit from the VHDL description. The output of logic synthesis is an optimized gate level circuit. Logic synthesis consists of two steps [3].

1. **Translation.** - Conversion of the given VHDL description into a non-optimal gate level description.
2. **Optimization.** - Both technology independent and technology dependent transformation of the initial gate level description obtained in translation phase. Output is a gate level description implementable in a target technology. The resulting implementation should meet area, speed, power, and testability requirements. Generated netlist can be used for custom layout or used for downloading to programmable logic gates such as FPGAs.

In this section, logic synthesis using 1.2 μm CMOSN standard cells and logic synthesis for a Xilinx 4003 FPGA are given as examples.

27.6.1 Synthesis to Standard Cells

In a standard cell architecture the layout consists of rectangular cells, all of the same height but varying width, placed in rows. In the standard cell approach, all the basic cells are already tested, analyzed, and specified and stored in the *cell library*. Cells are placed in rows and the space between two rows is called a *channel* which is used for routing as shown in Fig. 27.8 [30]. Using standard cells requires all the steps explained in Section 27.3. Therefore it takes a longer time for the complete design compared to using programmable logic devices like PLD (Programmable Logic Device)s or FPGAs but it is appropriate for large scale production.

As an example we will synthesize the multiplier shown in Fig. 27.4 for the 4 tap FIR filter. The logic synthesized gate level circuit is shown in Fig. 27.9 and requires 7 full adders, 11 NAND gates, 3 XOR, 13 NAND/AND, 3 INV/OR, 2 3-input NOR, 2 XNOR,2 INV/AND, 1 OR, and 1 inverter. Using the Autologic logic synthesizer, it is possible to optimize this design for speed or area. In this example, we optimized for area. Thus the multiplier is implemented using a ripple carry adder structure.

27.6.2 Synthesis to FPGAs

Programmable logic devices, including PLDs and field programmable gate arrays (FPGAs), provide more flexibility than semi-custom or full-custom application specific integrated circuits (ASICs). They can be designed and used in a short period of time compared to ASICs, for low volume production, and for test runs for high-volume products. A FPGA architecture is an array of logic cells that communicate with one another and with I/O via wires within routing channels as shown in Fig. 27.10. Like a semi-custom gate array, which consists of an array of transistors, an FPGA consists of an array of logic cells. In a gate array, routing is customized without programmable elements. In an FPGA, existing wire resources that run in horizontal and vertical columns (routing channels) are connected via programmable elements. These routing wires also connect logic to I/Os. Logic cell and routing architectures differ from vendor to vendor. FPGA programming

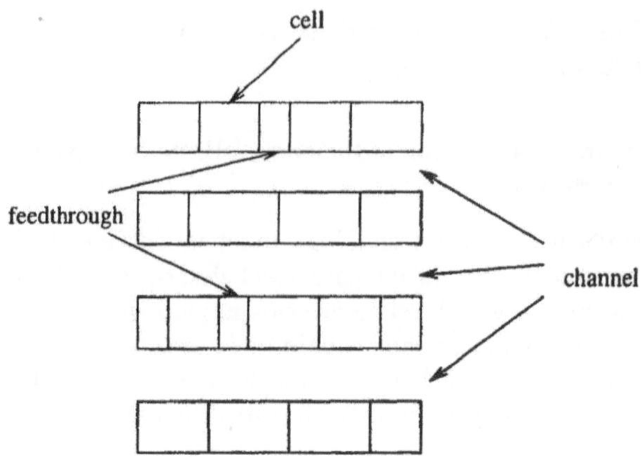

Figure 27.8 Standard cell architecture.

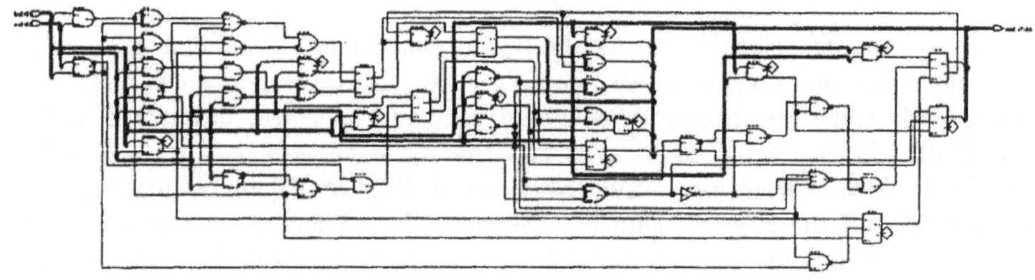

Figure 27.9 Synthesized 4 bit multiplier using CMOS standard cells.

methods include anti-fuse and SRAM-based technologies although only the latter allow true reconfigurability since anti-fuse can only be programmed once. In this chapter the Xilinx 4003 FPGA chip is used [31].

For the case of the 4 bit multiplier, we synthesized to the Xilinx 4003 using area optimization. The schematic of Fig. 27.11 is the result. The resulting circuit uses 39 AND, 1 3-input AND, 53 INV, 2 NAND, 1 3-input NAND, 1 3-input INV/NOR, 38 OR, 1 3-input OR, 2 3-input XNOR, 2 XOR, and 3 3-input XOR gates are used. Although the logic primitives for the Xilinx FPGA are slightly different than those used in the standard cell version, it is clear that the implementation is comparable in terms of gate count.

27.7 STRUCTURAL SIMULATION

Structural logic simulation is used to verify the correctness of the design with timing parameters. A structural simulation requires a physical gate level model of the circuit and is typically performed after the VHDL model has been synthesized

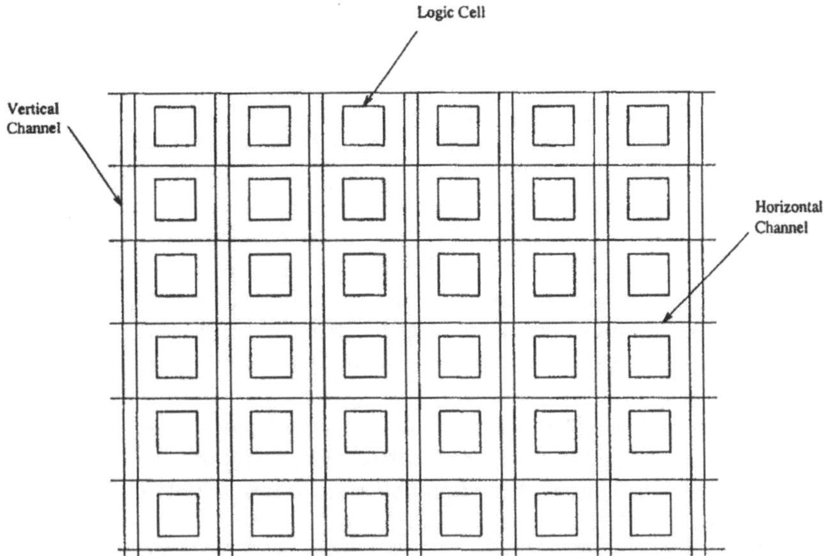

Figure 27.10 Generic FPGA architecture.

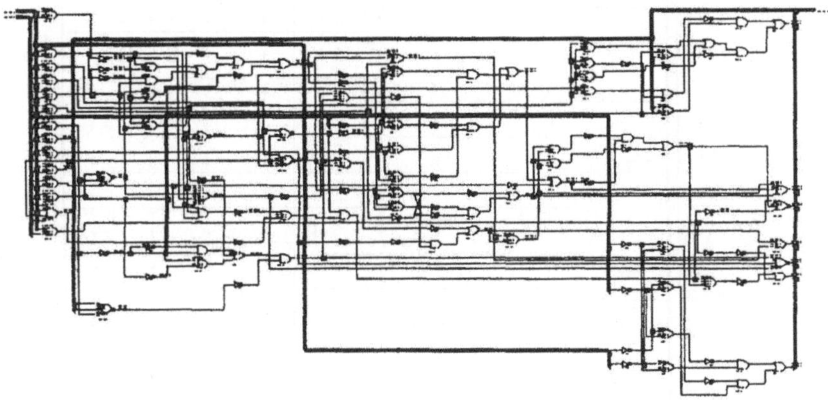

Figure 27.11 Synthesized 4 bit multiplier using FPGA.

to a logic gate level description. During structural simulation it is possible to check for logic hazards such as race conditions and setup and hold time violations.

The underlying logic models must contain complete timing models of the underlying technology. The logic models must either store or compute each of the following timing parameters: (1) the propagation delay for a rising output; (2) the propagation delay for a falling output; (3) the setup time for a flip-flop; (4) the hold time for a flip-flop. Each of these timing parameters is dependent, not only on the underlying technology, but also on the loading of the signals. The timing parameters are stored in a range of values including the minimum, maximum, and nominal values. A maximum time simulation verifies the correct operation of the clock and the setup time of the flip-flops. A minimum time simulation verifies hold times of the flip-flops. A nominal time simulation checks for logic glitches which may cause inappropriate behavior or higher power consumption. A set of structural logic test vectors must not only verify the correctness of the design but also exercise the inherent timing within the circuit. Thus it is necessary to provide a sequence of input transitions which exercise the low-to-high and high-to-low transitions internally and externally in the logic circuit.

Typically the designer begins with the functional test vector set and adds enough additional input vectors to toggle all the output pins. This verifies the correct operation of the circuit and the low-to-high and high-to-low transitions.

An example of a portion of the post-synthesis simulation for the 4 bit multiplier is shown in Fig. 27.12 for the Xilinx 4003 FPGA and in Fig. 27.13 for the standard cell implementation. It is clear that the propagation time of the multiplier using standard cells is much less than half that of the FPGA implementation.

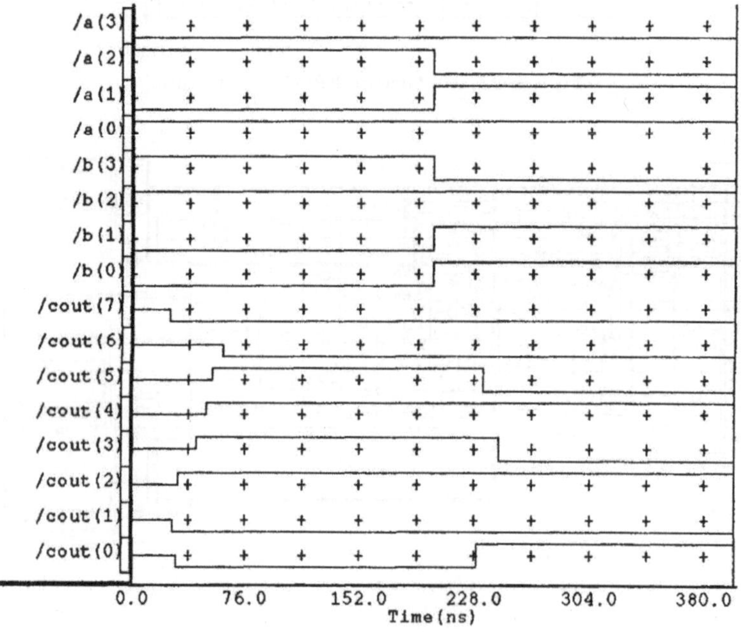

Figure 27.12 Simulation result of 4 bit multiplier after synthesis to FPGA.

27.8 DESIGN ANALYSIS

Once the circuit has been mapped to a particular technology it is possible to analyze it to see if it meets the design constraints. From synthesis, it is possible to obtain an immediate count of the gates needed to implement the circuit. This gives a good indication of the area required by the circuit. For the FPGA design, the number of gates determines how many FPGAs will be needed. For the standard

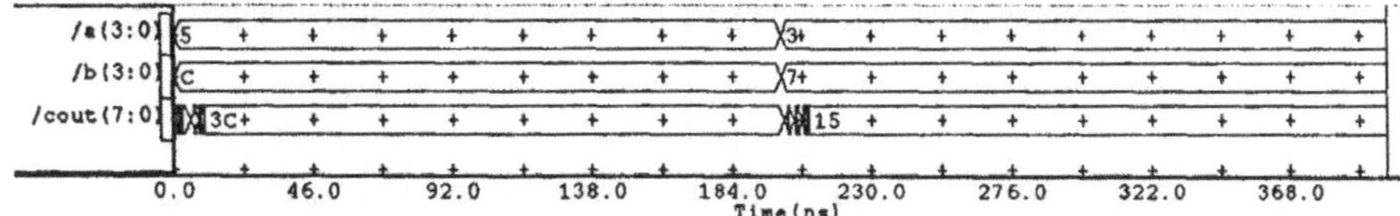

Figure 27.13 Simulation result of 4 bit multiplier after synthesis to standard cells.

cell design, the number of gates indicates the physical area necessary to layout the chip.

Additional design analysis includes critical path analysis, power estimation, and fault grading. Critical path analysis measures the speed of the circuit. Power estimation measures the average power consumption. Fault grading measures the testability of the circuit. In the following we will apply both critical path analysis and power estimation to our examples. Although fault grading is equally important, it is beyond the scope of this chapter.

27.8.1 Critical Path Analysis

Critical path analysis determines the speed of the design. In the case of clocked circuit design, the minimum clock period is determined by the longest path delay in the design. This critical path can occur in one of three places. Either from an external input to the input of a register, or from the output of a register to an external output, or between two registers. For a combinational circuit, the critical path is the longest path delay in the circuit. The critical path of a combinational circuit determines the propagation delay of that module.

For the FPGA design, the critical path within the 4 bit multiplier is 71.40 ns and occurs on the path from input b(0) to output cout(2). The multiplier has the longest critical path of any element in the design. When using standard cells, the critical path delay is 14.90 ns for the entire FIR filter. Thus the clock frequency for the standard cell circuit is 67 MHz compared to 13 MHz for the FPGA implementation. Remember a new output is generated every two clocks.

27.9 POWER ESTIMATION AND LOW POWER DESIGN

The continued increase in chip density and operating frequency have made power dissipation a major factor in VLSI design [32, 33]. Designing for low power has become important for all types of VLSI circuits and essential for applications such as portable computers and personal communications. To meet this requirement, many low power design techniques have been developed [34, 33, 35]. In CMOS circuit design, power consumption is caused mainly from switching transitions at the gate output. Many techniques have been developed to calculate circuit switching activity [36]-[39]. Power estimation techniques utilize the switching activity to calculate the power consumption. Statistical power estimation is one technique [40]. Another technique is to estimate the power based on circuit simulation [41]-[45]. Simulation based algorithms use simulation tools or extract energy consumption data beforehand and use this database to estimate the power consumption. In the following subsection, one simulation-based power estimation tool is introduced.

27.9.1 Power Estimation Using Input/Output Transition Analysis

As with other techniques, our estimation technique is based on a cell library. First, we discuss a new method to model the energy consumed by elements in the cell library. We compare our model to the state transition graph model. Next we discuss how the model is used to estimate power dissipation for a large circuit composed of library elements.

27.9.1.1 Library cell modeling

One method to estimate power utilizes an input/output transition energy database [46, 43, 47, 48]. Consider the NOR gate shown in Fig. 27.14. It will be used to explain the basic concept of this power estimation method which is called IOTA.

A library cell is modeled by storing the energy consumption for every possible pair of input transitions I_{t-1}, I_t, where I_{t-1} and I_t are input vectors at time $t-1$ and t. That is, energy consumption on every possible input transition is stored in the Input/Output Transition energy Table (IOTT). This table lists all possible input/output transitions and energy consumption data corresponding to each transition. Each entry in the IOTT lists a specific transition at the input, the outputs for each of the input vectors, and the corresponding energy consumption. Table 27.1 shows IOTT for the NOR gate.

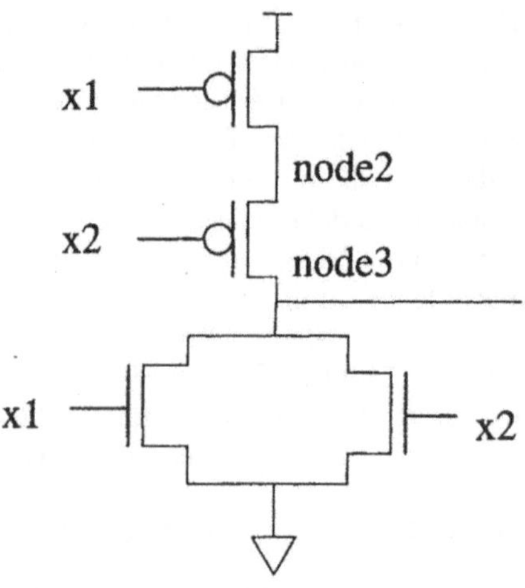

Figure 27.14 NOR gate.

27.9.1.2 Modeling sequential elements

For a complete system, sequential circuits should be analyzed by modeling energy consumption in memory elements. Energy consumption in sequential elements can be modeled using IOTT. However, in a sequential element, the current state of the element itself (not the internal nodes) has a significant effect on energy consumption. Hence, in addition to the previous input, the model also stores the previous state of the sequential element.

27.9.1.3 Relation to complete FSM model

For Comparison with the Finite State Machine (FSM) model, Fig. 27.15 shows the State Transition Diagram (STD) and Table 27.2 shows the state transitions of

Table 27.1 Example of Input/Output Transition Table (IOTT)

NOR INPUT/OUTPUT TRANSITION						
Previous Inputs		Current inputs		Prev_out	Cur_out	energy (nJ)
0	0	0	0	1	1	$W_{0000} = 0.000000111$
0	0	0	1	1	0	$W_{0001} = 0.003835260$
0	0	1	0	1	0	$W_{0010} = 0.004451810$
0	0	1	1	1	0	$W_{0011} = 0.004215160$
0	1	0	0	0	1	$W_{0100} = 0.003874200$
0	1	0	1	0	0	$W_{0101} = 0.000000099$
0	1	1	0	0	0	$W_{0110} = 0.003080660$
0	1	1	1	0	0	$W_{0111} = 0.000057917$
1	0	0	0	0	1	$W_{1000} = 0.004056090$
1	0	0	1	0	0	$W_{1001} = 0.002372670$
1	0	1	0	0	0	$W_{1010} = 0.000000015$
1	0	1	1	0	0	$W_{1011} = 0.000001490$
1	1	0	0	0	1	$W_{1100} = 0.003426430$
1	1	0	1	0	0	$W_{1101} = 0.000025406$
1	1	1	0	0	0	$W_{1110} = 0.000002445$
1	1	1	1	0	0	$W_{1111} = 0.000000000$

Table 27.2 State Transition Table

NOR STATE TRANSITION					
Pre_state		Inputs		Cur_state	
0	0	0	0	1	1
1	0	0	0	1	1
1	1	0	0	1	1
0	0	1	0	0	0
1	0	1	0	0	0
1	1	1	0	0	0
0	0	0	1	1	0
1	0	0	1	1	0
1	1	0	1	1	0
0	0	1	1	0	0
1	0	1	1	1	0
1	1	1	1	1	0

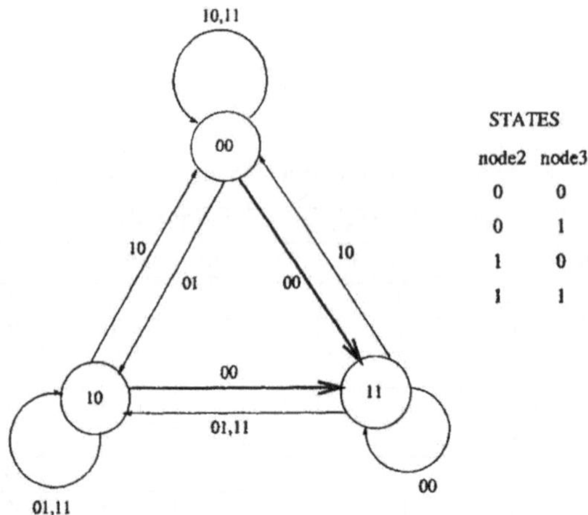

Figure 27.15 State transition diagram of NOR gate.

the NOR gate. The IOTT does not track the state of internal nodes in an entity; instead it relies on the input transition. It would appear that this model does not track the energy consumed at internal nodes in a library element. In fact, this simple model covers many of the transition energies represented in the *complete FSM model.* Hence, energy consumed at internal nodes is not ignored. Not tracking the state of an entity does not lead to a significant error. Consider the state and IOTT representations of a NOR gate in Tables 27.1 and 27.2 respectively. Consider the individual vectors 00, 01, and 10. Each of these vectors is a synchronizing sequence. For example, the vector 00 sets the state to 11. The input transition pair $\{00, i_1i_2\}$ (where i_1i_2 is a vector at the NOR gate inputs) models the edge energy in the complete FSM model for the corresponding triplet $\{00, 11, i_1i_2\}$. Though the state of the element is not tracked, all the energies on the transitions from the state 11 are represented in this model. For the states 00 and 10, all energies corresponding to triplets with previous inputs of 10 and 01, respectively, are represented in the IO model. When all single vectors are synchronizing sequences, there is a 1-to-1 map between the IOTT and the complete state table. Such a map is not always possible. The input 11 is not a synchronizing sequence. The destination state on an input of 11 may either be 00 or 10. Note that one of the two state elements is synchronized. Thus, a single input transition $\{11, i_1i_2\}$ represents two triplets $\{11, 00, i_1i_2\}$ and $\{11, 10, i_1i_2\}$ in the complete FSM model.

27.9.1.4 Computing transition energy

Since the IOTT method cannot accommodate all the internal node state transitions, average energy consumption is used for non-synchronizing sequences. This lets the IOTT method minimize the error in energy consumption caused by internal node state transitions. The errors from not-covered internal node state transitions are very small so the designer can assume all the internal nodes are covered using this method.

27.9.1.5 Circuit level power estimation

Once library elements have been characterized, given a sequence of N inputs, logic simulation is required to estimate the energy consumed by the circuit. The function of the simulation is to generate transition information at the inputs for each entity in the circuit. At each entity in the circuit, the number of occurrences of each input transition (in the IOTT) is counted. For a logic gate G_i with m inputs, if the energy consumption and the number of occurrences of each input transition pair are known, energy consumption of G_i can be estimated as follows.

$$Energy(G_i) = \sum_j TE_j \times W_j \tag{1}$$

where TE_j = Number of transitions(jth transition from IOTT)
W_j = Energy consumption of jth transition
i = gate number
j = corresponding input/output transition from IOTT

Assume that a combinational network *CN* has M logic gates. Assume the N patterns in the input sequence are input with clock cycle time T_{cycle}. Then, the average power dissipation of *CN* is:

$$P_{avg}(CN) = \frac{\sum_{i=1}^{M} Energy(G_i)}{N \times T_{cycle}} \tag{2}$$

27.9.1.4 Implementation

The power estimation technique described so far has been implemented in a tool called IOTA (Input Output Transition Analysis). Given a circuit, and an input sequence, the power dissipated by the circuit for that sequence is estimated using a two-phase process. In the first phase an IOTT is built for each element in the cell library. To build IOTTs for library elements, the Boolean function of the target gate is used to build the entries in the IOTT, but without transition energy consumption data. From input/output data, transition energy is calculated using SPICE. The transition energy data computed is added to the IOTT. In the second phase, a logic simulator is used to count the number of occurrences of each transition in the IOTT, at each entity's inputs. For each entity, the energy consumption is computed using the transition information and energy consumption information in its IOTT. The total energy consumption is the sum of the energy consumption over all the entities in the circuit.

27.9.1.7 Delay representation

IOTA can be used for zero delay simulation, unit delay simulation, or variable delay simulation. To use variable delay information, the smallest unit time is the degree of accuracy desired in modeling circuit delays. The digital CMOS circuits under consideration are assumed to be operating in a synchronous environment, i.e., they are being controlled by a global clock. Let T_{clk} denote the clock period, and T_{gd} denote the smallest gate delay in the circuit. To capture the glitches in the circuit, the clock period is assumed to be divided into S time slots [44, 45, 46, 48], where:

$$S \equiv \frac{T_{clk}}{T_{gd}} \tag{3}$$

The accuracy of glitch identification depends on the accuracy of the delay models.

Table 27.3 Basic Gate Power Estimation

	Power consumption				
Circuit	SPICE	IOTA	CPU time (SPICE)	CPU time (IOTA)	Error
INV	0.08319	0.083268	24	less than 0.1	0.09%
NOR	0.072471	0.071609	25.2	less than 0.1	1.19%
NAND	0.081323	0.081331	34.9	less than 0.1	0.01%
AND	0.12958	0.13063	45.5	less than 0.1	0.81%
XOR	0.1241	0.12799	45.2	less than 0.1	0.46%
MUX_1bit	0.023496	0.023569	39.3	less than 0.1	0.31%
*F-A	0.2865	0.2812	360	less than 0.1	1.85%
F-A	0.43615	0.45850	398	1	2.83%
D-FF	0.2120	0.2220	195	less than 0.1	2.79%

Units: power: mW, time: second. Full adder without inverters at the output)

Table 27.4 Power Consumption for Several Circuits

Circuit	SPICE (mW)	IOTA0 (mW)	IOTAV (mW)	IOTAVI (mW)	IOTAU (mW)	CPU time SPICE(sec)	CPU time IOTA(sec)	Error (%)
NOR2	0.1440	0.1324	0.1445	0.1399	0.1399	54.1	< 0.1	2.84
NOR3	0.2443	0.2354	0.2816	0.2367	0.2367	74.1	2	3.11
RCA_4bit	1.8350	1.4978	1.9842	1.8583	1.8493	61.8	35	0.77
RCA_8bit	4.2600	3.2847	5.1392	4.0760	4.0560	2054	59	4.79
CSA_8bit	7.2370	5.6898	8.0986	8.0333	7.8696	2721	102	8.74
4x4Mult	6.5790	5.1783	7.5230	7.4355	6.9198	12973	117	5.18
8x8Mult	42.070	25.874	64.323	52.388	48.388	39856	534	15.01
8x8mult_p	39.960	26.957	46.893	45.721	39.280	42827	637	1.70

RCA: ripple carry adder, CSA: carry select adder, Mult: multiplier, Multp: pipelined multiplier.

27.9.1.4 Results

Table 27.3 shows estimation results for individual gates and Table 27.4 represents estimation results for relatively large circuits. In Table 27.4, IOTA0 represents zero delay simulation, IOTAV for variable delay simulation, IOTAVI for variable delay simulation with inertial delay, and IOTAU for unit delay simulation. Results from Table 27.4 shows unit delay simulation has minimum errors compared to SPICE. This is because of the inaccuracy of using time slots when computing glitches. The error in the last column in Table 27.4 is for unit delay simulation.

Using IOTA the power consumption for the FIR filter implemented using standard cells has been estimated as shown in Table 27.5.

27.10 LAYOUT

The input to the physical design phase is a circuit diagram and the output is the layout of the circuit. Physical design is accomplished in several stages such

Table 27.5 Power Estimation of Example FIR Filter Using IOTA

Functional Unit	Power Consumption (mW)
Multiplier-Adder Unit	14.83
Example FIR filter	35.74

as partitioning, floorplanning, placement, routing, and compaction as explained below.

1. **Partitioning.** - When the design is too large to fit into one chip, the design should be split into small blocks so that each block can be fit into one chip.
2. **Floorplanning and placement.** - Circuit blocks should be exactly positioned at this stage of physical layout.
3. **Routing.** - Exactly positioned blocks should be connected to each other at this stage.
4. **Compaction.** - This is simply the task of compressing the layout in all directions such that the total area is reduced as much as possible.

The layout has been done for the complete 4 tap FIR filter and is shown in Fig. 27.16 for the standard cell implementation. After the layout, design rules should be checked to minimize the possibility of defects and malfunctioning of the chip prior to fabrication. The layout should also be checked against the original schematic description of the circuit to make sure the layout correctly represents the original design. Using backannotation from the layout it is possible to include additional timing parameters such as wire delays in the synthesized circuit.

27.11 STRUCTURAL SIMULATION

After layout of the logic gates, it is possible to do another structural simulation of the circuit with the additional timing parameters. The structural logic simulation not only includes the complete physical model of the logic gates of the circuit but also additional timing parameters such as input/output delays and wire delays. The structural simulation performed at this point is a repeat of the simulation performed prior to layout. If the structural simulation fails, the design must be modified and laid out again to remove the timing errors. Fig. 27.17 shows the final simulation of the standard cell based

FIR filter circuit after layout including all wire delays.

27.12 CONCLUSION

In this chapter we have studied the automated design process necessary to map a DSP algorithm to a completely functional FPGA or standard cell based

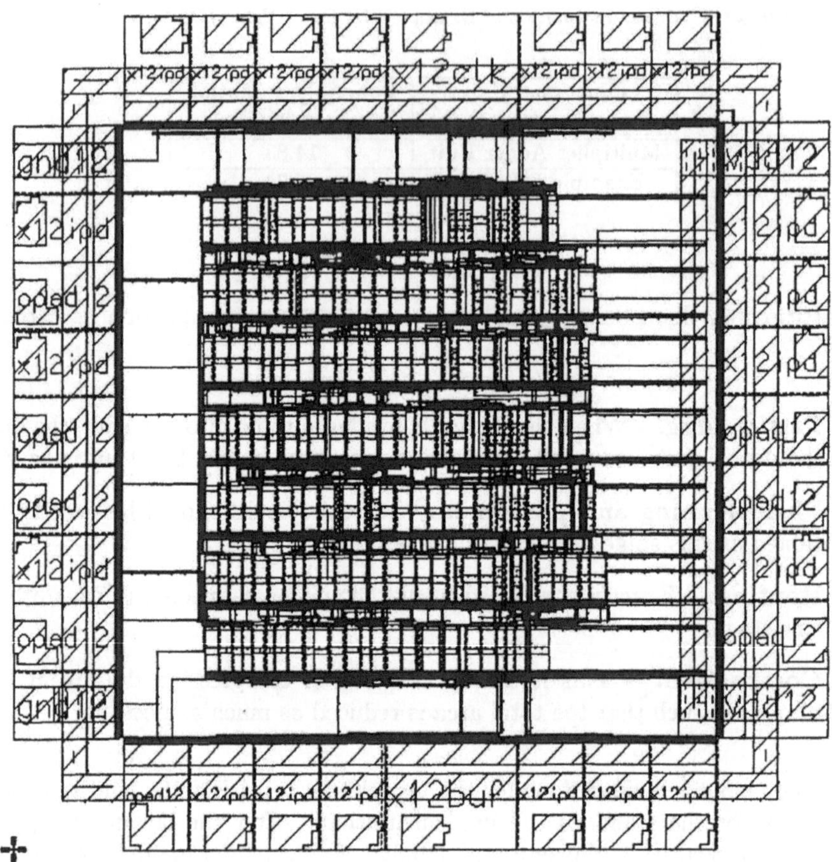

Figure 27.16 Layout of 4 tap FIR filter using standard cells.

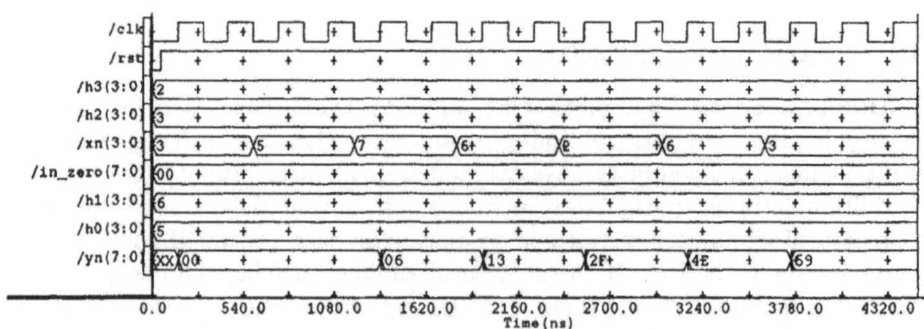

Figure 27.17 Simulation of 4 tap FIR filter after standard cells layout.

implementation. All steps in the design process have been demonstrated including high level synthesis, design entry, functional verification, logic synthesis, structural verification, design analysis, and final layout and verification. We have emphasized the VHDL necessary for proper logic synthesis. We have introduced a simulation based power estimation tool.

27.13 APPENDIX: VHDL CODE FOR 4 TAP FIR FILTER

27.13.1 Registers

Below is a VHDL code for a 4 bit register.

```
ENTITY reg_4bit IS
        PORT( four_in: IN std_logic_vector(3 downto 0);
                rst, clk: IN std_logic;
                four_out: BUFFER std_logic_vector(3 downto 0));
END reg_4bit;

ARCHITECTURE bhvrl OF reg_4bit IS
        SIGNAL pre_out: std_logic_vector(3 downto 0);
BEGIN
        clkp: PROCESS(clk, rst)
        BEGIN
                IF (rst = '0') THEN
                        pre_out <= "0000";
                ELSIF (CLK'event AND CLK = '1' AND
                CLK'last_value = '0') THEN
                        pre_out <= four_in;
                END IF;
        END PROCESS clkp;
        four_out <= pre_out;
END bhvrl;
```

Eight bit registers can be built using exactly same way except the length of the logic vectors.

27.13.2 Multiplexers

The 4 bit 2 to 1 multiplexer VHDL code is as follows.

```
ENTITY mux2_1_4bit IS
        PORT(in1, in2: IN std_logic_vector(3 downto 0);
                mux_out: OUT std_logic_vector(3 downto 0);
                ctrl: IN std_logic);
END mux2_1_4bit;

ARCHITECTURE bhvrl OF mux2_1_4bit IS
BEGIN
        PROCESS(ctrl, in1, in2)
        BEGIN
```

```
                IF (ctrl = '0') THEN
                    mux_out <= in1;
                ELSE
                    mux_out <= in2;
END IF;
END PROCESS;
END bhvrl;
```

27.13.3 Multiplier-Adder(MA unit)

For two levels of pipelining we put two registers, one between the 4 bit multiplier and the 8 bit adder and one register at the output of the adder. Below is the VHDL code for 4 bit multiplier.

```
ENTITY mult4 IS
        PORT(a,b: IN std_logic_vector(3 DOWNTO 0);
        cout: OUT std_logic_vector(7 DOWNTO 0) );
END mult4;

ARCHITECTURE behave OF mult4 IS
SIGNAL result : std_logic_vector(7 DOWNTO 0);
BEGIN
        arith_process: PROCESS(a,b)
        VARIABLE p: integer range 0 TO 255;
        BEGIN
                p := TO_INTEGER('0' & a) * TO_INTEGER('0' & b);
                result <= TO_STDLOGICVECTOR(p, 8 );
        END PROCESS arith_process;
        cout <= result;
END behave;
```

Below is the VHDL code for an 8 bit adder.

```
ENTITY add8 IS
        PORT(a: IN std_logic_vector(7 downto 0);
                b: IN std_logic_vector(7 downto 0);
                sum: BUFFER std_logic_vector(7 downto 0));
END add8;

ARCHITECTURE bhvrl of add8 IS
BEGIN
        sum <= a + b;
END;
```

Below is the VHDL code for the complete MA unit.

```
ENTITY mult_add IS
        PORT(rst, clk: IN std_logic;
                mul_in1, mul_in2: IN std_logic_vector(3 downto 0);
                add_in: IN std_logic_vector(7 downto 0);
                ma_out: BUFFER std_logic_vector(7 downto 0));
END mult_add;

ARCHITECTURE structural OF mult_add IS

- component initiation
        COMPONENT mult4
                PORT(a: IN std_logic_vector(3 downto 0);
                        b: IN std_logic_vector(3 downto 0);
                        cout: OUT std_logic_vector(7 downto 0));
        END COMPONENT;

        COMPONENT add8
                PORT(a: IN std_logic_vector(7 downto 0);
                        b: IN std_logic_vector(7 downto 0);
                        sum: BUFFER std_logic_vector(7 downto 0));
        END COMPONENT;

        COMPONENT reg_8bit
                PORT(eight_in: IN std_logic_vector(7 downto 0);
                        rst: IN std_logic;
                        clk: IN std_logic;
                        eight_out: BUFFER std_logic_vector(7 downto 0));
        END COMPONENT;

- map the instantiated components to the component models
- described at the top of this file
FOR M1: mult4 USE ENTITY work.mult4(behave);
FOR R1: reg_8bit USE ENTITY work.reg_8bit(bhvrl);
FOR R2: reg_8bit USE ENTITY work.reg_8bit(bhvrl);
FOR R3: reg_8bit USE ENTITY work.reg_8bit(bhvrl);
FOR A1: add8 USE ENTITY work.add8(bhvrl);

- temporary signal declaration
SIGNAL temp_mul, temp_reg1, temp_reg2, temp_reg3:
std_logic_vector(7 downto 0);

- description of component connections
BEGIN
        M1: mult4 PORT MAP (mul_in1, mul_in2, temp_mul);
        R1: reg_8bit PORT MAP (temp_mul, rst, clk, temp_reg1);
```

```
        R2: reg_8bit PORT MAP (add_in, rst, clk, temp_reg3);
        R3: reg_8bit PORT MAP (temp_reg2, rst, clk, ma_out);
        A1: add8 PORT MAP (temp_reg1, temp_reg3, temp_reg2);
END structural;
```

27.13.4 Switches

Below is a VHDL description of the output switch.

```
ENTITY switch_4 IS
        PORT( ctrl: IN std_logic;
                sw_in: IN std_logic_vector(3 downto 0);
                sw_out: OUT std_logic_vector(3 downto 0));
END switch_4;

ARCHITECTURE bhvrl OF switch_4 IS
BEGIN
        PROCESS(ctrl,sw_in)
        BEGIN
                IF (ctrl = '1') THEN
                        sw_out <= sw_in;
                END IF;
        END PROCESS;
END bhvrl;
```

27.13.5 Control Unit

Below is the VHDL code for the controller in the FIR filter.

```
ENTITY f_4fir_cont IS
        PORT(rst, clk: IN std_logic;
                ctrl: OUT std_logic);
END f_4fir_cont;

ARCHITECTURE bhvrl OF f_4fir_cont IS
TYPE state_enum IS (s0, s1);
SIGNAL state: state_enum := s0;
SIGNAL nexts: state_enum := s1;
BEGIN
        clkp: PROCESS(clk, rst)
        BEGIN
                IF (rst = '0') THEN
                state <= s0;
                ELSIF (clk'EVENT and clk = '1' and
                clk'LAST_VALUE = '0') THEN
```

```
                        state <= nexts;
                END IF;
        END PROCESS clkp;

        transp: PROCESS(state)
        BEGIN
                CASE state IS
                        WHEN s0 => nexts <= s1;
                                ctrl <= '0';
                        WHEN s1 => nexts <= s0;
                                ctrl <= '1';
                END CASE;
        END PROCESS transp;
END bhvrl;
```

REFERENCES

[1] M.C. McFarland, A.C. Parker, and R. Composano, "The high-level synthesis of digital systems," *Proceedings of the IEEE*, pp. 301–318, Feb. 1990.

[2] R. Camposano and W. Wolf, *High-level VLSI Design*. Kluwer Academic Publishers, Boston Mass., 1991.

[3] S. Devadas, A. Ghosh, and K. Keutzer, *Logic Synthesis*. McGraw-Hill, Inc., New York, NY, 1994.

[4] K. Parhi, "High-level synthesis algorithm and architecture transforms for DSP synthesis," *Journal of VLSI Signal Processing*, pp. 121–143, Jan. 1995.

[5] H. DeMan, F. Catthoor, G. Goossens, J. Vanhoof, J. Meerbergen, S. Note, and J. Huisken, "Architecture-driven synthesis techniques for VLSI implementation of DSP algorithms," *Proceedings of the IEEE*, pp. 319–335, Feb. 1990.

[6] L. Lucke, A. Brown, and K. Parhi, "Unfolding and retiming for high-level DSP synthesis," in *Proceedings of the IEEE International Symposium on Circuits and Systems*, pp. 2351–2354, June 1991.

[7] L. Lucke and K. Parhi, "Generalized ILP scheduling and allocation for high level VLSI synthesis," in *Proceedings of the IEEE Custom Integrated Circuits Conference*, pp. 5.4.1–5.4.4, May 1993.

[8] L. Lucke and K. Parhi, "Data-flow transformations for critical path time reduction in high-level DSP synthesis," *IEEE Transactions on Computer-Aided Design*, pp. 1063–1068, June 1993.

[9] K. Ito, L. Lucke, and K. Parhi, "Module selection and data format conversion for cost-optimal DSP synthesis," in *Proceedings of the IEEE/ACM International Conference on Computer-Aided Design*, Nov. 1994.

[10] V. Sundararajan and K. Parhi, "Synthesis of folded, pipelined architectures for multi-dimensional multirate systems," in *Proc. of IEEE International Conference on Acoustics, Speech, and Signal Processing*, May 1998.

[11] K. Parhi, C. Wang, and A. Brown, "Synthesis of control circuits in folded pipelined DSP architectures," *IEEE Journal of Solid State Circuits*, vol. 27, no. 1, pp. 29–43, Jan. 1992.

[12] E.A. Lee and D.G. Messerschmitt, "Synchronous data flow," *Proceedings of the IEEE*, vol. 75, no. 9, Sept. 1987.

[13] E.A. Lee and D.G. Messerschmitt, "Static scheduling of synchronous data flow programs for digital signal processing," *IEEE Transactions on Computer*, vol. 36, no. 1, pp. 24–35, Jan. 1987.

[14] C. Gebotys and M. Elmasry, "Optimal synthesis of high-performance architectures," *IEEE Journal of Solid-State Circuits*, vol. 27, no. 3, pp. 389–397, Mar. 1992.

[15] C.E. Leiserson and et.al., "Optimizing synchronous circuitry by retiming," in *3rd Caltech Conference on VLSI*, pp. 87–116, Mar. 1983.

[16] N. Park and A.C. Parker, "Sehwa: A software package for synthesis of pipelines from behavioral specifications," *IEEE Transactions on Computer-Aided Design*, vol. 7, no. 3, Mar. 1988.

[17] K.K. Parhi and D.G. Messerschmitt, "Static rate-optimal scheduling of iterative data-flow programs via optimum unfolding," *IEEE Transactions on Computer-Aided Design*, vol. 40, no. 2, Feb. 1991.

[18] C-T. Hwang, J-H. Lee, and Y-C. Hsu, "A formal approach to the scheduling problem in high level synthesis," *IEEE Transactions on Computer-Aided Design*, vol. 10, no. 4, pp. 464–475, Apr. 1991.

[19] T.C. Hu, "Parallel sequencing and assembly line problems," *Operations Research*, vol. 9, pp. 841–848, 1961.

[20] D.A. Schwartz and T.P. Barnwell, III, "Cyclo-static solutions: Optimal multiprocessor realizations of recursive algorithms," in *VLSI Signal Processing, II*, 1986.

[21] P.G. Paulin and J.P. Knight, "Force-directed scheduling for the behavioral synthesis of ASICs," *IEEE Transactions on Computer-Aided Design*, vol. 8, no. 6, June 1989.

[22] S.M. Heemstra de Groot, S. H. Gerez, and O.E. Herrmann, "Range-chart-guided iterative data-flow-graph scheduling," *IEEE Transactions on Circuits and Systems-I: Fund. Theory & Appl.*, vol. 39, no. 5, May 1992.

[23] M. Potkonjak and W. Wolf, "Heuristic techniques for synthesis of hard real-time DSP application specific systems," in *IEEE International Conference on Acoustics, Speech and Signal Processing*, pp. 1240–1243, 1996.

[24] K.K. Parhi and F. Catthoor, "Design of high-performance DSP systems," in *IEEE International Symposium on Circuits and Systems*, pp. 447–507, 1996.

[25] C.H. Gebotys, "Optimal methodology for synthesis of DSP multichip architectures," *Journal of VLSI Signal Processing*, vol. 11, pp. 9–19, Oct-Nov. 1995.

[26] L-F. Chao, E. Sha, "Static scheduling for synthesis of DSP algorithms on various models," *Journal of VLSI Signal Processing*, vol. 10, no. 3, pp. 207–223, Aug. 1995.

[27] W. Verhaegh and et. al., "Improved force-directed scheduling in high-throughput digital processing," *IEEE Transactions on Computer-Aided Design*, vol. 14, no. 8, pp. 945–960, Aug. 1995.

[28] K. Skahill, *VHDL for Programmable Logic*. Addison Wesley Publishing Company, Inc., Menlo Park, CA, 1996.

[29] R. Lipsett, C. Schaefer, and C. Ussery, *VHDL: Hardware Description and Design*. Kluwer Academic Publishers, Boston, Mass., 1989.

[30] N. Sherwani, *Algorithms for VLSI Physical Design Automation*. Kluwer Academic Publishers, Boston Mass., 1993.

[31] Xilinx, *The Programmable Logic Data Book*. Xilinx, San Jose, CA, 1994.

[32] R. Broderson, A. Chandrakasan, and S. Sheng, "Technologies for personal communications," in *Symposium on VLSI Circuits*, 1991.

[33] A. Chandrakasan, S. Sheng, and R. Broderson, "Low-power CMOS digital design," *IEEE Journal of Solid State Circuits*, vol. 27, no. 4, pp. 473–484, Apr. 1992.

[34] J. Rabaey and M. Pedram, *Low Power Design Methodologies*. Kluwer Academic Publishers, Boston Mass., 1996.

[35] A. Chandrakasan, S. Sheng, and R. Broderson, "Low-power techniques for portable real time DSP applications," *VLSI Design*, 1992.

[36] F. Najm, "Transition density: a new measure of activity in digital circuits," *IEEE Transactions on Computer-Aided Design*, vol. 12, no. 2, pp. 310–323, Feb. 1993.

[37] A. Ghosh, S. Devadas, K. Keutzer, and J. White, "Estimation of average switching activity in combinational and sequential circuits," in *29th ACM/IEEE Design Automation Conference*, pp. 253–259, June 1992.

[38] M. Xakellis and F. Najm, "Statistical estimation of the switching activity in digital circuits," in *31th ACM/IEEE Design Automation Conference*, pp. 728–733, June 1994.

[39] Y. Lim, K. Son, H. Park, and M. Soma, "A statistical approach to the estimation of delay-dependent switching activities in CMOS combinational circuits," in *33th ACM/IEEE Design Automation Conference*, pp. 445–450, June 1996.

[40] R. Burch, "A Monte Carlo approach for power estimation," *IEEE Transactions on VLSI Systems*, vol. 1, no. 1, pp. 63–71, Mar. 1993.

[41] S. Kang, "Accurate simulation of power dissipation in VLSI circuits," *IEEE Journal of Solid-State Circuits*, vol. 21, no. 5, pp. 889–891, Oct. 1986.

[42] C. Huang, B. Zhang, A. Deng, and B. Swirski, "The design and implementation of PowerMill," in *ACM/IEEE International Symposium on Low Power Design*, pp. 105–109, Apr. 1995.

[43] B. George, G. Yeap, M. Wuloka, and D. Gossain, "Power analysis for semi-custom design," in *IEEE Custom Integrated Circuits Conference*, pp. 249–252, 1994.

[44] J. Lin and W. Shen, "A cell-based power estimation in CMOS combinational circuits," in *Proceedings of International Conference on CAD*, pp. 304–309, 1994.

[45] J. Satyanarayana and K. Parhi, "Heat: Hierarchical energy analysis tool," in *33th ACM/IEEE Design Automation Conference*, pp. 9–14, June 1996.

[46] J. Lee and L. Lucke, "Library based hierarchical power estimation method using in/out transition activity," in *Proceedings of the IEEE Region 3 Technical Conference*, pp. 295–297, Apr. 1997.

[47] A. Bogliolo, L. Benini, and B. Ricco, "Power estimation of cell-based CMOS circuits," in *33th ACM/IEEE Design Automation Conference*, pp. 433–438, June 1996.

[48] J. Lee, L. Lucke, and B. Vinnakota, "Power estimation using input/output transition analysis," in *Proceedings of the IEEE International Symposium on Circuits and Systems*, June 1998.

Index

T - #0294 - 101024 - C0 - 254/178/47 [49] - CB - 9780824719241 - Gloss Lamination